Contents

MATHEMATICAL ANALYSIS

MATHEMATICAL ANALYSIS

Real, Complex, and Metric Spaces

Dr Ajay Kr Chaudhuri

Ex-Reader and Head of the Department of Mathematics [UG & PG]
J K College, Purulia, West Bengal

Pratikshan Mondal MSc

Assistant Professor, Department of Mathematics
Durgapur Government College

New Central Book Agency (P) Ltd

LONDON
HYDERABAD ERNAKULAM BHUBANESWAR
NEW DELHI KOLKATA PUNE GUWAHATI

NCBA

REGD OFFICE

8/1 Chintamoni Das Lane, Kolkata 700 009, India
email: ncbapvtltd@ncbapvtltd.com

OVERSEAS

NCBA (UK) Ltd, 149 Park Avenue North, Northampton, NN3 2HY, UK
email: ncbauk@yahoo.com

EXPORT

NCBA Exports Pvt. Ltd, 212 Shahpur Jat, New Delhi 110 049
email: ncbaexports@ncbapvtltd.com

BRANCHES

2/27 Ansari Road, 1st Floor
Daryaganj, **New Delhi** 110 002
email: ncbadel@ncbapvtltd.com

House No. 3-1-315, 1st Floor
Nimboliadda, Kachiguda, **Hyderabad** 500 027
email: ncbahydb@ncbapvtltd.com

Shop Nos. 3 & 4, Vinayak Towers
681/B Budhwar Peth, Appa Balwant Chowk
Pune 411 002
email: ncbapune@ncbapvtltd.com

GSS Shopping Complex, 1st Floor
Opposite College Post Office, Convent Road
Ernakulam 682 035
0484-4021054 / 53

Shop No. 15, Bharati Towers, Ground Floor
Forest Park
Bhubaneswar 751 001
email: ncbabub@ncbapvtltd.com

Radhamayee Complex [Second Floor]
Opp. Axis Bank, Panbazar, Jaswanta Road
Guwahati 781 001
email: ncbaguwahati@ncbapvtltd.com

MATHEMATICAL ANALYSIS • Ajay Kr Chaudhuri, Pratikshan Mondal

First Published: July 2012
Reprinted: April 2013
Reprinted: January 2016
Reprinted: January 2017
Thoroughly Revised and Enlarged Second Edition: January 2021

PUBLISHER AND TYPESETTER
New Central Book Agency (P) Ltd
8/1 Chintamoni Das Lane, Kolkata 700 009

PRINTER
New Central Book Agency (P) Ltd
Web-Offset Division, Dhulagarh, Sankrail, Howrah

TECHNICAL EDITOR
Dipan Roy

PROJECT TEAM
Subir Pal and Koushik Samanta

ISBN: 978 81 7381 780 9

Price ₹ 735.00

Dedication

To the memory of my Parents
Shri N C Chaudhuri and Smt. Sabitri Chaudhuri

Ajay Kumar Chaudhuri

AND

Dedicated to my mother Smt. Bhagyabati Mondal
and my respected teachers

Pratikshan Mondal

PART II
COMPLEX ANALYSIS

Preface to the Second Edition

Mathematical analysis (Real, Complex and Metric Space) was first published in July 2012. The book contains good number of theorems and solved problems in each part. We have tried our best to develop the basic concepts of number system, limit, continuity differentiation, bounds of a function which are the tools of analysis. After introducing these concepts we have entered into the main syllabus for Part-III (Hons) in Fifth Paper. Without expanding the volume of the book we have emphasized mainly on the corrections of the mistakes in each chapter which were overlooked in the first edition. In some cases the statements of the theorems and proofs of some of the theorems have been changed for better representation of the book. We hope we shall get good response from students and teachers to this second edition.

We offer our heartiest thanks to New Central Book Agency (P) Ltd for their best possible cooperation in bringing out the second edition of the book.

Ajay Kumar Chaudhuri

2 May 2020 **Pratikshan Mondal**

Preface to the First Edition

Mathematical analysis is a pillar of higher mathematics. Calculus is based on the concepts of limit (limit of a function, limit of a sequence). Real analysis is the rigorous extension of calculus. Mathematician Newton and Leibnitz have invented calculus and introduced ϵ as an arbitrary small positive number (however small may be) and a positive number δ depending on ϵ. (ϵ, δ) closely related positive numbers play an important role in real analysis for defining limit, continuity, uniform continuity, differentiability of a function, Riemann integration etc. The parallel matters are also developed in the field of complex analysis.

There are many textbooks on Real analysis, Complex analysis and Metric spaces. We have confined them in a single textbook by covering the present syllabi of BSc/BA (Hons.) in Mathematics of Calcutta University, Burdwan University, North Bengal University and Universities of all other states also. This book may also be used in MSc/MA courses in Mathematics.

The purpose of writing this book is to place the matters of Real analysis, Complex analysis and Metric spaces before the students in logical, systematical and simple ways so that the students may grasp the matters easily and feel interest on this subject.

The first two parts of this book provides a transaction of elementary calculus to advanced courses in real and complex analysis and the third part deals metric space, introduced by M Frechet to generalize the notion of distance in an arbitrary non-empty set. Here we have a platform to assign, more stress on the concept of convergent sequence of points, limit points, continuity, uniform continuity of a function at a point in their domain.

This book contains 15 chapters in the first part (Real analysis), 5 chapters in the second part (Complex analysis) and 6 chapters in the third part (Metric spaces). In each chapter of every part we cited good number of examples and solved harder problems by heading "Problems". The last chapter [Ch 15] of first part deals with some new definitions, theorems (which are not discussed earlier) and large number of extraordinary examples for helping the students in competitive examinations.

The book starts with a depth foundation of the real number system. The concepts of limit, continuity, differentiability, bounds of a set, and mapping are presented thoroughly. The foundation supports the subsequent chapters. Function of bounded variation, Riemann theory of integration, improper integrals, sequence of function and infinite series, Fourier series, power series, functions of several variables, implicit function, line, surface, double and triple integrals, beta function and gamma functions are presented in details with large number of examples.

In Part II, the basic definitions and properties of complex numbers, limit, continuity, differentiability of a complex function $w = f(z) = u + iv$, analytic function, sequence in a complex domain, power series, bilinear transformation are dealt with sufficient number of examples.

In Part III, the notion of metric spaces is discussed in details, duly exemplified by a large number of solved problems. In this part of the book the various properties of metric spaces such as completeness, compactness and connectedness etc. are given special emphasis. Each chapter is supported by remarks, notes along with solved and unsolved problems.

For preparation of the real analysis and complex analysis section we have taken the help of the textbook's of Apostol, Shanti Narayan, S C Malik, Phillips, Rudin, Brown and Churchill.

We consulted with the textbooks of Basudeva and Shirali, M O Searcöid, Jain and Ahmed, M N Mukherjee and J Sarkar to prepare the section on metric spaces.

Our sincere thanks are due to New Central Book Agency (P) Ltd for their excellent cooperation in bringing out this book. Any correction and suggestion towards the betterment of the book from teachers and students will be gratefully received by the authors.

Good Friday
6 April 2012

Ajay Kumar Chaudhuri
Pratikshan Mondal

Part I

REAL ANALYSIS

CHAPTER 1

SETS

Definition [1.1] : A collection of well-defined objects is called a set.

Well-defined objects means that all the objects are not coming to a particular collection.

Example. Collection of boys whose ages are less than 15 years.

Example. Collection of rivers which are flowing in West Bengal.

Example. Collection of numbers which are multiple of 7.

Sets are represented by capital letters $A, X, S, \ldots$.

Objects are called the elements or members of a set. These are represented by small letters $a, b, c, x, y, \ldots$.

Thus we define a set $X = \{x : P(x) \text{ is a statement for } x\}$. That is, X consists of all objects x's for which $P(x)$ is true and no other object.

A set X can be expressed in two forms

(i) Tabular form : Listing all the members of X.

(ii) Set-builder form : Stating the property of the element.

Example. $X = \{0, \pm 2, \pm 4, \pm 6, \ldots\}$, a tabular form.

Example. $X = \{x : x \text{ is an even number}\}$, a set-builder form, where x represents the element of X and the property of x is 'an even number'.

If x is an element or member of a set X, we write it by $x \in X$.

If x is not a member of X, we write it by $x \notin X$.

Problem [1.1] : Express the set $X = \{x : x \text{ is a root of } x^2 - 3x + 2 = 0\}$ in tabular form.

Solution. $x^2 - 3x + 2 = 0 \Rightarrow x = 1, 2$. Hence the tabular form of X is $\{1, 2\}$.

Definition [1.2] : [**Null set or Void set**] A set containing no element is called a null set and is denoted by ϕ.

Example. $X = \{x : x \text{ is a real number satisfying } x^2 + 1 = 0\}$. No such real number exists. So X has no element. Hence $X = \phi$.

Example. $X = \{x : x \text{ is an integer such that } 1 < x < 2\}$, $X = \phi$, as there is no integer in $(1, 2)$.

Definition [1.3] : A set S is said to be a subset of a set X, if $x \in S \Rightarrow x \in X \; \forall \, x \in S$.

We write it by $S \subseteq X$. X is itself a subset of X.

S is called a proper subset of X, if $S \neq X$. That is, there exists at least one element $x \in X$ such that $x \notin S$.

We write it by $S \subset X$.

Example. $X = \{x : x \text{ is an integer such that } 0 \leq x \leq 10\}$.
$S = \{x : x \text{ is an integer such that } 1 \leq x \leq 5\}$.

Then $S \subset X$. That is, S is a proper subset of X.

1.1 Equality of Two Sets

Two sets A and B are said to be equal, if and only if $A \subseteq B$ and $B \subseteq A$.

That is, $A \subseteq B, B \subseteq A \Leftrightarrow A = B$.

Definition [1.4] : **[Universal set]** Every set A is a subset of some fixed set. This fixed set is called the universal set and is denoted by U.

Let $A = \{x : x \text{ is an even integer}\}$
$U = \{x : x \text{ is an integer}\}$.

Then $A \subset U$. So U is called the universal set.

Theorem [1.1] : Null set is a subset of every set A. That is, $\phi \subset A$.

Proof. Since ϕ contains no element, there is no possibility that $x \in \phi \Rightarrow x \notin A$.

Hence $\phi \subset A$.

Alternative. If possible, let ϕ is not a subset of set A. Then $\exists \, x \in \phi$ such that $x \notin A$. But ϕ being a null set $x \notin \phi$. Thus $x \in \phi$ and $x \notin \phi$ arise a contradiction.

Hence $\phi \subset A$.

Definition [1.5] : **[Power set of a set A]** Collection of all subsets of a given set A is called the power set of A and is denoted by $P(A)$.

Thus $P(A) = \{\phi, A, A_i; A_i \subseteq A\}$.

Example. Let $A = \{1, 2, 3\}$.

Then $P(A) = \{\phi, A, \{1\}, \{2\}, \{3\}, \{1,2\}, \{2,3\}, \{1,3\}\}$.

Note. The elements of $P(A)$ are sets.

Theorem [1.2] : If A contains n elements, then number of elements in $P(A)$ is 2^n.

Proof. Let $A = \{a_1, a_2, \cdots, a_n\}$.

Number of subsets formed by single element a_i out of n is ${}^nC_1$.

Number of subsets formed by two elements a_i, a_j out of n is ${}^nC_2$.

....

Number of subsets formed by r elements $a_i, a_{i+1}, \cdots, a_{i+r-1}$ out of n is ${}^nC_r$.

Number of subsets formed by n elements $a_1, a_2, \cdots, a_n$ is ${}^nC_n = 1$.

Besides these, ϕ is a subset of A.

Total number of subsets are ${}^nC_1 + {}^nC_2 + \cdots + {}^nC_r + \cdots + {}^nC_n + 1 = (1+1)^n = 2^n$.

Hence number of elements of $P(A)$ is 2^n.

1.2 Fundamental Operations on Sets

Union (join) and meet (intersection) are two fundamental operations on sets. We get a new set if we apply these operations on any two sets.

Definition [1.6] : **Union** (or join) of two sets A and B, denoted by $A \cup B$, is defined by

$$A \cup B = \{x : x \in A \text{ or } x \in B\}.$$

Meet (or intersection) of any two sets A and B, denoted by $A \cap B$, is defined by

$$A \cap B = \{x : x \in A \text{ and } x \in B\}.$$

Besides these two operations we define an operation 'subtraction' of two sets.

Subtraction of two sets A and B, denoted by $A - B$, is defined by

$$A - B = \{x : x \in A \text{ but } x \notin B\}.$$

Definition [1.7] : **[Complement of a set A]** Let U be a universal set. Then the complement of A, denoted by A^c, is defined by $A^c = \{x : x \in U \text{ but } x \notin A\}$.

Note. $x \in A \Leftrightarrow x \notin A^c$ and $x \in A^c \Leftrightarrow x \notin A$.

1.3 Subtraction of Two Sets in Terms of Complement

Theorem [1.3] : If A and B be any two sets, then $A - B = A \cap B^c$.

Proof. Let $x \in A - B$. Then $x \in A$ but $x \notin B$

$$\Rightarrow x \in A \text{ and } x \in B^c$$
$$\Rightarrow x \in A \cap B^c$$
$$\Rightarrow A - B \subseteq A \cap B^c \qquad \text{(i)}$$

Again, let $y \in A \cap B^c$. Then $y \in A$ and $y \in B^c$

$$\Rightarrow y \in A \text{ and } y \notin B$$
$$\Rightarrow y \in A - B$$
$$\Rightarrow A \cap B^c \subseteq A - B \qquad \text{(ii)}$$

So from (i) and (ii), $A - B = A \cap B^c$.

Definition [1.8] : Symmetric difference of two sets A and B, denoted by $A \triangle B$, is defined by $A \triangle B = (A \cup B) - (A \cap B) = (A - B) \cup (B - A)$.

1.4 Venn Diagram

Sets and subsets can be represented by diagram which is known as Venn diagram. That is, diagrammatic representation of sets, union of sets, intersection of sets, symmetric difference of two sets, is called Venn diagram (Fig. 1.1).

1.5 Fundamental Laws

I. *Commutative law:* (i) $A \cup B = B \cup A$

(ii) $A \cap B = B \cap A$

(iii) $A \triangle B = B \triangle A$

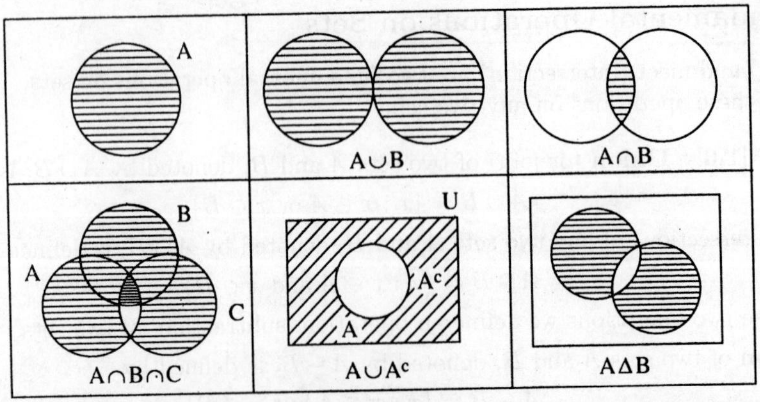

Fig. 1.1

II. *Associative law:* (i) $A \cup (B \cup C) = (A \cup B) \cup C$

(ii) $A \cap (B \cap C) = (A \cap B) \cap C$

(iii) $A \Delta (B \Delta C) = (A \Delta B) \Delta C$

III. *Distributive law:* (i) $A \cup (B \cap C) = (A \cup B) \cap (A \cup C)$

(ii) $A \cap (B \cup C) = (A \cap B) \cup (A \cap C)$

IV. *Complementary law:* (i) $A \cup A^c = U$

(ii) $A \cap A^c = \phi$

V. *Identity law:* (i) $A \cup \phi = A$

(ii) $A \cap U = A$

VI. *Idempotent law:* (i) $A \cup A = A$

(ii) $A \cap A = A$

VII. *De Morgan's law:* (i) $(A \cup B)^c = A^c \cap B^c$

(ii) $(A \cap B)^c = A^c \cup B^c$

VIII. (i) $A \cap \phi = \phi$ follows from the relation $\phi \subset A$.

(ii) $A \cup U = U$ follows from the relation $A \subset U$.

Definition [1.9] : **[Disjoint of two sets]** Two sets A and B are said to be disjoint, if

$$A \cap B = \phi.$$

Definition [1.10] : Let $\mathcal{A} = \{A_1, A_2, \ldots, A_n, \cdots\}$ be an arbitrary collection of sets A_i.

Then the **union** of all sets in $\mathcal{A}$, denoted by $\bigcup\limits_{A_i \in \mathcal{A}} A_i$, is defined by

$$\bigcup_{A_i \in \mathcal{A}} A_i = \{x : x \in A_i \text{ for at least one } A_i \in \mathcal{A}\}$$

It can be also written as

$$\bigcup_{i \in N} A_i = \{x : x \in A_i \text{ for at least one } i \in N\},$$

where N is called an index set.

The **meet** of all sets in $\mathcal{A}$, denoted by $\bigcap_{A_i \in \mathcal{A}} A_i$, is defined by

$$\bigcap_{A_i \in \mathcal{A}} A_i = \{x;\ x \in A_i \text{ for all } A_i \in \mathcal{A}\}$$

$$\text{or, } \bigcap_{i \in N} A_i = \{x;\ x \in A_i \text{ for all } i \in N\}.$$

Remark

1. If $\mathcal{A}$ is finite collection, we write

$$\bigcup_{A_i \in \mathcal{A}} A_i = \bigcup_{i=1}^{n} A_i = A_1 \cup A_2 \cup \cdots \cup A_n$$

$$\bigcap_{A_i \in A} A_i = \bigcap_{i=1}^{n} A_i = A_1 \cap A_2 \cap \cdots \cap A_n$$

2. If A is a countable collection, we write

$$\bigcup_{A_i \in A} A_i = \bigcup_{i=1}^{\infty} A_i = A_1 \cup A_2 \cup \cdots$$

$$\text{and } \bigcap_{A_i \in \mathcal{A}} A_i = \bigcap_{i=1}^{\infty} A_i = A_1 \cap A_2 \cap \cdots$$

Theorem [1.4] : Let $\mathcal{A} = \{A_1, A_2, \dots\}$ be a collection of sets and let B be any set. Then

(i) $B - \bigcup_{A_i \in \mathcal{A}} A_i = \bigcap_{A_i \in \mathcal{A}} (B - A)$

(ii) $B - \bigcap_{A_i \in \mathcal{A}} A_i = \bigcup_{A_i \in \mathcal{A}} (B - A)$

Proof.

(i) Let $x \in B - \bigcup_{A_i \in \mathcal{A}} A_i$. That is, $x \in B - \bigcup_{i \in N} A_i$

$$\Leftrightarrow x \in B \text{ but } x \notin \bigcup_{i \in N} A_i$$

$$\Leftrightarrow x \in B \text{ but } x \notin A_i\ \forall\ i \in N$$

$$\Leftrightarrow x \in B - A_i\ \forall\ i \in N$$

$$\Leftrightarrow x \in \bigcap_{i \in N} (B - A_i)$$

Hence $B - \bigcup_{A_i \in \mathcal{A}} A_i = \bigcap_{A_i \in \mathcal{A}} (B - A_i)$

(ii) Left to you.

Theorem [1.5] : **[De Morgan's laws]** Let $\mathcal{A}$ be a collection of sets A_i, then

(i) $\left[\bigcup_{A_i \in \mathcal{A}} A_i\right]^c = \bigcap_{A_i \in \mathcal{A}} A_i^c$

(ii) $\left[\bigcap_{A_i \in \mathcal{A}} A_i\right]^c = \bigcup_{A_i \in \mathcal{A}} A_i^c$

Proof.

(i) Let $x \in \left[\bigcup_{A_i \in \mathcal{A}} A_i\right]^c$. That is, $x \in \left[\bigcup_{i \in N} A_i\right]^c$

$$\Leftrightarrow x \notin \bigcup_{i \in N} A_i$$

$$\Leftrightarrow x \notin A_i \ \forall\, i \in N$$

$$\Leftrightarrow x \in A_i^c \ \forall\, i \in N$$

$$\Leftrightarrow x \in \bigcap_{i \in N} A_i^c$$

Hence, $\left[\bigcup_{i \in N} A_i\right]^c = \bigcap_{i \in N} A_i^c$. That is, $\left[\bigcup_{A_i \in \mathcal{A}} A_i\right]^c = \bigcap_{A_i \in \mathcal{A}} A_i^c$.

(ii) Left to you.

Definition [1.11] : [Finite and Infinite set] A set X is called finite, if X contains finite number of elements. Otherwise it is called an infinite set.

The number of elements of a finite set X is called the cardinal number of X or cardinality of X.

That is, if X contains m elements, we write card(X) or $n(X) = m$.

Theorem [1.6] :

(i) $n(A \cup B) = n(A) + n(B)$, when $A \cap B = \phi$.

(ii) $n(A \cup B) = n(A) + n(B) - n(A \cap B)$, when $A \cap B \neq \phi$.

Proof.

(i) When $A \cap B = \phi$, it immediately follows $n(A \cup B) = n(A) + n(B)$.

(ii) When $A \cap B \neq \phi$, in this case, $A - A \cap B, A \cap B, B - A \cap B$ are pairwise disjoint.

Also, $A = [A - (A \cap B)] \cup (A \cap B)$

$$\Rightarrow n(A) = n[A - (A \cap B)] + n(A \cap B) \qquad \text{(i)}$$

$$B = [B - (A \cap B)] \cup (A \cap B)$$

$$\Rightarrow n(B) = n[B - (A \cap B)] + n(A \cap B) \qquad \text{(ii)}$$

and $A \cup B = [A - (A \cap B)] \cup (A \cap B) \cup [B - (A \cap B)]$

$$\Rightarrow n(A \cup B) = n[A - (A \cap B)] + n(A \cap B) + n[B - (A \cap B)]$$

$$\Rightarrow n(A \cup B) = n(A) - n(A \cap B) + n(A \cap B) + n(B) - n(A \cap B) \ \text{[by (i) and (ii)]}$$

$$= n(A) + n(B) - n(A \cap B).$$

1.6 Cartesian Product of Two Sets

Definition [1.12] : Cartesian product (or simply product) of two sets A and B, denoted by $A \times B$, is defined by

$$A \times B = \{(x, y), x \in A, y \in B\}.$$

That is, product of two sets is a set containing of all ordered pairs of elements (x, y), where $x \in A$ and $y \in B$.

Example. Let $A = \{-1, 0\}$ and $B = \{1, 2, 3\}$.

Then $A \times B = \{(-1,1),\ (-1,2),\ (-1,3),\ (0,1),\ (0,2),\ (0,3)\}$

$B \times A = \{(1,-1),\ (1,0),\ (2,-1),\ (2,0),\ (3,-1),\ (3,0)\}$.

Remark

1. $A \times B \neq B \times A$
2. $A \times B = B \times A \Leftrightarrow A = B$
3. Product of finite number of sets $A_1, A_2, \dots, A_n$, denoted by $A_1 \times A_2 \times \cdots \times A_n$, is defined by $A_1 \times A_2 \times \cdots \times A_n = \{(x_1, x_2, \dots, x_n\}; x_1 \in A_1, \dots, x_n \in A_n\}$.
4. If $n(A) = m_1, n(B) = m_2$, then $n(A \times B) = m_1 \cdot m_2$.

Example. If $A = \{x;\ x \in R,\ 1 \leq x \leq 2\}$

$B = \{y;\ y \in R,\ 2 \leq y \leq 3\}$.

Here A and B are both infinite.

So $A \times B$ represents a rectangular region bounded by

$\{1 \leq x \leq 2, 2 \leq y \leq 3\}$

Diagrammatic representation is given by (Fig. 1.2).

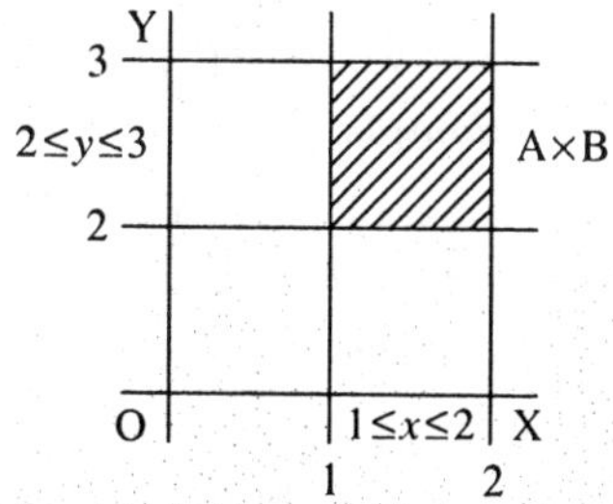

Fig. 1.2

1.7 Mapping (or Function)

Definition [1.13] : Let X and Y be two non-empty sets. If for each element $x \in X$, by some rule or assignment, there exists a unique element $y \in Y$, then this rule or assignment is called a mapping f of X into Y and is denoted by $f : X \to Y$. X is called the domain of f and Y is called the co-domain of f.

y is called the image of x under f and is denoted by $f(x) = y$.

Remark

1. The mapping $f : X \to Y$ is called a subset of $X \times Y$ so that $f = \{(x, y), y = f(x)\}$, where no two of which have the same first element.
   That is, if $(x, y), (x, z) \in f$, then $y = z$.
2. More than one element in X may have the same image in Y.
   That is, $(x_1, y), (x_2, y), \dots, (x_n, y)$ may belong to f.
3. Every element in Y may or may not be an image of an element of X.
   That is, $f(X) \subseteq Y$, where $f(X)$ is a set of all image points. $f(X)$ is called the range of f.
   Thus range of $f = f(X) = \{y, y = f(x)\} \subseteq Y$.

Example. Let $X = \{0, 1, 2, 3\}$ and $Y = \{a, b, c, d, e, f\}$.

1. $f = \{(0, a), (1, a), (2, b), (3, f)\} \subseteq X \times Y$
   Then f is a mapping of X into Y.
2. $f = \{(0, a)(1, b), (1, c), (2, d), (3, f)\} \subseteq X \times Y$
   Then f is not a mapping as the two pairs $(1, b)$ and $(1, c)$ have the same first element.
3. Let $f = \{(0, a), (1, a), (2, a), (3, f)\}$.
   Then range of $f = f(X) = \{a, f\}$, determined by the 2nd element of each pair of elements in f.

1.8 Classification of Mappings

A mapping $f : X \to Y$ is classified by

(i) Injective or one-one

(ii) Surjective or onto

(iii) Bijective or one-one-onto

Definition [1.14] :

(i) A mapping $f : X \to Y$ is called injective if $x_1, x_2 \in X$, $x_1 \neq x_2 \Rightarrow f(x_1) \neq f(x_2)$. In other words; $f(x_1) = f(x_2) \Rightarrow x_1 = x_2$.

Example. $f : R \to R$, defined by $f(x) = 2x + 3$, is an injective mapping.

(ii) A mapping $f : X \to Y$ is called surjective, if $f(X) = Y$. That is, for each $y \in Y$, $\exists\, x \in X$ such that $f(x) = y$.

Example. $f : R \to R$, defined by $f(x) = x^3$, is surjective.

(iii) A mapping $f : X \to Y$ is called bijective, if f is both injective and surjective.

Example. $f : R \to R$, defined by $f(x) = 2x + 3$, is bijective.

Definition [1.15] : **[Constant mapping]** A mapping $f : X \to Y$ is said to be constant if $f(x) = c$, $c \in Y$, $\forall\, x \in X$.

Example. Let $X = \{-1, 0, 1, 2\}, Y = \{a, b, c, d, e, f\}$, $f : X \to Y$.
Then $f = \{(-1, d), (0, d), (1, d), (2, d)\}$ is a constant mapping.

Definition [1.16] : **[Identity mapping]** A mapping $f : X \to X$ is said to be an identity mapping if $f(x) = x\ \forall\, x \in X$ and f is denoted by I_X. That is, $I_X : X \to X$, where $I_X(x) = x\ \forall\, x \in X$ is an identity mapping.

Example. Let $I_N : N \to N$ be a mapping such that $I_N(n) = n\ \forall\, n \in N$.

Then I_N is an identity mapping.

Example. Let $A = \{-1, 0, 1, 2\}$.

$I_A : A \to A$

Then $I_A = \{(-1, -1), (0, 0), (1, 1), (2, 2)\}$ is an identity mapping.

Definition [1.17] : **[Inverse mapping]** If $f : X \to Y$ is bijective, then $f^{-1} : Y \to X$ can be defined by $f^{-1}(y) = x$ such that $f(x) = y$ for each element $x \in X$ and $y \in Y$, and f^{-1} is called the inverse of f.

1.9 Composition of Mappings

Definition [1.18] : Let $f : X \to Y$ and $g : H \to Z$ be two mappings such that $f(X) \subseteq H$, i.e., the range of f is a subset of the domain of g. Let $x \in X$. Then $f(x) \in f(X) \Rightarrow f(x) \in H$. Again under g, $g(f(x)) \in Z$. Thus to $x \in X$, $\exists$ a unique element $g(f(x)) = z \in Z$. So $\exists$ a mapping $h : X \to Z$ so that $h(x) = g(f(x))$. We write $g(f(x)) = (g \circ f)(x)$.

Thus $h(x) = (g \circ f)(x)$. $h = g \circ f$ is called the compositions or product of two mappings f and g. Hence the product mapping $g \circ f : X \to Z$ can be defined by the rule $(g \circ f)(x) = g(f(x))\ \forall\, x \in X$.

Problem [1.2]: Let $f : Z \to Z$, $g : Z \to Z$ be two mappings such that $f(x) = x^2$ and $g(x) = |x|$. Then find $g \circ f$ and $f \circ g$. Here $Z = \{0, \pm1, \pm2, \dots\}$, set of all integers.

Solution. Let $x = 0$.

Then $(g \circ f)(x) = g(f(x)) = g(0) = 0$
$(f \circ g)(x) = f(g(x)) = f(0) = 0$

Let $x > 0$.

Then $(g \circ f)(x) = g(f(x)) = g(x^2) = x^2$ $[\because\ g(x^2) = |x^2| = x^2, x^2 > 0]$
$(f \circ g)(x) = f(g(x)) = f(x) = x^2$

Let $x < 0$.

Then $(g \circ f)(x) = g(f(x)) = g(x^2) = x^2$
$(f \circ g)(x) = g(g(x)) = f(-x) = (-x)^2 = x^2$ $[\because\ g(x) = |x| = -x, x < 0]$

Also, $f \circ g = g \circ f$.

Theorem [1.7]:

(i) If $f : X \to Y$, $I_X : X \to X$, and $I_Y : Y \to Y$, then $f \circ I_X = f$ and $I_Y \circ f = f$.
(ii) If $f : X \to Y$, then $f \circ f^{-1} = I_Y$ and $f^{-1} \circ f = I_X$, where f^{-1} is the inverse of f.
(iii) If f be bijective, then f^{-1} is also bijective.
(vi) Product of two bijective mappings is a bijective mapping.
(v) Commutative law in case of product of two mappings may not hold.
That is, $f \circ g \neq g \circ f\ \forall\ f, g$.
(vi) Associative law holds for product of mappings. That is, if $f : X \to Y$, $g : Y \to Z$, $h : Z \to K$, then $h \circ (g \circ f) = (h \circ g) \circ f$.

[Proofs are left.]

Definition [1.19]: Let $f : X \to Y$ be a mapping and $A \subseteq X$ and $B \subseteq Y$.

Then we define $f(A) = \{y; y = f(x) \text{ for some } x \in A\}$ and $f^{-1}(B) = \{x, f(x) \in B\}$.
Here $f(A) \subseteq Y$ and $f^{-1}(B) \subseteq X$.

Theorem [1.8]: Let $f : X \to Y$ be a mapping and let $A, B \subseteq X$. Then

(i) $f(A \cup B) = f(A) \cup f(B)$.
(ii) $f(A \cap B) \subseteq f(A) \cap f(B)$, but $f(A) \cap f(B) \not\subseteq f(A \cap B)$.
(iii) $f(A \cap B) = f(A) \cap f(B)$, if f is one-one.
(iv) Let $C, D \subseteq Y$. Then $f^{-1}[C \cup D] = f^{-1}(C) \cup f^{-1}(D)$.
(v) $f^{-1}(C \cap D) = f^{-1}(C) \cap f^{-1}(D)$.
(vi) $A \subseteq f^{-1}[f(A)]\ \forall\ A \subseteq X$.
(vii) $A = f^{-1}[f(A)]$, if f is one-one.
(viii) $f[f^{-1}(C) \subseteq C\ \forall\ C \subseteq Y$.
(ix) $f[f^{-1}(C)] = C\ \forall\ C \subseteq Y$, if and only if f is onto.

[Proofs are left.]

EXERCISE

(1) Prove that $A - B = A - (A \cap B) = (A \cup B) - B$.

(2) Show that $(A - B) - C = A - B \cup C = (A - B) \cap (A - C)$.

(3) Prove that

(i) $\{x; |x - 3| < 5\} = \{x; x < 8\} \cap \{x; x > -2\}$

(ii) $\{x; |x| \geq 5\} \cap \{y; |y| \leq 5\} = \{-5, 5\}$.

(4) Let $f : R \to R$ be defined by $f(x) = x^2$.
Find (i) $f^{-1}(16)$, (ii) $f^{-1}(5)$, (iii) $f[8, 13], f^{-1}[y, 16 \leq y \leq 25$.

(5) Let $f : R \to R$ be defined by $f(x) = \log x$. If $A = [1/2, 1], B = [1, 3]$, then find (i) $f(A)$, (ii) $f^{-1}(A)$, (iii) $f^{-1}(B)$, (iv) $f^{-1}(A \cup B)$, (v) $f^{-1}(A \cap B)$.

(6) If $f : A \to R, g : A \to R, A \subseteq R$ and $f \pm g, fg$, where $f \pm g : A \to R$ is defined by

$$f + g = \{x;\ f(x) \pm g(x),\ x \in A\}$$
$$(f \cdot g) = \{x;\ f(x) \cdot g(x),\ x \in A\}$$

Now if $A = \{x, y, z\}$ and $f = \{(x, 1), (y, 2), (z, 3)\}$, $g = \{(x, -1), (y, -2), (z, -3)\}$, then find (i) $f \pm g$, (ii) $4f + 5g$, (iii) $|2f|^2$, (iv) f^3, (v) $f \cdot g$.
[*Note.* $f = \{(x, 1), (y, 2), (z, 3)\} \Rightarrow f(x) = 1, f(y) = 2, f(z) = 3\}$]

(7) Let $f : R \to R$ be defined by $f(x) = \dfrac{2x - 2}{3x + 3}$. Determine whether f is bijective.
Find the range of f. Find f^{-1}, if it exists.

(8) Let $f : X \to Y$ be a mapping, then for $A, B \subseteq Y$, show that
(i) $f^{-1}(A - B) = f^{-1}(A) - f^{-1}(B)$ (ii) $f^{-1}(Y - B) = X - f^{-1}(B)$.

(9) Let $A = \{a, b, c, d\}$ and let $f : A \to A$ satisfy $f(a) = b, f(b) = c, f(c) = d, f(d) = a$. Suppose $g : A \to A$ satisfy $g(a) = c$ and that $f \circ g = g \circ f$. Find $g(b), g(c)$ and $g(d)$. Is it true that $g = f \circ f$?

(10) Let $f : R \to R$, defined by $f(x) = x^2$ and $g : R \to R$ defined by $g(x) = 2x + 3$. Show that $f \circ g \neq g \circ f$. Also find $(f \circ g)\sqrt{2}$.

(11) If A and B are two non-empty sets having m elements in common, then show that $A \times B$ and $B \times A$ have m^2 elements in common.

(12) If A and B be two sets such that $A \times B$ consists of 6 elements. If 3 elements of $A \times B$ be $(1, 3), (2, 5)$ and $(3, 3)$, find the remaining elements of $A \times B$.

CHAPTER 2

NUMBER SYSTEM

2.1 Concepts of Real Numbers

$N = \{1, 2, 3, \dots\}$ denotes the set of natural numbers. N is closed with respect to addition $(+)$ and multiplication $(\cdot)$ only. That is, $a, b \in N \Rightarrow a + b = c \in N$ and $a \cdot b = x \in N$. But N is not closed with respect to subtraction $(-)$ and division $(\div)$, since $1 - 2 = -1 \notin N$ and $1 \div 2 = \frac{1}{2} \notin N$.

To make it closed with respect to subtraction and division we extend N by introducing the set of all integers both $(+)$ve and $(-)$ve and zero. Thus the extended set $Z = \{0, \pm 1, \pm 2, \dots\}$ is closed with respect to $+, -$ and $\cdot$, but not closed with respect to division. So we further extend the set Z by introducing the set of all numbers of the form $\frac{a}{b}, a, b \in Z, b \neq 0$ and then we have a set $Q = \left\{\frac{a}{b}, a, b \in Z, b \neq 0, (a, b) = 1\right\}$, known as the set of all rational numbers. Here $\frac{a}{b}$ is either terminating or recurring decimals. $\frac{4}{5} = 0.8, \frac{2}{3} = 0.666 \cdots$. But there is a number which is neither terminating nor recurring.

$$\sqrt{2} = 1.414 \cdots$$

We call this number as an irrational number.

Thus $I_R = \left\{x \pm a^{1/m};\ \text{where } a \neq t^{rm}, a, t, r, m \in N, x \in Q\right\}$ is a set of irrational numbers.

Finally we have a set $\mathbb{R} = Q \cup I_R$, which is called the set of all real numbers. Here it is to be noted that $N \subset Z \subset Q \subset \mathbb{R}$. Also $I_R \subset \mathbb{R}$. Thus the set of all rational and irrational numbers form a set of real numbers, denoted by $\mathbb{R}$.

Theorem [2.1] : π^2 is an irrational number.

Proof. Let $f(x) = \dfrac{x^n(1-x)^n}{n!}$. Then for $0 < x < 1,\ 0 < f(x) < \dfrac{1}{n!}$.

Since $0 < x < 1 \Rightarrow 0 < 1 - x < 1 \Rightarrow 0 < x(1-x) < 1$

$$\Rightarrow 0 < x^n(1-x)^n < 1 \Rightarrow 0 < \frac{x^n(1-x)^n}{n!} < \frac{1}{n!}, \text{ i.e., } 0 < f(x) < \frac{1}{n!} \tag{i}$$

Now

$$\begin{aligned} f(x) &= \frac{1}{n!}\left[x^n - {}^nc_1 x^{n+1} + {}^nc_2 x^{n+2} - \cdots + (-1)^n x^{2n}\right] \\ &= \frac{1}{n!}\left[x^n - nx^{n+1} + \frac{n(n-1)}{2!}x^{n+2} - \cdots + (-1)^n x^{2n}\right] \end{aligned}$$

Therefore,

$$f'(x) = \frac{1}{n!}\left[nx^{n-1} - n(n+1)x^n + \frac{n(n-1)(n+2)}{2!}x^{n+1} - \cdots + (-1)^n 2nx^{2n-1}\right]$$

$$\Rightarrow f'(0) = 0$$

$$f''(x) = \frac{1}{n!}\left[n(n-1)x^{n-2} - n(n+1)nx^{n-1} + \frac{n(n-1)(n+2)(n+1)}{2!}x^n - \cdots \right.$$

$$\left. +(-1)^n 2n(2n-1)x^{2n-2}\right]$$

$$\Rightarrow f''(0) = 0$$

Differentiating $(n-1)$ times and putting $x = 0$, we get $f^{n-1}(0) = 0$. But

$$f^n(0) = \frac{1}{n!}[n!] = 1$$

as the other terms in $f^n(x)$ contains x,

$$f^{n+1}(0) = \frac{-n\cdot(n+1)!}{n!} = -n(n+1)$$

as the other terms in $f^{n+1}(x)$ contains x and $f^{n+1}(1) =$ integer.

$$f^{n+2}(0) = \frac{(n+2)!}{n!}\cdot\frac{n(n-1)}{2} = \frac{(n-1)n(n+1)(n+2)}{2}$$

as the other terms in $f^{n+2}(x)$ contains x and $f^{n+2}(1) =$ integer.

Proceeding in this way, we have

$$f^{2n}(x) = \frac{2n!}{n!}(-1)^n \ \forall\ x \text{ and } f^{2n+k}(x) = 0\ \forall\ x \text{ and } \forall\ k$$

Thus it follows that $f(0) = 0, f'(0) = 0, \cdots, f^{n-1}(0) = 0$.

$$f^n(0) = 1 = \text{integer and also } f^{n+k}(1) = \text{integer.}$$

Also $f(x) = 0$ for $x = 0$ and $x = 1$.

If possible, let π^2 be rational.

Then $\pi^2 = \frac{a}{b}$, where $a, b \in Z, (a,b) = 1, b \neq 0$.

We now consider the integral

$$F(x) = b^n \pi^{2n+1}\int_0^1 \sin\pi x f(x)dx$$

Integrating by parts, we get

$$F(x) = b^n\pi^{2n+1}\left[\left(\frac{-\cos\pi x}{\pi}f(x)\right)_0^1 + \frac{1}{\pi}\int_0^1 f'(x)\cos\pi x dx\right]$$

$$= b^n\pi^{2n+1}\left[-\frac{\cos\pi x}{\pi}f(x) + \frac{f'(x)\sin\pi x}{\pi^2} + \frac{1}{\pi^3}\cos\pi x f''(x) - \cdots + \frac{\cos\pi x}{\pi^{2n+1}}\cdot f^{2n}(x)\right]_0^1$$

Since the terms after $f^{2n}(x)$ vanishes as $f^{2n+k}(x) = 0\ \forall\ x,\ k > 0$ and the terms containing $\sin\pi x$ are zero for $x = 0$ and $x = 1$

$$\therefore\ F(x) = b^n\pi^{2n+1}\left[\frac{-\cos\pi x}{\pi}f(x) + \frac{1}{\pi^3}\cos\pi x f''(x) - \cdots + (-1)^n\frac{\cos\pi x}{\pi^{2n+1}}f^{2n}(x)\right]_0^1 \qquad \text{(ii)}$$

Again, we have $f(x) = f(1-x)$ for some values of $x, 0 < x < 1$.

$$x = 0 \Rightarrow f^r(0) = f^r(1) \;\forall\; r \geq 0$$

which proves that $f^r(1)$ is an integer $\forall\; r$.

So expression (ii) is numerically of the form

$$\begin{aligned} F(x) &= b^n \pi^{2n+1} \left[\frac{\cos \pi x}{\pi^{2r+1}} f^{2r}(x)\right]_0^1, \; 0 \leq r \leq 2n \\ &= b^n \pi^{(2n+1)-(2r+1)} \left[\cos \pi \cdot f^{2r}(1) - \cos 0 \cdot f^{2r}(0)\right] \\ &= b^n \pi^{2n-2r} \left[-f^{2r}(1) - f^{2r}(0)\right] \\ &= b^n \left(\frac{a}{b}\right)^{n-r} \cdot -2f^{2r}(1) \text{ [since } f^{2r}(0) = f^{2r}(1)] \\ &= -2b^r a^{n-r} f^{2r}(1) = \text{integer.} \end{aligned} \quad \text{(iii)}$$

Thus $F(x)$ is an integer $\forall\; x$. Also $0 < x < 1 \Rightarrow 0 < f(x) < \frac{1}{n!}$ [by (i)]

$$\Rightarrow \; 0 < b^n \pi^{2n+1} f(x) < \frac{b^n \cdot \pi^{2n+1}}{n!}$$

$$\Rightarrow \; 0 < b^n \pi^{2n+1} f(x) \sin \pi x < \frac{b^n \pi^{2n+1}}{n!} \quad \text{[since } 0 < \sin \pi x < 1 \;\forall x]$$

$$\Rightarrow \; 0 < b^n \pi^{2n+1} \int_0^1 f(x) \sin \pi x dx < \frac{b^n \pi^{2n+1}}{n!} \int_0^1 1 \cdot dx = \frac{\pi a^n}{n!} \quad \text{[for } \pi^2 = \tfrac{a}{b}]$$

Now $\frac{\pi a^n}{n!} \to 0$ as $n \to \infty$. So $\frac{\pi a^n}{n!} < 1$.

Therefore, $0 < F(x) < 1$, which arises a contradiction that $F(x)$ is an integer.

Hence π^2 is irrational.

Corollary [2.1] : π is an irrational number.

Proof. If π is rational, then π^2 is rational. Which contradicts that π^2 is irrational (Proved in above theorem). Hence π is irrational.

Theorem [2.2] : Exponential number e is irrational.

Proof. We have

$$e^x = 1 + \frac{x}{1!} + \frac{x^2}{2!} + \frac{x^3}{3!} + \cdots + \frac{x^k}{k!} + \cdots + \frac{x^{2k-1}}{(2k-1)!} + \frac{2^{2k}}{2k!} + \cdots \;\forall\; x \in \mathbb{R}.$$

Then $e^{-1} = 1 - \frac{1}{1!} + \frac{1}{2!} - \frac{1}{3!} + \cdots - \frac{1}{(2k-1)!} + \frac{1}{2k!} \cdots$.

Let $S_n = \sum\limits_{k=0}^{n} (-1)^k \frac{1}{k!}$.

Then

$$S_{2k-1} = 1 - 1 + \frac{1}{2!} - \frac{1}{3!} + \cdots - \frac{1}{(2k-1)!}, \; k \geq 1. \quad \text{(i)}$$

Therefore,

$$\begin{aligned} e^{-1} - S_{2k-1} &= \frac{1}{2k!} - \frac{1}{(2k+1)!} + \frac{1}{(2k+2)!} - \cdots \\ &= \frac{1}{2k!} - \left\{\frac{1}{(2k+1)!} - \frac{1}{(2k+2)!}\right\} - \cdots < \frac{1}{2k!} \end{aligned}$$

and so $0 < e^{-1} - S_{2k-1} < \frac{1}{2k!}$

$$\Rightarrow 0 < (2k-1)!\,(e^{-1} - S_{2k-1}) < \frac{1}{2k} \le \frac{1}{2}\ \forall\ k \ge 1$$

Now $(2k-1)!\ S_{2k-1}$ is an integer $\forall\ k \ge 1$. Since from (i),

when $k = 1$, $1!\ S_1 = 0$

when $k = 2$, $3!\ S_3 = 3!\left(1 - 1 + \frac{1}{2!}\right) =$ integer

when $k = 3$, $5!\ S_5 = 5!\left(1 - \frac{1}{1!} + \frac{1}{2!} - \frac{1}{3!} + \frac{1}{4!}\right) =$ integer

and so on.

Now if e^{-1} is rational, then $(2k-1)!e^{-1}$ would be an integer when k chosen so large and hence $(2k-1)!\,(e^{-1} - S_{2k-1})$ becomes an integer satisfying $0 < (2k-1)!\,(e^{-1} - S_{2k-1}) \le \frac{1}{2}$, which is not possible.

Therefore, e^{-1} is not rational and hence e^{-1} is irrational and consequently e is irrational.

Theorem [2.3]: If a and b are two (+)ve integers such that one of them has a prime factor p, (say) which does not divide the other, then $\log b^a$ is irrational.

Proof. If possible, let $\log b^a$ be rational.

Then $\log b^a = \frac{x}{y}$, where $x, y \in Z, (x, y) = 1, y \neq 0$.

$$\Rightarrow\ b^{\frac{x}{y}} = a \text{ [from definition of logarithm]}$$

$$\Rightarrow\ b^x = a^y \qquad \text{(i)}$$

which is not possible for any $x, y \in Z$.

For, if $p \mid a$ and $p \nmid b$, where p is prime.

Then $a = pt$ for some $t \in Z \Rightarrow a^y = p^y t^y = b^x$, from (i).

That is, $b^x = p^y t^y \Rightarrow b$ has a prime factor p, which contradicts that $p \nmid b$. So $\log b^a$ cannot be expressed in the form of $\frac{x}{y}$ and hence it is irrational.

Theorem [2.4]: If n is a (+)ve integer which is not a square number, then $\sqrt{n}$ is irrational.

Proof. We suppose that $n \neq x^2, x > 1$. If possible, let $\sqrt{n}$ be rational.

Then $\sqrt{n} = \frac{a}{b}$, $a, b \in Z, b \neq 0$ and $(a, b) = 1$

$$\Rightarrow\ n = \frac{a^2}{b^2} \qquad \text{(i)}$$

$$\Rightarrow\ a^2 = b^2 n \Rightarrow a^2 \text{ is multiple of } n$$

$$\Rightarrow\ a \text{ is multiple of } n. \qquad \text{(ii)}$$

Thus $a = \lambda n, \lambda \in Z^+$

$$\Rightarrow\ a^2 = \lambda^2 n^2$$

$$\Rightarrow\ b^2 n = \lambda^2 n^2 \text{ [by (i)]}$$

$$\Rightarrow\ b^2 = \lambda^2 n \Rightarrow b \text{ is multiple of } n \qquad \text{(iii)}$$

Consequently, $(a, b) \neq 1$, by (ii) and (iii). So $\sqrt{n}$ is not rational.

Next, let n have a square factor such that $n = x^2 k, k > 1$, k is prime.

Then $\sqrt{n} = x\sqrt{k}$.

Now if $\sqrt{n}$ is rational, then $\sqrt{k}$ is rational which contradicts the result what we have proved as above.

Hence $\sqrt{n}$ is irrational.

Theorem [2.5] : A number $x(>1) \neq y^n, y \in Z$, then $x^{\frac{1}{n}}$ is irrational.

Proof. If possible, let $x^{\frac{1}{n}}$ be rational.

Then $x^{\frac{1}{n}} = \frac{a}{b}$, $a, b \in Z, b \neq 0, (a, b) = 1$

$$\Rightarrow b^n x = a^n \qquad \text{(i)}$$

If $b = 1$, then $x = a^n$, contradicts the given hypotheses.

If $b \geq 2$, then $\exists$ a prime number p such that $p \mid b$ and hence by (i) $p \mid a^n$.

$p \mid a^n \Rightarrow p \mid a \Rightarrow (a, b) \neq 1$, a contradiction.

Therefore, $x^{\frac{1}{n}}$ is irrational.

Theorem [2.6] : Between any two rational numbers a and b, $\exists$ an infinite number of rational numbers.

Proof. Let a and b be two rational numbers such that $a < b$. Then $a < b \Rightarrow a + b < 2b$

$$\Rightarrow \frac{a+b}{2} < b \qquad \text{(i)}$$

Also $a + a < a + b$

$$\Rightarrow a < \frac{a+b}{2} \qquad \text{(ii)}$$

So by (i) and (ii), $a < \frac{a+b}{2} < b$. Again, $\frac{a+b}{2} + b < 2b$

$$\Rightarrow \frac{a+3b}{4} < b \qquad \text{(iii)}$$

Also $a < b \Rightarrow 3a < 3b \Rightarrow 4a < 3b + a$

$$\Rightarrow a < \frac{a+3b}{4} \qquad \text{(iv)}$$

So by (iii) and (iv), $a < \frac{a+3b}{4} < b$.

By this way we can construct an infinite number of rationals of the form $\frac{a+kb}{1+k}$, k is any (+)ve integer ≥ 1.

Hence the theorem.

Fundamental Properties of Real Numbers [2.7]

Let $\mathbb{R}$ be the set of all real numbers. Then $\langle \mathbb{R}, +, \cdot \rangle$ is an algebraic structure, having a non-empty set $\mathbb{R}$ with two binary compositions '+' and '$\cdot$'.

P(1) *Commutative law*:

(i) $a + b = b + a$,

(ii) $a \cdot b = b \cdot a \; \forall \; a, b \in \mathbb{R}$.

P(2) *Associative law:*

(i) $(a + b) + c = a + (b + c)$,

(ii) $(a \cdot b) \cdot c = a \cdot (b \cdot c) \; \forall \; a, b, c \in \mathbb{R}$.

P(3) *Identity law:*

(i) There exists an element $0 \in \mathbb{R}$ such that $0 + a = a + 0 = a \ \forall \ a \in \mathbb{R}$. Then '0' is called the zero element (additive identity) of $\mathbb{R}$.

(ii) There exists an element $1 \in \mathbb{R}$ such that $1 \cdot a = a \cdot 1 = a \ \forall \ a \in \mathbb{R}$, then 1 is called the unity element of $\mathbb{R}$ (or identity element with respect to '$\cdot$')

P(4) *Inverse law:*

(i) For each element $a \in \mathbb{R}$, $\exists$ an element $-a \in \mathbb{R}$ such that $a + (-a) = (-a) + a = 0 \cdot -a$ is called the additive inverse of a.

(ii) For each $a(\neq 0) \in \mathbb{R}$, $\exists$ an element $\frac{1}{a} \in \mathbb{R}$ such that $a \cdot \frac{1}{a} = 1$. $\frac{1}{a}$ is called the multiplicative inverse of a.

P(5) *Distributive law:*

$a \cdot (b + c) = a \cdot b + a \cdot c \ \forall \ a, b, c \in \mathbb{R}$.

All the above laws hold in $\mathbb{R}$, so $\langle \mathbb{R}, +, \cdot \rangle$ forms a field.

Theorem [2.8] : 0 and 1 exist uniquely in $\mathbb{R}$, followed by

(i) $x, a \in \mathbb{R}$, then $x + a = a \Rightarrow x = 0$.

(ii) $u, b \in \mathbb{R}, b \neq 0$, then $u \cdot b = b \Rightarrow u = 1$.

Proof. (i) $x = x + 0$ [by P(3)]
$= x + \{a + (-a)\}$ [by P(4)]
$= (x + a) + (-a)$
$= a + (-a)$ [by the given condition]
$= 0$ [by P(3)].

Thus $x + a = a \Rightarrow x = 0$. So '0' is unique.

(ii) $u = u \cdot 1$ [by P(3)]
$= u\left(b \cdot \frac{1}{b}\right)$ [by P(4)]
$= (u \cdot b) \cdot \frac{1}{b}$
$= b \cdot \frac{1}{b}$ [by the given condition]
$= 1$ [by P(3)].

Thus $u \cdot b = b \Rightarrow u = 1$. So '1' is unique.

The Order Properties of $\mathbb{R}$ [2.9]

The non-empty subset $\mathbb{R}^+$ of $\mathbb{R}$, called the set of (+)ve real numbers, has the following properties:

(1) $x, y \in \mathbb{R}^+ \Rightarrow x + y \in \mathbb{R}^+$

(2) $x, y \in \mathbb{R}^+ \Rightarrow x \cdot y \in \mathbb{R}^+$

(3) If $x \in \mathbb{R}$, then exactly one of the following holds: $x \in \mathbb{R}^+$, $x = 0$, $-x \in \mathbb{R}^+$, which is known as the **Trichotomy property** (Proof is left).

Definition [2.1] : Let $x, y \in \mathbb{R}$.

(i) If $x - y \in \mathbb{R}^+$, then we write $x > y$ or $y < x$.

(ii) If $x - y \in \mathbb{R}^+ \cup \{0\}$, then we write $x \geq y$ or $y \leq x$.

Theorem [2.10] :

(i) For any $x, y \in \mathbb{R}$, there exists exactly one of the relations $x > y, x = y, x < y$.

(ii) $x < y \Rightarrow x + z < y + z$ for any $z \in \mathbb{R}$

(iii) $x > y, y > z \Rightarrow x > z$ for any $x, y, z \in \mathbb{R}$.

(iv) If $x, a \in \mathbb{R}$, then $x + a = a \Rightarrow x = 0$.

(v) $x \cdot 0 = 0 \ \forall \ x \in \mathbb{R}$.

(vi) If $x \neq 0, y \in \mathbb{R}$, then $x \cdot y = 1 \Rightarrow y = \frac{1}{x}$.

(vii) $x \cdot y = 0 \Rightarrow$ either $x = 0$ or $y = 0$ for $\forall \ x, y \in \mathbb{R}$.

(viii) $x \cdot y > 0 \Rightarrow$ either $x > 0, y > 0$ or $x < 0, y < 0$.

(ix) If $0 \leq x < \epsilon$, then $x = 0$ for every arbitrary small (+)ve number ϵ.

(x) $x \leq y + \epsilon \Rightarrow x \leq y$ for every ϵ and $\forall \ x, y \in \mathbb{R}$.

(xi) If $x \in \mathbb{R}$ and $x \neq 0$, then $x^2 > 0$.

(xii) $1 > 0$.

(xiii) If $n \in N$, then $n > 0$.

Proof.

(i) From the Trichotomy Property [2.9](3), it follows that for $x, y \in \mathbb{R}$, either $x > y$ or $x = y$ or $x < y$.

(ii) $x < y \Rightarrow y - x \in \mathbb{R}^+$ and $(y + z) - (x + z) = y - x \in \mathbb{R}^+ \Rightarrow x + z < y + z$ for any $z \in \mathbb{R}$.

(iii) $x > y, y > z \Rightarrow x - y \in \mathbb{R}^+, y - z \in \mathbb{R}^+$

$\Rightarrow (x - y) + (y - z) \in \mathbb{R}^+$ [by Property [2.9](1)]

$\Rightarrow x - z \in \mathbb{R}^+ \Rightarrow x > z$

(iv) $x = x + 0 = x + \{a + (-a)\} = (x + a) + (-a) = a + (-a) = 0$.

(v) $x \cdot 0 + x = x \cdot 0 + x \cdot 1 [x \in \mathbb{R}, 0 \in \mathbb{R} \Rightarrow x \cdot 0 \in \mathbb{R}]$

$= x \cdot (0 + 1)$ [by Distributive law]

$= x \cdot 1$ [by Identity law] $= x$

$\Rightarrow x \cdot 0 = 0$ [by (iv)].

(vi) Let $x \neq 0$.
Then $y = 1 \cdot y = \left(\frac{1}{x} \cdot x\right) \cdot y = \frac{1}{x}(x \cdot y) = \frac{1}{x}$, since $xy = 1$.

(vii) Let $x \neq 0$.
Then $\frac{1}{x}(x \cdot y) = \left(\frac{1}{x} \cdot x\right) \cdot y = 1 \cdot y = y$
$\Rightarrow \frac{1}{x} \cdot 0 = y$ [since $x \cdot y = 0$]
$\Rightarrow 0 = y$, as $\frac{1}{x} \in \mathbb{R} \Rightarrow \frac{1}{x} \cdot 0 = 0$, i.e., $y = 0$.
So when $y \neq 0, x = 0$.
Hence $x \cdot y = 0 \Rightarrow$ either $x = 0$ or $y = 0$.

(viii) For any real number x, $\exists$ exactly one of the properties $x > 0, x = 0, x < 0$, by law of Trichotomy.
Then $x \cdot y > 0 \Rightarrow x \neq 0, y \neq 0$.

Again, by the law of Trichotomy $x \neq 0 \Rightarrow$ either $x > 0$ or $x < 0$.
Now, if $x > 0$, then $\frac{1}{x} > 0$ and so $y = 1 \cdot y = \left(\frac{1}{x} \cdot x\right) \cdot y = \frac{1}{x}(x \cdot y) > 0$.
Again, if $x < 0$, then $\frac{1}{x} < 0$ and hence $y = 1 \cdot y = \left(\frac{1}{x} \cdot x\right) \cdot y = \frac{1}{x}(x \cdot y) < 0$ [as $x \cdot y > 0$]
Thus $x \cdot y > 0 \Rightarrow x > 0, y > 0$ and $x < 0, y < 0$.
In the same way we can prove that $x \cdot y < 0 \Rightarrow x > 0, y < 0$ and $x < 0, y > 0$.

(ix) Let $x > 0$ and we choose $\epsilon = \frac{x}{n}, n \in Z^+, \epsilon > 0$. Then $0 < \frac{x}{n} < x \; \forall \; n \in Z^+$
$\Rightarrow 0 < \epsilon < x$
$\Rightarrow x \not< \epsilon$ for every $\epsilon > 0$, a contradiction.
Hence $x = 0$.

(x) Let $x > y$ and let $\epsilon = \frac{x-y}{2} (> 0)$.
Then $y + \epsilon = y + \frac{x-y}{2} = \frac{x+y}{2} < \frac{x+x}{2} = x$, by our consumption. So $x > y + \epsilon$, which contradicts that $x \leq y + \epsilon$.
Therefore, $x \leq y + \epsilon \Rightarrow x \leq y$.

(xi) $x \in \mathbb{R}$ and $x \neq 0 \Rightarrow x \in \mathbb{R}^+$ or $-x \in \mathbb{R}^+$ by property of Trichonity.
When $x \in \mathbb{R}^+, x^2 = x \cdot x \in \mathbb{R}^+ \Rightarrow x^2 > 0$ [by (2) of The Order Properties of $\mathbb{R}$ [2.9]].
When $-x \in \mathbb{R}^+, x^2 = (-x) \cdot (-x) \in \mathbb{R}^+ \Rightarrow x^2 > 0$
[by (2) of The Order Properties of $\mathbb{R}$ [2.9]].

(xii) $1 \in \mathbb{R}$ and $1 \neq 0$
Hence by (xi) $1^2 \in \mathbb{R}^+ \Rightarrow 1 = 1^2 > 0$, i.e., $1 > 0$.

(xiii) Let $n = 1$. By above result, $1 > 0$. So $1 \in \mathbb{R}^+$.
So the result is true for $n = 1$.
Let the result be true for $n = k$.
Then $k \in \mathbb{R}^+$. Also $1 \in \mathbb{R}^+$.
So $1 + k \in \mathbb{R}^+$ [by (1) of The Order Properties of $\mathbb{R}$ [2.9]].
$\Rightarrow 1 + k > 0$
Hence it is true $\forall \; n \in N$.
Thus $n \in N \Rightarrow n > 0$.

Problem [2.1]: $a, b \in \mathbb{R}$, then $a^2 + b^2 = 0 \Leftrightarrow a = 0, b = 0$.

Solution. Let $a^2 + b^2 = 0$. If possible, let $a \neq 0, b \neq 0$.
Then $a^2 > 0, b^2 > 0$ and hence $a^2 \in \mathbb{R}^+, b^2 \in \mathbb{R}^+$.
So $a^2 + b^2 \in \mathbb{R}^+$ [by (1) of The Order Properties of $\mathbb{R}$ [2.9]].
$\Rightarrow a^2 + b^2 > 0$, a contradiction.
So $a = 0, b = 0$.
Next, let $a = 0, b = 0$.

$$\left.\begin{aligned} \text{Then } a^2 &= a \cdot a = 0 \cdot 0 = 0 \\ b^2 &= b \cdot b = 0 \cdot 0 = 0 \end{aligned}\right\} \text{[by Prop. (v)]}$$

Therefore, $a^2 + b^2 = a \cdot a + b \cdot b = 0 + 0 = 0$ [by P(3) of (2.7)]

Problem [2.2]: $a, b, c, d \in \mathbb{R}$ satisfying $0 < a < b$ and $c < d < 0$. Give an example that $ac < bd$ and $bd < ac$.

Solution. $a > 0, c < 0$ and $b > 0, d < 0$.

Let $a = \frac{1}{2}, c = -2$ and $b = 4, d = -\frac{1}{2}$ satisfying $a < b, c < d$.

Also $ac = -1, bd = -2$. Therefore, $bd < ac$.

Again, $a = 4, c = -2$ and $b = 8, d = -\frac{1}{2}$, satisfying the conditions.

Then $ac = -8$ and $bd = -4 \Rightarrow ac < bd$.

Theorem [2.11] : Let $f : N \to \mathbb{R}$ be a function defined by $f(n) = a^n$, $a \in \mathbb{R}$. Then the following results hold.

(i) If $a > 1$, then $a^n > 1 + n(a-1)$.

(ii) If $a > 0$, then $a^n \gtrless 1$ and $a^{\frac{1}{n}} \gtrless 1$ according as $a \gtrless 1$.

(iii) If $a > 0$ and p is (+)ve rational, then $a^p \gtrless 1$ according as $a \gtrless 1$.

(iv) If $a > 0$ and $m > n$, then $a^m \gtrless a^n$ according as $a \gtrless 1$.

(v) Assuming the existence of the roots for $c > 1$, then $c^{\frac{1}{m}} < c^{\frac{1}{n}}$, if and only if $m > n$.

Proof.

(i) For if $a > 1$, we have $a = 1 + b, b > 0$.
So $a^n = (1+b)^n > 1 + nb = 1 + n(a-1)$.

(ii) $a > 1 \Rightarrow a^n > 1 \ \forall \ n \in N$. Now if $a < 1$. Then $b = \frac{1}{a} > 1$.

$$\Rightarrow b^n > 1$$
$$\Rightarrow \left(\frac{1}{a}\right)^n > 1$$
$$\Rightarrow a^n < 1.$$

Also $a > 1 \Rightarrow a^{\frac{1}{n}} > 1$.
For if $a^{\frac{1}{n}} < 1$, then $a = \left(a^{\frac{1}{n}}\right)^n < 1$, a contradiction. Similarly, if $a < 1$, then $a^{\frac{1}{n}} < 1$.

(iii) Let $p = \frac{m}{n}, m, n \in N, (m, n) = 1$.
Then $a^p = a^{\frac{m}{n}} = \sqrt[n]{a^m} = b^{\frac{1}{n}}$, where $b = a^m > 0$.
$\gtrless 1$ according as $b \gtrless 1$.
Hence, $a^p \gtrless 1$ according as $a^m \gtrless 1$.
That is, $a^p \gtrless 1$ according as $a \gtrless 1$ [by (ii)].

(iv) $\dfrac{a^m}{a^n} = a^{m-n}, m, n \in N$ and $m > n \Rightarrow m - n \in N$.
Therefore, $a^{m-n} \gtrless 1$ according as $a \gtrless 1$.
That is, $\dfrac{a^m}{a^n} \gtrless 1$ or $a^m \gtrless a^n$ according as $a \gtrless 1$ as $a > 0 \Rightarrow a^n > 0 \ \forall \ n \in N$.

(v) For $c > 1$, $c^{\frac{1}{n} - \frac{1}{m}} = c^{\frac{m-n}{mn}} = c^p > 1$.
when $p = \frac{m-n}{mn}$ is a (+)ve rational [by (iii)].
That is, when $m - n \in N$ and hence $m > n$.
Conversely, let $m > n$.
Then $\frac{m-n}{mn}$ is a (+)ve rational.
Now $c > 1 \Rightarrow c^{\frac{m-n}{mn}} > 1$ [by (iii)].

$$\rightarrow c^{\frac{1}{n}} - c^{\frac{1}{m}} > 1$$
$$\Rightarrow c^{\frac{1}{n}} > c^{\frac{1}{m}}.$$

That is $c^{\frac{1}{m}} < c^{\frac{1}{n}}$.

Absolute Value of a Real Number

Definition [2.2] : Let $a \in \mathbb{R}$. Then the absolute value of a, denoted by $|a|$, is defined by

$$|a| = a, \quad a > 0$$
$$= 0, \quad a = 0$$
$$= -0, \quad a < 0.$$

In other words, $|a| = a, \quad a \geq 0$
$= -a, \quad a < 0.$

Example. $|5| = 5$, $|-6| = -(-6) = 6$ as $a = -6$.

Theorem [2.12] : (Some fundamental results in connection of absolute value of real numbers)

(1) $|a \cdot b| = |a|\ |b|\ \forall\ a, b \in \mathbb{R}$

(2) $|a|^2 = a^2\ \forall\ a \in \mathbb{R}$

(3) $|a| > c \Rightarrow a > c$ or $a < -c$

(4) If $c \geq 0$, then $|a| \leq c \Leftrightarrow -c \leq a \leq c$

(5) $-|a| \leq a \leq |a|\ \forall\ a \in \mathbb{R}$

(6) $|a \pm b| \leq |a| + |b|\ \forall\ a, b \in \mathbb{R}$, known as triangle inequality.

(7) $|a| - |b| \leq |a - b|$

(8) $\left|\frac{a}{b}\right| = \frac{|a|}{|b|}$.

Proof.

1. Let $a > 0, b > 0$. Then $|a| = a, |b| = b$.
Also $a \cdot b > 0$. Therefore, $|a \cdot b| = ab \Rightarrow |a \cdot b| = |a| \cdot |b|$.
Next, let $a > 0, b < 0$, then $|a| = a, |b| = -b$. Also $a \cdot b < 0$.
Therefore, $|a \cdot b| = -(a \cdot b) = a \cdot (-b) = |a||b|$.
The same is true when $a < 0, b > 0$.
Again, let $a < 0, b < 0$. Then $|a| = -a, |b| = -b$. Also $a \cdot b > 0$.
Therefore, $|a \cdot b| = a \cdot b = (-a)(-b) = |a||b|$.

2. $|a|^2 = |a||a| = |a^2| = a^2$ as $a^2 > 0\ \forall\ a \in \mathbb{R}$ [by (1)]

3. $|a| > c \Rightarrow a > c$, when $a > 0$
and $|a| > c \Rightarrow -a > c$, when $a < 0$
$\Rightarrow a < -c$
Therefore, $|a| > c \Rightarrow a > c$ or $a < -c$.

4. $|a| \leq c, c \geq 0$

$$\Rightarrow a \leq c\ (a > 0) \tag{i}$$

and $-a \leq c\ (a < 0)$

$$\Rightarrow a \geq -c \tag{ii}$$

Combining (i) and (ii), we get $-c \leq a \leq c$.
Conversely, $-c \leq a \leq c \Rightarrow a \leq c$ and $-a \leq c \Rightarrow |a| \leq c$.

5. $|a| \geq 0$
   Replacing c by $|a|$ in the above result, we get $-|a| \leq a \leq |a|$.
6. For any real number $a, b, |a| \geq 0 |b| \geq 0$
   $\Rightarrow -|a| \leq a \leq |a|$
   and $-|b| \leq b \leq |b|$ [by (5)].
   Then $-\{|a| + |b|\} \leq a + b \leq |a| + |b|$
   and hence $|a + b| \leq |a| + |b|$ $[-c \leq a \leq c, c \geq 0 \Rightarrow |a| \leq c$ by (4)].
   Now replacing b by $-b$, we get $|a - b| \leq |a| + |-b| = |a| + |b|$.
7. $|a| = |(a - b) + b| \leq |a - b| + |b|$ [by (6)]
   $\Rightarrow |a| - |b| \leq |a - b|$
   Thus
   $$|a - b| \geq |a| - |b| \tag{iii}$$
   Again, $|a| = |(a + b) + (-b)| \leq |a + b| + |-b| = |a + b| + |b|$
   $\Rightarrow |a| - |b| \leq |a + b|$,
   $$\text{i.e., } |a + b| \geq |a| - |b| \tag{iv}$$
   So from (iii) and (iv), $|a \pm b| \geq |a| - |b|$.
8. Let $\frac{1}{b} = x$. Then $x \in \mathbb{R}$ $(b \neq 0)$.
   Now $|x||b| = |xb| = \left|\frac{1}{b} \cdot b\right| = |1| = 1$
   $$\Rightarrow |x| = \frac{1}{|b|} \tag{v}$$
   Again, $\left|\frac{a}{b}\right| = |ax| = |a||x| = |a|\frac{1}{|b|} = \frac{|a|}{|b|}$.

Corollary [2.2] : If $a_r \in \mathbb{R}\ \forall\ r = 1, 2, \ldots, n$, then $\left|\sum\limits_{r=1}^{n} a_r\right| \leq \sum\limits_{r=1}^{n} |a_r|$.

That is, $|a_1 + a_2 + \cdots + a_n| \leq |a_1| + |a_2| + \cdots + |a_n|$ (Proof is left).

Problem [2.3] : Find the set A of $x \in \mathbb{R}$ such that $|3x - 2| \leq 7$.

Solution. $|3x - 2| \leq 7 \Rightarrow -7 \leq 3x - 2 \leq 7$
$\Rightarrow -\frac{5}{3} \leq x \leq 3$

$A = \left\{x, x \in \mathbb{R} \text{ and } -\frac{5}{3} \leq x \leq 3\right\}$.

Problem [2.4] : Determine the set B of $x \in \mathbb{R}$ such that $|x - 1| < |x|$.

Solution. To remove the absolute value symbol, we consider the cases
(i) $x \geq 1$, (ii) $0 \leq x < 1$, (iii) $x < 0$.

1. For, $x \geq 1, (x - 1) < x \Rightarrow 0 < 1$, which is true.
   So all $x \geq 1 \in B$, without further restriction. Let the set be B_1, where $B_1 = \{x, x \geq 1\}$.
2. For, $0 \leq x < 1$, $-(x - 1) < x \Rightarrow 2x > 1 \Rightarrow x > \frac{1}{2}$.
   So those $x \in B$, when $\frac{1}{2} < x < 1$.
   Let this set be B_2, where $B_2 = \left\{x, \frac{1}{2} < x < 1\right\}$.
3. For, $x < 0$, $-(x - 1) < -x \Rightarrow 1 < 0$, which is absurd.
   So no $x < 0$ belongs to B. Then $B_3 = \{x, x < 0\} = \phi$.
   Therefore, $B = B_1 \cup B_2 \cup B_3 = \left\{x, x > \frac{1}{2}\right\}$.

Geometrical Representation of Real Numbers on a Line [2.13]

Let $\mathbb{R} = \{x, x \text{ is any real number}\}$.

and $L = \{P, P \text{ is a point on a line } l \text{ of infinite length}\}$.

Also let $f : L \to \mathbb{R}$, a mapping defined by $f(P) = x$ so that f is bijective.

Then $f^{-1} : \mathbb{R} \to L$ is also bijective.

Thus there exists a one-one correspondence between the points on a line and the real numbers. That is, for every point $P \in L$, $\exists$ a real number x such that $f(P) = x$ and for every real number x $\exists$ a point P in L such that $f^{-1}(x) = P$.

It follows from the above notion that the geometrical representation of the real numbers is the real line.

In this context, if $x \in \mathbb{R}$, then the absolute value $|x|$ of x can be regarded as the distance of x from the origin O, where number zero (0) is taken as the origin on the real line $X'OX$ of infinite length.

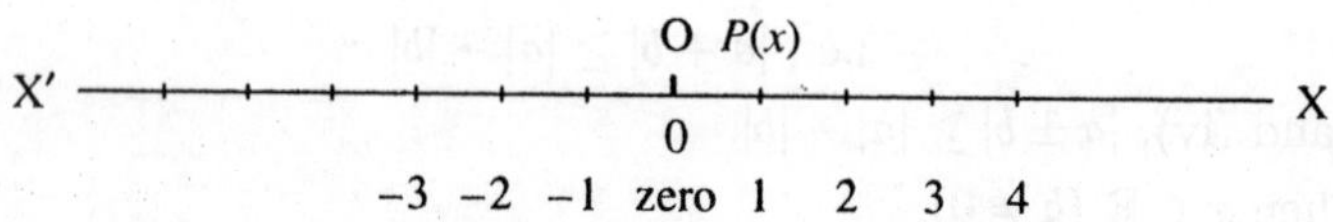

Fig. 2.1

So if P be a point representing the real number x, then $|x| = OP$ = distance of P from O.

Thus the distance of any two points P_1 and P_2 representing the real numbers a and b respectively, is $|a - b| = P_1P_2$.

The distance between $a = -2$ and $b = 4$ is

$$|a - b| = |-2 - 4| = 6 \text{ units.}$$

We represent the (+)ve integers by the points on the right of O (origin) with equal distance of length 1 unit and the equidistant points on the left of O respresent the (−)ve integers. Now to represent a (+)ve rational number $\frac{m}{n}$ $[m, n > 0, n \neq 0, m < n, (m, n) = 1]$ we move to the right of O and we divide the length OI (length of unity), I represents the real number 1 into n equal parts and taking m of these parts we get the point which will represent the rational number $\frac{m}{n}$.

To represent the negative rational $\frac{m}{n}$ [one of $m, n < 0$] we move to the left of O and follow the same rule as defined above.

We now consider the rational $\frac{m}{n}$, where $m > n$. Then $\frac{m}{n} = r + \frac{m_1}{n}$, where r is the integral part and $\frac{m_1}{n}$ is the rational part, where $m_1 < n$.

In this case we first represent the integral part r to the right (or left) of O according as r is (+)ve or (−)ve integer subject to the condition $m > 0$ or < 0, then we divide the next unit length [from r to $r + 1$] into n equal parts and taking m_1 of these parts we can finally represent the given rational. Thus we can represent every rational by a point on the line and conversely every point represents a rational. But still there remains points on the line which cannot be represented by a rational number. If we cut off a length OQ on the line from O so that Q is a point on the line and OQ is the diagonal of a square whose each side of length 1, then $OQ = \sqrt{2}$ and hence Q, a point on the line, represents an irrational number. By this way we have so many irrational numbers which are associated with a point on the line $X'OX$.

Hence we conclude that the geometrical interpretation of real numbers is the real line. We have introduced the concept of the distance of a real number x from O (origin) or the distances between the two points a and b by using the absolute value $|x|$ and $|a-b|$.

Now we need to discuss the notion that when a real number x is 'close to' or 'nearer to' another real number a.

Definition [2.3] : A real number x is said to be 'close to' or 'nearer to' another real number a if for every arbitrary small (+)ve number ϵ, $|x-a| < \epsilon$.

Neighbourhood of a Point

Definition [2.4] : Let $a \in \mathbb{R}$ and ϵ be an arbitrary small (+)ve number. Then the neighbourhood of a, denoted by $N_\epsilon(a)$, is defined by $N_\epsilon(a) = \{x, |x-a| < \epsilon\}$. We call it as ϵ-neighbourhood of a.

δ-neighbourhood of $a = N_\delta(a) = \{x, |x-a| < \delta\}$, where $\delta > 0$ and small.

Thus from definition, $N_\epsilon(a) = \{x, a-\epsilon < x < a+\epsilon\}$, so ϵ-neighbourhood of a is an open interval containing the point 'a'.

Theorem [2.13] : If $x \in N_\epsilon(a)$ for every $\epsilon > 0$, then $x = a$.

Proof.
$$\begin{aligned} x \in N_\epsilon(a) &\Rightarrow |x-a| < \epsilon \\ &\Rightarrow 0 < |x-a| < \epsilon \\ &\Rightarrow x - a = 0 \quad \text{[by Th. [2.10] (ix)]} \\ &\Rightarrow x = a \end{aligned}$$

Theorem [2.14] : $|x|$ is maximum of $(x, -x)$ and $|x| = |-x|$.

Proof. We have
$$\begin{aligned} |x| &= x, \text{ when } x \geq 0 \\ &= -x, \text{ when } x < 0 \end{aligned}$$

So
$$\begin{aligned} |x| &= x \geq -x \ \forall\, x \geq 0 \\ &= -x > x \ \forall\, x < 0 \end{aligned}$$

Thus $|x|$ is greater of two numbers x and $-x$.

That is, $|x| = \max(x, -x)$.

Also
$$\begin{aligned} |-x| &= \max(-x, -(-x)) \\ &= \max(-x, x) \\ &= \max(x, -x) = |x| \end{aligned}$$

Problem [2.5] : Show that, for any two real numbers a and b,

$$\max(a,b) = \frac{(a+b) + |a-b|}{2}$$

$$\min(a,b) = \frac{(a+b) - |a-b|}{2}$$

Solution.
$$\begin{aligned} |a-b| &= a-b \geq -(a-b), \ \forall\, a-b \geq 0 \\ &= -(a-b) > (a-b), \ \forall\, (a-b) < 0 \end{aligned}$$

$\Rightarrow |a-b|$ is greater of the two numbers $(a-b)$ and $-(a-b)$.

$\Rightarrow (a+b)+|a-b|$ is greater of two numbers $(a+b)+(a-b)$ and $(a+b)+(-a+b)$.

That is $(a+b)+|a-b|$ is greater of $2a$ and $2b$ or $\frac{(a+b)+|a-b|}{2}$ is greater of a and b.

Hence $\max(a,b)=\frac{1}{2}[(a+b)+|a-b|]$.

Again $\quad -|a-b|=-(a-b)<a-b$, when $a>b$

and $-|a-b|=(a-b)<-(a-b)$, when $a<b$.

Thus $-|a-b|$ is smaller than $(a-b)$ and $-(a-b)$.

$\Rightarrow (a+b)-|a-b|$ is smaller than $2a$ and $2b$.

$\Rightarrow \frac{1}{2}[(a+b)-|a-b|]$ is smaller than a and b.

Hence $\min(a,b)=\frac{1}{2}[(a-b)-|a-b|]$.

Corollary [2.3]: Show that if $a,b,c\in\mathbb{R}$, then $\min(a,b,c)=\min\{\min(a,b),c\}$.

Solution. Let $\min(a,b)=x$. Then

$$a\geq x,\ b\geq x. \tag{1}$$

Again, let $\min\{x,c\}=y$. Then

$$x\geq y, c\geq y \tag{2}$$

Hence $a\geq y, b\geq y, c\geq y$ by (1) and (2) which shows that $\min(a,b,c)=y$

$$=\min\{x,c\}=\min\{\min(a,b),c\}.$$

Problem [2.6]: Show that, if $a,b\in\mathbb{R}$, then $|a+b|=|a|+|b|$, if and only if $ab\geq 0$.

Solution. Let $a\cdot b>0$. Then

$$|a\cdot b|=a\cdot b \tag{1}$$

Now $\quad |a+b|^2=(a+b)^2=a^2+b^2+2a\cdot b=|a|^2+|b|^2+2|a\cdot b| \quad$ [by (1)]

$$=|a^2|+|b|^2+2|a||b|=[|a|+|b|]^2$$

$\Rightarrow |a+b|=|a|+|b|$

If $a\cdot b=0$, then either $a=0$ or $b=0$. Then the result is trivally true.

Again $\quad |a+b|=|a|+|b|$

$\Rightarrow |a||a+b|=|a|[|a|+|b|]$

$\Rightarrow |a^2+ab|=|a|^2+|ab|$

$\Rightarrow |a\cdot b|=|a^2+ab|-|a|^2=|a^2+ab|-a^2$

$\Rightarrow |a\cdot b|=ab$ if $a^2+ab>0$

$\Rightarrow a\cdot b>0$ and in that case $a^2+ab>0$ is true.

Also $|a\cdot b|=|a^2+ab|-a^2=0$, when $a\cdot b=0$.

Problem [2.7]: (i) Let $U=\{x, 0<x<1\}$ and $a\in U$ and ϵ is smaller of a and $1-a$. Then show that $V_\epsilon(a)\subset U$.

Solution. (i) Let $a = \frac{1}{m}$, $m > 1$. Then $a \in U$.

Also let smaller of $(a, 1-a) = p = \frac{1}{n}$ (say).

Where $p < a$ and $p < 1 - a$.

So $p < a \Rightarrow n > m$ and $p < 1 - a \Rightarrow a + p < 1$.

$$\begin{aligned} \text{Then } V_\epsilon(a) = V_p(a) &= \{x; a - p < x < a + p\} \\ &= \{x, \tfrac{1}{m} - \tfrac{1}{n} < x < 1\} \\ &= \{x, \tfrac{n-m}{mn} < x < 1\} \\ &= \{x, 0 < \tfrac{n-m}{mn} < x < 1\} \text{ as } n > m \subset U \end{aligned}$$

In particular, let $a = \frac{2}{5}$. Then $a \in U$.

Let $\epsilon = 0.33 = \frac{1}{3}$, then ϵ is smaller of

$$(a, 1-a) = \left(\frac{2}{5}, \frac{3}{5}\right) = (0.4, 0.6)$$

$$V_\epsilon(a) = \left\{x;\ \frac{2}{5} - \frac{1}{3} < x < \frac{2}{5} + \frac{1}{3}\right\} = \left\{x;\ \frac{1}{15} < x < \frac{11}{15}\right\} \subset U$$

(ii) If $U = \{x, 0 \le x \le 1\}$. Then show that $V_\epsilon(0) \not\subset U$.

Solution. $0 \in U$, $\quad V_\epsilon(0) = \{x;\ -\epsilon < x < \epsilon\}$

Then the number $x_\epsilon = -\frac{\epsilon}{2} \in V_\epsilon(0)$ but not in U.

So $V_\epsilon(0) \not\subset U$.

(iii) If x, y belong to the ϵ-neighbourhood of a and b, then $x + y$ belongs to the 2ϵ-neighbourhood of $a + b$ but not necessarily to the ϵ-neighbourhood of $a + b$.

Solution. Given that $x \in N_\epsilon(a)$ and $y \in N_\epsilon(b)$.

Then $|x - a| < \epsilon$ and $|y - b| < \epsilon$.

$$\begin{aligned} \text{Now } |(x+y) - (a+b)| &= |(x-a) + (y-b)| \\ &\le |x-a| + |y-b| < \epsilon + \epsilon = 2\epsilon \end{aligned}$$

Hence, $x + y \in N_{2\epsilon}(a + b)$.

We consider the following example for the necessity part.

Let $\epsilon = \frac{1}{10} = 0.1$, $2\epsilon = \frac{1}{5} = 0.2$ and let $a = \frac{1}{2}, b = 1$.

$$\left.\begin{aligned} \text{Then } \tfrac{1}{2} - 0.1 < x < \tfrac{1}{2} + 0.1 \Rightarrow 0.4 < x < 0.6 \\ 1 - 0.1 < y < 1 + 0.1 \Rightarrow 0.9 < y < 1.1 \end{aligned}\right\} \Rightarrow 1.3 < x + y < 1.7 \qquad (1)$$

Considering 2ϵ neighbourhood of $a + b$, we get $1.3 < x + y < 1.7$. (2)

Considering ϵ neighbourhood of $a + b$, we get $1.4 < x + y < 1.5$. (3)

From (1) and (2) it follows that $x + y \in N_{2\epsilon}(a + b)$. But from (1) and (3) it follows that

$$x + y \notin N_\epsilon(a + b)$$

Problem [2.8]: Let $\epsilon > 0$ and $\delta > 0$ and $a \in \mathbb{R}$. Then show that (i) $N_\epsilon(a) \cap N_\delta(a)$ and (ii) $N_\epsilon(a) \cup N_\delta(a)$ are γ-neighbourhood of a for appropriate value of γ.

Solution. (i) $N_\epsilon(a)$ is a neighbourhood of $a \Rightarrow a \in (a - \epsilon, a + \epsilon) \Rightarrow\ \exists\ \epsilon_1 (> 0)$ such that

$$a \in (a - \epsilon_1, a + \epsilon_1) \subseteq N_\epsilon(a).$$

Again, $N_\delta(a)$ is a neighbourhood of $a \Rightarrow \exists\ \delta_1(>0)$ such that

$$a \in (a-\delta_1, a+\delta_1) \subseteq N_\delta(a).$$

Let $\gamma = \min(\epsilon_1, \delta_1),\ \gamma > 0$.

Then $a \in (a-\gamma, a+\gamma) \subseteq (a-\epsilon_1, a+\epsilon_1) \subseteq N_\epsilon(a)$

and $a \in (a-\gamma, a+\gamma) \subseteq (a-\delta_1, a+\delta_1) \subseteq N_\delta(a)$

These together implies that $a \in (a-\gamma, a+\gamma) \subseteq N_\epsilon(a) \cap N_\delta(a)$, which shows that $N_\epsilon(a) \cap N_\delta(a)$ is a γ-neighbourhood of a.

(ii) It can be proved by taking $\gamma = \max(\epsilon_1, \delta_1)$.

EXERCISE

(1) Let $a(\neq 0), b, c \in \mathbb{R}$. Then show that the equation $ax + b = c$ determines a unique solution.

(2) Show that between any two distinct real numbers, there is always an irrational number and hence infinitely many irrational numbers.

(3) Show that the sum of two (+)ve real numbers is a (+)ve real number.

(4) Prove that between any two real numbers there lies at least one real number.

(5) If x is small, show that

(i) $\dfrac{1+ax}{1+bx}$ is nearby equal to $1+(a-b)x-b(a-b)x^2$,

(ii) if $a = 5, b = 2$ and $x < 0.1$, then error in this approximation is less than 0.01.

(6) Every open interval contains a rational number.

(7) Show that $\sqrt{8}$ is not a rational number.

(8) Show that there is no rational number whose square is 7.

(9) Prove that $\sqrt{2}+\sqrt{3}$ is irrational.

(10) If a, b, c, d are rational and x is irrational, show that $\dfrac{ax+b}{cx+d}$ is usually irrational, when does it become a rational?

(11) Let a and b be two (+)ve integers. Prove that $\sqrt{2}$ always lies between $\dfrac{a}{b}$ and $\dfrac{a+2b}{a+b}$. Which fraction is closer to $\sqrt{2}$?

(12) Prove that $\sqrt{n-1}+\sqrt{n+1}$ is irrational for ever integer $n \geq 1$.

(13) Show that the set $\mathbb{R}$ of real numbers is not countable.

(14) Show that the set Q of all rational numbers is countable.

(15) Find the rational number whose decimal expansion is $0.33444\cdots$.

(16) Prove that no rational number can have its cube equal to 3.

(17) If a (+)ve rational number m be close to $\sqrt{2}$, then show that $(4+3m)/(3+2m)$ is more closer to $\sqrt{2}$.

CHAPTER 3

BOUNDS OF A SET

Definition [3.1] : Let $S \subseteq \mathbb{R}$ be a given set.

If there exists a finite number $M \in \mathbb{R}$ such that $x \leq M\ \forall\ x \in S$, then we say that S is bounded above and M is called an **upper bound of S**.

If such M does not exist, then S is unbounded above.

Again any real number $M_1 > M$ is also an upper bound of S. So there exists an aggregate of upper bounds and let $A = \{M, M_1, M_2, \dots\}$, where $M < M_1 < M_2 \cdots$. Then the least of all upper bounds of S is called the least upper bound (or exact upper bound) of S. Here M is the **least upper bound of S**. We write

$$M = \text{l.u.b. of } S \text{ or } \textbf{supremum of } \boldsymbol{S}, \text{ i.e., } M = \sup S$$

If there exists a finite number $m \in \mathbb{R}$ such that $x \geq m\ \forall\ x \in S$, then we say S is bounded below and m is called a **lower bound of S**.

If such m does not exist, then S is unbounded below.

Any real number $m_1 < m$ is also a lower bound of S. Thus we have an aggregate of lower bounds $m, m_1, m_2, \dots$, where $m > m_1 > m_2 \cdots$.

Then the greatest of all lower bounds is called the **greatest lower bound** (g.l.b.). Here $m =$ g.l.b. or m is called an **infimum of S**. That is, $m = \inf S$.

Now S is called bounded, if it is both bounded above and bounded below. That is, if there are two finite numbers $M, m \in \mathbb{R}$ such that $m \leq x \leq M\ \forall\ x \in S$.

It is to be noted that M, m may or may not be a member of S.

Example. $S = \left\{1, \frac{1}{2}, \frac{1}{3}, \dots\right\}$.

Here $x \leq 1\ \forall\ x \in S$.

1 is an upper bound of S. So S is bounded above. Also 1 is the l.u.b. (or $\sup S$).

Here $1 \in S$. Again, $x \geq 0\ \forall\ x \in S$. So 0 is the lower bound of S. S is bounded below.

Also $0 =$ g.l.b. of S (or $\inf S$). Here $0 \notin S$. Thus S is bounded as $0 < x \leq 1\ \forall\ x \in S$.

Example. $S = \left\{-1, -\frac{1}{2}, -\frac{1}{3}, -\frac{1}{4}, \dots\right\}$

Here $x < 0\ \forall\ x \in S$. So '0' is an upper bound of S. Also $0 = \sup S$ but $0 \notin S$.

Again, $x \geq -1\ \forall\ x \in S$. So -1 is the lower bound of S.

Also $-1 = \inf S$ and $-1 \in S$. Here S is bounded.

Example. $N = \{1, 2, 3, \dots\}$, set of natural N is bounded below as $x \geq 1 \ \forall \ x \in S$ but not bounded above as there does not exist any real number M such that $x \leq M \ \forall \ x \in N$.

So S is unbounded above. In other words, S is not bounded.

Example. $Z = \{0, \pm 1, \pm 2, \dots\}$

$$Q = \{\tfrac{m}{n} : m, n \in Z, n \neq 0, (m, n) = 1\}$$

and $\mathbb{R} = \{x : x \text{ is any real number}\} = Q \cup I_{\mathbb{R}}$.

$Z, Q, \mathbb{R}$ has neither an upper bound M nor a lower bound m. Each of the set is unbounded.

Theorem [3.1]: [Properties of sup S] If M be the sup S, then it satisfies the following properties:

(i) $x \leq M \ \forall \ x \in S$

(ii) If M_r be an upper bound of S, then $M \leq M_r \ \forall \ r = 1, 2, 3, \dots$.

(iii) There exists an element $y \in S$ such that $y > M - \epsilon$ for every ϵ (> 0). Then $M - \epsilon$ is not an upper bound of S when ϵ is chosen however small.

(iv) M may or may not be a member of S.

(v) M is unique.

Proof. (iii) For, if $M - \epsilon$ is an upper bound of S, then it contradicts that M is the least upper bound of S as $M - \epsilon < M$ for every ϵ (> 0).

(v) If possible, let M and M' be the sup S. Then M, M' are both upper bounds of S. Then M being a l.u.b. of S and M' being an upper bound of S

$$M \leq M' \tag{i}$$

Again, M' being a l.u.b. of S and M being an upper bound of S,

$$M' \leq M \tag{ii}$$

Therefore, from (i) and (ii) it follows that $M = M'$.

Theorem [3.2]: [Properties of inf S] Let $m = \inf S$. Then

(i) $x \geq m \ \forall \ x \in S$.

(ii) If m_r be a lower bound of S, then $m \geq m_r \ \forall \ r = 1, 2, 3, \dots$.

(iii) For every ϵ (> 0), there exists $y \in S$ such that $y < m + \epsilon$. So $m + \epsilon$ cannot be a lower bound of S when ϵ is chosen however small.

(iv) m is unique.

Theorem [3.3]: [Archimedean property] If $x > 0$ and if $y \in \mathbb{R}$, then there exists a (+)ve integer n such that $nx > y$.

Proof. If possible, let $nx \leq y$, where $x, y \in \mathbb{R}$, $x > 0$ and $n \in Z^+$.

Let $S = \{nx : n \in Z^+\} \subseteq \mathbb{R}$.

S is bounded above as $nx \leq y$ for some $y \in \mathbb{R}$.

Hence S has a supremum, say M.

Then $nx \le M \ \forall\ n \in Z^+ \Rightarrow (n+1)x \le M \ \forall\ n \in Z^+$

$\Rightarrow nx \le M - x \ \forall\ n \in Z^+$

$\Rightarrow M - x$ is an upper bound of S.

$\Rightarrow M$ cannot be the $\sup S$ as $M - x < M, x > 0$.

Hence it arises a contradiction.

Therefore, $\exists$ a (+)ve integer n such that $nx > y$.

Corollary [3.1] : For any $x \in \mathbb{R}$, $\exists$ an $n \in Z^+$ such that $n > x$.

[*Hints.* Z^+ is unbounded above.]

Theorem [3.4] : The following statements are equivalent in a set $S \subseteq \mathbb{R}$, having a

(i) $\sup S = u$.

(ii) If $u \le v$, then v is an upper bound of S.

(iii) If $z < u, z \in S$, then z is not an upper bound of S.

(iv) If $z < u$, then $\exists\ s_z \in S$ such that $s_z > z$.

(v) For every $\epsilon\ (> 0)$, $\exists\ s_\epsilon \in S$ such that $u - \epsilon < s_\epsilon$.

Proof. Let $\sup S = u$.

Then $\sup S = u \Rightarrow u$ is an upper bound of S

$\Rightarrow x \le u \ \forall\ x \in S$

$\Rightarrow x \le u \le v$ as given in (ii)

$\Rightarrow x \le v \Rightarrow v$ is an upper bound of S.

Thus (i) $\Rightarrow$ (ii).

Next, let (ii) hold.

Then if z be an upper bound of S, $u \le z$, which contradicts the given condition in (iii).

Hence z is not an upper bound of S.

Thus (ii) $\Rightarrow$ (iii).

Now by (iii), $z < u \Rightarrow z$ is not an upper bound of $S \Rightarrow \exists\ s_z \in S$ such that $s_z > z$. That is, z cannot be an upper bound of S.

So (iii) $\Rightarrow$ (iv).

Again, $\epsilon > 0 \Rightarrow u - \epsilon < u$

$\Rightarrow u - \epsilon$ is not an upper bound of S, by (iv).

$\Rightarrow \exists\ s_\epsilon(\epsilon > 0) \in S$ such that $s_\epsilon > u - \epsilon$.

That is, $u - \epsilon < s_\epsilon$.

So (iv) $\Rightarrow$ (v).

Next let u is an upper bound of S and let $v < u$. We choose $\epsilon = u - v (> 0)$.

Then $v = u - \epsilon < s_\epsilon$, by (v).

$\Rightarrow v$ is not an upper bound of S.

$\Rightarrow u$ is the least upper bound of S.

That is, $\sup S = u$.

Problem [3.1]: Let $S \subseteq \mathbb{R}$ and let $S + a = \{x + a, x \in S, a \in \mathbb{R}\}$. Then show that $\sup(S + a) = \sup S + a$.

Solution. Let $\sup S = M$.

Then (i) $x \le M \ \forall \ x \in S$ and (ii) for every $\epsilon \ (> 0)$, $\exists \ s_\epsilon \in S$ such that $s_\epsilon > M - \epsilon$.

From (i)

$$x + a \le M + a \ \forall \ a \in \mathbb{R} \tag{1}$$

and from (ii) $\exists$ an $s_\epsilon + a \in S + a$ such that

$$s_\epsilon + a > M + a - \epsilon \tag{2}$$

Thus from (1) and (2) it follows that $M + a = \sup(S + a)$.

That is, $\sup S + a = \sup(S + a)$.

Theorem [3.5]: A lower bound ω of a non-empty set $S \subseteq \mathbb{R}$ is $\inf S$ if and only if, for every $\epsilon \ (> 0)$, there exists an element s_ϵ of S such that $s_\epsilon < \omega + \epsilon$.

Proof. Let ω be a lower bound of S and let $s_\epsilon \in S$ satisfies the condition that $s_\epsilon < \omega + \epsilon$.

Let $v > \omega$, then by taking $\epsilon = v - \omega (> 0)$, we have $v = \omega + \epsilon > s_\epsilon$.

$\Rightarrow$ v cannot be a lower bound of S.

$\Rightarrow$ ω is the greatest lower bound of S.

Hence $\inf S = \omega$.

Conversely, let $\inf S = \omega$.

Now for every $\epsilon \ (> 0)$, $\omega + \epsilon > \omega$, and hence $\omega + \epsilon$ cannot be a lower bound of S as ω is the greatest lower bound of S. So $\exists$ an $s_\epsilon \in S$ such that $\omega + \epsilon > s_\epsilon$.

That is, $s_\epsilon < \omega + \epsilon$.

Problem [3.2]: Let A and B be two non-empty subsets of $\mathbb{R}$ such that $a \le b \ \forall \ a, b$, where $a \in A$ and $b \in B$. Then show that $\sup A \le \inf B$.

Solution. $a \in A, b \in B$ and $a \le b \ \forall \ a, b$. Then $a \le b$

$\Rightarrow$ b is an upper bound of A

$$\Rightarrow \sup A \le b \tag{1}$$

Again, $b \ge \sup A \ \forall \ b \in B$

$\Rightarrow$ $\sup A$ is a lower bound of B

$\Rightarrow \inf B \ge \sup A$

Hence $\sup A \le \inf B$.

Problem [3.3]: Let S be a non-empty subset of $\mathbb{R}$ and let $b < 0$ and let $bS = \{bs, \ s \in S\}$. Then prove that (i) $\inf(bS) = b \sup S$ and (ii) $\sup(bS) = b \inf S$.

Solution. Let $\sup S = M$.

Then (i) $s \le M \ \forall \ s \in S$, (ii) for every $\epsilon \ (> 0)$, $\exists \ s_\epsilon \in S$ such that $s_\epsilon > M - \epsilon$.

When $b < 0$,

$$\text{(i)} \Rightarrow bs \ge bM \ \forall \ s \in S \tag{1}$$

$$\text{and (ii)} \Rightarrow bs_\epsilon < bM - b\epsilon$$

$b < 0 \Rightarrow -b > 0$.

So $(-b) \cdot \epsilon > 0$ as $\epsilon > 0$. Taking $(-b) \cdot \epsilon = \epsilon'$, we have

$$bs_\epsilon > bM + \epsilon' \tag{2}$$

So from (1) and (2) it follows that $bM = \inf(bS)$, i.e., $b(\sup S) = \inf bS$.

Similarly, proceeding as above, we can prove (ii) by taking $\inf S = m$.

Problem [3.4] : Show that 1 is $\sup S$, where $S = \{x, 0 \leq x \leq 1\}$.

Solution. S has an upper bound 1. We have to show that 1 is the least of all upper bounds of S. Let $v \in S$ and let $v < 1$. Then $\exists$ an element $u \in S$ such that $v < u$. For example, let $v = \frac{1}{5} < 1$. Then $\exists$ an element $u = \frac{2}{3}$ such that $v < u$. So v cannot be an upper bound of S and since $v(\nless 1)$ is chosen arbitrarily it follows that $\sup S = 1$.

Problem [3.5] : If A and B are bounded subsets of $\mathbb{R}$, then $A \cup B$ is bounded and sup $(A \cup B) = \sup(\sup A, \sup B)$.

Solution. Let M and N be the $\sup A$ and $\sup B$ respectively.

Then $x \leq M \ \forall \ x \in A$ and $x \leq N \ \forall \ x \in B$.

Now $A \cup B = \{x : x \in A \text{ or } x \in B\}$. Then

$$x \leq \sup(M, N) \ \forall \ x \in A \cup B \tag{1}$$

Again, $\exists$ an element $u \in A$ such that

$$u > M - \epsilon \ (\epsilon > 0) \tag{2}$$

and $\exists$ an element $v \in B$ such that

$$v > N - \epsilon \tag{3}$$

So from (2) and (3), it follows that $\exists \ y \in A \cup B$ such that

$$y > \sup(M, N) - \epsilon \tag{4}$$

Hence from (1) and (4), it follows that

$$\sup(A \cup B) = \sup(M, N) = \sup(\sup A, \sup B).$$

Definition [3.2] : **[Completeness property of Real Numbers according to Dedekind's form]** A set $\mathbb{R}$ of real numbers is divided into two classes L and U such that

(i) $L \neq \phi, U \neq \phi$

(ii) $L \cup U = \mathbb{R}$

(iii) $x \in L$ and $y \in U \Rightarrow x < y$.

Then either L has no greatest member or U has a least member.

Theorem [3.6] : **[The Completeness property of $\mathbb{R}$]** Every non-empty set $S \subseteq \mathbb{R}$, abounded above, has a supremum in $\mathbb{R}$.

Proof. Let $\mathbb{R}$ be divided into two classes L and U, where L and U are defined by

$$L = \{x, x \text{ is not an upper bound of } S\}.$$
$$U = \{x, x \text{ is an upper bound of } S\}.$$

Let M be an upper bound of S.

(i) Then $x \leq M \; \forall \; x \in S$.

Hence $x \in L$. That is, all the elements of S, except M are members of L. So $L \neq \phi$
$M \in U \Rightarrow U \neq \phi$.

(ii) $L \cup U = \mathbb{R}$.

(iii) Let $x \in L$ and $y \in U$. Then $x < y$.

For, if $x > y$, then x being greater than an upper bound of $S \in U$, a contradiction. Then either L has a greatest member or U has a least member according to Dedekind's property, we shall show that L has no greatest member. If possible, let ξ be the greatest member of L. Then $\xi \in L \Rightarrow \xi$ is not an upper bound of $S \Rightarrow \exists$ a member $p \in S$ such that $p > \xi$, i.e., $\xi < p$.

Now we have a rational number t such that $\xi < t < p$.

Now $t > \xi$, the greatest member of L.

Then $t \in U$. Hence t is an upper bound of S. (1)

Again, $t < p \Rightarrow t$ is less than a member of S and so $t \in L$. Consequently, t is not an upper bound of S. (2)

Thus a contradiction arises from (1) and (2). Hence L has no greatest member. Consequently, it follows that U has a least member. That is, U has a supremum.

The analogous property for infima can be deduced from the completeness property of $\mathbb{R}$ as follows:

Let S be a non-empty subset of $\mathbb{R}$ and it be bounded below.

Then $\bar{S} = \{-s, s \in S\}$ is bounded above and it has the supremum property which shows that $M = \sup \bar{S}$ in $\mathbb{R}$ and this implies that $-M = \inf S$ in $\mathbb{R}$.

Since, $-x \leq M \; \forall \; -x \in \bar{S}$ and for every $\epsilon \; (> 0)$, $\exists \; -y \in \bar{S}$ such that $-y > M - \epsilon$

$$\Rightarrow x \geq -M \; \forall \; x \in S$$

and for every $\epsilon \; (> 0)$, $\exists \; y \in S$ such that $y < -M + \epsilon$

$$\Rightarrow -M \text{ is the infimum of } S$$

i.e., $-M = \inf S$.

3.1 Bounded Functions

Definition [3.3] : Let $f : D \rightarrow \mathbb{R}$ be a real valued function whose domain D is a subset of $\mathbb{R}$.

Now f is said to be bounded above if the subset $f(D)(\subseteq \mathbb{R}) = \{f(x), x \in D\}$ is bounded above.

That is, if there exists a finite number $M \in \mathbb{R}$ such that $f(x) \leq M \; \forall \; x \in D$.

Then M is called the upper bound of $f(x)$ and f is said to be bounded above. Again, f is said to be bounded below, if $f(D)$ is bounded below.

That is, if there exists a finite number $m \in \mathbb{R}$ such that $f(x) \geq m \; \forall \; x \in D$.

So f is bounded, if $m \leq f(x) \leq M \; \forall \; x \in D$.

Equivalently, we can say, that if $|f(x)| \leq k \; \forall \; x \in D, k \in \mathbb{R}$, f is bounded.

Theorem [3.7] : If $f, g : D \rightarrow \mathbb{R}$ be two functions having the same domain D such that $f(x) \leq g(x) \; \forall \; x \in D$ and f, g are both bounded. Then $\sup\limits_{x \in D} f(x) \leq \sup\limits_{x \in D} g(x)$.

Proof. $f(x) \leq g(x) \ \forall \ x \in D \Rightarrow f(x) \leq g(x) \leq \sup g(x) \ \forall \ x \in D$

$\Rightarrow \sup g(D)$ is an upper bound of $f(D)$

$\Rightarrow \sup f(D) \leq \sup g(D)$

$\Rightarrow \sup\limits_{x \in D} f(x) \leq \sup\limits_{x \in D} g(x)$

It is to be noted that $f(x) \leq g(x) \not\Rightarrow$ any relation between $\sup f(D)$ and $\inf g(D)$.

Example. Let $f, g : D \to \mathbb{R}$, when $D = \{x, 1 \leq x \leq 3\}$ be defined by $f(x) = x - 1$ and $g(x) = x$ such that $f(x) \leq g(x) \ \forall \ x \in D$.

$$f(D) = \{0, 1, \dots, 2\}$$
$$g(D) = \{1, 1.5, \dots, 2, \dots, 3\}$$
$$\sup f(D) = 2 \text{ and } \sup g(D) = 3 \text{ and } \inf g(D) = 1.$$

Thus, $\sup\limits_{x \in D} f(x) < \sup\limits_{x \in D} g(x) = 3$.

Also $\sup f(D) > \inf g(D)$. (1)

Again, let f, g be defined by $f(x) = x - 1$ and $g(x) = 1/x$ and let $D = \{x, 0 < x \leq 1\}$ such that $f(x) \leq g(x) \ \forall \ x \in D$.

Also $f(D) = \{0, -\frac{1}{2}, -\frac{2}{3}, -\frac{3}{4}, \dots\}$

$g(D) = \{1, 2, 3, 4, \dots\}$.

$\sup f(D) = 0$ and $\inf g(D) = 1$.

So, $\sup f(D) < \inf g(D)$. (2)

From (1) and (2) it follows that $\nexists$ any relation between $\sup f(D)$ and $\inf g(D)$.

Theorem [3.8] : If $f, g : D \to \mathbb{R}$ be two functions such that $f(x) \leq g(y) \ \forall \ x, y \in D$, then $\sup f(D) \leq \inf g(D)$.

That is, $\sup\limits_{x \in D} f(x) \leq \inf\limits_{y \in D} g(y)$.

Proof. f, g are both bounded. Let $g(y) \leq M \ \forall \ y \in D, M \in \mathbb{R}$.

Then M is an upper bound of $g(D)$.

Now $f(x) \leq g(y) \ \forall \ x, y \in D$

$\Rightarrow f(x) \leq g(y) \leq M$, i.e., $f(x) \leq M \ \forall \ x \in D$

$\Rightarrow M$ is an upper bound of $f(D)$

$\Rightarrow \sup\limits_{x \in D} f(x) \leq M$ (i)

Since $g(y) \geq f(x) \ \forall \ y, \ x \in D$

$\Rightarrow g(y) \geq \sup f(D) \ \forall \ y \in D$

$\Rightarrow \sup f(D)$ is the lower bound of $g(D)$

$\Rightarrow \inf g(D) \geq \sup f(D)$

$\Rightarrow \inf\limits_{y \in D} g(y) \geq \sup\limits_{x \in D} f(x)$

That is $\sup\limits_{x \in D} f(x) \leq \inf\limits_{y \in D} g(y)$.

Theorem [3.9]: Let X be a non-empty subset of $\mathbb{R}$ and let $f, g : X \to \mathbb{R}$ be two functions such that $f(X)$ and $g(X)$ are the bounded ranges of f and g in $\mathbb{R}$. Then

$$\sup\{f(x) + g(x),\ x \in X\} \leq \sup\{f(x), x \in X\} + \sup\{g(x), x \in X\}$$

$$\text{and } \inf\{f(x) + g(x), x \in X\} \geq \inf\{f(x), x \in X\} + \inf\{g(x), x \in X\}$$

Proof. Let M, m be the sup and inf of $\{f(x), x \in X\}$ and P, p be the sup and inf of $\{g(x), x \in X\}$ since $f(X)$ and $g(X)$ are both bounded in $\mathbb{R}$.

Then $m \leq f(x) \leq M \;\forall\; x \in X$ and $p \leq g(x) \leq P \;\forall\; x \in X$.

$$\Rightarrow m + p \leq f(x) + g(x) \leq M + P \;\forall\; x \in X \tag{1}$$

Let $S = \{f(x) + g(x), x \in X\}$.

Then S is bounded in $\mathbb{R}$ by (1).

Also it follows that $M + P$ is an upper bound of S.

Hence $\sup S \leq M + P = \sup f(x) + \sup g(x) \;\forall\; x \in X$

$$\Rightarrow \sup\{f(x) + g(x), x \in X\} \leq \sup\{f(x), x \in X\} + \sup\{g(x), x \in X\}$$

Case of inf follows immediately from (1), when $m + p$ is a lower bound of S.

EXERCISE

(1) Prove that every finite set is bounded.

(2) Show that $S \subseteq T \Rightarrow \inf T \leq \inf S \leq \sup S \leq \sup T$.

(3) Let S_1, S_2 be two given sets and let $S = \{x_1 + x_2, x_1 \in S_1, x_2 \in S_2\}$. Then show that $\sup S = \sup S_1 + \sup S_2$ and $\inf S = \inf S_1 + \inf S_2$.

(4) Show that (i) supremum of a set S is unique and (ii) infimum of S is unique.

(5) Find the suprema and infima of the following sets:

(i) $\{-2, 2\} \cup \left\{1 + \frac{1}{n}, n \in N\right\} \cup \left\{-1 - \frac{1}{n}, n \in N\right\}$

(ii) $\left\{\frac{1}{n}, n \in N\right\} \cup \left\{1 + \frac{1}{n}, n \in N\right\}$.

(6) Let A and B be two bounded sets of (+)ve numbers and $C = \left\{\frac{y}{x}; x \in A, y \in B\right\}$. Find $\sup C$.

(7) If A and B be two bounded sets of real numbers, then

(i) show that $A \cup B$ is bounded,

(ii) find the $\sup(A \cup B)$ and $\inf(A \cup B)$.

(8) Let A and B be two sets of real numbers such that $A \subset B$ and B is bounded. Then show that A is bounded. Also show that $\sup A - \inf A \leq \sup B - \inf B$.
[i.e., oscillation of $A \leq$ oscillation of B].

(9) Let A and B be two bounded sets of real numbers and let $\sup A = M$ and $\sup B = N$. Let $C = \{xy, x \in A, y \in B\}$. Show that $\sup C = MN$.

(10) Find the sup and inf of the set:

$$S = \left\{\frac{1}{2^m} + \frac{1}{3^n} + \frac{1}{5^p},\ m, n, p \in N\right\}$$

(11) Let $S(\neq \phi) \subseteq \mathbb{R}$ and let $M = \sup S$. Then for every $a < M$, there is some $x \in S$ such that $a < x \leq M$. Prove it.

(12) Show that a real number r can be expressed as $r = a_0 + \frac{a_1}{10} + \frac{a_2}{10^2} + \cdots + \frac{a_n}{10^n}$, where a_0 is a non-negative integer and $a_1, a_2, \ldots, a_n$ are integers satisfying $0 \leq a_i \leq 9$.

CHAPTER 4

LIMIT OF A FUNCTION

'Mathematical Analysis' is generally based on the various concepts of limits, like 'the limit of a sequence of real numbers', 'the limit of a function'. Mathematician Isaac Newton (1642-1727) and Gottfried Leibniz (1646–1716) introduced the concept of limits of a function and struggled with the creation of Calculus. Leibniz introduced the term 'function', denoted by $f(x)$ depending on a variable x. He has also invented ϵ, an infinitesimally small or an arbitrary small (+)ve number, for developing concept of limit of a function.

Definition [4.1] : Let $f : D \to \mathbb{R}$, $D \subseteq \mathbb{R}$ be a function of x and let $a \in D$. $f(x)$ is said to have a limit L as $x \to a$ (i.e., $x \neq a$), which means that x becomes closer to a (or nearer to a), if the difference of $f(x)$ and L can be made as small as we like and it is written as $\lim_{x \to a} f(x) = L$ [symbolical notation of limit of $f(x)$ as x approaches to a].

(ϵ, δ)-definition

Definition [4.2] : A function $f(x)$ is said to have a limit L, if for every ϵ (> 0), there exists a (+)ve number δ (depending on ϵ) such that $|f(x) - L| < \epsilon$, whenever

$$0 < |x - a| < \delta \tag{1}$$

Limit Point

Definition [4.3] : A point $a \in \mathbb{R}$ is said to be a limit point of $D \subseteq \mathbb{R}$, if for every δ (> 0), there exists at least one point x of D, $x \neq a$ such that $|x - a| < \delta$.

In other words, every δ-neighbourhood $N_\delta(a)$ of a contains at least one point of D other than a, if $a \in D$.

Here $N_\delta(a) = (a - \delta, a + \delta)$, an open interval containing a.

$N'_\delta(a)$, known as the deleted neighbourhood of a. That is, a neighbourhood of a which does not contain a.

Thus 'a' is said to be a limit point of D, if $N'_\delta(a) \cap D \neq \phi$.

Also (1) can be written as $|f(x) - L| < \epsilon \;\forall\; x \in N'_\delta(a)$.

From definition of limit of $f(x)$, we have

$$\begin{aligned} & |f(x) - L| < \epsilon, \text{ whenever } 0 < |x - a| < \delta \\ \Rightarrow\; & L - \epsilon < f(x) < L + \epsilon \;\forall\; x \in (a - \delta, a + \delta), x \neq a \\ \Rightarrow\; & f(x) \in N_\epsilon(L) \;\forall\; x \in N'_\delta(a) \\ \Rightarrow\; & L \text{ is a limit point of } f(N'_\delta(a)) \text{ in } \mathbb{R}. \end{aligned}$$

Thus when $\lim_{x \to a} f(x) = L$, that is, when limit of $f(x)$ exists finitely when $x \to a$, we say $f(x)$ converges to L. If L does not exist finitely, we say $f(x)$ diverges to $\pm\infty$. In other words, $f(x)$ diverges to $\pm\infty$.

Theorem [4.1] : The limit of a function $f(x)$ is unique, if it exists.

Proof. Let the limit exist and, if possible, let L and L' be two limits of $f(x)$ as $x \to a$.

Then $|f(x) - L| < \frac{\epsilon}{2}$, whenever $0 < |x - a| < \delta_1$

and $|f(x) - L'| < \frac{\epsilon}{2}$, whenever $0 < |x - a| < \delta_2$

So $|L - L'| = |L - f(x) + f(x) - L'|$

$$\leq |f(x) - L| + |f(x) - L'| < \frac{\epsilon}{2} + \frac{\epsilon}{2} = \epsilon$$

whenever $0 < |x - a| < \delta$ where $\delta = \min(\delta_1, \delta_2)$.

Since ϵ is arbitrary, $|L - L'| = 0 \Rightarrow L = L'$.

Hence the theorem.

Problem [4.1] : Determine the following limits

(i) $\lim_{x \to a} f(x)$ where $f(x) = l \ \forall \ x \in D \subseteq \mathbb{R}$

(ii) $\lim_{x \to a} f(x)$ where $f(x) = x \ \forall \ x \in D \subseteq \mathbb{R}$

(iii) $\lim_{x \to a} f(x)$ where $f(x) = x^2 \ \forall \ x \in D \subseteq \mathbb{R}$

(iv) $\lim_{x \to a} f(x)$ where $f(x) = \frac{1}{x}$, $a > 0 \ \forall \ x \in D \subseteq \mathbb{R}$

Solution.

(i) Given that $0 < |x - a| < \delta$.

Now $|f(x) - l| = |l - l| = 0 < \epsilon$, whenever $0 < |x - a| < \delta$

$\Rightarrow \lim_{x \to a} f(x) = l$

(ii) $|f(x) - a| = |x - a| < \epsilon$ for $0 < |x - a| < \delta$ (taking $\delta < \epsilon$)

$\Rightarrow \lim_{x \to a} f(x) = a$

(iii) $|f(x) - a^2| = |x^2 - a^2| = |x - a||x + a|$ for $0 < |x - a| < \delta$

$$< \{|x| + |a|\}\delta \tag{1}$$

Again $|x| - |a| \leq |x - a| < \delta \Rightarrow |x| < |a| + \delta$

$\Rightarrow |x| + |a| < 2|a| + \delta.$

So from (1), $|f(x) - a^2| < \{2|a| + \delta\} \cdot \delta$

$$< 2|a| + \delta \cdot \frac{\epsilon}{2|a| + \delta} = \epsilon \text{ by choosing } \delta < \frac{\epsilon}{2|a| + \delta}$$

Hence $\lim_{x \to a} f(x) = a^2$.

(iv) Here $f(x) = \frac{1}{x}, a > 0$.

$$\left| f(x) - \frac{1}{a} \right| = \left| \frac{1}{x} - \frac{1}{a} \right| = \frac{|x - a|}{|x|a} < \frac{\delta}{|x|a} \tag{2}$$

Now taking $\delta = \frac{a}{2}(> 0)$

$$|a| - |x| < |x - a| < \frac{a}{2} \Rightarrow |x| > \frac{a}{2}$$

$$\Rightarrow a|x| > \frac{a^2}{2} \Rightarrow \frac{1}{a|x|} < \frac{2}{a^2}$$

So from (2),

$$\left|f(x) - \frac{1}{a}\right| < \frac{2}{a^2}\delta < \frac{2}{a^2} \cdot \frac{a^2\epsilon}{2} = \epsilon \text{ for } 0 < |x - a| < \frac{a^2\epsilon}{2}$$

That is,

$$\left|f(x) - \frac{1}{a}\right| < \epsilon \text{ for } 0 < |x - a| < \delta', \text{ where } \delta' = \min\left(\frac{a}{2}, \frac{a^2\epsilon}{2}\right)$$

$$\therefore \lim_{x \to a} f(x) = \frac{1}{a}$$

Problem [4.2]: Use (ϵ, δ) definition to establish $\lim_{x \to 1} \frac{x^2 - x + 1}{x + 1} = \frac{1}{2}$.

Solution.

$$\left|\frac{x^2 - x + 1}{x + 1} - \frac{1}{2}\right| = \left|\frac{2x^2 - 2x + 2 - x - 1}{2(x + 1)}\right| = \left|\frac{2x^2 - 3x + 1}{2(x + 1)}\right|$$

$$= \left|\frac{(2x - 1)(x - 1)}{2(x + 1)}\right| = \frac{|2x - 1||x - 1|}{2|x + 1|} \quad (1)$$

For $0 < |x - 1| < \delta$ (2)

$\Rightarrow 1 - \delta < x < 1 + \delta$

Now $x > 1 - \delta \Rightarrow x + 1 > 2 - \delta \Rightarrow |x + 1| > |2 - \delta|$

$$\Rightarrow \frac{1}{|x + 1|} < \frac{1}{|2 - \delta|} \Rightarrow \frac{1}{|x + 1|} < 1 \quad (3)$$

Also $2x - 1 < 1 + 2\delta$

$$\Rightarrow |2x - 1| < |1 + 2\delta| < 3, \text{ when } \delta = 1 \quad (4)$$

So from (1),

$$\left|\frac{x^2 - x + 1}{x + 1} - \frac{1}{2}\right| < \frac{1}{2} \cdot \frac{3}{1} \cdot \delta \text{ [by (2), (3) and (4)]}$$

$$< \epsilon, \text{ when } \delta < \frac{2\epsilon}{3}$$

$$\therefore \left|\frac{x^2 - x + 1}{x + 1} - \frac{1}{2}\right| < \epsilon, \text{ whenever } 0 < |x - 1| < \delta', \text{ where } \delta' = \min\left(1, \frac{2\epsilon}{3}\right)$$

Hence $\lim_{x \to 1} \frac{x^2 - x + 1}{x + 1} = \frac{1}{2}$.

4.1 Limit of a Function in Terms of a Sequence

Theorem [4.2]: $\lim_{x \to a} f(x) = L \Leftrightarrow \lim_{n \to \infty} f(x_n) = L$ for every sequence $\{x_n\}$ in the neighbourhood of a where f is defined, converging to a.

Proof. $\lim_{x \to a} f(x) = L$

$\Rightarrow$ for every ϵ (> 0), $\exists$ a (+)ve number δ such that

$$|f(x) - L| < \epsilon \text{ for } 0 < |x - a| < \delta \quad \text{(i)}$$

Here domain of f is $(a - \delta, a + \delta)$, a neighbourhood of a, $x \neq a$.

That is, $|f(x) - L| < \epsilon \; \forall \; x \in (a - \delta, a + \delta)$.

Thus, if we take $\delta = \frac{1}{n}$, then for each n, we have a point $x_n \in (a - \frac{1}{n}, a + \frac{1}{n})$.

Thus we have a sequence $\{x_n\}$ converging to a. Therefore, there exists a (+)ve integer m such that $|x_n - a| < \delta \; \forall \; n \geq m$.

Hence, for $x = x_n$, from (i) we have

$$|f(x_n) - L| < \epsilon \; \forall \; n \geq m \Rightarrow \lim_{n \to \infty} f(x_n) = L$$

Conversely, let $\lim_{n \to \infty} f(x_n) = L$, where $\{x_n\}$ converges to a we shall prove that

$$\lim_{x \to a} f(x) = L.$$

If possible, let $\lim_{x \to a} f(x) \neq L$. Then there exists at least one x in $0 < |x - a| < \frac{1}{n}$, where $\delta = \frac{1}{n}$, such that $|f(x) - L| \nless \epsilon$. That is $|f(x) - L| \geq \epsilon$, whenever $0 < |x_n - a| < \frac{1}{n}$, which shows that though $\{x_n\}$ converges to a, $\lim_{n \to \infty} f(x_n) \neq L$. Hence it contradicts to our hypothesis.

Therefore, $\lim_{x \to a} f(x) = L$.

Theorem [4.3] : The necessary and sufficient condition that $f(x)$ converges to L as $x \to a$ is that for ϵ (> 0) there exists a (+)ve number $\delta(\epsilon)$ such that $|f(x_2) - f(x_1)| < \epsilon \; \forall \; x_1, x_2 \in N'_\delta(a)$.

***Proof.* Condition is necessary**

Given that $|f(x) - L| < \epsilon \; \forall \; x \in N'_\delta(a)$. Let $x_1, x_2 \in N'_\delta(a)$.

Then $|f(x_1) - L| < \frac{\epsilon}{2}$, $|f(x_2) - L| < \epsilon_2$, whenever $0 < |x_1 - a| < \delta_1$, $0 < |x_2 - a| < \delta_2$

$$\begin{aligned} \Rightarrow |f(x_2) - f(x_1)| &= |f(x_2) - L + L - f(x_1)| \\ &\leq |f(x_2) - L| + |L - f(x_1)| \\ &< \frac{\epsilon}{2} + \frac{\epsilon}{2} = \epsilon \; \forall \; x_1, x_2 \in N'_\delta(a), \text{ where } \delta = \min(\delta_1, \delta_2). \end{aligned}$$

Condition is sufficient

Given that $|f(x_2) - f(x_1)| < \epsilon \; \forall \; x_1, x_2 \in N'_\delta(a)$ we have to show that $\lim_{x \to a} f(x) = L$.

Let $\{\epsilon_n\}$ be a monotonically decreasing sequence of (+)ve numbers such that $\epsilon_n \to 0$ as $n \to \infty$.

Now to each (+)ve number ϵ_n, $\exists$ a (+)ve integer δ_n (ϵ_n) such that

$$|f(x_2) - f(x_1)| < \epsilon_n \; \forall \; x_1, x_2 \in N'_{\delta_n}(a) \quad \text{(i)}$$

Thus we obtain a sequence $\{\delta_n\}$ of (+)ve numbers corresponding to $\{\epsilon_n\}$ and as δ_n depends on ϵ_n, δ_n is also monotonically decreasing sequence of (+)ve numbers.

Now taking $x_2 = x$, $x_1 = a + \delta_n$ we have from (i)

$$|f(x) - f(a+\delta_n)| < \epsilon_n \ \forall \ x \in N'_{\delta_n}(a)$$
$$\Rightarrow f(a+\delta_n) - \epsilon_n < f(x) < f(a+\delta_n) + \epsilon_n \ \forall \ x \in (a-\delta_n, a+\delta_n)$$
$$\Rightarrow f(x) \in (f(a+\delta_n) - \epsilon_n, \ f(a+\delta_n) + \epsilon_n) \text{ whose length is } 2\epsilon_n \quad \text{(ii)}$$
$$\Rightarrow |f(x) - f(a+\delta_n)| < \epsilon_n \text{ in the interval } (f(a+\delta_n) - \epsilon_n, f(a+\delta_n) + \epsilon_n) \quad \text{(iii)}$$

We write this interval as (a_n, b_n) so that $|b_n - a_n| = 2\epsilon_n$.

Thus we get a sequence of intervals (a_r, b_r) such that

$$(a_r, b_r) \supset (a_n, b_n) \ \forall \ r < n \text{ and } \lim_{n\to\infty} |b_n - a_n| = 0.$$

Therefore, we have $(a_1, b_1) \supset (a_2, b_2) \supset (a_3, b_3) \cdots \supset (a_n, b_n)$.

This shows that $a_1 < a_2 \cdots < a_n \cdots < b_1$ and $a_1 < b_n < b_{n-1} < \cdots < b_2 < b_1$.

Thus $\{a_n\}$ and $\{b_n\}$ are two sequences of which $\{a_n\}$ is monotonically increasing with bounded above by b_1 while $\{b_n\}$ is monotonically decreasing with bounded below by a_1. Then $\{a_n\}$ and $\{b_n\}$ are convergent.

Let $\lim_{n\to\infty} a_n = \lambda$ and $\lim_{n\to\infty} b_n = \mu$.

Then $|\lambda - \mu| = \lim_{n\to\infty} |a_n - b_n| = \lim_{n\to\infty} 2\epsilon_n = 0 \Rightarrow \lambda = \mu$.

Hence $\{a_n\}$ and $\{b_n\}$ converge to the same limit $L \cdot (\lambda = \mu = L)$

$$\Rightarrow L \in (a_n, b_n) \text{ as } n \to \infty$$

and ϵ (> 0) be so chosen that $2\epsilon_n < \epsilon$.

Therefore, $L \in (f(a+\delta_n) - \epsilon_n, f(a+\delta_n) + \epsilon_n)$ [by (ii)]

$$\Rightarrow f(a+\delta_n) - \epsilon_n < L < f(a+\delta_n) + \epsilon_n, \text{ whenever } 0 < |x-a| < \delta_n$$
$$\Rightarrow |f(a+\delta_n - L| < \epsilon_n \quad \text{(iv)}$$

$$\begin{aligned}\text{Then } |f(x) - L| &= |f(x) - f(a+\delta_n) + f(a+\delta_n) - L| \\ &\le |f(x) - f(a+\delta_n) + |f(a+\delta_n - L| \\ &< \epsilon_n + \epsilon_n = 2\epsilon_n < \epsilon \text{ for } 0 < |x-a| < \delta_n < \delta\end{aligned}$$

Hence $\lim_{n\to a} f(x) = L$.

Theorem [4.4]: If $\lim_{x\to a} f(x) = L$, $a \in \mathbb{R}$, then f is bounded on some neighbourhood of a, when $f : D \to \mathbb{R}$, $D \subseteq \mathbb{R}$.

Proof. Given that $|f(x) - L| < \epsilon$, $\forall \ x \in N'_\delta(a)$.

Then $|f(x)| - |L| \le |f(x) - L| < \epsilon$

$$\Rightarrow |f(x)| < |L| + \epsilon < |L| + 1 = M \ \forall \ x \in N'_\delta(a) \cap D$$

where ϵ is so chosen that $\epsilon < 1$.

Then it follows that $|f(x)| \le M \ \forall \ x \in N'_\delta(a)$

$\Rightarrow f$ is bounded on $N'_\delta(a)$ of a.

Corollary [4.1]: If $f : D \to \mathbb{R}, D \subseteq \mathbb{R}$ has a limit at c, where c is a limiting point of D, then f is bounded on some neighbourhood of c.

Proof. Let $f(x)$ has a limit L at $x = c$. That is, $|f(x) - L| < \epsilon \; \forall \; x \in N'_\delta(c)$

$$\Rightarrow |f(x)| - |L| \le |f(x) - L| < \epsilon \; \forall \; x \in N'_\delta(c)$$

$$\Rightarrow |f(x)| < |L| + 1, \text{ taking } \epsilon = 1, \; \forall \; x \in N'_\delta(c) \cap D.$$

Thus, $|f(x)| < |L| + 1 \; \forall \; x \in N'_\delta(c) \cap D$.

Now c being a limiting point of D, $c \in D$ or $c \notin D$, and

$$N'_\delta(c) \cap D \neq \phi \text{ or } N_\delta(c) \cap D \neq \phi$$

Thus, $|f(x)| < |L| + 1$, when $c \notin D$. (i)

When $c \in D$, we take $M = \sup\{|f(c)|, |L| + 1\}$, then

$$|f(x)| \le M \; \forall \; x \in N_\delta(c) \cap D \qquad \text{(ii)}$$

Hence by (i) and (ii), f is bounded on some neighbourhood of c.

Some Results on Limits

Theorem [4.5] : Let $f, g : D \to \mathbb{R}$ be two functions such that $\lim_{x \to a} f(x) = l_1$, $\lim_{x \to a} g(x) = l_2$, $a \in D \subseteq \mathbb{R}$, then

(i) $\lim_{x \to a} (f \pm g)x = l_1 \pm l_2$

(ii) $\lim_{x \to a} (f \cdot g)x = l_1 \cdot l_2$

(iii) $\lim_{x \to a} \frac{f}{g}(x) = \frac{l_1}{l_2}, l_2 \neq 0$

(iv) $f(x) < g(x) \Rightarrow l_1 \le l_2$

(v) $f(x) \le \phi(x) \le g(x) \Rightarrow \lim_{x \to a} \phi(x) = l$, when $l_1 = l_2 = l$

(vi) $\lim_{x \to a} f(x) > 0$ (or < 0) $\Rightarrow f(x) > 0$ (or < 0) $\forall \; x \in N'_\delta(x) \cap D$.

Proof.

(ii) Let $\{x_n\}$ be a sequence in D such that $\lim_{n \to \infty} x_n = a$.

Then $\lim_{n \to \infty} f(x_n) = \lim_{x \to a} f(x) = l_1$. [by Th. [4.2]].

So also $\lim_{n \to \infty} g(x_n) = \lim_{x \to \infty} g(x) = l_2$.

Then $\lim_{n \to \infty} (f \cdot g)(x_n) = \lim_{n \to \infty} [f(x_n) \cdot g(x_n)] = \lim_{n \to \infty} f(x_n) \cdot \lim_{n \to \infty} g(x_n) = l_1 \cdot l_2$.

So $\lim_{x \to a} (f \cdot g)(x) = l_1 \cdot l_2$.

(iii) We define $\left(\frac{f}{g}\right)(x) = \frac{f(x)}{g(x)}$

$$\lim_{x \to a} \left(\frac{f}{g}\right)(x) = \lim_{n \to \infty} \left(\frac{f}{g}\right)(x_n) = \lim_{n \to \infty} \frac{f(x_n)}{g(x_n)} = \frac{\lim_{n \to \infty} f(x_n)}{\lim_{n \to \infty} g(x_n)} = \frac{l_1}{l_2}, \; l_2 \neq 0, 1$$

If $l_2 = 0$, then $\lim_{x \to a} \left(\frac{f}{g}\right)(x)$ may or may not exist.

Let $\lim_{x \to a} g(x) = 0$ and let $(x - a)$ is a factor of $f(x)$, then

$$\lim_{x \to a} \frac{f(x)}{g(x)} = \lim_{x \to a} \frac{(x - a) f_1(x)}{(x - a) g_1(x)} \quad \left[\lim_{x \to a} g(x) = 0 \Rightarrow g(x) = (x - a) g_1(x)\right]$$

$$= \lim_{x \to a} \frac{f_1(x)}{g_1(x)} \quad [\text{as } x \neq a]$$

where $\lim_{x\to a} g_1(x) \neq 0$. So we proceed in this case as above.

But if $(x-a)$ is not a factor of $f(x)$, $\lim_{x\to a}\left(\frac{f}{g}\right)(x)$ does not exist.

[*Go through the examples.*]

Example. $\lim_{x\to 0}\frac{1}{x}$ does not exist.

Example. $\lim_{x\to 2}\frac{x^2-5}{3x-6}$ does not exist as $\lim_{x\to 2}(3x-6)=0$ and $(x-2)$ is not a factor of $f(x)=(x^2-5)$.

Example. $\lim_{x\to 2}\frac{x^2-4}{(3x-6)}$ exists. $\lim_{x\to 2} 3x-6=0$ and $(x-2)$ is a factor of $f(x)=x^2-4$.

So $\lim_{x\to 2}\frac{(x-2)(x+2)}{3(x-2)} = \lim_{x\to 2}\frac{x+2}{3}=\frac{4}{3}$.

(iv) We have $l_1-\epsilon < f(x) < g(x) < l_2+\epsilon$.
Then $l_1 < l_2 + 2\epsilon \Rightarrow l_1 \leq l_2$ [by (x) of Th. [2.10]]

(vi) Let $\lim_{x\to a} f(x)=L$ and let $L>0$.
Then $|f(x)-L|<\epsilon$, for $0<|x-a|<\delta$

$$\Rightarrow L-\epsilon < f(x) < L+\epsilon\ \forall\ x\in N'_\delta(a)\cap D$$

Taking $\epsilon=\frac{L}{2}$, we get

$$f(x) > L-\frac{L}{2}=\frac{L}{2}>0\ \forall\ x\in N'_\delta(a)\cap D$$

Again, let $L<0$.
Taking $\epsilon=-\frac{L}{2}$, we get

$$f(x) < L-\frac{L}{2}=\frac{L}{2}<0\ \forall\ x\in N'_\delta(a)\cap D$$

[Proofs of (i) and (v) are left.]

4.2 Continuity

Definition [4.4]: A function $f: D\to\mathbb{R}, D\subseteq\mathbb{R}$ is said to be continuous at a point $x=a\in D$, if $\lim_{x\to a} f(x)$ exists and equal to $f(a)$. That is, if $\lim_{x\to a+0} f(x) = \lim_{x\to a-0} f(x) = f(a)$, which may be written as $f(a+0)=f(a-0)=f(a)$. In other words, $\lim_{h\to 0} f(a+h)=f(a)$.

(ϵ,δ)-definition

Definition [4.5]: A function $f: D\to\mathbb{R}, D\subseteq\mathbb{R}$ is said to be continuous at $x=a\in D$, if for every ϵ (>0), there exists a (+)ve number δ such that $|f(x)-f(a)|<\epsilon$ whenever $|x-a|<\delta$. That is, if $|f(x)-f(a)|<\epsilon\ \forall\ x\in N_\delta(a)\cap D$.

Theorem [4.6]: The necessary and sufficient condition that $f: D\to\mathbb{R}$ to be continuous at $x=a\in D$ is $f(N_\delta(a)\cap D)\subseteq N_\epsilon(f(a))$.

Proof. Let f be continuous at $x=a\in D$. Then $|f(x)-f(a)|<\epsilon\ \forall\ x\in N_\delta(a)\cap D$

$$\Rightarrow f(a)-\epsilon < f(x) < f(a)+\epsilon\ \forall\ x\in N_\delta(a)\cap D$$
$$\Rightarrow f(x)\in N_\epsilon(f(a))\ \forall\ x\in N_\delta(a)\cap D$$
$$\Rightarrow f(N_\delta(a)\cap D)\subseteq N_\epsilon(f(a)).$$

Conversely, let $f(N_\delta(a) \cap D) \subseteq N_\epsilon f(a)$. Then $f(x) \in N_\epsilon f(a) \; \forall \; x \in \dot{N}_\delta(a) \cap D$

$$\Rightarrow |f(x) - f(a)| < \epsilon \; \forall \; x \in N_\delta(a) \cap D$$
$$\Rightarrow f \text{ is continuous at } x = a.$$

Continuity of f in an Interval

Definition [4.6]: A function $f : [a, b] \to \mathbb{R}, [a, b] \subseteq \mathbb{R}$ is said to be continuous in $[a, b]$, if for any $c \in [a, b]$, i.e., $a \leq c \leq b$, f is continuous at c.

That is f is continuous in $[a, b]$, if $\lim\limits_{x \to c} f(x) = f(c) \; \forall \; c \in [a, b]$.

Also f is said to be continuous at the left end point a, if $\lim\limits_{x \to a+0} f(x) = f(a)$ and f is said to be continuous at the right end point b, if $\lim\limits_{x \to b-0} f(x) = f(b)$.

Thus f is said to be continuous in $[a, b]$, if the following results hold:

1. $\lim\limits_{x \to c} f(x) = f(c) \; \forall \; c \in (a, b)$
2. $\lim\limits_{x \to a+0} f(x) = f(a)$
3. $\lim\limits_{x \to b-0} f(x) = f(b)$.

Example. Let $f(x) = x - \frac{1}{x}$ and $[a, b] = [1, 2]$. Let c be any point in $(1, 2)$ $(1 < c < 2)$

$$\lim_{x \to c} f(x) = \lim_{x \to c} \left(x - \frac{1}{x} \right) = f(c)$$

Also

$$\lim_{x \to 1+0} f(x) = \lim_{x \to 1+0} \left(x - \frac{1}{x} \right) = f(1)$$
$$\lim_{x \to 2-0} f(x) = \lim_{x \to 2-0} \left(x - \frac{1}{x} \right) = f(2)$$

So f is continuous in $[1, 2]$.

Example. Let $f(x) = x - \frac{1}{x}$ and $[a, b] = [0, 2]$. For any $c \in (0, 2)$, i.e., $0 < x < 2$

$$\lim_{x \to c} f(x) = f(c)$$

$\lim\limits_{x \to 0+0} f(x) = \lim\limits_{x \to 0+0} \left(x - \frac{1}{x}\right)$ does not exist as $\lim\limits_{x \to 0+0} \frac{1}{x}$ does not exist.

So f is not continuous in $[0, 2]$.

Types of Discontinuities

The cases arises when $f : D \to \mathbb{R}$ is not continuous at $x = a \in D$. If f is not continuous at $x = a$, we say f is discontinuous at $x = a$.

(i) **Ordinary discontinuity** at $x = a$.

If $\lim\limits_{x \to a-0} f(x) \neq \lim\limits_{x \to a+0} f(x)$, then f has an ordinary point of discontinuity at $x = a$.

In this case, $f(a)$ may or may not exist.

(ii) **Removable discontinuity.**

If $\lim\limits_{x \to a-0} f(x) = \lim\limits_{x \to a+0} f(x) \neq f(a)$ or $f(a)$ is undefined, then f has a removable discontinuity at $x = a$.

(iii) **Discontinuity of the 1st kind.**

f is said to have discontinuity of the 1st kind at $x = a$ if both $\lim_{x\to a+0} f(x)$ and $\lim_{x\to a-0} f(x)$ exist but are not equal.

1st kind discontinuity from the left if $\lim_{x\to a-0} f(x)$ exist but not equal to $f(a)$.

1st kind discontinuity from the right if $\lim_{x\to a+0} f(x)$ exists but not equal to $f(a)$.

(iv) **Discontinuity of 2nd kind.**

f has a 2nd kind of discontinuity at $x = a$, if neither $\lim_{x\to a+0} f(x)$ nor $\lim_{x\to a-0} f(x)$ exists.

(v) f has a 2nd kind of discontinuity at $x = a$ **from the left** if $\lim_{x\to a-0} f(x)$ does not exist and f has a 2nd kind of discontinuity at $x = a$ **from the right** if $\lim_{x\to a+0} f(x)$ does not exist.

Example. Let

$$f(x) = \begin{cases} (x+1)\sin\frac{1}{x}, & x \in (-1,1) \\ 0, & \text{otherwise,} \end{cases}$$

$$\lim_{x\to 0+0} f(x) = \lim_{x\to 0+0} x\sin\frac{1}{x} + \lim_{x\to 0+0} \sin\frac{1}{x} = 0 + \lim_{x\to 0+0} \sin\frac{1}{x} = \lim_{x\to 0+0} \sin\frac{1}{x}$$

which does not exist. Also $\lim_{x\to 0-0} f(x)$ does not exist.

So f has a second kind of discontinuity at $x = 0$.

$$\left.\begin{array}{l} \lim_{x\to 1+0}(x+1)\sin\frac{1}{x} = 2\sin 1 \\ \lim_{x\to 1-0}(x+1)\sin\frac{1}{x} = 2\sin 1 \end{array}\right\} \neq f(1) = 0$$

So f has a removable discontinuity at $x = 1$.

Example. $f(x) = x - [x]\ \forall\ x \geq 0$.

Then $\lim_{x\to 2+0} f(x) = 2 - 2 = 0$

$\lim_{x\to 2-0} f(x) = 2 - 1 = 1$

Therefore, $\lim_{x\to 2+0} f(x) \neq \lim_{x\to 2-0} f(x)$, when $f(2) = 0$.

So f has a 1st kind of discontinuity at $x = 2$.

Example. Let $f(x) = \begin{cases} \frac{\sin 3x}{x}, & x \neq 0 \\ 1, & x = 0 \end{cases}$

Then $\lim_{x\to 0} f(x) = \lim_{x\to 0} \dfrac{\sin 3x}{x} = 3\lim_{x\to 0} \dfrac{\sin 3x}{3x} = 3 \neq f(0)$.

So f has a removable discontinuity at $x = 0$.

Problem [4.3] : Find the nature of points of discontinuity of f where f is defined by

$$f(x) = \begin{cases} \dfrac{1}{2} - x, & 0 < x < \dfrac{1}{2} \\ \dfrac{3}{2} - x, & \dfrac{1}{2} < x < 1 \end{cases}$$

and $f(0) = 0,\ f\left(\dfrac{1}{2}\right) = \dfrac{1}{2},\ f(1) = 1$

Solution. $\lim_{x\to 0+0} f(x) = \lim_{x\to 0+0} \frac{1}{2} - x = \frac{1}{2} \neq f(0)$

So $x = 0$ is a point of discontinuity of 1st kind from the right.

$$\lim_{x\to \frac{1}{2}+0} f(x) = \lim_{x\to \frac{1}{2}+0} \frac{3}{2} - x = 1$$

$$\lim_{x\to \frac{1}{2}-0} f(x) = \lim_{x\to \frac{1}{2}-0} \frac{1}{2} - x = 0$$

Thus $\lim_{x\to \frac{1}{2}+0} f(x) \neq \lim_{x\to \frac{1}{2}-0} f(x)$.

So f has a point of discontinuity of first kind at $x = \frac{1}{2}$.

$$\lim_{x\to 1-0} f(x) = \lim_{x\to 1-0} \frac{3}{2} - x = \frac{1}{2} \neq f(1)$$

f has a point of discontinuity of 1st kind from the left at $x = 1$.

Problem [4.4]: Let $g : \mathbb{R} \to \mathbb{R}$ satisfy the relation $g(x + y) = g(x) \cdot g(y)\ \forall\ x, y \in \mathbb{R}$. If g is continuous at $x = 0$, then show that g is continuous at every point of $\mathbb{R}$. Also if we have $g(a) = 0$ for some $a \in \mathbb{R}$, then $g(x) = 0\ \forall\ x \in \mathbb{R}$.

Solution. Given that g is continuous at $x = 0$. So $\lim_{x\to 0} g(x) = g(0)$.

Again, $g(x + 0) = g(x)g(0)\ \forall\ x \in \mathbb{R},\ g(x) \neq 0$

$\Rightarrow g(x) = g(x)g(0)$

$\Rightarrow g(0) = 1$

Now let $c \in \mathbb{R}$, then $g(c + h) = g(c) \cdot g(h)$ for some $h \in \mathbb{R}$.

Therefore, $\lim_{h\to 0} g(c + h) = \lim_{h\to 0} g(c)g(h) = g(c) \lim_{h\to 0} g(h)$

$= g(c)g(0) = g(c)$ [as g is continuous at $x = 0$ and $g(0) = 1$]

$\Rightarrow g$ is continuous at $x = c$. So g is continuous in $\mathbb{R}$.

Let $g(a) = 0$ for some $a \in \mathbb{R}$.

Also g being continuous at every point $c \in \mathbb{R}$, $\lim_{x\to c} g(x) = g(c)$.

Now $g(x) = g(x - a + a) = g(x - a) \cdot g(a)$

$\lim_{x\to c} g(x) = g(a) \cdot \lim_{x\to c} g(x - a) = 0$ [as $g(a) = 0$]

$\Rightarrow g(c) = 0$

Since c is arbitrary, $g(x) = 0\ \forall\ x \in \mathbb{R}$.

Problem [4.5]: If $f : \mathbb{R} \to \mathbb{R}$ satisfies the relation $f(x + y) = f(x) + f(y)\ \forall\ x, y \in \mathbb{R}$ and if f is continuous at $x = 0$, then f is continuous at every point $c \in \mathbb{R}$. Also show that if f is continuous at a particular point $x = x_0$, then f is also continuous at every point $c \in \mathbb{R}$.

Solution. $x = 0,\ y = 0 \Rightarrow f(0 + 0) = f(0) + f(0)$

$\Rightarrow f(0) = f(0) + f(0)$

$\Rightarrow f(0) = 0$

Let f be continuous at $x = 0$. Then $\lim_{x\to 0} f(x) = f(0) = 0$.

Now let $c \in \mathbb{R}$.

Then $f(x) = f(x-c+c) = f(x-c) + f(c)$.

$$\Rightarrow \lim_{x\to c} f(x) = \lim_{x\to c} f(x-c) + f(c)$$
$$= \lim_{z\to 0} f(z) + f(c) \; [x - c = z \Rightarrow z \to 0 \text{ as } x \to c]$$
$$= f(0) + f(c) = f(c) \quad [\text{as } f \text{ is continuous at } x = 0]$$

Thus f is continuous at every point $c \in \mathbb{R}$.

Next, let f is continuous at $x = x_0 \in \mathbb{R}$.

Then $\lim_{x\to x_0} f(x) = f(x_0)$.

Now

$$f(x) = f(x - c + x_0 - x_0 + c)$$
$$= f(x - c + x_0) + f(-x_0) + f(c) \tag{1}$$

Again, $f(0) = f(x - x) = f(x) + f(-x)$

$$\Rightarrow 0 = f(x) + f(-x) \Rightarrow f(-x) = -f(x) \; \forall \; x \in \mathbb{R}$$

So from (1),

$$f(x) = f(x - c + x_0) - f(x_0) + f(c)$$
$$\Rightarrow \lim_{x\to c} f(x) = \lim_{x\to c} f(x - c + x_0) - f(x_0) + f(c)$$
$$= \lim_{z\to x_0} f(z) - f(x_0) + f(c)$$
$$= f(x_0) - f(x_0) + f(c) \; [\text{as } f \text{ is continuous at } x = x_0 \; \lim_{z\to x_0} f(z) = f(x_0)]$$
$$= f(c)$$

Thus f is continuous at every point $c \in \mathbb{R}$.

Problem [4.6] : Discuss the continuity of

(i) Polynomial function,

(ii) Rational function.

Solution.

1. Let $f(x) = a_0x^n + a_1x^{n-1} + \cdots + a_{n-1}x + a_n$ be a polynomial in x. Let $c \in \mathbb{R}$. Then $\lim_{x\to c} x^r$ exists $\forall \; r$ and the limit is equal to c^r. Thus

$$\lim_{x\to c} f(x) = \lim_{x\to c} \left[a_0x^n + a_1x^{n-1} + \cdots + a_{n-1}x + a_n\right]$$
$$= a_0 \lim_{x\to c} x^n + a_1 \lim_{x\to c} x^{n-1} + \cdots + a_{n-1} \lim_{x\to c} x + a_n$$
$$= a_0c^n + a_1c^{n-1} + \cdots + a_{n-1}c + a_n = f(c)$$

Hence $f(x)$ is continuous at every point in $\mathbb{R}$.

2. Let $p(x)$ and $q(x)$ be two polynomial on $\mathbb{R}$. Then $\frac{p(x)}{q(x)}$ is called a rational function of x. Let there are at most a finite number of real roots, say, $\alpha_1, \alpha_2, \ldots, \alpha_m$ of $q(x) = 0$. So if $x \notin \{\alpha_1, \alpha_2, \ldots, \alpha_m\}$, $q(x) \neq 0$.

Then we define the rational function $r(x)$ by

$$r(x) = \frac{p(x)}{q(x)} \text{ for } x \notin \{\alpha_1, \alpha_2, \ldots, \alpha_m\}$$

Let $x = c, c \notin \{\alpha_1, \alpha_2, \ldots, \alpha_m\}$, then $q(c) \neq 0$.

$$\lim_{x \to c} r(x) = \lim_{x \to c} \frac{p(x)}{q(x)} = \frac{p(c)}{q(c)} = r(c)$$

and hence $r(x)$ is continuous at $x = c$, when $c \notin \{\alpha_1, \alpha_2, \ldots, \alpha_m\}$.

Thus a rational function is continuous at every real number for which it is defined.

Problem [4.7]: Examine the continuity of (i) $\sin x$, (ii) $\cos x$ and (iii) $\tan x$ on $\mathbb{R}$.

Solution.

1. Let $f(x) = \sin x$ and let $c \in \mathbb{R}$.
   Then
   $$\begin{aligned} |f(x) - f(c)| &= |\sin x - \sin c| \\ &= \left|2\cos\tfrac{x+c}{2} \cdot \sin\tfrac{x-c}{2}\right| \\ &\leq 2\left|\cos\tfrac{x+c}{2}\right| \left|\sin\tfrac{x-c}{2}\right| \\ &< 2 \cdot 1 \cdot \tfrac{1}{2}|x - c|, \text{ whenever } |x - c| < \delta \\ &= |x - c| < \epsilon, \text{ taking } \delta = \epsilon \end{aligned}$$
   So $\sin x$ is continuous at $x = c$ and hence in $\mathbb{R}$ as c is an arbitrary point on $\mathbb{R}$.
2. Let $f(x) = \cos x$ and $c \in \mathbb{R}$.
   Then
   $$\begin{aligned} |f(x) - f(c)| &= |\cos x - \cos c| \\ &= 2\left|\sin\tfrac{x+c}{2}\right| \left|\sin\tfrac{x-c}{2}\right| \\ &\leq 2 \cdot 1 \cdot \tfrac{1}{2}|x - c| < \epsilon, \text{ whenever } |x - c| < \delta = \epsilon. \end{aligned}$$
   Hence $\cos x$ is continuous on $\mathbb{R}$.
3. $f(x) = \frac{\sin x}{\cos x}$ = rational function of x.
   So $f(x)$ is continuous for those values of x on $\mathbb{R}$ for which $\cos x \neq 0$, i.e., for those values of x, when $x \neq (2n+1)\frac{\pi}{2}$ [by Problem [4.6](ii)].

Properties of a Continuous Function

Theorem [4.7]:

(a) If $f, g : D \to \mathbb{R}$ be two continuous functions on $\mathbb{R}$, then (i) $f \pm g$, $f \cdot g$, $\frac{f}{g}$, $(g(x) \neq 0, x \in D)$ are continuous on D.

(b) If $f : D \to \mathbb{R}$ be the continuous on $\mathbb{R}$, then $|f|, \sqrt{f}, \frac{1}{f}$ (when $f(x) \neq 0, x \in D$) are continuous on $\mathbb{R}$.

(c) If $f : [a, b] \to \mathbb{R}$ be continuous at a point $c \in [a, b]$ and $f(c) \neq 0$, then f keeps the same sign as $f(c)$ in the neighbourhood $N_\delta(c)$ of $c\ \forall\ x \in N_\delta(c)$.

(d) If $f : [a, b] \to \mathbb{R}$ be continuous, then f is bounded and attains its bounds in $[a, b]$.

(e) If $f : [a, b] \to \mathbb{R}$ be continuous and $f(a) \cdot f(b) < 0$, then there exists at least one point $p \in (a, b)$ such that $f(p) = 0$.

(f) If $f : [a, b] \to \mathbb{R}$ be continuous on $[a, b]$ and $f(a) \neq f(b)$, then f assumes every value in between $f(a)$ and $f(b)$. [Intermediate value Theorem.]

(g) If $f : [a, b] \to \mathbb{R}$ be continuous and $f(x) \in [a, b] \; \forall \; x \in [a, b]$, then there exists a fixed point $c \in [a, b]$ such that $f(c) = c$. [Fixed point theorem.]

(h) The necessary and sufficient condition that a function $f : \mathbb{R} \to \mathbb{R}$ to be continuous at a point $c \in \mathbb{R}$ is that for every sequence $\{c_n\}$ converging to c [i.e., $\lim_{n\to\infty} c_n = c$], $\lim_{n\to\infty} f(c_n) = f(c)$.

Proof.

(b) Let $f : D \to \mathbb{R}$ be continuous on D and let us assume that

$$f(x) \geq 0 \; \forall \; x \in D \text{ and } \left(\sqrt{f}\right)(x) = \sqrt{f(x)}.$$

Then $\phi(x) = \sqrt{f(x)}$ can be defined $\forall \; x \in D$.
Let $c \in D$.
Now $|\phi(x) - \phi(c)| = \left|\sqrt{f(x)} - \sqrt{f(c)}\right|$

$$= \frac{|f(x) - f(c)|}{\left|\sqrt{f(x)} + \sqrt{f(c)}\right|} < \frac{|f(x) - f(c)|}{\sqrt{f(c)}} \qquad \text{(i)}$$

Since $f(x)$ is continuous, $f(x)$ assumes finite values $\forall \; x \in D$ and also

$$|f(x) - f(c)| < \sqrt{f(c)}\,\epsilon \; \forall \; x \in N_\delta(c) \cap D.$$

So from (i), $|\phi(x) - \phi(c)| < \epsilon \; \forall \; x \in N_\delta(c) \cap D$
$\Rightarrow \phi(x)$ is continuous at $x = c$.
$\Rightarrow \sqrt{f}$ is continuous on D.

(c) f is continuous on D,

$$\Rightarrow |f(x) - f(c)| < \epsilon \; \forall \; x \in N_\delta(c)$$
$$\Rightarrow f(c) - \epsilon < f(x) < f(c) + \epsilon \; \forall \; x \in N_\delta(c) \qquad \text{(i)}$$

$f(c) \neq 0 \Rightarrow f(c) > 0$ or $f(c) < 0$.
Let $f(c) > 0$, we choose $\epsilon < f(c)$. Then $f(c) - \epsilon > 0$, $f(c) + \epsilon > 0$.
So from (i), it follows that $f(x)$ lies between two (+)ve numbers and hence

$$f(x) > 0 \; \forall \; x \in N_\delta(c) \qquad \text{(ii)}$$

Again, let $f(c) < 0$. Then $-f(c) > 0$. We choose $\epsilon < -f(c)$.
Then $f(c) + \epsilon < 0$. Also $f(c) - \epsilon < 0$.
So from (i), it follows that $f(x)$ lies between two negative numbers and hence

$$f(x) < 0 \; \forall \; x \in N_\delta(c) \qquad \text{(iii)}$$

Thus we conclude from (ii) and (iii) that $f(x)$ keeps the same sign as $f(c)$ in $N_\delta(c)$.

(d) Let $f(x)$ be unbounded in $[a, b]$.
Then for each n, there exists a point x_n (say) in $[a, b]$ such that $|f(x_n)| > n$.
Thus we have a sequence $\{x_n\}$ in $[a, b]$.
Then $\{x_n\}$ is bounded and hence it has a limit point ξ. $\xi \in [a, b]$.
So f is continuous at $x = \xi$.
Therefore, $|f(x) - f(\xi)| < \epsilon \; \forall \; x \in N_\delta(\xi) \cap [a, b]$

$$\Rightarrow |f(x)| < |f(\xi)| + 1, \text{ taking } \epsilon = 1, \; \forall \; x \in N_\delta(\xi) \cap [a, b]$$
$$\Rightarrow n < |f(x_n)| < |f(\xi)| + 1 = \text{finite number} \qquad \text{(i)}$$

Now ξ being a limit point, there exist infinite number of points x_n in $N_\delta(\xi)$.

So (i) cannot be true, as n assumes infinite number of values and it cannot be less than a finite number.

Hence f is bounded in $[a, b]$.

Let M and m be the supremum and infimum of $f(x)$ in $[a, b]$. We have to show that there exist points $\alpha, \beta \in [a, b]$ such that $f(\alpha) = M$ and $f(\beta) = m$.

If possible, let $f(x) \neq M$ for any $x \in [a, b]$.

Then $M - f(x) \neq 0$.

We now consider a function $\phi(x) = \frac{1}{M-f(x)}$ in $[a, b]$. Then $\phi(x)$ is continuous in $[a, b]$ as $f(x)$ is continuous in $[a, b]$ and $M - f(x) \neq 0 \; \forall \; x \in [a, b]$.

Therefore, $\phi(x)$ is bounded in $[a, b]$.

So there exists a (+)ve number k such that $\phi(x) \leq k \; \forall \; x \in [a, b]$.

That is,

$$\frac{1}{M - f(x)} \leq k \Rightarrow M - f(x) \geq \frac{1}{k} \Rightarrow f(x) \leq M - \frac{1}{k} < M$$

$\Rightarrow M - \frac{1}{k}$ is an upper bound of $f(x)$ and it is less than M and hence it contradicts that M is the supremum (l.u.b.) of $f(x)$.

Therefore, $\exists \; \alpha \in [a, b]$ such that $f(\alpha) = M$.

The same can be proved in case of infimum of $f(x)$ by considering $\phi(x) = \frac{1}{f(x)-m}$, where $f(x) - m \neq 0$ for any $x \in [a, b]$.

(e) Let $f(a) > 0$ and $f(b) < 0$. Then

$$f(x) > 0 \; \forall \; x \in [a, a + \delta) \tag{i}$$

$$\text{and } f(x) < 0 \; \forall \; x \in (b - \delta, b] \tag{ii}$$

We consider the set $S = \{x, a \leq x \leq b \text{ and } f(x) > 0\}$

$$f(a) > 0 \Rightarrow a \in S \Rightarrow S \neq \phi$$

By order completeness property of real numbers, S has a supremum p, say, in $[a, b]$.

Then $a \leq p \leq b$.

We shall prove that $p \neq a$ and $p \neq b$ and $f(p) = 0$.

(i) $\Rightarrow [a, a + \delta) \subseteq S$

$\Rightarrow p \geq a + \delta \Rightarrow p \neq a$

Also (ii) $\Rightarrow b - \delta$ is an upper bound of S as $S = \{x, \; a < x < b - \delta, \text{ where } f(x) > 0\}$

$\Rightarrow p \leq b - \delta \Rightarrow p \neq b$

Now it is sufficient to show that $f(p) \ngtr 0$ and $f(p) \nless 0$.

If possible, let $f(p) > 0$.

Then $f(x) > 0 \; \forall \; x \in (p - \delta_1, p + \delta_1) \subseteq S$

$\Rightarrow \exists$ a member $p + \delta_1$ of S greater than the supremum p of S, which is not possible.

So $f(p) \ngtr 0$. (iii)

Again, if possible, let $f(p) < 0$.

Then $f(x) < 0 \; \forall \; x \in (p - \delta_2, p + \delta_2)$.

Therefore, p being supremum of S, $\exists\ \beta \in S$ such that $p - \delta_2 < \beta \leq p$ and consequently, $f(\beta) > 0$. But

$$\beta \in (p - \delta_2, p + \delta_2) \Rightarrow f(\beta) < 0$$

which arises a contradiction and hence $f(p) \not< 0$. (iv)

So from (iii) and (iv) it follows that $f(p) = 0$.

Thus, $\exists$ a point $p \in [a, b]$ such that $f(p) = 0$.

(f) It can be easily proved with the help of (e) by considering a function $\phi(x) = f(x) - c$, where $f(a) < c < f(b)$.

(g) $f : [a, b] \to \mathbb{R}$ is continuous and $f(x) \in [a, b] \subseteq \mathbb{R}\ \forall\ x \in [a, b]$.

If $f(a) = a$ or $f(b) = b$, then there is nothing to prove.

So let $f(a) \neq a$ and $f(b) \neq b$ and let us assume that $f(a) > a$ and $f(b) < b$.

We consider a function $\phi(x) = f(x) - x\ \forall\ x \in [a, b]$. $\phi(x)$ is continuous in $[a, b]$.

Also $\phi(a) = f(a) - a > 0$ and $\phi(b) = f(b) - b < 0$.

So $\phi(a) \cdot \phi(b) < 0$ and hence by (e), $\exists$ a point $c \in [a, b]$ such that $\phi(c) = 0$.

Therefore, $f(c) - c = 0 \Rightarrow f(c) = c$.

(h) Proof follows from Theorem [4.2].

[Proof of (a) and other parts of (b) are left.]

Problem [4.8]: Show that the function f defined by $f(x) = 2x + 1$ for $0 \leq x \leq 1$ satisfies the intermediate value theorem, while the function f defined by $f(x) = 2x + 1, 0 < x \leq 1$ and $f(0) = \frac{1}{2}$ does not satisfy the intermediate value theorem.

Solution. $f(x)$ is continuous in $[0, 1]$. Also $f(0) = 1$ and $f(1) = 3$.

Let c be any number lying between $f(0)$ and $f(1)$. That is, $1 \leq c \leq 3$. Then $f(x)$ assumes some value of $x \in [0, 1]$, as $f(x) = c \Rightarrow 2x + 1 = c \Rightarrow x = \frac{c-1}{2} \Rightarrow x \in [0, 1]$ for $1 \leq c \leq 3$.

But the function $f(x) = 2x + 1,\ 0 < x \leq 1$ and $f(0) = \frac{1}{2}$ is discontinuous in $[0, 1]$ as it is discontinuous at $x = 0$.

Also $f(0) = \frac{1}{2}, f(1) = 3$.

Let c be any real number such that $\frac{1}{2} \leq c \leq 3$. Then $f(x)$ does not assume all values between $f(0)$ and $f(1)$.

For $f(x) = c \Rightarrow x = \frac{c-1}{2}, \frac{1}{2} \leq c \leq 3$.

Let $c = \frac{2}{3}$, then $x = \frac{\frac{2}{3}-1}{2} = -\frac{1}{6} \notin [0, 1]$.

So there is no $x \in [0, 1]$ such that $f(x) = \frac{2}{3}$, a number lying between $f(0)$ and $f(1)$. Hence $f(x)$ fails to satisfy the result of intermediate value theorem.

Theorem [4.8]: Let $f : A \to \mathbb{R}$ and $g : B \to \mathbb{R}, A, B \subseteq \mathbb{R}$ be two functions such that f is continuous at a point $c \in A$ and g is continuous at $b = f(c) \in B$, and $f(A) \subseteq B$. Then $g \circ f : A \to \mathbb{R}$ is continuous at c.

Proof. f is continuous at $c \in A$.

$$\Rightarrow |f(x) - f(c)| < \epsilon \text{ for } |x - c| < \delta_1 \quad \text{(i)}$$

Also g being continuous at $b = f(c) \in B$

$$\Rightarrow |g(y) - g(b)| < \epsilon \text{ for } |y - b| < \delta_2 \quad \text{(ii)}$$

where $y = f(x) \in f(A) \subseteq B$

$\Rightarrow |g(f(x)) - g(f(c))| < \epsilon$ for $|f(x) - f(c)| < \delta_2$

$\Rightarrow |(g \circ f)(x) - (g \circ f)(c)| < \epsilon$ for $|f(x) - f(c)| < \epsilon$, taking $\delta_2 = \epsilon$

$\Rightarrow |(g \circ f)(x) - (g \circ f)(c)| < \epsilon$, whenever $|x - c| < \delta_1$ [by (i)]

$\Rightarrow g \circ f$ is continuous at $x = c \in A$.

Example. Dirichlet's function defined on $\mathbb{R}$ by

$$f(x) = \begin{cases} 1, & \text{when } x \text{ is irrational} \\ -1, & \text{when } x \text{ is rational,} \end{cases}$$

is discontinuous $\forall\, x \in \mathbb{R}$.

Example. A function $f : \mathbb{R} \to \mathbb{R}$ defined by

$$f(x) = \begin{cases} x, & \text{when } x \in I_{\mathbb{R}} \text{ [i.e., when } x \text{ is irrational]} \\ -x, & \text{when } x \in Q, \end{cases}$$

is continuous at $x = 0$ only.

Example. A function $f : [0, 1] \to \mathbb{R}$ defined by

$$f(x) = \begin{cases} -1, & \text{for } x \in Q \\ 1, & \text{for } x \in I_{\mathbb{R}}, \end{cases}$$

is discontinuous on $[0, 1]$, but $|f|$ is continuous on $[0.1]$.

Problem [4.9] : Let $f, g : \mathbb{R} \to \mathbb{R}$ be two functions and $c \in \mathbb{R}$. Also let $\lim_{x \to c} f(x) = b$ and that g is continuous at b. Then show that $\lim_{x \to c} (g \circ f)(x) = g(b)$.

Solution. $\lim_{x \to c} f(x) = b$

$$\Rightarrow |f(x) - b| < \epsilon \ \forall\, x \in N'\delta(c) \tag{1}$$

Again, g is continuous at b.

So $|g(y) - g(b)| < \epsilon$ for $|y - b| < \delta$.

$\Rightarrow |g(f(x)) - g(b)| < \epsilon$ for $|f(x) - b| < \delta$, $y = f(x)$ for some $x \in \mathbb{R}$.

$\Rightarrow |(g \circ f)(x) - g(b)| < \epsilon$ for $\forall\, x \in N'\delta(c)$, by (1) taking $\delta = \epsilon$.

$\Rightarrow \lim_{x \to c} (g \circ f)(x) = g(b)$.

Uniform Continuity

Let $f : [a, b] \to \mathbb{R}$ be a function such that it is continuous at every point $c \in [a, b]$, then we say f is uniformly continuous on $[a, b]$.

Definition [4.7] : A function $f : [a, b] \to \mathbb{R}$ is defined to be uniformly continuous on $[a, b]$, if for every $\epsilon\,(> 0)$, there exists a (+)ve number δ such that $|f(x_2) - f(x_1)| < \epsilon$, when any two points $x_1, x_2 \in [a, b]$ satisfies $|x_2 - x_1| < \delta$.

Theorem [4.9] : Uniform continuity of $f : [a, b] \to \mathbb{R}$ on $[a, b] \Rightarrow$ continuity of f on $[a, b]$, but the converse may not be true.

Proof. Let f be uniformly continuous on $[a, b]$.

Then by Definition [4.7], we have $|f(x_2) - f(x_1)| < \epsilon$, for any two points $x_1, x_2 \in [a, b]$ satisfying $|x_2 - x_1| < \delta$.

Let $x_2 = x$.

Then $|f(x) - f(x_1)| < \epsilon$, whenever $|x - x_1| < \delta$

$\Rightarrow f$ is continuous at $x = x_1$

$\Rightarrow f$ is continuous on $[a, b]$ as x_1 is chosen arbitrarily in $[a, b]$.

For the converse part, let $f(x) = \frac{1}{x}$ for $0 < x \leq 1$.

Then f is continuous on $(0, 1]$, since for any point $c \in (0, 1]$,

$$|f(x) - f(c)| = \left|\frac{1}{x} - \frac{1}{c}\right| = \left|\frac{x - c}{|x||c|}\right| = \frac{m|x - c|}{|c|} < \frac{m\delta}{|c|}$$

when $x = \frac{1}{m}$, such that $0 < \frac{1}{m} \leq 1$ and $|x - c| < \delta$.

$\Rightarrow |f(x) - f(c)| < \epsilon$, taking $\delta = \frac{|c|}{m} \cdot \epsilon$

Now let $x_2 = \frac{1}{n}$ and $x_1 = \frac{1}{n+1}$ be any points in $(0, 1]$.

Then $|x_2 - x_1| = \left|\frac{1}{n} - \frac{1}{n+1}\right| = \frac{1}{n(n+1)} < \frac{1}{n} < \delta$.

But $|f(x_2) - f(x_1)| = |n - (n + 1)| = 1 \not< \epsilon$, whenever $|x_2 - x_1| < \delta$.

So f is not uniformly continuous on $(0, 1]$.

Theorem [4.10] : If $f : [a, b] \to \mathbb{R}$ be continuous in the closed interval $[a, b]$, then f is uniformly continuous in $[a, b]$.

Proof. If not, then there exist two points x and u such that for every $\epsilon\,(> 0)$ there exists a (+)ve number δ so that $|f(x) - f(u)| \geq \epsilon$ for $|x - u| < \delta$.

Taking $\delta = \frac{1}{n}$, for each $n \in N$ we get two sequences $\{x_n\}$ and $\{u_n\}$ in $[a, b]$ such that $|f(x_n) - f(u_n)| \geq \epsilon$ for $|x_n - u_n| < \frac{1}{n}$, $\forall\, n \in N$.

Then the sequence $\{x_n\}$ and $\{u_n\}$ being both bounded in $[a, b]$, each of them has a limit point in $[a, b]$.

Let $\{x_n\}$ converge to ξ. Then there is a subsequence $\{x_{n_k}\}$ of $\{x_n\}$ that converges to ξ. Also $\xi \in [a, b]$ as $[a, b]$ is closed. The corresponding subsequence $\{u_{n_k}\}$ of $\{u_n\}$ also converges to the same limit ξ.

$$\begin{aligned} \text{Since} \qquad |u_{n_k} - \xi| &= |u_{n_k} - x_{n_k} + x_{n_k} - \xi| \\ &\leq |u_{n_k} - x_{n_k}| + |x_{n_k} - \xi| < \epsilon\ \forall\ n_k \in N \\ \text{as } |x_{n_k} - u_{n_k}| &< \frac{1}{n_k} = \epsilon/2 \text{ (say)} \\ \text{and } |x_{n_k} - \xi| &< \epsilon/2,\ \forall\ n_k \in N. \end{aligned}$$

Thus $u_{n_k} \to \xi$ as $k \to \infty$.

Now f is continuous at ξ. So both the sequences $\{f(x_{nk})\}$ and $\{f(u_{nk})\}$ must converge to $f(\xi)$. But this is not possible since $|f(x_n) - f(u_n)| \geq \epsilon\ \forall\ n \in N$ and hence f is not continuous at $x = \xi$.

So if f is not uniformly continuous, then f is not continuous at some point $\xi \in [a, b]$. Hence the Theorem.

Theorem [4.11] : If $f : D \to \mathbb{R}, D \subseteq \mathbb{R}$ be uniformly continuous on D and $\{x_n\}$ be a Cauchy sequence in D. Then $\{f(x_n)\}$ is also a Cauchy sequence in $\mathbb{R}$.

Proof. Let $\{x_n\}$ be a Cauchy sequence in D.

Then $|x_m - x_n| < \epsilon \ \forall \ m, n \geq N.$ (1)

Now f is uniformly continuous on $D \Rightarrow |f(x) - f(y)| < \epsilon, |x - y| < \delta, \ \forall \ x, y \in D.$

So for $x = x_m, y = y_n, |f(x_m) - f(y_n)| < \epsilon$, whenever $|x_m - y_n| < \delta = \epsilon \ \forall \ m, n \geq N.$

$\Rightarrow |f(x_m) - f(y_n)| < \epsilon \ \forall \ m, n \geq N \Rightarrow \{f(x_n)\}$ is a Cauchy sequence.

Theorem [4.12] : **[Continuous extension theorem]** A function $f : (a, b) \to \mathbb{R}$ is uniformly continuous on (a, b), if and only if f can be defined at $x = a$ and at $x = b$ such that the extended function $f_e : [a, b] \to \mathbb{R}$ is continuous on $[a, b]$.

Proof. Let $f_e : [a, b] \to \mathbb{R}$ be continuous so that $f(a)$ and $f(b)$ can be defined.

Thus continuity of f_e on $[a, b] \Rightarrow$ continuity of f on $(a, b) \Rightarrow$ uniform continuity of f on (a, b).

Now let f be uniformly continuous on (a, b). We have to extend f to f_e so that f_e is continuous on $[a, b]$.

Let $\{x_n\}$ be a sequence in (a, b) so that $\{x_n\}$ converges to a. That is $\lim\limits_{n\to\infty} x_n = a$. Then $\{x_n\}$ is a Cauchy sequence and hence by Theorem [4.11] $\{f(x_n)\}$ is a Cauchy sequence.

Again, if $\{u_n\}$ be another sequence on (a, b) such that $\lim\limits_{n\to\infty} u_n = a$.

Then $\lim\limits_{n\to\infty} (x_n - u_n) = a - a = 0.$

Also f is uniformly continuous on (a, b).

So $|f(x_n) - f(u_n)| < \epsilon$, whenever $|x_n - u_n| < \delta, \ \forall \ n \geq N \Rightarrow \lim\limits_{n\to\infty} \{f(x_n) - f(u_n)\} = 0.$

$$\begin{aligned}\text{Thus } \lim_{n\to\infty} f(u_n) &= \lim_{n\to\infty} \{f(u_n) - f(x_n)\} + \lim_{n\to\infty} f(x_n)\\ &= 0 + \lim_{n\to\infty} f(x_n) = \lim_{n\to\infty} f(x_n) = L \text{ (say)}.\end{aligned}$$

So we get the same value L for every sequence converging to a and this leads to the conclusion that f has a limit at a. If we define $L = f(a)$, then f is continuous at a.

With the similar argument f is considered to be continuous at $x = b$. Hence we conclude f has a continuous extension to the interval $[a, b]$.

Problem [4.10] : Show that $f(x) = x^2$ is uniformly continuous on $[-1, 1]$, but not uniformly continuous on $[0, \infty]$.

Solution. Let $f(x) = x^2$ and let x_1, x_2 be any points in $[-1, 1]$ such that $|x_2 - x_1| < \delta$.

$$\begin{aligned}\text{Then } |f(x_2) - f(x_1)| &= |x_2^2 - x_1^2 = |x_2 + x_1||x_2 - x_1|\\ &\leq [|x_2| + |x_1|]\delta\\ &\leq 2\delta \text{ as } |x_1|, |x_2| \leq 1\\ &< \epsilon, \text{ taking } \delta = \epsilon/2.\end{aligned}$$

So f is uniformly continuous on $[-1, 1]$.

Let $|x_2 - x_1| = \frac{\delta}{2} < \delta$.

By Archemedian Principle, $\exists$ a (+)ve integer n such that $n\delta^2 > \epsilon$.

Let $x_2 = n\delta + \delta/2$, $x_1 = n\delta \in [0, \infty)$.

Then $|f(x_2) - f(x_1)| = |x_2^2 - x_1^2| = |(x_2 - x_1)||x_2 + x_1|$

$$= \frac{\delta}{2}\left[2n\delta + \frac{\delta}{2}\right] = n\delta^2 + \frac{\delta^2}{4} > \epsilon$$

$\therefore f$ is not uniformly continuous in $[0, \infty)$.

4.3 Monotonic Functions

Definition [4.8] : A function $f : D \to \mathbb{R}, D \subseteq \mathbb{R}$ is said to be monotonically increasing on D, if $f(x_1) \leq f(x_2) \ \forall \ x_1, x_2 \in D$ and $x_1 \leq x_2$.

If $f(x_1) < f(x_2) \ \forall \ x_1, x_2 \in D$ and $x_1 < x_2$, then f is called strictly monotonically increasing.

f is said to be monotonically decreasing if $f(x_1) \geq f(x_2) \ \forall \ x_1, x_2 \in D$ and $x_1 \leq x_2$ and it is called strictly monotonically decreasing if $f(x_1) > f(x_2) \ \forall \ x_1, x_2 \in D$ and $x_1 < x_2$.

f is said to be monotone on D if it is either strictly monotonically increasing or strictly monotonically decreasing.

Example. Let $f : [0, 1] \to \mathbb{R}$ be defined by $f(x) = x$.

Then f is strictly monotonically increasing on $[0, 1]$ while f defined by $f(x) = \frac{1}{x^2}$ is strictly monotonically decreasing on $[0, 1]$.

Example. Let $f : [0, \pi/2] \to \mathbb{R}$ be defined by $f(x) = \sin x$.

Then f is strictly monotonically increasing on $[0, \pi/2]$, while $f : [0, \pi/2] \to \mathbb{R}$, defined by $f(x) = \cos x$ is strictly monotonically decreasing on $[0, \pi/2]$.

Unless or otherwise stated we shall use simply *strictly increasing* or *strictly decreasing.*

Example. Monotone functions are not necessarily continuous.

Let $f : [0, 3] \to \mathbb{R}$ be defined by

$$\begin{aligned} f(x) &= 1 \text{ for } 0 \leq x \leq 1 \\ &= 2 \text{ for } 1 < x \leq 2 \\ &= 3 \text{ for } 2 < x \leq 3. \end{aligned}$$

Then f is monotonically increasing on $[0, 3]$, but is discontinuous at $x = 1$ and at $x = 2$.

Locally increasing and locally decreasing

Definition [4.9] : $f : D \to \mathbb{R}$, $D \subseteq \mathbb{R}$ is said to be locally increasing at a point $c \in D$ if $\exists$ a (+)ve number δ such that $\forall \ x_1, x_2 \in (c - \delta, c + \delta)$ and $x_1 < x_2 \Rightarrow f(x_1) < f(x_2)$

That is, $c - \delta < x_1 < x_2 < x_3 < c + \delta \Rightarrow f(x_1) < f(x_2)$

i.e., $\forall \ x_1, x_2 \in N_\delta(c) \cap D, x_1 < x_2 \Rightarrow f(x_1) < f(x_2)$

Theorem [4.13] : If $f : D \to \mathbb{R}$ be locally increasing at every point of D, then f is increasing on D.

Proof. Let f be locally increasing at a point of $c \in D$.

Then by Def. [4.9],

$$f(x_1) < f(x_2) \ \forall \ x_1, x_2 \in (c - \delta, c + \delta) \text{ and } x_1 < x_2 \qquad \text{(i)}$$

Let $y_1, y_2 \in D$. If $y_1, y_2 \in (c-\delta, c+\delta) \cap D$ and $y_1 < y_2$, then by (i), $f(y_1) < f(y_2)$ and hence f is increasing on D.

Let α and β be any two points of D and $\delta_1, \delta_2 > 0$ such that $y_1 \in (\alpha-\delta_1, \alpha+\delta_1)$ and $y_2 \in (\beta-\delta_2, \beta+\delta_2)$ and f is locally increasing at α and β.

If the intervals $(\alpha-\delta_1, \alpha+\delta_1)$ and $(\beta-\delta_2, \beta+\delta_2)$ be overlapping, i.e., $\beta-\delta_2 < \alpha+\delta_1$, then $\alpha-\delta_1 < y_1 < r < \beta-\delta_2 < \alpha+\delta_1 < t < y_2 < \beta+\delta_2$

$\Rightarrow f(y_1) < f(\beta-\delta_2) < f(\alpha+\delta_1) < f(t) < f(y_2)$ [f is locally increasing at α and β] (ii)

$\Rightarrow f(y_1) < f(y_2)\ \forall\ y_1, y_2 \in D$ and $t < y_2$

$\Rightarrow f$ is increasing on D.

Next, let $(\alpha-\delta_1, \alpha+\delta_1)$ and $(\beta-\delta_2, \beta+\delta_2)$ be two non-overlapping intervals. Then we take a set of points $c_1, c_2, , \cdots, c_n$ in $(\alpha+\delta_1, \beta-\delta_2)$ so that we enclose them by overlapping intervals $(c_1-\delta', c_1+\delta'), (c_2-\delta'', c_2+\delta''), \cdots, \delta', \delta'', \cdots, > 0$ where $c_1-\delta' < \alpha+\delta_1, c_2-\delta'' < c_1+\delta'$ and so on. Then by case (ii) it follows that f is increasing on D.

Remark. Let $\exists$ (+)ve number δ_γ such that $(\gamma-\delta_\gamma, \gamma+\delta_\gamma) \supset (\beta-\delta_2, \alpha+\delta_1)$ and let f be locally increasing at γ, then $f(\beta-\delta_2) < f(\alpha+\delta_1)$ and (ii) is justified.

Corollary [4.2]: If $f : D \to \mathbb{R}$ be locally decreasing, then f is decreasing on D.

Theorem [4.14]: Let $f : [a,b] \to \mathbb{R}$, $[a,b] \subseteq \mathbb{R}$ be monotonically increasing on $[a,b]$ and let c be a point such that $a < c < b$. Then $f(c-0)$ and $f(c+0)$ both exist and

$$f(c-0) = \sup_{x\in(a,c)} f(x), \quad f(c+0) = \inf_{x\in(c,b)} f(x)$$

Proof. We consider the set $S = \{f(x), a < x < c\}$.

Then $S \neq \phi$. Also $x \in [a,b]$ and $x < c \Rightarrow f(x) \leq f(c) \Rightarrow f$ is bounded above by $f(c)$.

Let ξ be the supremum of S. Then $\xi \leq f(c)$.

Let $\epsilon\,(>0)$ be given. Then $f(x) \leq \xi < \xi+\epsilon$ for all $x \in (a,c)$.

Thus $f(x) < \xi + \epsilon$. (i)

We choose a (+)ve number δ such that $a < c-\delta < c$ and $f(c-\delta) > \xi - \epsilon$

When $x > c-\delta$, we have $f(x) > f(c-\delta) > \xi-\epsilon$ as f is monotonically increasing.

Thus for every $\epsilon\,(>0)$, $\exists$ a (+)ve number δ such that

$$\xi-\epsilon < f(x) < \xi+\epsilon \text{ for } c-\delta < x < c$$

$$\Rightarrow |f(x)-\xi| < \epsilon \text{ for } c-\delta < x < c \Rightarrow \lim_{x\to c-0} f(x) = \xi = \sup_{x\in(a,c)} f(x)$$

Hence $f(c-0)$ exists and it is equal to $\sup\limits_{a<x<c} f(x)$.

Next, considering the set $S = \{f(x), c < x < b\}$, which is non-empty and bounded below by $f(c)$ as f is monotonically increasing and proceeding as above by taking an $\inf S = \eta$, we can prove the result $\lim\limits_{x\to c+0} f(x) = \eta = \inf\limits_{c<x<b} f(x)$.

Theorem [4.15]: If $f : [a,b] \to \mathbb{R}$, $[a,b] \subseteq \mathbb{R}$ be monotonically decreasing on $[a,b]$ and c be a point such that $a < c < b$, then $f(c-0]$ and $f(c+0)$ both exist and

$$f(c-0] = \inf_{a<x<c} f(x) \text{ and } f(c+0) = \sup_{c<x<b} f(x).$$

[Proof is left]

Corollary [4.3] :

(i) $f(c-0) \le f(c) \le f(c+0)$

(ii) $f(c-0) \ge f(c) \ge f(c+0)$

[*Hints.* (i) can be proved when f is monotonically increasing on $[a, b]$ and (ii) can be proved when f is monotonically decreasing on $[a, b]$.]

Corollary [4.4] : If $f : [a, b] \to \mathbb{R}$ be continuous, then

$$\sup_{a<x<c} f(x) = f(c) = \inf_{c<x<b} f(x).$$

[*Hints.* f is continuous at $x = c \Rightarrow \lim_{x\to c-0} f(x) = f(c)$ and $\lim_{x\to c+0} f(x) = f(c)$.]

Definition [4.10] : If $f : [a, b] \to \mathbb{R}$, $[a, b] \subseteq \mathbb{R}$ be monotonically increasing on $[a, b]$ and $a < c < b$, then Jump of f at $x = c$ denoted by $J_f(c)$, is defined by

$$J_f(c) = \lim_{x\to c+0} f - \lim_{x\to c-0} f = \inf_{c<x<b} f - \sup_{a<x<c} f$$

Also $f(c) - \lim_{x\to c-0} f$ is called left-hand jump at $x = c$ and $\lim_{x\to c+0} f - f(c)$ is called right-hand jump at $x = c$.

Thus jump at $x = a$ is a right-hand jump and is written as $J_f(a) = \lim_{x\to a+0} f - f(a)$ and jump at $x = b$ is a left-hand jump and is written as $J_f(b) = f(b) - \lim_{x\to b-0} f$. LK is called the *height of the jump* of f at $x = c$.

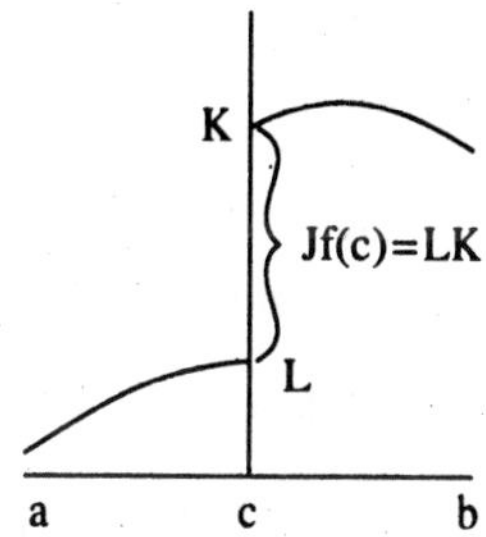

Fig. 4.1

Theorem [4.16] : If $f : [a, b] \to \mathbb{R}$ be continuous on $[a, b]$ and $c \in [a, b]$ such that $a < c < b$, then $J_f(c) = 0$.

Proof. $J_f(c) = \lim_{x\to c+0} f - \lim_{x\to c-0} f$. (i)

Now f is continuous at $x = c$, $a < c < b \Rightarrow \lim_{x\to c+0} f = \lim_{x\to c-0} f = f(c)$.

Hence, from (i) $J_f(c) = f(c) - f(c) = 0 = 0$.

Definition [4.11] : Let $f : [a, b] \to \mathbb{R}$ and c be a point in $[a, b]$ such that $a < c < b$. Then c is called a Jump discontinuity of f, if either the Jump of f at c or left-hand jump or right-hand jump of f at c is different from 0.

Problem [4.11] : Show that the function f defined by $f(x) = \frac{x}{|x|}$, if $x \ne 0$ and $f(0) = A$ has a jump discontinuity at 0, regardless to the value of A.

Solution. $\lim_{x\to 0+0} f(x) = \lim_{x\to 0+0} \dfrac{x}{|x|} = \lim_{x\to 0+0} 1 = 1$

$\lim_{x\to 0-0} f(x) = \lim_{x\to 0-0} \dfrac{x}{|x|} = \lim_{x\to 0-0} -1 = -1$

Then Jump of f at $x = 0$ is $\lim_{x\to 0+0} f(x) - \lim_{x+0-0} f(x) = 1 - (-1) = 2 \ne 0$.

Thus f has a Jump discontinuity at $x = 0$.

The heights of the Jump at $x = 0$ is 2.

Problem [4.12] : Show that the function f defined by $f(x) = 1$, if $x \neq 0$ and $f(0) = 0$, has a removable Jump discontinuity at $x = 0$.

Solution. $\lim_{x\to 0+0} f(x) = 1$ and $\lim_{x\to 0-0} f(x) = 1$.

$f(0+0)$ and $f(0-0)$ both exist and are equal but different from $f(0)$.
So $x = 0$ is removable discontinuity.

Again, $f(0) - \lim_{x\to 0-0} f(x) = 0 - 1 = -1 \neq 0$

or, $\lim_{x\to 0+0} f(x) - f(0) = 1 - 0 = 1 \neq 0$.

Hence f has a removable jump discontinuity at $x = 0$.

Problem [4.13] : Show that the function f, defined by $f(x) = [x]$ has a jump discontinuity for any integral values of x and the heights of jump is 1.

Solution. Let $x = k$, where k is a positive integer.

$$\text{Then} \quad \lim_{x\to k-0} f(x) = \lim_{x\to k-0} [x] = \lim_{h\to 0} [k-h]$$
$$= \lim_{h\to 0} [(k-1) + (1-h)] = k - 1$$
$$\text{and} \quad \lim_{x\to k+0} f(x) = \lim_{x\to k+0} [x] = \lim_{h\to 0} [k+h] = k$$
$$\text{Therefore} \quad \lim_{x\to k+0} f(x) - \lim_{x\to k-0} f(x) = k - (k-1) = 1 \neq 0.$$

Hence f has a jump discontinuity at $x = k$ and the height of jump is 1.

Again, let $x = -k$, k is a positive integer

$$\text{Then} \quad \lim_{x\to -k-0} f(x) = \lim_{x\to -k-0} [x] = \lim_{h\to 0} [-k-h]$$
$$= \lim_{h\to 0} [(-k-1) + (1-h)] = -(k+1)$$
$$\text{and} \quad \lim_{x\to -k+0} f(x) = \lim_{x\to -k+0} [x] = \lim_{h\to 0} [-k+h] = -k$$
$$\text{Therefore,} \quad \lim_{x\to -k+0} f(x) - \lim_{x\to -k-0} f(x) = -k + k + 1 = 1 \neq 0.$$

Thus f has a jump discontinuity at $x = -k$ and the height of jump is 1.

Problem [4.14] : Show that the function f defined by $f(x) = \sin\frac{1}{x}, x \neq 0$ and $f(0) = A$ has an irremovable discontinuity at $x = 0$, while the function f, defined by $f(x) = x\sin\frac{1}{x}, x \neq 0$, $f(0) = 1$ has a removable jump discontinuity at $x = 0$.

Solution. For the function $f(x) = \sin\frac{1}{x}$, neither of $\lim_{x\to 0+0} f(x)$ nor $\lim_{x\to 0-0} f(x)$ exists.

Hence $x = 0$ is an irremovable discontinuity.

Now for the function $f(x) = x\sin\frac{1}{x}$,

$$\lim_{x\to 0+0} x\sin\frac{1}{x} = 0 \text{ and } \lim_{x\to 0-0} x\sin\frac{1}{x} = 0$$

So $f(0+0)$ and $f(0-0)$ both exist and equal but different from $f(0)$. So $x = 0$ is a removable discontinuity.

Also $f(0) - \lim_{x\to 0-0} f(x) = 1 \neq 0$. So $x = 0$ is a removable jump discontinuity.

Problem [4.15] : Let $f : [a, b] \to \mathbb{R}, [a, b] \subseteq \mathbb{R}$ be increasing on $[a, b]$ and c is a point such that $a < c < b$. Then show that $J_f(c) = \inf[f(y) - f(x), x < c < y, x, y \in [a, b]]$.

Solution. $J_f(c) = \lim\limits_{x \to c+0} f(x) - \lim\limits_{x \to c-0} f(x) = \inf\limits_{c<x<b} f(x) - \sup\limits_{a<x<c} f(x)$

$$= \inf\{f(x), x > c\} - \sup\{f(x), x < c\}$$
$$= \inf\{f(y), y > c\} + \inf\{-f(x), x < c\}$$
$$= \inf\{f(y) - f(x), x < c < y\}, f(y) - f(x) \geq 0$$

as f is increasing, $y > x \Rightarrow f(y) \geq f(x)$.

Inverse Function

A function $f : [a, b] \to [c, d], [a, b], [c, d] \subseteq \mathbb{R}$ has an inverse, if and only if (i) f is injective and surjective, i.e., (ii) $f[a, b] = [c, d]$.

Theorem [4.17] : Every strictly monotone function is injective.

Proof. Let $f : [a, b] \to \mathbb{R}$ be a strictly increasing function. Then $x_1 < x_2 \Rightarrow f(x_1) < f(x_2)$.

Also if f be strictly decreasing, then $x_1 < x_2 \Rightarrow f(x_1) > f(x_2)$.

So $x_1 \neq x_2 \Rightarrow f(x_1) \neq f(x_2)$ and hence f is injective.

Theorem [4.18] : Inverse of strictly monotone and continuous function $f : [a, b] \to \mathbb{R}$ is monotone and continuous on $[c, d]$, where $f[a, b] = [c, d] \subseteq \mathbb{R}$.

Proof. By Theorem [4.17] f is injective. Also $f : [a, b] \to [c, d]$ is onto as $f[a, b] = [c, d] \subseteq \mathbb{R}$.

We have to show that $g : [c, d] \to [a, b]$, inverse of f is strictly monotone.

We suppose that f is strictly increasing on $[a, b]$.

Let $y_1, y_2 \in [c, d]$ with $y_1 < y_2$.

Then $\exists\ x_1, x_2 \in [a, b]$ such that $y_1 = f(x_1)$ and $y_2 = f(x_2)$.

We must have $x_1 < x_2$. For, if $x_1 \geq x_2$, then $y_1 = f(x_1) \geq f(x_2) = y_2$.

That is, $y_1 \geq y_2$, which contradicts to our assumption.

Hence $x_1 < x_2 \Rightarrow g(y_1) = x_1 < x_2 = g(y_2) \Rightarrow g$ is strictly increasing as y_1, y_2 are two arbitrary points on $[c, d]$ and $y_1 < y_2$.

Next, we have to show that g is continuous on $[c, d]$. If not, then g is discontinuous at some point $\alpha \in [c, d]$. Then $J_g(\alpha) \neq 0 \to \lim\limits_{y \to \alpha-0} g(y) < \lim\limits_{y \to \alpha+0} g(y)$ as g is strictly increasing.

Let $\exists$ a number $x \neq g(\alpha)$ such that $\lim\limits_{y \to \alpha-0} g(y) < x < \lim\limits_{y \to \alpha+0} g(y)$. Then $x \neq g(y)$ for any $y \in [c, d]$ and consequently $x \notin [a, b]$, which contradicts that $[a, b]$ is an interval.

Hence g is continuous on $[c, d]$.

Problem [4.16] :

(1) Show that the function $f : \mathbb{R} \to \mathbb{R}$ defined by $f(x) = \lim\limits_{n \to \infty} \frac{(1+\sin \pi x)^n - 1}{(1+\sin \pi x)^n + 1}$ is discontinuous at $x = 0, 1, 2, \ldots, n, \ldots$.

(2) Show that $f : \mathbb{R} \to \mathbb{R}$ defined by

$$f(x) = \lim_{n \to \infty} \left[\lim_{r \to 0} \frac{\sin^2(n!\,\pi x)}{\sin^2(n!\,\pi x) + r^2} \right] = \begin{cases} 0, & \text{when } x = \frac{p}{q},\ (p, q) = 1,\ q \neq 0, \text{ i.e., } x \in Q \\ 1, & \text{when } x \in I_{\mathbb{R}}, \text{ i.e., when } x \text{ is irrational.} \end{cases}$$

(3) If $\phi(t) = \lim_{x\to\infty} \frac{f(t)+\cos^{2x}\pi t\{2g(t)-f(t)\}}{1+\cos^{2x}\pi t}$

Then show that $\phi(t) = \begin{cases} g(t), & \text{when } t \text{ is an integer, i.e., } t \in Z \\ f(t), & \text{when } t \notin Z \end{cases}$

Solution.

(1) $\sin\pi x = 0 \;\forall\; x \in Z$. Then $\lim_{n\to\infty} \frac{(1+\sin\pi x)^n - 1}{(1+\sin\pi x)^n + 1} = \lim_{n\to\infty} \frac{1-1}{1+1} = 0.$

Let $x \notin Z$ and $2m < x < 2m+1$, m is an integer. Then $\sin\pi x > 0$ since πx is lying in 1st and 2nd quadrant $\Rightarrow 1+\sin\pi x > 1 \Rightarrow \frac{1}{1+\sin\pi x} < 1$.

Therefore,

$$\lim_{n\to\infty} \frac{1+\sin\pi x)^n - 1}{(1+\sin\pi x)^n + 1} = \lim_{n\to\infty} \frac{1 - \frac{1}{(1+\sin\pi x)^n}}{1 + \frac{1}{(1+\sin\pi x)^n}}$$

$$= \frac{1-0}{1+0} = 1 \left[\lim_{n\to\infty} z^n = 0, \text{ when } z < 1\right] \qquad (1)$$

Now, let $2m+1 < x < (2m+2)$, m is an integer $\sin\pi x < 0$ as πx is lying in the 3rd quadrant and 4th quadrant.

$\Rightarrow 1+\sin\pi x < 1$.

Therefore, $\lim_{n\to\infty} \frac{(1+\sin\pi x)^n - 1}{(1+\sin\pi x)^n + 1} = \frac{0-1}{0+1} = -1$ [by (1)]

Now, if $x = 2m$, $f(x) = 0$, $f(x+0) = 1$, $f(x-0) = -1$ and if $x = (2m+1)$, $f(x) = 0$, $f(x+0) = -1$, $f(x-0) = 1$.

Hence f is discontinuous at $x = 0, 1, 2, \ldots, n, \ldots$.

(2) Let $x = \frac{p}{q} \in Q$, $q \neq 0$.

Then, $n!\,\pi x$ is an integral multiple of π by taking n sufficiently large ($\because n \to \infty$).

For example, let $x = \frac{2}{97}$, then $n!\,\pi x$ is an integral multiple of π if $n = 98$ [$n > 97$].

$n!\,\pi x = 98!\,\pi \cdot \frac{2}{97} = m\pi$, where m is an integer and hence $\sin n!\,\pi x = 0$,

i.e., $\sin^2 n!\,\pi x = 0$.

$$\therefore f(x) = \lim_{n\to\infty}\left[\lim_{r\to 0} \frac{\sin^2(n!\,\pi x)}{\sin^2(n!\,\pi x) + r^2}\right] = \lim_{n\to\infty}\left[\lim_{r\to 0} \frac{0}{r^2}\right] = 0.$$

Now if $x \in I_{\mathbb{R}}$, i.e., when x is an irrational number, we have $0 < \sin^2 n!\,\pi x < 1$

$$\Rightarrow \lim_{n\to\infty}\left[\lim_{r\to 0} \frac{\sin^2 n!\,\pi x}{\sin^2 n\pi x + r^2}\right] = \lim_{n\to\infty} \frac{\sin^2 n!\,\pi x}{\sin^2 n!\,\pi x} = 1.$$

Since rational numbers and irrational numbers are dense in $\mathbb{R}$, f is discontinuous $\forall\, x \in \mathbb{R}$.

(3) Let t be an integer. Then $\cos\pi t = \pm 1$.

$\Rightarrow \cos^{2x}\pi t = 1$

$\Rightarrow \lim_{x\to\infty} \cos^{2x}\pi t = 1$

Therefore,

$$\phi(t) = \lim_{x\to\infty} \frac{f(t) + \cos^{2x}\pi t(2g(t) - f(t))}{1+\cos^{2x}\pi t} = \frac{f(t) + 1(2g(t) - f(t))}{1+1} = \frac{2g(t)}{2} = g(t).$$

Let t is not an integer. Then $|\cos \pi t| < 1$.

$$\Rightarrow \lim_{x\to\infty} \cos^{2x} \pi t = 0 \quad \left[\begin{array}{c} \because \ |z| < 1 \\ \lim_{n\to\infty} z^n = 0 \end{array}\right]$$

$$\therefore \phi(t) = \lim_{x\to\infty} \frac{f(t) + \cos^{2x} \pi t(2g(t) - f(t))}{1 + \cos^{2x} \pi t} = \frac{f(t)}{1} = f(t).$$

EXERCISE

(1) If $f : D \to \mathbb{R}, D \subseteq \mathbb{R}$ and $\lim_{x\to a} f(x) = l, a \in D$ exists, then show that the limit is unique.

(2) Does the $\lim_{x\to 0} \sin\left(\frac{1}{x}\right)$ exist? Justify your answer.

(3) If f is defined by $f(x) = \begin{cases} \dfrac{\sin[x]}{[x]}, & x \neq 0 \\ 0, & x = 0 \end{cases}$

where $[x]$ denotes the greatest integer not exceeding x, then find $\lim_{x\to 0} \dfrac{\sin[x]}{[x]}$.

(4) Prove that $\lim_{x\to 0+} \dfrac{x}{a}\left[\dfrac{b}{x}\right] = \dfrac{[b]}{a}$ and $\lim_{x\to 0+} \left[\dfrac{x}{a}\right] \cdot \dfrac{b}{x} = 0$.

(5) Show that $\lim_{x\to\infty} a^x \cdot \sin\dfrac{b}{a^x} = \begin{cases} 0, & \text{if } 0 < a < 1 \\ b, & \text{if } a > 1 \end{cases}$

(6) If $\lim_{x\to c} f(x) = l$ and $\{x_n\}$ converges to c, then show that $\lim_{n\to\infty} \{f(x_n)\} = l$ and conversely, show that $\lim_{x\to c} f(x) = l$, if for every sequence $\{x_n\}$ converging to c, $\lim_{n\to\infty} \{f(x_n)\} = l$.

(7) If f be defined by $f(x) = \begin{cases} 3 - x, & \text{when } x \text{ is irrational} \\ 2x, & \text{when } x \text{ is rational} \end{cases}$

prove that $\lim_{x\to a} f(x)$ exists, when $a = 1$ and the limit is equal to 2.

(8) Let $f : (a, b) \to \mathbb{R}$, $(a, b) \subseteq \mathbb{R}$ be monotone increasing and $M = \sup_{x\in(a,b)} f$, $m = \inf_{x\in(a,b)} f$, then show that $f(b - 0) = M$ and $f(a + 0) = m$.

(9) Let $f[0, 1] \to \mathbb{R}$ be defined by $f(x) = \begin{cases} 1, & \text{when } x \text{ is rational} \\ 0, & \text{when } x \text{ is irrational} \end{cases}$

there show that f is not continuous on $[0, 1]$.

(10) Give an example of a function f defined over an interval $I = [-1, 1]$, in each case such that

(i) f has a jump discontinuity at a point $c \in [-1, 1]$.

(ii) f has a removable discontinuity at $c \in [-1, 1]$.

(iii) f hs an infinite discontinuity at $c \in [-1, 1]$.

(11) If f be continuous on $[a, b]$ and assumes each value between $f(a)$ and $f(b)$ just once, then show that it is strictly monotone in $[a, b]$.

(12) Show that

(i) the function $\cos\dfrac{1}{x}$ defined in $(0, 1)$ is not uniformly continuous there,

(ii) the function $\tan x$ is uniformly continuous on $[a, b]$, where $-\dfrac{\pi}{2} < a < b < \dfrac{\pi}{2}$.

(13) If $f(x+y) = f(x)+f(y) \; \forall \; x, y \in \mathbb{R}$ and f is continuous at a point of $\mathbb{R}$, prove that f is uniformly continuous on $\mathbb{R}$.

(14) If f is continuous on $\mathbb{R}$ and satisfies the relation $f(x+y) = f(x) \cdot f(y) \; \forall \; x, y \in \mathbb{R}$, then show that either $f(x) = 0 \; \forall \; x \in \mathbb{R}$ or there exists $a > 0$ such that $f(x) = a^x \; \forall \; x \in \mathbb{R}$.

(15) (i) If f and g are both uniformly continuous on $[a, b]$, then show that $f \pm g$ are also uniformly continuous on $[a, b]$.

(ii) If in addition each function is bounded on $[a, b]$, then the product function fg is uniformly continuous on $[a, b]$. Is boundedness of each function on $[a, b]$ necessary for the uniform continuity of fg?

CHAPTER 5

THE DERIVATIVE AND EXPANSION OF FUNCTIONS

Definition [5.1] : Let $f : [a,b] \to \mathbb{R}$, $[a,b] \subseteq \mathbb{R}$ be a function. Let $y = f(x)$ for some $x \in [a,b]$. Therefore, y is dependent on x. Also let Δy be the increment of y corresponding to the increment Δx of x so that $y + \Delta y = f(x + \Delta x)$.

So $\Delta y = f(x + \Delta x) - f(x)$.

Then $\lim\limits_{\Delta x \to 0} \frac{\Delta y}{\Delta x} = \lim\limits_{\Delta x \to 0} \frac{f(x+\Delta x)-f(x)}{\Delta x}$ is called the derivative of $f(x)$ provided the limit exists.

Further, $\lim\limits_{\Delta x \to 0} \frac{\Delta y}{\Delta x}$ is written as $\frac{dy}{dx}$, called the derivative of y w.r.t. x. Thus we write

$$\frac{dy}{dx} = \lim_{h \to 0} \frac{f(x+h) - f(x)}{h}, \text{ taking } \Delta x = h$$

$$\text{or, } \frac{d}{dx} f(x) = f'(x) = \lim_{h \to 0} \frac{f(x+h) - f(x)}{h}$$

Let $c \in [a,b]$. Then the derivative of f at $x = c$ is given by

$$f'(c) = \left.\frac{dy}{dx}\right]_{x=c} = \lim_{h \to 0} \frac{f(c+h) - f(c)}{h}$$

$$= \lim_{x \to c} \frac{f(x) - f(c)}{x - c}, \text{ writing } x - c = h \Rightarrow x \to c \text{ as } h \to 0.$$

So $f'(0) = \lim\limits_{x \to 0} \frac{f(x)-f(0)}{x}$

$$Rf'(c) = \lim_{h \to 0+0} \frac{f(c+h) - f(c)}{h} \textit{ is called right-hand derivative}$$

$$\text{and } Lf'(c) = \lim_{h \to 0-} \frac{f(c+h) - f(c)}{h} \textit{ is called left-hand drivative.}$$

$f'(c)$ exists if $Rf'(c) = Lf'(c)$.

Thus a real number L is said to be the derivative of $f(x)$ at $x = c$, if for every $\epsilon > 0$, $\exists$ a (+)ve number δ such that $\left| \frac{f(x) - f(c)}{x - c} - L \right| < \epsilon$ for $0 < |x - c| < \delta$.

We write $L = f'(c)$.

$f'(c)$ is also called the *differential co-efficient of* $f(x)$ *at* $x - c$.

Theorem [5.1]: If $f : [a, b] \to \mathbb{R}$ and $g : [a, b] \to \mathbb{R}$ be both differentiable $\forall\, x \in [a, b]$, then

(i) f is continuous at $x = c \in [a, b]$. But converse is not true.

(ii) $\frac{d}{dx}[kf(x)] = k\frac{d}{dx}[f(x)] = kf'(x),\ k \in \mathbb{R}$.

(iii) $\frac{d}{dx}[f(x) + g(x)] = \frac{d}{dx}f(x) + \frac{d}{dx}g(x) = f'(x) + g'(x)$.

(iv) $\frac{d}{dx}[(fg)x] = f\frac{d}{dx}g(x) + g\frac{d}{dx}f(x) = fg'(x) + gf'(x)$.

(v) $\frac{d}{dx}\left[\left(\frac{f}{g}\right)x\right] = \frac{gf'(x) - fg'(x)}{[g(x)]^2}, g(x) \neq 0\ \forall\, x \in [a, b]$.

(vi) $\frac{d}{dx}\left[\left(\frac{1}{f}\right)(x)\right] = -\frac{f'(x)}{[f(x)]^2},\ f(x) \neq 0\ \forall\, x \in [a, b]$.

(vii) f is strictly increasing or strictly decreasing according as $f'(x) > 0$ or $f'(x) < 0\ \forall\, x \in [a, b]$ [Sign of derivative].

(viii) Let c be a point such that $a < c < b$ and let f has an extremum at $x = c$. If $f'(c)$ exists, then $f'(c) = 0$.

(ix) If f be differentiable in $[a, b]$ and $f'(a) \cdot f'(b) < 0$, then $\exists$ a point $c \in (a, b)$ such that $f'(c) = 0$ [known as Darboux's Theorem].

(x) If f be differentiable in $[a, b]$ and $f'(a) \neq f'(b)$ and $k \in \mathbb{R}$ lying between $f'(a)$ and $f'(b)$, then $\exists$ a point $c \in (a, b)$ such that $f'(c) = k$ [Intermediate Value Theorem for derivatives].

Proof.

(i) Let $c \in [a, b]$. Then $f'(c)$ exists.

Now $\lim_{x \to c}[f(x) - f(c)] = \lim_{x \to c}\left[\frac{f(x) - f(c)}{x - c}\right] \lim_{x \to c}(x - c) = f'(c) \cdot 0 = 0$ [$\because f'(c)$ is finite].

$\Rightarrow \lim_{x \to c} f(x) = f(c)$

$\Rightarrow f$ is continuous at $x = c$.

Next, let $f(x) = |x|$.

Then $f(x)$ is continuous at $x = 0$, since

$$\lim_{x \to 0+0} f(x) = \lim_{x \to 0+0} x = 0 \text{ and } \lim_{x \to 0-0} f(x) = \lim_{x \to 0-0} -x = 0 \text{ and } f(0) = 0$$

So $f(0 - 0) = f(0 + 0) = f(0)$.

Now $Rf'(0) = \lim_{h \to 0+0} \frac{f(h) - f(0)}{h} = \lim_{h \to 0+0} \frac{h}{h} = 1$

$Lf'(0) = \lim_{h \to 0-0} \frac{f(h) - f(0)}{h} = \lim_{h \to 0-} \frac{-h}{h} = -1$

$Rf'(0) \neq Lf'(0)$

Hence $f'(x)$ does not exist at $x = 0$.

Example. Let $f(x) = |x| + |x - 1|\ \forall\, x \in \mathbb{R}$.

Then f is continuous at $x = 0$ and 1, but not differentiable at these points.

(iv) Let $\phi = f \cdot g$.

Then $\phi(x) = (fg)(x) = f(x) \cdot g(x)$.

Let $c \in [a, b]$.

Then

$$\begin{aligned}\phi'(c) &= \lim_{x\to c}\frac{\phi(x)-\phi(c)}{x-c} = \lim_{x\to c}\frac{f(x)g(x)-f(c)g(c)}{x-c}\\ &= \lim_{x\to c}\frac{g(x)\{f(x)-f(c)\}+f(c)\{g(x)-g(c)\}}{x-c}\\ &= \lim_{x\to c}\frac{f(x)-f(c)}{x-c}\cdot g(x)+\lim_{x\to c}\frac{g(x)-g(c)}{x-c}\cdot f(c)\\ &= f'(c)\cdot g(c)+g'(c)f(c)\ [\text{as } g \text{ is continuous at } x=c,\ \lim_{x\to c} g(x)=g(c)]\end{aligned}$$

Hence $\phi'(x) = f'(x)g(x)+g'(x)f(x)\ \forall\ x \in [a,b]$, as c is chosen arbitrarily.

(v) Let $\phi = f/g$.

Then $\phi(x) = \left(\frac{f}{g}\right)(x) = \frac{f(x)}{g(x)}, g(x) \neq 0\ \forall\ x \in [a,b]$

Let $c \in [a,b]$.

Then

$$\begin{aligned}\phi'(c) &= \lim_{x\to c}\frac{\phi(x)-\phi(c)}{x-c} = \lim_{x\to c}\frac{\frac{f(x)}{g(x)}-\frac{f(c)}{g(c)}}{x-c} = \lim_{x\to c}\frac{f(x)g(c)-f(c)g(x)}{g(x)g(c)\cdot(x-c)}\\ &\lim_{x\to c}\frac{1}{g(x)g(c)}\cdot\lim_{x\to c}\cdot\left[\frac{f(x)-f(c)}{x-c}\cdot g(c)-f(c)\cdot\frac{g(x)-g(c)}{x-c}\right]\\ &= \frac{1}{[g(c)]^2}[f'(c)g(c)-g'(c)f(c)],\ g(c)\neq 0\ [\text{since } g \text{ is continuous at } x=c].\end{aligned}$$

Hence $\phi'(x) = \frac{f'(x)g(x)-g'(x)f(x)}{[g(x)]^2}$.

(vii) Let $c \in [a,b]$ and let $f'(c)$ exists and $f'(c) > 0$.

Then $\lim\limits_{x\to c}\frac{f(x)-f(c)}{x-c} = f'(c)$

$\Rightarrow \left|\frac{f(x)-f(c)}{x-c} - f'(c)\right| < \epsilon$, whenever $0 < |x-c| < \delta$

$\Rightarrow f'(c)-\epsilon < \frac{f(x)-f(c)}{x-c} < f'(c)+\epsilon\ \forall\ x \in (c-\delta, c+\delta)$

Now if $\epsilon(>0)$ be so chosen that $\epsilon < f'(c)$, then $\frac{f(x)-f(c)}{x-c}$ lies between two (+)ve numbers and hence $\frac{f(x)-f(c)}{x-c} > 0, x \neq c$.

$\Rightarrow f(x)-f(c) > 0$, if $c < x < c+\delta$

$\Rightarrow f(x) > f(c)\ \forall\ x > c$ satisfying $c < x < c+\delta$. (1)

Also $f(x)-f(c) < 0$, if $c-\delta < x < c$

$\Rightarrow f(x) < f(c)\ \forall\ x < c$ satisfying $c-\delta < x < c$. (2)

From (1) and (2), it follows that f is strictly increasing at $x = c$.

If $f'(c) < 0$, then proceeding as above we can prove that f is strictly decreasing.

(viii) Let $f(x)$ be maximum at $x = c$, $a < c < b$.

That is, $f(c)$ is the maximum value of $f(x)$.

Hence $\exists$ a (+) number δ such that $f(x) < f(c)\ \forall\ x \in (c-\delta, c+\delta), x \neq c$.

When $f'(c)$ exists, there arises three possibilities $f'(c) > 0, f'(c) < 0, f'(c) = 0$.

$f'(c) > 0 \Rightarrow \exists$ an open interval $(c, c+\delta_1), 0 < \delta_1 < \delta$ such that

$$f(x) > f(c)\ \forall\ x \in (c, c+\delta_1),$$

which arises a contradiction.

Again $f'(c) < 0 \Rightarrow \exists$ an open interval $(c - \delta_2, c)$, $0 < \delta_2 < \delta$ such that

$$f(x) > f(c) \; \forall \; x \in (c - \delta_2, c),$$

which again arises a contradiction.
Hence the only possibility is $f'(c) = 0$.

(ix) $f'(a) \cdot f'(b) < 0 \Rightarrow$ either $f'(a) < 0, f'(b) > 0$ or $f'(a) > 0, f'(b) < 0$.
Let us take, $f'(a) < 0$ and $f'(b) > 0$.

$$f'(a) < 0 \Rightarrow f(x) < f(a) \; \forall \; x \in (a, a + \delta_1), \delta_1 > 0 \quad (1)$$

$$\text{and } f'(b) > 0 \Rightarrow f(x) < f(b) \; \forall \; x \in (b - \delta_2, b), \delta_2 > 0 \quad (2)$$

Again, f being differentiable is continuous in $[a, b]$ and hence it attains its bounds in $[a, b]$.
Let $\exists \; c \in [a, b]$ such that $f(c) = m$, the g.l.b. of $f(x)$ in $[a, b]$. Then $c \neq a$.
For, if $c = a$, then $f(a) = m \Rightarrow f(x) < m$ by (1) $\forall \; x \in (a, a + \delta_1)$.
Also if $c = b$, then $f(b) = m \Rightarrow f(x) < m$ by (2) $\forall \; x \in (b - \delta_2, b)$ which are not possible.
Hence c is an interior point of $[a, b]$. That is, $a < c < b$.
Now we have to show that $f'(c) = 0$.

$$\left.\begin{array}{r} f'(c) > 0 \Rightarrow f(x) < f(c) = m \; \forall \; x \in (c - \delta, c) \\ \text{and } f'(c) < 0 \Rightarrow f(x) < f(c) = m \; \forall x \in (c, c + \delta) \end{array}\right\} \text{by Prop. (vii)}$$

Both arises a contradiction that m is the g.l.b. of $f(x)$.
Hence $f'(c) = 0$.

(x) [*Hints.* Consider a function $\phi(x) = f(x) - kx \; \forall \; x \in [a, b]$, $\phi'(x)$ exists. Apply Prop (ix).]
Proofs of (ii), (iii) and (vi) are left.

Remarks

(1) In Theorem [5.1], property (vii), we have seen that if $f'(x) > 0$ on an interval $[a, b]$, then f is strictly increasing there. But the converse is not true. Let $f : \mathbb{R} \to \mathbb{R}$ and let $f(x) = x^3$. Then f is strictly increasing on $\mathbb{R}$, but $f'(x) \not> 0 \; \forall \; x \in \mathbb{R}$, since $f'(x) = 3x^2 \Rightarrow f'(x) = 0$ for $x = 0$. The same is true for strictly decreasing function $f : \mathbb{R} \to \mathbb{R}$, defined by $f(x) = -x^3$.

(2) If $f'(x) > 0$ at a point $x = c \in \mathbb{R}$, then f is increasing at $x = c$. But this is not always true. We consider a function $f : \mathbb{R} \to \mathbb{R}$, defined by

$$f(x) = \begin{cases} x + 2x^2 \sin \frac{1}{x}, & \text{if } x \neq 0 \\ 0, & \text{if } x = 0. \end{cases}$$

$$f'(x) = 1 + 4x \sin \frac{1}{x} - 2 \cos \frac{1}{x}$$

Now

$$f'(0) = \lim_{x \to 0} \frac{f(x) - f(0)}{x - \theta} = \lim_{x \to 0} \frac{x + 2x^2 \sin \frac{1}{x}}{x} = \lim_{x \to 0} \left[1 + 2x \sin \frac{1}{x}\right] = 1 > 0$$

$f'(x) > 0$ at $x = 0$.

But yet we shall see that f is not increasing in the neighbourhood of 0.

For, $x = \frac{1}{\pi}, \frac{1}{2\pi}, \frac{1}{3\pi}, \frac{1}{4\pi}, \cdots$.

$$f'\left(\frac{1}{\pi}\right) = 1 - 2\cos\pi = 3 > 0$$

$$f'\left(\frac{1}{2\pi}\right) = 1 - 2\cdot\cos 2\pi = 1 - 2 = -1 < 0$$

$$f'\left(\frac{1}{3\pi}\right) > 0, f'\left(\frac{1}{4\pi}\right) < 0$$

So

$$f'\left(\frac{1}{n\pi}\right) > 0 \text{ and } f'\left(\frac{1}{(n+1)\pi}\right) < 0, \text{ if } n \text{ is odd}$$

$$\text{and } f'\left(\frac{1}{n\pi}\right) < 0 \text{ and } f'\left(\frac{1}{(n+1)\pi}\right) > 0, \text{ if } n \text{ is even.}$$

Thus f assumes both (+)ve and (−)ve values and hence when $n \to \infty$, f assumes both (+)ve and (−)ve values in the neighbourhood of 0 and hence f is not increasing.

Theorem [5.1A] Let f be defined in a domain $[a, b]$ and c be an interior point of $[a, b]$ [i.e., $a < c < b$] such that $f'(c) = 0$. The f has a maximum or minimum according as $f''(c) < 0$ or $f''(c) > 0$.

Proof. Existence of $f''(c)$.

$\Rightarrow f, f'$ exist and continuous at $x = c$.

$\Rightarrow f'$ exist in the neighbourhood $N_\delta(c) = (c - \delta, c + \delta)$ of c, where $N_\delta(c) \subset [a, b]$.

CASE (1) Let $f''(c) > 0$

Then $f''(c) > 0 \Rightarrow f'(x)$ is increasing in $N_\delta(c)$

$\Rightarrow f'(x) > f'(c)\ \forall\ x \in (c, c + \delta'), \delta < \delta$ and $f'(x) < f'(c)\ \forall\ x \in (c - \delta', c)$.

Now $f'(c) = 0$,

$\Rightarrow f'(x) > 0\ \forall\ x \in (c, c + \delta')$ and $f'(x) < 0\ \forall\ x \in (c - \delta', c)$.

$f'(x) > 0 \Rightarrow f$ is a increasing function in $(c, c + \delta')$

$$\Rightarrow f(x) > f(c) \text{ in } (c, c + \delta') \tag{1}$$

While $f'(x) < 0 \Rightarrow f$ is a decreasing function in $(c - \delta', c)$

$$\Rightarrow f(x) > f(c) \text{ in } (c - \delta', c) \tag{2}$$

(1) and (2) $\Rightarrow f(x) > f(c)\ \forall\ x \in\ (c - \delta', c + \delta'), x \neq c$

$\Rightarrow f$ has minimum at $x = c$.

CASE (2) Now $f''(c) < 0 \Rightarrow f(x) < f(c)\ \forall\ x \in (c - \delta', c + \delta')$, proceeding as in case (1).

$\Rightarrow f$ has maximum at $x = c$.

Generalisation of above theorem

Theorem [5.1B] Let f be defined in a domain $[a, b]$ and $a < c < b$ such that (i) $f'(c) = f''(c) = \cdots = f^{n-1}(c) - 0$ and (ii) $f^n(c)$ exists and $\not= 0$. Then f has neither maximum nor minimum when n is odd, while f has maximum or minimum according as $f^n(c) < 0$ or $f^n(c) > 0$ when n is even.

Proof. Existence of $f^n(c)$

$\Rightarrow$ Existence and continuity of $f, f', \ldots, f^{n-1}$ at $x = c$ and hence continuity of them in neighbourhood $N_\delta(c) = (c-\delta, c+\delta)$, $a < c < b$.

Let $f^n(c) > 0$.

Then $f^n(c) > 0 \Rightarrow f^{n-1}(x)$ s a increasing function in $N_\delta(c)$.

$\Rightarrow f^{n-1}(x) < f^{n-1}(c) \; \forall \; x \in (c-\delta', c), \delta' < \delta$

and $f^{n-1}(x) > f^{n-1}(c) \; \forall \; x \in (c, c+\delta')$

Now $f^{n-1}(c) = 0$

$$\left.\begin{aligned} \Rightarrow f^{n-1}(x) < 0 \text{ in } (c-\delta', c) \\ \text{and } f^{n-1}(x) > 0 \text{ in } (c, c+\delta') \end{aligned}\right\} \tag{1}$$

When $f^n(c) < 0$

$$\left.\begin{aligned} \Rightarrow f^{n-1}(x) > 0 \; \forall \; x \in (c-\delta', c) \\ \text{and } f^{n-1}(x) < 0 \; \forall \; x \in (c, c+\delta') \end{aligned}\right\} \tag{2}$$

By Taylor's Theorem,

$$f(c+h) = f(c) + hf'(c) + \frac{h^2}{2!}f''(c) + \cdots + \frac{h^{n-1}}{(n-1)!}f^{n-1}(c+\theta h), 0 < \theta < 1, |h| < \delta$$

$$\Rightarrow f(c+h) - f(c) = \frac{h^{n-1}}{(n-1)!}f^{n-1}(c+\theta h), 0 < \theta < 1, |h| < \delta, \text{ [by (i)]} \tag{3}$$

Let $f^n(c) > 0$.

CASE (i) When n is odd, $h^{n-1} > 0$ for any h, $|h| < \delta$.

Now, if $h < 0$, then $c + \theta h \in (c-\delta', c)$ and consequently $f^{n-1}(c+\theta h) < 0$ and $h > 0$, $c + \theta h \in (c, c+\delta')$ and so $f^{n-1}(c+\theta h) > 0$ by (1).

Thus from (3), $f(c+h) - f(c) < 0$ when $(c - \delta' < c + h < c)$ and $f(c+h) - f(c) > 0$,

$$c < c + h < c + \delta'.$$

That is, $f(c+h) < f(c)$ in $c - \delta' < c + h < c$ and $f(c+h) > f(c)$ in $c < c + h < c + \delta'$.

$\Rightarrow f(c)$ is not extremum at $x = c$ in $(c - \delta', c + \delta')$

The same can be shown when $f^n(c) < 0$.

CASE (ii) When n is even, $h^{n-1} > 0$ or < 0 according as $h > 0$ or $h < 0$ [$\because (n-1)$ is odd].

Now let $f^n(c) > 0$.

Then proceeding as in case (i) we have $f(c+h) > f(c) \; \forall \; x \in (c-\delta', c+\delta')$.

$\Rightarrow f$ has minimum at $x = c$.

Again when $f^n(c) < 0$, then proceeding as above, we have

$$f(c+h) < f(c) \; \forall \; x \in (c-\delta', c+\delta').$$

$\Rightarrow f$ has maximum at $x = c$.

Remarks.

(1) For extremum $f'(c) = 0$, if it exists, is a necessary condition but not a sufficient condition. Let $f(x) = x^3, f'(x) = 3x^2, f(0) = 0$

Now $f'(0) = 0$ at $x = 0$

$f(x) > f(0)\ \forall\ x > 0$

and $f(x) < f(0)\ \forall\ x < 0$

$\Rightarrow f$ has no extremum at $x = 0$.

(2) Theorem [5.1A] fails if $f'(x)$ is non existent at $x = c$.

In such case f has a maximum at $x = c$ if $f'(x)$ changes sign from (+)ve on the left of c to negative the (−)ve on the right of c (i.e., from $c - h$ to $c + h$) and f has a minimum at $x = c$ if $f'(x)$ changes sign (−)ve to the left of c to (+)ve on the right of c (i.e., from $c - h$ to $c + h$).

Example. Let $f(x) = 5 - (x-1)^{2/3}$.

$$f'(x) = -\frac{2}{3}\cdot\frac{1}{(x-1)^{1/3}}$$

$f'(x)$ does not exist at $x = 1$. Now

$$f'(1-h) = -\frac{2}{3}\cdot\frac{1}{(1-h-1)^{1/3}} = -\frac{2}{3}\cdot\frac{1}{(-h)^{1/3}} > 0$$

$$\text{and } f'(1+h) = -\frac{2}{3}\frac{1}{(1+h-1)^{1/3}} = -\frac{2}{3}\cdot\frac{1}{h^{1/3}} < 0.$$

So $f'(x)$ changes sign from (+)ve to (−)ve in $(1-h, 1+h)$.

$\therefore f$ has maximum at $x = 1$ and maxmum value of f is 5.

While $f(x) = 5 + (x-1)^{2/3}$

$$f'(x) = \frac{2}{3}\cdot\frac{1}{(x-1)^{1/3}}$$

Then $f'(1-h) < 0$ and $f'(1+h) > 0$.

$f'(x)$ changes sign from (−)ve to (+)ve in $(1-h, 1+h)$

So f' has minimum at $x = 1$.

Theorem [5.2] **[Chain Rule]** Let $f : D \to \mathbb{R}$ and $g : D' \to \mathbb{R}$, $D, D' \subseteq \mathbb{R}$ and $f(D) \subseteq D'$. Let $c \in D$. If f is differentiable at c and g is differentiable at $f(c)$, then the composition function $g \circ f : D \to \mathbb{R}$ is differentiable at c and $(g \circ f)'c = g'(f(c)) \cdot f'(c)$.

Therefore, $\forall\ x \in D, (g \circ f)'(x) = g'(f(x))f'(x)$.

Proof. f is differentiable on $D \Rightarrow f$ is continuous on D. Also $c \in D$. So f is continuous at $x = c$.

Hence $\lim\limits_{x\to c} f(x) = f(c)$.

Now

$$(g\circ f)'(c) = \lim_{x\to c}\frac{(g\circ f)(x) - (g\circ f)(c)}{x-c} = \lim_{x\to c}\frac{g(f(x)) - g(f(c))}{f(x)-f(c)}\cdot\frac{f(x)-f(c)}{x-c}$$

$$= \lim_{f(x)\to f(c)}\frac{g(f(x)) - g(f(c))}{f(x)-f(c)}\lim_{x\to c}\frac{f(x)-f(c)}{x-c} = g'(f(c))\cdot f'(c)$$

Since c is an arbitrary point in D, $(g\circ f)'(x) = g'(f(x))f'(x)\ \forall\ x \in D$.

Extension of Theorem [5.2]

Theorem [5.3] Let $f : \mathbb{R} \to \mathbb{R}$, $g : \mathbb{R} \to \mathbb{R}$, $h : \mathbb{R} \to \mathbb{R}$ be three differentiable functions on $\mathbb{R}$. Let $c \in \mathbb{R}$. Then the composite functions $h \circ g \circ f$ is differentiable at c and

$$(h \circ g \circ f)'(c) = h'(g(f(c)) \cdot g'(f(c)) \cdot f'(c).$$

Hence $(h \circ g \circ f)'(x) = h'(g(f(x)) \cdot g'(f(x)) \cdot f'(x)\ \forall\, x \in \mathbb{R}$.

Proof. $c \in \mathbb{R} \Rightarrow f(c) \in \mathbb{R} \Rightarrow g(f(c)) \in \mathbb{R}$.

Then $(h \circ g \circ f)'(c)$

$$= \lim_{x \to c} \frac{(h \circ g \circ f)(x) - (h \circ g \circ f)(c)}{x - c}$$

$$= \lim_{x \to c} \frac{h(g(f(x)) - h(g(f(c))}{g(f(x)) - g(f(c))} \cdot \frac{g(f(x)) - g(f(c))}{x - c}$$

$$= \lim_{g(f(x)) \to g(f(c))} \frac{h(g(f(x)) - h(g(f(c))}{g(f(x)) - g(f(c))} \cdot \lim_{f(x) \to f(c)} \frac{g(f(x)) - g(f(c))}{f(x) - f(c)} \cdot \lim_{x \to c} \frac{f(x) - f(c)}{x - c}$$

$$= h'(g(f(c)) \cdot g'(f(c)) \cdot f'(c)$$

As g is differentiable on $\mathbb{R}$ and $f(c) \in \mathbb{R}$, $g(f(x))$ is continuous at $f(c)$ and hence $\lim_{x \to c} g(f(x)) = g(f(c))$ and f is continuous at $c \Rightarrow \lim_{x \to c} f(x) = f(c)$.

Example. Let $f(x) = (\sin x^k)^m, m, k \in N$.

We define f, g, h by $f(x) = x^k$

$$g(f(x)) = \sin(f(x)) \text{ and } h(g(f(x)) = [\sin(f(x))]^m$$

So,

$$(h \circ g \circ f)'(x) = h'(g(f(x))g'(f(x))f'(x)$$
$$= \frac{d}{d(\sin x^k)}[\sin x^k]^m \cdot \frac{d}{d(x^k)}[\sin(x^k)] \cdot \frac{d}{dx}(x^k)$$
$$= m(\sin x^k)^{m-1} \cdot \cos(x^k) \cdot kx^{k-1} = mk(\sin x^k)^{m-1} \cdot \cos x^k \cdot x^{k-1}$$

Problem [5.1]: Let $g : \mathbb{R} \to \mathbb{R}$ be defined by $g(x) = x^2 \sin \frac{1}{x}$, $x \neq 0$ and $g(0) = 0$. Show that g is differentiable $\forall\, x \in \mathbb{R}$. Also show that g' is not bounded on $[-1, 1]$.

Solution.

$$g'(x) = 2x \sin \frac{1}{x} - \cos \frac{1}{x} \tag{1}$$

$\Rightarrow g'(x)$ exists $\forall\, x (\neq 0) \in \mathbb{R}$.

Again, $g'(0) = \lim_{h \to 0} \dfrac{g(h) - g(0)}{h} = \lim_{h \to 0} \dfrac{h^2 \sin \frac{1}{h} - 0}{h} = 0.$

So $g'(x)$ exists $\forall\, x \in \mathbb{R}$.

From (1), $\lim_{x \to 0} g'(x)$ does not exist. Therefore, g' is unbounded at $x = 0$.

That is g' is unbounded in $[-1, 1]$.

Definition [5.2]: A function $f : \mathbb{R} \to \mathbb{R}$ is called an even function if $f(-x) = f(x)\ \forall\, x \in \mathbb{R}$ and is called an odd function if $f(-x) = -f(x)\ \forall\, x \in \mathbb{R}$.

Theorem [5.4] If $f : \mathbb{R} \to \mathbb{R}$ be an even function and if $f'(x)$ exists, then f' is an odd function and vice versa.

Proof. Let f be an even function.

Then $f(-x) = f(x) \ \forall\ x \in \mathbb{R}$.

$$\text{Now } \frac{d}{dx}[f(-x)] = \frac{d}{dx}[f(x)] \Rightarrow f'(-x)(-1) = f'(x)$$
$$\Rightarrow f'(-x) = -f'(x) \ \forall\ x \in \mathbb{R}$$
$$\Rightarrow f' \text{ is an odd function.}$$

Next, let f be an odd function.

Then $f(-x) = -f(x) \ \forall\ x \in \mathbb{R}$.

$$\text{Now, } \frac{d}{dx}[f(-x)] = \frac{d}{dx}[-f(x)] \Rightarrow f'(-x)(-1) = -f'(x)$$
$$\Rightarrow f'(-x) = f'(x) \ \forall\ x \in \mathbb{R}$$
$$\Rightarrow f' \text{ is an even function.}$$

Problem [5.2] : If $f : \mathbb{R} \to \mathbb{R}$ is differentiable at $c \in \mathbb{R}$, show that

$$f'(c) = \lim \left[n \left\{ f\left(c + \frac{1}{n}\right) - f(c) \right\} \right].$$

However, show by an example that the existence of the limit of the sequence does not imply the existence of $f'(c)$.

Solution. By definition, $f'(c) = \lim\limits_{h \to 0} \dfrac{f(c+h) - f(c)}{h}$.

Now, let $h = \frac{1}{n}$ so that $h \to 0$ as $n \to \infty$.

$$\therefore\ f'(c) = \lim_{n\to\infty} \frac{f\left(c + \frac{1}{n}\right) - f(c)}{1/n} = \lim_{n\to\infty} \left[n \left\{ f\left(c + \frac{1}{n}\right) - f(c) \right\} \right].$$

We consider a function $f(x) = |x|$, $x \neq 0$ and $f(0) = 0$. Then $f'(0)$ does not exist.

$$\text{However, } \lim_{n\to\infty} \left[n \left\{ f\left(0 + \frac{1}{n}\right) - f(0) \right\} \right] = \lim_{n\to\infty} \left\{ n \cdot \frac{1}{n} \right\} = 1, \text{ exists.}$$

5.1 Differentiation of Inverse Function

Definition [5.3] : Let $f : [a, b] \to \mathbb{R}, [a, b] \subseteq \mathbb{R}$ be continuous and strictly monotone function on $[a, b]$. Then f^{-1}, inverse of f, is defined on the interval $[c, d] = f[a, b]$ such that $(f^{-1} \circ f)(x) = x \ \forall\ x \in [a, b]$.

That is $f^{-1}(f(x)) = x$

$$\Rightarrow (f^{-1})'(f(x))f'(x) = 1$$
$$\Rightarrow (f^{-1})'(f(x)) = \frac{1}{f'(x)} \ \forall\ x \in [a, b], \text{ provided } f'(x) \neq 0.$$

So for a point $c \in [a, b]$,

$$(f^{-1})'(f(c)) = \frac{1}{f'(c)} \text{ when } f'(c) \neq 0 \tag{1}$$

Also

$$f(x) = y \Rightarrow (f^{-1})'(y) = \frac{1}{f'(f^{-1}(y))}. \tag{2}$$

when $(f^{-1})'(f(x)) = \frac{1}{f'(x)}, f'(x) \neq 0 \ \forall\ x \in [a, b]$.

Problem [5.3]: Let $f : \mathbb{R} \to \mathbb{R}$ defined by $f(x) = x^3 + 2x + 1 \ \forall\ x \in \mathbb{R}$. Then find the value of $(f^{-1})'(f(x))$ at $x = 0, 1, -1$.

Solution. Here f, being a polynomial in x, is continuous.

Also $f'(x) = 3x^2 + 2 > 0 \ \forall\ x \in \mathbb{R}$. So f is strictly increasing in $\mathbb{R}$. Hence f^{-1} exists.

For $x = 0, f'(0) = 2 \neq 0$. Therefore, $(f^{-1})'(f(0)) = \dfrac{1}{f'(0)} = \dfrac{1}{2}$ or $(f^{-1})'(1) = \frac{1}{2}$.

For $x = 1, f'(1) = 5 \neq 0$. So $(f^{-1})'(f(1)) = \dfrac{1}{f'(1)} = \dfrac{1}{5}$ or, $(f^{-1})'(4) = \dfrac{1}{5}$.

For $x = -1, f'(-1) = 5 \neq 0$. So $(f^{-1})'f(-1) = \dfrac{1}{f'(-1)} = \dfrac{1}{5}$ or, $(f^{-1})'(-2) = \dfrac{1}{5}$.

Problem [5.4]: Find $(f^{-1})'$ when $f(x) = x^n$, when $n \in N$ and $x \in [0, \infty)$. Also discuss its existence.

Solution. $f(x) = x^n$. Let $f(x) = y$. Then $y = x^n \Rightarrow x = y^{1/n} = g(y)$, say.

f is continuous and strictly increasing in $[0, \infty)$. So g is also continuous and strictly increasing in $[0, \infty)$.

Now, $(g \circ f)(x) = g(f(x)) = g(y) = g(x^n) = (x^n)^{1/n} = x \Rightarrow g = f^{-1}$.

Thus $(f^{-1})'(f(x)) = \dfrac{1}{f'(x)} = \dfrac{1}{nx^{n-1}}$, for $x \neq 0$

$$\Rightarrow (f^{-1})'(y) = \frac{1}{n \cdot (y^{1/n})^{n-1}} = \frac{1}{n}\frac{y^{1/n}}{y} \Rightarrow (g')(y) = \frac{1}{n}\frac{y^{1/n}}{y}$$

which exists, when $y > 0$ and $g'(0)$ does not exist.

Hence $g' = f^{-1}$ is differentiable $\forall\ y > 0$.

Theorem [5.5] [Rolle's Theorem] If a function $f : [a, b] \to \mathbb{R}$, $[a, b] \subseteq \mathbb{R}$ be continuous in $[a, b]$, $f'(x)$ exists in the open interval (a, b), and if $f(a) = f(b)$, then there exists a point $c \in (a, b)$ such that $f'(c) = 0$.

Proof. If possible, let $f'(x) \neq 0$ for any x in (a, b). Since f is continuous in $[a, b]$, it attains its maximum M and minimum m somewhere in $[a, b]$. But f does not attain M and m at any interior point of $[a, b]$, otherwise $f'(x)$ vanishes at that point. So both are attained at the end points of $[a, b]$. Also $f(a) = f(b) \Rightarrow M = m \Rightarrow f$ is constant in $[a, b]$ and hence $f'(x)$ vanishes on (a, b). This contradicts the assumption that f' is never zero on (a, b). Therefore, $\exists$ a point c on (a, b) such that $f'(c) = 0$.

Remark. *Geometrical interpretation of Rolle's Theorem.*

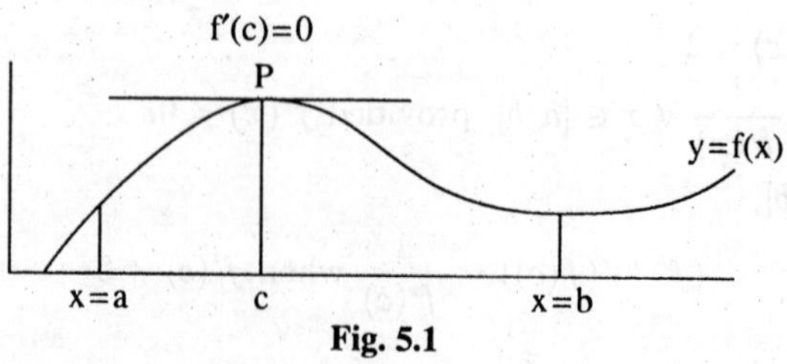

Fig. 5.1

$\dfrac{dy}{dx} = 0$ at a point $x = c$.

$\Rightarrow$ tangent at a point $P(c, f(c))$ is parallel to X-axis, where $a < c < b$.

Theorem [5.6] [Lagrange's Mean Value Theorem] If $f : [a,b] \to \mathbb{R}$, $[a,b] \subseteq \mathbb{R}$ be continuous in $[a,b]$ and f' exists in (a,b), then there exists a point c in (a,b) such that

$$f(b) - f(a) = (b-a)f'(c).$$

Proof. We consider a function $\phi(x) = f(x) + \lambda x \; \forall \, x \in [a,b]$, λ is a constant which is so chosen that $\phi(a) = \phi(b)$. Then ϕ is continuous on $[a,b]$. Existence of f' on $(a,b) \Rightarrow$ the existence of ϕ' on (a,b). Also $\phi(a) = \phi(b)$. So ϕ satisfies the conditions of Rolle's Theorem. Hence there exists a point c on (a,b) such that $\phi'(c) = 0$.

Now $\phi'(x) = f'(x) + \lambda$.

$$\phi'(c) = 0 \Rightarrow f'(c) = -\lambda \tag{i}$$

Again

$$\phi(a) = \phi(b) \Rightarrow \frac{f(b) - f(a)}{b-a} = -\lambda \tag{ii}$$

Hence from (i) and (ii), $f(b) - f(a) = (b-a)f'(c)$.

Remark. *Geometrical interpretation of Mean Value Theorem*

$$f(b) - f(a) = (b-a)f'(c)$$

$$\Rightarrow f'(c) = \frac{f(b) - f(a)}{b-a}$$

$$\Rightarrow \left.\frac{dy}{dx}\right|_{x=c} = \frac{f(b) - f(a)}{b-a}$$

$\Rightarrow$ Slope of the tangent at a point $(c, f(c))$ to the curve $y = f(x)$ is $\dfrac{f(b) - f(a)}{b-a}$.

Also slope of the line AB (i.e., a chord to the curve $y = f(x)$) joining the points $A(a, f(a))$ and $B(b, f(b))$ is $\frac{f(b)-f(a)}{b-a}$.

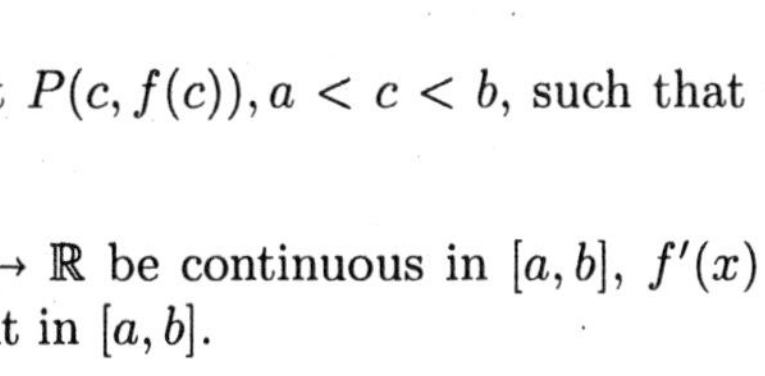

Fig. 5.2

Hence there exists a point $P(c, f(c)), a < c < b$, such that tangent at P is parallel to the chord AB.

Theorem [5.7] If $f : [a,b] \to \mathbb{R}$ be continuous in $[a,b]$, $f'(x)$ exists on (a,b) and $f'(x) = 0$ $\forall \, x \in (a,b)$, then f is constant in $[a,b]$.

Proof. Let x_1, x_2 be any two points in $[a,b]$ satisfying $x_1 < x_2$. Then f satisfies the conditions of Mean Value Theorem in $[x_1, x_2] \subseteq [a,b]$.

So $\quad f(x_2) - f(x_1) = (x_2 - x_1)f'(c), \; x_1 < c < x_2$

$\Rightarrow f(x_2) - f(x_1) = 0, \; f'(x) = 0 \; \forall \, x \in (a,b)$

$\Rightarrow f(x_2) = f(x_1) \Rightarrow f(x) = f(x_1), x \in [a,b]$ and $x > x_1$

$\Rightarrow f$ is constant in $[a,b]$.

Problem [5.5]: If $f, g : [a, b] \to \mathbb{R}$ be continuous on $[a, b]$, $f'(x), g'(x)$ exist on (a, b) and $f'(x) = g'(x)\ \forall\, x \in (a, b)$, then $f = g + \lambda$, where λ is constant.

Solution. Let $x \in [a, b]$ and $x > a$

Applying Lagrange's Mean Value Theorem to f and g in $[x, a]$, we have

$$f(x) - f(a) = (x - a)f'(c)$$
$$g(x) - g(a) = (x - a)g'(c),\ a < c < b$$

Now $f'(c) = g'(c)$ as $f'(x) = g'(x)\ \forall\, x \in (a, b)$.

$$\therefore\ f(x) - f(a) = g(x) - g(a)$$
$$\Rightarrow f(x) = g(x) - g(a) + f(a) = g(x) + \lambda$$
$$\Rightarrow f = g + \lambda, \text{ where } \lambda = f(a) - g(a) = \text{constant.}$$

Theorem [5.8] If $f, g : [a, b] \to \mathbb{R}$ be two functions satisfying the conditions,

(i) f, g are both continuous in $[a, b]$
(ii) $f'(x), g'(x)$ exists $\forall\, x \in (a, b]$
(iii) $g'(x) \neq 0$ for any x in (a, b), then there exists a point $c \in (a, b)$ such that

$$\frac{f(b) - f(a)}{g(b) - g(a)} = \frac{f'(c)}{g'(c)} \quad \text{[Cauchy's Mean Value Theorem].}$$

Proof. [*Hints.* Consider a function $\phi : [a, b] \to \mathbb{R}$ defined by $\phi(x) = f(x) + \lambda g(x)\ \forall\, x \in [a, b]$, where λ is a constant which is so chosen that $\phi(a) = \phi(b)$. Then proceed as in Theorem [5.6] and to show that $g(b) - g(a) \neq 0$. For, if $g(b) = g(a)$, then g satisfies the conditions of Rolle's Theorem and consequently $\exists$ a point $\xi \in (a, b)$ such that $g'(\xi) = 0$, contradicts the condition (iii).]

Remarks

(1) If we take $b - a = h$. The $b = a + h$. Also $c \in (a, b) \Rightarrow c = a + \theta h,\ 0 < \theta < 1$.
Thus Lagrange's Mean Value Theorem can be put in the form

$$f(a + h) = f(a) + hf'(a + \theta h),\ 0 < \theta < 1.$$

Replacing a by x, we get $f(x + h) = f(x) + hf'(x + \theta h),\ 0 < \theta < 1$.

(2) Cauchy's Mean Value Theorem can be written in the form

$$\frac{f(x + h) - f(x)}{g(x + h) - g(x)} = \frac{f'(x + \theta h)}{g'(x + \theta h)},\ 0 < \theta < 1.$$

(3) Lagrange's Mean Value Theorem can be deduced from Cauchy's Mean Value Theorem, by taking $g(x) = x$.
For $g(x) = x \Rightarrow g'(x) = 1 \Rightarrow g'(x + \theta h) = 1$ and hence

$$\frac{f(x + h) - f(x)}{x + h - x} = f'(x + \theta h) \Rightarrow f(x + h) = f(x) + hf'(x + \theta h)$$

That is, $f(b) - f(a) = (b - a)f'(c),\ a < c < b$.

5.2 Taylor's Theorem

Taylor's Theorem can be regarded as the extension of Lagrange's Mean Value Theorem. In Mean Value Theorem there is a relation between the values of a function and its first order derivative, while Taylor's Theorem establishes a relation between the values of a function and its higher order derivatives.

Theorem [5.9] [Taylor's Theorem] If a function $f : [a, b] \to \mathbb{R}$ possesses up to nth order derivatives satisfying

(i) $f, f', f'', \cdots, f^{n-1}$ are all continuous in $[a, b]$

(ii) $f^n(x)$ exists $\forall\ x \in (a, b)$,

then there exists a point $c \in (a, b)$ such that

$$\begin{aligned} f(b) &= f(a) + (b-a)f'(a) + \frac{(b-a)^2}{2!}f''(a) + \cdots + \frac{(b-a)^{n-1}}{(n-1)!}f^{n-1}(a) + \frac{(b-a)^n}{n!}f^n(c) \\ &= f'(a) + \sum_{k=1}^{n-1} \frac{(b-a)^k}{k!}f^k(a) + \frac{(b-a)^n}{n!}f^n(c) \end{aligned}$$

Proof. We consider a function

$$F(x) = G(x) - \left(\frac{b-x}{b-a}\right)^n G(a), \tag{i}$$

where

$$G(x) = f(b) - f(x) - (b-x)f'(x) \cdots \frac{(b-x)^{n-1}}{n-1!} f^{n-1}(x) \tag{ii}$$

Then $F(a) = G(a) - G(a) = 0$

$$\begin{aligned} F(b) = G(b) &= f(b) - f(b) - (b-b)f'(b) + \cdots + \frac{(b-b)^{n-1}}{(n-1)!}f^{n-1}(b) \\ &= 0 \end{aligned}$$

So

$$F(a) = F(b) \tag{iii}$$

$$\begin{aligned} F'(x) &= G'(x) + \frac{n(b-x)^{n-1}}{(b-a)^n}G(a) \\ &= -f'(x) + f'(x) - (b-x)f''(x) + \frac{2(b-x)}{2!}f''(x) - \frac{(b-x)^2}{2!}f'''(x) + \cdots \\ &\qquad \cdots + \frac{(n-1)(b-x)^{n-2}}{(n-1)!}f^{n-1}(x) - \frac{(b-x)^{n-1}}{(n-1)!}f^n(x) + \frac{n(b-x)^{n-1}}{(b-a)^n}G(a) \\ &= 0 - \frac{(b-x)^{n-1}}{(n-1)!}f^n(x) + \frac{n(b-x)^{n-1}}{(b-a)^n}G(a) \\ &= -\frac{(b-x)^{n-1}}{(n-1)!}f^n(x) + \frac{n(b-x)^{n-1}}{(b-a)^n}G(a) \end{aligned}$$

$\Rightarrow F'(x)$ exists as $f^n(x)$ exists in (a, b) (iv)

$F(x)$ is continuous in $[a, b]$. (v)

So F satisfies the conditions of Rolle's Theorem by (iii), (iv) and (v).

Hence, $\exists$ a point $c \in (a, b)$ such that $F'(c) = 0$.

$$\Rightarrow -\frac{(b-c)^{n-1}}{(n-1)!}f^n(c) + \frac{n(b-c)^{n-1}}{(b-a)^n}G(a) = 0 \text{ [by (iv)]}$$

$$\Rightarrow G(a) = \frac{(b-a)^n}{n!}f^n(c)$$

So from (ii),

$$G(a) = f(b) - f(a) - (b-a)f'(a) \cdots \frac{-(b-a)^{n-1}}{(n-1)!} f^{n-1}(a)$$

$$\Rightarrow f(b) = f(a) + (b-a)f'(a) + \frac{(b-a)^2}{2!} f''(a) + \cdots + \frac{(b-a)^{n-1}}{(n-1)!} f^{n-1}(a) + \frac{(b-a)^n}{n!} f^n(c)$$

$$= f(a) + \sum_{k=1}^{n-1} \frac{(b-a)^k}{k!} f^k(a) + \frac{(b-a)^n}{n!} f^n(c) \qquad \text{(vi)}$$

Remarks

(1) $b - a = h$ and $a < c < b \Rightarrow c = a + \theta h$, $0 < \theta < 1$. Then (vi) reduces to the form

$$f(a+h) = f(a) + \sum_{k=1}^{n-1} \frac{h^k}{k!} f^k(a) + \frac{h^n}{n!} f^n(a + \theta h), \ 0 < \theta < 1 \qquad \text{(vii)}$$

Replacing a by x,

$$f(x+h) = f(x) + \sum_{k=1}^{n-1} \frac{h^k}{k!} f^k(x) + \frac{h^n}{n!} f^n(x + \theta h) \qquad \text{(viii)}$$

(2) For $n = 1$, we have $f(a+h) = f(a) + hf'(a + \theta h)$, $0 < \theta < 1$ which is the form of Mean Value Theorem of 1st order.

For $n = 2$, we have $f(a+h) = f(a) + hf'(a) + \frac{h^2}{2!} f''(a + \theta h)$, $0 < \theta < 1$, is called the Mean Value Theorem of 2nd order. So Taylor's Theorem is the Mean Value Theorem of nth order.

(3) Putting $b = x$ in (vi) we have

$$f(x) = f(a) + \sum_{k=1}^{n-1} \frac{(x-a)^k f^k(x)}{k!} + \frac{(x-a)^n}{n!} f^n(x + \theta h), 0 < \theta < 1$$

is the expansion of $f(x)$ about the point $x = a$. It is also called the Taylor's series for $f(x)$.

(4) $\frac{(b-a)^n}{n!} f^n(c)$ or $\frac{h^n}{n!} f^n(a + \theta h)$ or $\frac{h^n}{n!} f^n(x + \theta h)$, $0 < \theta < 1$ are called the Lagrange's form of remainder, the remainder after n terms in (vi) or (vii) or (viii) and it is denoted by R_n.

5.3 Maclaurin's Series in Finite Form

Theorem [5.10] If $f : [a, b] \to \mathbb{R}$ possesses up to nth order derivatives satisfying (i) $f, f', \cdots, f^{n-1}$ are continuous in $[0, h]$ (ii) $f^n(x)$ exists in $(0, h)$, there exists a number θ, $0 < \theta < 1$ such that

$$f(x) = f(0) + xf'(0) + \frac{x^2}{2!} f''(0) + \cdots + \frac{x^n}{n!} f^n(\theta x), \ 0 < \theta < 1$$

for all $x \in [0, h]$.

Proof. It can be proved by taking $a = 0$ and $h = x$ in (vii) of Theorem [5.9].

The corresponding form of the remainder R_n is $\frac{x^n}{n!} f^n(\theta x)$.

Remark. $f(x) = f(0) + xf'(0) + \cdots + \frac{x^n}{n!} f^n(\theta x)$, $0 < \theta < 1$, is known as Maclaurin's series for $f(x)$ and $f(x)$ is expanded in the neighbourhood of $x = 0$.

Problem [5.6] : Let $f : [a, b] \to \mathbb{R}$, $[a, b] \subseteq \mathbb{R}$ be differentiable twice in the neighbourhood of $c \in [a, b]$ and $f''(x)$ be continuous at $x = c$. Then show that

$$f''(c) = \lim_{h\to 0} \frac{f(c+h) - 2f(c) + f(c-h)}{h^2}.$$

Give an example, where the R.H.S limit exists but $f''(c)$ does not exist.

Solution.

$$f(c+h) = f(c) + hf'(c) + \frac{h^2}{2!} f''(c+\theta h),\ 0 < \theta < 1,$$

by Taylor's Theorem in $[c-h,\ c+h]$ as $f'(x)$ and $f''(x)$ both exist at $x = c$ in the neighbourhood $(c-h, c+h)$ of c.

Also $f(c-h) = f(c) - hf'(c) + \dfrac{h^2}{2!} f''(c+\theta' h),\ 0 < \theta' < 1.$

Then

$$f(c+h) + f(c-h) = 2f(c) + \frac{h^2}{2}[f''(c+\theta h) + f''(c+\theta' h)]$$

$$\text{or, } \frac{f(c+h) - 2f(c) + f(c-h)}{h^2} = \frac{1}{2}[f''(c+\theta h) + f''(c+\theta' h)]\ 0 < \theta, \theta' < 1$$

$$\begin{aligned}\text{or, } \lim_{h\to 0} \frac{f(c+h) - 2f(c) + f(c-h)}{h^2} &= \frac{1}{2} \lim_{h\to 0}[f''(c+\theta h) + f''(c+\theta' h)] \\ &= \frac{1}{2} \cdot 2f''(c)\ [\because f''(x) \text{ is continuous at } x = c] \\ &= f''(c)\end{aligned}$$

For the next part of the problem, we consider function f defined by

$$f(x) = \begin{cases} x^2 + 1, & \text{when } 0 \le x < 1 \\ 2x, & \text{when } 1 \le x < 2 \end{cases}$$

Then,

$$f'(x) = \begin{cases} 2x, & \text{when } 0 \le x < 1 \\ 2, & \text{when } 1 \le x < 2 \end{cases}$$

We consider $c = 1 \in (0, 2)$.

Then

$$\begin{aligned}\lim_{h\to 0} \frac{f(1+h) - 2f(1) + f(1-h)}{h^2} &= \lim_{h\to 0} \frac{2(1+h) - 2\cdot 2 + (1-h)^2 + 1}{h^2} \\ &= \lim_{h\to 0} \frac{2 + 2h - 4 + 1 - 2h + h^2 + 1}{h^2} \\ &= \lim_{h\to 0} \frac{h^2}{h^2} = 1\end{aligned}$$

Therefore, the limit exists,

$$Rf''(1) = \lim_{x\to 1+0} \frac{f'(x) - f'(1)}{x-1} = \lim_{x\to 1+0} \frac{2-2}{x-1} = 0$$

$$Lf''(1) = \lim_{x\to 1-0} \frac{f'(x) - f'(1)}{x-1} = \lim_{x\to 1-0} \frac{2x-2}{x-1} = 2.$$

$\therefore Rf''(1) \neq Lf''(1) \Rightarrow f''(1)$ does not exist.

Problem [5.7] :

(i) If for a function f, f' and f'' exist and satisfies the inequalities $|f(x)| < A$, $|f''(x)| < B$ $\forall\, x > a$, where A, B are constants. Then show that $|f'(x)| < 2\sqrt{AB}$.

(ii) Apply Mean Value Theorem to prove that $\phi'(x) = F'\{f(x)\}f'(x)$, when $\phi(x) = F\{f(x)\}$, assuming the derivatives occur are continuous.

(iii) Let f be defined in $(-\pi/2, \pi/2)$ by $f(x) = x + \tan x$. Show that f has an inverse function g and that $g'(y) = \dfrac{1}{2 + (y - g(y))^2}$.

(iv) Let $f : \mathbb{R} \to \mathbb{R}$ be differentiable at c and $f(c) = 0$. Show that $g(x) = |f(x)|$ is differentiable at c, if and only if $f'(c) = 0$. Also prove the same if $f'(x)$ is continuous at $x = c$.

(v) Use Lagrange's Mean Value Theorem to prove that $1 + x < e^x < 1 + xe^x \;\forall\, x > 0$.

(vi) If f and g are defined on $[0, \infty)$ by

$$f(x) = \lim_{n\to\infty} \frac{x^n - 1}{x^n + 1} \text{ and } g(x) = \int_0^x f(x)dx,$$

then prove that g is continuous but not differentiable at $x = 1$.

(vii) Show that there exists one real value of x in $(0, 1)$ such that $a_0x^n + a_1x^{n-1} + \cdots + a_n = 0$, if $\dfrac{a_0}{n+1} + \dfrac{a_1}{n} + \cdots + \dfrac{a_{n-1}}{2} + a_n = 0$.

(viii) If $f''(x)$ exists $\forall\, x \in [a, b]$ and $\dfrac{f(c) - f(a)}{c - a} = \dfrac{f(b) - f(c)}{b - c}$, where $a < c < b$, then there exists a number ξ, $a < \xi < b$ such that $f''(\xi) = 0$.

(ix) If $f''(x)$ exists and $> 0 \;\forall\, x \in \mathbb{R}$. Then show that $f\left(\dfrac{x_1 + x_2}{2}\right) < \dfrac{f(x_1) + f(x_2)}{2}$.

(x) Let $f : [a, b] \to \mathbb{R}$ be continuous on $[a, b]$ and differentiable on (a, b). Show that if $\lim\limits_{x\to a} f'(x) = A$, then $f'(a)$ exists and is equal to A.

(xi) Find the value of θ in the Lagrange's form of R_n for expansion of $\frac{1}{1-x}$ in powers of x.

Solution.

(i) Let $x > a$ and $h > 0$.

Then $f(x + h) = f(x) + hf'(x) + \dfrac{h^2}{2!}f''(x + \theta h)$, $0 < \theta < 1$.

$$\Rightarrow hf'(x) = f(x + h) - f(x) - \frac{h^2}{2}f''(x + \theta h)$$

$$\Rightarrow |hf'(x)| = \left|f(x + h) - f(x) - \frac{h^2}{2}f''(x + \theta h)\right|$$

$$\leq |f(x + h)| + |f(x)| + \frac{h^2}{2}|f''(x + \theta h)| < A + A + \frac{h^2}{2}B$$

since $|f(x)| < A$ and $|f''(x)| < B \;\forall\, x > a$.

Thus $|f'(x)| < \dfrac{2A}{h} + \dfrac{hB}{2} = g(h)$, say.

So $|f'(x)|$ is a function of h and is less than $g(h)$ $\forall\, h$. Consequently $|f'(x)|$ is less than the least value of $g(h)$.

Now

$$g'(h) = 0 \Rightarrow -\frac{2A}{h^2} + \frac{B}{2} = 0 \Rightarrow h = \pm 2\sqrt{\frac{A}{B}}$$

$$\text{and } g''(h) = \frac{4A}{h^3} > 0 \text{ for } h = 2\sqrt{\frac{A}{B}}$$

Therefore, $g(h)$ is minimum, where $h = 2\sqrt{\frac{A}{B}}$.

$\therefore$ least value of $g(h) = 2 \times \frac{1}{2}A\sqrt{\frac{B}{A}} + \frac{B}{2}\cdot 2\sqrt{\frac{A}{B}} = \sqrt{AB} + \sqrt{AB} = 2\sqrt{AB}$

Hence $|f'(x)| < 2\sqrt{AB}$.

(ii) Given, $\phi(x) = F\{f(x)\}$. Let $f(x) = u$, then $\phi(x) = F(u)$.

Now

$$\begin{aligned}
\phi'(x) &= \lim_{h\to 0}\frac{\phi(x+h)-\phi(x)}{h} = \lim_{h\to 0}\frac{F\{f(x+h)\} - F\{f(x)\}}{h} \\
&= \lim_{h\to 0}\frac{F\{f(x) + hf'(x+\theta h)\} - F\{f(x)\}}{h}, \quad 0<\theta<1 \\
&= \lim_{h\to 0}\frac{F(u+H)-F(u)}{h}, \text{ taking } H = hf'(x+\theta h) \\
&= \lim_{h\to 0}\frac{F(u) + HF'(u+\theta' H) - F(u)}{h}, \text{by M.V. Theorem, } 0<\theta'<1 \\
&= \lim_{h\to 0}\frac{HF'(u+\theta' H)}{h} = \lim_{h\to 0}\frac{hf'(x+\theta h)F'(u+\theta' hf'(x+\theta h))}{h} \\
&= \lim_{h\to 0} f'(x+\theta h)F'(u+\theta' hf'(x+\theta h)) \\
&= f'(x)F'(u) = F'(u)f'(x) \quad [\because f' \text{ is continuous}] \\
&= F'\{f(x)\}f'(x)
\end{aligned}$$

(iii) Let $f(x) = x + \tan x$ in $(-\pi/2, \pi/2)$

f is continuus and strictly increasing in $(-\pi/2, \pi/2)$. So f has an inverse function g, say. That is $f^{-1} = g$, so that $(g\circ f)(x) = x$.

$$\Rightarrow g(f(x)) = x \tag{1}$$

$$\Rightarrow g'(f(x))f'(x) = 1$$

$$\Rightarrow g'(y) = \frac{1}{f'(x)}, \text{ when } f(x) = y \tag{2}$$

Also $f(x) = y \Rightarrow y = x + \tan x \Rightarrow \tan x = y - x = y - g(y)$ by (1).

So from (2), $g'(y) = \dfrac{1}{1+\sec^2 x} = \dfrac{1}{1+1+\tan^2 x} = \dfrac{1}{2+(y-g(y))^2}$

(iv) Let $f(x) > 0\ \forall\ x > c$ and $f(x) < 0\ \forall\ x < c$.

Then $g(x) = |f(x)| = \begin{cases} f(x), & x > c \\ -f(x), & x < c \end{cases}$

Now

$$Rg'(c) = \lim_{x\to c+0}\frac{g(x)-g(c)}{x-c} = \lim_{x\to c+0}\frac{f(x)-|f(c)|}{x-c} \quad [\because g(c) = |f(c)| = 0 \text{ as } f(c) = 0]$$

$$= \lim_{x\to c+0}\frac{f(x)}{x-c} = \lim_{h\to 0}\frac{f(c+h)}{h} \tag{i}$$

Also

$$Lg'(c) = \lim_{x\to c-0} \frac{g(x)-g(c)}{x-c} = \lim_{x\to c-0} \frac{-f(x)}{x-c} = -\lim_{h\to 0} \frac{f(c-h)}{-h} \tag{ii}$$

Therefore, $g'(c)$ exists, if and only if $f'(c) = -f'(c)$, i.e., iff $f'(c) = 0$ [by (i) and (ii)].
Again, from (i)

$$Rg'(c) = \lim_{h\to 0} \frac{f(c+h)}{h} = \lim_{h\to 0} \frac{f(c)+hf'(c+\theta h)}{h},\ 0<\theta<1$$

$$= \lim_{h\to 0} f'(c+\theta h) = f'(c) \tag{iii}$$

Also,

$$Lg'(c) = -\lim_{h\to 0} \frac{f(c-h)}{-h} \text{ [from (ii)]}$$

$$-\lim_{h\to 0} \frac{f(c)-hf'(c-\theta' h)}{-h}\ 0<\theta'<1$$

$$= -\lim_{h\to 0} f'(c-\theta' h) = -f'(c) \tag{iv}$$

$\therefore$ By (iii) and (iv), $g'(c)$ exists if and only if $f'(c) = -f'(c)$, i.e., $f'(c) = 0$.

Go through the examples

Example. Let $f(x) = x^3$ such that $f(0) = 0$. Also $f'(0) = 0$.
Again, $g(x) = |f(x)| \Rightarrow g(x) = x^3,\ \forall\ x > 0$ and $g(x) = -x^3,\ \forall\ x < 0$
and $g(0) = 0$

$$Rg'(0) = \lim_{x\to 0+0} \frac{g(x)-g(0)}{x} = \lim_{x\to 0+0} \frac{x^3}{x} = 0$$

$$Lg'(0) = \lim_{x\to 0-0} \frac{g(x)-g(0)}{x} = \lim_{x\to 0-0} \frac{-x^3}{x} = 0$$

Hence $g'(0)$ exists as $f'(0) = 0$.

Example. Let $f(x) = x^3 - 1$ such that $f(1) = 0$, but $f'(1) = 3 \neq 0$.

Also $g(x) = |f(x)| = \begin{cases} x^3-1, & \forall\ x > 1 \\ -(x^3-1), & \forall\ x < 1 \end{cases}$

and $g(1) = |f(1)| = 0$

Now

$$Rg'(1) = \lim_{x\to 1+0} \frac{g(x)-g(1)}{x-1} = \lim_{x\to 1+0} \frac{x^3-1}{x-1} = \lim_{x\to 1+0} (x^2+x+1) = 3$$

$$Lg'(1) = \lim_{x\to 1-0} \frac{g(x)-g(1)}{x-1} = \lim_{x\to 1-0} \frac{-(x^3-1)}{x-1} = -3$$

So $Rg'(1) \neq Lg'(1)$ and hence $g'(1)$ does not exist.

(v) Let $f(x) = e^x$ in $[0, x]$.
Then f is continuous in $[0, x]$ and $f'(x)$ exists in $(0, x)$. Therefore, by Lagrange's Mean Value Theorem, $f(x) - f(0) = (x-0)f'(c)$, for some $c \in (0, x)$

$$\Rightarrow \frac{e^x - 1}{x} = f'(c) = e^c \tag{1}$$

Again, $0 < c < x \Rightarrow e^0 < e^c < e^x$.
That is, $1 < e^c < c^x \Rightarrow 1 < \frac{e^x-1}{x} < e^x$, by (1)

$$\Rightarrow x+1 < e^x < 1 + xe^x$$

(vi) $f(x) = \lim_{n\to\infty} \frac{x^n - 1}{x^n + 1} = \begin{cases} -1, & 0 \le x < 1 \\ 1, & x > 1 \\ 0, & x = 1. \end{cases}$

f is bounded on $[0, \infty)$, though it is discontinuous at $x = 1$.
So f is integrable in $[0, x)\ \forall\ x > 0$, and let $|f(x)| \le \mu$.
Now

$$|g(x+h) - g(x)| = \left| \int_0^{x+h} f(t)dt - \int_0^x f(t)dt \right|$$
$$= \left| \int_x^{x+h} f(t)dt \right| \le |h|\mu < \frac{\epsilon}{\mu} \cdot \mu = \epsilon,\ 0 < |h| < \delta.$$

$\Rightarrow g$ is continuous on $[0, x)$.
Now

$$\frac{g(1+h) - g(1)}{h} = \frac{1}{h} \int_1^{1+h} f(t)dt$$
$$= \frac{1}{h} \cdot h\,\mu \quad \left[\because \int_a^b f(x)dx = \mu(b-a),\ \text{where } m \le \mu \le M.\right]$$
$$= \mu$$

$\therefore \lim_{h\to 0} \frac{g(1+h)-g(1)}{h} = \mu$, which does not exist.
Hence $g'(1)$ does not exist.

(vii) Let $f(x) = a_0 \frac{x^{n+1}}{n+1} + a_1 \frac{x^n}{n} + \cdots + \frac{a_{n-1}}{2} x^2 + a_n x$.
Then $f'(x) = a_0 x^n + a_1 x^{n-1} + \cdots + a_{n-1} x + a_n$.
So f being a polynomial in x, is continuous and differentiable $\forall\ x$.
Hence (a) f is continuous in $[0, 1]$
(b) f' exists in $(0, 1)$
(c) $f(0) = 0,\ f(1) = \frac{a_0}{n+1} + \frac{a_1}{n} + \cdots + \frac{a_{n-1}}{2} + a_n = 0 \Rightarrow f(0) = f(1)$.
So f satisfies all the conditions of Rolle's Theorem in $[0, 1]$.
Therefore, $\exists$ a real value c in $(0, 1)$ such that $f'(c) = 0$.
Thus $a_0 x^n + a_1 x^{n-1} + \cdots + a_{n-1} x + a_n = 0$ for some real value of x in $(0, 1)$.

(viii) $f''(x)$ exists in $[a, b]$.
$\Rightarrow f$ is continuous in $[a, b]$ and $f'(x)$ exists in (a, b)
$\Rightarrow f$ is continuous in $[a, c], [c, b]$ and $f'(x)$ exists in $(a, c), (c, b)$, where $a < c < b$.
So, by Lagrange's Mean Value Theorem,

$$\frac{f(c) - f(a)}{c - a} = f'(\xi_1),\ a < \xi_1 < c$$
$$\text{and } \frac{f(b) - f(c)}{b - c} = f'(\xi_2),\ c < \xi_2 < b$$

$\Rightarrow f'(\xi_1) = f'(\xi_2)$, by the given condition.
Hence f' satisfies all the conditions if Rolle's Theorem in $[\xi_1, \xi_2], a < \xi_1 < c < \xi_2 < b$.

Therefore, $\exists$ a point ξ, $\xi_1 < \xi < \xi_2$ such that $f''(\xi) = 0$.

That is, $\exists\ \xi \in (a, b)$ such that $f''(\xi) = 0$.

(ix) $f''(x) > 0 \Rightarrow f'$ is an increasing function of x.

Let $x_1, x_2 \in \mathbb{R}$ and $x_2 > x_1$.

Then $\dfrac{x_1 + x_2}{2} \in \mathbb{R}$ such that $x_1 < \dfrac{x_1 + x_2}{2} < x_2$.

$f''(x)$ exists. So we can apply Lagrange's Mean Value Theorem in

$$\left[x_1, \frac{x_1 + x_2}{2}\right] \text{ and } \left[\frac{x_1 + x_2}{2}, x_2\right].$$

Therefore,

$$f\left(\frac{x_1 + x_2}{2}\right) - f(x_1) = \frac{x_2 - x_1}{2} f'(\xi_1),\ x_1 < \xi_1 < \frac{x_1 + x_2}{2} \quad (1)$$

Also

$$f(x_2) - f\left(\frac{x_1 + x_2}{2}\right) = \frac{x_2 - x_1}{2} f'(\xi_2),\ \frac{x_1 + x_2}{2} < \xi_2 < x_2 \quad (2)$$

(1) – (2) gives

$$2f\left(\frac{x_1 + x_2}{2}\right) - f(x_1) - f(x_2) = \frac{x_2 - x_1}{2}[f'(\xi_1) - f'(\xi_2)] < 0$$

as $x_2 - x_1 > 0$ and $f'(\xi_1) < f'(\xi_2)$

$$\Rightarrow 2f\left(\frac{x_1 + x_2}{2}\right) < f(x_1) + f(x_2)$$

$$\Rightarrow f\left(\frac{x_1 + x_2}{2}\right) < \frac{f(x_1) + f(x_2)}{2}$$

(x) We have $f(a + h) = f(a) + hf'(a + \theta h),\ 0 < \theta < 1$

$$\Rightarrow \frac{f(a + h) - f(a)}{h} = f'(a + \theta h)$$

$$\Rightarrow \lim_{h \to 0} \frac{f(a + h) - f(a)}{h} = \lim_{h \to 0} f'(a + \theta h)$$

$$= \lim_{x \to a} f'(x)\ [a + \theta h = x \Rightarrow x \to a \text{ as } h \to 0]$$

$$= A \text{ [Given]}$$

$$\Rightarrow f'(a) = A$$

which shows that $f'(a)$ exists and equal to A.

(xi) Let $f(x) = \frac{1}{1-x} = (1 - x)^{-1}$

$$f'(x) = (1 - x)^{-2}, f''(x) = 2!(1 - x)^{-3}, \cdots, f^n(x) = n!(1 - x)^{-(n+1)}$$
$$f(0) = 1, f'(0) = 1, f''(0) = 2!, \cdots, f^n(\theta x) = n!(1 - \theta x)^{-(n+1)}, 0 < \theta < 1$$

Therefore, $f(x) = f(0) + xf'(0) + \frac{x^2}{2!}f''(0) + \cdots + \frac{x^n}{n!}f^n(\theta x)$

$$\Rightarrow \frac{1}{1-x} = 1 + x + x^2 + \cdots + x^{n-1} + \frac{x^n}{n!}n!(1-\theta x)^{-(n+1)}$$

$$\Rightarrow 1 = (1-x)(1 + x + \cdots + x^{n-1}) + x^n(1-x)(1-\theta x)^{-(n+1)}$$

$$\Rightarrow 1 = 1 - x^n + x^n(1-x)(1-\theta x)^{-(n+1)}$$

$$\Rightarrow 1 = (1-x)(1-\theta x)^{-(n+1)}$$

$$\Rightarrow \frac{1}{(1-x)} = (1-\theta x)^{-(n+1)} = \frac{1}{(1-\theta x)^{n+1}}$$

$$\Rightarrow \frac{1}{(1-x)^{\frac{1}{n+1}}} = \frac{1}{1-\theta x} \Rightarrow 1 - \theta x = (1-x)^{\frac{1}{n+1}}$$

$$\Rightarrow \theta x = 1 - (1-x)^{\frac{1}{n+1}} \Rightarrow \theta = \frac{1-(1-x)^{\frac{1}{n+1}}}{x}$$

Taylor's Theorem with Cauchy's form of Remainder R_n

Theorem [5.11] If function $f : [a, b] \to \mathbb{R}$, $[a, b] \subseteq \mathbb{R}$ possesses the following properties

(i) $f^{n-1}(x)$ is continuous $\forall\ x \in [a, b]$,

(ii) $f^n(x)$ exists in (a, b), then there exists a number θ, $0 < \theta < 1$, such that

$$f(a+h) = f(a) + hf'(a) + \frac{h^2}{2!}f''(a) + \cdots + \frac{h^{n-1}}{(n-1)!}f^{n-1}(a)$$
$$+ \frac{h^n(1-\theta)^{n-1}}{(n-1)!}f^n(a+\theta h),\ 0 < \theta < 1,$$

where $R_n = \dfrac{h^n(1-\theta)^{n-1}}{(n-1)!}f^n(a+\theta h)$.

Proof. We consider a function

$$F(x) = \phi(x) - \frac{b-x}{b-a}\phi(a), \tag{i}$$

where $\phi(x) = f(b) - f(x) - (b-x)f'(x) - \cdots - \dfrac{(b-x)^{n-1}}{(n-1)!}f^{n-1}(x)$.

Here $\phi(b) = 0 \Rightarrow F(b) = 0$

$F(a) = \phi(a) - \phi(a) = 0$.

So $F(a) = F(b)$.

F satisfies all the conditions of Rolle's Theorem. So $\exists\ \xi \in (a, b)$ such that $F'(\xi) = 0$.

$$F'(x) = \phi'(x) + \frac{\phi(a)}{(b-a)}$$
$$= -f'(x) + f'(x) - (b-x)f''(x) + \frac{2(b-x)}{2!}f''(x) - \cdots$$
$$+ \frac{(n-1)(b-x)^{n-2}}{(n-1)!}f^{n-1}(x) - \frac{(b-x)^{n-1}}{(n-1)!}f^n(x) + \frac{\phi(a)}{b-a}$$

$$\Rightarrow F'(x) = -\frac{(b-x)^{n-1}}{(n-1)!}f^n(x) + \frac{\phi(a)}{(b-a)}$$

So $F'(\xi) = 0$

$$\Rightarrow \phi(a) = \frac{(b-a)(b-\xi)^{n-1}}{(n-1)!} f^n(\xi)$$

$$\Rightarrow \phi(a) = h\frac{(a+h-a-\theta h)^{n-1}}{(n-1)!} f^n(a+\theta h) \begin{bmatrix} b-a=h \\ \xi = a+\theta h,\ 0<\theta<1 \end{bmatrix}$$

$$= \frac{h^n(1-\theta)^{n-1}}{(n-1)!} f^n(a+\theta h)$$

$$\phi(a) = f(b) - f(a) - (b-a)f'(a) - \cdots - \frac{(b-a)^{n-1}}{(n-1)!} f^{n-1}(a)$$

$$\Rightarrow \frac{h^n(1-\theta)^{n-1}}{(n-1)!} f^n(a+\theta h) = f(a+h) - f(a) - hf'(a) - \cdots - \frac{h^{n-1}}{(n-1)!} f^{n-1}(a)$$

$$\Rightarrow f(a+h) = f(a) + hf'(a) + \cdots + \frac{h^{n-1}}{(n-1)!} f^{n-1}(a)$$

$$+ \frac{h^n(1-\theta)^{n-1}}{n-1!} \cdot f^n(a+\theta h),\ 0<\theta<1$$

Problem [5.8]: Deduce Taylor's Theorem from Cauchy's Mean Value Theorem.

Solution. Let

$$\phi(x) = f(b) - f(x) - (b-x)f'(x) - \cdots - \frac{(b-x)^{n-1}}{(n-1)!} f^{n-1}(x)$$

and $\psi(x) = (b-x)^n$.

Then $\psi'(x) = -n(b-x)^{n-1}$ and $\psi'(x) \neq 0$ for any $x \in (a,b)$.

From Cauchy's Mean Value Theorem, we have

$$\frac{\phi(b)-\phi(a)}{\psi(b)-\psi(a)} = \frac{\phi'(\xi)}{\psi'(\xi)}, a<\xi<b$$

$$\Rightarrow \frac{-f(b)+f(a)+(b-a)f'(a)+\cdots+\frac{(b-a)^{n-1}}{(n-1)!}f^{n-1}(a)}{-(b-a)^n} = \frac{-\frac{(b-\xi)^{n-1}}{(n-1)!}f^n(\xi)}{-n(b-\xi)^{n-1}}$$

$$[\because \phi(b)=0, \psi(b)=0]$$

$$\Rightarrow -f(b)+f(a)+(b-a)f'(a)+\cdots+\frac{(b-a)^{n-1}}{(n-1)!}f^{n-1}(a) = -\frac{(b-a)^n}{n!}f^n(\xi)$$

$$\Rightarrow f(b) = f(a)+(b-a)f'(a)+\cdots+\frac{(b-a)^{n-1}}{(n-1)!}f^{n-1}(a)+\frac{(b-a)^n}{n!}f^n(\xi)$$

Problem [5.9]: If $f'(x)$ and $g'(x)$ exist $\forall\ x \in [a,b]$ and $g'(x) \neq 0$ for $a \leq x \leq b$ and if $f'(x) = 0\ \forall\ x \in (a,b)$, then show that f is constant in (a,b).

Proof. f, g satisfy the conditions of Cauchy's Mean Value Theorem.

Then we have $\dfrac{f(x)-f(a)}{g(x)-g(a)} = \dfrac{f'(\xi)}{g'(\xi)}, a<\xi<x<b$

$\Rightarrow f(x)-f(a) = 0$ as $f'(x) = 0$ and $g'(x) \neq 0\ \forall\ x \in (a,b)$

$\Rightarrow f(x) = f(a)$

$\Rightarrow f$ is constant in (a,b).

Problem [5.10] : Use Taylor's Theorem to find out the approximate value of $\sqrt{3}$.

Solution. $\sqrt{3} = \frac{7}{4}\left(1 - \frac{1}{49}\right)^{1/2}$. Let $f(x) = \sqrt{x}$.

Then $f'(x) = \frac{1}{2}\frac{1}{\sqrt{x}}, f''(x) = -\frac{1}{4}x^{-3/2}, f'''(x) = \frac{3}{8}x^{-5/2}, f^{IV}(x) = -\frac{15}{16}x^{-7/2}$.

We apply Taylor's Theorem for $n = 4$.

Then

$$f(x+h) = f(x) + hf'(x) + \frac{h^2}{2!}f''(x) + \frac{h^3}{3!}f'''(x) + \frac{h^4}{4!}f^{IV}(x+\theta h),\ 0 < \theta < 1, \tag{1}$$

where $R_4(x)$ = remainder after 4 terms.

Now we take $x = 1, h = -\frac{1}{49}$. Then from (1), with $R_3(x)$

$$f\left(1 - \frac{1}{49}\right) = f(1) - \frac{1}{49}\cdot\frac{1}{2\sqrt{1}} + \frac{1}{2\cdot 49^2}\left(-\frac{1}{4}\right) - \frac{1}{6}\cdot\frac{1}{(49)^3}\cdot\frac{3}{8}\left(1 - \frac{\theta}{49}\right)^{-5/12}$$

$$\text{or, } \left(1 - \frac{1}{49}\right)^{1/2} = 1 - \frac{1}{98} - \frac{1}{8\cdot(49)^2} - \frac{1}{16\cdot(49)^3}\frac{1}{\left(1 - \frac{\theta}{49}\right)^{5/12}} \tag{2}$$

Now

$$1 - \frac{\theta}{49} < 1 \Rightarrow \frac{1}{1 - \frac{\theta}{49}} > 1 \Rightarrow -\frac{1}{16\cdot(49)^3}\frac{1}{\left(1 - \frac{\theta}{49}\right)^{5/2}} < -\frac{1}{16(49)^3}$$

So from (2), the approximate value of $\left(1 - \frac{1}{49}\right)^{1/2}$ is given by

$$\begin{aligned}\left(1 - \frac{1}{49}\right)^{1/2} &= 1 - \frac{1}{98} - \frac{1}{8\cdot(49)^2}\\ &= 1 - 0.01 - 0.12 \times 0.0004 = 1 - 0.01 - 0.000048\\ &= 1 - 0.01 - 0.00005 = 1 - 0.01005 = 0.98995\end{aligned}$$

$$\therefore\ \frac{7}{4}\left(1 - \frac{1}{49}\right)^{1/2} = \frac{7 \times 0.98995}{4} = \frac{7 \times 0.99}{4} = 1.732$$

5.4 Taylor's Series in Infinite Form

Let f possesses continuous derivatives of every order in $[a, a+h]$.

Then, we have from Taylor's Theorem,

$$f(a+h) = f(a) + hf'(a) + \cdots + \frac{h^{n-1}}{(n-1)!}f^{n-1}(a) + R_n,$$

where R_n denotes the remainder after n terms.

Now let $S_n = f(a) + hf'(a) + \cdots + \frac{h^{n-1}}{(n-1)!}f^{n-1}(a)$. Then

$$f(a+h) = S_n + R_n \tag{1}$$

If $R_n \to 0$ as $n \to \infty$, $f(a+h) = \lim_{n\to\infty} S_n$, so that the infinite series

$$f(a) + hf'(a) + \frac{h^2}{2!}f''(a) + \cdots + \frac{h^n}{n!}f^n(a) + \cdots$$

converges and its sum is equal to $f(a+h)$.

Thus we have shown that if

(i) f possesses continuous derivatives of every order in $[a, a+h]$

(ii) $R_n \to 0$ as $n \to \infty$, then $f(a+h) = f(a) + hf'(a) + \frac{h^2}{2!}f''(a) + \cdots + \frac{h^n}{n!}f^n(a) + \cdots$ is known as Taylor's series in infinite form.

5.5 Maclaurin's Series in Infinite Form

If f possesses continuous derivatives of every order in $[0, h]$ and $R_n \to 0$ as $n \to \infty$, then for $x \in [0, h]$, we have

$$f(x) = f(0) + xf'(0) + \frac{x^2}{2!}f''(0) + \cdots + \frac{x^n}{n!}f^n(0) + \cdots$$

which can be deduced from Taylor's infinite series by replacing $a = 0$ and $h = x$.

Remarks

(1) $R_n \to 0$ as $n \to \infty$ is both a necessary and sufficient condition that $f(a+h)$ can be expanded in an infinite series. Sufficient part is established in the above result. Now, if $f(a+h) = \lim_{n\to\infty} S_n$, then $f(a+h) - S_n = R_n$, from (1)

$$\Rightarrow f(a+h) - \lim_{n\to\infty} S_n = \lim_{n\to\infty} R_n \Rightarrow \lim_{n\to\infty} R_n = 0$$

(2) Taylor's infinite series is also written as

$$f(x+h) = f(x) + hf'(x) + \frac{h^2}{2!}f''(x) + \cdots + \frac{h^n}{n!}f^n(x) + \cdots$$

This formula is used to expand a function of the form $f(x+h)$ in powers of h. For example, a function $\cos(x+h)$ can be expressed in powers of h, by taking $f(x) = \cos x$.

(3) Maclaurin's infinite series is used for expansion of functions in powers of x.

Example. Expand $\sin x$ in an infinite series in powers of x.

Let $f(x) = \sin x$. Then $f^n(x) = \sin\left(\frac{n\pi}{2} + x\right)$

$R_n = \frac{x^n}{n!}f^n(\theta x)$ in Maclaurin's expansion of finite form.

$$\therefore R_n = \frac{x^n}{n!}\sin\left(\frac{n\pi}{2} + \theta x\right)$$

$\Rightarrow |R_n| = \frac{|x^n|}{n!}\left|\sin\left(\frac{n\pi}{2} + \theta x\right)\right| \le \frac{|x^n|}{n!}\cdot 1$, since $\left|\sin\left(\frac{n\pi}{2} + \theta x\right)\right| \le 1$.

$\Rightarrow |R_n| \to 0$ as $n \to \infty$ for all values of x, i.e., $R_n \to 0$ as $n \to \infty$.

Thus f possesses continuous derivatives of every order $\forall\, x$ and $R_n \to 0$ as $n \to \infty$.

Hence by Maclaurin's infinite form

$$f(x) = f(0) + xf'(0) + \frac{x^2}{2!}f''(0) + \cdots + \frac{x^n}{n!}f^n(0) + \cdots$$

$$\Rightarrow \sin x = x - \frac{x^3}{3!} + \frac{x^5}{5!} - \cdots \text{ to } \infty$$

EXERCISE

(1) Let $f : I \to \mathbb{R}$, $I \subseteq \mathbb{R}$ be differentiable at $c \in I$ such that $f'(c) \neq 0$. Then prove that $|f|$ is differentiable at c and determine $|f|'(c)$.

(2) Let f be defined in the neighbourhood $(x-h, x+h)$ of x and let $f''(x)$ exists. Then prove that

$$\lim_{h\to 0} \frac{f(x+h)+f(x-h)-2f(x)}{h^2} = f''(x).$$

(3) Show that the function $f(x) = x^2 cos\frac{1}{x}$, $x \neq 0$ and $f(0) = 0$, is derivable at $x = 0$ but the derivative is not continuous at $x = 0$.

(4) If $a = -1$, $b \geq 1$ and $f(x) = \frac{1}{|x|}$, show that the conditions of Lagrange's Mean Value Theorem are not satisfied by the function f in $[a, b]$, but the result $\frac{f(b)-f(a)}{b-a} = f'(c)$ is true, if and only if $b > 1+\sqrt{2}$.

(5) Show that $a^x > x^a$, if $x > a \geq c$.

(6) If $\phi''(x) \geq 0 \ \forall \ x \in \mathbb{R}$, then show that

(i) $\phi\left[\frac{x_1+x_2}{2}\right] \leq \frac{1}{2}[\phi(x_1)+\phi(x_2)]$.

(ii) $\phi\left[\frac{x_1+x_2+\cdots+x_n}{n}\right] \leq \frac{1}{n}[\phi(x_1)+\phi(x_2)]$, where $x_1, x_2, \ldots, x_n \in [a, b] \subseteq \mathbb{R}$.

(7) If $f(x) = x^2 \sin\frac{1}{x}$, $x \neq 0$, $f(0) = 0$ and $g(x) = x \ \forall \ x \in \mathbb{R}$, then show that $\lim_{x\to 0} \frac{f'(x)}{g'(x)}$ does not exist but $\lim_{x\to 0}\left[\frac{f(x)}{g(x)}\right]$ exists and is equal to $\frac{f'(0)}{g'(0)}$.

(8) Let f be defined by $f(x) = \begin{cases} x^m \sin\frac{1}{x}, & x \neq 0 \\ 0, & x = 0 \end{cases}$

Examine the differentiability of f at $x = 0$. Find m, when f' is continuous at $x = 0$.

(9) Using Mean Value Theorem deduce the following:

(i) $|\sin x - \sin y| < |x - y|$.

(ii) $ny^{n-1}(x-y) < x^n - y^n < nx^{n-1}(x-y)$, if $0 < y < x$, $n \in N$.

(iii) $0 < \frac{1}{x}\log\left(\frac{e^x-1}{x}\right) < 1$.

(10) If $f : \mathbb{R} \to \mathbb{R}$ be a real valued function such that $|f(x) - f(y)| < |x-y|^p \ \forall \ x, y \in \mathbb{R}$ and $p > 1$, then show that f is constant.

(11) Verify Rolle's Theorem in $[a, b]$ for $f(x) = (x-a)^m(x-b)^n$, m, n are (+)ve integers.

(12) If f is differentiable at a point $c \in [a, b]$ and $\lim_{x\to c} f'(x) = l$ exists, then f' is continuous at $x = c$.

(13) Show that $\frac{\tan x}{x} > \frac{x}{\sin x}$ in $0 < x < \frac{\pi}{2}$.

(14) If f be defined by $f(x) = 0$, when $x \neq 0$ and $f(x) = 1$, when $x = 0$, then show that there does not exist any function $g(x)$ such that $g'(x) = f(x)$.

(15) Let $f(x+y) = f(x) + f(y) \ \forall \ x, y \in \mathbb{R}$. Prove that f is derivable on $\mathbb{R}$, if f is derivable at one point of $\mathbb{R}$.

(16) Find Maclaurin's infinite series of the following functions of the intervals indicated:

(i) $\log(1+2x)$ in $\left(-\frac{1}{2}, \frac{1}{2}\right]$

(ii) $e^x \sin x \ \forall\ x \in \mathbb{R}$

(iii) $(1+x)^m$, where $m \in \mathbb{R}$.

(17) Examine for maxima and minima of $f(x) = \sin nx \sin^n x,\ n \in N$.

(18) Show that the triangle of maximum area that can be inscribed in a circle is an equilateral triangle.

(19) Find the maximum and minimum values of $x + \sin 2x$ for $0 < x < 2\pi$.

(20) f is continuous on $[a, b]$, $f'(x)$ exists in (a, b) and $f(a) = f(b) = 0$. Prove that for every real λ, there is some $c \in (a, b)$ such that $f'(c) = \lambda f(c)$.

CHAPTER 6

FUNCTIONS OF BOUNDED VARIATION

Definition [6.1] : Let a function f be defined on $[a, b]$. A set of points $P_f = \{x_0, x_1, \cdots, x_n\}$ with the property $a = x_0 < x_1 < x_2 \cdots < x_n = b$ is called a partition (or division) of $[a, b]$. $[x_0, x_1], [x_1, x_2], \cdots, [x_{r-1}, x_r] \cdots [x_{n-1}, x_n]$ are called the closed subintervals of $[a, b]$, where $[x_{r-1}, x_r]$ is called the rth subinterval and $\delta_r = x_r - x_{r-1}$ is the length of the rth subinterval so that

$$\sum_{r=1}^{n}(x_r - x_{r-1}) = b - a.$$

$\delta f_r = f(x_r) - f(x_{r-1})$ is called the oscillation of f in the rth subinterval. So for the partition P_f of $[a, b]$ we have a sum

$$S_{P_f} = \sum_{r=1}^{n} |f(x_r) - f(x_{r-1})|,$$

the sum of the absolute values of oscillations of f in each subinterval of $[a, b]$.

Let $S_f = \{S_{P_f}, S_{Q_f}, \cdots\}$ be the set of all such sums for the partitions $P_f, Q_f, \cdots$ of $[a, b]$. If S_f is bounded above, then there exists a (+)ve number M such that

$$\sum_{r=1}^{n} |f(x_r) - f(x_{r-1}| \leq M \ \forall \ P_f \text{ of } [a, b]$$

then f is said to be a function of bounded variation on $[a, b]$ and henceforth we shall write f.b.v. on $[a, b]$.

$\sup S_f$ is known as the total variation of f on $[a, b]$ and is denoted by $V_f[a, b]$.

Thus $V_f[a, b] = \sup S_f \Rightarrow \sum |f(x_r) - f(x_{r-1})| \leq V_f[a, b]$.

Remark

(i) $V_f[a, b]$ is finite, since f is of bounded variation on $[a, b]$.

(ii) $V_f[a, b] \geq 0$ since each $S_{P_f} \geq 0$.

(iii) $V_f[a, b] = 0 \Rightarrow \sup S_f = 0 \Rightarrow f(x_r) - f(x_{r-1}) = 0 \ \forall \ r \Rightarrow f$ is constant.

Also, if f be constant, then $f(x_r) - f(x_{r-1}) = 0$ and hence $V_f[a, b] = 0$.

Hence $V_f[a, b] = 0$, if and only if f is constant.

6.1 Properties of a Function of Bounded Variation

(1) A bounded variation function f on $[a, b]$ is necessarily bounded on $[a, b]$.

(2) If f be monotonic on $[a, b]$, then f is bounded variation on $[a, b]$ and $V_f[a, b] = f(b) - f(a)$.

(3) If f, g be two functions of bounded variation on $[a, b]$, then $f \pm g, f \cdot g$ are also bounded variation on $[a, b]$. Further, if $g(x) \leq \lambda, \lambda > 0 \ \forall \ x \in [a, b]$, then f/g is a function of bounded variation on $[a, b]$.

(4) If f be bounded variation on $[a, b]$ and c be a point on $[a, b]$ such that $a < c < b$, then f is bounded variation on $[a, c]$ and $[c, b]$ and $V_f[a, b] = V_f[a, c] + V_f[c, b]$.
(Known as additive property of total variations)

Proof. (1) We consider a special partition $P_f = \{a, x, b\}$, where $a < x < b$. Since f is a function of bounded variation on $[a, b]$, for this partition P_f,

$$|f(x) - f(a)| + |f(b) - f(x)| \leq M \text{ [by Def. 6.1]}$$
$$\Rightarrow |f(x) - f(a)| \leq M$$
$$\Rightarrow |f(x)| - |f(a)| \leq |f(x) - f(a)| \leq M$$
$$\Rightarrow |f(x)| \leq |f(a)| + M \quad \text{(i)}$$

If $x = a$,

$$|f(b) - f(x)| \leq M \Rightarrow |f(x)| \leq |f(b)| + M \quad \text{(ii)}$$

If $x = b$,

$$|f(x) - f(a)| \leq M \Rightarrow |f(x)| \leq |f(a)| + M \quad \text{(iii)}$$

So from (i), (ii), (iii) it follows that f is bounded.

(2) Let f be monotonically increasing on $[a, b]$. Then $\delta f_r \geq 0$ in each subinterval for any partition P_f of $[a, b]$.

So $\sum\limits_{r=1}^{n} |f(x_r) - f(x_{r-1})| = \sum\limits_{r=1}^{n} |\delta f_r| = \sum\limits_{r=1}^{n} \delta f_r = f(b) - f(a)$.

$\Rightarrow f$ is bounded variation on $[a, b]$.

Also $\sup\{\sum |f(x_r) - f(x_{r-1})|\} = f(b) - f(a) \Rightarrow V_f[a, b] = f(b) - f(a)$.

The same is true, if f is monotonically decreasing on $[a, b]$ and $V_f[a, b] = f(a) - f(b)$.

(3) (i) Let $S_f = \{S_{P_f}, P_f \text{ a partition of } [a, b]\}$

$S_g = \{S_{P_g}, P_g \text{ a partition of } [a, b]\}$,

where $P_f = \{a = x_0 < x_1 \cdots < x_{r-1} < x_r \cdots x_n = b\} = P_g$.

f, g are functions of bounded variations on $[a, b]$. So

$$\sum |f(x_r) - f(x_{r-1}) \leq M \text{ and } \sum |g(x_r) - g(x_{r-1})| \leq M' \quad \text{(i)}$$

where M, M' are the bounds of S_f and S_g respectively.

Therefore, $\sum |(f \pm g)(x_r) - (f \pm g)(x_{r-1})|$

$$= \sum |\{f(x_r) - f(x_{r-1})\} \pm \{g(x_r) - g(x_{r-1})\}| \quad \text{(ii)}$$
$$\leq \sum |f(x_r) - f(x_{r-1})| + \sum |g(x_r) - g(x_{r-1})| \leq M + M'$$

$\Rightarrow f \pm g$ are functions of bounded variation on $[a, b]$.

Also $\{\sum |(f \pm g)(x_r) - (f \pm g)(x_{r-1})|\} \le V_f[a,b] + V_g[a,b]$ [by (ii)]

$$\Rightarrow \sup\left\{\sum |(f \pm g)(x_r) - (f \pm g)(x_{r-1})|\right\} \le V_f[a,b] + V_g[a,b]$$

$$\Rightarrow V_{f \pm g}[a,b] \le V_f[a,b] + V_g[a,b]$$

(ii) f, g are functions of bounded variations on $[a,b]$. So f, g are bounded on $[a,b]$. Let

$$\left.\begin{aligned} A &= \sup\{f(x),\ x \in [a,b]\} \\ B &= \sup\{g(x),\ x \in [a,b]\} \end{aligned}\right\} \quad \text{(iii)}$$

Also let $h = f \cdot g$ so that $h(x) = f(x)g(x)\ \forall\ x \in [a,b]$.

Now $\sum\limits_{r=1}^{n} |h(x_r) - h(x_{r-1})|$

$$\begin{aligned}
&= \sum |f(x_r)g(x_r) - f(x_{r-1})g(x_{r-1})| \\
&= \sum |f(x_r)\{g(x_r) - g(x_{r-1})\} + g(x_{r-1})\{f(x_r) - f(x_{r-1})\}| \\
&\le \sum \{|f(x_r)||g(x_r) - g(x_{r-1})| + |g(x_{r-1})||f(x_r) - f(x_{r-1})|\} \\
&\le A\sum |g(x_r) - g(x_{r-1})| + B\sum |f(x_r) - f(x_{r-1})| \qquad \text{(iv)} \\
&\le A \cdot M' + B \cdot M \ \text{[by (i) and (iii)]}
\end{aligned}$$

$\Rightarrow$ h is a function of bounded variation on $[a,b]$.

Also from (iv),

$$\left\{\sum_{r=1}^{n} |h(x_r) - h(x_{r-1})|\right\} \le AV_g[a,b] + BV_f[a,b]$$

$$\Rightarrow \sup\left\{\sum_{r=1}^{n} |h(x_r) - h(x_{r-1})|\right\} \le AV_g[a,b] + BV_f[a,b]$$

$$\Rightarrow V_{fg}[a,b] \le AV_g[a,b] + BV_f[a,b]$$

(iii) Let $h = \frac{f}{g}$ so that $h(x) = \frac{f(x)}{g(x)}\ \forall\ x \in [a,b]$, where

$$g(x) \ge \lambda > 0\ \forall\ x \in [a,b] \Rightarrow \frac{1}{g(x)} \le \frac{1}{\lambda}$$

$$\begin{aligned}
\sum |h(x_r) - h(x_{r-1})| &= \sum \left|\frac{f(x_r)}{g(x_r)} - \frac{f(x_{r-1})}{g(x_{r-1})}\right| \\
&= \sum \left|\frac{f(x_r)g(x_{r-1}) - g(x_r)f(x_{r-1})}{g(x_r)g(x_{r-1})}\right| \\
&= \sum \frac{|g(x_{r-1})\{f(x_r) - f(x_{r-1})\} - f(x_{r-1})\{g(x_r) - g(x_{r-1})\}|}{|g(x_r)||g(x_{r-1})|} \\
&\le \sum \frac{\{|g(x_{r-1})||f(x_r) - f(x_{r-1})| + |f(x_{r-1})||g(x_r) - g(x_{r-1})|}{\lambda^2} \\
&\le \sum \frac{B|f(x_r) - f(x_{r-1})| + A|g(x_r) - g(x_{r-1})|}{\lambda^2} \\
&\le \frac{BM + AM'}{\lambda^2} \ \text{[by (i) and (iii)]}
\end{aligned}$$

$\Rightarrow$ h is a function of bounded variation on $[a,b]$.

Also

$$\left\{\sum |h(x_r) - h(x_{r-1})|\right\} \le \{BV_f[a,b] + AV_g[a,b]\}/\lambda^2$$

$$\sup\left\{\sum |h(x_r) - h(x_{r-1})|\right\} \le \{BV_f[a,b] + AV_g[a,b]\}/\lambda^2$$

$$\therefore\ V_h[a,b] \le \frac{B}{\lambda^2} V_f[a,b] + \frac{A}{\lambda^2} V_g[a,b]$$

(4) Let $P_{1f} = \{a = x_0 < x_1 < \cdots < x_n = c\}$ and

$$P_{2f} = \{c = y_0 < y_1 \cdots < y_m = b\}$$

be the partitions of $[a,c]$ and $[c,b]$ so that

$$P_f = P_1 \cup P_2 = \{a = x_0 < x_1 \cdots < x_n = y_0 < y_1 \cdots < y_m = b\}$$

forming the partition of $[a,b]$.

Also $S_{P_{1f}} + S_{P_{2f}} = S_{P_f}$

$$\Rightarrow \sum_{r=1}^{n} |f(x_r) - f(x_{r-1})| + \sum_{r=1}^{m} |f(y_r) - f(y_{r-1})| = S_{P_f} \le V_f[a,b]$$

$$\Rightarrow \sum_{r=1}^{n} |f(x_r) - f(x_{r-1})| \le V_f[a,b] \quad \text{(i)}$$

$$\text{and } \sum_{r=1}^{m} |f(y_r) - f(y_{r-1})| \le V_f[a,b] \quad \text{(ii)}$$

(i) $\Rightarrow f$ is a f.v.b. on $[a,c]$ and

(ii) $\Rightarrow f$ is a f.v.b. on $[c,b]$.

Conversely, let $P = \{a = t_0 < t_1 \cdots < t_n = b\}$ be a partition of $[a,b]$ and let c be a point in $[a,b]$ s.t. $a < c < b$. We consider a new partition P_1 of $[a,b]$ so that $P_0 = P \cup \{c\}$, where $c \in [t_{r-1}, t_r]$, $r < n$. That is $P_{0f} = \{a = t_0 < t_1 \cdots < t_{r-1} < c < t_r < \cdots t_n = b\}$.

Then $\sum_{i=1}^{n} |f(t_i) - f(t_{i-1})|$

$$= \sum_{i=1}^{r-1} |f(t_i) - f(t_{i-1})| + |f(t_r) - f(t_{r-1})| + \sum_{i=r+1}^{n} |f(t_i) - f(t_{i-1})|$$

$$= \sum_{i=1}^{r-1} |f(t_i) - f(t_{i-1})| + |f(t_r) - f(c) + f(c) - f(t_{r-1})| + \sum_{i=r+1}^{n} |f(t_i) - f(t_{i-1})|$$

$$\le \sum_{i=1}^{r-1} |f(t_i) - f(t_{i-1})| + |f(t_r) - f(c)| + |f(c) - f(t_{r-1})| + \sum_{i=r+1}^{n} |f(t_i) - f(t_{i-1})| \quad \text{(iii)}$$

Now the points of P_0 in $[a,c]$ determine a partition

$$P_{1f} = \{a = t_0 < t_1 < \cdots < t_{r-1} = c\}$$

and those in $[c,b]$ determine a partition

$$P_{2f} = \{c < t_r < t_{r+1} < \cdots < t_n = b\}.$$

Hence from (iii) it follows that

$$S_{P_f}[a,b] \le S_{P_{0f}} = S_{P_{1f}} + S_{P_{2f}} \le V_f[a,c] + V_f[c,b]$$

$$\Rightarrow \sup S_f[a,b] \le V_f[a,c] + V_f[c,b] \Rightarrow V_f[a,b] \le V_f[a,c] + V_f[c,b] \qquad \text{(iv)}$$

Again, $V_f[a,c]$ and $V_f[c,b]$ being the total variations of f (i.e., the $\sup S_{P_f}$ over $[a,c]$ and $\sup S_{P_f}$ over $[c,b]$), there exist partitions

$$P_{1f} = \{a = x_1 < x_1 \cdots < x_m = c\}$$

$$\text{and } P_{2f} = \{c = y_0 < y_1 \cdots < y_n = b\}$$

on $[a,c]$ and $[c,b]$ respectively such that

$$\sum_{r=1}^{m} |f(x_r) - f(x_{r-1})| > V_f[a,c] - \frac{\epsilon}{2} \qquad \text{(v)}$$

$$\text{and } \sum_{i=1}^{n} |f(y_i) - f(y_{i-1})| > V_f[c,b] - \frac{\epsilon}{2} \text{ [by the property of } \sup S] \qquad \text{(vi)}$$

So from (v) and (vi) we have

$$\sum_{r=1}^{m} |f(x_r) - f(x_{r-1})| + \sum_{i=1}^{n} |f(y_i) - f(y_{i-1})| > V_f[a,c] + V_f[c,b] - \epsilon$$

$$\Rightarrow V_f[a,b] \ge V_f[a,c] + V_f[c,b] - \epsilon \qquad \text{(vii)}$$

As ϵ is arbitrary (+)ve number, it follows from (vii) that

$$V_f[a,b] \ge V_f[a,c] + V_f[c,b] \qquad \text{(viii)}$$

Hence from (iv) and (viii), we have $V_f[a,b] = V_f[a,c] + V_f[c,b]$.

Theorem [6.1] Let a function f defined on $[a,b]$ be such that (i) f is continuous in $[a,b]$, (ii) f' exists in (a,b), (iii) f' is bounded on $[a,b]$, then f is f.v.b. on $[a,b]$.

Proof. Let $P_f = \{a = x_0 < x_1 < \cdots < x_{r-1} < x_r \cdots < x_n = b\}$ be a partition of $[a,b]$.

f' is bounded on $[a,b] \Rightarrow |f'(x)| \le k,\ \forall\, x \in [a,b]$.

Again, by Mean Value Theorem, we have

$$f(x_r) - f(x_{r-1}) = (x_r - x_{r-1})f'(\xi_r),\ x_{r-1} < \xi_r < x_r$$

$$\Rightarrow |f(x_r) - f(x_{r-1})| = |f'(\xi_r)|\delta_r$$

$$\Rightarrow \sum_{r=1}^{n} |f(x_r) - f(x_{r-1})| = \sum_{r=1}^{n} |f'(\xi_r)|\delta_r \le k\sum_{r=1}^{n} \delta_r = k(b-a)$$

$\Rightarrow f$ is a f.b.v. on $[a,b]$.

Example. Let f be defined by $f(x) = \begin{cases} x^2 \cos\frac{1}{x}, & x \ne 0 \\ 0, & x = 0 \end{cases}$

Then (i) f is continuous in $[0,1]$

(ii) $f'(x) = 2x\cos\frac{1}{x} + \sin\frac{1}{x} \Rightarrow f'$ exists in $(0,1)$

(iii) $|f'(x)| = \left|2x\cos\frac{1}{x} + \sin\frac{1}{x}\right| \le 2|x|\left|\cos\frac{1}{x}\right| + \left|\sin\frac{1}{x}\right| \le 2\cdot1\cdot1 + 1 = 3$ and $f'(0) = 0$.

$\Rightarrow f'$ is bounded on $[0,1]$.

Therefore, by Theorem [6.1], f is a f.b.v. on $[0,1]$.

Remarks

(1) *Boundness of f' is not necessary for f to be a f.b.v.*
Let $f(x) = x^{\frac{1}{2}}$ in $[0, 1]$
$f'(x) = \frac{1}{2\sqrt{x}}$
$f'(x) \to +\infty$ as $x \to 0$. So f' is not bounded.
However, f being a monotonically increasing function, is a f.b.v. on $[0, 1]$.

(2) *A continuous function may not be f.b.v.*
We define a function f on $[0, 1]$ by

$$f(x) = \begin{cases} x\cos\left(\dfrac{\pi}{2x}\right), & \text{if } x \neq 0 \\ 0, & \text{if } x = 0. \end{cases}$$

Then f is continuous on $[0, 1]$.
We consider a partition $P_f = \left\{0, \dfrac{1}{2n}, \dfrac{1}{2n-1}, \cdots, \dfrac{1}{3}, \dfrac{1}{2}, 1\right\}$ having $2n$ subintervals.

$$\sum_{r=1}^{2n} |f(x_r) - f(x_{r-1})|$$
$$= |f(x_1) - f(x_0)| + |f(x_2) - f(x_1)| + \cdots + |f(x_{2n} - f(x_{2n-1})|$$
$$= \left|\frac{1}{2n}\cdot\cos\left(\frac{\pi\cdot 2n}{2}\right) - 0\right| + \left|\frac{1}{2n-1}\cos\left(\frac{\pi(2n-1)}{2}\right) - \frac{1}{2n}\cos\left(\frac{\pi\cdot 2n}{2}\right)\right|$$
$$+ \left|\frac{1}{2n-2}\cos\left(\frac{\pi}{2}\right)(2n-2) - \frac{1}{2n-1}\cos\left(\frac{\pi}{2}\right)(2n-1)\right| + \cdots$$
$$+ \left|1\cdot\cos\frac{\pi}{2} - \frac{1}{2}\cos\left(\frac{\pi}{2}\cdot 2\right)\right|$$
$$= \frac{1}{2n} + \frac{1}{2n} + \frac{1}{2n-2} + \frac{1}{2n-2} + \cdots + \frac{1}{4} + \frac{1}{4} + \frac{1}{2} + \frac{1}{2}$$
$$= \frac{1}{n} + \frac{1}{n-1} + \cdots + \frac{1}{2} + 1 = 1 + \frac{1}{2} + \frac{1}{3} + \cdots + \frac{1}{n}$$

which is not bounded $\forall\, n$, since $\sum_{1}^{\infty} \frac{1}{n}$ is a divergent series.
Hence, $V_f[0, 1] \to \infty$ and so f is not a f.b.v. on $[0, 1]$.

(3) *A function of bounded variation is not necessarily continuous.*
Let $f(x) = [x]$.
Then f is monotonically increasing in $[0, 3]$. So f is a f.b.v. on $[0, 3]$. But f is not continuous on $[0, 3]$ as it is discontinuous at $x = 1, 2$ and 3.

6.2 Total Variation on $[a, x]$ as a Function of x

Let f be a f.b.v. on $[a, b]$. We want to introduce the total variation as a function of x, the right end point, by keeping the left-hand point 'a' fixed so that $a < x \leq b$. We denote it by $V_f[a, x]$, or simply $V(x)$. Also $V(a) = V_f[a, a] = 0$.

Theorem [6.2] If f be a f.b.v. on $[a, b]$, then

(i) V is an increasing function on $[a, b]$.
(ii) $V - f$ is an increasing function on $[a, b]$.

Proof.

(i) We have $V(x) = V_f[a, x],\ a < x \leq b$.
Let x, y be any two points on $[a, b]$ such that $a < x < y \leq b$.
Then $V_f[a, y] = V_f[a, x] + V_f[x, y]$ [by (4) of [6.2]]
$\Rightarrow V(y) - V(x) = V_f[x, y] \geq 0$
$\Rightarrow V(y) \geq V(x)\ \forall\ y > x$
$\Rightarrow V$ is an increasing function on $[a, b]$.

(ii) Let $U = V - f$.
Then $U(x) = V(x) - f(x)\ \forall\ x \in [a, b]$
$U(y) = V(y) - f(y),\ a < x < y \leq b$
Therefore,

$$\begin{aligned} U(y) - U(x) &= \{V(y) - V(x)\} - \{f(y) - f(x)\} \\ &= V_f[x, y] - \{f(y) - f(x)\}. \qquad \text{(i)} \end{aligned}$$

Since

$$V_f[a, b] = \sup\left\{\sum_{r=1}^{n} |f(x_r) - f(x_{r-1})|\right\}$$

$$V_f[x, y] = \sup\{|f(y) - f(x)|, x, y \in [a, b]\} \Rightarrow V_f[x, y] \geq |f(y) - f(x)|$$

Hence from (i), it follows that $U(y) - U(x) \geq 0$
$\Rightarrow\ U(y) \geq U(x),\ \forall\ x, y$ satisfying $a < x < y \leq b$
$\Rightarrow\ U$ is an increasing function on $[a, b]$
$\Rightarrow\ V - f$ is an increasing function on $[a, b]$.

Theorem [6.3] Let f be defined on $[a, b]$. Then f is a f.b.v. on $[a, b]$, if and only if, f can be expressed as the difference of two increasing functions.

Proof. Let $U = V - f$.

Then by Theorem [6.2], U and V are both increasing on $[a, b]$ and consequently U and V are f.b.v. on $[a, b]$ [by (2) of 6.1].

Therefore $f = V - U$, difference of two increasing functions on $[a, b]$.

$\Rightarrow f$ is the difference of two functions of bounded variations on $[a, b]$

$\Rightarrow f$ is a f.b.v. on $[a, b]$ [by (3) of 6.1].

Next, let f be a f.b.v. on $[a, b]$, then by Theorem [6.2], $U = V - f$ is an increasing function on $[a, b] \Rightarrow f = V - U$ is the difference of two increasing functions of $[a, b]$.

Remark. The representation of f, a f.b.v. on $[a, b]$, as a difference of two increasing functions is not unique.

For, if $f = f_1 - f_2$, where f_1, f_2 are two increasing functions, then we can also write

$$f = (f_1 + g) - (f_2 + g)$$

where g is any increasing function, as the difference of two increasing functions $f_1' = f_1 + g$ and $f_2' = f_2 + g$ and consequently the representation of $f = f_1' - f_2'$ is not same as that of $f = f_1 - f_2$. Hence the result.

Theorem [6.4] Let f be a f.b.v. on $[a,b]$. Then $V(x) = V_f[a,x]$ is continuous on $[a,b]$, if and only if f is continuous on $[a,b]$.

Proof. Let V be continuous at a point $x = c \in [a,b]$ such that $a < c < x \le b$.

Then $|V(x) - V(c)| < \epsilon$ whenever $|x - c| < \delta$

$$\text{i.e., } \forall\, x \in (c-\delta, c+\delta) \tag{i}$$

Also

$$V_f[a,x] = V_f[a,c] + V_f[c,x] \tag{ii}$$

$|f(x) - f(c)| \le V_f[c,x] = V_f[a,x] - V_f[a,c]$ [by (ii)]

$\Rightarrow |f(x) - f(c)| \le V(x) - V(c) < \epsilon\ \forall\, x \in (c-\delta, c+\delta)$

$\Rightarrow f$ is continuous on $[a,b]$ as c is an arbitrary point on $[a,b]$.

Next, let f be continuous at point $c \in (a,b)$.

Then $|f(x) - f(c)| < \frac{\epsilon}{2}$ for $|x - c| < \delta$. That is,

$$|f(x) - f(c)| < \frac{\epsilon}{2}\ \forall\, x \in (c-\delta, c+\delta) \tag{iii}$$

Now for every $\epsilon(>0)$, there exists a partition P of $[c,b]$, where

$$P_f = \{x_0 = c < x_1 < x_2 \cdots < x_n = b\}$$

such that $\sum_{r=1}^{n} |f(x_r) - f(x_{r-1})| \le V_f[c,b]$.

$$V_f[c,b] = \sup\left\{\sum_{r=1}^{n} |f(x_r) - f(x_{r-1})|\right\} \Rightarrow \sum_{r=1}^{n} |f(x_r) - f(x_{r-1})| > V_f[c,b] - \frac{\epsilon}{2} \tag{iv}$$

We assume that $0 < x_1 - c < \delta$, for otherwise we have to add more points to P so that it can fulfil the condition $0 < x_1 - c < \delta$. Also the addition of more points to P, the sum

$$\sum_{r=1}^{n} |f(x_r) - f(x_{r-1})|$$

cannot decrease and consequently (iv) remains uneffected.

So from (iii), it follows that $|f(x_1) - f(c)| < \frac{\epsilon}{2}$ as

$$0 < x_1 - c < \delta \Rightarrow x_1 < c + \delta \Rightarrow x_1 \in (c-\delta, c+\delta)$$

Thus

$$\sum_{r=1}^{n} |f(x_r) - f(x_{r-1})| = |f(x_1) - f(x_0)| + \sum_{r=2}^{n} |f(x_r) - f(x_{r-1})|$$

$$= |f(x_1) - f(c)| + \sum_{r=2}^{n} |f(x_r) - f(x_{r-1})| < \frac{\epsilon}{2} + V_f[x,b] \tag{v}$$

Since $\{x_1, x_2, \cdots, x_n\}$ forms a partition of $[x_1, b]$, therefore by (iv) and (v)

$$V_f[c,b] - \frac{\epsilon}{2} < \frac{\epsilon}{2} + V_f[x_1,b] \Rightarrow V_f[c,b] - V_f[x_1,b] < \epsilon \tag{vi}$$

But

$$\begin{aligned} 0 \le V(x_1) - V(c) &= V_f[a,x_1] - V_f[a,c] \\ &= V_f[c,x_1]\ [\text{for } a < c < x_1 \le b,\ V_f[a,c] + V_f[c,x_1] = V_f[a,x_1]] \\ &= V_f[c,b] - V_f[x_1,b] < \epsilon\ [\text{by (vi)}] \end{aligned}$$

Thus, for $0 < x_1 - c < \delta$,

$$0 \le V(x_1) - V(c) < \epsilon \Rightarrow \lim_{x \to c+0} V(x) = V(c) \tag{vii}$$

Similarly, taking $0 < c - x_1 < \delta$, we can show that

$$\lim_{x \to c-0} V(x) = V(c) \tag{viii}$$

(vii) and (viii) $\Rightarrow V$ is continuous on $[a, b]$.

Theorem [6.5] [Jordan's Theorem] Let f be continuous on $[a, b]$. Then f is a f.b.v. on $[a, b]$, if and only if f can be expressed as the difference of two increasing continuous functions.

Proof. Let

$$U = V - f \tag{i}$$

If f be continuous then V is continuous on $[a, b]$, by Theorem [6.4].

From (i) it follows that U is continuous. Thus $f = V - U$, where V, U are both continuous.

Also by Theorem [6.2] V and U are both increasing functions and f is the difference of two increasing functions. Hence f is of bounded variation on $[a, b]$. Thus f is a f.b.v. on $[a, b]$, if and only if f is the difference of two increasing continuous functions on $[a, b]$.

Problem [6.1]: Show that f defined on $[0, 1]$ by

$$f(x) = \begin{cases} x^2 \sin \dfrac{1}{x^2}, & x \neq 0 \\ 0, & x = 0 \end{cases}$$

is not a f.b.v. on $[0, 1]$.

Solution.

(i) f is continuous on $[0, 1]$.

(ii) $f'(x) = 2x \sin \dfrac{1}{x^2} + x^2 \cos \left(\dfrac{1}{x^2}\right) \cdot \left(-\dfrac{2}{x^3}\right)$

$$= 2x \sin \left(\frac{1}{x^2}\right) - \frac{2}{x} \cos \left(\frac{1}{x^2}\right)$$

$$f'(0) = \lim_{x \to 0} \frac{f(x) - f(0)}{x} = \lim_{x \to 0} \frac{x^2 \sin \frac{1}{x^2}}{x} = 0$$

So $f'(x)$ exists on $[0, 1]$. Now

$$|f'(x)| = \left|2x \sin \frac{1}{x^2} - \frac{2}{x} \cos \frac{1}{x^2}\right|$$

$$\le 2|x| \sin \frac{1}{x^2} + \frac{2}{|x|} \left|\cos \frac{1}{x^2}\right| \le 2|x| + \frac{2}{|x|} \not\le k \;\forall\; x \in [0, 1]$$

That is x decreases, f' becomes unbounded.

Hence f is not a f.b.v. on $[0, 1]$.

EXERCISE

(1) Examine which of the following functions are of bounded variation on $[0,1]$:

(a) (i) $f(x) = x^3 \sin\left(\frac{1}{x}\right)$, if $x \neq 0, f(0) = 0$.

(ii) $f(x) = \sqrt{x} \sin\left(\frac{1}{x}\right)$, if $x \neq 0$, $f(0) = 0$.

(iii) $f(x) = x^2 \sin\left(\frac{1}{x^2}\right)$, if $x \neq 0, f(0) = 0$.

(b) Show that $f(x) = [x]$ is of bounded variation on $[0,3]$.

(2) Show that a polynomial f is of bounded variation on every compact interval $[a,b]$. Describe a method for finding the total variation of f on $[a,b]$, if the zeros of f' are known.

(3) Let f be a function of bounded variation on $[a,b]$ and $P = \{a = x_0 < x_1 \cdots < x_n = b\}$ be a partition of $[a,b]$. We define, $\Delta f_k = f(x_k) - f(x_{k-1}), k = 1, 2, \ldots, n$.

$$A(P) = \{k, \Delta f_k > 0\},\ B(P) = \{k, \Delta f_k < 0\}$$

$$P_f(a,b) = \sup\left\{\sum_{k\in A(P)} \Delta f_k : P \in \rho[a,b]\right\}$$

$$n_f(a,b) = \sup\left\{\sum_{k\in B(P)} |\Delta f_k| : P \in \rho[a,b]\right\}$$

as (+)ve and (−)ve variation of f on $[a,b]$ and $V(x) = V_f(a,x)$, $p(x) = p_f(a,x)$, $n(x) = n_f(a,x)$ $\forall\, x \in [a,b]$ and $V(a) = p(a) = n(a) = 0$. Then show that

(i) $V(x) = p(x) + n(x)$

(ii) $0 \leq p(x) \leq V(x)$ and $0 \leq n(x) \leq V(x)$

(iii) p and n are increasing on $[a,b]$

(iv) $f(x) = f(a) + p(x) - n(x)$

(v) $2p(x) = V(x) + f(x) - f(a), 2n(x) = V(x) - f(x) + f(a)$

(4) Prove that every absolutely continuous function f on $[a,b]$ is continuous and of bounded variation on $[a,b]$. Give an example that converse is not true.

(5) Find the total variation, (+)ve and (−)ve variations, for $f(x) = 2x - x^2, -1 \leq x \leq 2$.

(6) Find the total variation of $\sin x$ in $[0, 2\pi]$.

(7) Let $f : [a,b] \to \mathbb{R}$ be a function with bounded variation on $[a,b]$ and let $V : [a,b] \to \mathbb{R}$ be a function defined by $V(x) = V_a^b(f)\ \forall\, x \in [a,b]$. Then show that V and $V - f$ are increasing functions satisfying the inequality

$$\int_a^b V^2(x)dx - \frac{1}{b-a}\left\{\int_a^b V(x)dx\right\}^2$$
$$\geq \int_a^b V(x)f(x)dx - \frac{1}{b-a}\left\{\int_a^b V(x)dx \times \int_a^b f(x)dx\right\}$$

(8) If $\alpha : [a,b] \to \mathbb{R}$ is continuous on $[a,b]$ and differentiable on (a,b) with $\sup\limits_{a<x<b} |\alpha'(x)| \leq M$, then show that α is bounded variation and $V_\alpha[a,b] \leq M(b-a)$.

[*Hints.* By Mean Value Theorem,

$$\sum_{i=1}^{n} |\alpha(x_i) - \alpha(x_{i-1})| = \sum |\alpha'(t_i)|(x_i - x_{i-1})| \leq \sum_{i=1}^{n} M(x_i - x_{i-1}) = M(b-a)$$

$\Rightarrow \sup \sum_{i=1}^{n} |\alpha(x_i) - \alpha(k_{i-1})| \leq M(b-a)$

$\Rightarrow V_\alpha[a,b] \leq M(b-a)$

$\Rightarrow \alpha$ is bounded variation and $V_\alpha[a,b] \leq M(b-a)$.]

(9) If $\alpha, \beta : [a,b] \to \mathbb{R}$ are bounded variations, $c, d \in \mathbb{R}$, then show that $c\alpha + d\beta$ is bounded variation and $V_{c\alpha+d\beta}[a,b] \leq |c|V_\alpha[a,b] + |d|V_\beta[a,b]$.

(10) A function $f : [a,b] \to \mathbb{R}$ is constant, iff f is bounded variation and $V_f[a,b] = 0$.

(11) Show that a differentiable function $f : [a,b] \to \mathbb{R}$ with integral derivative is with bounded variation and $V_f[a,b] = \int_a^b |f'(x)|dx$.

(12) Let $\phi : [a,b] \to \mathbb{R}$ be $\mathbb{R}$-integrable. Then show that $f : [a,b] \to \mathbb{R}$ defined by $f(x) = \int_a^x \phi(t)dt$ $\forall\, x \in [a,b]$ is with bounded variation.

(13) A Lipschitz function of order 1 is with bounded variation. Prove it.

CHAPTER 7

THE RIEMANN INTEGRAL

A German mathematician Bernhard Riemann (1826–1866) introduced the concept of definite integral from the notion of limit of a sum of which term tends to zero when the number of terms tending to ∞. Literal meaning of integration is 'summation'. It can be also considered as the inverse process of differentiation. That is, if $f : [a, b] \to \mathbb{R}$ be a function and if there exists a function ϕ such that $\phi'(x) = f(x) \ \forall \ x \in [a, b]$, then ϕ is called the integral of f.

Riemann had defined the integrability of a real valued bounded function on a closed interval $[a, b] \subseteq \mathbb{R}$, using the limit of a sum. Further works have been done in this field. Among these, the theory of integration by Lebesgue (1902) is most noteworthy.

Definition [7.1]: Let $f : [a, b] \to \mathbb{R}$ be a bounded function on $[a, b]$ and let M, m be the bounds of f on $[a, b]$.

Let $P = \{a = x_0 < x_1 \cdots < x_{r-1} < x_r \cdots < x_n = b\}$ be a partition of $[a, b]$, where $x_r, r = 0, \cdots, n$ are called the points of division of $[a, b]$, $[x_{r-1}, x_r]$ is called the rth closed subinterval of $[a, b]$. $\delta_r = x_r - x_{r-1}$ is the length of the rth subinterval so that

$$\sum_{r=1}^{n} \delta_r = b - a.$$

Let M_r, m_r be the bounds of f in the rth subinterval $[x_{r-1}, x_r]$.

Then $O_r = M_r - m_r$ is called the oscillation of f in $[x_{r-1}, x_r]$.

$\omega(p, f) = \sum\limits_{r=1}^{n} O_r\delta_r = \sum\limits_{r=1}^{n} \{M_r - m_r\}\delta_r$ is called the oscillatory sum for the partition P.

The norm of the partion P, denoted by $||P||$, is defined by the length of the greatest of all subintervals $[x_{r-1}, x_r]$.

That is, $||P|| = \max\{x_1 - x_0, x_2 - x_1, \cdots, x_r - x_{r-1}, \cdots, x_n - x_{n-1}\}$.

The upper and lower sums, denoted by $S(P, f)$ and $s(P, f)$ respectively, are defined by

$$S(P, f) = \sum_{r=1}^{n} M_r\delta_r \text{ and } s(P, f) = \sum_{r=1}^{n} m_r\delta_r$$

Also $S(P, f) - s(P, f) = \omega(P, f)$.

Definition [7.2]: **[Refinement of a Partition of $[a, b]$]** A partition P^* is said to be a refinement of a partition $P\{a = x_0 < x_1 \cdots < x_{r-1} < x_r \cdots < x_n = b\}$ of $[a, b]$ if $P^* \supset P$.

That is, P^* forms a new partition of $[a,b]$ containing the points of P as well as some more points of division of $[a,b]$.

Theorem [7.1] If P^* is a refinement of $P\{a = x_0 < x_1 \cdots x_{r-1} < x_r \cdots < x_n = b\}$, then for a bounded function f on $[a,b]$.

(i) $S(P^*, f) \leq S(P, f)$ and (ii) $s(P^*, f) \geq s(P, f)$.

Proof. Let P^* contains just one point ξ more than P and let $x_{r-1} < \xi < x_r$.

Then $P^* = \{a = x_0 < x_1 \cdots < x_{r-1} < \xi < x_r \cdots < x_n = b\}$. Let ω_1, ω_2 and M_r be the upper bounds of f in $[x_{r-1}, \xi]$, $[\xi, x_r]$ and $[x_{r-1}, x_r]$ respectively.

Then $S(P^*, f) - S(P, f)$

$$= \sum_{k=1}^{r-1} \omega_k'(x_k - x_{k-1}) + \omega_1(\xi - x_{r-1}) + \omega_2(x_r - \xi) + \sum_{k=r}^{n-1} \omega_k'(x_{k+1} - x_k)$$

$$- \left\{ \sum_{k=1}^{r-1} \omega_k'(x_k - x_{k-1}) + M_r(x_r - x_{r-1}) + \sum_{k=r}^{n-1} \omega_k'(x_{k+1} - x_k) \right\}$$

$$= (\omega_1 - M_r)(\xi - x_{r-1}) + (\omega_2 - M_r)(x_r - \xi)$$

$$\leq 0 \ [\because \omega_1, \omega_2 \leq M_r] \Rightarrow S(P^*, f) \leq S(P, f)$$

In the same way, we can show that $s(P^*, f) \geq s(P, f)$.

Note. By the refinement of a partition P of $[a,b]$, the upper sums decreases while the lower sums increases.

Theorem [7.2] Let f be a bounded function on $[a,b]$ and let M and m be supremum and infimum of f on $[a,b]$. Then $m(b-a) \leq s(P,f) \leq S(P,f) \leq M(b-a)$.

Proof. Let M_r and m_r be the supremum and infimum of f in $[x_{r-1}, x_r]$

$$S(P,f) = \sum_{r=1}^{n} M_r \delta_r \text{ and } s(P,f) = \sum_{r=1}^{n} m_r \delta_r$$

We have $m \leq m_r \leq M_r \leq M \ \forall \ r = 1, 2, \cdots, n$

$$\Rightarrow m\delta_r \leq m_r\delta_r \leq M_r\delta_r \leq M\delta_r$$

$$\Rightarrow m\sum_{r=1}^{n} \delta_r \leq \sum_{r=1}^{n} m_r\delta_r \leq \sum_{r=1}^{n} M_r\delta_r \leq M\sum_{r=1}^{n} \delta_r$$

$$\Rightarrow m(b-a) \leq s(P,f) \leq S(P,f) \leq M(b-a)$$

which shows that each of lower sum and upper sum is bounded and also for each partition P, $s(P,f)$ is less than or equal to $S(P,f)$.

Definition [7.3] : **[Definition of Riemann integrability]** Let $f : [a,b] \to \mathbb{R}$, $[a,b] \subseteq \mathbb{R}$ be a bounded function on $[a,b]$. Then we have shown that $m(b-a) \leq s(P,f) \leq S(P,f) \leq M(b-a)$. Further we have observed that for a refinement P^* of P of $[a,b]$, $S(P^*, f) \leq S(P,f)$ and $s(P^*, f) \geq s(P,f)$.

Thus for all possible partitions P of $[a,b]$ we have

$$\left.\begin{array}{r} m(b-a) \leq \cdots S'(P^2, f) \leq S(P^1, f) \leq S(P, f) \\ \text{and } s(P,f) \leq s(P^1, f) \leq s(P^2, f) \cdots \leq M(b-a) \end{array}\right\} \quad \text{(i)}$$

where $P^1, P^2, \cdots$ are the refinement of P. Hence it follows that the set $S = \{S(P, f),\ P$ is any partition of $[a, b]\}$ of upper sums bounded below by $m(b-a)$ has an infimum and the set $s = \{s(P, f),\ P$ is any partition of $[a, b]\}$ of lower sums bounded above by $M(b-a)$ has a supremum, followed by (i).

The infimum of $S = \{S(P, f),\ P$ is any partition of $[a, b]\}$ is known as the upper Riemann integral of f on $[a, b]$ and is denoted by $\overline{\int_a^b} f(x)dx$.

The supremum of $s = \{s(P, f),\ P$ is any partition of $[a, b]\}$ is known as the lower Riemann integral of f on $[a, b]$ and is denoted by $\underline{\int_a^b} f(x)dx$.

Now f is said to be Riemann integrable on $[a, b]$ if $\overline{\int_a^b} f(x)dx = \underline{\int_a^b} f(x)dx$ and is denoted by

$$R\int_a^b f(x)dx \text{ or } f \in \mathbb{R}\ [a, b].$$

Problem [7.1]: Let $f : [0, 1] \to \mathbb{R}$ be a function defined by

$$f(x) = \begin{cases} 0, & \text{when } x \text{ is irrational} \\ 1, & \text{when } x \text{ is rational.} \end{cases}$$

Solution. We have $0 \le f(x) \le 1\ \forall\ x \in [0, 1]$, $m_r = 0$ and $M_r = 1$, $r = 1, 2, \cdots, n$ for any partition $P\{0 = x_0 < x_1 \cdots < x_n = 1\}$, which follows from the fact that rational and irrational points are everywhere dense.

$$\begin{aligned} S(P, f) &= \sum_{r=1}^{n} M_r\delta_r = M_1\delta_1 + \cdots + M_n\delta_n = 1 \cdot \delta_1 + \cdots + 1 \cdot \delta_n \\ &= \sum_{r=1}^{n} \delta_r = 1 - 0 = 1 \\ s(P, f) &= \sum_{r=1}^{n} m_r\delta_r = m_1\delta_1 + \cdots + m_n\delta_n = 0 \cdot \delta_1 + \cdots + 0 \cdot \delta_n \\ &= 0 \sum \delta_r = 0 \end{aligned}$$

Infimum of $\{S(P, f),\ P$ is a partition of $[a, b]\} = 1$.

$$\therefore \inf S = 1 = \int_0^{\overline{1}} f(x)dx$$

Also supremum of $\{s(P, f),\ P$ is a partition of $[a, b]\} = 0$.

$$\therefore \sup s = 0 = \int_{\underline{0}}^{1} f(x)dx$$

$$\therefore \int_0^{\overline{1}} f(x)dx \ne \int_{\underline{0}}^{1} f(x)dx$$

So f is not R-integrable.

Note. f is bounded but not R-integrable. So every bounded function is not R-integrable.

Problem [7.2]: Let $f : [0, a] \to \mathbb{R}, a > 0$ be defined by $f(x) = x^3\ \forall\ x \in [0, a]$.

Solution. f is bounded $\forall\ x \in [0, a]$.

Let $P = \left\{0, \frac{a}{n}, \frac{2a}{n}, \cdots, \frac{(n-1)a}{n}, \frac{na}{n} = a\right\}$ be a partition of $[0, a]$.

Then

$$S(P, f) = \sum_{r=1}^{n} M_r \delta_r = M_1 \left(\frac{a}{n} - 0\right) + M_2 \left(\frac{2a}{n} - \frac{a}{n}\right) + \cdots + M_n \left(a - \frac{(n-1)a}{n}\right)$$

$$= \left(\frac{a}{n}\right)^3 \cdot \frac{a}{n} + \left(\frac{2a}{n}\right)^3 \cdot \frac{a}{n} + \cdots + \left(\frac{na}{n}\right)^3 \cdot \frac{a}{n} = \frac{a^4}{n^4}\left[1^3 + 2^3 + \cdots + n^3\right]$$

$$= \frac{a^4}{n^4}\left\{\frac{n(n+1)}{2}\right\}^2 = \frac{a^4}{4}\left\{\frac{n^4 + 2n^3 + n^2}{n^4}\right\} = \frac{a^4}{4}\left\{1 + \frac{2}{n} + \frac{1}{n^2}\right\}$$

$$\int_0^{\overline{a}} f(x)dx = \inf S = \lim_{n\to\infty}\left\{\frac{a^4}{4}\left(1 + \frac{2}{n} + \frac{1}{n^2}\right)\right\} = \frac{a^4}{4}$$

Now

$$s(P, f) = \sum_{r=1}^{n} m_r \delta_r = m_1 \left(\frac{a}{n} - 0\right) + m_2 \left(\frac{2a}{n} - \frac{a}{n}\right) + \cdots + m_n \left(a - \frac{(n-1)a}{n}\right)$$

$$= 0 \cdot \frac{a}{n} + \left(\frac{a}{n}\right)^3 \cdot \frac{a}{n} + \left(\frac{2a}{n}\right)^3 \cdot \frac{a}{n} + \cdots + \left(\frac{(n-1)a}{n}\right)^3 \cdot \frac{a}{n}$$

$$= \frac{a^4}{n^4}[1^3 + 2^3 + \cdots + (n-1)^3]$$

$$= \frac{a^4}{n^4}\left\{\frac{n(n-1)}{2}\right\}^2 = \frac{a^4}{4}\left[1 - \frac{2}{n} + \frac{1}{n^2}\right]$$

$$\therefore \int_{\underline{0}}^{a} f(x)dx = \sup s = \lim_{n\to\infty} \frac{a^4}{4}\left\{1 - \frac{2}{n} + \frac{1}{n^2}\right\} = \frac{a^4}{4}$$

So $\int_0^{\overline{a}} f(x)dx = \int_{\underline{0}}^{a} f(x)dx = \frac{a^4}{4}$.

Hence f is R-integrable and $R\int_0^1 f(x)dx = \frac{a^4}{4}$.

Problem [7.3] : Show that a function $f : [0, \pi/4] \to \mathbb{R}$, defined by

$$f(x) = \begin{cases} \cos x, & \text{when } x \text{ is rational} \\ \sin x, & \text{when } x \text{ is irrational} \end{cases}$$

is not R-integrable on $[0, \pi/4]$.

Solution. Let $P\left\{0, \frac{\pi}{4n}, \frac{2\pi}{4n}, \cdots, \frac{(n-1)\pi}{4n}, \frac{n\pi}{4n} = \frac{\pi}{4}\right\}$ be a partition of $[0, \pi/4]$.

Since $\cos x$ is a decreasing function of x, $m_r = \cos\frac{r\pi}{4n}$ and $M_r = \cos\frac{(r-1)\pi}{4n}$ for any subinterval $[x_{r-1}, x_r] = \left[\frac{(r-1)\pi}{4n}, \frac{r\pi}{n}\right]$ of $[0, \pi/4]$ and $\sin x$ is an increasing function of x,

$$m_r = \sin\frac{(r-1)\pi}{4n} \text{ and } M_r = \sin\frac{r\pi}{4n}.$$

Now, rational and irrational points are dense on $[0, \pi/4]$.

Again, if $\phi(x) = \cos x - \sin x$. Then $\phi'(x) = -(\sin x + \cos x) < 0 \ \forall\ x \in [0, \pi/4]$

$\Rightarrow \phi(x)$ is decreasing function of x.

$\Rightarrow \phi(x) > \phi(\pi/4) \Rightarrow \phi(x) > 0 \Rightarrow \cos x > \sin x \ \forall\ x \in [0, \pi/4]$.

So in every subinterval upper bound is attained by $\cos x$ and lower bound is attained by $\sin x$ as rational and irrational points are dense in every subinterval.

Hence $M_r = \cos\frac{(r-1)\pi}{4n}$ and $m_r = \sin\frac{r\pi}{4n}$

$$S(P,f) = \sum_{r=1}^{n} M_r\delta_r = \sum_{r=1}^{n} \cos\frac{(r-1)\pi}{4n}\cdot\frac{\pi}{4n}$$

$$= \frac{\pi}{4n}\left[\cos 0 + \cos\frac{\pi}{4n} + \cos\frac{2\pi}{4n} + \cdots + \cos\frac{(n-1)\pi}{4n}\right]$$

$$= \frac{\pi}{4n}\cdot\frac{\cos\frac{n-1}{2}\cdot\frac{\pi}{4n}\cdot\sin\frac{n\pi}{8n}}{\sin\frac{\pi}{8n}} = \frac{\frac{\pi}{8n}}{\sin\frac{\pi}{8n}} 2\cos\left(\frac{\pi}{8}-\frac{\pi}{8n}\right)\sin\frac{\pi}{8}$$

$$\lim_{\|P\|\to 0} S(P,f) = \lim_{n\to\infty}\frac{\frac{\pi}{8n}}{\sin\frac{\pi}{8n}}\cdot 2\cos\left(\frac{\pi}{8}-\frac{\pi}{8n}\right)\cdot\sin\frac{\pi}{8}$$

$$= 1\cdot 2\cos\frac{\pi}{8}\cdot\sin\frac{\pi}{8} = \sin\frac{\pi}{4} = \frac{1}{\sqrt{2}}$$

Hence $\displaystyle\int_0^{\frac{\pi}{4}} f(x)dx = \frac{1}{\sqrt{2}}$. (1)

$$s(P,f) = \sum m_r\delta_r = \sum_{r=1}^{n}\sin\frac{r\pi}{4n}\cdot\frac{\pi}{4n} = \frac{\pi}{4n}\left[\sin\frac{\pi}{4n} + \sin\frac{2\pi}{4n} + \cdots + \sin\frac{n\pi}{4n}\right]$$

$$= \frac{\pi}{4n}\cdot\frac{\sin\left(\frac{\pi}{4n}+\frac{n-2}{2}\frac{\pi}{4n}\right)\sin\frac{n\pi}{8n}}{\sin\frac{\pi}{8n}} = \frac{\frac{\pi}{8n}}{\sin\frac{\pi}{8n}}\cdot 2\sin\left[\frac{\pi}{4n}+\frac{\pi}{8}-\frac{\pi}{4n}\right]\sin\frac{\pi}{8}$$

$$\lim_{\|P\|\to 0} s(P,f) = \lim_{n\to\infty}\frac{\frac{\pi}{8n}}{\sin\frac{\pi}{8n}}\cdot 2\sin\frac{\pi}{8}\sin\frac{\pi}{8} = 1\cdot 2\sin^2\frac{\pi}{8}$$

$$= 1-\cos\frac{\pi}{4} = 1-\frac{1}{\sqrt{2}}$$

So $\displaystyle\int_0^{\frac{\pi}{4}} f(x)dx = 1-\frac{1}{\sqrt{2}}$. (2)

Therefore, f is not R-integrable on $\left[0,\frac{\pi}{4}\right]$, followed by (1) and (2).

Problem [7.4]: Examine the R-integrability of f which is defined by on $[0,2]$

$$f(x) = x + x^2 \text{ for rational values of } x \text{ in } [0,2]$$
$$= x^2 + x^3 \text{ for irrational values of } x \text{ in } [0,2].$$

Solution. Let $\phi(x) = (x+x^2)-(x^2+x^3) = x - x^3$

$$\phi'(x) = 1-3x^2 < 0, \quad \text{when } 1 \le x \le 2$$
$$> 0, \quad \text{when } 0 \le x < 1.$$

So ϕ increases on $[0,1)$ and decreases on $[1,2]$.

For the first part, $\phi(x) > \phi(0) = 0$

$$\Rightarrow x + x^2 > x^2 + x^3 \;\forall\; x \in [0,1)$$

and in the second part, $\phi(x) < \phi(1) = 0$

$$\Rightarrow x + x^2 < x^2 + x^3. \text{ That is, } x^2 + x^3 > x + x^2.$$

Thus the upper bound attained by f in $[0,1]$ by $x+x^2$ and in $[1,2]$ by x^2+x^3.

Therefore,

$$\overline{\int_0^2} f(x)dx = \overline{\int_0^1} f(x)dx + \overline{\int_1^2} f(x)dx = \int_0^1 (x+x^2)dx + \int_1^2 (x^2+x^3)dx$$
$$= \frac{1}{2}+\frac{1}{3}+\frac{7}{3}+\frac{15}{4} = \frac{1}{2}+\frac{8}{3}+\frac{15}{4} = \frac{6+32+45}{12} = \frac{83}{12}$$

Again, the lower bound attained by f in $[0,1]$ by x^2+x^3 and in $[1,2]$ by $x+x^2$.

Therefore,

$$\underline{\int_0^2} f(x)dx = \underline{\int_0^1} f(x)dx + \underline{\int_1^2} f(x)dx = \int_0^1 (x^2+x^3)dx + \int_1^2 (x+x^2)dx$$
$$= \frac{1}{3}+\frac{1}{4}+\left(\frac{4}{2}-\frac{1}{2}\right)+\left(\frac{8}{3}-\frac{1}{3}\right)$$
$$= \frac{1}{3}+\frac{1}{4}+\frac{3}{2}+\frac{7}{3} = \frac{1}{4}+\frac{3}{2}+\frac{8}{3} = \frac{3+18+32}{12} = \frac{53}{12}$$

Thus, $\overline{\int_0^2} f(x)dx \neq \underline{\int_0^2} f(x)dx$.

Therefore, f is not R-integrable in $[0,2]$.

Problem [7.5] : If $f:[a,b]\to\mathbb{R}$ be bounded on $[a,b]$ and K be a (+)ve constant, then

$$\overline{\int_a^b} Kf(x)dx = K\overline{\int_a^b} f(x)dx \text{ and } \underline{\int_a^b} Kf(x)dx = K\underline{\int_a^b} f(x)dx.$$

Further, if K be a (−)ve constant, then

$$\overline{\int_a^b} Kf(x)dx = K\underline{\int_a^b} f(x)dx \text{ and } \underline{\int_a^b} Kf(x)dx = K\overline{\int_a^b} f(x)dx$$

Also deduce that if f is bounded and integrable on $[a,b]$, so is Kf, where K is any constant and that $\int_a^b Kf(x)dx = K\int_a^b f(x)dx$.

Solution. Let $P\{a=x_0<x_1\cdots<x_{r-1}<x_r\cdots<x_n=b\}$ be a partition of $[a,b]$ and M_r, m_r be the sup and inf of f in $[x_{r-1},x_r]$ and $\delta_r = x_r - x_{r-1}$.

When K is a (+)ve constant, KM_r, Km_r be the sup and inf of Kf in $[x_{r-1},x_r]$.

Then

$$\overline{\int_a^b} f(x)dx = \inf\left\{\sum_{r=1}^n M_r\delta_r\right\} \text{ and } \underline{\int_a^b} f(x)dx = \sup\left\{\sum_{r=1}^n m_r\delta_r\right\}.$$

Therefore,

$$\overline{\int_a^b} Kf(x)dx = \inf\left\{\sum_{r=1}^n KM_r\delta_r\right\} = K\inf\left\{\sum_{r=1}^n M_r\delta_r\right\} = K.\overline{\int_a^b} f(x)dx \tag{1}$$

Also

$$\underline{\int_a^b} Kf(x)dx = \sup\left\{\sum_{r=1}^n Km_r\delta_r\right\} = K\sup\left\{\sum_{r=1}^n m_r\delta_r\right\} - K\underline{\int_a^b} f(x)dx \tag{2}$$

when K is a (−)ve constant, KM_r, Km_r be the inf and sup of f in $[x_{r-1},x_r]$.

Hence

$$\begin{aligned}\int_a^{\overline{b}} Kf(x)dx &= \inf\left\{\sum_{r=1}^{n} Km_r\delta_r\right\} = \inf\{Km_1\delta_1 + Km_2\delta_2 + \cdots + Km_n\delta_n\} \\ &= \inf\{-\lambda m_1\delta_1 - \lambda m_2\delta_2 - \cdots - \lambda m_n\delta_n\}, \text{ where } K = -\lambda,\ \lambda > 0 \\ &= (-\lambda)\sup\left\{\sum_{r=1}^{n} m_r\delta_r\right\} = K\int_{\underline{a}}^{b} f(x)dx \qquad (3)\end{aligned}$$

Also

$$\int_{\underline{a}}^{b} Kf(x)dx = \sup\left\{\sum_{r=1}^{n} KM_r\delta_r\right\} = (-\lambda)\inf\left\{\sum_{r=1}^{n} M_r\delta_r\right\} = K\int_a^{\overline{b}} f(x)dx \qquad (4)$$

If f is bounded, then Kf is also bounded as

$$|f(x)| \le M \Rightarrow |Kf(x)| = |K||f(x)| \le |K|M\ \forall\ x \in [a,b].$$

Again, if f be integrable, then

$$\int_a^{\overline{b}} f(x)dx = \int_{\underline{a}}^{b} f(x)dx = \int_a^b f(x)dx \Rightarrow \int_a^{\overline{b}} Kf(x)dx = \int_{\underline{a}}^{b} Kf(x)dx$$

$\Rightarrow Kf$ is integrable on $[a,b]$ and $\int_a^b Kf(x)dx = K\int_a^b f(x)dx$ [by (1), (2), (3) and (4)].

Theorem [7.3] [Darboux's theorem] For every $\epsilon(>0)$, there exists a $(+)$ve number δ such that $S(P,f) < \int_a^{\overline{b}} f(x)dx + \epsilon$ and $s(P,f) > \int_{\underline{a}}^{b} f(x)dx - \epsilon\ \forall\ P$ with $||P|| \le \delta$.

Proof. Let $S = \{S(P,f),\ P$ is a partition of $[a,b]\}$ with $||P|| \le \delta$.

$\inf S = \int_a^{\overline{b}} f(x)dx$. So by property of $\inf S$, $\exists$ a partition P_1 of $[a,b]$ such that

$$S(P_1, f) < \int_a^{\overline{b}} f(x)dx + \frac{\epsilon}{2}. \qquad \text{(i)}$$

$||P_1|| \le \delta$, δ is a $(+)$ve number.

Here f is bounded. So $|f(x)| \le k\ \forall\ x \in [a,b]$.

Let $P_1\{a = x_0 < x_1 \cdots < x_p = b\}$ be a partition of $[a,b]$ containing $(p-1)$ points in (a,b), such that $||P_1|| \le \delta$, δ is a $(+)$ve number.

Let P_2 is a refinement of P_1 so that $P_2 \supset P_1 \Rightarrow S(P_2,f) \le S(P_1,f)$, by Theorem [7.1].

We first suppose that P_2 contains one more point than that of P_1, so one of the subintervals, say, $[x_{r-1}, x_r]$ of P_1 is divided into two subintervals $[x_{r-1}, \xi]$ and $[\xi, x_r]$ such that $x_{r-1} < \xi < x_r$, $r \le p$.

Let M_r, M_r', M_r'' be the sup of f is $[x_{r-1}, x_r]$, $[x_{r-1}, \xi]$ and $[\xi, x_r]$ respectively such that $\delta_r = \delta_r' + \delta_r''$, where $\delta_r' = \xi - x_{r-1}$ and $\delta_r'' = x_r - \xi$.

Then

$$\begin{aligned}S(P_1,f) - S(P_2,f) &= M_r(x_r - x_{r-1}) - M_r'(\xi - x_{r-1}) - M_r''(x_r - \xi) \\ &= (M_r - M_r')\delta_r' + (M_r - M_r'')\delta_r'' \qquad \text{(ii)}\end{aligned}$$

Also $\quad |f(x)| \le k \Rightarrow -k \le f(x) \le k$

$$\Rightarrow -k \le M_r' \le M_r \le k \Rightarrow 0 \le M_r - M_r' \le 2k$$

and $-k \le M_r'' \le M_r \le k \Rightarrow 0 \le M_r - M_r'' \le 2k$

Therefore, from (ii) follows that

$$\begin{aligned} S(P_1, f) - S(P_2, f) &\leq 2k\delta_r \; [\because \delta_r = \delta_r' + \delta_r''] \\ &\leq 2k\delta \; [\because ||P_1|| \leq \delta] \end{aligned} \quad \text{(iii)}$$

Now if P_2 contains at most $(p-1)$ more points that of P_1, then from (iii) it follows that

$$S(P_1, f) - S(P_2, f) \leq 2k(p-1)\delta. \quad \text{(iv)}$$

We choose $\delta > 0$ such that $\delta < \frac{\epsilon}{4k(p-1)}$ and let P be a division of $[a, b]$ such that $||P|| \leq \delta$ and let $P_2 = P_1 \cup P$. Then $P_2 \supset P$. Thus P_2 is a refinement of P, containing at most $(p-1)$ more points than P. Therefore, from (iii)

$$S(P, f) - 2k(p-1)\delta \leq S(P_2, f) \leq S(P_1, f) \Rightarrow S(P, f) \leq 2k(p-1)\delta + S(P_1, f)$$

$$< 2k(p-1)\frac{\epsilon}{4k(p-1)} + \int_a^{\bar{b}} f(x)dx + \frac{\epsilon}{2} = \frac{\epsilon}{2} + \frac{\epsilon}{2} + \int_a^{\bar{b}} f(x)dx = \int_a^{\bar{b}} f(x)dx + \epsilon$$

$$\Rightarrow S(P, f) < \int_a^{\bar{b}} f(x)dx + \epsilon$$

Thus the result (i) is true $\forall$ P with $||P|| \leq \delta$. The proof of $s(P, f) > \int_{\underline{a}}^b f(x)dx - \epsilon$ is similar to that of the first part.

Theorem [7.4] For every bounded function f in $[a, b]$, $\int_a^{\bar{b}} f(x)dx \geq \int_{\underline{a}}^b f(x)dx$. That is, upper R-integral is greater than or equal to the lower R-integral.

Proof. If possible, let $\int_a^{\bar{b}} f(x)dx < \int_{\underline{a}}^b f(x)dx$.

Then $\exists$ a real number k such that $\int_a^{\bar{b}} f(x)dx < k < \int_{\underline{a}}^b f(x)dx$.

We choose $\epsilon (> 0)$ such that

$$\int_a^{\bar{b}} f(x)dx = k - \epsilon \Rightarrow \int_a^{\bar{b}} f(x)dx + \epsilon = k \quad \text{(i)}$$

Then for every $\epsilon (> 0)$, there exists a (+)ve number δ such that for every partition P with $||P|| < \delta$, $S(P, f) < \int_a^{\bar{b}} f(x)dx + \epsilon$, by Theorem [7.3].

Thus

$$S(P, f) < k, \text{ by (i)} \quad \text{(ii)}$$

Also

$$\int_{\underline{a}}^b f(x)dx > k \Rightarrow \int_{\underline{a}}^b f(x)dx = k + \epsilon \Rightarrow \int_{\underline{a}}^b f(x)dx - \epsilon = k \quad \text{(iii)}$$

Again, by Theorem [7.3],

$$s(P, f) > \int_{\underline{a}}^b f(x)dx - \epsilon \Rightarrow s(P, f) > k, \text{ by (iii)} \quad \text{(iv)}$$

Thus from (ii) and (iv),

$$S(P, f) < k < s(P, f) \Rightarrow S(P, f) < s(P, f)$$

which arises a contradiction.

Hence, $\int_a^{\bar{b}} f(x)dx \geq \int_{\underline{a}}^b f(x)dx$.

Theorem [7.5] If a function $f : [a, b] \to \mathbb{R}$ be bounded on $[a, b]$, then for any partition $P\{a = x_0 < x_1 \cdots < x_n = b\}$ of $[a, b]$ with norm $||P|| \le \delta$, $\delta > 0$,

(i) $\lim_{||P|| \to 0} S(P, f) = \lim_{||P|| \to 0} \sum_{r=1}^{n} M_r \delta_r = \int_a^{\bar{b}} f(x)dx$.

(ii) $\lim_{||P|| \to 0} s(P, f) = \lim_{||P|| \to 0} \sum m_r \delta_r = \int_{\underline{a}}^{b} f(x)dx$.

Proof. Let $S_1 = \{S(P, f),\ P \text{ is a partition of } [a, b]\}$.

Then $\int_a^{\bar{b}} f(x)dx$ being an infimum of S_1, $S(P, f) \ge \int_a^{\bar{b}} f(x)dx$.

Now from Darboux Theorem [7.3]

$$S(P, f) < \int_a^{\bar{b}} f(x)dx + \epsilon \Rightarrow S(P, f) - \int_a^{\bar{b}} f(x)dx < \epsilon$$

$$\Rightarrow \left| S(P, f) - \int_a^{\bar{b}} f(x)dx \right| < \epsilon \quad \left[\because\ S(P, f) - \int_a^{\bar{b}} f(x)dx \ge 0\right]$$

$$\Rightarrow \lim_{||P|| \to 0} S(P, f) = \int_a^{\bar{b}} f(x)dx$$

Again, let $S_2 = \{s(P, f),\ P \text{ is a partition of } [a, b]\}$. Then $\int_{\underline{a}}^{b} f(x)dx$ being the supremum of S_2, $s(P, f) \le \int_{\underline{a}}^{b} f(x)dx$.

Now, by Darboux Theorem [7.3],

$$s(P, f) > \int_{\underline{a}}^{b} f(x)dx - \epsilon \Rightarrow \int_{\underline{a}}^{b} f(x)dx - s(P, f) < \epsilon$$

$$\Rightarrow \left| \int_{\underline{a}}^{b} f(x)dx - s(P, f) \right| < \epsilon \quad \left[\because\ \int_{\underline{a}}^{b} f(x)dx - s(P, f) \ge 0\right]$$

$$\Rightarrow \lim_{||P|| \to 0} s(P, f) = \int_{\underline{a}}^{b} f(x)dx$$

Theorem [7.6] Let a function $f : [a, b] \to \mathbb{R}$ be bounded on $[a, b]$. Then f is R-integrable on $[a, b]$, if and only if, for every $\epsilon (> 0)$, there exists a (+)ve number δ with $||P|| \le \delta$, where $P\{a = x_0 < x_1 \cdots < x_n = b\}$ is a partition of $[a, b]$, and every choice of ξ_r in $[x_{r-1}, x_r]$, such that $\left| \sum_{r=1}^{n} f(\xi_r)\delta_r - \int_a^b f(x)dx \right| < \epsilon$.

Proof. Let f be R-integrable on $[a, b]$. Then

$$\int_a^{\bar{b}} f(x)dx = \int_{\underline{a}}^{b} f(x)dx = \int_a^b f(x)dx \tag{i}$$

From Darboux's Theorem [7.3],

$$\left.\begin{aligned} S(P, f) &< \int_a^{\bar{b}} f(x)dx + \epsilon = \int_a^b f(x)dx + \epsilon \text{ with } ||P|| \le \delta \\ s(P, f) &> \int_{\underline{a}}^{b} f(x)dx - \epsilon = \int_a^b f(x)dx - \epsilon \text{ with } ||P|| \le \delta \end{aligned}\right\} \tag{ii}$$

We have $m_r \leq f(\xi_r) \leq M_r \ \forall \ \xi_r \epsilon [x_{r-1}, x_r]$

$$\Rightarrow \sum_{r=1}^{n} m_r \delta_r \leq \sum_{r=1}^{n} f(\xi_r)\delta_r \leq \sum_{r=1}^{n} M_r \delta_r \Rightarrow s(P,f) \leq \sum_{r=1}^{n} f(\xi_r)\delta_r \leq S(P,f)$$

$$\Rightarrow \int_a^b f(x)dx - \epsilon < s(P,f) \leq \sum_{r=1}^{n} f(\xi_r)\delta_r \leq S(P,f) < \int_a^b f(x)dx + \epsilon \text{ [by (ii)]}$$

$$\Rightarrow \int_a^b f(x)dx - \epsilon < \sum_{r=1}^{n} f(\xi_r)\delta_r < \int_a^b f(x)dx + \epsilon$$

$$\Rightarrow \left| \sum f(\xi_r)\delta_r - \int_a^b f(x)dx \right| < \epsilon \text{ [with } ||P|| \leq \delta]$$

$$\Rightarrow \lim_{||P|| \to 0} \sum f(\xi_r)\delta_r = \int_a^b f(x)dx$$

Conversely, let $\left|\sum f(\xi_r)\delta_r - \int_a^b f(x)dx\right| < \epsilon$ [with $||P|| \leq \delta$].

Then $\lim\limits_{||P|| \to 0} \sum f(\xi_r)\delta_r = \int_a^b f(x)dx$, holds for all values of $f(\xi_r)$.

If, in particular, f attains M_r at $\xi_r \in [x_{r-1}, x_r]$, i.e., if $f(\xi_r) = M_r$, then

$$\lim_{||P|| \to 0} \sum M_r \delta_r = \int_a^b f(x)dx \Rightarrow \int_a^{\overline{b}} f(x)dx = \int_a^b f(x)dx \text{ [by Th. [7.5]} \qquad \text{(iii)}$$

Again, if f attains m_r at $\xi_r \in [x_{r-1}, x_r]$, i.e., if $f(\xi_r) = m_r$, then

$$\lim_{||P|| \to 0} \sum m_r \delta_r = \int_{\underline{a}}^b f(x)dx \Rightarrow \int_{\underline{a}}^b f(x)dx = \int_a^b f(x)dx \qquad \text{(iv)}$$

Thus from (iii) and (iv),

$$\int_a^{\overline{b}} f(x)dx = \int_{\underline{a}}^b f(x)dx = \int_a^b f(x)dx.$$

Hence, f is R-integrable on $[a,b]$.

Note. The above theorem leads to an alternative definition of R-integrability.

Definition [7.4] : A function $f : [a,b] \to \mathbb{R}$ is said to be R-integrable, if f for every partition $P\{a = x_0 < x_1 \cdots < x_{r-1} < x_r < x_n = b\}$ of $[a,b]$ and every choice of ξ_r in $[x_{r-1}, x_r]$ such that $\lim\limits_{||P|| \to 0} \sum\limits_{r=1}^{n} f(\xi_r)\delta_r$ exists, then the limit is called the integral of f on $[a,b]$ and is denoted by $R\int_a^b f(x)dx$. That is, $R\int_a^b f(x)dx = \lim\limits_{||P|| \to 0} \sum\limits_{r=1}^{n} f(\xi_r)\delta_r$.

Problem [7.6] : Evaluate $\int_1^2 f(x)dx$, when $f(x) = x^2$, from the definition $\lim\limits_{||P|| \to 0} \sum f(\xi_r)\delta_r$.

Solution. Let $P\{1 = x_0, x_1, x_2, \cdots, x_n = 2\}$ be a division of $[1,2]$ so that the points of division x_r are given by

$$x_0 = 1, x_1 = 1 + \frac{1}{n}, x_2 = 1 + \frac{2}{n}, \cdots, x_r = 1 + \frac{r}{n}, \cdots, x_n = 1 + \frac{n}{n} = 2.$$

Here $\xi_r \in [x_{r-1}, x_r]$, we assume $\xi_r = x_r$.

Then

$$\int_1^2 f(x)dx = \sum_{||P||\to 0} f(x_r)\delta_r,\ \delta_r = \frac{1}{n} \text{ and } ||P|| \to 0 \Rightarrow n \to \infty.$$

Therefore, $\int_1^2 f(x)dx$

$$= \lim_{n\to\infty} \sum \left(1+\frac{r}{n}\right)^2 \delta_r = \lim_{n\to\infty} \sum \left(1+\frac{2r}{n}+\frac{r^2}{n^2}\right)\delta_r$$

$$= \lim_{n\to\infty}\left[\sum \delta_r + \frac{2}{n}\sum r\delta_r + \frac{1}{n^2}\sum r^2\delta_r\right]$$

$$= \lim_{n\to\infty}\left[1+\frac{2}{n}\left(1\cdot\frac{1}{n}+2\cdot\frac{1}{n}+\cdots+n\cdot\frac{1}{n}\right)+\frac{1}{n^2}\left(1^2\cdot\frac{1}{n}+2^2\cdot\frac{1}{n}+\cdots+n^2\cdot\frac{1}{n}\right)\right]$$

$$= 1 + \lim_{n\to\infty}\left[\frac{2}{n^2}\cdot\frac{n(n+1)}{2}+\frac{1}{n^3}\cdot\frac{n(n+1)(2n+1)}{6}\right]$$

$$= 1+1+\frac{1}{3} = \frac{7}{3}$$

Problem [7.7]: Evaluate $\int_{-1}^1 f(x)dx$ from the definition $\lim\limits_{||P||\to 0} \Sigma f(\xi_r)\delta_r$, when $f(x) = x|x|$.

Solution. $f(x) = \begin{cases} x^2, & x \geq 0 \\ -x^2, & x < 0 \end{cases}$

We consider a partition $P\{-1 = x_0 < x_1 \cdots < x_{r-1} < x_r \cdots < x_n = 0 = y_0 < y_1 \cdots y_{i-1} < y_i \cdots < y_n = 1\}$ where $\delta_r = x_r - x_{r-1}$ and $\delta_i' = y_i - y_{i-1}$

$$x_r = -1 + \frac{r}{n} \text{ and } y_i = \frac{i}{n} \text{ and } \xi_r \in [x_{r-1}, x_r] \text{ and } \xi_i' \in [y_{i-1}, y_i]$$

Taking $\xi_r = x_r$ and $\xi_i' = y_i$, we have, from the definition,

$$\int_{-1}^1 f(x)dx = \lim_{||P||\to 0}\left[\sum_{r=1}^n f(\xi_r)\delta_r + \sum_{i=1}^n f(\xi_i')\delta_i'\right] = \lim_{||P||\to 0}\left[\sum_{r=1}^n f(x_r)\delta_r + \sum_{i=1}^n f(y_i)\delta_i'\right]$$

$$= \lim_{||P||\to 0}\left[\sum -x_r^2\delta_r + \sum y_i^2\delta_i'\right] = \lim_{||P||\to 0}\left[\sum -\left(-1+\frac{r}{n}\right)^2\delta_r + \sum_{i=1}^n \left(\frac{i}{n}\right)^2\delta_i'\right]$$

$$= \lim_{n\to\infty}\sum -\left(1-\frac{2r}{n}+\frac{r^2}{n^2}\right)\delta_r + \lim_{n\to\infty}\sum \frac{i^2}{n^2}\delta_i'$$

$$= \lim_{n\to\infty}\sum\left\{-1\cdot\delta_r + \frac{2}{n}\sum r\delta_r - \frac{1}{n^2}\sum r^2\delta_r\right\} + \lim_{n\to\infty}\frac{1}{n^2}\sum i^2\delta_i'$$

$$= \lim_{n\to\infty}\left[(-1)\cdot 1 + \frac{2}{n^2}\frac{n(n+1)}{2} - \frac{1}{n^3}\cdot\frac{n(n+1)(2n+1)}{6}\right]$$

$$+ \lim_{n\to\infty}\frac{1}{n^3}\frac{n(n+1)(2n+1)}{6}$$

$$= \left(-1+1-\frac{1}{3}\right)+\frac{1}{3} = -\frac{1}{3}+\frac{1}{3} = 0$$

Theorem [7.7] The necessary and sufficient condition that a bounded function $f : [a, b] \to \mathbb{R}$ to be R-integrable on $[a, b]$ is for every $\epsilon(> 0)$, there exists a (+)ve number δ with $||P|| \leq \delta$ for every partition P of $[a, b]$, the oscillatory sum $\omega(P, f) < \epsilon$.

Proof. The condition is necessary. Let f be R-integrable on $[a, b]$. Then

$$\overline{\int_a^b} f(x)dx = \underline{\int_a^b} f(x)dx = \int_a^b f(x)dx \tag{i}$$

Now, from Darboux Theorem [7.3],

$$\left.\begin{aligned} S(P, f) &< \overline{\int_a^b} f(x)dx + \frac{\epsilon}{2} = \int_a^b f(x)dx + \frac{\epsilon}{2} \\ \text{and } s(P, f) &> \underline{\int_a^b} f(x)dx - \frac{\epsilon}{2} = \int_a^b f(x)dx - \frac{\epsilon}{2} \end{aligned}\right\} \quad \text{[by (i)]}$$

$$\Rightarrow -s(P, f) < -\int_a^b f(x)dx + \frac{\epsilon}{2}$$

Therefore, $S(P, f) - s(P, f) < \frac{\epsilon}{2} + \frac{\epsilon}{2} = \epsilon \Rightarrow \omega(P, f) < \epsilon$.

The condition is sufficient. Let the condition $\omega(P, f) < \epsilon$ hold.

$$\omega(P, f) = S(P, f) - s(P, f)$$

$$= \left\{S(P, f) - \overline{\int_a^b} f(x)dx\right\} + \left\{\overline{\int_a^b} f(x)dx - \underline{\int_a^b} f(x)dx\right\} + \left\{\underline{\int_a^b} f(x)dx - s(P, f)\right\} \tag{ii}$$

Each of the numbers on the brackets of the right-hand side of (ii) is non-negative and their sum is less than $\epsilon \Rightarrow$ each of the numbers on the bracket is less than ϵ

$$\Rightarrow 0 \le \overline{\int_a^b} f(x)dx - \underline{\int_a^b} f(x)dx < \epsilon$$

Now ϵ is an arbitrary small (+)ve number and a non-negative number

$$\overline{\int_a^b} f(x)dx - \underline{\int_a^b} f(x)dx < \epsilon$$

$$\Rightarrow \overline{\int_a^b} f(x)dx - \underline{\int_a^b} f(x)dx = 0 \Rightarrow \overline{\int_a^b} f(x)dx = \underline{\int_a^b} f(x)dx$$

$\Rightarrow f$ is R-integrable on $[a, b]$.

Some classes of integrable functions

Theorem [7.8] (1) If f is continuous on $[a, b]$, then f is R-integrable on $[a, b]$.

(2) If f is monotonic on $[a, b]$, then f is R-integrable on $[a, b]$.

(3) If a bounded function f on $[a, b]$ has a set of finite points of discontinuities on $[a, b]$, then f is R-integrable on $[a, b]$.

(4) If a bounded function f on $[a, b]$ has a set of points of discontinuities having finite number of limit points on $[a, b]$, then f is R-integrable on $[a, b]$.

Proof. (1) Continuity of f on a closed interval $[a, b]$

$\Rightarrow$ uniform continuity of f on $[a, b]$

$\Rightarrow$ for every $\epsilon(> 0)$, $\exists$ a (+)ve number δ such that

$$|f(x_2) - f(x_1)| < \frac{\epsilon}{b - a}, \ \forall \ x_1, x_2 \in [a, b] \text{ satisfying } |x_2 - x_1| < \delta \tag{i}$$

Let $P\{a = x_0 < x_1 \cdots < x_{r-1} < x_r \cdots < x_n = b\}$ be a partition of $[a, b]$ with the norm $||P|| \leq \delta$. Every continuous functions attains its bounds. Let $f(\alpha_r) = m_r$ and $f(\beta_r) = M_r$, where $\alpha_r, \beta_r \in [x_{r-1}, x_r]$, $f(\beta_r) \geq f(\alpha_r)$.

Then by (i), $|f(\beta_r) - f(\alpha_r)| < \frac{\epsilon}{b-a}$, since $|\beta_r - \alpha_r| \leq [x_r - x_{r-1}| < \delta$

$$\Rightarrow f(\beta_r) - f(\alpha_r) < \frac{\epsilon}{b-a} \Rightarrow M_r - m_r < \frac{\epsilon}{b-a}, \text{ for } n = 1, 2, \cdots, n.$$

Therefore,

$$\omega(P, f) = S(P, f) - s(P, f) = \sum_{r=1}^{n}(M_r - m_r)\delta_r < \sum_{r=1}^{n} \frac{\epsilon}{b-a}\delta_r = \frac{\epsilon}{b-a}(b-a) = \epsilon$$

$\Rightarrow f$ is R-integrable on $[a, b]$.

(2) Let f be monotonically increasing on $[a, b]$. We consider a partition

$$P\{a = x_0 < x_1 \cdots < x_n = b\}$$

of $[a, b]$ with the norm $||P|| < \frac{\epsilon}{f(b)-f(a)+1}$ where $f(b) - f(a) + 1 > 0$ as $f(b) \geq f(a)$.

In the rth subinterval $[x_{r-1}, x_r]$, $f(x_r) = M_r$ and $f(x_{r-1}) = m_r$.

Therefore, $\omega(P, f)$

$$= \sum_{r=1}^{n}(M_r - m_r)\delta_r < \sum_{r=1}^{n}[f(x_r) - f(x_{r-1})]\frac{\epsilon}{f(b) - f(a) + 1}$$

$$= \frac{\epsilon}{f(b) - f(a) + 1}[f(x_1) - f(x_0) + f(x_2) - f(x_1) + f(x_3) - f(x_2) + \cdots + f(x_n) - f(x_{n-1})]$$

$$= \frac{\epsilon}{f(b) - f(a) + 1}[f(x_n) - f(x_0)]$$

$$= \frac{\epsilon}{f(b) - f(a) + 1}[f(b) - f(a)] < \epsilon \left[\because \frac{f(b) - f(a)}{f(b) - f(a) + 1} < 1\right]$$

$\Rightarrow f$ is R-integrable on $[a, b]$.

(3) Let $S = \{a_1, a_2, \cdots, a_p\}$ be a set of finite points, where each a_i is a point of discontinuity of f on $[a, b]$ for $i = 1, 2, \cdots, p$.

We consider p-non-overlapping closed subintervals $[a_1', a_1''], [a_2', a_2''], \cdots, [a_p', a_p'']$ of $[a, b]$ such that $a_i \in [a_i', a_i''] \ \forall\, i = 1, 2, \cdots, p, a_i'' > a_i'$ and $\sum_{i=1}^{p}(a_i'' - a_i') < \frac{\epsilon}{2(M-m)}$, where M, m are sup and inf of f on $[a, b]$.

Consequently, the oscillatory sum arising from these subintervals

$$= \sum_{i=1}^{p}(M_i - m_i)(a_i'' - a_i') < \sum_{i=1}^{p}(M - m) \cdot (a_i'' - a_i')$$

$$= (M - m)\sum_{i=1}^{p}(a_i'' - a_i') < (M - m)\frac{\epsilon}{2(M - m)} = \frac{\epsilon}{2}$$

Now, f is continuous in the remaining $(p + 1)$ closed subintervals $[a, a_1'], [a_1'', a_2'], \cdots, [a_p'', b]$ of $[a, b]$. We can now form a partition of each of these $(p+1)$ subintervals so that the oscillatory sum arising from each of the $(p + 1)$ subintervals $< \frac{\epsilon}{2(p+1)}$, separately by (1) of Theorem [7.8].

Thus there exists a partition P of $[a,b]$ such that the resultant oscillatory sum

$$\omega(P,f) < \frac{\epsilon}{2} + \frac{\epsilon}{2(p+1)} \cdot (p+1) = \frac{\epsilon}{2} + \frac{\epsilon}{2} = \epsilon$$

Therefore, f is R-integrable on $[a,b]$.

(4) Let $S = \{a_1, a_2, \cdots, a_p\}$ be a finite set where each a_i is a limit point of the set of points of discontinuities of f on $[a,b]$, for $i = 1, 2, \cdots, p$. Also $a_i \in [a,b]$ as $[a,b]$ is closed.

We consider non-overlapping subintervals $[a_i', a_i'']$ such that

$$a_i \in [a_i', a_i''], a_i'' > a_i' \ \forall\, i = 1, 2, \cdots, p \text{ and } \sum_{i=1}^{p}(a_i'' - a_i') < \frac{\epsilon}{2(M-m)}$$

Then the oscillator sum arising from these subintervals

$$= \sum_{i=1}^{p}(M_i - m_i)(a_i'' - a_i') < \sum_{i=1}^{p}(M - m)(a_i'' - a_i')$$

$$= (M-m)\sum_{i=1}^{p}(a_i'' - a_i') < (M-m) \cdot \frac{\epsilon}{2(M-m)} = \frac{\epsilon}{2}$$

Every neighbourhood (a_i', a_i'') of a_i contains infinite number of points of discontinuities of f on $[a,b]$ as being the limit point on $[a,b]$ and, therefore, each of $(p+1)$ subintervals $[a, a_1']$, $[a_1'', a_2'], \cdots, [a_p'', b]$ contains only finite number of points of discontinuities of f on $[a,b]$.

We can now form a partition of each of $(p+1)$ subintervals so that the oscillatory sum arising from each of $(p+1)$ subintervals $< \frac{\epsilon}{2(p+1)}$, separately by (3) of Th. [7.8].

Hence the there exists a partition P of $[a,b]$ such that the resultant oscillatory sum

$$\omega(P,f) < \frac{\epsilon}{2} + \frac{\epsilon}{(p+1)} \cdot (p+1) = \frac{\epsilon}{2} + \frac{\epsilon}{2} = \epsilon.$$

Therefore, f is R-integrable on $[a,b]$.

Problem [7.8] : A function $f : [0,3] \to \mathbb{R}$ is defined by $f(x) = x[x]\ \forall\ x \in [0,3]$. Is f R-integrable on $[0,3]$ and if so, evaluate $\int_0^3 f(x)dx$.

Solution. $f(x) = x[x] \Rightarrow f(x) = \begin{cases} 0, & 0 \le x < 1 \\ x, & 1 \le x < 2 \\ 2x, & 2 \le x < 3 \\ 9, & x = 3 \end{cases}$

Then f has finite number of points discontinuities $1, 2$ and 3.

Hence f is R-integrable by (3) of Th. [7.8]. Now

$$\int_0^3 f(x)dx = \int_0^1 f(x)dx + \int_1^2 f(x)dx + \int_2^3 f(x)dx = 0 + \int_1^2 xdx + \int_2^3 2xdx$$

$$= \left[\frac{x^2}{2}\right]_1^2 + \left[x^2\right]_2^3 = \left(2 - \frac{1}{2}\right) + [9-4] = \frac{3}{2} + 5 = \frac{13}{2}$$

Problem [7.9] : A function $f : [0,1] \to \mathbb{R}$ is defined by

$f(x) = (-1)^{r-1}$, when $\frac{1}{r+1} < x \le \frac{1}{r}$, $r = 1, 2, 3, \cdots$

$f(0) = 0.$

Is f integrable on $[0,1]$. If so, evaluate $\int_0^1 f(x)dx$.

Solution. We have

$$f(x)=\begin{cases}1, \frac{1}{2}<x\leq 1, & \text{for } r=1\\ -1, \frac{1}{3}<r\leq \frac{1}{2}, & \text{for } r=2\\ 1, \frac{1}{4}<x\leq \frac{1}{3}, & \text{for } r=3\end{cases}$$

and so on. Also $f(0)=0$.

Points of discontinuities of f are $0, \frac{1}{2}, \frac{1}{3}, \frac{1}{4}, \cdots$ which are infinite in numbers.

Therefore, the set $S=\{0, \frac{1}{2}, \frac{1}{3}, \frac{1}{4}, \cdots\}$ of points of discontinuities has a finite limit point 0.

So f is R-integrable by (4) of Th. [7.8].

$$\int_{\frac{1}{n}}^{1} f(x)dx = \int_{\frac{1}{2}}^{1} f(x)dx + \int_{\frac{1}{3}}^{\frac{1}{2}} f(x)dx + \cdots + \int_{\frac{1}{n}}^{\frac{1}{n-1}} f(x)dx$$

$$= \int_{\frac{1}{2}}^{1} 1\cdot dx + \int_{\frac{1}{3}}^{\frac{1}{2}} -1\cdot dx + \int_{\frac{1}{4}}^{\frac{1}{3}} 1\cdot dx + \int_{\frac{1}{5}}^{\frac{1}{4}} -1\cdot dx + \cdots + \int_{\frac{1}{n}}^{\frac{1}{n-1}} (-1)^{n-2}\cdot dx$$

$$= \left(1-\frac{1}{2}\right) - \left(\frac{1}{2}-\frac{1}{3}\right) + \left(\frac{1}{3}-\frac{1}{4}\right) - \left(\frac{1}{4}-\frac{1}{5}\right) + \cdots + (-1)^{n-2}\left[\frac{1}{n-1}-\frac{1}{n}\right]$$

$$= \left(1-\frac{1}{2}+\frac{1}{3}-\frac{1}{4}-\cdots+(-1)^{n-2}\frac{1}{n-1}\right)$$

$$+\left(1-\frac{1}{2}+\frac{1}{3}-\frac{1}{4}+\cdots+(-1)^{n-2}\frac{1}{n}\right)-1$$

Then

$$\lim_{n\to\infty}\int_{\frac{1}{n}}^{1} f(x)dx = \left(1-\frac{1}{2}+\frac{1}{3}-\frac{1}{4}+\cdots \text{ to } \infty\right) + \left(1-\frac{1}{2}+\frac{1}{3}-\frac{1}{4}+\cdots \text{ to } \infty\right) - 1$$

$$\Rightarrow \int_{0}^{1} f(x)dx = \log(1+1)+\log(1+1)-1 = 2\log 2 - 1 = \log 4 - 1$$

$$\text{or,} \quad \int_{0}^{1} f(x)dx = \log\frac{4}{e}$$

Problem [7.10]: A function $f:[0,1]\to\mathbb{R}$ is defined by $f(x)=2rx$, when

$$\frac{1}{r+1}<x\leq\frac{1}{r}, \quad r=1,2,3,\cdots.$$

Show that f is R-integrable and $\int_0^1 f(x)dx=\frac{\pi^2}{6}$.

Solution.

$$f(x)=\begin{cases}2x, \frac{1}{2}<x\leq 1, & \text{for } r=1\\ 4x, \frac{1}{3}<x\leq \frac{1}{2}, & \text{for } r=2\\ 6x, \frac{1}{4}<x\leq \frac{1}{3}, & \text{for } r=3\end{cases}$$

and so on. Also $f(0)=0$.

So f has infinite number of points of discontinuities given by $S=\{0, \frac{1}{2}, \frac{1}{3}, \cdots\}$ and also S has a only one limit point 0.

Therefore, f is R-integrable.

Now

$$\begin{aligned}
\int_{\frac{1}{n}}^{1} f(x)dx &= \int_{\frac{1}{2}}^{1} f(x)dx + \int_{\frac{1}{3}}^{\frac{1}{2}} f(x)dx + \cdots + \int_{\frac{1}{n}}^{\frac{1}{n-1}} f(x)dx \\
&= \int_{\frac{1}{2}}^{1} 2xdx + \int_{\frac{1}{3}}^{\frac{1}{2}} 4xdx + \int_{\frac{1}{4}}^{\frac{1}{3}} 6xdx + \cdots + \int_{\frac{1}{n}}^{\frac{1}{n-1}} 2(n-1)x \\
&= \Big[x^2\Big]_{\frac{1}{2}}^{1} + 2\Big[x^2\Big]_{\frac{1}{3}}^{\frac{1}{2}} + 3\Big[x^2\Big]_{\frac{1}{4}}^{\frac{1}{3}} + (n-1)\Big[x^2\Big]_{\frac{1}{n}}^{\frac{1}{n-1}} \\
&= 1 - \left(\frac{1}{2}\right)^2 + 2\left(\frac{1}{2}\right)^2 - 2\left(\frac{1}{3}\right)^2 + 3\left(\frac{1}{3}\right)^2 - 3\left(\frac{1}{4}\right)^2 \\
&\qquad + \cdots + (n-1)\left(\frac{1}{n-1}\right)^2 - (n-1)\left(\frac{1}{n}\right)^2 \\
&= 1 + \left(\frac{1}{2}\right)^2 + \left(\frac{1}{3}\right)^2 + \left(\frac{1}{4}\right)^2 + \cdots + \frac{1}{n^2}
\end{aligned}$$

$$\begin{aligned}
\lim_{n\to\infty} \int_{\frac{1}{n}}^{1} f(x)dx &= \lim_{n\to\infty} \left[1 + \frac{1}{2^2} + \frac{1}{3^2} + \cdots + \frac{1}{n^2}\right] \\
\Rightarrow \int_{0}^{1} f(x)dx &= 1 + \frac{1}{2^2} + \frac{1}{3^2} + \cdots \infty \\
&= \frac{\pi^2}{6}
\end{aligned}$$

Note. [Please see Ch. 9]

Problem [7.11] : A function $f : [0,1] \to \mathbb{R}$ is defined by

$$f(x) = \frac{1}{a^{r-1}}, \text{ where } \frac{1}{a^r} < x \leq \frac{1}{a^{r-1}}, \; r = 1, 2, 3, \cdots \text{ and } a > 2.$$

Is f R-integrable on $[0,1]$. If so, find $\int_0^1 f(x)dx$.

Solution. We have

$$f(x) = \begin{cases} 1, \text{ when } \dfrac{1}{a} < x \leq 1, & \text{for } r = 1 \\ \dfrac{1}{a}, \text{ when } \dfrac{1}{a^2} < x \leq \dfrac{1}{a}, & \text{for } r = 2 \\ \dfrac{1}{a^2}, \text{ when } \dfrac{1}{a^3} < x \leq \dfrac{1}{a^2}, & \text{for } r = 3 \end{cases}$$

and so on.

$a > 2 \Rightarrow f$ is monotonically increasing on $[0,1]$

$\Rightarrow f$ is R-integrable on $[0,1]$, by (2) of Th. [7.8].

In other words, f has infinite number of points of discontinuities given by

$$S = \left\{\frac{1}{a}, \frac{1}{a^2}, \frac{1}{a^3}, \cdots, a > 2\right\}$$

when S has a finite limit point 0 and, therefore, f is R-integrable on $[0,1]$.

Now

$$\int_{\frac{1}{a^n}}^{1} f(x)dx = \int_{\frac{1}{a}}^{1} f(x)dx + \int_{\frac{1}{a^2}}^{\frac{1}{a}} f(x)dx + \cdots + \int_{\frac{1}{a^n}}^{\frac{1}{a^{n-1}}} f(x)dx$$

$$= \int_{\frac{1}{a}}^{1} 1dx + \int_{\frac{1}{a^2}}^{\frac{1}{a}} \frac{1}{a}dx + \int_{\frac{1}{a^3}}^{\frac{1}{a^2}} \frac{1}{a^2}dx + \cdots + \int_{\frac{1}{a^n}}^{\frac{1}{a^{n-1}}} \frac{1}{a^{n-1}}dx$$

$$= \left(1-\frac{1}{a}\right) + \frac{1}{a}\left(\frac{1}{a}-\frac{1}{a^2}\right) + \frac{1}{a^2}\left(\frac{1}{a^2}-\frac{1}{a^3}\right) + \cdots + \frac{1}{a^{n-1}}\left(\frac{1}{a^{n-1}}-\frac{1}{a^n}\right)$$

$$= \frac{a-1}{a} + \frac{1}{a^3}(a-1) + \frac{1}{a^5}(a-1) + \cdots + \frac{1}{a^{2n-1}}(a-1)$$

$$= \frac{(a-1)}{a}\left[1 + \frac{1}{a^2} + \frac{1}{a^4} + \cdots + \frac{1}{a^{2n-2}}\right]$$

$$= \frac{a-1}{a}\frac{1-\left(\frac{1}{a^2}\right)^n}{1-\frac{1}{a^2}} = \frac{a-1}{a}\frac{a^2}{(a^2-1)}\left[1-\left(\frac{1}{a^2}\right)^n\right]$$

$$\lim_{n\to\infty}\int_{\frac{1}{a^n}}^{1} f(x)dx = \frac{a}{a+1}\lim_{n\to\infty}\left[1-\left(\frac{1}{a^2}\right)^n\right], \quad a > 2$$

$$\Rightarrow \int_0^1 f(x)dx = \frac{a}{a+1}\cdot 1 = \frac{a}{a+1}$$

Problem [7.12] : Evaluate $\int_0^{\overline{2}} f(x)dx$, when f is defined by

$$f(x) = 0 \text{ for } x = \frac{n}{n+1}, \frac{n+1}{n}, \ n = 1, 2, 3 \text{ and } f(x) = 1 \text{ for elsewhere.}$$

Is f R-integrable on $[0, 2]$? Is f is continuous at $x = 1$.

Solution. $f(x) = 0$ for $x = \frac{1}{2}, 2, \frac{2}{3}, \frac{3}{2}, \frac{3}{4}, \frac{4}{3}, \cdots$.

So f has infinite number of points of discontinuities having finite limit point 1.

Therefore, f is R-integrable on $[0, 2]$.

Also

$$\int_0^{\overline{2}} f(x)dx = \inf\left\{\sum_{r=1}^{n} M_r\delta_r\right\}$$

$$= \inf\{1\cdot\delta_1 + 1\cdot\delta_2 + \cdots + 1\cdot\delta_n\}$$

$$= \inf\left\{1\cdot\sum_{r=1}^{n}\delta_r\right\} = 2$$

$\lim_{x\to 1-0} f(x) = 0$ and $\lim_{x\to 1+0} f(x) = 0$, but $f(1) = 1$.

Since

$$1 - 0.0001 = 0.9999 = \frac{9999}{10000} = \frac{9999}{9999+1}$$

$$1 + 0.0001 = 1.0001 = \frac{10000+1}{10000}$$

which are of form $\frac{n}{n+1}$ or $\frac{n+1}{n}$.

So f is discontinuous at $x = 1$.

Problem [7.13] : A function $f : [0, 1] \to \mathbb{R}$ is defined by

$$f(0) = 0$$

$$\text{and } f(x) = 0, \text{ if } x \text{ is irrational}$$

$$= \tfrac{1}{q}, \text{ if } x = \tfrac{p}{q}, \text{ where } p, q \text{ are (+)ve integers prime to each other.}$$

Show that f is R-integrable on $[0, 1]$ and $\int_0^1 f(x)dx = 0$.

Solution. The function f is evidently bounded on $[0, 1]$. The oscillation of f is at most 1.

Let $q > 0$ be given. There exist only finite number of (+)ve integers q such that $\frac{1}{q} > \frac{\epsilon}{2}$ and let these finite number of values of q be $q_1, q_2, \cdots, q_m$. So there are only finite number of points $\frac{p}{q_i}$, $i = 1, 2, \cdots, m$, at which $\frac{1}{q_i} > \frac{\epsilon}{2}$.

We call these points as exceptional points. We consider non-overlapping closed subintervals $[a_1, b_1], [a_2, b_2], \cdots, [a_m, b_m]$ of $[0, 1]$ such that

$$\frac{p}{q_i} \in [a_i, b_i], b_i > a_i, \ \forall\, i = 1, 2, \cdots, m$$

$$\text{and } \sum_{i=1}^{m}(b_i - a_i) < \frac{\epsilon}{2} \tag{1}$$

Since each of these subintervals contains rational and irrational points, the oscillation of f in each of these subinterval is less than 1.

Consequently, at each point $\frac{p}{q}$ in the remaining $(m + 1)$ subintervals

$$[0, a_1], [b_1, a_2], \cdots, [b_m, 1], \ \frac{1}{q} < \frac{\epsilon}{2}$$

so that the oscillation of $f = \left|f(x) - f\left(\frac{p}{q}\right)\right| = \frac{1}{q} < \frac{\epsilon}{2}$.

Thus we have a partition $P\{0, a_1, b_1, a_2, b_2, \cdots, a_m, b_m, 1\}$ of $[0, 1]$ such that the oscillatory sum

$$\begin{aligned} \omega(P, f) &= \sum_{i=1}^{m}(M_i - m_i)(b_i - a_i) + \sum_{r=0}^{m}(M_r - m_r)(a_{r+1} - b_r), \text{ where } b_0 = 0 \text{ and } a_{m+1} = 1 \\ &< \sum_{i=1}^{m} 1 \cdot (b_i - a_i) + \sum_{r=0}^{m} \frac{\epsilon}{2}(a_{r+1} - b_r) \\ &< \frac{\epsilon}{2} + \frac{\epsilon}{2} \cdot 1 = \epsilon \cdot \left[\sum_{r=0}^{m}(a_{r+1} - b_r) < 1\right] \end{aligned}$$

$\therefore f$ is R-integrable on $[0, 1]$. Also $\underline{\int_0^1} f(x)dx = \sup \sum\limits_{r=1}^{n} m_r\delta_r = 0$ $[m_r = 0 \ \forall\, r]$

$\Rightarrow \int_0^1 f(x)dx = 0$.

Problem [7.14] : A constant function $f : [a, b] \to \mathbb{R}$ is R-integral, prove it.

Solution. Let $f : [a, b] \to \mathbb{R}$ be defined by $f(x) = K \ \forall\, x \in [a, b]$, K is constant.

Let $P\{a = x_0 < x_1 \cdots < x_n = b\}$ be a partition of $[a, b]$ so that

$$M_r = m_r = K \ \forall\, [x_{r-1}, x_r], \ r = 1, 2, \cdots, n.$$

Then

$$\int_a^{\overline{b}} f(x)dx = \inf\{S(P,f),\ P \text{ is a partition of } [a,b]\}$$

$$= \inf\left[\sum_{r=1}^{n} M_r\delta_r\right] = \inf[K\delta_1 + K\delta_2 + \cdots + K\delta_n]$$

$$= \inf\left[K\sum_{r=1}^{n}\delta_r\right] = K(b-a)$$

$$\int_{\underline{a}}^{b} f(x)dx = \sup\{s(P,f),\ P \text{ is a partition of } [a,b]\}$$

$$= \sup\left\{\sum_{r=1}^{n} m_r\delta_r\right\} = \sup\left\{K\sum_{r=1}^{n}\delta_r\right\} = K(b-a)$$

$$\therefore \quad \int_a^{\overline{b}} f(x)dx = \int_{\underline{a}}^{b} f(x)dx$$

Hence, f is R-integrable on $[a,b]$.

Problem [7.15]: A function $f : [0,1] \to \mathbb{R}$ is defined by $f(x) = \frac{1}{n}$ for $\frac{1}{n+1} < x \le \frac{1}{n}$, $n = 1, 2, 3, \cdots$.

Show that f is R-integrable on $[0,1]$ and evaluate $\int_0^1 f(x)dx$.

Solution.

$$f(x) = \begin{cases} 1, \text{ when } \frac{1}{2} < x \le 1, & \text{for } n = 1 \\ \frac{1}{2}, \text{ when } \frac{1}{3} < x \le \frac{1}{2}, & \text{for } n = 2 \\ \frac{1}{3}, \text{ when } \frac{1}{4} < x \le \frac{1}{3}, & \text{for } n = 3 \end{cases}$$

and so on.

The points of discontinuities of f are $\frac{1}{3}, \frac{1}{3}, \frac{1}{4}, \cdots$

The set $\left\{\frac{1}{2}, \frac{1}{3}, \frac{1}{4}, \cdots\right\}$ has only one limit point 0. So f is R-integrable on $[0,1]$.

Now

$$\int_{\frac{1}{n}}^{1} f(x)dx = \int_{\frac{1}{2}}^{1} f(x)dx + \int_{\frac{1}{3}}^{\frac{1}{2}} f(x)dx + \cdots + \int_{\frac{1}{n}}^{\frac{1}{(n-1)}} f(x)dx$$

$$= \int_{\frac{1}{2}}^{1} 1dx + \int_{\frac{1}{3}}^{\frac{1}{2}} \frac{1}{2}dx + \int_{\frac{1}{4}}^{\frac{1}{3}} \frac{1}{3}dx + \cdots + \int_{\frac{1}{n}}^{\frac{1}{n-1}} \frac{1}{n-1}dx$$

$$= \left(1 - \frac{1}{2}\right) + \frac{1}{2}\left(\frac{1}{2} - \frac{1}{3}\right) + \frac{1}{3}\left(\frac{1}{3} - \frac{1}{4}\right) + \cdots + \frac{1}{n-1}\left(\frac{1}{n-1} - \frac{1}{n}\right)$$

$$= \left\{1 + \frac{1}{2^2} + \frac{1}{3^2} + \cdots + \frac{1}{(n-1)^2}\right\} - \left\{\frac{1}{2} + \frac{1}{2}\cdot\frac{1}{3} + \frac{1}{3}\cdot\frac{1}{4} + \cdots + \frac{1}{(n-1)}\frac{1}{n}\right\}$$

$$= \left\{1 + \frac{1}{2^2} + \frac{1}{3^2} + \cdots + \frac{1}{(n-1)^2}\right\} - \left\{1 - \frac{1}{n}\right\}$$

$$\lim_{n\to\infty} \int_{\frac{1}{n}}^{1} f(x)dx = \left\{1 + \frac{1}{2^2} + \frac{1}{3^2} + \cdots \text{ to } \infty\right\} - 1 \Rightarrow \int_0^1 f(x)dx = \frac{\pi^2}{6} - 1.$$

Note. $t_{n-1} = \frac{1}{(n-1)n} = \frac{1}{n-1} - \frac{1}{n}$, $n \geq 2$.

$\sum t_{n-1} = S_n = \left(1 - \frac{1}{n}\right)$, where $S_n = \left\{\frac{1}{2} + \frac{1}{2} \cdot \frac{1}{3} + \cdots + \frac{1}{(n-1)} \cdot \frac{1}{n}\right\}$.

Properties of integrable functions

Theorem [7.9] (1) If a function $f : [a, b] \to \mathbb{R}$ be bounded and integrable on $[a, b]$ and c is a point of $[a, b]$, then f is also integrable on $[a, c]$ and $[c, b]$.

Conversely, if f is integrable on $[a, c]$ and $[c, b]$, then it is also integrable on $[a, b]$.

In both cases, $\int_a^b f(x)dx = \int_a^c f(x)dx + \int_c^b f(x)dx$.

(2) If M, m be the sup and inf of f on $[a, b]$, then

(i) $m(b-a) \leq \int_a^b f(x)dx \leq M(b-a), b \geq a$,

(ii) $m(b-a) \geq \int_a^b f(x)dx \geq M(b-a), b \leq a$,

provided f is integrable on $[a, b]$.

(iii) There exists a number μ lying betweeen m and M such that $\int_a^b f(x)dx = \mu(b-a)$.

(iv) If f is continuous, then there exists a number ξ, $a < \xi < b$ such that

$$\int_a^b f(x)dx = (b-a)f(\xi) = hf(a + \theta h),\ 0 < \theta < 1.$$

(3) If $|f(x)| \leq k\ \forall\ x \in [a, b]$, then $\left|\int_a^b f(x)dx\right| \leq k|b-a|$.

(4) If $f(x) \geq 0\ \forall\ x \in [a, b]$, then $\int_a^b f(x)dx \geq 0$, when $b \geq a$
≤ 0, when $b \leq a$.

Proof. (1) Let f be bounded and integrable on $[a, b]$. Then for every $\epsilon(> 0)$, $\exists$ a (+)ve number δ such that for each partition P of $[a, b]$ with norm $||P|| \leq \delta$, $\omega(P, f) < \epsilon$. We consider the partition P of $[a, b]$, defined by

$$P\{a = x_0 < x_1 \cdots < x_k = c = x_{k+1} < x_{k+2} < \cdots < x_n = b\}$$

so that $c \in P$ and $\sum\limits_{r=0}^{k} \delta_r + \sum\limits_{r=k+1}^{n} \delta_r = c - a + b - c = b - a$, with $x_k = c$.

Then $P_1\{a = x_0 < x_1 \cdots < x_k = c\}$ forms a partition of $[a, c]$.

and $P_2\{c = x_{k+1} < x_{k+2} < \cdots < x_n = b\}$ forms a partition of $[c, b]$ so that $P_1 \cup P_2 = P$.

$$S(P_1, f) = \sum_{r=1}^{k} M_r\delta_r \text{ and } S(P_2, f) = \sum_{r=k+1}^{n} M_r\delta_r$$

$$s(P_1, f) = \sum_{r=1}^{k} m_r\delta_r \text{ and } s(P_2, f) = \sum_{r=k+1}^{n} m_r\delta_r$$

Also

$$S(P, f) = \sum_{r=1}^{n} M_r\delta_r = \sum_{r=1}^{k} M_r\delta_r + \sum_{r=k+1}^{n} M_r\delta_r \quad \text{(i)}$$

$$s(P, f) = \sum_{r=1}^{n} m_r\delta_r = \sum_{r=1}^{k} m_r\delta_r + \sum_{r=k+1}^{n} m_r\delta_r \quad \text{(ii)}$$

From (i) $S(P, f) = S(P_1, f) + S(P_2, f)$ and from (ii) $s(P, f) = s(P_1, f) + s(P_2, f)$.

$$\text{Hence } \omega(P,f) = S(P,f) - s(P,f)$$
$$= \{S(P_1,f) - s(P_1,f)\} + \{S(P_2,f) - s(P_2,f)\}$$
$$= \omega(P_1,f) + \omega(P_2,f)$$

Therefore, $\omega(P_1,f) + \omega(P_2,f) < \epsilon$ by the given condition. Since each member on the left side is non-negative and their sum $< \epsilon$, it follows that $\omega(P_1,f) < \epsilon$ and $\omega(P_2,f) < \epsilon$.

Hence f is R-integrable on $[a,c]$ and $[c,b]$.

Conversely, let f be integrable on $[a,c]$ and $[c,b]$, where $a < c < b$.

Then $\omega(P_1,f) < \frac{\epsilon}{2}$ and $\omega(P_2,f) < \frac{\epsilon}{2}$, where P_1 and P_2 are the partitions of $[a,c]$ and $[c,b]$ respectively.

Hence $P = P_1 \cup P_2$ forms a partition of $[a,b]$.

Therefore, $\omega(P,f) = \omega(P_1,f)+\omega(P_2,f) < \frac{\epsilon}{2}+\frac{\epsilon}{2} = \epsilon$ which shows that f is R-integrable on $[a,b]$.

Now

$$S(P,f) = S(P_1,f) + S(P_2,f) \tag{iii}$$

$\int_a^{\overline{c}} f(x)dx$ and $\int_c^{\overline{b}} f(x)dx$ being the inf of the sets $\{S(P_1,f),\ P_1$ is a partition of $[a,c]\}$ and $\{S(P_2,f),\ P_2$ is a partition of $[c,b]$ respectively, it follows that $S(P_1,f) \geq \int_a^{\overline{c}} f(x)dx$ and $S(P_2,f) \geq \int_c^{\overline{b}} f(x)dx$.

Therefore, from (iii) $S(P,f) \geq \int_a^{\overline{c}} f(x)dx + \int_c^{\overline{b}} f(x)dx$.

Now $\int_a^{\overline{c}} f(x)dx + \int_c^{\overline{b}} f(x)dx$ forms a lower bound of $\{S(P,f),\ P$ is a partition of $[a,b]\}$ having an infimum $\int_a^{\overline{b}} f(x)dx$.

Hence,

$$\int_a^{\overline{b}} f(x)dx \geq \int_a^{\overline{c}} f(x)dx + \int_c^{\overline{b}} f(x)dx \Rightarrow \int_a^b f(x)dx \geq \int_a^c f(x)dx + \int_c^b f(x)dx \tag{iv}$$

since

$$\int_a^{\overline{b}} f(x)dx = \int_a^b f(x)dx,\ \int_a^{\overline{c}} f(x)dx = \int_a^c f(x)dx \text{ and } \int_c^{\overline{b}} f(x)dx = \int_c^b f(x)dx$$

as f is simultaneously integrable on $[a,b]$, $[a,c]$ and $[c,b]$.

Similarly considering, $s(P,f) = s(P_1,f)+s(P_2,f)$ and taking the supremum values, we have

$$\int_{\underline{a}}^b f(x)dx \leq \int_{\underline{a}}^c f(x)dx + \int_{\underline{c}}^b f(x)dx \Rightarrow \int_a^b f(x)dx \leq \int_a^c f(x)dx + \int_c^b f(x)dx \tag{v}$$

So, from (iv) and (v)

$$\int_a^b f(x)dx = \int_a^c f(x)dx + \int_c^b f(x)dx.$$

Alternative proof of

$$\int_a^b f(x)dx = \int_a^c f(x)dx + \int_c^b f(x)dx$$

when f is simultaneously integrable on $[a,b]$, $[a,c]$ and $[c,b]$.

By Theorem [7.6], we have

$$\left|\sum_{r=1}^{n} f(\xi_r)\delta_r - \int_a^b f(x)dx\right| < \frac{\epsilon}{3}, \quad \left|\sum_{r=1}^{k} f(\xi_r)\delta_r - \int_a^c f(x)dx\right| < \frac{\epsilon}{3}$$

$$\text{and} \quad \left|\sum_{r=k+1}^{n} f(\xi_r)\delta_r - \int_c^b f(x)dx\right| < \frac{\epsilon}{3} \text{ with } ||P|| \le \delta.$$

Also $\sum_{r=1}^{n} f(\xi_r)\delta_r = \sum_{r=1}^{k} f(\xi_r)\delta_r + \sum_{r=k+1}^{n} f(\xi_r)\delta_r$.

Therefore,

$$\left|\int_a^b f(x)dx - \int_a^c f(x)dx - \int_c^b f(x)dx\right|$$

$$= \left|\int_a^b f(x)dx - \sum_{r=1}^{n} f(\xi_r)\delta_r + \sum_{r=1}^{k} f(\xi_r)\delta_r + \sum_{r=k+1}^{n} f(\xi_r)\delta_r - \int_a^c f(x)dx - \int_c^b f(x)dx\right|$$

$$\le \left|\int_a^b f(x)dx - \sum_{r=1}^{n} f(\xi_r)\delta_r\right| + \left|\int_a^c f(x)dx - \sum_{r=1}^{k} f(\xi_r)\delta_r\right| + \left|\int_c^b f(x)dx - \sum_{r=k+1}^{n} f(\xi_r)\delta_r\right|$$

$$< \frac{\epsilon}{3} + \frac{\epsilon}{3} + \frac{\epsilon}{3} = \epsilon \text{ with } ||P|| \le \delta$$

$$\Rightarrow \int_a^b f(x)dx = \int_a^c f(x)dx + \int_c^b f(x)dx$$

(2) (i) Let $\xi_r \in [x_{r-1}, x_r]$. Let $b > a$. Then $m \le m_r \le f(\xi_r) \le M_r \le M$

$$\Rightarrow \sum_{r=1}^{n} m\delta_r \le \sum_{r=1}^{n} f(\xi_r)\delta_r \le \sum_{r=1}^{n} M\delta_r$$

$$\Rightarrow m(b-a) \le \sum_{r=1}^{n} f(\xi_r)\delta_r \le M(b-a) \quad \left[\because \sum_{r=1}^{n} \delta_r = b-a\right]$$

$$\Rightarrow m(b-a) \le \lim_{||P||\to 0} \sum_{r=1}^{n} f(\xi_r)\delta_r \le M(b-a)$$

$$\Rightarrow m(b-a) \le \int_a^b f(x)dx \le M(b-a) \tag{i}$$

(ii) Next, let $b < a$. Then $a > b$. Therefore from (i)

$$m(a-b) \le \int_b^a f(x)dx \le M(a-b)$$

$$\Rightarrow -m(b-a) \le -\int_a^b f(x)dx \le -M(a-b)$$

$$\Rightarrow m(b-a) \ge \int_a^b f(x)dx \ge M(b-a)$$

For $a = b$, the result is trivial.

(iii) Let μ be a number lying between m and M. That is, $m \le \mu \le M$. Then from

$$m(b-a) \le \int_a^b f(x)dx \le M(b-a) \Rightarrow \int_a^b f(x)dx = \mu(b-a)$$

Also from $m(a-b) \le \int_b^a f(x)dx \le M(a-b),\ a > b$

$$\Rightarrow \int_b^a f(x)dx = \mu(a-b) \Rightarrow -\int_a^b f(x)dx = -\mu(b-a) \Rightarrow \int_a^b f(x)dx = \mu(b-a)$$

(iv) Let f be continuous on $[a,b]$ and $m \le \mu \le M$ since f attains every values between m and M, $\exists$ a point $\xi \in [a,b]$ such that $f(\xi) = \mu$.

Therefore, from (2) (iii), $\int_a^b f(x))dx = (b-a)f(\xi)$.

Also $b - a = h \Rightarrow \int_a^b f(x)dx = hf(a+\theta h),\ 0 < \theta < 1$.

(3) $|f(x)| \le K \Rightarrow -K \le f(x) \le K$.

m, M being the inf and sup of f on $[a,b]$ we have $-K \le m \le f(x) \le M \le K\ \forall\ x \in [a,b]$

$$\Rightarrow -K(b-a) \le m(b-a) \le \int_a^b f(x)dx \le M(b-a) \le K(b-a) \text{ where } b > a$$

$$\Rightarrow -K(b-a) \le \int_a^b f(x)dx \le K(b-a)$$

$$\Rightarrow \left|\int_a^b f(x)dx\right| \le K(b-a),\ b > a \qquad \text{(i)}$$

So, when $b < a$, i.e., $a > b$, $-K(a-b) \le \int_b^a f(x)dx \le K(a-b)$

$$\Rightarrow \left|\int_b^a f(x)dx\right| \le K(a-b) \Rightarrow \left|\int_a^b f(x)dx\right| \le K(a-b),\ a > b \qquad \text{(ii)}$$

From (i) and (ii) it follows that $\left|\int_a^b f(x)dx\right| \le K|b-a|$.

(4) Let $b > a$. Now $f(x) \ge 0\ \forall\ x \in [a,b] \Rightarrow m \ge 0$.

Therefore by (i) and (ii) of (2)

$$\int_a^b f(x)dx \ge m(b-a) \ge 0\ [\because b - a > 0]$$

Again, $b < a \Rightarrow a > b$

$$\Rightarrow \int_b^a f(x)dx \ge m(a-b) \ge 0\ [\because a - b > 0]$$

$$\Rightarrow -\int_a^b f(x)dx \ge 0 \Rightarrow \int_a^b f(x)dx \le 0$$

The results are trivial when $a = b$.

Lemma [7.1]: If M and m be the sup and inf of a bounded function f on $[a,b]$, then $M - m$, the oscillation of f on $[a,b]$ is the supremum of the set $S = \{|f(\alpha) - f(\beta)|,\ \alpha, \beta \in [a,b]\}$.

Proof. Let $\alpha, \beta \in [a,b]$. Then $f(\alpha) \ge m, f(\beta) \le M$ or $f(\beta) \ge m, f(\alpha) \le M$.

Now $f(\alpha) \ge m \Rightarrow -f(\alpha) \le -m$ and $f(\beta) \le M$

$$\Rightarrow f(\beta) - f(\alpha) \le M - m \qquad \text{(i)}$$

Also $f(\beta) \geq m \Rightarrow -f(\beta) \leq -m$ and $f(\alpha) \leq M$

$$\Rightarrow f(\alpha) - f(\beta) \leq M - m \tag{ii}$$

Thus $|f(\alpha) - f(\beta)| \leq M - m$ [by (i) and (ii)]

Therefore $(M - m)$ is an upper bound of S.

Now M is the $\sup f$ on $[a, b]$. Hence for every $\epsilon(> 0)$, $\exists\ \alpha_1 \in [a, b]$ such that

$$f(\alpha_1) > M - \epsilon/2 \tag{iii}$$

Again, m is the inf of f in $[a, b]$. Hence for every $\epsilon(> 0)$, $\exists\ \beta_1 \in [a, b]$ such that $f(\beta_1) < m + \epsilon/2$.

That is

$$-f(\beta_1) > -m - \epsilon/2 \tag{iv}$$

Therefore, $f(\alpha_1) - f(\beta_1) > M - m - \epsilon$ [by (iii) and (iv)]

$$\Rightarrow |f(\alpha_1) - f(\beta_1)| > M - m - \epsilon\ [\because |f(\alpha_1) - f(\beta_1)| \geq f(\alpha_1) - f(\beta_1)]$$

So, $\exists\ \alpha_1, \beta_1 \in [a, b]$ such that $|f(\alpha_1) - f(\beta_1)| > M - m - \epsilon$.

It follows that $M - m$ is the $\sup S$.

Theorem [7.10] If $f : [a, b] \to \mathbb{R}$ be bounded and R-integrable on $[a, b]$, then

(1) (i) $|f|$ is bounded and R-integrable on $[a, b]$. Converse is not true.

(ii) $\left|\int_a^b f(x)dx\right| \leq \left|\int_a^b |f(x)|dx\right|$

(iii) cf, where c is a constant, is bounded and R-integrable on $[a, b]$.
Also $\int_a^b cf(x)dx = c\int_a^b f(x)dx$.

(iv) $\frac{1}{f}$, when $f(x) \geq \lambda, \lambda > 0\ \forall\ x \in [a, b]$, is bounded and R-integrable on $[a, b]$.

(2) If $f, g : [a, b] \to \mathbb{R}$ be both bounded and R-integrable on $[a, b]$, then

(i) $f \pm g$ is bounded and R-integrable on $[a, b]$.
Also $\int_a^b (f \pm g)(x)dx = \int_a^b f(x)dx \pm \int_a^b g(x)dx$.

(ii) $f \cdot g$ is bounded and R-integrable on $[a, b]$.

(iii) f/g when $g(x) \geq \lambda,\ \lambda > 0\ \forall\ x \in [a, b]$, is bounded and R-integrable on $[a, b]$.

(iv) $f \geq g \Rightarrow \int_a^b f(x)dx \geq \int_a^b g(x)dx$ and $f \leq g \Rightarrow \int_a^b f(x)dx \leq \int_a^b g(x)dx$.

Proof. (1) (i) f is bounded $\Rightarrow |f(x)| \leq k\ \forall\ x \in [a, b] \Rightarrow\ |f|$ is bounded.

Since f is R-integrable on $[a, b]$, for every $\epsilon(> 0)$, $\exists$ a (+)ve number δ such that for a partition $P\{a = x_0 < x_1 \cdots x_{r-1} < x_r \cdots < x_n = b\}$ of $[a, b]$ with $||P|| \leq \delta, \omega(P, f) < \epsilon$.

Now, let M_r, m_r and M_r', m_r' be the sup and inf of f and $|f|$ respectively in $[x_{r-1}, x_r]$.

Let $\alpha_1, \alpha_2 \in [x_{r-1}, x_r]$.

$$\left.\begin{array}{ll}\text{Then} & |f(\alpha_1)| - |f(\alpha_2)| \leq |f(\alpha_1) - f(\alpha_2)| \\ \text{Also} & |f(\alpha_2)| - |f(\alpha_1)| \leq |f(\alpha_1) - f(\alpha_2)|\end{array}\right\} \tag{i}$$

Hence from (i),

$$\begin{aligned}
&||f(\alpha_1)|-|f(\alpha_2)|| \le |f(\alpha_1)-f(\alpha_2)| \ \forall\ \alpha_1,\alpha_2 \in [x_{r-1},x_r]\\
\Rightarrow\ &||f(\alpha_1)|-|f(\alpha_2)|| \le M_r - m_r\\
\Rightarrow\ &\sup\{||f(\alpha_1)|-|f(\alpha_2)||, \alpha_1,\alpha_2 \in [x_{r-1},x_r]\} \le M_r - m_r\\
\Rightarrow\ &M_r' - m_r' \le M_r - m_r\\
\Rightarrow\ &\sum_{r=1}^{n}(M_r'-m_r')\delta_r \le \sum_{r=1}^{n}(M_r-m_r)\delta_r = \omega(P,f) < \epsilon\\
\Rightarrow\ &\omega(P,|f|) < \epsilon\\
\Rightarrow\ &|f| \text{ is } R\text{-integrable on } [a,b].
\end{aligned}$$

Converse of the theorem is not true, follows from the example:
Let $f : [a,b] \to \mathbb{R}$ be defined by

$$\begin{aligned}
f(x) &= -1\ \forall \text{ irrational values of } x \text{ on } [a,b]\\
&= 1\ \forall \text{ rational values of } x \text{ on } [a,b].
\end{aligned}$$

Then $|f(x)| = 1\ \forall\ x \in [a,b]$

$$\overline{\int_a^b} |f(x)|dx = \underline{\int_a^b} |f(x)|dx = (b-a)$$

$\Rightarrow |f|$ is R-integrable on $[a,b]$. But f is not R-integrable on $[a,b]$, since

$$\overline{\int_a^b} f(x)dx = (b-a) \text{ and } \underline{\int_a^b} f(x)dx = -(b-a)$$

$$\Rightarrow \quad \overline{\int_a^b} f(x)dx \ne \underline{\int_a^b} f(x)dx.$$

Proof. (1) (ii) $\int_a^b |f(x)|dx$ exists. So $\int_a^b |f(x)|dx = \lim\limits_{||P||\to 0} \sum\limits_{r=1}^{n} |f(\xi_r)|\delta_r$.

Now

$$\begin{aligned}
\left|\int_a^b f(x)dx\right| &= \lim_{||P||\to 0} |\Sigma f(\xi_r)\delta_r| \le \lim_{||P||\to 0} [|f(\xi_1)|\delta_1 + |f(\xi_2)|\delta_2 + \cdots + |f(\xi_n)|\delta_n]\\
&= \lim_{||P||\to 0} \sum_{r=1}^{n} |f(\xi_r)|\delta_r = \int_a^b |f(x)|dx
\end{aligned}$$

$$\therefore \left|\int_a^b f(x)dx\right| \le \int_a^b |f(x)|dx = \left|\int_a^b |f(x)|dx\right|$$

Alternative proof of $-|f(x)| \le f(x) \le |f(x)|\ \forall\ x \in [a,b]$

$$\Rightarrow -\int_a^b |f(x)|dx \le \int_a^b f(x)dx \le \int_a^b |f(x)|dx, \text{ when } b \ge a$$

$$\Rightarrow \left|\int_a^b f(x)dx\right| \le \int_a^b |f(x)|dx.$$

Remark. When $b \geq a$,

$$\left|\int_a^b f(x)dx\right| \leq \int_a^b |f(x)|dx = \left|\int_a^b |f(x)|dx\right|.$$

When $b \leq a$, i.e., $a \geq b$,

$$\left|\int_b^a f(x)dx\right| \leq \int_b^a |f(x)|dx$$
$$\Rightarrow \left|-\int_a^b f(x)dx\right| \leq \int_b^a |f(x)|dx \Rightarrow \left|\int_a^b f(x)dx\right| \leq \left|\int_a^b |f(x)|dx\right|$$

Proof. (1) (iii) Let M_r', m_r' and M_r, m_r be the sup and inf of cf and f on $[x_{r-1}, x_r]$ respectively.

Let $c > 0$ and let $\alpha_1, \alpha_2 \in [x_{r-1}, x_r]$.

Then $|cf(\alpha_1) - cf(\alpha_2)| = |c(f(\alpha_1) - f(\alpha_2))| = |c|\ |f(\alpha_1) - f(\alpha_2)| < c(M_r - m_r)$.

$$\Rightarrow \sup\{|cf(\alpha_1) - cf(\alpha_2)|, \alpha_1, \alpha_2 \in [x_{r-1}, x_r]\} < c(M_r - m_r)$$
$$\Rightarrow M_r' - m_r' < c(M_r - m_r)$$
$$\Rightarrow \sum_{r=1}^{n}(M_r' - m_r')\delta_r < c\sum(M_r - m_r)\delta_r < c \cdot \frac{\epsilon}{c} = \epsilon$$
$$\left[\because \sum(M_r - m_r)\delta_r = \omega(P, f) < \frac{\epsilon}{c}\right]$$
$$\Rightarrow \omega(P, cf) < \epsilon$$
$$\Rightarrow cf \text{ is } R\text{-integrable on } [a, b].$$

Let $c < 0$, we take $c = -\lambda, \lambda > 0$.

Then $|cf(\alpha_1) - cf(\alpha_2)| = |-\lambda|\ |f(\alpha_1) - f(\alpha_2)| = \lambda|f(\alpha_1) - f(\alpha_2)|$

$$\Rightarrow \sum(M_r' - m_r')\delta_r < \lambda\sum(M_r - m_r)\delta_r < \lambda \cdot \frac{\epsilon}{\lambda} = \epsilon$$
$$\Rightarrow \omega(P, cf) \text{ is } R\text{-integrable on } [a, b].$$

The result is trivial, when $c = 0$.

Also

$$\int_a^b cf(x)dx = \lim_{||P|| \to 0} \sum cf(\xi_r)\delta_r,\ \xi_r \in [x_{r-1}, x_r]$$
$$= c \lim_{||P|| \to 0} \sum f(\xi_r)\delta_r = c\int_a^b f(x)dx$$

Proof. (1) (iv) $f(x) \geq \lambda, \lambda > 0 \Rightarrow \left|\frac{1}{f(x)}\right| \leq \frac{1}{\lambda}\ \forall\ x \in [a, b] \Rightarrow \frac{1}{f}$ is bounded on $[a, b]$.

Let M_r', m_r' and $M_r; m_r$ be the sup and inf of $\frac{1}{f}$ and f respectively on $[x_{r-1}, x_r]$.

Also let $\alpha_1, \alpha_2 \in [x_{r-1}, x_r]$

$$\left|\frac{1}{f(\alpha_1)} - \frac{1}{f(\alpha_2)}\right| = \frac{|f(\alpha_2) - f(\alpha_1)|}{|f(\alpha_1)|\,|f(\alpha_2)|} \leq \frac{M_r - m_r}{\lambda^2}$$

$$\Rightarrow \sup\left\{\left|\frac{1}{f(\alpha_1)} - \frac{1}{f(\alpha_2)}\right|, \alpha_1, \alpha_2 \in [x_{r-1}, x_r]\right\} \leq \frac{M_r - m_r}{\lambda^2}$$

$$\Rightarrow M'_r - m'_r \leq \frac{M_r - m_r}{\lambda^2}$$

$$\Rightarrow \sum_{r=1}^{n}(M'_r - m'_r)\delta_r \leq \frac{\Sigma(M_r - m_r)\delta_r}{\lambda^2} = \frac{\omega(P, f)}{\lambda^2} < \frac{\epsilon}{\lambda^2} < \epsilon$$

$$\Rightarrow \omega\left(P, \frac{1}{f}\right) < \epsilon$$

$\Rightarrow \frac{1}{f}$ is R-integrable on $[a, b]$.

Proof. (2) (i) f, g are both bounded on $[a, b]$

$\Rightarrow \exists$ a (+)ve number k such that $|f(x)| \leq k, |g(x)| \leq k \; \forall \; x \in [a, b]$.

Then $|f \pm g(x)| = |f(x) \pm g(x)| \leq |f(x)| + |g(x)| \leq 2k \; \forall \; x \in [a, b] \Rightarrow f \pm g$ is bounded on $[a, b]$.

Let $P\{a = x_0 < x_1 \cdots < x_n = b\}$ be a partition of $[a, b]$.

f, g are R-integrable on $[a, b]$

$$\Rightarrow \omega(P, f) < \frac{\epsilon}{2} \text{ and } \omega(P, g) < \frac{\epsilon}{2} \qquad \text{(i)}$$

Let $M''_r, m''_r, M_r, m_r, M'_r, m'_r$ be the sup and inf of $f + g$, f and g on $[x_{r-1}, x_r]$ respectively. Also let $\alpha_1, \alpha_2 \in [x_{r-1}, x_r]$.

$$\begin{aligned} \text{Then } |(f+g)\alpha_1 - (f+g)\alpha_2| &= |\{f(\alpha_1) - f(\alpha_2)\} + \{g(\alpha_1) - g(\alpha_2)\}| \\ &\leq |f(\alpha_1) - f(\alpha_2)| + |g(\alpha_1) - g(\alpha_2)| \\ &\leq (M_r - m_r) + (M'_r - m'_r) \text{ [by Lemma 7.1]} \end{aligned}$$

$\Rightarrow \sup\{|(f+g)\alpha_1 - (f+g)\alpha_2|, \alpha_1, \alpha_2 \in [x_{r-1}, x_r]\} \leq (M_r - m_r) + (M'_r - m'_r)$

$\Rightarrow (M''_r - m''_r) \leq (M_r - m_r) + (M'_r - m'_r)$

$\Rightarrow \Sigma(M''_r - m''_r)\,\delta_r \leq \sum_{r=1}^{n}(M_r - m_r)\delta_r + \sum_{r=1}^{n}(M'_r - m'_r)\,\delta_r = \omega(P, f) + \omega(P, g) < \frac{\epsilon}{2} + \frac{\epsilon}{2} = \epsilon$

$\Rightarrow \omega(P, f + g) < \epsilon$

$\Rightarrow (f + g)$ is R-integrable on $[a, b]$.

The same case be shown for $f - g$, choosing M''_r, m''_r be the sup and inf of $f - g$ on $[x_{r-1}, x_r]$.

Since f, g are integrable on $[a, b]$, for every $\epsilon > 0$, $\exists$ a (+)ve number δ, with the norm $||P|| \leq \delta$,

$$\left|\sum f(\xi_r)\delta_r - \int_a^b f(x)dx\right| < \frac{\epsilon}{2}$$

$$\text{and } \left|\sum g(\xi_r)\delta_r - \int_a^b g(x)dx\right| < \frac{\epsilon}{2} \forall \; \xi_r \in [x_{r-1}, x_r].$$

Then

$$\left|\sum (f+g)(\xi_r)\delta_r - \left\{\int_a^b f(x)dx + \int_a^b g(x)dx\right\}\right|$$

$$= \left|\left\{\sum f(\xi_r)\delta_r - \int_a^b f(x)dx\right\} + \left\{\sum g(\xi_r)\delta_r - \int_a^b g(x)dx\right\}\right|$$

$$\leq \left|\sum f(\xi_r)\delta_r - \int_a^b f(x)dx\right| + \left|\sum g(\xi_r)\delta_r - \int_a^b g(x)dx\right| < \frac{\epsilon}{2} + \frac{\epsilon}{2} = \epsilon, ||P|| \leq \delta$$

$$\Rightarrow \lim_{||P||\to 0} \sum (f+g)(\xi_r)\delta_r = \int_a^b f(x)dx + \int_a^b g(x)dx$$

$$\Rightarrow \int_a^b (f+g)(x)dx = \int_a^b f(x)dx + \int_a^b g(x)dx \text{ [Since } (f+g) \text{ is integrable on } [a,b]]$$

Corollary [7.1] : If f, g be bounded and R-integrable on $[a, b]$ and λ, μ be any constants, then $\lambda f + \mu g$ is also bounded and R-integrable on $[a, b]$. Also

$$\int_a^b (\lambda f + \mu g)(x)dx = \lambda \int_a^b f(x)dx + \mu \int_a^b g(x)dx$$

[*Hints.* The result follows from (1) (iii) and (2) (i) of Theorem [7.10].]

Proof. (2) (ii) f, g are bounded $\Rightarrow \exists$ a (+)ve number K such that $|f(x)| \leq K,\ |g(x)| \leq K\ \forall\ x \in [a, b]$

$$\Rightarrow\ |(fg)(x)| = |f(x)\cdot g(x)| = |f(x)||g(x)| \leq K^2\ \forall\ x \in [a,b]$$

$$\Rightarrow\ f\cdot g \text{ is bounded on } [a,b].$$

f, g are R-integrable on $[a, b]$.

$\Rightarrow$ for a partition $P\{a = x_0 < x_1 \cdots < x_n = b\}$ with the norm

$$||P|| \leq \delta,\ \omega(P, f) < \frac{\epsilon}{2K} \text{ and } \omega(P, g) < \frac{\epsilon}{2K}.$$

Now let M_r'', m_r'', M_r, m_r and M_r', m_r' be the sup and inf of $f\cdot g$, f and g respectively on $[x_{r-1}, x_r]$.

Let $\alpha_1, \alpha_2 \in [x_{r-1}, x_r]$.

Then

$$\begin{aligned}|(f\cdot g)(\alpha_1) - (f\cdot g)(\alpha_2)| &= |f(\alpha_1)g(\alpha_1) - f(\alpha_2)g(\alpha_2)|\\ &= |f(\alpha_1)\{g(\alpha_1) - g(\alpha_2)\} - g(\alpha_2)\{f(\alpha_2) - f(\alpha_1)\}|\\ &\leq |f(\alpha_1)|\ |g(\alpha_1) - g(\alpha_2)| + |g(\alpha_2)|\ |f(\alpha_2) - f(\alpha_1)|\\ &\leq K(M_r' - m_r') + K(M_r - m_r) \quad \text{[by Lemma [7.1]]}\end{aligned}$$

$$\Rightarrow \sup\{|fg)\alpha_1 - (f(g)\alpha_2|, \alpha_1, \alpha_2 \in [x_{r-1}, x_r]\} \leq K[(M_r - m_r) + (M_r' - m_r')]$$

$$\Rightarrow M_r'' - m_r'' \leq K[(M_r - m_r) + (M_r' - m_r')]$$

$$\Rightarrow \sum_{r=1}^{n} (M_r'' - m_r'')\delta_r \leq K\left[\sum (M_r - m_r)\delta_r + \sum (M_r' - m_r')\delta_r\right]$$

$$\Rightarrow \omega(P, fg) \leq K[\omega(P, f) + \omega(P, g)] < K\left[\frac{\epsilon}{2K} + \frac{\epsilon}{2K}\right] = \epsilon$$

$$\Rightarrow f\cdot g \text{ is } R\text{-integrable on } [a,b].$$

Proof. (2) (iii) f is bounded on $[a,b] \Rightarrow \exists$ a (+)ve number K such that $|f(x)| \leq K$ and $|g(x)| \leq K \ \forall \ x \in [a,b]$.

Also $g(x) \geq \lambda \Rightarrow \frac{1}{g(x)} \leq \frac{1}{\lambda} \ \forall \ x \in [a,b], \ \lambda > 0$.

Therefore, $\left|\left(\frac{f}{g}\right)(x)\right| = \left|\frac{f(x)}{g(x)}\right| \leq \frac{K}{\lambda} \ \forall \ x \in [a,b]$.

$\Rightarrow \frac{f}{g}$ is bounded on $[a,b]$. f, g are R-integrable on $[a,b]$.

$\Rightarrow$ for a partition $P\{a = x_0 < x_1 \cdots < x_n = b\}$ of $[a,b]$ with the norm $||P|| \leq \delta$, $\omega(P,f) < \frac{\epsilon}{2K}$ and $\omega(P,g) < \frac{\epsilon}{2K}$.

Let $\alpha_1, \alpha_2 \in [x_{r-1}, x_r]$.

Then

$$\begin{aligned}
\left|\left(\frac{f}{g}\right)(\alpha_1) - \left(\frac{f}{g}\right)(\alpha_2)\right| &= \left|\frac{f(\alpha_1)}{g(\alpha_1)} - \frac{f(\alpha_2)}{g(\alpha_2)}\right| \\
&= \left|\frac{f(\alpha_1)\{g(\alpha_2) - g(\alpha_1)\} - g(\alpha_1)\{f(\alpha_2) - f(\alpha_1)\}}{g(\alpha_1) \cdot g(\alpha_2)}\right| \\
&\leq \frac{|f(\alpha_1)|\ |g(\alpha_2) - g(\alpha_1)| + |g(\alpha_1)|\ |f(\alpha_2) - f(\alpha_1)|}{|g(\alpha_1)|\ |g(\alpha_2)|} \\
&\leq \frac{1}{\lambda^2}[K(M_r' - m_r') + K(M_r - m_r)]
\end{aligned}$$

[$\because$ $|g(x)| \leq K \ \forall \ x \in [a,b]$ and by Lemma [7.1].]

$$\Rightarrow \sup\left|\left(\frac{f}{g}\right)(\alpha_1) - \left(\frac{f}{g}\right)(\alpha_2)\right|, \alpha_1, \alpha_2 \in [x_{r-1}, x_r] \leq \frac{1}{\lambda^2}[K(M_r' - m_r') + K(M_r - m_r)]$$

$$\Rightarrow (M_r'' - m_r'') \leq \frac{K}{\lambda^2}[(M_r - m_r) + (M_r' - m_r')]$$

$$\Rightarrow \sum_{r=1}^{n}(M_r'' - m_r'')\delta_r \leq \frac{K}{\lambda^2}\left[\sum_{r=1}^{n}(M_r - m_r)\delta_r + \sum_{r=1}^{n}(M_r' - m_r')\delta_r \leq \frac{K}{\lambda^2}[\omega(P,f) + \omega(P,g)\right]$$

$$\Rightarrow \omega\left(P, \frac{f}{g}\right) < \frac{K}{\lambda^2}\left[\frac{\epsilon}{2K} + \frac{\epsilon}{2K}\right] = \frac{\epsilon}{\lambda^2} < \epsilon$$

$\Rightarrow \frac{f}{g}$ is R-integrable on $[a,b]$.

Note. Here M_r'', m_r'' are the sup and inf of $\frac{f}{g}$ on $[x_{r-1}, x_r]$.

Proof. (2) (iv) Let $\phi(x) = f(x) - g(x)$. Then $f(x) \geq g(x) \Rightarrow \phi(x) \geq 0 \ \forall \ x \in [a,b]$.

Therefore, when $b \geq a$ by (4) of Theorem [7.9],

$$\begin{aligned}
\int_a^b \phi(x)dx \geq 0 &\Rightarrow \int_a^b \{f(x) - g(x)\}dx \geq 0 \\
&\Rightarrow \int_a^b f(x)dx - \int_a^b g(x)dx \geq 0 \quad \text{[by (2) (i) of Theorem [7.10].]} \\
&\Rightarrow \int_a^b f(x)dx \geq \int_a^b g(x)dx
\end{aligned}$$

Also $b \leq a$

$$\Rightarrow \int_a^b \phi(x)dx \leq 0 \Rightarrow \int_a^b \{f(x) - g(x)\}dx \leq 0 \Rightarrow \int_a^b f(x)dx \leq \int_a^b g(x)dx.$$

Corollary [7.2] : If $f : [a,b] \to \mathbb{R}$ be bounded and R-integrable on $[a,b]$, then f^2 is also bounded and R-integrable on $[a,b]$.

Proof. [*Hints.* $|f(x)| \leq K \ \forall\, x \in [a,b] \Rightarrow |f^2(x)| = |f(x) \cdot f(x)| = |f(x)|\,|f(x)| \leq K^2 \ \forall\, x \in [a,b]$.]

Let $\alpha_1, \alpha_2 \in [x_{r-1}, x_r]$.

$$\begin{aligned}\text{Then } |f^2(\alpha_1) - f^2(\alpha_2)| &= |\{f(\alpha_1) - f(\alpha_2)\}\{f(\alpha_1) + f(\alpha_2)\}| \\ &\leq |f(\alpha_1) - f(\alpha_2)|\{|f(\alpha_1)| + |f(\alpha_2)|\} \\ &\leq 2K|f(\alpha_1) - f(\alpha_2)|\end{aligned}$$

$\Rightarrow (M'_r - m'_r) \leq 2K(M_r - m_r)$, where M'_r, m'_r and M_r, m_r are sup and inf of f^2 and f on $[x_{r-1}, x_r]$ respectively

$$\Rightarrow \sum (M'_r - m'_r)\delta_r \leq 2K \sum_{r=1}^{n} (M_r - m_r)\delta_r$$

$$\Rightarrow \omega(P, f^2) \leq 2K\omega(P, f) < 2K\frac{\epsilon}{2K} = \epsilon$$

$\Rightarrow f^2$ is R-integrable on $[a,b]$.

Corollary [7.3] : If f is R-integrable on $[a,b]$ and $[c,d] \subseteq [a,b]$, then f is R-integrable on $[c,d]$.

[*Hints.* Use (1) of Theorem [7.9]. f is R-integrable on $[a,b]$, $c < d < b \Rightarrow f$ is R-integrable on $[c,d]$.]

Theorem [7.11] If $f : [a,b] \to \mathbb{R}$ be bounded and integrable on $[a,b]$, then f^+ and f^-, the positive and negative part of f, are bounded and integrable on $[a,b]$.

Proof. The positive part f^+ of f and the negative part f^- of f are defined by

$$f^+(x) = \begin{cases} f(x), & \text{if } f(x) \geq 0 \\ 0, & \text{if } f(x) < 0 \end{cases}$$

$$f^-(x) = \begin{cases} 0, & \text{if } f(x) > 0 \\ -f(x), & \text{if } f(x) \leq 0 \end{cases}$$

Also $|f|$ is defined by

$$|f|(x) = \begin{cases} f(x), & \text{if } f(x) \geq 0 \\ -f(x), & \text{if } f(x) < 0 \end{cases}$$

So it follows that

$$|f| = f^+ + f^- \qquad \text{(i)}$$

$$\text{and } f = f^+ - f^- \qquad \text{(ii)}$$

From definition of f^+ and f^- it follows that they are bounded, if f is bounded on $[a,b]$.

$$\text{(i)} + \text{(ii)} \Rightarrow f^+ = \frac{1}{2}[f + |f|]$$

$$\text{(i)} - \text{(ii)} \Rightarrow f^- = \frac{1}{2}[|f| - f]$$

Hence f^+ and f^- are integrable on $[a,b]$ by (1) (i) and (2) (i) of Theorem [7.10].

Problem [7.16] : (i) If f is a non-negative continuous function on $[a,b]$ and if $f(c) > 0$ for some $c \in [a,b]$, then $\int_a^b f(x)dx > 0$. Prove it.

(ii) If $f(x) \geq 0 \ \forall \ x \in [a, b]$ and be continuous on $[a, b]$, then $\int_a^b f(x)dx = 0 \Rightarrow f(x) = 0 \ \forall \ x \in [a, b]$. Prove it.

(iii) If f be bounded and integrable on $[a, b]$, then show that $\int_a^b [f(x)]^2 dx = 0$, if and only if $f(c) = 0$, at every point c, of continuity of f, on $[a, b]$.

(iv) Give an example of a non-negative function f which is R-integrable on $[a, b]$ such that $f(x_0) > 0$ for some $x_0 \in [a, b]$ but $\int_a^b f(x)dx = 0$.

(v) Show that, if $f(x) = g(x)$ at most finite number of points on $[a, b]$, then either both are R-integrable or neither is.

Solution. (i) $f(x) \geq 0 \ \forall \ x \in [a, b]$. f is continuous at $x = c$ and $f(c) > 0 \Rightarrow f(x) > 0 \ \forall \ x \in (c - \delta, c + \delta), \delta > 0$, when $a < c < b$.

Again, continuity of f at $x = c$

$$\Rightarrow |f(x) - f(c)| < \epsilon \ \forall \ x \in (c - \delta, c + \delta)$$

$$\Rightarrow f(x) > f(c) - \epsilon = \frac{1}{2} f(c), \text{ taking } \epsilon = \frac{1}{2} f(c), f(c) > 0.$$

Hence

$$\int_a^b f(x)dx = \int_a^{c-\delta} f(x)dx + \int_{c-\delta}^{c+\delta} f(x)dx + \int_{c+\delta}^b f(x)dx$$

$$\geq \int_{c-\delta}^{c+\delta} f(x)dx > \frac{1}{2} f(c) \int_{c-\delta}^{c+\delta} dx = \delta f(c) > 0$$

(ii) From (i), $a < c < b$, $f(c) > 0 \Rightarrow \int_a^b f(x)dx > 0$, contradiction arises and so $\int_a^b f(x)dx = 0 \Rightarrow f(x) = 0 \ \forall \ x \in [a, b]$. Similarly it can be proved when $f(a) > 0$ and $f(b) > 0$.

Hence the result.

(iii) Let f be continuous at a point $x = c$, $a < c < b$ and let $\phi(x) = [f(x)]^2$.

Then $\phi(x) \geq 0 \ \forall \ x \in [a, b]$.

Also let $f(c) \neq 0$. Then either $f(c) > 0$ or $f(c) < 0$. In each case, $\phi(c) > 0$.

$$\Rightarrow \int_a^b \phi(x)dx > 0 \text{ by (i)}$$

$$\Rightarrow \int_a^b [f(x)]^2 dx > 0$$

Hence, $\int_a^b [f(x)]^2 dx = 0$, if and only if $[f(x)]^2 = 0 \forall x \in [a, b]$ by (ii)

$$\Rightarrow [f(c)]^2 = 0 \Rightarrow f(c) = 0.$$

(iv) Let us consider a function f on $[a, b]$ such that

$$f(x) = \begin{cases} 1, & \text{if } x = a, b \\ 0, & \text{if } x \in (a, b) \end{cases}$$

Then, if $x_0 = a$ or b, $f(x_0) = 1 > 0$.

Hence, $f(x)$ is discontinuous at $x = a$ and $x = b$. f is continuous in (a, b).

So f has finite number of points of discontinuities and hence f is R-integrable on $[a, b]$.

Also $\int_a^b f(x)dx = 0$, since $f(x) = 0 \ \forall \ x \in (a, b)$.

(v) Let $\phi = f - g$. Then $\phi(x) = f(x) - g(x) = 0$ for finite number of points on $[a, b]$. Consequently, ϕ has finite number of points discontinuities and hence ϕ is R-integrable on $[a, b]$.

Now $f = \phi + g \Rightarrow f$ is R-integrable, if g is R-integrable.

Also $g = f - \phi \Rightarrow g$ is R-integrable, if f is R-integrable.

Again, if g is not R-integrable, i.e., if $\int_a^b g(x)dx$ does not exist, $\int_a^b f(x)dx$ does not exist and vice versa. Hence the result.

Integral function of a function f

Definition [7.5] : Let f be R-integrable on $[a, b]$. Then a function $\phi(t)$, $a < t \leq b$, defined by $\phi(t) = \int_a^t f(x)dx$ or $\int_t^b f(x)dx$, $a \leq t < b$ with $\phi(a) = 0$ or $\phi(b) = 0$ is called an integrable function of a function f.

Theorem [7.12] (i) If f be an integable function on $[a, b]$, then the integral function $\phi(t) = \int_a^t f(x)dx$, $a < t \leq b$ is continuous on $[a, b]$.

(ii) If f is continuous on $[a, b]$ and $\phi(t) = \int_a^t f(x)dx$, then $\phi'(t) = f(t)\ \forall\ t \in [a, b]$.

(iii) If f is R-integrable on $[a, b]$ and if $\phi(t) = \int_a^t f(x)dx$, $a < t \leq b$ and if f is continuous at a point $t = t_0 \in [a, b]$, then $\phi'(t_0) = f(t_0)$.

Proof. (i) Here f is bounded. So $|f(x)| \leq K,\ \forall\ x \in [a, b],\ K > 0$.

Let $c \in [a, b]$ and $h > 0$ so that $c + h \in [a, b]$.

We have

$$\phi(c) = \int_a^c f(x)dx \text{ and } \phi(c+h) = \int_a^{c+h} f(x)dx.$$

Therefore,

$$\begin{aligned}\phi(c+h) - \phi(c) &= \int_a^{c+h} f(x)dx - \int_a^c f(x)dx \\ &= \int_c^{c+h} f(x)dx\end{aligned}$$

$$\begin{aligned}\Rightarrow |\phi(c+h) - \phi(c)| - \left|\int_c^{c+h} f(x)dx\right| &\leq \left|\int_c^{c+h} |f(x)|dx\right| \quad \text{[by (ii) of Theorem [7.10]]} \\ &\leq K\left|\int_c^{c+h} dx\right| = K|h|\end{aligned}$$

Taking $|h| < \frac{\epsilon}{K}$

$$|\phi(c+h) - \phi(c)| = K|h| < K\frac{\epsilon}{K} = \epsilon.$$

$\Rightarrow \phi(t)$ is continuous at c.

Since c is arbitrary, ϕ is continuous on $[a, b]$.

Corollary [7.4] : $\phi(t) = \int_t^b f(x)dx$, $a \leq t < b$ is continuous on $[a, b]$.

Proof. (ii) Let $c \in [a, b]$ and $h > 0$ so that $c + h \in [a, b]$.

Then $\phi(c)=\int_a^c f(x)dx$ and $\phi(c+h)=\int_a^{c+h} f(x)dx$.

$$\Rightarrow \phi(c+h)-\phi(c)=\int_c^{c+h} f(x)dx = hf(c+\theta h), 0<\theta<1 \text{ [by (2) (iv) of Th. [7.9].]}$$

$$\Rightarrow \frac{\phi(c+h)-\phi(c)}{h}=f(c+\theta h)$$

$$\Rightarrow \lim_{h\to 0}\frac{\phi(c+h)-\phi(c)}{h}=\lim_{h\to 0} f(c+\theta h)=f(c), \text{ since } f \text{ is continuous at } x=c\in[a,b]$$

$$\Rightarrow \phi'(c)=f(c)$$

$$\Rightarrow \phi'(t)=f(t)\ \forall\ t\in[a,b] \text{ as } c \text{ is an arbitrary point of } [a,b].$$

The same can be established by taking $h<0$ so that $c-h\in[a,b]$.

Remark. The continuity of f on $[a,b]$ is only a sufficient condition for a function to be a derivative of a function f on $[a,b]$ but is not a necessary condition.

We consider the following example.

Example. We define a function $f:[0,1]\to\mathbb{R}$ by

$$f(x)=\begin{cases}3x^2\sin\frac{1}{x^2}-2\cos\frac{1}{x^2}, & x\neq 0\\ 0, & x=0\end{cases}$$

Also let $F(x)=x^3\sin\frac{1}{x^2}, x\neq 0$, $F(x)=0, x=0$. Then $F'(x)=f(x)$ for $x\neq 0$.

Again,

$$F'(0)=\lim_{x\to 0}\frac{F(x)-F(0)}{x}=\lim_{x\to 0}\frac{x^3\sin\frac{1}{x^2}}{x}$$

$$=\lim_{x\to 0}x^2\sin\frac{1}{x^2}=0 \quad \left[\because \sin\frac{1}{x^2} \text{ is bounded}\right]$$

So $F'(0)$ exists and $F'(0)=f(0)$. Thus $F'(x)=f(x)\ \forall\ x\in[0,1]$.

But f is not continuous at $x=0$. Hence the condition is not necessary.

Proof. (iii) Let $h>0$ and h is so chosen that $t_0+h\in[a,b]$. We denote $[t_0,t_0+h]$ by δh and let M_0, m_0 be the sup and inf of f on $[t_0,t_0+h]$.

Then $|f(x)-f(t_0)|\leq M_0-m_0=0(f,\delta h)=$ oscillation of f on δh.

$$\Rightarrow f(t_0)-0(f,\delta h)\leq f(x)\leq f(t_0)+0(f,\delta h)$$

$$\Rightarrow h[f(t_0)-0(f,\delta h)]\leq\int_{t_0}^{t_0+h} f(x)dx\leq h[f(t_0)+0(f,\delta h)] \text{ [by 2(i) of Th. [7.9].]}$$

$$\Rightarrow f(t_0)-0(f,\delta h)\leq\frac{\phi(t_0+h)-\phi(t_0)}{h}\leq f(t_0)+0(f,\delta h)\ [h\neq 0] \qquad \text{(i)}$$

Now, since f is continuous at $x=t_0$, $\lim_{h\to 0}(f,\delta h)=0$.

So taking limit as $h\to 0$ in (1), we get

$$f(t_0)\leq\phi'(t_0)\leq f(t_0)\Rightarrow\phi'(t_0)=f(t_0).$$

The same can be established if $h<0$ proceeding as above where $\delta_1=[t_0-h,t_0]$.

Remark. It follows from (ii) and (iii) that $\phi'(t)=f(t)\Rightarrow\phi'=f\Rightarrow$ continuity of f on $[a,b]$ $\Rightarrow$ differentiability of ϕ on $[a,b]$.

Definition [7.6] : [Primitive of a function] Let f be a function defined on $[a, b]$. If there exists a differentiable function ϕ such that $\phi'(x) = f(x) \ \forall \ x \in [a, b]$, then ϕ is called the primitive of f.

Remark. It follows from Th. [7.12] that every continuous function f has a primitive $\int_a^t f(x)dx$.

Example. Let $f : [0, 1] \to \mathbb{R}$ be defined by

$$f(x) = \begin{cases} 2x\cos\frac{1}{x} + \sin\frac{1}{x}, & x \neq 0 \\ 0, & x = 0. \end{cases}$$

Then f has a primitive ϕ, where

$$\phi(x) = \begin{cases} x^2\cos\frac{1}{x}, & x \neq 0 \\ 0, & x = 0 \end{cases}$$

Since $\phi'(x) = f(x) \ \forall \ x \in [0, 1]$.

Remarks

(1) The above example shows that though f is not continuous on $[0, 1]$, $\phi'(x)$ exists. So it is to be noted that the derivative of a function is not necessarily continuous.

(2) Primitive of a function is not unique. For if ϕ be primitive of f, $\phi + c$ where c is constant is a primitive of f.

Theorem [7.13] [Fundamental theorem of integral calculus] If a function $f : [a, b] \to \mathbb{R}$ be bounded and integrable on $[a, b]$ and if there exists a differentiable function ϕ such that $\phi'(x) = f(x) \ \forall \ x \in [a, b]$, then $\int_a^b f(x)dx = \phi(b) - \phi(a)$.

Proof. We consider a partition $P\{a = x_0 < x_1 \cdots < x_{r-1} < x_r \cdots < x_n = b\}$ of $[a, b]$, with the norm $||P|| \leq \delta, \delta > 0$.

ϕ is differentiable on $[a, b] \Rightarrow \phi$ is differentiable on every subinterval $[x_{r-1}, x_r]$ of $[a, b]$.

Also ϕ, being a differentiable function, is continuous on $[a, b]$ and hence it is continuous on $[x_{r-1}, x_r]$.

Therefore, applying Lagrange's Mean Value theorem to ϕ on $[x_{r-1}, x_r]$, we have

$$\phi(x_r) - \phi(x_{r-1}) = (x_r - x_{r-1})\phi'(\xi_r), \ x_{r-1} < \xi_r < x_r \qquad \text{(i)}$$

$$\Rightarrow \sum_{r=1}^{n}\{\phi(x_r) - \phi(x_{r-1})\} = \sum_{r=1}^{n}\phi'(\xi_r)\delta_r = \sum_{r=1}^{n} f(\xi_r)\delta_r$$

$$\Rightarrow \phi(x_1) - \phi(x_0) + \phi(x_2) - \phi(x_1) + \cdots + \phi(x_n) - \phi(x_{n-1}) = \sum_{r=1}^{n} f(\xi_r)\delta_r$$

$$\Rightarrow \phi(x_n) - \phi(x_0) = \sum_{r=1}^{n} f(\xi_r)\delta_r$$

$$\Rightarrow \phi(b) - \phi(a) = \lim_{||P|| \to 0}\sum_{r=1}^{n} f(\xi_r)\delta_r = \int_a^b f(x)dx \text{ [by Th. [7.6].]}$$

Example. From the example of Definition [7.6], f is not continuous at $x = 0$ but is continuous on $(0, 1)$. Thus f, having a finite number of points of discontinuities on $[0, 1]$, is R-integrable on $[0, 1]$.

Also $\int_0^1 f(x)dx = \phi(1) - \phi(0) = \cos 1$.

Remarks

(1) A function may have a primitive but not an integral.

For example, $f(x) = 2x \sin \frac{1}{x^2} - \frac{2}{x} \cos \frac{1}{x^2}$, $x \neq 0$ and $f(0) = 0$ has a primitive $\phi(x) = x^2 \sin \frac{1}{x^2}$, but f is not R-integrable on $[-1, 1]$ as it is unbounded at $x = 0$.

(2) A function f may be R-integrable but has no primitive.

For example, we define a function f, by $f(x) = \begin{cases} x \sin \frac{1}{x}, & x \neq 0 \\ 0, & x = 0 \end{cases}$

f is continuous on $[0, 1]$ and hence it is R-integrable on $[0, 1]$.

But f has no primitive. That is, there does not exist a function ϕ such that $\phi' = f$.

Problem [7.17] : If $f(x) = x \cos x$ on $[0, \pi/2]$, then find the primitive of f.

Solution. Here f is continuous on $[0, \pi/2]$ and, therefore, f is R-integrable on $[0, \pi/2]$.

Hence primitive of f is given by

$$\phi(t) = \int_0^t f(x)dx = \int_0^t x \cos x dx$$

$$= (x \sin x)_0^t - (-\cos x)_0^t = t \sin t + \cos t - 1$$

$\therefore \phi(x) = x \sin x + \cos x - 1$ is the primitive of $f(x)$.

Problem [7.18] : If $f(x) = 2 \cos x + e^x$, then find the primitive of f on a closed interval $[a, b]$ and hence evaluate $\int_a^b f(x)dx$.

Solution. f is continuous on $[a, b]$ and hence $\int_a^b f(x)dx$ exists. Primitive of f is given by

$$\phi(t) = \int_a^t f(x)dx = \int_a^t (2 \cos x + e^x)dx$$

$$= [2 \sin x + e^x]_a^t = (2 \sin x + e^t) - (2 \sin a + e^a)$$

That is, $\phi(x) = (2 \sin x + e^x) - (2 \sin a + e^a)$ [or simply $(2 \sin x + e^x]$

$$\int_a^b f(x)dx = \phi(b) - \phi(a) = (2 \sin b + e^b) - (2 \sin a + e^a)$$

$$= 2(\sin b - \sin a) + (e^b - e^a)$$

Theorem [7.14] [First Mean Value theorem] If $f : [a, b] \to \mathbb{R}$ and $g : [a, b] \to \mathbb{R}$ be both bounded and R-integrable on $[a, b]$ and g preserves the same sign on $[a, b]$, then there exists a number μ lying between m and M, the inf and sup of f on $[a, b]$ such that

$$\int_a^b (fg)(x)dx = \mu \int_a^b g(x)dx.$$

If further f be continuous on $[a, b]$, then there exists a number $\xi \in [a, b]$ such that

$$\int_a^b (f \cdot g)(x)dx = f(\xi) \int_a^b g(x)dx.$$

Proof. The existence of $\int_a^b f(x)dx$ and $\int_a^b g(x)dx \Rightarrow$ the existence of

$$\int_a^b (fg)(x)dx, \text{ [by Th. [7.10), 2(ii)]}$$

Now let $g(x) > 0 \ \forall \ x \in [a, b]$. We have $m \leq f(x) \leq M$

$$\Rightarrow \ mg(x) \leq f(x)g(x) \leq Mg(x)$$

$$\Rightarrow \ m \int_a^b g(x)dx \leq \int_a^b f(x)g(x)dx \leq M \int_a^b g(x)dx, \ b \geq a$$

$$\Rightarrow \int_a^b f(x)g(x)dx = \mu \int_a^b g(x)dx, \text{ when } m \leq \mu \leq M \tag{i}$$

Now if $b \leq a$, i.e., if $a \geq b$, we have from (i)

$$\int_b^a f(x)g(x)dx = \mu \int_b^a g(x)dx \Rightarrow \int_a^b f(x)g(x)dx = \mu \int_a^b g(x)dx \tag{ii}$$

So from (i) and (ii), it follows that $\int_a^b f(x)g(x)dx = \mu \int_a^b g(x)dx$

$$\therefore \int_a^b (f \cdot g)(x)dx = \mu \int_a^b g(x)dx$$

Next, let $g(x) < 0$. Then $\phi(x) = -g(x) > 0$. Therefore, by above result

$$\int_a^b f(x)[\phi(x)]dx = \mu \int_a^b \phi(x)dx$$

$$\Rightarrow \int_a^b f(x)\{-g(x)\}dx = \mu \int_a^b -g(x)dx$$

$$\Rightarrow \int_a^b f(x)g(x)dx = \mu \int_a^b g(x)dx$$

When $g(x) = 0$, the case is trivial.

If further, f is continuous on $[a, b] \Rightarrow f$ assumes every value between m and $M \Rightarrow$ there exists a number $\xi \in [a, b]$ such that $f(\xi) = \mu$.

Hence, $\int_a^b (f \cdot g)(x)dx = f(\xi) \int_a^b g(x)dx$.

Problem [7.19] : Verify the First Mean Value Theorem for

(i) $f(x) = x, g(x) = x^2 \ \forall \ x \in [-1, 1]$

(ii) $f(x) = x, g(x) = e^x \ \forall \ x \in [-1, 1]$.

Solution. (i) f, g are both continuous on $[-1, 1]$. So f, g are both R-integrable on $[-1, 1]$ and hence $f \cdot g$ is R-integrable on $[-1, 1]$ and $g(x) > 0 \ \forall \ x \in [-1, 1]$.

Also $f(-1) = -1 = m$, inf of f on $[-1, 1]$

$f(1) = 1 = M$, sup of f on $[-1, 1]$.

Now $\int_{-1}^1 (f \cdot g)(x)dx = \int_{-1}^1 x^3 dx = 0$, while $\int_{-1}^1 g(x)dx = \int_{-1}^1 x^2 dx = \frac{2}{3}$.

Then $\int_{-1}^1 (fg)(x)dx = f(\xi) \int_{-1}^1 g(x)dx$, when $\xi = 0 \in [-1, 1]$, since $f(0) = 0$.

(ii) f, g are both continuous on $[-1, 1]$. So f, g are both R-integrable on $[-1, 1]$ and hence $(f \cdot g)$ is R-integrable on $[-1, 1]$.

Also $g(x) = e^x > 0 \ \forall \ x \in [-1, 1]$

$f(-1) = -1, \ m = -1$, inf of f on $[-1, 1]$

$f(1) = 1, \ M = 1$, sup of f on $[-1, 1]$.

Now

$$\int_{-1}^{1}(fg)(x)dx = \int_{-1}^{1} f(x)g(x)dx = \int_{-1}^{1} xe^x dx$$

$$= \left[xe^x - e^x\right]_{-1}^{1} = \left[(e-e) - \left(-\frac{1}{e} - \frac{1}{e}\right)\right] = \frac{2}{e}$$

Also

$$\int_{-1}^{1}(g)(x)dx = \int_{-1}^{1} e^x dx = \left[e^x\right]_{-1}^{1} = e - \frac{1}{e} = \frac{e^2-1}{e}.$$

Then $\int_{-1}^{1}(fg)(x)dx = f(\xi)\int_{-1}^{1} g(x)dx$

$$\Rightarrow \frac{2}{e} = \xi \cdot \frac{e^2-1}{e}$$

$$\Rightarrow \xi = \frac{2}{e^2-1} \in [-1,1], \text{ since } e > 2 \Rightarrow e^2 > 4 \Rightarrow e^2 - 1 > 3 \Rightarrow \xi < \frac{2}{3} \in [-1,1].$$

Hence, in both cases, First Mean Value Theorem is applicable.

Remark. If $f(x) = e^x$ and $g(x) = x \;\forall\; x \in [-1,1]$. Then First Mean Value Theorem is not applicable on $[-1,1]$, since $g(x)$ does not preserve the same sign on $[-1,1]$ as $g(x) < 0$ on $[-1,0]$ and $g(x) > 0$ on $[0,1]$.

Lemma [7.2]: [Abel's lemma] If

(i) $\{a_1, a_2, \cdots, a_n\}$ is a monotonically decreasing set of n (+)ve numbers

(ii) $v_1, v_2, \cdots, v_n$ are any n-numbers

(iii) k and K are two numbers such that $k < \sum_{i=1}^{p} v_i < K,\; 1 \le p \le n$, and $s_p = \sum_{i=1}^{p} v_i$, then

$$a_1 k < \sum_{r=1}^{n} a_r v_r < a_1 K.$$

Proof.

$$\sum_{r=1}^{n} a_r v_r = a_1 v_1 + a_2 v_2 + \cdots + a_n v_n$$

$$= a_1 s_1 + a_2(s_2 - s_1) + \cdots + a_n(s_n - s_{n-1}) \;[\because\; s_p = v_1 + v_2 + \cdots + v_p]$$

$$= (a_1 - a_2)s_1 + (a_2 - a_3)s_2 + \cdots + (a_{n-1} - a_n)s_{n-1} + a_n s_n$$

By (i) $a_{i-1} - a_i > 0$ for $i = 2, 3, \ldots, n, a_n > 0$ and by (iii) $\forall\; p \le n,\; k < s_p < K$, it follows that

$$\sum_{r=1}^{n} a_r v_r < (a_1 - a_2)K + (a_2 - a_3)K + \cdots + (a_{n-1} - a_n)K + a_n K = a_1 K$$

Also $\sum_{r=1}^{n} a_r v_r > (a_1 - a_2)k + (a_2 - a_3)k + \cdots + (a_{n-1} - a_n)k + a_n k = a_1 k$.

Hence $a_1 k < \sum_{r=1}^{n} a_r v_r < a_1 K$.

Theorem [7.15] [Second Mean Value Theorem, Bonnet's form] If f and ψ be both bounded and R-integrable on $[a,b]$, ψ is monotonicaly decreasing and $\psi(x) > 0 \;\forall\; x \in [a,b]$. Then there exists a number $\xi \in [a,b]$ such that $\int_a^b f(x)\psi(x)dx = \psi(a)\int_a^{\xi} f(x)dx$.

Proof. Let $P\{a = x_0 < x_1 \cdots < x_{r-1} < x_r \cdots < x_n = b\}$ be a partition of $[a, b]$ with the norm $||P|| \le \delta$.

Let $\xi_1 = a$ and $\xi_r \in [x_{r-1}, x_r]$, when $r \ne 1$. Then applying the result 2(i) of Theorem [7.9] in $[x_{r-1}, x_r]$

$$m_r\delta_r \le \int_{x_{r-1}}^{x_r} f(x)dx \le M_r\delta_r \tag{i}$$

Also

$$m_r\delta_r \le f(\xi_r)\delta_r \le M_r\delta_r \tag{ii}$$

From (i)

$$\sum_{r=1}^{p} m_r\delta_r \le \sum \int_{x_{r-1}}^{x_r} f(x)dx \le \sum_{r=1}^{p} M_r\delta_r,\ p \le n$$

$$\Rightarrow \sum_{r=1}^{p} m_r\delta_r \le \left[\int_{x_0}^{x_1} f(x)dx + \int_{x_1}^{x_2} f(x)dx + \cdots + \int_{x_{p-1}}^{x_p} f(x)dx\right] \le \sum_{r=1}^{p} M_r\delta_r$$

$$\Rightarrow \sum_{r=1}^{p} m_r\delta_r \le \int_a^{x_p} f(x)dx \le \sum_{r=1}^{p} M_r\delta_r \tag{iii}$$

From (ii)

$$\sum_{r=1}^{p} m_r\delta_r \le \sum_{r=1}^{p} f(\xi_r)\delta_r \le \sum_{r=1}^{p} M_r\delta_r \tag{iv}$$

Hence from (iii) and (iv), we have

$$\left|\int_a^{x_p} f(x)dx - \sum_{r=1}^{p} f(\xi_r)\delta_r\right| \le \sum_{r=1}^{p}(M_r - m_r)\delta_r \le \sum_{r=1}^{n}(M_r - m_r)\delta_r = \sum_{r=1}^{n} O_r\delta_r$$

where $O_r = M_r - m_r =$ oscillation of f in $[x_{r-1}, x_r]$

$$\Rightarrow \int_a^{x_p} f(x)dx - \sum_{r=1}^{n} O_r\delta_r \le \sum_{r=1}^{p} f(\xi_r)\delta_r \le \int_a^{x_p} f(x)dx + \sum_{r=1}^{n} O_r\delta_r \tag{v}$$

Now $\phi(x_p) = \int_a^{x_p} f(x)dx$ is continuous on $[a, b]$ by (1) of Theorem [7.12].

Therefore, it is bounded and let A and B be the sup and inf of $\phi(x_p)$ on $[a, b]$, so that

$$B \le \int_a^{x_p} f(x)dx \le A$$

$$\Rightarrow B - \sum_{r=1}^{n} O_r\delta_r \le \int_a^{x_p} f(x)dx - \sum_{r=1}^{n} O_r\delta_r \le A + \sum_{r=1}^{n} O_r\delta_r \tag{vi}$$

Combining (v) and (vi), we get

$$B - \sum_{r=1}^{n} O_r\delta_r \le \sum_{r=1}^{p} f(\xi_r)\delta_r \le A + \sum_{r=1}^{n} O_r\delta_r \tag{vii}$$

Now to apply Abel's Lemma [7.2], we take

$$a_r = \psi(\xi_r), v_r = f(\xi_r)\delta_r, k = B - \sum_{r=1}^{n} O_r\delta_r \text{ and } K = A + \sum_{r=1}^{n} O_r\delta_r$$

and then we have

$$\psi(\xi_1)\left(B-\sum_{r=1}^{n}O_r\delta_r\right)\le\sum_{r=1}^{n}f(\xi_r)\psi(\xi_r)\delta_r\le\psi(\xi_1)\left(A+\sum_{r=1}^{n}O_r\delta_r\right)$$

Taking $||P||\to 0$,

$$\psi(a)\cdot B\le\lim_{||P||\to 0}\sum f(\xi_r)\psi(\xi_r)\delta_r\le\psi(a)\cdot A\quad\left[\begin{array}{l}\because \xi_1=a \text{ and } \sum O_r\delta_r\to 0\\ \text{as } ||P||\to 0\end{array}\right]$$

$\Rightarrow \int_a^b f(x)\psi(x)dx=\mu\psi(a)$ when $B\le\mu\le A$.

Here $f\psi$ is R-integrable since $f\psi$ is the product of two R-integrable functions f and ψ, as ψ being a monotonically decreasing function is R-integrable.

The continuous function $\int_a^{x_p} f(x)dx$ assumes every value between its bounds A and B.

Hence there exists a number $\xi\in[a,b]$ such that $\int_a^{\xi} f(x)dx=\mu$.

Therefore, $\int_a^b f(x)\psi(x)dx=\psi(a)\int_a^{\xi} f(x)dx$.

Note. $\lim_{||P||\to 0}\sum M_r\delta_r=\overline{\int_a^b} f(x)dx$ and $\lim_{||P||\to 0}\sum m_r\delta_r=\underline{\int_a^b} f(x)dx$.

Thus $\lim_{||P||\to 0}\sum(M_r-m)\delta_r=\overline{\int_a^b} f(x)dx-\underline{\int_a^b} f(x)dx=\int_a^b f(x)dx-\int_a^b f(x)dx=0$ as f is R-integrable on $[a,b]$.

Theorem [7.16] [Second Mean Value Theorem, Weierstrass's form] If f and g be both bounded and R-integrable on $[a,b]$, g is monotonic on $[a,b]$, then there exists a number $\xi\in[a,b]$ such that $\int_a^b f(x)g(x)dx=g(a)\int_a^{\xi} f(x)dx+g(b)\int_{\xi}^b f(x)dx$.

Proof. Let g be monotonically decreasing on $[a,b]$.

Then $\psi(x)=g(x)-g(b)$ is monotonically decreasing and $\psi(x)>0\ \forall\ x\in[a,b]$.

Therefore, by Theorem [7.15], there exists a number $\xi\in[a,b]$ such that

$$\int_a^b f(x)\psi(x)dx=\psi(a)\int_a^{\xi} f(x)dx$$

$$\Rightarrow \int_a^b f(x)\{g(x)-g(b)\}dx=\{g(a)-g(b)\}\int_a^{\xi} f(x)dx$$

$$\Rightarrow \int_a^b f(x)g(x)dx=g(b)\int_a^b f(x)dx-g(b)\int_a^{\xi} f(x)dx+g(a)\int_a^{\xi} f(x)dx$$

$$=g(a)\int_a^{\xi} f(x)dx+g(b)\left\{\int_a^b f(x)dx+\int_{\xi}^a f(x)dx\right\}$$

$$=g(a)\int_a^{\xi} f(x)dx+g(b)\int_{\xi}^b f(x)dx.$$

Next, let g be monotonically increasing on $[a,b]$.

Then $\psi(x)=g(b)-g(x)$ is monotonically decreasing and $\psi(x)>0\ \forall\ x\in[a,b]$.

Therefore by Theorem [7.15], there exists a number $\xi \in [a, b]$ such that

$$\int_a^b f(x)\psi(x)dx = \psi(a)\int_a^\xi f(x)dx$$

$$\Rightarrow \int_a^b f(x)\{g(b) - g(x)\}dx = \{g(b) - g(a)\}\int_a^\xi f(x)dx$$

$$\Rightarrow \int_a^b f(x)g(x)dx = g(b)\int_a^b f(x)dx - g(b)\int_a^\xi f(x)dx + g(a)\int_a^\xi f(x)dx$$

$$= g(a)\int_a^\xi f(x)dx + g(b)\left\{\int_a^b f(x)dx + \int_\xi^a f(x)dx\right\}$$

$$= g(a)\int_a^\xi f(x)dx + g(b)\int_\xi^b f(x)dx$$

Note. (1) $\psi(x) = g(x) - g(b)$. Let $x_1, x_2 \in [a, b]$ and $x_1 < x_2$.

Then $\psi(x_1) = g(x_1) - g(b)$, $\psi(x_2) = g(x_2) - g(b) \Rightarrow \psi(x_2) - \psi(x_1) = g(x_2) - g(x_1) < 0$, if g is monotonically decreasing and hence $\psi(x_2) < \psi(x_1) \Rightarrow \psi$ is decreasing.

(2) When $\psi(x) = g(b) - g(x)$, we can show in the same way that $\psi(x)$ is decreasing, if $g(x)$ is increasing.

Theorem [7.17] If g is monotonic and g', f are all continuous on $[a, b]$ and $g'(x) > 0 \ \forall \ x \in [a, b]$, then there exists a number $\xi \in [a, b]$ such that

$$\int_a^b f(x)g(x)dx = g(a)\int_a^\xi f(x)dx + g(b)\int_\xi^b f(x)dx$$

(A particular case of Weierstrass's form of Second Mean Value Theorem)

Proof. g is monotonic and f is continuous $\Rightarrow g$ and f are both R-integrable on $[a, b] \Rightarrow g \cdot f$ is R-integrable on $[a, b]$. So $\int_a^b (f \cdot g)(x)dx$ exists.

Let $F(x) = \int_a^x f(t)dt$.

Then $F(a) = 0$ and since f is continuous, $F(x)$ is differentiable and $F'(x) = f(x)$.

Therefore,

$$\int_a^b (fg)(x)dx = \int_a^b f(x)g(x)dx = \int_a^b g(x)F'(x)dx$$

$$= \left[g(x)F(x)\right]_a^b - \int_a^b F(x)g'(x)dx \quad \text{(i)}$$

on integration by parts.

$F(x)$ is continuous, g' is continuous and $g'(x) > 0 \ \forall \ x \in [a, b]$. Therefore, by First Mean Value Theorem, $\exists$ a number $\xi \in [a, b]$ such that

$$\int_a^b F(x)g'(x)dx = F(\xi)\int_a^b g'(x)dx$$

$$= F(\xi)\left[g(x)\right]_a^b = F(\xi)[g(b) - g(a)] \quad \text{(ii)}$$

Hence from (i) by (ii)

$$\begin{aligned}\int_a^b f(x)g(x)dx &= g(b)F(b) - F(\xi)[g(b) - g(a)]\\ &= g(a)F(\xi) + g(b)[F(b) - F(\xi)]\\ &= g(a)\int_a^{\xi} f(x)dx + g(b)\left[\int_a^b f(x)dx - \int_a^{\xi} f(x)dx\right]\\ &= g(a)\int_a^{\xi} f(x)dx + g(b)\int_{\xi}^b f(x)dx\end{aligned}$$

Problem [7.20]: Verify Second Mean Value Theorem of Bonnet's form on $[0,1]$ for $f(x) = x$ and $g(x) = \frac{1}{1+x^2}$.

Solution. f, g are both bounded and R-integrable on $[0,1]$. g is monotonically decreasing on $[0,1]$ and $g(x) > 0\ \forall\ x \in [0,1]$. Also $\int_0^1 (f \cdot g)(x)dx$ exists.

Now

$$\int_0^1 (f \cdot g)(x)dx = \int_0^1 \frac{x}{1+x^2}dx = \frac{1}{2}\left[\log(1+x^2)\right]_0^1 = \frac{1}{2}\log 2$$

Again, by Second Mean Value Theorem of Bonnet's form, there exists a number ξ such that

$$\int_0^1 \frac{x}{1+x^2}dx = g(0)\int_0^{\xi} f(x)dx = 1\int_0^{\xi} xdx = \frac{\xi^2}{2}$$

$$\Rightarrow \frac{1}{2}\log 2 = \frac{\xi^2}{2} \Rightarrow \xi^2 = \log 2 \Rightarrow \xi = \sqrt{\log 2} = \sqrt{0.3010} \Rightarrow \xi = 0.55 \text{ (approx)} \in [0,1]$$

Problem [7.21]: Show that for the validity of the Second Mean Value Theorem, g must be necessarily monotonic by showing that the theorem does not hold if

$$f(x) = x^2, g(x) = \cos x; \text{ on } \left[-\frac{\pi}{2}, \frac{\pi}{2}\right]$$

Solution. Here f, g are both bounded and R-integrable on $[-\pi/2, \pi/2]$, $g(x) > 0\ \forall\ x \in [-\pi/2, \pi/2]$, but g is not monotonic on $[-\pi/2, \pi/2]$ as g is increasing on $[-\pi/2, 0]$ and g is decreasing on $[0, \pi/2]$. Consequently, Second Mean Value Theorem is not applicable to the given functions on $[-\pi/2, \pi/2]$.

Also

$$\int_{-\pi/2}^{\pi/2} f(x)g(x)dx = \int_{-\pi/2}^{\pi/2} x^2 \cos x dx = 2\int_0^{\pi/2} x^2 \cos x dx > 0 \qquad (1)$$

as $x^2 \cos x > 0\ \forall\ x \in [0, \pi/2]$

$$\text{and } g(-\pi/2)\int_{-\pi/2}^{\xi} f(x)dx + g(\pi/2)\int_{\xi}^{\pi/2} f(x)dx$$

$$= \cos(-\pi/2)\int_{-\pi/2}^{\xi} x^2 dx + \cos(\pi/2)\int_{\xi}^{\pi/2} x^2 dx = 0 \qquad (2)$$

So from (1) and (2), it follows that the theorem holds only when g is monotonic on $[a, b]$.

Remark. The monotnic condition for the function g cannot be dropped.

Problem [7.22] : Show that

(i) $\left|\int_a^b \frac{\sin x}{x}dx\right| \leq \frac{2}{a}$, if $b > a > 0$. Also show that $\left|\int_a^b \frac{\sin x}{x}dx\right| < \frac{4}{a}$. (ii) $\left|\int_a^b \sin x^2 dx\right| \leq \frac{1}{a}$.

Solution. (i) Let $f(x) = \sin x$ and $g(x) = \frac{1}{x}$. Both are bounded and R-integrable on $[a, b]$, where $b > a > 0$.

g is monotonically decreasing on $[a, b]$ and $g(x) > 0\ \forall\ x \in [a, b]$.

So applying Second Mean Value Theorem of Bonnet's form,

$$\int_a^b f(x)g(x)dx = g(a)\int_a^\xi f(x)dx = \frac{1}{a}\int_a^\xi \sin x dx = \frac{1}{a}[-\cos\xi + \cos a]$$

$$\therefore \left|\int_a^b f(x)g(x)dx\right| \leq \frac{1}{a}[|\cos\xi| + |\cos a|] \leq \frac{1}{a}[1+1] = \frac{2}{a}$$

Now applying Second Mean Value Theorem of Weierstrass's from, we have

$$\begin{aligned}\left|\int_a^b f(x)g(x)dx\right| &= \left|g(a)\int_a^\xi f(x)dx + g(b)\int_\xi^b f(x)dx\right| \\ &= \left|\frac{1}{a}[-\cos\xi + \cos a] + \frac{1}{b}[-\cos b + \cos\xi]\right| \\ &\leq \frac{1}{a}\{|\cos\xi| + |\cos a|\} + \frac{1}{b}\{|\cos b| + |\cos\xi|\} \\ &\leq \frac{2}{a} + \frac{2}{b} < \frac{2}{a} + \frac{2}{a} = \frac{4}{a}\left[\because\ b > a \Rightarrow \frac{1}{b} < \frac{1}{a}\right]\end{aligned}$$

(ii) [*Hints.* Consider $\int_a^b \frac{2x\sin x^2}{2x}dx$ and proceed as in (1) by taking $f(x) = 2x\sin x^2$ and $g(x) = \frac{1}{2x}$.]

Problem [7.23] :

(i) If a function f be continuous on $[0, 1]$, show that $\lim\limits_{n\to\infty} \int_0^1 \frac{nf(x)}{1+n^2x^2}dx = \frac{\pi}{2}f(0)$.

(ii) if f be continuous on $[-1, 1]$, then show that $\lim\limits_{h\to 0} \int_{-1}^1 \frac{hf(x)}{1+h^2x^2}dx = 2\pi f(0)$.

(iii) if f be continuous on $[a, b]$, then prove that $\int_a^b [f(x)]^2 dx = f(\alpha)f(\beta)(b-a)$ for some points $\alpha, \beta \in [a, b]$.

(iv) Prove that $\lim\limits_{x\to\infty} e^{-x^2}\int_0^x e^{t^2}dt = 0$.

(v) Prove that $\frac{2\pi^2}{9} \leq \int_{\pi/6}^{\pi/2} \frac{2x}{\sin x}dx \leq \frac{4\pi^2}{9}$.

(vi) Prove that $\frac{\pi}{6} > \int_0^{1/2} \frac{dx}{\sqrt{1-x^{2n}}} > \frac{1}{2}\ \forall\ n > 1$.

Solution. (i) Let M be the sup of f on $[0, 1]$. So $f(x) \leq M\ \forall\ x \in [0, 1]$.

We divide $[0, 1]$ into two parts $\left[0, \frac{1}{\sqrt{n}}\right]$ and $\left[\frac{1}{\sqrt{n}}, 1\right]$.

Therefore,

$$\int_0^1 \frac{nf(x)dx}{1+n^2x^2} = \int_0^{\frac{1}{\sqrt{n}}} \frac{nf(x)dx}{1+n^2x^2} + \int_{\frac{1}{\sqrt{n}}}^1 \frac{nf(x)dx}{1+n^2x^2}.$$

Applying First Mean Value Theorem to both parts, we have

$$\int_0^1 \frac{nf(x)dx}{1+n^2x^2} = f(\alpha_n)\int_0^{\frac{1}{\sqrt{n}}} \frac{ndx}{1+n^2x^2} + f(\beta_n)\int_{\frac{1}{\sqrt{n}}}^1 \frac{ndx}{1+n^2x^2}$$

where $0 \le \alpha_n \le \frac{1}{\sqrt{n}}, \frac{1}{\sqrt{n}} \le \beta_n \le 1$

$$= f(\alpha_n)\big[\tan^{-1} nx\big]_0^{1/\sqrt{n}} + f(\beta_n)\big[\tan^{-1} nx\big]_{1/\sqrt{n}}^1 \qquad (1)$$

Now $n \to \infty \Rightarrow f(\alpha_n) \to f(0)$.

So the first part of (1) $f(\alpha_n)[\tan^{-1}\sqrt{n}] \Rightarrow f(0)\frac{\pi}{2}$ as $n \to \infty$.

Now, the second part of (1) $= f(\beta_n)[\tan^{-1} n - \tan^{-1}\sqrt{n}]$

$$\le M[\tan^{-1} n - \tan^{-1}\sqrt{n}] \Rightarrow 0 \text{ as } n \to \infty$$

Hence, $\lim_{n\to\infty}\int_0^1 \frac{nf(x)dx}{1+n^2x^2} = \frac{\pi}{2}f(0)$.

(ii) We divide $[-1,1]$ into 3 subintervals $[-1,-\sqrt{h}], [-\sqrt{h},\sqrt{h}], [\sqrt{h},1]$. Then

$$\int_{-1}^1 \frac{hf(x)dx}{1+h^2x^2} = \int_{-1}^{-\sqrt{h}} \frac{hf(x)dx}{1+h^2x^2} + \int_{-\sqrt{h}}^{\sqrt{h}} \frac{hf(x)dx}{1+h^2x^2} + \int_{\sqrt{h}}^1 \frac{hf(x)}{1+h^2x^2}dx.$$

Applying First Mean Value theorem to each part, we have

$$\int_{-1}^1 \frac{hf(x)dx}{1+h^2x^2} = f(\alpha_h)\int_{-1}^{-\sqrt{h}} \frac{hdx}{1+h^2x^2} + f(\beta_h)\int_{-\sqrt{h}}^{\sqrt{h}} \frac{hdx}{1+h^2x^2} + f(\gamma_h)\int_{\sqrt{h}}^1 \frac{hdx}{1+h^2x^2}$$

where $-1 \le \alpha_h \le -\sqrt{h}, -\sqrt{h} \le \beta_h \le \sqrt{h}, \sqrt{h} \le \gamma_h \le 1$

$$= f(\alpha_h)\big[\tan^{-1} hx\big]_{-1}^{-\sqrt{h}} + f(\beta_h)\big[\tan^{-1} hx\big]_{-\sqrt{h}}^{\sqrt{h}} + (f(\gamma_h)\big[\tan^{-1} hx\big]_{\sqrt{h}}^1$$

$$M = \sup f \Rightarrow f(x) \le M \;\forall\; x \in [-1,1].$$

Also $h \to 0 \Rightarrow f(\beta_h) \to f(0)$.

First part, $\lim_{h\to 0} f(\alpha_h)\int_{-1}^{-\sqrt{h}} \frac{hdx}{1+h^2x^2} \le \lim_{h\to 0}\{M[-\tan^{-1} h^{3/2} + \tan^{-1} h]\} = 0$.

Third part, $\lim_{h\to 0} f(\gamma_h)\big[\tan^{-1} hx\big]_{\sqrt{h}}^1 \le M\big[\tan^{-1} h - \tan^{-1} h^{3/2}\big] = 0$.

Therefore, $\lim_{h\to 0}\int_{-1}^1 \frac{hf(x)dx}{1+h^2x^2} = \lim_{h\to 0} f(\beta_h)[2\cdot\tan^{-1} h^{3/2}] = 2\pi f(0)$.

(iii) Continuity of f on $[a,b] \Rightarrow R$-integrability of f on $[a,b] \Rightarrow R$-integrability of $f \cdot f = f^2$ on $[a,b]$.

So $\int_a^b f(x)\cdot f(x)dx = \int_a^b [f(x)]^2 dx$ exists.

Applying first Mean Value Theorem to $\int_a^b f(x)\cdot f(x)dx$, we have

$$\int_a^b [f(x)]^2 dx = f(\alpha)\int_a^b f(x)dx \text{ for same } \alpha \in [a,b] \qquad (1)$$

Again, applying the result 2 (iv) of Theorem [7.9], $\exists$ a point $\beta \in [a,b]$ such that $\int_a^b f(x)dx = f(\beta)(b-a)$, since f is continuous on $[a,b]$.

Hence from (1), $\int_a^b [f(x)]^2 dx = f(\alpha)f(\beta)(b-a)$, for some $\alpha, \beta \in [a,b]$.

(iv) $\lim_{x\to\infty} e^{-x^2}\int_0^x e^{t^2}dt = \lim_{x\to\infty}\frac{\int_0^x e^{t^2}dt}{e^{x^2}} = \lim_{x\to\infty}\frac{F(x)}{e^{x^2}}\cdot\left(\frac{\infty}{\infty}\right)$, where $F(x)=\int_0^x e^{t^2}dt$.

Here $\lim_{x\to\infty} F(x)=\infty$. e^{t^2} is continuous $\forall\ t\geq 0$. So $F'(x)=e^{x^2}$.

Therefore, by L. Hospital's rule,

$$\lim_{x\to\infty} e^{-x^2}\int_0^x e^{t^2}dt = \lim_{x\to\infty}\frac{F'(x)}{2xe^{x^2}} = \lim_{x\to\infty}\frac{e^{x^2}}{2xe^{x^2}} = \lim_{x\to\infty}\frac{1}{2x}=0$$

(v) $\sin x$ is an increasing function in $\left[\frac{\pi}{6},\frac{\pi}{2}\right]$. So the minimum value of $\sin x$ is $\sin\frac{\pi}{6}=\frac{1}{2}$ and the maximum value of $\sin x$ is $\sin\frac{\pi}{2}=1$.

Thus $\frac{1}{2}\leq \sin x\leq 1$

$$\Rightarrow\ 2\geq\frac{1}{\sin x}\geq 1\Rightarrow 1\leq\frac{1}{\sin x}\leq 2\ \Rightarrow\ 2x\leq\frac{2x}{\sin x}\leq 4x$$

$$\Rightarrow\ \int_{\pi/6}^{\pi/2} 2x\leq\int_{\pi/6}^{\pi/2}\frac{2x}{\sin x}dx\leq 4\int_{\pi/6}^{\pi/2} xdx\Rightarrow\left[x^2\right]_{\pi/6}^{\pi/2}\leq\int_{\pi/6}^{\pi/2}\frac{2x}{\sin x}dx\leq 2\left[x^2\right]_{\pi/6}^{\pi/2}$$

$$\Rightarrow\ \left[\frac{\pi^2}{4}-\frac{\pi^2}{36}\right]\leq\int_{\pi/6}^{\pi/2}\frac{2x}{\sin x}dx\leq 2\left[\frac{\pi^2}{4}-\frac{\pi^2}{36}\right]\Rightarrow\frac{2\pi^2}{9}\leq\int_{\pi/6}^{\pi/2}\frac{2x}{\sin x}dx\leq\frac{4\pi^2}{9}$$

(vi) At $x=0$, the inequality reduces to equality. So we consider x for $0<x\leq\frac{1}{2}$

$$x^2\leq\frac{1}{4}\Rightarrow x^{2n}<x^2\ \forall\ n>1\Rightarrow 1-x^{2n}>1-x^2\ \forall\ n>1\ \Rightarrow\ \sqrt{1-x^{2n}}>\sqrt{1-x^2}$$

$$\Rightarrow\ \frac{1}{\sqrt{1-x^{2n}}}<\frac{1}{\sqrt{1-x^2}} \qquad (1)$$

Also

$$x^{2n}<x^2\Rightarrow x^{2n}<1\Rightarrow 1-x^{2n}<1\Rightarrow\frac{1}{\sqrt{1-x^{2n}}}>1\ [\because x^2<1] \qquad (2)$$

So from (1) and (2), $1<\frac{1}{\sqrt{1-x^{2n}}}<\frac{1}{\sqrt{1-x^2}}$.

Each function being continuous on $[0,1/2]$ is R-integrable on $[0,1/2]$

$$\Rightarrow\ 1\int_0^{1/2} 1dx<\int_0^{1/2}\frac{1}{\sqrt{1-x^{2n}}}dx<\int_0^{1/2}\frac{1}{\sqrt{1-x^2}}dx$$

$$\Rightarrow\ \frac{1}{2}<\int_0^{1/2}\frac{1}{\sqrt{1-x^{2n}}}dx<\left[\sin^{-1}x\right]_0^{1/2}=\frac{\pi}{6}$$

Thus $\frac{1}{2}<\int_0^{1/2}\frac{1}{\sqrt{1-x^{2n}}}dx<\frac{\pi}{6}$.

Note. It follows that $0.5<\int_0^{1/2}\frac{1}{\sqrt{1-x^{2n}}}<0.524$.

Problem [7.24] : If f be bounded and R-integrable on $[a,b]$, then show that

$$\lim_{n\to\infty}\int_a^b f(x)\sin nxdx=0.$$

Solution. Let $I_n=\int_a^b f(x)\sin nxdx$.

Since f is R-integrable on $[a,b]$, there exists a partition

$$P\{a=x_0<x_1\cdots<x_{r-1}<x_r\cdots<x_n=b\}$$

of $[a, b]$ with norm $||P|| \leq \delta$ such that $\omega(P, f) = \sum_{r=1}^{n} O_r \delta_r < \frac{\epsilon}{2}$

$$I_n = \int_a^b f(x) \sin nx dx = \sum_{r=1}^{n} \int_{x_{r-1}}^{x_r} f(x) \sin nx dx$$

$$= \sum_{r=1}^{n} \int_{x_{r-1}}^{x_r} [\{f(x) - f(x_{r-1})\} + f(x_{r-1})] \sin nx dx$$

$$= \sum_{r=1}^{n} \int_{x_{r-1}}^{x_r} \{f(x) - f(x_{r-1})\} \sin nx dx + \sum f(x_{r-1}) \int_{x_{r-1}}^{x_r} \sin nx dx$$

Then

$$|I_n| \leq \sum_{r=1}^{n} \int_{x_{r-1}}^{x_r} |f(x) - f(x_{r-1}) \sin nx| dx + \sum |f(x_{r-1})| \left| \int_{x_{r-1}}^{x_r} \sin nx dx \right|$$

$$\leq \sum_{r=1}^{n} \int_{x_{r-1}}^{x_r} |f(x) - f(x_{r-1})| \, |\sin nx| dx + \sum_{r=1}^{n} |f(x_{r-1})| \left| -\frac{\cos nx_r}{n} + \frac{\cos nx_{r-1}}{n} \right|$$

$$\leq \sum_{r=1}^{n} \int_{x_{r-1}}^{x_r} O_r dx + \sum_{r=1}^{n} |f(x_{r-1})| \left| \frac{1}{n} + \frac{1}{n} \right|$$

$$= \sum_{r=1}^{n} O_r \delta_r + \frac{2}{n} \sum_{r=1}^{n} |f(x_{r-1})| < \frac{\epsilon}{2} + \frac{2}{n} \sum_{r=1}^{n} |f(x_{r-1})| \quad (1)$$

If the partition P be kept fixed, then $\sum_{r=1}^{n} |f(x_{r-1})|$ remains fixed and

$$\lim_{n \to \alpha} \frac{2}{n} \sum_{r=1}^{n} |f(x_{r-1})| \to 0.$$

That is $\frac{2}{n} \sum_{r=1}^{n} |f(x_{r-1})| < \frac{\epsilon}{2} \; \forall \, n \geq m$.

Hence, from (1), $|I_n| < \frac{\epsilon}{2} + \frac{\epsilon}{2} = \epsilon \; \forall \, n \geq m \Rightarrow I_n \to 0$ as $n \to \infty$.

That is $\lim_{n \to \infty} \int_a^b f(x) \sin nx dx = 0$.

Corollary [7.5]: If f be bounded and R-integrable on $[a, b]$, then $\lim_{n \to \infty} \int_a^b f(x) \cos nx dx = 0$.

[*Hints.* Proceed as in the above problem.]

Note. The Problem [7.24] is known as Riemann Lebesgue Lemma.

Problem [7.25]: The function f and g are defined on $[0, \infty]$ by $f(x) = \lim_{n \to \infty} \frac{x^n + 2}{x^n + 1}$ and $g(x) = \int_0^x f(t) dt$. Then prove that g is continuous but not derivable at $x = 1$.

Solution.

$$f(x) = \begin{cases} 2, & \text{for } x = 0 \\ 2, & \text{for } 0 < x < 1 \\ 1, & \text{for } x > 1 \\ 3/2, & \text{for } x = 1 \end{cases}$$

So f is bounded on $[0, \infty)$. Also f has a point of discontinuity at $x = 1$. That is, f has a finite point of discontinuity on $[0, \infty)$. So f is R-integrable on $[0, \infty)$.

Then $g(x) = \int_0^x f(t)dt$ is an integral function of f.

Let $c \in [0, \infty)$ and $h(> 0)$ is so chosen that $c + h \subset [0, \infty)$.

Then $|g(c+h) - g(c)| = \left|\int_c^{c+h} f(t)dt\right| \leq k|h|$.

$$< k \cdot \frac{\epsilon}{k} = \epsilon \text{ when } |h| < \delta \text{ and taking } \delta = \frac{\epsilon}{k}.$$

$\Rightarrow g$ is continuous at $x = c$.

$\Rightarrow g$ is continuous on $[0, \infty)$ as c is an arbitrary point.

Again

$$\frac{g(c+h) - g(c)}{h} = \frac{1}{h}\int_c^{c+h} f(t)dt$$

$$\lim_{h \to 0+0} \frac{g(1+h) - g(1)}{h} = \lim_{h \to 0+0} \frac{1}{h}\int_1^{1+h} f(t)dt = \lim_{h \to 0+0} \frac{1}{h} \cdot \int_1^{1+h} 1 \cdot dt$$
$$= \lim_{h \to 0+0} \frac{1}{h} \cdot h = 1 \; [\because f(x) = 1, \text{ for } x > 1]$$

Again

$$\lim_{h \to 0-0} \frac{g(1+h) - g(1)}{h} = \lim_{h \to 0-0} \frac{1}{h}\int_1^{1+h} 2 \cdot dt \; [\because f(x) = 2, \text{ for } x < 1]$$
$$= 2$$

$\therefore Rg'(1) \neq Lg'(1)$

$\therefore$ $g(x)$ is not derivable at $x = 1$.

Problem [7.26] :

(1) Applying First Mean Value theorem of integral calculas show that

$$\tfrac{\pi}{6} < \int_0^{1/2} \frac{dx}{\sqrt{(1-x^2)(1-k^2x^2)}} < \frac{\pi}{6}\frac{1}{\sqrt{1-\frac{k^2}{4}}}, \; 0 < k < 1.$$

(2) If f and ϕ be both bounded and R-integrable in $[a, b]$, then show that

$$\left[\int_a^b f(x)\phi(x)dx\right]^2 \leq \int_a^b [f(x)]^2 dx \cdot \int_a^b [f(x)]^2 dx$$

when does the sign of equality hold?

(3) If f and g both integrable in $[a, b]$ and $\int_a^b f(x)dx = \int_a^b g(x)dx$, then $f(\xi) = g(\xi)$ for some $\xi \in [a, b]$.

Solution. (1) Let $f(x) = \dfrac{1}{\sqrt{1 - k^2x^2}}$ and $\phi(x) = \frac{1}{\sqrt{1-x^2}}$.

$\phi(x) > 0 \; \forall \; x \in \left[0, \frac{1}{2}\right]$ or $\phi(x) < 0 \; \forall \; x \in \left[0, \frac{1}{2}\right]$ if the negative sign is taken in $\sqrt{1 - x^2}$.

So ϕ keeps the same sign in $\left[0, \frac{1}{2}\right]$

$$m = \inf \text{ of } f = 1$$

$$M = \sup f = \frac{1}{\sqrt{1 - \frac{k^2}{4}}}, 0 < k < 1.$$

Applying First Mean Value Theorem of integral calculus

$$m\int_0^{\frac{1}{2}}\phi(x)dx < \int_0^{\frac{1}{2}} f(x)\phi(x)dx < M\int_0^{\frac{1}{2}}\phi(x)dx$$

$$\Rightarrow 1\int_0^{\frac{1}{2}}\frac{1}{\sqrt{1-x^2}}dx < \int_0^{\frac{1}{2}}\frac{1}{\sqrt{1-x^2}\sqrt{1-k^2x^2}}dx < \frac{1}{\sqrt{1-\frac{k^2}{4}}}\int_0^{\frac{1}{2}}\frac{1}{\sqrt{1-x^2}}dx$$

$$\Rightarrow [\sin^{-1}x]_0^{\frac{1}{2}} < \int_0^{\frac{1}{2}}\frac{1}{\sqrt{1-x^2}\sqrt{1-k^2x^2}}dx < \frac{1}{\sqrt{1-\frac{k^2}{4}}}[\sin^{-1}x]_0^{\frac{1}{2}}$$

$$\Rightarrow \frac{\pi}{6} < \int_0^{\frac{1}{2}}\frac{dx}{\sqrt{1-x^2}\sqrt{1-k^2x^2}} < \frac{\pi}{6}\cdot\frac{1}{\sqrt{1-\frac{k^2}{4}}}$$

(2) Let $P\{a = x_0 < x_1 < x_2 \cdots < x_{r-1} < x_r \cdots < x_n = b\}$ be a partition of $[a, b]$ where $\delta_r = x_r - x_{r-1}$ and $\xi_r \in [x_{r-1}, x_r]$ and $||P|| < \delta$.

$\int_a^b f(x)dx$, $\int_a^b \phi(x)dx$ and $\int_a^b f(x)\phi(x)dx$ all exist in $[a, b]$.

By definition,

$$\int_a^b f(x)dx = \lim_{||P||\to 0}\sum f(\xi_r)\delta_r$$

$$\text{and } \int_a^b \phi(x)dx = \lim_{||P||\to 0}\sum \phi(\xi_r)\delta_r$$

$$\text{and } \int_a^b f(x)\phi(x)dx = \lim_{||P||\to 0}\sum f(\xi_r)\phi(\xi_r)\delta_r$$

Then

$$\int_a^b [f(x)]^2dx = \lim_{||P||\to 0}\sum [f(\xi_r)]^2\delta_r = \lim_{||P||\to 0}\sum\left[f(\xi_r)\sqrt{\delta_r}\right]^2$$

$$\int_a^b [\phi(x)]^2dx = \lim_{||P||\to 0}\sum\left[\phi(\xi_r)\sqrt{\delta_r}\right]^2$$

$$\text{and } \left[\int_a^b f(x)\phi(x)dx\right]^2 = \left[\lim_{||P||\to 0}\sum f(\xi_r)\phi(\xi_r)\delta_r\right]^2$$

Let

$$a_r = f(\xi_r)\sqrt{\delta_r},\ b_r = \phi(\xi_r)\sqrt{\delta_r} \tag{1}$$

Then by Cauchy's inequality

$$\left(\sum a_r b_r\right)^2 \le \sum a_r^2 \cdot \sum b_r^2$$

$$\Rightarrow \left[\lim_{||P||\to 0}\sum f(\xi_r)\phi(\xi_r)\delta_r\right]^2 \le \lim_{||P||\to 0}\sum [f(\xi_r)]^2\delta_r \cdot \lim_{||P||\to 0}\sum [\phi(\xi_r)]^2\delta_r$$

$$\Rightarrow \left[\int_a^b f(x)\phi(x)dx\right]^2 \le \int_a^b [f(x)]^2dx \cdot \int_a^b [\phi(x)]^2dx$$

Equality sign holds when

$$\frac{a_1}{b_1} = \frac{a_2}{b_2} = \cdots$$

$$\Rightarrow \frac{f(\xi_1)}{\phi(\xi_1)} = \frac{f(\xi_2)}{\phi(\xi_2)} = \cdots \text{ [by (1)]}$$

$$\Rightarrow f, \phi \text{ are both constant functions.}$$

(3) f, g are both integrable in $[a, b]$
$\Rightarrow (f - g)$ is integrable in $[a, b]$
$\Rightarrow \int_a^b (f - g)(x)dx$ exists.
By First Mean Value Theorem of integral calculus

$$\int_a^b (f - g)(x)dx = (b - a)(f - g)\xi \text{ for some } \xi \in [a, b]$$

$$\Rightarrow \int_a^b f(x)dx - \int_a^b g(x)dx = (b - a)[f(\xi) - g(\xi)], \; b > a$$

$$\Rightarrow 0 = (b - a)[f(\xi) - g(\xi)]$$

$$\Rightarrow f(\xi) = g(\xi) \text{ for some } \xi \in [a, b].$$

EXERCISE

(1) Let $f : [a, b] \to \mathbb{R}$ be defined by

$$f(x) = \begin{cases} \dfrac{1}{q^2}, & \text{when } x = \dfrac{p}{q} \\ \dfrac{1}{q^3}, & \text{when } x = \sqrt{\dfrac{p}{q}}, \; (p, q) = 1 \end{cases}$$

and $f(x) = 0$ for all other values of $x \in [a, b]$. Show that f is R-integrable on $[a, b]$.

(2) If $f(x, y) = 1 + 2y$ for rational values of x and $f(x, y) = 0$ for irrational values of x, then examine $R \int_0^1 f(x, y)dy$ and evaluate it.

(3) If f be bounded and R-integrable on $[a, b]$, then show that $\int_a^b [f(x)]^2 dx = 0 \Leftrightarrow f(c) = 0$ for every point c of continuity of f on $[a, b]$.

(4) If $f(x) = 0 \; \forall \; x \in [0, 1]$ except at the set of points $x_1, x_2, \ldots, x_n, \ldots, \; n \in N$ and $f(x_n) = \dfrac{1}{\sqrt{n}}$, then show that f is R-integrable on $[0, 1]$.

(5) Prove that $\lim\limits_{x \to \infty} e^{-x^2} \int_0^x e^{t^2} dt = 0$.

(6) Show that

(i) $\dfrac{1}{2} < \displaystyle\int_0^1 \frac{dx}{\sqrt{4 - x^2 + x^3}} < \frac{\pi}{6}$.

(ii) $\dfrac{1}{2} < \displaystyle\int_0^{1/2} \frac{dx}{\sqrt{(1 - x^{2n})}} < \frac{\pi}{6}, \; \forall \, n > 1$.

(iii) $\dfrac{2\pi^2}{9} \le \displaystyle\int_{\pi/6}^{\pi/2} \frac{2x}{\sin x} \le \frac{4\pi^2}{9}$.

(7) Estimate $\int_{\pi/4}^{\pi/3} \frac{\sin x}{x} dx$ from $m(b-a) \le \int_a^b f(x)dx \le M(b-a)$ to show that

$$\frac{\sqrt{3}}{8} \le \int_{\pi/4}^{\pi/3} \frac{\sin x}{x} dx \le \frac{\sqrt{2}}{6}.$$

From the same inequality show that $\int_0^{\pi/2} e^{-R\sin x} dx < \int_0^{\pi/2} e^{-\frac{2Rx}{\pi}} dx = \frac{\pi}{2R}\left(1-e^{-R}\right).$

(8) Let f be defined on $[1, 3]$ by $f(x) = \begin{cases} 1, & 1 \le x < 2 \\ 2, & 2 \le x \le 3 \end{cases}$

Then state with reasons,

(i) whether $\int_1^3 f(x)dx$ exists;

(ii) whether $\int_a^b f(x)dx = (b-a)f(\xi)$, for some $\xi \in [a, b]$ holds here;

(iii) whether Fundamental Theorem of integral calculus is applicable to $f(x)$ on $[1, 3]$.

(9) Let $f : [0, 2] \to \mathbb{R}$ be defined by $f(x) = \begin{cases} x + x^2, & \text{when } x \text{ is rational} \\ x^2 + x^3, & \text{when } x \text{ is irrational.} \end{cases}$

Examine whether f is R-integrable on $[0, 2]$.

(10) Let $f : \mathbb{R} \to \mathbb{R}$ be continuous on $\mathbb{R}$ and $G : \mathbb{R} \to \mathbb{R}$ be defined by $G(x) = \int_0^{\sin x} f(t)dt$ for $x \in \mathbb{R}$. Show that G is differentiable on $\mathbb{R}$ and find $G'(x)$.

(11) Let f be continuous and let $F(x) = \frac{1}{2}\int_0^x (x-t)^2 f(t)dt$. Show that

$$F'(x) = x\int_0^x f(t)dt - \int_0^x tf(t)dt.$$

(12) Let $f : [a, b] \to \mathbb{R}$ be bounded and f^2 is R-integrable on $[a, b]$. Does it follow that f is R-integrable on $[a, b]$? Does the answer change if f^3 is R-integrable on $[a, b]$?

(13) If $f, g : [a, b] \to \mathbb{R}$ be both bounded and R-integrable on $[a, b]$, then show that

(i) $\lambda f(x) \pm \mu g(x)$, λ, μ are constants.

(ii) $\lambda\mu f(x) \cdot g(x)$ are R-integrable on $[a, b]$. Deduce that $5f(x), 2f^3(x)$ are R-integrable on $[a, b]$.

(14) If $f(x) = 2n$, for $x = \frac{1}{n}, n \in N$ and $f(x) = 0$, otherwise. Show that f is not R-integrable on $[0, 1]$.

CHAPTER 8

IMPROPER INTEGRALS

The theory of R-integration is developed on the basis of (i) when the domain of integration is finite and (ii) when the function f is bounded on $[a,b]$. The R-integral $\int_a^b f(x)dx$ is called proper integral.

Now the question arises (i) when the domain of integration is finite but f is unbounded (ii) when f is bounded but the domain of integration is infinite.

In both cases we assign the integrals as *improper integrals.*

Definition [8.1] : A function $f : [a,b] \to \mathbb{R}$ is said to be unbounded on $[a,b]$, if there exists a point c, $a \leq c \leq b$ such that f is not bounded in any neighbourhood of c.

That is, $|f(x)| \geq k,\ \forall\ x \in [c-\delta, c+\delta]$.

In other words, $f(x) \to +\infty$ or $-\infty$ as $x \to c$. c is called a point of infinite discontinuity of f on $[a,b]$. f may have a point of infinite discontinuity at $x=a, x=b$ or at $x=c, a<c<b$.

Definition [8.2] : An integral $\int_a^b f(x)dx$, when f is unbounded on $[a,b]$, is called an improper integral of *first kind.*

An integral of the type $\int_a^\infty f(x)dx$ or $\int_{-\infty}^b f(x)dx$ or $\int_{-\infty}^\infty f(x)dx$, i.e., when the domain of integration is infinite, is called an improper integral of *second kind.*

Example. (1) $\int_{-1}^1 \frac{1}{x}dx$, $\int_1^2 \frac{1}{x-1}dx$, $\int_2^3 \frac{1}{x-3}dx$ are improper integrals of first kind.

The first one has a point of infinite discontinuity at $x=0$, the second one has a point of infinite discontinuity at $x=1$ and the third one has a point of infinite discontinuity at $x=3$.

(2) $\int_0^\infty \frac{1}{1+x^2}dx$, $\int_{-\infty}^1 \frac{1}{x}dx$, $\int_{-\infty}^\infty \frac{xdx}{(1+x^2)^3}$ are improper integrals of second kind.

(3) $\int_0^\infty \frac{1}{1-x}dx$, $\int_{-\infty}^1 \frac{1}{x+1}dx$, $\int_{-\infty}^\infty \frac{dx}{x^2}$ are the examples of improper integrals of mixed types.

The first one has a point of infinite discontinuity at $x=1 \in [0,\infty)$ and at the same time the domain of integration is infinite. In this case, we break up it into two parts

$$\int_0^1 \frac{1}{1-x}dx + \int_1^\infty \frac{1}{1-x}dx.$$

The first part is an improper integral of first kind while the second part is an improper integral of second kind. The same is true for the second one and third one also.

Existence or convergence of an improper integral of first kind

Definition [8.3]: (1) If in an improper integral $\int_a^b f(x)dx$, f has only point of infinite discontinuity at $x = a$, then $\int_a^b f(x)dx$ exists or converges at $x = a$, if $\lim_{\epsilon \to 0} \int_{a+\epsilon}^b f(x)dx$ exists finitely, $0 < \epsilon \leq b - a$. Otherwise, we say that the improper integral $\int_a^b f(x)dx$ does not exist or it does not converge at $x = a$.

(2) If f has a point of infinite discontinuity at $x = b$ only, then $\int_a^b f(x)dx$ exists or converges at $x = b$, if $\lim_{\epsilon \to 0} \int_a^{b-\epsilon} f(x)dx$ exists finitely, otherwise, it does not exist or does not converge at $x = b$.

Example. $\int_1^2 \frac{1}{x-1}dx$.

$f(x) = \frac{1}{x-1}$, has a point of infinite discontinuity at $x = 1$ only.

Now, we write

$$\int_1^2 \frac{1}{x-1}dx = \lim_{\epsilon \to 0} \int_{1+\epsilon}^2 \frac{1}{x-1}dx = \lim_{\epsilon \to 0} \left[\log(x-1)\right]_{1+\epsilon}^2$$
$$= \lim_{\epsilon \to 0} [\log 1 - \log \epsilon] = -\lim_{\epsilon \to 0} \log \epsilon$$

which does not exist.

So $\int_1^2 \frac{1}{x-1}dx$ does not converge at $x = 1$.

Example. $\int_0^1 \frac{1}{\sqrt{1-x^2}}dx$.

$f(x) = \frac{1}{\sqrt{1-x^2}}$ has a point of infinite discontinuity at $x = 1$ only. Then

$$\int_0^1 \frac{dx}{\sqrt{1-x^2}}dx = \lim_{\epsilon \to 0} \int_0^{1-\epsilon} \frac{dx}{\sqrt{1-x^2}} = \lim_{\epsilon \to 0} \left[\sin^{-1} x\right]_0^{1-\epsilon}$$
$$= \lim_{\epsilon \to 0} [\sin^{-1}(1-\epsilon)] = \sin^{-1} 1 = \frac{\pi}{2}$$

exists finitely.

Hence $\int_0^1 \frac{1}{\sqrt{1-x^2}}dx$ exists. That is, it converges at $x = 1$.

(3) If in an improper integral $\int_a^b f(x)dx$, f has points of infinite discontinuity at $x = a$ and $x = b$ only, we choose a point c, such that $a < c < b$. Now if the improper integrals $\int_a^c f(x)dx$ and $\int_c^b f(x)dx$ both exist, i.e., if they converge at $x = a$ and $x = b$ respectively, we say that the improper integral $\int_a^b f(x)dx$ converges.

Otherwise, it does not exist. In this case, we write

$$\int_a^b f(x)dx = \int_a^c f(x)dx + \int_c^b f(x)dx$$

For the first part $x = a$ is a point of infinite discontinuity and for the second part $x = b$ is a point of infinite discontinuity.

So $\int_a^b f(x)dx = \lim_{\epsilon \to 0} \int_{a+\epsilon}^c f(x)dx + \lim_{\epsilon' \to 0} \int_c^{b-\epsilon'} f(x)dx$.

Now $\int_a^b f(x)dx$ exists provided both the limits exist finitely.

Example. $\int_0^2 \frac{dx}{(2x-x^2)} = \int_0^2 \frac{dx}{x(2-x)}$.

$f(x) = \frac{1}{x(2-x)}$ has points of infinite discontinuity at $x = 0$ and $x = 2$ only.

We choose point $x = 1$ in $[0, 2]$. Then

$$\begin{aligned}\int_0^2 \frac{dx}{x(2-x)}dx &= \int_0^1 \frac{dx}{x(2-x)} + \int_1^2 \frac{dx}{x(2-x)} \\ &= \lim_{\epsilon\to 0}\int_\epsilon^1 \frac{dx}{x(2-x)} + \lim_{\epsilon'\to 0}\int_1^{2-\epsilon'} \frac{dx}{x(2-x)} \\ &= \frac{1}{2}\lim_{\epsilon\to 0}\left[\log x - \log(2-x)\right]_\epsilon^1 + \frac{1}{2}\lim_{\epsilon'\to 0}\left[\log x - \log(2-x)\right]_1^{2-\epsilon'} \\ &= \frac{1}{2}\lim_{\epsilon\to 0}[-\log\epsilon + \log(2-\epsilon)] + \frac{1}{2}\lim_{\epsilon'\to 0}[\log(2-\epsilon') - \log\epsilon']\end{aligned}$$

Both the limits do not exist. Hence the given integral $\int_0^2 \frac{dx}{x(2-x)}$ does not exist.

Example. $\int_0^1 \frac{dx}{\sqrt{x(1-x)}} = \int_0^1 \frac{dx}{\sqrt{\left(\frac{1}{2}\right)^2 - \left(x-\frac{1}{2}\right)^2}}$.

$f(x) = \frac{1}{\sqrt{x(1-x)}}$ has points of infinite discontinuity at $x = 0$ and $x = 1$ only.

We choose a point $\frac{1}{2}$ in $[0, 1]$.

Then $\int_0^1 \frac{dx}{\sqrt{x(1-x)}} = \int_0^{1/2} \frac{dx}{\sqrt{x(1-x)}} + \int_{1/2}^1 \frac{dx}{\sqrt{x(1-x)}}$.

For the first part, 0 is a point of infinite discontinuity and for the second part, 1 is a point of infinite discontinuity.

So

$$\begin{aligned}\int_0^1 \frac{dx}{\sqrt{x\cdot(1-x)}} &= \lim_{\epsilon\to 0}\int_{0+\epsilon}^{1/2} \frac{dx}{\sqrt{\left(\frac{1}{2}\right)^2 - \left(x-\frac{1}{2}\right)^2}} + \lim_{\epsilon'\to 0}\int_{1/2}^{1-\epsilon'} \frac{dx}{\sqrt{\left(\frac{1}{2}\right)^2 - \left(x-\frac{1}{2}\right)^2}} \\ &= \lim_{\epsilon\to 0}\left[\sin^{-1}\frac{x-\frac{1}{2}}{\frac{1}{2}}\right]_\epsilon^{1/2} + \lim_{\epsilon'\to 0}\left[\sin^{-1}\frac{x-\frac{1}{2}}{\frac{1}{2}}\right]_{1/2}^{1-\epsilon'} \\ &= \lim_{\epsilon\to 0}[-\sin^{-1}(2\epsilon-1)] + \lim_{\epsilon'\to 0}[\sin^{-1}(1-\epsilon')] = \frac{\pi}{2} + \frac{\pi}{2} = \pi\end{aligned}$$

Both the limits exists finitely. Hence $\int_0^1 \frac{dx}{\sqrt{x(1-x)}}$ exists and is equal to π.

(4) If f has a point of infinite discontinuity at $x = c$ only, where $a < c < b$, i.e., at an interior point c of $[a, b]$, then we write,

$$\int_a^b f(x)dx = \int_a^c f(x)dx + \int_c^b f(x)dx,$$

in which c is only point of infinite discontinuity in both parts. So $\int_a^b f(x)dx$ exists, if both converges at $x = c$.

Similarly, if f has finite number of points of infinite discontinuities at $c_1, c_2, \cdots, c_n$ where $a < c_i < b, i = 1, 2, \cdots, n$, then we write,

$$\int_a^b f(x)dx = \int_a^{c_1} f(x)dx + \int_{c_1}^{c_2} f(x)dx + \cdots + \int_{c_n}^b f(x)dx$$

and $\int_a^b f(x)dx$ exists if all the integrals on the R.H.S. are convergent, otherwise it does not exist.

Example. $\int_{-1}^1 \frac{dx}{x^{2/3}}$.

$f(x) = \frac{1}{x^{2/3}}$ has only a point of infinite discontinuity at $x = 0 \in [-1, 1]$.

So

$$\begin{aligned}\int_{-1}^{1}\frac{dx}{x^{2/3}} &= \int_{-1}^{0}\frac{dx}{x^{2/3}}+\int_{0}^{1}\frac{dx}{x^{2/3}} = \lim_{\epsilon\to 0}\int_{-1}^{0-\epsilon}\frac{dx}{x^{2/3}}+\lim_{\epsilon'\to 0}\int_{0+\epsilon'}^{1}\frac{dx}{x^{2/3}}\\ &= \lim_{\epsilon\to 0}\left[3x^{1/3}\right]_{-1}^{-\epsilon}+\lim_{\epsilon'\to 0}\left[3x^{1/3}\right]_{\epsilon'}^{1}\\ &= 3\left[\lim_{\epsilon\to 0}(-\epsilon)^{1/3}+1\right]+3\lim_{\epsilon'\to 0}[1-\epsilon'^{1/3}]\\ &= 3+3=6\end{aligned}$$

Both the limit exists independently.

So $\int_{-1}^{1}\frac{dx}{x^{2/3}}$ exists and its value is equal to 6.

Existence or convergence of an improper integral of second kind

Definition [8.4] : (1) For an improper integral $\int_a^\infty f(x)dx$, let f be bounded and integrable in $[a,t]$, $\forall\ t\geq a$ so that the proper integral $\int_a^t f(x)dx$ exists. If $\lim\limits_{t\to\infty}\int_a^t f(x)dx$ exist finitely, the improper integral $\int_a^\infty f(x)dx$ exists or it converges at ∞. Otherwise, $\int_a^\infty f(x)dx$ does not exist.

(2) For an improper integral $\int_{-\infty}^b f(x)dx$, let f be bounded and integrable in $[t,b]$, $\forall\ t\leq b$ so that $\int_t^b f(x)dx$ exists. If $\lim\limits_{t\to-\infty}\int_t^b f(x)dx$ exists finitely, the improper integral $\int_{-\infty}^b f(x)dx$ exists or that it converges at $-\infty$. Otherwise, $\int_{-\infty}^b f(x)dx$ does not exist.

Example. $\int_0^\infty \frac{dx}{1+x^2}$.

f is bounded and integrable in $[0,t]$, $\forall\ t\geq 0$.

Now

$$\int_0^\infty\frac{dx}{1+x^2}=\lim_{t\to\infty}\int_0^t\frac{dx}{1+x^2}=\lim_{t\to\infty}\left[\tan^{-1}x\right]_0^\infty=\frac{\pi}{2},\ \text{exists finitely.}$$

So $\int_0^\infty\frac{dx}{1+x^2}$ exists or it converges at ∞ and its value is equal to $\frac{\pi}{2}$.

Example. $\int_{-\infty}^1\frac{dx}{1+x^2}$.

Here f is bounded and integrable on $[t,1]$, $\forall\ t\leq 1$.

Now

$$\begin{aligned}\int_{-\infty}^{1}\frac{dx}{1+x^2} &= \lim_{t\to-\infty}\int_t^1\frac{dx}{1+x^2}=\lim_{t\to-\infty}[\tan^{-1}1-\tan^{-1}t]\\ &= \frac{\pi}{4}-\left(-\frac{\pi}{2}\right)=\frac{3\pi}{4},\ \text{exists finitely.}\end{aligned}$$

So $\int_{-\infty}^1\frac{dx}{1+x^2}$ exists or it converges at $-\infty$ and its value is equal to $\frac{3\pi}{4}$.

Example. $\int_1^\infty\frac{1}{\sqrt{x}}dx$.

f is bounded and integrable on $[1,t]$, $\forall\ t\geq 1$.

Now

$$\int_1^\infty\frac{1}{\sqrt{x}}dx=\lim_{t\to\infty}\int_1^t\frac{dx}{\sqrt{x}}=\lim_{t\to\infty}2\left[\sqrt{x}\right]_1^t=\lim_{t\to\infty}2[\sqrt{t}-1]\to\infty$$

So limit does not exist finitely. Hence, $\int_1^\infty\frac{1}{\sqrt{x}}dx$ does not exist or it does not converge at $x=\infty$.

(3) For an improper integral $\int_{-\infty}^{\infty} f(x)dx$, we choose a number c in $(-\infty, \infty)$ so that we write

$$\int_{-\infty}^{\infty} f(x)dx = \int_{-\infty}^{c} f(x)dx + \int_{c}^{\infty} f(x)dx$$

Now for the first part, let f be bounded and integrable in $[t, c]$ $\forall\ t \leq c$, and for the second part, let f be bounded and integrable on $[c, t']$, $\forall\ t' \geq c$. Then $\int_{-\infty}^{\infty} f(x)dx$ exists, if

$$\lim_{t \to -\infty} \int_{t}^{c} f(x)dx \text{ and } \lim_{t' \to \infty} \int_{c}^{t} f(x)dx, \text{ exists finitely.}$$

Example. $\int_{-\infty}^{\infty} \frac{2x dx}{(1+x^2)^3}$.

We choose a number 0 in $(-\infty, \infty)$ so that

$$\int_{-\infty}^{\infty} \frac{2x dx}{1 + x^2)^3} = \int_{-\infty}^{0} \frac{2x dx}{(1 + x^2)^3} + \int_{0}^{\infty} \frac{2x dx}{(1 + x^2)^3}$$

In the first part, f is bounded and integrable in $[t, 0]$, $\forall\ t \leq 0$ and, in the second part, f is bounded and integrable in $[0, t']$, $\forall\ t' \geq 0$.

So

$$\int_{-\infty}^{\infty} \frac{2x dx}{(1 + x^2)^3} = \lim_{t \to -\infty} \int_{t}^{0} \frac{2x dx}{(1 + x^2)^3} + \lim_{t' \to \infty} \int_{0}^{t'} \frac{2x dx}{(1 + x^2)^3}$$

$$= \lim_{t \to -\infty} \left[-\frac{1}{2(1 + x^2)^2} \right]_{t}^{0} + \lim_{t' \to \infty} \left[-\frac{1}{2(1 + x^2)^2} \right]_{0}^{t'}$$

$$= \lim_{t \to -\infty} \left[-\frac{1}{2} + \frac{1}{2(1 + t^2)^2} \right] + \lim_{t' \to \infty} \left[-\frac{1}{2\left(1 + t'^2\right)^2} + \frac{1}{2} \right] = -\frac{1}{2} + \frac{1}{2} = 0$$

Both the limits exist finitely. So $\int_{-\infty}^{\infty} \frac{2x dx}{(1+x^2)^3}$ exists and its value is equal to 0.

General types of improper integrals

In case of second kind of improper integrals, if f also possesses finite number of points of infinite discontinuities, then these types of improper integrals are treated as general types of improper integrals.

In this case, we choose an interval $[a, b]$ which includes all the points of infinite discontinuities and then for the existence of the improper integral we consider the existence of the integrals

$$\int_{-\infty}^{a} f(x)dx, \int_{a}^{b} f(x)dx \text{ and } \int_{a}^{\infty} f(x)dx$$

in accordance with the definition given above. Now, if all the above integrals exist, we say that the integral $\int_{-\infty}^{\infty} f(x)dx$ exist. We write,

$$\int_{-\infty}^{\infty} f(x)dx = \int_{-\infty}^{a} f(x)dx + \int_{a}^{b} f(x)dx + \int_{b}^{\infty} f(x)dx.$$

Test for convergence

Example. $\int_{-\infty}^{\infty} \frac{dx}{(1+x)^3}$.

Here domain of integration is infinite and at the same time f is unbounded at $x = -1$ in $(-\infty, \infty)$.

We write,

$$\int_{-\infty}^{\infty} \frac{dx}{(1+x)^3} = \int_{-\infty}^{-2} \frac{dx}{(1+x)^3} + \int_{-2}^{2} \frac{dx}{(1+x)^3} + \int_{2}^{\infty} \frac{dx}{(1+x)^3}. \tag{1}$$

The 2nd part is a first kind of improper integral. f has a point of infinite discontinuity at $x = -1$ only. So

$$\begin{aligned}\int_{-2}^{2} \frac{dx}{(1+x)^3} &= \int_{-2}^{-1} \frac{dx}{(1+x)^3} + \int_{-1}^{2} \frac{dx}{1+x)^3} \\ &= \lim_{\epsilon\to 0} \int_{-2}^{-1-\epsilon} \frac{dx}{(1+x)^3} + \lim_{\epsilon'\to 0} \int_{-1+\epsilon'}^{2} \frac{dx}{(1+x)^3} \\ &= \lim_{\epsilon\to 0} \left[-\frac{1}{2(1+x)^2}\right]_{-2}^{-1-\epsilon} + \lim_{\epsilon'\to 0} \left[-\frac{1}{2(1+x)^2}\right]_{-1+\epsilon'}^{2} \\ &= \lim_{\epsilon\to 0} \left[-\frac{1}{2\epsilon^2} + \frac{1}{2}\right] + \lim_{\epsilon'\to 0} \left[-\frac{1}{18} + \frac{1}{2\epsilon'^2}\right] \to \infty\end{aligned}$$

Limit does not exist. So the given integral $\int_{-\infty}^{\infty} \frac{dx}{(1+x)^3}$ does not exist though the 1st and the last integrals of (1) exist.

Test of convergence of first kind of improper integrals by comparison test

Theorem [8.1] The necessary and sufficient condition that the improper integral $\int_a^b f(x)dx$ converges at $x = a$, where $f(x) > 0 \ \forall \ x \in [a, b]$ is, there exists a (+)ve number K such that $\int_{a+\epsilon}^b f(x)dx < K$, for $0 < \epsilon \leq b - a$.

Proof. Let $\phi(\epsilon) = \int_{a+\epsilon}^b f(x)dx$, $0 < \epsilon \leq b - a$. The improper integral $\int_a^b f(x)dx$ converges at $x = a$, if $\lim\limits_{\epsilon\to 0} \int_{a+\epsilon}^b f(x)dx$ exists finitely.

$f(x) > 0 \ \forall \ x \in [a, b] \Rightarrow \phi(\epsilon) > 0$.

Now if $\epsilon_2 < \epsilon_1$, then $\phi(\epsilon_2) - \phi(\epsilon_1) = \int_{a+\epsilon_2}^{a+\epsilon_1} f(x)dx > 0 \Rightarrow \phi(\epsilon_2) > \phi(\epsilon_1)$.

So $\phi(\epsilon)$ monotonically increases as ϵ decreases. Hence, $\lim\limits_{\epsilon\to 0} \phi(\epsilon)$ exists finitely, if and only if $\phi(\epsilon)$ is bounded above.

That is, if $\exists$ a (+)ve number K such that $\phi(\epsilon) < K$, for $0 < \epsilon \leq b - a$

$$\Rightarrow \int_{a+\epsilon}^b f(x)dx < K, \text{ for } 0 < \epsilon \leq b - a$$

Theorem [8.2] [Comparison test] If f, g be two functions such that $f(x) \leq g(x)$, and $f(x)$, $g(x) \geq 0 \ \forall \ x \in [a, b]$, then (1) $\int_a^b g(x)dx$ converges $\Rightarrow \int_a^b f(x)dx$ converges. (2) $\int_a^b f(x)dx$ does not converge $\Rightarrow \int_a^b g(x)dx$ does not converge.

Proof. Let $x = a$ is the only point of infinite discontinuity of f and g on $[a, b]$. f, g are both bounded and integrable on $[a + \epsilon, b], 0 < \epsilon \leq b - a$. Also $f(x) \leq g(x) \ \forall \ x \in [a, b]$

$$\Rightarrow \int_{a+\epsilon}^b f(x)dx \leq \int_{a+\epsilon}^b g(x)dx \tag{i}$$

Let $\int_a^b g(x)dx$ converges at $x = a$. Then by Theorem [8.1], $\exists$ a (+)ve number K such that

$$\int_{a+\epsilon}^b g(x)dx < K \text{ for } 0 < \epsilon \leq b - a.$$

Therefore, from (i) it follows that $\int_{a+\epsilon}^{b} f(x)dx \leq \int_{a+\epsilon}^{b} g(x)dx < K$ for $0 < \epsilon \leq b-a$

$$\Rightarrow \int_{a}^{b} f(x)dx \text{ converges at } x = a.$$

Next, let $\int_{a}^{b} f(x)dx$ does not converge at $x = a$.

Then $\int_{a+\epsilon}^{b} f(x)dx$ is not bounded above. So $\int_{a+\epsilon}^{b} f(x)dx > M$ for $0 < \epsilon \leq b-a$

$$\Rightarrow \int_{a+\epsilon}^{b} g(x)dx \geq \int_{a+\epsilon}^{b} f(x)dx > M, \text{ for } 0 < \epsilon \geq b-a$$

$$\Rightarrow \int_{a+\epsilon}^{b} g(x)dx \text{ is unbounded above.}$$

$$\Rightarrow \lim_{\epsilon \to 0} \int_{a+\epsilon}^{b} g(x)dx \text{ does not exist finitely.}$$

$$\Rightarrow \int_{a}^{b} g(x)dx \text{ does not converge at } x = a.$$

Theorem [8.3] [Comparison test] If $f(x)$ and $g(x)$ be both positive for all $x \in [a, b]$ and $\lim_{x \to a+0} \frac{f(x)}{g(x)} = l$, where l is neither zero nor infinite and a is only point of infinite discontinuity of f and g on $[a, b]$, then the two integrals $\int_{a}^{b} f(x)dx$ and $\int_{a}^{b} g(x)dx$ either both converge or both do not converge.

Proof. $f(x), g(x) > 0 \Rightarrow l > 0$. We choose $\epsilon(> 0) < l$.

Now $\lim_{x \to a+0} \frac{f(x)}{g(x)} = l \Rightarrow \exists$ a number c, $a < c < b$ such that $\left|\frac{f(x)}{g(x)} - l\right| < \epsilon, \ \forall \ x \in (a, c)$

$$\Rightarrow \ (l-\epsilon)g(x) < f(x) < (l+\epsilon)g(x) \ \forall \ x \in (a, c) \qquad \text{(i)}$$

Now $\int_{a}^{b} f(x)dx$ converges at $x = a \Rightarrow \int_{a}^{c} f(x)dx$ converges.

$$\Rightarrow (l-\epsilon) \int_{a}^{c} g(x)dx \text{ converges [by (i)]}$$

$$\Rightarrow \int_{a}^{c} g(x)dx \text{ converges at } x = a$$

$$\Rightarrow \int_{a}^{b} g(x) \text{ converges at } x = a.$$

It can be similarly proved from the relation $(l-\epsilon)g(x) < f(x)$, i.e., $f(x) > (l-\epsilon)g(x)$, that if $\int_{a}^{b} g(x)dx$ does not converge, then $\int_{a}^{b} f(x)dx$ does not converge.

Again, let $\int_{a}^{b} g(x)dx$ converges at $x = a$.

Then $\int_{a}^{c} g(x)dx$ converges and so $(l+\epsilon) \int_{a}^{c} g(x)dx$ converges at $x = a$.

Then from (i), $f(x) < (l+\epsilon)g(x) \Rightarrow \int_{a}^{c} f(x)dx$ converges at $x = a$

$$\Rightarrow \int_{a}^{b} f(x)dx \text{ converges.}$$

Similarly, we can show that if $\int_{a}^{b} f(x)dx$ does not converge, then $\int_{a}^{b} g(x)dx$ does not converge.

A useful comparison integral

Theorem [8.4] [μ-test of convergence] (1) The improper integrals $\int_a^b \frac{dx}{(x-a)^\mu}$ having $x = a$, the only point of infinite discontinuity, converges at $x = a$, if and only if $\mu < 1$.

Proof. Here $\mu \geq 0$, otherwise it becomes a proper integral.

Let $\mu \neq 1$. Then

$$\int_a^b \frac{dx}{(x-a)^\mu} = \lim_{\epsilon \to 0} \int_{a+\epsilon}^b \frac{dx}{(x-a)^\mu} = \lim_{\epsilon \to 0} \left[\frac{(x-a)^{-\mu+1}}{-\mu+1}\right]_{a+\epsilon}^b$$

$$= \lim_{\epsilon \to 0} \frac{1}{1-\mu}\left[\frac{1}{(b-a)^{\mu-1}} - \frac{1}{\epsilon^{\mu-1}}\right]$$

$$= \frac{1}{1-\mu}\frac{1}{(b-a)^{\mu-1}} - \frac{1}{1-\mu}\lim_{\epsilon \to 0}\frac{1}{\epsilon^{\mu-1}}$$

Now, if $\mu < 1$, then $\lim_{\epsilon \to 0} \frac{1}{\epsilon^{\mu-1}} = 0$ and hence

$$\lim_{\epsilon \to 0} \int_{a+\epsilon}^b \frac{dx}{(x-a)^\mu} = \frac{1}{1-\mu}\frac{1}{(b-a)^{\mu+1}}$$

exists finitely and consequently the given integral converges at $x = a$.

If $\mu > 1$, then $\lim_{\epsilon \to 0} \frac{1}{\epsilon^{\mu-1}} \to \infty$ as $\mu - 1 > 0$ and hence $\lim_{\epsilon \to 0} \int_{a+\epsilon}^b \frac{dx}{(x-a)^\mu}$ does not exist and so $\int_a^b \frac{dx}{(x-a)^\mu}$ does not converge at $x = a$ for $\mu > 1$.

If $\mu = 1$, then

$$\int_a^b \frac{dx}{(x-a)^\mu} = \int_a^b \frac{dx}{(x-a)} = \lim_{\epsilon \to 0}\left[\log(x-a)\right]_{a+\epsilon}^b = \lim_{\epsilon \to 0}[\log(b-a) - \log\epsilon] \to \infty$$

Therefore, limit does not exist and so $\int_a^b \frac{dx}{(x-a)^\mu}$ does not converge when $\mu = 1$. Therefore, the given improper integral converges at $x = a$, if and only if $\mu < 1$, otherwise it does not converge or we say it diverges.

(2) The improper integral $\int_a^b \frac{dx}{(b-x)^\mu}$ having $x = b$, the only point of infinite discontinuity converges at $x = b$, if and only if $\mu < 1$.

[*Hints.* The proof is similar to (1).]

Remark. The statements (1) and (2) of Theorem [8.4] may be given as follows:

(1) If $\lim_{x \to a+0} [f(x)(x-a)^\mu] = l$ which is neither zero nor infinite, then $\int_a^b f(x)dx$ converges at $x = a$, if $\mu < 1$ and diverges, if $\mu \geq 1$, when f is unbounded at $x = a$ only.

(2) If $\lim_{x \to b-0} [f(x)(b-x)^\mu] = l$ which is neither zero nor infinite, then $\int_a^b f(x)dx$ converges at $x = b$, if $\mu < 1$ and diverges, if $\mu \geq 1$, when f is unbounded at $x = b$ only.

Example. Let us examine the convergence of $\int_0^1 \frac{dx}{\sqrt{x}(1+x)}$.

Here $f(x) = \frac{1}{\sqrt{x}(1+x)}$. $x = 0$ is the only point of infinite discontinuity of f.

Let $g(x) = \frac{1}{\sqrt{x}} = \frac{1}{x^{1/2}}$.

(i) $f, g \geq 0 \ \forall \ x \in [0, 1]$.

(ii) $\lim_{x \to 0+} \frac{f(x)}{g(x)} = \lim_{x \to 0+} \frac{1}{\sqrt{x}(1+x)} \Big/ \frac{1}{\sqrt{x}} = \lim_{x \to 0+} \frac{1}{1+x} = 1$, which is neither zero nor infinite.

Also by μ-Test of convergence, $\mu = \frac{1}{2} < 1$. Therefore, $\int_0^1 g(x)dx = \int_0^1 \frac{1}{x^{1/2}}dx$ converges at $x = 0$.

Hence by Comparison Test, $\int_0^1 \frac{dx}{\sqrt{x}(1+x)}$ converges at $x = 1$.

Example. $\int_0^1 \frac{dx}{\sqrt{x(1-x)}}$.

$f(x) = \frac{1}{\sqrt{x(1-x)}}$. $x = 0$ and 1, which are both points of infinite discontinuities of f. So we write,

$$\int_0^1 f(x)dx = \int_0^{1/2} f(x)dx + \int_{1/2}^1 f(x)dx \tag{1}$$

In the first part, $x = 0$ is the only point of infinite discontinuity, and in the 2nd part, $x = 1$ is the only point of infinite discontinuity.

For the first part, we take $g(x) = \frac{1}{\sqrt{x}} = \frac{1}{x^{1/2}}$.

Then (i) $f, g \geq 0 \ \forall \ x \in [0, 1]$

(ii) $\lim_{x \to 0+0} \frac{f(x)}{g(x)} = 1$ ($\neq 0$ or infinity).

Also $\int_0^{1/2} g(x)dx$ converges at $x = 0$ as $\mu = \frac{1}{2} < 1$. So $\int_0^{1/2} f(x)dx$ converges at $x = 0$. Again, for the 2nd part we take $g(x) = \frac{1}{\sqrt{1-x}} = \frac{1}{(1-x)^{1/2}}$.

Then (i) $f, g \geq 0 \ \forall \ x \in [0, 1]$

(ii) $\lim_{x \to 1-} \frac{f(x)}{g(x)} = \lim_{x \to 1-} \frac{1}{\sqrt{x}} = 1$ ($\neq 0$ or infinity).

Also $\int_{1/2}^1 g(x)dx$ converges at $x = 1$ as $\mu = \frac{1}{2} < 1$. So $\int_{1/2}^1 f(x)dx$ converges at $x = 1$. Since both the parts on the R.H.S. of (1) converge, the given integral $\int_0^1 f(x)dx$ converges.

Theorem [8.5] [Beta function] Beta function, denoted by $\beta_{m,n}$ and defined by

$$\beta_{m,n} = \int_0^1 x^{m-1}(1-x)^{n-1}dx$$

exists, if and only if m, n are both positive.

Proof. If $m, n \geq 1$, then $\int_0^1 x^{m-1}(1-x)^{n-1}dx$ is a proper integral. Now $x = 0$ is a point of infinite discontinuity, when $m < 1$ and $x = 1$ is a point of infinite discontinuity when $n < 1$.

Let $m < 1$ and $n < 1$ and let c be a number such that $0 < c < 1$.

Then $\beta_{m,n} = \int_0^c x^{m-1}(1-x)^{n-1}dx + \int_c^1 x^{m-1}(1-x)^{n-1}dx$.

For the 1st part, $x = 0$ is a point of infinite discontinuity.

Let $f(x) = \frac{(1-x)^{n-1}}{x^{1-m}}$ and $g(x) = \frac{1}{x^{1-m}}$.

(i) $f, g > 0 \ \forall \ x \in [0, c]$

(ii) $\lim_{x \to 0+} \frac{f(x)}{g(x)} = \lim_{x \to 0+} (1-x)^{n-1} = 1$ [$\neq 0$ nor infinity]

Also $\int_0^c g(x)dx = \int_0^c \frac{1}{x^{1-m}}dx$, converges at $x = 0$, if and only if $1 - m < 1$, i.e., $m > 0$. Therefore, 1st part converges at $x = 0$, if and only if $m > 0$. For the 2nd part, $x = 1$ is a point of infinite discontinuity.

Let $f(x) = \frac{x^{m-1}}{(1-x)^{1-n}}$ and $g(x) = \frac{1}{(1-x)^{1-n}}$.

(i) $f, g > 0 \; \forall \; x \in [c, 1]$

(ii) $\lim_{x \to 1-} \frac{f(x)}{g(x)} = \lim_{x \to 1-} x^{m-1} = 1$ ($\neq 0$ nor infinity)

Also $\int_c^1 g(x)dx = \int_c^1 \frac{1}{(1-x)^{1-n}}dx$ converges at $x = 1$, if and only if $1 - n < 1$, i.e., $n > 0$.

Hence $\int_c^1 x^{m-1}(1-x)^{n-1}dx$ converges at $x = 1$, if and only if $n > 0$.

Therefore, $\beta_{m,n}$ converges if and only if $m > 0, n > 0$. That is, $\beta_{m,n}$ exists if and only if $m > 0, n > 0$.

Note. $\beta_{m,n} = \int_0^1 x^{m-1}(1-x)^{n-1}dx$ is also known as *First Eulerian Integral.*

Problem [8.1]: (1) Show that $\int_0^{\pi/2} x^m \operatorname{cosec}^n x dx$ exists, if and only if $n < m + 1$.

(2) Examine the convergence of

(i) $\int_0^{\pi} \frac{\sqrt{x}}{\sin x}dx$

(ii) $\int_0^{\pi/2} \sin^{p-1} x \cos^{q-1} x dx$

(iii) $\int_0^{\pi} \frac{\sin x}{x^p}dx$

(iv) $\int_0^1 x^{n-1} \log x dx$

(v) $\int_0^1 x^{m-1}(1-x)^{n-1} \log x dx$

(3) Show that the integral $\int_0^{\pi/2} \log \sin x dx$ is convergent and hence evaluate it.

(4) Prove that $\int_0^1 \frac{dx}{x^2 + x^{1/2}}$ is convergent.

(5) Show that $\int_0^1 \frac{\log x}{\sqrt{x}}dx$ is convergent but $\int_1^2 \frac{\sqrt{x}}{\log x}dx$ is divergent.

(6) Show that $\int_0^1 \frac{x^p \log x}{(1+x)^2}dx$ converges, where $p > -1$.

Solution. (1) $\int_0^{\pi/2} x^m \operatorname{cosec}^n x dx = \int_0^{\pi/2} \left(\frac{x}{\sin x}\right)^n \frac{1}{x^{n-m}}dx.$

The integral is proper if $n - m \leq 0$, since $\lim_{x \to 0} \left(\frac{x}{\sin x}\right) = 1$.

So we consider the case, when $n - m > 0$. In this case, 0 is the only point of infinite discontinuity of the integrand $f(x) = \left(\frac{x}{\sin x}\right)^n \frac{1}{x^{n-m}}$.

Let $g(x) = \frac{1}{x^{n-m}}$.

Then (i) $f, g > 0 \; \forall \; x \in [0, \pi/2]$

(ii) $\lim_{x \to 0+} \frac{f(x)}{g(x)} = \lim_{x \to 0+} \left(\frac{x}{\sin x}\right)^n = 1$ ($\neq 0$ nor infinity).

Now $\int_0^{\pi/2} g(x)dx = \int_0^{\pi/2} \frac{1}{x^{n-m}}dx$, converges at $x = 0$, if and only if $n - m < 1$, i.e., $n < m + 1$.

So the given integral $\int_0^{\pi/2} f(x)dx = \int_0^{\pi} x^m \operatorname{cosec}^n x dx$. Converges at $x = 0$, if and only if $n < m + 1$.

(2) (i) $\int_0^{\pi} \frac{\sqrt{x}}{\sin x}dx$. Here 0 and π are both points of infinite discontinuity of $f(x) = \frac{\sqrt{x}}{\sin x}$ in $[0, \pi]$. So we write,

$$\int_0^{\pi} \frac{\sqrt{x}}{\sin x}dx = \int_0^{\pi/2} \frac{\sqrt{x}}{\sin x}dx + \int_{\pi/2}^{\pi} \frac{\sqrt{x}}{\sin x}dx \tag{1}$$

In the 1st part, $\int_0^{\pi/2} \frac{\sqrt{x}}{\sin x}dx = \int_0^{\pi/2} \left(\frac{x}{\sin x}\right) \frac{1}{x^{1/2}}dx$, 0 is the only point of infinite discontinuity.

Let $g(x) = \frac{1}{x^{1/2}}$.

Then (i) $f, g > 0 \ \forall\, x \in [0, \pi/2]$

(ii) $\lim_{x\to 0+} \frac{f(x)}{g(x)} = \lim_{x\to 0+} \left(\frac{x}{\sin x}\right) = 1$ ($\neq 0$ nor infinity).

Also $\int_0^{\pi/2} g(x)dx = \int_0^{\pi/2} \frac{1}{x^{1/2}} dx$ converges at $x = 0$, since $\mu = \frac{1}{2} < 1$.

Therefore, $\int_0^{\pi/2} f(x)dx = \int_0^{\pi/2} \left(\frac{x}{\sin x}\right) \frac{1}{\sqrt{x}} dx$ converges at $x = 0$.

The 2nd part, $\int_{\pi/2}^{\pi} \frac{\sqrt{x}}{\sin x} dx$ has a point of infinite discontinuity at $x = \pi$.

Let $g(x) = \frac{1}{\sin x}$.

Then (i) $f, g > 0 \ \forall\, x \in [\pi/2, \pi]$

(ii) $\lim_{x\to\pi-} \frac{f(x)}{g(x)} = \lim_{x\to\pi} \sqrt{x} = \sqrt{\pi}$ exists finitely.

Now

$$\int_{\pi/2}^{\pi} \frac{1}{\sin x} dx = \lim_{\epsilon\to 0} \int_{\pi/2}^{\pi-\epsilon} \operatorname{cosec} x dx = \lim_{\epsilon\to 0} \left[\log \tan \frac{x}{2}\right]_{\pi/2}^{\pi-\epsilon}$$
$$= \lim_{\epsilon\to 0} \left[\log \tan\left(\frac{\pi}{2} - \frac{\epsilon}{2}\right) - \log \tan \frac{\pi}{4}\right] = \lim_{\epsilon\to 0} \log \cot \frac{\epsilon}{2} = \infty$$

$\int_{\pi/2}^{\pi} \frac{1}{\sin x} dx$ does not converge at $x = \pi$. So $\int_{\pi/2}^{\pi} \frac{\sqrt{x}}{\sin x} dx$ does not converge at $x = \pi$.

From (1), it follows that $\int_0^{\pi} \frac{\sqrt{x}}{\sin x} dx$ does not converge, since the 2nd part does not converge.

(2) (ii) $\int_0^{\pi/2} \sin^{p-1} x \cos^{q-1} x dx$. We put $\sin^2 x = z$. Then $2 \sin x \cos x dx = dz$.

Also $x \in \left[0, \frac{\pi}{2}\right] \Rightarrow z \in [0, 1]$. Then the given integral

$$\int_0^{\pi/2} \sin^{p-1} x \cos^{q-1} x dx = \int_0^{\pi/2} \sin^{p-2} x \cos^{q-2} x (\sin x \cdot \cos x) dx$$
$$= \frac{1}{2} \int_0^1 z^{\frac{p-2}{2}} \cdot (1-z)^{\frac{q-2}{2}} dz = \frac{1}{2} \int_0^1 z^{\frac{p}{2}-1} \cdot (1-z)^{\frac{q}{2}-1} dz$$
$$= \frac{1}{2} \beta\left(\frac{p}{2}, \frac{q}{2}\right)$$

$\beta\left(\frac{p}{2}, \frac{q}{2}\right)$ converges only when $\frac{p}{2} > 0$ and $\frac{q}{2} > 0$ by Theorem [8.5]. Hence the given integral converges, if and only if $p > 0$ and $q > 0$.

(2) (iii) $\int_0^{\pi} \frac{\sin x}{x^p} dx = \int_0^{\pi} \left(\frac{\sin x}{x}\right) \cdot \frac{1}{x^{p-1}} dx$. Then integral is proper, when $p \leq 0$.

Let $p - 1 > 0$. Then $x = 0$ is the only point of infinite discontinuity.

Let $g(x) = \frac{1}{x^{p-1}}$.

Then (i) $f, g > 0 \ \forall\, x \in [0, \pi]$

(ii) $\lim_{x\to 0+} \frac{f(x)}{g(x)} = \lim_{x\to 0+} \frac{\sin x}{x} = 1$ ($\neq 0$ nor infinity).

Also $\int_0^{\pi} g(x)dx = \int_0^{\pi} \frac{1}{x^{p-1}} dx$ converges at $x = 0$, if $p - 1 < 1$, i.e., if $p < 2$. Hence the given integral converges, only when $p < 2$.

(2) (iv) $\int_0^1 x^{n-1} \log x dx$.

When $n - 1 > 0$, the integral is proper, as $\lim_{x\to 0} x^{n-1} \log x = 0$ consequently. 0 is not a point of infinite discontinuity. We consider the case *when* $n - 1 \leq 0$.

$f(x) = x^{n-1}\log x < 0$ in $[0,1]$. We consider the integrand $f(x) = -x^{n-1}\log x$ so that $f(x) > 0$ in $[0,1]$. '0' is the only point of infinite discontinuity. There exists a (+)ve integer m such that $m+n-1>0$.

Now $\lim_{x\to 0}\left[-x^{m+n-1}\log x\right] = 0$

$$\Rightarrow \left|-x^{m+n-1}\log x\right| < \epsilon$$

$$\Rightarrow -x^{m+n-1}\log x < \epsilon$$

$$\Rightarrow -x^{n-1}\log x < \frac{\epsilon}{x^m}$$

$$\Rightarrow \int_0^1 -x^{n-1}\log x dx < \epsilon\int_0^1 \frac{1}{x^m}dx$$

Now $\int_0^1 \frac{1}{x^m}dx$ converges at $x=0$, if and only if $m<1$. So it can be possible to choose a (+)ve integer $m<1$ such that $m+n-1>0$.

$$\Rightarrow 1-n<m \Rightarrow 1-n<1 \Rightarrow n>0$$

Thus the integral converges at $x=0$, if and only if $n>0$.

When $n=0$, the integral becomes $\int_0^1 x^{-1}\log x dx = \int_0^1 \frac{\log x}{x}dx$.

Here 0 is a point of infinite discontinuity.

Then $\int_0^1 \frac{\log x}{x}dx = \lim_{\epsilon\to 0}\int_\epsilon^1 \frac{\log x}{x}dx = \lim_{\epsilon\to 0}\left[-\frac{(\log\epsilon)^2}{2}\right] \to -\infty$.

So $\int_0^1 \frac{\log x}{x}dx$ does not converge at $x=0$.

When $n<0$, $x^{-1}<x^{n-1}$, $\forall\ x\in[0,1]$.

$$\Rightarrow \int_0^1 -x^{n-1}\log x dx > \int_0^1 x^{-1}\log x dx \quad [\because \log x < 0\ \forall\ x\in(0,1)]$$

$$\Rightarrow \int_0^1 -x^{n-1}\log x dx \text{ does not converge at } x=0 \text{ as}$$

$$\int_0^1 -x^{-1}\log x dx \text{ does not converge at } x=0.$$

$$\Rightarrow \int_0^1 x^{n-1}\log x dx \text{ diverges for } n<0.$$

(2) (v) $\int_0^1 x^{m-1}(1-x)^{n-1}\log x dx$.

$$f(x) = x^{m-1}(1-x)^{n-1}\log x < 0\ \forall\ x\in[0,1].$$

So we consider the integrand $-x^{m-1}(1-x)^{n-1}\log x$ so that $f(x)>0\ \forall\ x\in[0,1]$.

Also 0 and 1 are possible points of infinite discontinuity.

We consider a number c, $0<c<1$ so that

$$\int_0^1 -x^{m-1}(1-x)^{n-1}\log x dx$$

$$= \int_0^c -x^{m-1}(1-x)^{n-1}\log x dx + \int_c^1 -x^{m-1}(1-x)^{n-1}\log x dx.$$

We now examine the convergence of the first part, $\int_0^c -x^{m-1}(1-x)^{n-1}\log x dx$.

The integral is proper, when $m-1\geq 0$.

Since $\lim\limits_{x\to 0} -x^{m-1}(1-x)^{n-1}\log x = -\lim\limits_{x\to 0}(x^{m-1}\log x)(1-x)^{n-1} = 0$

$\Rightarrow$ 0 is not a point of infinite discontinuity.

So the integral is improper, when $m-1<0$ and 0 is the only point of infinite discontinuity.

We now choose a (+)ve integer p such that $p+m-1>0$.

Also $\lim\limits_{x\to 0+}(x^{p+m-1}\log x)(1-x)^{n-1} = 0$

$$\begin{aligned}
&\Rightarrow -x^{p+m-1}\log x(1-x)^{n-1} < \epsilon\\
&\Rightarrow -x^{m-1}(1-x)^{n-1}\log x < \frac{\epsilon}{x^p}\\
&\Rightarrow \int_0^c -x^{m-1}(1-x)^{n-1}\log x dx < \int_0^c \frac{1}{x^p}dx
\end{aligned}$$

Now $\int_0^c \frac{1}{x^p}dx$ converges, if and only if $p<1$. Hence, $\int_0^c -x^{m-1}(1-x)^{n-1}\log x dx$ convrges when $p<1$, where $m+p-1>0$.

$$\therefore m+p-1>0 \Rightarrow 1-m<p<1 \Rightarrow m>0.$$

Therefore, 1st part converges at $x=0$, if and only if $m>0$.

Next, we examine the convergence of the 2nd part, $\int_c^1 -x^{m-1}(1-x)^{n-1}\log x dx$.

Here

$$\begin{aligned}
\lim_{x\to 1-} -(1-x)^{n-1}\log x &= -\lim_{x\to 1-}\frac{\log x}{(1-x)^{1-n}}\left(\frac{0}{0}\right)\\
&= +\lim_{x\to 1-}\frac{1}{(1-n)(1-x)^{-n}\cdot x}\\
&= \lim_{x\to 1-}\frac{(1-x)^n}{(1-n)\cdot x} \text{ exists finitely, if } n\geq 0.
\end{aligned}$$

So $x=1$ is not a point of infinite discontinuity when $n\geq 0$ and consequently the integral becomes a proper integral. When $n<0, f(x)\to\infty$ as $x\to 1$.

So 1 is a point of infinite discontinuity of f. Now $f(x) = -x^{m-1}(1-x)^{n-1}\log x$.

Let $g(x) = \frac{1}{(1-x)^q}$. Then $\int_c^1 g(x)dx$ converges at $x=1$, if and only if $q<1$.

Now $\lim\limits_{x\to 1}\frac{f(x)}{g(x)} = -\lim\limits_{x\to 1-}\cdot\frac{x^{m-1}\log x}{(1-x)^{1-n-q}}$, exists finitely, if and only if $n+q\geq 0$.

Therefore, $\int_c^1 f(x)dx$ converges at $x=1$, when $q<1$.

Now $n+q\geq 0 \Rightarrow n\geq -q>-1$, i.e., $n>-1$. So the 2nd part converges, if and only if $n>-1$. Therefore, the given integral converges when $m>0$ and $n>-1$.

(3) $\int_0^{\pi/2}\log\sin x dx = \int_0^{\pi/2}\log\left(\frac{\sin x}{x}\right)dx + \int_0^{\pi/2}\log x dx$.

1st part is a proper integral, 0 is not a point of infinite discontinuity, since $\lim\limits_{x\to 0}\log\frac{\sin x}{x} = 0$.

2nd part, $\int_0^{\pi/2}\log x dx = \lim\limits_{\epsilon\to 0}\int_\epsilon^{\pi/2}\log x dx$, as 0 is the only point of infinite discontinuity.

Now

$$\lim_{\epsilon\to 0}\int_0^{\pi/2}\log x dx = \lim_{\epsilon\to 0}\left[x\log x - x\right]_\epsilon^{\pi/2}$$
$$= \lim_{\epsilon\to 0}\left[\frac{\pi}{2}\log\frac{\pi}{2} - \frac{\pi}{2} - \epsilon\log\epsilon - \epsilon\right]$$
$$= \frac{\pi}{2}\log\frac{\pi}{2} - \frac{\pi}{2}, \text{ a finite number since } \lim_{\epsilon\to 0}\epsilon\log\epsilon = 0.$$

Therefore, the 2nd part converges at $x = 0$. Hence the given integral converges at $x = 0$.

Now, let $I = \int_0^{\pi/2}\log\sin x dx$. Then $I = \int_0^{\pi/2}\log\cos x dx$.

Therefore,

$$I + I = \int_0^{\pi/2}\log\frac{\sin 2x}{2}dx \Rightarrow 2I = \int_0^{\pi/2}\log\sin 2x dx - \frac{\pi}{2}\log 2 \tag{1}$$

Let $I' = \int_0^{\pi/2}\log\sin 2x dx$. Put $2x = z, 2dx = dz$, $z : 0 \to \pi$ as $x : 0 \to \frac{\pi}{2}$

$$\therefore\ I' = \frac{1}{2}\int_0^{\pi}\log\sin z dz$$
$$= \int_0^{\pi/2}\log\sin x dx \left[\text{Since } \frac{1}{2}\int_0^{2a}f(x)dx = \int_0^a f(x)dx, \text{ if } f(2a-x) = f(x)\right]$$
$$= \int_0^{\pi/2}\log\sin x dx = I \text{ [Here } \log\sin(\pi - x) = \log\sin x]$$

From (1), $2I = I - \frac{\pi}{2}\log 2 \Rightarrow I = -\frac{\pi}{2}\log 2 = \frac{\pi}{2}\log\left(\frac{1}{2}\right)$.

(4) $\int_0^1\frac{dx}{x^2+x^{1/2}} = \int_0^1\frac{dx}{x^{1/2}[x^{3/2}+1]}$

Here 0 is the only point of infinite discontinuity $f(x) = \frac{1}{x^{1/2}[x^{3/2}+1]}$. Let $g(x) = \frac{1}{x^{1/2}}$.

Then (i) $f, g > 0\ \forall\ x \in [0, 1]$

(ii) $\lim\limits_{x\to 0+}\frac{f(x)}{g(x)} = \lim\limits_{x\to 0+}\frac{1}{x^{3/2}+1} = 1$ [$\neq 0$ nor infinity]

Also $\int_0^1 g(x)dx = \int_0^1\frac{dx}{x^{1/2}}$ converges at $x = 0$, since $\mu = 1/2 < 1$.

Therefore, by comparison test, the given integral converges at $x = 0$.

(5) $\int_0^1\frac{\log x}{\sqrt{x}}$. Here $\frac{\log x}{\sqrt{x}} < 0$ in $[0, 1]$.

Thus we take $f(x) = -\frac{\log x}{\sqrt{x}}$ so that $f(x) > 0\ \forall\ x \in [0, 1]$.

$$f(x) = -x^{-\frac{1}{2}}\log x.$$

0 is the only point of infinite discontinuity. We choose a (+)ve number μ such that $\mu - \frac{1}{2} > 0$.

Then $\lim\limits_{x\to 0+} x^\mu f(x) = \lim\limits_{x\to 0+} -x^{\mu-1/2}\log x = 0.$

Hence, $\int_0^1 f(x)dx$ converges at $x = 0$, if and only if $\mu < 1$.

Also $\mu - \frac{1}{2} > 0 \Rightarrow \mu > \frac{1}{2}$.

So there is a number μ satisfying $\frac{1}{2} < \mu < 1$.

Hence, the given integral $\int_0^1\frac{\log x}{\sqrt{x}}dx$ converges at $x = 0$.

Now, in $\int_1^2\frac{\sqrt{x}}{\log x}dx$, $f(x) \to \infty$ as $x \to 1$.

So I is the only point of infinite discontinuity $f(x) = \frac{\sqrt{x}}{\log x}$.

Let $g(x) = \frac{1}{x-1}$.

Then (i) $f, g > 0 \ \forall \ x \in [1,2]$

$$\text{(ii)} \lim_{x\to 1+} \frac{f(x)}{g(x)} = \lim_{x\to 1+} \frac{\sqrt{x}(x-1)}{\log x} = \lim_{x\to 1+} \frac{x^{3/2}-x^{1/2}}{\log x} \left(\frac{0}{0} \text{ form}\right)$$

$$= \lim_{x\to 1+} \cdot \frac{\frac{3}{2}x^{1/2} - \frac{1}{2}x^{-1/2}}{\frac{1}{x}}$$

$$= \lim_{x\to 1+} \frac{3}{2}x^{3/2} - \frac{1}{2}x^{1/2} = 1 \ (\neq 0 \text{ nor infinity}).$$

Also $\int_1^2 g(x)dx = \int_1^2 \frac{dx}{(x-1)}$ diverges as $\mu = 1$.

Therefore, by comparison list, the given integral diverges.

(6) $\int_0^1 \frac{x^p \log x}{(1+x)^2} dx$

The integral is proper when $p > 0$, since $\lim_{x\to 0+} \frac{x^p \log x}{(1+x)^2} = 0$. When $p < 0$, $x = 0$ is a point of infinite discontinuity as $f(x) \to \infty$ as $x \to 0+$.

Now $\frac{x^p \log x}{(1+x)^2} < 0$ in $[0,1]$.

So we consider $f(x) = -\frac{x^p \log x}{(1+x)^2}$ so that $f(x) > 0 \ \forall \ x \in [0,1]$.

We choose a (+)ve number μ such that $\mu + p > 0$. Then

$$\lim_{x\to 0+} -x^{\mu+p} \frac{\log x}{(1+x)^2} = 0$$

$$\Rightarrow \ -x^{\mu+p} \frac{\log x}{(1+x)^2} < \epsilon \Rightarrow -\frac{x^p \log x}{(1+x)^2} < \frac{\epsilon}{x^\mu}$$

$$\Rightarrow \ \int_0^1 -\frac{x^p \log x}{(1+x)^2} dx < \epsilon \int_0^1 \frac{1}{x^\mu} dx$$

Now $\int_0^1 \frac{1}{x^\mu} dx$, converges only when $\mu < 1$. Therefore, the given integral converges, when $\mu < 1$.

Therefore, $\mu + p > 0 \Rightarrow p > -\mu > -1, \therefore p > -1$.

Hence the given integral converges, when $p > -1$.

General test for convergence at a of an improper integral $\int_a^b f(x)dx$, where $f(x)$ is not necessarily positive

Theorem [8.6] The necessary and sufficient condition that an improper integral $\int_a^b f(x)dx$ converges at $x = a$ is that for every $\epsilon(> 0)$, there exists a (+)ve number δ, such that

$$\left|\int_{a+\eta_1}^{a+\eta_2} f(x)dx\right| < \epsilon, \text{ for } 0 < \eta_1, \eta_2 < \delta.$$

Proof. Let $F(\eta) = \displaystyle\int_{a+\eta}^{b} f(x)dx.$ (1)

The improper integral $\int_a^b f(x)dx$, having $x = a$, a point of infinite discontinuity, converges at $x = a$, if and only if $\lim_{\eta\to 0} F(\eta)$ exists finitely. Now the necessary and sufficient condition for the existence of $\lim\limits_{\eta \to 0} F(\eta)$ is for every $\epsilon (> 0)$, there exists a (+)ve number δ such that

$$|F(\eta_1) - F(\eta_2)| < \epsilon, \text{ for } 0 < \eta_1, \eta_2 < \delta$$

$$\Rightarrow \left| \int_{a+\eta_1}^{b} f(x)dx - \int_{a+\eta_2}^{b} f(x)dx \right| < \epsilon, \text{ for } 0 < \eta_1, \eta_2 < \delta, \text{ by (i)}$$

$$\Rightarrow \left| \int_{a+\eta_1}^{a+\eta_2} f(x)dx \right| < \epsilon, \text{ for } 0 < \eta_1, \eta_2 < \delta$$

Corollary [8.1]: The necessary and sufficient condition for the convergence of the improper integral $\int_a^b f(x)dx$ at $x = b$, is that for every $\epsilon (> 0)$, there exists a (+)ve number δ such that $\left| \int_{b-\eta_2}^{b-\eta_1} f(x)dx \right| < \epsilon$, for $0 < \eta_1, \eta_2 < \delta$.

[*Hints.* Take $F(\eta) = \int_a^{b-\eta} f(x)dx$ and proceed as in the above theorem.]

Definition [8.5]: [**Absolute convergence**] An improper integral $\int_a^b f(x)dx$ is said to be absolutely convergent, if $\int_a^b |f(x)|dx$ is convergent.

Example. $\int_0^1 \frac{\sin(1/x)}{\sqrt{x}} dx$ is absolutely convergent. Since $\left| \frac{\sin \frac{1}{x}}{\sqrt{x}} \right| \leq \frac{1}{\sqrt{x}}$

$$\Rightarrow \int_0^1 \left| \frac{\sin \frac{1}{x}}{\sqrt{x}} \right| dx \leq \int_0^1 \frac{1}{\sqrt{x}} dx$$

$$\Rightarrow \int_0^1 \left| \frac{\sin \frac{1}{x}}{\sqrt{x}} \right| dx \text{ is convergent as } \int_0^1 \frac{dx}{\sqrt{x}} \text{ is convergent.}$$

$$\Rightarrow \int_0^1 \frac{\sin \frac{1}{x}}{\sqrt{x}} dx \text{ is absolutely convergent.}$$

Theorem [8.7] Every absolutely convergent integral is convergent. That is, the existence of $\int_a^b |f(x)|dx \Rightarrow$ the existence of $\int_a^b f(x)dx$.

Proof. Let $\int_a^b f(x)dx$ be absolutely convergent. Therefore, $\int_a^b |f(x)|dx$ is convergent.

Then

$$\left| \int_{a+\eta_1}^{a+\eta_2} |f(x)|dx \right| < \epsilon, \text{ for } 0 < \eta_1, \eta_2 < \delta \text{ by Th [8.6]} \qquad \text{(i)}$$

Now

$$\left| \int_{a+\eta_1}^{a+\eta_2} f(x)dx \right| \leq \left| \int_{a+\eta_1}^{a+\eta_2} |f(x)|dx \right| \text{ [by (ii) of Th. [7.10]} \qquad \text{(ii)}$$

Hence from (ii) by (i),

$$\left| \int_{a+\eta_1}^{a+\eta_2} f(x)dx \right| < \epsilon \text{ for } 0 < \eta_1, \eta_2 < \delta \Rightarrow \int_a^b f(x)dx \text{ is convergent.}$$

Problem [8.2]: Show that $\int_0^1 \frac{\sin \frac{1}{x}}{x^p} dx, p > 0$ converges absolutely for $p < 1$.

Solution. $\left| \frac{\sin \frac{1}{x}}{x^p} \right| \leq \frac{1}{x^p} \Rightarrow \int_0^1 \left| \frac{\sin \frac{1}{x}}{x^p} \right| dx \leq \int_0^1 \frac{1}{x^p} dx.$

Now $\int_0^1 \frac{1}{x^p} dx$ converges at $x = 0$, if and only if $p < 1$.

Hence $\int_0^1 \left|\frac{\sin\frac{1}{x}}{x^p}\right| dx$ converges for $p < 1$.

$$\Rightarrow \int_0^1 \frac{\sin\frac{1}{x}}{x^p} dx \text{ converges absolutely for } p < 1.$$

Test for convergence of second kind of improper integral

Convergence of $\int_a^\infty f(x)dx$ at ∞. From definition, $\int_a^\infty f(x)dx = \lim\limits_{t\to\infty} \int_a^t f(x)dx$ when f is bounded and integrable in $[a,t]$, $\forall\, t \geq a$.

Theorem [8.8] $f(x)$ is positive for all x in $[a,\infty]$.
The necessary and sufficient condition for convergence of $\int_a^\infty f(x)dx$ is that there exists a (+)ve number K such that $\int_a^t f(x)dx \leq K$, $\forall\, t \geq a$.

Proof. Let $F(t) = \int_a^t f(x)dx$. From definition, $\int_a^\infty f(x)dx = \lim\limits_{t\to\infty} \int_a^t f(x)dx$.

$\int_a^\infty f(x)dx$ converges at ∞, if $\lim\limits_{t\to\infty} \int_a^t f(x)dx$ exists finitely.

That is, if $\lim\limits_{t\to\infty} F(t)$ exists finitely. Let $t_1, t_2 \geq a$ and $t_2 \geq t_1$.

Then $F(t_2) - F(t_1) = \int_{t_1}^{t_2} f(x)dx > 0$ as $f > 0$ in $[a,t]$ $\forall\, t \geq a$

$$\Rightarrow F(t_2) > F(t_1) \;\forall\, t_2 \geq t_1.$$

So F increases as t increases and therefore, $\lim\limits_{t\to 0} F(t)$ exists finitely, if and only if $F(t)$ is bounded above. Hence $\exists$ a (+)ve number K such that $F(t) \leq K$, $\forall\, t \geq a$.

That is, $\int_a^t f(x)dx \leq K$, $\forall\, t \geq a$.

Theorem [8.9] [Comparison test]

(1) If f, g be two functions such that

  (i) $f(x), g(x) > 0 \;\forall\, x \in [a,\infty)$,

  (ii) $\lim\limits_{x\to\infty} \frac{f(x)}{g(x)} = l$, where l is neither zero nor infininity. Then the improper integrals $\int_a^\infty f(x)dx$ and $\int_a^\infty g(x)dx$ converge or diverge together.

(2) (i) $f(x), g(x) > 0 \;\forall\, x \in [a,\infty]$,

  (ii) $\lim\limits_{x\to\infty} \frac{f(x)}{g(x)} = 0$ and if $\int_a^\infty g(x)dx$ converges, then $\int_a^\infty f(x)dx$ converges.

(3) (i) $f(x), g(x) > 0 \;\forall\, x \in [a,\infty]$,

  (ii) $\lim\limits_{x\to\infty} \frac{f(x)}{g(x)} = \infty$ and if $\int_a^\infty g(x)dx$ diverges, then $\int_a^\infty f(x)dx$ diverges.

Proof. (1) $f(x), g(x) > 0$ and $\lim\limits_{x\to\infty} \frac{f(x)}{g(x)} = l \Rightarrow l > 0$.

There exists a (+)ve number $K > a$ such that $\left|\frac{f(x)}{g(x)} - l\right| < \epsilon$, $\forall\, x > K$

$$\Rightarrow (l-\epsilon)g(x) < f(x) < (l+\epsilon)g(x),\; \forall\, x > K > a \tag{i}$$

where $\epsilon(> 0) < l$.

Now let $\int_a^\infty g(x)dx$ converges at ∞. Then there exists a (+)ve number M such that

$$\int_a^t g(x)dx \leq M \;\forall\, t \geq a \text{ [by Th. [8.8]]}.$$

Then from (i), $f(x) < (l+\epsilon)g(x)$

$$\Rightarrow \int_a^t f(x)dx < (l+\epsilon)\int_a^t g(x)dx \le (l+\epsilon)M = M_1 \text{ (say)}$$

where $M_1 > 0$, $\forall\, t \ge a$

$$\Rightarrow \int_a^t f(x)dx \le M_1 \;\forall\, t \ge a \Rightarrow \int_a^\infty f(x)dx \text{ converges at } \infty.$$

Next, the $\int_a^\infty f(x)dx$ converges at ∞. Then there exists a (+)ve number P such that

$$\int_a^t f(x)dx \le P, \forall\, t \ge a \text{ [by Th. [8.8]]}.$$

Then from (i), $(l-\epsilon)g(x) < f(x)$

$$\Rightarrow (l-\epsilon)\int_a^t g(x)dx < \int_a^t f(x)dx \le P, \forall\, t \ge a \Rightarrow (l-\epsilon)\int_a^t g(x)dx \le P, \forall\, t \ge a$$

$$\Rightarrow (l-\epsilon)\int_a^\infty g(x)dx \text{ converges at } \infty$$

$$\Rightarrow \int_a^\infty g(x)dx \text{ converges at } \infty.$$

(2) $\lim_{x\to\infty}\dfrac{f(x)}{g(x)} = 0 \Rightarrow \dfrac{f(x)}{g(x)} < \epsilon,\ \forall\, x > K > a \Rightarrow f(x) < \epsilon g(x),\ \forall\, x > K\ [\because\ g(x) > 0]$

$$\Rightarrow \int_a^t f(x)dx < \epsilon\int_a^t g(x)dx \tag{ii}$$

Now if $\int_a^\infty g(x)dx$ converges at ∞, there exists a (+)ve number K_1 such that $\int_a^t g(x)dx \le K_1\ \forall\, t \ge a$.

So from (ii), $\int_a^t f(x)dx \le \epsilon K_1,\ \epsilon > 0,\ \forall\, t \ge a \Rightarrow \int_a^\infty f(x)dx$ converges at ∞.

(3) $\lim_{x\to\infty}\dfrac{f(x)}{g(x)} = \infty$

$$\Rightarrow \frac{f(x)}{g(x)} > M,\ \forall\, x > K \Rightarrow f(x) > Mg(x) \tag{iii}$$

Now, if $\int_a^\infty g(x)dx$ diverges, then $\int_a^t g(x)dx$ is unbounded above $\forall\, t \ge a$. So there exists a (+)ve number K_1 such that $\int_a^t g(x)dx > K_1,\ \forall\, t \ge a$. Hence from (iii),

$$\int_a^t f(x)dx > M\int_a^t g(x)dx \Rightarrow \int_a^t f(x)dx > MK_1 = K' \text{ (say)}, \forall\, t \ge a$$

$$\Rightarrow \int_a^\infty f(x)dx \text{ diverges.}$$

A useful comparison integral

Theorem [8.10] An improper integral $\int_a^\infty \frac{dx}{x^\mu}$, $a > 0$ converges at ∞, if and only if $\mu > 1$.

Proof. From definition, $\displaystyle\int_a^\infty \frac{dx}{x^\mu} = \lim_{t\to\infty}\int_a^t \frac{dx}{x^\mu}$ (i)

Now $f(x) = \frac{1}{x^\mu}$ is bounded and integrable in $[a, t]$. $\forall\, t \ge a$.

Also

$$\int_a^t \frac{dx}{x^\mu} = \left[\frac{x^{-\mu+1}}{-\mu+1}\right]_a^t, \ \mu \neq 1 = \frac{1}{1-\mu}\left[\frac{1}{t^{\mu-1}} - \frac{1}{a^{\mu-1}}\right]$$

Therefore,

$$\lim_{t\to\infty}\int_a^t \frac{dx}{x^\mu} = \lim_{t\to\infty}\frac{1}{1-\mu}\left[\frac{1}{t^{\mu-1}} - \frac{1}{a^{\mu-1}}\right] = \frac{1}{(\mu-1)}\frac{1}{a^{\mu-1}}$$

if $\mu - 1 > 0$, i.e., if $\mu > 1$.

So when $\mu > 1$, the limit exists finitely. Therefore, from (i) it follows that $\int_a^\infty \frac{dx}{x^\mu}$ converges at ∞.

Now, if $\mu - 1 < 0$, $\lim\limits_{t\to\infty} \frac{1}{1-\mu}\left[\frac{1}{t^{\mu-1}} - \frac{1}{a^{\mu-1}}\right] \to \infty$. That is, when $\mu < 1$, the limit does not exist.

So $\int_a^\infty \frac{dx}{x^\mu}$ diverges. Again, when $\mu = 1$, $\lim\limits_{t\to\infty}\int_a^t \frac{dx}{x} = \lim\limits_{t\to\infty}[\log t - \log a] \to \infty$.

So the limit does not exist and hence $\int_a^\infty \frac{dx}{x^\mu}$ diverges.

Therefore, $\int_a^\infty \frac{dx}{x^\mu}$ converges at ∞, if and only if $\mu > 1$.

Remark. The same is true for an improper integral $\int_{-\infty}^b \frac{dx}{x^\mu}$, $b < 0$.

Put $x = -y$. Then $\int_{-\infty}^b \frac{dx}{x^\mu} = \int_{-b}^\infty \frac{dy}{(-y)^\mu}$, $-b > 0$, which is of the form $\int_a^\infty \frac{dx}{x^\mu}, a > 0$.

Test for convergence

Example. $$\int_0^\infty \frac{x^{1/2}}{(1+x)^2} = \int_0^1 \frac{x^{1/2}dx}{(1+x)^2} + \int_1^\infty \frac{x^{1/2}dx}{(1+x)^2} \qquad (1)$$

1st part is a proper integral.

$$\text{2nd part} = \int_1^\infty \frac{dx}{x^{2-1/2}\left(\frac{1}{x}+1\right)^2} = \int_1^\infty \frac{dx}{x^{3/2}\left(1+\frac{1}{x}\right)^2}$$

Here $f(x) = \dfrac{1}{x^{3/2}\left(1+\frac{1}{x}\right)^2}$. Let $g(x) = \dfrac{1}{x^{3/2}}$

(i) $f, g > 0 \ \forall \ x \in [1, \infty)$

(ii) $\lim\limits_{x\to\infty}\frac{f(x)}{g(x)} = \lim\limits_{x\to\infty}\frac{1}{\left(1+\frac{1}{x}\right)^2} = 1$ ($\neq 0$ nor infinity).

Also $\int_1^\infty \frac{dx}{x^{3/2}}$ converges at ∞, since $\mu = \frac{3}{2} > 1$.

$\therefore \int_1^\infty f(x)dx$ converges at ∞. [by Th. [8.10]].

Hence from (1) it follows that $\int_0^\infty \frac{x^{1/2}dx}{(1+x)^2}$ converges at ∞.

Example. $\displaystyle\int_{-\infty}^{-2} \frac{dx}{1-x}$. Put $x = -y$. Then

$$\int_{-\infty}^{-2} \frac{dx}{1-x} = \int_2^\infty \frac{dy}{1+y} = \lim_{t\to\infty}\int_2^t \frac{dy}{1+y} = \lim_{t\to\infty}[\log(1+t) - \log 3] \to \infty.$$

So the limit does not exist. Therefore, $\int_{-\infty}^{-2} \frac{dx}{1-x}$ diverges at $-\infty$.

Alternative method (By comparison test)

$$\int_2^\infty \frac{dy}{1+y} = \int_2^\infty \frac{dy}{y\left(1+\frac{1}{y}\right)}.\ f(y) = \frac{1}{y\left(1+\frac{1}{y}\right)}.\ \text{Let } g(y) = \frac{1}{y}.$$

Then (i) $f, g > 0\ \forall\ y \in [2, \infty)$

(ii) $\lim\limits_{t\to\infty} \frac{f(y)}{g(y)} = \lim\limits_{t\to\infty} \frac{1}{\frac{1}{y}+1} = 1$ ($\neq 0$ nor infinity).

Also $\int_2^\infty g(y)dy = \int_2^\infty \frac{1}{y}dy$ diverges as $\mu = 1$.

$\therefore \int_2^\infty f(y)dy$ diverges. So the given integral diverges.

Example. $\displaystyle\int_0^\infty \frac{x^{p-1}}{1+x}dx = \int_0^1 \frac{x^{p-1}}{1+x}dx + \int_1^\infty \frac{x^{p-1}}{1+x}dx$

1st part is proper, if $p-1 \geq 0$. When $p-1<0$, $x=0$ is a point of infinite discontinuity.

$$f(x) = \frac{1}{x^{1-p}(1+x)}.\ \text{Let } g(x) = \frac{1}{x^{1-p}}$$

Then (i) $f, g > 0\ \forall\ x \in [0, 1]$

(ii) $\lim\limits_{x\to 0} \frac{f(x)}{g(x)} = \lim\limits_{x\to 0} \frac{1}{1+x} = 1$ [$\neq 0$ nor infinity].

Also $\int_0^1 g(x)dx = \int_0^1 \frac{1}{x^{1-p}}dx$ converges when $1-p<1$. That is, when

$$p > 0 \tag{1}$$

2nd part is an improper integral, where $f(x) = \frac{x^{p-1}}{1+x} = \frac{1}{x^{2-p}\left(\frac{1}{x}+1\right)}$. Let $g(x) = \frac{1}{x^{2-p}}$.

Then (i) $f, g > 0\ \forall\ x \in [1, \infty)$

(ii) $\lim\limits_{x\to\infty} \frac{f(x)}{g(x)} = \lim\limits_{x\to\infty} \frac{1}{\left(\frac{1}{x}+1\right)} = 1$ [$\neq 0$ nor infinity].

Also $\int_1^\infty g(x)dx = \int_1^\infty \frac{1}{x^{2-p}}dx$ converges when $2-p>1$. That is, when

$$1 > p \tag{2}$$

So from (1) and (2) it follows that the given integral converges, when $0<p<1$.

Example. $\displaystyle\int_0^\infty \frac{\sin^2 x}{x^2}dx = \int_0^1 \frac{\sin^2 x}{x^2}dx + \int_1^\infty \frac{\sin^2 x}{x^2}dx.$

First part is a proper integral as $\lim\limits_{x\to 0}\left(\frac{\sin x}{x}\right)^2 = 1$.

2nd part is an improper integral, $\int_1^\infty \frac{\sin^2 x}{x^2}dx \leq \int_1^\infty \frac{1}{x^2}dx$.

Now $\int_1^\infty \frac{1}{x^2}dx$ converges at ∞ as $\mu = 2 > 1$.

$\therefore$ by comparison test, $\int_1^\infty \frac{\sin^2 x}{x^2}dx$ converges at ∞.

Hence the given integral converges at ∞.

Theorem [8.11] [Gamma function] Gamma function, denoted by $\Gamma(n)$ defined by $\Gamma(n) = \int_0^\infty e^{-x}x^{n-1}dx$ is convergent, if and only if $n > 0$.

Proof. $\int_0^\infty e^{-x}x^{n-1}dx = \int_0^1 e^{-x}x^{n-1}dx + \int_1^\infty e^{-x}x^{n-1}dx.$

1st part is a proper integral, when $n-1 \geq 0$. When $n-1<0$, $x=0$ is a point of infinite discontinuity of f, where $f(x) = e^{-x}x^{n-1} = \frac{e^{-x}}{x^{1-n}}$.

We examine the convergence of the 1st part at $x = 0$. Let $g(x) = \frac{1}{x^{1-n}}$.

Then (i) $f, g > 0 \ \forall \ x \in [0, 1]$

(ii) $\lim_{x \to 0+} \frac{f(x)}{g(x)} = \lim_{x \to 0} e^{-x} = 1$ [$\neq 0$ nor infinity].

Also $\int_0^1 g(x)dx = \int_0^1 \frac{1}{x^{1-n}}dx$ converges at $x = 0$, if and only if $1 - n < 1$, i.e., if and only if

$$n > 0 \quad \text{(i)}$$

2nd part is an improper integral

$$e^x = 1 + x + \frac{x^2}{2!} + \cdots + \frac{x^{n+1}}{(n+1)!} + \cdots \Rightarrow e^x > \frac{x^{n+1}}{(n+1)!} \ \forall \ x \in [1, \infty)$$

$$\Rightarrow e^{-x} < (n+1)! \frac{1}{x^{n+1}} \Rightarrow e^{-x} x^{n-1} < (n+1)! \frac{x^{n-1}}{x^{n+1}} \Rightarrow e^{-x} x^{n-1} < (n+1)! \frac{1}{x^2}$$

$$\Rightarrow \int_1^\infty e^{-x} x^{n-1} dx < (n+1)! \int_1^\infty \frac{1}{x^2} dx$$

Now $\int_1^\infty \frac{1}{x^2}dx$ converges at ∞ as $\mu = 2 > 1$.

Therefore, by comparison test, $\int_1^\infty e^{-x}x^{n-1}dx$ converges at ∞, for all values of n. (ii)

So from (i) and (ii) it follows that $\Gamma(n)$ converges, if and only if $n > 0$.

Note. $\Gamma(n) = \int_0^\infty e^{-x}x^{n-1}dx$ is also known as *Second Eulerian integral.*

General test for convergence at ∞ when the integrand is not necessarily positive

Theorem [8.12] The necessary and sufficient condition that an improper integral $\int_a^\infty f(x)dx$, when $f(x) \not> 0 \ \forall \ x \in [a, \infty)$, to be convergent at ∞, is that for every $\epsilon > 0$, there exists a (+)ve number K such that $\left|\int_{t_1}^{t_2} f(x)dx\right| < \epsilon \ \forall \ t_1, t_2 \geq K$.

Proof. Let $F(t) = \int_a^t f(x)dx$. From definition $\int_a^\infty f(x)dx = \lim_{t \to \infty} \int_a^t f(x)dx$, when $f(x)$ is bounded and integrable in $[a, t] \ \forall \ t \geq a$.

Now $\int_a^\infty f(x)dx$ convergence at ∞, if $\lim_{t \to \infty} \int_a^t f(x)dx = \lim_{t \to \infty} F(t)$ exists finitely.

The necessary and sufficient condition for the existence of the limit $\lim_{t \to \infty} F(t)$ is that for every $\epsilon(> 0)$, there exists a (+)ve number K such that $|F(t_2) - F(t_1)| < \epsilon, \ \forall \ t_1, t_2 \geq K$

$$\Rightarrow \left| \int_a^{t_2} f(x)dx - \int_a^{t_1} f(x)dx \right| < \epsilon, \ \forall \ t_1, t_2 \geq K$$

$$\Rightarrow \left| \int_{t_1}^{t_2} f(x)dx \right| < \epsilon, \ \forall \ t_1, t_2 \geq K$$

Definition [8.6] : An improper integral $\int_a^\infty f(x)dx$ is said to be absolutely convergent, if $\int_a^\infty |f(x)|dx$ converges at ∞.

Example. The improper integral $\int_1^\infty \frac{\sin x}{x^3}dx$ converges absolutely at ∞.

Since $\int_1^\infty \left|\frac{\sin x}{x^3}\right| dx \leq \int_1^\infty \frac{1}{x^3}dx$ and since $\int_1^\infty \frac{1}{x^3}dx$ converges at ∞ as $\mu = 3 > 1$, $\int_1^\infty \left|\frac{\sin x}{x^3}\right| dx$ converges at ∞ by comparison test.

Theorem [8.13] Every absolutely convergent integral at ∞ is convergent. That is, if $\int_a^\infty |f(x)|dx$ converges at ∞, then $\int_a^\infty f(x)dx$ converges at ∞.

Proof. Let $\int_a^\infty |f(x)|dx$ converge at ∞. Then by Th. [8.12],

$$\left|\int_{t_1}^{t_2} |f(x)|dx\right| < \epsilon, \ \forall \ t_1, t_2 \geq K \qquad \text{(i)}$$

Now $\left|\int_{t_1}^{t_2} f(x)dx\right| \leq \left|\int_{t_1}^{t_2} |f(x)|dx\right| \Rightarrow \left|\int_{t_1}^{t_2} f(x)dx\right| < \epsilon, \ \forall \ t_1, t_2 \geq K$, by (i)

$$\Rightarrow \int_a^\infty f(x)dx \text{ converges at } \infty.$$

Theorem [8.14] If a function ϕ is bounded in $[a, \infty)$ and integrable in $[a, t]$, $\forall \ t \geq a$ and $\int_a^\infty f(x)dx$ converges absolutely at ∞, then $\int_a^\infty f(x)\phi(x)dx$ is absolutely convergent at ∞.

Proof. If ϕ is bounded in $[a, \infty)$, there exists a (+)ve number K such that

$$|\phi(x)| \leq K, \ \forall \ x \geq a \qquad \text{(i)}$$

Again, if $\int_a^\infty f(x)dx$ converges absolutely at ∞, then there exists a (+)ve number M such that

$$\int_a^t |f(x)|dx < M, \ \forall \ t \geq a, \text{ by Th. [8.8]} \qquad \text{(ii)}$$

Then $\displaystyle\int_a^t |f(x)\phi(x)|dx = \int_a^t |f(x)||\phi(x)|dx \leq K \int_a^t |f(x)|dx$, by (i) $\forall \ t \geq a$

$$\leq K \cdot M \ \forall \ t \geq a$$

$$\Rightarrow \int_a^t |f(x)\phi(x)|dx \leq K_1, \ \forall \ t \geq a, \text{ where } K_1 = K \cdot M, \text{ a (+)ve number}$$

$$\Rightarrow \int_a^\infty |f(x)\phi(x)|dx \text{ is convergent at } \infty$$

$$\Rightarrow \int_a^\infty f(x)\phi(x)dx \text{ is absolutely convergent at } \infty.$$

Test for conditional convergence

Theorem [8.15] (1) **[Abel's Test for convergence]** If a function ϕ is bounded and monotonic in $[a, \infty)$ and $\int_a^\infty f(x)dx$ converges at ∞, then $\int_a^\infty f(x)\phi(x)dx$ converges at ∞.

Proof. ϕ is bounded in $[a, \infty)$

$$\Rightarrow |\phi(x)| \leq K \ \forall \ x \geq a \qquad \text{(i)}$$

ϕ is bounded and monotonic in $[a, \infty)$.

$\Rightarrow \phi$ is R-integrable in $[a, t) \ \forall \ t \geq a$. Also $\int_a^\infty f(x)dx$ converges at $\infty \Rightarrow f$ is R-integrable in $[a, t) \ \forall \ t \geq a$.

Therefore, applying 2nd Mean Value Theorem of Weirstrass's form, we get

$$\int_{t_1}^{t_2} f(x)\phi(x)dx = \phi(t_1)\int_{t_1}^{\xi} f(x)dx + \phi(t_2)\int_{\xi}^{t_2} f(x)dx \text{ when } a < t_1 < \xi \leq t_2$$

$$\Rightarrow \left|\int_{t_1}^{t_2} f(x)\phi(x)dx\right| \leq |\phi(t_1)|\left|\int_{t_1}^{\xi} f(x)dx\right| + |\phi(t_2)|\left|\int_{\xi}^{t_2} f(x)dx\right| \qquad \text{(ii)}$$

Again, $\int_a^\infty f(x)dx$ converges at ∞

$$\Rightarrow \left|\int_{t_1}^{t_2} f(x)dx\right| < \frac{\epsilon}{2K} \ \forall\, t_1, t_2 \geq K \text{ [by Th. [8.12]]} \tag{iii}$$

Now $t_1, t_2 \geq K \Rightarrow \xi$ lying between t_1, t_2 is also $\geq K$.

Therefore, from (iii),

$$\left|\int_{t_1}^{\xi} f(x)dx\right| < \frac{\epsilon}{2K} \text{ and } \left|\int_{\xi}^{t_2} f(x)dx\right| < \frac{\epsilon}{2K},\ \forall\, t_1, t_2 \geq K \tag{iv}$$

From (ii), we get $\left|\int_{t_1}^{t_2} f(x)\phi(x)dx\right| \leq K \cdot \frac{\epsilon}{2K} + K \cdot \frac{\epsilon}{2K} = \epsilon$ [by (i) and (iv)]

$$\Rightarrow \int_a^\infty f(x)\phi(x)dx \text{ converges at } \infty.$$

(2) [**Dirichlet's test for convergence**] If a function ϕ is bounded and monotonic in $[a, \infty)$ and $\phi \to 0$ as $x \to \infty$ and if $\int_a^t f(x)dx$ is bounded $\forall\, t \geq a$, then $\int_a^\infty f(x)\phi(x)dx$ converges at ∞.

Proof. ϕ is R-integrable on $[a, t]$, $\forall\, t \geq a$ as ϕ is monotonic on $[a, \infty)$. f is also R-integrable on $[a, t]$, $\forall\, t \geq a$ as $\int_a^t f(x)dx$ is bounded.

Therefore, applying 2nd Mean Value Theorem of Weierstrass's form,

$$\int_{t_1}^{t_2} f(x)\phi(x)dx = \phi(t_1)\int_{t_1}^{\xi} f(x)dx + \phi(t_2)\int_{\xi}^{t_2} f(x)dx \tag{i}$$

for $a < t_1 \leq \xi \leq t_2$.

Again, $\int_a^t f(x)dx$ is bounded $\Rightarrow \exists$ a (+)ve number K such that

$$\left|\int_a^t f(x)dx\right| \leq K \ \forall\, t \geq a \tag{ii}$$

$$\text{Now } \left|\int_{t_1}^{\xi} f(x)dx\right| = \left|\int_a^{t_1} f(x)dx - \int_a^{\xi} f(x)dx\right|$$

$$\leq \left|\int_a^{t_1} f(x)dx\right| + \left|\int_a^{\xi} f(x)dx\right| \leq 2K \text{ [by (ii)], as } \xi > a \tag{iii}$$

$$\text{Also } \left|\int_{\xi}^{t_2} f(x)dx\right| = \left|\int_a^{t_2} f(x)dx - \int_a^{\xi} f(x)dx\right|$$

$$\leq \left|\int_a^{t_2} f(x)dx\right| + \left|\int_a^{\xi} f(x)dx\right| \leq 2K \text{ [by (ii)]} \tag{iv}$$

From (i),

$$\left|\int_{t_1}^{t_2} f(x)dx\right| \leq |\phi(t_1)|\left|\int_{t_1}^{\xi} f(x)dx\right| + |\phi(t_2)|\left|\int_{\xi}^{t_2} f(x)dx\right| \tag{v}$$

Now, since $\phi(x) \to 0$ as $x \to \infty$, $\exists$ a (+)ve number d such that $|\phi(x)| \leq \frac{\epsilon}{4K} \ \forall\, x \geq d$.

In particular,

$$|\phi(t_1)| \le \frac{\epsilon}{4K} \text{ and } |\phi(t_2)| \le \frac{\epsilon}{4K}, \ \forall\, t_1, t_2 \ge d \tag{vi}$$

From (v), it follows that $\left|\int_{t_1}^{t_2} f(x)\phi(x)dx\right| \le \frac{\epsilon}{4K}\cdot 2K + \frac{\epsilon}{4K}\cdot 2K = \epsilon,\ \forall\, t_1, t_2 \ge d$ by (iii), (iv) and (vi)

$$\Rightarrow \int_a^\infty f(x)\phi(x)dx \text{ converges at } \infty.$$

Test for convergence of

Example. (1) $\int_1^\infty e^{-ax}\cdot\frac{1}{x^2}dx, a \ge 0$ (2) $\int_1^\infty \frac{\cos x}{x}dx$

(3) $\int_0^\infty \frac{\sin x}{x}dx$ (4) $\int_0^\infty e^{-ax}\frac{\sin x}{x}dx,\ a \ge 0$

(1) Let $\phi(x) = e^{-ax}$ and $f(x) = \frac{1}{x^2}$. Now ϕ is bounded and monotonic in $[1, \infty]$. $\int_1^\infty f(x)dx = \int_1^\infty \frac{1}{x^2}dx$ is convergent at ∞ as $\mu = 2 > 1$.

Therefore, by Abel's test $\int_1^\infty f(x)\phi(x)dx$ converges at ∞. That is, $\int_1^\infty e^{-ax}.\frac{1}{x^2}dx$ converges at ∞.

(2) Let $\phi(x) = \frac{1}{x}$ and $f(x) = \cos x$. Then ϕ is bounded and monotonically decreasing on $[1, \infty]$ and $\phi \to 0$ as $x \to \infty$.

Also $\left|\int_1^t f(x)dx\right| = \left|\int_1^t \cos x dx\right| \le |\sin t| + |\sin 1| \le 2.$

Therefore, $\int_1^t f(x)dx$ is bounded and hence by Dirichlet's test $\int_1^\infty f(x)\phi(x)dx = \int_1^\infty \frac{\cos x}{x}dx$ converges at ∞.

$$(3)\quad \int_0^\infty \frac{\sin x}{x}dx = \int_0^1 \frac{\sin x}{x}dx + \int_1^\infty \frac{\sin x}{x}dx \tag{1}$$

1st part of (1) is a proper integral as $\lim_{x\to 0}\frac{\sin x}{x} = 1$. That is, 0 is not a point of infinite discontinuity. For the 2nd part of (1) let $\phi(x) = \frac{1}{x}$ and $f(x) = \sin x$. Then ϕ is bounded and monotonically decreasing on $[1, \infty]$ and $\phi \to 0$ as $x \to \infty$.

Also $\left|\int_1^t f(x)dx\right| = \left|\int_1^t \sin x dx\right| \le |\cos t| + |\cos 1| \le 2 \Rightarrow \int_1^t f(x)dx$ is bounded.

Therefore, by Dirichlet's test $\int_1^\infty \frac{\sin x}{x}dx$ is convergent.

Hence, $\int_0^\infty \frac{\sin x}{x}dx$ is convergent.

$$(4)\quad \int_0^\infty e^{-ax}\frac{\sin x}{x}dx = \int_0^1 e^{-ax}\frac{\sin x}{x}dx + \int_1^\infty e^{-ax}\frac{\sin x}{x}dx,\ a \ge 0 \tag{1}$$

1st part is a proper integral as $\lim_{x\to 0}\frac{\sin x}{x} = 1$. For the 2nd part of (1), let $\phi(x) = e^{-ax}$ and $f(x) = \frac{\sin x}{x}$. Then ϕ is bounded, monotonic on $[1, \infty)$. Also $\int_1^\infty \frac{\sin x}{x}dx$ is convergent by example (3). Therefore, by Abel's test, $\int_1^\infty e^{-ax}\frac{\sin x}{x}dx$ is convergent and hence $\int_0^\infty e^{-ax}\frac{\sin x}{x}dx$ is convergent.

Problem [8.3]: Show that the improper integral $\int_0^\infty \frac{\sin x}{x}dx$ is not absolutely convergent.

Solution. It is sufficient to show that $\int_0^\infty \left|\frac{\sin x}{x}\right| dx = \int_0^\infty \frac{|\sin x|}{x}dx$ is not convergent.

$\int_0^{n\pi} \frac{|\sin x|}{x}dx$ is a proper integral.

Also $\displaystyle\int_0^{n\pi}\frac{|\sin x|}{x}dx=\sum_{r=1}^{n}\int_{(r-1)\pi}^{r\pi}\frac{|\sin x|}{x}dx.$

Putting $x=(r-1)\pi+y$, y ranges from 0 to π as x ranges from $(r-1)\pi$ to $r\pi$.

$$\begin{aligned}\int_0^{n\pi}\frac{|\sin x|}{x}dx&=\sum_{r=1}^{n}\int_0^{\pi}\frac{|\sin\{(r-1)\pi+y\}|}{(r-1)\pi+y}dy\\&=\sum_{r=1}^{n}\int_0^{\pi}\frac{|(-1)^{r-1}\sin y|}{(r-1)\pi+y}dy\\&=\sum_{r=1}^{n}\int_0^{\pi}\frac{\sin y}{(r-1)\pi+y}dy\geq\sum_{r=1}^{n}\int_0^{\pi}\frac{\sin y}{r\pi}dy\end{aligned}$$

Since $r\pi$ is the maximum value of $(r-1)\pi+y$, when $y\in[0,\pi]$

$$\Rightarrow\int_0^{n\pi}\frac{|\sin x|}{x}dx\geq\sum_{r=1}^{n}\frac{1}{r\pi}\big[-\cos y\big]_0^{\pi}=\sum_{r=1}^{n}\frac{2}{r\pi}$$

$$\Rightarrow\int_0^{n\pi}\frac{|\sin x|}{x}dx\geq\frac{2}{\pi}\sum_{r=1}^{n}\frac{1}{r}$$

Now $\sum_{r=1}^{n}\frac{1}{r}$ is a divergent series when $n\to\infty$.

That is $\sum_{r=1}^{n}\frac{1}{r}\to\infty$ as $n\to\infty$.

$$\Rightarrow\int_0^{n\pi}\frac{|\sin x|}{x}dx\to\infty\text{ as }n\to\infty.$$

Let t be a real number. Then $\exists$ a (+)ve integer n such that $n\pi\leq t\leq(n+1)\pi$

$$\Rightarrow\int_0^{t}\frac{|\sin x|}{x}dx\geq\int_0^{n\pi}\frac{|\sin x|}{x}dx$$

Let $t\to\infty$ so that $n\to\infty$.

Thus we have $\int_0^t\frac{|\sin x|}{x}dx\to\infty\Rightarrow\int_0^\infty\frac{|\sin x|}{x}$ is not convergent.

Therefore, $\int_0^\infty\frac{\sin x}{x}dx$ is not absolutely convergent.

Problem [8.4] : Show that the improper integral $\int_2^\infty\frac{\cos x}{\log x}dx$ is convergent but not absolutely convergent.

Solution. Let $f(x)=\cos x$ and $\phi(x)=\frac{1}{\log x}$.

Then $\phi'(x)=-\frac{1}{(\log x)^2}\cdot\frac{1}{x}<0\ \forall\ x\in[2,\infty)$.

So ϕ is bounded and monotonically decreasing on $[2,\infty)$ and $\phi(x)\to0$ as $x\to\infty$.

Also $\left|\int_2^t\cos x\,dx\right|\leq2$.

Therefore, by Dirichlet's test

$$\int_2^\infty f(x)\phi(x)dx=\int_2^\infty\frac{\cos x}{\log x}dx\text{ converges at }\infty.$$

Now $\displaystyle\int_2^{(2n+1)\frac{\pi}{2}} \frac{|\cos x|}{\log x}dx, (\log x > 0 \ \forall\ x \in [2,\infty))$

$$= \int_2^{\frac{3\pi}{2}} \frac{|\cos x|}{\log x}dx + \int_{\frac{3\pi}{2}}^{\frac{5\pi}{2}} \frac{|\cos x|}{\log x}dx + \cdots + \int_{(2n-1)\frac{\pi}{2}}^{(2n+1)\frac{\pi}{2}} \frac{|\cos x|}{\log x}dx$$

$$= \int_{\frac{\pi}{2}}^{2} \frac{|\cos x|}{\log x}dx + \int_2^{\frac{3\pi}{2}} \frac{|\cos x|}{\log x}dx + \int_{\frac{3\pi}{2}}^{\frac{5\pi}{2}} \frac{|\cos x|}{\log x}dx + \cdots + \int_{(2n-1)\frac{\pi}{2}}^{(2n+1)\frac{\pi}{2}} \frac{|\cos x|}{\log x}dx - \int_{\frac{\pi}{2}}^{2} \frac{|\cos x|}{\log x}dx$$

$$= \int_{\frac{\pi}{2}}^{\frac{3\pi}{2}} \frac{|\cos x|}{\log x} + \int_{\frac{3\pi}{2}}^{\frac{5\pi}{2}} \frac{|\cos x|}{\log x}dx + \cdots + \int_{(2n-1)\frac{\pi}{2}}^{(2n+1)\frac{\pi}{2}} \frac{|\cos x|}{\log x}dx - \int_{\frac{\pi}{2}}^{2} \frac{|\cos x|}{\log x}dx$$

$$= \sum_{r=1}^{n} \int_{(2r-1)\frac{\pi}{2}}^{(2r+1)\frac{\pi}{2}} \frac{|\cos x|}{\log x}dx - \int_{\frac{\pi}{2}}^{2} \frac{|\cos x|}{\log x}dx \qquad (1)$$

2nd part of (1) is a proper integral. We put $x = (2r-1)\frac{\pi}{2} + y$ for the 1st part. Then $y \in [0,\pi]$.

$$\text{1st part} = \sum_{r=1}^{n} \int_0^{\pi} \frac{|\cos\{(2r-1)\frac{\pi}{2}+y\}|}{\log\{(2r-1)\frac{\pi}{2}+y\}}dy$$

$$= \sum_{r=1}^{n} \int_0^{\pi} \frac{\sin y}{\log\{(2r-1)\frac{\pi}{2}+y\}}dy \quad \left[\because \cos\left\{(2n-1)\frac{\pi}{2}+y\right\} = (-1)^{2n-1}\sin y\right]$$

Now $(2r-1)\frac{\pi}{2} + y$ is maximum when $y = \pi$ and maximum value is $(2r+1)\frac{\pi}{2}$.
So $(2r-1)\frac{\pi}{2} + y < (2r+1)\frac{\pi}{2}$

$$\Rightarrow \log\left\{(2r-1)\frac{\pi}{2}+y\right\} < \log(2r+1)\frac{\pi}{2}$$

$$\Rightarrow \frac{1}{\log\left\{(2r-1)\frac{\pi}{2}+y\right\}} > \frac{1}{\log(2r+1)\frac{\pi}{2}}$$

$$\text{So 1st part} \geq \sum_{r=1}^{n} \int_0^{\pi} \frac{\sin y}{\log(2r+1)\frac{\pi}{2}}dy = \sum_{r=1}^{n} \frac{1}{\log(2r+1)\frac{\pi}{2}} \int_0^{\pi} \sin y dy = \sum_{r=1}^{n} \frac{2}{\log(2r+1)\frac{\pi}{2}}$$

which is a divergent series as $n \to \infty$.

Therefore, from (1) it follows that $\int_2^{\infty} \frac{|\cos x|}{\log x}dx$ does not converges at ∞.

Hence, $\int_2^{\infty} \frac{\cos x}{\log x}dx$ is not absolutely convergent.

Problem [8.5]: (1) Examine the convergence of the following improper integrals

(i) $\int_0^{\infty} \frac{\cos x dx}{\sqrt{1+x^3}}$ (ii) $\int_0^{\infty} \frac{\cos x}{\sqrt{x^2+x}}dx$ (iii) $\int_1^{\infty} \frac{\sin x}{x^p}dx$

(iv) $\int_0^{\infty} e^{-a^2x^2} \frac{\sin 2bx}{x}dx$ (v) $\int_0^{\infty} \left(\frac{1}{1+x} - e^{-x}\right)\frac{dx}{x}$ (vi) $\int_{-\infty}^{\infty} e^{-x^2}dx$

(vii) $\int_{-\infty}^{\infty} e^{-\left(x-\frac{a}{x}\right)^2}dx$

(2) Show that $\int_0^\infty \frac{dx}{1+x^4\sin^2 x}dx$ converges but $\int_0^\infty \frac{dx}{1+x^2\sin^2 x}$ does not converge.

(3) Discuss the convergence of $\int_0^{\pi/2} \cos 2nx \log \sin x dx$ and evaluate it.

(4) Show that $\int_0^1 \log \Gamma(x)dx$ is convergent and hence evaluate it.

(5) Show that $\int_0^\infty \frac{\sin(x+x^2)}{x^p}dx$ converges for $-1 < p < 2$.

Solution. (1) (i) $$\int_0^\infty \frac{\cos x}{\sqrt{1+x^3}}dx = \int_0^1 \frac{\cos x}{\sqrt{1+x^3}}dx + \int_1^\infty \frac{\cos x}{\sqrt{1+x^3}}. \qquad (1)$$

1st part of (1) is a proper integral.

For the 2nd part, let $f(x) = \cos x$ and $\phi(x) = \frac{1}{\sqrt{1+x^3}}$

$$\phi'(x) = -\frac{1}{2}(1+x^3)^{-3/2}\cdot 3x^2 = -\frac{3x^2}{2\cdot\sqrt{(1+x^3)^3}} < 0 \;\forall\; x \in [1,\infty]$$

ϕ is bounded and monotonically decreasing in $[1,\infty]$ and $\phi(x) \to 0$ as $x \to \infty$.

Also $\left|\int_1^t \cos x dx\right| \le 2$.

Therefore, by Dirichlet's test $\int_1^\infty \frac{\cos x}{\sqrt{1+x^3}}dx$ is converges. Hence the given integral converges at ∞.

(ii) $$\int_0^\infty \frac{\cos x}{\sqrt{x+x^2}}dx = \int_0^1 \frac{\cos x}{\sqrt{x+x^2}}dx + \int_1^\infty \frac{\cos x}{\sqrt{x+x^2}}dx \qquad (1)$$

For the 1st part $x = 0$ is a point of infinite discontinuity.

Let $f(x) = \cos x$ and $\phi(x) = \frac{1}{\sqrt{x+x^2}}$; $f, \phi > 0$ in $[0,1]$.

$$\lim_{x\to 0}\frac{f(x)}{\phi(x)} = 1.$$

Also $\phi(x) = \frac{1}{x^{1/2}\sqrt{1+x}} \Rightarrow \int_0^1 \phi(x)dx$ converges at $x = 0$ by comparison test. So the 1st part of (1) converges at $x = 0$.

2nd part $\int_1^\infty \frac{\cos x}{\sqrt{x+x^2}}dx$ also converges at ∞ by Dirichlet's test, proceeding as in (1) (i).

Therefore, the given integral is convergent.

(iii) $\int_1^\infty \frac{\sin x}{x^p}dx$. Let $f(x) = \sin x$ and $\phi(x) = \frac{1}{x^p}$.

Then ϕ is bounded, monotonically decreasing in $[1,\infty)$ and $\to 0$ as $x \to \infty$, if $p > 0$.

Also $\left|\int_1^t \sin x dx\right| \le 2$.

Therefore, by Dirichlet's test the given integral converges at ∞, if $p > 0$.

(iv) $$\int_0^\infty e^{-a^2x^2}\frac{\sin 2bx}{x}dx$$
$$= 2b\int_0^\infty e^{-a^2x^2}\frac{\sin 2bx}{2bx}dx$$
$$= 2b\left[\int_0^1 e^{-a^2x^2}\frac{\sin 2bx}{2bx}dx + \int_1^\infty e^{-a^2x^2}\frac{\sin 2bx}{2bx}dx\right] \qquad (1)$$

1st part is a proper integral as $\lim_{x\to 0}\frac{\sin 2bx}{2bx} = 1$.

For the 2nd part, let $f(x) = \frac{\sin 2bx}{2bx}$ and $\phi(x) = e^{-a^2x^2}$.

Then $\int_1^\infty f(x)dx = \int_1^\infty \frac{\sin 2bx}{2bx}dx$ is convergent by Dirichlet's test. Also $\phi(x)$ is bounded and monotonically decreasing in $[1,\infty)$.

Therefore, by Abel's test $\int_1^\infty f(x)\phi(x)dx = \int_1^\infty e^{-a^2x^2}\frac{\sin 2bx}{2bx}dx$ is convergent.

Hence from (1) it follows that the given integral is convergent.

(v) $$\int_0^\infty \left(\frac{1}{1+x}-e^{-x}\right)\frac{dx}{x} = \int_0^1 \left(\frac{1}{1+x}-e^{-x}\right)\frac{dx}{x} + \int_1^\infty \left(\frac{1}{1+x}-e^{-x}\right)\frac{dx}{x}.$$

In the first part, $x=0$ is not a point of infinite discontinuity since

$$\lim_{x\to 0}\left(\frac{1}{1+x}-e^{-x}\right)\frac{dx}{x} = \lim_{x\to 0}\left(\frac{1-e^{-x}(1+x)}{x(1+x)}\right)\left(\frac{0}{0}\right)$$

$$= \lim_{x\to 0}\frac{(e^{-x}(1+x)-e^{-x})}{(1+x)+x} = \lim_{x\to 0}\left(\frac{x}{e^x(1+2x)}\right) = 0$$

$$\text{2nd part} = \int_1^\infty \left(\frac{1}{1+x}-e^{-x}\right)\frac{dx}{x} = \int_1^\infty \frac{1-(1+x)e^{-x}}{x(1+x)}dx = \int_1^\infty \frac{1-(1+x)e^{-x}}{x^2(1+1/x)}dx$$

Let $f(x) = \frac{1-(1+x)e^{-x}}{x^2(1+1/x)}$ and $\phi(x) = \frac{1}{x^2}$

(i) $f,\phi > 0 \;\forall\; x\in[1,\infty)$.

(ii) $\lim_{x\to\infty}\frac{f(x)}{\phi(x)} = \lim_{x\to\infty}\frac{1-(1+x)e^{-x}}{(1+\frac{1}{x})} = 1.$

Also $\int_1^\infty \phi(x)dx = \int_1^\infty \frac{1}{x^2}dx$ converges at ∞. So the integral of the 2nd part is convergent. Hence the given integral is convergent.

(vi) $I = \int_{-\infty}^\infty e^{-x^2}dx = 2\int_0^\infty e^{-x^2}dx$. Put $x^2 = z$. Then $dx = \frac{dz}{2\sqrt{z}}$

$x: 0\to\infty \Rightarrow z: 0\to\infty$.

$$I = \int_0^\infty e^{-z}\frac{dz}{2\sqrt{z}} = \frac{1}{2}\int z^{-\frac{1}{2}}e^{-z}dz = \frac{1}{2}\int_0^\infty z^{\frac{1}{2}-1}e^{-z}dz = \frac{1}{2}\Gamma\left(\frac{1}{2}\right),\; n=\frac{1}{2}>0$$

Therefore, the given integral converges by Th [8.11].

(vii) $$\int_{-\infty}^\infty e^{-(x-\frac{a}{x})^2}dx = 2\int_0^\infty e^{-(x-\frac{a}{x})^2}dx$$

$$= 2\int_0^\infty e^{-\left\{x^2-2a+\frac{a^2}{x^2}\right\}}dx = 2e^{2a}\int_0^\infty e^{-x^2}\cdot e^{-\frac{a^2}{x^2}}dx \tag{1}$$

Now

$$\int_0^\infty e^{-x^2}e^{-\frac{a^2}{x^2}}dx = \int_0^1 e^{-x^2}e^{-\frac{a^2}{x^2}}dx + \int_1^\infty e^{-x^2}e^{-\frac{a^2}{x^2}}dx$$

1st part is a proper integral as 0 is not a point of infinite discontinuity. For the 2nd part, let $f(x) = e^{-x^2}$ and $\phi(x) = e^{-\frac{a^2}{x^2}}$.

Then $\int_1^\infty e^{-x^2}dx$ is convergent, since $\int_0^\infty e^{-x^2}dx$ is convergent by (vi) and

$$\int_0^\infty e^{-x^2}dx = \int_0^1 e^{-x^2}dx + \int_1^\infty e^{-x^2}dx$$

where $\int_0^1 e^{-x^2}dx$ is a proper integral.

Also $\phi(x) = e^{-\frac{a^2}{x^2}} \Rightarrow \phi'(x) = e^{-\frac{a^2}{x^2}}\left(\frac{2a^2}{x^3}\right) > 0 \; \forall \; x \in [1, \infty)$. So ϕ is monotonically increasing and ϕ is bounded in $[1, \infty)$.

Therefore, by Abel's theorem,

$$\int_1^\infty f(x)\phi(x)dx = \int_1^\infty e^{-x^2} e^{-\frac{a^2}{x^2}} dx \text{ is convergent.}$$

Hence from (1) it follows that the given integral is convergent.

(2) $\displaystyle\int_0^\infty \frac{dx}{1 + x^4 \sin^2 x}$.

We consider the proper integral $\int_0^{n\pi} \frac{dx}{1+x^4\sin^2 x}$.

$$\int_0^{n\pi} \frac{dx}{1 + x^4 \sin^2 x} = \sum_{r=1}^{n} \int_{(r-1)\pi}^{r\pi} \frac{dx}{1 + x^4 \sin^2 x}$$

Now $x \geq (r-1)\pi \; \forall \; x \in [(r-1)\pi, r\pi]$

$$\Rightarrow x^4 \geq (r-1)^4\pi^4 \Rightarrow 1 + x^4 \sin^2 x \geq 1 + (r-1)^4\pi^4 \sin^2 x$$

Hence,

$$\sum_{r=1}^{n} \int_{(r-1)\pi}^{r\pi} \frac{dx}{1 + x^4 \sin^2 x} \leq \sum_{r=1}^{n} \int_{(r-1)\pi}^{r\pi} \frac{dx}{1 + (r-1)^4\pi^4 \sin^2 x}.$$

We put $x = (r-1)\pi + y$. Then as $x \in [(r-1)\pi, r\pi], y \in [0, \pi]$ and so

$$\sum_{r=1}^{n} \int_{(r-1)\pi}^{r\pi} \frac{dx}{1 + x^4 \sin^2 x} \leq \sum_{r=1}^{n} \int_0^{\pi} \frac{dy}{1 + (r-1)^4\pi^4 \sin^2 y}$$

$$= \sum_{r=1}^{n} \int_0^{\pi} \frac{\sec^2 y dy}{1 + \tan^2 y + (r-1)^4\pi^4 \tan^2 y}$$

$$[\because \{\sin(r-1)\pi + y\}^2 = \{(-1)^{r-1} \sin y\}^2 = \sin^2 y]$$

$$= \sum_{r=1}^{n} 2 \cdot \int_0^{\pi/2} \frac{\sec^2 y dy}{1 + \tan^2 y + (r-1)^4\pi^4 \tan^2 y}$$

$$= \sum_{r=1}^{n} 2 \cdot \int_0^{\infty} \frac{dz}{1 + \{1 + (r-1)^4\pi^4\}z^2}, \text{ where } \tan y = z$$

$$= \sum_{r=1}^{n} \frac{2 \times \sqrt{1 + (r-1)^4\pi^4}}{1 + (r-1)^4\pi^4} \left[\tan^{-1} z\sqrt{1 + (r-1)^4\pi^4}\right]_0^\infty$$

$$= \sum_{r=1}^{n} \frac{\pi}{\sqrt{1 + (r-1)^4\pi^4}} = \sum_{r=1}^{n} \frac{\pi}{r^2} \frac{1}{\sqrt{\frac{1}{r^4} + \left(1 - \frac{1}{r}\right)^4 \pi^4}}$$

So, let $n \to \infty$. Then $r \to \infty$ and consequently,

$$\int_0^\infty \frac{dx}{1 + x^4 \sin^2 x} dx \leq \sum_1^\infty \frac{1}{r^2} \left[\because \sum_{r=1}^{\infty} \frac{\pi}{r^2\sqrt{\frac{1}{r^4} - \left(1-\frac{1}{r}\right)^4 \pi^4}} \sim \sum_{r=1}^{\infty} \frac{1}{r^2}\right]$$

The series $\sum\limits_1^\infty \frac{1}{r^2}$ is a convergent series and hence the given integral $\int_0^\infty \frac{dx}{1+x^4\sin^2 x}$ is convergent.

Let us now consider the integral $\int_0^{n\pi} \frac{dx}{1+x^2\sin^2 x}$

$$\int_0^{n\pi} \frac{dx}{1+x^2\sin^2 x} = \sum_{r=1}^{n} \int_{(r-1)\pi}^{r\pi} \frac{dx}{1+x^2\sin^2 x}$$

$$[\because x \le r\pi \Rightarrow 1+x^2\sin^2 x \le 1+r^2\pi^2\sin^2 x]$$

$$\Rightarrow \int_0^{n\pi} \frac{dx}{1+x^2\sin^2 x} \ge \sum_{r=1}^{n} \int_{(r-1)\pi}^{r\pi} \frac{dx}{1+r^2\pi^2\sin^2 x}$$

$$= \sum_{r=1}^{n} \int_0^{\pi} \frac{dy}{1+r^2\pi^2\sin^2 y} \quad [\text{Put } x=(r-1)\pi+y]$$

$$= \sum_{r=1}^{n} \int_0^{\pi} \frac{\sec^2 y dx}{(1+\tan^2 y)+r^2\pi^2\tan^2 y}$$

$$= \sum_{r=1}^{n} 2\int_0^{\pi/2} \frac{\sec^2 y dy}{(1+\tan^2 y)+r^2\pi^2\tan^2 y}$$

$$= \sum_{r=1}^{n} 2\int_0^{\infty} \frac{dz}{1+(1+r^2\pi^2)z^2} \quad [\text{Put } \tan y = z]$$

$$= \sum_{r=1}^{n} 2\frac{1}{\sqrt{1+r^2\pi^2}}\frac{\pi}{2} = \sum_{r=1}^{n} \frac{\pi}{r\sqrt{\frac{1}{r^2}+\pi^2}}$$

Let $n \to \infty$. Then $r \to \infty$ and hence

$$\int_0^{\infty} \frac{dx}{1+x^2\sin^2 x} \ge \sum_{r=1}^{\infty} \frac{1}{r} \left[\sum_{r=1}^{n} \frac{\pi}{r\sqrt{\frac{1}{r^2}+\pi^2}} \sim \sum_{r=1}^{\infty} \frac{1}{r} \right] \tag{1}$$

Now the series $\sum_{r=1}^{\infty} \frac{1}{r}$ is a divergent series and hence from (1) it follows that $\int_0^{\infty} \frac{dx}{1+x^2\sin^2 x}$ does not converge.

(3) $\int_0^{\pi/2} \cos 2nx \log \sin x dx$, $x=0$ is the only point of infinite discontinuity.

Now $\lim_{x\to 0} x^{\mu} \cos 2nx \log \sin x = 0$, when $\mu > 0$ (1)

$$\Rightarrow |\cos 2nx \log \sin x| < \frac{\epsilon}{x^{\mu}}$$

$$\Rightarrow \int_0^{\pi/2} |\cos 2nx \log \sin x| dx < \epsilon \int_0^{\pi/2} \frac{dx}{x^{\mu}}$$

$$\Rightarrow \int_0^{\pi/2} |\cos 2nx \log \sin x| dx \text{ is convergent, if and only if } \mu < 1.$$

So for $0 < \mu < 1$, the given integral converges absolutely. Hence the given integral is convergent for $0 < \mu < 1$.

Let $I = \int_0^{\pi/2} \cos 2nx \log \sin x dx$. Integrating by parts,

$$= \left[\log \sin x \cdot \frac{\sin 2nx}{2n}\right]_0^{\pi/2} - \int_0^{\pi/2} \frac{\cos x}{\sin x} \cdot \frac{\sin 2nx}{2n} dx$$

$$= -\frac{1}{2n} \int_0^{\pi/2} \frac{\sin 2nx \cos x}{\sin x} dx \tag{2}$$

Let $I_n = \int_0^{\pi/2} \frac{\sin 2nx \cos x}{\sin x} dx$

$$\sin(2n+1)x = \sin 2nx \cos x + \cos 2nx \sin x$$

$$\Rightarrow \frac{\sin(2n+1)x}{\sin x} = \frac{\sin 2nx \cos x}{\sin x} + \cos 2nx$$

$$\Rightarrow \int_0^{\pi/2} \frac{\sin 2nx \cos x}{\sin x} = \int_0^{\pi/2} \frac{\sin(2n+1)x}{\sin x} dx - \int_0^{\pi/2} \cos 2nx dx$$

$$= \int_0^{\pi/2} \frac{\sin(2n+1)x}{\sin x} dx \quad \left[\because \int_0^{\pi/2} \cos 2nx dx = 0\right] \tag{3}$$

Also

$$\frac{\sin(2n-1)x}{\sin x} = \frac{\sin 2nx \cos x}{\sin x} - \cos 2nx$$

$$\Rightarrow \int_0^{\pi/2} \frac{\sin 2nx \cos x}{\sin x} dx = \int_0^{\pi/2} \frac{\sin(2n-1)x}{\sin x} dx \tag{4}$$

Thus from (3) and (4) it follows that

$$I_n = \int_0^{\pi/2} \frac{\sin(2n+1)x}{\sin x} dx = \int_0^{\pi/2} \frac{\sin(2n-1)x}{\sin x} dx$$

Hence it follows, $I_n = \displaystyle\int_0^{\pi/2} \frac{\sin x}{\sin x} dx = \frac{\pi}{2}$.

Therefore from (2), $I = -\dfrac{1}{2n} \cdot \dfrac{\pi}{2} = -\dfrac{\pi}{4n}$.

(4) $\displaystyle\int_0^1 \log \Gamma(x)\, dx$.

$\Gamma(x) = \displaystyle\int_0^\infty e^{-p} p^{x-1} dp$ converges for $x > 0$.

Then $\Gamma(x+1) = \displaystyle\int_0^\infty e^{-p} p^x dp$ is a proper integral for

$$x > 0 \tag{1}$$

Now $\Gamma(x+1) = x\Gamma(x)$

$$\Rightarrow \int_0^1 \log \Gamma(x+1) dx = \int_0^1 \log x dx + \int_0^1 \log \Gamma(x) dx$$

$$\Rightarrow \int_0^1 \log \Gamma(x) dx = \int_0^1 \log \Gamma(x+1) dx - \int_0^1 \log x dx \tag{2}$$

1st part is a proper integral by (1).

For the 2nd part $x = 0$ is a point of infinite discontinuity.

$$\int_0^1 \log x dx = \lim_{\epsilon \to 0} \int_\epsilon^1 \log x dx = \lim_{\epsilon \to 0} \left[x \log x - x\right]_\epsilon^1$$

$$= \lim_{\epsilon \to 0} [-1 - \epsilon \log \epsilon + \epsilon] = -1 \quad \left[\because \lim_{\epsilon \to 0} \epsilon \log \epsilon = 0\right]$$

So 2nd part converges at $x = 0$.

From (2) it follows that $\int_0^1 \log \Gamma(x) dx$ converges at $x = 0$.

Let $I = \int_0^1 \log\Gamma(x)dx$. Then $I = \int_0^1 \log\Gamma(1-x)dx$.

$$2I = \int_0^1 \log\Gamma x \cdot \Gamma(1-x)dx \quad \left[\text{we have } \Gamma m \cdot \Gamma(1-m) = \frac{\pi}{\sin \pi m}\right]$$

$$= \int_0^1 \log\frac{\pi}{\sin \pi x}dx = \frac{1}{\pi}\int_0^\pi \log\frac{\pi}{\sin z}dz, \text{ putting } \pi x = z$$

$$= \frac{1}{\pi}\int_0^\pi \log \pi dz - \frac{1}{\pi}\cdot 2\int_0^{\pi/2} \log \sin z dz$$

$$= \log \pi - \frac{1}{\pi}\cdot 2 \cdot \frac{\pi}{2}\log\frac{1}{2} \text{ [see Problem [8.1]—(3)]}$$

$$= \log \pi + \log 2 = \log 2\pi$$

$$\therefore\ I = \frac{1}{2}\log 2\pi$$

(5) $$\int_0^\infty \frac{\sin(x+x^2)}{x^p}dx = \int_0^1 \frac{\sin(x+x^2)}{x+x^2}\cdot\frac{x+x^2}{x^p} + \int_1^\infty \frac{\sin(x+x^2)}{x^p}dx$$

$$\text{1st part} = \int_0^1 \frac{\sin(x+x^2)}{x+x^2}\cdot\frac{(1+x)}{x^{p-1}}dx.$$

$x = 0$ is a point of infinite discontinuity, if $p - 1 > 0$.

$$f(x) = \frac{\sin(x+x^2)}{x+x^2}\cdot\frac{(1+x)}{x^{p-1}}$$

Let $\phi(x) = \dfrac{1}{x^{p-1}}$.

Then $\displaystyle\lim_{x\to 0+}\frac{f(x)}{\phi(x)} = \lim_{x\to 0+}\frac{\sin(x+x^2)}{x+x^2}\cdot(1+x) = 1\ (\neq 0, \infty)$.

Also $\displaystyle\int_0^1 \phi(x)dx = \int_0^1 \frac{1}{x^{p-1}}dx$ converges for $p - 1 < 1$. That is, when $p < 2$.

So first part converges when $p < 2$

$$\text{2nd part} = \int_1^\infty \frac{\sin(x+x^2)}{x^p} = \int_1^\infty \frac{(1+2x)\sin(x+x^2)}{(1+2x)x^p}dx$$

Let $f(x) = (1+2x)\sin(x+x^2)$ and $\phi(x) = \dfrac{1}{(1+2x)x^p} = \dfrac{1}{x^{p+1}\left(\frac{1}{x}+2\right)}$.

Here $\phi(x)$ is bounded and monotonically decreases in $[1, \infty)$ and $\phi \to 0$ as $x \to \infty$, if and only if

$$p + 1 > 0 \tag{1}$$

Also

$$\left|\int_1^t f(x)dx\right| = \left|\int_1^t (1+2x)\sin(x+x^2)dx\right| = \left|\left[-\cos\left(x+x^2\right)\right]_1^t\right| \leq 2 \tag{2}$$

So from (1) and (2) it follows that the 2nd part converges at ∞, if $p + 1 > 0$.

That is, if $p > -1$. Hence the given integral converges, when $-1 < p < 2$.

Problem [8.6]: If f is continuous on $[0, 1]$, then show that $\int_0^1 \frac{f(x)}{\sqrt{1-x^2}}dx$ is convergent and hence show that $\int_0^{\pi/2} f(\sin u)du$ is a proper integral.

Solution. f is continuous and therefore, it is bounded on $[0, 1]$. (1)

If $f(1) = 0$, then 1 is not a point of infinite discontinuity of the integrand.

For, if $f(1) = 0$, then

$$\lim_{x\to 1-} \frac{f(x)}{\sqrt{1-x^2}} \left(\frac{0}{0}\right) = \lim_{x\to 1-} \frac{f'(x)}{\frac{1}{2\sqrt{1-x^2}}\cdot(-2x)} = \lim_{x\to 1} \frac{f'(x)\sqrt{1-x^2}}{-x} = 0$$

Therefore, if $f(1) = 0$, the integral is proper.

Let $f(1) \neq 0$ (2)

Then $x = 1$ is a point of infinite discontinuity of $F(x) = \frac{f(x)}{\sqrt{1-x}\sqrt{1+x}}$.

Let $\phi(x) = \dfrac{1}{\sqrt{1-x}}$.

Now $\displaystyle\lim_{x\to 1-} \frac{F(x)}{\phi(x)} = \lim_{x\to 1-} \frac{f(x)}{\sqrt{1+x}} = \frac{f(1)}{\sqrt{2}}$ ($\neq 0$ or ∞) by (1) and (2).

Also $\displaystyle\int_0^1 \phi(x)dx = \int_0^1 \frac{dx}{\sqrt{1-x}}$ converges as $\mu = \dfrac{1}{2} < 1$.

Therefore, $\displaystyle\int_0^1 F(x)dx$ converges at $x = 1$. Hence, $\displaystyle\int_0^1 \frac{f(x)}{\sqrt{1-x^2}}dx$ is convergent.

Put $x = \sin u$. Then $x \in [0, 1] \Rightarrow u \in \left[0, \frac{\pi}{2}\right]$.

So, $\displaystyle\int_0^1 \frac{f(x)dx}{\sqrt{1-x^2}} = \int_0^{\pi/2} \frac{f(\sin u)}{\sqrt{1-\sin^2 u}} \cdot \cos u du = \int_0^{\pi/2} f(\sin u)du.$

$f(\sin u)$ is continuous in $[0, \pi/2]$ as f is continuous in $[0, 1]$ and $\sin u$ is continuous on $[0, \pi/2]$.

Therefore, $f(\sin u)$ is bounded on $[0, \pi/2]$ and consequently, $\int_0^{\pi/2} f(\sin u)du$ is a proper integral.

Alternative process

$f(x)$ being continuous on $[0, 1]$ is bounded on $[0, 1]$. Also $\phi(x) = \frac{1}{\sqrt{1-x^2}}$ keeps the same sign on $[0, 1]$. Therefore, by First Mean Value theorem, there exists a number $\xi \in [0, 1]$ such that

$$\int_0^1 \frac{f(x)}{\sqrt{1-x^2}}dx = f(\xi)\int_0^1 \frac{1}{\sqrt{1-x^2}}dx \qquad (1)$$

Now $x = 1$ is a point of infinite discontinuity of $\phi(x)$.

So $\displaystyle\int_0^1 \frac{1}{\sqrt{1-x^2}}dx = \lim_{\epsilon\to 0}\int_0^{1-\epsilon} \frac{dx}{\sqrt{1-x^2}} = \lim_{\epsilon\to 0}\left[\sin^{-1}(1-\epsilon)\right] = \frac{\pi}{2}$ exists finitely.

Therefore, $\displaystyle\int_0^1 \frac{dx}{\sqrt{1-x^2}}$ is convergent.

So from (1),

$$\int_0^1 \frac{f(x)}{\sqrt{1-x^2}}dx \leq M\int_0^1 \frac{dx}{\sqrt{1-x^2}} \quad \left[\begin{array}{l}\text{since } f \text{ is bounded,}\\ f(x) \leq M \ \forall\ x \in [0,1]\end{array}\right]$$

and hence, $\displaystyle\int_0^1 \frac{f(x)}{\sqrt{1-x^2}}dx$ is convergent.

Theorem [8.16] [Frullani's Integral] If ϕ is continuous in $[0, \infty)$ and $\lim\limits_{x \to 0^+} \phi(x) = \phi_0$, $\lim\limits_{x \to \infty} \phi(x) = \phi_1$. Then $\int\limits_0^\infty \frac{\phi(ax)-\phi(bx)}{x} dx = (\phi_0 - \phi_1) \log \frac{b}{a}$, when $a, b > 0$.

Proof. We consider the proper integral $I = \displaystyle\int_\epsilon^t \frac{\phi(ax) - \phi(bx)}{x} dx$.

Then $I = \displaystyle\int_\epsilon^t \frac{\phi(ax)}{x} dx - \int_\epsilon^t \frac{\phi(bx)}{x} dx$. Putting $ax = u$, $bx = v$, we get

$$\begin{aligned} I &= \int_{a\epsilon}^{at} \frac{\phi(u)}{u/a} \frac{du}{a} - \int_{b\epsilon}^{bt} \frac{\phi(v)}{v/b} \frac{du}{b} = \int_{a\epsilon}^{at} \frac{\phi(u)}{u} du - \int_{b\epsilon}^{bt} \frac{\phi(v)}{v} dv \\ &= \int_{a\epsilon}^{at} \frac{\phi(u)}{u} du - \int_{b\epsilon}^{bt} \frac{\phi(u)}{u} du = \int_{a\epsilon}^{b\epsilon} \frac{\phi(u)}{u} du - \int_{at}^{bt} \frac{\phi(u)}{u} du \qquad \text{(i)} \end{aligned}$$

$\phi(x)$ is continuous in $[a\epsilon, b\epsilon]$ and $[at, bt]$.

So applying First Mean Value Theorem to both integrals, $\exists\, \xi, n$ belonging to $[a\epsilon, b\epsilon]$ and $[at, bt]$ respectively such that

$$\begin{aligned} \int_{a\epsilon}^{b\epsilon} \frac{\phi(u)}{u} du &= \phi(\xi) \int_{a\epsilon}^{b\epsilon} \frac{1}{u} du = \phi(\xi) \left[\log u \right]_{a\epsilon}^{b\epsilon} \\ &= \phi(\xi) \log \frac{b}{a}, \quad a\epsilon \le \xi \le b\epsilon \qquad \text{(ii)} \end{aligned}$$

Similarly,

$$\int_{at}^{bt} \frac{\phi(u)}{u} du = \phi(\eta) \log \frac{b}{a}, \quad at \le \eta \le bt \qquad \text{(iii)}$$

Therefore, from (i) $I = [\phi(\xi) - \phi(\eta)] \log \frac{b}{a}$ by (ii) and (iii)

$$\text{i.e.,} \quad \int_\epsilon^t \frac{\phi(ax) - \phi(bx)}{x} dx = [\phi(\xi) - \phi(\eta) \log \frac{b}{a}$$

Taking $\epsilon \to 0$ and $t \to \infty$, we have $\lim\limits_{\epsilon \to 0} \phi(\xi) = \lim\limits_{\xi \to 0} \phi(\xi) = \phi_0 \quad [\because \epsilon \to 0 \Rightarrow \xi \to 0]$.

Also $\lim\limits_{t \to \infty} \phi(\eta) = \lim\limits_{\eta \to \infty} \phi(\eta) = \phi_1$ $[\because t \to \infty \Rightarrow \eta \to \infty]$ and hence

$$\int_0^\infty \frac{\phi(ax) - \phi(bx)}{x} dx = [\phi_0 - \phi_1] \log \frac{b}{a}.$$

Remark.

(1) If $\int_a^b f(x) dr$ converges at $x = 0$, then by Th [8.6] we have

$$\left| \int_{\eta_1}^{\eta_2} f(x) dx \right| < \epsilon \text{ for } 0 < \eta_1, \eta_2 < \delta$$

$\Rightarrow \lim\limits_{\epsilon \to 0} \int\limits_{\eta_1}^{\eta_2} f(x) dx = 0$ when $F(\epsilon) = \int\limits_\epsilon^b f(x) dx$.

So, if $\int\limits_0^b \frac{\phi(x)}{x} dx$ converges at $x = 0$, then $\lim\limits_{\epsilon \to 0} \int\limits_{a\epsilon}^{b\epsilon} \frac{\phi(x)}{x} dx = 0$.

(2) If $\int_a^\infty \frac{\phi(x)}{x}dx$ converges at $x=\infty$, then by Th [8.12]

$$\left|\int_{t_1}^{t_2} \frac{\phi(x)}{x}dx\right| < \epsilon \ \forall \ t_1, t_2 \geq k \text{ and } F(t) = \int_a^t \frac{\phi(x)}{x}dx$$

$\Rightarrow t \to \infty \int_{t_1}^{t_2} \frac{\phi(x)}{x}dx = 0$

In other words, $\lim_{t\to\infty} \int_{at}^{bt} \frac{\phi(x)}{dx} = 0$

(3) Therefore, if $\int_0^\infty \frac{\phi(x)}{x}dx$ converges at $x=0$ and $\lim_{x\to\infty} \phi(x) = \phi_1$, then by remark (1),

$$\int_0^\infty \frac{\phi(ax)-\phi(bx)}{x}dx = -\phi_1 \log\frac{b}{a} = \phi_1 \log\frac{a}{b}$$

(4) Again if $\int_a^\infty \frac{\phi(x)}{x}dx$ converges at ∞ and $\lim_{x\to 0} \phi(x) = \phi_0$, then by remark (2) we have

$$\int_0^\infty \frac{\phi(ax)-\phi(bx)}{x}dx = \phi_0 \log\frac{b}{a}$$

(5) Now, if both $\int_0^a \frac{\phi(x)}{x}dx$ and $\int_a^\infty \frac{\phi(x)}{x}dx$ converges at $x=0$ and $x=\infty$ respectively, then

$$\int_0^\infty \frac{\phi(ax)-\phi(bx)}{x}dx = 0.$$

Problem [8.7]: Show that

(i) $\int_0^\infty \frac{e^{-ax}-e^{-bx}}{x}dx = \log\frac{b}{a}, a, b > 0$, (ii) $\int_0^\infty \frac{e^{-ax}\cos ax - e^{-bx}\cos bx}{x}dx = \log\frac{b}{a}$

(iii) $\int_0^\infty \frac{e^{-ax}\sin ax - e^{-bx}\sin bx}{x}dx = 0$, (iv) $\int_0^\infty \frac{\tan^{-1} ax - \tan^{-1} bx}{x}dx = \frac{\pi}{2}\log\frac{a}{b}$, $a, b > 0$

(v) $\int_0^\infty \frac{\cos ax - \cos bx}{x}dx = \log\frac{b}{a}$ and hence deduce that

$$\int_0^\infty \frac{\sin ax \sin bx}{x}dx = \frac{1}{2}\log\frac{a+b}{a-b}, \ a > b > 0$$

Solution. [**Hints.**

(i) $\phi(x) = e^{-x}$

(ii) $\phi(x) = e^{-x}\cos x$

(iii) $\phi(x) = e^{-x}\sin x$

(iv) $\phi(x) = \tan^{-1} x$ and apply Frullani's integral.

(v) Let $\phi(x) = \cos x$. Then $\int_a^\infty \frac{\phi(x)}{x}dx = \int_a^\infty \frac{\cos x}{x}dx$ is convergent at ∞

[See Example (2), Page-172] and $\lim_{x\to 0} \phi(x) = \phi_0 = 1$

Therefore, by Remark (2) we have $\int_0^\infty \frac{\cos ax - \cos bx}{x}dx = \phi_0 \log\frac{b}{\imath} = \log\frac{b}{a}$

Now $\sin ax \sin bx = \frac{1}{2}(2\sin ax \sin bx) = \frac{1}{2}[\cos(a-b)x - \cos(a+b)x]$

$\therefore \int_0^\infty \frac{\sin ax \sin bx}{x}dx = \frac{1}{2}\int_0^\infty \frac{\cos(a-b)x - \cos(a+b)x}{x}dx = \frac{1}{2}\log\frac{a+b}{a-b}$.]

EXERCISE

(1) Let f be continuous on $I = \{x; 2 \le x < \infty\}$ and $\lim_{x\to\infty} x(\log x)^2 f(x) = A(\ne 0)$, then prove that $\int_2^\infty f(x)dx$ is convergent.

(2) Show that $\int_0^1 (x^m + x^{-m}) \log(1+x)\frac{dx}{x}$ is convergent, if $-1 < m < 1$.

(3) If f is defined on $[0,1]$ by $f(x) = (-1)^{n+1} \cdot n(n+1), \frac{1}{n+1} \le x < \frac{1}{n}, n \in N$, then show that $\int_0^1 f(x)dx$ does not converge.

(4) Examine the convergence of $\int_0^\infty f(x)dx$ when f is defined by $f(x) = (-1)^{n-1}$, for $n-1 \le x < n, n \in N$ for $0 \le x < \infty$.

(5) Show that $\int_a^\infty \frac{1}{x^\mu}dx (a > 0)$ exists, if $\mu > 1$ and does not exist, if $\mu \le 1$.

Use the above result to test the convergence of $\int_{e^2}^\infty \frac{dx}{x\log(\log x)}dx$.

(6) Apply Dirichlet's test to establish the convergence of

(i) $\int_0^1 \frac{1}{x^{3/2}} \sin\left(\frac{1}{x}\right) dx$, (ii) $\int_0^\infty \sin(x^2)dx$.

(7) Apply Abel's test to show that

(i) $\int_a^\infty e^{-ax}\frac{\sin x}{x}dx$ is convergent and (ii) $\int_a^\infty e^{-a^2x^2}\frac{\sin 2bx}{x}dx$ is convergent, $a > 0$.

(8) Show that the integrals $\int_1^\infty \frac{\sin x}{x^p}dx$ and $\int_1^\infty \cos\left(x^{1+p}\right) dx$ are convergent, if $p > 0$.

(9) Discuss the convergence of $\int_0^{\pi/2} \cos 2nx \log \sin x dx$ and evaluate it when n is a (+)ve integer.

Also deduce $\int_0^{\pi/2} \cos 2nx \log \cos x dx = (-1)^{n+1}\frac{\pi}{4n}$.

(10) Show that $\int_0^\infty \frac{\sin ax}{x}dx = \frac{\pi}{2}$ or 0 or $-\frac{\pi}{2}$ according as $a > 0$ or $a = 0$ or $a < 0$. Deduce that $\int_0^\infty \frac{\cos ax - \cos bx}{x^2}dx = \frac{\pi}{2}(b-a),\ 0 < a < b$.

(11) Show that $\int_0^\infty \frac{x\,dx}{1 + x^4\sin^2 x}$ is divergent.

(12) Examine the convergence of

(i) $\int_0^\infty \frac{x^{m-1}+x^{n-1}}{1-x}dx$, (ii) $\int_0^\infty x^m(\log x)^n dx$, (iii) $\int_0^{\pi/2} \sin^{p-1} x \cos^{q-1} x\, dx$.

(13) A function $f(x)$ is unbounded on $[0,1]$ but is R-integrable on $[\epsilon, 1]$, $0 < \epsilon < 1$ so that $\int_0^1 f(x)dx$ diverges. Is it true that $\int_0^1 [f(x)]^2$ will also diverge? Give reasons.

(14) If $\int_1^\infty f(x)dx$ converges and $\lim_{n\to\infty} f(x) = b$, then show that $b = 0$.

CHAPTER 9

SEQUENCE OF FUNCTIONS

Definition [9.1] : Sequence of functions defines a sequence whose members are all real valued functions defined on an interval I as domain. Members are denoted by $f_n(x)$, $\forall\ x \in I, n \in N$ and the sequence is represented by $\{f_n(x)\}$.

Example. $\{x, x^2, x^3, \cdots, x^n, \cdots\}$ where $f_n(x) = x^n$ and it is represented by $\{x^n\}$.

Example. $\left\{\frac{1}{1^x}, \frac{1}{2^x}, \frac{1}{3^x}, \cdots, \frac{1}{n^x}, \cdots\right\}$ where $f_n(x) = \frac{1}{n^x}$ and is represented by $\left\{\frac{1}{n^x}\right\}$.

Example. $\left\{\frac{x}{1+1^2x^2}, \frac{2x}{1+2^2x^2}, \cdots, \frac{nx}{1+n^2x^2}, \cdots,\right\}$ where $f_n(x) = \frac{1}{1+n^2x^2}$ and it is represented by $\left\{\frac{nx}{1+n^2x^2}\right\}$.

Note. Each member of the sequence $\{f_n(x)\}$ is a real valued functions depend on the variable x.

9.1 Pointwise Convergence

Let f_n be a real valued function defined on $I = [a, b]$, for each $n \in N$.

Now to each point $c \in I$, there generates a real number sequence $\{f_n(c)\}$ with the members $f_1(c)$, $f_2(c), \cdots, f_n(c), \cdots$.

We suppose that this sequence is convergent. In fact, we suppose that each of the sequences arising for different members of I is convergent. Naturally, there exists a real valued function $f(c)$ with domain I such that $\lim\limits_{n\to\infty} f_n(c) = f(c)$.

Thus for different members $d, e, h, \cdots$ of I, the corresponding sequences $\{f_n(d)\}$, $\{f_n(e)\}$, $\{f_n(h)\}, \cdots$ converge to $f(d), f(e), f(h), \cdots$ respectively.

The function f thus defined above, is called the *pointwise limit of the pointwise convergent sequences* $\{f_n\}$ of functions defined on I, to each $c \in I$.

That is, $\lim\limits_{n\to\infty} f_n(x) = f(x),\ x \in I$.

Therefore, if f is the pointwise limit of a sequences of function $\{f_n\}$ defined on $I = [a, b]$, then for every $\epsilon > 0$, and to each $x \in I$, there exists a (+)ve integer m such that

$$|f_n(x) - f(x)| < \epsilon,\ \forall\ n \geq m. \qquad (1)$$

Note. Here m is depending on both x and ϵ.

Example. The sequence of functions $\{f_n(x)\}$, where $f_n(x) = x^n$ is pointwise convergent for $0 \le x \le 1$.

$$\lim_{n\to\infty} f_n(x) = \begin{cases} \lim_{n\to\infty} x^n = 0, & \text{for } 0 < x < 1 \\ 0, & \text{for } x = 0 \\ 1, & \text{for } x = 1. \end{cases}$$

Pointwise limit f is given by

$$f(x) = \begin{cases} 0, & 0 \le x < 1 \\ 1, & x = 1 \end{cases}$$

and the sequence $\{f_n(x)\}$ is pointwise convergent on $[0, 1]$.

Example. The sequence of functions $\{f_n(x)\}$ defined by $f_n(x) = \frac{nx}{1+n^2x^2}$ is pointwise convergent $\forall\ x \in [a, b]$ and the pointwise limit is 0.

$$\lim_{n\to\infty} f_n(x) = \lim_{n\to\infty} \frac{nx}{1+n^2x^2} = \lim_{n\to\infty} \frac{\frac{x}{n}}{\frac{1}{n^2}+x^2} = 0,\ \forall\ x \in [a, b]$$

So the pointwise limit is 0 and the sequences $\{f_n(x)\}$ converges pointwise $\forall\ x \in [a, b]$.

9.2 Uniform Convergence

Definition [9.2] : A sequence of functions $\{f_n(x)\}$ is said to converge uniformly on a domain $I = [a, b]$, if there exists a function $f(x)$ with domain I and to each $\epsilon(> 0)$, there corresponds a (+)ve integer m such that $|f_n(x) - f(x)| < \epsilon,\ \forall\ n \ge m$ and $\forall\ x \in [a, b]$.

Note. Here m is depending on ϵ only.

Observations

Example. The sequence $\{f_n(x)\} = \{x^n\}$ with domain $[0, 1]$ is a pointwise convergent with pointwise limit $f(x) = 0$ for $0 \le x < 1$ and $f(x) = 1$, for $x = 1$.

Let (i) $x = 3/4$ and $\epsilon = 1/4$, then the least (+)ve integer $m = 5$ to hold (1)
(ii) $x = 3/4$ and $\epsilon = 1/2$, then the least (+)ve integer $m = 3$ to hold (1)
(iii) $x = 2/5$ and $\epsilon = 1/4$, then the least (+)ve integer $m = 2$ to hold (1)
(iv) $x = 2/5$ and $\epsilon = 1/2$, then the least (+)ve integer $m = 1$ to hold (1).

Thus we observe that (1) is satisfied for (i), (ii), (iii) and (iv) respectively for $n \ge 5$, $n \ge 3$, $n \ge 2$, $n \ge 1$. It has been found that m is depending on both x and ϵ.

Example. The sequence $\{f_n(x)\} = \left\{\frac{nx}{1+n^2x^2}\right\}$ with domain $I = [a, b]$ [That is, with domain of real numbers] is a pointwise convergent with pointwise limit $f(x) = 0,\ \forall\ x \in [a, b]$.

Let $x = 1/5$ and $\epsilon = 1/3$, then the least (+)ve integer $m = 14$ to hold (1)
$x = 1/5$ and $\epsilon = 1/4$, then the least (+)ve integer $m = 19$ to hold (1)
$x = 3/4$ and $\epsilon = 1/4$, then the least (+)ve integer $m = 5$ to hold (1).

In this case also, m is depending on both x and ϵ.

Example. We now consider a sequence of functions $\{f_n(x)\}$, where $f_n(x) = \frac{x}{1+nx}$ for $0 \le x < \infty$.

Here $f(x) = \lim_{n\to\infty} f_n(x) = \lim_{n\to\infty} \frac{\frac{x}{n}}{\frac{1}{n}+x} = 0,\ \forall\ x \in [0, \infty)$.

Then

$$|f_n(x) - f(x)| < \epsilon \Rightarrow \frac{x}{1+nx} < \epsilon, \ \forall\, x \in [0, \infty) \tag{2}$$

Let $\epsilon = 1/4$ and

$x = 1$, then the least (+)ve integer m for which (2) holds is $m = 4$.

$$x = 2, \ \frac{2}{1+2\cdot 4} = \frac{2}{9} = 0.22 < \epsilon$$

$$x = 3, \ \frac{3}{1+3\cdot 4} = \frac{3}{13} = 0.23 < \epsilon$$

$$x = \frac{3}{2}, \ \frac{3/2}{1+3/2\cdot 4} = \frac{3}{14} = 0.21 < \epsilon$$

Here it can be shown that for a given $\epsilon(> 0)$, it is possible to find one least (+)ve integer $m = 4$ so that (2) holds for all $x \in [0, \infty)$. That is, for a given $\epsilon = 1/4$,

$$|f_n(x) - f(x)| < \epsilon, \ \forall\, x \in [0, \infty] \text{ and } \forall\, n \geq 4.$$

If we take $\epsilon = 1/2$ and then for any value of x in $[0, \infty)$, (2) holds $\forall\, n \geq 2$.

So in this case there exists one least (+)ve integer, depending on ϵ only and we say that $\{f_n(x)\}$ in uniformly convergent.

Note. (i) For pointwise convergence m is depending on both x and ϵ. Here m is not fixed. For uniform convergence m is depending on ϵ. Here m is fixed.

(ii) From the first two examples it follows that when a sequence of functions $\{f_n(x)\}$ converges pointwise to $f(x)$, then for a given $\epsilon(> 0)$, each point $x_i \in [a, b]$ determines an integer m_i such that $|f_n(x_i) - f(x_i)| < \epsilon$ for $n \geq m_i$ and consequently all points x_i of $[a, b]$ determines a sequence $\{m_1, m_2, m_3, \cdots\}$.

Now, if $\{m_i\}$ is bounded above with supremum m (say), then $|f_n(x) - f(x)| < \epsilon$ hold $\forall\, x \in [a, b]$ and $\forall\, n \geq m$ and hence $\{f_n(x)\}$ converges uniformly on $[a, b]$. Uniform convergence $\Rightarrow$ pointwise convergence but not the converse.

Cauchy's criterion for uniform convergence

Theorem [9.1] The necessary and sufficient condition for uniform convergence of a sequence of functions $\{f_n(x)\}$ on a domain $[a, b]$ is that for every $\epsilon(> 0)$, there exists a (+)ve integer m such that $|f_{n+p}(x) - f_n(x)| < \epsilon$, $\forall\, n \geq m$, $\forall\, x \in [a, b]$ and $\forall\, p \in N$.

Proof. *The condition is necessary.*

Let $\{f_n(x)\}$ converge uniformly on $[a, b]$ to the limit function $f(x)$. Then for a given $\epsilon > 0$, there exists a (+)ve integer m such that

$$|f_n(x) - f(x)| < \frac{\epsilon}{2}, \ \forall\, n \geq m \text{ and } \forall\, x \in [a, b]. \tag{i}$$

Since $\forall\, p \in N$ and $\forall\, n \geq m \Rightarrow n + p \geq m$ and hence from (1),

$$|f_{n+p}(x) - f(x)| < \frac{\epsilon}{2}, \ \forall\, x \in [a, b], \ \forall\, n \geq m \text{ and } \forall\, p \in N. \tag{ii}$$

Therefore, $|f_{n+p}(x) - f_n(x)| = |f_{n+p}(x) - f(x) + f(x) - f_n(x)|$

$$\leq |f_{n+p}(x) - f(x)| + |f_n(x) - f(x)| < \frac{\epsilon}{2} + \frac{\epsilon}{2} = \epsilon, \text{ by (i) and (ii)}.$$

Thus, $|f_{n+p}(x) - f_n(x)| < \epsilon$, $\forall\, x \in [a, b]$, $\forall\, n \geq m$ and $\forall\, p \in N$.

The condition is sufficient.

Let the given condition hold.

Then by Cauchy's general principle of convergence, the sequence $\{f_n(x)\}$ converges to the limit f, for each $x \in [a,b]$. Thus the sequence $\{f_n(x)\}$ is pointwise convergent.

Now we have to show that $\{f_n(x)\}$ converges uniformly on $[a,b]$.

For a given $\epsilon(>0)$, we choose a (+)ve integer m such that

$$|f_{n+p}(x) - f_n(x)| < \epsilon/2, \ \forall\, n \geq m, \ \forall\, x \in [a,b] \text{ and } \forall\, p \in N.$$

$$\Rightarrow f_n(x) - \epsilon/2 < f_{n+p}(x) < f_n(x) + \epsilon/2.$$

Now keeping n fixed and letting $p \to \infty$, we have

$$f_n(x) - \epsilon/2 \leq f(x) \leq f_n(x) + \epsilon/2, \ \forall\, n \geq m \text{ and } \forall\, x \in [a,b]$$

$$\Rightarrow |f_n(x) - f(x)| < \epsilon/2 < \epsilon, \ \forall\, n \geq m \text{ and } \forall\, x \in [a,b].$$

Thus the convergence is uniform.

Note. (i) n fixed, $p \to \infty$, $\lim\limits_{n\to\infty} f_{n+p}(x) = f(x)$, due to the pointwise convergence of $\{f_n(x)\}$ to $f(x)$.

(ii) The above theorem may be stated as, for each $\epsilon > 0$, $\exists$ a (+)ve integer k such that $|f_m(x) - f_n(x)| < \epsilon$, $\forall\, m, n \geq k$ and $\forall\, x \in [a,b]$.

Problem [9.1]: Apply Cauchy's criterion of uniform convergence to show that the following sequences of functions $\{f_n(x)\}$ converge uniformly on the given domains,

(1) $f_n(x) = \frac{n}{x+n}$ on $[0,1]$

(2) $f_n(x) = \frac{x}{1+nx}$ on $[0,\infty]$.

Solution. (1) $f_n(x) = \frac{n}{x+n}$, $f_{n+p}(x) = \frac{n+p}{x+n+p}$, $p \in N$

$$|f_{n+p}(x) - f_n(x)| = \left|\frac{n+p}{x+n+p} - \frac{n}{x+n}\right| = \frac{|(x+n)(n+p) - n(x+n+p)|}{(x+n+p)(x+n)}$$

$$= \frac{px}{(x+n)(x+n+p)} = \frac{px}{(x+n)^2 + px + np} \leq \frac{px}{px+np} \leq \frac{px}{np}$$

$$= \frac{x}{n} \leq \frac{1}{n} < \epsilon, \ \forall\, n \geq m \text{ and } \forall\, x \in [0,1] \text{ and } \forall\, p \in N$$

$\Rightarrow \{f_n(x)\}$ converges uniformly on $[0,1]$.

(2) $f_n(x) = \frac{x}{1+nx}$, $f_{n+p}(x) = \frac{x}{1+(n+p)x}$, $p \in N$

$$|f_{n+p}(x) - f_n(x)| = \left|\frac{x}{1+(n+p)x} - \frac{x}{1+nx}\right| = \left|\frac{x\{1+nx-1-nx-px\}}{(1+nx)(1+(n+p)x)}\right|$$

$$= \frac{xpx}{(1+nx)^2 + (1+nx)px} < \frac{xpx}{(1+nx)px}$$

$$= \frac{x}{1+nx} \leq \frac{x}{nx} \ [\text{for } x \neq 0]$$

$$= \frac{1}{n} < \epsilon, \ \forall\, n \geq m \text{ and } \forall\, x \in (0,\infty) \text{ and } \forall\, p \in N.$$

The same is true for $x = 0$. Hence $\{f_n(x)\}$ converges uniformly on $[0,\infty)$.

Problem [9.2]: (i) A sequence of functions $\{f_n(x)\}$ is defined on $[0,a]$, $0 < a < 1$, by $f_n(x) = x^n$, $x \in [0,a]$. Show that the sequence $\{f_n(x)\}$ converges uniformly on $[0,a]$.

Solution. Let $\epsilon(>0)$ be given. Then $\forall\ x \in [0, a]$,

$$|f_m(x) - f_n(x)| = |x^m - x^n| \le |x^m| + |x^n|$$
$$\le a^m + a^n \le 2a^m, \text{ if } n \ge m\ [\because 0 < a < 1 \Rightarrow a^n \le a^m]$$

Now $|f_m(x) - f_n(x)| < \epsilon$, if $a^m < \epsilon/2$

$$\text{i.e., if } \frac{1}{a^m} > \frac{2}{\epsilon}$$
$$\text{i.e., if } m \log(1/a) > \log\frac{2}{\epsilon}$$
$$\text{i.e., if } m > \frac{\log\frac{2}{\epsilon}}{\log\frac{1}{a}}$$

Let k be the integral part of $\log\frac{2}{\epsilon}/\log\frac{1}{a}$. Then k is a natural number.

Therefore, $|f_m(x) - f_n(x)| < \epsilon$, $\forall\ m, n$ satisfying $n \ge m \ge k$ and $\forall\ x \in [0, a], 0 < a < 1$ and hence by Cauchy's criterion the given sequence is uniformly convergent on $[0, a]$.

(ii) Let $D \subseteq R$ and for each $n \in N$, $f_n : D \to \mathbb{R}$ is bounded on D. If $\{f_n(x)\}$ be a sequence of pointwise convergent on D, then the limit function f may not be bounded on D.

Solution. Let $f_n(x) = 1 + x + x^2 + \cdots + x^{n-1}$, $x \in [0, 1)$.

Then $\lim_{n\to\infty} f_n(x) = \frac{1}{1-x}, x \in [0, 1)$.

The sequence $\{f_n(x)\}$ converges on $[0, 1)$ to the function f, given by $f(x) = \frac{1}{1-x}, x \in [0, 1)$.

$f(x) \to \infty$ as $x \to 1 - 0$. So f is unbounded on $[0, 1]$.

$$|f_n(x)| = |1 + x + \cdots + x^{n-1}| \le 1 + |x| + |x^2| + \cdots + |x^{n-1}| < n,\ \forall\ x \in [0, 1)$$

Thus each f_n is bounded on $[0, 1]$ but f is unbounded on $[0, 1)$.

Theorem [9.2] Let $D \subseteq R$ and for each $n \in N$, $f_n : D \to \mathbb{R}$ is bounded on D. If the sequence $\{f_n(x)\}$ be uniformly convergent on D, then the limit function f is bounded on D.

Proof. Let $\epsilon(>0)$ be given. Since $\{f_n(x)\}$ converges uniformly on D, there exists a natural number m such that $|f_n(x) - f(x)| < \epsilon$, $\forall\ n \ge m$, $\forall\ x \in D$.

Let $\epsilon = 1$. Then there exists a natural number k (say), such that $|f_n(x) - f(x)| < 1$, $\forall\ n \ge k$ and $\forall\ x \in D$.

Now $|f(x)| = |f(x) - f_k(x) + f_k(x)| \le |f(x) - f_k(x)| + |f_k(x)| < 1 + |f_k(x)|$.

Since each f_n is bounded on D, $\exists$ a (+)ve number B such that $|f_k(x)| \le B$, $\forall\ x \in D$.

Hence, $|f(x)| < 1 + B$, $\forall\ x \in D \Rightarrow f$ is bounded on D.

Theorem [9.3] If $\{f_n(x)\}$ and $\{g_n(x)\}$ converge uniformly on $[a, b]$, then $\{(f_n \pm g_n\}(x)\}$ converge uniformly on $[a, b]$.

Proof. Let $\epsilon(>0)$ be given.

Since $\{f_n(x)\}$ and $\{g_n(x)\}$ converge uniformly on $[a, b]$, there exist $m_1, m_2 \in N$ such that

$$|f_m(x) - f_n(x)| < \epsilon/2;\ \forall\ x \in [a, b] \text{ and } \forall\ n \ge m_1$$
$$\text{and } |g_m(x) - g_n(x)| < \epsilon/2;\ \forall\ x \in [a, b] \text{ and } \forall\ n \ge m_2.$$

[Using Cauchy's criterion of uniform convergence]

Then $|(f_m \pm g_m)(x) - (f_n \pm g_n)(x)| = |\{f_m(x) - f_n(x)\} \pm \{g_m(x) - g_n(x)\}|$
$\leq |f_m(x) - f_n(x)| + |g_m(x) - g_n(x)|$
$< \epsilon/2 + \epsilon/2 = \epsilon,\ \forall\ x \in [a, b]$ and $\forall\ n \geq m$, where $m = \max(m_1, m_2)$

$\Rightarrow \{(f_n \pm g_n)(x)\}$ converges uniformly on $[a, b]$, according to Cauchy's criterion for uniform convergence.

Problem [9.3] : If a sequence $\{f_n(x)\}$ of continuous real valued functions converges uniformly on $[a, b]$ and for each $n \in N$, $F_n(x) = \int_a^x f_n(t)dt$, $a < x < b$, then prove that $\{F_n(x)\}$ converges uniformly on $[a, b]$.

Solution. Since $\{f_n(x)\}$ converges uniformly on $[a, b]$, for a given $\epsilon(> 0)$, $\exists$ a (+)ve integer k such that

$$|f_m(x) - f_n(x)| < \frac{\epsilon}{b-a},\ \forall\ m, n \geq k \text{ and } \forall\ x \in [a, b], \tag{1}$$

by Cauchy's criterion of uniform convergence.

Now

$$F_m(x) - F_n(x) = \int_a^x \{f_m(t) - f_n(t)\}dt$$

$$\Rightarrow |F_m(x) - F_n(x)| = \left|\int_a^x \{f_m(t) - f_n(t)\}dt\right| < \int_a^x |f_m(t) - f_n(t)|dt$$

$$< \int_a^x \frac{\epsilon}{b-a}dt \text{ [by (1)]}$$

$$= \frac{\epsilon}{b-a}(x-a) < \frac{\epsilon}{b-a}(b-a) = \epsilon,\ \forall\ m, n \geq k \text{ and } \forall\ x \in [a, b]$$

$\Rightarrow \{F_n(x)\}$ converges uniformly on $[a, b]$.

Problem [9.4] : Define two sequences $\{f_n(x)\}$ and $\{g_n(x)\}$ as follows:

$$f_n(x) = x\left(1 + \tfrac{1}{n}\right), \text{ if } x \in R,\ n \in N$$

$$g_n(x) = \begin{cases} \frac{1}{n}, & \text{if } x = 0, \text{ or if irrational} \\ b + \frac{1}{n}, & \text{if } x \text{ is rational, say, } x = \frac{a}{b},\ b > 0. \end{cases}$$

Let $h_n(x) = f_n(x)g_n(x)$. Then prove that

(1) $\{f_n(x)\}, \{g_n(x)\}$ both converge uniformly on every bounded interval.

(2) $\{h_n(x)\}$ does not converge uniformly on any bounded interval.

Solution. Let $f_n : I \to R$ and $g_n : I \to R$, where I is a closed bounded interval, $I \subseteq R$.

Now $f(x) = \lim_{n\to\infty} f_n(x) = x$.

Then for a given ϵ,

$$|f_n(x) - f(x)| = \left|x + \frac{x}{n} - x\right| = \frac{|x|}{n} < \epsilon \Rightarrow n > \frac{|x|}{\epsilon} = m(\epsilon, x),\ \forall\ x \in I,$$

$m(\epsilon, x) \to$ finite number. So $\exists$ a (+)ve integer m such that $|f_n(x) - f(x)| < \epsilon$, $\forall\ n \geq m$ and $\forall\ x \in I$. Hence $\{f_n(x)\}$ converges uniformly on I.

Now

$$g(x) = \begin{cases} \lim_{n\to\infty} g_n(x) = 0, & \text{if } x = 0 \text{ or irrational} \\ b,\ b > 0, & \text{if } x = \frac{a}{b},\ b > 0. \end{cases}$$

Then for a given ϵ,

$$|g_n(x) - g(x)| = \left|\frac{1}{n} - 0\right| < \epsilon \Rightarrow \frac{1}{n} < \epsilon \Rightarrow n > \frac{1}{\epsilon} = m \text{ (say)}.$$

So $\exists$ a (+)ve integer m such that $|g_n(x) - g(x)| < \epsilon$,

$$\forall\ n \geq m \text{ and for } x = 0 \text{ or irrational points} \in I \qquad (1)$$

Again, when $x = \frac{a}{b}, b > 0 \in I$,

$$|g_n(x) - g(x)| = \left|b + \frac{1}{n} - b\right| = \left|\frac{1}{n}\right| < \epsilon \Rightarrow n > \frac{1}{\epsilon} = m \text{ (say)}.$$

So in this case also, $\exists$ a (+)ve integer m such that

$$|g_n(x) - g(x)| < \epsilon,\ \forall\ n \geq m \text{ and } \forall\ x = \frac{a}{b} \in I \qquad (2)$$

So from (1) and (2) it follows that $\{g_n(x)\}$ converges uniformly on I.

Given that $h_n(x) = f_n(x)g_n(x)$.

So $h_n(x)$ is defined by: $h_n(x) = x\left(1 + \frac{1}{n}\right)\frac{1}{n}$; for $x = 0$ or irrational

$$\text{and } h_n(x) = x\left(1 + \frac{1}{n}\right)\left(b + \frac{1}{n}\right);\ \text{for } x = \frac{a}{b},\ b > 0.$$

Then $h(x) = \lim_{n\to\infty} f_n(x)g_n(x) = \lim_{n\to\infty} x\left(1 + \frac{1}{n}\right)\frac{1}{n} = 0,\ x = 0$ or irrational

$$\text{and } h(x) = \lim_{n\to\infty} f_n(x)g_n(x) = \lim_{n\to\infty} x\left(1 + \frac{1}{n}\right)\left(b + \frac{1}{n}\right) = bx,\ x = \frac{a}{b},\ b > 0.$$

Now

$$|h_n(x) - h(x)| = \left|x\left(1 + \frac{1}{n}\right)\frac{1}{n} - 0\right| = \left|\frac{x}{n} + \frac{x}{n^2}\right|.$$

Now $\frac{x}{n} + \frac{x}{n^2} \to 0$ as $n \to \infty$ for $x = 0$ or any irrational and so $\left|\frac{x}{n} + \frac{x}{n^2}\right| < \epsilon,\ \forall\ n \geq m$.

Thus $|h_n(x) - h(x)| < \epsilon,\ \forall\ n \geq m$ and for $x = 0$ or irrational.

Again, when $x = \frac{a}{b}, b > 0$. Let $\epsilon(> 0)$ be given. Say, $\epsilon = 1/2$.

$$|h_n(x) - h(x)| = \left|x\left(1 + \frac{1}{n}\right)\left(b + \frac{1}{n}\right) - bx\right| = \left|\frac{x}{n} + \frac{bx}{n} + \frac{x}{n^2}\right|$$

Let $n = m, x = \frac{m}{b} \in I$. Then $\frac{x}{n} + \frac{bx}{n} + \frac{x}{n^2} = \frac{1}{b} + 1 + \frac{1}{bm} > 1$ since $b > 0$.

$$\therefore\ |h_n(x) - h(x)| > 1.$$

That is, $|h_n(x) - h(x) \not< \epsilon,\ \forall$ rational points $\in I$.

Hence, $|h_n(x) - h(x)| \not< \epsilon,\ \forall\ x \in I$.

Therefore, $\{h_n(x)\}$ does not converge uniformly on I.

Remark. In the Problem [9.4], it has been found that $\{h_n(x)\}$, where $h_n(x) = f_n(x)g_n(x)$ and both $f_n(x)$ and $g_n(x)$ are unbounded, does not converge uniformly on any bounded closed interval I.

But if each $f_n(x)$ and $g_n(x)$ are bounded on I, then $\{h_n(x)\}$ converges uniformly on I, when both $\{f_n(x)\}$ and $\{g_n(x)\}$ converge uniformly on I.

Proof. Let $f_n \to f$ and $g_n \to g$ uniformly on I. Then by Cauchy's criterion of uniform convergence,

$$\left.\begin{aligned} &|f_m(x) - f_n(x)| < \epsilon/2M_1,\ \forall\ m,n \geq k_1,\ \forall\ x \in I \\ \text{and } &|g_m(x) - g_n(x)| < \epsilon/2M_2,\ \forall\ m,n \geq k_2,\ \forall\ x \in I \end{aligned}\right\} \qquad \text{(i)}$$

Also, since f_n, g_n are bounded,

$$|f_n(x)| \leq M_1 \text{ and } |g_n(x)| \leq M_2,\ \forall\ x \in I \qquad \text{(ii)}$$

$$\begin{aligned} \text{Now } |h_m(x) - h_n(x)| &= |f_m(x)g_m(x) - f_n(x)g_n(x)| \\ &= |f_m(x)\{g_m(x) - g_n(x)\} + g_n(x)\{f_m(x) - f_n(x)\}| \\ &\leq |f_m(x)||g_m(x) - g_n(x)| + |g_n(x)||f_m(x) - f_n(x)| \\ &\leq M_1 \frac{\epsilon}{2M_1} + M_2 \cdot \frac{\epsilon}{2M_2} = \epsilon \text{ [by (i) and (ii)]} \\ &\forall\ m,n \geq k, \forall\ x \in I, \text{ where } k = \max(k_1, k_2). \end{aligned}$$

$\Rightarrow \{h_n(x)\}$ converges uniformly on I.

Problem [9.5] : Show that the sequence of functions $\{f_n(x)\}$, where $f_n(x) = e^{-nx}$ is pointwise but not uniformly convergent in $[0, \infty)$. Also show that the convergence is uniform in $[k, \infty)$, $k > 0$.

Solution. $\lim\limits_{n\to\infty} f_n(x) = \lim\limits_{n\to\infty} e^{-nx} = 0$ for $0 < x < \infty$.

For $x = 0$, $\lim\limits_{n\to\infty} f_n(x) = 1$.

$$\text{So } f(x) = \begin{cases} 0, & 0 < x < \infty \\ 1, & x = 0. \end{cases}$$

So the sequence $\{f_n(x)\}$ is pointwise convergent to $f(x)$ in $[0, \infty)$. Now, if possible, let $\{f_n(x)\}$ converge uniformly on $[0, \infty)$. Then for a given $\epsilon(> 0)$,

$$|f_n(x) - f(x)| < \epsilon,\ \forall\ x \in [0, \infty) \text{ and } \forall\ n \geq m \qquad (1)$$

Now $|f_n(x) - f(x) < \epsilon$,

$$\Rightarrow e^{-nx} < \epsilon \Rightarrow \frac{1}{e^{-nx}} > \frac{1}{\epsilon} \Rightarrow e^{nx} > \frac{1}{\epsilon}$$

$$\Rightarrow nx > \log\frac{1}{\epsilon} \Rightarrow n > \frac{\log\frac{1}{\epsilon}}{x} = m(\epsilon, x) \text{ (say)}$$

Now $m(\epsilon, x) \to \infty$ as $x \to 0$, so it is not possible to find m such that $|f_n(x) - f(x)| < \epsilon$, $\forall\ x \in [0, \infty)$ and $\forall\ n \geq m$. Hence the sequence $\{f_n(x)\}$ is not uniformly convergent on $[0, \infty)$.

But $m(\epsilon, x) \to \dfrac{\log\frac{1}{\epsilon}}{k}$ as $x \to k$ and $m(\epsilon, x) \to 0$ as $x \to \infty$.

The maximum value of $m(\epsilon, k) = \dfrac{\log\frac{1}{\epsilon}}{k}$ in $[k, \infty)$, $k > 0$. So it is always possible to find a (+)ve integer $m \geq$ the integral part of $\dfrac{\log\frac{1}{\epsilon}}{k}$ such that

$$|f_n(x) - f(x)| < \epsilon,\ \forall\ n \geq m,\ \forall\ x \in [k, \infty),\ k > 0.$$

Hence, $\{f_n(x)\}$ converges uniformly on $[k, \infty)$, $k > 0$.

Problem [9.6] : Show that the sequence of function $\{f_n(x)\}$, where $f_n(x) = x^n$ is uniformly convergent on $[0, k]$, $k < 1$ and only pointwise convergent on $[0, 1]$.

Solution. $\lim\limits_{n\to\infty} f_n(x) = \lim\limits_{n\to\infty} x^n = \begin{cases} 0, & \text{for } 0 \le x < 1 \\ 1, & \text{for } x = 1. \end{cases}$

So the given sequence is pointwise convergent with pointwise limit f given by

$$f(x) = 0, \ 0 \le x < 1 \text{ and } f(x) = 1 \text{ for } x = 1.$$

Now for a given ϵ and for $0 < x < k < 1$, $|f_n(x) - f(x)| < \epsilon$,

$$\text{if } |x^n - 0| < \epsilon,$$

$$\text{i.e., if } x^n < \epsilon,$$

$$\text{i.e., if } n \log \frac{1}{x} > \log \frac{1}{\epsilon},$$

$$\text{i.e., if } n > \frac{\log \frac{1}{\epsilon}}{\log \frac{1}{x}} = m(\epsilon, x) \text{ (say)}.$$

Now $x \to 0$, $m(\epsilon, x) \to 0$ and $x \to k$, $m(\epsilon, x) \to \dfrac{\log \frac{1}{\epsilon}}{\log \frac{1}{k}}$, the maximum value of $m(\epsilon, x)$ on $[0, k]$, $0 < k < 1$.

So it is always possible to find a (+)ve integer $m \ge$ the integral part of $\dfrac{\log \frac{1}{\epsilon}}{\log \frac{1}{k}}$ such that

$$|f_n(x) - f(x)| < \epsilon, \ \forall \, x \in (0, k] \text{ and } \forall \, n \ge m.$$

Also for $x = 0$, $|f_n(x) - f(x)| = 0 < \epsilon$, $\forall \, n \ge 1$. Thus for any $\epsilon > 0$, there exists a (+)ve integer m such that $|f_n(x) - f(x)| < \epsilon$, $\forall \, x \in [0, k]$, $k < 1$, $\forall \, n \ge m$.

Hence the sequence is uniformly convergent on $[0, k]$.

But if $x \to 1$, $m(\epsilon, x) \to \infty$. So it is not possible to find an integer m such that

$$|f_n(x) - f(x)| < \epsilon, \ \forall \, n \ge m \text{ and } \forall \, x \in [0, 1].$$

Hence the sequence is not uniformly convergent on $[0, 1]$. It is also true for any interval containing I.

Problem [9.7] : Let $f_n(x) = x^n$. Then the sequence $\{f_n(x)\}$ converges pointwise but not uniformly on $[0, 1]$. Let g be continuous on $[0, 1]$ with $g(1) = 0$. Prove that the sequence $\{g(x)x^n\}$ converges uniformly on $[0, 1]$.

Solution. The 1st part of the problem is already shown in Problem [9.6].

Now, let $h_n(x) = g(x)x^n$, $0 \le x \le 1$.

Then, $h(x) = \lim\limits_{n\to\infty} g(x)x^n = 0$ for $0 \le x < 1$ [$\because g$ being continuous is bounded on $[0, 1]$].

Also, at $x = 1$, $h(x) = \lim\limits_{n\to\infty} g(1) = 0$, since $g(1) = 0$.

Thus $h(x) = 0$, $\forall \, x \in [0, 1]$.

Now $|h_n(x) - h(x)| = |g(x)x^n|$.

Here, $\lim\limits_{n\to\infty} g(x)x^n = 0$, $\forall \, x \in [0, 1] \Rightarrow |g(x)x^n| < \epsilon$, $\forall \, n \ge m$, $\forall \, x \in [0, 1]$.

Therefore, $|h_n(x) - h(x)| < \epsilon$, $\forall \, n \ge m$, $\forall \, x \in [0, 1]$

$\Rightarrow \{h_n(x)\}$ converges uniformly on $[0, 1]$.

Problem [9.8]: Show that the sequence of function $\{f_n(x)\}$, where $f_n(x) = \tan^{-1}(nx)$, is uniformly convergent on $[a, k]$, $a > 0$ but is only pointwise convergent on $[0, k]$, where $k > a$.

Solution. $\lim_{n\to\infty} f_n(x) = \begin{cases} \lim_{n\to\infty} \tan^{-1}(nx) = \frac{\pi}{2}, & \text{for } 0 < a < x < k. \\ 0, & \text{for } x = 0 \end{cases}$

The pointwise limit $f(x) = \begin{cases} \frac{\pi}{2}, & 0 < a < x < k \\ 0, & x = 0. \end{cases}$

The sequence is pointwise convergent to $f(x)$ on $[0, k]$.

Now for a given $\epsilon > 0$, and for $0 < a < x < k$, $|f_n(x) - f(x)| < \epsilon$,

$$\text{if } |\tan^{-1}(nx) - \pi/2| < \epsilon,$$
$$\text{i.e., if } \cot^{-1}(nx) < \epsilon,$$
$$\text{i.e., if } \cot(\cot^{-1} nx) > \cot\epsilon \; [\because \alpha < \beta \Rightarrow \cot\alpha > \cot\beta]$$
$$\text{i.e., if } nx > \cot\epsilon,$$
$$\text{i.e., if } n > \frac{\cot\epsilon}{x} = m(\epsilon, x).$$

Now $x \to a$, $m(\epsilon, x) \to \cot\frac{\epsilon}{a}$ and $x \to k$, $k > a$, $m(\epsilon, x) \to \cot\frac{\epsilon}{k}$. But x increases $m(\epsilon, x)$ decreases and so the maximum value of $m(\epsilon, x)$ is $\cot\frac{\epsilon}{a}$. So it is always possible to find a (+)ve integer $m \geq$ integral part of $\cot\frac{\epsilon}{a}$ such that

$$|f_n(x) - f(x)| < \epsilon, \; \forall\, n \geq m \text{ and } \forall\, x \in [a, k], \; a > 0.$$

Hence the sequence is uniformly convergent on $[a, k]$, $a > 0$.

But as $x \to 0$, $m(\epsilon, x) \to \infty$. So there does not exist m such that $|f_n(x) - f(x)| < \epsilon$ for all $n \geq m$ and $\forall\, x \in [0, k]$.

Hence the given sequence is pointwise convergent $[0, k]$ but not uniformly convergent.

Problem [9.9]: Show that every pointwise convergent is not necessarily convergent.

Solution. We shall show it by illustrating an example.

Let $f_n(x) = \dfrac{nx}{1 + n^2x^2}$, $x \in \mathbb{R}$.

Then $\lim_{n\to\infty} f_n(x) = \lim_{n\to\infty} \dfrac{nx}{1 + n^2x^2} = \lim_{n\to\infty} \dfrac{\frac{x}{n}}{\frac{1}{n^2} + x^2} = 0$, $\forall\, x \in R$.

Thus the limit function $f(x) = 0$, $\forall\, x \in \mathbb{R}$. So the sequence $\{f_n(x)\}$ converges pointwise to

$$f(x) = 0, \; \forall\, x \in \mathbb{R}.$$

Now, we shall show that the convergence of $\{f_n(x)\}$ is not uniform in any interval $[a, b] \subseteq \mathbb{R}$, containing 0 as an interior point. If possible, let $\{f_n(x)\}$ converge uniformly to the pointwise limit $f(x)$.

Let $\epsilon > 0$ be given. Then there exists m such that $\forall\, x \in [a, b]$ and $\forall\, n \geq m$.

$$|f_n(x) - f(x)| < \epsilon \Rightarrow \frac{nx}{1 + n^2x^2} < \epsilon.$$

Let $\epsilon = 1/3$. Then $\exists$ an integer k such that $k \geq m$ and $\frac{1}{k} \in [a, b]$. Taking $n = k$ and $x = \frac{1}{k}$,

$$\frac{nx}{1 + n^2x^2} = \frac{k \cdot \frac{1}{k}}{1 + k^2\frac{1}{k^2}} = \frac{1}{2} \not< \epsilon = \frac{1}{3}$$

which arises a contradiction and consequently the sequence $\{f_n(x)\}$ is not uniformly convergent in any interval $[a, b]$ containing 0 as an interior point.

Note. For large value of k, $\frac{1}{k} \to 0$ and so 0 to be an interior point of $[a, b] \subseteq \mathbb{R}$.

Problem [9.10] : Show that 0 is a point of non-uniform convergence of the sequence of function $\{f_n(x)\}$, where $f_n(x) = 1 - (1 - x^2)^n$.

Solution.
$$\begin{aligned}\lim_{n\to\infty} f_n(x) &= \lim_{n\to\infty} \{1 - (1 - x^2)^n\}\\ &= 0, \text{ when } |1 - x^2| = 1, \text{ i.e., when } x = 0, \pm\sqrt{2}\\ &= 1, \text{ when } |1 - x^2| < 1, \text{ i.e., when } 0 < |x| < \sqrt{2}.\end{aligned}$$

Thus $f(x) = 0$ for $x = 0, \pm\sqrt{2}$ and $f(x) = 1$ for $0 < |x| < \sqrt{2}$.

Here domain is $[-\sqrt{2}, \sqrt{2}]$ and 0 is a point of the domain. If possible, let the sequence be uniformly convergent in $(-k, k)$, a neighbourhood of 0, where $0 < |k| < \sqrt{2}$.

Then for a given $\epsilon(> 0)$, there exists a (+)ve integer m such that
$$|f_n(x) - f(x)| < \epsilon, \ \forall\, n \geq m \text{ and } \forall\, x \in (-k, k).$$

Let $\epsilon = \frac{1}{2}$ and $n = m$.

Then
$$\begin{aligned}&|f_m(x) - f(x)| < 1/2, \ \forall\, x \in (-k, k)\\ &\Rightarrow |1 - (1 - x^2)^m - 1| < 1/2\\ &\Rightarrow (1 - x^2)^m < 1/2, \ \forall\, x \in (-k, k)\end{aligned}$$

Now $(1 - x^2)^m \Rightarrow 1$ as $x \to 0$ and then $1 < 1/2$, which is not possible. We thus arrive a contradiction. Hence 0 is a point of non-uniform convergence.

Test for uniform convergence of $\{f_n(x)\}$

Theorem [9.4] If $\{f_n(x)\}$ be a sequence of function defined on $[a, b]$ such that
$$\lim_{n\to\infty} f_n(x) = f(x), \ \forall\, x \in [a, b] \text{ and if } M_n = \sup_{x\in[a,b]} |f_n(x) - f(x)|$$
then $\{f_n(x)\}$ converges uniformly to $f(x)$ on $[a, b]$, if and only if $M_n \to 0$ as $n \to \infty$.

Proof. Let $\{f_n(x)\}$ converge uniformly to $f(x)$ on $[a, b]$. Then for given $\epsilon(> 0)$, there exists a (+)ve integer m such that $|f_n(x) - f(x)| < \epsilon, \forall\, n \geq m, \forall\, x \in [a, b]$.

Therefore,
$$\begin{aligned}&|f_n(x) - f(x)| < \epsilon\\ &\Rightarrow \sup |f_n(x) - f(x)| < \epsilon, \ \forall\, n \geq m, \ x \in [a, b]\\ &\Rightarrow M_n < \epsilon, \ \forall\, n \geq m\\ &\Rightarrow M_n \to 0 \text{ as } n \to \infty.\end{aligned}$$

Conversely, let $M_n \to 0$ as $n \to \infty$. Then $\exists$ a (+)ve integer m such that
$$M_n < \epsilon, \ \forall\, n \geq m \tag{i}$$

Also from definition of M_n,
$$|f_n(x) - f(x)| < M_n, \ \forall\, x \in [a, b] \tag{ii}$$

$\Rightarrow |f_n(x) - f(x)| < \epsilon, \ \forall\, x \in [a, b], \ \forall\, n \geq m$ [by (i) and (ii)]

$\Rightarrow \{f_n(x)\}$ converges uniformly to $f(x)$ on $[a, b]$.

Problem [9.11] : Show that $\{f_n(x)\}$ where $f_n(x) = nxe^{-nx^2}$ is pointwise convergent but not uniformly convergent on $[0, \infty)$.

Solution. $\lim_{n\to\infty} f_n(x) = 0,\ \forall\, x \in [0, \infty],\ n \in N.$

Thus the pointwise limit $f(x) = 0,\ \forall\, x \in [0, \infty)$. So $\{f_n(x)\}$ is a pointwise convergent sequence. Now $|f_n(x) - f(x)| = nxe^{-nx^2}$.

Then

$$\frac{d}{dx}|f_n(x) - f(x)| = \frac{d}{dx}\left[nxe^{-nx^2}\right] = ne^{-nx^2} - 2n^2x^2e^{-nx^2} = ne^{-nx^2}\{1 - 2nx^2\}$$

For extremum, $ne^{-nx^2}\{1 - 2nx^2\} = 0 \Rightarrow x = \frac{1}{\sqrt{2n}}$ as $x \in [0, \infty)$.

Now

$$\begin{aligned}\frac{d^2}{dx^2}|f_n(x) - f(x)| &= \frac{d}{dx}\left\{ne^{-nx^2}(1 - 2nx^2)\right\}\\ &= -2n^2xe^{-nx^2}(1 - 2nx^2) - 4n^2xe^{-nx^2}\\ &= 2n^2xe^{-nx^2}\{-1 + 2nx^2 - 2\}\\ &= \frac{-2\sqrt{2}\,n^{3/2}}{e^{1/2}} < 0 \text{ for } x = \frac{1}{\sqrt{2n}}\end{aligned}$$

$\therefore |f_n(x) - f(x)|$ is maximum at $x = \frac{1}{\sqrt{2n}}$.

That is, $$\sup_{x\in[0,\infty)} |f_n(x) - f(x)| = n\frac{1}{\sqrt{2n}}e^{-n\cdot\frac{1}{2n}} = \frac{\sqrt{n}}{\sqrt{2}\,e^{1/2}}.$$

$$\Rightarrow M_n = \frac{\sqrt{n}}{\sqrt{2e}} \text{ and } M_n \not\to 0 \text{ as } n \to \infty.$$

So $\{f_n(x)\}$ is not uniformly convergent on $[0, \infty)$. Also as $n \to \infty$, $x \to 0$. So 0 is a point of non-uniform convergence of $\{f_n(x)\}$.

Problem [9.12] : Test the following sequences for uniform convergence:

(1) $\left\{\frac{\sin nx}{\sqrt{n}}\right\}, 0 \le x \le 2\pi$ (2) $\left\{\frac{n^2x}{1 + n^3x^2}\right\},\ 0 \le x \le 1$

(3) $\left\{\frac{2n^2x^2}{e^{n^2x^2}} - 1\right\},\ \forall\, x \in R$ (4) $\left\{1 - \frac{x^n}{n}\right\},\ x \in [0, 1]$

(5) $\{nx(1 - x)^n\},\ x \in [0, 1]$

Solution. (1) Here $f(x) = \lim_{n\to\infty} f_n(x) = \lim_{n\to\infty} \frac{\sin nx}{\sqrt{n}} = 0$, since $\sin nx$ is bounded.

Therefore, $|f_n(x) - f(x)| = \frac{|\sin nx|}{\sqrt{n}} \le \frac{1}{\sqrt{n}},\ \forall\, x \in [0, 2\pi]$.

$$M_n = \sup_{x\in[0,2\pi]} |f_n(x) - f(x)| = \frac{1}{\sqrt{n}}.\quad M_n \to 0 \text{ as } n \to \infty.$$

Therefore, $\left\{\frac{\sin nx}{\sqrt{n}}\right\}$ converges uniformly on $[0, 2\pi]$.

(2) $f_n(x) = \frac{n^2x}{1 + n^3x^2},\ 0 \le x \le 1$

$$f(x) = \lim_{n\to\infty} f_n(x) = 0,\ \forall\, x \in [0, 1]$$

Now

$$|f_n(x) - f(x)| = \left|\frac{n^2x}{1+n^3x^2} - 0\right| = \frac{n^2x}{(1+n^3x^2)}$$

$$\frac{d}{dx}|f_n(x) - f(x)| = \frac{d}{dx}\cdot\frac{n^2x}{1+n^3x^2} = \frac{n^2(1+n^3x^2) - n^2x\cdot 2n^3x}{(1+n^3x^2)^2}$$

$$= \frac{n^2+n^5x^2-2n^5x^2}{(1+n^3x^2)^2} = \frac{n^2-n^5x^2}{(1+n^3x^2)^2} = 0 \Rightarrow x^2 = \frac{1}{n^3} \Rightarrow x = \frac{1}{n^{3/2}}$$

$$\frac{d^2}{dx^2}|f_n(x) - f(x)| = \frac{d}{dx}\left[\frac{n^2-n^5x^2}{(1+n^3x^2)^2}\right]$$

$$= \frac{-2n^5x(1+n^3x^2)^2 - (n^2-n^5x^2)\cdot 2(1+n^3x^2)\cdot 2n^3x}{(1+n^3x^2)^4}$$

$$= \frac{-2n^5x(1+n^3x^2) - 4(n^2-n^5x^2)n^3x}{(1+n^3x^2)^3} < 0 \text{ for } x = \frac{1}{n^{3/2}}$$

$$M_n = \sup_{x\in[0,1]} |f_n(x) - f(x)| = \frac{n^2\cdot\frac{1}{n^{3/2}}}{1+n^3\cdot\frac{1}{n^3}} = \frac{1}{2}n^{1/2}.$$

$M_n \to \infty$ as $n \to \infty$. That is, $M_n \not\to 0$ as $n \to \infty$.

Therefore, $\left\{\dfrac{n^2x}{1+n^3x^2}\right\}$ is not uniformly convergent on $[0,1]$.

(3) $f_n(x) = \left(\dfrac{2n^2x^2}{e^{n^2x^2}} - 1\right)$.

Now $f(x) = \lim\limits_{n\to\infty} f_n(x) = \lim\limits_{n\to\infty}\left(\dfrac{2n^2x^2}{e^{n^2x^2}} - 1\right) = -1,\ \forall\, x \in R.$

Since

$$\lim_{n\to\infty}\frac{2n^2x^2}{e^{n^2x^2}}\left(\frac{\infty}{\infty}\right) = \lim_{n\to\infty}\frac{4n^2x}{e^{n^2x^2}\cdot 2n^2x} = \lim_{n\to\infty}\frac{2}{e^{n^2x^2}} = 0$$

$$|f_n(x) - f(x)| = \left|\frac{2n^2x^2}{e^{n^2x^2}} - 1 + 1\right| = \frac{2n^2x^2}{e^{n^2x^2}}$$

$$\frac{d}{dx}|f_n(x) - f(x)| = \frac{d}{dx}\left[\frac{2n^2x^2}{e^{n^2x^2}}\right]$$

$$= \frac{(4n^2x - 4n^4x^3)}{e^{n^2x^2}} = 0 \Rightarrow\ x^2 = \frac{1}{n^2} \Rightarrow x = \pm\frac{1}{n},\ x \in R$$

$$\frac{d^2}{dx^2}|f_n(x) - f(x)| = \frac{d}{dx}\left[\frac{4n^2x - 4n^4x^3}{e^{n^2x^2}}\right]$$

$$= \frac{4n^2 - 20n^4x^2 + 8n^6x^4}{e^{n^2x^2}} < 0,\ \text{for } x = \pm\frac{1}{n}$$

$$M_n = \sup_{n\in R}|f_n(x) - f(x)| = \frac{2n^2\frac{1}{n^2}}{e^{n^2\frac{1}{n^2}}} = \frac{2}{e},\ \text{for } x = \frac{1}{n} \text{ or, } x = -\frac{1}{n}$$

$\Rightarrow M_n \not\to 0$ as $n \to \infty$.

Hence, $\left\{\dfrac{2n^2x^2}{e^{n^2x^2}} - 1\right\}$ is not uniformly convergent on $\mathbb{R}$.

(4) $f_n(x) = 1 - \dfrac{x^n}{n}$, $0 \le x \le 1$.

For $x = 0$, $f(x) = \lim_{n\to\infty} f_n(x) = 1$.

$x = 1$, $f(x) = \lim_{n\to\infty} f_n(x) = \lim_{n\to\infty} (1 - \frac{1}{n}) = 1$.

$0 < x < 1$, $f(x) = \lim_{n\to\infty} f_n(x) = 1$. $\left[\lim_{n\to\infty} x^n = 0\right]$

So $f(x) = 1$, $\forall\, x \in [0,1]$.

Then $M_n = \sup_{x\in[0,1]} |f_n(x) - f(x)| = \sup_{x\in[0,1]} \left|1 - \dfrac{x^n}{n} - 1\right| = \sup_{x\in[0,1]} \dfrac{x^n}{n} = \dfrac{1}{n}$

$\Rightarrow M_n \to 0$ as $n \to \infty$.

$\Rightarrow \left\{1 - \dfrac{x^n}{n}\right\}$ converges uniformly to 1 on $[0,1]$.

(5) $f_n(x) = nx(1-x)^n$, $0 \le x \le 1$.

For $x = 0$, $f(x) = 0$

$x = 1$, $f(x) = 0$.

When $0 < x < 1$, then $0 < 1 - x < 1$.

Let $1 - x = \frac{1}{y}$, $y > 1$. Also let $y = 1 + a$, $a > 0$.

Then $f_n(x) = nx(1-x)^n = \dfrac{nx}{(1+a)^n} < \dfrac{2nx}{n(n-1)a^2}$

$$\left[\begin{aligned} &\because (1+a)^n = 1 + na + \frac{n(n-1)a^2}{2} + \cdots + a^n \\ &\Rightarrow (1+a)^n > \frac{n(n-1)}{2}\cdot a^2 \Rightarrow \frac{1}{(1+a)^n} < \frac{2}{n(n-1)a^2}. \end{aligned}\right]$$

$$\Rightarrow 0 < f_n(x) < \frac{2nx}{n(n-1)a^2},\ \forall\, x \in (0,1).$$

By Sandwich theorem on limit, $\lim_{n\to\infty} f_n(x) = 0 = f(x)$.

Hence, $f(x) = 0$, $\forall\, x \in [0,1]$.

Now $M_n = \sup_{x\in[0,1]} |f_n(x) - f(x)| = \sup_{x\in[0,1]} nx(1-x)^n$.

Let $g_n(x) = nx(1-x)^n$

$$g_n'(x) = n(1-x)^n + n^2x(1-x)^{n-1}\cdot(-1)$$

$$= n(1-x)^{n-1}\{(1-x) - nx\} = 0 \Rightarrow 1 - x - nx = 0 \Rightarrow x = \frac{1}{1+n}$$

$$g_n''(x) = -n(n-1)(1-x)^{n-2}(1-x-nx) - n(1-x)^{n-1}(1+n)$$

$$\Rightarrow g_n''\left(\frac{1}{n+1}\right) = -n(n+1)\frac{n^{n-1}}{(n+1)^{n-1}} = \frac{-n^n}{(n+1)^{n-2}} < 0.$$

$\therefore\ g_n(x)$ is maximum at $x = \frac{1}{n+1}$.

$$M_n = \sup_{x\in[0,1]} |f_n(x) - f(x)| = \frac{n}{n+1}\left(1 - \frac{1}{n+1}\right)^n$$
$$= \frac{n^{n+1}}{(n+1)^{n+1}} = \frac{1}{\left(1+\frac{1}{n}\right)} \frac{1}{\left(1+\frac{1}{n}\right)^n}$$
$$\lim_{n\to\infty} M_n = 1 \cdot \frac{1}{e} = e^{-1}$$

Thus $M_n \not\to 0$ as $n \to \infty$.

Therefore, $\{nx(1-x)^n\}$ does not converge uniformly on $[0,1]$.

Properties of uniformly convergent sequence of functions $\{f_n(x)\}$

Theorem [9.5] (1) If a sequence of continuous functions $\{f_n(x)\}$ converges uniformly to $f(x)$ on $[a,b]$, then $f(x)$ is continuous on $[a,b]$.

(2) If $\{f_n(x)\}$ be a sequence of integrable functions and converges uniformly to $f(x)$ on $[a,b]$. Then f is integrable on $[a,b]$ and $\int_a^b f(x)dx = \lim_{n\to\infty} \int_a^b f_n(x)dx$, i.e., $\left\{\int_a^x f_n(t)dt\right\}$ converges uniformly to $\int_a^x f(t)dt$.

(3) If $\{f_n(x)\}$ be a sequence of differentiable functions with pointwise limit $f(x)$ on $[a,b]$ and if $\{f_n'(x)\}$, sequence of differentials, converge uniformly to $\phi(x)$ on $[a,b]$, where each $f_n'(x)$ is continuous on $[a,b]$, then

(i) $\{f_n(x)\}$ converges uniformly to f on $[a,b]$

(ii) $f'(x) = \phi(x)$, i.e., $\left[\lim_{n\to\infty} f_n(x)\right]' = \phi(x)$.

Proof. (1) Since $\{f_n(x)\}$ converges uniformly to $f(x)$ on $[a,b]$, for a given $\epsilon(>0)$,

$$|f_n(x) - f(x)| < \frac{\epsilon}{3},\ \forall\ n \geq m \text{ and } \forall\ x \in [a,b]. \tag{i}$$

Let $c \in [a,b]$.

Then since $f_n(x)$ is continuous on $[a,b]$, for every $\epsilon > 0$, $\exists$ a (+)ve number δ, such that

$$|f_n(x) - f_n(c)| < \frac{\epsilon}{3},\ \text{for } |x-c| < \delta,\ \forall\ n \in N \tag{ii}$$

In particular, $n = m$, from (i)

$$|f_m(x) - f(x)| < \frac{\epsilon}{3},\ \forall\ x \in [a,b] \Rightarrow |f_m(c) - f(c)| < \frac{\epsilon}{3}$$

Also from (ii) $|f_m(x) - f_m(c)| < \frac{\epsilon}{3}$ for $|x-c| < \delta$.

Now
$$|f(x) - f(c)| = |f(x) - f_m(x) + f_m(x) - f_m(c) + f_m(c) - f(c)|$$
$$\leq |f_m(x) - f(x)| + |f_m(x) - f_m(c)| + |f_m(c) - f(c)|$$
$$< \tfrac{\epsilon}{3} + \tfrac{\epsilon}{3} + \tfrac{\epsilon}{3} = \epsilon,\ \text{for } |x-c| < \delta$$

$\Rightarrow$ for every $\epsilon(>0)$, $\exists$ a (+)ve number δ such that $|f(x) - f(c)| < \epsilon$, for $|x-c| < \delta$.

$\Rightarrow$ $f(x)$ is continuous at $x = c \in [a,b]$.

$\Rightarrow$ $f(x)$ is continuous on $[a,b]$ as c is an arbitrary point on $[a,b]$.

(2) Since $\{f_n(x)\}$ converges uniformly to $f(x)$, for given $\epsilon(>0)$, there exists a (+)ve integer m such that

$$|f_n(x) - f(x)| < \frac{\epsilon}{2(b-a)}, \ \forall \ n \geq m, \ \forall \ x \in [a, b] \tag{i}$$

In particular, $n = m$,

$$|f_m(x) - f(x)| < \frac{\epsilon}{2(b-a)}, \ \forall \ x \in [a, b]. \tag{ii}$$

Let

$$f(x) - f_m(x) = R_m(x) \tag{iii}$$

Since $f_m(x)$ is R-integrable on $[a, b]$, there exists a partition P of $[a, b]$ such that the oscillatory sum $\omega(P, f_m) < \epsilon/2$.

Again, from (ii), $|R_m(x)| < \frac{\epsilon}{2(b-a)}$ by (iii)

$\Rightarrow$ oscillation of R_m in each subinterval is less than $\frac{\epsilon}{2(b-a)}$.

Therefore,

$$\omega(P, R_m) = \sum_{r=1}^{n} (M_r - m_r)\delta_r < \frac{\epsilon}{2(b-a)} \sum_{r=1}^{n} \delta_r = \frac{\epsilon}{2(b-a)}(b-a) = \frac{\epsilon}{2}.$$

Also from (iii),

$$f(x) = f_m(x) + R_m(x) \Rightarrow \omega(P, f) \leq \omega(P, f_m) + \omega(P, R_m) < \frac{\epsilon}{2} + \frac{\epsilon}{2} = \epsilon.$$

Thus for every $\epsilon(>0)$, $\exists$ a partition P of $[a, b]$ such that $\omega(P, f) < \epsilon$ and consequently, f is R-integrable on $[a, b]$.

Now f_n is R-integrable and f has also been proved R-integrable on $[a, b]$, so $f_n - f$ is R-integrable on $[a, b]$.

Now $\left|\int_a^b f_n(x)dx - \int_a^b f(x)dx\right| = \left|\int_a^b \{f_n(x) - f(x)\}dx\right|$

$$\leq \int_a^b |f_n(x) - f(x)|dx < \int_a^b \frac{\epsilon}{2(b-a)}dx, \ \forall \ n \geq m \ [\text{by (i)}]$$

$$< \frac{\epsilon}{2(b-a)} \cdot (b-a) = \frac{\epsilon}{2} < \epsilon, \ \forall \ n \geq m \Rightarrow \lim_{n\to\infty} \int_a^b f_n(x)dx = \int_a^b f(x)dx$$

It immediately follows for $a < x < b$, that

$$\left|\int_a^x f_n(t)dt - \int_a^x f(t)dt\right| = \left|\int_a^x \{f_n(t) - f(t)\}dt\right|$$

$$\leq \int_a^x |f_n(t) - f(t)|dt < \frac{\epsilon}{b-a}(x-a) < \epsilon, \ \forall \ n \geq m \text{ and } \forall \ x \in [a, b]$$

$\Rightarrow \left\{\int_a^x f_n(t)dt\right\}$ converges uniformly to $\int_a^x f(t)dt$ on $[a, b]$.

(3) $\{f_n(x)\}$ converges pointwise to $f(x)$ on $[a, b]$

$\Rightarrow$ to every $\epsilon(>0)$ and to each point $c \in [a, b]$, $\exists$ a (+)ve integer m such that

$|f_n(c) - f(c)| < \epsilon, \ \forall \ n \geq m$

$$\Rightarrow \ |f_{n+p}(c) - f_n(c)| < \epsilon, \ \forall \ n \geq m, \ p \in N \tag{i}$$

Again, uniform convergence of $\{f_n'(x)\}$ to $\phi(x)$ on $[a,b]$

$$\Rightarrow |f_{n+p}'(x) - f_n'(x)| < \frac{\epsilon}{2(b-a)},\ \forall\, n \geq m,\ \forall\, x \in [a,b],\ p \in N \qquad \text{(ii)}$$

[By Cauchy's criterion]

Applying Lagranges Mean Value Theorem to the function $\{f_{n+p}(x) - f_n(x)\}$ for any two points x and t, $a \leq t < x \leq b$, we get

$$\{f_{n+p}(x) - f_n(x)\} - \{f_{n+p}(t) - f_n(t)\}$$

$$= (x-t)\{f_{n+p}'(\xi) - f_n'(\xi)\}, \text{ where } t < \xi < x \Rightarrow |\{f_{n+p}(x) - f_n(x)\} - \{f_{n+p}(t) - f_n(t)\}|$$

$$\leq |x-t|\ |f_{n+p}'(\xi) - f_n'(\xi)\}| < |x-t| \frac{\epsilon}{2(b-a)} \text{ [by (ii)]}$$

$$\leq \frac{(b-a)\xi}{2(b-a)} = \frac{\epsilon}{2} \qquad \text{(iii)}$$

Now $|f_{n+p}(x) - f_n(x)| = |f_{n+p}(x) - f_n(x) - f_{n+p}(t) + f_n(t) + f_{n+p}(t) - f_n(t)|$

$$\leq |\{f_{n+p}(x) - f_n(x)\} - \{f_{n+p}(t) - f_n(t)\}| + |f_{n+p}(t) - f_n(t)|$$

$$< \frac{\epsilon}{2} + \frac{\epsilon}{2} = \epsilon \text{ [by (iii) and (i)] } \forall\, n \geq m,\ \forall\, x \in [a,b],\ p \in N$$

$\Rightarrow \{f_n(x)\}$ converges uniformly to $f(x)$ on $[a,b]$.

$\{f_n'(x)\}$ converges uniformly to $\phi(x)$ on $[a,b]$ and each $\{f_n'(x)\}$ is continuous. Then by Prop. (1), Theorem [9.5], ϕ is continuous on $[a,b]$ and hence ϕ is integrable on $[a,b]$.

Therefore, for any $t \in [a,b]$,

$$\int_a^t \phi(x)dx = \lim_{n\to\infty} \int_a^t f_n'(x)dx = \lim_{n\to\infty}[f_n(t) - f_n(a)] = \lim_{n\to\infty} f_n(t) - \lim_{n\to\infty} f_n(a)$$

$$= f(t) - f(a) \ [\because \{f_n(x)\} \text{ converges pointwise on } [a,b]]$$

Since $\phi(x)$ is continuous on $[a,t]$,

$$\phi(t) = \frac{d}{dt}[f(t) - f(a)]$$

$$\Rightarrow \phi(t) = f'(t) \Rightarrow \phi(x) = f'(x),\ \forall\, x \in [a,b]$$

$$\Rightarrow \phi(x) = \left[\lim_{n\to\infty} f_n(x)\right]'$$

Example. A sequence of continuous functions with a discontinuous limit function.

Let $f_n(x) = \frac{x^{2n}}{1+x^{2n}}$, $\forall\, x \in R,\ n \in N$.

Then $f_n(x)$ is continuous $\forall\, x \in R$. For, if $c \in R$, then $\lim_{x\to c+0} f(x) = \lim_{x\to c-0} f(x) = f(c)$.

$$\text{Now } f(x) = \lim_{n\to\infty} f_n(x) = \begin{cases} 0, & \text{if } |x| < 1 \\ \frac{1}{2}, & \text{if } |x| = 1 \\ 1, & \text{if } |x| > 1, \end{cases}$$

which shows that the limit function f is discontinuous at $x = 1$ and $x = -1$. So by virtue of Prop. (1), Th. [9.5], $\{f_n(x)\}$ is not uniformly convergent on any bounded interval.

Example. A sequence of functions for which $\lim_{n\to\infty} \int_0^1 f_n(x)dx \neq \int_0^1 f(x)dx$.

Let $f_n(x) = n^2 x(1-x)^n$, $\forall\, x \in \mathbb{R}$. We consider a bounded interval $[0,1]$.

For $x = 0,\ \ f(x) = 0$
$\quad x = 1,\ \ f(x) = 0$

For $0 < x < 1$, $1 - x < 1$. Let $\frac{1}{y} = 1 - x$. Then $y > 1$.

Let $y = 1 + a,\ a > 0$. Then $f_n(x) = \dfrac{n^2 x}{(1+a)^n} < \dfrac{6n^2 x}{n(n-1)(n-2)a^3}$

$$0 < f_n(x) < \frac{6n^2 x}{n(n-1)(n-2)a^3},\ \forall\, x \in [0,1]$$

$\Rightarrow \lim\limits_{n\to\infty} f_n(x) = 0$, since $\lim\limits_{n\to\infty} \dfrac{6n^2 x}{n(n-1)(n-2)a^3} = 0$, i.e., $f(x) = 0$ for $0 < x < 1$.

Thus $f(x) = 0,\ \forall\, x \in [0,1]$. Therefore, $\int_0^1 f(x)dx = 0$.
But

$$\lim_{n\to\infty} \int_0^1 f_n(x)dx = \lim_{n\to\infty} \int_0^1 n^2 x(1-x)^n dx$$

$$= \lim_{n\to\infty} n^2 \int_0^1 (1-t)t^n dt, \text{ by putting } 1 - x = t.$$

$$= \lim_{n\to\infty} n^2 \left[\frac{t^{n+1}}{n+1} - \frac{t^{n+2}}{n+2}\right]_0^1 = \lim_{n\to\infty} \left[\frac{n^2}{n+1} - \frac{n^2}{n+2}\right]$$

$$= \lim_{n\to\infty} \cdot \frac{n^2}{(n+1)(n+2)} = \lim_{n\to\infty} \frac{1}{\left(1+\frac{1}{n}\right)\left(1+\frac{2}{n}\right)} = 1$$

Hence, $\lim\limits_{n\to\infty} \int_0^1 f_n(x)dx \neq \int_0^1 f(x)dx$.

So by virtue of Prop. (2) of Th. [9.5], $\{f_n(x)\}$ is not uniformly convergent.

Example. A sequence of differentiable functions $\{f_n(x)\}$ with limit 0 for which $\{f_n'(x)\}$ diverges. That is, $\lim\limits_{n\to\infty} f_n'(x) \neq f'(x)$ for any $x \in R$.

Let $f_n(x) = \dfrac{\sin nx}{\sqrt{n}},\ \forall\, x \in R$

$$|f_n(x)| = \frac{|\sin nx|}{\sqrt{n}} \le \frac{1}{\sqrt{n}} \Rightarrow \lim_{n\to\infty} |f_n(x)| = 0 \Rightarrow \lim_{n\to\infty} f_n(x) = 0,\ \forall\, n \in N,\ x \in R$$

$\Rightarrow f(x) = 0$

Now $f_n'(x) = \sqrt{n}\cos nx \Rightarrow \lim\limits_{n\to\infty} f_n'(x)$ does not exist for any $x \Rightarrow \{f_n'\}$ diverges and

$$\lim_{n\to\infty} f_n'(x) \neq f'(x).$$

Thus we see that the limit of the sequence of derivatives is not equal to the derivative of the limit of the sequence of functions. So by virtue of Prop. (3), Theorem [9.5], the sequence $\{f_n'(x)\}$ cannot converge uniformly in this case.

Problem [9.13]:

(1) Let $\{f_n(x)\}$ be a sequence of real valued continuous functions defined on $[0,1]$ and let $f_n \to f$ uniformly on $[0,1]$. Then prove or disprove $\lim\limits_{n\to\infty} \int_0^{1-\frac{1}{n}} f_n(x)dx = \int_0^1 f(x)dx$.

(2) Let $f_n(x) = \frac{1}{n}e^{-n^2x^2}$, if $x \in \mathbb{R}$, $n \in N$. Prove that $f_n \to 0$ uniformly on R, while $f_n' \to 0$ pointwise on R, and the convergence of $\{f_n'\}$ is not uniform on any interval containing the origin.

Solution. (1) Each $f_n(x)$ is continuous on $[0, 1]$.

$\Rightarrow f_n(x)$ is integrable on $[0, 1]$.

Also $f_n \to f$ uniformly on $[0, 1]$

$$\Rightarrow f \text{ is integrable on } [0,1], \text{ by Prop. (2), Th. [9.5].}$$

$$\Rightarrow \lim_{n\to\infty} \int_0^t f_n(x)dx = \int_0^t f(x)dx \text{ for any } t \in [0,1] \qquad (1)$$

Now, for $t = 1 - \frac{1}{n} \Rightarrow t \in [0, 1]$, $\forall\, n \in N$.

Also $n \to \infty \Rightarrow t = 1$.

Therefore, from (1), $\lim\limits_{n\to\infty} \int_0^{1-\frac{1}{n}} f_n(x)dx = \int_0^1 f(x)dx$.

Note. [Go through the example.]

Let $f_n(x) = \frac{n}{x+n}$, $0 \le x \le 1$.

Then $f_n(x)$ is continuous on $[0, 1]$, $\forall\, n \in N$.

$$f(x) = \lim_{n\to\infty} f_n(x) = 1, \ \forall\, x \in [0,1]$$

Then $|f_n(x) - f(x)| = \left|\frac{n}{x+n} - 1\right| = \left|\frac{-x}{x+n}\right|$

$= \frac{x}{x+n} < \epsilon$ if $\frac{x+n}{x} > \frac{1}{\epsilon}$, i.e., if $n > x\left(\frac{1}{\epsilon} - 1\right) = m(\epsilon, x)$.

Now $m(\epsilon, x) \to 0$ as $x \to 0$ and $m(\epsilon, x) \to \left(\frac{1}{\epsilon} - 1\right)$ as $x \to 1$.

So $m(\epsilon, x) \to$ finite number. Hence $\exists$ a (+)ve integer m such that

$$|f_n(x) - f(x)| < \epsilon, \ \forall\, n \ge m, \ \forall\, x \in [0,1].$$

Therefore, $\{f_n(x)\}$ converges uniformly to 1 on $[0, 1]$.

$$\int_0^{1-\frac{1}{n}} \frac{n}{x+n}dx = n\big[\log(x+n)\big]_0^{1-\frac{1}{n}} = n\left[\log\left(1+n-\frac{1}{n}\right) - \log n\right]$$

$$= n\left[\log\left(1 + \frac{1}{n} - \frac{1}{n^2}\right)\right]$$

$$\lim_{n\to\infty} \int_0^{1-\frac{1}{n}} \frac{n}{x+n}dx = \lim_{n\to\infty} n\left[\log\left(1+\frac{1}{n} - \frac{1}{n^2}\right)\right] = 1$$

Also $\displaystyle\int_0^1 f(x)dx = \int_0^1 1 \cdot dx = [x]_0^1 = 1.$

Thus $\displaystyle\lim_{n\to\infty} \int_0^{1-\frac{1}{n}} f_n(x)dx = \int_0^1 f(x)dx.$

(2) $f_n(x) = \frac{1}{n}e^{-n^2x^2}$

$$\therefore f(x) = \lim_{n\to\infty} f_n(x) = 0, \ \forall\, x \in \mathbb{R}$$

$$|f_n(x) - f(x)| = \frac{1}{n}e^{-n^2x^2}$$

$$\frac{d}{dx}|f_n(x) - f(x)| = \frac{1}{n}e^{-n^2x^2} \cdot (-2n^2x) = 0 \Rightarrow x = 0$$

Also $\frac{d^2}{dx^2}|f_n(x) - f(x)| < 0$ for $x = 0$.

$$M_n = \sup_{x\in R}[f_n(x) - f(x)] = \frac{1}{n}$$

$M_n \to 0$ as $n \to \infty$. So $\{f_n\} \to 0$ uniformly on R. Now $f_n'(x) = -2nxe^{-n^2x^2}$.

$$\begin{aligned}\lim_{n\to\infty} f_n'(x) &= \lim_{n\to\infty} -2nx \cdot e^{-n^2x^2} = \lim_{n\to\infty} \frac{-2n}{e^{n^2x^2}} = \left(\frac{\infty}{\infty}\right)\\ &= \lim_{n\to\infty} \frac{-2}{e^{n^2x^2} \cdot 2nx^2} = 0 \Rightarrow \phi(x) = 0,\ \forall\ x \in R\end{aligned}$$

$$|f_n'(x) - \phi(x)| = \left|-2nxe^{-n^2x^2}\right| = 2nxe^{-n^2x^2}$$

$$\frac{d}{dx}|f_n'(x) - \phi(x)| = 2ne^{-n^2x^2} - 4e^{-n^2x^2}n^3x^2 = 0 \Rightarrow x = \pm\frac{1}{\sqrt{2}\,n}$$

$$\frac{d^2}{dx^2}|f_n'(x) - \phi(x)| < 0 \text{ for } x = -\frac{1}{\sqrt{2}\,n}.$$

$$M_n = \sup_{x\in R}|f_n'(x) - \phi(x)| = 2n\frac{1}{-\sqrt{2}\,n}e^{-n^2\cdot\frac{1}{2n^2}} = -\sqrt{2}e^{-\frac{1}{2}}.$$

$\Rightarrow M_n \not\to 0$ as $n \to \infty$. So $\{f_n'\}$ does not converge uniformly to $\phi(x)$.

Also $n \to \infty \Rightarrow x \to 0$. So 0 is a point of non-convergence on R.

So the convergence of $\{f_n'(x)\}$ is not uniform in any interval $(-k, k)$ containing 0.

Thus $\{f_n'\}$ converges pointwise on R.

Note. $\lim_{n\to\infty} \int_0^{1-\frac{1}{n}} \frac{1}{n}e^{-n^2x^2}\,dx = \int_0^1 f(x)dx.$

9.3 Infinite Series of Functions

A series of the form $\sum_1^\infty f_n(x)$ in which each term of the series is a function of x is called an infinite series of the functions $f_n(x)$.

We write, $S_n(x) = \sum_{k=1}^{n} f_k(x) = f_1(x) + f_2(x) + \cdots + f_n(x)$ so that $\{S_n(x)\}$ forms a sequence of functions.

Definition [9.3] : The series $\sum_1^\infty f_n(x)$ is said to converge pointwise on $[a, b]$ to $S(x)$, if the sequence $\{S_n(x)\}$ converges pointwise to $S(x)$ on $[a, b]$.

That is, to each $\epsilon(> 0)$ and to each $x \in [a, b]$, there exists a (+)ve integer m such that $|S_n(x) - S(x)| < \epsilon,\ \forall\ n \geq m$.

Here, $S(x)$ is called the pointwise sum of the infinite series.

Definition [9.4] : The series $\sum_1^\infty f_n(x)$ is said to converge uniformly on $[a, b]$ to $S(x)$, if the sequence $\{S_n(x)\}$ converges uniformly to $S(x)$ on $[a, b]$.

That is, for every $\epsilon > 0$, there exists a (+)ve integer m such that $|S_n(x) - S(x)| < \epsilon,\ \forall\ n \geq m$ and $\forall\ x \in [a, b]$.

Here $S(x)$ is called the uniform sum of the series.

Example. We consider an infinite series of functions of the form $\sum\limits_{k=1}^{\infty} \frac{x}{n(n+1)}$.

$$\text{We write, } S_n(x) = \sum_{k=1}^{n} \frac{x}{k(k+1)} = \frac{x}{1\cdot 2} + \frac{x}{2\cdot 3} + \cdots + \frac{x}{n(n+1)}$$

$$f_1(x) = \frac{x}{1\cdot 2} = x\left\{1 - \frac{1}{2}\right\}$$

$$f_2(x) = \frac{x}{2\cdot 3} = x\left\{\frac{1}{2} - \frac{1}{3}\right\}$$

$$\cdots\cdots\cdots\cdots\cdots\cdots\cdots\cdots$$

$$f_n(x) = \frac{x}{n(n+1)} = x\left\{\frac{1}{n} - \frac{1}{n+1}\right\}$$

Then $S_n(x) = f_1(x) + \cdots + f_n(x) = x\left\{1 - \frac{1}{n+1}\right\} = \frac{nx}{n+1}$, we consider both pointwise and uniform convergence of the given series in $[0, k]$

$$\lim_{n\to\infty} S_n(x) = \lim_{n\to\infty} \frac{nx}{n+1} = \lim_{n\to\infty} \frac{x}{1+\frac{1}{n}} = x, \ \forall\, x \in [0, k].$$

$S(x) = 0$, for $x = 0$.

Hence, $\{S_n(x)\}$ converges pointwise to x, to each $x \in [0, k]$. (1)

Again, $|S_n(x) - S(x)| = \left|\frac{nx}{n+1} - x\right| = \left|\frac{-x}{n+1}\right|$

$= \frac{x}{n+1} < \epsilon$, if $n + 1 > \frac{x}{\epsilon}$, i.e., if $n > \left(\frac{x}{\epsilon} - 1\right) = m(\epsilon, x)$.

Now $m(\epsilon, x) \to -1$ as $x \to 0$ and $m(\epsilon, x) \to \frac{k}{\epsilon} - 1$ as $x \to k$.

So $m(\epsilon, x) \to$ finite number for all $x \in [0, k]$.

Hence, $\exists$ a (+)ve integer m such that $|S_n(x) - S(x)| < \epsilon$, $\forall\, n \geq m$ and $\forall\, x \in [0, k]$

$\Rightarrow \{S_n(x)\}$ converges uniformly to x on $[0, k]$ (2)

So from (1) and (2) it follows that the series $\sum\limits_{n=1}^{\infty} \frac{x}{n(n+1)}$ converges pointwise and also converges uniformly to x on $[0, k]$.

Properties of uniform convergence of infinite series of functions

Theorem [9.6] (1) Let $f_n : [a, b] \to \mathbb{R}$, $[a, b] \subseteq \mathbb{R}$ be continuous on $[a, b]$, $\forall\, n \in N$ and let $\sum\limits_{1}^{\infty} f_n(x)$ be uniformly convergent on $[a, b]$, then the sum function $S(x)$ is continuous on $[a, b]$.

(2) If $f_n : [a, b] \to \mathbb{R}$, $[a, b] \subseteq \mathbb{R}$ be integrable on $[a, b]$, $\forall\, n \in N$ and let $\sum\limits_{1}^{\infty} f_n(x)$ be uniformly convergent on $[a, b]$ to the sum function $S(x)$, then

(i) $S(x)$ is integrable on $[a, b]$, (ii) $\int_a^b \sum f_n(x)dx = \int_a^b S(x)dx = \sum \int_a^b f_n(x)dx$.

(3) If $\sum\limits_{1}^{\infty} f_n(x)$ is a pointwise convergent series of differentiable functions on $[a, b]$ and each $f_n'(x)$ is continuous on $[a, b]$ and further $\sum\limits_{1}^{\infty} f_n'(x)$, the series of differentials, is uniformly convergent on $[a, b]$, then $\frac{d}{dx}\left(\sum f_n(x)\right) = \sum f_n'(x)$. That is, the derivative of the sum is equal to sum of their derivatives.

Proof. (1) $S_n(x) = f_1(x) + f_2(x) + \cdots + f_n(x)$.

So $S_n(x)$ is continuous on $[a, b]$, since each $f_n(x)$ is continuous on $[a, b]$.

Again, the uniform convergence of $\sum_1^\infty f_n(x)$ to $S(x)$ on $[a, b] \Rightarrow \{S_n(x)\}$ converges uniformly to $S(x)$ on $[a, b]$. Then for a given $\epsilon(> 0)$, there exists a (+)ve integer m such that

$$|S_n(x) - S(x)| < \frac{\epsilon}{3},\ \forall\, n \geq m,\ \forall\, x \in [a, b] \quad \text{(i)}$$

In particular, $n = m$,

$$|S_m(x) - S(x)| < \frac{\epsilon}{3},\ \forall\, x \in [a, b]. \quad \text{(ii)}$$

Let $c \in [a, b]$. $S_m(x)$ is continuous on $[a, b]$

$$\Rightarrow |S_m(x) - S_m(c)| < \frac{\epsilon}{3}, \text{ whenever } |x - c| < \delta. \quad \text{(iii)}$$

$$\begin{aligned}
\text{Therefore, } |S(x) - S(c)| &= |S(x) - S_m(x) + S_m(x) - S_m(c) + S_m(c) - S(c)| \\
&\leq |S_m(x) - S(x)| + |S_m(x) - S_m(c)| + |S_m(c) - S(c)| \\
&< \frac{\epsilon}{3} + \frac{\epsilon}{3} + \frac{\epsilon}{3} = \epsilon, \text{ whenever } |x - c| < \delta \text{ [by (ii) and (iii)]}
\end{aligned}$$

$\Rightarrow S(x)$ is continuous at $x = c$

$\Rightarrow S(x)$ is continuous on $[a, b]$ as c is any point on $[a, b]$.

(2) $S_n(x) = \sum_1^n f_k(x)$

$\Rightarrow S_n(x)$ is integrable on $[a, b]$ as each $f_n(x)$ is integrable on $[a, b]$.

Also uniform convergence of $\sum_1^\infty f_n(x)$ to $S(x)$ on $[a, b] \Rightarrow \{S_n(x)\}$ converges uniformly to $S(x)$ on $[a, b]$.

Hence, by Prop. (2), Th. [9.5], $S(x)$ is integrable on $[a, b]$.

$$\text{Also} \quad \lim_{n\to\infty} \int_a^b S_n(x)dx = \int_a^b S(x)dx$$

$$\Rightarrow \lim_{n\to\infty} \int_a^b \{f_1(x) + \cdots + f_n(x)\}dx = \int_a^b \{f_1(x) + f_2(x) + \cdots\}$$

$$\Rightarrow \lim_{n\to\infty} \left[\int_a^b f_1(x)dx + \cdots + \int_a^b f_n(x)dx\right] = \int_a^b \sum f_n(x)dx$$

$$\Rightarrow \int_a^b f_1(x)dx + \cdots + \int_a^b f_n(x)dx + \cdots = \int_a^b \sum f_n(x)dx$$

$$\Rightarrow \sum_1^\infty \int_a^b f_n(x)dx = \int_a^b \sum f_n(x)dx$$

(3) Let $S_n(x) = f_1(x) + f_2(x) + \cdots + f_n(x)$. (i)

Then $\sum_1^\infty f_n(x)$ converge pointwise to $S(x)$ on $[a, b]$.

$\Rightarrow \{S_n(x)\}$ converges pointwise to $S(x)$ on $[a, b]$.

From (i), $S_n'(x) = f_1'(x) + f_2'(x) + \cdots + f_n'(x)$.

Let $\sum_1^\infty f_n'(x)$ converge uniformly to $\phi(x)$ on $[a,b]$. Then $\{S_n'(x)\}$ converges uniformly to $\phi(x)$ on $[a,b]$. Each $f_n'(x)$ is continuous $\Rightarrow S_n'(x)$ is continuous $\forall\, n \in N \Rightarrow \phi(x)$ is continuous on $[a,b]$.

So for any $t \in [a,b]$,

$$\begin{aligned}\int_a^t \phi(x)dx &= \lim_{n\to\infty}\int_a^t S_n'(x)dx\\ &= \lim_{n\to\infty}\left[S_n(x)\right]_a^t = \lim_{n\to\infty}[S_n(t)-S_n(a)]\\ &= S(t)-S(a) \quad [\because \{S_n(x)\} \text{ converges pointwise to } S(x)].\end{aligned}$$

Since $\phi(x)$ is continuous,

$$\frac{d}{dt}[S(t)-S(a)] = \phi(t)$$

$$\Rightarrow\ S'(t) = \phi(t) \text{ for } t\in[a,b] \Rightarrow S'(x)=\phi(x),\ \forall\, x\in[a,b]$$

$$\Rightarrow\ \frac{d}{dx}[S(x)] = \sum_1^\infty f_n'(x) \Rightarrow \frac{d}{dx}\sum f_n(x) = \sum_1^\infty\left\{\frac{d}{dx}(f_n(x))\right\}.$$

Test for the uniform convergence of an infinite series of functions

Theorem [9.7] [General principle of convergence] The necessary and sufficient condition for the uniform convergence of an infinite series $\sum_1^\infty f_n(x)$ on $[a,b]$ is that, for every $\epsilon(>0)$, there exists a (+)ve integer m such that $|f_{n+1}(x)+f_{n+2}(x)+\cdots+f_{n+p}(x)|<\epsilon$, $\forall\, n\ge m$ and $\forall\, x\in[a,b]$ and $\forall\, p\in N$.

Proof. Let $S_n(x) = f_1(x)+f_2(x)+\cdots+f_n(x)$.

Then $\sum_1^\infty f_n(x)$ converges uniformly on $[a,b]$.

$\Rightarrow \{S_n(x)\}$ converges uniformly on $[a,b]$.

The necessary and sufficient condition that a sequence of function $\{S_n(x)\}$ converges uniformly on $[a,b]$ is that, for every $\epsilon>0$, there exists a (+)ve m such that $|S_{n+p}(x)-S_n(x)|<\epsilon$, $\forall\, n\ge m$ and $\forall\, x\in[a,b]$ and $\forall\, p\in N$ [by Prop. (3), Th. [9.1]]

$$\begin{aligned}&\Rightarrow |\{f_1(x)+f_2(x)+\cdots+f_n(x)+f_{n+1}(x)+\cdots+f_{n+p}(x)\}-\{f_1(x)+f_2(x)\\&\qquad+\cdots+f_n(x)\}|<\epsilon,\ \forall\, n\ge m \text{ and } \forall\, x\in[a,b],\ \forall\, p\in N\\&\Rightarrow |f_{n+1}(x)+\cdots+f_{n+p}(x)|<\epsilon,\ \forall\, n\ge m,\ \forall\, x\in[a,b] \text{ and } \forall\, p\in N.\end{aligned}$$

Theorem [9.8] [Weierstrass's M-Test for uniform convergence] An infinite series of functions $\sum_1^\infty f_n(x)$ converges uniformly on $[a,b]$, if there exists a convergent series $\sum M_n$ of (+)ve numbers such that $|f_n(x)|\le M_n$, $\forall\, n\in N$ and $\forall\, x\in[a,b]$.

Proof. Since $\sum M_n$ is a convergent series of (+)ve numbers, by Cauchy's general principle of convergence, for every $\epsilon(>0)$, there exists a (+)ve number m such that

$$|M_{n+1}+M_{n+2}+\cdots+M_{n+p}|<\epsilon,\ \forall\, n\ge m \text{ and } \forall\, p\in N$$

$$\Rightarrow M_{n+1}+M_{n+2}+\cdots+M_{n+p}<\epsilon,\ \forall\, n\ge m \text{ and } \forall\, p\in N \qquad \text{(i)}$$

$$[\because M_n>0,\ \forall\, n\in N]$$

Now, $|f_{n+1}(x)+f_{n+2}(x)+\cdots+f_{n+p}(x)| \le |f_{n+1}(x)|+|f_{n+2}(x)|+\cdots+|f_{n+p}(x)|$
$\le M_{n+1}+M_{n+2}+\cdots+M_{n+p}$, since $|f_n(x)| \le M_n,\ \forall\ x \in [a,b],\ \forall\ n \in N.$
$< \epsilon\ \forall\ n \ge m,\ \forall\ x \in [a,b]$ and $\forall\ P \in N$ [by (i)]

$\Rightarrow \sum_1^\infty f_n(x)$ converges uniformly on $[a,b]$ by Th. [9.7].

Example. The series $\sum_{n=1}^{\infty} \frac{1}{n^p}\cos nx$ converges uniformly on R for $p > 1$.

Here $f_n(x) = \frac{1}{n^p}\cos nx \Rightarrow |f_n(x)| = |\frac{1}{n^p}\cos nx| \le \frac{1}{n^p} = M_n,\ \forall\ n \in N,\ x \in \mathbb{R}.$

Now $\sum M_n = \sum_1^\infty \frac{1}{n^p}$, infinite series of (+)ve numbers and also convergent, if $p > 1$.

Therefore, by Th. [9.8], $\sum \frac{1}{n^p}\cos nx$ converges uniformly on $\mathbb{R}$, if $p > 1$.

Example. The series $\sum_0^\infty \frac{x^n}{n!}$ is uniformly convergent on $[-1,1]$.

Here $f_n(x) = \frac{x^n}{n!}$.
Then $|f_n(x)| = |\frac{x^n}{n!}| \le \frac{1}{n!}\ \forall\ x \in [-1,1]$.
Let $M_n = \frac{1}{n!}$.
Then $\sum_1^\infty M_n = \sum_1^\infty \frac{1}{n!}$ is a convergent series as

$$\sum_1^\infty \frac{1}{n!} = 1 + \frac{1}{2!} + \frac{1}{3!} + \frac{1}{4!} + \cdots + \frac{1}{n!} + \cdots$$

$$< 1 + \frac{1}{2} + \frac{1}{2^2} + \frac{1}{2^3} + \cdots + \frac{1}{2^{n-1}} + \cdots,$$

which is a convergent series as being an infinite G.P. series with common ratio $\frac{1}{2} < 1$.

Therefore, $|f_n(x)| \le M_n,\ \forall\ n,\ \forall\ x \in [-1,1]$ and $\sum_1^\infty M_n$ is a convergent series and hence by Weierstrass Test $\sum_1^\infty \frac{x^n}{n!}$ is uniformly convergent on $[-1,1]$.

Example. The series $\sum_1^\infty a_n \sin nx$ and $\sum_1^\infty a_n \cos nx$ are uniformly convergent on $\mathbb{R}$, if $\sum_1^\infty |a_n|$ converges.

Here $f_n(x) = a_n \sin nx$ or $a_n \cos nx$
$|f_n(x)| = |a_n \sin nx| \le |a_n|,\ \forall\ x \in \mathbb{R}.$

Let $M_n = |a_n|$. Then, if $\sum M_n = \sum_1^\infty |a_n|$ converges then $\sum a_n \sin nx$ and $\sum_1^\infty a_n \cos nx$ uniformly converges on R.

Theorem [9.9] (1) If $\sum_1^\infty |f_n(x)|$ converges uniformly on $[a,b]$, then $\sum f_n(x)$ also converges uniformly on $[a,b]$.

(2) If $\sum_{n=1}^\infty f_n(x)$ converges uniformly on $[a,b]$ and g is bounded on $[a,b]$, then the series $\sum_1^\infty g(x) f_n(x)$ is uniformly convergent on $[a,b]$.

Proof. (1) Since $\sum_1^\infty |f_n(x)|$ converges uniformly on $[a,b]$, for every $\epsilon > 0$, there exists a (+)ve integer m such that $||f_{n+1}(x)| + |f_{n+2}(x)| + \cdots + |f_{n+p}(x)|| < \epsilon\ \forall\ x \in [a,b]$ and $\forall\ n \geq m$, $\forall\ p \in N$, by Th. [9.7]

$$\Rightarrow |f_{n+1}(x)| + |f_{n+2}(x)| + \cdots + |f_{n+p}(x)| < \epsilon,\ \forall\ x \in [a,b],\ \forall\ n \geq m,\ \forall\ p \in N \quad \text{(i)}$$

Now $|f_{n+1}(x) + f_{n+2}(x) + \cdots + f_{n+p}(x)| \leq |f_{n+1}(x)| + |f_{n+2}(x)| + \cdots + |f_{n+p}(x)|$
$< \epsilon,\ \forall\ x \in [a,b],\ \forall\ n \geq m$ and $\forall\ p \in N$, by (i).

$\Rightarrow \sum_1^\infty f_n(x)$ converges uniformly on $[a,b]$.

(2) Since $\sum_1^\infty f_n(x)$ converges uniformly on $[a,b]$, for every $\epsilon > 0$, there exists a (+)ve integer m such that $|f_{n+1}(x) + f_{n+2}(x) + \cdots + f_{n+p}(x)| < \epsilon/k,\ \forall\ x \in [a,b]\ \forall\ n \geq m$ and $\forall\ p \in N$.

Also $g(x)$ is bounded on $[a,b] \Rightarrow |g(x)| \leq k,\ \forall\ x \in [a,b]$.

Now $|g(x)f_{n+1}(x) + g(x)f_{n+2}(x) + \cdots + g(x)f_{n+p}(x)|$
$\leq |g(x)|\,|f_{n+1}(x) + f_{n+2}(x) + \cdots + f_{n+p}(x)|$
$< k \cdot \dfrac{\epsilon}{k} = \epsilon,\ \forall\ x \in [a,b],\ \forall\ n \geq m$ and $\forall\ p \in N$.

$\Rightarrow \sum_1^\infty g(x)f_n(x)$ converges uniformly on $[a,b]$.

Problem [9.14]: (1) If $\sum a_n$ converges absolutely, then prove that $\sum \frac{a_n x^n}{1+x^{2n}}$ and $\sum \frac{a_n x^{2n}}{1+x^{2n}}$ converges uniformly for all x.

(2) Prove that $\sum \frac{1}{n^3+n^4x^2}$ is uniformly convergent for all x and that the derivative of the sum is given by term by term differentiation.

(3) Show that the series $(1-x)^2 + x(1-x)^2 + x^2(1-x)^2 + \cdots$ is not uniformly convergent in $[0,1]$.

(4) Show that $\sum_0^\infty \frac{1}{1+n^2+n^4x^2}$ converges uniformly $\forall\ x$ and examine where $f_n'(0)$ can be found by term by term differentiation.

(5) Show that the series $\frac{x}{x+1} + \frac{x}{(x+1)(2x+1)} + \frac{x}{(2x+1)(3x+1)} + \cdots$ is uniformly convergent in $[k,\infty), k > 0$. Show also that the series is not uniformly convergent in $[0,\infty)$.

(6) Show that the series $\frac{x}{1+x^2} + \left(\frac{2^2x}{1+2^3x^2} - \frac{x}{1+x^2}\right) + \left(\frac{3^2x}{1+3^3x^2} - \frac{2^2x}{1+2^3x^2}\right) + \cdots$ does not converge uniformly on $[0,1]$.

Solution. (1) For the series $\displaystyle\sum \frac{a_n x^n}{1+x^{2n}}$, $f_n(x) = \dfrac{a_n x^n}{1+x^{2n}}$.

The maximum value of $\dfrac{x^n}{1+x^{2n}}$ as given by $x = 1$.

So, $|f_n(x)| = \left|\dfrac{a_n x^n}{1+x^{2n}}\right| = |a_n| \left|\dfrac{x^n}{1+x^{2n}}\right| \leq |a_n| \cdot \dfrac{1}{2} < |a_n|\ \forall x$.

Let $M_n = |a_n|$. Then $|f_n(x)| < M_n$, $\forall\ x$ and $\forall\ n \in N$. Since $\sum a_n$ is absolutely convergent, $\sum |a_n|$ is convergent and so $\sum M_n$ is a convergent series of (+)ve numbers

Therefore, by Weierstrass's M-Test, $\displaystyle\sum \frac{a_n x^n}{1+x^{2n}}$ is uniformly convergent for all x.

Again, for the series $\sum a_n \dfrac{x^{2n}}{1+x^{2n}}$,

$$f_n(x) = a_n \frac{x^{2n}}{1+x^{2n}} \Rightarrow |f_n(x)| = |a_n| \left|\frac{x^{2n}}{1+x^{2n}}\right| \le |a_n|,\ \forall\ x$$

Then $|f_n(x)| \le M_n$, $\forall\ x$ and $\forall\ n \in N$, when $|a_n| = M_n$.

$\sum M_n = \sum |a_n|$ is a convergent series of (+)ve numbers.

Therefore, by Weierstrass's M-Test $\sum \dfrac{a_n x^{2n}}{1+x^{2n}}$ is uniformly convergent.

(2) Here $f_n(x) = \dfrac{1}{n^3+n^4x^2}$. So, $|f_n(x)| = \left|\dfrac{1}{n^3+n^4x^2}\right| \le \dfrac{1}{n^3}$, $\forall\ x$ and $\forall\ n \in N$.

Let $M_n = \dfrac{1}{n^3}$. Then $|f_n(x)| \le M_n$, $\forall\ n \in N$, $\forall\ x$.

Now $\sum M_n = \sum \dfrac{1}{n^3}$ is a convergent series of (+)ve numbers.

Therefore, by Weierstrass's M-Test, $\sum \dfrac{1}{n^3+n^4x^2}$ is uniformly convergent $\forall\ x$.

Now $\lim\limits_{n\to\infty} f_n(x) = \lim\limits_{n\to\infty} \dfrac{1}{n^3+n^4x^2} = 0$, $\forall\ x \Rightarrow f(x) = 0$ and so $f'(x) = 0$.

Again, $f_n'(x) = -\dfrac{1\cdot 2n^4x}{(n^3+n^4x^2)^2} = -\dfrac{\frac{2}{n^4}x}{\left(\frac{1}{n}+x^2\right)^2}$.

$\lim\limits_{n\to\infty} f_n'(x) = 0$, $\forall\ x$ [Since $f_n'(0) = 0$].

Thus $\lim\limits_{n\to\infty} f_n'(x) = f'(x) \Rightarrow f_1'(x) + f_2'(x) + \cdots = \dfrac{d}{dx}[f(x)]$

$$\Rightarrow \sum \frac{d}{dx} f_n(x) = \frac{d}{dx}\left(\sum f_n(x)\right)$$

which shows that the derivative of sum (i.e., $\frac{d}{dx}\sum f_n(x)$) is equal to the term by term differentiation (i.e., $\sum \frac{d}{dx} f_n(x)$).

(3) Here $S_n(x) = (1-x)^2 + x(1-x)^2 + x^2(1-x)^2 + \cdots + x^{n-1}(1-x)^2$

$$= (1-x)^2\left\{1+x+x^2+\cdots+x^{n-1}\right\}$$

$$= (1-x)^2 \cdot \frac{1-x^n}{1-x},\ \text{for } 0<x<1 = (1-x)(1-x^n)$$

$\lim\limits_{n\to\infty} S_n(x) = 1-x$, for $0<x<1$

$S_n(0) = 1$

$S_n(1) = 0$

$$\text{Thus } S(x) = \begin{cases} 1-x, & 0<x<1 \\ 1, & x=0 \\ 0, & x=1 \end{cases}$$

So, $S(x)$ is continuous on $[0,1]$.

Now $|S_n(x) - S(x)| = |(1-x)(1-x^n) - (1-x)| = (1-x)x^n < \epsilon,\ 0 < x < 1$

$$\text{if } x^n < \frac{\epsilon}{(1-x)}$$

$$\text{i.e., if } n \log \frac{1}{x} > \log\left(\frac{1-x}{\epsilon}\right)$$

$$\text{i.e., if } n > \log\left(\frac{1-x}{\epsilon}\right) \Big/ \log\frac{1}{x} = m(\epsilon, x)$$

$m(\epsilon, x) \to 0$ as $x \to 0$ but $m(\epsilon, x) \to \infty$ as $x \to 1$. So there does not exist an m such that $|S_n(x) - S(x)| < \epsilon$, $\forall\ n \geq m$ and $\forall\ x \in [0,1]$. Therefore, $\{S_n(x)\}$ is not uniformly convergent on $[0,1]$ and hence the series given is not uniformly convergent on $[0,1]$.

(4) Here $f_n(x) = \dfrac{1}{1+n^2+n^4x^2}$. Then $|f_n(x)| = \left|\dfrac{1}{1+n^2+n^4x^2}\right| \leq \dfrac{1}{n^2}$, $\forall\ x \in \mathbb{R}$.

Let $M_n = \dfrac{1}{n^2}$. Then, $|f_n(x)| \leq M_n$, $\forall\ n$ and $\forall\ x \in R$.

Also $\sum M_n = \sum \dfrac{1}{n^2}$ is a convergent series.

Therefore, $\sum \dfrac{1}{1+n^2+n^4x^2}$ is uniformly convergent $\forall\ x$, by Weierstrass's M-Test.

Also $\lim\limits_{n\to\infty} f_n(x) = \lim\limits_{n\to\infty} \dfrac{1}{1+n^2+n^4x^2} = 0,\ \forall\ x.$

$\therefore\ f(x) = 0$ and hence $f'(x) = 0$.

Now $f_n'(x) = -\dfrac{1 \cdot 2n^4x}{(1+n^2+n^4x^2)^2} = -\dfrac{\frac{2x}{n^4}}{\left(\frac{1}{n^4} + \frac{1}{n^2} + x^2\right)^2}$

$\lim\limits_{n\to\infty} f_n'(x) = 0,\ \forall\ x$. [Since $f_n'(0) = 0$]

Therefore, $\lim\limits_{n\to\infty} f_n'(x) = f'(x),\ \forall\ x \Rightarrow \sum\limits_1^\infty \dfrac{d}{dx} f_n(x) = \dfrac{d}{dx}\sum f_n(x)$ and hence $f'(0) = 0$.

(5) $S_n(x) = \dfrac{x}{x+1} + \dfrac{x}{(x+1)(2x+1)} + \cdots + \dfrac{x}{((n-1)x+1)(nx+1)}$

$$f_n(x) = \frac{x}{((n-1)x+1)(nx+1)} = \frac{1}{(n-1)x+1} - \frac{1}{nx+1}$$

$$f_1(x) = 1 - \frac{1}{x+1}$$

$$f_2(x) = \frac{1}{x+1} - \frac{1}{2x+1}$$

....

$$f_n(x) = \frac{1}{(n-1)x+1} - \frac{1}{nx+1}$$

$$\therefore\ S_n(x) = \sum_1^n f_k(x) = 1 - \frac{1}{nx+1}$$

$S(x) = \lim\limits_{n\to\infty} S_n(x) = 1,\ \forall\ x \in (0,\infty)$ and $S(x) = 0$, for $x = 0$.

Now $|S_n(x) - S(x)| = \left|1 - \frac{1}{nx+1} - 1\right| = \frac{1}{nx+1} < \epsilon,$

if $nx + 1 > \frac{1}{\epsilon}$, i.e., if $n > \left(\frac{1}{\epsilon} - 1\right)/x = m(\epsilon, x)$.

$m(\epsilon, x) \to \dfrac{(\frac{1}{\epsilon}-1)}{k}$ as $x \to k(>0)$ and $m(\epsilon, x) \to 0$ as $x \to \infty$.

So $m(\epsilon, x) \to$ finite number $\forall\ x \in [k, \infty)$. Therefore, there exists a (+)ve integer m such that $|S_n(x) - S(x)| < \epsilon$, $\forall\ n \geq m$ and $\forall\ x \in [k, \infty)$.

Therefore, $\{S_n(x)\}$ is uniformly convergent on $[k, \infty)$ and hence the given series is uniformly convergent on $[k, \infty)$.

But $m(\epsilon, x) \to \infty$ as $x \to 0$ and consequently, it is not possible to find a (+)ve integer m such that $|S_n(x) - S(x)| < \epsilon$, $\forall\ n \geq m$ and $\forall\ x \in [0, \infty)$.

So $\{S_n(x)\}$ is not uniformly convergent on $[0, \infty)$ and hence the given series is not uniformly convergent on $[0, \infty)$.

(6) Here
$$f_1(x) = \frac{x}{1+x^2}$$
$$f_2(x) = \frac{2^2x}{1+2^3x^2} - \frac{x}{1+x^2}$$
$$f_3(x) = \frac{3^2x}{1+3^3x^2} - \frac{2^2x}{1+2^3x^2}$$
....
$$f_n(x) = \frac{n^2x}{1+n^3x^2} - \frac{(n-1)^2x}{1+(n-1)^3x^2}$$

Then $S_n(x) = \sum_{k=1}^{n} f_k(x) = \dfrac{n^2x}{1+n^3x^2}$.

Now $S(x) = \lim_{n\to\infty} S_n(x) = \lim_{n\to\infty} \dfrac{n^2x}{1+n^3x^2} = 0$, $\forall\ x \in [0, 1]$.

Let $M_n = \sup_{x\in[0,1]} |S_n(x) - S(x)| = \sup_{x\in[0,1]} \left|\dfrac{n^2x}{1+n^3x^2}\right| = \sup_{x\in[0,1]} \dfrac{n^2x}{1+n^3x^2}$

$$\frac{d}{dx}\left[\frac{n^2x}{1+n^3x^2}\right] = 0 \Rightarrow n^2 - n^5x^2 = 0 \Rightarrow x = \frac{1}{n^{3/2}}$$

Also $\dfrac{d^2}{dx^2}\left[\dfrac{n^2x}{1+n^3x^2}\right] < 0$, for $x = \frac{1}{n^{3/2}}$.

So $\sup_{x\in[0,1]} \dfrac{n^2x}{1+n^3x^2} = \dfrac{n^2 \cdot \frac{1}{n^{3/2}}}{1+n^3\cdot\frac{1}{n^3}} = \dfrac{1}{2}\sqrt{n}$.

$\therefore M_n = \frac{1}{2}\sqrt{n}$. $M_n \to \infty$ as $n \to \infty$. So $M_n \not\to 0$ as $n \to \infty$.

Therefore, the given series is not uniformly convergent on $[0, 1]$.

Sufficient conditions for uniform convergence of a series

Theorem [9.10] [Dirichlet's Test] Let $F_n(x) = \sum_{k=1}^{n} f_k(x)$ be the nth partial sum of the series $\sum_{1}^{\infty} f_n(x)$, where $f_n(x)$ is a real valued function defined on $[a, b]$ such that $|F_n(x)| \leq k$, $\forall\ n$ and $\forall\ x \in [a, b]$. Also let $\{g_n(x)\}$ be a sequence of positive real valued function on $[a, b]$ such that $g_{n+1}(x) \leq g_n(x)$ for each x in $[a, b]$ and for each $n \in N$ and assume that $\{g_n\} \to 0$ uniformly on $[a, b]$. Then the series $\sum_{n=1}^{n} f_n(x)g_n(x)$ converges uniformly on $[a, b]$.

Proof. $\{g_n\} \to 0$ uniformly on $[a, b] \Rightarrow$ for every $\epsilon(>0)$, $\exists$ a (+)ve integer m such that

$$|g_n(x)| < \frac{\epsilon}{2k},\ \forall\ n \geq m \text{ and } \forall\ x \in [a, b] \qquad \text{(i)}$$

Also $F_n(x) = \sum\limits_{k=1}^{n} f_k(x)$. (ii)

Let $S_n(x) = \sum\limits_{k=1}^{n} f_k(x) g_k(x)$

Then
$$\begin{aligned} S_n(x) &= f_1(x)g_1(x) + f_2(x)g_2(x) + \cdots + f_n(x)g_n(x) \\ &= F_1(x)g_1(x) + \{F_2(x) - F_1(x)\}g_2(x) + \{F_3(x) - F_2(x)\}g_3(x) \\ &\qquad + \cdots + \{F_n(x) - F_{n-1}(x)\}g_n(x) \quad \text{[by (ii)]} \\ &= F_1(x)\{g_1(x) - g_2(x)\} + F_2(x)\{g_2(x) - g_3(x)\} \\ &\qquad + \cdots + F_n(x)\{g_n(x) - g_{n+1}(x)\} + F_n(x)g_{n+1}(x) \\ &= \sum_{k=1}^{n} F_k(x)\{g_k(x) - g_{k+1}(x)\} + F_n(x)g_{n+1}(x). \end{aligned}$$

For $n > m$,

$$S_n(x) - S_m(x) = \sum_{k=m+1}^{n} F_k(x)\{g_k(x) - g_{k+1}(x)\} + F_n(x)g_{n+1}(x) - F_m(x)g_{m+1}(x)$$

$$\begin{aligned} \Rightarrow |S_n(x) - S_m(x)| &\leq \sum_{k=m+1}^{n} |F_k(x)||g_k(x) - g_{k+1}(x)| + |F_n(x)||g_{n+1}(x)| + |F_m(x)||g_{m+1}(x)| \\ &\leq k \sum_{k=m+1}^{n} (g_k(x) - g_{k+1}(x)) + k(g_{n+1}(x)) + k(g_{m+1}(x)) \\ &\qquad \text{(since } g_{k+1} \leq g_k(x), g_n(x) > 0 \text{ and } |F_n(x)| \leq k,\ \forall\ x \in [a, b]\ \&\ \forall\ n \in N] \\ &= k[g_{m+1}(x) - g_{m+2}(x) + g_{m+2}(x) - g_{m+3}(x) \\ &\qquad + \cdots + g_n(x) - g_{n+1}(x)] + kg_{n+1}(x) + kg_{m+1}(x) \\ &= k[g_{m+1}(x) - g_{n+1}(x)] + kg_{n+1}(x) + kg_{m+1}(x) \\ &= 2kg_{m+1}(x) < 2k\frac{\epsilon}{2k} = \epsilon,\ \forall\ n > m,\ \forall\ x \in [a, b] \text{ [by (i)]} \end{aligned}$$

$\Rightarrow \{S_n(x)\}$ converges uniformly on $[a, b] \Rightarrow \sum\limits_{1}^{\infty} f_n(x)g_n(x)$ converges uniformly on $[a, b]$.

Note. The same is true, if $\{g_n(x)\}$ is monotonically increasing on $[a, b]$.

Definition [9.5] : A sequence of functions $\{f_n(x)\}$ is said to uniformly bounded on $[a, b]$, if there exists a (+)ve number k such that $|f_n(x)| < k$ for each $x \in [a, b]$ and for each $n \in N$.

Theorem [9.11] [Abel's test] If (i) $\sum\limits_{1}^{\infty} f_n(x)$ converges unformly on $[a, b]$, (ii) the sequence $\{g_n(x)\}$ is a sequence of non-negative monotonic decreasing function for each $n \in N$ and for each $x \in [a, b]$ and (iii) $\{g_n(x)\}$ is uniformly bounded on $[a, b]$, then the series $\sum\limits_{1}^{\infty} f_n(x)g_n(x)$ is uniformly convergent on $[a, b]$.

Proof. Since $\{g_n(x)\}$ is uniformly bounded on $[a, b]$, $\exists$ a (+)ve number k such that $|g_n(x)| < k$, for each $x \in [a, b]$ and for each $n \in N \Rightarrow g_n(x) < k,\ \forall\ x \in [a, b]$ and $\forall\ n \in N$.

Also since $\sum_{1}^{\infty} f_n(x)$ converges uniformly on $[a,b]$, $|F_n(x)-F_m(x)|<\epsilon$, $\forall\, n\geq m$, $\forall\, x\in[a,b]$, where

$$F_n(x)=\sum_{k=1}^{n} f_k(x) \Rightarrow \left|\sum_{k=m+1}^{n} f_k(x)\right| < \frac{\epsilon}{k},\ \forall\, n\geq m,\ \forall\, x\in[a,b]$$

Now using Abel's lemma (Lemma [7.2], Ch-7), we get

$$\left|\sum_{k=m+1}^{n} g_n(x)f_n(x)\right| \leq g_{m+1}(x)\cdot \max_{m+1,m+2,\cdots,n}\left|\sum_{k=m+1}^{p} f_k(x)\right| \quad (1\leq p\leq n)$$

$$< k\cdot\frac{\epsilon}{k}=\epsilon,\ \forall\, x\in[a,b],\ \forall\, n\geq m$$

$$\Rightarrow |S_n(x)-S_m(x)|<\epsilon,\ \forall\, x\in[a,b],\ \forall\, n\geq m,$$

where $S_n(x)=\sum_{k=1}^{n} f_k(x)g_k(x)$. Therefore, $\sum_{1}^{\infty} f_n(x)g_n(x)$ converges uniformly on $[a,b]$.

Example. $\sum \dfrac{\cos n\theta}{n^p}$ converges uniformly for $0<p\leq 1$ in any interval $[\alpha, 2\pi-\alpha]$, $\alpha>0$.

We shall apply Dirichlet's test for its uniform convergence on $[\alpha,2\pi-\alpha]$, $\alpha>0$.

Let $g_n(\theta)=\dfrac{1}{n^p}$ and $f_n(\theta)=\cos n\theta$.

For $0<p\leq 1$, $\{g_n(\theta)\}$ is monotonically decreasing and $g_n(\theta)$ is (+)ve $\forall\,\theta\in[\alpha,2\pi-\alpha]$.

Also $\lim_{n\to\infty} g_n(\theta)=\lim_{n\to\infty}\dfrac{1}{n^p}=0,\ 0<p\leq 1$

$$=g(\theta).$$

$|g_n(\theta)-g(\theta)|=\left|\dfrac{1}{n^p}\right|<\epsilon$, if $n^p>\dfrac{1}{\epsilon}$, i.e., if $n>\left(\dfrac{1}{\epsilon}\right)^{\frac{1}{p}}=m$ (integral part), say.

So $\exists$ a (+)ve integer m such that $|g_n(\theta)-g(\theta)|<\epsilon$, $\forall\, n\geq m$.

Hence $\{g_n(\theta)\}$ converges uniformly to zero in $[\alpha,2\pi-\alpha]$.

Now $\left|\sum_{k=1}^{n} f_k(\theta)\right| = |\cos\theta+\cos 2\theta+\cdots+\cos n\theta| = \dfrac{\left|\cos\left(\frac{n+1}{2}\theta\right)\cdot\sin\left(\frac{n}{2}\theta\right)\right|}{\sin\frac{\theta}{2}} \leq \dfrac{1}{\sin\frac{\theta}{2}}$

$$=\operatorname{cosec}\theta/2\leq \operatorname{cosec}\frac{1}{2}(2\pi-\alpha)=\operatorname{cosec}\alpha/2,\ \forall\,\theta\in[\alpha,2\pi-\alpha]$$

So $\{F_n(\theta)\}$, where $F_n(\theta)=\sum_{k=1}^{n} f_k(\theta)$ is uniformly bounded on $[\alpha,2\pi-\alpha]$.

So all the conditions of Dirichlet's test are satisfied.

Hence, $\sum f_n(\theta)g_n(\theta)=\sum\frac{\cos n\theta}{n^p}$ is uniformly convergent.

Note. The same is true for the series $\sum\frac{\sin n\theta}{n^p}$.

Example. The series $\sum a_n n^{-x}$ converges uniformly on $[0,1]$, if $\sum a_n$ converges uniformly on $[0,1]$.

Let $g_n(x)=n^{-x}=\frac{1}{n^x}$ and $f_n(x)=a_n$. Then $g_n(x)$ is (+)ve in $[0,1]$.

$\{g_n(x)\}$ is monotonically decreasing in $[0,1]$.

Also $|g_n(x)|=\left|\dfrac{1}{n^x}\right|\leq 1$ for $0\leq x\leq 1$.

So $\{g_n(x)\}$ is uniformly bounded on $[0,1]$. Again, $\sum\limits_1^\infty f_n(x) = \sum\limits_1^\infty a_n$ converges uniformly on $[0,1]$ by the given condition.

Therefore, by Abel's test, the series $\sum\limits_1^\infty f_n(x)g_n(x) = \sum a_n n^{-x}$ converges uniformly on $[0,1]$.

Problem [9.15] : (1) Assume that $g_{n+1}(x) \le g_n(x)$ and $g_n(x) > 0$ for each x in T and each $n = 1, 2, \cdots$ and suppose that $g_n \to 0$ uniformly on T. Prove that $\sum\limits_1^\infty (-1)^{n+1} g_n(x)$ converges uniformly on T.

(2) Show that the series $\sum\limits_1^\infty \frac{(-1)^{n-1}}{n} x^n$ converges uniformly on $[0,1]$.

(3) Prove that if δ is any fixed (+)ve number less than 1, the series $\sum\limits_1^\infty \frac{x^n}{n+1}$ is uniformly convergent in $[-\delta, \delta]$.

(4) Show that the series $\sum\limits_1^\infty \frac{nx^2}{n^3+x^3}$ converges uniformly on $[0,k]$, $k > 0$ and hence show that $\lim\limits_{x \to 1} \left[\sum\limits_{n=1}^\infty \frac{nx^2}{n^3+x^3}\right] = \sum\limits_{n=1}^\infty \frac{n}{n^3+1}$.

(5) Apply Dirichlet's test to show that $\sum\limits_1^\infty \frac{\sin nx}{n}$ converges uniformly on $[a,b]$, where $0 < a \le x \le b < 2\pi$, but $\sum\limits_1^\infty \frac{\sin nx}{n}$ is not uniformly convergent on $(-1,1)$.

Solution. (1) Let $f_n(x) = (-1)^{n+1}$

Then $F_n(x) = \sum\limits_{k=1}^n f_k(x) = 1 - 1 + 1 - 1 + \cdots + (-1)^{n+1} = \begin{cases} 0, & \text{when } n \text{ is even} \\ 1, & \text{when } n \text{ is odd.} \end{cases}$

So $|F_n(x)| \le 1$, $\forall\, x \in T$.

Now $\{g_n(x)\}$ is monotonically decreasing and $g_n(x)$ is (+)ve for each $x \in T$ and for each $n \in N$. Also $\{g_n(x)\}$ converges uniformly to zero on T.

Therefore, the conditions of Dirichlet's test are satisfied and hence the series

$$\sum_1^\infty f_n(x)g_n(x) = \sum_1^\infty (-1)^{n+1} g_n(x) \text{ converges uniformly on } T.$$

(2) Let $g_n(x) = x^n$ and $f_n(x) = \dfrac{(-1)^{n-1}}{n}$.

Then $\{g_n(x)\}$ is monotonically decreasing, $g_n(x) \ge 0$ for each x in $[0,1]$ and for each $n \in N$. Also $|g_n(x)| = |x^n| \le 1$ for all $x \in [0,1]$. That is, $\{g_n(x)\}$ is uniformly bounded on $[0,1]$.

Also the series $\sum\limits_1^\infty f_n(x) = \sum_1^\infty \frac{(-1)^{n-1}}{n} = 1 - \frac{1}{2} + \frac{1}{3} - \frac{1}{4} + \cdots$ an alternating infinite series, where each $f_n > 0$, $f(n+1) < f(n)\ \forall\, n \in N$ and $f_n \to 0$ as $n \to \infty$, is convergent. Hence by Abel's test the series $\sum\limits_1^\infty f_n(x)g_n(x) = \sum\limits_1^\infty \frac{(-1)^{n-1}}{n} \cdot x^n$ is uniformly convergent on $[0,1]$.

(3) Let $g_n(x) = \dfrac{1}{n+1}$ and $f_n(x) = x^n$. Since for every value of x in $[-\delta, \delta]$, we have $|x| \le \delta < 1$.

Therefore,

$$|F_n(x)| = |x + x^2 + \cdots + x^n| \le |x| + |x|^2 + \cdots + |x|^n \le \delta + \delta^2 + \cdots + \delta^n$$
$$= \frac{\delta(1-\delta^n)}{1-\delta} < \frac{\delta}{1-\delta}$$

Also $\{g_n(x)\}$ is (+)ve monotonic decreasing sequence and converges uniformly to zero on $[-\delta, \delta]$.

Therefore, by Dirichlet's test, the series $\sum\limits_{1}^{\infty} f_n(x)g_n(x) = \sum\limits_{1}^{\infty} \frac{x^n}{n+1}$ is uniformly convergent on $[-\delta, \delta]$.

(4) Let $g_n(x) = nx^2$ and $f_n(x) = \frac{1}{n^3+x^3}$.

Here $\{g_n(x)\}$ is positive monotonic increasing sequence for each $x \in [0, k]$, $k > 0$ and $\{g_n(x)\}$ is uniformly bounded on $[0, k]$, i.e., $|g_n(x)| \le M$ for each $x \in [0, k]$ and for each $n \in N$, where M is a (+)ve number.

$$|f_n(x)| = \left|\frac{1}{n^3+x^3}\right| \le \frac{1}{n^3},\ \forall\ x \in [0, k].$$

$\sum\limits_{1}^{\infty} M_n = \sum\limits_{1}^{\infty} \frac{1}{n^3}$ is a convergent series. So by Weierstrass's M-test, $\sum\limits_{1}^{\infty} f_n(x)$ is uniformly convergent on $[0, k]$.

Therefore, by Abel's test, the series $\sum\limits_{1}^{\infty} f_n(x)g_n(x) = \sum\limits_{1}^{\infty} \frac{nx^2}{n^3+x^3}$ is uniformly convergent on $[0, k]$.

Let $F_n(x) = u_1(x) + u_2(x) + \cdots + u_n(x) = \sum\limits_{k=1}^{n} u_k(x)$, where $u_k(x) = g_k(x)f_k(x)$ and $\sum\limits_{1}^{\infty} u_k(x)$ converges uniformly on $[0, k]$. Then

$$\lim_{x\to 1}\left(\sum_{n=1}^{\infty} \frac{nx^2}{n^3+x^3}\right) = \lim_{x\to 1}\lim_{n\to\infty}\sum_{k=1}^{n} u_k(x) = \lim_{x\to 1}\lim_{n\to\infty} F_n(x) \tag{1}$$

Again,

$$\sum_{n=1}^{\infty} \lim_{x\to 1} \frac{nx^2}{n^3+x^3} = \sum_{n=1}^{\infty} \frac{n}{n^3+1} \tag{2}$$

But

$$\sum_{n=1}^{\infty} \lim_{x\to 1} u_n(x) = \lim_{n\to\infty}\sum_{k=1}^{n}\left[\lim_{x\to 1} u_k(x)\right] = \lim_{n\to\infty}\lim_{x\to 1}\sum_{k=1}^{n} u_k(x)$$
$$= \lim_{n\to\infty}\lim_{x\to 1} F_n(x) = \lim_{x\to 1}\left(\sum_{n=1}^{\infty} \frac{nx^2}{n^3+x^3}\right) \quad \text{[by (1)]} \tag{3}$$

Hence it follows from (2) and (3), that

$$\lim_{x\to 1}\left(\sum_{n=1}^{\infty} \frac{nx^2}{n^3+x^3}\right) = \sum_{n=1}^{\infty} \frac{n}{n^3+1}$$

since

$$\lim_{x\to 1}\left(\sum_{n=1}^{\infty}\frac{nx^2}{n^3+x^3}\right)=\lim_{x\to 1}\lim_{n\to\infty}\sum_{k=1}^{n}u_k(x)=\lim_{n\to\infty}\lim_{x\to 1}\left(\sum_{k=1}^{n}u_k(x)\right)$$

$$=\lim_{n\to\infty}\left(\lim_{x\to 1}\sum_{1}^{n}\frac{nx^2}{n^3+x^3}\right)=\lim_{n\to\infty}\left(\sum_{k=1}^{n}\frac{n}{n^3+1}\right)=\sum_{n=1}^{\infty}\frac{n}{n^3+1}$$

[See Remark]

Remark. If $\{f_n(x)\}$ converges uniformly on $[a,b]$ and c is a point on $[a,b]$ such that $\lim\limits_{x\to c} f_n(x)=a_n$, then

(i) $\{a_n\}$ converges and

(ii) $\lim\limits_{x\to c} f(x)=\lim\limits_{n\to\infty} a_n \Rightarrow \lim\limits_{x\to c}\lim\limits_{n\to\infty} f_n(x)=\lim\limits_{n\to\infty}\lim\limits_{x\to c} f_n(x)$.

Since $\{f_n(x)\}$ converges uniformly, $|f_{n+p}(x)-f_n(x)|<\epsilon,\ \forall\, n\geq m,\ \forall\, x\in[a,b]$ and $\forall\, p\in N$. Keeping n,p fixed and letting $x\to c$, we have $|a_{n+p}-a_n|<\epsilon,\ \forall\, n\geq m$ and $\forall\, p\in N\Rightarrow\{a_n\}$ converges. Let it converge to L. Then

$$|a_n-L|<\frac{\epsilon}{3},\ \forall\, n\geq m_1. \tag{1}$$

Also $\{f_n(x)\}$ converges uniformly to $f(x)$

$$\Rightarrow |f_n(x)-f(x)|<\frac{\epsilon}{3},\ \forall\, n\geq m_2,\ \forall\, x\in[a,b] \tag{2}$$

Again, $\lim\limits_{x\to c} f_n(x)=a_n\Rightarrow|f_n(x)-a_n|<\dfrac{\epsilon}{3}$, whenever $|x-c|<\delta$. (3)

Hence, by (1), (2) and (3), $|f(x)-L|=|f(x)-f_n(x)+f_n(x)-a_n+a_n-L|<\epsilon$, whenever $|x-c|<\delta$

$$\Rightarrow \lim_{x\to c} f(x)=L=\lim_{n\to\infty} a_n$$

$$\Rightarrow \lim_{x\to c}\lim_{n\to\infty} f_n(x)=\lim_{n\to\infty}\lim_{x+c} f_n(x)$$

(5) Let $g_n(x)=\dfrac{1}{n}$ and $f_n(x)=\sin nx$.

Then $\{g_n(x)\}$ is a (+)ve monotonic decreasing sequence on $[a,b]$, where $0<a\leq x\leq b<2\pi$ and converges to zero $\forall\, x\in[a,b]$.

Now $\displaystyle F_n(x)=\sum_{k=1}^{n}f_k(x)=\sin x+\sin 2x+\cdots+\sin nx=\frac{\sin\frac{n+1}{2}x\cdot\sin\frac{nx}{2}}{\sin\frac{x}{2}}$

$$|F_n(x)|=\frac{\left|\sin\frac{n+1}{2}x\right|\left|\sin\frac{nx}{2}\right|}{\left|\sin\frac{x}{2}\right|}\leq\frac{1}{\sin\frac{x}{2}}=\operatorname{cosec}\frac{x}{2}$$

which is bounded for all $x\in[a,b]$, $0<a\leq x\leq b<2\pi$.

Therefore, $|F_n(x)|\leq k,\ \forall\, x\in[a,b],\ k=\sup\limits_{x\to[a,b]}\left\{\operatorname{cosec}\dfrac{x}{2}\right\}$.

Hence, by Dirichlet's test, $\sum\limits_{1}^{\infty} f_n(x)g_n(x)=\sum\limits_{1}^{\infty}\frac{\sin nx}{n}$ converges uniformly on $[a,b]$.

Since $\operatorname{cosec}\frac{x}{2}$ is unbounded at $x=0\in(-1,1)$, $|F_n(x)|\nleq k,\ \forall\, x\in(-1,1)$ and consequently, $\sum\limits_{1}^{\infty}\frac{\sin nx}{n}$ is not uniformly convergent on $[-1,1]$.

EXERCISE

(1) Let f_n be defined by $f_n(x) = 1 - |1 - x^2|^n$ in $[-\sqrt{2}, \sqrt{2}]$. Show that $\{f_n\}$ is pointwise convergent but not uniformly convergent in $[-\sqrt{2}, \sqrt{2}]$.

(2) Prove that $\sum a_n x^{-n}$ is uniformly convergent on $[0, 1]$, if $\sum a_n$ converges uniformly in $[0, 1]$.

(3) Show that $\delta(x) = \sum_{n=0}^{\infty} \frac{1}{1 + n^2 + n^4 x^2}$ converges uniformly for all values of x. Examine whether $\delta'(0)$ can be obtained by term by term differentiation.

(4) Discuss the uniform convergence of $1 + \frac{e^{-2x}}{2^2 - 1} + \frac{e^{-4x}}{4^2 - 1} + \frac{e^{-6x}}{6^2 - 1} + \cdots$ in $(0, \infty)$.

(5) Let f_n be defined by $f_n(x) = \begin{cases} n^2 x, & \text{for } 0 \le x \le \frac{1}{n} \\ n(2 - nx), & \text{for } \frac{1}{n} < x \le \frac{2}{n} \\ 0, & \text{for } \frac{2}{n} < x \le 1, \end{cases}$

$\forall\, n \ge 2$.

Then show that

(i) each f_n is R-integrable in $[0, 1]$.

(ii) $\{f_n\}$ has a pointwise limit f which is R-integrable in $[0, 1]$.

(iii) $\lim_{n \to \infty} \int_0^1 f_n(x)dx \ne \int_0^1 f(x)dx$.

(6) Let $\{f_n(x)\}$ be a sequence of function, $n \in N$, where $f_n(x) = x^n$, $0 \le x \le 1$. Then prove that

(i) $\{f_n(x)\}$ converges pointwise but not uniformly on $[0, 1]$.

(ii) $\{f_n(x)\}$ is uniformly convergent on $[0, k]$, $0 < k < 1$.

(iii) $f_n(x)$ is continuous but $\{f_n(x)\}$ converges pointwise to a discontinuous function.

(7) Show that the series $\sum_{n=0}^{\infty} x^n$ converges uniformly to $\frac{1}{1 - x}$ on $|x| < \delta, 0 < \delta < 1$.

(8) Show that the series $\sum_{1}^{\infty} (-1)^n \frac{x^n}{n}$ converges uniformly but not absolutely on $[0, 1]$.

(9) Show that the sum function $S(x)$ of the series $\sum_{0}^{\infty} \frac{x^2}{(1 + x^2)^n}$ is not continuus at $x = 0$ although each term of the series is continuous at $x = 0$.

(10) Let $f(x) = e^{-x} + 2e^{-2x} + 3 \cdot e^{-3x} + \cdots + ne^{-nx} + \cdots$, $x > 0$. Show that $f(x)$ is continuous for all real $x > 0$ and the series can be integrated term by term on $[a, b]$, $a > 0$. Prove also that $\int_{\log 2}^{\log 3} f(x)dx = \frac{1}{2}$.

(11) Let $Q = \{x_1, x_2, \ldots, x_n, \cdots\}$ be the countable set of rational numbers on $[0, 1]$ and each f_n, $n \in N$ be defined on $[0, 1]$ by

$$f_n(x) = \begin{cases} 1, & \text{for } x = x_i,\ i = 1, 2, \ldots, n \\ 0, & \text{otherwise.} \end{cases}$$

Also let $\{f_n(x)\}$, $n \in N$, converge pointwise to f. Then state with reasons,

(i) whether each f_n is R-integral on $[0, 1]$;

(ii) whether f is R-integrable on $[0, 1]$;

(iii) whether $\{f_n\}$, $n \in N$, converges uniformly to f on $[0, 1]$.

CHAPTER 10

POWER SERIES

Definition [10.1] : An infinite series of the form $\sum_{0}^{\infty} a_n(x - x_0)^n$, where x, x_0, a_n are real numbers is called a Power Series.

In particular, if $x_0 = 0$, the series $\sum_{0}^{\infty} a_n(x - x_0)^n$ takes the form $\sum_{0}^{\infty} a_n x^n$.

We consider $\sum_{0}^{\infty} a_n x^n$ as a general form of a power series in x.

Note. We shall use P.S. in place of power series.

Example. $1 + 2x + 3x^2 + \cdots \infty$ is a power series, where $a_n = (n+1)$. That is, the power series is of the form $\sum_{0}^{\infty}(n+1)x^n$.

Example. $1 - \dfrac{x}{2} + \dfrac{x^2}{3} - \dfrac{x^4}{4} + \cdots \infty$, is a power series of the form

$$\sum_{0}^{\infty} \frac{(-1)^n}{n+1} x^n, \text{ where } a_n = \frac{(-1)^n}{n+1}.$$

Observations

A power series $\sum_{0}^{\infty} a_n x^n$ is an infinite series, so the question of convergence and non-convergence (divergence) arises.

(i) A P.S. $\sum_{0}^{\infty} a_n x^n$ always converges at $x = 0$.

(ii) A P.S. may not converge for any value of x except $x = 0$. In this case, we say it is nowhere convergent.

Example. $\sum_{0}^{\infty} n!\, x^n$ is a nowhere convergent series.

(iii) A P.S. may converge for all $x \in \mathbb{R}$. Then it is called everywhere convergent.

Example. $\sum_{0}^{\infty} \frac{x^n}{n!}$, $\sum_{0}^{\infty}(-1)^n \cdot \frac{x^n}{n!}$ converges $\forall\, x$.

(iv) A P.S. may converge for some values of x and diverges for other values of x.

Example. $\sum x^n$ converges for $|x| < 1$ and diverges for $|x| \geq 1$.

Definition [10.2] : If a P.S. $\sum_0^\infty a_n x^n$ converges at $x = x_0$, then $\lim_{n\to\infty} a_n x_0^n = 0$.

So $|a_n x_0^n| < \epsilon \; \forall \; n \geq m$.

We can choose a finite (+)ve number M such that $|a_n x_0^n| < M, \; \forall \; n$.

Theorem [10.1] If a P.S. $\sum_0^\infty a_n x^n$ converges for $x = x_0$, then it converges absolutely for all x satisfying $|x| < |x_0|$.

Proof. From Definition [10.2], we have $|a_n x_0^n| < M, \; \forall \; n$.

Also $|x| < |x_0| \Rightarrow \left|\dfrac{x}{x_0}\right| < 1$.

Now

$$\begin{aligned}\sum_0^\infty |a_n x^n| &= |a_0| + |a_1 x| + |a_2 x^2| + \cdots + |a_n x^n| + \cdots \infty \\ &= |a_0| + |a_1 x_0| \left|\frac{x}{x_0}\right| + |a_0 x_0^2| \left|\frac{x}{x_0}\right|^2 + \cdots \infty \\ &< M + M\left|\frac{x}{x_0}\right| + M\left|\frac{x}{x_0}\right|^2 + \cdots \infty \\ &= M[1 + r + r^2 + \cdots \infty], \text{ where } r = \left|\frac{x}{x_0}\right|.\end{aligned}$$

Series in the bracket being an infinite G.P. series with common ratio $r < 1$ is convergent.

Therefore, $\sum_1^\infty |a_n x^n|$ is less than a convergent series and hence $\sum_0^\infty |a_n x^n|$ is convergent and consequently, $\sum_0^\infty a_n x^n$ is absolutely convergent $\forall \; x$ satisfying $|x| < x_0$.

Theorem [10.2] If the P.S. $\sum_0^\infty a_n x^n$ diverges for $x = x_0$, then it diverges for all x satisfying $|x| > |x_0|$.

Proof. Let the P.S. $\sum_0^\infty a_n x^n$ converge at $x = x_1$. Then by above theorem it converges for all x satisfying $|x| < |x_1|$. So for $x = x_0$, $|x_0| < |x_1|$ which contradicts our hypothesis $|x| < |x_0| \; \forall \; x$.

Hence the theorem.

10.1 Radius of Convergence and Interval of Convergence

Definition [10.3] : If there exists a real (+)ve number R such that the P.S. $\sum a_n x^n$ converges absolutely for $|x| < R$ and diverges for $|x| > R$, then R is called the radius of convergence and $(-R, R)$ is called the interval of convergence.

Note. A P.S. may converge or diverge at $x = R$.

Example. The P.S. $\sum_0^\infty \dfrac{x^n}{n^2}$ converges for $|x| < 1$ and also at $x = 1$. So it converges for $|x| \leq 1$ and diverges for $|x| > 1$. But the P.S. $\sum \dfrac{x^n}{n}$ converges for $|x| < 1$ and diverges at $x = 1$.

The radius of convergence of the series is 1 and interval of convergence is $(-1, 1)$.

Theorem [10.3] (1) If R be the radius of convergence of the P.S.

$$\sum_{0}^{\infty} a_n x^n \text{ and } \lim_{n\to\infty}\left|\frac{a_{n+1}}{a_n}\right| = a, \text{ then } R = \frac{1}{a}.$$

(2) If R be the radius of convergence of the P.S. $\sum a_n x^n$ and $\overline{\lim}_{n\to\infty} |a_n|^{1/n} = a$, then $R = \frac{1}{a}$.

In other words, $\dfrac{1}{R} = \lim_{n\to\infty}\left|\dfrac{a_{n+1}}{a_n}\right|$ or $\dfrac{1}{R} = \lim_{n\to\infty} |a_n|^{1/n}$.

Proof. (1) By D. Alembert's ratio test, $\lim_{n\to\infty}\left|\dfrac{a_{n+1}x^{n+1}}{a_n x^n}\right| = \lim_{n\to\infty}\left|\dfrac{a_{n+1}}{a_n}\right| |x| = a|x|$.

Then $\sum_{0}^{\infty} a_n x^n$ converges, if $a|x| < 1$, i.e., if $|x| < \frac{1}{a}$ and diverges, if $|x| > \frac{1}{a}$.

So radius of convergence $R = \frac{1}{a}$.

(2) By Cauchy's root test, $\overline{\lim}_{n\to\infty} |a_n x^n|^{1/n} = \overline{\lim}_{n\to\infty} |a_n|^{1/n}|x| = a|x|$.

So the series converges for $a|x| < 1$, i.e., if $|x| < \frac{1}{a}$ and diverges for a $|x| > 1$, i.e., if $|x| > \frac{1}{a}$.

Hence $R = \frac{1}{a}$.

So

$$\frac{1}{R} = \lim_{n\to\infty}\left|\frac{a_{n+1}}{a_n}\right| \text{ or } \frac{1}{R} = \lim_{n\to\infty} |a_n|^{1/n} \tag{i}$$

Note. (i) (1) and (2) of Th. [10.3] stand as the working rule for finding the radius convergence. We apply the rule which is applicable for a given series.

(ii) If $a = 0$, $R = \frac{1}{a} \Rightarrow R \to \infty$. We say the P.S. is everywhere convergent.

(iii) If $a \to \infty$, then $R = 0$. We say that the P.S. is nowhere convergent.

Problem [10.1]: Determine the radius of convergence and interval of convergence for the following P.S.

(1) $\displaystyle\sum_{0}^{\infty} \frac{(2n)!}{(n!)^2} x^n$ (2) $\displaystyle\sum \frac{(n+1)}{(n+2)(n+3)} x^n$

(3) $\displaystyle\sum_{n=2}^{\infty} \frac{(x+2)^n}{\log n}$ (4) $1 + \dfrac{a\cdot b}{1\cdot c}x + \dfrac{a(a+1)b(b+1)}{1\cdot 2c(c+1)}x^2 + \cdots$ (5) $\displaystyle\sum_{1}^{\infty} \frac{x^n}{n^n}$

Solution. (1)

$$\lim_{n\to\infty}\left|\frac{a_{n+1}}{a_n}\right| = \lim_{n\to\infty}\left|\frac{(2n+2)!}{(n+1)!^2}\cdot\frac{(n!)^2}{2n!}\right| = \lim_{n\to\infty}\frac{(2n+2)(2n+1)}{(n+1)^2}$$

$$= \lim_{n\to\infty}\frac{4n^2+6n+2}{n^2+2n+1} = \lim_{n\to\infty}\frac{4+6/n+2/n^2}{1+2/n+1/n^2} = 4$$

$\therefore \frac{1}{R} = 4 \Rightarrow R = \frac{1}{4}$ = radius of convergence and interval of convergence is $\left(-\frac{1}{4}, \frac{1}{4}\right)$.

(2) $$\lim_{n\to\infty}\left|\frac{a_{n+1}}{a_n}\right| = \lim_{n\to\infty}\left|\frac{(n+2)}{(n+3)(n+4)}\cdot\frac{(n+2)(n+3)}{n+1}\right|$$

$$\lim_{n\to\infty}\frac{(n+2)^2}{(n+4)(n+1)} = \lim_{n\to\infty}\frac{n^2+4n+1}{n^2+5n+1} = 1.$$

So $R = 1$ and interval of convergence is $(-1, 1)$.

(3) Here $a_n = \dfrac{(x+2)^n}{\log n}$, $n \geq 2$.

Then

$$\lim_{n\to\infty}\left|\frac{a_{n+1}}{a_n}\right| = \lim_{n\to\infty}\left|\frac{(x+2)^{n+1}}{\log(n+1)}\cdot\frac{\log n}{(x+2)^n}\right|$$

$$= \lim_{n\to\infty}|x+2|\cdot\left|\frac{\log n}{\log(n+1)}\right| = |x+2|\lim_{n\to\infty}\left|\frac{\log n}{\log(n+1)}\right|$$

$$= |x+2| \left[\because \lim_{n\to\infty}\frac{\log n}{\log(n+1)}\left(\frac{\infty}{\infty}\right) = \lim_{n\to\infty}\frac{n+1}{n} = 1\right]$$

Given series converges, if $|x+2| < 1$, i.e., if $-1 < x+2 < 1$, i.e., if $-3 < x < -1$.
So the interval of convergence is $(-3,-1)$ and the radius of convergence is $\frac{1}{2}(-1+3) = 1$.

(4) Here $a_n = \dfrac{\{a(a+1)\cdots(a+n-1)\}\{b(b+1)\cdots|b+n-1)\}}{\{1\cdot 2\cdots(n-1)\}\{c(c+1)\cdots(c+n-1)\}}$.

Then

$$\lim_{n\to\infty}\left|\frac{a_{n+1}}{a_n}\right| = \lim_{n\to\infty}\left|\frac{a(a+1)\cdots(a+n)\cdot b(b+1)\cdots(b+n)}{n!\;c(c+1)\cdots(c+n)}\cdot\frac{(n-1)!\;c(c+1)\cdots(c+n-1)}{a(a+1)\cdots(a+n-1)\cdot b(b+1)\cdots(b+n-1)}\right|$$

$$= \lim_{n\to\infty}\left|\frac{(a+n)(b+n)}{n(c+n)}\right|$$

$$= \lim_{n\to\infty}\left|\frac{n^2+(a+b)n+ab}{n^2+nc}\right| = 1$$

So $\frac{1}{R} = 1 \Rightarrow R = 1$ and interval of convergence is $(-1,1)$.

(5) Here we apply Cauchy's root test

$$\overline{\lim_{n\to\infty}}\,|a_n|^{1/n} = \overline{\lim_{n\to\infty}}\left|\frac{1}{n^n}\right|^{1/n} = \lim_{n\to\infty}\frac{1}{n} = 0$$

So $\frac{1}{R} = 0 \Rightarrow R \to \infty$. Interval of convergence is $(-\infty,\infty)$.
So the series is everywhere convergent.

10.2 Uniform Convergence

Definition [10.4]: Let $S(x) = \sum\limits_{0}^{\infty} a_n x^n$ and $S_n(x) = \sum\limits_{k=1}^{n} a_k x^k$. The P.S. $\sum a_n x^n$ is said to converge uniformly on $[a,b]$, if for every $\epsilon(>0)$, there exists a (+)ve integer m such that

$$|S(x) - S_n(x)| < \epsilon,\ \forall\, n \geq m \text{ and } \forall\, x \in [a,b].$$

$$\Rightarrow |R_n(x)| < \epsilon,\ \forall\, n \geq m \text{ and } \forall\, x \in [a,b]$$

where $R_n(x) = \sum\limits_{k=n+1}^{\infty} a_k x^k$.

Theorem [10.4] If a P.S. $\sum\limits_{0}^{\infty} a_n x^n$ converges absolutely for $|x| < R$, then it converges uniformly on $|x| \leq R_1$, where $R_1 < R$.

Solution. $\sum\limits_{n=0}^{\infty} |a_n x^n| = |a_0| + |a_1 x| + |a_2 x^2| + \cdots + |a_n x^n| + R_n(x)$, where

$$R_n(x) = \sum_{k=n+1}^{\infty} |a_k x^k|$$

If the series converges absolutely on $|x| < R$, then for every $\epsilon(> 0)$, there exists a (+)ve integer m such that

$$|R_n(x)| < \epsilon, \ \forall\, n \geq m. \tag{1}$$

Now if x lies in the interval $|x| \leq R_1 < R$, then each term of $\sum\limits_{k=n+1}^{\infty} |a_k x^k|$ will attain the maximum value at $x = R_1$.

Hence, from (1) it follows that $\sum\limits_{k=n+1}^{\infty} |a_k R_1^n| < \epsilon, \ \forall\, n \geq m$

$$\Rightarrow \sum_{k=n+1}^{\infty} |a_k x^k| < \epsilon, \ \forall\, x \in [-R_1, R_1]$$

$$\Rightarrow |R_n(x)| < \epsilon, \ \forall\, x \in [-R_1, R_1] \text{ and } \forall\, n \geq m$$

$$\Rightarrow |S(x) - S_n(x)| < \epsilon, \ \forall\, x \in [-R_1, R_1] \text{ and } \forall\, n \geq m$$

$$\Rightarrow \sum_{0}^{\infty} a_n x^n \text{ converges uniformly on } [-R_1, R_1]$$

i.e., $\sum\limits_{0}^{\infty} a_n x^n$ converges uniformly on $|x| \leq R_1$.

Corollary [10.1]: Let $R(\neq 0)$ be the radius of convergence of the P.S. $\sum\limits_{0}^{\infty} a_n x^n$. Then the series converges absolutely and uniformly on $[a, b]$, where $-R < a < b < R$.

Theorem [10.5] [Cauchy-Hadamard theorem] Every P.S. $\sum\limits_{0}^{\infty} a_n x^n$ having a radius of convergence R, $0 \leq R < \infty$ has the following properties.

(1) The series converges absolutely for $|x| < R$.

(2) The series converges uniformly for $|x| \leq P$, where $0 \leq P < R$.

(3) The series diverges for $|x| > R$, i.e., the terms of the series are unbounded.

Proof. (1) We define R by

$$\frac{1}{R} = \overline{\lim_{n\to\infty}}\, |a_n|^{1/n} \tag{i}$$

We choose a real number P such that $|x| < P < R$. Then $\frac{1}{P} > \frac{1}{R}$.

Then from (i), there exists a (+)ve integer m such that $|a_n|^{1/n} < \frac{1}{P}, \ \forall\, n \geq m$

$$\Rightarrow |a_n| < \frac{1}{P^n}, \ \forall\, n \geq m$$

$$\Rightarrow |a_n x^n| < \left|\frac{x}{P}\right|^n = \left\{\frac{|x|}{P}\right\}^n \ [P > 0].$$

Now $\sum_{0}^{\infty}\left(\frac{|x|}{P}\right)^n$ is a G.P. series with common ratio < 1 $\left[\text{as } |x| < P \Rightarrow \frac{|x|}{P} < 1\right]$ and so

$$\sum_{0}^{\infty}\left(\frac{|x|}{P}\right)^n \text{ converges.}$$

Therefore, the given series $\sum_{0}^{\infty} a_n x^n$ converges absolutely.

(2) Let $0 \le P < R$. We have to show that the series converges uniformly for $|x| \le P$. We choose real number P' such that $P < P' < R$. It follows from case (i), that

$$|a_n| < \frac{1}{(P')^n}, \ \forall\, n \ge m$$

$$\Rightarrow |a_n x^n| < \left(\frac{|x|}{P'}\right)^n \le \left(\frac{P}{P'}\right)^n \text{ as } |x| \le P.$$

Now the series $\sum\left(\frac{P}{P'}\right)^n$ of positive constants is convergent as being a G.P. series with common ratio < 1.

Therefore, by Weierstrass's test, the P.S. $\sum_{0}^{\infty} a_n x^n$ converges absolutely for $|x| < R$ and hence it converges uniformly for $|x| \le P < R$.

(3) Let $|x| > R$. We choose P such that $R < P < |x|$. Then $\frac{1}{R} > \frac{1}{P}$.

Now from $\frac{1}{R} = \overline{\lim_{n\to\infty}}\ |a_n|^{1/n}$, we have $|a_n|^{1/n} > \frac{1}{P}$, for infinitely many values of n

$$\Rightarrow |a_n x^n| > \left(\frac{|x|}{P}\right)^n, \text{ for infinitely many values of } n \tag{ii}$$

Here $\frac{|x|}{P} > 1$.

So $\sum_{0}^{\infty}|a_n x^n|$ diverges, as the $\sum_{0}^{\infty}\left(\frac{|x|}{P}\right)^n$ being a G.P. series with common ratio > 1, diverges.

Each term of the series is unbounded since $\lim_{n\to\infty}\left(\frac{|x|}{P}\right)^n \to \infty$ as $\frac{|x|}{P} > 1$.

Theorem [10.6] (1) If $S(x) = \sum_{0}^{\infty} a_n x^n$, then $S(x)$ is continuous within its interval of convergence $(-R, R)$. That is, $S(x)$ is continuous on $[a, b]$, where $-R < a < b < R$.

(2) A P.S. $\sum_{0}^{\infty} a_n x^n$ can be integrated term by term within its interval of convergence $(-R, R)$.

That is $\int_a^b S(x)dx = \sum_{n=0}^{\infty}\int_a^b a_n x^n dx = \int_a^b a_0 dx + \int_a^b a_1 x dx + \cdots$, where $-R < a < b < R$.

(3) A P.S. $\sum_{0}^{\infty} a_n x^n$ can be differentiated term by term within its interval of convergence $(-R, R)$. That is, $S'(x) = \sum_{n=0}^{\infty}\frac{d}{dx}(a_n x^n)$.

Proof. (1) Let $S_n(x) = \sum\limits_{k=0}^{n} a_n x^k$.

Then $S_n(x)$, being a polynomial in x, is continuous on $[a, b]$, where $-R < a < b < R$. Let $c \in [a, b]$. Then $S_n(x)$ is continuous at $x = c$

$$\Rightarrow |S_n(x) - S_n(c)| < \frac{\epsilon}{3},\ |x - c| < \delta \qquad \text{(i)}$$

Again, the P.S. $\sum\limits_{0}^{\infty} a_n x^n$ converges uniformly to $S(x)$ on $[a, b]$ [by Corollary of Th. [10.4]].

$$\Rightarrow |S(x) - S_n(x)| < \frac{\epsilon}{3},\ \forall\, n \geq m,\ \forall\, x \in [a, b] \qquad \text{(ii)}$$

In particular, $n = m$,

$$|S(x) - S_m(x)| < \frac{\epsilon}{3},\ \forall\, x \in [a, b] \qquad \text{(iii)}$$

$$\begin{aligned}\text{Then } |S(x) - S(c)| &= |S(x) - S_m(x) + S_m(x) + S_m(c) + S_m(c) - S(c)| \\ &\leq |S(x) - S_m(x)| + |S_m(x) - S_m(c)| + |S_m(c) - S(c)| \\ &< \frac{\epsilon}{3} + \frac{\epsilon}{3} + \frac{\epsilon}{3} = \epsilon, \text{ whenever } |x - c| < \delta \text{ [by (i), (ii) and (iii)]}\end{aligned}$$

$\Rightarrow$ $S(x)$ is continuous at $x = c$.

$\Rightarrow$ $S(x)$ is continuous on $[a, b]$ as c is an arbitrary point on $[a, b]$.

(2) $S(x) = \sum\limits_{0}^{\infty} a_n x^n$. Then $S(x)$ is continuous on $[a, b]$ and hence $S(x)$ is integrable on $[a, b]$.

Since the P.S. $\sum\limits_{0}^{\infty} a_n x^n$ converges uniformly on $[a, b]$, we have $|S(x) - S_n(x)| < \frac{\epsilon}{b-a}$, $\forall\, n \geq m$ and $\forall\, x \in [a, b]$, where

$$S_n(x) = \sum_{k=0}^{n} a_k x^k,\ b > a \qquad \text{(i)}$$

$$\begin{aligned}\text{Now } \left|\int_a^b S_n(x)dx - \int_a^b S(x)dx\right| = \left|\int_a^b \{S_n(x) - S(x)\}dx\right| &\leq \int_a^b |S_n(x) - S(x)|dx \\ &< \int_a^b \frac{\epsilon}{b-a}dx \text{ [by (i)]} \\ &= \epsilon,\ \forall\, n \geq m\end{aligned}$$

$$\Rightarrow \lim_{n\to\infty} \int_a^b S_n(x)dx = \int_a^b S(x)dx$$

That is

$$\int_a^b S(x)dx = \lim_{n\to\infty} \int_a^b \sum_0^n (a_k x^k)dx = \sum_0^\infty \int_a^b a_n x^n dx = \int_a^b a_0 dx + 1 \int_a^b a_1 x dx + \cdots$$

(3) Let

$$S(x) = \sum_0^\infty a_n x^n = a_0 + a_1 x + a_2 x^2 + \cdots + a_n x^n + \cdots \qquad \text{(i)}$$

Let $\phi(x) = a_1 + 2a_2 x + 3a_3 x^2 + \cdots + na_n x^{n-1} + \cdots = \sum\limits_0^\infty na_n x^{n-1}$ which is also a P.S.

Let R' be its radius of convergence.

$$\text{Then } \frac{1}{R'} = \lim_{n\to\infty}\left|\frac{(n+1)a_{n+1}}{na_n}\right| = \lim_{n\to\infty}\left|\left(1+\frac{1}{n}\right)\right|\left|\frac{a_{n+1}}{a_n}\right|$$

$$= \lim_{n\to\infty}\left|\frac{a_{n+1}}{a_n}\right| = \frac{1}{R}$$

where R is the radius of convergence of $\sum_0^\infty a_n x^n$. Then $R' = R$.

So the P.S. $\sum_0^\infty n a_n x^{n-1}$ has the same interval of convergence $(-R, R)$.

Therefore, by Prop. (2), $\sum_0^\infty n a_n x^{n-1}$ can be integrated term by term within $(-R, R)$. Then

$$\int_0^x \phi(x)dx = \int_0^x (a_1 + 2a_2x + 3a_3x^2 + \cdots + na_nx^{n-1} + \cdots)dx$$

$$= \int_0^x a_1 dx + \int_0^x 2a_2 x dx + \cdots + \int_0^x na_nx^{n-1}dx + \cdots$$

where $[0, x] \subset (-R, R)$.

$$\Rightarrow \int_0^x \phi(x)dx = a_1x + a_2x^2 + a_3x^3 + \cdots + a_nx^n + \cdots = S(x) - a_0 \quad \text{[by (i)]}$$

$\phi(x)$ is continuous on $[0, x] \subset (-R, R)$.

$$\Rightarrow \frac{d}{dx}[S(x) - a_0] = \phi(x)$$

$$\Rightarrow \frac{d}{dx}[S(x)] = a_1 + 2a_2x + 3a_3x^2 + \cdots + na_nx^{n-1} + \cdots$$

$$= \frac{d}{dx}[a_0] + \frac{d}{dx}[a_1x] + \frac{d}{dx}[a_2x^2] + \cdots$$

$$\Rightarrow \frac{d}{dx}\left[\sum_{n=0}^\infty a_nx^n\right] = \sum_0^\infty \frac{d}{dx}[a_nx^n].$$

Theorem [10.7] A function can be represented uniquely in the form of a P.S. on a given interval of convergence (uniqueness of power series).

Proof. If possible, let a function $f(x)$ can be expressed in the forms

$$\sum_0^\infty a_nx^n \text{ and } \sum_0^\infty b_nx^n \text{ within } (-R, R).$$

That is, $f(x) = \sum_0^\infty a_nx^n$ and $f(x) = \sum_0^\infty b_nx^n$ for $|x| < R$.

We have to show that $a_n = b_n$, $\forall\, n = 0, 1, 2, \cdots$.

Now $f(x) = a_0 + a_1x + a_2x^2 + \cdots + a_nx^n + \cdots$

$f(x) = b_0 + b_1x + b_2x^2 + \cdots + b_nx^n + \cdots$

Then

$$0 = (a_0 - b_0) + (a_1 - b_1)x + (a_2 - b_2)x^2 + \cdots + (a_n - b_n)x^n + \cdots \quad \text{(i)}$$

since two P.S. can be added or subtracted term by term within $(-R, R)$. Therefore, for $x = 0$, $a_0 = b_0$ from (i).

Again $f'(x) = a_1 + 2a_2x + \cdots + na_nx^{n-1} + \cdots$

$f'(x) = b_1 + 2b_2x + \cdots + nb_nx^{n-1} + \cdots$

since a P.S. can be differentiated term by term within the interval of convergence.

Therefore, $0 = (a_1 - b_1) + 2(a_2 - b_2)x + \cdots + n(a_n - b_n)x^{n-1} + \cdots$

$\Rightarrow a_1 - b_1 = 0, \quad \text{for } x = 0$

$\Rightarrow a_1 = b_1.$

We now assume that $a_m = b_m$ $(m < n)$.

Then $f^{m+1}(x) = \sum\limits_{n=m+1}^{\infty} \{n(n-1) - (n-m)\}a_nx^{n-m-1}$ and also

$$f^{m+1}(x) = \sum_{n=m+1}^{\infty} \{n(n-1) - (n-m)\}b_nx^{n-m-1}$$

obtained by successive differentiation of $f(x)$ up to $(m+1)$ times and using $a_m = b_m$ for $|x| < R$.

So $0 = \sum\limits_{n=m+1}^{\infty} \{n(n-1)\cdots(n-m)\}\{a_n - b_n\}x^{n-m-1}.$

Now putting $x = 0$, we get $a_{m+1} = b_{m+1}$.

Hence, by mathematical induction, $a_n = b_n$, $\forall\, n = 0, 1, 2, \cdots$.

Corollary [10.2] : If $f(x) = \sum\limits_{0}^{\infty} a_nx^n$ possesses derivatives of all orders on $[a, b]$, where $-R < a < b < R$, then $f^k(0) = k!\, a_k$ for $k = 0, 1, 2, \cdots$.

Proof. $f(x) = a_0 + a_1x + a_2x^2 + a_3x^3 + \cdots + a_nx^n + a_{n+1}x^{n+1} + \cdots$ differentiating successively w.r.t., x we get,

$$\begin{aligned}
f'(x) &= a_1 + 2a_2x + 3a_3x^2 + \cdots + a_nx^{n-1} + (n+1)a_{n+1}x^n + \cdots \\
f''(x) &= 2a_2 + 3\cdot 2a_3x + \cdots + n(n-1)a_nx^{n-2} + (n+1)na_{n+1}x^{n-1} + \cdots
\end{aligned}$$

....

$$\begin{aligned}
f^n(x) &= n(n-1)(n-2)\cdot\left(n - \overline{n-1}\right)a_nx^{n-n} + (n+1)n\cdots(n+1-n)x^{(n+1)-n}\cdot a_{n+1} + \cdots \\
&= n!\, a_n + (n+1)!\, a_{n+1}x + (n+2)!\, a_{n+2}x^2 + \cdots \\
&= n!\, a_n + \text{ terms containing } x.
\end{aligned}$$

Thus for $x = 0$, $f^n(0) = n!\, a_n$, which is true $\forall\, n = 1, 2, \cdots, k \cdots k \leq n$.

Hence $f^k(0) = k!\, a_k$ [since $f^{k+r}(x)$ contains x for all $r \geq 1$].

Theorem [10.8] If $\sum\limits_{n=0}^{\infty} a_n$ converges, then $\sum\limits_{n=0}^{\infty} a_nx^n$ converges uniformly for $0 \leq x \leq 1$.

Proof. Since $\sum\limits_{n=0}^{\infty} a_n$ converges, by Cauchy's criterion of convergence for every $\epsilon(> 0)$, there exists a (+)ve integer m_1 such that

$$\left|\sum_{k=m+1}^{n} a_n\right| < \epsilon,\ \forall\, m, n \geq m_1 \Rightarrow -\epsilon < a_{m+1} + a_{m+2} + \cdots + a_n < \epsilon,\ \forall\, m, n \geq m_1.$$

Now if $0 \leq k \leq 1$, then applying Abel's Lemma [7.2] to $\{a_k\}_{m+1}^{n}$ and $\{x^k\}_{m+1}^{n}$ we have $-\epsilon x^{m+1} < a_{m+1}x^{m+1} + a_{m+2}x^{m+2} + \cdots + a_n x^n < \epsilon x^{m+1}$

$$\Rightarrow \left|\sum_{k=m+1}^{n} a_k x^k\right| < \epsilon \cdot x^{m+1} < \epsilon \ (\forall\ m, n \geq m_1, 0 \leq x \leq 1)$$

$$\Rightarrow |S_n(x) - S_m(x)| < \epsilon,\ \forall\ m, n \geq m_1, \text{ where } S_n = \sum_{k=0}^{n} a_k x^k$$

$$\Rightarrow \sum_{0}^{\infty} a_n x^n \text{ converges uniformly on } [0, 1].$$

Alternative Proof

The P.S. $\sum_{1}^{\infty} a_n x^n$ converges at $x = 1$, since $\sum_{0}^{\infty} a_n$ converges.

Therefore, by Th. [10.1], $\sum_{0}^{\infty} a_n x^n$ converges absolutely for $|x| < 1$ and hence it converges uniformly for $|x| < 1$. But since it converges at $x = 1$ also, it converges uniformly on $[0, 1]$.

Corollary [10.3] : If $\sum_{0}^{\infty} |a_n|$ be convergent, then show that $\sum_{1}^{\infty} a_n x^n$ is uniformly convergent on $[-1, 1]$.

Theorem [10.9] [Abel's Theorem on limit] If the P.S. $\sum_{0}^{\infty} a_n x^n$ converges with finite radius of convergence R and the series also converges at $x = R$, then $\lim\limits_{x \to R-0} f(x) = \sum_{n=0}^{\infty} a_n R^n$, where $f(x) = \sum_{0}^{\infty} a_n x^n$ for $-R < x < R$.

Proof. Without any loss of generality we can take $R = 1$. To justify it, we put $x = Ry$.

Then $\sum_{0}^{\infty} a_n x^n = \sum a_n R^n y^n = \sum_{0}^{\infty} b_n y^n$, where $b_n = a_n R^n$.

Let the P.S. $\sum_{0}^{\infty} b_n y^n$ converge for $|y| < R'$.

$$\text{Then } \frac{1}{R'} = \overline{\lim}\, |b_n|^{1/n} = \overline{\lim}\, |a_n R^n|^{1/n} = \overline{\lim}\, |a_n|^{1/n} \cdot R$$

$$= \frac{1}{R} \cdot R = 1 \Rightarrow R' = 1.$$

So let $\sum_{0}^{\infty} a_n x^n$ be a P.S. with radius of convergence 1.

Thus we have $f(x) = \sum a_n x^n$ for $-1 < x < 1$.

Also $f(1) = \sum_{0}^{\infty} a_n$.

Now

$$\frac{1}{1-x}f(x) = \frac{1}{(1-x)}(a_0 + a_1x + a_2x^2 + a_3x^3 + \cdots + a_nx^n + \cdots)$$
$$= (1 + x + x^2 + \cdots + x^n + \cdots)(a_0 + a_1x + a_2x^2 + \cdots + a_nx^n + \cdots)$$
$$= a_0 + (a_0 + a_1)x + |a_0 + a_1 + a_2)x^2 + \cdots = \sum_0^{\infty} C_nx^n$$

where $C_n = a_0 + a_1 + a_2 + \cdots + a_n = \sum_0^n a_k \Rightarrow f(x) = (1-x)\sum_0^{\infty} C_nx^n$

$$\Rightarrow f(x) - f(1) = (1-x)\sum_0^{\infty} C_nx^n - f(1) \quad \text{(i)}$$

Also

$$(1-x)\sum_0^{\infty} x^n = 1 \text{ for } -1 < x < 1 \quad \text{(ii)}$$

since $(1-x)\lim_{n\to\infty}\frac{1-x^n}{1-x} = (1-x)\cdot\frac{1}{1-x} = 1$ as $\lim_{n\to\infty} x^n = 0$.

Then using this result, (i) can be written as

$$f(x) - f(1) = (1-x)\sum_0^{\infty} C_nx^n - f(1)(1-x)\sum_0^{\infty} x^n$$
$$= (1-x)\sum_0^{\infty}\{C_n - f(1)\}x^n \text{ for } -1 < x < 1 \quad \text{(iii)}$$

Now $C_n = \sum_{k=0}^n a_k \to f(1)$ as $n \to \infty$

$$\Rightarrow C_n - f(1) \to 0 \text{ as } n \to \infty$$
$$\Rightarrow |C_n - f(1)| < \frac{\epsilon}{2},\ \forall\, n \geq m \quad \text{(iv)}$$

From (iii),

$$|f(x) - f(1)| = \left|(1-x)\sum_0^n\{C_n - f(1)\}x^n + (1-x)\sum_{n+1}^{\infty}\{C_n - f(1)\}x^n\right|$$
$$\leq (1-x)\sum_0^n |C_n - f(1)|x^n + (1-x)\sum_{n+1}^{\infty}|C_n - f(1)|x^n$$
$$\leq (1-x)\sum_0^n |C_n - f(1)|x^n + \frac{\epsilon}{2}\cdot 1 \text{ [by (ii) and (iv)]} \quad \text{(v)}$$

So for a fixed n,

$$(1-x)\sum_0^n |C_n - f(1)|x^n = (1-x)\,[|C_0 - f(1)| + |C_1 - f(1)|x + \cdots + |C_n - f(1)|x^n]$$

is a (+)ve continuous function of x and takes zero value at $x = 1$

$$\Rightarrow |(1-x)| \sum_{0}^{n} |C_n - f(1)| x^n < \frac{\epsilon}{2}, \ \forall\, x \in (1-\delta, 1)$$

$$\Rightarrow (1-x) \sum_{0}^{n} |C_n - f(1)| x^n < \frac{\epsilon}{2}, \ \forall\, x \in (1-\delta, 1)$$

Thus from (v), $|f(x) - f(1)| < \frac{\epsilon}{2} + \frac{\epsilon}{2} = \epsilon, \ \forall\, x \in (1-\delta, 1)$

$$\Rightarrow \lim_{x \to 1-0} f(x) = f(1) = \sum_{0}^{\infty} a_n$$

Problem [10.2]: (1) Show that $\log(1+x) = x - \frac{x^2}{2} + \frac{x^3}{3} - \frac{x^4}{4} + \cdots$ for $-1 < x < 1$ and hence show that $\log 2 = 1 - \frac{1}{2} + \frac{1}{3} - \frac{1}{4} + \cdots$ by using Abel's theorem.

(2) Find the P.S. of $\tan^{-1} x$ and deduce the sum of $1 - \frac{1}{3} + \frac{1}{5} - \frac{1}{7} + \cdots$.

(3) Show that $\log(1-x) = -x - \frac{x^2}{2} - \frac{x^3}{3} - \cdots -1 \leq x < 1$.

(4) Show that $\sin^{-1} x = x + \frac{1}{2 \cdot 3} x^2 + \frac{1 \cdot 3}{2 \cdot 4 \cdot 5} x^5 + \frac{1 \cdot 3 \cdot 5}{2 \cdot 4 \cdot 6} \cdot \frac{x^7}{7} + \cdots (-1 < x \leq 1)$

and hence show that $\frac{\pi}{2} = 1 + \frac{1}{2 \cdot 3} + \frac{1 \cdot 3}{2 \cdot 4} \cdot \frac{1}{5} + \cdots$.

Solution. (1) Let $f(x) = \log(1+x)$.

Then

$$f'(x) = \frac{1}{1+x} = (1+x)^{-1} = 1 - x + x^2 - x^3 + x^4 - \cdots$$
$$= \sum_{0}^{\infty} (-1)^n x^n \tag{1}$$

Now $\frac{1}{R} = \lim_{n \to \infty} \left| \frac{u_{n+1}}{u_n} \right| = \lim_{n \to \infty} |-1| = 1 \Rightarrow R = 1$.

Therefore, the P.S. converges absolutely for $|x| < 1$ and hence it converges uniformly for $|x| < 1$. That is, for $-1 < x < 1$.

Therefore, the P.S. in (1) can be integrated term by term in $-1 < x < 1$.

So

$$\int_0^x \frac{1}{1+x} dx = \int_0^x \sum_{0}^{\infty} (-1)^n x^n dx = \int_0^x (1 - x + x^2 - x^3 + \cdots) dx$$
$$= \int_0^x 1 dx - \int_0^x x dx + \int_0^x x^2 dx + \cdots$$
$$\log(1+x) = x - \frac{x^2}{2} + \frac{x^3}{3} - \frac{x^4}{4} + \cdots$$

$f(x)$ is continuous on $-1 < x < 1$.

By Abel's theorem, $\lim_{x \to 1-} f(x) = f(1) \Rightarrow \lim_{x \to 1-} \log(1+x) = \log 2$

$$\Rightarrow \lim_{x \to 1-} \left\{ x - \frac{x^2}{2} + \frac{x^3}{3} - \frac{x^4}{4} + \cdots \right\} = \log 2$$

$$\Rightarrow 1 - \frac{1}{2} + \frac{1}{3} - \frac{1}{4} + \cdots = \log 2$$

(2) Let $f(x) = \tan^{-1} x$. Then $f'(x) = \dfrac{1}{1+x^2} = (1+x^2)^{-1} = 1 - x^2 + x^4 - x^6 + \cdots$ having a radius of convergence 1.

Therefore, the P.S. $1 - x^2 + x^4 - x^6 + \cdots$ converges uniformly for $-1 < x < 1$. Therefore, integrating term by term in $-1 < x < 1$, we get

$$\int_0^x \frac{1}{1+x^2} dx = \int_0^x (1 - x^2 + x^4 - x^6 + \cdots) dx$$

$$= \int_0^x 1 dx - \int_0^x x^2 dx + \int_0^x x^4 dx - \cdots$$

$$\tan^{-1} x = x - \frac{x^3}{3} + \frac{x^5}{5} - \frac{x^7}{7} + \cdots \text{ for } -1 < x < 1$$

Also at $x = 1$,

$$1 - \frac{1}{3} + \frac{1}{5} - \frac{1}{7} + \cdots \text{ is convergent}$$

and at $x = -1$,

$$-\left(1 - \frac{1}{3} + \frac{1}{5} - \frac{1}{7} + \cdots\right) \text{ is convergent.}$$

Therefore, the series $x - \dfrac{x^3}{3} + \dfrac{x^5}{5} - \dfrac{x^7}{7} + \cdots$ converges uniformly on $-1 \le x \le 1$.

Hence, $\tan^{-1} x = x - \dfrac{x^3}{3} + \dfrac{x^5}{5} - \dfrac{x^7}{7} + \cdots$ for $-1 \le x \le 1$.

Also by Abel's theorem $\lim_{x \to 1-0} f(x) = f(1) \Rightarrow 1 - \dfrac{1}{3} + \dfrac{1}{5} - \dfrac{1}{7} + \cdots = \dfrac{\pi}{4}$.

(3) Let $f(x) = \log(1-x)$

We have $(1-x)^{-1} = 1 + x + x^2 + \cdots + x^n + \cdots = \sum_0^\infty x^n$.

$$\frac{1}{R} = \lim_{n \to \infty} \left| \frac{u_{n+1}}{u_n} \right| = 1$$

So $\sum_0^\infty x^n$ converges absolutely for $|x| < 1$.

Also at $x = -1$, $\sum_0^\infty x^n = \frac{1}{2}$. So it converges at $x = -1$

But it diverges at $x = 1$.

So, the P.S. $\sum_0^\infty x^n$ converges uniformly for $-1 \le x < 1$.

Hence the P.S. can be integrated term by term in $[0, x]$

$$\therefore \int_0^x (1-x)^{-1} dx = \int_0^x 1 dx + \int_0^x x dx + \int_0^x x^2 dx + \cdots$$

$$\Rightarrow \int_0^x \frac{1}{1-x} dx = x + \frac{x^2}{2} + \frac{x^3}{3} + \cdots$$

$$\Rightarrow -\log(1-x) = x + \frac{x^2}{2} + \frac{x^3}{3} + \cdots$$

$$\Rightarrow \log(1-x) = -x - \frac{x^2}{2} - \frac{x^3}{3} - \cdots$$

$$\Rightarrow f(x) = \log(1-x) = -x - \frac{x^2}{2} - \frac{x^3}{3} - \cdots \text{ for } -1 \le x < 1.$$

(4) Let $f(x) = \sin^{-1} x$

$$f'(x) = \frac{1}{\sqrt{1-x^2}} = (1-x^2)^{-1/2} = 1 + \frac{1}{2}x^2 + \frac{\frac{1}{2}\left(\frac{1}{2}+1\right)}{2!}x^4 + \frac{\frac{1}{2}\left(\frac{1}{2}+1\right)\left(\frac{1}{2}+2\right)}{3!}x^6 + \cdots$$

$$\text{or, } \frac{1}{\sqrt{1-x^2}} = 1 + \frac{1}{2}x^2 + \frac{1\cdot 3}{2\cdot 4}x^4 + \frac{1\cdot 3\cdot 5}{2\cdot 4\cdot 6}x^6 + \cdots \tag{1}$$

Now

$$\frac{1}{R} = \lim_{n\to\infty}\left|\frac{u_{n+1}}{u_n}\right| = \lim_{n\to\infty}\left|\frac{1\cdot 3\cdot 5\cdots(2n+1)}{2\cdot 4\cdot 6\cdots(2n+2)}\cdot\frac{2\cdot 4\cdot 6\cdots 2n}{1\cdot 3\cdot 5\cdots(2n-1)}\right|$$

$$= \lim_{n\to\infty}\left|\frac{2n+1}{2n+2}\right| = 1$$

It converges uniformly on $(-1, 1)$.

So the series in (1) can be integrated term by term. Hence we get

$$\int_0^x \frac{1}{\sqrt{1-x^2}} dx = \int_0^x 1 dx + \int_0^x \frac{x^2}{2} + \int_0^x \frac{1\cdot 3}{2\cdot 4}x^4 + \cdots$$

$$\Rightarrow \sin^{-1} x = x + \frac{x^3}{2\cdot 3} + \frac{1\cdot 3}{2\cdot 4}\cdot\frac{x^5}{5} + \frac{1\cdot 3\cdot 5}{2\cdot 4\cdot 6}\cdot\frac{x^7}{7} + \cdots \text{ for } -1 < x < 1 \tag{2}$$

Again, at $x = 1$, the P.S. $x + \frac{x^3}{2\cdot 3} + \frac{1\cdot 3}{2\cdot 4}\frac{x^5}{5} + \cdots$ becomes

$$1 + \frac{1}{2\cdot 3} + \frac{1\cdot 3}{2\cdot 4}\cdot\frac{1}{5} + \frac{1\cdot 3\cdot 5}{2\cdot 4\cdot 6}\frac{1}{7} + \cdots$$

which is convergent, since the series

$$< 1 + \frac{1}{2^2} + \frac{1}{2^3} + \frac{1}{2^4} + \cdots = 1 + \frac{1}{2^2}\left(1 + \frac{1}{2} + \frac{1}{2^2} + \cdots\right)$$

where $1 + \frac{1}{2} + \frac{1}{2^2} + \cdots$ being an infinite G.P. series having a common ratio < 1 is convergent.

Therefore, the series in (2) converges uniformly on $-1 < x \le 1$. Hence,

$$\sin^{-1} x = x + \frac{x^3}{2\cdot 3} + \frac{1\cdot 3}{2\cdot 4}\frac{x^5}{5} + \frac{1\cdot 3\cdot 5}{2\cdot 4\cdot 6}\frac{x^7}{7} + \cdots \text{ for } -1 < x \le 1.$$

Also $f(x)$ is continuous at $x = 1$, so by Abel's theorem,

$$f(1) = \lim_{x\to 1-0} f(x) \Rightarrow \frac{\pi}{2} = 1 + \frac{1}{2\cdot 3} + \frac{1\cdot 3}{2\cdot 4}\cdot\frac{1}{5} + \frac{1\cdot 3\cdot 5}{2\cdot 4\cdot 6}\cdot\frac{1}{7} + \cdots.$$

EXERCISE

(1) Show that $\sum_{1}^{\infty} 2^{-n}x^n$ represents a power series with radius of convergence $(-2, 2)$. Does it converge at $x = \pm 2$. Show that the given series converges uniformly in $[-a, a]$ for $0 < a < 2$.

(2) Let $f(x) = |x| \ \forall\ x \in R$. Is there a power series $\sum a_n x^n$ such that $f(x) = \sum_{0}^{\infty} a_n x^n$?

(3) Taking $f(x) = \sum_{n=1}^{\infty} \frac{x^n}{(n+1)^2 n!}$, $-\infty < x < \infty$ show that $f(x) = \frac{1}{x}\int_0^x \frac{e^x - 1}{x} dx$, with proper justification.

(4) Prove that the power series $\sum a_n(x-a)^n$ with radius of convergence R, converges absolutely, if $|x - a| < R$ and diverges if $|x - a| > R$.

(5) If $\sum_{n=\delta}^{\infty} |a_n|$ be convergent, then show that $\sum_{0}^{\infty} a_n x^n$ converges uniformly on $[-1, 1]$.

(6) If $f(x) = \sum a_n x^n$, $|x| < R$ and $g(x) = \sum b_n x^n$, $|x| < r$, then show that $f(x) + g(x) = \sum c_n x^n$ and $f(x) \cdot g(x) = \sum d_n x^n$, converge for $|x| < R'$, where $R' = R \cap r$, $c_n = a_n + b_n$ and $d_n = a_0 b_n + a_1 b_{n-1} + \cdots + a_n b_0$.

(7) If R be the radius of convergence of $\sum a_n x^n$, then construct a power series other than $\sum x^n$ so that $\sum a_n b_n x^n$ has also the same radius of convergence.

(8) If R be the radius of convergence of the power series $\sum a_n x^n$, then prove that $R^{1/k}$, $k \in N$, is the radius of convergence of $\sum a_n x^{kn}$.

(9) Let ρ be the radius of convergence of a power series $\sum a_n(x - x_0)^n$ and the series converges at $x_0 + \rho$. Then show that the series is uniformly convergent in $[x_0 - r, x_0 + r]$ for any r, $0 < r < \rho$ [Abel's Theorem].

(10) Find the interval of uniform convergence of

(a) $\sum \left(\frac{n}{n+1}\right)^{n^2} x^n$

(b) $\sum \frac{1}{(n+1)^2} x^n$.

(11) Show that the series $\frac{1}{4} + \sum \frac{(-1)^n \cos nx}{a^2 - n^2}$, where a is a real number but $a \neq 0, \pm 1, \pm 2, \ldots$, converges uniformly on every finite interval of R, the set of real numbers.

(12) If $|x| = 1$, then show that $\int_0^x \log(1+x)dx = \frac{x^2}{1 \cdot 2} - \frac{x^3}{2 \cdot 3} + \frac{x^4}{3 \cdot 4} - \cdots$. Does the result hold for $x = \pm 1$? State with reasons.

CHAPTER 11

FOURIER SERIES

Definition [11.1]: **[Periodic function]** A function $f : \mathbb{R} \to \mathbb{R}$ is said to be periodic if for any $\lambda \neq 0$, $f(x \pm \lambda) = f(x)$, $\forall\, x \in \mathbb{R}$. λ is called periodicity of f.

Example. $\sin x, \cos x$ are periodic functions with periodicity 2π.
For, if $f(x) = \sin x$ or $\cos x$, then $f(x \pm 2\pi) = f(x)$.

11.1 Some Fundamental Results

Theorem [11.1] If $f : \mathbb{R} \to \mathbb{R}$ be a periodic function with periodicity λ, then

(1) $\int_a^b f(x)dx = \int_{a+\lambda}^{b+\lambda} f(x)dx$.

(2) $f(x + n\lambda) = f(x)$, $\forall\, x \in \mathbb{R}$ and $\forall\, n \in Z$.

(3) $\int_{-\lambda/2}^{\lambda/2} f(x)dx = \int_{-\lambda/2}^{\lambda/2} f(a+x)dx$ for any $a \in \mathbb{R}$.

(4) $\int_{-\pi}^{\pi} f(x)dx = \int_{-\pi}^{\pi} f(\xi + x)dx$ for any ξ, where periodicity of f is 2π.

Proof. (1) Put $x = y - \lambda$. Then

$$\int_a^b f(x)dx = \int_{a+\lambda}^{b+\lambda} f(y-\lambda)dy = \int_{a+\lambda}^{b+\lambda} f(y)dy = \int_{a+\lambda}^{b+\lambda} f(x)dx \;[\because f(y-\lambda) = f(y)]$$

(2) For $n = \pm 1$, $f(x \pm \lambda) = f(x)$. So result is true for $n = \pm 1$.

Let it be true for any $n \in Z$. Then $f(x + n\lambda) = f(x)$, $\forall\, x \in \mathbb{R}$.

$$\begin{aligned} \text{Now } f\left(x + \overline{n+1}\lambda\right) &= f(x + n\lambda + \lambda) = f(y + \lambda), \text{ say, } y = x + n\lambda \\ &= f(y),\ \forall\, y \in \mathbb{R} \\ &= f(x + n\lambda) = f(x),\ \forall\, x \in \mathbb{R}. \end{aligned}$$

So the result is true for $n + 1$. Hence it is true $\forall\, n \in Z$.

(3) Put $a + x = y$. Then we have

$$\begin{aligned} \int_{-\lambda/2}^{\lambda/2} f(a+x)dx &= \int_{a-\lambda/2}^{a+\lambda/2} f(y)dy \\ &= \int_{a-\lambda/2}^{-\lambda/2} f(y)dy + \int_{-\lambda/2}^{\lambda/2} f(y)dy + \int_{\lambda/2}^{a+\lambda/2} f(y)dy \qquad \text{(i)} \end{aligned}$$

Now putting $y = z - \lambda$, we see that

$$\int_{a-\lambda/2}^{-\lambda/2} f(y)dy = \int_{a+\lambda/2}^{\lambda/2} f(z-\lambda)dz = -\int_{\lambda/2}^{a+\lambda/2} f(z)dz$$

$$= -\int_{\lambda/2}^{a+\lambda/2} f(y)dy \; [\because f(z-\lambda) = f(z)] \qquad \text{(ii)}$$

So from (i)

$$\int_{-\lambda/2}^{\lambda/2} f(a+x)dx = \int_{-\lambda/2}^{\lambda/2} f(y)dy = \int_{-\lambda/2}^{\lambda/2} f(x)dx \text{ [by (ii)]}$$

(4) We have

$$\int_{-\pi}^{\pi} f(x+\xi)dx = \int_{-\pi+\xi}^{\pi+\xi} f(y)dy$$

$$= \int_{-\pi+\xi}^{-\pi} f(y)dy + \int_{-\pi}^{\pi} f(y)dy + \int_{\pi}^{\pi+\xi} f(y)dy$$

Putting $y = z - 2\pi$ in the 1st part, we get

$$\int_{-\pi}^{\pi} f(x+\xi)dx = -\int_{\pi}^{\pi+\xi} f(z-2\pi)dz + \int_{-\pi}^{\pi} f(y)dy + \int_{\pi}^{\pi+\xi} f(y)dy$$

$$= -\int_{\pi}^{\pi+\xi} f(z)dz + \int_{-\pi}^{\pi} f(y)dy + \int_{\pi}^{\pi+\xi} f(y)dy$$

$$= \int_{-\pi}^{\pi} f(x)dx$$

11.2 Periodic Extension

Let f be a function defined in the domain $[-\pi, \pi]$ and let f be a periodic function with periodicity 2π. Then $f(x \pm 2\pi) = f(x)$, $\forall\, x \in (-\pi, \pi)$, we can define the periodic extension of f by shifting the graph of f in $(-\pi, \pi)$ by $2\pi, 4\pi, \cdots$, to the right or left as $f(x + n2\pi) = f(x)$, $\forall\, x \in (-\pi, \pi)$ and $\forall\, n \in Z$ [by (2) Th. [11.1]].

Thus we can define f outside its domain $(-\pi, \pi)$ by using its periodicity formula.

For, $-3\pi < x < -\pi$, we use $f(x + 2\pi) = f(x)$ and we have

$$f(-3\pi + 2\pi) = f(-3\pi) \Rightarrow f(-\pi) = f(-3\pi).$$

For $-5\pi < x < -3\pi$, $f(-5\pi + 2\pi) = f(-5\pi)$

$$\Rightarrow f(-3\pi) = f(-5\pi) \Rightarrow f(-5\pi) = f(-3\pi) = f(-\pi)$$

Also for $\pi < x < 3\pi$, we use $f(x - 2\pi) = f(x)$ and we have

$$f(3\pi - 2\pi) = f(3\pi) \Rightarrow f(\pi) = f(3\pi)$$

For $3\pi < x < 5\pi$, $f(5\pi - 2\pi) = f(5\pi) \Rightarrow f(3\pi) = f(5\pi)$

$$\therefore\; f(5\pi) = f(3\pi) = f(\pi).$$

So proceeding as above the domain of f can be extended outside $(-\pi, \pi)$. The execution domains are $\cdots (-5\pi, -3\pi), (-3\pi, -\pi), (\pi, 3\pi), (3\pi, 5\pi), \cdots$.

The graphical representation of f with domain $(-\pi, \pi)$ is given by Fig. 11.1,

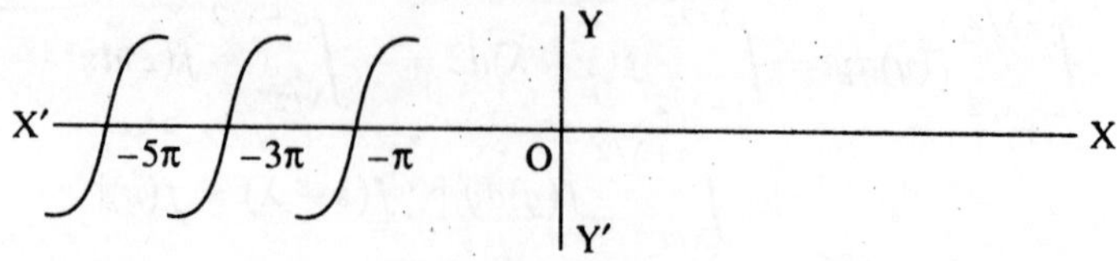

Fig. 11.1

which shows that the values of $f(x)$ are undetermined at odd multiples of π.

11.3 Some Useful Results of Integrals Associated with m, n, where m, n are (+)ve Integers or Zero

(1) (i) $\displaystyle\int_{-\pi}^{\pi} \cos nx dx = 2\int_0^{\pi} \cos nx dx = 2\left[\frac{\sin nx}{n}\right]_0^{\pi} = 0$

(ii) $\displaystyle\int_{-\pi}^{\pi} \sin nx dx = 0$ [$\because \sin nx$ is an odd function]

(2) (i) $$\int_{-\pi}^{\pi} \cos mx \cos nx dx = \begin{cases} 0, & \text{for } m \neq n \\ \pi, & \text{for } m = n \neq 0 \\ 2\pi, & \text{for } m = n = 0 \end{cases}$$

Since

$$\int_{-\pi}^{\pi} \cos mx \cos nx dx = 2\int_0^{\pi} \cos mx \cos nx dx$$
$$= \int_0^{\pi} [\cos(m+n)x + \cos(m-n)x]dx = 0 \text{ for } m \neq n$$
$$= \int_0^{\pi} (\cos 2mx + 1)dx \text{ for } m = n \neq 0$$
$$= \int_0^{\pi} dx = \pi$$

Also $\displaystyle\int_{-\pi}^{\pi} \cos mx \cos nx dx = \int_{-\pi}^{\pi} 1 \cdot dx$, for $m = n = 0$

$$= 2\pi$$

(ii) $$\int_{-\pi}^{\pi} \sin mx \sin nx = \begin{cases} 0, & \text{when } m \neq n \text{ and } m = n = 0 \\ \pi, & \text{when } m = n \neq 0 \end{cases}$$

(3) $\displaystyle\int_{-\pi}^{\pi} \cos mx \sin nx dx = 0$ [$\because \cos mx \sin nx$ is an odd function].

Fourier Series

Definition [11.2]: A trigonometrical series of the form $\dfrac{a_0}{2} + \displaystyle\sum_{n=1}^{n}(a_n \cos nx + b_n \sin nx) = \dfrac{a_0}{2} + (a_1 \cos x + a_2 \cos 2x + \cdots) + (b_1 \sin x + b_2 \sin 2x + \cdots)$, where a_0, a_n, b_n are constants $\forall\, n \in N$ and every term of the series which is periodic with period 2π is called a 'Fourier

Series' if the co-efficients $a_0, a_n, b_n, n \in N$ can be derived from a periodic function f with periodicity 2π by

$$\left.\begin{aligned} a_0 &= \frac{1}{\pi}\int_{-\pi}^{\pi} f(x)dx \\ a_n &= \frac{1}{\pi}\int_{-\pi}^{\pi} f(x)\cos nx dx \\ \text{and } b_n &= \frac{1}{\pi}\int_{-\pi}^{\pi} f(x)\sin nx dx \end{aligned}\right\} \tag{A}$$

a_0, a_n, b_n are called Fourier co-efficients of f.

We write $f(x) = \frac{a_0}{2} + \sum\limits_{1}^{\infty}(a_n \cos nx + b_n \sin nx)$.

Note. (1) Though we write $f(x) = \frac{a_0}{2} + \sum(a_n \cos nx + b_n \sin nx)$, $f(x)$ is not necessarily the sum of the series.

(2) The series $\frac{a_0}{2} + \sum(a_n \cos nx + b_n \sin nx)$, where a_0, a_n, b_n are determined from a function f by the formulae in (A) is called a Fourier series generated by f and we write it $f(x) \sim \frac{a_0}{2} + \sum(a_n \cos nx + b_n \sin nx)$.

(3) The formulae in (A) may be replaced by

$$a_n = \frac{1}{\pi}\int_{-\pi}^{\pi} f(x)\cos nx dx, \text{ for } n = 0, 1, 2, \ldots$$

$$\text{and } b_n = \frac{1}{n}\int_{-\pi}^{\pi} f(x)\sin nx dx, \text{ for } n = 1, 2, \cdots$$

It is also to be noted that to cover $a_0 = \frac{1}{\pi}\int_{-\pi}^{\pi} f(x)dx$, we use $\frac{a_0}{2}$ rather than a_0 in the given series.

(4) It is to be noted that the Fourier co-efficients a_0, a_n, b_n have been obtained under the assumption that f is bounded and integrable on $[-\pi, \pi]$.

Theorem [11.2] The series $\frac{a_0}{2} + \sum\limits_{1}^{\infty}(a_n \cos nx + b_n \sin nx)$ is a Fourier series if it converges uniformly to the sum function $f(x)$ on $[-\pi, \pi]$.

Proof. Let

$$f(x) = \frac{a_0}{2} + \sum_{1}^{\infty}(a_n \cos nx + b_n \sin nx) \tag{i}$$

Here the sum function f is bounded and integrable on $[-\pi, \pi]$.

Since the series (i) converges uniformly to f, it can be integrated term by term on $[-\pi, \pi]$. So,

$$\begin{aligned} \int_{-\pi}^{\pi} f(x)dx &= \frac{a_0}{2}\int_{-\pi}^{\pi} dx + \sum_{n=1}^{\infty} a_n \int_{-\pi}^{\pi} \cos nx dx + \sum_{n=1}^{\infty} b_n \int_{-\pi}^{\pi} \sin nx dx \\ &= a_0\pi + 0 + 0 \text{ [by (1) (i), (ii) of Art. 11.3]} \end{aligned}$$

$$\Rightarrow a_0 = \frac{1}{\pi}\int_{-\pi}^{\pi} f(x)dx \tag{ii}$$

$\cos nx$ is bounded on $[-\pi, \pi]$ and so the uniform convergence of (i) remains uneffected if it is multiplied by $\cos nx$.

Therefore, multiplying (i) by $\cos nx$ and integrating term by term, we have

$$\int_{-\pi}^{\pi} f(x)\cos nx dx = \frac{a_0}{2}\int_{-\pi}^{\pi} \cos nx dx + \sum_{\substack{m=1\\ m\neq n}}^{\infty} a_m \int_{-\pi}^{\pi} \cos mx \cos nx dx$$

$$+a_n \int_{-\pi}^{\pi} \cos^2 nx dx + \sum_{m=1}^{\infty} b_m \int_{-\pi}^{\pi} \cos nx \sin mx dx$$

$$= \frac{a_0}{2}\cdot 0 + \sum_{1}^{\infty} a_m \cdot 0 + a_n\pi + \sum_{n=1}^{\infty} b_n \cdot 0 \quad \left[\begin{array}{l}\text{by (1), (2) and (3)}\\ \text{of Art. 11.3}\end{array}\right]$$

$$= a_n \cdot \pi \Rightarrow a_n = \frac{1}{\pi}\int_{-\pi}^{\pi} f(x)\cos nx dx \quad \text{(iii)}$$

$\sin nx$ is bounded on $[-\pi, \pi]$ and so the same as above is true on $[-\pi, \pi]$. Thus, after multiplying (i) by $\sin nx$ and integrating term by term, we get

$$\int_{-\pi}^{\pi} f(x)\sin nx dx = \frac{a_0}{2}\int_{-\pi}^{\pi} \sin nx dx + \sum_{m=1}^{\infty} a_m \int_{-\pi}^{\pi} \cos mx \sin nx dx$$

$$+\sum_{\substack{m=1\\ m\neq n}}^{\infty} b_m \int_{-\pi}^{\pi} \sin mx \sin nx dx + b_n \int_{-\pi}^{\pi} \sin^2 nx dx$$

$$\int_{-\pi}^{\pi} f(x)\sin nx dx = a_0\cdot 0 + \sum_{m=1}^{\infty} a_m \cdot 0 + \sum_{\substack{m=1\\ m\neq n}}^{\infty} b_m \cdot 0 + b_n \cdot \pi \quad \left[\begin{array}{l}\text{by (1), (2) and (3)}\\ \text{of Art. 11.3}\end{array}\right]$$

$$\Rightarrow b_n = \frac{1}{\pi}\int_{-\pi}^{\pi} f(x)\sin nx dx \quad \text{(iv)}$$

By (1), (2) and (3) of Art. 11.3 the Fourier co-efficients $a_0, a_n, b_n, n \in N$, follow the formulae in (A) and hence the given trigonometrical series converging uniform to f is a Fourier series.

Definition [11.3] : [Piecewise monotonic] A function $f : [a, b] \to \mathbb{R}$ is said to be piecewise monotonic if there exists a partition $P\{a = x_0, x_1, \cdots, x_{r-1}, x_r \cdots x_n = b\}$ of $[a, b]$ into finite number of subintervals $[a, x_1], [x_1, x_2], \cdots [x_{n-1}, b]$ such that the function f is monotonic on each of the subintervals.

Note. Every monotonic function on $[a, b]$ is piecewise monotone but every piecewise monotone function is not monotone on $[a, b]$.

For example: Let f be defined on $[0, 1]$ by

$$f(x) = x \text{ in } \left[0, \frac{1}{2}\right]$$

$$= 1 - x \text{ in } \left[\frac{1}{2}, 1\right],$$

where f is monotonically increasing on $\left[0, \frac{1}{2}\right]$ and f is monotonically decreasing on $\left[\frac{1}{2}, 1\right]$. Therefore, f is piecewise monotone on $[0, 1]$ but not monotone on $[0, 1]$.

11.4 Dirichlet's Condition

A function $f : [-\pi, \pi] \to \mathbb{R}$ will be said to satisfy Dirichlet's conditions on $[-\pi, \pi]$, the domain of f, when one of the two following conditions are imposed on f.

(i) f is bounded, periodic with period 2π, integrable on $[-\pi, \pi]$ and the interval can be broken up into finite number of open subintervals in each of which f is monotonic.

(ii) The finite number of points infinite discontinuity of f, if it possesses on $[-\pi, \pi]$, are excluded by arbitrary small neighbourhoods so that f remains bounded in the remaining part of the interval and this can be broken into finite number of open subintervals, in each of which f is monotonic.

Further, the improper integral $\int_{-\pi}^{\pi} |f(x)|dx$ is convergent.

Theorem [11.3] If f is bounded and integrable on $[0, a]$, $a > 0$ and monotonic in $[0, \alpha]$, $0 < \alpha < a$, then

$$\lim_{n\to\infty} \int_0^a f(x)\frac{\sin nx}{x}dx = f(0+) \int_0^\infty \frac{\sin x}{x}dx.$$

Proof. Since f is bounded and monotonic in $[0, \alpha]$, $f(0+)$ exists.

We first suppose that $f(0+) = 0$. Without any loss of generality we may assume that $f(0) = 0$. Applying 2nd Mean Value theorem of Weierstrass's form in $[0, \alpha]$, we have

$$\int_0^\alpha f(x)\frac{\sin nx}{x}dx = f(0)\int_0^\xi \frac{\sin nx}{x}dx + f(\alpha)\int_\xi^\alpha \frac{\sin nx}{x}dx$$

where $0 \le \xi \le \alpha$

$$= f(\alpha)\int_\xi^\alpha \frac{\sin nx}{x}dx = f(\alpha)\int_{n\xi}^{n\alpha} \frac{\sin y}{y}dy = f(\alpha)\int_{n\xi}^{n\alpha} \frac{\sin x}{x}dx \quad \text{(i)}$$

Since $\displaystyle\int_0^\infty \frac{\sin x}{x}dx$ is convergent, $\exists$ a (+)ve number k such that

$$\left|\int_0^t \frac{\sin x}{x}dx\right| \le k, \ \forall\, t \ge 0 \quad \text{(ii)}$$

Then $$\left|\int_{n\xi}^{n\alpha} \frac{\sin x}{x}dx\right| = \left|\int_0^{n\alpha} \frac{\sin x}{x}dx - \int_0^{n\xi} \frac{\sin x}{x}dx\right|$$
$$\le \left|\int_0^{n\alpha} \frac{\sin x}{x}dx\right| + \left|\int_0^{n\xi} \frac{\sin x}{x}dx\right| \le 2k, \text{ by (ii)}$$

Also $f(\alpha) \to 0$ as $\alpha \to 0 + 0$. Therefore, for every $\epsilon > 0$, $\exists$ a (+)ve number δ such that

$$|f(\alpha)| < \frac{\epsilon}{4k} \text{ for } 0 < \alpha < \delta \quad \text{(iii)}$$

Hence, from (i)

$$\left|\int_0^\alpha f(x)\frac{\sin nx}{x}dx\right| = |f(\alpha)|\left|\int_{n\xi}^{n\alpha} \frac{\sin x}{x}dx\right| < \frac{\epsilon}{4k}\cdot 2k$$
$$= \frac{\epsilon}{2} \text{ for } 0 < \alpha < \delta \quad \text{(iv)}$$

Now $$\int_0^a f(x)\frac{\sin nx}{x}dx = \int_0^\alpha f(x)\frac{\sin nx}{x}dx + \int_\alpha^a f(x)\frac{\sin nx}{x}dx.$$

The 2nd integral on the right-hand side $\to 0$ as $n \to \infty$. Since $\phi(x) = \frac{f(x)}{x}$ is bounded and integrable on $[\alpha, a]$,

$$x > \alpha > 0 \Rightarrow \lim_{n\to\infty} \int_\alpha^a \phi(x)\sin nx dx = 0$$

by Riemann Lebsgue integral [See Problem 7.19, Ch.-7]. So $\exists$ a (+)ve integer m such that

$$\left|\int_{\alpha}^{a} f(x)\frac{\sin nx}{x}dx\right| < \frac{\epsilon}{2},\ \forall\ n \geq m \tag{v}$$

Hence

$$\left|\int_{0}^{a} f(x)\frac{\sin nx}{x}dx\right| \leq \left|\int_{0}^{\alpha} f(x)\frac{\sin nx}{x}dx\right| + \left|\int_{\alpha}^{a} f(x)\frac{\sin nx}{x}dx\right|$$
$$< \frac{\epsilon}{2} + \frac{\epsilon}{2} = \epsilon,\ \forall\ n \geq m \text{ [by (iv) and (v)]}$$

$$\Rightarrow \lim_{n\to\infty}\int_0^a f(x)\frac{\sin nx}{x}dx = 0$$

In general case, let $f(0+) \neq 0$. Since $\lim\limits_{x\to 0+}[f(x)-f(0+)] = 0$, replacing $f(x)$ by $f(x)-f(0+)$ in the above result, we get

$$\lim_{n\to\infty}\int_0^a \{f(x)-f(0+)\}\frac{\sin nx}{x}dx = 0$$

$$\Rightarrow \lim_{n\to\infty}\int_0^a f(x)\frac{\sin nx}{x}dx = \lim_{n\to\infty} f(0+)\int_0^a \frac{\sin nx}{x}dx$$
$$= f(0+)\lim_{n\to\infty}\int_0^{na}\frac{\sin y}{y}dy = f(0+)\int_0^{\infty}\frac{\sin x}{x}dx$$

Corollary [11.1] : If f satisfies the condition of the above theorem on $[0, a]$, $0 < a < \pi$, then

$$\lim_{n\to\infty}\int_0^a f(x)\frac{\sin nx}{\sin x}dx = f(0+)\int_0^{\infty}\frac{\sin x}{x}dx.$$

Proof. We write,

$$f(x)\cdot\frac{\sin nx}{\sin x} = f(x)\frac{x}{\sin x}\cdot\frac{\sin nx}{x} = \phi(x)\frac{\sin nx}{x}$$

where $\phi(x) = f(x)\frac{x}{\sin x}$.

Now $x/\sin x$ is (+)ve and monotonically increasing on $[0, \pi/2]$. It follows that ϕ is monotonically increasing on $[0, \alpha], \alpha < \pi/2$.

$$\lim_{n\to\infty}\int_0^a f(x)\frac{\sin nx}{\sin x}dx = \lim_{n\to\infty}\int_0^a \phi(x)\frac{\sin nx}{x}dx$$
$$= \phi(0+)\int_0^{\infty}\frac{\sin x}{x}dx$$
$$= f(0+)\int_0^{\infty}\frac{\sin x}{x}dx$$

Since $\lim\limits_{x\to 0+0}\phi(x) = \lim\limits_{x\to 0+} f(x)\cdot\dfrac{x}{\sin x} = f(0+)$, i.e., $\phi(0+) = f(0+)$.

Also if f be decreasing, $-f$ is increasing and then by above

$$\lim_{n\to\infty}\int_0^a [-f(x)]\frac{\sin nx}{\sin x}dx = -f(0+)\int_0^{\infty}\frac{\sin x}{x}dx$$

$$\Rightarrow \lim_{n\to\infty}\int_0^a f(x)\frac{\sin nx}{\sin x}dx = f(0+)\int_0^{\infty}\frac{\sin x}{x}dx$$

Theorem [11.4] If $f : [-\pi, \pi] \to \mathbb{R}$ be bounded and integrable on $[-\pi, \pi]$ and piecewise monotonic on $[-\alpha, 0]$ and $[0, \alpha]$ (not necessarily in the same sense) where $0 < \alpha < \pi$, then

$$\frac{a_0}{2} + \sum_{n=1}^{\infty} a_n = \frac{f(0+) + f(0-)}{\pi} \int_0^{\infty} \frac{\sin x}{x} dx$$

where $a_0, a_n, n \in N$, are the Fourier co-efficient of f.

Proof. We have

$$\frac{a_0}{2} + \sum_{1}^{m} a_n = \frac{1}{2\pi} \int_{-\pi}^{\pi} f(x)dx + \frac{1}{\pi} \sum_{n=1}^{m} \int_{-\pi}^{\pi} f(x) \cos nx dx$$

$$= \frac{1}{2\pi} \int_{-\pi}^{\pi} f(x) \left[1 + 2\sum_{n=1}^{m} \cos nx\right] dx = \frac{1}{2\pi} \int_{-\pi}^{\pi} f(x) \frac{\sin\left(m + \frac{1}{2}\right)x}{\sin\frac{x}{2}} dx$$

$$= \frac{1}{2\pi} \int_{-\pi}^{0} f(x) \frac{\sin\left(m + \frac{1}{2}\right)x}{\sin\frac{x}{2}} dx + \frac{1}{2\pi} \int_{0}^{\pi} f(x) \frac{\sin\left(m + \frac{1}{2}\right)x}{\sin\frac{x}{2}} dx$$

$$= \frac{1}{2\pi} \int_{0}^{\pi} f(-x) \frac{\sin\left(m + \frac{1}{2}\right)x}{\sin\frac{x}{2}} dx + \frac{1}{2\pi} \int_{0}^{\pi} f(x) \frac{\sin\left(m + \frac{1}{2}\right)x}{\sin\frac{x}{2}} dx$$

$$= \frac{1}{2\pi} \cdot 2 \int_{0}^{\pi/2} f(-2x) \frac{\sin(2m+1)x}{\sin x} dx + \frac{1}{2\pi} \cdot 2 \int_{0}^{\pi/2} f(2x) \frac{\sin(2m+1)x}{\sin x} dy$$

[putting $x = 2z$ and changing z to x]

Then

$$\frac{a_0}{2} + \sum_{n=1}^{\infty} a_n = \frac{1}{\pi}\left[f(0-) \int_0^{\infty} \frac{\sin x}{x} dx\right] + \frac{1}{\pi}\left[f(0+) \int_0^{\infty} \frac{\sin x}{x} dx\right]$$

when $m \to \infty$ by Th. [11.3]

$$= \frac{1}{\pi}[f(0-) + f(0+)] \int_0^{\infty} \frac{\sin x}{x} dx$$

Corollary [11.2]: Deduce that $\int_0^{\infty} \frac{\sin x}{x} dx = \frac{\pi}{2}$. Let $f(x) = 1,\ \forall\ x \in [-\pi, \pi]$. Then

$$\frac{a_0}{2} = \frac{1}{2\pi} \int_{-\pi}^{\pi} 1 \cdot dx = 1 \text{ and } a_n = \frac{1}{\pi} \int_{-\pi}^{\pi} 1 \cdot \cos nx dx = 0.$$

Therefore, from above result, $1 = \frac{1}{\pi}[1 + 1] \int_0^{\infty} \frac{\sin x}{x} dx \Rightarrow \int_0^{\infty} \frac{\sin x}{x} dx = \frac{\pi}{2}.$

Corollary [11.3]: Show that $\frac{a_0}{2} + \sum_{1}^{\infty} a_n = \frac{f(0+) + f(0-)}{2}$.

Main Theorem

Theorem [11.5] If $f : [-\pi, \pi] \to \mathbb{R}$ be bounded, integrable and piecewise monotonic on $[-\pi, \pi]$ and f is periodic with period 2π, then

$$\frac{a_0}{2} + \sum_{n=1}^{\infty} (a_n \cos n\xi + b_n \sin n\xi) = \frac{1}{2}[f(\xi - 0) + f(\xi + 0)] \text{ for } -\pi < \xi < \pi$$

$$= \frac{1}{2}[f(\pi - 0) + f(-\pi + 0)] \text{ for } \xi = \pm\pi,$$

where $a_0, a_n, b_n, n \in N$, are the Fourier co-efficient of f.

Proof. We have

$$\frac{a_0}{2}+\sum_{n=1}^{m}(a_n\cos n\xi+b_n\sin n\xi) \text{ for any } \xi\in(-\pi,\pi)$$

$$=\frac{1}{2\pi}\int_{-\pi}^{\pi}f(x)dx+\sum_{n=1}^{m}\frac{1}{\pi}\int_{-\pi}^{\pi}f(x)\{\cos nx\cos n\xi+\sin nx\sin n\xi)dx$$

$$=\frac{1}{2\pi}\int_{-\pi}^{\pi}f(x)\left\{1+\sum_{n=1}^{m}2\cos(nx-n\xi)\right\}dx$$

$$=\frac{1}{2\pi}\int_{-\pi}^{\pi}f(x+\xi)\left[1+2\sum_{n=1}^{m}\cos nx\right]dx \quad \left[\begin{array}{l}\text{by (4) of [11.1], since } f \text{ and cosine}\\ \text{are both periodic in } [-\pi,\pi]\end{array}\right]$$

$$=\frac{1}{2\pi}\int_{-\pi}^{\pi}f(x+\xi)\frac{\sin\left(m+\frac{1}{2}\right)}{\sin\frac{x}{2}}dx$$

$$=\frac{1}{2\pi}\int_{-\pi}^{0}f(x+\xi)\frac{\sin\left(m+\frac{1}{2}\right)x}{\sin\frac{x}{2}}dx+\frac{1}{2\pi}\int_{0}^{\pi}f(x+\xi)\frac{\sin\left(m+\frac{1}{2}\right)x}{\sin\frac{x}{2}}dx$$

Putting $x=-2y$ for the 1st part and $x=2z$ in the 2nd part,

$$=\frac{1}{\pi}\int_{0}^{\pi/2}f(-2y+\xi)\frac{\sin(2m+1)y}{\sin y}dy+\frac{1}{2\pi}\int_{0}^{\pi/2}f(2z+\xi)\frac{\sin(2m+1)z}{\sin z}dz$$

$$=\frac{1}{\pi}[f(\xi-0)+f(\xi+0)]\int_{0}^{\infty}\frac{\sin x}{x}dx, \text{ as } m\to\infty$$

$$=\frac{1}{\pi}[f(\xi-0)+f(\xi+0)\frac{\pi}{2} \text{ [by Cor. of Th. [11.4]]}$$

$$=\frac{1}{2}[f(\xi-0)+f(\xi+0)] \text{ for } -\pi<\xi<\pi$$

At $\xi=\pi$,

$$\frac{1}{2}[f(\xi-0)+f(\xi+0)]=\frac{1}{2}[f(\pi-0)+f(\pi+0)]$$

$$=\frac{1}{2}[f(\pi-0)+f(\pi+0-2\pi)] \ [\because f(x)=f(x-2\pi)]$$

$$=\frac{1}{2}[f(\pi-0)+f(-\pi+0)]$$

Also at $\xi=-\pi$,

$$\frac{1}{2}[f(\xi-0)+f(\xi+0)]=\frac{1}{2}[f(\pi-0)+f(-\pi+0)] \ [\because f(x)=f(x+2\pi)]$$

Hence the theorem.

Corollary [11.4] : If f is continuous at a point $x=\xi$, $-\pi<\xi<\pi$, then

$$\frac{a_0}{2}+\sum_{n=1}^{\infty}(a_n\cos n\xi+b_n\sin n\xi)=f(\xi).$$

Proof. Since f is continuous at $x=\xi$, we have

$$\lim_{x\to\xi+0}f(x)=\lim_{x\to\xi-0}f(x)=f(\xi)\Rightarrow f(\xi+0)=f(\xi-0)=f(\xi)$$

Therefore, from above theorem,

$$\frac{a_0}{2}+\sum_{n=1}^{\infty}(a_n\cos n\xi+b_n\sin n\xi)=\frac{f(\xi+0)+f(\xi-0)}{2}=\frac{2f(\xi)}{2}=f(\xi).$$

Thus we say that, if f is continuous on $[-\pi,\pi]$, the series converges to $f(x)$, $\forall\, x\in[-\pi,\pi]$.

Note. Thus we have

$$S(x)=\frac{1}{2}[f(x+0)+f(x-0)]\ \forall\, x\in[-\pi,\pi]$$

$$\text{and }\ S(x)=\frac{1}{2}[f(-\pi+0)+f(\pi-0)]\ \text{at } x=\pm\pi$$

$$\text{and }\ S(x)=f(x),\ \text{when } f \text{ is continuous at } x\in[-\pi,\pi]$$

Here $S(x)$ denotes the sum function of the Fourier series $\frac{a_0}{2}+\sum\limits_{n=1}^{\infty}(a_n\cos nx+b_n\sin nx)$ corresponding to the function f defined on $[-\pi,\pi]$.

Theorem [11.6] If f be bounded and integrable on $[-\pi,\pi]$ and if a_n, b_n are Fourier coefficients, then $\sum\limits_1^{\infty}(a_n^2+b_n^2)$ is convergent.

Proof. We have $\int\limits_{-\pi}^{\pi}\left\{f-\sum\limits_1^m(a_n\cos nx+b_n\sin nx)\right\}^2dx\geq 0$ as $\left\{f-\sum\limits_1^m(a_n\cos nx+b_n\sin nx)\right\}^2$ is non-negative.

Therefore,

$$\int_{-\pi}^{\pi}\left\{f^2+\left[\sum_1^m(a_n\cos nx+b_n\sin nx)\right]^2-2f\sum_1^m(a_n\cos nx+b_n\sin nx)\right\}dx\geq 0$$

$$\Rightarrow \int_{-\pi}^{\pi}f^2dx+\sum_1^m a_n^2\int_{-\pi}^{\pi}\cos^2 nx dx+\sum_{n=1}^m b_n^2\int_{-\pi}^{\pi}\sin^2 nx dx+\sum_{n=1}^m a_nb_n\int_{-\pi}^{\pi}\sin 2nx dx$$

$$-2\sum_{n=1}^m a_n\int_{-\pi}^{\pi}f\cos nx dx-2\sum_{n=1}^m b_n\int_{-\pi}^{\pi}f\sin nx dx\geq 0$$

$$\Rightarrow \int_{-\pi}^{\pi}f^2dx+\sum_{n=1}^m a_n^2\cdot\pi+\sum_{n=1}^m b_n^2\cdot\pi+\sum_{n=1}^m a_nb_n\cdot 0-2\sum_{n=1}^m\pi a_n^2-2\sum_{n=1}^m\pi bn^2\geq 0$$

[by (1), (2) and (3) of Art. [11.3]]

[since $a_n=\frac{1}{\pi}\int_{-\pi}^{\pi}f\cos nx dx\Rightarrow \pi a_n=\int_{-\pi}^{\pi}f\cos nx dx$ and $b_n=\frac{1}{\pi}\int_{-\pi}^{\pi}f\sin nx dx$ $\Rightarrow \pi b_n=\int_{-\pi}^{\pi}f\sin nx dx$]

$$\Rightarrow \int_{-\pi}^{\pi}f^2dx-\pi\sum_1^{\infty}a_n^2-\pi\sum_1^{\infty}b_n^2\geq 0,\ \text{taking } m\to\infty$$

$$\Rightarrow \sum_1^{\infty}(a_n^2+b_n^2)\leq\frac{1}{\pi}\int_{-\pi}^{\pi}f^2dx\ \text{[Known as Bessel's inequality]}$$

Again, f is integrable $\Rightarrow f^2$ is integrable $\Rightarrow \int_{-\pi}^{\pi}f^2dx$ exists on $[-\pi,\pi]\Rightarrow \sum\limits_1^{\infty}(a_n^2+b_n^2)$ is convergent.

Corollary [11.5] : Fourier co-efficients a_n, b_n of an integrable function f form a null sequence.

Proof. Since $\sum\limits_{1}^{\infty}(a_n^2+b_n^2)$ is convergent, $\lim\limits_{n\to\infty}(a_n^2+b_n^2)=0$

$$\Rightarrow \lim_{n\to\infty} a_n = 0 \text{ and } \lim_{n\to\infty} b_n = 0$$
$$\Rightarrow \{a_n\} \text{ and } \{b_n\} \text{ form a null sequence.}$$

Corollary [11.6] : Let ϕ be bounded and integrable on $[a,b]$, then

$$\lim_{n\to\infty}\int_a^b \phi(x)\cos nx dx = 0 \text{ and } \lim_{n\to\infty}\int_a^b \phi(x)\sin nx dx = 0.$$

Proof. If $a, b \in [-m\pi, m\pi]$ for a fixed $m, m \in N$, then we define

$$\phi(x) = f(x)\ \forall\ x \in [a,b] \subset [-m\pi, m\pi]$$
$$= 0 \text{ when } x \text{ is not a point of an interval of the form } [-m\pi, m\pi].$$

Then f becomes a periodic function with period 2π.
Therefore,

$$\int_a^b \phi(x)\cos nx dx = \int_{-m\pi}^{m\pi} f(x)\cos nx dx \text{ [for a fixed } m \in N]$$
$$= \int_{-m\pi}^{-\pi} f(x)\cos nx dx + \int_{-\pi}^{\pi} f(x)\cos nx dx + \int_{\pi}^{m\pi} f(x)\cos nx dx$$
$$= 0 + \int_{-\pi}^{\pi} f(x)\cos nx dx + 0 \text{ [by definition]}$$
$$= \pi a_n$$

Similarly, $\int_a^b \phi(x)\sin nx dx = \pi b_n$, where a_n, b_n are Fourier co-efficients. Then by Cor. (11.5) of Theorem [11.6], $a_n, b_n \to 0$ as $n \to \infty$.

Hence, $\lim\limits_{n\to\infty}\int_a^b \phi(x)\cos nx dx = 0$ and $\lim\limits_{n\to\infty}\int_a^b \phi(x)\sin nx dx = 0$.

Now if a and b do not belong to the same interval $[-m\pi, m\pi], m \in N$, we divide the interval $[a,b]$ into k finite number of subintervals $[a, x_1], [x_1, x_2], \dots, [x_{n-1}, x_n], \dots [x_{k-1}, b]$ so that some of the subinterval $[x_{r-1}, x_r] \subset [-m\pi, m\pi]$ for some $m \in N$.

We define $\phi(x) = f(x)\ \forall\ x \in [x_{r-1}, x_r] \subset [-m\pi, m\pi]$ for some $m \in N$
$= 0$, otherwise.

Then
$$\int_a^b \phi(x)\cos nx dx = \int_a^{x_1}\phi(x)\cos nx dx + \int_{x_1}^{x_2}\phi(x)\cos nx dx + \cdots + \int_{x_{n-1}}^{b}\phi(x)\cos nx dx$$
$$= \int_{-m_1\pi}^{m_1\pi} f(x)\cos nx dx + \int_{-m_2\pi}^{m_2\pi} f(x)\cos nx dx + \cdots$$
$$+ \int_{-m_k\pi}^{m_k\pi} f(x)\cos nx dx,\ m_r \in N,\ r = 1, 2, \cdots, k$$
$$= \sum_{r=1}^{k}\int_{-m_r\pi}^{m_r\pi} f(x)\cos nx dx = k\int_{-\pi}^{\pi} f(x)\cos nx dx$$

proceeding as above, each of $\int_{-m_r\pi}^{m_r\pi} f(x)\cos nx dx = \int_{-\pi}^{\pi} f(x)\cos nx dx$
$= k\pi a_n.$

Similarly, $\int_a^b \phi(x)\sin nx dx = k\pi b_n$.
Now $\lim_{n\to\infty} a_n = 0$ and $\lim_{n\to\infty} b_n = 0$

$$\Rightarrow \lim_{n\to\infty}\int_a^b \phi(x)\cos nx dx = 0 \text{ and } \lim_{n\to\infty}\int_a^b \phi(x)\sin nx dx = 0.$$

Problem [11.1]: (1) Find the F.S. of the following function defined by

$$f(x) = \begin{cases} -x, & \text{for } -\pi < x < 0 \\ x, & \text{for } 0 < x < \pi \end{cases}$$

Hence deduce that $\frac{\pi^2}{8} = \frac{1}{1^2} + \frac{1}{3^2} + \frac{1}{5^2} + \cdots$
(2) Obtain F.S. for the function f defined by

$$f(x) = \begin{cases} -\pi - x, & \text{for } -\pi \le x < -\frac{\pi}{2} \\ x, & \text{for } -\frac{\pi}{2} \le x < \frac{\pi}{2} \\ \pi - x, & \text{for } \frac{\pi}{2} \le x \le \pi \end{cases}.$$

Find the graph of the given function.
(3) If $f(x) = cx$ for $-\pi < x < \pi, f(-\pi) = f(\pi) = 0$ and $f(x + 2\pi) = f(x)$, $\forall\, x$.
Show that $f(x) = 2c\left[\sin x - \frac{1}{2}\sin 2x + \frac{1}{3}\sin 3x - \frac{1}{4}\sin 4x + \cdots\right]$ $\forall\, x$.
Draw the graph of $f(x)$. $f(x)$ satisfies all the Dirichlet's conditions in $[-\pi, \pi]$.

Solution. (1) Let the F.S. corresponding to the given function $f(x)$ be $\frac{a_0}{2} + \sum_{n=1}^{\infty}(a_n \cos nx + b_n \sin nx)$, where

$$a_0 = \frac{1}{\pi}\int_{-\pi}^{\pi} f(x)dx = \frac{1}{\pi}\left[\int_{-\pi}^{0} -xdx + \int_0^{\pi} xdx\right]$$
$$= \frac{1}{\pi}\left[\int_0^{\pi} xdx + \int_0^{\pi} xdx\right] = \frac{2}{\pi}\left[\frac{\pi^2}{2}\right] = \pi$$
$$a_n = \frac{1}{\pi}\int_{-\pi}^{\pi} f(x)\cos nxdx = \frac{1}{\pi}\left[\int_{-\pi}^{0} -x\cos nxdx + \int_0^{\pi} x\cos nxdx\right]$$
$$= \frac{2}{\pi}\int_0^{\pi} x\cos nxdx = \frac{2}{\pi}\left[\left(\frac{x\sin nx}{n}\right)_0^{\pi} - \int_0^{\pi}\frac{\sin nx}{n}dx\right] = \frac{2}{\pi}\cdot\frac{1}{n^2}[(-1)^n - 1]$$
$$b_n = \frac{1}{\pi}\int_{-\pi}^{\pi} f(x)\sin nxdx = \frac{1}{\pi}\left[\int_{-\pi}^{0} -x\sin nxdx + \int_0^{\pi} x\sin nxdx\right]$$
$$= \frac{1}{\pi}\left[-\int_0^{\pi} x\sin nxdx + \int_0^{\pi} x\sin nxdx\right] = 0$$

Therefore, the F.S. for $f(x)$ is given by

$$f(x) = \frac{\pi}{2} + \sum_{n=1}^{\infty}\frac{2}{\pi n^2}\{(-1)^n - 1)\}\cos nx = \frac{\pi}{2} + \frac{2}{\pi}\sum_{n=1}^{\infty}\frac{(-1)^n - 1}{n^2}\cos nx$$
$$= \frac{\pi}{2} + \frac{2}{\pi}\left[-\frac{2}{1^2}\cos x - \frac{2}{3^2}\cos 3x - \frac{2}{5^2}\cos 5x - \cdots\right]$$
$$= \frac{\pi}{2} - \frac{4}{\pi}\left[\frac{\cos x}{1^2} + \frac{\cos 3x}{3^2} + \frac{\cos 5x}{5^2} + \cdots\right]$$

At $x=\pi$, sum of the F.S. is given by

$$S(\pi)=\frac{1}{2}[f(-\pi+0)+f(\pi-0)]=\frac{1}{2}[\pi+\pi]=\pi$$

$$\therefore\ \pi=\frac{\pi}{2}+\frac{4}{\pi}\left[\frac{1}{1^2}+\frac{1}{3^2}+\frac{1}{5^2}+\cdots\right]\text{ or, }\frac{\pi^2}{8}=\frac{1}{1^2}+\frac{1}{3^2}+\frac{1}{5^2}+\cdots$$

(2) Let the F.S. for the given function f is $\frac{a_0}{2}+\sum_1^\infty(a_n\cos nx+b_n\sin nx)$, where

$$a_0=\frac{1}{\pi}\int_{-\pi}^{\pi}f(x)dx=\frac{1}{\pi}\left[\int_{-\pi}^{-\pi/2}(-\pi-x)dx+\int_{-\pi/2}^{\pi/2}xdx+\int_{\pi/2}^{\pi}(\pi-x)dx\right]$$

$$=\frac{1}{\pi}\left[\int_{-\pi}^{-\pi/2}-\pi dx+\int_{\pi/2}^{\pi}xdx+\int_{\pi/2}^{\pi}\pi dx-\int_{\pi/2}^{\pi}xdx\right]\left[\because\int_{-\pi/2}^{\pi/2}xdx=0\right]$$

$$=\frac{1}{\pi}\left[-\int_{\pi/2}^{\pi}\pi dx+\int_{\pi/2}^{\pi}\pi dx\right]=0$$

$$a_n=\frac{1}{\pi}\int_{-\pi}^{\pi}f(x)\cos nxdx$$

$$=\frac{1}{\pi}\left[\int_{-\pi}^{-\pi/2}-(\pi+x)\cos nxdx+\int_{-\pi/2}^{\pi/2}x\cos nxdx+\int_{\pi/2}^{\pi}(\pi-x)\cos nxdx\right]$$

$$=\frac{1}{\pi}\left[\int_{-\pi}^{-\pi/2}-\pi\cos nxdx-\int_{-\pi}^{-\pi/2}x\cos nxdx+0+\int_{\pi/2}^{\pi}\pi\cos nxdx\right.$$

$$\left.-\int_{\pi/2}^{\pi}x\cos nxdx\right]=0$$

$$b_n=\frac{1}{\pi}\int_{-\pi}^{\pi}f(x)\sin nxdx$$

$$=\frac{1}{\pi}\left[\int_{-\pi}^{-\pi/2}-(\pi+x)\sin nxdx+\int_{-\pi/2}^{\pi/2}x\sin nxdx+\int_{\pi/2}^{\pi}(\pi-x)\sin nxdx\right]$$

$$=\frac{1}{\pi}\left[\int_{\pi/2}^{\pi}\pi\sin nxdx-\int_{\pi/2}^{\pi}x\sin nxdx+2\int_{0}^{\pi/2}x\sin nxdx\right.$$

$$\left.+\int_{\pi/2}^{\pi}\pi\sin nxdx-\int_{\pi/2}^{\pi}x\sin nxdx\right]$$

$$=\frac{2}{\pi}\left[\int_{\pi/2}^{\pi}\pi\sin nxdx-\int_{\pi/2}^{\pi}x\sin nxdx+\int_{0}^{\pi/2}x\sin nxdx\right]$$

$$=\frac{2}{\pi}\left[-\frac{\pi\cos n\pi}{n}+\frac{\pi\cos n\frac{\pi}{2}}{n}+\frac{\pi\cos n\pi}{n}-\frac{\pi}{2n}\cos n\frac{\pi}{2}+\frac{\sin n\frac{\pi}{2}}{n^2}\right.$$

$$\left.-\frac{\pi}{2n}\cos n\frac{\pi}{2}+\frac{\sin n\frac{\pi}{2}}{n^2}\right]$$

$$=\frac{2}{\pi}\left[\frac{2\sin n\frac{\pi}{2}}{n^2}\right]=\frac{4\sin n\frac{\pi}{2}}{\pi n^2}$$

Hence the F.S. for the given function f is

$$f(x) = \sum_{1}^{\infty} \frac{4}{\pi n^2} \sin n\frac{\pi}{2} \sin nx = \frac{4}{\pi}\sum_{1}^{\infty} \frac{\sin n\frac{\pi}{2} \sin nx}{n^2}$$

Graph of the given function

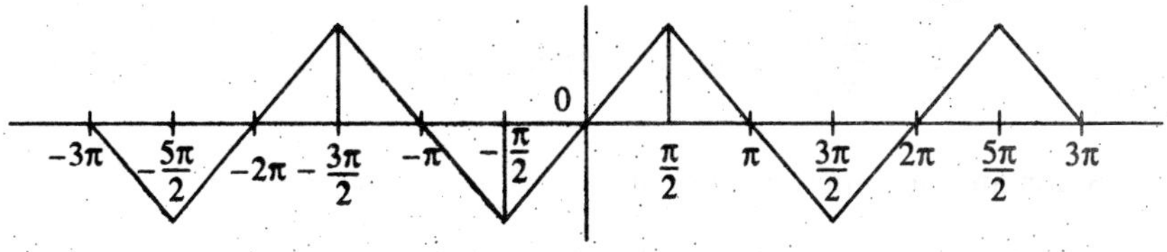

Fig. 11.2

(3) Let the F.S. for the given function be $\frac{a_0}{2} + \sum_{1}^{\infty}(a_n \cos nx + b_n \sin nx)$, where

$$a_0 = \frac{1}{\pi}\int_{-\pi}^{\pi} cx dx = 0$$

$$a_n = \frac{1}{\pi}\int_{-\pi}^{\pi} cx \cos nx dx = 0$$

$$b_n = \frac{1}{n}\int_{-\pi}^{\pi} cx \sin nx dx = \frac{2c}{\pi}\int_{0}^{\pi} x \sin nx dx$$

$$= \frac{2c}{\pi}\left[\left(-\frac{x \cos nx}{n}\right)_0^{\pi} + \left(\frac{\sin nx}{n^2}\right)_0^{\pi}\right]$$

$$= \frac{2c}{\pi}\left[\frac{-\pi(-1)^n}{n}\right] = 2c\frac{(-1)^{n+1}}{n}$$

Therefore,

$$f(x) = \sum_{1}^{\infty} b_n \sin nx = 2c\sum_{1}^{\infty}(-1)^{n+1}\frac{\sin nx}{n}$$

$$= 2c\left[+\sin x - \frac{\sin 2x}{2} + \frac{\sin 3x}{3} - \cdots\right] \; \forall\, x \in (-\pi, \pi)$$

Graph of $f(x)$

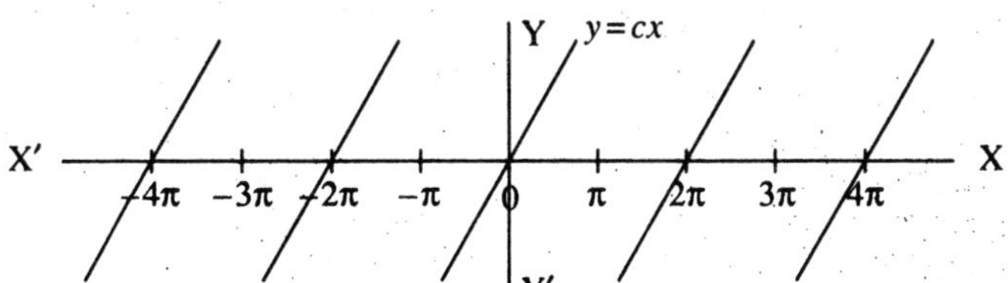

Fig. 11.3

11.4.1 Fourier Series for Odd and Even Functions

Let f be an odd function. Then $f(-x) = -f(x)\ \forall\ x$. Therefore, the F.S. corresponding to $f(x)$ in $[-\pi, \pi]$

$$= \frac{a_0}{2} + \sum a_n \cos nx + b_n \sin nx$$

where

$$a_0 = \frac{1}{\pi}\int_{-\pi}^{\pi} f(x)dx = 0 \text{ as } f \text{ is an odd function}$$

$$a_n = \frac{1}{\pi}\int_{-\pi}^{\pi} f(x)\cos nx dx = 0 \text{ as } f(x)\cos nx \text{ becomes an odd function}$$

$$b_n = \frac{1}{\pi}\int_{-\pi}^{\pi} f(x)\sin x dx = \frac{2}{\pi}\int_{0}^{\pi} f(x)\sin nx dx \text{ as } f(x)\sin nx \text{ becomes an even function.}$$

Thus the F.S. for an odd function $f(x)$ in $[-\pi, \pi]$ is $\sum_1^\infty b_n \sin nx$, where

$$b_n = \frac{2}{\pi}\int_0^\pi f(x)\sin nx dx$$

which represents a *sine series only.*

Next, let f be an even function. Then $f(-x) = f(x)\ \forall\ x$.

Therefore,

$$a_0 = \frac{1}{\pi}\int_{-\pi}^{\pi} f(x)dx = \frac{2}{\pi}\int_0^\pi f(x)dx$$

$$a_n = \frac{1}{\pi}\int_{-\pi}^{\pi} f(x)\cos nx dx = \frac{2}{\pi}\int_0^\pi f(x)\cos nx dx$$

$$\text{and } b_n = \frac{1}{\pi}\int_{-\pi}^{\pi} f(x)\sin nx dx = 0 \text{ as } f(x)\sin nx \text{ becomes an odd function.}$$

Thus the F.S. for an even function f in $[-\pi, \pi]$ is $\frac{a_0}{2} + \sum_1^\infty a_n \cos nx$, where

$$a_0 = \frac{2}{\pi}\int_0^\pi f(x)dx \text{ and } a_n = \frac{2}{\pi}\int_0^\pi f(x)\cos nx dx$$

which represents a *cosine series only.*

11.4.2 Fourier Series in the Half-Range Interval $[0, \pi]$

It will be seen that the F.S. in the half-range interval $[0, \pi]$ is either a sine series or cosine series.

(i) Let f be bounded, integrable and piecewise monotonic on $[0, \pi]$ and let the Fourier series be a sine series $\sum_1^\infty b_n \sin nx$; then the sum of the sine series, where

$$b_n = \frac{2}{\pi}\int_0^\pi f(x)\sin nx dx$$

is $S(x) = \frac{1}{2}[f(x+0) + f(x-0)]$ for $0 < x < \pi$ and $S(x) = 0$ for $x = 0$ and $x = \pi$.

Now to have a sine series only we define an odd function $F(x)$ in $[-\pi, \pi]$ such that

$$F(x) = f(x)\ \forall\ x \in [0, \pi].$$

Then $F(x) = -f(-x)$ in $[-\pi, 0]$.

Clearly F will be bounded, integrable and piecewise monotonic on $[-\pi, \pi]$. Thus F satisfies all the conditions of the main theorem. Then the F.S. corresponding to $F(x)$, where $F(x)$ is are odd function, is $\sum_{1}^{\infty} b_n \sin nx$, a sine series only, where

$$b_n = \frac{2}{\pi}\int_0^\pi F(x)\sin nx dx = \frac{2}{\pi}\int_0^\pi f(x)\sin nx dx \; [\because F(x) = f(x) \text{ in } [0,\pi]]$$

$$S(x) = \frac{1}{2}[F(x+0) + F(x-0)] = \frac{1}{2}[f(x+0) + f(x-0)] \text{ for } 0 < x < \pi$$

$$S(0) = \frac{1}{2}[F(+0) + F(-0)] = \frac{1}{2}[F(+0) - F(+0)] = 0$$

$$S(\pi) = \frac{1}{2}[F(\pi - 0) + F(-\pi + 0)] = \frac{1}{2}[F(\pi - 0) - F(\pi - 0)] = 0$$

(ii) Let f be bounded and integrable and piecewise monotonic on $[0, \pi]$. Then the F.S. of f is a cosine series of the form $a_0 + \sum_{1}^{\infty} a_n \cos nx$, where

$$a_0 = \frac{2}{\pi}\int_0^\pi f(x)dx \text{ and } a_n = \frac{2}{\pi}\int_0^\pi f(x)\cos nx dx$$

Sum of the series is given by

$$S(x) = \frac{1}{2}[f(x+0) + f(x-0) \text{ for } 0 < x < \pi$$

$$S(0) = f(+0) \text{ and } S(\pi) = f(\pi - 0)$$

We define an even function $F(x)$ on $[-\pi, \pi]$ such that

$$F(x) = f(x) \;\forall\; x \in [0,\pi] \text{ and } F(x) = f(-x) \;\forall\; x \in [-\pi, 0].$$

F is periodic with period 2π. Clearly F will be bounded, integrable and piecewise monotonic on $[-\pi, \pi]$. Thus F satisfies all the conditions of the main Theorem [11.5].

Then the F.S. corresponding to an even function F is of the form $\frac{a_0}{2} + \sum_{1}^{\infty} a_n \cos nx$, where

$$a_0 = \frac{1}{\pi}\int_{-\pi}^{\pi} F(x)dx = \frac{1}{\pi}\left[\int_{-\pi}^{0} F(x)dx + \int_0^\pi F(x)dx\right]$$

$$= \frac{2}{\pi}\int_0^\pi f(x)dx \text{ [By definition of } F]$$

$$a_n = \frac{1}{\pi}\int_{-\pi}^{\pi} F(x)\cos nx dx = \frac{2}{\pi}\int_0^\pi F(x)\cos nx dx$$

$$= \frac{2}{\pi}\int_0^\pi f(x)\cos nx dx$$

$$S(x) = \text{sum of the series}$$

$$= \frac{1}{2}[F(x+0) + F(x-0] \text{ for } -\pi < x < \pi$$

$$= \frac{1}{2}[f(x+0) + f(x-0)] \text{ for } 0 < x < \pi$$

$$S(0) = \frac{1}{2}[F(+0) + F(-0)] = \frac{1}{2}[F(+0) + F(+0)] = F(+0) = f(+0)$$

$$S(\pi) = \frac{1}{2}[F(-\pi + 0) + F(\pi - 0)] = \frac{1}{2}[F(\pi - 0) + F(\pi - 0)]$$

$$= F(\pi - 0) = f(\pi - 0)$$

Problem [11.2]: (1) Find (a) the Fourier sine series and (b) the Fourier cosine series for $f(x) = \pi - x, 0 < x < \pi$. Also find the sum of the series at $x = 0$ and $x = \pi$ in both cases.

(2) Let $f(x) = \begin{cases} \pi/3, & \text{for } 0 < x < \pi/3 \\ 0, & \text{for } \pi/3 < x < 2\pi/3 \\ -\pi/3, & \text{for } 2\pi/3 < x < \pi. \end{cases}$

Then obtain the F.S. for f on $[0, \pi]$ and the sum of the series at $x = \pi, 2\pi/3$.

(3) Obtain a sine series which will be equal to x^2 for $0 \leq x \leq \pi$.

Solution. (1) (a) We define

$$f(x) = \pi - x \text{ for } 0 \leq x \leq \pi$$
$$f'(x) = -1 < 0 \ \forall \ x \in [0, \pi]$$

$\Rightarrow f$ is monotonically decreasing on $[0, \pi]$.

Thus f is bounded, integrable and monotonic on $[0, \pi]$. Here f is periodic with period 2π. Then f satisfies all the Dirichlet's conditions on $[0, \pi]$, a half-range interval of $[-\pi, \pi]$.

The Fourier Sine series of f on $[0, \pi]$ is of the form $\sum b_n \sin nx$, where

$$b_n = \frac{2}{\pi} \int_0^\pi f(x) \sin nx dx.$$

Here $b_n = \dfrac{2}{\pi} \displaystyle\int_0^\pi (\pi - x) \sin nx dx$

$$= 2\left[\left(-\frac{\cos nx}{n}\right)_0^\pi\right] - \frac{2}{\pi}\left[\left(-\frac{x \cos nx}{n}\right)_0^\pi + \left(\frac{\sin nx}{n^2}\right)_0^\pi\right]$$

$$= \frac{2}{n}[(-1)^{n+1} + 1] - \frac{2}{n}(-1)^{n+1} = \frac{2}{n}$$

Hence, $\pi - x = \sum\limits_1^\infty \frac{2}{n} \sin nx$.

Also $S(0) = S(\pi) = 0$. This can be directly verified from

$$S(x) = 2\left[\sin x + \frac{\sin 2x}{2} + \frac{\sin 3x}{3} + \cdots\right].$$

(b) The Fourier cosine series is of the form $\frac{a_0}{2} + \sum\limits_1^\infty a_n \cos nx$, where

$$a_0 = \frac{2}{\pi} \int_0^\pi f(x) dx.$$

Here $f(x) = \pi - x$.

$$\therefore \ a_0 = \frac{2}{\pi} \int_0^\pi (\pi - x) dx = \frac{2}{\pi}\left[\pi^2 - \frac{\pi^2}{2}\right] = \pi$$

$$a_n = \frac{2}{\pi} \int_0^\pi f(x) \cos nx dx = \frac{2}{\pi} \int_0^\pi (\pi - x) \cos nx dx$$

$$= \frac{2}{\pi}\left[\left(\frac{\pi \sin nx}{n}\right)_0^\pi\right] - \left[\left(\frac{x \sin nx}{n}\right)_0^\pi + \left(\frac{\cos nx}{n^2}\right)_0^\pi\right]$$

$$= -\frac{2}{\pi n^2}[(-1)^n - 1] = \frac{2[1 - (-1)^n]}{\pi n^2}$$

Therefore,

$$\pi - x = \frac{a_0}{2} + \sum a_n \cos nx = \frac{\pi}{2} + \frac{2}{\pi}\sum_{1}^{\infty} \frac{1-(-1)^n}{n^2}\cos nx$$

$$= \frac{\pi}{2} + \frac{4}{\pi}\left[\frac{\cos x}{1^2} + \frac{\cos 3x}{3^2} + \frac{\cos 5x}{5^2} + \cdots\right]$$

$S(0) = f(+0) = \pi$

and $S(\pi) = f(\pi - 0) = 0$, since $f(x) = \pi - x \Rightarrow f(+0) = \pi$ and $f(\pi - 0) = 0$.

(2) We define f by

$$f(x) = \begin{cases} \pi/3, & \text{for } 0 \le x < \pi/3 \\ 0, & \text{for } \pi/3 \le x < 2\pi/3 \\ -\pi/3, & \text{for } 2\pi/3 \le x \le \pi \end{cases}$$

so that f remains bounded. f has finite numbers of points of discontinuities at $x = \pi/3$ and $x = 2\pi/3$. So f is integrable on $[0, \pi]$. Also f is piecewise monotonic on $[0, \pi/3]$, $[\pi/3, 2\pi/3]$ and $[2\pi/3, \pi]$. Here f is supposed to be periodic with period 2π. Then $[0, \pi]$ is a half-range of $[-\pi, \pi]$.

Let the F.S. be a sine series. Then $f(x) = \sum_{1}^{\infty} b_n \sin nx$, where

$$b_n = \frac{2}{\pi}\int_0^{\pi} f(x)\sin nx dx$$

$$= \frac{2}{\pi}\left[\int_0^{\pi/3} f(x)\sin nx dx + \int_{\pi/3}^{2\pi/3} f(x)\sin nx dx + \int_{2\pi/3}^{\pi} f(x)\sin nx dx\right]$$

$$= \frac{2}{\pi}\left[\int_0^{\pi/3} \frac{\pi}{3}\sin nx dx + 0 + \int_{2\pi/3}^{\pi} -\frac{\pi}{3}\sin nx dx\right]$$

$$= \frac{2}{3}\left[-\frac{\cos n\frac{\pi}{3} - 1}{n}\right] + \frac{2}{3}\left[\frac{\cos n\pi - \cos 2n\frac{\pi}{3}}{n}\right] = \frac{2}{3}\left[\frac{1+(-1)^n}{n} - \frac{\cos n\frac{\pi}{3} + \cos 2n\frac{\pi}{3}}{n}\right]$$

$$= \frac{2}{3}\left[\frac{1+(-1)^n}{n} - \frac{2}{n}\cos n\frac{\pi}{2}\cdot\cos n\frac{\pi}{6}\right]$$

Then $b_n = 0$ for any odd integer n.

$\therefore b_1 = b_3 = b_5 = \cdots = 0$, when n is odd.

When n is even

$$b_2 = \frac{2}{3}\left[1 + \frac{1}{2}\right] = 1$$

$$b_4 = \frac{2}{3}\left[\frac{1}{2} - \frac{2}{4}\left(-\frac{1}{2}\right)\right] = \frac{2}{3}\cdot\left[\frac{1}{2} + \frac{1}{4}\right] = \frac{2}{3}\cdot\frac{3}{4} = \frac{1}{2}$$

$$b_6 = \frac{2}{3}\left[\frac{1}{3} - \frac{1}{3}\cos 3\pi \cos \pi\right] = \frac{2}{3}\left[\frac{1}{3} - \frac{1}{3}\right] = 0$$

$$b_8 = \frac{2}{3}\left[\frac{1}{4} - \frac{1}{4}\cos 4\pi \cos\frac{2\pi}{3}\right] = \frac{2}{3}\left[\frac{3}{8}\right] = \frac{1}{4}$$

and so on.

Then $f(x) = b_2 \sin 2x + b_4 \sin 4x + b_6 \sin 6x + b_8 \sin 8x + \cdots$

$$= \sin 2x + \frac{\sin 4x}{2} + \frac{\sin 8x}{4} + \cdots$$

$S(\pi) = 0$. This can be obtained directly from the series.

At $x = \frac{2\pi}{3}$, $0 < x < \pi$,

$$S(x) = \frac{1}{2}[f(x+0) + f(x-0)], 0 < x < \pi$$

$$= \frac{1}{2}\left[f\left(\frac{2\pi}{3}+0\right) + f\left(\frac{2\pi}{3}-0\right)\right] = \frac{1}{2}\left[-\frac{\pi}{3}+0\right] = -\frac{\pi}{6}$$

(3) $f(x) = x^2$, $0 \le x \le \pi$.

Then f is bounded on $[0, \pi]$, $f'(x) = 2x \ge 0 \;\forall\; x \in [0, \pi]$.

$\Rightarrow f$ is monotonically increasing on $[0, \pi]$. Thus f is bounded, integrable, monotonic and periodic with period 2π. So f satisfies all the Dirichlet's conditions.

The sine series of f is $\sum b_n \sin nx$, where

$$b_n = \frac{2}{\pi}\int_0^{\pi} f(x)\sin nx dx = \frac{2}{\pi}\int_0^{\pi} x^2 \sin nx dx$$

$$= \frac{2}{\pi}\left[\left(\frac{-x^2\cos nx}{n}\right)_0^{\pi} + \frac{2}{n}\int_0^{\pi} x\cos nx dx\right]$$

$$= \frac{2}{\pi}\left[-\frac{\pi^2(-1)^n}{n}\right] + \frac{4}{n\pi}\left[\left(\frac{x\sin nx}{n}\right)_0^{\pi} + \left(\frac{\cos nx}{n^2}\right)_0^{\pi}\right]$$

$$= \frac{2\pi}{n}(-1)^{n+1} + \frac{4}{\pi n^3}[(-1)^n - 1]$$

Therefore, $x^2 = \dfrac{2}{\pi}\displaystyle\sum_1^{\infty}\left[\frac{\pi^2(-1)^{n+1}}{n} + \frac{2}{n^3}\{(-1)^n - 1\}\right]\sin nx.$

11.5 Fourier Series in the Interval $[0, 2\pi]$ for a Function f with Periodicity 2π

Let f satisfy the conditions of Main Theorem [11.5] in $[0, 2\pi]$.

Now if we put $x = y + \pi$, then $f(x) = f(y+\pi) = F(y)$, say. Then $F(y)$ satisfies the conditions of Th. [11.5] in $[-\pi, \pi]$ as $x \in [0, 2\pi] \Rightarrow y \in [-\pi, \pi]$.

Then the F.S. corresponding to $F(y) = \dfrac{a_0'}{2} + \displaystyle\sum_1^{\infty}(a_n' \cos ny + b_n' \sin ny)$, where

$$a_0' = \frac{1}{\pi}\int_{-\pi}^{\pi} F(y)dy = \frac{1}{\pi}\int_0^{2\pi} F(x-\pi)d(x-\pi) = \frac{1}{\pi}\int_0^{2\pi} f(x)dx$$

$$a_n' = \frac{1}{\pi}\int_{-\pi}^{\pi} F(y)\cos ny dy = \frac{1}{\pi}\int_0^{2\pi} F(x-\pi)\cos n(x-\pi)dx$$

$$= \frac{(-1)^n}{\pi}\int_0^{2\pi} f(x)\cos nx dx \;[\because F(x-\pi) = f(x-\pi+\pi) = f(x)]$$

$$b_n' = \frac{1}{\pi}\int_{-\pi}^{\pi} F(y)\sin ny dy = \frac{(-1)^n}{\pi}\int_0^{2\pi} f(x)\sin nx dx$$

Therefore, the F.S. corresponding to $f(x)$ on $[0, 2\pi]$ will be of the form

$$\frac{a_0'}{2} + \sum_{n=1}^{\infty}\{a_n' \cos n(x-\pi) + b_n' \sin n(x-\pi)\}$$
$$= \frac{a_0'}{2} + \sum_{n=1}^{\infty}\{(-1)^n a_n' \cos nx + (-1)^n b_n' \sin nx\}$$

Now from above,

$$(-1)^n a_n' = \frac{1}{\pi}\int_0^{2\pi} f(x)\cos nx dx = a_n \text{ (say)}$$
$$(-1)^n b_n' = \frac{1}{\pi}\int_0^{2\pi} f(x)\sin nx dx = b_n \text{ (say)}$$
$$\text{and } a_0' = a_0.$$

Then the F.S. reduces to the form $\frac{a_0}{2} + \sum\limits_{n=1}^{\infty}\{a_n \cos nx + b_n \sin nx\}$, where

$$a_0 = \frac{1}{\pi}\int_0^{2\pi} f(x)dx$$
$$a_n = \frac{1}{\pi}\int_0^{2\pi} f(x)\cos nx dx$$
$$\text{and } b_n = \frac{1}{\pi}\int_0^{2\pi} f(x)\sin nx dx$$

Sum of the series corresponding to the F.S. represented by $F(y)$ on $[-\pi, \pi]$

$$S(y) = \frac{1}{2}[F(y+0) + F(y-0)] = \frac{1}{2}f(y+\pi+0) + f(y+\pi-0)]$$
$$\Rightarrow S(x) = \frac{1}{2}[f(x+0) + f(x-0) \text{ for } 0 < x < 2\pi$$
$$S(2\pi) = \frac{1}{2}[f(2\pi-0) + f(+0)] \text{ for } x = 2\pi$$
$$S(0) = \frac{1}{2}[f(2\pi-0) + f(+0)] \text{ for } x = 0$$

Thus the sum of the F.S. corresponding to $f(x)$ on $[0, 2\pi]$ is given by

$$S(x) = \frac{1}{2}[f(x+0) + f(x-0)],\ 0 < x < 2\pi$$
$$S(2\pi) = S(0) = \frac{1}{2}[f(2\pi-0) + f(+0)].$$

11.6 F.S. in the Interval $(-l, l)$ with Period $2l$, where l is any Real Number

Considering the function $f(x)$ and putting $y = \pi x/l$, we have

$$f(x) = f\left(\frac{ly}{\pi}\right) = F(y), \text{ say,}$$

so that $y \in [-\pi, \pi]$ as $x \in [-l, l]$. Also if f satisfies the conditions of the main theorem [11.5] on $[-l, l]$, then F satisfies the conditions of the main theorem on $[-\pi, \pi]$. So the F.S. corresponding to $F(y)$ on $[-\pi, \pi]$ is of the form $\frac{a_0}{2} + \sum_1^{\infty}\{a_n \cos ny + b_n \sin xy\}$, where

$$a_0 = \frac{1}{\pi}\int_{-\pi}^{\pi} F(y)dy = \frac{1}{\pi}\int_{-l}^{l} F\left(\frac{\pi x}{l}\right)\frac{\pi}{l}\cdot dx$$

$$= \frac{1}{l}\int_{-l}^{l} f(x)dx \left[\because F(y) = f\left(\tfrac{ly}{\pi}\right) \Rightarrow F\left(\tfrac{\pi x}{l}\right) = f(x)\right]$$

$$a_n = \frac{1}{\pi}\int_{-\pi}^{\pi} F(y)\cos nydy = \frac{1}{\pi}\cdot\frac{\pi}{l}\int_{-l}^{l} F\left(\frac{\pi x}{l}\right)\cos\frac{n\pi x}{l}dx = \frac{1}{l}\int_{-l}^{l} f(x)\cos\frac{n\pi x}{l}dx$$

$$b_n = \frac{1}{\pi}\int_{-\pi}^{\pi} F(y)\sin nydy = \frac{1}{l}\int_{-l}^{l} f(x)\sin\frac{n\pi x}{l}dx$$

Hence, the F.S. corresponding to $f(x)$ on $[-l, l]$ is $\frac{a_0}{2} + \sum_1^{\infty}\left(a_n \cos\frac{n\pi x}{l} + b_n \sin\frac{n\pi x}{l}\right)$, where a_0, a_n, b_n are given above.

Sum of the F.S. for $-\pi < y < \pi$

$$S(y) = \frac{1}{2}[F(y+0) + F(y-0)]$$

$$\Rightarrow S(x) = \frac{1}{2}[f(x+0) + f(x-0)] \text{ for } -l < x < l$$

$$S(+\pi) = \frac{1}{2}[F(-\pi+0) + F(\pi-0)]$$

$$\Rightarrow S(l) = \frac{1}{2}[f(-l+0) + f(l-0)] \left[\because F(y) = f\left(\tfrac{ly}{\pi}\right) \Rightarrow F(-\pi+0) = f(-l+0)\right]$$

Also $S(-l) = \frac{1}{2}[f(-l+0) + f(l-0)]$.

11.7 F.S. in the Interval $[a, b]$, a, b are Real Numbers, with Period $b - a$

Let f satisfies all the conditions of the main theorem [11.5] on $[a, b]$. Then the F.S. corresponding to f on $[a, b]$ is of the form $\frac{a_0}{2} + \sum_1^{\infty}\left\{a_n \cos\left(\frac{2n\pi x}{b-a}\right) + b_n \sin\left(\frac{2n\pi x}{b-a}\right)\right\}$, where

$$a_0 = \frac{2}{b-a}\int_a^b f(x)dx$$

$$a_n = \frac{2}{b-a}\int_a^b f(x)\cos\frac{2n\pi x}{b-a}dx$$

$$b_n = \frac{2}{b-a}\int_a^b f(x)\sin\frac{2n\pi x}{b-a}dx$$

$$S(x) = \frac{1}{2}[f(x+0) + f(x-0)],\ a < x < b$$

$$S(a) = \frac{1}{2}[f(a+0) + f(b-0)] = S(b)$$

Let $y = \frac{2\pi}{b-a}x - \frac{b+a}{b-a}\pi$.

This can be obtained by considering a transformation $y = lx + m$ so that $y = -\pi$ when $x = a$, and $y = \pi$ when $x = b$.

Thus we have a function $F(y)$ which will satisfy the conditions of the main theorem on $[-\pi, \pi]$.

Proceeding as in [11.6] we can obtain the F.S. of $f(x)$ on $[a, b]$, as given above.

Problem [11.3] : (1) Prove that for $0 < \theta < 2\pi$, $\frac{1}{2}(\pi - \theta) = \sum_{n=1}^{\infty} \frac{\sin n\theta}{n}$. Find $S(0)$ and $S(2\pi)$.

(2) Show that $x + x^2 = \frac{1}{3} + \frac{2}{\pi} \sum_{n=1}^{\infty} (-1)^n \left[\frac{2\cos n\pi x}{n^2\pi} - \frac{\sin n\pi x}{n}\right]$, when $-1 < x < 1$.

(3) If $f(x) = \begin{cases} 1, & \text{when } 0 < x < 1 \\ 2, & \text{when } 1 < x < 3 \\ \frac{3}{4}, & \text{when } x = 0, 1, 3 \end{cases}$

and $f(x+3) = f(x)\ \forall\ x$, show that $f(x) = \frac{5}{3} - \frac{2}{\pi} \sum_{n=1}^{\infty} \frac{\sin \frac{n\pi}{3}}{n} \left\{\cos \frac{n\pi}{3}(2x-1)\right\}$.

Note. When f is not defined at the end points of the interval in which f is defined, we define f at the end points arbitrarily so that f remains bounded.

Solution. (1) Here $f(\theta) = \frac{1}{2}[\pi - \theta]$ for $0 < \theta < 2\pi$.

We define $f(0) = \frac{\pi}{2}$ and $f(2\pi) = -\frac{\pi}{2}$. Then f is continuous on $[0, 2\pi]$. So f is bounded and integrable on $[0, 2\pi]$.

Also $f'(\theta) = -\frac{1}{2} < 0$. So f is monotonically decreasing on $[0, 2\pi]$. Therefore, f satisfies all the Dirichlet's conditions on $[0, 2\pi]$.

Let the F.S. for the function $f(\theta) = \frac{1}{2}(\pi - \theta)$ be $\frac{a_0}{2} + \sum_{1}^{\infty}(a_n \cos n\theta + b_n \sin n\theta)$, where

$$a_0 = \frac{1}{\pi}\int_0^{2\pi} f(\theta)d\theta = \frac{1}{\pi}\int_0^{2\pi} \frac{1}{2}(\pi - \theta)d\theta = \frac{1}{2}[2\pi] - \frac{1}{4\pi}\cdot 4\pi^2 = \pi - \pi = 0$$

$$a_n = \frac{1}{\pi}\int_0^{2\pi} f(\theta)\cos n\theta d\theta = \frac{1}{\pi}\int_0^{2\pi} \frac{1}{2}(\pi - \theta)\cos n\theta d\theta$$

$$= \frac{1}{2}\left[\frac{\sin n\theta}{n}\right]_0^{2\pi} - \frac{1}{2\pi}\left[\left(\frac{\theta \sin n\theta}{n}\right)_0^{2\pi} + \left(\frac{\cos n\theta}{n^2}\right)_0^{2\pi}\right] = 0$$

$$b_n = \frac{1}{\pi}\int_0^{2\pi} f(\theta)\sin n\theta d\theta = \frac{1}{\pi}\int_0^{2\pi} \frac{1}{2}(\pi - \theta)\sin n\theta d\theta$$

$$= \frac{1}{2}\left[-\frac{\cos n\theta}{n}\right]_0^{2\pi} - \frac{1}{2\pi}\left\{\left(-\frac{\theta\cos n\theta}{n}\right)_0^{2\pi} + \left(\frac{\sin n\theta}{n^2}\right)_0^{2\pi}\right\} = \frac{1}{2\pi n}\cdot 2\pi = \frac{1}{n}$$

Hence, $\frac{1}{2}(\pi - \theta) = \sum_{1}^{\infty} \frac{\sin n\theta}{n}$.

Also $S(0) = \frac{1}{2}[f(0+0) + f(2\pi - 0)] = \frac{1}{2}\left[\frac{\pi}{2} - \frac{\pi}{2}\right] = 0 = S(2\pi)$.

(2) $f(x) = x + x^2$ for $-1 < x < 1$.

We define $f(x) = x + x^2$ for $-1 \le x \le 1$. Then f is continuous on $[-1, 1]$. So f is bounded and integrable on $[-1, 1]$.

Also $f'(x) = 1 + 2x > 0$ for $x > -\frac{1}{2}$ and $f'(x) < 0$ for $x < -\frac{1}{2}$. So f is monotonically increasing on $-1 \le x < -\frac{1}{2}$ and monotonically decreasing on $-\frac{1}{2} < x \le 1$.

Thus f is piecewise monotonic on $[-1, 1]$. Therefore, f satisfies all the Dirichlet's conditions on $[-1, 1]$. Let the F.S. of $f(x) = \frac{a_0}{2} + \sum(a_n \cos n\pi x + b_n \sin n\pi x)$ on $[-1, 1]$, where

$$a_0 = \frac{1}{1}\int_{-1}^{1} f(x)dx = \int_{-1}^{1}(x + x^2)dx = \frac{2}{3}$$

$$a_n = \frac{1}{1}\int_{-1}^{1} f(x)\cos n\pi x dx = \int_{-1}^{1}(x + x^2)\cos n\pi x dx$$

$$= 2\int_0^{\pi} x^2 \cos n\pi x dx = 2\left[\left(x^2\frac{\sin n\pi x}{n\pi}\right)_0^1 - \int_0^1 2x\frac{\sin n\pi x}{n\pi}dx\right]$$

$$= -\frac{4}{n\pi}\left[\left(\frac{-x\cos n\pi x}{n\pi}\right)_0^1 + \left(\frac{\sin n\pi x}{n^2\pi^2}\right)_0^1\right]$$

$$= -\frac{4}{n\pi}\cdot\frac{-(-1)^n}{n\pi} = \frac{4(-1)^n}{n^2\pi^2}$$

$$b_n = \frac{1}{1}\int_{-1}^{1} f(x)\sin n\pi x dx = \int_{-1}^{1}(x + x^2)\sin n\pi x dx$$

$$= 2\int_0^1 x\sin n\pi x dx = 2\left[\left(-\frac{x\cos n\pi x}{n\pi}\right)_0^1 + \left(\frac{\sin n\pi x}{n^2\pi^2}\right)_0^1\right]$$

$$= 2\frac{(-1)\cdot(-1)^n}{n\pi} = -\frac{2(-1)^n}{n\pi}$$

Therefore, $x + x^2 = \frac{a_0}{2} + \sum_1^{\infty}(a_n \cos n\pi x + b_n \sin n\pi x)$

$$= \frac{1}{2}\cdot\frac{2}{3} + \frac{2}{\pi}\sum_1^{\infty}\left\{\frac{(-1)^n 2\cos n\pi x}{n^2\pi.} - \frac{(-1)^n \sin n\pi x}{n}\right\}$$

(3) Here $f(0) = f(1) = f(3) = \frac{3}{4}$.

f has number of points of discontinuities on $[0, 3]$ at $x = 0, 1$ and 3. So f is integrable on $[0, 3]$ and f is also bounded on $[0, 3]$. Again, f is piecewise monotonic on $[0, 1]$ and on $[1, 3]$. So f satisfies all the Dirichlet's conditions on $[0, 3]$. Also f is periodic with periodicticity 3. Therefore, by Art. 11.7, $a = 0, b = 3$, the F.S. for $f(x)$ is

$$\frac{a_0}{2} + \sum_1^{\infty}\left(a_n \cos\frac{2n\pi x}{3} + b_n \sin\frac{2n\pi x}{3}\right),$$

where

$$a_0 = \frac{2}{3}\int_0^3 f(x)dx = \frac{2}{3}\left[\int_0^1 1dx + \int_1^3 2dx\right] = \frac{2}{3}[1 + 2\cdot 2] = \frac{10}{3}$$

$$a_n = \frac{2}{3}\int_0^3 f(x)\cos\frac{2n\pi x}{3}dx = \frac{2}{3}\left[\int_0^1 \cos\frac{2n\pi x}{3} + 2\int_1^3 \cos\frac{2n\pi x}{3}dx\right]$$

$$= \frac{2}{3}\left[\left(\frac{\sin\frac{2n\pi x}{3}}{\frac{2n\pi}{3}}\right)_0^1 + 2\left(\frac{\sin\frac{2n\pi x}{3}}{\frac{2n\pi}{3}}\right)_1^3\right]$$

$$= \frac{1}{n\pi}\left[\sin\frac{2n\pi}{3} + 2\left(-\sin\frac{2n\pi}{3}\right)\right] = \frac{1}{n\pi}\left(-\sin\frac{2n\pi}{3}\right)$$

$$
\begin{aligned}
b_n &= \frac{2}{3}\int_0^3 f(x)\sin\frac{2n\pi x}{3}dx \\
&= \frac{2}{3}\left[\int_0^1 \sin\frac{2n\pi x}{3} + 2\int_1^3 \sin\frac{2n\pi x}{3}dx\right] \\
&= \frac{2}{3}\left[-\frac{\cos\frac{2n\pi x}{3}}{\frac{2n\pi}{3}}\right]_0^1 + \frac{4}{3}\left[-\frac{\cos\frac{2n\pi x}{3}}{\frac{2n\pi}{3}}\right]_1^3 \\
&= \frac{1}{n\pi}\left[1 - \cos 2n\frac{\pi}{3} + 2\left(-1 + \cos 2n\frac{\pi}{3}\right)\right] \\
&= \frac{1}{n\pi}\left[-1 + \cos 2n\frac{\pi}{3}\right]
\end{aligned}
$$

Therefore,

$$
\begin{aligned}
f(x) &= \frac{1}{2}\cdot\frac{10}{3} + \sum_1^\infty \left\{\frac{1}{n\pi}\left(-\sin\frac{2n\pi}{3}\right)\cos\frac{2n\pi x}{3} + \frac{1}{n\pi}\left(-1+\cos\frac{2n\pi}{3}\right)\sin\frac{2n\pi x}{3}\right\} \\
&= \frac{5}{3} + \frac{1}{\pi}\sum_1^\infty -\frac{1}{n}\sin\frac{2n\pi x}{3} + \frac{1}{n}\left\{\cos\frac{2n\pi}{3}\sin\frac{2n\pi x}{3} - \sin\frac{2n\pi}{3}\cos\frac{2n\pi x}{3}\right\} \\
&= \frac{5}{3} + \frac{1}{\pi}\sum_1^\infty -\frac{1}{n}\sin\frac{2n\pi x}{3} + \frac{1}{n}\sin\left\{\frac{2n\pi x}{3} - \frac{2n\pi}{3}\right\} \\
&= \frac{5}{3} - \frac{1}{\pi}\sum_1^\infty \frac{1}{n}\left\{\sin\frac{2n\pi x}{3} - \sin(x-1)\frac{2n\pi}{3}\right\} \\
&= \frac{5}{3} - \frac{1}{\pi}\sum_1^\infty \frac{1}{n}\cdot 2\cos\left(\frac{2n\pi x}{3} - \frac{n\pi}{3}\right)\cdot\sin\frac{n\pi}{3} \\
&= \frac{5}{3} - \frac{2}{\pi}\sum_1^\infty \frac{1}{n}\sin\frac{n\pi}{3}\cos(2x-1)\frac{n\pi}{3}
\end{aligned}
$$

11.8 Half-Range Interval $[0, l]$

We obtain either sine series or cosine series on $[0, l]$ with period $2l$. So the interval $[0, l]$ is considered a half-range interval of $[-l, l]$.

Let f satisfy all the conditions of the main theorem [11.5] on $[0, l]$.

Introducing an odd function $F(x)$ on $[-l, l]$, defined by

$$
\begin{aligned}
F(x) &= f(x) \ \forall\ x \in [0, l] \\
&= -f(x) \text{ in } [-l, 0]
\end{aligned}
$$

so that F satisfies all the conditions of the Main Theorem on $[-l, l]$ and the F.S. corresponding to $F(x)$ is a sine series $\sum_1^\infty b_n \sin\frac{n\pi x}{l}$, where

$$
b_n = \frac{2}{l}\int_0^l F(x)\sin\frac{n\pi x}{l}dx = \frac{2}{l}\int_0^l f(x)\sin\frac{n\pi x}{l}dx
$$

$$
\therefore f(x) \sim \sum_1^\infty b_n \sin\frac{n\pi x}{l}, \text{ where } b_n = \frac{2}{l}\int_0^l f(x)\sin\frac{n\pi x}{l}dx.
$$

Sum of the series is given by

$$S(x) = \frac{1}{2}[F(x+0) + F(x-0)] \;\forall\; x, \;\; -l < x < l$$
$$= \frac{1}{2}[f(x+0) + f(x-0)] \text{ for } 0 < x < l$$
$$S(0) = \frac{1}{2}[F(+0) + F(-0)] = 0$$
$$S(l) = \frac{1}{2}[F(l-0) + F(-l+0)] = 0$$

Similarly, introducing an even function $F(x)$ on $[-l, l]$ defined by

$$F(x) = \begin{cases} f(x), & \text{for } 0 < x < l \\ -f(x), & \text{for } -l < x < 0 \end{cases}$$

so that F satisfies all the conditions of Main Theorem [11.5] on $[-l, l]$ and hence the F.S. corresponding to $F(x)$ on $[-l, l]$ is of the form $\frac{a_0}{2} + \sum_1^\infty a_n \cos\frac{n\pi x}{l}$ a cosine series, where

$$a_0 = \frac{2}{l}\int_0^l f(x)dx$$
$$a_n = \frac{2}{l}\int_0^l f(x)\cos\frac{n\pi x}{l}dx$$
$$S(x) = \text{sum of the F.S. of } f(x)$$
$$= \frac{1}{2}[f(x+0) + f(x-0)], \; 0 < x < l$$
$$S(0) = f(+0)$$
$$\text{and } S(l) = f(l-0)$$

Problem [11.4] : Find Fourier sine and cosine series of $f(x)$ when $f(x) = \pi - x$, in $0 < x < \pi$ with period 2π. Also find the F.S. of $f(x) = \pi - x$ in $0 < x < \pi$ with period π.

Solution. We define $f(0) = \pi$, $f(\pi) = 0$. f satisfies all the Dirichlet's conditions on $[0, \pi]$.

$f(x) = \pi - x$, $0 < x < \pi$ with period 2π.

In this case, the interval is a half-range interval of $[-\pi, \pi]$.

So the sine series corresponding to $f(x) = \pi - x$ is $\sum_1^\infty b_n \sin nx$, where

$$b_n = \frac{2}{\pi}\int_0^\pi f(x)\sin nx dx$$
$$= \frac{2}{\pi}\int_0^\pi (\pi - x)\sin nx dx$$
$$= \frac{2}{\pi}\left[\pi\left(-\frac{\cos nx}{n}\right)_0^\pi - \left\{\left(-\frac{x\cos nx}{n}\right)_0^\pi + \left(\frac{\sin nx}{n^2}\right)_0^\pi\right\}\right]$$
$$= \frac{2}{n}\left\{1 - (-1)^n\right\} + \frac{2}{n}(-1)^n\Big\} = \frac{2}{n}$$

$$\therefore \pi - x = \sum_1^\infty \frac{2}{n}\sin nx = 2\left(\frac{\sin x}{1} + \frac{\sin 2x}{2} + \frac{\sin 3x}{3} + \cdots\right) \quad (1)$$

The corresponding cosine series is $\frac{a_0}{2} + \sum_1^\infty a_n \cos nx$, where

$$a_0 = \frac{2}{\pi}\int_0^\pi f(x)dx = \frac{2}{\pi}\int_0^\pi (\pi - x)dx = 2\pi - \pi = \pi$$

$$\begin{aligned} a_n &= \frac{2}{\pi}\int_0^\pi f(x)\cos nx dx = \frac{2}{\pi}\int_0^\pi (\pi - x)\cos nx dx \\ &= \frac{2}{\pi}\left[\pi\left(\frac{\sin nx}{n}\right)_0^\pi - \left\{\left(\frac{x\sin nx}{n}\right)_0^\pi + \left(\frac{\cos nx}{n^2}\right)_0^\pi\right\}\right] \\ &= -\frac{2}{\pi}\left[\frac{(-1)^n - 1}{n^2}\right] = \frac{2}{\pi}\frac{[1-(-1)^n]}{n^2} \end{aligned}$$

$$\begin{aligned} \therefore\ \pi - x &= \frac{\pi}{2} + \frac{2}{\pi}\sum_1^\infty \frac{1-(-1)^n}{n^2}\cos nx \\ &= \frac{\pi}{2} + \frac{4}{\pi}\left[\frac{\cos x}{1^2} + \frac{\cos 3x}{3^2} + \frac{\cos 5x}{5^2} + \cdots\right] \end{aligned}$$

Now F.S. of $f(x) = \pi - x$ in $[0, \pi]$ with period π is of the form

$$\frac{a_0}{2} + \sum_1^\infty [a_n \cos 2nx + b_n \sin 2nx]$$

by Art. 11.7, when $a = 0, b = \pi$, where

$$a_0 = \frac{2}{\pi}\int_0^\pi f(x)dx = \frac{2}{\pi}\int_0^\pi (\pi - x)dx = \pi$$

$$\begin{aligned} a_n &= \frac{2}{\pi}\int_0^\pi f(x)\cos 2nx dx = \frac{2}{\pi}\int_0^\pi (\pi - x)\cos 2nx dx \\ &= \frac{2}{\pi}\left[\pi\left(\frac{\sin 2nx}{2n}\right)_0^\pi - \left\{\left(\frac{x\sin 2nx}{2n}\right)_0^\pi + \left(\frac{\cos 2nx}{4n^2}\right)_0^\pi\right\}\right] \\ &= \frac{2}{4\pi n^2}[1-1] = 0 \end{aligned}$$

$$\begin{aligned} b_n &= \frac{2}{\pi}\int_0^\pi f(x)\sin 2nx dx = \frac{2}{\pi}\int_0^\pi (\pi - x)\sin 2nx dx \\ &= \frac{2}{\pi}\left[\left(-\frac{\cos 2nx}{2n}\right)_0^\pi - \left\{\left(-\frac{x\cos 2nx}{2n}\right)_0^\pi + \left(\frac{\sin 2nx}{4n^2}\right)_0^\pi\right\}\right] \\ &= \frac{2}{\pi}\left(\frac{\pi}{2n}\right) = \frac{1}{n} \end{aligned}$$

$$\begin{aligned} \therefore\ \pi - x &= \frac{\pi}{2} + \sum_1^\infty \frac{\sin 2nx}{n} \\ &= \frac{\pi}{2} + \left\{\frac{\sin 2x}{1} + \frac{\sin 4x}{2} + \frac{\sin 6x}{3} + \cdots\right\} \qquad (2) \end{aligned}$$

Note. It is to be noted that the sine series for $f(x) = \pi - x$, in $[0, \pi]$ with period 2π and the sine series in $[0, \pi]$ with period π, obtained in (1) and (2) respectively are different. The interval $[0, \pi]$ with period π is treated as a complete interval of the form $[a, b]$.

Problem [11.5] : If $f(x) = \begin{cases} \frac{a}{4} - x, & \text{for } 0 \le x \le \frac{a}{2} \\ x - \frac{3}{4}a, & \text{for } \frac{a}{2} < x \le a \end{cases}$

then show that $\forall\, x \in [0, a]$

$$f(x) = \frac{2a}{\pi^2} \sum_{n=1}^{\infty} \frac{1}{(2n-1)^2} \cos \frac{(4n-2)}{a} \pi x \text{ so that } f(x+a) = f(x).$$

Solution. Here f satisfies all the Dirichlet's conditions on $[0, a]$. f is periodic with period a. The Fourier Series corresponding f on $[0, a]$ is of the form $\frac{a_0}{2} + \sum\limits_{1}^{\infty} \{a_n \cos \frac{2n\pi x}{a} + b_n \sin \frac{2n\pi x}{a}\}$, where

$$\begin{aligned} a_0 &= \frac{2}{a} \int_0^a f(x)dx \\ &= \frac{2}{a} \left[\int_0^{a/2} \left(\frac{a}{4} - x\right) dx + \int_{a/2}^a \left(x - \frac{3a}{4}\right) dx \right] \\ &= \frac{2}{a} \left[\int_0^{a/2} \left(\frac{a}{4} - x\right) dx + \int_0^{a/2} \left(\frac{a}{4} - x\right) dx \right] \end{aligned}$$

$$\left[\begin{array}{l} \text{Putting } x = a - z \text{ in the 2nd part, we get} \\ \displaystyle\int_{a/2}^a \left(x - \frac{3a}{4}\right) dx = \int_0^{a/2} \left(\frac{a}{4} - z\right) dz = \int_0^{a/2} \left(\frac{a}{4} - x\right) dx \end{array} \right]$$

$$\begin{aligned} &= \frac{2}{a} \cdot 2 \cdot \int_0^{a/2} \left(\frac{a}{4} - x\right) dx = \frac{4}{a} \left[\frac{a}{4} \cdot x - \frac{x^2}{2} \right]_0^{a/2} = \frac{4}{a} \left[\frac{a^2}{8} - \frac{a^2}{8} \right] = 0 \\ a_n &= \frac{2}{a} \int_0^a f(x) \cos \frac{2n\pi x}{a} dx \\ &= \frac{2}{a} \left[\int_0^{a/2} \left(\frac{a}{4} - x\right) \cos \frac{2n\pi x}{a} dx + \int_{a/2}^a \left(x - \frac{3a}{4}\right) \cos \frac{2n\pi x}{a} dx \right] \end{aligned}$$

Putting $x = a - z$ in the 2nd part, we get

$$\begin{aligned} a_n &= \frac{2}{a} \left[\int_0^{a/2} \left(\frac{a}{4} - x\right) \cos \frac{2n\pi x}{a} + \int_0^{a/2} \left(\frac{a}{4} - x\right) \cos \frac{2n\pi x}{a} dx \right] \\ &= \frac{4}{a} \left[\int_0^{a/2} \left(\frac{a}{4} - x\right) \cos \frac{2n\pi x}{a} dx \right] \\ &= \frac{4}{a} \left[\frac{a}{4} \left(\frac{\sin \frac{2n\pi x}{a}}{\frac{2n\pi}{a}} \right)_0^{a/2} - \left\{ \left(\frac{x \sin \frac{2n\pi x}{a}}{\frac{2n\pi}{a}} \right)_0^{a/2} + \left(\frac{\cos \frac{2n\pi x}{a}}{\left(\frac{2n\pi}{a}\right)^2} \right)_0^{a/2} \right\} \right] \\ &= \frac{4}{a} \left[\frac{a^2}{8n\pi} \cdot 0 - 0 - \frac{a^2}{4n^2\pi^2} \{(-1)^n - 1\} \right] \\ &= \frac{a}{n^2\pi^2} \{1 - (-1)^n\}. \end{aligned}$$

$a_n = 0$, when n is even.

When n is odd, say $2m-1$, then

$$a_n = \frac{a \cdot 2}{(2m-1)^2\pi^2} = \frac{2a}{(2m-1)^2\pi^2}$$

$$b_n = \frac{2}{a}\int_0^a f(x)\sin\frac{2n\pi x}{a}dx$$

$$= \frac{2}{a}\left[\int_0^{a/2}\left(\frac{a}{4}-x\right)\sin\frac{2n\pi x}{a}dx + \int_{a/2}^{a}\left(x-\frac{3a}{4}\right)\sin\frac{2n\pi x}{a}dx\right]$$

$$= \frac{2}{a}\left[\int_0^{a/2}\left(\frac{a}{4}-x\right)\sin\frac{2n\pi x}{a}dx - \int_0^{a/2}\left(\frac{a}{4}-x\right)\sin\frac{2n\pi x}{a}dx\right] = 0$$

Therefore, the F.S. is

$$\sum_{m=1}^{\infty}\frac{2a}{(2m-1)^2\pi^2}\cos\frac{(4m-2)\pi x}{a} = \frac{2a}{\pi^2}\sum_{m=1}^{\infty}\frac{1}{(2m-1)^2}\cos\frac{(4m-2)\pi x}{a}$$

Problem [11.6] :

(1) Prove that the even function $f(x) = |x|$ on $-\pi < x < \pi$ has a cosine F.S. of the form $\frac{\pi}{2} - \frac{4}{\pi}\left\{\cos x + \frac{\cos 3x}{3^2} + \frac{\cos 5x}{5^2} + \cdots\right\}$.
Apply Dirichlet's conditions of convergence to show that the series converges to $|x|\ \forall\, x \in [-\pi, \pi]$.
Also show that $\frac{\pi^2}{8} = 1 + \frac{1}{3^2} + \frac{1}{5^2} + \cdots$.

(2) Find the F.S. which represents $|\sin x|$ in $-\pi \le x \le \pi$.

(3) Show that for $f(x) = \cos kx$ on $[-\pi, \pi]$, where k is not an integer,

$$\cos kx = \frac{\sin k\pi}{\pi}\left[\frac{1}{k} - \frac{2k\cos x}{k^2-1^2} + \frac{2k\cos 2x}{k^2-2^2} + \cdots\right]$$

Deduce that

$$\pi\cot k\pi = \frac{1}{k} + 2k\sum_{1}^{\infty}\frac{1}{k^2-n^2} \text{ and } \frac{\pi}{\sin k\pi} = \sum(-1)^n\left(\frac{1}{n+k} + \frac{1}{n+1-k}\right).$$

(4) Find the F.S. of $f(x)$ with period 2π, where $f(x) = 0$, $-\pi < x < a$, $f(x) = 1$, $a \le x \le b$, $f(x) = 0, b < x < \pi$. Find the sum of the series for $x = 4\pi + a$ and deduce that $\sum_{1}^{\infty}\frac{\sin n(b-a)}{n} = \frac{\pi-b+a}{2}$.

(5) Let $f(x) = \begin{cases} 1, & -\frac{\pi}{2} \le x \le \frac{\pi}{2} \\ 0, & x \in \left(-\pi, -\frac{\pi}{2}\right) \cup \left(\frac{\pi}{2}, \pi\right) \end{cases}$
and $f(x+2\pi) = f(x)\ \forall\, x$. Find the F.S. of $f(x)$ and examine the convergence of the series in $\left(-\frac{\pi}{2}, \frac{\pi}{2}\right)$ and at the end point $x = \frac{\pi}{2}$.

(6) Show that $\log\left|\sin\frac{x}{2}\right| = -\log 2 - \sum_{n=1}^{\infty}\frac{\cos nx}{n}$, if $x \ne 2k\pi$ for any integer k.

(7) Let $f(x) = \begin{cases} 0, & \text{for } -\pi \le x < 0 \\ 1, & \text{for } 0 \le x \le \pi \end{cases}$
and $f(x+2\pi) = f(x)\ \forall\, x \in [-\pi, \pi]$. Show that the F.S. is $\frac{1}{2} + \frac{2}{\pi}\left[\frac{\sin x}{1} + \frac{\sin 3x}{3} + \frac{\sin 5x}{5} + \cdots\right]$ which does converge to $\frac{1}{2}$ at $x = \pi$.
Also show that $\frac{\pi}{4} = 1 - \frac{1}{3} + \frac{1}{5} - \frac{1}{7} + \cdots$.

Solution. (1) We define $f(x) = |x|$ for $-\pi \le x \le \pi$ so that f remains bounded.

f is also integrable; piecewise monotonic and periodic with period 2π. So f satisfies all the Dirichlet's conditions on $[-\pi, \pi]$. Here $f(x)$ is an even function. Then the F.S. of f is of the form $\frac{a_0}{2} + \sum_1^{\infty} a_n \cos nx$, where

$$a_0 = \frac{2}{\pi}\int_0^{\pi} f(x)dx \text{ and } a_n = \frac{2}{\pi}\int_0^{\pi} f(x)\cos nx dx.$$

Here $f(x) = x,\ 0 \le x \le \pi$.

Therefore,

$$a_0 = \frac{2}{\pi}\int_0^{\pi} x dx = \pi$$

$$a_n = \frac{2}{\pi}\int_0^{\pi} x\cos nx dx = \frac{2}{\pi}\left[\left(\frac{x\sin nx}{n}\right)_0^{\pi} + \left(\frac{\cos nx}{n^2}\right)_0^{\pi}\right] = \frac{2}{\pi}\left[\frac{(-1)^n - 1}{n^2}\right]$$

Then the F.S. of $f(x)$ is

$$\frac{\pi}{2} + \frac{2}{\pi}\sum_{n=1}^{\infty}\frac{(-1)^n - 1}{n^2}\cos nx = \frac{\pi}{2} - \frac{4}{\pi}\left[\frac{\cos x}{1^2} + \frac{\cos 3x}{3^2} + \frac{\cos 5x}{5^2} + \cdots\right] \tag{1}$$

The function f is continuous at $x = 0$.

$$S(0) = f(0) = 0$$

Therefore, from (1) for $x = 0$,

$$S(0) = \frac{\pi}{2} - \frac{4}{\pi}\left[\frac{1}{1^2} + \frac{1}{3^2} + \frac{1}{5^2} + \cdots\right] \text{ or, } \frac{\pi^2}{8} = \frac{1}{1^2} + \frac{1}{3^2} + \frac{1}{5^2} + \cdots$$

$f(x) = |x|$ for $-\pi \le x \le \pi$.

f is continuous in $[-\pi, \pi]$. So if c be any point in $[-\pi, \pi]$, then $S(c) = f(c)$.

$\Rightarrow$ the series converges to $f(x) = |x|$ for all $x \in [-\pi, \pi]$.

(2) $f(x) = |\sin x|,\ -\pi \le x \le \pi$.

f is an even function. f satisfies all the Dirichlet's conditions on $[-\pi, \pi]$. So the F.S. of $f(x)$ is of the form $\frac{a_0}{2} + \sum a_n \cos nx$

$$a_0 = \frac{2}{\pi}\int_0^{\pi} f(x)dx = \frac{2}{\pi}\int_0^{\pi}\sin x dx = \frac{4}{\pi}$$

$$\begin{aligned}
a_n &= \frac{2}{\pi}\int_0^{\pi}\sin x\cos nx dx = \frac{1}{\pi}\int_0^{\pi} 2\cos nx\sin x dx \\
&= \frac{1}{\pi}\int_0^{\pi}\{\sin(n+1)x - \sin(n-1)x\}dx \\
&= \frac{1}{\pi}\left[-\frac{\cos(n+1)x}{n+1} + \frac{\cos(n-1)x}{n-1}\right]_0^{\pi} \\
&= \frac{1}{\pi}\left[(-1)^n\left\{\frac{1}{n+1} - \frac{1}{n-1}\right\} + \left\{\frac{1}{n+1} - \frac{1}{n-1}\right\}\right] \\
&= \frac{1}{\pi}\left[\{(-1)^n + 1\}\frac{-2}{n^2 - 1}\right] = -\frac{2}{\pi}\frac{\{(-1)^n + 1\}}{n^2 - 1} \text{ for } n \ne 1
\end{aligned}$$

$$\text{So } a_1 = \frac{2}{\pi}\int_0^{\pi} \sin x \cos x dx = \frac{1}{\pi}\left[-\frac{\cos 2x}{2}\right]_0^{\pi} = 0.$$

$$\text{Therefore, the F.S. of } f(x) = \frac{1}{2}\cdot\frac{4}{\pi} - \frac{2}{\pi}\sum_{n=2}^{\infty}\frac{(-1)^n+1}{n^2-1}\cos nx$$

$$= \frac{2}{\pi} - \frac{2}{\pi}\sum_{m=1}^{\infty}\frac{2}{(2m)^2-1}\cos 2mx,$$

since for $n = 2m-3$, $a_n = 0$, for $m = 2, 3, \ldots$.

(3) $f(x) = \cos kx$ in $-\pi \le x \le \pi$, $k \notin Z$.

f is bounded, integrable and periodic with period 2π. It satisfies all the Dirichlet's conditions on $[-\pi, \pi]$.

f is an even function. Therefore, the F.S. of $f(x)$ represents a cosine series of the form $\frac{a_0}{2} + \sum\limits_1^{\infty} a_n \cos nx$, where

$$a_0 = \frac{2}{\pi}\int_0^{\pi} f(x)dx = \frac{2}{\pi}\int_0^{\pi}\cos kx = \frac{2}{\pi}\left[\frac{\sin kx}{k}\right]_0^{\pi} = \frac{2}{k\pi}\sin k\pi$$

$$a_n = \frac{2}{\pi}\int_0^{\pi} f(x)\cos nx dx = \frac{2}{\pi}\int_0^{\pi}\cos kx \cos nx dx = \frac{1}{\pi}\int_0^{\pi}[\cos(n+k)x + \cos(n-k)x]dx$$

$$= \frac{1}{\pi}\left[\frac{\sin(n+k)\pi}{n+k} + \frac{\sin(n-k)\pi}{n-k}\right] = \frac{1}{\pi}\left[\frac{(-1)^n \sin k\pi}{n+k} + \frac{(-1)^{n+1}\sin k\pi}{n-k}\right]$$

$$= \frac{(-1)^n \sin k\pi}{\pi}\left[\frac{1}{n+k} - \frac{1}{n-k}\right] = \frac{(-1)^{n+1}2k\sin k\pi}{\pi(n^2-k^2)} = \frac{2(-1)^n k \sin k\pi}{\pi(k^2-n^2)}$$

Therefore, F.S. of $f(x)$ is $\frac{a_0}{2} + \sum\limits_1^{\infty} a_n \cos nx$.

$$\therefore \cos kx = \frac{1}{2}\cdot\frac{2}{k\pi}\sin k\pi + \sum_1^{\infty}\frac{2k(-1)^n \sin k\pi}{\pi(k^2-n^2)}\cos nx$$

$$= \frac{1}{k\pi}\sin k\pi + \frac{2k}{\pi}\sin k\pi\sum_1^{\infty}\frac{(-1)^n\cdot\cos nx}{k^2-n^2}$$

$$= \frac{1}{k\pi}\sin k\pi + \frac{2k\sin k\pi}{\pi}\left\{-\frac{\cos x}{k^2-1^2} + \frac{\cos 2x}{k^2-2^2} - \frac{\cos 3x}{k^2-3^2} + \cdots\right\}$$

$$= \frac{\sin k\pi}{\pi}\left\{\frac{1}{k} - \frac{2k\cos x}{k^2-1^2} + \frac{2k\cos 2x}{k^2-2^2} - \frac{2k\cos 3x}{k^2-3^2} + \cdots\right\} \quad (1)$$

f is continuous at $x = \pi$. So it converges to $f(x)$ at $x = \pi$

$$\therefore S(\pi) = f(\pi) = \cos k\pi$$

$$\therefore \cos k\pi = \frac{\sin k\pi}{\pi}\left\{\frac{1}{k} + \frac{2k}{k^2-1^2} + \frac{2k}{k^2-2^2} + \frac{2k}{k^2-3^2} + \cdots\right\}$$

$$\text{or } \pi\cot k\pi = \frac{1}{k} + 2k\sum_1^{\infty}\frac{1}{k^2-n^2}$$

f is also continuous at $x = 0$. So it converges to $f(x)$ at $x = 0$.

$$S(0) = f(0) = 1$$

$$\therefore 1 = \frac{\sin k\pi}{\pi}\left\{\frac{1}{k} - \frac{2k}{k^2 - 1^2} + \frac{2k}{k^2 - 2^2} - \frac{2k}{k^2 - 3^2} + \cdots\right\} \text{ [From (1)]}$$

$$\text{or, } \frac{\pi}{\sin k\pi} = \left\{\frac{1}{k} + \frac{2k}{1^2 - k^2} - \frac{2k}{2^2 - k^2} + \frac{2k}{3^2 - k^2} + \cdots\right\}$$

$$= \left\{\frac{1}{k} + \left(\frac{1}{1-k} - \frac{1}{1+k}\right) + \left(\frac{1}{2+k} - \frac{1}{2-k}\right) + \left(\frac{1}{3-k} - \frac{1}{3+k}\right) + \cdots\right\}$$

$$= \left\{\left(\frac{1}{k} + \frac{1}{1-k}\right) - \left(\frac{1}{1+k} + \frac{1}{2-k}\right) + \left(\frac{1}{2+k} + \frac{1}{3-k}\right) + \cdots\right\}$$

$$= \sum_{0}^{\infty}(-1)^n\left(\frac{1}{n+k} + \frac{1}{n+1-k}\right)$$

(4) We define $f(x) = 0, -\pi \le x < a, f(x) = 1, a \le x \le b, f(x) = 0, b < x \le \pi$.

Then f satisfies all the conditions of Dirichlet.

Let the F.S. of $f(x)$ be $\frac{a_0}{2} + \sum_{1}^{\infty}(a_n \cos nx + b_n \sin nx)$ on $[-\pi, \pi]$, where

$$a_0 = \frac{1}{\pi}\int_{-\pi}^{\pi} f(x)dx = \frac{1}{\pi}\left[\int_{-\pi}^{a} 0 \cdot dx + \int_{a}^{b} 1 \cdot dx + \int_{b}^{\pi} 0 \cdot dx\right] = \frac{b-a}{\pi}$$

$$a_n = \frac{1}{\pi}\int_{-\pi}^{\pi} f(x)\cos nx dx = \frac{1}{\pi}\left[\int_{-\pi}^{a} 0 \cdot dx + \int_{a}^{b} 1 \cdot \cos nx dx + \int_{b}^{\pi} 0 \cdot \cos nx dx\right]$$

$$= \frac{1}{\pi}\left[\frac{\sin nb - \sin na}{n}\right]$$

$$b_n = \int_{a}^{b} \sin nx dx = \left[\frac{\cos na - \cos nb}{n}\right]$$

Therefore,

$$f(x) = \frac{b-a}{2\pi} + \frac{1}{\pi}\sum_{1}^{\infty}\left\{\frac{\sin nb - \sin na}{n} \cdot \cos nx + \frac{\cos na - \cos nb}{n} \cdot \sin nx\right\}$$

$$= \frac{b-a}{2\pi} + \frac{1}{\pi}\sum_{1}^{\infty}\frac{\sin(nb - nx) - \sin(na - nx)}{n} \quad (1)$$

f is periodic with period 2π

$$f(x) = f(x + 2\pi) = f(x + 4\pi) \Rightarrow f(a) = f(a + 2\pi) = f(a + 4\pi)$$

$\therefore$ at $x = a$, f is also periodic with period 2π.

So the F.S. of $f(x)$ converges to $\frac{1}{2}[f(a+0) + f(a-0)] = \frac{1}{2}[1 + 0] = \frac{1}{2}$.

$\therefore S(a) = \frac{1}{2}$. Also $S(4\pi + a) = S(a) = \frac{1}{2}$.

From (1),

$$\frac{1}{2} = \frac{b-a}{2\pi} + \frac{1}{\pi}\sum_{1}^{\infty}\frac{\sin(nb - na)}{n} \text{ or, } \frac{\pi - b + a}{2\pi} = \frac{1}{\pi}\sum_{1}^{\infty}\frac{\sin n(b-a)}{n}.$$

(5) $f(x)=\begin{cases}0, & \text{for } -\pi<x<-\frac{\pi}{2}\\ 1, & \text{for } -\frac{\pi}{2}\le x\le\frac{\pi}{2}\\ 0, & \text{for } \frac{\pi}{2}<x<\pi\end{cases}$

At $x=\pm\pi$, f is not defined. We define f at $x=\pm\pi$ by $f(x)=0$ so that f remains bounded on $[-\pi,\pi]$. f has finite points of infinite discontinuity at $x=-\frac{\pi}{2}$ and $x=\frac{\pi}{2}$. So f is bounded, integrable and priodic with period 2π. Thus f satisfies all the Dirichlet's conditions.

Let the F.S. of f on $[-\pi,\pi]$ be $\frac{a_0}{2}+\sum(a_n\cos nx+b_n\sin nx)$, where

$$a_0=\frac{1}{\pi}\int_{-\pi/2}^{\pi/2}1\cdot dx=\frac{\pi}{\pi}=1$$

$$a_n=\frac{1}{\pi}\int_{-\pi/2}^{\pi/2}1\cdot\cos nx dx=\frac{1}{\pi}\left[\frac{2\sin\frac{n\pi}{2}}{n}\right]=\frac{2}{\pi n}\sin\frac{n\pi}{2}$$

$$b_n=\frac{1}{\pi}\int_{-\pi/2}^{\pi/2}\sin nx dx=0 \text{ [}\sin nx \text{ is an odd function]}$$

$$\therefore\ f(x)=\frac{1}{2}+\frac{2}{\pi}\sum_1^\infty\frac{1}{n}\sin\frac{n\pi}{2}\cos nx$$

When n is even, $\sin\frac{n\pi}{2}=0$.

When n is odd, $\sin\frac{n\pi}{2}=\pm1$. So taking $n=2m-1$,

$$f(x)=\frac{1}{2}+\frac{2}{\pi}\sum_1^\infty\frac{1}{(2m-1)}\sin(2m-1)\frac{\pi}{2}\cos(2m-1)x$$

$$=\frac{1}{2}+\frac{2}{\pi}\left[\frac{\cos x}{1}-\frac{1}{3}\cos 3x+\frac{1}{5}\cos 5x-\cdots\right]$$

f is continuous in $\left(-\frac{\pi}{2},\frac{\pi}{2}\right)$. So it converges in $\left(-\frac{\pi}{2},\frac{\pi}{2}\right)$ to 1.

For $x=0$, $S(0)=f(0)=1$.

$$1=\frac{1}{2}+\frac{2}{\pi}\left[1-\frac{1}{3}+\frac{1}{5}-\frac{1}{7}+\cdots\right] \text{ or, } \frac{\pi}{4}=1-\frac{1}{3}+\frac{1}{5}-\frac{1}{7}+\cdots$$

At $x=\frac{\pi}{2}$, f converges to

$$\frac{1}{2}\left[f\left(\frac{\pi}{2}+0\right)+f\left(\frac{\pi}{2}-0\right)\right]=\frac{1}{2}[0+1]=\frac{1}{2}$$

$\therefore\ S\left(\frac{\pi}{2}\right)=\frac{1}{2}$, which can be verified directly from the series putting $x=\frac{\pi}{2}$.

(6) $f(x)=\log\left|\sin\frac{x}{2}\right|, x\neq 2k\pi$ for any $k\in Z$.

So f is bounded, integrable and periodic with period 2π in $[-\pi,\pi]$, since $2k\pi\notin[-\pi,\pi]$ for any integer k.

So f satisfies all the Dirichlet's conditions. Also f is an even function. So the Fourier Series of $f(x)$ is of the form $\frac{a_0}{2}+\sum_1^\infty a_n\cos nx$, where

$$a_0=\frac{2}{\pi}\int_0^\pi\log\left|\sin\frac{x}{2}\right|dx=\frac{4}{\pi}\int_0^{\pi/2}\log\sin xdx \left[\text{Putting } \frac{x}{2}=z\right]$$

$$=\frac{4}{\pi}\cdot\frac{\pi}{2}\log\frac{1}{2}=2\log\frac{1}{2}$$

$$a_n = \frac{2}{\pi}\int_0^{\pi} \log\left|\sin\frac{x}{2}\right| \cos nx dx$$

$$= \frac{4}{\pi}\int_0^{\pi/2} \log\sin x \cos 2nx dx \quad \left[\text{Putting } \frac{x}{2} = z\right]$$

$$= \frac{4}{\pi}\left[\left(\frac{\sin 2nx}{2n}\log\sin x\right)_0^{\pi/2} - \int_0^{\pi/2}\frac{\sin 2nx \cdot \cos x}{2n\sin x}dx\right]$$

$$= -\frac{2}{n\pi}\int_0^{\pi/2}\frac{\sin 2nx \cos x}{\sin x}dx \tag{1}$$

Now

$$\int_0^{\pi/2}\frac{\sin(2n+1)x}{\sin x}dx = \int_0^{\pi/2}\frac{\sin 2nx\cos x}{\sin x}dx + \int_0^{\pi/2}\cos 2nx dx$$

$$\left[\begin{array}{l} \because \sin(2x+1)x = \sin 2nx\cos x + \cos 2nx\sin x \\ \text{or, } \dfrac{\sin(2n+1)x}{\sin x} = \dfrac{\sin 2nx\cos x}{\sin x} + \cos 2nx \end{array}\right]$$

Thus

$$\int_0^{\pi/2}\frac{\sin(2n+1)x}{\sin x}dx = \int_0^{\pi/2}\frac{\sin 2nx\cos x}{\sin x}dx \quad \left[\because \int_0^{\pi/2}\cos 2nx dx = 0\right]$$

Also

$$\int_0^{\pi/2}\frac{\sin(2n-1)dx}{\sin x} = \int_0^{\pi/2}\frac{\sin 2nx\cos x}{\sin x}dx$$

$$\therefore \int_0^{\pi/2}\frac{\sin(2n+1)x}{\sin x}dx = \int_0^{\pi/2}\frac{\sin(2n-1)x}{\sin x} = \cdots = \int_0^{\pi/2}\frac{\sin x}{\sin x}dx = \frac{\pi}{2}$$

Hence, $\displaystyle\int_0^{\pi/2}\frac{\sin 2nx\cos x}{\sin x}dx = \frac{\pi}{2}$.

Therefore, from (1), $a_n = -\frac{2}{n\pi}\cdot\frac{\pi}{2} = -\frac{1}{n}$.

Hence, the Fourier Series of $f(x) = \log\left|\sin\frac{x}{2}\right|$

$$= \frac{1}{2}\cdot 2\log\frac{1}{2} - \sum_1^{\infty}\frac{1}{n}\cos nx = -\log 2 - \sum_1^{\infty}\frac{\cos nx}{n}.$$

(7) f has a point of discontinuity at $x = 0$ only. So f is integrable on $[-\pi, \pi]$. f is also bounded and periodic with period 2π.

Thus f satisfies all the Dirichlet's conditions on $[-\pi, \pi]$.

Let the F.S. of $f(x)$ be $\frac{a_0}{2} + \sum(a_n\cos nx + b_n\sin nx)$, where

$$a_0 = \frac{1}{\pi}\int_{-\pi}^{\pi} f(x)dx = \frac{1}{\pi}\left[\int_{-\pi}^{0} 0\cdot dx + \int_0^{\pi} 1\cdot dx\right] = 1$$

$$a_n = \frac{1}{\pi}\int_{-\pi}^{\pi} f(x)\cos nx dx = \frac{1}{\pi}\int_0^{\pi}\cos nx dx = \frac{1}{\pi}\left[\frac{\sin nx}{n}\right]_0^{\pi} = 0$$

$$b_n = \frac{1}{\pi}\int_0^{\pi}\sin nx dx = \frac{1}{\pi}\left[-\frac{\cos nx}{n}\right]_0^{\pi} = \frac{1}{n\pi}[1-(-1)^n]$$

So the F.S. of $f(x)$

$$= \frac{1}{2} + \frac{1}{\pi}\sum_{1}^{\infty} \frac{\{1-(-1)^n\}}{n} \sin nx$$

$$= \frac{1}{2} + \frac{2}{\pi}\left[\sin x + \frac{\sin 3x}{3} + \frac{\sin 5x}{5} + \cdots\right] \quad [\because \text{ even term vanishes}]$$

f is continuous at $x = \frac{\pi}{2}$. So it converges to $f(x)$ at $x = \frac{\pi}{2}$.
Therefore, $S\left(\frac{\pi}{2}\right) = f\left(\frac{\pi}{2}\right) = 1$

$$\therefore 1 = \frac{1}{2} + \frac{2}{\pi}\left[1 - \frac{1}{3} + \frac{1}{5} - \frac{1}{7} - \cdots\right] \text{ or, } \frac{\pi}{4} = 1 - \frac{1}{3} + \frac{1}{5} - \frac{1}{7} + \cdots.$$

Theorem [11.7] [Parseval's Formula] If f be bounded integrable and periodic with 2π in $[-\pi, \pi]$, then $\displaystyle\int_{-\pi}^{\pi} [f(x)]^2 dx = \pi\left[\frac{a_0^2}{2} + \sum_{1}^{\infty}(a_n^2 + b_n^2)\right]$.

Proof. Let f satisfy all the Dirichlet's conditions in $[-\pi, \pi]$.

Also let f be continuous on $[-\pi, \pi]$. Then the Fourier Series $\frac{a_0}{2} + \sum\limits_{1}^{\infty}(a_n \cos nx + b_n \sin nx)$ of f has the sum $f(x)\ \forall\ x \in [-\pi, \pi]$.

So, $f(x) = \frac{a_0}{2} + \sum\limits_{1}^{\infty}(a_n \cos nx + b_n \sin nx)$.

Now multiplying by f and integrating term by term in $[-\pi, \pi]$, we get

$$\int_{-\pi}^{\pi} [f(x)]^2 dx = \frac{a_0}{2}\int_{-\pi}^{\pi} f(x)dx + \sum_{1}^{\infty} a_n \int_{-\pi}^{\pi} f(x)\cos nx dx + \sum_{1}^{\infty} b_n \int_{-\pi}^{\pi} f(x)\sin nx dx$$

$$= \frac{a_0}{2}\cdot \pi a_0 + \sum_{1}^{\infty} a_n \cdot \pi a_n + \sum_{1}^{\infty} b_n \cdot \pi b_n$$

[Since the series being a F.S. a_0, a_n, b_n are Fourier coefficients.]

$$= \pi\left[\frac{a_0^2}{2} + \sum_{1}^{\infty}(a_n^2 + b_n^2)\right],$$

which is known as Paseval's formula.

Note. (1) In $0 < x < \pi$, (Half-range with cosine series), the Parseval's formula stands as

$$\int_{0}^{\pi} [f(x)]^2 dx = \frac{\pi}{2}\left[\frac{a_0^2}{2} + \sum_{1}^{\infty} a_n^2\right]$$

(2) In $0 < x < \pi$ (Half-range with sine series), the Parseval's formula stands as

$$\int_{0}^{\pi} [f(x)]^2 dx = \frac{\pi}{2}\sum_{1}^{\infty} b_n^2.$$

(3) In $-l < x < l$, the Parseval's formula is $\displaystyle\int_{-l}^{l} [f(x)]^2 dx = l\left[\frac{a_0^2}{2} + \sum(a_n^2 + b_n^2)\right]$.

(4) In $0 < x < l$, Parseval's formula is

$$\int_{0}^{l} [f(x)]^2 dx = \frac{l}{2}\left[\frac{a_0^2}{2} + \sum_{1}^{\infty} a_n^2\right] \text{ or, } \int_{0}^{l} [f(x)]^2 dx = \frac{l}{2}\sum b_n^2.$$

Problem [11.7]: Show that for $0 < x < \pi$,

(i) $x(\pi - x) = \dfrac{\pi^2}{6} - 4\sum_1^\infty \dfrac{\cos 2mx}{(2m)^2}$, (ii) $x(\pi - x) = \dfrac{8}{\pi}\sum_1^\infty \dfrac{\sin(2m-1)x}{(2m-1)^3}$.

Deduce from (i) and (ii) respectively that

(iii) $\sum_1^\infty \dfrac{1}{n^4} = \dfrac{\pi^4}{90}$, (iv) $\sum_1^\infty \dfrac{1}{n^6} = \dfrac{\pi^6}{945}$.

Solution. (i) We define, $f(0) = 0 = f(\pi)$. So f is continuous $[0, \pi]$.

In the half-range $0 < x < \pi$, cosine series is $\frac{a_0}{2} + \sum a_n \cos nx$, where

$$a_0 = \frac{2}{\pi}\int_0^\pi f(x)dx = \frac{2}{\pi}\int_0^\pi x(\pi - x)dx = \frac{2}{\pi}\left[\frac{\pi^3}{2} - \frac{\pi^3}{3}\right] = \frac{\pi^2}{3} \quad (1)$$

$$\begin{aligned}
a_n &= \frac{2}{\pi}\int_0^\pi f(x)\cos nx dx = \frac{2}{\pi}\int_0^\pi x(\pi - x)\cos nx dx \\
&= \frac{2}{\pi}\left[\pi\int_0^\pi x\cos nx dx - \int_0^\pi x^2 \cos nx dx\right] \\
&= \frac{2}{\pi}\left[\pi\left(\frac{x\sin nx}{n} + \frac{\cos nx}{n^2}\right)_0^\pi - \left\{\left(\frac{x^2 \sin nx}{n}\right)_0^\pi - \int_0^\pi 2x\frac{\sin nx}{n}dx\right\}\right] \\
&= \frac{2}{n^2}\{(-1)^n - 1\} + \frac{2}{\pi}\left\{-\frac{2x\cos nx}{n^2} + \frac{2\sin nx}{n^3}\right\}_0^\pi = \frac{2}{n^2}\{(-1)^n - 1\} - \frac{4}{n^2}(-1)^n \\
&= -\frac{4}{n^2}, \text{ when } n \text{ is even} \\
&= 0, \text{ when } n \text{ is odd}
\end{aligned} \quad (2)$$

Hence $x(\pi - x) = \dfrac{\pi^2}{6} - 4\sum_1^\infty \dfrac{\cos 2mx}{(2m)^2}$.

(ii) In half-range $[0, \pi]$, sine series is of the form $\sum_1^\infty b_n \sin nx$, where

$$\begin{aligned}
b_n &= \frac{2}{\pi}\int_0^\pi x(\pi - x)\sin nx dx \\
&= \frac{2}{\pi}\left[\pi\left(-\frac{x\cos nx}{n} + \frac{\sin nx}{n^2}\right)_0^\pi - \int_0^\pi x^2 \sin nx dx\right] \\
&= -\frac{2\pi(-1)^n}{n} - \frac{2}{\pi}\left[\left(-\frac{x^2\cos nx}{n}\right)_0^\pi + \int_0^\pi 2x\frac{\cos nx}{n}dx\right] \\
&= (-1)^{n+1}\frac{2\pi}{n} + \frac{2\pi}{n}(-1)^n - \frac{2}{\pi}\left\{\frac{2x\sin nx}{n^2} + \frac{2\cos nx}{n^3}\right\}_0^\pi \\
&= -\frac{4}{\pi n^3}\{(-1)^n - 1\} \\
&= \frac{8}{\pi n^3}, \text{ when } n \text{ is odd} \\
&= 0, \text{ when } n \text{ is even.}
\end{aligned}$$

Thus, $x(\pi - x) = \dfrac{8}{\pi}\sum_1^\infty \dfrac{\sin(2m-1)x}{(2m-1)3}$.

(iii) In $0 < x < \pi$, $\int_0^\pi [f(x)]^2 dx = \frac{\pi}{2}\left[\frac{a_0^2}{2} + \sum a_n^2\right]$

$$\Rightarrow \int_0^\pi \{x(\pi - x)\}^2 dx = \frac{\pi}{2}\left[\frac{\pi^4}{18} + 16\sum \frac{1}{(2m)^4}\right] \text{ [by (1) and (2)]}$$

$$\Rightarrow \int_0^\pi x^2\{\pi^2 - 2\pi x + x^2\}\, dx = \frac{\pi^5}{36} + 8\pi \sum_1^\infty \frac{1}{(2m)^4}$$

$$\Rightarrow \left[\pi^2 \cdot \frac{\pi^3}{3} - 2\pi\frac{\pi^4}{4} + \frac{\pi^5}{5}\right] = \frac{\pi^5}{36} + \frac{8\pi}{2^4}\sum_1^\infty \frac{1}{m^4}$$

$$\Rightarrow \pi^5\left[\frac{1}{3} - \frac{1}{2} + \frac{1}{5}\right] = \frac{\pi^5}{36} + \frac{8\pi}{16}\sum_1^\infty \frac{1}{m^4}$$

$$\Rightarrow \pi^5\left[\frac{10 - 15 + 6}{30}\right] = \frac{\pi^5}{36} + \frac{\pi}{2}\sum_1^\infty \frac{1}{m^4}$$

$$\Rightarrow \pi^5\left[\frac{1}{30} - \frac{1}{36}\right] = \frac{\pi}{2}\sum_1^\infty \frac{1}{m^4}$$

$$\Rightarrow \pi^5 \cdot \frac{1}{180} \times \frac{2}{\pi} = \sum_1^\infty \frac{1}{m^4} \Rightarrow \frac{\pi^4}{90} = \sum_1^\infty \frac{1}{m^4}$$

That is, $\sum_1^\infty \frac{1}{n^4} = \frac{\pi^4}{90}$.

(iv) $\int_0^\pi [f(x)]^2 dx = \frac{\pi}{2}\sum_1^\infty b_n^2 \Rightarrow \frac{\pi^5}{30} = \frac{\pi}{2}\sum_1^\infty \frac{64}{\pi^2(2m-1)^6}$ for $n = 2m - 1$

$$\Rightarrow \frac{\pi^4}{15} \times \frac{\pi^2}{64} = \sum_1^\infty \frac{1}{(2m-1)^6} = \frac{1}{1^6} + \frac{1}{3^6} + \frac{1}{5^6} + \cdots$$

$$\Rightarrow \frac{\pi^6}{960} = \frac{1}{1^6} + \frac{1}{3^6} + \frac{1}{5^6} + \cdots \tag{3}$$

Now let $S = \sum_1^\infty \frac{1}{n^6} = \left(\frac{1}{1^6} + \frac{1}{3^6} + \frac{1}{5^6} + \cdots\right) + \left(\frac{1}{2^6} + \frac{1}{4^6} + \frac{1}{6^6} + \cdots\right)$.

Then

$$S = \frac{\pi^6}{960} + \frac{1}{2^6}\left(\frac{1}{1^6} + \frac{1}{2^6} + \frac{1}{3^6} + \frac{1}{4^6} + \cdots\right) = \frac{\pi^6}{960} + \frac{1}{2^6} \cdot S = \frac{\pi^6}{960} + \frac{S}{64}$$

$$\Rightarrow S - \frac{S}{64} = \frac{\pi^6}{960} \Rightarrow \frac{63S}{64} = \frac{\pi^6}{960} \Rightarrow S = \frac{\pi^6}{960} \times \frac{64}{63} = \frac{\pi^6}{945} \Rightarrow \sum_1^\infty \frac{1}{n^6} = \frac{\pi^6}{945}$$

Problem [11.8]: Prove that in $0 < x < c$,

$$x = \frac{c}{2} - \frac{4c}{\pi^2}\left(\cos\frac{\pi x}{c} + \frac{1}{3^2}\cos\frac{3\pi x}{c} + \frac{1}{5^2}\cos\frac{5\pi^2}{c} + \cdots\right\}$$

and hence deduce that

(1) $\frac{1}{1^4} + \frac{1}{3^4} + \frac{1}{5^4} + \cdots = \frac{\pi^4}{96}$

(2) $\frac{1}{1^4} + \frac{1}{2^4} + \frac{1}{3^4} + \frac{1}{4^4} + \cdots = \frac{\pi^4}{90}$.

Solution. We consider the half-range $[0, c]$ with period $2c$. We define, $f(0) = 0$ and $f(c) = c$ so that f is bounded. Thus f is bounded integrable and with period $2c$. f satisfies all the Dirichlet's conditions in $[0, c]$: Let the F.S. of $f(x) = x$ represent a cosine series of the form $\frac{a_0}{2} + \sum_1^\infty a_n \cos\frac{n\pi x}{c}$, where

$$a_0 = \frac{2}{c}\int_0^c f(x)dx = \frac{2}{c}\int_0^c x dx = c$$

$$a_n = \frac{2}{c}\int_0^c f(x)\cos\frac{n\pi x}{c}dx = \frac{2}{c}\int_0^c x\cos\frac{n\pi x}{c}dx = \frac{2}{c}\left[\left(\frac{x\sin\frac{n\pi x}{c}}{\frac{n\pi}{c}}\right)_0^c + \left(\frac{\cos\frac{n\pi x}{c}}{\left(\frac{n\pi}{c}\right)^2}\right)_0^c\right]$$

$$= \frac{2}{c}\left[\frac{c^2}{n^2\pi^2}\{(-1)^n - 1\}\right] = \frac{2c}{\pi^2}\frac{\{(-1)^n - 1\}}{n^2}$$

Therefore, the F.S. of $f(x) = x$ is

$$\frac{c}{2} - \frac{2c}{\pi^2}\sum_1^\infty \frac{1-(-1)^n}{n^2}\cos\frac{n\pi x}{c}$$

$$= \frac{c}{2} - \frac{4c}{\pi^2}\left[\cos\frac{\pi x}{c} + \frac{1}{3^2}\cos\frac{3\pi x}{c} + \frac{1}{5^2}\cos\frac{5\pi x}{c} + \cdots\right] \quad [\because \text{ even term vanishes}]$$

Deductions (1) By Parseval's formula, $\int_0^c [f(x)]^2 dx = \frac{c}{2}\left[\frac{a_0^2}{2} + \sum_1^\infty a_n^2\right]$

$$\Rightarrow \int_0^c x^2 dx = \frac{c}{2}\left[\frac{c^2}{2} + \frac{4c^2}{\pi^4}\sum\frac{4}{n^4}\right], \ n \text{ is odd.}$$

$$\Rightarrow \frac{c^3}{3} = \frac{c^3}{4} + \frac{8c^3}{\pi^4}\sum\frac{1}{n^4}, \ n \text{ is odd.}$$

$$\Rightarrow \frac{c^3}{12}\cdot\frac{\pi^4}{8c^3} = \sum\frac{1}{n^4}, \ n \text{ is odd}$$

$$\Rightarrow \frac{\pi^4}{96} = \frac{1}{1^4} + \frac{1}{3^4} + \frac{1}{5^4} + \cdots$$

(2) Let $S = \sum_1^\infty \frac{1}{n^4} = \frac{1}{1^4} + \frac{1}{3^4} + \frac{1}{5^4} + \cdots + \frac{1}{2^4} + \frac{1}{4^4} + \frac{1}{6^4} + \cdots$

$$= \frac{\pi^4}{96} + \frac{1}{2^4}\left(\frac{1}{1^4} + \frac{1}{2^4} + \frac{1}{3^4} + \frac{1}{4^4} + \cdots\right)$$

$$\Rightarrow S = \frac{\pi^4}{96} + \frac{1}{16}\cdot S \Rightarrow \frac{15S}{16} = \frac{\pi^4}{96} \Rightarrow S = \frac{\pi^4}{90}$$

That is, $\sum\frac{1}{n^4} = \frac{\pi^4}{90}$.

Problem [11.9]: (1) Obtain a sine series which will be equal to x^2 for $0 \le x \le \pi$. Draw the graph of the sum function of the series for $-3\pi \le x \le 3\pi$.

(2) Obtain the Fourier cosine series for the periodic function f defined by $f(t) = \sin\left(\frac{\pi t}{l}\right)$, $0 < t < l$.

(3) If $f(x) = \begin{cases} -\frac{\pi}{4}, & \text{when } -\pi < x < 0 \\ \frac{\pi}{4}, & \text{when } 0 < x < \pi \end{cases}$

and $f(-\pi) = f(0) = f(\pi) = 0$ and $f(x+2\pi) = f(x)$, $\forall\, x \in [-\pi, \pi]$. Then show that

$$f(x) = \sin x + \tfrac{1}{3}\sin 3x + \tfrac{1}{5}\sin 5x + \cdots.$$

Deduce that $\frac{\pi}{4} = 1 - \frac{1}{3} + \frac{1}{5} - \frac{1}{7} + \cdots$.

(4) Show that for $0 < x < \pi$, $e^x = \frac{2}{\pi}\sum\limits_{n=1}^{\infty}\{1-(-1)^n e^\pi\}\frac{n\sin nx}{n^2+1}$. What are the sum of the series for $-\pi \le x \le 0$?

(5) Obtain F.S. for the function $f(x) = x\sin x$ in $(-\pi, \pi)$. Hence deduce that

$$\tfrac{\pi}{4} = \tfrac{1}{2} + \tfrac{1}{1\cdot 3} - \tfrac{1}{3\cdot 5} + \cdots.$$

(6) Let f be periodic with period 2π and let $S_n(f,x)$ be the nth partial sum of the F.S.

$$\tfrac{a_0}{2} + \sum_{n=1}^{\infty}(a_n \cos nt + b_n \sin nt)$$

generated by f in $[-\pi, \pi]$. Then $S_n(f,x) = \frac{1}{\pi}\int_{-\pi}^{\pi} f(t)D_n(t-x)dt$, where

$$D_n(t-x) = \tfrac{1}{2} + \sum_{k=1}^{n}\cos k(t-x).$$

(7) For any h, $0 < h < \pi$ define $\phi_h(x) = 1$ for $x \in (-h, h)$ and $\phi_n(x) = 0$ for $x \in [-\pi - h] \cup [h, \pi]$ and $\phi_h(x+2\pi) = \phi_h(x)\ \forall\ x$.

Find the F.S. of $\phi_h(x)$ and deduce that $\frac{1}{2} + \sum\limits_{n=1}^{\infty}\frac{\sin nh}{nh} = \frac{\pi}{2h}$.

(8) Find the F.S. of $\sin x$ in $[-\pi, \pi]$.

Solution. (1) Let $f(x) = x^2$, defined on $[0, \pi]$. Then f satisfies all the Dirichlet's conditions in the half-range interval $[0, \pi]$ with period 2π.

The half-range sine series for f is of the form $\sum\limits_{1}^{\infty} b_n \sin nx$, where

$$b_n = \frac{2}{\pi}\int_0^\pi f(x)\sin nx dx = \frac{2}{\pi}\int_0^\pi x^2 \sin nx dx = \frac{2}{\pi}\left[\left(-\frac{x^2\cos nx}{n}\right)_0^\pi + \frac{2}{\pi}\cdot\frac{1}{n}\int_0^\pi x\cos nx dx\right]$$

$$= \frac{2}{\pi}\frac{\pi^2(-1)^{n+1}}{n} + \frac{4}{n\pi}\left[\left(\frac{x\sin nx}{n}\right)_0^\pi + \left(\frac{\cos nx}{n^2}\right)_0^\pi\right] = \frac{2\pi(-1)^{n+1}}{n} + \frac{4}{\pi}\left[\frac{(-1)^n - 1}{n^3}\right]$$

f is also continuous on $[0, \pi]$. Hence, $S(x) = f(x)$. So,

$$f(x) = \sum_{1}^{\infty}\left[\frac{2\pi}{n}(-1)^{n+1} + \frac{4}{\pi n^3}\{(-1)^n - 1\}\right]\sin nx.$$

Graph of sum function $f(x) = x^2$:

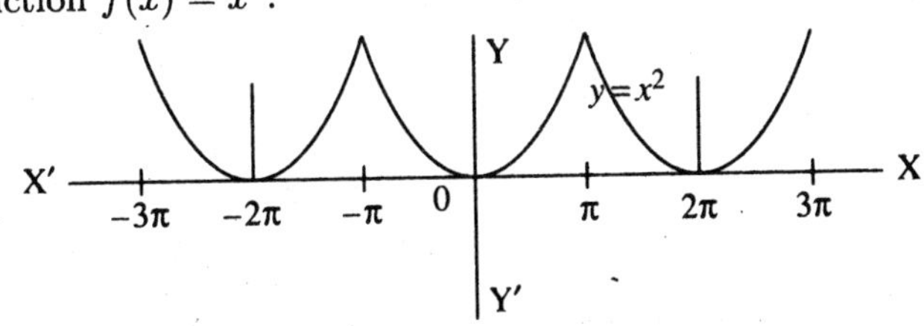

Fig. 11.4

$$f(x+2\pi) = f(x) \Rightarrow f(0) = f(2\pi) = f(4\pi)\cdots$$
$$f(x-2\pi) = f(x) \Rightarrow f(0) = f(-2\pi) = f(-4\pi)\cdots$$

(2) $f(t) = \sin\frac{\pi t}{l}$ for $0 < t < l$.

We define, $f(0) = f(l) = 0$.

Then f is bounded. f is also integrable and periodic with period $2l$.

The half-range Fourier cosine series of f is of the form $\frac{a_0}{2} + \sum_1^\infty a_n \cos\frac{n\pi t}{l}$, where

$$a_0 = \frac{2}{l}\int_0^l f(t)dt = \frac{2}{l}\int_0^l \sin\frac{\pi t}{l}dt = \frac{2}{l}\left[-\frac{\cos\frac{\pi t}{l}}{\frac{\pi}{l}}\right]_0^l = \frac{2}{l}\times\frac{l}{\pi}\cdot 2 = \frac{4}{\pi}$$

$$a_n = \frac{2}{l}\int_0^l \sin\frac{\pi t}{l}\cos\frac{n\pi t}{l}dt = \frac{1}{l}\int_0^l\left[\sin(n+1)\frac{\pi t}{l} + \sin(1-n)\frac{\pi t}{l}\right]dt$$

$$= \frac{1}{l}\left[-\frac{\cos(n+1)\frac{\pi t}{l}}{(n+1)\frac{\pi}{l}} + \frac{\cos(n-1)\frac{\pi t}{l}}{(n-1)\frac{\pi}{l}}\right]_0^l$$

$$= \frac{1}{l}\left[\frac{l}{(n+1)\pi}\{(-1)^{n+2}+1\} + \frac{l}{(n-1)\pi}\{(-1)^{n-1}-1\}\right]$$

$$= \frac{1}{\pi}\left[+\frac{(-1)^n}{n+1} - \frac{(-1)^n}{(n-1)} + \frac{1}{n+1} - \frac{1}{(n-1)}\right]$$

$$= \frac{1}{\pi}\left[-\frac{2}{n^2-1}\{(-1)^n+1\}\right] \text{ for } n \neq 1.$$

$$= -\frac{2}{\pi(n^2-1)}\{1+(-1)^n\}$$

$a_n = 0$, when n is odd, $n \neq 1$.

$a_n = \frac{-4}{\pi[4m^2-1]}$, when n is even, say $n = 2m$.

$$\text{Now } a_1 = \frac{2}{l}\int_0^l \sin\frac{\pi t}{l}\cos\frac{\pi t}{l}dt = \frac{1}{l}\left[-\frac{\cos\frac{2\pi t}{l}}{\frac{2\pi}{l}}\right]_0^l = \frac{l}{l\times 2\pi}[-1+1] = 0.$$

Therefore, the F.S. is $\frac{2}{\pi} - \frac{4}{\pi}\sum_{m=1}^\infty \frac{1}{4m^2-1}\cos\frac{2m\pi t}{l}$.

(3) f is bounded, integrable and periodic with period 2π. f is also piecewise monotonic in $[-\pi, 0]$ and $[0, \pi]$. So f satisfies all the Dirichlet's conditions in $[-\pi, \pi]$.

The F.S. of f in $[-\pi, \pi]$ is of the form $\frac{a_0}{2} + \sum(a_n \cos nx + b_n \sin nx)$, where

$$a_0 = \frac{1}{\pi}\int_{-\pi}^{\pi} f(x)dx = \frac{1}{\pi}\left[\int_{-\pi}^0 -\frac{\pi}{4}dx + \int_0^\pi \frac{\pi}{4}dx\right] = \frac{1}{\pi}\left[-\int_0^\pi \frac{\pi}{4}dx + \int_0^\pi \frac{\pi}{4}dx\right] = 0$$

$$a_n = \frac{1}{\pi}\int_{-\pi}^{\pi} f(x)\cos nx dx = \frac{1}{\pi}\left[\int_{-\pi}^0 -\frac{\pi}{4}\cos nx dx + \int_0^\pi \frac{\pi}{4}\cos nx dx\right]$$

$$= \frac{1}{\pi}\left[-\int_0^\pi \frac{\pi}{4}\cos nx dx + \int_0^\pi \frac{\pi}{4}\cos nx dx\right] = 0$$

$$b_n = \frac{1}{\pi}\int_{-\pi}^{\pi} f(x)\sin nx dx = \frac{1}{\pi}\left[\int_{-\pi}^0 -\frac{\pi}{4}\sin nx dx + \int_0^\pi \frac{\pi}{4}\sin nx dx\right]$$

$$= \frac{1}{\pi}\cdot 2\int_0^\pi \frac{\pi}{4}\sin nx dx = \frac{2}{\pi}\cdot\frac{\pi}{4}\left[-\frac{\cos nx}{n}\right]_0^\pi = \frac{1}{2}\frac{[1-(-1)^n]}{n}$$

Therefore, F.S. of $f(x)$ is $\sum_1^\infty \frac{1}{2}\frac{\{1-(-1)^n\}}{n}\sin nx$

$$\therefore\ f(x) = \sin x + \frac{1}{3}\sin 3x + \frac{1}{5}\sin 5x \quad \text{[even term vanishes]} \tag{1}$$

f is continuous at $x = \frac{\pi}{2}$. So $S\left(\frac{\pi}{2}\right) = f\left(\frac{\pi}{2}\right) = \frac{\pi}{4}$.

Therefore, from (1) for $x = \frac{\pi}{2}$, $\frac{\pi}{4} = 1 - \frac{1}{3} + \frac{1}{5} - \frac{1}{7} + \cdots$.

(4) $f(x) = e^x$, $0 < x < \pi$, half-range interval of $[-\pi, \pi]$, we define, $f(0) = 1$ and $f(\pi) = e^\pi$, so that f remains bounded on $[0, \pi]$. f is also integrable and periodic with period 2π.

Let the F.S. for f in the half-range interval $[0, \pi]$ be a sine series of the form $\sum_1^\infty b_n \sin nx$, where

$$b_n = \frac{2}{\pi}\int_0^\pi e^x \sin nx dx = \frac{2}{\pi}\left[e^x \frac{[\sin nx - n\cos nx]}{n^2+1}\right]_0^\pi$$

$$= \frac{2}{\pi}\left[\frac{e^\pi\{0 - n(-1)^n\} + n}{n^2+1}\right] = \frac{2}{\pi}\left[\frac{n\{1-(-1)^n e^\pi\}}{n^2+1}\right]$$

f is continuous on $[0, \pi]$,

$$S(x) = f(x)\ \forall\ x \in [0, \pi]$$

$$f(x) = \frac{2}{\pi}\sum_1^\infty \left\{\frac{1-(-1)^n e^\pi}{n^2+1}\right\} n \sin nx$$

$$S(x) = \frac{1}{2}[-f(-x+0) - f(-x-0)] \text{ for } -\pi < x < 0$$

$$= \frac{1}{2}\left[-e^{-x} - e^{-x}\right] = -e^{-x}, \quad -\pi < x < 0$$

and $S(0) = 0 = S(-\pi)$ [by Art. 11.4.2]

(5) $f(x) = x\sin x$, $-\pi < x < \pi$. We define, $f(-\pi) = f(\pi) = 0$.

Then f is bounded, integrable, periodic 2π. Also f is piecewise monotonic in $(-\pi, \pi)$. f satisfies all the Dirichlet's conditions. Here f is an even function. So the F.S. of f is a cosine series of the form $\frac{a_0}{2} + \sum_1^\infty a_n \cos nx$, where

$$a_0 = \frac{2}{\pi}\int_0^\pi x\sin x dx = \frac{2}{\pi}\big[-x\cos x + \sin x\big]_0^\pi = 2$$

$$a_n = \frac{2}{\pi}\int_0^\pi x\sin x\cos nx dx = \frac{1}{\pi}\int_0^\pi x\{\sin(n+1)x - \sin(n-1)x\}dx$$

$$= \frac{1}{\pi}\left[-\frac{x\cos(n+1)x}{n+1} + \frac{\sin(n+1)x}{(n+1)^2} + \frac{x\cos(n-1)x}{(n-1)} - \frac{\sin(n-1)x}{(n-1)^2}\right]_0^\pi$$

$$= \frac{1}{\pi}\left[\frac{-\pi(-1)^{n+1}}{n+1} + \frac{\pi(-1)^{n-1}}{n-1}\right]$$

$$= (-1)^n\left[\frac{1}{n+1} - \frac{1}{n-1}\right] = \frac{-2(-1)^n}{n^2-1}, \quad n \neq 1$$

$$a_1 = \frac{2}{\pi}\int_0^\pi x\sin x\cos x dx = \frac{1}{\pi}\int_0^\pi x\sin 2x dx = \frac{1}{\pi}\left[-\frac{x\cos 2x}{2} + \frac{\sin 2x}{2^2}\right]_0^\pi = -\frac{1}{2}$$

Therefore, the F.S. for $f(x) = x\sin x$

$$= \frac{1}{2}\cdot 2 + a_1\cos x + \sum_{n=2}^{\infty}\frac{2(-1)^{n+1}}{n^2-1}\cos nx$$

$$= 1 - \frac{1}{2}\cos x + \left[-\frac{2}{3}\cos 2x + \frac{2}{8}\cos 3x - \frac{2}{15}\cos 4x + \cdots\right] \tag{1}$$

Here f is continuous at $x = \frac{\pi}{2}$.

$$\therefore\ S\left(\frac{\pi}{2}\right) = f\left(\frac{\pi}{2}\right) = \frac{\pi}{2}$$

So from (1) for $x = \frac{\pi}{2}$,

$$S\left(\frac{\pi}{2}\right) = 1 + \frac{2}{1\cdot 3} - \frac{2}{3\cdot 5} + \frac{2}{5\cdot 7} - \cdots$$

$$\frac{\pi}{2} = 1 + \frac{2}{1\cdot 3} - \frac{2}{3\cdot 5} + \frac{2}{5\cdot 7} - \cdots \text{ or } \frac{\pi}{4} = \frac{1}{2} + \frac{1}{1\cdot 3} - \frac{1}{3\cdot 5} + \frac{1}{5\cdot 7} - \cdots$$

(6) We have $a_0 = \frac{1}{\pi}\int_{-\pi}^{\pi} f(t)dt,\ a_k = \frac{1}{\pi}\int_{-\pi}^{\pi} f(t)\cos ktdt,\ b_k = \frac{1}{\pi}\int_{-\pi}^{\pi} f(t)\sin ktdt.$

Now

$$S_n(f,x) = \frac{a_0}{2} + \sum_{k=1}^{n}(a_k\cos kx + b_k\sin kx)$$

$$= \frac{1}{2\pi}\int_{-\pi}^{\pi} f(t)dt + \sum_{k=1}^{n}\left\{\frac{1}{\pi}\int_{-\pi}^{\pi} f(t)\cos ktdt\right\}\cos kx$$

$$+ \sum_{k=1}^{n}\left\{\frac{1}{\pi}\int_{-\pi}^{\pi} f(t)\sin ktdt\right\}\sin kx$$

$$= \frac{1}{\pi}\int_{-\pi}^{\pi} f(t)\left[\frac{1}{2} + \sum_{k=1}^{n}\{\cos kt\,\cos kx + \sin kt\,\sin kx\}\right]dt$$

$$= \frac{1}{\pi}\int_{-\pi}^{\pi} f(t)\left[\frac{1}{2} + \sum_{k=1}^{n}\cos k(t-x)\right]dt = \frac{1}{\pi}\int_{-\pi}^{\pi} f(t)D_n(t-x)dt$$

where

$$D_n(t-x) = \frac{1}{2} + \sum_{k=1}^{n}\cos k(t-x)$$

(7) Clearly $\phi_h(x)$ is periodic, bounded and integrable and piecewise monotonic on $[-\pi,\pi]$. Hence ϕ_h satisfies Dirichlet's conditions. Let $\frac{a_0}{2} + \sum\limits_{n=1}^{\infty}(a_n\cos nx + b_n\sin nx)$ be the F.S. of ϕ_h in $[-\pi,\pi]$.

$$a_0 = \frac{1}{\pi}\int_{-\pi}^{\pi}\phi_h(x)dx = \frac{1}{\pi}\int_{-h}^{h}dx = \frac{2h}{\pi}$$

$$a_n = \frac{1}{\pi}\int_{-\pi}^{\pi}\phi_h(x)\cos nxdx = \frac{1}{\pi}\int_{-h}^{h}1\cdot\cos nxdx = \frac{2}{\pi}\int_{0}^{h}\cos nxdx = \frac{2\cdot\sin nh}{n\pi}$$

$$b_n = \frac{1}{\pi}\int_{-\pi}^{\pi}\phi_h(x)\sin nxdx = \frac{1}{\pi}\int_{-h}^{h}\sin nxdx = 0$$

Therefore, F.S. is

$$\frac{h}{\pi}+\sum_{n=1}^{\infty}\frac{2\sin nh}{n\pi}\cos nx=\frac{h}{\pi}+\frac{2}{\pi}\sum_{n=1}^{\infty}\frac{\sin nh\cos nx}{n} \tag{1}$$

Here 0 is a point of continuity. So $S(0)=\phi_h(0)=1$.

$\therefore$ from (1) for $x=0$,

$$\phi_h(0)=\frac{h}{\pi}+\frac{2}{\pi}\sum_{n=1}^{\infty}\frac{\sin nh}{n}\Rightarrow 1=\frac{h}{\pi}+\frac{2}{\pi}\sum_{n=1}^{\infty}\frac{\sin nh}{n}\Rightarrow\frac{\pi}{2h}=\frac{1}{2}+\sum_{n=1}^{\infty}\frac{\sin nh}{nh}$$

(8) Here $f(x)=\sin x$ is bounded, periodic and piecewise monotonic on $[-\pi,\pi]$. f is integrable on $[-\pi,\pi]$. Thus f satisfies Dirichlet's conditions on $[-\pi,\pi]$. The F.S. of $\sin x$ is of the form $\sum b_n\sin nx$, as $\sin x$ is an odd function, where

$$\begin{aligned}b_n&=\frac{2}{\pi}\int_0^{\pi}\sin x\sin nx\,dx=\frac{1}{\pi}\int_0^{\pi}[\cos(n-1)x-\cos(n+1)x]dx\\&=\frac{1}{\pi}\left[\frac{\sin(n-1)x}{n-1}-\frac{\sin(n+1)x}{n+1}\right]_0^{\pi}=0\text{ for }n\neq 1\\b_1&=\frac{2}{\pi}\int_0^{\pi}\sin^2 x\,dx=\frac{1}{\pi}\int_0^{\pi}(1-\cos 2x)dx=1\end{aligned}$$

Therefore, the F.S. of $\sin x$ is $b_1\sin x+\sum_2^{\infty}b_n\sin nx=\sin x$.

EXERCISE

(1) Show that (i) $\sum_{n=1}^{\infty}\frac{\cos nx}{n}$, (ii) $\sum_{n=1}^{\infty}\left(\frac{\sin nx}{n^2}+\frac{\cos nx}{n}\right)$ are Fourier series while $\sum_1^{\infty}\frac{\sin nx}{\sqrt{n}}$ and $\sum_1^{\infty}\sin nx$ are not Fourier series.

(2) Find the Fourier series of $f(x)=\sin x$.

(3) Obtain the Fourier series of $f(x)=x\sin x$ on $[-\pi,\pi]$, where $f(x\pm 2\pi)=f(x)$. Hence deduce that $\frac{\pi}{4}=\frac{1}{2}+\frac{1}{1\cdot 3}-\frac{1}{3\cdot 5}+\frac{1}{5\cdot 7}-\cdots$.

(4) Expanding x and x^2 in $[0,\pi]$ as cosine multiples of x show that

$$\sum_1^{\infty}\frac{\cos nx}{n^2}=\frac{3x^2-6\pi x+2\pi^2}{12}\text{ for }0\leq x\leq\pi.$$

(5) Find the Fourier series of $|x|$ in $[-6,6]$.

(6) Let $f:[0,\pi]\to\mathbb{R}$ be defined as follows:

$$f(x)=\begin{cases}\frac{\pi}{3}, & 0\leq x<\frac{\pi}{3}\\ 0, & \frac{\pi}{3}\leq x<\frac{2\pi}{3}\\ -\frac{\pi}{3}, & \frac{2\pi}{3}\leq x\leq\pi.\end{cases}$$

Also $f\left(\frac{\pi}{3}\right)=\frac{\pi}{2}$, $f\left(\frac{2\pi}{3}\right)=-\frac{\pi}{12}$.

Find a Fourier series of $f(x)$ in $[0, \pi]$.

Deduce that $\dfrac{\pi}{2\sqrt{3}} = 1 - \dfrac{1}{5} + \dfrac{1}{7} - \dfrac{1}{11} + \dfrac{1}{13} - \dfrac{1}{17} + \cdots$.

(7) Expand $x - \dfrac{1}{3}$ in Fourier series in $[0, 1]$.

(8) Let $f : [0, 3] \to \mathbb{R}$ be defined by

$$f(x) = \begin{cases} x, & 0 \le x < 1 \\ 0, & 1 \le x < 2 \\ 3 - x, & 2 \le x \le 3. \end{cases}$$

Find the Fourier series of $f(x)$ on $[0, 3]$ as a (i) series of sines, (ii) series of cosines.

(9) Show that $\displaystyle\sum_{1}^{\infty} \frac{\sin nx}{n}$ is the Fourier series of $f(x) = \dfrac{1}{2}(\pi - x)$ in $[0, 2\pi]$. What is the value to which the series converges at $x = \dfrac{\pi}{2}$?

(10) Find the Fourier series of f with period 2π defined as

$$f(x) = \begin{cases} 0, & \text{for } -\pi < x < a \\ 1, & \text{for } a \le x \le b \\ 0, & \text{for } b < x < \pi. \end{cases}$$

Find the sum of the series at $x = 4\pi + a$ and deduce that $\displaystyle\sum_{1}^{\infty} \frac{\sin n(b-a)}{n} = \frac{\pi - b + a}{2}$.

(11) Find the Fourier series f with period 2π defined as follows:

$$f(x) = \begin{cases} 0, & \text{for } -\pi \le x < 0 \\ x, & \text{for } 0 < x \le \pi. \end{cases}$$

What is the sum of the series at $x = -5\pi$?

(12) Let f, a periodic function with period 2π, be defined by

$$f(x) = \begin{cases} x^2, & \text{for } 0 \le x < \dfrac{\pi}{2} \\ 0, & \text{for } \dfrac{\pi}{2} \le x \le \pi. \end{cases}$$

Then show that f can be represented by a series sine multiples of x such that co-efficient of $\sin nx$ is $\left(\dfrac{4}{\pi n^3} - \dfrac{\pi}{2n}\right) \cos \dfrac{n\pi}{2} + \dfrac{2}{n^2} \sin \dfrac{n\pi}{2} - \dfrac{4}{\pi n^3}$.

(13) Find (i) the series of sines, (ii) the series of cosines which will represent $f(x)$ in $(0, l)$, where

$$f(x) = \begin{cases} kx, & \text{when } 0 < x < \dfrac{l}{2} \\ k(l - x), & \dfrac{l}{2} \le x < l. \end{cases}$$

CHAPTER 12

FUNCTION OF SEVERAL VARIABLES

Definition [12.1] : If the three variables u, x, y be so related that for every pair of values of x and y in a domain D, u has a definite value, then u is said to a function of x and y and is written as $u = f(x, y)$, where x and y are independent variables and u is a depending variable.

It is to be noted that if u has a single definite value for $(x, y) \in D$, then u is called a *single-valued function* and if u has more than one definite value, then u is called a *multiple-valued function*. $u = f(x, y)$ is a function of two variables.

Example. $u = x + y$ and let $D \subseteq \mathbb{R}$. Then u is a single-valued function.

Example. $u^2 = 2x + y$ and let $D = N \times N$.

Then u has two values for every $(x, y) \in N \times N$, u is a multiple-valued function.

Example. Let u be the area of a triangle ABC having base $= x$ unit and altitude $= y$ unit, then $u = \frac{1}{2}xy$ square unit, has a single value for every $(x, y) \in N \times N$. So u is a single-valued function of two independent variables x and y.

12.1 Geometrical Interpretation of $u = f(x, y)$

Let $z = f(x, y)$ be a single-valued function. Let OX and OY be rectangular axes of co-ordinates. We cut a length $OA = x$ from OX and a length $OB = y$ from OY. Then for the pair of values of $(OA, OB) = (x, y)$, we have a point P on XY-plane.

We now draw a perpendicular of length $|\overrightarrow{PQ}| = z$ at P, on XY-plane. Thus the values of z obtained from the relation $z = f(x, y)$ describe points like Q. Then the locus of Q will describe a surface generated by x, y, z. This surface is known as three dimensional surface. Hence $z = f(x, y)$ represents a surface referred to axes OX, OY, OZ in space.

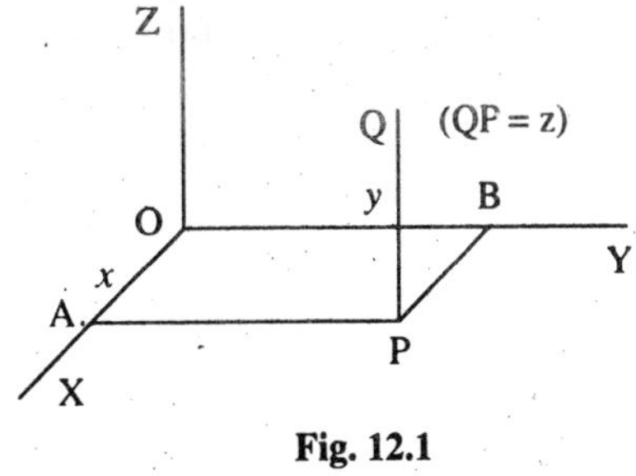

Fig. 12.1

Definition [12.2] : If we consider 3 independent variables x, y, z and one dependent variable u, such that for every set of values $(x, y, z) \in D$, u can be determined uniquely, then u is called a function of three independent variables x, y, z and is denoted by $u = f(x, y, z)$.

In a similar way, considering n independent variables $x_1, x_2, \cdots, x_n$ and a dependent variable u such that for every set of values of $(x_1, x_2, \cdots, x_n) \in D$, u has a definite value, then u is called a function of n independent variables and is denoted by $u = f(x_1, x_2, \cdots, x_n)$. D is called the domain of f.

Example. $u = 2x + 3y + 5z$, a function of 3 independent variables $x, y, z \in z^+$, where u is a dependent variable.

Example. $u = x_1^2 + x_2^2 + \cdots + x_n^2$, a function of n independent variables $x_1, x_2, \cdots, x_n \in R^n$, u is a dependent variable.

Explicit and Implicit functions

Explicit function. The above functional relations $u = f(x, y), u = f(x, y, z), u = f(x_1, x_2, \cdots, x_n)$ define explicit functions.

Examples given above are all explicit functions.

Implicit function. From a functional relation

$$f(x_1, x_2, \cdots, x_n) = 0. \tag{1}$$

Connecting n variables $x_1, x_2, \cdots, x_n$ it is not always possible to decide whether this equation determines x_i as a function of $x_1, x_2, \cdots, x_{i-1}, x_{i+1}, \cdots, x_n$ for any $i = 1, 2, \cdots, n$. If so, we have $x_i = g(x_1, x_2, \cdots, x_{i-1}, x_{i+1}, \cdots, x_n)$ for some function g. We say that g is defined implicitly by (1).

In other words, g is called an implicit function defined by $f(x_1, x_2, \cdots, x_i, \cdots, x_n) = 0$, $1 \leq i \leq n$.

Example. If $f(x, y, z) = (2x - y + 3z) = 0,$ (1)
then $y = 2x + 3z = g(x, z)$, is an implicit function defined by (1).

Example. If $f(x, y, z) = x^2 + y^2 + z^2 = 0$, then it determines no implicit function, while $f(x, y, z) = x^2 + y^2 - z^2 = 0$ determines implicit functions $z = \pm\sqrt{x^2 + y^2} = g(x, y)$.

Example. From the functional relation $f(x, y) = xy \sin x + \cos y = 0$, it is not easy to write y explicitly in terms of x, but we cannot conclude that implicit function does not exist from the given relation.

So the question of existence of implicit function from a functional relation to be discussed.

Definition [12.3] : [Neighbourhood of a point $(a, b) \in \mathbb{R}^2$] The neighbourhood (*nhd*) of a point $(a, b) \in \mathbb{R}^2$ is a subset S of $\mathbb{R}^2$ such that $S = \{(x, y) : |x - a| < \delta, |y - b| < \delta\} \delta > 0$, is called a square *nhd* of (a, b) and is denoted by $S_\delta(a, b)$.

$S = \{(x, y) : (x - a)^2 + (y - b)^2 < \delta^2\} \subseteq \mathbb{R}^2$ is called a circular *nhd* of (a, b) and is denoted by $C_\delta(a, b)$.

If $p, q, r, s \in \mathbb{R}$ such that $p < a < r, q < y < s$.

Then the subset $s = [(p, q) \times (r, s)] = \{(x, y) : p < x < r, q < y < s\} \subseteq \mathbb{R}^2$ is called a rectangular *nhd* of (a, b) and is denoted by $R_\delta(a, b)$.

Generally *nhd* of a point (a, b) is denoted by $N_\delta(a, b)$. If $(a, b) \notin N_\delta(a, b)$, then $N_\delta(a, b)$ is called deleted *nhd* of (a, b) and is denoted by $N'_\delta(a, b)$.

Definition [12.4] : [Limits of a function $f(x, y)$ as $(x, y) \to (a, b)$] Let $f : D \to \mathbb{R}$, $D \subseteq \mathbb{R}^2$. The function f is said to have a limit 'L' as $(x, y) \to (a, b)$, if for $\epsilon(> 0)$, $\exists$ a (+)ve number δ such that $|f(x, y) - L| < \epsilon$ whenever $0 < |x - a| < \delta$, $0 < |y - b| < \delta$.

That is, $|f(x,y) - L| < \epsilon$ whenever $(x,y) \in N'_\delta(a,b) \cap D$.

Symbolically, we write $\lim\limits_{(x,y)\to(a,b)} f(x,y) = L$. '$L$' is called double limit or simultaneous limit.

Theorem [12.1] [Necessary condition for the existence of double limit]
If $\lim\limits_{(x,y)\to(a,b)} f(x,y) = L$, then $f(x,\phi(x)) \to L$ as $x \to a$ where $\phi(x)$ is a real valued function of one variable x and $\phi(x) \to b$ as $x \to a$.

Proof. $\lim\limits_{(x,y)\to(a,b)} f(x,y) = L$.

$\Rightarrow$ for every $\epsilon(>0), \exists$ a (+)ve number δ such that

$$|f(x,y) - L| < \epsilon \text{ whenever } 0 < |x-a| < \delta, 0 < |y-b| < \delta \qquad \text{(i)}$$

Given that $\phi(x) \to b$ as $x \to a$

$\Rightarrow$ $\exists$ a (+)ve number λ corresponding δ such that

$$|\phi(x) - b| < \delta \text{ whenever } 0 < |x-a| < \lambda \qquad \text{(ii)}$$

Let $\mu = \min(\delta, \lambda)$.

Then by (i) and (ii), $|f(x,\phi(x)) - L| < \epsilon$ whenever $0 < |x-a| < \mu$ and hence

$$\lim_{x\to a} f(x,\phi(x)) = L.$$

Note. Non existence of double limit

If $\exists$ two functions $y = \phi_1(x)$ and $y = \phi_2(x)$ such that $x \to a$, $\phi_1(x)$, $\phi_2(x) \to b$ and if $\lim\limits_{x\to a} f(x,\phi_1(x)) \neq \lim\limits_{x\to a} f(x,\phi_2(x))$, we say that double limit does not exist.

Example (1). Find $\lim\limits_{(x,y)\to(0,0)} xy \cdot \dfrac{x^2-y^2}{x^2+y^2}$

Method (1): (ϵ, δ) definition

$$\left|xy \cdot \frac{x^2-y^2}{x^2+y^2} - 0\right| = \left|xy \cdot \frac{x^2-y^2}{x^2+y^2}\right| = |x|\,|y| \frac{|x^2-y^2|}{|x^2+y^2|}$$

$$\leq |x|\,|y| \quad [\because |x^2-y^2| \leq |x^2+y^2|]$$

$$\leq \delta^2 \text{ for } 0 < |x| < \delta, 0 < |y| < \delta < \epsilon \text{ choosing } \delta = \sqrt{\epsilon}, \epsilon > 0$$

$$\Rightarrow \lim_{(x,y)\to(0,0)} xy \cdot \frac{x^2-y^2}{x^2+y^2} = 0$$

Method (2): By substitution in polar form.

Let $x = r\cos\theta, y = r\sin\theta$

$x^2+y^2 = r^2 \Rightarrow r \to 0$ as $(x,y) \to (0,0)$

$$|f(x,y) - 0| = \left|xy \cdot \frac{x^2-y^2}{x^2+y^2}\right| = \left|r^2 \sin\theta\cos\theta \cdot \frac{r^2\cos 2\theta}{r^2}\right|$$

$$= \left|\frac{r^2}{2}\sin 2\theta \cos 2\theta\right| = \frac{r^2}{4}|\sin 4\theta| \leq \frac{r^2}{4} < \epsilon, \text{ if } r^2 < 4\epsilon$$

$$\text{i.e., if } 0 < x^2+y^2 < 4\epsilon = \delta^2, \text{ i.e., if } 0 < x^2+y^2 < \delta^2$$

So $|f(x,y) - 0| < \epsilon$ whenever $0 < x^2+y^2 < \delta^2$ $(C_\delta(0,0))$.

$\Rightarrow \lim\limits_{(x,y)\to(0,0)} f(x,y) = 0$

Example (2). Let $f(x, y) = \begin{cases} x \sin \dfrac{1}{y} + y \sin \dfrac{1}{x}, & xy \neq 0 \\ 0, & xy = 0 \end{cases}$

$$\begin{aligned} |f(x, y) - 0| &= \left| x \sin \frac{1}{y} + y \sin \frac{1}{x} \right| \leq |x| \left| \sin \frac{1}{y} \right| + |y| \left| \sin \frac{1}{x} \right| \leq |x| + |y| \\ &< \delta + \delta \text{ whenever } 0 < |x| < \delta, 0 < |y| < \delta \\ &= 2\delta = \epsilon, \text{ choosing } \delta = \epsilon/2, \epsilon > 0 \end{aligned}$$

$\therefore |f(x, y) - 0| < \epsilon$ whenever $0 < |x| < \delta, 0 < |y| < \delta$.

$\therefore \lim_{(x,y)\to(0,0)} f(x, y) = 0.$

Example (3). Non existence of double limit

Let $f(x, y) = \begin{cases} \dfrac{x^2 y}{x^4 + y^2}, & \text{if } x^4 + y^2 \neq 0 \\ 0, & \text{if } x = y = 0 \end{cases}$

Let $(x, y) \to (0, 0)$ along a line $y = mx$. Then $f(x, y) = f(x, mx) = \dfrac{mx^3}{x^4 + m^2 x^2}$

$$\therefore \lim_{(x,y)\to(0,0)} f(x, y) = \lim_{x\to 0} \frac{mx^3}{x^4 + m^2 x^2} = \lim_{x\to 0} \frac{mx}{x^2 + m^2} = 0$$

But, let $(x, y) \to (0, 0)$ along a curve $y = mx^2$ so that $y \to 0$ as $x \to 0$

$$\therefore \lim_{(x,y)\to(0,0)} \frac{x^2 y}{x^4 + y^2} = \lim_{x\to 0} \cdot \frac{x^2 \cdot mx^2}{x^4 + m^2 x^4} = \lim_{x\to 0} \cdot \frac{m}{1 + m^2} = \frac{m}{1 + m^2}$$

which is different for different values of m.

Hence $\lim_{(x,y)\to(0,0)} f(x, y)$ does not exist.

Definition [12.5] : [Repeated limit (or Iterated limit)] Let f be defined in some $N_\delta(a, b)$. Then $\lim_{y\to b} f(x, y)$, if it exists, is a function of x say $\phi(x)$. If then $\lim_{x\to a} \phi(x)$ exists and say equal to λ. We write it by $\lim_{x\to a} \lim_{y\to b} f(x, y) = \lambda$. λ is called the repeated limit or iterated limit of f as $y \to b$, $x \to a$.

Other repeated limit $\lim_{y\to b} \lim_{x\to a} f(x, y)$ can be defined by changing the order of the limit by taking first $x \to a$ and then $y \to b$.

Example (4). Find $\lim_{x\to 0} \lim_{y\to 0} xy \cdot \dfrac{x^2 - y^2}{x^2 + y^2}$ and $\lim_{y\to 0} \lim_{x\to 0} xy \cdot \dfrac{x^2 - y^2}{x^2 + y^2}$

$$\lim_{x\to 0} \lim_{y\to 0} xy \cdot \frac{x^2 - y^2}{x^2 + y^2} = \lim_{x\to 0} 0 = 0$$

Also $\lim_{y\to 0} \lim_{x\to 0} xy \cdot \dfrac{x^2 - y^2}{x^2 + y^2} = \lim_{y\to 0} 0 = 0$

Repeated limit exists and are equal.

Example (5). Find $\lim_{x\to 0} \lim_{y\to 0} \dfrac{x - y}{x + y}$ and $\lim_{y\to 0} \lim_{x\to 0} \dfrac{x - y}{x + y}$.

Now

$$\lim_{x\to 0}\lim_{y\to 0}\frac{x-y}{x+y}=\lim_{x\to 0}\frac{x}{x}=1$$

$$\lim_{y\to 0}\lim_{x\to 0}\frac{x-y}{x+}=\lim_{y\to 0}\frac{-y}{y}=\lim_{y\to 0}\cdot -1=-1$$

Both the repeated limits exists but unequal.

Example (6). If $f(x,y)=\begin{cases} x\sin\dfrac{1}{y}+y\sin\dfrac{1}{x}, & xy\neq 0\\ 0, & xy=0\end{cases}$

Show that the repeated limits at $(0,0)$ do not exist.

$$\lim_{x\to 0}\lim_{y\to 0}\left(x\sin\frac{1}{y}+y\sin\frac{1}{x}\right)=\lim_{x\to 0}\left[x\lim_{y\to 0}\sin\frac{1}{y}+\sin\frac{1}{x}\lim_{y\to 0}y\right]$$

$$=\lim_{x\to 0}\left[x\lim_{y\to 0}\sin\frac{1}{y}+0\right]=\lim_{x\to 0}\left[x_1\lim_{y\to 0}\sin\frac{1}{y}\right]$$

does not exist for any constant value $x=x_1(\neq 0)$.

Also

$$\lim_{y\to 0}\lim_{x\to 0}\left[x\sin\frac{1}{y}+y\sin\frac{1}{x}\right]=\lim_{y\to 0}\left[y_1\lim_{x\to 0}\sin\frac{1}{x}\right]$$

does not exist for any constant value $y=y_1(\neq 0)$.

Remarks.

(1) If repeated limits exist and equal, the double limit exist. It follows from example (1) and (4)

(2) If the double limit exists it does not imply that the repeated limits will exist. It follows from example (2) and (6).

(3) If the repeated limits exist and unequal, the double limit does not exist.

In example (5), repeated limits exists but unequal while $\lim\limits_{(x,y)\to(0,0)}\dfrac{x-y}{x+y}$ does not exist.

For $y=mx$, $y\to 0$ as $x\to 0$, we have $\lim\limits_{x\to 0}\dfrac{x-mx}{x+mx}=\dfrac{1-m}{1+m}$ which is different for different values of m.

Hence the double limit does not exist.

Continuity of $f(x,y)$ at a point $(a,b)\in D$

Definition [12.6] : Let $f:D\to\mathbb{R}$ and $(a,b)\in D$, $D\subseteq\mathbb{R}$. Then $f(x,y)$ is said to be continuous at a point (a,b) if for every $\epsilon(>0)$, $\exists$ a (+)ve number δ such that

$$|f(x,y)-f(a,b)|<\epsilon\ \forall\ (x,y)\in N_\delta(a,b)\cap D$$

$$\text{i.e., } |f(x,y)-f(a,b)|<\epsilon \text{ whenever } |x-a|<\delta, |y-b|<\delta.$$

In other words: $\lim\limits_{(x,y)\to(a,b)} f(x,y)$ exists and equal to $f(a,b)$, $f(a,b)$ is defined.

Note. (1) Not only at (a,b) but also in the *nhd* of (a,b), f is to be defined for continuity of it at (a,b). A function f which is not continuous at (a,b) is said to be discontinuous at (a,b).

(2) If $f(x,y)$ is continuous at (a,b), then $f(x,b)$, a function single variable x is continuous at $x=a$. Similarly $f(a,y)$ a function singler variable y, is continuous at $y=b$.

Example **(1). Show that the function $f(x,y) = \dfrac{x^3 - y^3}{x^2 + y^2}$, $(x,y) \neq (0,0)$ and $f(0,0) = 0$ is continuous at $(0,0)$.**

Let $x = r\cos\theta, y = r\sin\theta$. Then $x^2 + y^2 = r^2$.

Therefore,

$$\left|\frac{x^3 - y^3}{x^2 + y^2} - 0\right| = \left|\frac{r^3(\cos^3\theta - \sin^3\theta)}{r^2}\right|$$

$$= r\left|\frac{(\cos 3\theta + 3\cos\theta)}{4} - \frac{(3\sin\theta - \sin 3\theta)}{4}\right|$$

$$= \frac{r}{4}|(\cos 3\theta + \sin 3\theta) + 3(\cos\theta - \sin\theta)|$$

$$\leq \frac{r}{4}[|\cos 3\theta| + |\sin 3\theta| + 3|\cos\theta| + 3|\sin\theta|]$$

$$\leq \frac{r}{4}[1 + 1 + 3 + 3] = \frac{r}{4}\cdot 8 = 2r < \epsilon \text{ if } r < \frac{\epsilon}{2}$$

$$\therefore \left|\frac{x^3 - y^3}{x^2 + y^2} - f(0,0)\right| < \epsilon \text{ whenever } 0 < x^2 + y^2 < \frac{\epsilon^2}{4} = \delta^2.$$

$\therefore f(x,y)$ is continuous at $(0,0)$.

Example **(2). Show that $f(x,y) = \dfrac{xy}{\sqrt{x^2 + y^2}}$ $(x,y) \neq (0,0)$ and $f(0,0) = 0$. Examine the continuity of $f(x,y)$ at $(0,0)$.**

Let $x = r\cos\theta, y = r\sin\theta$.

Then $|f(x,y) - f(0,0)| = \left|\dfrac{xy}{\sqrt{x^2 + y^2}}\right| = \left|\dfrac{r^2\cos\theta\sin\theta}{r}\right| = \dfrac{r}{2}|\sin 2\theta| \leq \dfrac{r}{2} < \epsilon$

$\Rightarrow |f(x,y) - 0| < \epsilon$ whenever $0 < x^2 + y^2 < 4\epsilon^2 = \delta^2$

$\Rightarrow f(x,y)$ is continuous at $(0,0)$.

Alternative. $x^2 < x^2 + y^2 \Rightarrow |x| < \left|\sqrt{x^2 + y^2}\right|$.

Similarly $|y| < \left|\sqrt{x^2 + y^2}\right|$.

Therefore

$$|f(x,y) - f(0,0)| = \left|\frac{xy}{\sqrt{x^2 + y^2}}\right|$$

$$= \frac{|x||y|}{\left|\sqrt{x^2 + y^2}\right|} \leq \frac{\left|\sqrt{x^2 + y^2}\right|\left|\sqrt{x^2 + y^2}\right|}{\left|\sqrt{x^2 + y^2}\right|} \leq \left|\sqrt{x^2 + y^2}\right| = \epsilon \text{ (say)}$$

$\therefore |f(x,y) - f(0,0) < \epsilon$ for $0 < x^2 + y^2 < \epsilon^2 = \delta^2$ $[\delta = \epsilon]$

$\Rightarrow f(x,y)$ is continuous at $(0,0)$.

Example **(3). Show that $f(x,y) = \dfrac{x^3 + y^3}{x - y}$, when $x \neq y$ and $f(x,y) = 0$ when $x = y$ is discontinuous at $(0,0)$.**

Let $(x, y) \to (0,0)$ along $y = x - mx^3$, $y \to 0$ as $x \to 0$

$$\lim_{(x,y)\to(0,0)} f(x,y) = \lim_{x\to 0} \frac{x^3 + (x - mx^3)^3}{x - x + mx^3} = \lim_{x\to 0} x^3 \frac{[1 + (1 - mx^2)]^3}{mx^3}$$
$$= \lim_{x\to 0} \frac{1 + (1 - mx^2)^3}{m} = \frac{2}{m}$$

which is different for different values of m. Hence limit does not exist

So $f(x, y)$ is not continuous at $(0, 0)$.

Partial derivatives

Let $u = f(x, y)$ be a function of two independent variables x and y.

Here derivative of u with respect to x, keeping other independent variable y constant is called the partial derivative of u with respect to x, denoted by $\frac{\partial u}{\partial x}$, is defined by

$$\frac{\partial u}{\partial x} = \lim_{\Delta x\to 0} \frac{f(x + \Delta x, y) - f(x, y)}{\Delta x}$$
$$= \lim_{h\to 0} \frac{(x + h, y) - f(x, y)}{h} \quad [\text{Taking } \Delta x = \text{increment of } x = h]$$

Similarly, derivative of u with respect to y, keeping other independent variable x constant, is called the partial derivative of u with respect to y, denoted by $\frac{\partial u}{\partial y}$, is defined by

$$\frac{\partial u}{\partial y} = \lim_{\Delta y\to 0} \frac{f(x, y + \Delta y) - f(x, y)}{\Delta y}$$
$$= \lim_{k\to 0} \frac{f(x, y + k) - f(x, y)}{k} \quad [\text{Taking } \Delta y = \text{increment of } y = k]$$

Note.

(1) $\frac{\partial u}{\partial x}, \frac{\partial u}{\partial y}$ are also represented by, $u_x or fx(x, y), u_y or f_y(x, y)$ respectively.

(2) $\left[\frac{\partial u}{\partial x}\right]_{(a,b)} = \lim_{h\to 0} \frac{f(a+h,b)-f(a,b)}{h}$

$\left[\frac{\partial u}{\partial y}\right]_{(a,b)} = \lim_{k\to 0} \frac{f(a,b+k)-f(a,b)}{k}$

Example (1). Let $f(x, y) = \begin{cases} \dfrac{xy}{x^2 + y^2}, & (x, y) \neq (0, 0) \\ 0, & (x, y) = (0, 0) \end{cases}$

Then, $f_x(0, 0) = 0$ and $f_y(0, 0) = 0$

$$f_x(0,0) = \lim_{h\to 0} \frac{f(0 + h, 0) - f(0, 0)}{h} = \lim_{h\to 0} \frac{f(h, 0) - h(0, 0)}{h} = \lim_{h\to 0} \frac{0}{h} = 0$$
$$f_y(0,0) = \lim_{k\to 0} \frac{f(0, 0 + k) - f(0, 0)}{k} = \lim_{k\to 0} \frac{f(0, k) - f(0, 0)}{k} = \lim_{k\to 0} \frac{0}{k} = 0$$

Example (2). If $f(x, y) = \dfrac{xy(x^2 - y^2)}{x^2 + y^2}$ where $x^2 + y^2 \neq 0$ and $f(0, 0) = 0$ show that

$$f_x(x, 0) = 0 = f_y(0, y)$$
$$f_x(0, y) = -y, \ f_y(x, 0) - x$$

$$f_x(x,0) = \lim_{h\to 0} \frac{f(x + h, 0) - f(x, 0)}{h} = \lim_{h\to 0} \frac{(x + h) \cdot 0 \cdot (x^2) - 0}{x^2} = \lim_{h\to 0} \frac{0}{x^2} = 0$$

Similarly $f_y(0,y)=0$.

Now

$$f_x(0,y)=\lim_{h\to 0}\frac{(h,y)-f(0,y)}{h}=\lim_{h\to 0}\frac{\frac{hy(h^2-y^2)}{h^2+y^2}-0}{h}$$
$$=\lim_{h\to 0}\frac{y(h^2-y^2)}{h^2+y^2}=\lim_{h\to 0}\frac{h^2y-y^3}{h^2+y^2}=\frac{-y^3}{y^2}=-y$$

Also

$$f_y(x,0)=\lim_{k\to 0}\frac{f(x,k)-f(x,0)}{k}=\lim_{k\to 0}\left(\frac{\frac{xk(x^2-k^2)}{x^2+k^2}-0}{k}\right)=\lim_{k\to 0}\frac{x^3-k^2x}{x^2+k^2}=\frac{x^3}{x^2}=x.$$

Differentiability

Let (x,y), $(x+h,y+k)$ be the two neighbouring points in the domain D of $f(x,y)$.

$df=f(x+dx,y+dy)-f(x,y)=$ change of f when (x,y) is changed to $(x+dx,y+dy)$.

f is said to be differentiable at a point (x,y) if df can be expressed in the form

$$df=Adx+Bdy+dx\phi(dx,dy)+dy\psi(dx,dy) \tag{1}$$

When A,B are constants independent of dx,dy, and $\phi,\psi\to 0$ as $dx,dy\to 0$ simultaneously.

Also $Adx+Bdy$ is called the differential of f.

Thus $df=Adx+Bdy\Rightarrow f(x+dx,y+dy)-f(x,y)=Adx+Bdy$

From (1), when $(dx,dy)\to(0,0)$ we get $f(x+dx,y+dy)-f(x,y)\to 0$

$\Rightarrow \lim\limits_{(dx,dy)\to(0,0)} f(x+dx,y+dy)=f(x,y)$

$\Rightarrow f(x,y)$ is continuous at (x,y)

$\Rightarrow$ Every differentiable function is continuous.

Again from (1), taking $dy=0$, we get $f(x+dx,y)-f(x,y)=Adx+\phi(dx,0)dx$

$$\Rightarrow \lim_{dx\to 0}\frac{f(x+dx,y)-f(x,y)}{dx}=\lim_{dx\to 0}[A+\phi(dx,0)]$$

$$\Rightarrow f_x(x,y)=\frac{\partial f}{\partial x}=A \tag{2}$$

Similarly, taking $dx=0$ and taking $\lim\limits_{dy\to 0}$, we get

$$f_y(x,y)=\frac{\partial f}{\partial y}=B \tag{3}$$

From (2) and (3) it follows that the constants A and B are partial derivatives of f with respect to x and y respectively

$\Rightarrow$ differentiable function possesses the partial derivatives fx and fy.

Hence $df=\dfrac{\partial f}{\partial x}dx+\dfrac{\partial f}{\partial y}dy$.

Also a function differentiale at a point (a,b) in D domain of f if $df=Ah+Bk+h\phi(h,k)+k\psi(h,k)$ where $A=f_x$, $B=f_y$ and $\phi,\psi\to 0$ as $(h,k)\to(0,0)$.

Example (1). $f(x,y) = \begin{cases} x\sin\frac{1}{x} + y\sin\frac{1}{y}, & xy \neq 0 \\ x\sin\frac{1}{x}, & x \neq 0, y = 0 \\ y\sin\frac{1}{y}, & y \neq 0, x = 0 \\ 0, & x = 0, y = 0 \end{cases}$

is continuous but not differentiable at $(0,0)$.

We have $|f(x,y) - f(0,0)| = \left|x\sin\frac{1}{x} + y\sin\frac{1}{y}\right|$
$\leq |x|\left|\sin\frac{1}{x}\right| + |y|\left|\sin\frac{1}{y}\right|$
$\leq |x| + |y| < \epsilon$ whenever $|x| < \delta, |y| < \delta, \delta = \frac{\epsilon}{2}$

$\therefore f(x,y)$ is continuous at $(0,0)$.

Now $f_x(0,0) = \lim\limits_{h\to 0} \frac{f(h,0)-f(0,0)}{h} = \lim\limits_{h\to 0} \frac{h\sin\frac{1}{h}}{h} = \lim\limits_{h\to 0} \sin\frac{1}{h}$ which does not exist.

Similarly, $f_y(0,0)$ does not exist.

Therefore $f(x,y)$ is not differentiable at $(0,0)$.

Example (2) $f(x,y) = \sqrt{|xy|}$ is not differentiable at $(0,0)$ though f_x, f_y both exist at $(0,0)$.

We have $f_x(0,0) = \lim\limits_{h\to 0} \frac{f(h,0)-f(0,0)}{h} = \lim\limits_{h\to 0} \frac{0-0}{h} = 0$

Also $f_y(0,0) = 0$. So both f_x, f_y exist at $(0,0)$.

Now if f is differentiable at $(0,0)$, then we have $df = Ah + Bk + h\phi + k\psi$, where $A = f_x$, $B = f_y$ and $\phi, \psi \to 0$ as $(h,k) \to (0,0)$.

$\therefore df = f(h,k) - f(0,0) = \sqrt{hk}$.

$\therefore \sqrt{hk} = h\phi + k\psi$ $[\because f_x(0,0) = 0, f_y(0,0) = 0]$

Putting $h = r\cos\theta, k = r\sin\theta$, $r \to 0 \Rightarrow (h,k) \to (0,0)$ as θ is arbitrary.

Then $\sqrt{r\cos\theta \cdot r\sin\theta} = r\phi\sin\theta + r\psi\cos\theta$

$\Rightarrow \sqrt{\cos\theta\sin\theta} = \phi\sin\theta + \psi\cos\theta$

Now taking $\lim\limits_{r\to 0}$ where θ is arbitrary.

R.H.S $\to 0$ as $(h,k) \to (0,0)$ but $\sqrt{\cos\theta\sin\theta} \neq 0 \;\forall\; \theta$.

Hence f is not differentiable at $(0,0)$.

Theorem [12.2] [Existence theorem] Let $u = f(x,y)$ be a function of two variables x and y and let $(a,b) \in D$, the domain of definition of f such that

(i) $f(a,b) = 0$

(ii) the function f possesses continuous partial order derivatives f_x and f_y in the neighbourhood of (a,b) and

(iii) $f_y(a,b) \neq 0$, then there exists a rectangle $[a-h,\ a+h,\ b-k,\ b+k]$ about (a,b) such that for all $x \in [a-h,\ a+h]$ the function $f(x,y) = 0$, determines one and only one solution $y = \phi(x)$ in $[b-k, b+k]$ with the following properties:

(1) $b = \phi(a)$

(2) $f[x,\phi(x)] = 0 \;\forall\; x \in\ [a-h, a+h]$

(3) $\phi(x)$ is derivable and both $\phi(x)$ and $\phi'(x)$ are continuous on $[a-h, a+h]$.

(4) $\phi'(x) = -\frac{f_x}{f_y}$.

Proof. Let $f_y(a,b) > 0$.

Let f_x, f_y be continuous in a neighbourhood $R_1[a-h_1, a+h_1; b-k_1, b+k_1]$ of (a,b).

Continuity of f_x, f_y in $R_1 \Rightarrow$ continuity of f in R_1. Since f_y is continuous and $f_y(a,b) > 0$, there exists a neighbourhood $R_2[a-h_2, a+h_2, b-k_2, b+k_2]$, $h_2 < h_1, k_2 < k_1$ (i.e., $R_2 \subset R_1$) such that $f_y(x,y) > 0 \ \forall \ (x,y) \in R_2$.

Now since $f_y(x,y) > 0 \ \forall \ (x,y) \in R_2$, it follows that $\forall \ x \in [a-h_2, a+h_2]$ the function f of y strictly increases from $b-k_2$ to $b+k_2$.

In particular, $f(a,b) = 0$, we have $f(a, b-k) < 0$ and $f(a, b+k) > 0$, $k < k_2$, for f is strictly increasing function of y, when $y = b$.

Again, since f is continuous, there exists a (+)ve number $h < h_2$ such that $\forall \ x \in [a-h, a+h]$, $f(x, b-k) < 0$ and $f(x, b+k) > 0$.

Now for every fixed value of x in $[a-h, a+h]$, the continuous function of y strictly increases from negative to positive value as y increases from $b-k$ to $b+k$, therefore, there exists one value of y for which $f(x,y) = 0$. Thus for each x in $[a-h, a+h]$, there is a y in $[b-k, b+k]$ for which $f(x,y) = 0$, which concludes that y is a function of x, say $y = \phi(x)$, satisfying the properties (1) and (2).

Now we shall prove the *uniqueness of the solution* $y = \phi(x)$ for which $f(x,y) = 0$.

If possible, let y_1, y_2 be two values of y in $[b-k, b+k]$ so that $f(x, y_1) = 0, f(x, y_2) = 0$.

Now $f(x,y)$ is a function of single variable y for $x \in [a-h, a+h]$. Also $f_y(x,y)$ exists and continuous in $[a-h, a+h, b-k, b+k]$. Therefore, by Rolle's theorem, $f_y(x,y) = 0$ for some value of y in $[y_1, y_2]$, which contradicts that the fact $f_y(x,y) \neq 0$ in $[a-h, a+h, b-k, b+k]$.

Hence the solution is unique.

Next, let (x,y) and $(x+\Delta x, y+\Delta y) \in [a-h, a+h, b-k, b+k]$ such that $y = \phi(x)$ and $y + \Delta y = \phi(x+\Delta x)$, $f(x,y) = 0$, $f(x+\Delta x, y+\Delta y) = 0$.

$$\therefore \ f(x+\Delta x, y+\Delta y) - f(x,y) = 0$$

$$\Rightarrow \ f(x+\Delta x, y+\Delta y) - f(x+\Delta x, y) + f(x+\Delta x, y) - f(x,y) = 0$$

$$\Rightarrow \ \Delta y f_y(x+\Delta x, y+\theta\Delta y) + \Delta x f_x(x+\theta'\Delta x, y) = 0.$$

[by M.V. Theorem and $0 < \theta, \theta' < 1$]

Since $f_y(x,y) \neq 0$ in $[a-h, a+h, b-k, b+k]$, we have $\frac{\Delta y}{\Delta x} = -\frac{f_x(x+\theta'\Delta x, y)}{f_y(x+\Delta x, y+\theta\Delta y)}$.

f_x, f_y are continuous and $(\Delta x, \Delta y) \to (0,0)$

$$\Rightarrow \ \frac{dy}{dx} = -\frac{f_x(x,y)}{f_y(x,y)} \Rightarrow \phi'(x) = -\frac{f_x(x,y)}{f_y(x,y)}$$

$\Rightarrow \ \phi$ is derivable and so $\phi(x)$ is continuous in $[a-h, a+h, b-k, b+k]$.

Again, $\phi'(x)$ being the quotient of two continuous functions with $f_y(x,y) \neq 0$ is continuous in $[a-h, a+h, b-k, b+k]$.

Hence the theorem.

Example. Let $f(x,y) = x^2 + y^2 - 1$ and the point be $(0,1)$.

Then (i) $f(0,1) = 0$, (ii) $f_x = 2x$, $f_y = 2y$.

Both f_x, f_y are continuous in the neighbourhood $[0,1]$. Also (iii) $f_y(0,1) = 2 \neq 0$.

So all the conditions of the existence of the unique solution $y = \phi(x)$ are satisfied. Now $f(x,y) = 0 \Rightarrow y = \sqrt{1-x^2}$ and $y = -\sqrt{1-x^2}$.

(1) The solution is the implicit function $y = \sqrt{1-x^2}$ defined in the neighbourhood of $(0, 1)$, where $|x| < 1$ and $y > 0$.

(2) The solution $y = -\sqrt{1-x^2}$ is the implicit function defined in the neighbourhood of $(0, -1)$, where $|x| < 1, y < 0$.

Hence $y = \sqrt{1-x^2}$ is the unique implicit function near $(0, 1)$.

Problem [12.1] : Examine the existence of the unique solutions near the point $(1, -1)$ for $f(x, y) = y^4 + y^2x^2 - 2x^5 = 0$.

Solution. Here $f(1, -1) = 1 + 1 - 2 = 0$.

$$f_x = 2xy^2 - 10x^4$$
$$f_y = 4y^3 + 2yx^2.$$

f_x, f_y are both continuous about $(1, -1)$. Also $f_y(1, -1) = -4 - 2 = -6 \neq 0$.

All the conditions of the existence of the unique solution $y = \phi(x)$ are satisfied.

Now $f(x, y) = 0 \Rightarrow y^4 + y^2x^2 - 2x^5 = 0 \Rightarrow y^2 = \dfrac{-x^2 \pm \sqrt{x^4 + 8x^5}}{2} = \dfrac{-x^2 \pm x^2\sqrt{1+8x}}{2}$ of which $y^2 = \dfrac{-x^2 + x^2\sqrt{1+8x}}{2} = \dfrac{x^2\left\{\sqrt{1+8x} - 1\right\}}{2}$ determines the unique solution $y = \phi(x)$.

$$y = \pm\frac{x}{\sqrt{2}}\left\{\sqrt{1+8x} - 1\right\}^{1/2} \text{ for } x \geq 0$$

$$\Rightarrow y = \frac{x}{\sqrt{2}}\left\{\sqrt{1+8x} - 1\right\}^{1/2} \text{ and } y = -\frac{x}{\sqrt{2}}\left\{\sqrt{1+8x} - 1\right\}^{1/2}$$

are two implicit functions defined in the region $x \geq 0, y > 0$ and $x \geq 0, y < 0$.

But $\phi(1) = \frac{1}{\sqrt{2}} \cdot \sqrt{2} = 1 \neq -1$ for $y = \frac{x}{\sqrt{2}}\left\{\sqrt{1+8x} - 1\right\}$.

That is, $y = \frac{x}{\sqrt{2}}\left\{\sqrt{1+8x} - 1\right\}^{1/2}$ does not satisfy $b = \phi(a)$.

Therefore, $y = \frac{-x}{\sqrt{2}}\left\{\sqrt{1+8x} - 1\right\}, x \geq 0, y < 0$ is the unique implicit function near $(1, -1)$.

Problem [12.2] : Examine the existence of unique implicit function for $f(x, y) = 2xy - \log xy - 2 = 0$ near $(1, 1)$ and also find the first derivative.

Solution. $f(x, y) = 2xy - \log xy - 2 = 2xy - \log x - \log y - 2$

$f(1, 1) = 2 - 2 = 0$

$f_x(x, y) = 2y - \frac{1}{x}$

$f_y(x, y) = 2x - \frac{1}{y}$.

Both are continuous in $[1-h, 1+h, 1-k, 1+k]$, $h, k > 0$.

Also $f_y(1, 1) = 2 - 1 = 1 \neq 0$.

So $\exists$ a unique solution though it is not possible to find $y = \phi(x)$ directly from the given relation.

$$\phi'(x) = -\frac{f_x(x, y)}{f_y(x, y)}.$$

$$\therefore \phi'(x) = -\frac{2-1}{2-1} = -1 \text{ at } (1, 1).$$

Problem [12.3] : Show that $f(x,y) = x^3 + y^3 - 3xy + y = 0$ has a unique solution $y = \phi(x)$ near $(0,0)$ and find $\phi'(0)$.

Solution. Here $f_x(x,y) = 3x^2 - 3y$

$$f_y(x,y) = 3y^2 - 3x + 1.$$

$f(0,0) = 0$. Also $f_y(0,0) = 1 \neq 0$.

$f_x(x,y), f_y(x,y)$ are continuous in the neighbourhood of $(0,0)$.

Therefore, $\exists$ a unique solution $y = \phi(x)$ in the neighbourhood of $(0,0)$.

$$\phi'(0) = -\frac{f_x(0,0)}{f_y(0,0)} = \frac{0}{1} = 0.$$

12.2 Jacobians

Definition [12.7] : Let u and v be functions of two independent variables x and y. Then the determinant $\begin{vmatrix} \frac{\partial u}{\partial x} & \frac{\partial u}{\partial y} \\ \frac{\partial v}{\partial x} & \frac{\partial v}{\partial y} \end{vmatrix}$ is called the Jacobian of u and v with respect to x and y and is denoted by

$$\frac{\partial(u,v)}{\partial(x,y)} \text{ or } J\left(\frac{u,v}{x,y}\right).$$

Thus, if u,v,w are functions of three independent variables x,y,z, then the Jacobian of u,v,w with respect to x,y,z is defined by $\dfrac{\partial(u,v,w)}{\partial(x,y,z)} = \begin{vmatrix} \frac{\partial u}{\partial x} & \frac{\partial u}{\partial y} & \frac{\partial u}{\partial z} \\ \frac{\partial v}{\partial x} & \frac{\partial v}{\partial y} & \frac{\partial v}{\partial z} \\ \frac{\partial w}{\partial x} & \frac{\partial w}{\partial y} & \frac{\partial w}{\partial z} \end{vmatrix}$.

Example. Let $x = r\cos\theta; y = r\sin\theta$.

Here x,y are functions of two independent variables r and θ.

$$\text{Then } \frac{\partial(x,y)}{\partial(r,\theta)} = \begin{vmatrix} \frac{\partial x}{\partial r} & \frac{\partial x}{\partial \theta} \\ \frac{\partial y}{\partial r} & \frac{\partial y}{\partial \theta} \end{vmatrix} = \begin{vmatrix} \cos\theta & -r\sin\theta \\ \sin\theta & r\cos\theta \end{vmatrix} = r(\cos^2\theta + \sin^2\theta) = r.$$

Again, to find $\frac{\partial(r,\theta)}{\partial(x,y)}$, we express r and θ in terms of x and y.

So $r^2 = x^2 + y^2$ and $\theta = \tan^{-1}\frac{y}{x}$.

$$\frac{\partial r}{\partial x} = \frac{x}{r}, \frac{\partial \theta}{\partial x} = \frac{-y}{x^2+y^2} = -\frac{y}{r^2}.$$

$$\frac{\partial r}{\partial y} = \frac{y}{r}, \frac{\partial \theta}{\partial y} = \frac{x}{x^2+y^2} = \frac{x}{r^2}.$$

$$\text{Therefore, } \frac{\partial(r,\theta)}{\partial(x,y)} = \begin{vmatrix} \frac{\partial r}{\partial x} & \frac{\partial r}{\partial y} \\ \frac{\partial \theta}{\partial x} & \frac{\partial \theta}{\partial y} \end{vmatrix} = \begin{vmatrix} \frac{x}{r} & \frac{y}{r} \\ -\frac{y}{r^2} & \frac{x}{r^2} \end{vmatrix} = \frac{x^2+y^2}{r^3} = \frac{r^2}{r^3} = \frac{1}{r}.$$

Example. Let $x = r\sin\theta\cos\phi$, $y = r\sin\theta\sin\phi$ and $z = r\cos\theta$. Then x,y,z are functions of r,θ and ϕ.

Let us find out $\dfrac{\partial(x,y,z)}{\partial(r,\theta,\phi)}$.

We determine,

$$\frac{\partial x}{\partial r} = \sin\theta\cos\phi, \quad \frac{\partial x}{\partial \theta} = r\cos\theta\cos\phi, \quad \frac{\partial x}{\partial \phi} = -r\sin\theta\sin\phi$$
$$\frac{\partial y}{\partial r} = \sin\theta\sin\phi, \quad \frac{\partial y}{\partial \theta} = r\cos\theta\sin\phi, \quad \frac{\partial y}{\partial \phi} = r\sin\theta\cos\phi$$
$$\frac{\partial z}{\partial r} = \cos\theta, \quad \frac{\partial z}{\partial \theta} = -r\sin\theta, \quad \frac{\partial z}{\partial \phi} = 0.$$

Then

$$\frac{\partial(x,y,z)}{\partial(r,\theta,\phi)} = \begin{vmatrix} \frac{\partial x}{\partial r} & \frac{\partial x}{\partial \theta} & \frac{\partial x}{\partial \phi} \\ \frac{\partial y}{\partial r} & \frac{\partial y}{\partial \theta} & \frac{\partial y}{\partial \phi} \\ \frac{\partial z}{\partial r} & \frac{\partial z}{\partial \theta} & \frac{\partial z}{\partial \phi} \end{vmatrix} = \begin{vmatrix} \sin\theta\cos\phi & r\cos\theta\cos\phi & -r\sin\theta\sin\phi \\ \sin\theta\sin\phi & r\cos\theta\sin\phi & r\sin\theta\cos\phi \\ \cos\theta & -r\sin\theta & 0 \end{vmatrix}$$
$$= r^2 \begin{vmatrix} \sin\theta\sin\phi & \cos\theta\cos\phi & -\sin\theta\sin\phi \\ \sin\theta\sin\phi & \cos\theta\sin\phi & \sin\theta\cos\phi \\ \cos\theta & -\sin\theta & 0 \end{vmatrix}$$
$$= r^2 \cdot \sin\theta \text{ [expanding the determinant].}$$

Properties

Theorem [12.3] (1) If u and v are functions of x and y, then $\dfrac{\partial(u,v)}{\partial(x,y)} \times \dfrac{\partial(x,y)}{\partial(u,v)} = 1$.

(2) If the functions u, v, w of three independent variables x, y, z are not independent, then $\dfrac{\partial(u,v,w)}{\partial(x,y,z)} = 0$.

(3) If the variables x, y, u, v are connected by the relations $f_1(x,y,u,v) = 0$ and $f_2(x,y,u,v) = 0$, then $\dfrac{\partial(f_1,f_2)}{\partial(u,v)} \times \dfrac{\partial(u,v)}{\partial(x,y)} = (-1)^2 \dfrac{\partial(f_1,f_2)}{\partial(x,y)}$.

(4) If $J = \dfrac{\partial(f_1,f_2)}{\partial(u,v)} \neq 0$, then $\dfrac{\partial u}{\partial x} = -\dfrac{\frac{\partial(f_1,f_2)}{\partial(x,v)}}{J}$, $\dfrac{\partial u}{\partial y} = -\dfrac{\frac{\partial(f_1,f_2)}{\partial(y,v)}}{J}$, $\dfrac{\partial v}{\partial x} = -\dfrac{\frac{\partial(f_1,f_2)}{\partial(u,x)}}{J}$ and $\dfrac{\partial v}{\partial y} = -\dfrac{\frac{\partial(f_1,f_2)}{\partial(u,y)}}{J}$.

Proof. (1) $\dfrac{\partial(u,v)}{\partial(x,y)} \times \dfrac{\partial(x,y)}{\partial(u,v)}$

$$= \begin{vmatrix} \frac{\partial u}{\partial x} & \frac{\partial u}{\partial y} \\ \frac{\partial v}{\partial x} & \frac{\partial v}{\partial y} \end{vmatrix} \times \begin{vmatrix} \frac{\partial x}{\partial u} & \frac{\partial x}{\partial v} \\ \frac{\partial y}{\partial u} & \frac{\partial y}{\partial v} \end{vmatrix} = \begin{vmatrix} \frac{\partial u}{\partial x}\cdot\frac{\partial x}{\partial u} + \frac{\partial u}{\partial y}\cdot\frac{\partial y}{\partial u} & \frac{\partial u}{\partial x}\cdot\frac{\partial x}{\partial v} + \frac{\partial u}{\partial y}\cdot\frac{\partial y}{\partial v} \\ \frac{\partial v}{\partial x}\cdot\frac{\partial x}{\partial u} + \frac{\partial v}{\partial y}\cdot\frac{\partial y}{\partial u} & \frac{\partial v}{\partial x}\cdot\frac{\partial x}{\partial v} + \frac{\partial v}{\partial y}\cdot\frac{\partial y}{\partial v} \end{vmatrix} \quad \text{(i)}$$

$$u = f(x,y) \qquad v = g(x,y)$$
$$\Rightarrow \frac{\partial u}{\partial u} = 1 = \frac{\partial u}{\partial x}\cdot\frac{\partial x}{\partial u} + \frac{\partial u}{\partial y}\cdot\frac{\partial y}{\partial u} \qquad \frac{\partial v}{\partial v} = 1 = \frac{\partial v}{\partial x}\cdot\frac{\partial x}{\partial v} + \frac{\partial v}{\partial y}\cdot\frac{\partial y}{\partial v} \quad \text{(ii)}$$
$$\frac{\partial u}{\partial v} = 0 = \frac{\partial u}{\partial x}\cdot\frac{\partial x}{\partial v} + \frac{\partial u}{\partial y}\cdot\frac{\partial y}{\partial v} \qquad \frac{\partial v}{\partial u} = 0 = \frac{\partial v}{\partial x}\cdot\frac{\partial x}{\partial u} + \frac{\partial v}{\partial y}\cdot\frac{\partial y}{\partial u}$$

From (i) by (ii) $\dfrac{\partial(u,v)}{\partial(x,y)} \times \dfrac{\partial(x,y)}{\partial(u,v)} = \begin{vmatrix} 1 & 0 \\ 0 & 1 \end{vmatrix} = 1.$

(2) If u, v, w are not independent, then $f(u,v,w) = 0$. (i)

Then differentiating (i) w.r.t. x, y and z respectively, we get the following results:

$$\frac{\partial f}{\partial u}\cdot\frac{\partial u}{\partial x} + \frac{\partial f}{\partial v}\cdot\frac{\partial v}{\partial x} + \frac{\partial f}{\partial w}\cdot\frac{\partial w}{\partial x} = 0 \quad \text{(ii)}$$

$$\frac{\partial f}{\partial u}\cdot\frac{\partial u}{\partial y} + \frac{\partial f}{\partial v}\cdot\frac{\partial v}{\partial y} + \frac{\partial f}{\partial w}\cdot\frac{\partial w}{\partial y} = 0 \quad \text{(iii)}$$

$$\frac{\partial f}{\partial u}\cdot\frac{\partial u}{\partial z} + \frac{\partial f}{\partial v}\cdot\frac{\partial v}{\partial z} + \frac{\partial f}{\partial w}\cdot\frac{\partial w}{\partial z} = 0. \quad \text{(iv)}$$

Eliminating $\dfrac{\partial f}{\partial u}, \dfrac{\partial f}{\partial v}, \dfrac{\partial f}{\partial w}$ from (ii), (iii) and (iv), we have

$$\begin{vmatrix} \frac{\partial u}{\partial x} & \frac{\partial v}{\partial x} & \frac{\partial w}{\partial x} \\ \frac{\partial u}{\partial y} & \frac{\partial v}{\partial y} & \frac{\partial w}{\partial y} \\ \frac{\partial u}{\partial z} & \frac{\partial v}{\partial z} & \frac{\partial w}{\partial z} \end{vmatrix} = 0 \Rightarrow \begin{vmatrix} \frac{\partial u}{\partial x} & \frac{\partial u}{\partial y} & \frac{\partial u}{\partial z} \\ \frac{\partial v}{\partial x} & \frac{\partial v}{\partial y} & \frac{\partial v}{\partial z} \\ \frac{\partial w}{\partial x} & \frac{\partial w}{\partial y} & \frac{\partial w}{\partial z} \end{vmatrix} = 0 \Rightarrow \frac{\partial(u,v,w)}{\partial(x,y,z)} = 0.$$

(3) $f_1(x,y,u,v) = 0$ and $f_2(x,y,u,v) = 0$, where u, v are implicit functions of x and y.

Differentiating them w.r.t. x and y, we get

$$\left.\begin{aligned} \frac{\partial f_1}{\partial x} + \frac{\partial f_1}{\partial u}\cdot\frac{\partial u}{\partial x} + \frac{\partial f_1}{\partial v}\cdot\frac{\partial v}{\partial x} &= 0 \\ \frac{\partial f_1}{\partial y} + \frac{\partial f_1}{\partial u}\cdot\frac{\partial u}{\partial y} + \frac{\partial f_1}{\partial v}\cdot\frac{\partial v}{\partial y} &= 0 \\ \frac{\partial f_2}{\partial x} + \frac{\partial f_2}{\partial u}\cdot\frac{\partial u}{\partial x} + \frac{\partial f_2}{\partial v}\cdot\frac{\partial v}{\partial x} &= 0 \\ \frac{\partial f_2}{\partial y} + \frac{\partial f_2}{\partial u}\cdot\frac{\partial u}{\partial y} + \frac{\partial f_2}{\partial v}\cdot\frac{\partial v}{\partial y} &= 0 \end{aligned}\right\} \quad \text{(i)}$$

Now

$$\frac{\partial(f_1,f_2)}{\partial(u,v)} \times \frac{\partial(u,v)}{\partial(x,y)} = \begin{vmatrix} \frac{\partial f_1}{\partial u} & \frac{\partial f_1}{\partial v} \\ \frac{\partial f_2}{\partial u} & \frac{\partial f_2}{\partial v} \end{vmatrix} \times \begin{vmatrix} \frac{\partial u}{\partial x} & \frac{\partial u}{\partial y} \\ \frac{\partial v}{\partial x} & \frac{\partial v}{\partial y} \end{vmatrix}$$

$$= \begin{vmatrix} \frac{\partial f_1}{\partial u}\cdot\frac{\partial u}{\partial x} + \frac{\partial f_1}{\partial v}\cdot\frac{\partial v}{\partial x} & \frac{\partial f_1}{\partial u}\cdot\frac{\partial u}{\partial y} + \frac{\partial f_1}{\partial v}\cdot\frac{\partial v}{\partial y} \\ \frac{\partial f_2}{\partial u}\cdot\frac{\partial u}{\partial x} + \frac{\partial f_2}{\partial v}\cdot\frac{\partial v}{\partial x} & \frac{\partial f_2}{\partial u}\cdot\frac{\partial u}{\partial y} + \frac{\partial f_2}{\partial v}\cdot\frac{\partial v}{\partial y} \end{vmatrix}$$

$$= \begin{vmatrix} -\frac{\partial f_1}{\partial x} & -\frac{\partial f_1}{\partial y} \\ -\frac{\partial f_2}{\partial x} & -\frac{\partial f_2}{\partial y} \end{vmatrix} \quad \text{[by (i)]}$$

$$= (-1)^2 \frac{\partial(f_1,f_2)}{\partial(x,y)}.$$

(4) Taking total differential of $f_1(x, y, u, v) = 0$ and $f_2(x, y, u, v) = 0$, we get

$$df_1 = \frac{\partial f_1}{\partial x}dx + \frac{\partial f_1}{\partial y}dy + \frac{\partial f_1}{\partial u}du + \frac{\partial f_1}{\partial v}dv = 0 \qquad \text{(i)}$$

$$df_2 = \frac{\partial f_2}{\partial x}dx + \frac{\partial f_2}{\partial y}dy + \frac{\partial f_2}{\partial u}du + \frac{\partial f_2}{\partial v}dv = 0. \qquad \text{(ii)}$$

We rewrite (i) and (ii) by

$$\frac{\partial f_1}{\partial u}du + \frac{\partial f_1}{\partial v}dv = -\left(\frac{\partial f_1}{\partial x}dx + \frac{\partial f_1}{\partial y}dy\right) \qquad \text{(iii)}$$

$$\frac{\partial f_2}{\partial u}du + \frac{\partial f_2}{\partial v}dv = -\left(\frac{\partial f_2}{\partial x}dx + \frac{\partial f_2}{\partial y}dy\right). \qquad \text{(iv)}$$

We use Cramer's rule in (iii) and (iv) for finding du and dv.

Then

$$du = \frac{\begin{vmatrix} -\left(\frac{\partial f_1}{\partial x}dx + \frac{\partial f_1}{\partial y}dy\right) & \frac{\partial f_1}{\partial v} \\ -\left(\frac{\partial f_2}{\partial x}dx + \frac{\partial f_2}{\partial y}dy\right) & \frac{\partial f_2}{\partial v} \end{vmatrix}}{\begin{vmatrix} \frac{\partial f_1}{\partial u} & \frac{\partial f_1}{\partial v} \\ \frac{\partial f_2}{\partial u} & \frac{\partial f_2}{\partial v} \end{vmatrix}} \Rightarrow du = \frac{-\begin{vmatrix} \frac{\partial f_1}{\partial x} & \frac{\partial f_1}{\partial v} \\ \frac{\partial f_2}{\partial x} & \frac{\partial f_2}{\partial v} \end{vmatrix} dx - \begin{vmatrix} \frac{\partial f_1}{\partial y} & \frac{\partial f_1}{\partial v} \\ \frac{\partial f_2}{\partial y} & \frac{\partial f_2}{\partial v} \end{vmatrix} dy}{J},$$

where $J = \dfrac{\partial(f_1, f_2)}{\partial(u, v)} \neq 0$.

$$\Rightarrow \frac{\partial u}{\partial x}dx + \frac{\partial u}{\partial y}dy = -\frac{\frac{\partial(f_1, f_2)}{\partial(x, v)}}{J}dx - \frac{\frac{\partial(f_1, f_2)}{\partial(y, v)}}{J}dy \Rightarrow \frac{\partial u}{\partial x} = -\frac{\frac{\partial(f_1, f_2)}{\partial(x, v)}}{J}, \frac{\partial u}{\partial y} = -\frac{\frac{\partial(f_1, f_2)}{\partial(y, v)}}{J}.$$

Similarly,

$$dv = \frac{-\begin{vmatrix} \frac{\partial f_1}{\partial u} & \frac{\partial f_1}{\partial x} \\ \frac{\partial f_2}{\partial u} & \frac{\partial f_2}{\partial x} \end{vmatrix} dx - \begin{vmatrix} \frac{\partial f_1}{\partial u} & \frac{\partial f_1}{\partial y} \\ \frac{\partial f_2}{\partial u} & \frac{\partial f_2}{\partial y} \end{vmatrix} dy}{J} \Rightarrow \frac{\partial v}{\partial x} = -\frac{\frac{\partial(f_1, f_2)}{\partial(u, x)}}{J} \text{ and } \frac{\partial v}{\partial y} = -\frac{\frac{\partial(f_1, f_2)}{\partial(u, y)}}{J}.$$

Example. Let $u = x^2 - y^2$ and $v = 2xy$.

To find $\dfrac{\partial(x, y)}{\partial(u, v)}$, we apply the rule $\dfrac{\partial(u, v)}{\partial(x, y)} \cdot \dfrac{\partial(x, y)}{\partial(u, v)} = 1$ [without expressing x and y as a function of u and v].

Here $\dfrac{\partial u}{\partial x} = 2x$, $\dfrac{\partial u}{\partial y} = -2y$, $\dfrac{\partial v}{\partial x} = 2y$, $\dfrac{\partial v}{\partial y} = 2x$.

$$\frac{\partial(u, v)}{\partial(x, y)} = \begin{vmatrix} 2x & -2y \\ 2y & 2x \end{vmatrix} = 4(x^2 + y^2).$$

Therefore, $\dfrac{\partial(x, y)}{\partial(u, v)} = \dfrac{1}{\frac{\partial(u, v)}{\partial(x, y)}} = \dfrac{1}{4(x^2 + y^2)}$.

Example. Let $u = e^x \cos y,\ v = e^x \sin y$, where $x = lr + sm$ and $y = mr - sl$.
Then verify

$$\frac{\partial(u,v)}{\partial(x,y)} \times \frac{\partial(x,y)}{\partial(r,s)} = \frac{\partial(u,v)}{\partial(r,s)} \tag{1}$$

Here $\dfrac{\partial u}{\partial x} = e^x \cos y,\quad \dfrac{\partial u}{\partial y} = -e^x \sin y,\quad \dfrac{\partial v}{\partial x} = e^x \sin y,\quad \dfrac{\partial v}{\partial y} = e^x \cos y$

$\dfrac{\partial x}{\partial r} = l,\quad \dfrac{\partial x}{\partial s} = m\quad \dfrac{\partial y}{\partial r} = m.\quad \dfrac{\partial y}{\partial s} = -l.$

Then L.H.S. of (1)

$$= \begin{vmatrix} e^x \cos y & -e^x \sin y \\ e^x \sin y & e^x \cos y \end{vmatrix} \times \begin{vmatrix} l & m \\ m & -l \end{vmatrix} = -e^{2x} \cdot (l^2 + m^2) \tag{2}$$

Again

$u = e^{lr+sm} \cos(mr - sl),\ v = e^{lr+sm} \sin(mr - sl)$

$\dfrac{\partial u}{\partial r} = le^{lr+sm} \cos(mr - sl) - me^{lr+sm} \sin(mr - sl) = e^{lr+sm}[l \cos(mr - sl) - m \sin(mr - sl]$

$\dfrac{\partial u}{\partial s} = e^{lr+sm}[m \cos(mr - sl) + l \sin(mr - sl)]$

$\dfrac{\partial v}{\partial r} = e^{lr+sm}[l \sin(mr - sl) + m \cos(mr - sl)]$

$\dfrac{\partial v}{\partial s} = e^{lr+sm}[m \sin(mr - sl) - l \cos(mr - sl)].$

R.H.S. of (1),

$$\frac{\partial(u,v)}{\partial(r,s)} = e^{2(lr+sm)} \begin{vmatrix} l \cos y - m \sin y & m \cos y + l \sin y \\ l \sin y + m \cos y & m \sin y - l \cos y \end{vmatrix}$$

$$= e^{2x}\left[-(m \sin y - l \cos y)^2 - (m \cos y + l \sin y)^2\right] = -e^{2x}\,(l^2 + m^2). \tag{3}$$

Result follows from (2) and (3).

Problem [12.4]: Are $x + y - z,\ x - y + z,\ x^2 + y^2 + z^2 - 2yz$ functionally dependent? If so, find a relation between them.

Solution. Let $u = x + y - z,\ v = x - y + z,\ w = x^2 + y^2 + z^2 - 2yz$

$$\frac{\partial u}{\partial x} = 1,\quad \frac{\partial u}{\partial y} = 1,\quad \frac{\partial u}{\partial z} = -1$$

$$\frac{\partial v}{\partial x} = 1,\quad \frac{\partial v}{\partial y} = -1,\quad \frac{\partial v}{\partial z} = 1$$

$$\frac{\partial w}{\partial x} = 2x,\quad \frac{\partial w}{\partial y} = 2y - 2z,\quad \frac{\partial w}{\partial z} = 2z - 2y.$$

$$\frac{\partial(u,v,w)}{\partial(x,y,z)} = \begin{vmatrix} \frac{\partial u}{\partial x} & \frac{\partial u}{\partial y} & \frac{\partial u}{\partial z} \\ \frac{\partial v}{\partial x} & \frac{\partial v}{\partial y} & \frac{\partial v}{\partial z} \\ \frac{\partial w}{\partial x} & \frac{\partial w}{\partial y} & \frac{\partial w}{\partial z} \end{vmatrix} = \begin{vmatrix} 1 & 1 & -1 \\ 1 & -1 & 1 \\ 2x & 2(y-z) & 2(z-y) \end{vmatrix} = \begin{vmatrix} 2 & 0 & 0 \\ 1 & -1 & 1 \\ 2x & 2(y-z) & 2(z-y) \end{vmatrix}$$

$$= 2\{-2(z - y) - 2(y - z)\} = 4 \times 0 = 0.$$

Therefore, u, v, w are dependent.

Again, $u^2 = \Sigma x^2 + 2xy - 2yz - 2zx,\ v^2 = \Sigma x^2 - 2xy - 2yz + 2zx.$

$\therefore u^2 + v^2 = 2[\Sigma x^2 - 2yz] = 2w$ which is a relation between u, v and w.

12.3 Jacobians (General Form)

Definition [12.8] : If $u_1, u_2, \cdots, u_n$ be differentiable functions of n-variables $x_1, x_2, \cdots, x_n$, then the determinant

$$\begin{vmatrix} \frac{\partial u_1}{\partial x_1} & \frac{\partial u_1}{\partial x_2} & \cdots & \frac{\partial u_1}{\partial x_n} \\ \frac{\partial u_2}{\partial x_1} & \frac{\partial u_2}{\partial x_2} & \cdots & \frac{\partial u_2}{\partial x_n} \\ \cdots & \cdots & \cdots & \cdots \\ \frac{\partial u_n}{\partial x_1} & \frac{\partial u_n}{\partial x_2} & \cdots & \frac{\partial u_n}{\partial x_n} \end{vmatrix}$$

is called the Jacobian of $u_1, u_2, \cdots, u_n$ with respect to $x_1, x_2, \cdots, u_n$ and is denoted by

$$\frac{\partial(u_1, u_2, \cdots, u_n)}{\partial(x_1, x_2, \cdots, x_n)} \text{ or } J\left(\frac{u_1, u_2, \cdots, u_n}{x_1, x_2, \cdots, x_n}\right).$$

If the functions $u_1, u_2, \cdots, u_n$ are defined by $u_1 = f_1(x_1), u_2 = f_2(x_1, x_2), \cdots, u_n = f_n(x_1, x_2, \cdots, x_n)$, then

$$\frac{\partial(u_1, u_2, \cdots, u_n)}{\partial(x_1, x_2, \cdots, x_n)} = \begin{vmatrix} \frac{\partial u_1}{\partial x_1} & 0 & 0 & \cdots & 0 \\ \frac{\partial u_2}{\partial x_1} & \frac{\partial u_2}{\partial x_2} & 0 & \cdots & 0 \\ \cdots & \cdots & \cdots & \cdots & \cdots \\ \frac{\partial u_n}{\partial x_1} & \frac{\partial u_n}{\partial x_2} & \frac{\partial u_n}{\partial x_3} & \cdots & \frac{\partial u_n}{\partial x_n} \end{vmatrix} = \frac{\partial u_1}{\partial x_1} \cdot \frac{\partial u_2}{\partial x_2} \cdot \frac{\partial u_3}{\partial x_3} \cdots \frac{\partial u_n}{\partial x_n}.$$

Example. If $u_1, u_2, \cdots, u_n$ are defined by $u_1 = 1 - x_1, u_2 = x_1(1 - x_2), u_3 = x_1x_2(1 - x_3), \cdots, u_n = x_1 \cdot x_2 \cdot x_3 \cdots x_{n-1}(1 - x_n)$, then

$$\frac{\partial(u_1, u_2, \cdots, u_n)}{\partial(x_1, x_2, \cdots, x_n)} = (-1)^n x_1^{n-1} x_2^{n-2} \cdots x_{n-1}.$$

Here $u_1 = f_1(x_1), u_2 = f_2(x_1, x_2), \cdots, u_n = f_n(x_1, x_2, \cdots, x_n)$.

$$\therefore \frac{\partial(u_1, u_2, \cdots, u_n)}{\partial(x_1, x_2, \cdots, x_n)} = \frac{\partial u_1}{\partial x_1} \cdot \frac{\partial u_2}{\partial x_2} \cdots \frac{\partial u_n}{\partial x_n} = (-1)(-x_1)(-x_1 \cdot x_2) \cdots (-x_1x_2 \cdots x_{n-1})$$
$$= (-1)^n x_1^{n-1} x_2^{n-2} \cdots x_{n-1}.$$

Corollary [12.1] : If $u_1 = \cos x_1, u_2 = \sin x_1 \cos x_2, y_3 = \sin x_1 \sin x_2 \cos x_3, \cdots, u_n = \sin x_1 \sin x_2 \cdots \sin x_{n-1} \cos x_n$, then

$$\frac{\partial(u_1, u_2, \cdots, u_n)}{\partial(x_1, x_2, \cdots, x_n)} = (-1)^n \sin^n x_1 \sin^{n-1} x_2 \cdots \sin x_n.$$

Theorem [12.4] [**Chain rule**] If $u_1, u_2, \cdots, u_n$ are functions of $y_1, y_2, \cdots, y_n$ and $y_1, y_2, \cdots, y_n$ are functions of $x_1, x_2, \cdots, x_n$, then

$$\frac{\partial(u_1, u_2, \cdots, u_n)}{\partial(x_1, x_2, \cdots, x_n)} = \frac{\partial(u_1, u_2, \cdots, u_n)}{\partial(y_1, y_2, \cdots, y_n)} \cdot \frac{\partial(y_1, y_2, \cdots, y_n)}{\partial(x_1, x_2, \cdots, x_n)}.$$

Proof. We have

$$u_1 = f_1(y_1, y_2, \cdots, y_n), u_2 = f_2(y_1, y_2, \cdots, y_n) \text{ and } u_n = f_n(y_1, y_2, \cdots, y_n).$$

Then

$$\frac{\partial u_1}{\partial x_1} = \frac{\partial u_1}{\partial y_1}\cdot\frac{\partial y_1}{\partial x_1} + \frac{\partial u_1}{\partial y_2}\cdot\frac{\partial y_2}{\partial x_1} + \cdots + \frac{\partial u_1}{\partial y_n}\cdot\frac{\partial y_n}{\partial x_1} = \sum_{r=1}^{n}\frac{\partial u_1}{\partial y_r}\cdot\frac{\partial y_r}{\partial x_1}$$

$$\frac{\partial u_1}{\partial x_2} = \frac{\partial u_1}{\partial y_1}\cdot\frac{\partial y_1}{\partial x_2} + \frac{\partial u_1}{\partial y_2}\cdot\frac{\partial y_2}{\partial x_2} + \cdots + \frac{\partial u_1}{\partial y_n}\cdot\frac{\partial y_n}{\partial x_2} = \sum\frac{\partial u_1}{\partial y_r}\cdot\frac{\partial y_r}{\partial x_2}$$

$$\cdots\cdots\cdots\cdots\cdots\cdots\cdots\cdots\cdots\cdots\cdots\cdots$$

$$\frac{\partial u_1}{\partial x_n} = \sum\frac{\partial u_1}{\partial y_r}\cdot\frac{\partial y_r}{\partial x_n}.$$

Proceeding as above,

$$\frac{\partial u_2}{\partial x_1} = \sum\frac{\partial u_2}{\partial y_r}\cdot\frac{\partial y_r}{\partial x_1}, \ldots, \frac{\partial u_2}{\partial x_n} = \sum\frac{\partial u_2}{\partial y_r}\cdot\frac{\partial y_r}{\partial x_n}$$

$$\frac{\partial u_n}{\partial x_1} = \sum\frac{\partial u_n}{\partial y_r}\cdot\frac{\partial y_r}{\partial x_1}, \ldots, \frac{\partial u_n}{\partial x_n} = \sum\frac{\partial u_n}{\partial y_r}\cdot\frac{\partial y_r}{\partial x_n}. \qquad \text{(i)}$$

Now $\dfrac{\partial(u_1, u_2, \cdots, u_n)}{\partial(y_1, y_2, \cdots, y_n)}\cdot\dfrac{\partial(y_1, y_2, \cdots, y_n)}{\partial(x_1, x_2, \cdots, x_n)}$

$$= \begin{vmatrix} \frac{\partial u_1}{\partial y_1} & \frac{\partial u_1}{\partial y_2} & \cdots & \frac{\partial u_1}{\partial y_n} \\ \frac{\partial u_2}{\partial y_1} & \frac{\partial u_2}{\partial y_2} & \cdots & \frac{\partial u_2}{\partial y_n} \\ \cdots & \cdots & \cdots & \cdots \\ \frac{\partial u_n}{\partial y_1} & \frac{\partial u_n}{\partial y_2} & \cdots & \frac{\partial u_n}{\partial y_n} \end{vmatrix} \times \begin{vmatrix} \frac{\partial y_1}{\partial x_1} & \frac{\partial y_1}{\partial x_2} & \cdots & \frac{\partial y_1}{\partial x_n} \\ \frac{\partial y_2}{\partial x_1} & \frac{\partial y_2}{\partial x_2} & \cdots & \frac{\partial y_2}{\partial x_n} \\ \cdots & \cdots & \cdots & \cdots \\ \frac{\partial y_n}{\partial x_1} & \frac{\partial y_n}{\partial x_2} & \cdots & \frac{\partial y_n}{\partial x_n} \end{vmatrix}$$

$$= \begin{vmatrix} \sum\frac{\partial u_1}{\partial y_r}\cdot\frac{\partial y_r}{\partial x_1} & \sum\frac{\partial u_1}{\partial y_r}\cdot\frac{\partial y_r}{\partial x_2} & \cdots & \sum\frac{\partial u_1}{\partial y_r}\cdot\frac{\partial y_r}{\partial x_n} \\ \sum\frac{\partial u_2}{\partial y_r}\cdot\frac{\partial y_r}{\partial x_1} & \sum\frac{\partial u_2}{\partial y_r}\cdot\frac{\partial y_r}{\partial x_2} & \cdots & \sum\frac{\partial u_2}{\partial y_r}\cdot\frac{\partial y_r}{\partial x_r} \\ \cdots\cdots & \cdots\cdots & \cdots & \cdots\cdots \\ \sum\frac{\partial u_n}{\partial y_r}\cdot\frac{\partial y_r}{\partial x_1} & \sum\frac{\partial u_n}{\partial y_r}\cdot\frac{\partial y_r}{\partial x_2} & \cdots & \sum\frac{\partial u_n}{\partial y_r}\cdot\frac{\partial y_r}{\partial x_n} \end{vmatrix} \left[\begin{array}{l}\text{Multiplying row by}\\ \text{column operations}\end{array}\right]$$

$$= \begin{vmatrix} \frac{\partial u_1}{\partial x_1} & \frac{\partial u_1}{\partial x_2} & \cdots & \frac{\partial u_1}{\partial x_n} \\ \frac{\partial u_2}{\partial x_1} & \frac{\partial u_2}{\partial x_2} & \cdots & \frac{\partial u_2}{\partial x_n} \\ \cdots & \cdots & \cdots & \cdots \\ \frac{\partial u_n}{\partial x_1} & \frac{\partial u_n}{\partial x_2} & \cdots & \frac{\partial u_n}{\partial x_n} \end{vmatrix} \quad \text{[by using results in (i)]}$$

$$= \frac{\partial(u_1, u_2, \cdots, u_n)}{\partial(x_1, x_2, \cdots, x_n)}.$$

Hence the result.

Theorem [12.5] Show that $\dfrac{\partial(y_1, y_2, \cdots, y_n)}{\partial(x_1, x_2, \cdots, x_n)}\cdot\dfrac{\partial(x_1, x_2, \cdots, x_n)}{\partial(y_1, y_2, \cdots, y_n)} = 1.$

Proof. Let $y_1 = f_1(x_1, x_2, \cdots, x_n), y_2 = f_2(x_1, x_2, \cdots, x_n), \cdots, y_n = f_n(x_1, x_2, \cdots, x_n)$, which can be put in the following forms:

$$x_1 = F_1(y_1, y_2, \cdots, y_n), x_2 = F_2(y_1, y_2, \cdots, y_n), \cdots, x_n = F_n(y_1, y_2, \cdots, y_n).$$

Now

$$1 = \frac{\partial f_1}{\partial x_1} \cdot \frac{\partial x_1}{\partial y_1} + \frac{\partial f_1}{\partial x_2} \cdot \frac{\partial x_2}{\partial y_1} + \cdots + \frac{\partial f_1}{\partial x_n} \cdot \frac{\partial x_n}{\partial y_1} = \sum \frac{\partial f_1}{\partial x_r} \cdot \frac{\partial x_r}{\partial y_1} = \sum \frac{\partial y_1}{\partial x_r} \cdot \frac{\partial x_r}{\partial y_1}$$

$$0 = \frac{\partial f_1}{\partial x_1} \cdot \frac{\partial x_1}{\partial y_2} + \frac{\partial f_1}{\partial x_2} \cdot \frac{\partial x_2}{\partial y_2} + \cdots + \frac{\partial f_1}{\partial x_n} \cdot \frac{\partial x_n}{\partial y_2} = \sum \frac{\partial f_1}{\partial x_r} \cdot \frac{\partial x_r}{\partial y_2} = \sum \frac{\partial y_1}{\partial x_r} \cdot \frac{\partial x_r}{\partial y_2}$$

$$0 = \sum \frac{\partial f_1}{\partial x_r} \cdot \frac{\partial x_r}{\partial y_n} = \sum \frac{\partial y_1}{\partial x_r} \cdot \frac{\partial x_r}{\partial y_n}.$$

Thus in general

$$\sum \frac{\partial y_p}{\partial x_r} \cdot \frac{\partial x_r}{\partial y_p} = 1 \text{ for } p = 1, 2, \ldots, n \text{ and } \sum \frac{\partial y_p}{\partial x_r} \cdot \frac{\partial x_r}{\partial y_q} = 0 \text{ for } p \neq q$$

Now $\dfrac{\partial(y_1, y_2, \cdots, y_n)}{\partial(x_1, x_2, \cdots, x_n)} \cdot \dfrac{\partial(x_1, x_2, \cdots, x_n)}{\partial(y_1, y_2, \cdots, y_n)}$

$$= \begin{vmatrix} \frac{\partial y_1}{\partial x_1} & \frac{\partial y_1}{\partial x_2} & \cdots & \frac{\partial y_1}{\partial x_n} \\ \frac{\partial y_2}{\partial x_1} & \frac{\partial y_2}{\partial x_2} & \cdots & \frac{\partial y_2}{\partial x_n} \\ \cdots & \cdots & \cdots & \cdots \\ \frac{\partial y_n}{\partial x_1} & \frac{\partial y_n}{\partial x_2} & \cdots & \frac{\partial y_n}{\partial x_n} \end{vmatrix} \times \begin{vmatrix} \frac{\partial x_1}{\partial y_1} & \frac{\partial x_1}{\partial y_2} & \cdots & \frac{\partial x_1}{\partial y_n} \\ \frac{\partial x_2}{\partial y_1} & \frac{\partial x_2}{\partial y_2} & \cdots & \frac{\partial x_2}{\partial y_n} \\ \cdots & \cdots & \cdots & \cdots \\ \frac{\partial x_n}{\partial y_1} & \frac{\partial x_n}{\partial y_2} & \cdots & \frac{\partial x_n}{\partial y_n} \end{vmatrix}$$

$$= \begin{vmatrix} \sum \frac{\partial y_1}{\partial x_r} \cdot \frac{\partial x_r}{\partial y_1} & \sum \frac{\partial y_1}{\partial x_r} \cdot \frac{\partial x_r}{\partial y_2} & \cdots & \sum \frac{\partial y_1}{\partial x_r} \cdot \frac{\partial x_r}{\partial y_n} \\ \sum \frac{\partial y_2}{\partial x_r} \cdot \frac{\partial x_r}{\partial y_1} & \sum \frac{\partial y_2}{\partial x_r} \cdot \frac{\partial x_r}{\partial y_2} & \cdots & \sum \frac{\partial y_2}{\partial x_r} \cdot \frac{\partial x_r}{\partial y_n} \\ \cdots\cdots\cdots & \cdots\cdots\cdots & \cdots & \cdots\cdots\cdots \\ \sum \frac{\partial y_n}{\partial x_r} \cdot \frac{\partial x_r}{\partial y_1} & \sum \frac{\partial y_n}{\partial x_r} \cdot \frac{\partial x_r}{\partial y_2} & \cdots & \sum \frac{\partial y_n}{\partial x_r} \cdot \frac{\partial x_r}{\partial y_n} \end{vmatrix}$$

$$= \begin{vmatrix} 1 & 0 & \cdots & 0 \\ 0 & 1 & \cdots & 0 \\ \cdots & \cdots & \cdots & \cdots \\ 0 & 0 & \cdots & 1 \end{vmatrix} = 1$$

12.4 Jacobian of Implicit Functions

Let $u_1, u_2, \cdots, u_n,\ x_1, x_2, \cdots, x_n$ be connected by the functional relations

$$\begin{aligned} F_1(u_1, u_2, \cdots, u_n,\ x_1, x_2, \cdots, x_n) &= 0 \\ F_2(u_1, u_2, \cdots, u_n,\ x_1, x_2, \cdots, x_n) &= 0 \\ \cdots\cdots\cdots\cdots\cdots\cdots\cdots\cdots\cdots\cdots \\ F_n(u_1, u_2, \cdots, u_n, x_1, x_2, \cdots, x_n) &= 0 \end{aligned}$$

Here $u_1, u_2, \cdots, u_n$ may be regarded as implicit functions of $x_1, x_2, \cdots, x_n$.

We have, by differentiation,

$$\frac{\partial F_1}{\partial x_1}+\frac{\partial F_1}{\partial u_1}\cdot\frac{\partial u_1}{\partial x_1}+\frac{\partial F_1}{\partial u_2}\cdot\frac{\partial u_2}{\partial x_1}+\cdots+\frac{\partial F_1}{\partial u_n}\cdot\frac{\partial u_n}{\partial x_1}=0$$

$$\frac{\partial F_1}{\partial x_2}+\frac{\partial F_1}{\partial u_1}\cdot\frac{\partial u_1}{\partial x_2}+\frac{\partial F_1}{\partial u_2}\cdot\frac{\partial u_2}{\partial x_2}+\cdots+\frac{\partial F_1}{\partial u_n}\cdot\frac{\partial u_n}{\partial x_2}=0$$

$$\cdots\cdots\cdots\cdots\cdots\cdots\cdots\cdots\cdots\cdots$$

$$\frac{\partial F_1}{\partial x_n}+\frac{\partial F_1}{\partial u_1}\cdot\frac{\partial u_1}{\partial x_n}+\frac{\partial F_1}{\partial u_2}\cdot\frac{\partial u_2}{\partial x_n}+\cdots+\frac{\partial F_1}{\partial u_n}\cdot\frac{\partial u_n}{\partial x_n}=0$$

The same can be obtained from $F_2,\cdots,F_n$.

Now $\dfrac{\partial(F_1,F_2,\cdots,F_n)}{\partial(u_1,u_2,\cdots,u_n)}\cdot\dfrac{\partial(u_1,u_2,\cdots,u_n)}{\partial(x_1,x_2,\cdots,x_n)}$

$$=\begin{vmatrix}\frac{\partial F_1}{\partial u_1} & \frac{\partial F_1}{\partial u_2} & \cdots & \frac{\partial F_1}{\partial u_n}\\ \frac{\partial F_2}{\partial u_1} & \frac{\partial F_2}{\partial u_2} & \cdots & \frac{\partial F_2}{\partial u_n}\\ \cdots & \cdots & \cdots & \cdots\\ \frac{\partial F_n}{\partial u_1} & \frac{\partial F_n}{\partial u_2} & \cdots & \frac{\partial F_n}{\partial u_n}\end{vmatrix}\times\begin{vmatrix}\frac{\partial u_1}{\partial x_1} & \frac{\partial u_1}{\partial x_2} & \cdots & \frac{\partial u_1}{\partial x_n}\\ \frac{\partial u_2}{\partial x_1} & \frac{\partial u_2}{\partial x_2} & \cdots & \frac{\partial u_2}{\partial x_n}\\ \cdots & \cdots & \cdots & \cdots\\ \frac{\partial u_n}{\partial x_1} & \frac{\partial u_n}{\partial x_2} & \cdots & \frac{\partial u_n}{\partial x_n}\end{vmatrix}$$

$$=\begin{vmatrix}\sum\frac{\partial F_1}{\partial u_r}\cdot\frac{\partial u_r}{\partial x_1} & \sum\frac{\partial F_1}{\partial u_r}\cdot\frac{\partial u_r}{\partial x_2} & \cdots & \sum\frac{\partial F_1}{\partial u_r}\cdot\frac{\partial u_r}{\partial x_n}\\ \sum\frac{\partial F_2}{\partial u_r}\cdot\frac{\partial u_r}{\partial x_1} & \sum\frac{\partial F_2}{\partial u_r}\cdot\frac{\partial u_r}{\partial x_2} & \cdots & \sum\frac{\partial F_2}{\partial u_r}\cdot\frac{\partial u_r}{\partial x_n}\\ \cdots\cdots & \cdots\cdots & \cdots & \cdots\cdots\\ \sum\frac{\partial F_n}{\partial u_r}\cdot\frac{\partial u_r}{\partial x_1} & \sum\frac{\partial F_n}{\partial u_r}\cdot\frac{\partial u_r}{\partial x_2} & \cdots & \sum\frac{\partial F_n}{\partial u_r}\cdot\frac{\partial u_r}{\partial x_n}\end{vmatrix}$$

$$=\begin{vmatrix}-\frac{\partial F_1}{\partial x_1} & -\frac{\partial F_1}{\partial x_2} & \cdots & -\frac{\partial F_1}{\partial x_n}\\ -\frac{\partial F_2}{\partial x_1} & -\frac{\partial F_2}{\partial x_2} & \cdots & -\frac{\partial F_2}{\partial x_n}\\ \cdots & \cdots & \cdots & \cdots\\ -\frac{\partial F_n}{\partial x_1} & -\frac{\partial F_n}{\partial x_2} & \cdots & -\frac{\partial F_n}{\partial x_n}\end{vmatrix}=(-1)^n\frac{\partial(F_1,F_2,\cdots,F_n)}{\partial(x_1,x_2,\cdots,x_n)}$$

$$\Rightarrow\frac{\partial(u_1,u_2,\cdots,u_n)}{\partial(x_1,x_2,\cdots,x_n)}=(-1)^n\cdot\frac{\partial(F_1,F_2,\cdots,F_n)}{\partial(x_1,x_2,\cdots,x_n)}\Big/\frac{\partial(F_1,F_2,\cdots,F_n)}{\partial(u_1,u_2,\cdots,u_n)}$$

Note. This result may be regarded as the generalisation of the result $\frac{dy}{dx}=-\frac{\partial f}{\partial x}/\frac{\partial f}{\partial y}$, where x,y are connected by the relation $f(x,y)=0$.

Problem [12.5] : Show that the functions $u=x+y-z$, $v=x-y+z$, $w=x^2+y^2+z^2-2yz$ are not independent of one another. Also find the relation between them.

Solution.

$$\frac{\partial(u,v,w)}{\partial(x,y,z)}=\begin{vmatrix}\frac{\partial u}{\partial x} & \frac{\partial u}{\partial y} & \frac{\partial u}{\partial z}\\ \frac{\partial v}{\partial x} & \frac{\partial v}{\partial y} & \frac{\partial v}{\partial z}\\ \frac{\partial w}{\partial x} & \frac{\partial w}{\partial y} & \frac{\partial w}{\partial z}\end{vmatrix}=\begin{vmatrix}1 & 1 & -1\\ 1 & -1 & 1\\ 2x & 2y-2z & 2z-2y\end{vmatrix}=\begin{vmatrix}1 & 1 & 0\\ 1 & -1 & 0\\ 1 & 2y-2z & 0\end{vmatrix}(c_2+c_3)$$

$$=0.$$

Since the Jacobian is zero, the functions are not independent.

Again, $u+v=2x$ and $u-v=2(y-z)$.

Therefore, $(u+v)^2+(u-v)^2=4(x^2+y^2+z^2-2yz)=4w$.

Problem [12.6] : The roots of the equation in λ, $(\lambda-x)^3+(\lambda-y)^3+(\lambda-z)^3=0$ are u,v,w. Prove that $\dfrac{\partial(u,v,w)}{\partial(x,y,z)}=-\dfrac{2(y-z)(z-x)(x-y)}{(v-w)(w-u)(u-v)}$.

Solution. Given equation in λ, is of the form

$$3\lambda^3-3\lambda^2(x+y+z)+3\lambda(x^2+y^2+z^2)-(x^3+y^3+z^3)=0.$$

Since u,v,w are its roots, it follows that

$$\begin{aligned}u+v+w&=+(x+y+z)\\ uv+vw+uw&=x^2+y^2+z^2\\ uvw&=\frac{1}{3}(x^3+y^3+z^3).\end{aligned}$$

Now let

$$\begin{aligned}F_1&=u+v+w-x-y-z=0\\ F_2&=uv+vw+uw-x^2-y^2-z^2=0\\ F_3&=uvw-\frac{1}{3}(x^3+y^3+z^3)=0.\end{aligned}$$

Now

$$\begin{aligned}\frac{\partial(F_1,F_2,F_3)}{\partial(x,y,z)}&=\begin{vmatrix}-1&-1&-1\\-2x&-2y&-2z\\-x^2&-y^2&-z^2\end{vmatrix}=-2\begin{vmatrix}1&1&1\\x&y&z\\x^2&y^2&z^2\end{vmatrix}\\ &=-2(x-y)(y-z)(z-x).\end{aligned}$$

Again,

$$\begin{aligned}\frac{\partial(F_1,F_2,F_3)}{\partial(u,v,w)}&=\begin{vmatrix}1&1&1\\v+w&u+w&u+v\\vw&uw&vu\end{vmatrix}=\begin{vmatrix}1&0&1\\v+w&w-v&u+v\\vw&u(w-v)&uv\end{vmatrix}(c_2-c_3)\\ &=(w-v)\begin{vmatrix}1&0&0\\v+w&1&u-w\\vw&u&v(u-w)\end{vmatrix}(c_3-c_1)\\ &=(w-v)(u-w)\begin{vmatrix}1&0&0\\v+w&1&1\\vw&u&v\end{vmatrix}\\ &=(w-v)(u-w)(v-u)\\ &=-(u-v)(v-w)(w-u).\end{aligned}$$

$\therefore$ from Art (12.4)

$$\frac{\partial(u,v,w)}{\partial(x,y,z)}=-\frac{\partial(F_1,F_2,F_3)}{\partial(x,y,z)}\Bigg/\frac{\partial(F_1,F_2,F_3)}{\partial(u,v,w)}=-\frac{2(x-y)(y-z)(z-x)}{(u-v)(v-w)(w-u)}.$$

Problem [12.7] : Find the Jacobian $\frac{\partial(x,y)}{\partial(r,\theta)}$ of the transformations $x = re^{\alpha\theta}$, $y = r^2e^{-\alpha\theta}$, α is constant.

Solution.

$$\frac{\partial(x,y)}{\partial(r,\theta)} = \begin{vmatrix} \frac{\partial x}{\partial r} & \frac{\partial x}{\partial \theta} \\ \frac{\partial y}{\partial r} & \frac{\partial y}{\partial \theta} \end{vmatrix} = \begin{vmatrix} e^{\alpha\theta} & \alpha re^{\alpha\theta} \\ 2re^{-\alpha\theta} & -\alpha r^2 e^{-\alpha\theta} \end{vmatrix}$$

$$= e^{\alpha\theta}e^{-\alpha\theta}\begin{vmatrix} 1 & \alpha r \\ 2r & -\alpha r^2 \end{vmatrix} = -\alpha r^2 - 2\alpha r^2 = -3\alpha r^2.$$

12.5 Maxima and Minima

Let $f : D \subseteq \mathbb{R}^2 \to \mathbb{R}$ be a function and let $(a,b) \in D$. Then $f(a,b)$ is an extreme value of $f(x,y)$, if for any point $(x,y) \neq (a,b)$ in some neighbourhood of (a,b), $f(x,y) - f(a,b)$ keeps the same sign.

The extreme value $f(a,b)$ is called maximum, if $f(x,y) - f(a,b) < 0$ or minimum, if $f(x,y) - f(a,b) > 0$.

Theorem [12.6] A necessary condition for $f(x,y)$ to have an extreme value at (a,b) is that $f_x(a,b) = 0$, $f_y(a,b) = 0$, provided f_x, f_y exist at (a,b).

Proof. If $f(a,b)$ is an extreme value of $f(x,y)$, a function of two variables, at (a,b), then it is also an extreme value of $f(x,b)$ at $x = a$ when $f(x,b)$ is considered as a function of single variable x. We shall show that $f_x(a,b) = 0$ at $x = a$ for an extreme value provided $f_x(x,b)$ exists at $x = a$. Let $f(x,b)$ has a local maximum at $x = a$.

Then $f(x,b) \leq f(a,b)$.

Hence, for $a < x < a+\delta$, $\frac{f(x,b)-f(a,b)}{x-a} \leq 0$ and for $a - \delta < x < a$, $\frac{f(x,b)-f(a,b)}{x-a} \geq 0$

$$\Rightarrow \lim_{x\to a+0}\frac{f(x,b)-f(a,b)}{x-a} \leq 0 \text{ and } \lim_{x\to a-0}\frac{f(x,b)-f(a,b)}{x-a} \geq 0$$

$$\Rightarrow f_x(a,b) \leq 0 \text{ and } f_x(a,b) \geq 0 \Rightarrow f_x(a,b) = 0, \text{ since } f_x(a,b) \text{ exists.}$$

Similarly, considering $f(a,y)$ as a function of single variable y and assuming that $f_y(a,y)$ exists at $y = b$, we can show that $f_y(a,b) = 0$ for an extreme value at $y = b$.

Note. (1) Considering the local minimum at $x = a$, we have $f(x,b) \geq f(a,b)$. Then proceeding as above, we obtain $f_x(a,b) = 0$ and $f_y(a,b) = 0$.

(2) The conditions for existance of extreme value of $f(x,y)$ at (a,b) are necessary but not sufficient. It follows from the example:

Example. Let $f(x,y) = (y - x^2)(y - 2x^2)$. Then $f(0,0) = 0$.

Also

$$f_x(0,0) = \lim_{h\to 0}\frac{f(h,0)-f(0,0)}{h} = \lim_{h\to 0}\frac{(-h^2)(-2h^2)}{h} = 0$$

$$f_y(0,0) = \lim_{k\to 0}\frac{f(0,k)-f(0,0)}{k} = \lim_{k\to 0}\frac{k^2}{k} = 0.$$

So both $f_x(0,0)$ and $f_y(0,0)$ exist.

Now $f(0,0) = 0$, but the function assumes both (+)ve and (−)ve values in every neighbourhood of $(0,0)$, so f has neither a local maximum nor a local minimum at $(0,0)$.

(3) The points given by $f_x(x,y) = 0$ and $f_y(x,y) = 0$ are called *Stationary Points.*

Theorem [12.7] [Sufficient conditions] Let $f_x(a,b) = 0$, $f_y(a,b) = 0$ and let $f(x,y)$ possess continuous second order partial derivatives in a certain neighbourhood N of (a,b) and $f_{xx}(a,b)$, $f_{xy}(a,b)$, $f_{yy}(a,b)$ are not all zero. Then obtain the sufficient conditions so that $f(x,y)$ has an extremum at (a,b).

Proof. We write, $f_{xx}(a,b) = A$, $f_{xy}(a,b) = B$, $f_{yy}(a,b) = C$.

Let $(a+h, b+k)$ be a point in the neighbourhood N of (a,b).

Using Taylor's theorem, we have

$$\begin{aligned} f(a+h,b+k) &= f(a,b) + [hf_x(a,b) + kf_y(a,b)] \\ &+\frac{1}{2!}\left[h^2 f_{xx}(a+\theta h, b+\theta k) + 2hkf_{xy}(a+\theta h, b+\theta k)\right. \\ &\left.+k^2 f_{yy}(a+\theta h, b+\theta k)\right],\ 0<\theta<1. \end{aligned} \tag{i}$$

Let us take, $f_{xx}(a+\theta h, b+\theta k) - f_{xx}(a,b) = \alpha$

$$f_{xy}(a+\theta h, b+\theta k) - f_{xy}(a,b) = \beta$$
$$f_{yy}(a+\theta h, b+\theta k) - f_{yy}(a,b) = \gamma.$$

Since 2nd order partial derivatives are continuous at $(a,b), \alpha, \beta, \gamma \to 0$ as $(h,k) \to (0,0)$.

From (i),

$$\begin{aligned} f(a+h,b+k) - f(a,b) &= \frac{1}{2!}\left[h^2(\alpha + f_{xx}(a,b)) + 2hk(\beta + f_{xy}(a,b)) + k^2(\gamma + f_{yy}(a,b))\right] \\ &= \frac{1}{2!}[Ah^2 + 2Bhk + Ck^2 + (h^2\alpha + 2hk\beta + k^2\gamma)] = \frac{1}{2}[\phi + \rho], \end{aligned}$$

where $\phi = Ah^2 + 2Bhk + Ck^2$ and $\rho = h^2\alpha + 2hk\beta + k^2\gamma$.

Then $\rho \to 0$ as $(h,k) \to (0,0)$, since $\alpha, \beta, \gamma \to 0$. When $(h,k) \to (0,0)$, sign of ρ is unknown.

Several cases arise:

(1) ϕ never vanishes and keeps a constant sign.

$\rho \to 0$ as $(h,k) \to (0,0)$.

$\Rightarrow$ ρ is very small

$\Rightarrow$ sign of $\phi + \rho$ depends on ϕ

$\Rightarrow$ $f(a+h,b+k) - f(a,b) \lesseqgtr 0$ according as $\phi \lesseqgtr 0$

$\Rightarrow$ $f(a,b)$ is maximum or minimum value, if $\phi < 0$ or $\phi > 0$ [by Definition 12.5]

(2) ϕ changes sign: In this case $f(a,b)$ will not be extremum, since $f(a+h,b+k) - f(a,b)$ and ϕ have same sign when ρ is small.

(3) ϕ does not change sign but vanishes for some values of (h,k).

In this case, the sign of $f(a+h,b+k) - f(a,b)$ will depend on ρ whose sign is not known. So we cannot draw any result in this case. Further investigations are required. We have

$$\phi = Ah^2 + 2Bhk + Ck^2 = \frac{(Ah+Bk)^2 + k^2(AC - B^2)}{A}.$$

We assume that $A \neq 0$.

(a) First let $A > 0$.

Now if $AC - B^2 > 0$, then $(Ah+Bk)^2 + k^2(AC - B^2)$ being the sum of two (+)ve numbers can never vanish unless $h = 0$, $k = 0$ simultaneously and this is not possible.

Therefore, $\phi > 0$ and consequently, $f(a+h, b+k) - f(a,b) > 0 \Rightarrow f(x,y)$ has minimum at (a, b).

Next, let $A < 0$.

Then $AC - B^2 > 0 \Rightarrow \phi < 0$ and consequently, $f(a+h, b+k) - f(a,b) < 0 \Rightarrow f(x,y)$ has maximum at (a, b).

Thus we conclude that $f(x, y)$ has (i) maximum when $AC - B^2 > 0$ and $A < 0$ (ii) minimum when $AC - B^2 > 0$ and $A > 0$.

(b) Now $\phi = \dfrac{(Ah+Bk)^2 - k^2(B^2 - AC)}{A}$.

If $AC - B^2 < 0$, the sign of $\phi >$ or < 0 according as $(Ah + Bk)^2 >$ or $< k^2(B^2 - AC)$, which shows that ϕ does not keep the same sign for all values of (h, k). Thus $f(a, b)$ is neither maximum nor minimum of $f(x, y)$ at (a, b).

(c) If $AC - B^2 = 0$, then $\phi = \dfrac{(Ah+Bk)^2}{A}$.

Here $\phi > 0$, when $A > 0$ and $\phi < 0$, when $A < 0$.

Also ϕ may be zero for some values of (h, k) for which $Ah + Bk = 0$.

This shows that ϕ, without changing sign, may vanish for certain values of (h, k). In this case the sign of $f(a+h, b+k) - f(a, b)$ depends on ρ and it requires further investigation.

We now consider that $A = 0$.

Then $\phi = Ah^2 + 2Bhk + Ck^2$ becomes $2Bhk + Ck^2$.

$$\therefore \phi = k(2Bh + Ck).$$

(d) If $B \neq 0$, then ϕ changes sign with k. So there is no extreme value.

(e) If $B = 0$, then $\phi = Ck^2$, where $C = f_{yy}(a, b)$ has a fixed sign. Consequently, ϕ does not change its sign but may be zero for $k = 0$ (without $h = 0$) . So further investigation is required.

Note. In case of further investigation for doubtful cases, we need to consider the terms of higher order than the second order of Taylor's expansion of $f(a+h, b+k)$. This is very complicated.

Example. Let $f(x, y) = 2x^2 - xy + 2y^2 - 20x$. Then $f_x = 4x - y - 20$, $f_y = -x + 4y$, $f_{xx} = 4$, $f_{xy} = -1$, $f_{yy} = 4$.

For extremum, $f_x(xy) = 0, f_y(x, y) = 0$

$$\Rightarrow 4x - y - 20 = 0 \tag{1}$$

$$\text{and } -x + 4y = 0. \tag{2}$$

From (2), $x = 4y$. Therefore, from (1), $15y = 20 \Rightarrow y = 4/3, x = 16/3$.

Stationary point is $(16/3, 4/3)$

$$A = f_{xx} = 4 > 0,\ B = f_{xy} = -1, f_{yy} = 4 \text{ at } (16/3, 4/3).$$

Then $AC - B^2 = f_{xx} \cdot f_{yy} - (f_{xy})^2 = 4 \cdot 4 - (-1)^2 = 15 > 0$. Also $A > 0$.

Therefore, $f(x, y)$ has minimum at $(16/3, 4/3)$.

Problem [12.8]: (1) Prove that $f(x, y) = x^2 - 2xy + y^2 + x^3 - y^3 + x^5$ has neither maximum nor a minimum at $(0, 0)$.

(2) Show that the function $f(x,y) = (y-x)^4 + (x-2)^4$ has a minimum at $(2,2)$.

(3) Examine the extreme values for the function $f(x,y) = (x^2+y^2)e^{6x+2x^2}$.

(4) Discuss maximum or minimum of $f(x,y) = 2\sin\frac{x-y}{2}\cos\frac{x-y}{2} + \cos(x+y)$.

(5) Show that the maximum and minimum of the function

$$u(x,y) = \frac{ax^2+by^2+2hxy+2gx+2fy+c}{a'x^2+b'y^2+2h'xy+2g'x+2f'y+c'}$$

are given by the roots of the equation

$$\begin{vmatrix} a-a'u & h-h'u & g-g'u \\ h-h'u & b-b'u & f-f'u \\ g-g'u & f-f'u & c-c'u \end{vmatrix} = 0.$$

Solution. (1) $f_x = 2x - 2y + 3x^2 + 5x^4,\ f_y = -2x + 2y - 3y^2$

$$f_{xx} = 2 + 6x + 20x^3,\ f_{xy} = -2,\ f_{yy} = 2 - 6y.$$

For extremum, $f_x(x,y) = 0, f_y(x,y) = 0$

$$\Rightarrow 2x - 2y + 3x^2 + 5x^4 = 0 \quad (1)$$

$$-2x + 2y - 3y^2 = 0. \quad (2)$$

From (2), $2x = 2y - 3y^2$ or, $x = \dfrac{2y-3y^2}{2} = y\left(\dfrac{2-3y}{2}\right)$.

From (1),

$$2y - 3y^2 - 2y + 3y^2\left(\frac{2-3y}{2}\right)^2 + 5y^4\left(\frac{2-3y}{2}\right)^4 = 0$$

$$\Rightarrow y^2 = 0 \text{ and } -3 + 3\left(\frac{2-3y}{2}\right)^2 + 5y^2\left(\frac{2-3y}{2}\right)^4 = 0.$$

$y = 0$. Then $x = 0$. So $(0,0)$ is a point of extremum.

$$A = f_{xx}(0,0) = 2, B = f_{xy}(0,0) = -2, C = f_{yy}(0,0) = 2.$$

$$AC - B^2 = 4 - (-2)^2 = 4 - 4 = 0.$$

Test fails.

Now $f(x,y) = (x-y)^2 + (x-y)(x^2+xy+y^2) + x^5$. Then points along the line $y - x = 0$, it gives $f(x,y) = x^5$.

In every neighbourhood of $(0,0)$, there are points, where $f(x,y) > 0$ and there are points, where $f(x,y) < 0$.

So $f(x,y) - f(0,0)$ does not keep the same sign and consequently $f(0,0)$ is neither a maximum nor a minimum.

(2) $f(x,y) = (y-x)^4 + (x-2)^4$,

$$f_x = -4(y-x)^3 + 4(x-2)^3,\ f_y = 4(y-x)^3,$$

$$f_{xx} = 12(y-x)^2 + 12(x-2)^2,$$

$$f_{xy} = -12(y-x)^2, f_{yy} = 12(y-x)^2.$$

For extremum, $f_x(x,y) = 0, f_y(x,y) = 0$

$\Rightarrow$ (i) $(x-2)^3 - (y-x)^3 = 0$

(ii) $(y-x)^3 = 0 \Rightarrow y = x$.

For $y = x, x = 2$ and so $y = 2$.

Therefore, $(2, 2)$ is a stationary point.

$$A = f_{xx}(2,2) = 0, C = f_{yy}(2,2) = 0, B = f_{xy}(2,2) = 0$$
$$AC - B^2 = 0.$$

Test fails.

$f(x, y) = (y - x)^4 + (x - 2)^4$.

Then points along the line $y - x = 0$, it gives $f(x, y) = (x - 2)^4$.

In every neighbourhood of $(2, 2)$, $f(x, y) > 0$.

Therefore, $f(x, y) - f(2, 2) > 0$ for every neighbourhood of $(2, 2)$ and hence $f(2, 2)$ is minimum.

Thus $f(x, y)$ has minimum at $(2, 2)$.

(3) $f(x, y) = (x^2 + y^2)e^{6x+2x^2}$.

$f_x = 2x \cdot e^{6x+2x^2} + (x^2 + y^2)(6 + 4x)e^{6x+2x^2} = \{2x + (x^2 + y^2)(6 + 4x)\}e^{6x+2x^2}$.

$f_y = 2ye^{6x+2x^2}$.

$f_{xx} = \{2 + 2x(6 + 4x) + 4(x^2 + y^2)\}e^{6x+2x^2} + \{2x + (x^2 + y^2)(6 + 4x)\}(6 + 4x)e^{6x+2x^2}$.

$f_{xy} = 2y(6 + 4x)e^{6x+2x^2}$.

$f_{yy} = 2e^{6x+2x^2}$.

For extremum, $f_x(x, y) = 0$, $f_y(x, y) = 0$

$$\Rightarrow \{2x + (x^2 + y^2)(6 + 4x)\} = 0 \tag{1}$$

$$\text{and} \quad y = 0, \tag{2}$$

since $e^{6x+2x^2} \neq 0$.

Therefore, $y = 0 \Rightarrow 2x + 6x^2 + 4x^3 = 0$ [from (1)]

$\Rightarrow x = 0$ and $4x^2 + 6x + 2 = 0 \Rightarrow 2x^2 + 3x + 1 = 0 \Rightarrow (2x + 1)(x + 1) = 0$

$\Rightarrow x = -\dfrac{1}{2}, x = -1.$

So the stationary points are $(0, 0), \left(-\frac{1}{2}, 0\right), (-1, 0)$.

At the point $(0, 0)$,

$$A = f_{xx}(0,0) = 2, B = f_{xy}(0,0) = 0, C = f_{yy}(0,0) = 2.$$
$$AC - B^2 = 2 \cdot 2 - 0 = 4 > 0.$$

Therefore, $f(x, y)$ has minimum value at $(0, 0)$.

At the point $\left(-\frac{1}{2}, 0\right)$,

$$A = f_{xx}\left(-\frac{1}{2}, 0\right) = \{2 - (6 - 2) + 1\} \cdot e^{-3+\frac{1}{2}} = -e^{-5/2} = -\frac{1}{e^{5/2}} < 0$$

$$B = f_{xy}\left(-\frac{1}{2}, 0\right) = 0, C = f_{yy}\left(-\frac{1}{2}, 0\right) = \frac{2}{e^{5/2}}.$$

Then $AC - B^2 = -\dfrac{2}{e^5} - 0 = -\dfrac{2}{e^5} < 0$.

Therefore, $f(x, y)$ has neither a maximum nor a minimum at $\left(-\frac{1}{2}, 0\right)$.

At the point $(-1,0)$,

$$A = f_{xx}(-1,0) = \{2 - 2(6-4) + 4\}e^{-6+2} = 2e^{-4} = \frac{2}{e^4}$$

$$B = f_{xy}(-1,0) = 0, C = f_{yy}(-1,0) = 2e^{-4} = \frac{2}{e^4}.$$

$$\therefore AC - B^2 = \frac{4}{e^8} > 0.$$

Also $A = \frac{2}{e^4} > 0$. Therefore, $f(x,y)$ has minimum at $(-1,0)$.

(4) $f(x,y) = 2\sin\frac{x+y}{2}\cos\frac{x-y}{2} + \cos(x+y),\ x,y > 0 = \sin x + \sin y + \cos(x+y)$.

$$f_x = \cos x - \sin(x+y),\ f_y = \cos y - \sin(x+y),$$
$$f_{xx} = -\sin x - \cos(x+y),\ f_{yy} = -\sin y - \cos(x+y),$$
$$f_{xy} = -\cos(x+y).$$

For extremum, $f_x(x,y) = 0, f_y(x,y) = 0$

$$\Rightarrow \cos x = \sin(x+y), \quad (1)$$
$$\cos y = \sin(x+y). \quad (2)$$

So from (1) and (2), $\cos x = \cos y \Rightarrow x = y$.

Then from (1), $\cos x = \sin 2x$ or, $\cos x(1 - 2\sin x) = 0$.

Then $\cos x = 0 \Rightarrow x = (2n+1)\frac{\pi}{2},\ n = 0,1,2,\cdots$ and $\sin x = \frac{1}{2} \Rightarrow x = n\pi + (-1)^n\frac{\pi}{6}$.

Thus the stationary points are given by $x = y = (2n+1)\frac{\pi}{2}$ and $x = y = n\pi + (-1)^n\frac{\pi}{6}$.

Let $x = \frac{\pi}{2}$. Then $y = \frac{\pi}{2}$.

$$A = f_{xx}\left(\frac{\pi}{2},\frac{\pi}{2}\right) = -1+1 = 0, C = f_{yy}\left(\frac{\pi}{2},\frac{\pi}{2}\right) = -1+1 = 0, B = f_{xy}\left(\frac{\pi}{2},\frac{\pi}{2}\right) = 1.$$

Then $AC - B^2 = -1 < 0$.

So $f(x,y)$ has neither a maximum nor a minimum at $x = y = \frac{\pi}{2}$.

Now, for $x = y = \frac{3\pi}{2}$,

$$A = f_{xx}\left(\frac{3\pi}{2},\frac{3\pi}{2}\right) = 2, C = f_{yy}\left(\frac{3\pi}{2},\frac{3\pi}{2}\right) = 2, B = f_{xy}\left(\frac{3\pi}{2},\frac{3\pi}{2}\right) = 1.$$

$$\therefore AC - B^2 = 4 - 1 > 0$$

Also $A = 2 > 0$.

Then $f(x,y)$ has minimum at $x = \frac{3\pi}{2}$. It can be observed that $f(x,y)$ has neither a maximum nor a minimum at $5\pi/2$ and $f(x,y)$ has a minimum at $7\pi/2$.

The same is true with every 2π interval.

Now, for $x=y=n\pi+(-1)^n\frac{\pi}{6}$, $n=0,1,2,\cdots$

$$A=f_{xx}(x,y)=-\sin\left\{n\pi+(-1)^n\frac{\pi}{6}\right\}-\cos\left\{2n\pi+(-1)^n\frac{\pi}{3}\right\}$$

$$=-(-1)^{2n}\sin\frac{\pi}{6}-\cos\frac{\pi}{3}=-\frac{1}{2}-\frac{1}{2}=-1,$$

$$C=f_{yy}(x,y)=-\sin\left\{n\pi+(-1)^n\frac{\pi}{6}\right\}-\cos\left\{2n\pi+(-1)^n\frac{\pi}{3}\right\}=-1,$$

$$B=f_{xy}(x,y)=-\cos\left\{2n\pi+(-1)^n\frac{\pi}{3}\right\}=-\frac{1}{2}.$$

$$\therefore AC-B^2=1-\left(-\frac{1}{2}\right)^2=1-\frac{1}{4}=\frac{3}{4}>0.$$

Also $A<0$.

$\therefore f(x,y)$ has maximum $\forall\ x=y=n\pi+(-1)^n\frac{\pi}{6}$.

(5) $$u=\frac{ax^2+by^2+2hxy+2gx+2fy+c}{a'x^2+b'y^2+2h'xy+2g'x+2f'y+c'}$$

$$\Rightarrow u(a'x^2+b'y^2+2h'xy+2g'x+2f'y+c')=ax^2+by^2+2hxy+2gx+2fy+c.$$

Now differentiating partially w.r.t. x and y respectively, we get

$$u_x(a'x^2+b'y^2+2h'xy+2g'x+2f'y+c)+2u(a'x+h'y+g')=2(ax+by+g) \quad (1)$$

$$\text{and } u_y(a'x^2+b'y^2+2h'xy+2g'x+2f'y+c)+2u(b'y+h'x+f')=2(by+hx+f). \quad (2)$$

For extremum, $u_x=0, u_y=0$. Then from (1) and (2), we get

$$u(a'x+h'y+g')=ax+hy+g \quad (3)$$

$$\text{and } u(b'y+h'x+f')=by+hx+f. \quad (4)$$

Now (3) x + (4) y gives

$$u(a'x^2+2h'xy+b'y^2+g'x+f'y)=(ax^2+2hxy+by^2+gx+fy). \quad (5)$$

Again, from the given relation

$$u(a'x^2+2h'xy+b'y^2+2g'x+2f'y+c')$$
$$=ax^2+2hxy+by^2+2gx+2fy+c$$
$$\Rightarrow u(a'x^2+2h'xy+b'y^2+g'x+f'y)+u(g'x+f'y+c')$$
$$=(ax^2+2hxy+by^2+gx+fy)+(gx+fy+c)$$
$$\Rightarrow u(g'x+f'y+c')=gx+fy+c \text{ [by (5)]}. \quad (6)$$

Thus from (3), (4) and (6), we get

$$(a-a'u)x+(h-h'u)y+g-g'u=0, \quad (7)$$
$$(h-h'u)x+(b-b'u)y+f-f'u=0, \quad (8)$$
$$(g-g'u)x+(f-f'u)y+c-c'u=0. \quad (9)$$

Eliminating x,y from (7), (8) and (9), we get

$$\begin{vmatrix} a-a'u & h-h'u & g-g'u \\ h-h'u & b-b'u & f-f'u \\ g-g'u & f-f'u & c-c'u \end{vmatrix}=0.$$

which gives the maximum and minimum values of u.

Problem [12.9] : A rectangular box, open at the top is to have a volume of 32 cc. Find the dimensions of the box requiring least material for its construction.

Solution. Let the length, breadth, height of the rectangular box be a cm, b cm, c cm respectively.

Then

$$V = abc = 32 \text{ cc.} \quad (1)$$

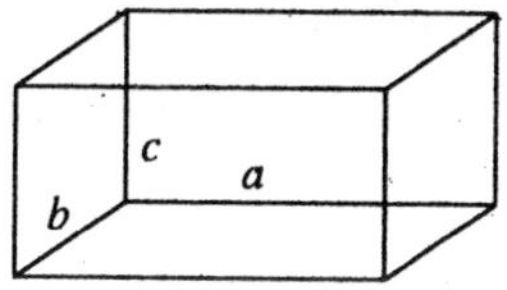

Fig. 12.2

Total surface area

$$S = 2(ac + bc) + ab. \quad (2)$$

Putting the value of $b = \frac{32}{ac}$in (2), we get

$$S = 2\left(a + \frac{32}{ac}\right) \cdot c + \frac{32a}{ac} = 2ac + \frac{64}{a} + \frac{32}{c}$$

$$\frac{\partial S}{\partial a} = 2c - \frac{64}{a^2}, \ \frac{\partial S}{\partial c} = 2a - \frac{32}{c^2}.$$

For extremum, $\frac{\partial S}{\partial a} = 0$, $\frac{\partial S}{\partial c} = 0$.

$$\therefore \ 2c - \frac{64}{a^2} = 0 \text{ and } 2a - \frac{32}{c^2} = 0 \Rightarrow \ c = \frac{32}{a^2} \text{ and } a = \frac{16}{c^2}.$$

$$\therefore \ a^2 = \frac{256}{c^4} \Rightarrow \frac{32}{c} = \frac{256}{c^4} \Rightarrow c^3 = 8 \Rightarrow c = 2 \ [c \neq 0] \text{ and hence } a = \frac{16}{4} = 4.$$

Therefore, stationary point is $(4, 2)$.

$$\frac{\partial^2 S}{\partial a^2} = \frac{128}{a^3}, \ \frac{\partial^2 S}{\partial c^2} = \frac{64}{c^3}, \ \frac{\partial^2 S}{\partial a \partial c} = 2.$$

$$\text{Now } A = \left.\frac{\partial^2 S}{\partial a^2}\right]_{(4,2)} = 2, \ \ C = \left.\frac{\partial^2 S}{\partial c^2}\right]_{(4,2)} = 8, \ \ B = \left.\frac{\partial^2 S}{\partial a \partial c}\right]_{(4,2)} = 2.$$

$$\therefore AC - B^2 = 16 - 4 = 12 > 0.$$

Also $A > 0$. Therefore, S is minimum at $a = 4, c = 2$ and $b = \frac{32}{8} = 4$.

Dimensions are 4 cm, 4 cm and 2 cm.

12.6 Working Rule of Finding the Extremum for a Function of Three Independent Variables

Let $f(x, y, z)$ be a function of three independent variables x, y, z.

$$df = \frac{\partial f}{\partial x}dx + \frac{\partial f}{\partial y}dy + \frac{\partial f}{\partial z}dz.$$

For extremum,

$$df = 0 \Rightarrow \frac{\partial f}{\partial x} = 0, \ \frac{\partial f}{\partial y} = 0, \ \frac{\partial f}{\partial z} = 0. \quad (1)$$

$$d^2 f = \frac{\partial^2 f}{\partial x^2}(dx)^2 + \frac{\partial^2 f}{\partial y^2}(dy)^2 + \frac{\partial^2 f}{\partial z^2}(dz)^2 + 2\frac{\partial^2 f}{\partial x \partial y}dxdy + 2\frac{\partial^2 f}{\partial y \partial z}dydz + 2\frac{\partial^2 f}{\partial z \partial x}dzdx$$

$$= f_{xx}(dx)^2 + f_{yy}(dy)^2 + f_{zz}(dz)^2 + 2f_{xy}dxdy + 2f_{yz}dydz + 2f_{zx}dzdx. \quad (2)$$

The extremum point will be a maximum point or a minimum point according as $d^2 f < 0$ or $d^2 f > 0$.

The point will be neither a maximum nor a minimum if d^2f does not keep the same sign in the neighbourhood of the stationary point (x, y, z).

Further investigation is required if d^2f keeps the same sign but vanishes at some points in the neighbourhood of a stationary point (x, y, z). The matrix of the quadratic form given in (2) is

$$\begin{bmatrix} f_{xx} & f_{xy} & f_{xz} \\ f_{yx} & f_{yy} & f_{yz} \\ f_{zx} & f_{zy} & f_{zz} \end{bmatrix}.$$

From the theory of quadratic forms that this form, i.e., d^2f is *positive definite*, if and only if the three principal minors

$$f_{xx}, \begin{vmatrix} f_{xx} & f_{xy} \\ f_{yx} & f_{yy} \end{vmatrix}, \begin{vmatrix} f_{xx} & f_{xy} & f_{xz} \\ f_{yx} & f_{yy} & f_{yz} \\ f_{zx} & f_{zy} & f_{zz} \end{vmatrix}$$

are all *positive*. d^2f will be always *negative*, if and only if the signs of above three minors are alternatively negative and positive.

Example. Let $f(x, y, z) = 2xyz + x^2 + y^2 + z^2$.

$$\frac{\partial f}{\partial x} = 2yz + 2x, \frac{\partial f}{\partial y} = 2xz + 2y, \frac{\partial f}{\partial z} = 2xy + 2z$$

$$df = (2yz + zx)dx + (2xz + 2y)dy + (2xy + 2z)dz$$

$$d^2f = 2(dx)^2 + 2(dy)^2 + 2(dz)^2 + 2z\,dxdy + 2x\,dydz + 2y\,dzdx.$$

Now, for extremum, $f_x = 0, f_y = 0, f_z = 0$

$$\Rightarrow yz + x = 0 \quad \cdots \text{(i)} \quad \Rightarrow xyz + x^2 = 0 \Rightarrow x^2 = -xyz$$

$$zx + y = 0 \quad \cdots \text{(ii)} \quad \Rightarrow xyz + y^2 = 0 \Rightarrow y^2 = -xyz$$

$$xy + z = 0 \quad \cdots \text{(iii)} \quad \Rightarrow xyz + z^2 = 0 \Rightarrow z^2 = -xyz.$$

$$\therefore \quad x^2 = y^2 = z^2 \Rightarrow x = y = z \text{ or, } x = -y, \ y = -z,$$

$$x = y = z \Rightarrow x = 0, \ x = -1, \ \therefore (x, y, z) = (0, 0, 0), (-1, -1, -1)$$

$$y = -x, \ z = x \Rightarrow -x^2 + x = 0 \Rightarrow x = 0, \ x = 1.$$

$$\therefore \quad x = 1, \ y = -1, \ z = 1, \ \therefore (x, y, z) = (1, -1, 1).$$

Stationary points are $(0, 0, 0), (-1, -1, -1)$ and $(1, -1, 1), (-1, 1, 1), (1, 1, -1)$.

Three principal minors at $(0, 0, 0)$

$$f_{xx} = 2 > 0, \begin{vmatrix} f_{xx} & f_{xy} \\ f_{yx} & f_{yy} \end{vmatrix} = \begin{vmatrix} 2 & 0 \\ 0 & 2 \end{vmatrix} = 4 > 0, \begin{vmatrix} f_{xx} & f_{xy} & f_{xz} \\ f_{yx} & f_{yy} & f_{yz} \\ f_{zx} & f_{zy} & f_{zz} \end{vmatrix} = \begin{vmatrix} 2 & 0 & 0 \\ 0 & 2 & 0 \\ 0 & 0 & 2 \end{vmatrix} = 8 > 0.$$

Thus the three minors are all positive.

$$\therefore d^2f > 0.$$

Hence, $f(x, y, z)$ has minimum at $(0, 0, 0)$.

Now at the points $(-1, -1, -1)$

$$[f_{xx} = 2 > 0, \begin{vmatrix} f_{xx} & f_{xy} \\ f_{yx} & f_{yy} \end{vmatrix} = 0, \begin{vmatrix} f_{xx} & f_{xy} & f_{xz} \\ f_{yx} & f_{yy} & f_{yz} \\ f_{zx} & f_{zy} & f_{zz} \end{vmatrix} = \begin{vmatrix} 2 & -2 & -2 \\ -2 & 2 & -2 \\ -2 & -2 & 2 \end{vmatrix} = -32.$$

Consequently, it does not follow any rule of (+)ve definite or (−)ve definite of d^2f. Therefore, $f(x,y,z)$ has neither a maximum nor a minimum at $(-1,-1,-1)$.

The same is true at $(-1,1,1)$ and $(1,1,-1)$ and $(1,-1,1)$.

Problem [12.10] : (1) Show that the function $f(x,y,z) = 3\log(x^2+y^2+z^2) - 2x^3 - 2y^3 - 2z^3$ has only one extreme value, viz., $\log 3/e^2$

(2) Show that the maximum value of $xy(z-h)\left(\frac{x^2}{a^2}+\frac{y^2}{b^2}-\frac{z^2}{c^2}\right)$ is $\left(\frac{2h}{5}\right)^5\frac{ab}{c^4}$, where $x,y,z,h,a,b,c>0$.

Solution. (1) $f(x,y,z) = 3\log(x^2+y^2+z^2) - 2x^3 - 2y^3 - 2z^3$

$$f_x = \frac{6x}{(x^2+y^2+z^2)} - 6x^2,\ f_y = \frac{6y}{x^2+y^2+z^2} - 6y^2,\ f_z = \frac{6z}{x^2+y^2+z^2} - 6z^2$$

For extremum, $df = 0 \Rightarrow f_x = 0, f_y = 0, f_z = 0$

$$\Rightarrow x\left(\frac{1}{x^2+y^2+z^2} - x\right) = 0 \tag{1}$$

$$y\left(\frac{1}{x^2+y^2+z^2} - y\right) = 0 \tag{2}$$

$$z\left(\frac{1}{x^2+y^2+z^2} - z\right) = 0 \tag{3}$$

$$\Rightarrow x = y = z = 0.$$

But $(x,y,z) \neq (0,0,0)$.

For other values of x,y,z, $1 - x(x^2+y^2+z^2) = 0 \Rightarrow x^2+y^2+z^2 = 1/x$.

Similarly, $x^2+y^2+z^2 = 1/y$ and $x^2+y^2+z^2 = 1/z$.

Thus $\frac{1}{x} = \frac{1}{y} = \frac{1}{z} \Rightarrow x = y = z$.

Therefore, $x^2+y^2+z^2 = \frac{1}{x} \Rightarrow 3x^3 = 1 \Rightarrow x = \frac{1}{3^{1/3}}$

$\therefore$ stationary point is $(x,y,z) = \left(\frac{1}{3^{1/3}}, \frac{1}{3^{1/3}}, \frac{1}{3^{1/3}}\right)$

$$f_{xx} = \frac{6}{(x^2+y^2+z^2)} - \frac{12x^2}{(x^2+y^2+z^2)^2} - 12x$$

$$f_{yy} = \frac{6}{(x^2+y^2+z^2)} - \frac{12y^2}{(x^2+y^2+z^2)^2} - 12y$$

$$f_{yx} = f_{xy} = \frac{-12xy}{(x^2+y^2+z^2)^2},\ f_{yz} = f_{zy} = \frac{-12zy}{(x^2+y^2+z^2)^2}$$

$$f_{zx} = f_{xz} = \frac{-12xz}{(x^2+y^2+z^2)^2}.$$

Now at the point $\left(\frac{1}{3^{1/3}}, \frac{1}{3^{1/3}}, \frac{1}{3^{1/3}}\right)$, $x^2+y^2+z^2 = \frac{3}{3^{2/3}} = 3^{1/3}$

$$f_{xx} = \frac{6}{3^{1/3}} - \frac{12}{3^{2/3}\,.\,3^{2/3}} - \frac{12}{3^{1/3}} = \frac{6}{3^{1/3}} - \frac{4}{3^{1/3}} - \frac{12}{3^{1/3}} = -\frac{10}{3^{1/3}}$$

$$f_{yy} = -\frac{10}{3^{1/3}},\ f_{zz} = -\frac{10}{3^{1/3}},\ f_{xy} = f_{yx} = -\frac{12}{3^{2/3}\,.\,3^{2/3}} = -\frac{4}{3^{1/3}}$$

$$f_{yz} = f_{zy} = -\frac{4}{3^{1/3}},\ f_{zx} = f_{xz} = -\frac{4}{3^{1/3}}.$$

Now three principal minors,

$$f_{xx} = -\frac{10}{3^{1/3}} < 0$$

$$\begin{vmatrix} f_{xx} & f_{xy} \\ f_{yx} & f_{yy} \end{vmatrix} = \frac{1}{3^{2/3}} \begin{vmatrix} 10 & 4 \\ 4 & 10 \end{vmatrix} = \frac{84}{3^{2/3}} > 0$$

$$\begin{vmatrix} f_{xx} & f_{xy} & f_{xz} \\ f_{yx} & f_{yy} & f_{yz} \\ f_{zx} & f_{zy} & f_{zz} \end{vmatrix} = -\frac{1}{3} \begin{vmatrix} 10 & 4 & 4 \\ 4 & 10 & 4 \\ 4 & 4 & 10 \end{vmatrix} = -\frac{1}{3}[840 - 96 - 96] < 0.$$

The signs of the three principal minors are of (−)ve, (+)ve and (−)ve respectively. Therefore, $d^2f < 0$.

Hence, f has maximum at $\left(\frac{1}{3^{1/3}}, \frac{1}{3^{1/3}}, \frac{1}{3^{1/3}}\right)$.

The maximum value of

$$f(x,y,z) = 3\log\left(\frac{3}{3^{2/3}}\right) - 2\left(\frac{1}{3} + \frac{1}{3} + \frac{1}{3}\right) = 3\log 3^{1/3} - 2 = \log 3 - \log e^2 = \log\frac{3}{e^2}.$$

(2) Let $\quad u = f(x,y,z) = xy(z-h)\left(\frac{x^2}{a^2} + \frac{y^2}{b^2} - \frac{z^2}{c^2}\right).$

$$\log u = \log x + \log y + \log(z-h) + \log\left(\frac{x^2}{a^2} + \frac{y^2}{b^2} - \frac{z^2}{c^2}\right).$$

Then

$$\frac{1}{u}u_x = \frac{1}{x} + \frac{2x/a^2}{\frac{x^2}{a^2} + \frac{y^2}{b^2} - \frac{z^2}{c^2}}, \frac{1}{u}u_y = \frac{1}{y} + \frac{2y/b^2}{\frac{x^2}{a^2} + \frac{y^2}{b^2} - \frac{z^2}{c^2}}, \frac{1}{u}u_z = \frac{1}{z-h} + \frac{-2z/c^2}{\frac{x^2}{a^2} + \frac{y^2}{b^2} - \frac{z^2}{c^2}}.$$

For extremum, $u_x = 0, u_y = 0, u_z = 0$

$$\Rightarrow \frac{3x^2}{a^2} + \frac{y^2}{b^2} - \frac{z^2}{c^2} = 0 \qquad (1)$$

$$\frac{x^2}{a^2} + \frac{3y^2}{b^2} - \frac{z^2}{c^2} = 0 \qquad (2)$$

$$\frac{x^2}{a^2} + \frac{y^2}{b^2} - \frac{3z^2}{c^2} + \frac{2zh}{c^2} = 0. \qquad (3)$$

$$(1) - (2) \Rightarrow \frac{x^2}{a^2} = \frac{y^2}{b^2} \Rightarrow x = \frac{ay}{b} \; (x, y, a, b > 0)$$

$$(2) - (3) \Rightarrow \frac{2y^2}{b^2} + \frac{2z^2}{c^2} - \frac{2zh}{c^2} = 0 \qquad (4)$$

(1) + (2) + (3) gives

$$\frac{5x^2}{a^2} + \frac{5y^2}{b^2} - \frac{5z^2}{c^2} = -\frac{2zh}{c^2}.$$

$$\Rightarrow 5\left(\frac{x^2}{a^2} + \frac{y^2}{b^2} - \frac{z^2}{c^2}\right) = -\frac{2zh}{c^2} \Rightarrow 5\left(\frac{3z^2}{c^2} - \frac{2zh}{c^2} - \frac{z^2}{c^2}\right) = -\frac{2zh}{c^2} \text{ [by (3)]}$$

$$\Rightarrow 5 \cdot \frac{2z^2}{c^2} = -\frac{2zh}{c^2} + \frac{10zh}{c^2} = \frac{8zh}{c^2} \Rightarrow z = \frac{4h}{5}.$$

From (4),

$$\frac{2y^2}{b^2}+\frac{2}{c^2}\cdot\frac{16h^2}{25}-\frac{8h^2}{5c^2}=0 \text{ or, } \frac{y^2}{b^2}+\frac{16h^2}{25c^2}-\frac{4h^2}{5c^2}=0$$

$$\text{or, } \frac{y^2}{b^2}=\frac{4h^2}{c^2}\left[\frac{1}{5}-\frac{4}{25}\right]=\frac{4h^2}{25c^2} \text{ or, } \frac{y}{b}=\frac{2h}{5c}, \quad \therefore y=\frac{2bh}{5c}.$$

Now $\frac{x}{a}=\frac{y}{b}\Rightarrow x=\frac{2ah}{5c}$. Stationary point is $\left(\frac{2ah}{5c},\frac{2bh}{5c},\frac{4h}{5}\right)$.

$$\frac{1}{u}u_{xx}-\frac{1}{u^2}(u_x)^2=-\frac{1}{x^2}+\frac{2/a^2}{\frac{x^2}{a^2}+\frac{y^2}{b^2}-\frac{z^2}{c^2}}-\frac{(2x/a^2)\cdot(2x/a^2)}{\left(\frac{x^2}{a^2}+\frac{y^2}{b^2}-\frac{z^2}{c^2}\right)^2}$$

$$u_x=0\Rightarrow u_{xx}=u\left[-\frac{1}{x^2}+\frac{2/a^2}{\left(\frac{x^2}{a^2}+\frac{y^2}{b^2}-\frac{z^2}{c^2}\right)}-\frac{4x^2/a^4}{\left(\frac{x^2}{a^2}+\frac{y^2}{b^2}-\frac{z^2}{c^2}\right)^2}\right].$$

For $(x,y,z)=\left(\frac{2ah}{5c},\frac{2bh}{5c},\frac{4h}{5}\right)$, $\frac{x^2}{a^2}+\frac{y^2}{b^2}-\frac{z^2}{c^2}=-\frac{8h^2}{25c^2}$.

$$\therefore u_{xx}=u\left[-\frac{25c^2}{4a^2h^2}-\frac{2}{a^2}\times\frac{25c^2}{8h^2}-\frac{\frac{4}{a^4}\cdot\frac{4a^2h^2}{25c^2}}{\frac{64h^4}{625c^4}}\right]$$

$$=-u\left[\frac{25c^2}{4a^2h^2}+\frac{25c^2}{4a^2h^2}+\frac{25c^2}{4a^2h^2}\right]<0\ [\because u>0].$$

We can verify that the other two principal minors, $\begin{vmatrix} u_{xx} & u_{xy}\\ u_{yx} & u_{yy}\end{vmatrix}$ and $\begin{vmatrix} u_{xx} & u_{xy} & u_{xz}\\ u_{yx} & u_{yy} & u_{yz}\\ u_{zx} & u_{zy} & u_{zz}\end{vmatrix}$ are respectively >0 and <0. Hence, $d^2f<0$.

So u has maximum at $\left(\frac{2ah}{5c},\frac{2bh}{5c},\frac{4h}{5}\right)$ and the maximum value of u is

$$\frac{2ah}{5c}\cdot\frac{2bh}{5c}\left(\frac{4h}{5}-h\right)\left(\frac{4a^2h^2}{a^2\cdot 25c^2}+\frac{4b^2h^2}{b^2\cdot 25c^2}-\frac{16h^2}{25c^2}\right)=-\frac{4abh^3}{125c^2}\left(-\frac{8h^2}{25c^2}\right)=\left(\frac{2h}{5}\right)^5\frac{ab}{c^4}.$$

12.7 Lagrange's Method of Multipliers

Here we shall find out the stationary points by using undetermined multipliers on the view of Lagrange's method of multipliers.

CASE (1) : Let $u=f(x,y,z)$ be a given function.

We wish to find out the stationary point of u subject to the condition $\phi(x,y,z)=0$.

Now

$$df=\frac{\partial f}{\partial x}dx+\frac{\partial f}{\partial y}dy+\frac{\partial f}{\partial z}dz \tag{1}$$

$$0=\frac{\partial \phi}{\partial x}dx+\frac{\partial \phi}{\partial y}dy+\frac{\partial \phi}{\partial z}dz. \tag{2}$$

Multiplying (2) by λ, where λ is a constant and adding with (1), we get

$$df=\left(\frac{\partial f}{\partial x}+\lambda\frac{\partial \phi}{\partial x}\right)dx+\left(\frac{\partial f}{\partial y}+\lambda\frac{\partial \phi}{\partial y}\right)dy+\left(\frac{\partial f}{\partial z}+\lambda\frac{\partial \phi}{\partial z}\right)dz.$$

For extremum, $df = 0$,

$$\Rightarrow \left.\begin{aligned} \frac{\partial f}{\partial x} + \lambda \frac{\partial \phi}{\partial x} &= 0 \\ \frac{\partial f}{\partial y} + \lambda \frac{\partial \phi}{\partial y} &= 0 \\ \frac{\partial f}{\partial z} + \lambda \frac{\partial \phi}{\partial z} &= 0 \end{aligned}\right\}, \tag{A}$$

since x, y, z are three independent variables.

From (A), we can determine the multiplier λ and the three unknowns x, y, z. So the equation (A) will determine the stationary point of $u = f(x, y, z)$.

$$d^2 f = \frac{\partial^2 f}{\partial x^2}(dx)^2 + \frac{\partial^2 f}{\partial y^2}(dy)^2 + \frac{\partial^2 f}{\partial z^2}(dz)^2 + 2\frac{\partial^2 f}{\partial x \partial y} dxdy + 2\frac{\partial^2 f}{\partial y \partial z} dydz + 2\frac{\partial^2 f}{\partial z \partial x} dxdz. \tag{B}$$

If for a stationary point (x, y, z) determined from (A), $d^2 f < 0$, then f is maximum at (x, y, z) and $d^2 f > 0$, then f is minimum at (x, y, z).

CASE (2) : Let $u = f(x, y, z)$.

We now want to determine the stationary points of u subject to the following conditions $\phi_1(x, y, z) = 0$, $\phi_2(x, y, z) = 0$.

Then

$$df = \frac{\partial f}{\partial x} dx + \frac{\partial f}{\partial y} dy + \frac{\partial f}{\partial z} dz \tag{1}$$

$$0 = \frac{\partial \phi_1}{\partial x} dx + \frac{\partial \phi_1}{\partial y} dy + \frac{\partial \phi_1}{\partial z} dz \tag{2}$$

$$0 = \frac{\partial \phi_2}{\partial x} dx + \frac{\partial \phi_2}{\partial y} dy + \frac{\partial \phi_2}{\partial z} dz. \tag{3}$$

Now multiplying (2) by λ_1, (3) by (λ_2) and then adding to (1), where λ_1, λ_2 are constants, we get

$$df = \left(\frac{\partial f}{\partial x} + \lambda_1 \frac{\partial \phi_1}{\partial x} + \lambda_2 \frac{\partial \phi_2}{\partial x}\right) dx + \left(\frac{\partial f}{\partial y} + \lambda_1 \frac{\partial \phi_1}{\partial y} + \lambda_2 \frac{\partial \phi_2}{\partial y}\right) dy + \left(\frac{\partial f}{\partial z} + \lambda_1 \frac{\partial \phi_1}{\partial z} + \lambda_2 \frac{\partial \phi_2}{\partial z}\right) dz.$$

For extremum, $df = 0$,

$$\Rightarrow \left.\begin{aligned} \frac{\partial f}{\partial x} + \lambda_1 + \frac{\partial \phi_1}{\partial x} + \lambda_2 \frac{\partial \phi_2}{\partial x} &= 0 \\ \frac{\partial f}{\partial y} + \lambda_1 \frac{\partial \phi_1}{\partial y} + \lambda_2 \frac{\partial \phi_2}{\partial y} &= 0 \\ \frac{\partial f}{\partial z} + \lambda_1 \frac{\partial \phi_1}{\partial z} + \lambda_2 \frac{\partial \phi_2}{\partial z} &= 0 \end{aligned}\right\}. \tag{C}$$

The equations (C), involving three unknowns x, y, z and two multipliers λ_1, λ_2, will determine their values.

So from (C), we get the stationary point (x, y, z) of $u = f(x, y, z)$.

Again, from (B), f has maximum at (x, y, z), if $d^2 f < 0$ and f has minimum at (x, y, z), if $d^2 f > 0$.

An Alternative Rule (Working Rule)

We consider a function $F(x, y, z) = f(x, y, z) + \lambda_1\phi_1(x, y, z) + \lambda_2\phi_2(x, y, z)$ for finding the extremum of $f(x, y, z)$ subject to the conditions $\phi_1(x, y, z) = 0$, $\phi_2(x, y, z) = 0$. λ_1, λ_2 are two undetermined multipliers.

$$\begin{aligned} dF &= \frac{\partial F}{\partial x}dx + \frac{\partial F}{\partial y}dy + \frac{\partial F}{\partial z}dz \\ &= \left(\frac{\partial f}{\partial x} + \lambda_1\frac{\partial \phi_1}{\partial x} + \lambda_2\frac{\partial \phi_2}{\partial x}\right)dx + \left(\frac{\partial f}{\partial y} + \lambda_1\frac{\partial \phi_1}{\partial y} + \lambda_2\frac{\partial \phi_2}{\partial y}\right)dy \\ &\quad + \left(\frac{\partial f}{\partial z} + \lambda_1\frac{\partial \phi_1}{\partial z} + \lambda_2\frac{\partial \phi_2}{\partial z}\right)dz \end{aligned}$$

$$\Rightarrow \quad \begin{aligned} \frac{\partial F}{\partial x} &= \frac{\partial f}{\partial x} + \lambda_1\frac{\partial \phi_1}{\partial x} + \lambda_2\frac{\partial \phi_2}{\partial x} \\ \frac{\partial F}{\partial y} &= \frac{\partial f}{\partial y} + \lambda_1\frac{\partial \phi_1}{\partial y} + \lambda_2\frac{\partial \phi_2}{\partial y} \\ \frac{\partial F}{\partial z} &= \frac{\partial f}{\partial z} + \lambda_1\frac{\partial \phi_1}{\partial z} + \lambda_2\frac{\partial \phi_2}{\partial z} \end{aligned}$$

For extremum, $dF = 0 \Rightarrow \frac{\partial F}{\partial x} = 0, \frac{\partial F}{\partial y} = 0, \frac{\partial F}{\partial z} = 0$, since x, y, z are three linearly independent variables.

Therefore,

$$\left.\begin{aligned} \frac{\partial f}{\partial x} + \lambda_1\frac{\partial \phi_1}{\partial x} + \lambda_2\frac{\partial \phi_2}{\partial x} &= 0 \\ \frac{\partial f}{\partial y} + \lambda_1\frac{\partial \phi_1}{\partial y} + \lambda_2\frac{\partial \phi_2}{\partial y} &= 0 \\ \frac{\partial f}{\partial z} + \lambda_1\frac{\partial \phi_1}{\partial z} + \lambda_2\frac{\partial \phi_2}{\partial z} &= 0 \end{aligned}\right\} \tag{D}$$

which will determine the stationary points of $F(x, y, z)$.

From (D) and (C) it follows that the stationary points of F determines the stationary points of f. Also a stationary point of f will be a point of maximum, if $d^2F < 0$ and a point of minimum, if $d^2F > 0$.

General Case

Let $u = f(x_1, x_2, \cdots, x_n)$ be a function of n independent variables. We want to determine the stationary points of u subject to the following conditions:

$$\phi_1(x_1, x_2, \cdots, x_n) = 0, \phi_2(x_1, x_2, \cdots, x_n) = 0, \cdots, \phi_m(x_1, x_2, \cdots, x_n) = 0.$$

We define $F(x_1, x_2, \cdots, x_n) = f(x_1, x_2, \cdots, x_n) + \lambda_1\phi_1(x_1, x_2, \cdots, x_n) + \lambda_2\phi_2(x_1, x_2, \cdots, x_n) + \cdots + \lambda_m\phi_m(x_1, x_2, \cdots, x_n)$, where $\lambda_1, \lambda_2, \cdots, \lambda_m$ are m undetermined multipliers.

$$\begin{aligned} dF &= \frac{\partial F}{\partial x_1}dx_1 + \frac{\partial F}{\partial x_2}dx_2 + \cdots + \frac{\partial F}{\partial x_n}dx_n \\ &= \left(\frac{\partial f}{\partial x_1} + \lambda_1\frac{\partial \phi_1}{\partial x_1} + \cdots + \lambda_m\frac{\partial \phi_m}{\partial x_1}\right)dx_1 + \left(\frac{\partial f}{\partial x_2} + \lambda_1\frac{\partial \phi_1}{\partial x_2} + \cdots + \lambda_m\frac{\partial \phi_m}{\partial x_2}\right)dx_2 \\ &\quad \cdots\cdots\cdots\cdots\cdots\cdots\cdots\cdots\cdots\cdots\cdots\cdots \\ &\quad + \left(\frac{\partial f}{\partial x_n} + \lambda_1\frac{\partial \phi_1}{\partial x_n} + \cdots + \lambda_m\frac{\partial \phi_m}{\partial x_n}\right)dx_n. \end{aligned}$$

For extremum, $dF = 0, \Rightarrow \dfrac{\partial F}{\partial x_1} = 0, \dfrac{\partial F}{\partial x_2} = 0, \cdots, \dfrac{\partial F}{\partial x_n} = 0$, since $x_1, x_2, \cdots, x_n$ are independent variables.

So

$$\left.\begin{aligned}
\frac{\partial f}{\partial x_1} + \lambda_1 \frac{\partial \phi_1}{\partial x_1} + \cdots + \lambda_m \frac{\partial \phi_m}{\partial x_1} &= 0 \\
\frac{\partial f}{\partial x_2} + \lambda_1 \frac{\partial \phi_1}{\partial x_2} + \cdots + \lambda_m \frac{\partial \phi_m}{\partial x_2} &= 0 \\
\cdots\cdots\cdots\cdots\cdots\cdots\cdots\cdots\cdots & \\
\frac{\partial f}{\partial x_n} + \lambda_1 \frac{\partial \phi_1}{\partial x_n} + \cdots + \lambda_m \frac{\partial \phi_m}{\partial x_n} &= 0
\end{aligned}\right\} \quad \text{(E)}$$

We solve this system of equations for '$n+m$' unknowns, $x_1, x_2, \cdots, x_n$ and $\lambda_1, \lambda_2, \cdots, \lambda_m$.

The point $(x_1, x_2, \cdots, x_n)$ thus obtained is a stationary point and this stationary point will be a point of maximum, if $d^2F < 0$ and a point of minimum, if $d^2F > 0$.

Note. $F(x, y, z)$ is called Lagrangian function.

Example. Let $f(x, y, z) = x^2 + y^2 + z^2 - yz - zx - xy$.

We shall find out the extremum of $f(x, y, z)$ subject to the condition $\phi(x, y, z) = x^2 + y^2 + z^2 - 2x + 2y + 6z + 9 = 0$.

We define,

$$F(x, y, z) = (x^2 + y^2 + z^2 - xy - yz - zx) + \lambda(x^2 + y^2 + z^2 - 2x + 2y + 6z + 9)$$

$$\frac{\partial F}{\partial x} = 2x - y - z + \lambda(2x - 2)$$

$$\frac{\partial F}{\partial y} = 2y - x - z + \lambda(2y + 2)$$

$$\frac{\partial F}{\partial z} = 2z - x - y + \lambda(2z + 6).$$

For extremum, $dF = 0, \Rightarrow \dfrac{\partial F}{\partial x} = 0, \dfrac{\partial F}{\partial y} = 0, \dfrac{\partial F}{\partial z} = 0$

$$\Rightarrow 2x - y - z + 2\lambda(x - 1) = 0 \quad (1)$$

$$2y - x - z + 2\lambda(y + 1) = 0 \quad (2)$$

$$2z - x - y + 2\lambda(z + 3) = 0. \quad (3)$$

From (1), (2) and (3), we have

$$-\lambda = \frac{2x - y - z}{2(x - 1)} = \frac{2y - x - z}{2(y + 1)} = \frac{2z - x - y}{2(z + 3)} = \frac{0}{2(x + y + z) + 6}$$

$$\Rightarrow \lambda = 0, x + y + z \neq -3 \cdots.$$

Also $2x - y - z = 0, 2y - x - z = 0, 2z - x - y = 0 \Rightarrow x = y = z$.

The system has infinite number of solutions of the form (c, c, c), $c \neq -1$.

Stationary point is (c, c, c), $c \neq -1$.

Therefore,

$$F_x = 2x - y - z,\ F_y = 2y - x - z,\ F_z = 2z - x - y,\ F_{xx} = 2,\ F_{yy} = 2,\ F_{zz} = 2,$$

$$F_{xy} = F_{yx} = -1,\ F_{yz} = F_{zy} = -1,\ F_{zx} = F_{xz} = -1$$

for all the stationary points $(c, c, c), c \neq -1$.

$$\begin{aligned}
\therefore d^2F &= \sum \frac{\partial^2 f}{\partial x^2}(dx)^2 + 2\sum \frac{\partial^2 f}{\partial x \partial y} dxdy \\
&= 2(dx)^2 + 2(dy)^2 + 2(dz)^2 - 2dxdy - 2dydz - 2dzdx \\
&= (dx - dy)^2 + (dy - dz)^2 + (dz - dx)^2 > 0.
\end{aligned}$$

$\therefore F$ is minimum at (c, c, c), $c \neq -1$

$\Rightarrow f$ is minimum at (c, c, c).

Problem [12.11]: (1) Show that the volume of the greatest rectangular parallelopiped that can be inscribed in the ellipsoid $\frac{x^2}{a^2} + \frac{y^2}{b^2} + \frac{z^2}{c^2} = 1$ is $\frac{8abc}{3\sqrt{3}}$.

(2) Find the shortest distance from the point (a_1, a_2, a_3) in R^3 to the plane whose equation is $b_1x_1 + b_2x_2 + b_3x_3 + b_0 = 0$.

(3) Find the maximum of $(xyz)^2$ subject to the condition $x^2 + y^2 + z^2 = 1$.

(4) Find the shorest distance from the point $(0, b)$ on the Y-axis to the parabola $x^2 - 4y = 0$. Use Lagrange's method.

(5) If $ax^2 + by^2 + cz^2 + 2fyz + 2gxz + 2hxy = k$ and $lx + my + nz = 0$, then find the extremum values of $x^2 + y^2 + z^2$.

(6) If $xyz = abc$, show that the minimum value of $bcx + cay + abz$ is $3abc$.

Solution. (1) Volume of the rectangular parallelopiped inscribed in the given elepsoid is $8xyz$.

Let $V = 8xyz$.

Subject to the condition

$$\frac{x^2}{a^2} + \frac{y^2}{b^2} + \frac{z^2}{c^2} - 1 = 0. \tag{1}$$

We define the Lagrangian function F by

$$F(x, y, z) = 8xyz + \lambda\left(\frac{x^2}{a^2} + \frac{y^2}{b^2} + \frac{z^2}{c^2} - 1\right)$$

$$dF = \left(8yz + \frac{2\lambda x}{a^2}\right)dx + \left(8xz + \frac{2\lambda y}{b^2}\right)dy + \left(8xy + \frac{2\lambda z}{c^2}\right)dz.$$

For extremum, $dF = 0 \Rightarrow \dfrac{\partial F}{\partial x} = 0, \dfrac{\partial F}{\partial y} = 0, \dfrac{\partial F}{\partial z} = 0$

$$\Rightarrow 8yz + 2\lambda\frac{x}{a^2} = 0 \Rightarrow 4yz + \frac{\lambda x}{a^2} = 0 \tag{2}$$

$$8zx + 2\lambda\frac{y}{b^2} = 0 \Rightarrow 4zx + \frac{\lambda y}{b^2} = 0 \tag{3}$$

$$8xy + 2\lambda\frac{z}{c^2} = 0 \Rightarrow 4xy + \frac{\lambda z}{c^2} = 0. \tag{4}$$

(2) x + (3) y + (4) z gives $12xyz + \lambda = 0$ by (1) $\Rightarrow \lambda = -12xyz$.

From (2),

$$4yz - \frac{12x^2yz}{a^2} = 0 \Rightarrow x^2 = \frac{a^2}{3} \Rightarrow x = \frac{a}{\sqrt{3}}.$$

Similarly, we get $y = \frac{b}{\sqrt{3}}, z = \frac{c}{\sqrt{3}}$ $[\because\ x, y, z > 0]$

$$\lambda = -\frac{12abc}{3\sqrt{3}} = -\frac{4abc}{\sqrt{3}}.$$

$$d^2F = \frac{2\lambda}{a^2}(dx)^2 + \frac{2\lambda}{b^2}(dy)^2 + \frac{2\lambda}{c^2}(dz)^2 + 16z\,dxdy + 16y\,dzdx + 16x\,dydz$$

$$= -\frac{8abc}{\sqrt{3}}\left[\frac{(dx)^2}{a^2} + \frac{(dy)^2}{b^2} + \frac{(dz)^2}{c^2}\right] + \frac{16}{\sqrt{3}}(c\,dxdy + b\,dzdx + a\,dxdz). \quad (5)$$

Now differentiating (1), we get $\frac{2x}{a^2}dx + \frac{2y}{b^2}dy + \frac{2z}{c^2}dz = 0.$

At the point $\left(\frac{a}{\sqrt{3}}, \frac{b}{\sqrt{3}}, \frac{c}{\sqrt{3}}\right)$, we get $\frac{dx}{a} + \frac{dy}{b} + \frac{dz}{c} = 0.$

Squaring it, we have

$$\frac{(dx)^2}{a^2} + \frac{(dy)^2}{b^2} + \frac{(dz)^2}{c^2} + \frac{2dxdy}{ab} + \frac{2dydz}{bc} + \frac{2dzdx}{ca} = 0$$

$$\text{or,} \quad -abc\left[\frac{(dx)^2}{a^2} + \frac{(dy)^2}{b^2} + \frac{(dz)^2}{c^2}\right] = 2c\,dxdy + 2a\,dydz + 2b\,dzdx. \quad (6)$$

From (5) by (6),

$$d^2F = -\frac{16abc}{3\sqrt{3}}\left[\frac{(dx)^2}{a^2} + \frac{(dy)^2}{b^2} + \frac{(dz)^2}{c^2}\right] < 0.$$

So $\left(\frac{a}{\sqrt{3}}, \frac{b}{\sqrt{3}}, \frac{c}{\sqrt{3}}\right)$ is a point of maximum and maximum value of $V = \frac{8abc}{3\sqrt{3}}$.

(2) The distance of the point (a_1, a_2, a_3) from a point (x_1, x_2, x_3) lying in the given plane is $d^2 = (x_1 - a_1)^2 + (x_2 - a_2)^2 + (x_3 - a_3)^2 = f(x_1, x_2, x_3)$.

Subject to the condition $b_1x_1 + b_2x_2 + b_3x_3 + b_0 = 0$, we consider the Lagrangian function

$$F(x_1, x_2, x_3) = (x_1 - a_1)^2 + (x_2 - a_2)^2 + (x_3 - a_3)^2 + \lambda(b_1x_1 + b_2x_2 + b_3x_3 + b_0)$$

$$dF = \{2(x_1 - a_1) + \lambda b_1\}dx_1 + \{2(x_2 - a_2) + \lambda b_2\}dx_2 + \{2(x_3 - a_3) + \lambda b_3\}dx_3.$$

For extremum, $dF = 0 \Rightarrow F_{x_1} = 0, F_{x_2} = 0, F_{x_3} = 0$

$$\Rightarrow 2(x_1 - a_1) + \lambda b_1 = 0 \quad (1)$$

$$2(x_2 - a_2) + \lambda b_2 = 0 \quad (2)$$

$$2(x_3 - a_3) + \lambda b_3 = 0. \quad (3)$$

Then

$$-\frac{\lambda}{2} = \frac{x_1 - a_1}{b_1} = \frac{x_2 - a_2}{b_2} = \frac{x_3 - a_3}{b_3}$$

$$= \frac{b_1x_1 + b_2x_2 + b_3x_3 - (a_1b_1 + a_2b_2 + a_3b_3)}{b_1^2 + b_2^2 + b_3^2}$$

$$= -\frac{[b_0 + a_1b_1 + a_2b_2 + a_3b_3]}{b_1^2 + b_2^2 + b_3^2} = -\frac{[b_0 + \Sigma a_1b_1]}{\Sigma b_1^2}.$$

Taking $k = \dfrac{(b_0 + \Sigma a_1 b_1)}{\Sigma b_1^2}$, we get $\lambda = 2k$.

$$\frac{x_1 - a_1}{b_1} = -k \Rightarrow x_1 = a_1 - b_1 k$$
$$\frac{x_2 - a_2}{b_2} = -k \Rightarrow x_2 = a_2 - b_2 k$$
$$\frac{x_3 - a_3}{b_3} = -k \Rightarrow x_3 = a_3 - b_3 k.$$

Stationary point is $(a_1 - b_1 k, a_2 - b_2 k, a_3 - b_3 k)$.

Now $F_{x_1x_1} = 2 = F_{x_2x_2} = F_{x_3x_3}$.

Also $F_{x_1x_2} = F_{x_2x_1} = F_{x_1x_3} = F_{x_3x_1} = F_{x_2x_3} = F_{x_3x_2} = 0$.

$$d^2F = 2\left[(dx_1)^2 + (dx_2)^2 + (dx_3)^2\right] > 0$$

$\therefore f$ is minimum at $(a_1 - b_1 k, a_2 - b_2 k, a_3 - b_3 k)$.

$\therefore$ shortest distance is given by

$$d = \sqrt{(a_1 - b_1 k - a_1)^2 + (a_2 - b_2 k - a_2)^2 + (a_3 - b_3 k - a_3)^2}$$
$$= k\sqrt{b_1^2 + b_2^2 + b_3^2} = \frac{b_0 + \Sigma a_1 b_1}{\Sigma b_1^2}\sqrt{\Sigma b_1^2} = \frac{b_0 + \Sigma a_1 b_1}{\sqrt{\Sigma b_1^2}}.$$

(3) Let $f(x, y, z) = x^2y^2z^2$.

Subject to the condition: $x^2 + y^2 + z^2 - 1 = 0$.

We define the Lagrangian function,

$$F(x, y, z) = x^2y^2z^2 + \lambda(x^2 + y^2 + z^2 - 1)$$
$$dF = (2xy^2z^2 + 2\lambda x)dx + (2yx^2z^2 + 2\lambda y)dy + (2zx^2y^2 + 2\lambda z)dz.$$

For extremum, $dF = 0 \Rightarrow F_x = 0, F_y = 0, F_z = 0$

$$\Rightarrow 2xy^2z^2 + 2\lambda x = 0 \quad (1)$$
$$2yx^2z^2 + 2\lambda y = 0 \quad (2)$$
$$2zx^2y^2 + 2\lambda z = 0. \quad (3)$$

(1) x + (2) y + (3) $z \Rightarrow 3x^2y^2z^2 + \lambda = 0 \Rightarrow \lambda = -3x^2y^2z^2$.

From (1), $xy^2z^2 - 3x^2y^2z^2 \cdot x = 0 \Rightarrow x^2 = \dfrac{1}{3}$ $(x \neq 0) \Rightarrow x = \pm\dfrac{1}{\sqrt{3}}$.

Similarly, $y = \pm\dfrac{1}{\sqrt{3}}$, $z = \pm\dfrac{1}{\sqrt{3}}$.

Stationary points are $\left(\frac{1}{\sqrt{3}}, \frac{1}{\sqrt{3}}, \frac{1}{\sqrt{3}}\right), \left(-\frac{1}{\sqrt{3}}, -\frac{1}{\sqrt{3}}, -\frac{1}{\sqrt{3}}\right)$

$$\left(-\frac{1}{\sqrt{3}}, \frac{1}{\sqrt{3}}, \frac{1}{\sqrt{3}}\right), \left(-\frac{1}{\sqrt{3}}, -\frac{1}{\sqrt{3}}, \frac{1}{\sqrt{3}}\right), \left(\frac{1}{\sqrt{3}}, -\frac{1}{\sqrt{3}}, \frac{1}{\sqrt{3}}\right),$$
$$\left(\frac{1}{\sqrt{3}}, -\frac{1}{\sqrt{3}}, -\frac{1}{\sqrt{3}}\right), \left(\frac{1}{\sqrt{3}}, \frac{1}{\sqrt{3}}, -\frac{1}{\sqrt{3}}\right), \left(-\frac{1}{\sqrt{3}}, \frac{1}{\sqrt{3}}, -\frac{1}{\sqrt{3}}\right).$$

We consider the stationary point $\left(\frac{1}{\sqrt{3}}, \frac{1}{\sqrt{3}}, \frac{1}{\sqrt{3}}\right)$.

$$\lambda = -\frac{3}{27} = -\frac{1}{9}$$

$$F_{xx} = 2y^2z^2 + 2\lambda = \frac{2}{9} - \frac{2}{9} = 0, F_{yy} = 0, F_{zz} = 0.$$

Also $F_{xy} = 4xyz^2 = \frac{4}{9}, F_{yz} = \frac{4}{9}, F_{zx} = \frac{4}{9}$.

$$\therefore d^2F = F_{xx}(dx)^2 + F_{yy}(dy)^2 + F_{zz}(dz)^2 + 2F_{xy}dxdy + 2F_{yz}dydz + 2F_{zx}dzdx$$

$$= 2 \cdot \frac{4}{9}[dxdy + dydz + dzdx]. \quad (4)$$

Now $x^2 + y^2 + z^2 = 1 \Rightarrow xdx + ydy + zdz = 0$

$$\Rightarrow dx + dy + dz = 0 \text{ at } \left(\frac{1}{\sqrt{3}}, \frac{1}{\sqrt{3}}, \frac{1}{\sqrt{3}}\right) \Rightarrow \sum(dx)^2 = -2\sum dxdy.$$

From (4), $d^2F = -\frac{4}{9}\sum(dx)^2 < 0$.

So f is maximum at $\left(\frac{1}{\sqrt{3}}, \frac{1}{\sqrt{3}}, \frac{1}{\sqrt{3}}\right)$. Maximum value of $f = \frac{1}{3} \cdot \frac{1}{3} \cdot \frac{1}{3} = \frac{1}{27}$.

(4) Let (α, β) be any point on the parabola $x^2 = 4y$. Then distance of the point $(0, b)$ from (α, β).

$$d^2 = \alpha^2 + (\beta - b)^2 = f(\alpha, \beta).$$

Subject to the condition $\alpha^2 - 4\beta = 0$, we define the Lagrangian function

$$F(\alpha, \beta) = \alpha^2 + (\beta - b)^2 + \lambda(\alpha^2 - 4\beta)$$

$$dF = (2\alpha + 2\alpha\lambda)d\alpha + \{2(\beta - b) - 4\lambda\}d\beta.$$

From extremum, $dF = 0 \Rightarrow F_\alpha = 0, F_\beta = 0$

$$\Rightarrow 2\alpha + 2\alpha\lambda = 0, \quad (1)$$

$$2(\beta - b) - 4\lambda = 0. \quad (2)$$

From (1), $\lambda = -1$ $(\alpha \neq 0)$ and from (2), $2(\beta - b) + 4 = 0 \Rightarrow \beta = b - 2$.

$$\alpha^2 = 4\beta = 4(b - 2) \Rightarrow \alpha = \pm 2\sqrt{b-2} \ (b \geq 2).$$

Stationary points are $\left(2\sqrt{b-2}, b-2\right), \left(-2\sqrt{b-2}, b-2\right)$.

$$F_{\alpha\alpha} = 2 + 2\lambda = 0, F_{\beta\beta} = 2, F_{\alpha\beta} = 0$$

$$d^2F = F_{\alpha\alpha}(d\alpha)^2 + 2F_{\alpha\beta}d\alpha d\beta + F_{\beta\beta}(d\beta)^2 = 2(d\beta)^2 > 0$$

for $\left(2\sqrt{b-2}, b-2\right)$ or $\left(-2\sqrt{b-2}, b-2\right)$.

$\therefore f$ is minimum at any of them.

Therefore, the shortest distance $d = \sqrt{4(b-2)+4} = 2\sqrt{b-1}$, $(b \geq 2)$.

(5) Let $f(x, y, z) = x^2 + y^2 + z^2 = u$ (say).

Subject to the conditions, $ax^2 + by^2 + cz^2 + 2fyz + 2gzx + 2hxy - k = 0$ and $lx + my + nz = 0$, we form the Lagrangian function,

$$F(x, y, z) = x^2 + y^2 + z^2 + \lambda_1(ax^2 + by^2 + cz^2 + 2fyz + 2gzx + 2hxy - k)$$
$$+\lambda_2(lx + my + nz)$$

$$dF = \{2x + \lambda_1(2ax + 2gz + 2hy) + \lambda_2 l\}dx + \{2y + \lambda_1(2by + 2fz + 2hx) + \lambda_2 m\}dy$$
$$+\{2z + \lambda_1(2cz + 2fy + 2gx) + \lambda_2 n\}dz.$$

For extremumm, $dF = 0 \Rightarrow F_x = F_y = F_z = 0$

$$\Rightarrow 2x + 2\lambda_1(ax + gz + hy) + \lambda_2 l = 0 \quad (1)$$

$$2y + 2\lambda_1(by + fz + hx) + \lambda_2 m = 0 \quad (2)$$

$$2z + 2\lambda_1(cz + fy + gx) + \lambda_2 n = 0. \quad (3)$$

(1) x + (2) y + (3) $z \Rightarrow 2u + 2\lambda_1 k = 0 \Rightarrow \lambda_1 = -\frac{u}{k}$ $(k \neq 0)$.

For $\lambda_1 = -\frac{u}{k}$, from (1),

$$-\lambda_2 = \frac{2x - 2\frac{u}{k}(ax + gz + hy)}{l} \text{ or, } -\frac{\lambda_2}{2} = \frac{kx - u(ax + gz + hy)}{lk} \quad (4)$$

Similarly,

$$-\frac{\lambda_2}{2} = \frac{ky - u(by + fz + hx)}{mk} \quad (5)$$

$$\text{and } -\frac{\lambda_2}{2} = \frac{kz - u(cz + fy + gx)}{nk}. \quad (6)$$

(4) and (5) gives $m\{kx - u(ax + gz + hy)\} = l\{ky - u(by + fz + hx)\}$

$$\Rightarrow (mk - mua + luh)x + (lub - lk - muh)y + (luf - mug)z = 0. \quad (7)$$

(5) and (6) gives $n\{ky - u(by + fz + hx)\} = m\{kz - u(cz + fy + gx)\}$

$$\Rightarrow (mug - nuh)x + \{(nk - nub + muf)y + (muc - mk - nuf)z = 0. \quad (8)$$

(4) and (6) gives $n\{kx - u(ax + gz + hy) = l\{kz - u(cz + fy + gx)\}$

$$\Rightarrow (nk - nua + lug)x + (luf - nuh)y + (luc - lk - nug)z = 0. \quad (9)$$

Eliminating x, y, z from (7), (8) and (9), we get an equation in u in the form of a determinant,

$$\begin{vmatrix} mk - mua + luh & lub - lk - muh & luf - mug \\ mug - nuh & nk - nub + muf & muc - mk - nuf \\ nk - nua + lug & luf - nuh & luc - lk - nug \end{vmatrix} = 0$$

$$\text{or, } \begin{vmatrix} mk - u(ma - lh) & -lk - u(mh - lb) & u(lf - mg) \\ u(mg - nh) & nk - u(nb - mf) & -mk - u(nf - mc) \\ nk - u(na - lg) & u(lf - nh) & -lk - u(ng - lc) \end{vmatrix} = 0,$$

which will give the extremum of u.

(6) Let $f(x, y, z) = bcx + cay + abz = u$ (say).

Subject to the condition $xyz - abc = 0$, we define the Lagrangian function

$$F(x, y, z) = f(x, y, z) + \lambda(xyz - abc) = bcx + cay + abz + \lambda(xyz - abc)$$

$$dF = (bc + \lambda yz)dx + (ca + \lambda xz)dy + (ab + \lambda xy)dz.$$

For extremum, $dF = 0$,

$$\Rightarrow F_x = bc + \lambda yz = 0 \quad (1)$$

$$F_y = ca + \lambda xz = 0 \quad (2)$$

$$F_z = ab + \lambda xy = 0. \quad (3)$$

From (1), $xbc + \lambda xyz = 0 \Rightarrow \frac{x}{a} = -\lambda \Rightarrow x = -a\lambda$ $[\because xyz = abc]$.

Similarly, $\frac{y}{b} = -\lambda$, $\frac{z}{c} = -\lambda \Rightarrow y = -b\lambda$, $z = -c\lambda$.

Then $xyz = -abc\lambda^3 \Rightarrow \lambda = -1$.

Hence $x = a, y = b, z = c$ is a stationary point.

$$F_x = bc - yz, F_y = ca - zx, F_z = ab - xy, F_{xx} = 0, \quad F_{yy} = 0$$
$$F_{zz} = 0, \quad F_{xy} = -z, \quad F_{yz} = -x, \quad F_{zx} = -y.$$

Then,

$$d^2F = 2F_{yz}dydz + 2F_{zx}dzdx + 2F_{xy}dxdy$$
$$= -2xdydz - 2ydzdx - 2zdxdy = -[2adydz + 2bdzdx + 2cdxdy] \quad (4)$$

at the stationary point (a, b, c).

Again, from $xyz = abc$, we have $yzdx + zxdy + xydz = 0 \Rightarrow bcdx + cady + abdz = 0$ at (a, b, c)

$$\Rightarrow \frac{dx}{a} + \frac{dy}{b} + \frac{dz}{c} = 0 \Rightarrow \sum \frac{(dx)^2}{a^2} = -2\left[\frac{dydz}{bc} + \frac{dzdx}{ac} + \frac{dxdy}{ab}\right]$$
$$\Rightarrow abc \sum \left(\frac{dx}{a}\right)^2 = -[2adydz + 2bdzdx + 2cdxdy]$$

$\therefore$ from (4), $d^2F = abc\sum\left(\frac{dx}{a}\right)^2 > 0$ $[a, b, c > 0]$.
$\therefore f$ has minimum at $(x, y, z) = (a, b, c)$ and minimum value of $f = 3abc$.

12.8 Integrals as Functions of a Parameter

We have $\mathbb{R}^2 = \{(x, y), x, y \in \mathbb{R}\}$.

Let $u = f(x, y)$ be defined in a rectangular domain $R\ [a, b, c, d] \subseteq R^2$.

$R\ [a, b, c, d]$ defines $a \le x \le b, c \le y \le d$.

We suppose that $f(x, y)$ is continuous in $R\ [a, b, c, d]$. Then for a fixed value of y in $[c, d]$, $f(x, y)$ becomes a function of single variable x and remains continuous in $[a, b]$.

Therefore, $\int_a^b f(x, y)dx$ exists and the integral is a function of y and we denote it by $\phi(y) = \int_a^b f(x, y)dx$. Here y is considered as a parameter.

Similarly, for a fixed value of x in $[a, b]$, $f(x, y)$ becomes a function of y and remains continuous in $[c, d]$ and hence the integral $\int_c^d f(x, y)dy$ exists and the integral is a function of x and we denote it by $\phi(x)$.

That is, $\phi(x) = \int_c^d f(x, y)dy$. Here x is considered as a parameter.

Example. Let $f(x, y) = x^2 + xy + y^3$, $R = [0, 1, 0, 1]$. Then $f(x, y)$ is continuous in R.

For a fixed y,

$$\int_0^1 (x^2 + xy + y^3)dx = \left[\frac{x^3}{3} + y\frac{x^2}{2} + y^3x\right]_0^1 = \frac{1}{3} + \frac{y}{2} + y^3$$

is a function of y.

For a fixed x,

$$\int_0^1 (x^2 + xy + y^3)dy = \left(x^2y + x\frac{y^2}{2} + \frac{y^4}{4}\right)_0^1 = x^2 + \frac{x}{2} + \frac{1}{4}$$

is a function of x.

It also follows that $f(x, y)$ remains continuous in $[0, 1]$ for fixed value of y and for a fixed value of x.

Properties of integral functions

Theorem [12.8] (1) Let $f(x,y)$ be continuous in $R\ [a,b;c,d]$. Then $\phi(y)=\int_a^b f(x,y)dx$ is continuous in $[c,d]$.

(2) If (i) $f(x,y)$ be continuous in $R\ [a,b;\ c,d]$, (ii) $f_y(x,y)$ exists and is continuous in $R\ [a,b;c,d]$, then $\phi(y)=\int_a^b f(x,y)dx$ is differentiable in $[c,d]$ and $\phi'(y)=\int_a^b f_y(x,y)dx$. [Differentiation under the integral sign].

(3) If (i) $f(x,y)$ be continuous in $R\ [a,b;c,d]$, (ii) $f_y(x,y)$ exists and is continuous in $R\ [a,b;c,d]$, (iii) $g_1(y), g_2(y)$ be two differentiable functions of y in $[c,d]$, (iv) $\forall\ y \in [c,d]$ the points $(g_1(y),y)$ and $(g_2(y),y) \in [c,d]$, then $\phi(y)=\int_{g_1(y)}^{g_2(y)} f(x,y)dx$ is differentiable in $[c,d]$ and $\phi'(y)=\int_{g_1(y)}^{g_2(y)} f_y(x,y)dx+g_2'(y)f(g_2(y),y)-g_1'(y)f(g_1(y),y)$.
[Differentiation under the integral sign when the limits of integration are functions of the parameter y.]

Note. Rules (2), (3) are known as Leibniz's rule (1) and (2) respectively.

Proof. (1) Let y, $y+k$ be any two points in $[c,d]$, $\phi(y)=\int_a^b f(x,y)dx$.

Then $\phi(y+k)=\int_a^b f(x,y+k)dx$.

Therefore,

$$\phi(y+k)-\phi(y)=\int_a^b \{f(x,y+k)-f(x,y)\}dx. \tag{i}$$

$f(x,y)$ is continuous in $R\ [a,b;c,d]$

$\Rightarrow f(x,y)$ is uniformly continuous in R.

$\Rightarrow$ for $\epsilon(>0)$, there exists a (+)ve number δ such that $|f(x_2,y_2)-f(x_1,y_1)|<\epsilon$, whenever $|x_2-x_1|\le\delta$, $|y_2-y_1|\le\delta$.

Applying this result to the given continuous function $f(x,y)$ in R, we have

$$|f(x,y+k)-f(x,y)|<\frac{\epsilon}{b-a},$$

whenever $|k|\le\delta$, and x is any point in $[a,b], b>a$.

From (i),

$$\begin{aligned}|\phi(y+k)-\phi(y)| &= \left|\int_a^b \{f(x,y+k)-f(x,y)\}dx\right| \\ &\le \int_a^b |f(x,y+k)-f(x,y)|dx < \int_a^b \frac{\epsilon}{b-a}dx=\epsilon, \text{ whenever } |k|\le\delta.\end{aligned}$$

$\Rightarrow \phi(y)$ is continuous in $[c,d]$.

(2) Let $y, y+k$ be any two points in $[c,d]$.

Then

$$\begin{aligned}\phi(y+k)-\phi(y) &= \int_a^b \{f(x,y+k)-f(x,y)\}dx \\ &= \int_a^b kf_y(x,y+\theta k)dx, 0<\theta<1 \text{ [by Lagrange's M.V. Th.]}\end{aligned}$$

$$\Rightarrow \frac{\phi(y+k)-\phi(y)}{k}=\int_a^b f_y(x,y+\theta k)dx.$$

Therefore, $\dfrac{\phi(y+k)-\phi(y)}{k}-\displaystyle\int_a^b f_y(x,y)dx=\int_a^b \{f_y(x,y+\theta k)-f_y(x,y)\}dx$

$$\Rightarrow \left|\frac{\phi(y+k)-\phi(y)}{k}-\int_a^b f_y(x,y)dx\right| \le \int_a^b |f_y(x,y+\theta k)-f_y(x,y)|dx. \qquad \text{(ii)}$$

$f_y(x,y)$ is continuous in $R\,[a,b;c,d]$

$\Rightarrow f_y(x,y)$ is uniformly continuous in R.

$\Rightarrow$ for every $\epsilon(>0)$, $\exists$ a (+)ve number δ, such that $|f_y(x,y+\theta k)-f_y(x,y)|<\frac{\epsilon}{b-a}$, whenever $|k|\le\delta$ for any point $x\in[a,b]$, $b>a$.

Therefore, from (1),

$$\left|\frac{\phi(y+k)-\phi(y)}{k}-\int_a^b f_y(x,y)dx\right| \le \int_a^b \frac{\epsilon}{b-a}dx=\epsilon, \text{ whenever } |k|\le\delta$$

$$\Rightarrow \lim_{k\to 0}\frac{\phi(y+k)-\phi(y)}{k}=\int_a^b f_y(x,y)dx \text{ for } \forall\, y\in[c,d].$$

Limit exists since $f_y(x,y)$ being continuous on $R\,[a,b,c,d]$ is R-integrable on $[a,b]$.
So $\phi(y)$ is differentiable on $[c,d]$.

$$\text{and } \phi'(y)=\int_a^b f_y(x,y)dx.$$

That is, $\dfrac{d}{dy}\phi(y)=\displaystyle\int_a^b \frac{\partial}{\partial y}(f(x,y))dx.$

(3) Let y, $y+k$ be any two points in $[c,d]$.

Then $\phi(y+k)-\phi(y)$

$$=\int_{g_1(y+k)}^{g_2(y+k)} f(x,y+k)dx-\int_{g_1(y)}^{g_2(y)} f(x,y)dx$$

$$=\int_{g_1(y)}^{g_2(y)} f(x,y+k)dx-\int_{g_1(y)}^{g_1(y+k)} f(x,y+k)dx+\int_{g_2(y)}^{g_2(y+k)} f(x,y+k)dx-\int_{g_1(y)}^{g_2(y)} f(x,y)dx$$

$$=\int_{g_1(y)}^{g_2(y)} \{f(x,y+k)-f(x,y)\}dx+\int_{g_2(y)}^{g_2(y+k)} f(x,y+k)dx-\int_{g_1(y)}^{g_1(y+k)} f(x,y+k)dx$$

$$=\int_{g_1(y)}^{g_2(y)} kf_y(x,y+\theta k)dx+\{g_2(y+k)-g_2(y)\}f(\xi,y+k)-\{g_1(y+k)-g_1(y)\}f(\eta,y+k),$$

[Applying M.V. Theorem in the first part and applying the property of R-integrability for $\int_a^b f(x)dx=(b-a)f(\xi)$, $\xi\in[a,b]$ in the 2nd and 3rd part.]

where ξ lies in between $g_2(y)$, $g_2(y+k)$ and η lies in between $g_1(y), g_1(y+k)$.

Also $\lim\limits_{k\to 0} f(\xi,y+k)=f(g_2(y),y)$ and $\lim\limits_{k\to 0} f(\eta,y+k)=f(g_1(y),y)$.

Therefore,

$$\frac{\phi(y+k)-\phi(y)}{k}=\int_{g_1(y)}^{g_2(y)} f_y(x,y+\theta k)dx+\frac{g_2(y+k)-g_2(y)}{k}f(\xi,y+k)$$
$$-\frac{g_1(y+k)-g_1(y)}{k}f(\eta,y+k).$$

Taking limit $k \to 0$, we get

$$\phi'(y) = \int_{g_1(y)}^{g_2(y)} f_y(x,y)dx + g_2'(y)f(g_2(y),y) - g_1'(y)f(g_1(y),y),$$

since $\lim_{k\to 0} \int_{g_1(y)}^{g_2(y)} f_y(x, y+\theta k)dx = \int_{g_1(y)}^{g_2(y)} f_y(x,y)dx$, which has been shown in (2).

Theorem [12.9] [Inversion of the order of integration]: Let $f(x,y)$ be continuous in $R\,[a,b;c,d]$. Then $\int_c^d \left\{\int_a^b f(x,y)dx\right\} dy = \int_a^b \left\{\int_c^d f(x,y)dy\right\} dx$.

Proof. $f(x,y)$ is continuous, so $\int_a^b f(x,y)dx$ exists.

Also $\phi(y) = \int_a^b f(x,y)dx$ is continuous in $[c,d]$ [by (1) of Th. (12.7)] and hence $\int_c^d \phi(y)dy$ exists. So the integral on the left-hand side exists. Similarly, continuity of $f(x,y) \Rightarrow$ the existence of $\int_c^d f(x,y)dy$. Also $\phi(x) = \int_c^d f(x,y)dy$ is continuous in $[a,b]$ and hence, $\int_a^b \phi(x)dx$ exists and therefore, the integral on the R.H.S. also exists.

We now define,

$$F(t) = \int_c^t \left\{\int_a^b f(x,y)dx\right\} dy \text{ and } G(t) = \int_a^b \left\{\int_c^t f(x,y)dy\right\} dx$$

so that $F(c) = G(c) = 0$ and $t \in [c,d]$.

We have $\phi(y) = \int_a^b f(x,y)dx$. Then $F(t) = \int_c^t \phi(y)dy$

$$\Rightarrow F'(t) = \phi(t) = \int_a^b f(x,t)dx. \tag{i}$$

Again

$$G'(t) = \int_a^b \left\{\frac{\delta}{\delta t}\int_c^t f(x,y)dy\right\} dx = \int_a^b f(x,t)dx \tag{ii}$$

Therefore, from (i) and (ii) $F'(t) = G'(t)$, $\forall\, t \in [c,d]$.

Integrating w.r.t. 't', we get $F(t) = G(t) + A$.

For $t = c$, $F(c) = G(c) = 0 \Rightarrow A = 0$. Hence, $F(t) = G(t)$ $\forall\, t \in [c,d]$.

Putting $t = d$, we get

$$\int_c^d \left\{\int_a^b f(x,y)dx\right\} dy = \int_a^b \left\{\int_c^d f(x,y)dy\right\} dx.$$

Note. The repeated integrals may be written as

$$\int_c^d dy \left\{\int_a^b f(x,y)dx\right\} \text{ and } \int_a^b dx \left\{\int_c^d f(x,y)dy\right\}.$$

Problem [12.12]: (1) If f, f_y be continuous in $R\,[a,b;c,d]$ and F be bounded and integrable on $[a,b]$, then show that $\phi(y) = \int_a^b f(x,y)F(x)dx$ is derivable in $[c,d]$ and $\phi'(y) = \int_a^b f_y(x,y)F(x)dx$.

(2) If f be continuous in $R\,[a,b;c,d]$, then show that $g(x) = \int_c^d f(x,y)dy$ is continuous on $[a,b]$. Further, if f_x exists and is continuous in R, then show that $g'(x)$ exists and $g'(x) = \int_c^d f_x(x,y)dy$ $\forall\, x \in [a,b]$.

(3) Let $F(x,y) = \int_c^y dv \int_a^x f(u,v)du$ and $f(x,y)$ is continuous in R $[a,b;c,d]$, then show that $F_{xy}(x,y) = F_{yx}(x,y) = f(x,y)\ \forall\ (x,y) \in R$. Deduce that $\int_a^b dx \left\{\int_c^d f(x,y)dy\right\} = F(a,c) + F(b,d) - F(a,d) - F(b,c) = \int_c^d dy \left\{\int_a^b f(x,y)dx\right\}$.

(4) Show that $\int_0^{\pi/2} \frac{\log(1+y\sin^2 x)}{\sin^2 x}dx = \pi\left[\sqrt{1+y}-1\right]$.

(5) Show that $\int_0^a \frac{\log(1+ax)}{1+x^2}dx = \frac{1}{2}\log(1+a^2)\tan^{-1}a$ and hence deduce that $\int_0^1 \frac{\log(1+x)}{1+x^2}dx = \frac{\pi\log 2}{8}$.

(6) Show that $y = \frac{1}{k}\int_0^x f(t)\sin k(x-t)dt$ satisfies the differential equation $\frac{d^2y}{dx^2} + k^2y = f(x)$, where k is a constant.

(7) Let $f(x,y) = \frac{y^2}{x^2+y^2}$ be defined in R $[0,1;0,1]$ and let $g(y) = \int_0^1 f(x,y)dx$. Then show that $g'(+0) \neq g'(-0)$ and each of them differs from $\int_0^1 f_y(x,0)dx$.

(8) Show that $\int_0^1 \frac{x^y-1}{\log x}dx = \log(1+y)$ and hence show that $\int_0^1 \frac{x-1-\log x}{(\log x)^2}dx = \log\frac{4}{e}$.

(9) Prove that $\int_{\frac{\pi}{2}-\alpha}^{\pi/2} \sin\theta\cos^{-1}(\cos\alpha\,\mathrm{cosec}\,\theta)d\theta = \frac{\pi}{2}(1-\cos\alpha)$.

(10) If $F(x) = \int_{\sin x}^{e^x} \sqrt{1+y^3}dy$, then evaluate $F'(x)$.

Solution. (1) Continuity of $f(x,y)$ in R $[a,b;c,d] \Rightarrow$ integrability of $f(x,y)$ in $R \Rightarrow f(x,y)$ is integrable on $[a,b]\ \forall\ y \in [c,d]$.

$F(x)$ is integrable on $[a,b]$. So $\int_a^b f(x,y)F(x)dx$ exists. Also $|F(x)| \leq M\ \forall\ x \in [a,b]$.

We have $\phi(y) = \int_a^b f(x,y)F(x)dx$. Let $y, y+k$ be any points in $[c,d]$.

Then

$$\phi(y+k)-\phi(y) = \int_a^b \{f(x,y+k)F(x) - f(x,y)F(x)\}dx = \int_a^b \{f(x,y+k)-f(x,y)\}F(x)dx$$

$$= \int_a^b kf_y(x,y+\theta k)F(x)dx,\ 0<\theta<1 \text{ [by M.V. Th.]}$$

or, $$\left|\frac{\phi(y+k)-\phi(y)}{k} - \int_a^b f_y(x,y)F(x)dx\right|$$

$$= \left|\int_a^b f_y(x,y+\theta k)F(x)dx - \int_a^b f_y f(x,y)F(x)dx\right|$$

$$\leq \int_a^b |f_y(x,y+\theta k)-f_y(x,y)|\,|F(x)|dx \leq M\int_a^b |f_y(x,y+\theta k)-f_y(x,y)|dx. \tag{1}$$

Since $f_y(x,y)$ is continuous on R $[a,b;c,d]$, it is uniformly continuous on R. So for any value of x, $|f_y(x,y+\theta k)-f_y(x,y)| < \frac{\epsilon}{M(b-a)}$, for $|k| \leq \delta$.

Therefore, from (1),

$$\left|\frac{\phi(y+k)-\phi(y)}{k} - \int_a^b f_y(x,y)F(x)dx\right| < M\int_a^b \frac{\epsilon}{M(b-a)}dx = \epsilon \text{ for } |k| \leq \delta$$

$$\Rightarrow \lim_{k\to 0}\frac{\phi(y+k)-\phi(y)}{k} = \int_a^b f_y(x,y)F(x)dx$$

Limit exists as $f_y(x,y)\,F(x)$ is integrable on $[a,b]$. So $\phi(y)$ is derivable and

$$\phi'(y)=\int_a^b f_y(x,y)F(x)dx.$$

(2) Proofs are similar as done in (1) and (2) of Theorem [12.7].

(3) If $\phi(x)=\int_a^x f(u,v)du$, then $\phi'(x)=\dfrac{\partial}{\partial x}\int_a^x f(u,v)du=f(x,v)$.

Now $F(x,y)=\int_c^y dv\int_a^x f(u,v)du$.

Then

$$F_x(x,y)=\int_c^y dv\left\{\frac{\partial}{\partial x}\int_a^x f(u,v)du\right\}=\int_c^y dv f(x,v)=\int_c^y f(x,v)dv$$

$$F_{xy}(x,y)=\frac{\partial}{\partial y}\int_c^y f(x,v)dv=f(x,y).$$

Again, $F(x,y)=\int_a^x du\int_c^y f(u,v)dv$ [by inversion order of integration]

Then,

$$F_y(x,y)=\int_a^x du\frac{\partial}{\partial y}\left\{\int_c^y f(u,v)dv\right\}=\int_a^x f(u,y)du$$

$$\Rightarrow F_{yx}(x,y)=\frac{\partial}{\partial x}\int_a^x f(u,y)du=f(x,y).$$

Thus we have proved $F_{xy}=F_{yx}=f(x,y)$.

Now

$$\int_a^b dx\int_c^d f(x,y)dy=\int_a^b dx\int_c^d F_{xy}dy=\int_a^b\frac{\partial}{\partial x}\left\{\int_c^d F_y dy\right\}dx=\int_a^b\frac{\partial}{\partial x}\left[F(x,y)\right]_c^d dx$$

$$=\int_a^b[F_x(x,d)-F_x(x,c)]dx=\left[F(x,d)-F(x,c)\right]_a^b$$

$$=F(b,d)-F(b,c)-F(a,d)+F(a,c)$$

$$=F(b,d)+F(a,c)-F(b,c)-F(a,d).$$

Since $\int_a^b dx\left\{\int_c^d f(x,y)dy\right\}=\int_c^d dy\left\{\int_a^b f(x,y)dx\right\}$, the result follows.

(4) Let

$$\phi(y)=\int_0^{\pi/2}\frac{\log(1+y\sin^2 x)}{\sin^2 x}dx,\ f(x,y)=\frac{\log(1+y\sin^2 x)}{\sin^2 x}$$

'0' is not a point of infinite discontinuity of $f(x,y)$, since

$$\lim_{x\to 0}\frac{\log(1+y\sin^2 x)}{\sin^2 x}\left(\frac{0}{0}\right)=\lim_{x\to 0}\frac{1}{1+y\sin^2 x}\cdot\frac{2y\sin x\cos x}{2\sin x\cos x}=y.$$

Also

$$f_y(x,y)=\frac{1}{1+y\sin^2 x}\cdot\frac{\sin^2 x}{\sin^2 x}=\frac{1}{1+y\sin^2 x}.$$

Therefore, $f(x,y)$ and $f_y(x,y)$ are both continuous on $\left[0,\frac{\pi}{2}\right]$.

Differentiation under the integral sign exists. So

$$\phi'(y) = \int_0^{\pi/2} \frac{\partial}{\partial y} \frac{\log(1+y\sin^2 x)}{\sin^2 x} dx.$$

$$= \int_0^{\pi/2} \frac{1}{1+y\sin^2 x} dx = \int_0^{\pi/2} \frac{\sec^2 x dx}{1+\tan^2 x + y\tan^2 x}$$

$$\phi'(y) = \frac{1}{1+y} \int_0^{\pi/2} \frac{\sec^2 x dx}{\tan^2 x + \frac{1}{1+y}}$$

$$= \frac{1}{1+y} \int_0^{\infty} \frac{dz}{z^2 + \left(\frac{1}{\sqrt{1+y}}\right)^2} \text{ [putting } \tan x = z]$$

$$= \frac{1}{\sqrt{1+y}} \left[\tan^{-1} z\sqrt{1+y}\right]_0^{\infty}$$

$$= \frac{1}{\sqrt{1+y}} \frac{\pi}{2}.$$

Integrating w.r.t. y, we get $\phi(y) = \frac{\pi}{2} \cdot 2\sqrt{1+y} + A = \pi\sqrt{1+y} + A$.

For $y = 0$, $\phi(0) = 0$. So $A = -\pi$.

$$\therefore \phi(y) = \pi[\sqrt{1+y} - 1].$$

(5) Let $\phi(a) = \displaystyle\int_0^a \frac{\log(1+ax)}{1+x^2} dx.$

Here, $f(a,x) = \dfrac{\log(1+ax)}{1+x^2}$ is continuous in $[0,a]$, since 0 is not a point of infinite discontinuity of $f(a,x)$.

Also, $f_a(a,x) = \dfrac{x}{(1+ax)(1+x^2)}$ is continuous in $[0,a]$. So, differentiation under the integral sign holds.

Therefore,

$$\phi'(a) = \int_0^a \frac{\partial}{\partial a} \cdot \frac{\log(1+ax)}{1+x^2} dx + 1 \cdot \frac{\log(1+a^2)}{1+a^2}$$

$$= \int_0^a \frac{x}{(1+ax)(1+x^2)} dx + \frac{\log(1+a^2)}{1+a^2}$$

$$= \int_0^a \left[\frac{\frac{1}{1+a^2}x + \frac{a}{1+a^2}}{1+x^2} + \frac{-\frac{a}{1+a^2}}{1+ax} \right] dx + \frac{\log(1+a^2)}{1+a^2}$$

$$= \int_0^a \frac{1}{(1+a^2)} \frac{x+a}{x^2+1} dx - \int_0^a \frac{1}{1+a^2} \cdot \frac{a}{1+ax} dx + \frac{\log(1+a^2)}{1+a^2}$$

$$= \frac{1}{2(1+a^2)} \int_0^a \frac{2x}{x^2+1} dx + \frac{a}{(1+a^2)} \int_0^a \frac{1}{x^2+1} dx - \frac{1}{1+a^2} \int_0^a \frac{a}{1+ax} dx + \frac{\log(1+a^2)}{1+a^2}$$

$$= \frac{1}{2(1+a^2)} \left[\log(x^2+1)\right]_0^a + \frac{a}{1+a^2} \left[\tan^{-1} x\right]_0^a - \frac{1}{1+a^2} \left[\log(1+ax)\right]_0^a + \frac{\log(1+a^2)}{1+a^2}$$

$$= \frac{1}{2(1+a^2)} \log(1+a^2) + \frac{a}{1+a^2} \tan^{-1} a - \frac{1}{1+a^2} \log(1+a^2) + \frac{\log(1+a^2)}{1+a^2}.$$

Now integrating w.r.t. a, we get

$$\begin{aligned}\phi(a) &= \frac{1}{2}\int \frac{1}{1+a^2}\log(1+a^2)da + \frac{1}{2}\int \frac{2a}{1+a^2}\tan^{-1} a da \\ &= \frac{1}{2}\int \frac{2a}{1+a^2}\tan^{-1} a da + \frac{1}{2}\int \frac{1}{1+a^2}\log(1+a^2)da \\ &= \frac{1}{2}\left[\log(1+a^2)\tan^{-1} a - \int \frac{1}{1+a^2}\log(1+a^2)da\right] \\ &\quad + \frac{1}{2}\int \frac{1}{1+a^2}\log(1+a^2)da + A \\ \phi(a) &= \frac{1}{2}\log(1+a^2)\tan^{-1} a + A\end{aligned}$$

For $a = 0, \phi(0) = 0 \Rightarrow A = 0$.

$$\therefore \phi(a) = \frac{1}{2}\log(1+a^2)\tan^{-1} a \Rightarrow \int_0^a \frac{\log(1+ax)}{1+x^2}dx = \frac{1}{2}\log(1+a^2)\tan^{-1} a.$$

Taking $a = 1$, we get

$$\int_0^1 \frac{\log(1+x)}{1+x^2}dx = \frac{1}{2}\log 2\tan^{-1} 1 = \frac{\pi}{8}\log 2.$$

(6) $y = \dfrac{1}{k}\displaystyle\int_0^x f(t)\sin k(x-t)dt$ (1)

Differentiating w.r.t. x,

$$\begin{aligned}\frac{dy}{dx} &= \frac{1}{k}\int_0^x \frac{\partial}{\partial x}\{f(t)\sin k(x-t)\}dt + \frac{1}{k}\cdot 1\cdot f(x)\sin k(x-x) \\ \Rightarrow \frac{dy}{dx} &= \frac{1}{k}\int_0^x kf(t)\cos k(x-t)dt + 0 = \int_0^x f(t)\cos k(x-t)dt.\end{aligned}$$

Again differentiating w.r.t. x, we get

$$\frac{d^2y}{dx^2} = -k\int_0^x f(t)\sin k(x-t)dt + 1\cdot f(x)\cos k(x-x) = -k^2y + f(x) \text{ [by (1)]}.$$

$$\therefore \frac{d^2y}{dx^2} + k^2y = f(x). \tag{2}$$

Hence (1) satisfies the differential equation (2).

Note. Here $f(t)\sin k(k-t)$ and $\frac{\partial}{\partial x}[f(t)\sin k(x-t)]$ are supposed to be continuous so that Leibniz's rule (2) holds.

(7) $f(x, y) = \dfrac{y^2}{x^2+y^2}$ defined in $R\ [0, 1; 0, 1]$

$$\begin{aligned}g(y) &= \int_0^1 f(x,y)dx = \int_0^1 \frac{y^2}{x^2+y^2}dx = y^2\cdot\frac{1}{y}\left[\tan^{-1}\frac{x}{y}\right]_0^1 = y\tan^{-1}\frac{1}{y} \\ g'(+0) &= \lim_{k\to 0+0}\frac{g(k)-g(0)}{k} = \lim_{k\to 0+0}\frac{k\tan^{-1}\frac{1}{k} - 0}{k} = \frac{\pi}{2} \\ g'(-0) &= \lim_{k\to 0-}\frac{g(k)-g(0)}{k} = \lim_{k\to 0-}\frac{k\tan^{-1}\frac{1}{k}}{k} = -\frac{\pi}{2}.\end{aligned}$$

Therefore, $g'(+0)$ and $g'(-0)$ both exist but $g'(+0) \neq g'(-0)$.

Also $f_y(x,y) = \dfrac{2y}{x^2+y^2} - y^2\dfrac{1}{(x^2+y^2)^2}\cdot 2y = 2y\cdot x^2/(x^2+y^2)^2.$

$f_y(x,0) = 0$

$\therefore \displaystyle\int_0^1 f_y(x,0)dx = 0.$

So they differ from $\displaystyle\int_0^1 f_y(x,0)dx$.

(8) Let $\phi(y) = \displaystyle\int_0^1 \frac{x^y-1}{\log x}dx$. Here $f(x,y) = \dfrac{x^y-1}{\log x}$ and $f_y(x,y) = \dfrac{x^y \log x}{\log x} = x^y$ are both continuous in $[0,1]$, since $x=0$ and $x=1$ are not points of infinite discontinuity as

$$\lim_{x\to 0+}\frac{x^y-1}{\log x} = 0 \text{ and } \lim_{x\to 1}\frac{x^y-1}{\log x} = y.$$

Differentiating under the integral sign

$$\phi'(y) = \int_0^1 \frac{\partial}{\partial y}\left(\frac{x^y-1}{\log x}\right)dx = \int_0^1 x^y dx = \left[\frac{x^{y+1}}{y+1}\right]_0^1 = \frac{1}{y+1}.$$

Then integrating w.r.t. y, we get $\phi(y) = \log(1+y) + A$.

For, $y=0$, $\phi(0) = 0 \Rightarrow A = 0$.

$$\therefore \phi(y) = \log(1+y), \text{ i.e., } \int_0^1 \frac{x^y-1}{\log x}dx = \log(1+y) \qquad (1)$$

Using repeated integrals in $R\,[0,1;0,1]$

$$\int_0^1 dy\int_0^1 \frac{x^y-1}{\log x}dx = \int_0^1 dx\int_0^1 \frac{x^y-1}{\log x}dy. \qquad (2)$$

$$\text{L.H.S.} = \int_0^1 \log(1+y)dy \text{ [by (1)]}$$

$$= \left[y\log(1+y)\right]_0^1 - \int_0^1 \frac{y}{1+y}dy = \log 2 - \left[y - \log(1+y)\right]_0^1$$

$$= \log 2 - 1 + \log 2 = 2\log 2 - 1 = \log\frac{4}{e}.$$

$$\text{R.H.S.} = \int_0^1 \frac{1}{\log x}\left[\frac{x^y}{\log x} - y\right]_0^1 dx = \int_0^1 \frac{1}{\log x}\left[\frac{x}{\log x} - 1 - \frac{1}{\log x}\right]dx$$

$$= \int_0^1 \frac{x-1-\log x}{(\log x)^2}dx.$$

Therefore, from (2), $\displaystyle\int_0^1 \frac{x-1-\log x}{(\log x)^2} = \text{L.H.S.} = \log\frac{4}{e}.$

(9) Let $\phi(\alpha) = \displaystyle\int_{\frac{\pi}{2}-\alpha}^{\pi/2} \sin\theta\cos^{-1}(\cos\alpha\,\text{cosec}\,\theta)d\theta$. Here $f(\alpha,\theta) = \sin\theta\cos^{-1}(\cos\alpha\,\text{cosec}\,\theta)$ and $f_\alpha(\alpha,\theta) = \sin\theta\dfrac{\sin\alpha\,\text{cosec}\,\theta}{\sqrt{1-\cos^2\alpha\,\text{cosec}^2\theta}}$ are both continuous in $\left[\dfrac{\pi}{2}-\alpha, \dfrac{\pi}{2}\right]$.

Also, $g(\alpha) = \frac{\pi}{2} - \alpha$, possesses derivative.

Differentiating under the integral sign, by Leibniz's Rule (2), we get

$$\phi'(\alpha) = \int_{\frac{\pi}{2}-\alpha}^{\pi/2} \frac{\partial}{\partial \alpha}\left[\sin\theta \cos^{-1}(\cos\alpha \operatorname{cosec}\theta)\right] d\theta + 0$$

$$-\frac{d}{d\alpha}\left(\frac{\pi}{2}-\alpha\right)\left[\sin\left(\frac{\pi}{2}-\alpha\right)\cos^{-1}\left\{\cos\alpha \operatorname{cosec}\left(\frac{\pi}{2}-\alpha\right)\right\}\right]$$

$$= \int_{\frac{\pi}{2}-\alpha}^{\pi/2} \frac{\sin\alpha}{\sqrt{1-\cos^2\alpha \operatorname{cosec}^2\theta}} d\theta + \cos\alpha \cos^{-1}\left(\frac{\cos\alpha}{\cos\alpha}\right)$$

$$= \int_{\frac{\pi}{2}-\alpha}^{\pi/2} \frac{\sin\alpha \sin\theta}{\sqrt{\sin^2\theta - \cos^2\alpha}} d\theta + \cos\alpha \cdot 0 = \int_{\frac{\pi}{2}-\alpha}^{\pi/2} \frac{\sin\alpha\sin\theta}{\sqrt{\sin^2\alpha - \cos^2\theta}} d\theta.$$

Putting $\cos\theta = z, -\sin\theta d\theta = dz, z : \sin\alpha$ to 0

$$= \sin\alpha \int_{\sin\alpha}^{0} -\frac{dz}{\sqrt{\sin^2\alpha - z^2}} = \sin\alpha \int_0^{\sin\alpha} \frac{dz}{\sqrt{\sin^2\alpha - z^2}}$$

$$= \sin\alpha \left[\sin^{-1}\frac{z}{\sin\alpha}\right]_0^{\sin\alpha}.$$

$$\therefore \phi'(\alpha) = \frac{\pi}{2}\sin\alpha.$$

Now integrating w.r.t. α, we get $\phi(\alpha) = -\frac{\pi}{2}\cos\alpha + A$.

For $\alpha = 0$, $\phi(0) = 0 \Rightarrow A = \frac{\pi}{2}$.

$$\therefore \phi(\alpha) = \frac{\pi}{2}(1-\cos\alpha).$$

(10) $F(x) = \displaystyle\int_{\sin x}^{e^x} \sqrt{1+y^3}dy.$

Here $f(x,y) = \sqrt{1+y^3}$ and $f_x(x,y) = 0$ are both continuous and e^x, $\sin x$ are differentiable functions of x.

Differentiating w.r.t. x under the integral sign by Leibniz's Rule (2), we get

$$F'(x) = \int_{\sin x}^{e^x} \frac{\partial}{\partial x}\sqrt{1+y^3}dy + \frac{d}{dx}(e^x)\sqrt{1+e^{3x}} - \frac{d}{dx}(\sin x)\sqrt{1+\sin^3 x}$$

$$= 0 + e^x\sqrt{1+e^{3x}} - \cos x\sqrt{1+\sin^3 x}$$

$$\therefore F'(x) = e^x\sqrt{1+e^{3x}} - \cos x\sqrt{1+\sin^3 x}.$$

12.9 Operation of Differentiation and Integration Under the Sign of Integration in Case of Improper Integrals

In the previous part Art. 12.8 of this Chapter we have defined $\phi(y) = \int_a^b f(x,y)dx$ as an integral function of the parameter y, when $\int_a^b f(x,y)dx$ is considered as a proper integral throughout the part Art. 12.8. Application of differentiation and integration to $\phi(y)$ [or to $\phi(x) = \int_a^b f(x,y)dy$] are done by stating some sufficient conditions.

Now the question arises for differentiation and integration under the integral sign when $\phi(y)$ is defined by an *improper integral* like (i) $\int_a^\infty f(x,y)dx$, (ii) $\int_a^b f(x,y)dx$ when $x = a$ is a point of infinite discontinuity. Here $f(x,y)$ is defined in $R\ [x \geq a; c, d]$ and $R\ [a, b; c, d]$, when $x = a$ is a point of infinite discontinuity respectively.

We shall observe that it is possible to state sufficient conditions for the operation of differentiation and integration under the sign of integration when the integral defined by $\phi(y)$ is an improper integral.

Definition [12.9] : Uniform convergence of an improper integral:
Let $\phi(y) = \int_a^\infty f(x,y)dx$ be convergent. Then $\phi(y)$ is said to be uniformly convergent in $[c,d]$, if for every $\epsilon(>0)$, $\exists$ a (+)ve number k such that $\left|\int_a^t f(x,y)dx - \phi(y)\right| < \epsilon,\ \forall\, t \geq k$.

That is, $\left|\int_t^\infty f(x,y)dx\right| < \epsilon,\ \forall\, t \geq k$.

Definition [12.10] : Let the integral $\phi(y) = \int_a^b f(x,y)dx$ be convergent at $x = a$, where a is a point of infinite discontinuity of $f(x,y)$. Then $\phi(y)$ is said to be uniformly convergent in the interval $[c,d]$, if for every $\epsilon(>0)$, $\exists$ a (+)ve number δ such that $\left|\int_{a+\eta}^b f(x,y)dx - \phi(y)\right| < \epsilon$, for $0 < \eta \leq \delta$.

Test for uniform convergence [12.6.1]

(1) **Weierstrass's M-test:** If there exists a function $M_n(x)$ with domain $[a,\infty)$ such that

(i) $M(x) > 0,\ \forall\, x \in [a,\infty)$,

(ii) $\int_a^\infty M(x)dx$ is convergent $\forall\, x \geq a$ and

(iii) $|f(x,y)| \leq M(x),\ \forall\, x,y \in R\ [a,\infty; c,d]$, then $\int_a^\infty f(x,y)dx$ is said to be uniformly convergent.

Note. Similar result is applicable to the improper integral $\int_a^b f(x,y)dx$. Here $M(x) > 0\ \forall\, x \in (a,b]$ or $x \in [a,b)$ or $x \in (a,b)$ according as a or b or both points of infinite discontinuity.

(2) **General test:** The improper integral $\phi(y) = \int_a^\infty f(x,y)dx$ is uniformly convergent in $[a,\infty; c,d]$, if and only if for every $\epsilon(>0)$, $\exists$ a (+)ve number k independent of x such that $\left|\int_{t_1}^{t_2} f(x,y)dx\right| < \epsilon,\ \forall\, t_1, t_2 \geq k$.

Example. Let $\phi(y) = \displaystyle\int_0^\infty e^{-x^2}\cos xy\,dx$.

Here $f(x,y) = e^{-x^2}\cos xy$. Now $|f(x,y)| \leq e^{-x^2} = M(x)$, say.

Then (i) $M(x) > 0\ \forall\, x \in [0,\infty)$

(ii) $\displaystyle\int_0^\infty M(x)dx = \int_0^\infty e^{-x^2}dx$ is convergent.

(iii) Also $|f(x,y)| \leq M(x)\ \forall\, x,y \in [0,\infty; c,d]$

Then $\phi(y)$ is uniformly convergent in $[0,\infty)$.

Example. $\displaystyle\int_0^1 \frac{\sin yx}{\sqrt{x(1-x)}}dx$.

Here $|f(x,y)| = \left|\dfrac{\sin yx}{\sqrt{x(1-x)}}\right| \leq \dfrac{1}{\sqrt{x(1-x)}} = M(x)$

(i) $M(x) > 0\ \forall\, x \in (0,1)$.

(ii) $\displaystyle\int_0^1 M(x)dx = \int_0^1 \frac{1}{\sqrt{x(1-x)}}dx$ is convergent.

Therefore, $\displaystyle\int_0^1 \frac{\sin yx}{\sqrt{x(1-x)}}dx$ is uniformly convergent.

Example. $\displaystyle\int_1^\infty \frac{x\cos xy}{1+x^2}dx.$

Here $\dfrac{x}{1+x^2}$ is (+)ve and monotonically decreasing in $[1,\infty)$ and $\to 0$ as $x\to\infty$.

Therefore,

$$\left|\int_{t_1}^{t_2} \frac{x\cos xy}{1+x^2}dx\right| = \left|\frac{t_1}{1+t_1^2}\int_{t_1}^{\xi}\cos xy dx\right| \quad \text{[by Bonnet's form]}$$
$$\leq \frac{2t_1}{k(1+t_1^2)} \;\forall\, y \geq k,$$

where k is a fixed (+)ve number independent of x.

$$< \epsilon, \text{ taking } \frac{t_1}{k(1+t_1^2)} < \frac{\epsilon}{2}.$$

Thus $\displaystyle\left|\int_{t_1}^{t_2} \frac{x\cos xy}{1+x^2}dx\right| < \epsilon \;\forall\, t_1, t_2 \geq k.$

Thus the integral is uniformly convergent in $[k,\infty]$, $k>0$.

Theorem [12.10] If $\phi(y) = \int_a^\infty f(x,y)dx$ be uniformly convergent in $R\,[x\geq a; c,d]$ and $f(x,y)$ is continuous in R, then $\phi(y)$ is continuous in $[c,d]$.

Proof. Since $\phi(y) = \int_a^\infty f(x,y)dx$ is uniformly convergent in $[c,d]$, we have

$$\left|\int_t^\infty f(x,y)dx\right| < \frac{\epsilon}{3},\; \forall\, t > a. \tag{i}$$

Now let $y+k\in[c,d]$.

Then

$$\begin{aligned}
\phi(y+k)-\phi(y) &= \int_a^\infty f(x,y+k)dx - \int_a^\infty f(x,y)dx \\
&= \int_a^t f(x,y+k)dx - \int_a^t f(x,y)dx + \int_t^\infty f(x,y+k)dx - \int_t^\infty f(x,y)dx \\
\Rightarrow |\phi(y+k)-\phi(y)| &\leq \int_a^t |f(x,y+k)-f(x,y)|dx + \left|\int_t^\infty f(x,y+k)dx\right| + \left|\int_t^\infty f(x,y)dx\right| \\
&< \int_a^t |f(x,y+k)-f(x,y)|dx + \frac{\epsilon}{3}+\frac{\epsilon}{3} \text{ [by (i)]}
\end{aligned}$$

Again, $[a,t]$ is a finite range and f being continuous in $R\,[a,t;c,d]$, it is uniformly continuous in R and hence $|f(x,y+k)-f(x,y)| < \frac{\epsilon}{3(t-a)}$ for $|k|<\delta$.

Therefore, $|\phi(y+k)-\phi(y)| < \int_a^t \frac{\epsilon}{3(t-a)}dx + \frac{2\epsilon}{3} = \frac{\epsilon}{3}+\frac{2\epsilon}{3} = \epsilon$ for $|k|<\delta$.

$\therefore\ \phi(y)$ is continuous in $[c,d]$.

Theorem [12.11] Let $f(x,y)$ be continuous in $R\,[x\geq a; c,d]$ and let $\phi(y) = \int_a^\infty f(x,y)dx$ be uniformly convergent in $[c,d]$. Then ϕ can be integrated under the sign of integration. That is,

$$\int_c^d \phi(y)dy = \int_c^d \left\{\int_a^\infty f(x,y)dx\right\}dy = \int_a^\infty \left\{\int_c^d f(x,y)dy\right\}dx.$$

Proof. Since $\phi(y) = \int_a^\infty f(x,y)dx$ is uniformly convergent, we have

$$\left|\int_t^\infty f(x,y)dx\right| < \frac{\epsilon}{d-c} \text{ for } t \geq k. \qquad \text{(i)}$$

Now

$$\int_c^d \phi(y)dy = \int_c^d \left\{\int_a^\infty f(x,y)dx\right\} dy$$

$$= \int_c^d \left\{\int_a^t f(x,y)dx\right\} + \left\{\int_t^\infty f(x,y)dx\right\} dy$$

$$\Rightarrow \int_c^d \phi(y)dy - \int_c^d \left\{\int_a^t f(x,y)dx\right\} dy = \int_c^d \left\{\int_t^\infty f(x,y)dx\right\} dy$$

$$\Rightarrow \left|\int_c^d \phi(y)dy - \int_a^t \left\{\int_c^d f(x,y)dy\right\} dx\right| = \left|\int_c^d \left\{\int_t^\infty f(x,y)dx\right\}\right| dy$$

$$\leq \int_c^d \left|\int_t^\infty f(x,y)dx\right| dy < \int_c^d \frac{\epsilon}{d-c} dy = \epsilon \ \forall \ t \geq k \text{ [by (i)]}$$

$$\Rightarrow \int_c^d \phi(y)dy = \int_a^\infty \left\{\int_c^d f(x,y)dy\right\} dx$$

$$\Rightarrow \int_c^d \phi(y)dy = \int_c^d \left\{\int_a^\infty f(x,y)dx\right\} dy = \int_a^\infty \left\{\int_c^d f(x,y)dy\right\} dx.$$

Theorem [12.12] Let $f(x,y)$ and $f_y(x,y)$ be continuous in R $[x \geq a; c,d]$. Also let $\phi(y) = \int_a^\infty f(x,y)dx$ and $\int_a^\infty f_y(x,y)dx$ converge uniformly in $[c,d]$. Then $\phi'(y) = \int_a^\infty f_y(x,y)dx$.

Proof. Let $\psi(y) = \int_a^\infty f_y(x,y)dx$.

Then by Theorem [12.10],

$$\int_c^y \psi(y)dy = \int_c^y \left\{\int_a^\infty f_y(x,y)dx\right\} dy = \int_a^\infty \left\{\int_c^y f_y(x,y)dy\right\} dx$$

$$= \int_a^\infty .[f(x,y) - f(x,c)]dx = \phi(y) - \phi(c).$$

Differentiating, we get $\psi(y) = \phi'(y)$ [since $\psi(y)$ is continuous by Th. [12.9]

$$\therefore \phi'(y) = \int_a^\infty f_y(x,y)dx.$$

Theorem [12.13] If $f(x,y)$ has a point of infinite discontinuity at $x = a$ for some or $\forall$ $y \in [c,d]$, then uniformly convergent improper integral $\phi(y) = \int_a^b f(x,y)dx$ is continuous, when $f(x,y)$ is continuous in R $[a+\eta, b; c,d]$, for some (+)ve integer η.

Proof. Improper integral $\phi(y) = \int_a^b f(x,y)dx$ is uniformly convergent.

Therefore, by Definition [12.6],

$$\left|\int_{a+\eta}^b f(x,y)dx - \phi(y)\right| < \frac{\epsilon}{3}, \text{ for } 0 < \eta \leq \delta \Rightarrow \left|\int_{a+\eta}^a f(x,y)dx\right| < \frac{\epsilon}{3}. \qquad \text{(i)}$$

Let $y + k \in [c,d]$.

Then

$$\phi(y+k)-\phi(y)=\int_a^b f(x,y+k)dx-\int_a^b f(x,y)dx$$

$$=\int_{a+\eta}^b f(x,y+k)-\int_{a+\eta}^a f(x,y+k)dx-\int_{a+\eta}^b f(x,y)dx$$

$$+\int_{a+\eta}^b f(x,y)dx-\int_a^b f(x,y)dx$$

$$\Rightarrow |\phi(y+k)-\phi(y)|=\left|\int_{a+\eta}^b \{f(x,y+k)-f(x,y)\}dx-\int_{a+\eta}^a f(x,y+k)dx\right.$$

$$\left.+\left\{\int_{a+\eta}^b f(x,y)dx-\phi(y)\right\}\right|$$

$$\leq \int_{a+\eta}^b |f(x,y+k)-f(x,y)|dx+\left|\int_{a+\eta}^a f(x,y+k)dx\right|$$

$$+\left|\int_{a+\eta}^b f(x,y)dx-\phi(y)\right|$$

Since $f(x,y)$ is continuous in R $[a+\eta,b;c,d]$, it is uniformly continuous in R and so $|f(x,y+k)-f(x,y)|<\frac{\epsilon}{b-(a+\eta)}$, $|k|\leq\delta$.

$$\therefore |\phi(y+k)-\phi(y)|<\int_{a+\eta}^b \frac{\epsilon}{3\{b-(a+\eta)\}}dx+\frac{\epsilon}{3}+\frac{\epsilon}{3} \text{ for } |k|\leq\delta \text{ [by (i)]}$$

$$=\frac{\epsilon}{3}+\frac{\epsilon}{3}+\frac{\epsilon}{3}=\epsilon \text{ for } |k|\leq\delta$$

$\Rightarrow \phi(y)$ is continuous in $[c,d]$.

Theorem [12.14] Let $f(x,y)$ be continuous in R $[a+\eta,b;c,d]$ and let the improper integral $\phi(y)=\int_a^b f(x,y)dx$, where $x=a$ is apoint of infinite discontinuity of $f(x,y)$, be uniformly convergent in $[c,d]$, then ϕ can be integrated under the sign of integration. That is,

$$\int_c^d \phi(y)dy=\int_c^d\left\{\int_a^b f(x,y)dx\right\}dy=\int_a^b\left\{\int_c^d f(x,y)dy\right\}dx.$$

Proof. $\phi(y)=\int_a^b f(x,y)dx$ is uniformly convergent in $[c,d]$

$$\Rightarrow \left|\int_{a+\eta}^b f(x,y)dx-\int_a^b f(x,y)dx\right|<\epsilon \text{ for } 0<\eta\leq\delta$$

$$\Rightarrow \left|\int_{a+\eta}^a f(x,y)dx\right|<\frac{\epsilon}{d-c} \text{ for } 0<\eta\leq\delta,\ d>c \text{ [by Def. 12.6]} \qquad \text{(i)}$$

Now

$$\int_c^d \phi(y)dy=\int_c^d\left\{\int_a^b f(x,y)dx\right\}dy=\int_c^d\left\{\int_{a+\eta}^b f(x,y)dx+\int_a^{a+\eta} f(x,y)dx\right\}dy$$

$$=\int_c^d\left\{\int_{a+\eta}^b f(x,y)dx\right\}dy+\int_c^d\left\{\int_a^{a+\eta} f(x,y)dx\right\}dy$$

$$\Rightarrow \left| \int_c^d \phi(y)dy - \int_c^d \left\{ \int_{a+\eta}^b f(x,y)dx \right\} dy \right| \le \int_c^d \left| \int_a^{a+\eta} f(x,y)dx \right| dy$$

$$< \int_c^d \frac{\epsilon}{d-c} dy = \epsilon \text{ for } 0 < \eta \le \delta \text{ [by (i)]} \tag{ii}$$

$$\Rightarrow \int_c^d \phi(y)dy = \int_c^d \left\{ \int_a^b f(x,y)dx \right\} dy.$$

Also from (ii),

$$\left| \int_c^d \phi(y)dy - \int_{a+\eta}^b \left\{ \int_c^d f(x,y)dy \right\} dx \right| < \epsilon, 0 < \eta \le \delta \text{ [by Th. [12.8]}$$

Therefore, $\int_c^d \phi(y)dy = \int_a^b \left\{ \int_c^d f(x,y)dy \right\} dx$.

Hence the result.

Theorem [12.15] If $f(x,y)$ be continuous in R $[a+\eta, b; c, d]$, $0 < \eta \le \delta$ and the improper integral $\phi(y) = \int_a^b f(x,y)dx$, where $x = a$ is a point of infinite discontinuity of $f(x,y)$, be uniformly convergent in $[c,d]$ and the improper integral $\int_a^b f_y(x,y)dx$ also converges uniformly in $[c,d]$, then $\phi'(y) = \int_a^b f_y(x,y)dx$.

Proof. Let $\psi(y) = \int_a^b f_y(x,y)dx$. Since $\psi(y)$ converges uniformly $\forall\ y \in [c,d]$, it can be integrated under sign of integration.

Therefore,

$$\int_c^y \psi(y)dy = \int_c^y \left\{ \int_a^b f_y(x,y)dx \right\} dy = \int_a^b \left\{ \int_c^y f_y(x,y)dy \right\} dx \text{ [by Th. [12.13]}$$

$$= \int_a^b [f(x,y)]_c^y dx = \int_a^b [f(x,y) - f(x,c)]dx = \phi(y) - \phi(c).$$

Differentiating, we get $\psi(y) = \phi'(y)$.

$$\therefore \phi'(y) = \int_a^b f_y(x,y)dx.$$

Example. Evaluate $\displaystyle\int_0^\infty e^{-\alpha x} \frac{\sin \beta x}{x}$, where $\alpha \ge 0$ and deduce that $\displaystyle\int_0^\infty \frac{\sin x}{x} dx = \frac{\pi}{2}$.

Let $\displaystyle\phi(\alpha,\beta) = \int_0^\infty e^{-\alpha x} \frac{\sin \beta x}{x} dx$, where $\displaystyle f(\alpha,\beta) = e^{-\alpha x} \frac{\sin \beta x}{x}$.

Here α, β are both parameters.

(i) Here $f(\alpha,\beta)$ is continuous $\forall\ \alpha, \beta$, where $\beta \in (-\infty, \infty)$.

(ii) $\displaystyle\left| e^{-\alpha x} \frac{\sin \beta x}{x} dx \right| \le \frac{e^{-\alpha x}}{x} dx.$

Now $\displaystyle\int_0^\infty \frac{e^{-\alpha x}}{x} dx$ is convergent by Abel's test.

Therefore, by Weierstrass's M-Test [Def. 12.6.1 – (1)], $\displaystyle\int_0^\infty e^{-\alpha x} \frac{\sin \beta x}{x} dx$ is uniformly convergent in $[-\infty, \infty)$. Again $f_\beta(\alpha,\beta) = e^{-\alpha x} \cos \beta x$.

Then $|f_\beta(\alpha,\beta)| = |e^{-\alpha x}\cos\beta x| \le e^{-\alpha x}$, $\alpha > 0$. Also $\int_0^\infty e^{-\alpha x}dx$ converges for $\alpha > 0$.

Hence, by Weierstrass's M-Test, $\int_0^\infty e^{-\alpha x}\cos\beta x dx$ is uniformly convergent in $(-\infty,\infty)$.

That is, $\int_0^\infty f_\beta(\alpha,\beta)$ is uniformly convergent in $(-\infty,\infty)$.

Therefore, differentiation under the sign of integration holds. We have

$$\phi_\beta(\alpha,\beta) = \int_0^\infty e^{-\alpha x}\cos\beta x dx = \left[e^{-\alpha x}\frac{(\beta\sin\beta x - \alpha\cos\beta x)}{\alpha^2+\beta^2}\right]_0^\infty = \frac{\alpha}{\alpha^2+\beta^2}.$$

Integrating w.r.t. β,

$$\phi(\alpha,\beta) = \int \frac{\alpha}{\alpha^2+\beta^2}d\beta = \tan^{-1}\frac{\beta}{\alpha} + c,$$

where c is constant. $\phi(\alpha,0) = 0$ for $\beta = 0$. So $0 = \tan^{-1}0 + c \Rightarrow c = 0$.

Therefore, $\phi(\alpha,\beta) = \tan^{-1}\frac{\beta}{\alpha}$. Now $\alpha \to 0 \Rightarrow \phi(\alpha,\beta) \to \frac{\pi}{2}$, when $\beta > 0$.

$$\therefore \int_0^\infty \frac{\sin\beta x}{x}dx = \frac{\pi}{2}.$$

In particular, $\beta = 1$, we get $\displaystyle\int_0^\infty \frac{\sin x}{x}dx = \frac{\pi}{2}$.

Problem [12.13]: (1) Show that $\displaystyle\int_0^\infty e^{-x^2}dx = \frac{\sqrt{\pi}}{2}$.

(2) Show that $\displaystyle\int_0^\infty e^{-x^2-\frac{a^2}{x^2}}dx = \frac{\sqrt{\pi}}{2}e^{-2|a|}$.

(3) Evaluate: $\displaystyle\int_0^\infty e^{-x^2}\cos\alpha x dx$ and hence show that $\displaystyle\int_0^\infty e^{-x^2}\cos x dx = \frac{\sqrt{\pi}}{2}e^{-\frac{1}{4}}$.

(4) Evaluate: $\displaystyle\int_0^\infty \frac{\cos xy}{1+x^2}dx$ and hence show that $\displaystyle\int_0^\infty \frac{\sin xy}{x(1+x^2)}dx = \frac{\pi}{2}(1-e^{-y})$.

(5) Show that $\displaystyle\int_0^\infty \frac{e^{-ax}-e^{-bx}}{x}\cos mx dx = \frac{1}{2}\log\frac{b^2+m^2}{a^2+m^2}$, $a, b > 0$.

(6) Show that $\displaystyle\int_0^\infty \frac{e^{-ax}\sin mx}{x}dx = \tan^{-1}\frac{m}{a}$, $a > 0$.

Solution. (1) Let $I = \int_0^\infty e^{-x^2}dx$.

Replacing x by αx, we get $I = \int_0^\infty e^{-\alpha^2x^2}\alpha dx$.

Now multiplying both sides by $e^{-\alpha^2}$ and integrating w.r.t. α from 0 to ∞, we have

$$I\cdot\int_0^\infty e^{-\alpha^2}d\alpha = \int_0^\infty e^{-\alpha^2}\left\{\int_0^\infty e^{-\alpha^2x^2}\alpha dx\right\}d\alpha$$

$$\Rightarrow I^2 = \int_0^\infty\left\{\int_0^\infty e^{-\alpha^2(1+x^2)}\alpha d\alpha\right\}dx = \int_0^\infty\left[-\frac{e^{-\alpha^2(1+x^2)}}{2(1+x^2)}\right]_0^\infty dx$$

$$= \frac{1}{2}\int_0^\infty\frac{dx}{1+x^2} = \frac{1}{2}\left[\tan^{-1}x\right]_0^\infty = \frac{\pi}{4}.$$

$$\therefore I = \frac{\sqrt{\pi}}{2}.$$

(2) (i) $f(a,x) = e^{-x^2-\frac{a^2}{x^2}}$ is continuous in $[0,\infty)$.

(ii) $\left|e^{-x^2-\frac{a^2}{x^2}}\right| = \left|\dfrac{e^{-x^2}}{e^{\frac{a^2}{x^2}}}\right| \le e^{-x^2}$.

$\int_0^\infty e^{-x^2}dx$ is convergent.

Therefore by Weierstrass's M-Test $\int_0^\infty e^{-x^2-\frac{a^2}{x^2}}dx$ is uniformly convergent.

(iii) $|f_a(a,x)| = \left|e^{-x^2-\frac{a^2}{x^2}} \cdot \dfrac{-2a}{x^2}\right| = \left|-2a \cdot \dfrac{e^{-x^2-\frac{a^2}{x^2}}}{x^2}\right|$

$$\le |-2a|\frac{e^{-x^2}}{x^2+a^2}\left[\because e^{a^2/x^2} \ge 1+\frac{a^2}{x^2}\right].$$

By Abel's Test, $\displaystyle\int_0^\infty \frac{e^{-x^2}}{x^2+a^2}dx$ is convergent.

Therefore, by Weierstrass's M-Test, $\int_0^\infty f_a(a,x)dx$ is uniformly convergent.
So differentiation under the integral sign is applicable.

Let $\phi(a) = \displaystyle\int_0^\infty e^{-x^2-\frac{a^2}{x^2}}dx$.

Then $\phi'(a) = \displaystyle\int_0^\infty \frac{\partial}{\partial a}\left\{e^{-x^2-\frac{a^2}{x^2}}\right\}dx = \int_0^\infty e^{-x^2-\frac{a^2}{x^2}} \cdot \left(-\frac{2a}{x^2}\right)dx.$

CASE 1: When $a > 0$. Put $\dfrac{a}{x} = z$. Then $-\dfrac{a}{x^2}dx = dz$ and $z : \infty$ to 0

$$= \int_\infty^0 e^{-\frac{a^2}{z^2}-z^2}2dz = -2\int_0^\infty e^{-z^2-\frac{a^2}{z^2}}dz = -2 \cdot \phi(a) \Rightarrow \frac{\phi'(a)}{\phi(a)} = -2.$$

Integrating w.r.t. a, we get $\log\phi(a) = -2a + c$.

For $a = 0, \phi(0) = \displaystyle\int_0^\infty e^{-x^2}dx = \frac{\sqrt{\pi}}{2}$ [by (1)] $\Rightarrow \log\dfrac{\sqrt{\pi}}{2} = 0 + c$.

$$\therefore c = \log\frac{\sqrt{\pi}}{2}.$$

Therefore,

$$\log\phi(a) = -2a + \log\frac{\sqrt{\pi}}{2} \text{ or, } \log\frac{\phi(a)}{\frac{\sqrt{\pi}}{2}} = -2a \text{ or, } \phi(a) = \frac{\sqrt{\pi}}{2}e^{-2a}. \quad (1)$$

CASE 2: When $a < 0$, and $a/x = z$, z varies from $-\infty$ to 0 as x varies from 0 to ∞. Then

$$\phi'(a) = \int_{-\infty}^0 e^{-\frac{a^2}{z^2}-z^2}2dz = -2\int_\infty^0 e^{-\frac{a^2}{t^2}-t^2}dt \text{ [for } z = -t]$$

$$= 2\int_0^\infty e^{-t^2-\frac{a^2}{t^2}}dt$$

$$= 2\phi(a) \Rightarrow \frac{\phi'(a)}{\phi(a)} = 2 \Rightarrow \phi(a) = \frac{\sqrt{\pi}}{2}e^{2a} \text{ [proceeding as above]} \quad (2)$$

Hence from (1) and (2), we get $\phi(a) = \frac{\sqrt{\pi}}{2}e^{-2|a|}$.

(3) Let $\phi(\alpha) = \int_0^{\infty} e^{-x^2} \cos \alpha x dx$, $\cos \alpha x$ is bounded in $[0, \infty]$.

(i) $f(\alpha, x) = e^{-x^2} \cos \alpha x$ is continuous in $[0, \infty)$.

(ii) $|e^{-x^2} \cos \alpha x| \le e^{-x^2}$ and $\int_0^{\infty} e^{-x^2} dx$ is convergent.

Therefore, by Weierstrass's M-Test, $\int_0^{\infty} e^{-x^2} \cos \alpha x dx$ is uniformly convergent.

(iii) $f_\alpha(\alpha, x) = -e^{-x^2} x \sin \alpha x$, $|f_\alpha(\alpha, x)| \le e^{-x^2} x$.

Therefore, $\int_0^{\infty} f_\alpha(\alpha, x) dx$ is uniformly convergent by Weierstrass's M-Test.

Differentiation under the integral sign holds to $\phi(\alpha) = \int_0^{\infty} e^{-x^2} \cos \alpha x dx$.

Differentiating,

$$\phi'(\alpha) = \int_0^{\infty} -xe^{-x^2} \sin \alpha x dx = \frac{1}{2} \int_0^{\infty} -2xe^{-x^2} \sin \alpha x dx$$

$$= \frac{1}{2} \left[\left(e^{-x^2} \sin \alpha x \right)_0^{\infty} - \alpha \int_0^{\infty} e^{-x^2} \cos \alpha x dx \right] = -\frac{\alpha}{2} \phi(\alpha) \Rightarrow \frac{\phi'(\alpha)}{\phi(\alpha)} = -\frac{\alpha}{2}.$$

Integrating w.r.t. α, we get $\log \phi(\alpha) = -\frac{\alpha^2}{4} + c$.

For $\alpha = 0$, $\phi(0) = \int_0^{\infty} e^{-x^2} dx = \frac{\sqrt{\pi}}{2} \Rightarrow \log \frac{\sqrt{\pi}}{2} = c$.

$$\therefore \log \phi(\alpha) = -\frac{\alpha^2}{4} + \log \frac{\sqrt{\pi}}{2} \Rightarrow \phi(\alpha) = \frac{\sqrt{\pi}}{2} e^{-\alpha^2/4}.$$

Therefore, $\int_0^{\infty} e^{-x^2} \cos \alpha x dx = \frac{\sqrt{\pi}}{2} e^{-\alpha^2/4}$.

Putting $\alpha = 1$, $\phi(1) = \frac{\sqrt{\pi}}{2} e^{-1/4} \Rightarrow \int_0^{\infty} e^{-x^2} \cos x dx = \frac{\sqrt{\pi}}{2} e^{-1/4}$.

(4) Let $\phi(y) = \int_0^{\infty} \frac{\cos xy}{1 + x^2} dx$, $\cos xy$ is bounded.

(i) $f(x, y) = \frac{\cos xy}{1 + x^2}$ is continuous on $[0, \infty)$.

(ii) $\left| \frac{\cos xy}{1 + x^2} \right| \le \frac{1}{1 + x^2}$ and $\int_0^{\infty} \frac{1}{1 + x^2} dx$ is convergent.

So by Weierstrass's M-Test, $\int_0^{\infty} f(x, y) dx$ is uniformly convergent.

(iii) $|f_y(x, y)| = \left| -\frac{x \sin xy}{1 + x^2} \right| = \frac{x \sin xy}{1 + x^2}$.

Now

$$\int_0^{\infty} f_y(x, y) dx = \int_0^1 \frac{x \sin xy}{1 + x^2} dx + \int_1^{\infty} \frac{x \sin xy}{1 + x^2} dx. \tag{1}$$

For the 2nd Part, $\frac{x}{1 + x^2}$ is monotonically decreasing $\forall\ x \ge 1$ and $\to 0$ as $x \to \infty$.

Also $\int_1^t \sin xy dx$ is bounded.

Therefore, by Dirichlet's Test, $\int_1^{\infty} \frac{x \sin xy}{1 + x^2} dx$ is convergent $\forall\ t \ge 1$.

So $\int_1^\infty f_y(x,y)dx$ is uniformly convergent in $[1,\infty]$ and hence $\int_0^\infty f_y(x,y)dx$ is uniformly convergent which follows from (1) as the 1st part is a proper integral.

Differentiation under the integral sign holds.

So,

$$\phi'(y) = -\int_0^\infty \frac{x\sin xy}{1+x^2}dx = -\int_0^\infty \frac{x^2\sin xy}{x(1+x^2)}dx = -\int_0^\infty \frac{\{(1+x^2)-1\}\sin xy}{x(1+x^2)}dx$$

$$= -\int_0^\infty \frac{\sin xy}{x}dx + \int_0^\infty \frac{\sin xy}{x(1+x^2)}dx. \tag{2}$$

Again

$$\int_0^y \phi(y)dy = \int_0^y dy\int_0^\infty \frac{\cos xy}{1+x^2}dx = \int_0^\infty \frac{dx}{1+x^2}\int_0^y \cos xy\,dy$$

$$= \int_0^\infty \left[\frac{\sin xy}{x}\right]_0^y \frac{dx}{1+x^2} = \int_0^\infty \frac{\sin xy}{x(1+x^2)}dx = \phi'(y) + \int_0^\infty \frac{\sin xy}{x}dx \text{ [by (2)]}$$

$$= \phi'(y) + \frac{\pi}{2}$$

$$\text{or, } \phi'(y) = -\frac{\pi}{2} + \int_0^y \phi(y)dy. \tag{3}$$

Here $\phi(y)$ is continuous by Th. [12.9].

So differentiating (3), we get

$$\phi''(y) = \phi(y) \Rightarrow \frac{d^2z}{dy^2} - z = 0 \text{ for } \phi(y) = z.$$

A.E. is $m^2 - 1 = 0$, $m = \pm 1$. Solution is $z = c_1e^{-y} + c_2e^y$.

That is,

$$\phi(y) = c_1e^{-y} + c_2e^y, \tag{4}$$

c_1, c_2 are constants.

For $y = 0$,

$$\phi(0) = \int_0^\infty \frac{dx}{1+x^2} = \frac{\pi}{2} \Rightarrow \frac{\pi}{2} = c_1 + c_2. \tag{5}$$

Again from (3), $\phi'(0) = -\frac{\pi}{2}$.

$$\phi'(y) = -c_1e^{-y} + c_2e^y \Rightarrow -\frac{\pi}{2} = -c_1 + c_2 \text{ [for } y = 0]. \tag{6}$$

From (5) and (6), $c_2 = 0$, $c_1 = \frac{\pi}{2}$.

So from (4), $\phi(y) = \frac{\pi}{2}e^{-y}$.

$$\therefore \int_0^\infty \frac{\cos xy}{1+x^2}dx = \frac{\pi}{2}e^{-y}.$$

From (2),

$$\phi'(y) = -\frac{\pi}{2} + \int_0^\infty \frac{\sin xy}{x(1+x^2)}dx$$

$$\text{or, } -\frac{\pi}{2}e^{-y} = -\frac{\pi}{2} + \int_0^\infty \frac{\sin xy}{x(1+x^2)}dx, \text{ or, } \int_0^\infty \frac{\sin xy}{x(1+x^2)}dx = \frac{\pi}{2}(1-e^{-y}).$$

(5) Let $f(x, y) = e^{-xy} \cos mx$ be defined in $R\ [0, \infty; a, b]$ with $0 < a < b$.

(i) Then $f(x, y)$ is continuous in R.

(ii) $|f(x, y)| = |e^{-xy} \cos mx| \leq e^{-xy}$. Then $\int_0^\infty e^{-xy} dx$ is convergent $\forall\ y \in [a, b]$.

Therefore, by Weierstrass's M-Test, $\int_0^\infty e^{-xy} \cos mx dx$ is uniformly convergent.

Therefore, $\phi(y) = \int_0^\infty e^{-xy} \cos mx dx$ can be integrated under the integral sign.

$$\int_a^b \phi(y) dy = \int_a^b \left\{ \int_0^\infty e^{-xy} \cos mx dx \right\} dy = \int_0^\infty \left\{ \int_a^b e^{-xy} \cos mx dy \right\} dx.$$

Now

$$\text{R.H.S.} = \int_0^\infty \left[\frac{e^{-xy}}{-x} \right]_a^b \cos mx dx = \int_0^\infty \frac{e^{-ax} - e^{-bx}}{x} \cos mx dx. \tag{7}$$

Also

$$\int_a^b \left\{ \int_0^\infty e^{-xy} \cos mx dx \right\} dy = \int_a^b \left[e^{-xy} \frac{(m \sin mx - y \cos mx)}{m^2 + y^2} \right]_0^\infty dy$$

$$= \int_a^b \frac{y}{m^2 + y^2} dy = \frac{1}{2} \log \frac{b^2 + m^2}{a^2 + m^2}. \tag{8}$$

$\therefore$ from (7) and (8), it follows that $\displaystyle\int_0^\infty \frac{e^{-ax} - e^{-bx}}{x} \cos mx dx = \frac{1}{2} \log \frac{b^2 + m^2}{a^2 + m^2}$.

(6) Follows from example in P-332.

EXERCISE

(1) If f is continuous of each variables x and y separately in a certain neighbourhood of (a, b), $f(a, b) = 0$, f_y is continuous at (a, b) and $f_y(a, b) \neq 0$, then show that the equation $f(x, y) = 0$ determines a unique continuous implicit solution $y = \phi(x)$ near (a, b).

(2) If $f(x, y) = y^4 - y^2 + x^2$, discuss the existence of the function $y = \phi(x)$ in the neighbourhood of $(0, 0)$.

(3) Examine the existence of the unique solution near the points indicated. Find also the derivative of the solution, if it exists.

(i) $xy \sin x + \cos y = 0$, $\left(0, \frac{\pi}{2}\right)$ (ii) $y^3 \cos x + y^2 \sin^2 x = 7$, $\left(\frac{\pi}{3}, 2\right)$

(iii) $x^4 + x^2y^2 + 2y^5 = 0$, $(-1, 1)$ (iv) $2x + 3y + 5x^2 + 2xy + 3y^2 = 0$, $(0, 0)$.

(4) Let $f(x, y) = x^2 + y^2 - 1$ and a point $(0, 1)$ so that $f(0, 1) = 0$ and $f_y(0, 1) \neq 0$. How do you reconcile the conclusion of the existence of unique solution with the fact that $x^2 + y^2 - 1 = 0$ defines two functions $y = +\sqrt{1 - x^2}$ and $y = -\sqrt{1 - x^2}$.

(5) Find the stationary points of

(i) $f(x, y) = x^3y^2(1 - x - y)$, (ii) $f(x, y) = xy + \dfrac{a^3}{x} + \dfrac{a^3}{y}$.

(6) Find the values of x and y for which $x^2 + y^2 + 6x = 12$ has a minimum value and find this minimum value.

(7) Find a point within a triangle such that the sum of the square of its distances from the three angular points is a minimum.

(8) A rectangular box, open at the lop, is to have a volume of 32 cc. Find the dimensions of the box requiring least material for its construction.

(9) Use the method of Lagrange's multipliers to find the volume of the largest rectangular parallelopiped that can be inscribed in the ellipsoid $\frac{x^2}{a^2}+\frac{y^2}{b^2}+\frac{z^2}{c^2}=1$.

(10) Show that the maximum volume of the rectangular solid that can be inscribed in a sphere $x^2+y^2+z^2=a^2$ is a cube.

(11) The temperature T at any point (x, y, z) in space is $T = 400xyz^2$. Find the highest temperature at the surface of a unit sphere $x^2+y^2+z^2=1$.

(12) (i) If $3x+2y+2z=\lambda$, then show that the maximum value of $x^3y^2z^2$ is $\left(\frac{\lambda}{7}\right)^7$.

(ii) If $xyz = abc$, then show that the minimum value of $\frac{x}{a}+\frac{y}{b}+\frac{z}{c}$ is 3.

(13) Using the method of Jacobian show that,

(i) $f(x)+f(y)=f\left(\frac{x+y}{1-xy}\right)$, $xy \neq 1$, if $f(0)=0$ and $f'(x)=\frac{1}{1+x^2}$.

(ii) $\tan^{-1}x+\tan^{-1}y=\tan^{-1}\frac{x+y}{1-xy}$.

(14) Let $f(x+y)=\frac{f(x)+f(y)}{1-f(x)\cdot f(y)}$, $f(x)\cdot f(y)\neq 1$. x, y are two independent variables and $f(z)$ is differentiable function of z and $f(0)=0$. Using Jacobian method, show that $f(z)=\tan \alpha z$, α is constant.

(15) Examine for extreme values of $x^2+y^2+z^2+2xyz$.

(16) Show that the function $(x+y)^2+(y+z)^2+xyz$ has no maximum or minimum values.

(17) If f and f_x be continuous in $[a, b;\ c, d]$ and G be bounded and intergrable in $[c, d]$, then show that $\int_c^d f(x,y)G(y)dy$ is derivable in $[a, b]$ with derivative $\int_c^d f_x(x,y)G(y)dy$.

(18) If $f(x,y)=\frac{y^2}{x^2+y^2}$ and $G(y)=\int_0^1 f(x,y)dx$, then show that $G'(+0)$ and $G'(-0)$ differ from each other and from $\int_0^1 f_y(x,0)dx$.

(19) If $\int_0^{\pi/2}\frac{dx}{1+\alpha\cos x}=\frac{\cos^{-1}\alpha}{\sqrt{1-\alpha^2}}$ for $0\le\alpha<1$, then show that

$$\int_0^{\pi/2}\sec x\log\left(\frac{1+a\cos x}{1+b\cos x}\right)dx=\frac{1}{2}\left[(\cos^{-1}b)^2-(\cos^{-1}a)^2\right]$$

for $0\le a<1$, $0\le b<1$.

Deduce that $\int_0^{\pi/2}\sec x\log\left(1+\frac{1}{2}\cos x\right)dx=\frac{5\pi^2}{72}$.

(20) Show that $\int_0^\infty e^{-ax^2}dx=\frac{1}{2}\sqrt{\frac{\pi}{a}}$.

Deduce that $\int_0^\infty e^{-ax^2}x^{2n}dx=\frac{1\cdot 3\cdot 5\cdots(2n-1)}{e^{2n+1}}\frac{\sqrt{\pi}}{a^{n+\frac{1}{2}}}$.

(21) Evaluate $\int_0^\infty e^{-x^2}\cos x\,dx$.

(22) If $u=\int_{x-\lambda t}^{x+\lambda t}\phi(y)dy$, where ϕ possesses continuous derivatives, then show that $\frac{\partial^2 u}{\partial t^2}=\lambda^2\frac{\partial^2 u}{\partial x^2}$.

(23) Evaluate $\frac{d}{d\alpha}\int_{\sin\alpha}^{\cos\alpha}\log(x+\alpha)dx$.

(24) Find $\frac{dy}{dx}$, if $\int_{\sqrt{x}}^{2y}e^{-xt^2}dx=-2$.

(25) Prove that the function $y=\int_0^\infty\frac{e^{-xz}}{1+x^2}dz$ satisfies the differential equation $\frac{d^2y}{dx^2}+y=\frac{1}{x}$.

CHAPTER 13

MULTIPLE INTEGRALS

13.1 The Concept of a Plane Curve

Let ϕ, ψ be two real valued functions defined in a domain $[\alpha, \beta]$.

That is, $\phi, \psi : [\alpha, \beta] \to \mathbb{R}^2$, where $\mathbb{R}^2 = \{(x, y), x, y \in \mathbb{R}\}$.

Then the set $S = \{(x, y);\ x = \phi(t),\ y = \psi(t),\ t \in [\alpha, \beta]\} \subseteq \mathbb{R}^2$ determines a plane curve. That is $x = \phi(t),\ y = \psi(t)$ forms this curve. The curve is said to be *closed*, if $\phi(\alpha) = \phi(\beta)$ and $\psi(\alpha) = \psi(\beta)$.

The points corresponding to α and β are called end points of the curve. Thus the curve is closed, if its end points coincide.

Now if $f : [a, b] \to \mathbb{R}^2$, then the set $S = \{(x, y),\ y = f(x),\ x \in [a, b]\} \subseteq \mathbb{R}^2$ is also a plane curve, as we may see on setting $x = t,\ y = f(t)$. Thus $x = t,\ y = f(t)$ also determines a plane curve.

Example. Let $\phi, \psi : [0, 2\pi] \to \mathbb{R}^2$.

Then the set $= \{(x, y),\ x = \cos t,\ y = \sin t,\ t \in [0, 2\pi]\}$ determines a plane curve.

Also $\cos 0 = 1 = \cos(2\pi)$ and $\sin 0 = 0 = \sin(2\pi)$.

The curve is closed.

Example. $x = t$ and $y = \sqrt{1 - t^2}$ determine a plane curve $x^2 + y^2 = 1$.

13.2 Line Integral

Let $x = \phi(t),\ y = \psi(t)$ be a curve C, where ϕ, ψ defined by $\phi, \psi : [\alpha, \beta] \to \mathbb{R}^2$, $t \in [\alpha, \beta]$.

Let $f(x, y)$ be a bounded function defined on every point of the curve C. [i.e., the domain of f contains the curve C.]

Suppose $[\alpha = t_0 < t_1 < t_2 \cdots < t_{r-1} < t_r \cdots < t_n = \beta]$ be a partition P of $[\alpha, \beta]$ and $x_r = \phi(\xi_r), y_r = \psi(\xi_r)$, where $\xi_r \in [t_{r-1}, t_r]$.

We form the sum,

$$S = \sum f[\phi(\xi_r),\ \psi(\xi_r)](x_r - x_{r-1}). \tag{1}$$

Then, if $\lim\limits_{||P|| \to 0} S = \lim\limits_{||P|| \to 0} \sum f(\phi(\xi_r), \psi(\xi_r))\delta_r$, where $\delta_r = (x_r - x_{r-1})$, the length of the rth sub-interval and $||P||$ is the norm of the partition P of $[\alpha, \beta]$, exists finitely, we call the limit as the line integral of $f(x, y)$ along the curve C and we denote it by the symbol $\int_C f(x, y)dx$. The limit is independent of the choice of ξ_r.

It means that for every $\epsilon > 0$, there corresponds a (+)ve number δ such that for every partition P of $[\alpha, \beta]$ with norm $||P|| < \delta$ and every choice of ξ_r in $[t_{r-1}, t_r]$, there exists a number L such that $|\sum f(\phi(\xi_r),\ \psi(\xi_r))\delta_r - L| < \epsilon$.

Thus, we have $L = \int_C f(x, y)dx$.

Theorem [13.1] [Sufficient conditions for the existence of the line integral]: If (i) f, ϕ and ψ are continuous (ii) ϕ possesses continuous derivative ϕ', then the line integral $\int_C f(x, y)dx$ exists and $\int_C f(x, y)dx = \int_\alpha^\beta f(\phi(t), \psi(t))\phi'(t)dt$.

Proof. Here the curve C is denoted by $x = \phi(t),\ y = \psi(t)$.

Let $P = \{\alpha = t_0 < t_1 \cdots < t_{r-1} < t_r \cdots < t_n = \beta\}$ be a partition of $[\alpha, \beta]$.

Now

$$\begin{aligned} x_r - x_{r-1} &= \phi(t_r) - \phi(t_{r-1}) = (t_r - t_{r-1})\phi'(\eta_r), \eta_r \in [t_{r-1}, t_r] \text{ [by M.V. Th. to } \phi] \\ &= \phi'(\eta_r)\delta r. \end{aligned} \tag{i}$$

$$\begin{aligned} S &= \sum f[\phi(\xi_r), \psi(\xi_r)][x_r - x_{r-1}] \\ &= \sum f[\phi(\xi_r), \psi(\xi_r)]\phi'(\eta_r)\delta_r \text{ [by (i)]} \\ &= \sum f[\phi(\xi_r), \psi(\xi_r)][\phi'(\xi_r) + \phi'(\eta_r) - \phi'(\xi_r)]\delta_r \\ &= \sum f[\phi(\xi_r), \psi(\xi_r)]\phi'(\xi_r)\delta r + \sum f[\phi(\xi_r), \psi(\xi_r)][\phi'(\eta_r) - \phi'(\xi_r)]\delta r \end{aligned} \tag{ii}$$

Since ϕ' and $f(\phi, \psi)$ are continuous, it follows that $\lim\limits_{||P||\to 0} \sum f[\phi(\xi_r), \psi(\xi_r)]\phi'(\xi_r)\delta r$ exists to a finite limit

$$\int_\alpha^\beta f[\phi(t), \psi(t)]\phi'(t)dt. \tag{iii}$$

Again, $[f(\phi, \psi)]$ is continuous; it is bounded. Then there exists a (+)ve number k such that

$$|f[\phi(t), \psi(t)]| \le k\ \forall\ t \in [\alpha, \beta].$$

Therefore,

$$\begin{aligned} \left|\sum f[\phi(\xi_r), \psi(\xi_r)][\phi'(\eta_r) - \phi'(\xi_r)]\delta_r\right| &\le \left|\sum f[\phi(\xi_r), \psi(\xi_r)]\right| |\phi'(\eta_r) - \phi'(\xi_r)|\delta_r \\ &\le k\sum(M_r - m_r)\delta_r < k\frac{\epsilon}{k} = \epsilon, \end{aligned}$$

where M_r, m_r are the bounds of ϕ' in $[t_{r-1}, t_r]$.

Thus

$$\lim_{||P||\to 0} \sum f[\phi(\xi_r), \psi(\xi_r)]\ [\phi'(\eta_r) - \phi'(\xi_r)]\delta_r = 0. \tag{iv}$$

Hence from (ii), it follows that

$$\int_C f(x, y)dx = \int_\alpha^\beta f[\phi(t), \psi(t)]\phi'(t)dt \text{ [by (iii), (iv)].}$$

Note. (1) Similarly, we can define and examine the existence of the line integral $\int_C g(x, y)dy$ and the line integral $\int_C [f(x, y)dx + g(x, y)dy]$.

That is

$$\int_C g(x,y)dy = \int_{\alpha}^{\beta} g[\phi(t),\psi(t)]\psi'(t)dt$$

$$\text{and } \int_C f(x,y)dx + g(x,y)dy = \int_{\alpha}^{\beta} [f(\phi(t),\psi(t))\phi'(t) + g(\phi(t),\psi(t))\psi'(t)]dt.$$

(2) Along a curve $y = \phi(x),\ a \le x \le b$,

$$\int_C g(x,y)dy = \int_a^b g(x,\phi(x))\phi'(x)dx \ [\because dy = \phi'(x)dx].$$

(3) $\displaystyle\int_{-C} f(x,y)dx = -\int_C f(x,y)dx.$

(4)

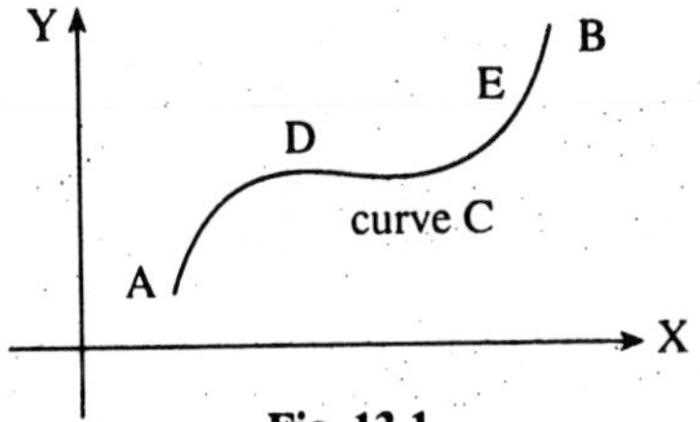

Fig. 13.1

$$\int_{AB} f(x,y)dx = \int_{AD} f(x,y)dx + \int_{DE} f(x,y)dx + \int_{EB} f(x,y)dx$$

the line integral $\int_C f(x,y)dx$ is the additive for the arcs $\widehat{AD}, \widehat{DE}, \widehat{EB}$ of the curve C represented by the arc $\widehat{AB}$.

Example. Evaluate the line integrals (i) $\int_C (x^2+y^2)dx$, (ii) $\int_C (x^2+y^2)dy$, where C is the arc of the parabola $y^2 = 4ax$ from $(0,0)$ to $(a, 2a)$.

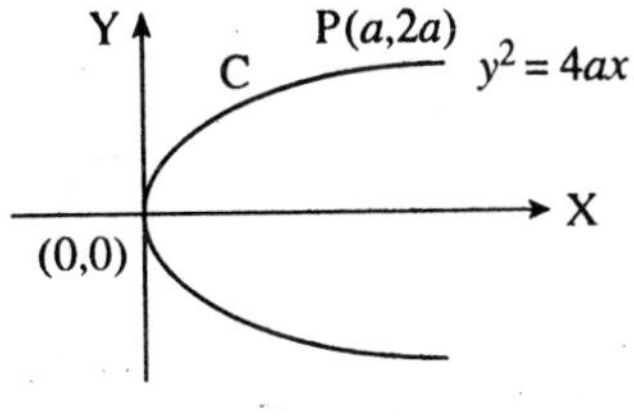

Fig. 13.2

Here

$$C = \text{arc } \widehat{OP}$$

$$\text{and } x = \phi(t) = at^2,$$

$$y = \psi(t) = 2at.$$

t varies from 0 to 1.

$$f(x,y) = x^2 + y^2;\ dx = 2at\,dt,\ dy = 2a\,dt.$$

(i) $\displaystyle\int_C (x^2+y^2)dx = \int_0^1 (a^2t^4 + 4a^2t^2)2at\,dt$ [by Th. [13.1]

$$= 2a^3\int_0^1 (t^5 + 4t^3)dt = 2a^3\left[\frac{1}{6}+1\right] = \frac{7a^3}{3}.$$

(ii) $\displaystyle\int_C (x^2+y^2)dy = \int_0^1 (a^2t^4 + 4a^2t^2)2a\,dt$

$$= 2a^3\left[\int_0^1 (t^4+4t^2)dt\right] = 2a^3\left[\frac{1}{5}+\frac{4}{3}\right] = \frac{46a^3}{15}.$$

Example. Show that $\int_C [(x-y)^3dx + (x-y)^3dy] = 3\pi a^4$ taken along the circle $x^2+y^2 = a^2$ in the counter-clockwise sense.

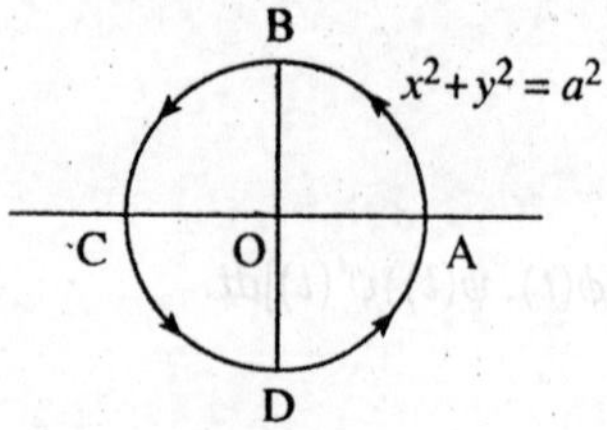

Fig. 13.3

Here C is the arc $ABCDA$ of the circle

$$x^2+y^2=a^2.$$

We put

$$x = a\cos t = \phi(t)$$
$$y = a\sin t = \psi(t)$$
$$t : 0 \to 2\pi.$$

Then $\displaystyle\int_C (x-y)^3 dx + (x-y)^3 dy$

$$= \int_0^{2\pi} [a^3(\cos t - \sin t)^3(-a\sin t) + a^3(\cos t - \sin t)^3 a\cos t]\, dt \text{ [by Note (1) of Th.(13.1)]}$$

$$= a^4 \int_0^{2\pi} (\cos t - \sin t)^4 dt = a^4 \int_0^{2\pi} [1-\sin 2t]^2 dt$$

$$= a^4 \int_0^{2\pi} [1 - 2\sin 2t + \sin^2 2t]\, dt = a^4 \int_0^{2\pi} \left[1 - 2\sin 2t + \frac{1-\cos 4t}{2}\right] dt$$

$$= a^4 \left[\frac{3}{2}t + \cos 2t - \frac{1}{8}\sin 4t\right]_0^{2\pi} = a^4 \cdot \frac{3}{2} \cdot 2\pi = 3\pi a^4$$

Example. Show that $\displaystyle\int_C (x^2y\,dx + xy^2dy) = \frac{38}{\sqrt{3}}a^4$, where the integration is taken in the clockwise sense along the hexagon whose vertices are $(\pm 3a, 0), (\pm 2a, \pm\sqrt{3}a)$.

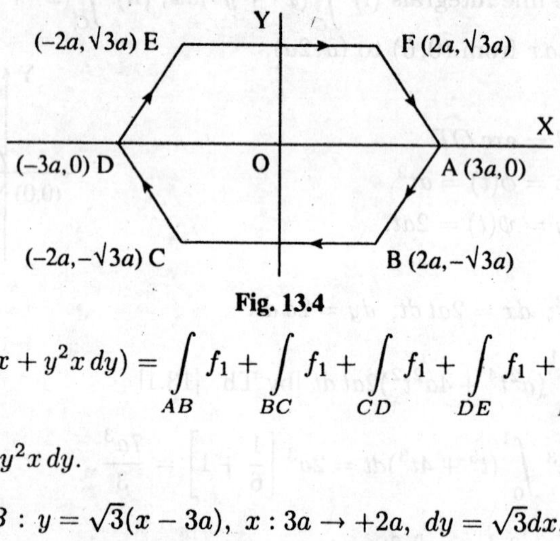

Fig. 13.4

$$\int_C (x^2y\,dx + y^2x\,dy) = \int_{AB} f_1 + \int_{BC} f_1 + \int_{CD} f_1 + \int_{DE} f_1 + \int_{EF} f_1 + \int_{FA} f_1,$$

where $f_1 = x^2y\,dx + y^2x\,dy$.

(i) Equation of AB : $y = \sqrt{3}(x-3a)$, $x : 3a \to +2a$, $dy = \sqrt{3}dx$.

(ii) Equation of BC : $y = -\sqrt{3}a$, $x : 2a \to -2a$, $dy = 0$.

(iii) Equation of CD : $y = -\sqrt{3}(x+3a)$, $x : -2a \to -3a$, $dy = -\sqrt{3}dx$.

(iv) Equation of DE : $y = \sqrt{3}(x+3a)$, $x : -3a \to -2a$, $dy = \sqrt{3}dx$.

(v) Equation of EF : $y = \sqrt{3}a$, $x : -2a \to 2a$, $dy = 0$.

(vi) Equation of FA : $y = -\sqrt{3}(x-3a)$, $x : 2a \to 3a$, $dy = -\sqrt{3}dx$.

Now (i) $\int_{3a}^{2a} x^2 \cdot \sqrt{3}(x-3a) + 3(x-3a)^2 x\sqrt{3}\, dx$

$$= \sqrt{3} \int_{3a}^{2a} (x-3a)(x^2+3x^2-9ax)dx$$

$$= \sqrt{3} \int_{3a}^{2a} (x-3a)(4x^2-9ax)dx$$

$$= \sqrt{3} \int_{3a}^{2a} (4x^3-21x^2a+27a^2x)dx.$$

(vi) $\int_{2a}^{3a} \left\{-x^2\sqrt{3}(x-3a) + 3(x-3a)^2 \cdot x \cdot \left(-\sqrt{3}\right)\right\} dx$

$$= \sqrt{3} \int_{3a}^{2a} (x-3a)(x^2+3x^2-9ax)dx = \sqrt{3} \int_{3a}^{2a} (4x^3-21ax^2+27a^2x)dx.$$

$$\text{(i)} + \text{(vi)} = 2\sqrt{3} \int_{3a}^{2a} (4x^3-21x^2a+27xa^2)dx = -2\sqrt{3} \int_{2a}^{3a} (4x^3-21x^2a+27xa^2)dx$$

$$= -2\sqrt{3} \left[x^4 - 7x^3a + \frac{27}{2}a^2x^2\right]_{2a}^{3a} = -2\sqrt{3}a^4 \left[65 - 133 + \frac{135}{2}\right] = \sqrt{3}a^4. \quad (1)$$

(ii) $\int_{2a}^{-2a} x^2 \left(-\sqrt{3}a\right) dx = \sqrt{3}a \int_{-2a}^{2a} x^2 dx = \sqrt{3}a \left[\frac{x^3}{3}\right]_{-2a}^{2a} = \frac{16}{3}\sqrt{3}a^4.$

(v) $\int_{-2a}^{2a} x^2\sqrt{3}a\, dx = \frac{16}{3}\sqrt{3}a^4.$

$$\text{(ii)} + \text{(v)} = \frac{32}{3}\sqrt{3}a^4. \quad (2)$$

(iii) $\int_{-2a}^{-3a} \left\{-x^2\sqrt{3}(x+3a) + 3(x+3a)^2 \cdot x \cdot \left(-\sqrt{3}\right)\right\} dx = \sqrt{3} \int_{-3a}^{-2a} (x+3a)(4x^2+9xa)dx.$

(iv) $\int_{-3a}^{-2a} \left\{x^2\sqrt{3}(x+3a) + 3(x+3a)^2 x\sqrt{3}\right\} dx = \sqrt{3} \int_{-3a}^{-2a} (x+3a)(4x^2+9xa)dx.$

$$\text{(iii)} + \text{(iv)} = 2\sqrt{3} \int_{-3a}^{-2a} (x+3a)(4x^2+9xa)dx = 2\sqrt{3} \int_{-3a}^{-2a} [4x^3+21x^2a+27a^2x]dx$$

$$= 2\sqrt{3} \left[x^4 + 7ax^3 + \frac{27}{2}a^2x^2\right]_{-3a}^{-2a} = 2\sqrt{3} \left[-65 + 133 - \frac{135}{2}\right] = \sqrt{3}a^4. \quad (3)$$

Therefore

$$\int_C x^2y\, dx + y^2x dy = \text{(i)} + \text{(ii)} + \text{(iii)} + \text{(iv)} + \text{(v)} + \text{(vi)} = (1) + (2) + (3)$$

$$= \sqrt{3}a^4 + \frac{32}{3}\sqrt{3}a^4 + \sqrt{3}a^4$$

$$= \sqrt{3}a^4 \left(2 + \frac{32}{3}\right) = \frac{38a^4}{\sqrt{3}}.$$

Example. Find the value of $\int (x+y^2)dx + (x^2-y)dy$ taken in the clockwise sense along the closed curve C formed by $y^2 = x$ and $y = x$.

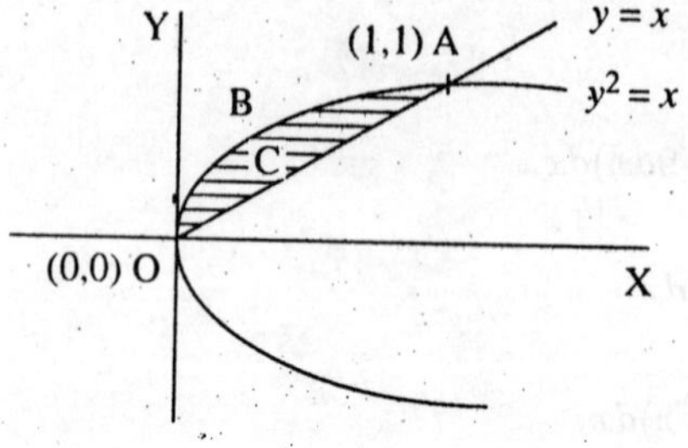

Fig. 13.5

$$y^2 = x \text{ and } y = x$$
$$\Rightarrow x^2 = x$$
$$\Rightarrow x = 0,\ x = 1$$
$$x = 0,\ y = 0 \text{ and } x = 1,\ y = 1.$$

Closed curve C is bounded by the arc $\widehat{OBA}$ and the line AO in the clockwise sense between the points $(0,0)$ and $(1,1)$.

$$\int_C (x+y^2)dx + (x^2-y)dy$$

$$= \int\limits_{\substack{\widehat{OBA} \\ y^2=x \\ y:0\to 1}} (x+y^2)dx + (x^2-y)dy + \int\limits_{\substack{\overline{AO} \\ y=x \\ x:1\to 0}} (x+y^2)dx + (x^2-y)dy. \tag{1}$$

$$\text{1st Part} = \int_0^1 \left[(y^2+y^2)\cdot 2ydy + (y^4-y)dy\right] = \int_0^1 (4y^3+y^4-y)dy = \left[y^4 + \frac{y^5}{5} - \frac{y^2}{2}\right]_0^1$$
$$= 1 + \frac{1}{5} - \frac{1}{2} = \frac{7}{10}.$$

$$\text{2nd Part} = \int_1^0 (x+x^2)dx + (x^2-x)dx = -\int_0^1 2x^2dx = -\frac{2}{3}.$$

$$\text{Therefore, from (1), } \int_C (x+y^2)dx + (x^2-y)dy = \frac{7}{10} - \frac{2}{3} = \frac{21-20}{30} = \frac{1}{30}.$$

13.3 Double Integrals

Definition [13.1]: Let $f(x,y)$ be a bounded function defined in a rectangle $E = \{(x,y),\ a \le x \le b, c \le y \le d\} \subseteq \mathbb{R}^2$, where $a < b$ and $c < d$.

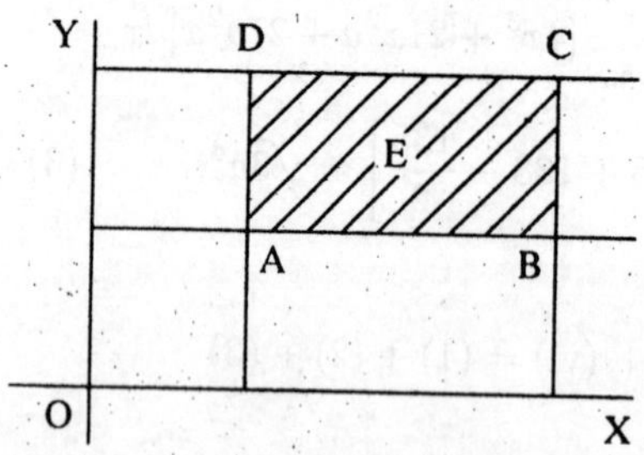

Fig. 13.6

E is the rectangle $\square ABCD$.

We form a partition P of E by drawing the lines through E parallel to the Y-axis $x = x_i, i = 0,1,\cdots,n$ and parallel to the X-axis $y = y_j,\ j = 0,1,\cdots,m$. With the points of division $a = x_0 < x_1 \cdots < x_{n-1} < x_n = b$ and $c = y_0 < y_1 < y_2 \cdots < y_{m-1} < y_m = d$.

$$E_{ij} = \{(x,y); x_{i-1} \le x \le x_i, y_{j-1} \le y \le y_j\}$$

denotes the ijth sub-rectangle on E and $A(E_{ij})$ = area of the sub-rectangle E_{ij}.

Let $M = \sup\limits_{(x,y)\in E} f(x,y)$ and $m = \inf\limits_{(x,y)\in E} f(x,y)$ and correspondingly we define,

$$M_{ij} = \sup_{(x,y)\in E_{ij}} f(x,y) \text{ and } m_{ij} = \inf_{(x,y)\in E_{ij}} f(x,y)$$

in the ijth sub-rectangle on E.

We define the upper sum and lower sum for the partition P of E by

$$U(f,P) = \sum_{ij} M_{ij}A(E_{ij}) \text{ and } L(f,P) = \sum_{ij} m_{ij}A(E_{ij}).$$

It follows from definition of $L(f,P)$ and $U(f,P)$ that

$$mA(E) \le L(f,p) \le U(f,p) \le MA(E), \tag{1}$$

where

$$A(E) = \text{area of the rectangle } E = (b-a)(d-c)$$
$$\text{and } A(E_{ij}) = \text{area of the rectangle } E_{ij} = (x_i - x_{i-1})(y_j - y_{j-1}).$$

Thus $U(f,p)$ and $L(f,P)$ are bounded.

The infimum of the set of upper sums $U(f,P)$ is called the *upper integral* and supremum of the lower sums $L(f,P)$ is called the *lower integral* of f over E and are denoted by the symbols

$$U = \overline{\iint_E} f(x,y)dx\,dy \text{ and } L = \underline{\iint_E} f(x,y)dx\,dy \text{ respectively.}$$

That is,

$$\inf_P U(f,P) = \overline{\iint_E} f(x,y)dx\,dy \text{ and } \sup_P L(f,P) = \underline{\iint_E} f(x,y)dx\,dy.$$

Both the integrals exist as $f(x,y)$ is bounded on E. The bounded function $f : E \to R$ is said to be integrable if

$$\underline{\iint_E} f(x,y)dx\,dy = \overline{\iint_E} f(x,y)dx\,dy.$$

In such case, we define the *double integral* of f over E by

$$\underline{\iint_E} f(x,y)dx\,dy = \overline{\iint_E} f(x,y)dx\,dy = \iint_E f(x,y)dx\,dy.$$

Theorem [13.2] It can be shown that $\overline{\iint_E} f(x,y)dx\,dy \ge \underline{\iint_E} f(x,y)dx\,dy$ for every partition P on E.

Proof. For, if P^* be the refinement of the partition P on E, then each of the sub-rectangle in P^* is contained in one of the sub-rectangles of P and hence it follows that

$$U(f,P^*) \le U(f,P) \text{ and } L(f,P^*) \ge L(f,P). \tag{i}$$

Next, let P_1 and P_2 be the two partitions on E and P be the common refinement of P_1 and P_2 so that $P = P_1 \cup P_2$. Then

$$L(f,P_1) \le L(f,P) \le U(f,P) \le U(f,P_2).$$

So for any partition P_1 on E, taking supremum over all P_1, we have

$$\underline{\iint_E} f(x,y)dx\,dy \le U(f,P_2) \text{ keeping } P_2 \text{ fixed.}$$

Since P_2 is also arbitrary, taking infimum over P_2, we get

$$\underline{\iint_E} f(x,y)dx\,dy \le \overline{\iint_E} f(x,y)dx\,dy.$$

That is,

$$\overline{\iint_E} f(x,y)dx\,dy \ge \underline{\iint_E} f(x,y)dx\,dy.$$

Theorem [13.3] [Darboux's theorem]: To every $\epsilon(>0)$, there exists a (+)ve number δ such that for every partition P of $E\{(x,y), a \le x \le b, c \le y \le d\} \subseteq \mathbb{R}^2$ with norm $||P|| \le \delta$, $U(f,P) < U + \epsilon$ and $L(f,P) > L - \epsilon$.

The proof is exactly similar to the corresponding proof of Th. [7.3] for a function of single variable.

Condition of Integrability : First form

Theorem [13.4] The necessary and sufficient condition that a bounded function $f(x,y)$ defined on E to be integrable on E is that for every $\epsilon(>0)$, there exists a (+)ve number δ such that $U(f,P) - L(f,P) < \epsilon$ for every partition P with norm $||P|| < \delta$.

Proof. *The condition is necessary.* Let $f(x,y)$ be integrable on E.

Then $U(f,P) = L(f,P) = I$, where $I = \iint_E f(x,y)dx\,dy$.

By Th. [13.3],

$$U(f,P) < U + \frac{\epsilon}{2} = I + \frac{\epsilon}{2} \tag{i}$$

$$\text{and } L(f,P) > L - \frac{\epsilon}{2} = I - \frac{\epsilon}{2}. \tag{ii}$$

Then $U(f,P) - L(f,P) < \epsilon$ [by (i) and (ii)].

The condition is sufficient. If $U(f,P), L(f,P)$ be the upper and lower sum corresponding to the partition P of E, then $U(f,P) - L(f,P) = (U(f,P) - U) + (U - L) + (L - L(f,P)) < \epsilon$ by the given condition.

Since each of the numbers $U(f,P) - U$, $U - L$, $L - L(f,P)$ is non-negative, it follows that $0 \le U - L < \epsilon$.

As ϵ is an arbitrary (+)ve number and a non-negative number $U - L < \epsilon$, it follows that

$$U - L = 0$$

$$\text{i.e., } U = L = \overline{\iint_E} f(x,y)dx\,dy = \underline{\iint_E} f(x,y)dx\,dy.$$

Hence $f(x,y)$ is integrable on E.

Second form

The necessary and sufficient condition that a bounded $f(x,y)$ defined on E to be integrable on E is that for $\epsilon(>0)$, there corresponds a partition for which the oscillator sum

$$U(f,P) - L(f,P) < \epsilon.$$

Theorem [13.5] Every continuous function $f(x,y)$ defined on $E : [a,b,c,d]$ is integrable on E.

Proof. $f(x,y)$ is continuous on $E : [a,b;c,d]$

$\Rightarrow$ $f(x,y)$ is uniformly continuous on E

$\Rightarrow$ for every $\epsilon(>0)$, there corresponds a partition P of E such that the oscillation $M_{ij} - m_{ij}$ of f on every sub-rectangle E_{ij} of $P < \dfrac{\epsilon}{(b-a)(d-c)}$

$$\Rightarrow U(f,P) - L(f,P) = \sum_{i,j}(M_{ij} - m_{ij})A(E_{ij}) < \frac{\epsilon}{(b-a)(d-c)} \sum_{i,j} A(E_{ij}) = \epsilon$$

$\Rightarrow$ $f(x,y)$ is integrable on E.

Theorem [13.6] A bounded function $f(x,y)$ defined on $E : [a,b;c,d]$ having finite number of points of discontinuities is integrable on E.

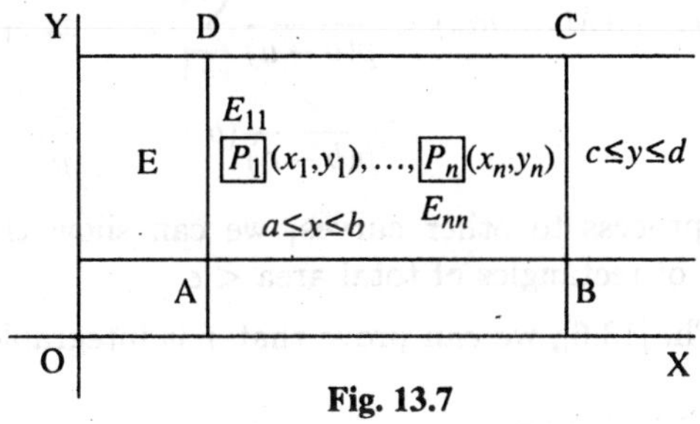

Fig. 13.7

Proof. Let $f(x,y)$ be bounded on $E = \square ABCD : [a,b;c,d]$.

Let $P_1(x_1,y_1)$, $P_2(x_2,y_2)\cdots P_n(x_n,y_n)$ be finite number of points of discontinuities, enclosed by the sub-rectangles E_{11}, $E_{22}, \cdots, E_{nn}$, so that the sum of the areas of these sub-rectangles is less that $\frac{\epsilon}{2(M-m)}$, where M, m be the bounds of f on E.

Then ω_1 = part of oscillator sum $U(f,P) - L(f,P)$ arising from these sub-rectangles

$$= \sum_{i,j}(M_{ij} - m_{ij})A(E_{ij}),\ i = j,\ i = 1,2,\cdots,n$$

$$< (M-m)\sum_{i,j} A(E_{ij}) < (M-m)\cdot\frac{\epsilon}{2(M-m)} = \frac{\epsilon}{2}. \qquad \text{(i)}$$

On producing the sides of these sub-rectangles $E_{11}, E_{22}, \cdots, E_{nn}$ enclosing the points of discontinuities $P_1, P_2, \ldots, P_n$, we get a partition of E into sub-rectangles E_{ij}. Then $f(x,y)$ is continuous in the remaining sub-rectangles E_{ij}, where $i \neq j$.

The sub-rectangles $E_{ij}, i \neq j$ can be further partitioned into sub-rectangles so that the oscillatory sum $\omega_2 = U(f,P) - L(f,P)$ arising from these partitions is less than $< \frac{\epsilon}{2}$ (by Th. [13.5]).

Thus we have a partition of E such that the resultant oscillatory sum

$$\omega_P < \omega_1 + \omega_2 = \frac{\epsilon}{2} + \frac{\epsilon}{2} = \epsilon.$$

Hence f is integrable on E.

Theorem [13.7] Let f be a bounded function on $E = [a,b;c,d] \subseteq \mathbb{R}^2$ and let its points of discontinuities lie on a finite number of curves of the form $y = \phi_1(x), x = \psi_1(y)$, etc., where $\phi_1, \psi_1, \cdots$ are continuous, then f is integrable on E.

Proof. Let there be p finite number of curves like

$$y = \phi_1(x), x = \psi_1(y), y = \phi_2(x), x = \psi_2(y), \cdots, y = \phi_p(x), x = \psi_p(y)$$

such that the points of discontinuities of $f(x, y)$ lie on these p-curves. Since ϕ_1 is continuous on $[a, b]$, there corresponds a partition $P\{a = x_0 < x_1 < \cdots < x_{r-1} < x_r < \cdots < x_n = b\}$ of $[a, b]$ such that the oscillation of ϕ_1 in every sub-interval $[x_{r-1}, x_r] < \frac{\epsilon}{p(b-a)}$.

That is, $M_r - m_r < \dfrac{\epsilon}{p(b-a)}$, where M_r, m_r be the bounds of ϕ_1 in $[x_{r-1},\ x_r]$. (i)

Then the points of $y = \phi_1(x)$ corresponding to the values of x in $[x_{r-1}, x_r]$ of the partition P belongs to the rectangle $[x_{r-1}, x_r; M_r, m_r]$ of area $(x_r - x_{r-1})(M_r - m_r)$.

Then all the points of the curve $y = \phi_1(x)$ have been enclosed in rectangles of total area $\sum(x_r - x_{r-1})(M_r - m_r)$, where

$$\sum_{r=1}^{n}(x_r - x_{r-1})(M_r - m_r) < \frac{\epsilon}{p(b-a)} \sum_{r=1}^{n}(x_r - x_{r-1}) \text{ [by (i)]}$$

$$= \frac{\epsilon}{p(b-a)}(b-a) = \frac{\epsilon}{p}.$$

Continuing the above process to other curves, we can show that all the curves can be enclosed in a finite number of rectangles of total area $< \epsilon$.

Now proceeding as in Th. [13.6], we can prove that f is integrable on E.

Some Properties of Double integrals

Let f, g be integrable functions of (x, y) on a rectangle $E \subseteq \mathbb{R}^2$. Then

(i) $\displaystyle\iint_E Kf(x, y)dx\,dy = K \iint_E f\,dx\,dy$, K is any constant.

(ii) $\displaystyle\iint_E (f + g)(x, y)dx\,dy = \iint_E f(x, y)dx\,dy + \iint_E g(x, y)dx\,dy.$

(iii) $f \cdot g$ is also integrable on E.

(iv) f/g is integrable on E, if $|g| \geq K > 0$ on E.

(v) $|f|$ is also integrable on E and $\displaystyle\left|\iint_E f\,dx\,dy\right| \leq \iint_E |f|dx\,dy.$

Example. Let $f(x, y) = \begin{cases} 1, & \forall \text{ rational values of } x, y \in E\ [0, 1; 0, 1] \\ 0, & \forall \text{ other values of } x, y \text{ in } E. \end{cases}$

Let $P < 0 = x_0 < x_1 \cdots < x_n = 1, 0 = y_0 < y_1 \cdots < y_m = 1]$ be a partition of E

$$E_{ij} = \{(x, y), x_{i-1} \leq x \leq x_i,\ y_{j-1} \leq y \leq y_j\}$$

be the ijth sub-rectangle of E, $i = 1, 2, \cdots, n;\ j = 1, 2, \cdots, m$.

Each sub-rectangle E_{ij} contains points (rational) where $f(x, y) = 1$ and also points (other than rational), where $f(x, y) = 0$.

Therefore, $M_{ij} = 1$ and $m_{ij} = 0$ in each E_{ij} for $i = 1, 2, \cdots, n;\ j = 1, 2, \cdots, m$.

Then

$$U(f,P)=\sum_{i,j} M_{ij}A(E_{ij})=\sum 1\cdot A(E_{ij})=A(E)=1$$

$$\text{and } L(f,P)=\sum_{ij} m_{ij}A(E_{ij})=\sum 0\cdot A(E_{ij})=0.$$

Since P is arbitrary,

$$\overline{\iint_E} f(x,y)dx\,dy=\inf[U(f,P),\ P \text{ is any partition of } E]=1$$

$$\text{and } \underline{\iint_E} f(x,y)dx\,dy=\sup[L(f,P),\ P \text{ is any partition of } E]=0.$$

$\therefore \overline{\iint_E} f(x,y)dx\,dy \neq \underline{\iint_E} f(x,y)dx\,dy$. So $\iint_E f(x,y)dx\,dy$ does not exist.

Example. Let $f(x,y)$ be defined on E $[0,1;0,1]$ by

$$f(x,y)=\begin{cases}1, & \text{for } x=1,\ 0\le y\le 1\\ 0, & \text{for } 0\le x<1,\ 0\le y\le 1.\end{cases}$$

Let $P\{0=x_0<x_1\cdots x_{n-1}<x_n=1,\ 0=y_0<y_1\cdots<y_{n-1}<y_n=1\}$ be a partition of E such that $x_n-x_{n-1}<\epsilon$.

Then $M_{nj}=1$ at $x_n=1,\ j=1,2,\cdots,n$.

That is, $M_{n1}=1, M_{n2}=1,\cdots,M_{nm}=1$ and

$$M_{ij}=0 \text{ for } i=1,\cdots,n-1;\ j=1,2,\cdots,m. \tag{1}$$

Therefore,

$$\begin{aligned}U(f,P)&=\sum_{i,j} M_{ij}AE(ij)=\sum_{j=1}^{m} M_{nj}AE(nj)\ [\text{by (1)}]\\ &=\sum_{j=1}^{m} 1\cdot(x_n-x_{n-1})(y_j-y_{j-1})\\ &=(x_n-x_{n-1})\sum_{j=1}^{m}(y_j-y_{j-1})=(x_n-x_{n-1})\cdot 1<\epsilon.\end{aligned} \tag{2}$$

Again, since $m_{ij}=0$ for $i=1,2,\cdots,n;\ j=1,2,\cdots,m$.

$$L(f,P)=0. \tag{3}$$

Therefore, $U(f,P)-L(f,P)<\epsilon$ by (2) and (3).

Thus $f(x,y)$ is integrable on E, since a bounded function $f(x,y)$ on E is integrable on E, if and only if for every $\epsilon(>0)$, there corresponds a partition P of E such that $U(f,P)-L(f,P)<\epsilon$ [by Th. [13.4, Second form]

It also follows that $\overline{\iint_E} f(x,y)dx\,dy=0 \Rightarrow \iint_E f(x,y)dx\,dy=0$.

Note. If the rectangle E is partitioned into finite number of sub-rectangles $E_1, E_2, \cdots, E_n$, then $\iint\limits_E f\,dx\,dy = \iint\limits_{E_1} f\,dx\,dy + \iint\limits_{E_2} f\,dx\,dy + \cdots + \iint\limits_{E_n} f\,dx\,dy.$

Theorem [13.8] Let $f(x, y)$ be continuous on $E = [a, b; c, d] \subset \mathbb{R}^2$, where E is closed and bounded region in $\mathbb{R}^2$. Then there exists a point $(x_0, y_0) \in E$ such that

$$\iint\limits_E f(x, y)dx\,dy = f(x_0, y_0)A(E).$$

Proof. Since f is continuous on E, there exist points $(u_1, v_1), (u_2, v_2) \in E$ such that

$$f(u_1, v_1) = M = \sup_{(x,y)\in E} f(x, y) \text{ and } f(u_2, v_2) = m = \inf_{(x,y)\in E} f(x, y).$$

$$\text{Then, } mA(E) = \iint\limits_E m\,dx\,dy \le \iint\limits_E f(x, y)dx\,dy \le \iint\limits_E M\,dx\,dy = MA(E)$$

$$\Rightarrow m \le \frac{1}{A(E)} \iint\limits_E f(x, y)dx\,dy \le M$$

$$\Rightarrow \frac{1}{A(E)} \iint\limits_E f(x, y)dx\,dy = \mu, \; m \le \mu \le M.$$

Since $f(x, y)$ is continuous on E, there exists a point $(x_0, y_0) \in E$ such that $\mu = f(x_0, y_0)$ and hence

$$\frac{1}{A(E)} \iint\limits_E f(x, y)dx\,dy = f(x_0, y_0) \Rightarrow \iint\limits_E f(x, y)dx\,dy = A(E)f(x_0, y_0).$$

Note. (1) Let $f(x, y)$ be bounded on $E = [a, b; c, d] \subset \mathbb{R}^2$. For a fixed value of x in $[a, b]$, $f(x, y)$ be considered as a function of single variable y on $[c, d]$, we then define a function

$$g(y) = \int_a^b f(x, y)dx \text{ on } [c, d].$$

If $g(y)$ is integrable on $[c, d]$, then $\int_c^d g(y)dy = \int_c^d \left\{\int_a^b f(x, y)dx\right\} dy$.

Hence the repeated integral $\int_c^d \left\{\int_a^b f(x, y)dx\right\} dy$ exists and equal to the double integral

$$\iint\limits_E f(x, y)dx\,dy = \int_a^b \int_c^d f(x, y)dx\,dy.$$

Similarly, if we define a function $g(x) = \int_c^d f(x, y)dy$ and if $g(x)$ is integrable on $[a, b]$, then

$$\int_a^b g(x)dx = \int_a^b \left\{\int_c^d f(x, y)dy\right\} dx$$

and in that case, the repeated integral $\int_a^b \left\{\int_c^d f(x, y)dy\right\} dx$ exists and equal to the double integral

$$\iint\limits_E f(x, y)dx\,dy = \int_a^b \int_c^d f(x, y)dx\,dy.$$

(2) We shall use the repeated integrals as defined above in finding out the double integral.

(3) *Repeated integrals* are also called as *iterated integrals.*

(4) The repeated integrals (iterated integrals) are defined by (i) $\int_a^b \left\{\int_c^d f(x,y)dy\right\} dx$, where c, d may be functions of x or constants or (ii) $\int_c^d \left\{\int_a^b f(x,y)dx\right\} dy$, where a, b may be functions of y or constants.

Theorem [13.9] **[Fubini's theorem]** Let f be a bounded function defined on $E = [a, b; c, d] \subset \mathbb{R}^2$ and the double integral $\iint\limits_E f(x,y)dx\,dy$ exist. If $\int_c^d f(x,y)dy$ $\left[\text{or } \int_a^b f(x,y)dx\right]$ exists for each fixed x in $[a, b]$ (or for each fixed y in $[c, d]$), then the repeated integral

$$\int_a^b \left\{\int_c^d f(x,y)dy\right\} dx \left(\text{or } \int_c^d \left\{\int_a^b f(x,y)dx\right\} dy\right)$$

exists and is equal to the double integral $\iint\limits_E f\,dx\,dy$. That is,

$$\iint\limits_E f\,dx\,dy = \int_a^b \left\{\int_c^d f(x,y)dy\right\} dx = \int_c^d \left\{\int_a^b f(x,y)dx\right\} dy.$$

Proof. Let $g(x) = \displaystyle\int_c^d f(x,y)dy$ for $x \in [a, b]$. Then we have to show that $\displaystyle\int_a^b g(x)dx$ exists and equal to $\displaystyle\iint\limits_E f\,dx\,dy$.

Let $a = x_0 < x_1 < x_2 \cdots x_{i-1} < x_i < \cdots < x_n = b$ be a partition of $[a, b]$ with $||P|| < \frac{\delta}{2}$ and let $\xi_i \in [x_{i-1}, x_i]$ for $i = 1, 2, \cdots, n$ and $c = y_0 < y_1 < y_2 < \cdots < y_{j-1} < y_j \cdots < y_m = d$ be a partition of $[c, d]$ with $||P|| < \frac{\delta}{2}$ and let $\eta_j \in [y_{j-1}, y_j]$ for $j = 1, 2, \cdots, m$.

Then $P = \{E_{ij}, E_{ij} = [x_{i-1}, x_i; y_{j-1}, y_j]\}$ forms the partition of $E = [a, b; c, d] \subset \mathbb{R}^2$, with $||P|| < \delta$ and $(\xi_i, \eta_j) \in E_{ij}$. Since the upper integral U is the infimum of the upper sums, we have for the partition P

$$\sum_{i,j} M_{ij} A(E_{ij}) < U + \epsilon,\ \epsilon > 0 \tag{i}$$

$$\text{or } \sum_{i,j} M_{ij}(x_i - x_{i-1})(y_j - y_{j-1}) < U + \epsilon.$$

Again,

$$\int_c^{\bar{d}} f(x,y)dy \le \sum_j M_{ij}(y_j - y_{j-1}), \tag{ii}$$

since for any value of $x \in [x_{i-1}, x_i]$, M_{ij} is an upper bound of f in $[y_{j-1}, y_j]$.

It follows from (ii) that $\sum\limits_j M_{ij}(y_j - y_{j-1})$; $x \in [x_{i-1}, x_i]$ is an upper bound of

$$\phi(x) = \int_c^{\bar{d}} f(x,y)dy.$$

Therefore,

$$\int_a^{\bar{b}} \phi(x)dx \le \sum_{i,j} [M_{ij}(y_j - y_{j-1})](x_i - x_{i-1})$$

$$\text{or,} \int_a^{\bar{b}} \left\{ \int_c^{\bar{d}} f(x,y)dy \right\} dx \le \sum_{i,j} M_{ij}(x_i - x_{i-1})(y_j - y_{j-1}) < U + \epsilon \text{ [by (i)].} \quad \text{(iii)}$$

Since by hypothesis, $\int_c^d f(x,y)dy$ exists, we have $\int_c^{\bar{d}} f dy = \int_{\underline{c}}^d f dy = \int_c^d f dy.$

Hence from (iii), $\int_a^{\bar{b}} \left\{ \int_c^d f dy \right\} dx < U + \epsilon$

$$\text{i.e.,} \int_a^{\bar{b}} \left\{ \int_c^d f dy \right\} dx \le U, \text{ as } \epsilon \text{ is an arbitrary (+)ve number.} \quad \text{(iv)}$$

By considering the lower integral L we can similarly show that

$$\int_{\underline{a}}^b \left\{ \int_c^d f dy \right\} dx \ge L. \quad \text{(v)}$$

Thus we get

$$\int_a^{\bar{b}} \left\{ \int_c^{\bar{d}} f(x,y)dy \right\} dx = \int_a^{\bar{b}} \left\{ \int_c^d f(x,y)dy \right\} dx \le U = \overline{\iint_E} f(x,y)dx\,dy \quad \text{(vi)}$$

and similarly,

$$\int_{\underline{a}}^b \left\{ \int_{\underline{c}}^d f(x,y)dy \right\} dx = \int_{\underline{a}}^b \left\{ \int_c^d f(x,y)dy \right\} dx \ge L = \iint_{\underline{E}} f(x,y)dx\,dy \text{ [by (iv) and (v)].}$$

Therefore,

$$\iint_{\underline{E}} f(x,y)dx\,dy \le \int_{\underline{a}}^b \left\{ \int_c^d f(x,y)dy \right\} dx$$

$$\le \int_a^{\bar{b}} \left\{ \int_c^d f(x,y)dy \right\} dx \le \overline{\iint_E} f(x,y)dx\,dy. \quad \text{(vii)}$$

But by hypothesis,

$$\overline{\iint_E} f(x,y)dx\,dy = \iint_{\underline{E}} f(x,y)dx\,dy = \iint_E f(x,y)dx\,dy.$$

Then from (vii),

$$\iint_E f(x,y)dx\,dy = \int_a^{\bar{b}} \left\{ \int_c^d f(x,y)dy \right\} dx$$

$$= \int_{\underline{a}}^b \left\{ \int_c^d f(x,y)dy \right\} dx = \int_a^b \left\{ \int_c^d f(x,y)dy \right\} dx.$$

Example. Let $f(x, y)$ be defined in $E = [0, 1; 0, 1]$ by

$$f(x,y) = \begin{cases} 1/2, & \text{when } y \text{ is rational} \\ x, & \text{when } y \text{ is irrational.} \end{cases}$$

Then the repeated integral $\int_0^1 \left\{\int_0^1 f(x,y)dx\right\} dy = \frac{1}{2}$, but $\int_0^1 \left\{\int_0^1 f(x,y)dy\right\} dx$ does not exist.

When y is rational, $\int_0^1 f(x,y)dx = \int_0^1 \frac{1}{2}dx = \frac{1}{2}$ and when y is irrational

$$\int_0^1 f(x,y)dx = \int_0^1 x\,dx = \left.\frac{x^2}{2}\right]_0^1 = \frac{1}{2}.$$

So $\int_0^1 f(x,y)dy$ exists and equal to $\frac{1}{2}$ for all rational and irrational values of y and hence the repeated integral

$$\int_0^1 \left\{\int_0^1 f(x,y)dx\right\} dy = \int_0^1 \frac{1}{2}dy = \frac{1}{2}.$$

Now we consider the existence of $\int_0^1 f(x,y)dy$. We divide $[0,1]$ into two halves:

$$\left[0, \frac{1}{2}\right] \text{ and } \left[\frac{1}{2}, 1\right].$$

In $\left[0, \frac{1}{2}\right]$, $f(0, y) = \frac{1}{2}$, y is rational.

$f(0, y) = 0$, y is irrational.

$f(a, y) = \frac{1}{2}$, $0 < a < \frac{1}{2}$, y is rational.

$f(a, y) = a$, $0 < a < \frac{1}{2}$, y is irrational.

Therefore, $M = \frac{1}{2}$, $m = 0$ of $f(x, y)$ in $\left[0, \frac{1}{2}\right]$.

In $\left[\frac{1}{2}, 1\right]$, $f\left(\frac{1}{2}, y\right) = \frac{1}{2}$, y is rational.

$f\left(\frac{1}{2}, y\right) = \frac{1}{2}$, y is irrational.

$f(a, y) = \frac{1}{2}$, $\frac{1}{2} < a < 1$, y is rational.

$f(a, y) = a$, $\frac{1}{2} < a < 1$, y is irrational.

$f(1, y) = \frac{1}{2}$, y is rational.

$f(1, y) = 1$, y is rational.

Therefore, $M = 1$, $m = \frac{1}{2}$ of $f(x, y)$ in $\left[\frac{1}{2}, 1\right]$.

Hence

$$U = \int_0^{\overline{1}} f(x,y)dy = \int_0^{\overline{\frac{1}{2}}} f\,dy + \int_{\frac{1}{2}}^{\overline{1}} f\,dy = \int_0^{\frac{1}{2}} \frac{1}{2}dy + \int_{\frac{1}{2}}^{1} 1dy$$

$$= \frac{1}{2}\cdot\frac{1}{2} + \frac{1}{2} = \frac{1}{4} + \frac{1}{2} = \frac{3}{4}.$$

$$L = \int_{\underline{0}}^{1} f(x,y)dy = \int_{\underline{0}}^{\frac{1}{2}} f\,dy + \int_{\underline{\frac{1}{2}}}^{1} f\,dy = 0 + \int_{\frac{1}{2}}^{1} \frac{1}{2}dy = \frac{1}{4}.$$

$$\therefore U \neq L.$$

Therefore, $\int_0^1 f(x,y)dy$ does not exist and hence the other repeated integral

$$\int_0^1 \left\{ \int_0^1 f(x,y)dy \right\} dx$$

does not exist.

Problem [13.1] : Let $f(x,y)$ be defined over $E = [0,1;0,1]$ by

$$f(x,y) = \begin{cases} 1, & \text{if } x \text{ be rational} \\ 3y^2, & \text{if } x \text{ be irrational.} \end{cases}$$

Then (i) $f(x,y)$ is not integrable over E.

(ii) the repeated integral $\int_0^1 \left\{ \int_0^1 f(x,y)dy \right\} dx$ exists.

(iii) the other repeated integral does not exist.

Solution.

$$f(x,0) = \begin{cases} 1, & \text{when } x \text{ is rational.} \\ 0, & \text{when } x \text{ is irrational.} \end{cases}$$

$$f(x,a) = \begin{cases} 1, & 0 < a \le 1, \text{ when } x \text{ is rational.} \\ 3a^2, & 0 < a < 1, \text{ when } x \text{ is irrational.} \end{cases}$$

$$f(x,1) = 3,\ a = 1, \text{ when } x \text{ is irrational.}$$

Also $0 < a < 1 \Rightarrow 0 < a^2 < 1 \Rightarrow 3a^2 < 3$.

$M = 3, m = 0$ are the bounds of $f(x,y)$ in $[0,1]$.

(i) Let $E_{ij} = [x_{i-1}, x_i, y_{j-1}, y_j]$

$A(E_{ij}) = (x_i - x_{i-1})(y_j - y_{j-1})$.

Let $(\xi, \eta) \in E_{ij}$.

Then by definition it follows that $M_{ij} = 1$ and $m_{ij} = 3\eta^2$ for $0 < \eta < 1$, are respectively upper bound and lower bound of $f(x,y)$ in E_{ij}.

Then

$$\overline{\iint_E} f(x,y)dx\,dy = \inf\left[\sum_{i,j} M_{ij} AE(ij)\right] = \inf\left[\sum_{i,j} 1 \cdot AE(E_{ij})\right] = 1.$$

$$\underline{\iint_E} f(x,y)dx\,dy = \sup\left[\sum_{i,j} m_{ij} A(E_{ij})\right] = \sup\left[\sum_{i,j} 3\eta^2 A(E_{ij})\right] = 3\eta^2.$$

Therefore, $\overline{\iint\limits_E} f(x,y)dx\,dy \neq \underline{\iint\limits_E} f(x,y)dx\,dy \;\forall\; \eta.$

Hence, $\iint\limits_E f(x,y)dx\,dy$ does not exist.

That is, $f(x,y)$ is not integrable in E.

(ii) Again

$$\int_0^1 f(x,y)dy = \int_0^1 1\cdot dy = 1 \text{ [when } x \text{ is rational]}$$

$$\text{and } \int_0^1 f(x,y)dy = \int_0^1 3y^2dy = \left[y^3\right]_0^1 = 1 \text{ [when } x \text{ is irrational]}.$$

Then $\int_0^1 \left\{\int_0^1 f(x,y)dy\right\} dx = \int_0^1 1\cdot dx = 1.$

So, the repeated integral exists.

(iii) Since $M = 3, m = 0$ are bounds of $f(x,y)$ in $[0,1]$, we have

$$\int_0^{\overline{1}} f(x,y)dx = \int_0^1 3dx = 3 \text{ and } \int_{\underline{0}}^1 f(x,y)dx = 0.$$

So $\int_0^1 f(x,y)dx$ does not exist and hence the other repeated integral $\int_0^1 \left\{\int_0^1 f(x,y)dx\right\} dy$ does not exist.

Example. Define the function f by

$$f(x,y) = \begin{cases} \frac{1}{y^2}, & 0<x<y<1 \\ -\frac{1}{x^2}, & 0<y<x<1 \\ 0, & \text{otherwise.} \end{cases}$$

Examine the existence of $\iint\limits_E f\,dx\,dy,\;\; E : [0,1;0,1].$

Let $0 < x < 1$.

Then

$$\int_0^1 f(x,y)dy = \int_0^x f(x,y)dy + \int_x^1 f(x,y)dy = -\int_0^x \frac{1}{x^2}dy + \int_x^1 \frac{1}{y^2}dy$$

$\left[\text{as by definition, } f(x,y) = -\frac{1}{x^2} \text{ for } 0<y<x \text{ and } f(x,y) = \frac{1}{y^2} \text{ for } x<y<1\right]$

$$= \left[-\frac{1}{x^2}\cdot x\right] + \left[-\frac{1}{y}\right]_x^1 = -\frac{1}{x} + \left[-1+\frac{1}{x}\right] = -1.$$

Therefore

$$\int_0^1 \left[\int_0^1 f(x,y)dy\right] dx = \int_0^1 -1\cdot dx = -1. \tag{1}$$

Again, let $0 < y < 1$.

Then

$$\int_0^1 f(x,y)dx = \int_0^y f(x,y)dx + \int_y^1 f(x,y)dx = \int_0^y \frac{1}{y^2}dx - \int_y^1 \frac{1}{x^2}dx \text{ [by definition]}$$

$$= \frac{1}{y} - \left[-\frac{1}{x}\right]_y^1 = \frac{1}{y} - \left[-1 + \frac{1}{y}\right] = 1.$$

Therefore,

$$\int_0^1 \left\{\int_0^1 f(x,y)dx\right\} dy = \int_0^1 1 \cdot dy = 1. \tag{2}$$

From (1) and (2), it follows that $\int_0^1 \left\{\int_0^1 f(x,y)dy\right\} dx \neq \int_0^1 \left\{\int_0^1 f(x,y)dx\right\} dy.$

Hence, $\iint_E f\,dx\,dy$ does not exist.

Example. Evaluate the integral $\iint_E \frac{x-y}{x+y}dx\,dy$, where $E = [0,1;0,1]$.

Here $f(x,y) = \frac{x-y}{x+y}$ is bounded in $E' = \{0 < x \le 1,\ 0 < y \le 1\}$.

$(0,0)$ is the only point of infinite discontinuity.

For $x \neq 0$,

$$g(x) = \int_0^1 \frac{x-y}{x+y}dy = -\int_0^1 \frac{y+x-2x}{x+y}dy = -\int_0^1 \left[1 - \frac{2x}{x+y}\right] dy$$

$$= -\left[y - 2x\log(x+y)\right]_0^1 = -\left[1 - 2x\log\left(\frac{1+x}{x}\right)\right] = -1 + 2x\log\left(\frac{1+x}{x}\right).$$

Here $\int_0^1 g(x)dx$ is a proper integral as $\lim_{x\to 0}\left[x\log\left(\frac{1+x}{x}\right) = 0\right]$.

Then $\int_0^1 \left\{\int_0^1 f(x,y)dy\right\} dx$

$$= \int_0^1 g(x)dx = \int_0^1 \left[2x\log\left(\frac{1+x}{x}\right) - 1\right] dx$$

$$= \int_0^1 [2x\log(1+x) - 2x\log x - 1]dx$$

$$= \left[\log(1+x)\cdot x^2\right]_0^1 - \int_0^1 \frac{x^2}{1+x}dx - \left[x^2\log x\right]_0^1 + \int_0^1 \frac{x^2}{x}dx - [x]_0^1$$

$$= \log 2 - \int_0^1 \frac{x(1+x) - x}{1+x}dx + \frac{1}{2} - 1$$

$$= \log 2 - \frac{1}{2} - \int_0^1 x\,dx + \int_0^1 \frac{1+x-1}{1+x}dx$$

$$= \log 2 - \frac{1}{2} - \frac{1}{2} + 1 - \left[\log(1+x)\right]_0^1$$

$$= \log 2 - \log 2 = 0.$$

Again, for $y \neq 0$, $g(y) = \int_0^1 \frac{x-y}{x+y}dx = 1 - 2y\log\left(\frac{1+y}{y}\right)$.

Here $\int_0^1 g(y)dy$ is a proper integral as $\lim_{y\to 0} y\log\left(\frac{1+y}{y}\right) = 0$.

Then proceeding as above, $\int_0^1 \left\{\int_0^1 f(x,y)dx\right\}dy = \int_0^1 g(y)dy = 0$.

Hence $\iint\limits_E f(x,y)dxdy = 0$.

Example. Let $f(x,y) = \frac{x-y}{(x+y)^3}$ and $E = [0,1;0,1]$.

Then show that $\int_0^1 \left\{\int_0^1 \frac{x-y}{(x+y)^3}dy\right\}dx = \frac{1}{2}$ and $\int_0^1 \left\{\int_0^1 \frac{x-y}{(x+y)^3}dx\right\}dy = -\frac{1}{2}$.

$f(x,y) = \frac{x-y}{(x+y)^3}$.

$f(x,y)$ is bounded in $0 < x \leq 1,\ 0 < y \leq 1$.

$(0,0)$ is a point of infinite discontinuity.

Also for a fixed $x \neq 0$, $f(x,y)$ is a bounded function of y.

The same is true for a fixed $y \neq 0$.

Let $x \neq 0$.

Then

$$g(x) = \int_0^1 \frac{x-y}{(x+y)^3}dy = \int_0^1 \left\{\frac{2x}{(x+y)^3} - \frac{1}{(x+y)^2}\right\}dy = \left[\frac{-x}{(x+y)^2} + \frac{1}{x+y}\right]_0^1$$

$$= \left[-\frac{x}{(x+1)^2} + \frac{1}{x+1} - \left(-\frac{1}{x} + \frac{1}{x}\right)\right] = \left[\frac{-x+x+1}{(x+1)^2}\right] = \frac{1}{(x+1)^2}.$$

$$g(0) = \int_0^1 -\frac{1}{y^2}dy, \text{ which does not exist.}$$

But

$$\int_0^1 \left\{\int_0^1 f(x,y)dy\right\}dx = \int_0^1 g(x)dx = \int_0^1 \frac{1}{(x+1)^2}dx$$

$$= \left[-\frac{1}{x+1}\right]_0^1 = -\frac{1}{2} + 1 = \frac{1}{2}. \qquad (1).$$

Again, let $y \neq 0$.

Then

$$\phi(y) = \int_0^1 f(x,y)dx = \int_0^1 \frac{x-y}{(x+y)^3}dx = \int_0^1 \left[\frac{1}{(x+y)^2} - \frac{2y}{(x+y)^3}\right]dx = -\frac{1}{(1+y)^2}$$

and $\phi(0) = \int_0^1 \frac{x}{x^3}dx = \int_0^1 \frac{1}{x^2}dx$ does not exist.

But

$$\int_0^1 \left\{\int_0^1 f(x,y)dx\right\}dy = \int_0^1 -\frac{1}{(1+y)^2}dy = -\frac{1}{2}. \qquad (2)$$

From (1) and (2), the integral $\iint_E f(x,y)dx\,dy$ does not exist, where $E = [0,1,;0,1]$, though the repeated integrals exist.

Example. Show that $\int_0^1 \left\{ \int_0^1 \frac{x^2-y^2}{x^2+y^2} dy \right\} dx = \int_0^1 \left\{ \int_0^1 \frac{x^2-y^2}{x^2+y^2} dx \right\} dy.$

Here $f(x,y) = \dfrac{x^2-y^2}{x^2+y^2}$ is bounded on $E : [0 < x \le 1,\ 0 < y \le 1]$.

$(0,0)$ is the point of infinite discontinuity. For any fixed value of $x \neq 0$, the function $f(x,y)$ is a bounded function of y.

Let $x \neq 0$.

$$\text{Then } g(x) = \int_0^1 \frac{x^2-y^2}{x^2+y^2} dy = -\int_0^1 \left(1 - \frac{2x^2}{x^2+y^2}\right) dy$$

$$= -1 + \frac{2x^2}{x}\left[\tan^{-1}\frac{y}{x}\right]_0^1 = -1 + 2x \cdot \tan^{-1}\frac{1}{x}.$$

Therefore,

$$\int_0^1 \left\{ \int_0^1 \frac{x^2-y^2}{x^2+y^2} dy \right\} dx = \int_0^1 g(x)dx = \int_0^1 \left[-1 + 2x\tan^{-1}\frac{1}{x} \right] dx$$

$$= -1 + \left[x^2 \tan^{-1}\frac{1}{x} \right]_0^1 - \int_0^1 x^2 \cdot \frac{1}{1+\frac{1}{x^2}} \left(-\frac{1}{x^2}\right) dx$$

$$= -1 + \frac{\pi}{4} + \int_0^1 \frac{x^2}{x^2+1} dx$$

$$= -1 + \frac{\pi}{4} + \int_0^1 dx - \int_0^1 \frac{1}{x^2+1} dx$$

$$= -1 + \frac{\pi}{4} + 1 - \frac{\pi}{4} = 0.$$

Let $y \neq 0$. $f(x,y)$ is a bounded function of x.

$$\text{Then } \phi(y) = \int_0^1 \frac{x^2-y^2}{x^2+y^2} dx = \int_0^1 \left(1 - \frac{2y^2}{x^2+y^2}\right) dx = 1 - 2y\tan^{-1}\frac{1}{y}.$$

Therefore, $\int_0^1 \left\{ \int_0^1 \frac{x^2-y^2}{x^2+y^2} dx \right\} = \int_0^1 \left(1 - 2y\tan^{-1}\frac{1}{y}\right) dy = 0$ [proceeding as above].

Hence, $\int_0^1 \left\{ \int_0^1 \frac{x^2-y^2}{x^2+y^2} dy \right\} dx = \int_0^1 \left\{ \int_0^1 \frac{x^2-y^2}{x^2+y^2} dx \right\} dy.$

In other words, $\iint_E f(x,y)dx\,dy$ is integrable over $E\ [0,1;0,1]$.

Example. Show that $\int_\lambda^1 \left\{ \int_\lambda^1 f\,dy \right\} dx = \int_\lambda^1 \left\{ \int_\lambda^1 f\,dx \right\} dy$ for $f(x,y) = \dfrac{y^2-x^2}{(y^2+x^2)^2}$, $0 < \lambda < 1$.

But $\int_0^1 \left\{ \int_0^1 f\,dy \right\} dx \neq \int_0^1 \left\{ \int_0^1 f\,dx \right\} dy.$

Here $f(x,y)$ is bounded on $E = [\lambda, 1; \lambda, 1]$ for $0 < \lambda < 1$.

$$g(x) = \int_\lambda^1 f(x,y)dy = \int_\lambda^1 \frac{y^2 - x^2}{(y^2+x^2)^2}dy = \int_\lambda^1 \frac{dy}{y^2+x^2} - \int_\lambda^1 \frac{2x^2}{(y^2+x^2)^2}dy$$

$$= \frac{1}{x}\left[\tan^{-1}\frac{y}{x}\right]_\lambda^1 - 2x^2\int_\lambda^1 \frac{1}{(y^2+x^2)^2}dy. \tag{1}$$

Put $y = x\tan\theta$ in the 2nd part.

Then $dy = x\sec^2\theta\, d\theta$.

$$\int \frac{dy}{(y^2+x^2)^2} = \int \frac{x\sec^2\theta\, d\theta}{x^4\sec^4\theta} = \frac{1}{x^3}\int \cos^2\theta\, d\theta = \frac{1}{2x^3}\int (1+\cos 2\theta)d\theta$$

$$= \frac{1}{2x^3}\left[\theta + \frac{1}{2}\sin 2\theta\right] = \frac{1}{2x^3}\left[\tan^{-1}\frac{y}{x} + \frac{xy}{x^2+y^2}\right].$$

From (1),

$$g(x) = \frac{1}{x}\left[\tan^{-1}\frac{1}{x} - \tan^{-1}\frac{\lambda}{x}\right] - \frac{1}{x}\left[\tan^{-1}\frac{y}{x} + \frac{xy}{x^2+y^2}\right]_\lambda^1$$

$$= \frac{1}{x}\left[\tan^{-1}\frac{1}{x} - \tan^{-1}\frac{\lambda}{x}\right] - \frac{1}{x}\left[\tan^{-1}\frac{1}{x} + \frac{x}{x^2+1} - \tan^{-1}\frac{\lambda}{x} - \frac{\lambda x}{x^2+\lambda^2}\right]$$

$$= \frac{\lambda}{x^2+\lambda^2} - \frac{1}{x^2+1}.$$

Therefore,

$$\int_\lambda^1 \left\{\int_\lambda^1 f(x,y)dy\right\}dx = \int_\lambda^1 g(x)dx = \int_\lambda^1 \left[\frac{\lambda}{x^2+\lambda^2} - \frac{1}{x^2+1}\right]dx$$

$$= \left[\tan^{-1}\frac{x}{\lambda} - \tan^{-1}x\right]_\lambda^1 = \tan^{-1}\frac{1}{\lambda} - \frac{\pi}{4} - \frac{\pi}{4} + \tan^{-1}\lambda$$

$$= \cot^{-1}\lambda + \tan^{-1}\lambda - \frac{\pi}{2} = \frac{\pi}{2} - \frac{\pi}{2}$$

$$= 0.$$

Proceeding in the same way we can show that $\displaystyle\int_\lambda^1 \left\{\int_\lambda^1 \frac{y^2-x^2}{(y^2+x^2)^2}dx\right\}dy = 0.$

$f(x,y)$ is bounded on $E[0 < x \le 1;\ 0 < x \le 1]$.

$(0,0)$ is a point of infinite discontinuity.

Also for a fixed $x \ne 0$, $f(x,y)$ is a bounded function of y.

Let $x \ne 0$.

Then $\displaystyle g(x) = \int_0^1 f(x,y)dy = \int_0^1 \frac{y^2-x^2}{(y^2+x^2)^2}dy$

$$= \frac{1}{x}\left[\tan^{-1}\frac{y}{x}\right]_0^1 - \frac{1}{x}\left[\tan^{-1}\frac{y}{x} + \frac{xy}{x^2+y^2}\right]_0^1 = -\left[\frac{y}{x^2+y^2}\right]_0^1 = -\frac{1}{x^2+1}.$$

Therefore,

$$\int_0^1 \left\{\int_0^1 f(x,y)dy\right\}dx = \int_0^1 g(x)dx = \int_0^1 -\frac{1}{x^2+1}dx = -\left[\tan^{-1}x\right]_0^1 = -\frac{\pi}{4} \tag{2}$$

Again, for a fixed $y \ne 0$, $f(x,y)$ is a bounded function of x.

Let $y \neq 0$.

$$\text{Then } g(y) = \int_0^1 f(x,y)dx = \int_0^1 \frac{y^2 - x^2}{(y^2+x^2)^2}dx$$

$$= -\left[\int_0^1 \frac{x^2+y^2-2y^2}{(x^2+y^2)^2}dx\right] = -\int_0^1 \frac{1}{x^2+y^2}dx + 2y^2\int_0^1 \frac{dx}{(x^2+y^2)^2}$$

$$= -\frac{1}{y}\left[\tan^{-1}\frac{x}{y}\right]_0^1 + \frac{1}{y}\left[\tan^{-1}\frac{x}{y} + \frac{xy}{x^2+y^2}\right]_0^1 \quad \text{[Proceeding as above]}$$

$$= \frac{1}{y^2+1}.$$

Therefore,

$$\int_0^1 \left\{\int_0^1 f(x,y)dx\right\} dy = \int_0^1 g(y)dy = \int_0^1 \frac{1}{y^2+1}dy = \left[\tan^{-1} y\right]_0^1 = \frac{\pi}{4}. \tag{3}$$

Hence from (2) and (3), it follows that

$$\int_0^1 \left\{\int_0^1 f(x,y)dy\right\} dx \neq \int_0^1 \left\{\int_0^1 f(x,y)dx\right\} dy.$$

13.4 Integrability and Integral of Bounded Function f over any Bounded Domain E

Definition [13.2] : Let E be a bounded domain and let R be a rectangle such that $E \subseteq R$. We define a function $F(x,y)$ over R as follows:

$$F(x,y) = \begin{cases} f(x,y), & \forall\ (x,y) \in E \\ 0, & \forall\ (x,y) \notin E. \end{cases}$$

Then f is said to be integrable over E, if F is integrable over R and

$$\iint\limits_E f(x,y)dx\,dy = \iint\limits_R F(x,y)dx\,dy.$$

Theorem [13.10] If f be continuous in a domain E which is bounded by the curves $y = \phi(x)$, $y = \psi(x)$, $x = a$, $x = b$, where ϕ, ψ are continuous $\forall\ x \in [a,b]$ and $\phi(x) \leq \psi(x)\ \forall\ x \in [a,b]$,

$$\text{then } \iint\limits_E f(x,y)dx\,dy = \int_a^b \left\{\int_{\phi(x)}^{\psi(x)} f(x,y)dy\right\} dx.$$

Proof. Let $R = [a,b;c,d]$ enclose $E = [a,b;\phi(x),\psi(x)]$. That is, $E \subseteq R$.

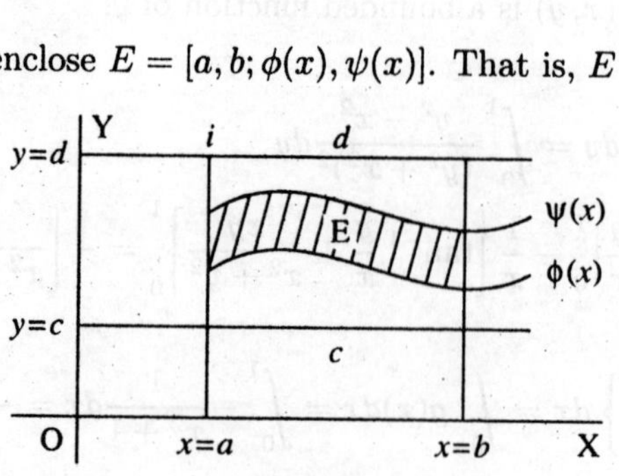

Fig. 13.8

Let $F(x, y)$ be defined as in Def. [13.2].

We have $$\iint_E f(x,y)dx\,dy = \iint_R F(x,y)dx\,dy = \int_a^b \left\{ \int_c^d F(x,y)dy \right\} dx$$
$$= \int_a^b \left\{ \int_c^{\phi(x)} F\,dy + \int_{\phi(x)}^{\psi(x)} F\,dy + \int_{\psi(x)}^{d} F\,dy \right\} dx$$
$$= \int_a^b \left\{ \int_{\phi(x)}^{\psi(x)} F\,dy \right\} dx$$
$$= \int_a^b \left\{ \int_{\phi(x)}^{\psi(x)} f(x,y)dy \right\} dx,$$

since by Def. [13.2], $F(x, y) = 0 \;\forall\; (x, y) \notin E$, the two integrals $\int_a^{\phi(x)} F\,dy$ and $\int_{\psi(x)}^{d} F\,dy$ are zero and $F(x, y) = f(x, y) \;\forall\; (x, y) \in E$.

Note. If f be continuous in a domain E bounded by the curves $x = \phi(y)$, $x = \psi(y)$, $y = c, y = d$, where ϕ, ψ are continuous and $\phi(y) \leq \psi(y) \;\forall\; y \in [c, d]$, then

$$\iint_E f(x,y)dx\,dy = \int_c^d \left\{ \int_{\phi(y)}^{\psi(y)} F(x,y)dx \right\} dy = \int_c^d \left\{ \int_{\phi(y)}^{\psi(y)} f(x,y)dx \right\} dy.$$

Example. The double integral $\iint_E f(x,y)dx\,dy$, where the domain of integration is $\{(x, y), x^2 + y^2 \leq 1\}$ is equivalent to the repeated integral $\int_{-1}^{1} \left\{ \int_{-\sqrt{1-x^2}}^{\sqrt{1-x^2}} f(x,y)dy \right\} dx.$

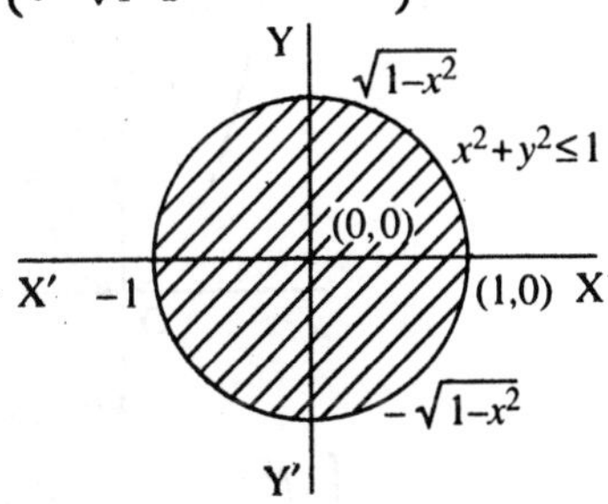

Fig. 13.9

Here $E = \{-1, 1, -\sqrt{1 - x^2}, \sqrt{1 - x^2}\}$.

Lines are moving $\|$ to Y-axis.

Changing the order of integration, we have

$$\int_{-1}^{1} \left\{ \int_{-\sqrt{1-x^2}}^{\sqrt{1-x^2}} f\,dy \right\} dx = \int_{-1}^{1} \left\{ \int_{-\sqrt{1-y^2}}^{\sqrt{1-y^2}} f\,dx \right\} dy$$

(Lines are moving $\|$ to X-axis.)

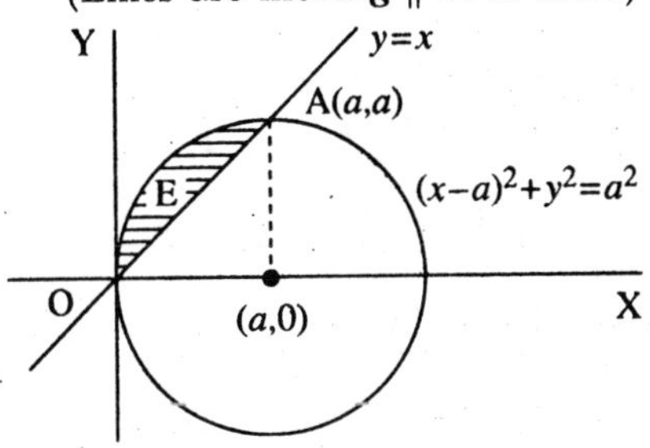

Fig. 13.10

Example. Evaluate $\iint y\,dx\,dy$ over the region bounded by the line $y = x$ and the circle $x^2 - 2ax + y^2 = 0$.

Here

$$x^2 + y^2 - 2ax = 0$$
$$\Rightarrow (x - a)^2 + y^2 = a^2,$$

a circle with centre at $(a, 0)$ and of radius a.

Here $E = \{0, a, x, \sqrt{2ax - x^2}\}$.

$$\iint_E y\,dx\,dy = \int_0^a \left\{ \int_x^{\sqrt{2ax-x^2}} y\,dy \right\} dx \tag{1}$$

$$= \int_0^a \left[\frac{y^2}{2} \right]_x^{\sqrt{2ax-x^2}} dx = \frac{1}{2} \int_0^a [2ax - x^2 - x^2] dx$$

$$= \int_0^a (ax - x^2) dx = \left[a\frac{x^2}{2} - \frac{x^3}{3} \right]_0^a = \left[\frac{a^3}{2} - \frac{a^3}{3} \right] = \frac{a^3}{6}.$$

Note. Changing the order of integration, we get

$$\int_0^a y \left\{ \int_{a-\sqrt{a^2-y^2}}^{y} dx \right\} dy$$

[In (1), lines are ∥ to Y-axis. Taking lines ∥ to X-axis, we have $x = a - \sqrt{a^2 - y^2}$ and $x = y$.
$\because x^2 + y^2 - 2ax = 0$
$\Rightarrow (x - a)^2 = a^2 - y^2 \Rightarrow x = a \pm \sqrt{a^2 - y^2}$.]

Here $x = a - \sqrt{a^2 - y^2}$ as the region lies in $x = 0$ to $x = a$.

Problem [13.2] : Change the order of integration in the following cases and evaluate:

(i) $\displaystyle\int_0^a \left\{ \int_{x^2/2}^{2a-x} xy\,dy \right\} dx$ (ii) $\displaystyle\int_{y=0}^{2} \left\{ \int_{x=1}^{\sqrt{4-y}} (x + y) dx \right\} dy$

(iii) $\displaystyle\int_0^\infty \left\{ \int_{y=x}^{\infty} \frac{e^{-y}}{y} dy \right\} dx$. (iv) $\displaystyle\int_0^a \int_0^x \frac{\phi'(y) dx\,dy}{\sqrt{(a - x)(x - y)}}$.

Solution. (i)

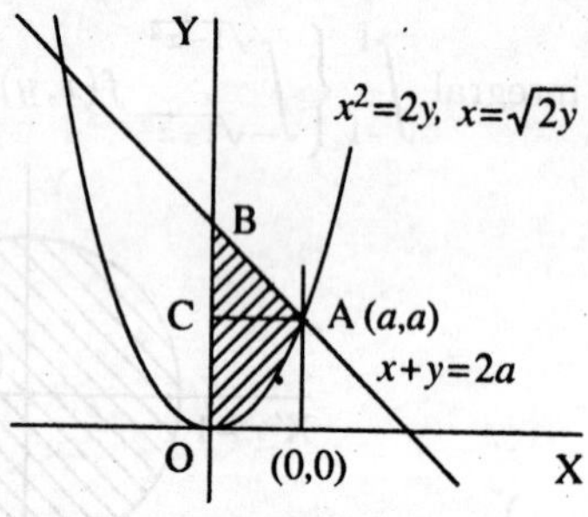

Fig. 13.11

Region $\left[0, a; \frac{x^2}{2}, 2a - x\right]$ is the portion $OABCO$. It consists of two parts $OACO$ and $CABC$.

$$OACO = \left[0, \sqrt{2y};\ 0, a\right]$$
$$CABC = [0, 2a - y;\ a, 2a].$$

The strips are taken ∥ to X-axis.

Then, $\displaystyle\int_0^a \left\{ \int_{x^2/2}^{2a-x} xy\,dy \right\} dx$

$$= \int_0^a \left\{ \int_{x=0}^{\sqrt{2y}} xy\,dx \right\} dy + \int_a^{2a} \left\{ \int_0^{2a-y} xy\,dx \right\} dy \quad \left[\text{changing the order of integration}\right]$$

$$= \int_0^a y \cdot \frac{2y}{2} dy + \int_a^{2a} y \frac{(2a - y)^2}{2} dy = \int_0^a y^2 dy + \frac{1}{2} \int_a^{2a} (4a^2 y - 4ay^2 + y^3) dy$$

$$= \frac{a^3}{3} + \frac{1}{2} \left[2a^2 \cdot 3a^2 - \frac{4a}{3} \cdot 7a^3 + \frac{15}{4} a^4 \right] = \frac{a^3}{3} + 3a^4 - \frac{14}{3} a^4 + \frac{15}{8} a^4$$

$$= \frac{a^3}{3} + \frac{72 - 112 + 45}{24} a^4 = \frac{a^3}{3} + \frac{5}{24} a^4 = \frac{a^3}{24}(5a + 8).$$

(ii) Region $\left[1, \sqrt{4-y};\ 0, 2\right]$ $\quad x = 1,\ x = \sqrt{4-y} \Rightarrow x^2 = -(y-4)$

$y = 0,\ y = 2$

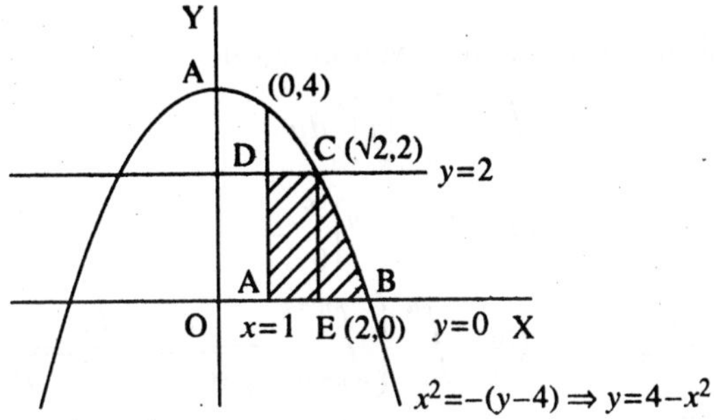

Fig. 13.12

For $y = 2, x = \sqrt{2}$.

Given region $= ABCDA$ consists of two parts $AECDA$ and $EBCE$.

$$AECDA = \left[1, \sqrt{2};\ 0, 2\right],\quad EBCE = \left[\sqrt{2}, 2;\ 0, 4 - x^2\right].$$

The strips are taken $\|$ to Y-axis.

Then, $\displaystyle\int_0^2 \left[\int_{x=1}^{\sqrt{4-y}} (x+y)dx\right] dy$

$$= \int_1^{\sqrt{2}} \left[\int_{y=0}^{2} (x+y)dy\right] dx + \int_{\sqrt{2}}^{2} \left[\int_{y=0}^{4-x^2} (x+y)dy\right] dx$$

$$= \int_1^{\sqrt{2}} \left(xy + \frac{y^2}{2}\right)_0^2 dx + \int_{\sqrt{2}}^{2} \left(xy + \frac{y^2}{2}\right)_0^{4-x^2} dx$$

$$= \int_1^{\sqrt{2}} (2x+2)dx + \int_{\sqrt{2}}^{2} \left[x(4-x^2) + \frac{1}{2}(4-x^2)^2\right] dx$$

$$= \left[x^2 + 2x\right]_1^{\sqrt{2}} + \int_{\sqrt{2}}^{2} \left[4x - x^3 + 8 - 4x^2 + \frac{x^4}{2}\right] dx$$

$$= \left[2 + 2\sqrt{2} - 3\right] + \left[2x^2 - \frac{x^4}{4} + 8x - \frac{4x^3}{3} + \frac{x^5}{10}\right]_{\sqrt{2}}^{2}$$

$$= \left(2\sqrt{2} - 1\right) + \left[\left(8 - 4 + 16 - \frac{32}{3} + \frac{32}{10}\right) - \left(4 - 1 + 8\sqrt{2} - \frac{8\sqrt{2}}{3} + \frac{4\sqrt{2}}{10}\right)\right]$$

$$= \left(2\sqrt{2} - 1\right) + \left[20 - \frac{32}{3} + \frac{16}{5}\right] - \left[3 + \left(8 - \frac{8}{3} + \frac{2}{5}\right)\sqrt{2}\right]$$

$$= \left(2\sqrt{2} - 1\right) + \left[\frac{300 - 160 + 48}{15}\right] - \left[3 + \frac{120 - 40 + 6}{15}\sqrt{2}\right]$$

$$= \left(2\sqrt{2} - 1\right) + \frac{188}{15} - \left[\frac{45 + 86\sqrt{2}}{15}\right] = 2\sqrt{2} - 1 + \frac{143}{15} - \frac{86\sqrt{2}}{15}$$

$$= \frac{128}{15} - \frac{56\sqrt{2}}{15} = \frac{8}{15}\left(16 - 7\sqrt{2}\right).$$

(iii) Region $[0, \infty;\ x, \infty]$. Here strips are vertical [i.e., $\parallel$ to Y-axis]

Now taking the strips $\parallel$ to X-axis we have $x = 0, x = y$ and $y = 0$ to ∞.

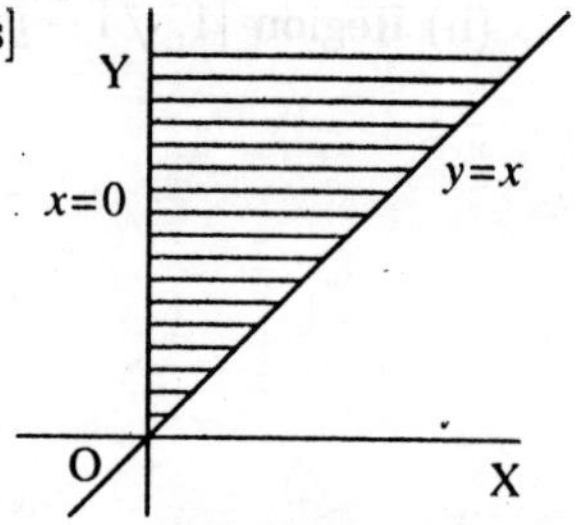

Fig. 13.13

Changing the order of integration, the given integral

$$= \int_0^\infty \left[\int_{x=0}^{y} \frac{e^{-y}}{y} dx\right] dy = \int_0^\infty \frac{e^{-y}}{y}\left[x\right]_0^y dy$$

$$= \int_0^\infty \frac{e^{-y}}{y} y\, dy = \int_0^\infty e^{-y} dy = \left[-e^{-y}\right]_0^\infty = 1.$$

(iv) $\displaystyle\int_0^a \left\{\int_{y=0}^{y=x} \frac{\phi'(y)dy}{\sqrt{(a-x)(x-y)}}\right\} dx$

$$= \int_{y=0}^{a} \left\{\int_{x=y}^{a} \frac{\phi'(y)dx}{\sqrt{(a-x)(x-y)}}\right\} dy$$

[taking the strips $\parallel$ to X-axis for the change of order of integration.]

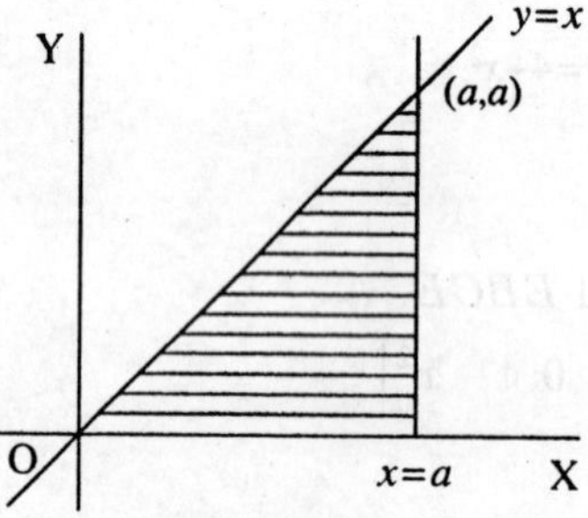

Fig. 13.14

Put

$$x = a\sin^2\theta + y\cos^2\theta \tag{1}$$

$$dx = (a\sin 2\theta - y\sin 2\theta)d\theta.$$

Now for $x = y, x(1-\cos^2\theta) = a\sin^2\theta \Rightarrow \sin^2\theta(x-a) = 0 \Rightarrow \sin\theta = 0 \Rightarrow \theta = 0.$

When $x = a$, $(a-y)\cos^2\theta = 0 \Rightarrow \theta = \pi/2$. Thus $\theta : 0 \to \pi/2$.

Therefore, given integral

$$= \int_0^a \left\{\int_y^a \frac{\phi'(y)dx}{\sqrt{(a-x)(x-y)}}\right\} dy$$

$$= \int_0^a \left[\phi'(y)\int_0^{\pi/2} \frac{2(a-y)(\sin\theta\cos\theta d\theta}{(a-y)\sin\theta\cos\theta}\right] dy \quad \left[\begin{array}{l}\because (a-x) = (a-y)\cos^2\theta \text{ and} \\ (x-y) = (a-y)\sin^2\theta\end{array}\right]$$

$$= \int_0^a \left[\phi'(y)2\int_0^{\pi/2} d\theta\right] dy = \int_0^a \phi'(y)\cdot\pi dy = \pi\left[\phi(y)\right]_0^a = \pi[\phi(a) - \phi(0)].$$

Problem [13.3]: (1) Evaluate $\displaystyle\int_0^\infty \frac{\sin rx}{x} dx$ from $\displaystyle\int_0^\infty\int_0^\infty e^{-xy}\sin rx\, dx\, dy$ with the help of the change of order of integration.

(2) Show that the value of $\displaystyle\iint \sqrt{4y - x^2}dx\, dy$ over the interior of the circle $x^2 + y^2 = 2y$ is $\pi + 8/3$.

(3) Change the order of integration in the integral $\displaystyle\int_0^1 \left\{\int_0^{\sqrt{1-x^2}} \frac{dy}{(1+e^y)\sqrt{1-x^2-y^2}}\right\} dx$ and hence evaluate.

(4) Evaluate $\displaystyle\iint_E (x^2+y^2)dx\, dy$ over the domain bounded by $xy = 1, y = 0, y = x, x = 2$.

(5) Evaluate $\displaystyle\iint_E (x+y+a)dx\, dy$ taken over $E = \{(x,y) : x^2 + y^2 \le a^2\}$.

Solution. (1) Let

$$I = \int_0^\infty \int_0^\infty e^{-xy} \sin rx \, dx \, dy = \int_0^\infty \sin rx \left\{ \int_0^\infty e^{-xy} dy \right\} dx$$

$$= \int_0^\infty \sin rx \left[\frac{e^{-xy}}{-x} \right]_0^\infty dx = \int_0^\infty \frac{\sin rx}{x} dx. \quad (1)$$

Again

$$I = \int_0^\infty \left\{ \int_0^\infty e^{-xy} \sin rx \, dx \right\} dy \quad \left[\begin{array}{l}\text{changing order of} \\ \text{integration}\end{array}\right]$$

$$= \int_0^\infty \left[-e^{-xy} \frac{(r \cos rx + y \sin rx)}{r^2 + y^2} \right]_0^\infty dy$$

$$= \int_0^\infty \frac{r}{r^2 + y^2} dy \quad \left[\begin{array}{l}\text{since } \cos rx, \sin rx \text{ are} \\ \text{bounded } \forall\ x \geq 0\end{array}\right]$$

$$= \left[\tan^{-1} \frac{y}{r} \right]_0^\infty = \frac{\pi}{2} (r > 0). \quad (2)$$

Thus from (1) and (2), $\displaystyle\int_0^\infty \frac{\sin rx}{x} dx = \frac{\pi}{2}.$

(2)

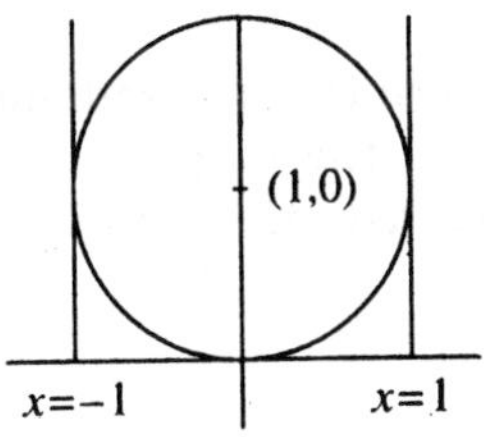

Fig. 13.15

$$x^2 + y^2 = 2y$$
$$\Rightarrow x^2 + (y-1)^2 = 1.$$

Also $x = \pm\sqrt{2y - y^2}$

$$E = \left[-\sqrt{2y - y^2}, \sqrt{2y - y^2}; 0, 2 \right]$$

$$\iint_E \sqrt{4y - x^2} dx \, dy = \int_0^2 \left[\int_{-\sqrt{2y-y^2}}^{\sqrt{2y-y^2}} \sqrt{4y - x^2} dx \right] dy$$

$$= \int_0^2 \left[\frac{x}{2} \sqrt{4y - x^2} + \frac{4y}{2} \sin^{-1} \frac{x}{2\sqrt{y}} \right]_{-\sqrt{2y-y^2}}^{\sqrt{2y-y^2}} dy$$

$$= \int_0^2 2 \left[\frac{\sqrt{2y - y^2}}{2} \sqrt{4y - 2y + y^2} + \frac{4y}{2} \sin^{-1} \frac{\sqrt{2y - y^2}}{2\sqrt{y}} \right] dy$$

$$= \int_0^2 \left[\sqrt{(2y - y^2)(2y + y^2)} + 4y \sin^{-1} \frac{\sqrt{2 - y}}{2} \right] dy$$

$$= \int_0^2 y\sqrt{4 - y^2} dy + 4 \int_0^2 y \sin^{-1} \frac{\sqrt{2 - y}}{2} dy. \quad \text{(i)}$$

1st Part. Put $4 - y^2 = t^2 \Rightarrow y \, dy = -t \, dt$

$$-\int_2^0 t^2 dt = \int_0^2 t^2 dt = \left[\frac{t^3}{3} \right]_0^2 = \frac{8}{3} \quad \text{(ii)}$$

2nd Part.

$$4\int_0^2 y\sin^{-1}\frac{\sqrt{2-y}}{2}dy = \left[4\cdot\frac{y^2}{2}\sin^{-1}\frac{\sqrt{2-y}}{2}\right]_0^2 - 4\int_0^2 \frac{y^2}{2}\frac{1}{\sqrt{1-\frac{2-y}{4}}}\times\frac{(-1)}{4\sqrt{2-y}}dy$$

$$= 0 + 2\int_0^2 \frac{2y^2}{\sqrt{2+y}}\frac{1}{4\sqrt{2-y}}dy = \int_0^2 \frac{y^2}{\sqrt{4-y^2}}dy$$

$$= -\int_0^2 \frac{4-y^2-4}{\sqrt{4-y^2}}dy = -\int_0^2 \sqrt{4-y^2}\,dy + 4\int_0^2 \frac{1}{\sqrt{4-y^2}}dy$$

$$= -\left[\frac{y\sqrt{4-y^2}}{2} + \frac{4}{2}\sin^{-1}\frac{y}{2}\right]_0^2 + 4\left[\sin^{-1}\frac{y}{2}\right]_0^2$$

$$= -2\cdot\frac{\pi}{2} + 4\cdot\frac{\pi}{2} = -\pi + 2\pi = \pi. \qquad \text{(iii)}$$

Therefore, from (i) by (ii) + (iii), given integral $= \frac{8}{3} + \pi$.

(3)

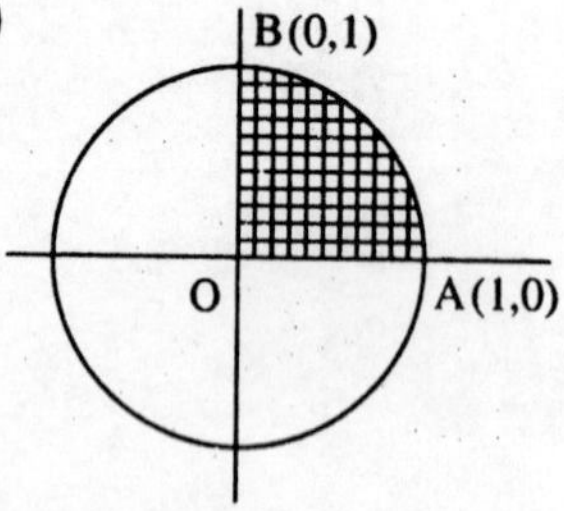

Fig. 13.16

$$y = \sqrt{1-x^2} \Rightarrow x^2 + y^2 = 1.$$

$$E = \left[0, 1;\ 0, \sqrt{1-x^2}\right],$$

region of the circle $x^2 + y^2 = 1$ in the 1st quadrant. Here strips are $\parallel$ to Y-axis. Taking the strips $\parallel$ to X-axis, change of order of integration takes the form

$$\int_0^1 \left\{\int_0^{\sqrt{1-y^2}} \frac{1}{(1+e^y)\sqrt{1-y^2-x^2}}dx\right\}dy$$

$$= \int_0^1 \frac{1}{1+e^y}\left[\sin^{-1}\frac{x}{\sqrt{1-y^2}}\right]_0^{\sqrt{1-y^2}} = \int_0^1 \frac{1}{1+e^y}\frac{\pi}{2}dy$$

$$= \frac{\pi}{2}\int_0^1 \frac{1+e^y-e^y}{1+e^y}dy = \frac{\pi}{2}\left[\int_0^1 dy - \int_0^1 \frac{e^y}{1+e^y}dy\right]$$

$$= \frac{\pi}{2}\left[1 - (\log(1+e^y))_0^1\right] = \frac{\pi}{2}\left[\log e - \log\frac{(1+e)}{2}\right] = \frac{\pi}{2}\log\frac{2e}{1+e}.$$

(4)

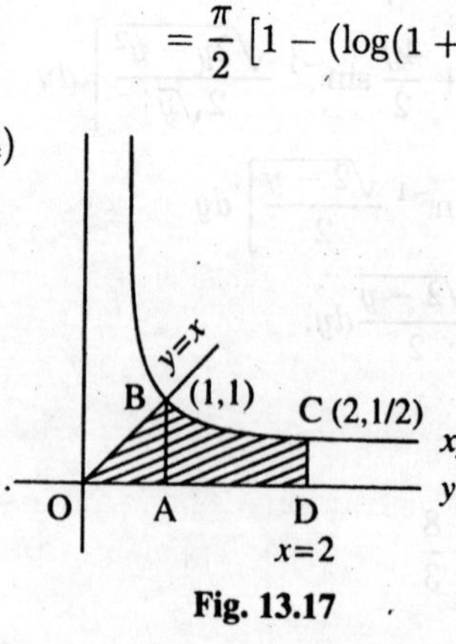

Fig. 13.17

Region of integration is

$$E = ODCBO,\ E_1 = OABO,\ E_2 = ADCBA.$$

Then $E = E_1 \cup E_2$, where

$$E_1 = \{0, 1; 0, x\} \text{ and } E_2 = \left\{1, 2; 0, \frac{1}{x}\right\}.$$

$$\iint_E (x^2+y^2)dydx = \iint_{E_1}(x^2+y^2)dy\,dx + \int_{E_2}(x^2+y^2)dydx$$

$$= \int_0^1 \left\{\int_0^x (x^2+y^2)dy\right\}dx + \int_1^2 \left\{\int_0^{1/x}(x^2+y^2)dy\right\}dx$$

$$= \int_0^1 \left(x^2y+\frac{y^3}{3}\right)_0^x dx + \int_1^2 \left(x^2y+\frac{y^3}{3}\right)_0^{1/x} dx$$

$$= \int_0^1 \frac{4}{3}x^3 dx + \int_1^2 \left(x+\frac{1}{3x^3}\right)dx$$

$$= \frac{1}{3} + \left[\frac{x^2}{2}-\frac{1}{6x^2}\right]_1^2 = \frac{1}{3}+\left[2-\frac{1}{24}-\frac{1}{2}+\frac{1}{6}\right]$$

$$= \frac{1}{3}+\left[\frac{48-1-12+4}{24}\right] = \frac{1}{3}+\frac{39}{24} = \frac{1}{3}+\frac{13}{8} = \frac{8+39}{24} = \frac{47}{24}.$$

(5)

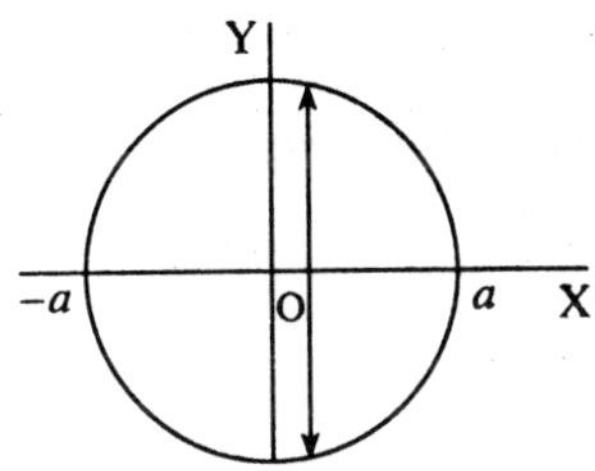

Fig. 13.18

Here

$$E = \{(x,y), x^2+y^2 \le a^2\}$$
$$y = \pm\sqrt{a^2-x^2}$$
$$E = \left\{-a, a; -\sqrt{a^2-x^2},\ \sqrt{a^2-x^2}\right\}.$$

$$\iint_E (x+y+a)dxdy = \int_{-a}^{a}\left\{\int_{y=-\sqrt{a^2-x^2}}^{\sqrt{a^2-x^2}}(x+y+a)dy\right\}dx = \int_{-a}^{a}\left[xy+\frac{y^2}{2}+ay\right]_{-\sqrt{a^2-x^2}}^{\sqrt{a^2-x^2}}dx$$

$$= \int_{-a}^{a}\left[x\sqrt{a^2-x^2}+\frac{a^2-x^2}{2}+a\sqrt{a^2-x^2}\right.$$
$$\left. +x\sqrt{a^2-x^2}-\frac{a^2-x^2}{2}+a\sqrt{a^2-x^2}\right]dx$$

$$= \int_{-a}^{a}\left\{2x\sqrt{a^2-x^2}+2a\sqrt{a^2-x^2}\right\}dx$$

$$= \int_{-a}^{a} 2x\sqrt{a^2-x^2}dx + 2a\int_{-a}^{a}\sqrt{a^2-x^2}dx$$

$$= 0+4a\left[\frac{x}{2}\sqrt{a^2-x^2}+\frac{a^2}{2}\sin^{-1}\frac{x}{a}\right]_0^a = 4a^3\cdot\frac{\pi}{4} = \pi a^3.$$

Change of variable in a double integral (or multiple integral)

In case of evaluation of a double integral it is sometimes necessary to make a change of variables.

We shall write, $\iint_E f(x,y)dx\,dy$ be terms of new variables u and v, if the transformation be $x=\phi(u,v)$, $y=\psi(u,v)$. We assume that the transformation be continuous one to one so that the closed region E on xy-plane is mapped on to a region E' in uv-plane. We also assume that $\phi(u,v)$ and $\psi(u,v)$ have continuous first order partial derivatives.

Then $dx\,dy$ will be replaced by $|J(u,v)|du\,dv$, where $J(u,v)$ is the Jacobian of the transformation.

Thus $\iint_E f(x,y)dx\,dy = \iint_{E'} f\{\phi(u,v), \psi(u,v)\}|J(u,v)|du\,dv$, where

$$J(u,v) = \frac{\partial(x,y)}{\partial(u,v)} = \begin{vmatrix} \dfrac{\partial x}{\partial u} & \dfrac{\partial x}{\partial v} \\ \dfrac{\partial y}{\partial u} & \dfrac{\partial y}{\partial v} \end{vmatrix}.$$

Transformation in a multiple integral

If the closed region E^n in the n-dimensional space of $x_1, x_2, \cdots, x_n$ is mapped onto E'^n in the n-dimensional space of $u_1, u_2, \cdots, u_n$, by a continuous one-one transformation whose Jacobian $J = \dfrac{\partial(x_1, x_2, \cdots, x_n)}{\partial(u_1, u_2, \cdots, u_n)}$, then we write the integral

$$\iint_{E^n} f(x_1, x_2, \cdots, x_n)dx_1 dx_2 \cdots dx_n = \iint_{E'^n} f(\phi_1, \phi_2, \cdots, \phi_n)|J|du_1 du_2 \cdots du_n,$$

where $x_1 = \phi_1(u_1, u_2, \cdots, u_n), x_2 = \phi_2(u_1, u_2, \cdots, u_n), \cdots, x_n = \phi_n(u_1, u_2, \cdots, u_n)$.

Some particular cases

(i) Let $x = r\cos\theta, y = r\sin\theta$, in two dimensional space.

Then $J = \dfrac{\partial(x,y)}{\partial(r,\theta)} = \begin{vmatrix} \dfrac{\partial x}{\partial r} & \dfrac{\partial y}{\partial r} \\ \dfrac{\partial x}{\partial \theta} & \dfrac{\partial y}{\partial \theta} \end{vmatrix} = \begin{vmatrix} \cos\theta & \sin\theta \\ -r\sin\theta & r\cos\theta \end{vmatrix} = r.$

Therefore, $\iint_E f(x,y)dx\,dy = \iint_{E'} f(r\cos\theta, r\sin\theta)|J|dr\,d\theta = \iint_{E'} f(r\cos\theta, r\sin\theta)r\,dr\,d\theta.$

(ii) Let $x = r\cos\phi\sin\theta,\ y = r\sin\phi\sin\theta,\ z = r\cos\theta$ in three dimensional space. Then

$$J = \frac{\partial(x,y,z)}{\partial(r,\theta,\phi)} = \begin{vmatrix} \dfrac{\partial x}{\partial r} & \dfrac{\partial y}{\partial r} & \dfrac{\partial z}{\partial r} \\ \dfrac{\partial x}{\partial \theta} & \dfrac{\partial y}{\partial \theta} & \dfrac{\partial z}{\partial \theta} \\ \dfrac{\partial x}{\partial \phi} & \dfrac{\partial y}{\partial \phi} & \dfrac{\partial z}{\partial \phi} \end{vmatrix} = \begin{vmatrix} \cos\phi\sin\theta & \sin\phi\sin\theta & \cos\theta \\ r\cos\phi\cos\theta & r\sin\phi\cos\theta & -r\sin\theta \\ -r\sin\phi\sin\theta & r\cos\phi\sin\theta & 0 \end{vmatrix} = r^2\sin\theta$$

Therefore,

$$\iiint_E f(x,y,z)dx\,dy\,dz = \iiint_{E'} f(r\cos\phi\sin\theta,\ r\sin\phi\sin\theta,\ r\cos\theta),\ |J|dr\,d\theta\,d\phi$$

$$= \iiint_{E'} f(r\cos\phi\sin\theta,\ r\sin\phi\sin\theta,\ r\cos\theta) \times r^2\sin\theta dr\,d\theta\,d\phi.$$

Example. Evaluate $\iint_E \left(1 - \dfrac{x^2}{a^2} - \dfrac{y^2}{b^2}\right)dx\,dy$, where E is the region in the (+)ve quadrant of the ellipse $\dfrac{x^2}{a^2} + \dfrac{y^2}{b^2} = 1$.

Changing ellipse to a circle, put $x = au$ and $y = bv$.

Then $J = \dfrac{\partial(x, y)}{\partial(u, v)} = \begin{vmatrix} \dfrac{\partial x}{\partial u} & \dfrac{\partial y}{\partial u} \\ \dfrac{\partial x}{\partial v} & \dfrac{\partial y}{\partial v} \end{vmatrix} = \begin{vmatrix} a & 0 \\ 0 & b \end{vmatrix} = ab.$

Then

$$\iint_E \left(1 - \frac{x^2}{a^2} - \frac{y^2}{b^2}\right) dx\, dy = \iint_{E':(u^2+v^2=1)} (1 - u^2 - v^2)\, |J|\, du\, dv$$

$$= \iint_{E':(u^2+v^2=1)} (1 - u^2 - v^2) ab\, du\, dv.$$

Again, putting $u = r\cos\theta,\ v = r\sin\theta,\ J = \frac{\partial(u,v)}{\partial(r,\theta)} = r$, we have

$$\iint_E \left(1 - \frac{x^2}{a^2} - \frac{y^2}{b^2}\right) dx\, dy = \int_{r=0}^{1}\int_{\theta=0}^{\pi/2} (1 - r^2) ab\ r\ dr\ d\theta$$

$$= ab\frac{\pi}{2}\int_{r=0}^{1} (r - r^3) dr = ab\frac{\pi}{2}\left[\frac{1}{2} - \frac{1}{4}\right]$$

$$= ab \cdot \frac{\pi}{2} \cdot \frac{1}{4} = ab\frac{\pi}{8}.$$

Example. Show that $\displaystyle\iint_E \frac{\sqrt{a^2b^2 - b^2x^2 - a^2y^2}}{\sqrt{a^2b^2 + b^2x^2 + a^2y^2}} dx\, dy = ab\frac{\pi}{4}\left(\frac{\pi}{2} - 1\right)$, where E is the region in the (+)ve quadrant of the ellipse $\dfrac{x^2}{a^2} + \dfrac{y^2}{b^2} = 1$.

Changing ellipse to circle we put $x = au,\ y = bv$. Then $J = ab$.

$$\iint_E \frac{\sqrt{a^2b^2 - b^2x^2 - a^2y^2}}{\sqrt{a^2b^2 + b^2x^2 + a^2y^2}} dx\, dy = \iint_{E':u^2+v^2=1} \frac{\sqrt{1 - u^2 - v^2}}{\sqrt{1 + u^2 + v^2}} ab\, du\, dv.$$

Now putting $u = r\cos\theta$ and $v = r\sin\theta$, we have

$$\iint_E \frac{\sqrt{a^2b^2 - b^2x^2 - a^2y^2}}{\sqrt{a^2b^2 + b^2x^2 + a^2y^2}} dx\, dy$$

$$= ab\int_{r=0}^{1}\int_{\theta=0}^{\pi/2} \sqrt{\frac{1 - r^2}{1 + r^2}}\, r\, dr\, d\theta = \frac{\pi}{2} \cdot ab\int_{r=0}^{1} \sqrt{\frac{1 - r^2}{1 + r^2}}\, r\, dr.$$

Put $r^2 = \cos 2\phi, 2r\, dr = -2\sin 2\phi\, d\phi \Rightarrow r\, dr = -\sin 2\phi\, d\phi, r = 0, \phi = \pi/4, r = 1, \phi = 0$

$$= \frac{\pi}{2}ab\int_0^{\pi/4} \frac{\sin\phi}{\cos\phi} \cdot \sin 2\phi d\phi = \frac{\pi}{2}ab\int_0^{\pi/4} 2\sin^2\phi\, d\phi$$

$$= \frac{\pi}{2}ab\int_0^{\pi/4} (1 - \cos 2\phi) d\phi = \frac{\pi}{2}ab\left[\phi - \frac{1}{2}\sin 2\phi\right]_0^{\pi/4}$$

$$= \frac{\pi}{2}ab\left[\frac{\pi}{4} - \frac{1}{2}\right] = \frac{\pi}{4}ab\left(\frac{\pi}{2} - 1\right).$$

Example. Using the transformation $x+y=u, y=uv$, find the value of the integral

$$\int_0^1 \int_{y=0}^{1-x} e^{\frac{y}{x+y}} \, dy \, dx.$$

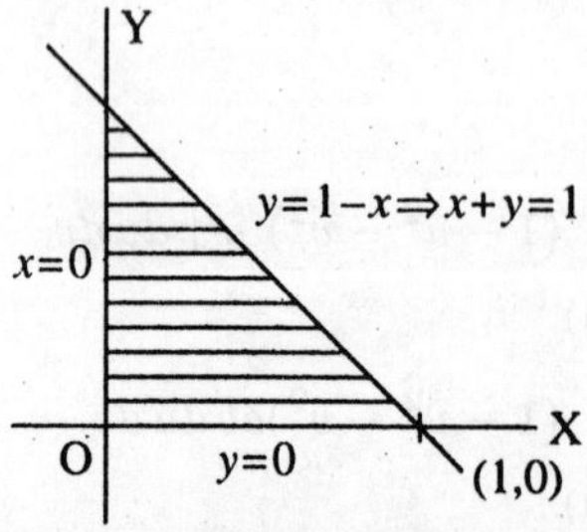

Fig. 13.19(i)

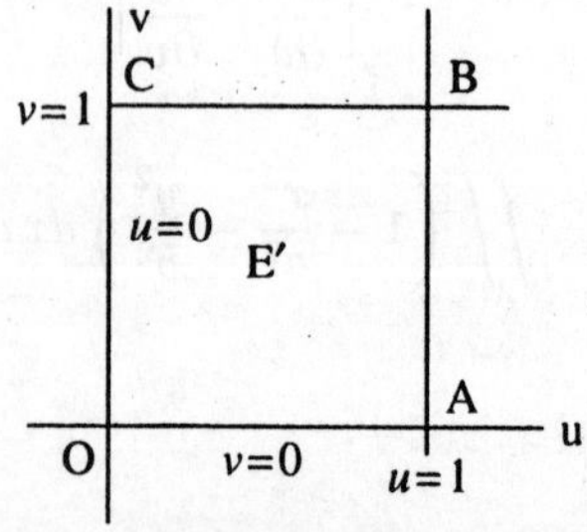

Fig. 13.19(ii)

$x+y=1 \Rightarrow u=1.$

$x=0$, then $y=u \Rightarrow v=1.$

The region $E=[0,1;0,x+y=1]$ is transformed to a rectangular region $E'=\{0,1;0,1\}$ by the given transformation.

Also $x=u(1-v)$ and $y=uv$ gives $J=\dfrac{\partial(x,y)}{\partial(u,v)}=\begin{vmatrix} 1-v & v \\ -u & u \end{vmatrix}=u.$

Therefore,

$$\int_0^1 \left\{\int_0^{1-x} e^{\frac{y}{x+y}} dy\right\} dx = \int_0^1 \int_0^1 e^{\frac{uv}{u}} |J| \, du \, dv = \int_0^1 u\left\{\int_0^1 e^v dv\right\}$$

$$= \int_0^1 u[e-1]du = \left[(e-1)\frac{u^2}{2}\right]_0^1 = \frac{1}{2}(e-1).$$

Example. Show that $\iiint x^2y^2z^2 dx\,dy\,dz$ taken throughout the tetrahedron bounded by the planes $x=0$, $y=0$, $z=0$ and $x+y+z=1$.

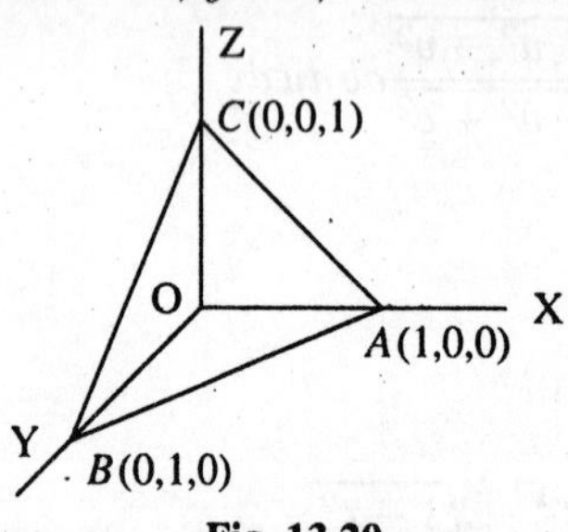

Fig. 13.20

Tetrahedron: $OABCA$ bounded by the planes

(i) BOC $[x=0]$

(ii) AOC $[y=0]$

(iii) AOB $[z=0]$

(iv) ABC $[x+y+z=1]$.

$$\iiint\limits_{OABCA} x^2y^2z^2 dx\,dy\,dz = \int_{x=0}^1 dx \int_{y=0}^{1-x} dy \int_{z=0}^{1-(x+y)} x^2y^2z^2 dz$$

$$= \int_{x=0}^1 dx \int_0^{1-x} x^2y^2 \frac{\{1-(x+y)\}^3}{3} dy$$

$$= \frac{1}{3}\int_0^1 x^2 dx \left[\int_0^{1-x} y^2\{(1-x)-y\}^3 dy\right]$$

[Integrating the integral in [ ], by parts and putting limits]

$$= \frac{1}{3}\int_0^1 x^2 \frac{1}{60}(1-x)^6 dx = \frac{1}{180}\int_0^1 x^2(1-x)^6 dx$$

$$= \frac{1}{180}\left[-\frac{2}{7\times 9\times 8}(1-x)^9\right]_0^1 \text{ [integrating by parts]}$$

$$= \frac{1}{90\times 7\times 8\times 9} = \frac{1}{45360}.$$

Example. Evaluate $\iiint\limits_E \sqrt{a^2b^2c^2 - a^2c^2x^2 - c^2a^2y^2 - a^2b^2z^2}\, dx\, dy\, dz$, where E is the region bounded by the ellipsoid $\frac{x^2}{a^2} + \frac{y^2}{b^2} + \frac{z^2}{c^2} = 1$.

$$\iiint\limits_E \sqrt{a^2b^2c^2 - a^2c^2x^2 - c^2a^2y^2 - a^2b^2z^2}\, dx\, dy\, dz$$

$$= abc \iiint\limits_E \sqrt{1 - \left(\frac{x^2}{a^2} + \frac{y^2}{b^2} + \frac{z^2}{c^2}\right)}\, dx\, dy\, dz.$$

We convert the ellipsoid to a sphere by the transformation $x = au, y = bv, z = cw$.

That is, E is reduced to the sphere $u^2 + v^2 + w^2 = 1$.

The given integral $= abc \iiint\limits_{E'} \sqrt{1 - (u^2 + v^2 + w^2)}\, |J|\, du\, dv\, dw$.

Here $E' = \left[(u, v, w), u^2 + v^2 + w^2 = 1\right]$

$$J = \frac{\partial(x, y, z)}{\partial(u, v, w)} = abc = a^2b^2c^2 \iiint\limits_{E'} \sqrt{1 - (u^2 + v^2 + w^2)}\, du\, dv\, dw$$

Again, putting $u = r\cos\phi\sin\theta,\ v = r\sin\phi\sin\theta,\ w = r\cos\theta$, where $\theta : 0 \to \pi,\ \phi : 0 \to 2\pi,\ r : 0 \to 1$.

$$J = \frac{\partial(u, v, w)}{\partial(r, \theta, \phi)} = r^2\sin\theta$$

$$= a^2b^2c^2 \int_{r=0}^1 \int_{\theta=0}^{\pi} \int_{\phi=0}^{2\pi} \sqrt{1 - r^2}\cdot |J|\, dr\, d\theta\, d\phi$$

$$= a^2b^2c^2 \int_0^1 \int_0^{\pi} \int_0^{2\pi} r^2\sqrt{1 - r^2}\, \sin\theta\, dr\, d\theta\, d\phi$$

$$= 2\pi a^2b^2c^2 \int_0^1 r^2\sqrt{1 - r^2}\Big[-\cos\theta\Big]_0^{\pi} dr$$

$$= 4\pi a^2b^2c^2 \int_0^1 r^2\sqrt{1 - r^2}\, dr.$$

Put $r = \sin\alpha,\ dr = \cos\alpha\, d\alpha$

$$= 4\pi a^2b^2c^2 \int_0^{\pi/2} \sin^2\alpha\, \cos^2\alpha\, d\alpha = 4\pi a^2b^2c^2 \cdot \frac{1}{4}\int_0^{\pi/2} \sin^2 2\alpha\, d\alpha$$

$$= \pi a^2b^2c^2 \int_0^{\pi/2} \left(\frac{1 - \cos 4\alpha}{2}\right) d\alpha = \pi a^2b^2c^2 \cdot \frac{1}{2}\cdot\frac{\pi}{2} = \frac{\pi^2}{4}a^2b^2c^2.$$

Example. Evaluate $\iiint\limits_E \frac{dx\,dy\,dz}{x^2+y^2+(z-2)^2}$, where the region $E = \{(x, y, z),$ $x^2+y^2+z^2=1\}$.

We put $x = r\cos\phi\sin\theta$, $y = r\sin\phi\sin\theta$, $z = r\cos\theta$, $r : 0 \to 1$, $\theta : 0 \to \pi$ and $\phi : 0 \to 2\pi$, $|J| = r^2\sin\theta$.

$$\text{Given integral} = \int_{r=0}^{1}\int_0^{\pi}\int_0^{2\pi} \frac{|J|\,dr\,d\theta\,d\phi}{r^2\sin^2\theta+(r\cos\theta-2)^2} = \int_0^1\int_0^{\pi}\int_0^{2\pi}\frac{r^2\sin\theta\,dr\,d\theta\,d\phi}{r^2-4r\cos\theta+4}$$

$$= 2\pi\int_{r=0}^{1} r^2\left[\int_0^{\pi}\frac{\sin\theta\,d\theta}{r^2-4r\cos\theta+4}\right]dr$$

$$= 2\pi\int_{r=0}^{1} r^2\left[\frac{1}{4r}\log(r^2-4r\cos\theta+4)\right]_0^{\pi}dr$$

$$= \frac{\pi}{2}\int_{r=0}^{1} r^2\left[\frac{1}{r}\log(2+r)^2-\frac{1}{r}\log(2-r)^2\right]dr = \frac{\pi}{2}\int_{r=0}^{1} 2r\log\frac{2+r}{2-r}dr$$

$$= \pi\left[\left(\frac{r^2}{2}\log\frac{2+r}{2-r}\right)_0^1 - \int_0^1\frac{r^2}{2}\cdot\frac{2-r}{2+r}\cdot\frac{1(2-r)+1(2+r)}{(2-r)^2}dr\right]$$

$$= \pi\left[\frac{1}{2}\log 3 - \int_0^1\frac{2r^2}{4-r^2}dr\right] = \pi\left[\frac{1}{2}\log 3 + 2\int_0^1\frac{4-r^2-4}{4-r^2}dr\right]$$

$$= \pi\left[\frac{1}{2}\log 3 + 2 - 8\cdot\frac{1}{4}\left(\log\frac{2+r}{2-r}\right)_0^1\right]$$

$$= \pi\left[\frac{1}{2}\log 3 + 2 - 2\log 3\right] = \pi\left[2-\frac{3}{2}\log 3\right].$$

Example. Show that $\iiint\limits_E z^2 dx\,dy\,dz$, where E is the region of the hemisphere $z \geq 0$, $x^2 + y^2 + z^2 \leq a^2$ is $\frac{2}{15}\pi a^5$.

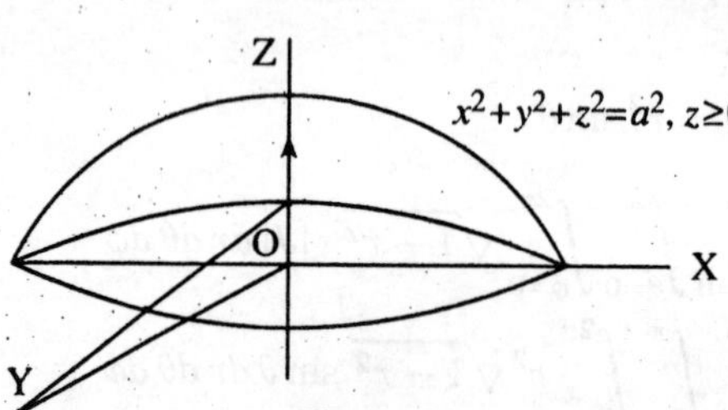

Fig. 13.21

$E = \{z \geq 0, x^2+y^2+z^2 \leq a^2\}$, upper half of the hemisphere $z = \sqrt{a^2-(x^2+y^2)}$.

$$\text{Given integral} = \iint\left\{\int_0^{\sqrt{a^2-(x^2+y^2)}} z^2dz\right\}dx\,dy = \frac{1}{3}\iint\{a^2-(x^2+y^2)\}^{3/2}\,dx\,dy.$$

Put $x = r\cos\theta, y = r\sin\theta$. Then $|J| = r$, $r : 0 \to a$ and $\theta : 0 \to 2\pi$

$$= \frac{1}{3}\int_{r=0}^{a}\int_{\theta=0}^{2\pi}(a^2-r^2)^{3/2}\,r\,dr\,d\theta = \frac{2\pi}{3}\left(-\frac{1}{2}\right)\int_{r=0}^{a}(a^2-r^2)^{3/2} - 2r\,dr$$

$$= \frac{2\pi}{3}\left(-\frac{1}{2}\right)\left[\frac{2}{5}(a^2-r^2)^{5/2}\right]_0^a = -\frac{2\pi}{15}[0-a^5] = \frac{2\pi a^5}{15}.$$

Example. Find the volume cut off from the sphere $x^2+y^2+z^2=a^2$ by the cylinder $x^2+y^2=ax$.

Hints.

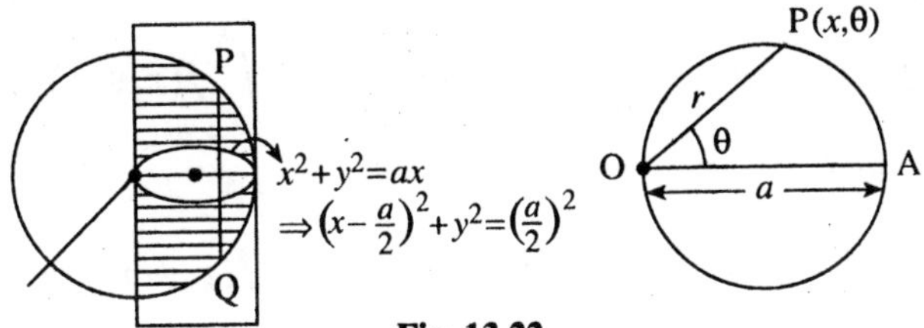

Fig. 13.22

$$V=2\iint_D\left[\int_0^{\sqrt{a^2-x^2-y^2}}dz\right]dx\,dy,\quad D: x^2+y^2\le ax$$

$$=2\cdot\int_0^{\pi}\int_0^{a\cos\theta}\sqrt{a^2-r^2}\cdot r\,dr\,d\theta=\frac{2}{3}a^3\left(\pi-\frac{4}{3}\right)\quad\left[\begin{array}{l}\text{Changing to polar:}\\ 0\le\theta\le\pi;0\le r\le a\cos\theta\end{array}\right]$$

Green's Theorem

Theorem [13.11] [Connection between double integral and line integral] Let a region E bounded by the contour C consisting of curves $x=a, x=b, y=\phi(x), y=\psi(x)$ such that $\phi(x)\le\psi(x)\ \forall\ x\in[a,b]$. Let $f(x,y)$ and $g(x,y)$ be real valued continuous functions defined on E, possessing continuous partial derivatives f_y and g_x on E. Then

$$\iint_E(g_x(x,y)-f_y(x,y))dx\,dy=\int_C(f\,dx+g\,dy),$$

where the line integral is taken in (+)ve sense.

Proof.

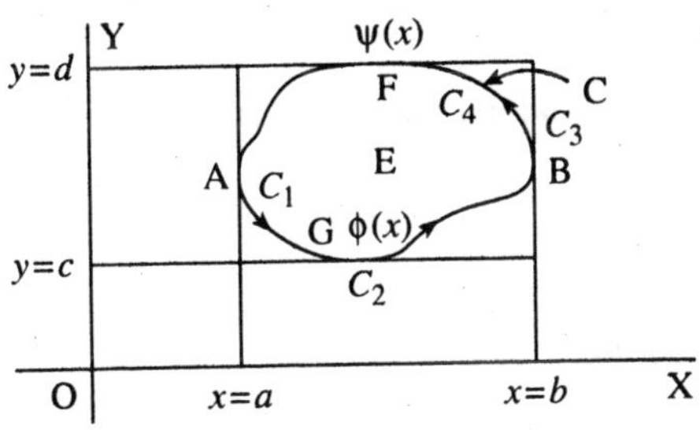

$y=\phi(x)$ represents $\widehat{AGB}$

$y=\psi(x)$ represents $\widehat{BFA}$

Fig. 13.23

Here E is the region bounded by the contour C consisting of curves $x=a$, $x=b$, $y=\phi(x)$, $y=\psi(x)$. Let $C_1: x=a$, $C_2: y=\phi(x)$, $C_3: x=b$, $C_4: y=\psi(x)$ denote the four parts of C taken in the (+)ve sense. E is piecewise regular with respect to Y-axis. Thus

$$\int_C f(x,y)dx=\int_{C_1}f\,dx+\int_{C_2}f\,dx+\int_{C_3}dx+\int_{C_4}f\,dx. \tag{i}$$

Now $C_1: x=a\Rightarrow\displaystyle\int_{C_1}f\,dx=0\quad[\because dx=0]$

$C_2: y=\phi(x),\ x: a\to b$

$$\therefore\int_{C_2}f(x,y)dx=\int_a^b f(x,\phi(x))dx$$

$C_3 : x = b \Rightarrow \int_{C_3} f\,dx = 0 \quad [\because dx = 0]$

$C_4 : y = \psi(x), x : b \to a.$

$$\therefore \int_{C_4} f(x,y)dx = \int_b^a f(x,\psi(x))dx = -\int_a^b f(x,\psi(x))dx.$$

Then from (i), we have

$$\int_C f(x,y)dx = \int_a^b f(x,\phi(x))dx - \int_a^b f(x,\psi(x))dx. \tag{ii}$$

Again

$$\iint_E f_y(x,y)dx\,dy = \int_a^b \left\{ \int_{y=\phi(x)}^{\psi(x)} f_y(x,y)dy \right\} dx = \int_a^b \Big[f(x,y)\Big]_{\phi(x)}^{\psi(x)} dx$$

$$= \int_a^b f(x,\psi(x))dx - \int_a^b f(x,\phi(x))dx = -\int_C f(x,y)dx \text{ [by (ii)]}. \tag{iii}$$

Now we consider the region E which is bounded by the contour C consisting of curves $C_1' : y = c$, $C_2' : x = \alpha(y)$, $C_3' : y = d$, $c_4' : x = \beta(y)$. $x = \alpha(y)$ represents the arc FAG and $x = \beta(y)$ represents the arc GBF, $\alpha(y) \le x \le \beta(y)\ \forall\ y \in [c,d]$. E is piecewise regular with respect to X-axis.

Then

$$\int_C g(x,y)dy = \int_{C_1'} g\,dy + \int_{C_2'} g\,dy + \int_{C_3'} g\,dy + \int_{C_4'} g\,dy. \tag{iv}$$

Here

$C_1' : y = c \Rightarrow \int_{C_1'} g\,dy = 0 \quad [\because dy = 0]$

$C_2' : x = \alpha(y) \Rightarrow \int_d^c g(\alpha(y),y)dy$ [along the arc FAG]

$C_3' : y = d \Rightarrow \int_{C_3'} g\,dy = 0$

$C_4' : x = \beta(y) \Rightarrow \int_c^d g(\beta(y),y)dy.$

Then

$$\int_c g(x,y)dy = \int_d^c g(\alpha(y),y)dy + \int_c^d g(\beta(y),y)dy \quad \text{[From (iv)]} \tag{v}$$

Again

$$\iint_E g_x(x,y)dy\,dx = \int_c^d \left[\int_{x=\alpha(y)}^{\beta(y)} g_x(x,y)dx \right] dy = \int_c^d \Big[g(x,y)\Big]_{\alpha(y)}^{\beta(y)} dy$$

$$= \int_c^d g(\beta(y),y)dy - \int_c^d g(\alpha(y),y)dy$$

$$= \int_d^c g(\alpha(y),y)dy + \int_c^d g(\beta(y),y)dy = \int_C g(x,y)dy. \tag{vi}$$

So from (v) and (vi), we get

$$\iint_E (g_x(x,y) - f_y(x,y))dx\,dy = \iint_E g_x(x,y)dx\,dy - \iint_E f_y(x,y)dx\,dy$$

$$= \int_C g(x,y)dy + \int_C f(x,y)dx \text{ [by (iii) and (vi)]}$$

$$= \int_C (g\,dy + f\,dx), \quad \text{(vii)}$$

when E is regular with respect to both the axes.

Remarks. (1) The above theorem can be generalised. That is, if the region E bounded by the contour C be divided into a finite number of sub-regions so that f, g satisfy the properties stated in the Theorem [13.11], then the theorem holds in each sub-region.

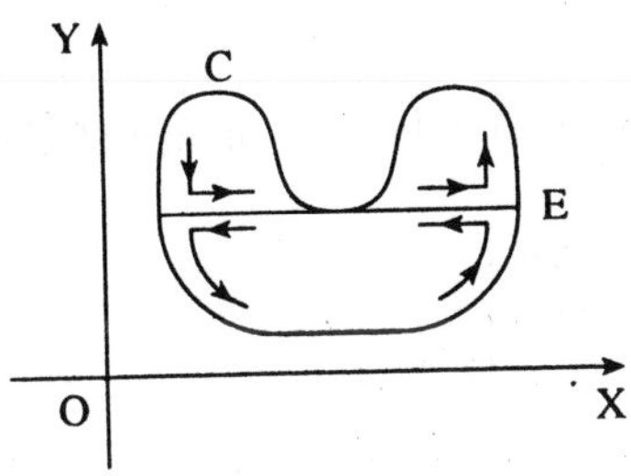

Fig. 13.24(i)

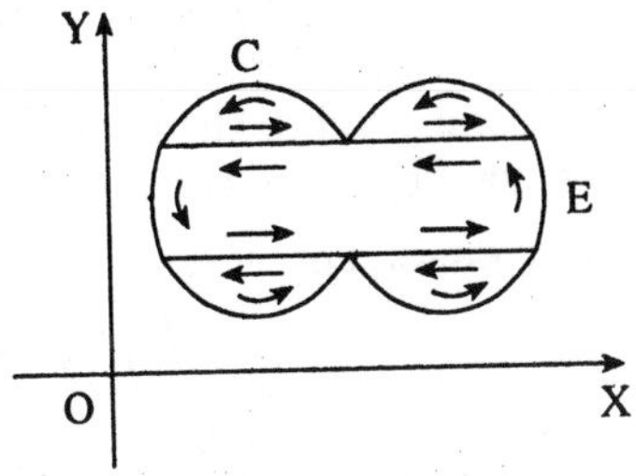

Fig. 13.24(ii)

Totality of the double integrals in the sub-regions gives the double integral over E, while on addition, the parts of the line integrals along the connecting lines cancel one another, since each of these described twice [as shown in Fig (13.24(ii))] in opposite direction. Finally we arrive at the theorem for the whole region E.

(2) Green's theorem in the plane can be applied to a multiple connected region E as shown in the following diagrams.

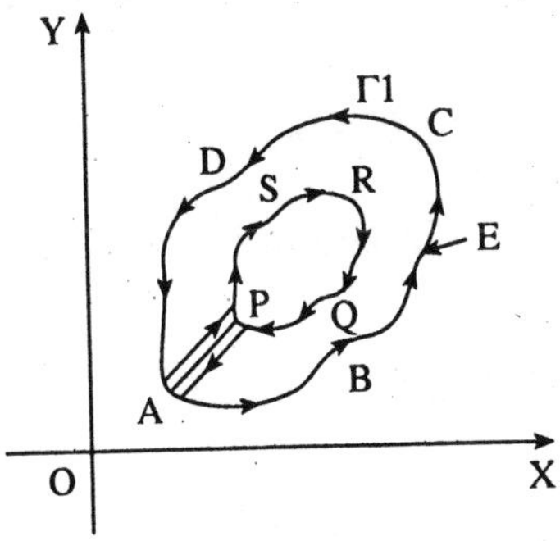

Fig. 13.25(i)

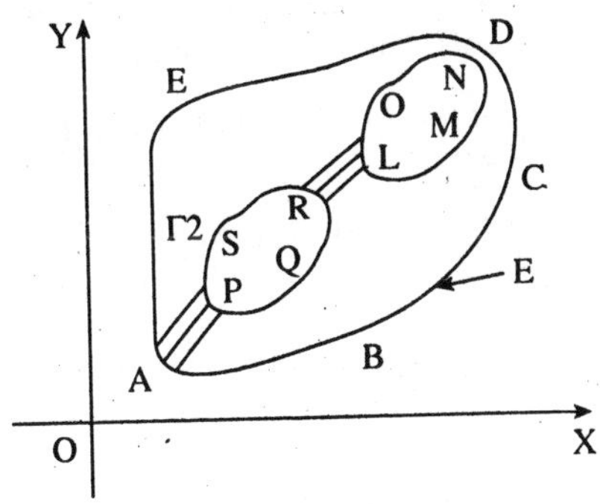

Fig. 13.25(ii)

In Fig (13.25(i)), the boundary of E consists of an exterior boundary $ABCDA\,(\Gamma 1)$ and an interior boundary $PSRQP\,(\Gamma 2)$, the boundary of E is taken in the (+)ve sense. We draw a line AP for cross connection of the exterior boundary to the interior boundary. Thus the region E is bounded by $APSRQPASBCDA\,(\Gamma)$ a simply connected region and so Green's theorem is valid.

Therefore,

$$\int_{APSRQPABCDA} (f\,dx + g\,dy) = \iint_E \left(\frac{\partial g}{\partial x} - \frac{\partial f}{\partial y}\right) dx\,dy.$$

The integral on the left side

$$= \int_{AP} + \int_{PSRQP} + \int_{PA} + \int_{ABCDA} = \int_{PSRQP} + \int_{ABCDA} = \int_{ABCDA} + \int_{PSRQP}$$

$$= \int_{\Gamma 1} + \int_{\Gamma 2} \left[\because \int_{PA} = -\int_{AP}\right]$$

and hecne $\displaystyle\int_\Gamma (f\,dx + g\,dy) = \iint_E \left(\frac{\partial g}{\partial x} - \frac{\partial f}{\partial y}\right) dx\,dy.$

The same can be applied to the region E in Fig 13.25(ii), where boundary of E is connected to two interior regions by cross cut AP and RL.

(3) **Particular case.** Let $f(x, y) = -y$. Then from (iii),

$$\iint_E f_y(x, y)dx\,dy = +\int_C y\,dx,$$

the area of the region E, which can be split up into finite number of domains regular with respect to Y-axis.

Now, if $g(x, y) = x$, then from (vi),

$$\iint_E g_x(x, y)dx\,dy = \int_C x\,dy$$

the area of the region E, which can be split up into finite number of domains regular with respect to X-axis.

Thus the area of the region E, which can be divided into finite number of domains regular with respect to both the axes, is given by

$$\frac{1}{2}\int_C (x\,dy - y\,dx), \text{ since } f_y(x, y) = -1 \text{ and } g_x(x, y) = 1$$

and $\displaystyle\iint_E [g_x(x, y) - f_y(x, y)]dx\,dy = \int_C (f\,dx + g\,dy)$ [from (vii)]

$$\Rightarrow \iint_E (1 + 1)dx\,dy = \int_C (-y\,dx + x\,dy) \Rightarrow \iint_E dx\,dy = \frac{1}{2}\int_C (x\,dy - y\,dx). \quad \text{(viii)}$$

Problem [13.4]: Verify Green's Theorem by evaluating in two ways for the following line integrals:

(1) $\displaystyle\int_C (x^2 y dx + xy^2 dy)$, where C is taken along the closed path formed by $y = x$ and $x^2 = y^3$ in the 1st quadrant.

(2) $\displaystyle\int_C [(1 - x^2)y\,dx + (1 + y^2)x\,dy]$, where C is $x^2 + y^2 = a^2$.

Solution. (i)

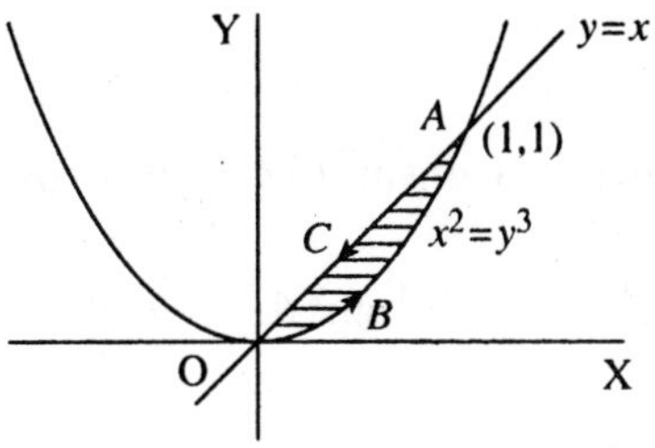

Fig. 13.26

Here contour C is $OBACO$ which encloses the region E, bounded by $y = x$ and $x^2 = y^3$.

$$\int_C (x^2y\,dx + y^2x\,dy) = \int_{\substack{OBA\\x^2=y^3}} (x^2y\,dx + y^2x\,dy) + \int_{\substack{ACO\\y=x}} (x^2y\,dx + y^2x\,dy)\,.$$

For the 1st part, $dx = \frac{3}{2}y^{1/2}dy,\ y : 0 \to 1$

$$\text{1st Part} = \int_0^1 \left[y^4\frac{3}{2}y^{1/2}dy + y^2y^{3/2}dy\right] = \int_0^1 \left[\frac{3}{2}y^{9/2} + y^{7/2}\right] dy$$
$$= \frac{3}{2}\cdot\frac{2}{11} + \frac{2}{9} = \frac{3}{11} + \frac{2}{9} = \frac{49}{99}.$$
$$\text{2nd Part} = \int_1^0 [x^3dx + x^3dx] = \int_1^0 2x^3dx = -\frac{1}{2}.$$

Therefore,

$$\int_C (x^2y\,dx + y^2x\,dy) = \frac{49}{99} - \frac{1}{2} = \frac{98-99}{198} = -\frac{1}{198}. \qquad (1)$$

Again, $\displaystyle\iint_E [g_x(x,y) - f_y(x,y)]dx\,dy$, where $g(x,y) = y^2x,\ f(x,y) = x^2y$

$$= \int_{x=0}^1 \int_{y=x^{2/3}}^{y=x} (y^2 - x^2)dx\,dy = \int_0^1 \left[\int_{y=x^{2/3}}^{y=x} (y^2-x^2)dy\right] dx$$
$$= \int_0^1 \left[\frac{y^3}{3} - x^2y\right]_{x^{2/3}}^{x} dx = \int_0^1 \left[\left(\frac{x^3}{3} - x^3\right) - \left(\frac{x^2}{3} - x^{8/3}\right)\right] dx$$
$$= \int_0^1 \left[-\frac{2x^3}{3} - \frac{x^2}{3} + x^{8/3}\right] dx = -\frac{2}{3}\cdot\frac{1}{4} - \frac{1}{9} + \frac{3}{11}$$
$$= -\frac{1}{6} - \frac{1}{9} + \frac{3}{11} = -\frac{1}{198}. \qquad ($$

From (1) and (2), Greens's Theorem is verified.

(ii)

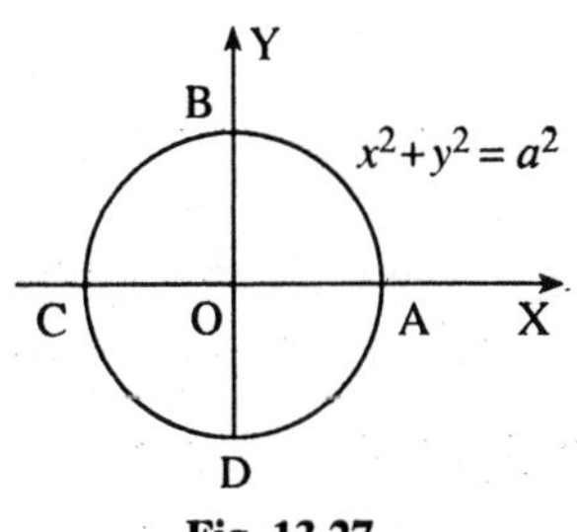

Fig. 13.27

Here contour C : $ABCDA$ representing the circle $x^2 + y^2 = a^2$.
Let $x = a\cos\theta,\ y = a\sin\theta,\ \theta : 0 \to 2\pi$.

$$\int_C (1-x^2)y\,dx + (1+y^2)x\,dy$$

$$= \int_0^{2\pi} [-(1-a^2\cos^2\theta)\,a\sin\theta\cdot a\sin\theta + (1+a^2\sin^2\theta)\,a\cos\theta\cdot a\cos\theta]\,d\theta$$

$$= \int_0^{2\pi} [-a^2\sin^2\theta + a^4\sin^2\theta\cos^2\theta + a^2\cos^2\theta + a^4\sin^2\theta\cos^2\theta]d\theta$$

$$= \int_0^{2\pi} a^2(\cos^2\theta - \sin^2\theta)d\theta + \frac{a^4}{2}\int_0^{2\pi} \sin^2 2\theta d\theta$$

$$= a^2\int_0^{2\pi} \cos 2\theta d\theta + \frac{a^4}{2}\int_0^{2\pi} \frac{1-\cos 4\theta}{2}d\theta = 0 + \frac{a^4}{2}\cdot\frac{2\pi}{2} = \frac{1}{2}\pi a^4. \qquad (1)$$

Here $f(x,y) = (1-x^2)y, g(x,y) = (1+y^2)x, f_y = 1-x^2, g_x = 1+y^2$.

Then $\iint_E (g_x - f_y)dx\,dy$, E is bounded by $c : x^2+y^2 = a^2$.

$$\therefore \iint_{x^2+y^2=a^2} (1+y^2-1+x^2)dx\,dy = \iint_{x^2+y^2=a^2} (y^2+x^2)dxdy.$$

For $x = r\cos\theta, y = r\sin\theta, r : 0 \to a, \theta : 0 \to 2\pi$

$$= \int_{r=0}^{a}\int_{\theta=0}^{2\pi} r^2\cdot r\,dr\,d\theta = 2\pi\left[\frac{r^4}{4}\right]_0^a = \frac{\pi a^4}{2}. \qquad (2)$$

Green's theorem is verified from (1) and (2).

Example. Prove that the line integral $\int_c \frac{x\,dy - y\,dx}{x^2+y^2}$ taken in the (+)ve direction over any closed contour C with the origin inside it, is equal to 2π.

Here the origin O is inside the closed contour C, we consider a circle (of contour Γ) with centre at O and of radius, say a, so that Γ does not intersect C.

Let E be the region bounded by C and Γ.

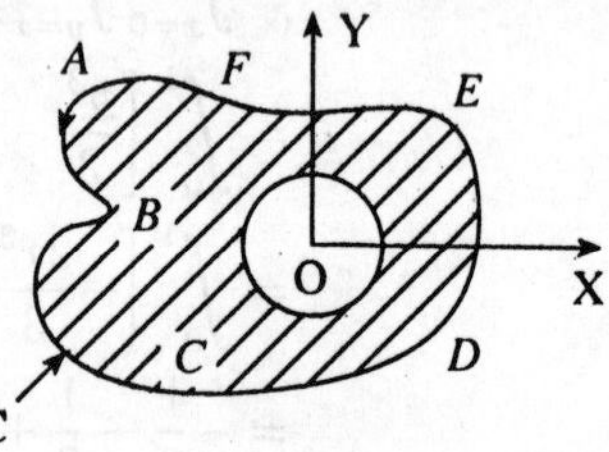

Fig. 13.28

$$f(x,y) = -\frac{y}{x^2+y^2},\; g(x,y) = \frac{x}{x^2+y^2}.$$

$$f_y = -\left[\frac{1\cdot(x^2+y^2) - 2y\cdot y}{(x^2+y^2)^2}\right].$$

$$g_x = \frac{1\cdot(x^2+y^2) - 2x^2}{(x^2+y^2)^2}.$$

By Green's Theorem,

$$\iint_E [g_x - f_y]dx\,dy = \iint_{C-\Gamma}\left[\frac{y^2-x^2}{(x^2+y^2)^2} - \frac{y^2-x^2}{x^2+y^2}\right]dx\,dy = \iint_{C-\Gamma} 0\cdot dx\,dy = 0$$

$$\Rightarrow \int_{C-\Gamma} \frac{x\,dy - y\,dx}{x^2+y^2} = 0 \Rightarrow \int_C \frac{x\,dy-y\,dx}{x^2+y^2} = \int_\Gamma \frac{x\,dy - y\,dx}{x^2+y^2}, \quad \text{where } \Gamma : x^2+y^2 = a^2,\; x = a\cos\theta,\; y = a\sin\theta$$

$$= \int_0^{2\pi} \frac{(a^2\cos^2\theta + a^2\sin^2\theta)}{a^2}d\theta = \int_0^{2\pi} d\theta = 2\pi.$$

Example. Find the area of the loop of Descarte's folium $x^3 + y^3 = 3axy$ with the help of line integral.

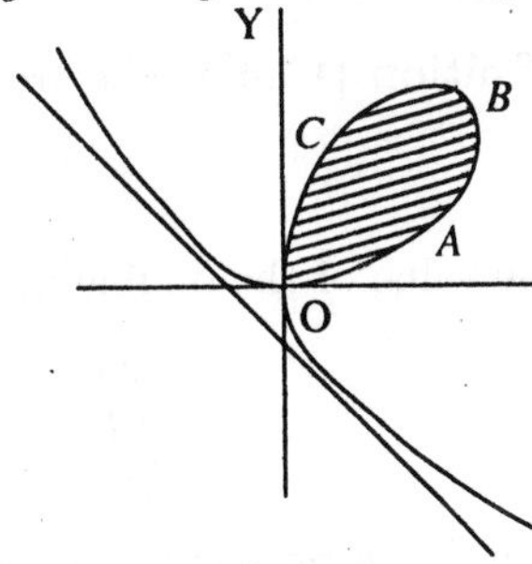

Fig. 13.29

Area of the loop $OABCO$ of the curve $x^3 + y^3 = 3axy$

$$= \frac{1}{2}\int_C (x\,dy - y\,dx) \text{ [by (viii), Remark (3)]}$$

To obtain the parametric equation of the contour C of the folium, we put $y = tx$. Then from $x^3 + y^3 = 3axy$

$$\Rightarrow x^3 + t^3x^3 = 3atx^2 \Rightarrow x(1+t^3) = 3at$$

$$\Rightarrow x = \frac{3at}{1+t^3} \text{ and } y = \frac{3at^2}{1+t^3}.$$

Here $\frac{y}{x} = t \Rightarrow t = \tan\theta$, where θ varies from 0 to $\frac{\pi}{2}$. Therefore, t varies from 0 to ∞.

Hence area of the loop $= \frac{1}{2}\int_C (x\,dy - y\,dx)$

$$= \frac{1}{2}\int_0^\infty \frac{3at}{1+t^3} d\left(\frac{3at^2}{1+t^3}\right) - \frac{3at^2}{1+t^3} d\left(\frac{3at}{1+t^3}\right)$$

$$= \frac{1}{2}\int_0^\infty \left[\frac{3at}{1+t^3}\cdot\frac{6at(1+t^3)-9at^4}{(1+t^3)^2} - \frac{3at^2}{1+t^3}\cdot\frac{3a(1+t^3)-9at^3}{(1+t^3)^2}\right] dt$$

$$= \frac{1}{2}\int_0^\infty \frac{9a^2t^2(2-t^3-1+2t^3)}{(1+t^3)^3} dt = \frac{3a^2}{2}\int_0^\infty \frac{3t^2}{(1+t^3)^2} dt$$

$$= \frac{3a^2}{2}\left[-\frac{1}{1+t^3}\right]_0^\infty = \frac{3a^2}{2}$$

13.5 Surfaces

Definition [13.3] : Let $\vec{f}$ be a vector valued function with $\vec{f} : D(\subseteq \mathbb{R}^2) \to \mathbb{R}^3$, then a surface in $\mathbb{R}^3 =$ Range $\vec{f} \subseteq \mathbb{R}^3$.

Thus if f, g, h be three real valued functions with domain D, and $(u, v) \in D \subseteq \mathbb{R}^2$, where $x = f(u, v)$, $y = g(u, v)$, $z = h(u, v)$, then the set

$$S = \{f(u,v),\ g(u,v), h(u,v);\ (u,v) \in D\} \subseteq \mathbb{R}^3$$

is a surface in $\mathbb{R}^3$.

Thus while a curve in $\mathbb{R}^3$ is a set with one degree of freedom, a surface in $\mathbb{R}^3$ is a set with two degrees of freedom.

That is, if a curve is represented by one parameter, the surface requires two parameters for its representation.

Therefore, if $x = f(t)$, $y = \phi(t)$, then $z = F(x, y) = F(x(t), y(t))$ represents a surface.

A surfaces of the forms $x = f_1(y, z)$, $y = f_2(x, z)$, $z = f_3(x, y)$ are known as regular with X-axis, Y-axis and Z-axis respectively.

S is called a *smooth surface*, if f, g, h possesses continuous partial derivatives at each point

$$(u, v) \in D \text{ and } \frac{\partial(x,y)}{\partial(u,v)}, \frac{\partial(y,z)}{\partial(u,v)}, \frac{\partial(z,x)}{\partial(u,v)}$$

do not all vanish simultaneously at any point.

Surface Area

Definition [13.4] : The area of a surface S given by

$$x = f(u,v), y = g(u,v), z = h(u,v), u, v \in D \subseteq \mathbb{R}^2$$

is the value of the double integral

$$\iint_D \sqrt{\left(\frac{\partial(y,z)}{\partial(u,v)}\right)^2 + \left(\frac{\partial(z,x)}{\partial(u,v)}\right)^2 + \left(\frac{\partial(x,y)}{\partial(u,v)}\right)^2}\, du\, dv,$$

when f, g, h possesses continuous first order partial derivatives in D. (1)

(1) Let the *surface be plane.*

We take $x = u, y = v, z = 0$, where $(u,v) \in D \subseteq \mathbb{R}^2$ [That is, D is the XY-plane].

$$\frac{\partial(x,y)}{\partial(u,v)} = \begin{vmatrix} \frac{\partial x}{\partial u} & \frac{\partial y}{\partial u} \\ \frac{\partial x}{\partial v} & \frac{\partial y}{\partial v} \end{vmatrix} = \begin{vmatrix} 1 & 0 \\ 0 & 1 \end{vmatrix} = 1, \quad \frac{\partial(y,z)}{\partial(u,v)} = 0, \quad \frac{\partial(z,x)}{\partial(u,v)} = 0.$$

Then the surface area of the plane $= \iint_D 1 \cdot du\, dv =$ area of D [by (1)].

In fact, the surface coincides with D.

(2) Let the surface be given by $z = h(x,y)$.

We take it as $x = u, y = v, z = h(u,v)$, where $(u,v) = (x,y) \in D \subset \mathbb{R}^2$.

We have

$$\frac{\partial(y,z)}{\partial(u,v)} = \begin{vmatrix} \frac{\partial y}{\partial u} & \frac{\partial z}{\partial u} \\ \frac{\partial y}{\partial v} & \frac{\partial z}{\partial v} \end{vmatrix} = \begin{vmatrix} 0 & \frac{\partial z}{\partial u} \\ 1 & \frac{\partial z}{\partial v} \end{vmatrix} = -\frac{\partial z}{\partial u} = -\frac{\partial z}{\partial x}.$$

$$\frac{\partial(z,x)}{\partial(u,v)} = \begin{vmatrix} \frac{\partial z}{\partial u} & \frac{\partial x}{\partial u} \\ \frac{\partial z}{\partial v} & \frac{\partial x}{\partial v} \end{vmatrix} = \begin{vmatrix} \frac{\partial z}{\partial u} & 1 \\ \frac{\partial z}{\partial v} & 0 \end{vmatrix} = -\frac{\partial z}{\partial v} = -\frac{\partial z}{\partial y}.$$

$$\frac{\partial(x,y)}{\partial(u,v)} = 1.$$

Then the surface area

$$= \iint_D \sqrt{\left(-\frac{\partial z}{\partial x}\right)^2 + \left(-\frac{\partial z}{\partial y}\right)^2 + 1}\, dx\, dy \quad \text{[from Def. 13.4]}$$

$$= \iint_D \sqrt{1 + \left(\frac{\partial z}{\partial x}\right)^2 + \left(\frac{\partial z}{\partial y}\right)^2}\, dx\, dy. \quad (2)$$

Note. (1) The result from Differential Geometry that $-\frac{\partial z}{\partial x}, -\frac{\partial z}{\partial y}, 1$ are the direction ratios of the normal to the surface at any point (x,y,z) on the surface.

Let δs be the elementary area of the surface lying on the tangent plane at (x, y, z). Then the projection δA of δs on XY-plane is given by

$$\delta A = \frac{\delta s}{\sqrt{1+\left(\frac{\partial z}{\partial x}\right)^2+\left(\frac{\partial z}{\partial y}\right)^2}}.$$

$$\therefore \delta s = \sqrt{1+\left(\frac{\partial z}{\partial x}\right)^2+\left(\frac{\partial z}{\partial y}\right)^2}\cdot \delta A = \sqrt{1+\left(\frac{\partial z}{\partial x}\right)^2+\left(\frac{\partial z}{\partial y}\right)^2}\,\delta x\,\delta y.$$

Then on integrating, $S = \iint\limits_D \sqrt{1+\left(\frac{\partial z}{\partial x}\right)^2+\left(\frac{\partial z}{\partial y}\right)^2}\,dx\,dy$, which justifies the formula (2).

(2) If

$$\left(\frac{\partial f}{\partial u}\right)^2+\left(\frac{\partial g}{\partial u}\right)^2+\left(\frac{\partial h}{\partial u}\right)^2 = A,\quad \frac{\partial f}{\partial u}\cdot\frac{\partial f}{\partial v}+\frac{\partial g}{\partial u}\cdot\frac{\partial g}{\partial v}+\frac{\partial h}{\partial u}\cdot\frac{\partial h}{\partial v} = B$$

and $\left(\frac{\partial f}{\partial v}\right)^2+\left(\frac{\partial g}{\partial v}\right)^2+\left(\frac{\partial h}{\partial v}\right)^2 = C.$

Then it can be easily verified that, $\left[\frac{\partial(y,z)}{\partial(u,v)}\right]^2+\left[\frac{\partial(z,x)}{\partial(u,v)}\right]^2+\left[\frac{\partial(x,y)}{\partial(u,v)}\right]^2 = AC-B^2.$

Hence surface area

$$S = \iint\limits_D \sqrt{AC-B^2}\,du\,dv \quad \text{[from (1)]}. \tag{3}$$

Example. Find the surface area of a sphere $x^2+y^2+z^2=a^2$.

For the given sphere, we have $x = a\sin\theta\cos\phi, y = a\sin\theta\sin\phi, z = a\cos\theta$, where

$$0 \le \theta \le \pi, 0 \le \phi \le 2\pi.$$

Here

$$A = \sum\left(\frac{\partial x}{\partial\theta}\right)^2 = (a\cos\theta\cos\phi)^2+(a\cos\theta\sin\phi)^2+(-a\sin\theta)^2$$
$$= a^2.$$
$$B = \sum\frac{\partial x}{\partial\theta}\cdot\frac{\partial x}{\partial\phi}$$
$$= -(a\cos\theta\cos\phi\cdot a\sin\theta\sin\phi)+(a\cos\theta\sin\phi\cdot a\sin\theta\cos\phi)+(-a\sin\theta\cdot 0)$$
$$= 0.$$
$$C = \sum\left(\frac{\partial x}{\partial\phi}\right)^2 = (-a\sin\theta\sin\phi)^2+(a\sin\theta\cos\phi)^2+0$$
$$= a^2\sin^2\theta.$$

Therefore, the surface area by (3)

$$S = \int_{\theta=0}^{\pi}\int_{\theta=0}^{2\pi}\sqrt{AC-B^2}\,d\theta\,d\phi = \int_{\theta=0}^{\pi}\int_{\theta=0}^{2\pi}\sqrt{a^4\sin^2\theta}d\theta\,d\phi$$
$$= 2\pi a^2\int_0^{\pi}\sin\theta d\theta = 2\pi a^2\big[-\cos\theta\big]_0^{\pi} = 4\pi a^2.$$

Example. Find the area of the part of the surface of the paraboloid $4ax = y^2 + z^2$ enclosed by the cylinder $y^2 = ax$ and the plane $x = 3a$.

$y^2 + z^2 = 4ax$ gives $\dfrac{\partial z}{\partial x} = \dfrac{2a}{z}, \dfrac{\partial z}{\partial y} = -\dfrac{y}{z}$.

Then $1 + \left(\dfrac{\partial z}{\partial x}\right)^2 + \left(\dfrac{\partial z}{\partial y}\right)^2 = 1 + \dfrac{4a^2}{z^2} + \dfrac{y^2}{z^2} = \dfrac{y^2 + z^2 + 4a^2}{z^2} = \dfrac{4ax + 4a^2}{\sqrt{4ax - y^2}}$.

Then the required surface area of the paraboloid bounded by $y^2 = ax$ and the plane $x = 3a$

$$= \int_0^{3a} \int_{-\sqrt{ax}}^{\sqrt{ax}} \sqrt{1 + \left(\frac{\partial z}{\partial x}\right)^2 + \left(\frac{\partial z}{\partial y}\right)^2}\, dx\, dy$$

$$= \int_0^{3a} \int_{-\sqrt{ax}}^{\sqrt{ax}} \sqrt{\frac{4ax + 4a^2}{4ax - y^2}}\, dx\, dy = 2\sqrt{a} \int_0^{3a} \int_{-\sqrt{ax}}^{\sqrt{ax}} \sqrt{\frac{x + a}{4ax - y^2}}\, dxdy$$

$$= 2\sqrt{a} \int_0^{3a} \sqrt{x + a} \left[\sin^{-1} \frac{y}{2\sqrt{ax}}\right]_{-\sqrt{ax}}^{\sqrt{ax}} dx = 2\sqrt{a} \int_0^{3a} \sqrt{x + a} \left[2 \sin^{-1} \frac{1}{2}\right] dx$$

$$= 2\sqrt{a} \cdot \frac{\pi}{3} \int_0^{3a} \sqrt{x + a}\, dx = \frac{2\pi}{3} \sqrt{a} \left[\frac{2}{3} (x + a)^{3/2}\right]_0^{3a}$$

$$= \frac{4}{9} \pi \sqrt{a} \left[(4a)^{3/2} - a^{3/2}\right] = \frac{4\pi}{9} \sqrt{a} \cdot 7a^{3/2} = \frac{28}{9} \pi a^2.$$

Example. Find the surface area of the sphere $x^2 + y^2 + z^2 = 9$ lying inside the cylinder $x^2 + y^2 = 3y$.

$$x^2 + y^2 + z^2 = 9$$

$$2x + 2z \frac{\partial z}{\partial x} = 0 \Rightarrow \frac{\partial z}{\partial x} = -\frac{x}{z}$$

$$2y + 2z \frac{\partial z}{\partial y} = 0 \Rightarrow \frac{\partial z}{\partial y} = -\frac{y}{z}.$$

Then

$$\sqrt{1 + \left(\frac{\partial z}{\partial x}\right)^2 + \left(\frac{\partial z}{\partial y}\right)^2} = \sqrt{1 + \frac{x^2}{z^2} + \frac{y^2}{z^2}} = \frac{\sqrt{x^2 + y^2 + z^2}}{z} = \frac{3}{z} = \frac{3}{\sqrt{9 - (x^2 + y^2)}}.$$

Bounding surface $x^2 + y^2 = 3y$. For $x = r\cos\theta, y = r\sin\theta \Rightarrow r = 3\sin\theta$.

Required surface area

$$= 4 \int_{\theta=0}^{\pi/2} \int_{r=\theta}^{3\sin\theta} \frac{3}{\sqrt{9 - r^2}} \cdot r \cdot dr\, d\theta = 12 \int_{\theta=0}^{\pi/2} \left[\int_0^{3\sin\theta} \frac{r\, dr}{\sqrt{9 - r^2}} dr\right] d\theta$$

$$= 12 \left(-\frac{1}{2}\right) \int_{\theta=0}^{\pi/2} \left[\int_0^{3\sin\theta} \frac{-2r\, dr}{\sqrt{9 - r^2}} dr\right] d\theta = -6 \int_{\theta=0}^{\pi/2} \left[\frac{(9 - r^2)^{-\frac{1}{2}+1}}{-\frac{1}{2} + 1}\right]_0^{3\sin\theta} d\theta$$

$$= -12 \int_{\theta=0}^{\pi/2} \left[(9 - 9\sin^2\theta)^{1/2} - 9^{1/2}\right] d\theta = -12 \int_{\theta=0}^{\pi/2} (3\cos\theta - 3) d\theta$$

$$= -36 \left[\sin\theta - \theta\right]_0^{\pi/2} = -36 \left[1 - \frac{\pi}{2}\right]$$

$$= 18[\pi - 2].$$

Example. Show that the surface area of the torus $x = (r - \cos v)\cos u$, $y = (r - \cos v)\sin u$, $z = \sin v$, where $-\pi \le u, v \le \pi, r > 1$ is $4\pi^2 r$.

$$S = \text{surface area of the torus}$$
$$= \iint_D \sqrt{\left(\frac{\partial(y,z)}{\partial(u,v)}\right)^2 + \left(\frac{\partial(z,x)}{\partial(u,v)}\right)^2 + \left(\frac{\partial(x,y)}{\partial(u,v)}\right)^2}\, du\, dv. \qquad (1)$$

Here $x = (r - \cos v)\cos u$, $y = (r - \cos v)\sin u$, $z = \sin v$

$$\frac{\partial(y,z)}{\partial(u,v)} = \begin{vmatrix} \frac{\partial y}{\partial u} & \frac{\partial z}{\partial u} \\ \frac{\partial y}{\partial v} & \frac{\partial z}{\partial v} \end{vmatrix} = \begin{vmatrix} (r - \cos v)\cos u & 0 \\ \sin v \sin u & \cos v \end{vmatrix} = (r - \cos v)\cos u \cos v,$$

$$\frac{\partial(z,x)}{\partial(u,v)} = (r - \cos v)\sin u \cos v \text{ and } \frac{\partial(x,y)}{\partial(u,v)} = -(r - \cos v)\sin v.$$

Then

$$\sum \left(\frac{\partial(y,z)}{\partial(u,v)}\right)^2 = (r - \cos v)^2[\cos^2 u \cos^2 v + \sin^2 u \cos^2 v + \sin^2 v]$$
$$= (r - \cos v)^2[\cos^2 v + \sin^2 v] = (r - \cos v)^2.$$
$$S = \int_{-\pi}^{\pi}\int_{-\pi}^{\pi} (r - \cos v)du\, dv \quad [\text{from (1)}]$$
$$= 2\pi \int_{-\pi}^{\pi} (r - \cos v)dv = 2\pi \Big[rv - \sin v\Big]_{-\pi}^{\pi}$$
$$= 2\pi \cdot 2\pi r = 4\pi^2 r.$$

Example. If a surface is given by the equations $x = c\sin u$, $y = c\cos v$, $z = c(\cos u + \cos v)$, prove that its area bounded by $u = 0, u = \frac{\pi}{2}, v = 0, v = \frac{\pi}{2}$ is $\frac{\pi c^2}{4}\left[1 - \sum_{n=1}^{\infty} \frac{P_{2n}^2}{2n-1}\right]$, where $P_{2n} = \frac{(2n-1)(2n-3)\cdots 5\cdot 3\cdot 1}{2n(2n-2)\cdots 4\cdot 2}$.

$$\frac{\partial(y,z)}{\partial(u,v)} = \begin{vmatrix} \frac{\partial y}{\partial u} & \frac{\partial z}{\partial u} \\ \frac{\partial y}{\partial v} & \frac{\partial z}{\partial v} \end{vmatrix} = \begin{vmatrix} 0 & -c\sin u \\ -c\sin v & -c\sin v \end{vmatrix} = -c^2 \sin u \sin v$$

$$\frac{\partial(z,x)}{\partial(u,v)} = \begin{vmatrix} \frac{\partial z}{\partial u} & \frac{\partial x}{\partial u} \\ \frac{\partial z}{\partial v} & \frac{\partial x}{\partial v} \end{vmatrix} = \begin{vmatrix} -c\sin u & c\cos u \\ -c\sin v & 0 \end{vmatrix} = c^2 \sin v \cos u$$

$$\frac{\partial(x,y)}{\partial(u,v)} = \begin{vmatrix} c\cos u & 0 \\ 0 & -c\sin v \end{vmatrix} = -c^2 \cos u \sin v$$

$$\sum \left(\frac{\partial(y,z)}{\partial(u,v)}\right)^2 = c^4[\sin^2 u \sin^2 v + \sin^2 v \cos^2 u + \cos^2 u \sin^2 v]$$
$$= c^4[\sin^2 v(1 + \cos^2 u)].$$

Required surface area

$$S=\int_{u=0}^{\pi/2}\int_{v=0}^{\pi/2}\sqrt{\left(\sum\frac{\partial(y,z)}{\partial(u,v)}\right)^2}\,du\,dv=c^2\int_0^{\pi/2}\int_0^{\pi/2}\sin v(1+\cos^2 u)^{1/2}dv\,du$$

$$=c^2\int_0^{\pi/2}\left\{\int_0^{\pi/2}\sin v\,dv\right\}(1+\cos^2 u)^{1/2}du$$

$$=c^2\int_0^{\pi/2}(1+\cos^2 u)^{1/2}du\qquad\left[\because\int_0^{\pi/2}\sin v\,dv=1\right]$$

$$=c^2\int_0^{\pi/2}\left[1+\frac{1}{2}\cos^2 u+\frac{\frac{1}{2}\left(\frac{1}{2}-1\right)}{2!}(\cos^2 u)^2+\frac{\frac{1}{2}\left(\frac{1}{2}-1\right)\left(\frac{1}{2}-2\right)}{3!}(\cos^2 u)^3+\cdots\right.$$
$$\left.+\frac{\frac{1}{2}\left(\frac{1}{2}-1\right)\left(\frac{1}{2}-2\right)\cdots\left(\frac{1}{2}-n+1\right)}{n!}(\cos^2 u)^n+\cdots\right]du$$

$$=c^2\int_0^{\pi/2}\left[1+\frac{1}{2}\cdot\cos^2 u-\frac{1}{2^2}\cdot\frac{1}{2!}(\cos^2 u)^2+\frac{1\cdot 3}{2^3\cdot 3!}(\cos^2 u)^3-\cdots\right.$$
$$\left.+(-1)^{n-1}\cdot\frac{1\cdot 3\cdot 5\cdots(2n-3)}{2^n\cdot n!}(\cos^2 u)^n+\cdots\right]du$$

$$=c^2\int_0^{\pi/2}\left[1+\frac{1}{2}\cos^2 u-\frac{1}{2^2}\cdot\frac{1}{2!}(\cos^2 u)^2+\frac{1\cdot 3}{2^3\cdot 3!}(\cos^2 u)^3+\cdots\right.$$
$$\left.+\frac{(-1)^{n-1}}{(2n-1)}\frac{(2n-1)(2n-3)\cdots 3\cdot 1}{2^n n(n-1)\cdots 2\cdot 1}(\cos^2 u)^n+\cdots\right]du$$

$$=c^2\int_0^{\pi/2}\left[1+\sum_1^{\infty}\frac{(-1)^{n-1}}{2n-1}\cdot\frac{(2n-1)(2n-3)\cdots 3\cdot 1}{2n\cdot(2n-2)\cdots 4\cdot 2}\cos^{2n}u\right]du$$

$$\left[\begin{array}{l}\because 2^n\cdot n(n-1)\cdots 2\cdot 1=(2n)\cdot 2(n-1)\cdots 2^2\cdot 2\\ \quad=2n(2n-2)\cdots 4\cdot 2\end{array}\right]$$

$$=c^2\left[\frac{\pi}{2}+\sum_1^{\infty}\frac{(-1)^{n-1}}{2n-1}\cdot\frac{(2n-1)(2n-3)\cdots 3\cdot 1}{2n\cdot(2n-2)\cdots 4\cdot 2}\cdot\frac{(2n-1)}{2n}\cdot\frac{(2n-3)}{2n-2}\cdots\frac{3}{4}\cdot\frac{1}{2}\cdot\frac{\pi}{2}\right]$$

$$\left[\because\int_0^{\pi/2}\cos^n u\,du=\frac{n-1}{n}\cdot\frac{n-3}{n-2}\cdots\frac{3}{4}\cdot\frac{1}{2}\cdot\frac{\pi}{2},\text{ when } n \text{ is even}\right]$$

$$=c^2\frac{\pi}{2}\left[1-\sum_1^{\infty}\frac{(-1)^n}{2n-1}\left\{\frac{(2n-1)(2n-3)\cdots 3\cdot 1}{2n(2n-2)\cdot 4\cdot 2}\right\}^2\right]$$

$$=c^2\frac{\pi}{2}\left[1-\sum_1^{\infty}\frac{(-1)^n P_{2n}^2}{2n-1}\right],\text{ when } P_{2n}=\frac{(2n-1)(2n-3)\cdots 3\cdot 1}{2n\cdot(2n-2)\cdots 4\cdot 2}.$$

Example. Show that area of the surface $cz=xy$ intercepted by the cylinder $x^2+y^2=b^2$ is

$$\iint_A\frac{1}{c}\sqrt{c^2+x^2+y^2}.$$

If A is the area of the circle $x^2+y^2=b^2$, $z=0$, then surface area $=\dfrac{2\pi}{3c}\left[(c^2+b^2)^{3/2}-c^3\right]$.

$$S = \text{surface area} = \iint_A \sqrt{1 + \left(\frac{\partial z}{\partial x}\right)^2 + \left(\frac{\partial z}{\partial y}\right)^2}\, dx\, dy.$$

Here $\dfrac{\partial z}{\partial x} = \dfrac{y}{c}$ and $\dfrac{\partial z}{\partial y} = \dfrac{x}{c}$.

Then $S = \iint_A \sqrt{1 + \dfrac{y^2}{c^2} + \dfrac{x^2}{c^2}}\, dx\, dy = \dfrac{1}{c} \iint \sqrt{c^2 + x^2 + y^2}\, dx\, dy.$

For the circle $x^2 + y^2 = b^2, z = 0$, let $x = r\cos\theta, y = r\sin\theta, \theta : \theta : 0 \to 2\pi, r : 0 \to b$.

$$S = \frac{1}{c} \int_0^{2\pi} \int_0^b \sqrt{c^2 + r^2} \cdot r\, dr\, d\theta = \frac{\pi}{c} \int_0^b \sqrt{c^2 + r^2}\, 2r \cdot dr$$

$$= \frac{\pi}{c} \left[\frac{2}{3}(c^2 + r^2)^{3/2}\right]_0^b = \frac{2\pi}{3c} \left[(c^2 + b^2)^{3/2} - c^3\right].$$

13.6 Volume Integrals (Triple Integrals)

Let $f(x, y, z)$ be a function of three independent variables. Then the integral of f is called volume integral, known as triple integral in $\mathbb{R}^3$. It is a simple and straight extension of the ideas in respect of double integrals.

13.6.1 Integrability and Integral of a Bounded Function $f(x, y, z)$ over a Rectangular Parallelopiped R

If region in $\mathbb{R}^3$, denoted by R, is defined by

$$R = [a, b;\ c, d;\ e, g] = [a \le x \le b,\ c \le y \le d,\ e \le z \le g].$$

Let f be bounded and defined in R.

Let $P_1 = \{a = x_0 < x_1 \cdots < x_{r-1} < x_r < \cdots < x_m = b\}$

$P_2 = \{c = y_0 < y_1 < \cdots < y_{s-1} < y_s < \cdots < y_n = d\}$

$P_3 = \{e = z_0 < z_1 < \cdots < z_{t-1} < z_t < \cdots < z_p = g\}$

be the partitions of $[a, b]$, $[c, d]$ and $[e, g]$ respectively. Then these three partitions P_1, P_2, P_3 together constitute a partition P of R into mnp sub-rectangular parallelopiped R_{rst} $[x_{r-1}, x_r;\ y_{s-1}, y_s;\ z_{t-1}, z_t]$, where $r : 1 \to m$, $s : 1 \to n$, $t : 1 \to p$.

We write, $V_{rst} = (x_r - x_{r-1}) \cdot (y_s - y_{s-1}) \cdot (z_t - z_{t-1})$ is the volume of R_{rst}.

Let M, m be the bounds of f in R.

Also, let M_{rst}, m_{rst} be the supremum and infimum of $f(x, y, z)$ in R_{rst}.

We write, $S(P, f) = \displaystyle\sum_{t=1}^{p}\sum_{s=1}^{n}\sum_{r=1}^{m} M_{rst}V_{rst}$ and $s(P, f) = \displaystyle\sum_{t=1}^{p}\sum_{s=1}^{n}\sum_{r=1}^{m} m_{rst}V_{rst}$.

It can be easily shown that

$$m(b - a)(d - c)(g - e) \le s(P, f) \le S(P, f) \le M(b - a)(d - c)(g - e).$$

This shows that the sets $s = \{s(P, f),\ P$ is a partition of $R\}$ and $S = \{S(P, f),\ P$ is a partition of $R\}$ has a supremum and an infimum respectively. The supremum of s [i.e., the supremum of the set of all lower sums] is known as the *upper integral* over R, denoted by

$$\overline{\iiint_R} f(x,y,z)dx\,dy\,dz$$

and the infimum of S [i.e., infimum of the set of all upper sums] is known as the *lower integral* over R, denoted by

$$\iiint_{\underline{R}} f(x,y,z)dx\,dy\,dz.$$

If these two integrals are equal, then we say, that f is *integrable over* R and the common value is called the *integral* of f over R and is denoted by

$$\iiint_R f(x,y,z)dx\,dy\,dz.$$

13.6.2 Darboux theorem

(1) Let $f(x,y,z)$ be bounded in $R\ [a,b;\ c,d;\ e,g]$. Then for every $\epsilon(>0)$, there exists a (+)ve number δ such that

$$S(P,f) < \overline{\iiint_R} f(x,y,z)dx\,dy\,dz + \epsilon \text{ and } s(P,f) > \iiint_{\underline{R}} f(x,y,z)dx\,dy\,dz - \epsilon$$

for every partition P of R with norm $\mu(P) < \delta$. [$\mu(P)$ = greatest volume of the sub-rectangular parallelopiped].

(2) The necessary and sufficient condition that a bounded function $f(x,y,z)$ to be integrable over a rectangular parallelopiped R is that for every $\epsilon > 0$, there is a partition P or R such that $S(P,f) - s(P,f) < \epsilon$.

(3) Every continuous function on R is integrable.

(4) A bounded function whose points of discontinuities (finite or infinite in numbers) can be enclosed in a finite number of parallelopiped, the sum of whose volumes is less than $\epsilon(>0)$, is integrable.

(5) A bounded function whose points of discontinuities lie on a finite number of surfaces of the form $z = \phi(x,y)$, etc., where ϕ is continuous is integrable.

(Proofs are similar to those for single integrals).

Calculation of triple integrals, reduction to ordinary integrals

Theorem [13.12] If (i) $\iiint_R f(x,y,z)dx\,dy\,dz$ and (ii) $\iint_S f(x,y,z)dx\,dy$, where $e \le z \le g$ and $S = [a,b;\ c,d]$ both exist, then $\int_e^g \left[\iint_S f(x,y,z)dx\,dy\right] dz$ also exists and

$$\int_e^g \left[\iint_S f(x,y,z)dx\,dy\right] dz = \iiint_R f(x,y,z)dx\,dy\,dz.$$

Proof. Let P be a partition of R into mnp sub-rectangular parallelpiped $[x_{r-1}, x_r, y_{s-1}, y_s; z_{t-1}, z_t]$; M_{rs}, m_{rs} be the upper and lower bound of $f(x,y,z)$ defined on R, in rstth sub-rectangular parallelopiped.

We write, the upper integral $\overline{\iiint\limits_R} f(x,y,z)dx\,dy\,dz = U$ and the lower integral

$$\underline{\iiint\limits_R} f(x,y,z)dx\,dy\,dz = L$$

Then from Darboux Theorem [13.6.2(1)],

$$S(P,f) = \sum_{t=1}^{p}\sum_{s=1}^{n}\sum_{r=1}^{m} M_{rst}V_{rst} < U + \epsilon. \tag{i}$$

For each fixed value of $z \in [z_{t-1}, z_t]$, we have

$$\iint\limits_S f(x,y,z)dx\,dy \leq \sum_{s=1}^{n}\sum_{r=1}^{m} M_{rst}(x_r - x_{r-1})(y_s - y_{s-1}) \tag{ii}$$

$$\Rightarrow \int_e^g \left[\iint\limits_S f(x,y,z)dx\,dy\right] dz \leq \sum_{t=1}^{p}\sum_{s=1}^{n}\sum_{r=1}^{m} M_{rst}V_{rst} < U + \epsilon \text{ [by (1)]}$$

$$\Rightarrow \int_e^g \left[\iint\limits_S f(x,y,z)dx\,dy\right] dz \leq U \text{ as } \epsilon \text{ is an arbitrary (+)ve number.}$$

Proceeding as above and by Darboux Theorem, we get $\int_e^g \left[\iint\limits_S f(x,y,z)dx\,dy\right] dz \geq L.$

Now since $\iiint\limits_R f(x,y,z)dx\,dy\,dz$ exists, $U = L$.

Hence, $\int_e^g \left[\iint\limits_S f(x,y,z)dx\,dy\right] dz = \iiint\limits_R f(x,y,z)dx\,dy\,dz.$

Corollary [13.1] : (i) If $f(x,y,z)$ be continuous over R. Then

$$\iiint\limits_R f(x,y,z)dx\,dy\,dz = \int_e^g \left[\int_c^d \left[\int_a^b f(x,y,z)dx\right] dy\right] dz,$$

where $R = [a,b;\ c,d;\ e,g]$. We can interchange the order of integration on the R.H.S. in any manner.

(ii) If f be continuous in a region E bounded by the surfaces $z = \phi(x,y)$, $z = \psi(x,y)$; $y = g(x)$, $y = h(x)$; $x = a$, $x = b$. Then

$$\iiint\limits_E f(x,y,z)dx\,dy\,dz = \int_a^b \left[\int_{g(x)}^{h(x)} \left[\int_{\phi(x,y)}^{\psi(x,y)} f(x,y,z)dz\right] dy\right] dx,$$

where $\phi(x,y) \leq \psi(x,y)$, $g(x) \leq h(x)$, $b \leq a\ \forall\ (x,y,z) \in \mathbb{R}^3$, $E \subseteq \mathbb{R}^3$.

Proof. Let D be the projection of the domain E which is enclosed by $R = [a,b;\ c,d;\ g,h]$. E is bounded by the surface $z = \phi(x,y)$ and $z = \psi(x,y)$, $\phi(x,y) \leq \psi(x,y)$.

Then for every fixed point $(x,y) \in D$, the single integral $\int_{\phi(x,y)}^{\psi(x,y)} f(x,y,z)dz$ exists.

Assuming the existence of $\iiint\limits_E f(x, y, z)dx\,dy\,dz$, we can write,

$$\iiint\limits_E f(x,y,z)dx\,dy\,dz = \iint\limits_D \left[\int_{\phi(x,y)}^{\psi(x,y)} f(x,y,z)dz\right] dx\,dy. \qquad \text{(i)}$$

Let $\int_{\phi(x,y)}^{\psi(x,y)} f(x,y,z)dz = F(x,y)$ and let the conditions for reduction of the double integral to the iterated integral be satisfied on D, where $D = [a, b;\ g(x), h(x)]$, then

$$\iint\limits_D F(x,y)dx\,dy = \int_a^b \left[\int_{g(x)}^{h(x)} f(x,y,z)dy\right] dx,\ g(x) \le h(x),\ b \le a.$$

Therefore, from (i),

$$\iiint\limits_E f(x,y,z)dx\,dy\,dz = \int_a^b \left[\left\{\int_{g(x)}^{h(x)} \int_{\phi(x,y)}^{\psi(x,y)} f(x,y,z)dz\right\} dy\right] dx$$

$$= \int_a^b dx \int_{g(x)}^{h(x)} dy \int_{\phi(x,y)}^{\psi(x,y)} f(x,y,z)dz.$$

Corollary [13.2] : Change of variables in Triple integrals. Let $x = X(u,v,w)$, $y = Y(u,v,w)$, $z = Z(u,v,w)$ be a one-one transformation $E \to E'$, where E, E' are domains on $\mathbb{R}^3$ of (x,y,z) and $\mathbb{R}^3$ of (u,v,w) respectively.

Then $f(x,y,z) = f(X(u,v,w), Y(u,v,w), Z(u,v,w)) = F(u,v,w)$ and $dxdy\,dz = |J|dudvdw$, where $J = \dfrac{\partial(x,y,z)}{\partial(u,v,w)}$ and hence $\iiint\limits_E f(x,y,z)dxdydz = \iiint\limits_{E'} F(u,v,w)|J|du\,dv\,dw$.

Theorem [13.13] [Gauss's Theorem (or **Divergence Theorem**)] Let S be any closed surface and E be a three dimensional region such that S encloses E and let f, g, h be three functions with continuous partial derivatives f_x, g_y, h_z at each point of E and S. Then

$$\iiint\limits_E (f_x + g_y + h_z)dx\,dy\,dz = \iint\limits_S (f\,dy\,dz + g\,dz\,dx + h\,dx\,dy),$$

where the surface integral is taken over the exterior of S.

Proof.

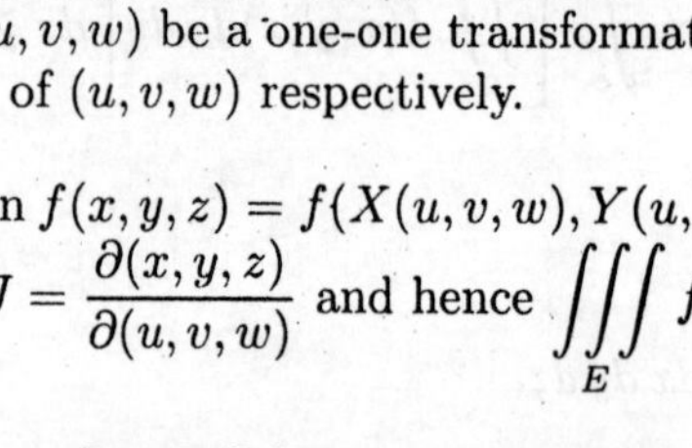

Fig. 13.30(i)

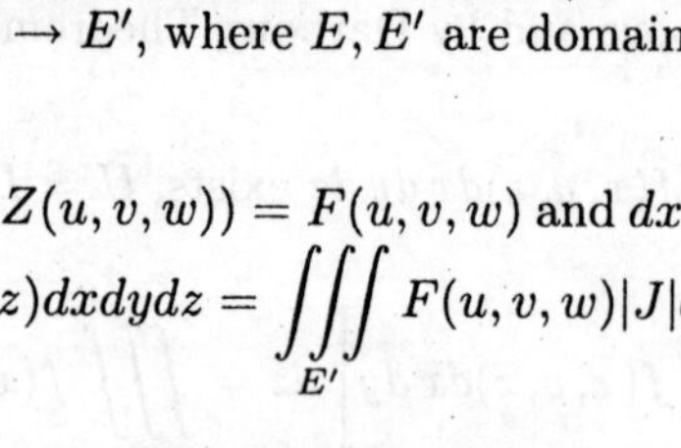

Fig. 13.30(ii)

Here E is a three dimensional region; $E \subseteq \mathbb{R}^3$.

E is bounded by

(i) two smooth surfaces, $S_1 : z = \psi(x, y)$, $S_2 : z = \phi(x, y)$ above and below.

(ii) S_3 : a lateral cylindrical surface

(iii) with generators parallel to Z-axis.

$S_1 \cup S_2 \cup S_3 = S$, the whole surface area enclosing E.

Let D be the projection of S on XY-plane. We are considering the exterior of the surface S. Outward normal to S_1 and S_2 are in opposite directions. So S_1, S_2 are of opposite orientation. Accordingly, if D denotes the region of projection of S_1 on XY-plane, then $-D$ denotes the region of projection of S_2 on XY-plane.

Therefore,

$$\iiint\limits_E h_z dx\,dy\,dz = \iint\limits_D \left[\int_{\phi(x,y)}^{\psi(x,y)} h_z dz\right] dx\,dy = \iint\limits_D \Big[h(x, y, z)\Big]_{\phi(x,y)}^{\psi(x,y)} dx\,dy$$

$$= \iint\limits_D [h(x, y, \psi(x, y)) - h(x, y, \phi(x, y))]dx\,dy$$

$$= \iint\limits_D h(x, y, \psi(x, y))dx\,dy + \iint\limits_{-D} h(x, y, \phi(x, y))dx\,dy$$

$$= \iint\limits_{S_1} h(x, y, z)dx\,dy + \int\limits_{S_2} h(x, y, z)dx\,dy$$

$$= \iint\limits_{S_1} h(x, y, z)dx\,dy + \iint\limits_{S_2} h(x, y, z)dx\,dy + \iint\limits_{S_3} h(x, y, z)dx\,dy,$$

where $\iint\limits_{S_3} h(x, y, z)dx\,dy = 0$, as it is taken over the outerside of the lateral surface S_3

$$= \iint\limits_S h(x, y, z)dx\,dy. \tag{i}$$

The relation (i) is also valid for any domain E which is piecewise regular with respect to Z-axis.

E is also regular with respect to all the co-ordinate axes. So considering regular with X-axis and Y-axis respectively, we can show that

$$\iiint\limits_E f_x dx\,dy\,dz = \iint\limits_S f\,dy\,dz \tag{ii}$$

$$\text{and} \quad \iiint\limits_E g_y dx\,dy\,dz = \iint\limits_S g\,dx\,dz \tag{iii}$$

by proceeding as in (i).

Thus (i) + (ii) + (iii) gives

$$\iiint\limits_E (f_x + g_y + h_z)dx\,dy\,dz = \iint\limits_S [f\,dy\,dz + g\,dz\,dx + h\,dx\,dy]. \tag{iv}$$

Note. (1) The result (iv) holds for any regular or piecewise regular domain E with a smooth boundary S.

(2) If $\cos\alpha, \cos\beta, \cos\gamma$ be the direction cosines of the outward unit normal of S

$$\text{Projection of } S \text{ on } YZ\text{-plane } dy\,dz = dS\cos\alpha$$
$$\text{Projection of } S \text{ on } ZX\text{-plane, } dz\,dx = dS\cos\beta$$
$$\text{and Projection of } S \text{ on } XY\text{-plane, } dx\,dy = dS\cos\gamma.$$

So the result (iv) can be written as

$$\iiint_E (f_x + g_y + hz)dx\,dy\,dz = \iint_S (f\cos\alpha + g\cos\beta + h\cos\gamma)dS. \quad \text{(v)}$$

(3) In case of vector function.

Let $\vec{F} = iP + jQ + kR$ be vector function defined on E enclosed by S.

Then

$$\text{div } F = \Delta \cdot F = \left(i\frac{\partial}{\partial x} + j\frac{\partial}{\partial y} + k\frac{\partial}{\partial z}\right)\cdot(iP + jQ + kR) = \frac{\partial P}{\partial x} + \frac{\partial Q}{\partial y} + \frac{\partial R}{\partial z}.$$

Let $\hat{n} = i\cos\alpha + j\cos\beta + k\cos\gamma$ be an outward unit normal on S.

$$F\cdot\hat{n} = P\cos\alpha + Q\cos\beta + R\cos\gamma.$$

Hence, Gauss's theorem may be expressed from (v) as

$$\iiint_E \text{div } \vec{F}dv = \iint_S F\cdot n dS \quad [\because dV = dx\,dy\,dz] \text{ taking } P = f, Q = g, R = h.$$

Example. Show using Gauss's Theorem that the surface integral

$$\iint_S [(x^3 - yz)dy\,dz - 2x^2y\,dz\,dx + 2dx\,dy]$$

taken over the outer surface of the cube bounded by the planes $x = 0$, $x = a$, $y = 0$, $y = a$, $z = 0$, $z = a$ is $\frac{1}{3}a^5$.

Here $f(x, y, z) = x^3 - yz \Rightarrow f_x = 3x^2$

$g(x, y, z) = -2x^2y \Rightarrow g_y = -2x^2$

$h(x, y, z) = 2 \Rightarrow h_z = 0.$

By Gauss's theorem,

$$\iint_S (f\,dy\,dz + g\,dz\,dx + h\,dx\,dy) = \int_0^a\int_0^a\int_0^a (f_x + g_y + h_z)dx\,dy\,dz$$
$$= \int_0^a\int_0^a\int_0^a (3x^2 - 2x^2 + 0)dx\,dy\,dz$$
$$= \int_0^a\int_0^a \left[\int_0^a x^2dx\right]dy\,dz = \frac{a^3}{3}\cdot a\cdot a = \frac{a^5}{3}.$$

Example. Evaluate $\iint_S (x\,dy\,dz + y\,dz\,dx + z\,dx\,dy)$ taken over the outer surface of the cube $[0, a; 0, a; 0, a]$ and verify Gauss's theorem.

Here $f(x, y, z) = x \Rightarrow f_x = 1$

$g(x, y, z) = y \Rightarrow g_y = 1$

$h(x, y, z) = z \Rightarrow h_z = 1.$

$$\iint_S (x\,dy\,dz + y\,dz\,dx + z\,dx\,dy) = \int_0^a \int_0^a \int_0^a (1 + 1 + 1)dx\,dy\,dz = 3a^3. \quad (1)$$

Also, $\iint_S (x\,dy\,dz + y\,dz\,dx + z\,dx\,dy)$.

Any point P on XY-plane has a fixed Z-coordinate.

That is, for the given cubic, $z = a$.

Similarly, $x = a, y = a$.

$$= \int_0^a \int_0^a [a\,dy\,dz + a\,dz\,dx + a\,dx\,dy] = a[3a^2] = 3a^3. \quad (2)$$

Gauss's theorem is verified by (1) and (2).

Example. Evaluate $\iint_S (y^2z^2dy\,dz + z^2x^2dx\,dz + x^2y^2dx\,dy)$, where S is the part of the surface of the sphere $x^2 + y^2 + z^2 = 1$ above XY-plane.

Projection of the surface S on XY-plane is $x^2 + y^2 = 1,\ z = 0$.

Therefore, the given integral is

$$= \iint_{\substack{S:x^2+y^2=1\\ z=0}} x^2y^2dx\,dy = \int_{-1}^1 x^2 \left[\int_{-\sqrt{1-x^2}}^{\sqrt{1-x^2}} y^2dy\right] = \int_{-1}^1 x^2 \left[\frac{y^3}{3}\right]_{-\sqrt{1-x^2}}^{\sqrt{1-x^2}} dx$$

$$= \frac{1}{3}\int_{-1}^1 2x^2 \sqrt[3]{1-x^2}dx = \frac{2}{3}\cdot 2\int_0^1 x^2 \left(\sqrt{1-x^2}\right)^{3/2} dx.$$

Putting $x = \sin\theta,\ dx = \cos\theta d\theta$

$$= \frac{4}{3}\int_0^{\pi/2} \sin^2\theta\cos^4\theta d\theta = \frac{4}{3}\left[\int_0^{\pi/2}\cos^4\theta d\theta - \int_0^{\pi/2}\cos^6\theta d\theta\right]$$

$$= \frac{4}{3}\left[\frac{3}{4}\cdot\frac{1}{2}\cdot\frac{\pi}{2} - \frac{5}{6}\cdot\frac{3}{4}\cdot\frac{1}{2}\cdot\frac{\pi}{2}\right] = \frac{4}{3}\cdot\frac{3\pi}{4}\left[\frac{1}{4} - \frac{5}{24}\right] = \pi\cdot\left[\frac{6-5}{24}\right] = \frac{\pi}{24}.$$

Example. Show that $\iint_S (ax\,dy\,dz + by\,dz\,dx + cz\,dx\,dy) = \frac{4}{3}\pi(a + b + c)$, where S is the surface of the sphere $x^2 + y^2 + z^2 = 1$.

Here $f = ax, g = by, h = cz$. $\quad\therefore f_x = a, g_y = b, h_z = c.$

Then given integral

$$= \iiint_{E:x^2+y^2+z^2=1} (f_x + g_y + h_z)dx\,dy\,dz = (a + b + c)\iiint_{E:x^2+y^2+z^2=1} dx\,dy\,dz.$$

Put $x = r\sin\theta\cos\phi, y = r\sin\theta\sin\phi, z = r\cos\theta$.

$r: 0 \to 1,\ \theta: 0 \to \pi$ and $\phi: 0 \to 2\pi$

$$= (a+b+c)\int_{r=0}^{1}\int_{0}^{\pi}\int_{0}^{2\pi} |J|\, dr\, d\theta\, d\phi,\ |J| = r^2 \sin\theta$$

$$= (a+b+c)\int_{r=0}^{1}\int_{0}^{\pi}\int_{0}^{2\pi} r^2 \sin\theta dr\, d\theta\, d\phi$$

$$= \frac{a+b+c}{3} 2\pi\Big[-\cos\theta\Big]_0^{\pi}$$

$$= \frac{4\pi}{3}(a+b+c).$$

Example. Evaluate $\iint_S (x\, dy\, dz + y\, dz\, dx + z^2 dx\, dy)$, where S denotes the closed surface bounded by the cone $x^2+y^2=z^2$ and the plane $z=1$.

Here $f=x,\ g=y,\ h=z^2$
$f_x = 1,\ g_y = 1,\ h_z = 2z$

$$\therefore z: \sqrt{x^2+y^2} \to 1$$

$$\text{Given integral} = \iiint_{\substack{E: x^2+y^2=z^2 \\ z=1}} (f_x + g_y + h_z)\, dx\, dy\, dz$$

$$= \iint_{D: x^2+y^2=1} \left[\int_{z=\sqrt{x^2+y^2}}^{1} (2+2z)dz\right] dx\, dy$$

$$= \iint_{D: x^2+y^2=1} \left[2\left(1-\sqrt{x^2+y^2}\right) + (1-(x^2+y^2)\right] dx\, dy$$

$$= \iint_{D: x^2+y^2=1} \left(3 - 2\sqrt{x^2+y^2} - (x^2+y^2)\right) dx\, dy$$

For $x = r\cos\theta,\ y = r\sin\theta,\ dx\, dy = |J| dr\, d\theta = r\, dr\, d\theta$

$$= \int_{r=0}^{1}\int_{\theta=0}^{2\pi} [3 - 2r - r^2]\, r\, dr\, d\theta = 2\pi\left[3\cdot\frac{r^2}{2} - 2\cdot\frac{r^3}{3} - \frac{r^4}{4}\right]_0^1$$

$$= 2\pi\left[\frac{3}{2} - \frac{2}{3} - \frac{1}{4}\right] = 2\pi\left[\frac{18-8-3}{12}\right] = \frac{7\pi}{6}.$$

Theorem [13.14] [Stoke's Theorem] Let S be a smooth oriented surface bounded by a curve C oriented in the same sense and let $f(x,y,z)$, $g(x,y,z)$, $h(x,y,z)$ be three continuous functions with continuous first order partial derivatives at each point (x,y,z) of the domain D containing S. Then

$$\int_C (f\, dx + g\, dy + h\, dz)$$

$$= \iint_S \left[\left(\frac{\partial h}{\partial y} - \frac{\partial g}{\partial z}\right) dy\, dz + \left(\frac{\partial f}{\partial z} - \frac{\partial h}{\partial x}\right) dz\, dx + \left(\frac{\partial g}{\partial x} - \frac{\partial f}{\partial y}\right) dx\, dy\right].$$

Proof. Let the oriented surface S be given by $x = x(u,v)$, $y = y(u,v)$, $z = z(u,v)$, where $(u,v) \in D$, D is an oriented domain in the uv-plane.

Let Γ denotes the oriented boundary curve of the domain D and Γ is given by

$$u = u(t), v = v(t), a \le t \le b.$$

Proof follows from the following steps

(i) The line integral along C is expressed as an ordinary integral I.

(ii) The ordinary integral I is expressed as a line integral along Γ.

(iii) Using Green's theorem the line integral along Γ is expressed as a double integral over D.

(iv) Finally, the double integral over D is expressed as a surface integral over S.

Thus, the line integral along C is reduced to the surface integral over S

$$x = x(u, v) \Rightarrow \frac{dx}{dt} = \frac{\partial x}{\partial u} \cdot \frac{du}{dt} + \frac{\partial x}{\partial v} \cdot \frac{dv}{dt}$$

$$y = y(u, v) \Rightarrow \frac{dy}{dt} = \frac{\partial y}{\partial u} \cdot \frac{du}{dt} + \frac{\partial y}{\partial v} \cdot \frac{dv}{dt}$$

$$z = z(u, v) \Rightarrow \frac{dz}{dt} = \frac{\partial z}{\partial u} \cdot \frac{du}{dt} + \frac{\partial z}{\partial v} \cdot \frac{dv}{dt}.$$

Then $\int_C (f\,dx + g\,dy + h\,dz)$

$$= \int_a^b \left[f\left(\frac{\partial x}{\partial u} \cdot \frac{du}{dt} + \frac{\partial x}{\partial v} \cdot \frac{dv}{dt}\right) + g\left(\frac{\partial y}{\partial u} \cdot \frac{du}{dt} + \frac{\partial y}{\partial v} \cdot \frac{dv}{dt}\right) + h\left(\frac{\partial z}{\partial u} \cdot \frac{du}{dt} + \frac{\partial z}{\partial v} \cdot \frac{dv}{dt}\right)\right] dt$$

$$= \int_\Gamma \left[\left(f\frac{\partial x}{\partial u} + g\frac{\partial y}{\partial u} + h\frac{\partial z}{\partial u}\right) du + \left(f\frac{\partial x}{\partial v} + g\frac{\partial y}{\partial v} + h\frac{\partial z}{\partial v}\right) dv\right]$$

$$= \iint_D \left[\frac{\partial}{\partial u}\left(f\frac{\partial x}{\partial v} + g\frac{\partial y}{\partial v} + h\frac{\partial z}{\partial v}\right) - \frac{\partial}{\partial v}\left(f\frac{\partial x}{\partial u} + g\frac{\partial y}{\partial u} + h\frac{\partial z}{\partial u}\right)\right] du\,dv \left[\begin{matrix}\text{by Green's} \\ \text{theorem}\end{matrix}\right]. \quad \text{(i)}$$

Now $\frac{\partial}{\partial u}\left(f\frac{\partial x}{\partial v} + g\frac{\partial y}{\partial v} + h\frac{\partial z}{\partial v}\right) - \frac{\partial}{\partial v}\left(f\frac{\partial x}{\partial u} + g\frac{\partial y}{\partial u} + h\frac{\partial z}{\partial u}\right)$

$$= \left[\left(\frac{\partial f}{\partial x} \cdot \frac{\partial x}{\partial u} + \frac{\partial f}{\partial y} \cdot \frac{\partial y}{\partial u} + \frac{\partial f}{\partial z} \cdot \frac{\partial z}{\partial u}\right)\frac{\partial x}{\partial v} + f\frac{\partial^2 x}{\partial u \partial v}\right.$$

$$+ \left(\frac{\partial g}{\partial x} \cdot \frac{\partial x}{\partial u} + \frac{\partial g}{\partial y} \cdot \frac{\partial y}{\partial u} + \frac{\partial g}{\partial z} \cdot \frac{\partial z}{\partial u}\right)\frac{\partial y}{\partial v} + g\frac{\partial^2 y}{\partial u \partial v}$$

$$\left. + \left(\frac{\partial h}{\partial x} \cdot \frac{\partial x}{\partial u} + \frac{\partial h}{\partial y} \cdot \frac{\partial y}{\partial u} + \frac{\partial h}{\partial z} \cdot \frac{\partial z}{\partial u}\right)\frac{\partial z}{\partial v} + h\frac{\partial^2 z}{\partial u \partial v}\right]$$

$$- \left[\left(\frac{\partial f}{\partial x} \cdot \frac{\partial x}{\partial v} + \frac{\partial f}{\partial y} \cdot \frac{\partial y}{\partial v} + \frac{\partial f}{\partial z} \cdot \frac{\partial z}{\partial v}\right)\frac{\partial x}{\partial u} + f\frac{\partial^2 x}{\partial v \partial u}\right.$$

$$+ \left(\frac{\partial g}{\partial x} \cdot \frac{\partial x}{\partial v} + \frac{\partial g}{\partial y} \cdot \frac{\partial y}{\partial v} + \frac{\partial g}{\partial z} \cdot \frac{\partial z}{\partial v}\right)\frac{\partial y}{\partial u} + g\frac{\partial^2 y}{\partial v \partial u}$$

$$\left. + \left(\frac{\partial h}{\partial x} \cdot \frac{\partial x}{\partial v} + \frac{\partial h}{\partial y} \cdot \frac{\partial y}{\partial v} + \frac{\partial h}{\partial z} \cdot \frac{\partial z}{\partial v}\right)\frac{\partial z}{\partial u} + h\frac{\partial^2 z}{\partial v \partial u}\right].$$

Here $\frac{\partial^2 x}{\partial u \partial v} = \frac{\partial^2 x}{\partial v \partial u}, \frac{\partial^2 y}{\partial u \partial v} = \frac{\partial^2 y}{\partial v \partial u}, \frac{\partial^2 z}{\partial u \partial v} = \frac{\partial^2 z}{\partial v \partial u}$ cancelling the common terms in the brackets we get

$$\frac{\partial}{\partial u}\left(f\frac{\partial x}{\partial v}+g\frac{\partial y}{\partial v}+h\frac{\partial x}{\partial v}\right)-\frac{\partial}{\partial v}\left(f\frac{\partial x}{\partial u}+g\frac{\partial y}{\partial u}+h\frac{\partial z}{\partial u}\right)$$

$$=\left[\left(\frac{\partial f}{\partial y}\cdot\frac{\partial y}{\partial u}+\frac{\partial f}{\partial z}\cdot\frac{\partial z}{\partial u}\right)\frac{\partial x}{\partial v}+\left(\frac{\partial g}{\partial x}\cdot\frac{\partial x}{\partial u}+\frac{\partial g}{\partial z}\cdot\frac{\partial z}{\partial u}\right)\frac{\partial y}{\partial v}+\left(\frac{\partial h}{\partial x}\cdot\frac{\partial x}{\partial u}+\frac{\partial h}{\partial y}\cdot\frac{\partial y}{\partial u}\right)\frac{\partial z}{\partial v}\right]$$

$$-\left[\left(\frac{\partial f}{\partial y}\cdot\frac{\partial y}{\partial v}+\frac{\partial f}{\partial z}\cdot\frac{\partial z}{\partial v}\right)\frac{\partial x}{\partial u}+\left(\frac{\partial g}{\partial x}\cdot\frac{\partial x}{\partial v}+\frac{\partial g}{\partial z}\cdot\frac{\partial z}{\partial v}\right)\frac{\partial y}{\partial u}+\left(\frac{\partial h}{\partial x}\cdot\frac{\partial x}{\partial v}+\frac{\partial h}{\partial y}\cdot\frac{\partial y}{\partial v}\right)\frac{\partial z}{\partial u}\right]$$

$$=\left(\frac{\partial h}{\partial y}-\frac{\partial g}{\partial z}\right)\left(\frac{\partial y}{\partial u}\cdot\frac{\partial z}{\partial v}-\frac{\partial y}{\partial v}\cdot\frac{\partial z}{\partial u}\right)+\left(\frac{\partial f}{\partial z}-\frac{\partial h}{\partial x}\right)\left(\frac{\partial z}{\partial u}\cdot\frac{\partial x}{\partial v}-\frac{\partial z}{\partial v}\cdot\frac{\partial x}{\partial u}\right)$$

$$+\left(\frac{\partial g}{\partial x}-\frac{\partial f}{\partial y}\right)\left(\frac{\partial x}{\partial u}\cdot\frac{\partial y}{\partial v}-\frac{\partial y}{\partial u}\cdot\frac{\partial x}{\partial v}\right)$$

$$=\left(\frac{\partial h}{\partial y}-\frac{\partial g}{\partial z}\right)\frac{\partial(y,z)}{\partial(u,v)}+\left(\frac{\partial f}{\partial z}-\frac{\partial h}{\partial x}\right)\frac{\partial(z,x)}{\partial(u,v)}+\left(\frac{\partial g}{\partial x}-\frac{\partial f}{\partial y}\right)\frac{\partial(x,y)}{\partial(u,v)}.$$

Therefore, from (1),

$$\int_C (f\,dx+g\,dy+h\,dz)$$

$$=\iint_D\left[\left(\frac{\partial h}{\partial y}-\frac{\partial g}{\partial z}\right)\frac{\partial(y,z)}{\partial(u,v)}+\left(\frac{\partial f}{\partial z}-\frac{\partial h}{\partial x}\right)\frac{\partial(z,x)}{\partial(u,v)}+\left(\frac{\partial g}{\partial x}-\frac{\partial f}{\partial y}\right)\frac{\partial(x,y)}{\partial(u,v)}\right]du\,dv$$

$$=\iint_S\left[\left(\frac{\partial h}{\partial y}-\frac{\partial g}{\partial z}\right)dy\,dz+\left(\frac{\partial f}{\partial z}-\frac{\partial h}{\partial x}\right)dz\,dx+\left(\frac{\partial g}{\partial x}-\frac{\partial f}{\partial y}\right)dx\,dy\right]$$

$$\left[\because \frac{\partial(y,z)}{\partial(u,v)}du\,dv=dy\,dz,\ \frac{\partial(z,x)}{\partial(u,v)}du\,dv=dz\,dx \text{ and } \frac{\partial(x,y)}{\partial(u,v)}du\,dv=dx\,dy\right]$$

Example. Verify Stoke's theorem for the line integral $\int_C (x^2dx+xy\,dy)$, where C denotes the square in the plane $z=0$ with the sides along the lines $x=0, y=0, x=a, y=a$.

$$\int_C (x^2dx+xy\,dy)=\underset{\substack{OA\\y=0}}{\int}+\underset{\substack{AB\\x=a}}{\int}+\underset{\substack{BC\\y=a}}{\int}+\underset{\substack{CO\\x=0}}{\int}=\int_0^a x^2dx+a\int_0^a y\,dy+\int_a^0 x^2dx+0$$

$$=a\cdot\frac{a^2}{2}=\frac{a^3}{2}. \qquad (1)$$

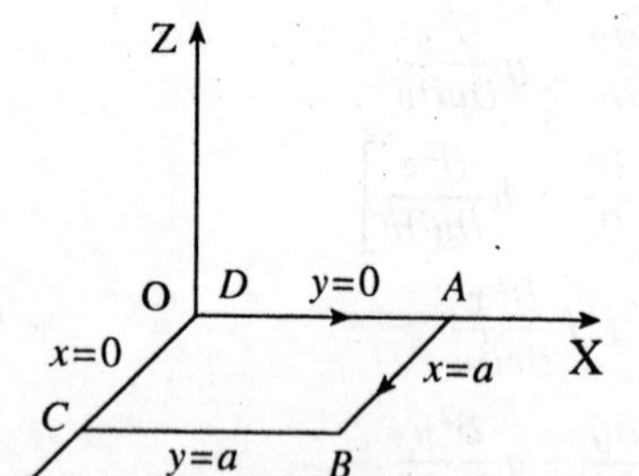

Fig. 13.31

Here $f=x^2, g=xy, h=0$.

$$\text{R.H.S.}=\iint_S (y-0)dx\,dy=\int_0^a\left[\int_0^a y\,dy\right]dx$$

$$=a\cdot\frac{a^2}{2}=\frac{a^3}{2}. \qquad (2)$$

Therefore, Stoke's theorem is verified by (1) and (2).

Example. Verify Stoke's Theorem in evaluating $\int_\Gamma 3y\,dx-zx\,dy+y\,z^2dz$, where S is the surface of the paraboloid $x^2+y^2=2z$ bounded by $z=2$ and Γ is its boundary.

Γ is the boundary of $x^2+y^2=2z$, $z=2$, $x^2+y^2=4$.

We take, $x = 2\cos\theta,\ y = 2\sin\theta,\ z = 2,\ dz = 0$

$$\int_{\Gamma}(3y\,dx - zx\,dy + y\,z^2dz) = \int_{-\pi}^{\pi}[6\sin\theta(-2\sin\theta) - 4\cos\theta(2\cos\theta)]d\theta$$

$$= -\int_{-\pi}^{\pi}(12\sin^2\theta + 8\cos^2\theta)d\theta$$

$$= -2\int_0^{\pi} 8d\theta - 2\int_0^{\pi} 2(1-\cos 2\theta)d\theta$$

$$= -16\pi - 4\pi$$

$$= -20\pi. \tag{1}$$

Again, for the surface integral,

$$f = 3y,\quad g = -zx,\quad h = yz^2$$
$$f_y = 3,\quad g_x = -z,\quad h_y = z^2,\ h_z = 2yz$$
$$g_z = -x,\ f_z = 0,\quad h_x = 0.$$

Then

$$\iint\limits_{\substack{x^2+y^2=2z\\ z=2}}\left[\left(\frac{\partial h}{\partial y} - \frac{\partial g}{\partial z}\right)dy\,dz + \left(\frac{\partial f}{\partial z} - \frac{\partial h}{\partial x}\right)dz\,dx + \left(\frac{\partial g}{\partial x} - \frac{\partial f}{\partial y}\right)dx\,dy\right]$$

$$= \iint\limits_{\substack{x^2+y^2=4\\ z=2}}\left[(z^2+x)dy\,dz + (0-0)dz\,dx + (-z-3)dx\,dy\right]$$

$$= -5\iint\limits_{x^2+y^2=4} dx\,dy\ [z=2, dz=0]$$

$$= -5\int_{r=0}^{2}\int_{-\pi}^{\pi} r\,dr\,d\theta$$

$$= -10\pi\left[\frac{r^2}{2}\right] = -20\pi. \tag{2}$$

Hence Stoke's theorem is verified from (1) and (2).

Example. Show that $\iint\limits_S (y-z)dy\,dz + (z-x)dz\,dx + (z-y)dx\,dy = a^3\pi$, where S is the part of the surface $x^2+y^2-2ax+az=0,\ z \geq 0$.

$$x^2+y^2-2ax+az = 0,\ z = 0$$
$$(x-a)^2 + y^2 = a^2,\ z \geq 0.$$

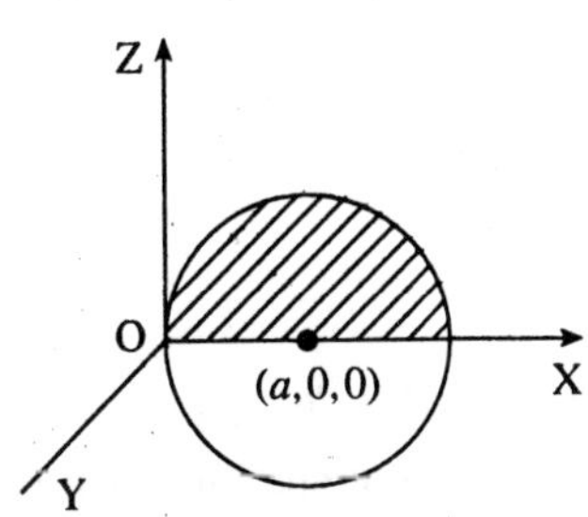

Fig. 13.32

Let $f = y^2+z^2,\ g = z^2+x^2,\ h = x^2+y^2$.

Then $$\frac{\partial h}{\partial y} - \frac{\partial g}{\partial z} = 2(y-z)$$
$$\frac{\partial f}{\partial z} - \frac{\partial h}{\partial x} = 2(z-x)$$
$$\frac{\partial g}{\partial x} - \frac{\partial f}{\partial y} = 2(x-y).$$

$$\therefore \iint_S \left[\left(\frac{\partial h}{\partial y}-\frac{\partial g}{\partial z}\right)dy\,dz+\left(\frac{\partial f}{\partial z}-\frac{\partial h}{\partial x}\right)dx\,dz+\left(\frac{\partial g}{\partial x}-\frac{\partial f}{\partial y}\right)dx\,dy\right]$$

$$= 2\iint_S [(y-z)dydz+(z-x)dx\,dz+(x-y)dx\,dy]$$

$$\Rightarrow \iint_S (y-z)dy\,dz+(z-x)dx\,dz+(x-y)dx\,dy$$

$$= \frac{1}{2}\int_\Gamma (f\,dx+g\,dy+h\,dz) \text{ [by Stoke's theorem]}$$

$$= \frac{1}{2}\int_{\substack{\Gamma:(x-a)^2+y^2=a^2\\ z=o}} (y^2+z^2)dx+(z^2+x^2)dy+(x^2+y^2)dz.$$

$$\text{Given integral} = \frac{1}{2}\int_{(x-a)^2+y^2=a^2} (y^2dx+x^2dy).$$

Putting $x=a+a\cos\theta, y=a\sin\theta, dx=-a\sin\theta d\theta,\ dy=a\cos\theta d\theta.$

$$\text{Given integral} = \frac{1}{2}\int_{-\pi}^{\pi}\{a^2\sin^2\theta(-a\sin\theta)+a^2(1+\cos\theta)^2a\cos\theta\}\,d\theta$$

$$= \frac{a^3}{2}\int_{-\pi}^{\pi}(-\sin^3\theta+\cos^3\theta+2\cos^2\theta+\cos\theta)\,d\theta$$

$$= \frac{a^3}{2}\left[0+2\int_0^\pi\frac{1}{4}(\cos3\theta+3\cos\theta)d\theta+4\int_0^\pi\frac{(1+\cos2\theta)d\theta}{2}+2\int_0^\pi\cos\theta d\theta\right]$$

$$= \frac{a^3}{2}\left[0+4\cdot\frac{\pi}{2}+0\right]=a^3\pi.$$

Example. Apply Stoke's theorem to prove that $\int_C (y\,dx+z\,dy+x\,dz)=-2\sqrt{2}\pi a^2$, where C is the curve given by $x^2+y^2+z^2-2ax-2ay=0,\ x+y=2a$ and begins at a point $(2a,0,0)$ and goes at first below Z-plane.

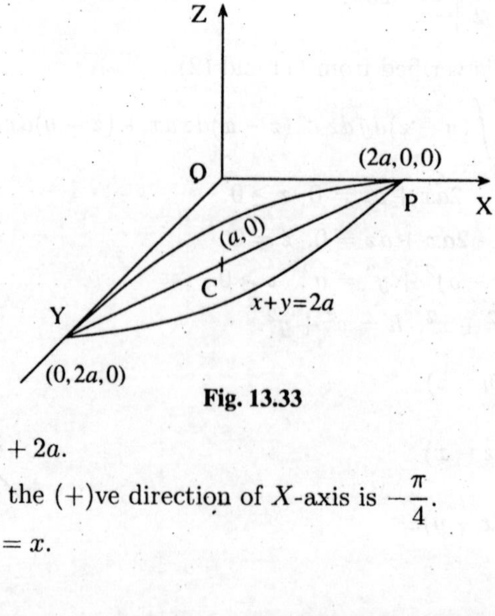

Fig. 13.33

$x+y=2a \Rightarrow y=-x+2a.$

Slope of the line with the (+)ve direction of X-axis is $-\dfrac{\pi}{4}$.

Now $f=y,\ g=z,\ h=x.$

$$\int\limits_{\substack{C:(x-a)^2+(y-a)^2=2a^2 \\ x+y=2a}} (y\,dx + z\,dy + x\,dz)$$

$$= \iint\limits_S \left(\frac{\partial h}{\partial y} - \frac{\partial g}{\partial z}\right) dy\,dz + \left(\frac{\partial f}{\partial z} - \frac{\partial h}{\partial x}\right) dz\,dx + \left(\frac{\partial g}{\partial x} - \frac{\partial f}{\partial y}\right) dx\,dy$$

$$= \iint\limits_S \left(\frac{\partial g}{\partial x} - \frac{\partial f}{\partial y}\right) dx\,dy \text{ [projection on } XY\text{-plane]}$$

$$= -\iint dx\,dy = \sqrt{2}\iint\limits_S -\sqrt{2}\frac{1}{\sqrt{2}}dx\,dy$$

$$= -\sqrt{2}\iint\limits_S dx\,dy\cos\left(\frac{\pi}{4}\right) = -\sqrt{2}\iint\limits_S dS$$

$$= -\sqrt{2}\,(\text{area of the circle of radius } \sqrt{2}\,a)$$

$$= -\sqrt{2}\cdot\pi\cdot 2a^2 = -2\sqrt{2}\,\pi a^2.$$

EXERCISE

(1) Evaluate $\int_C \left[(x^2 + 2y^2)\,dx + 3xy\,dy\right]$ around the triangle ABC, whose vertices are $A = (0,0)$, $B = (0,2)$, $C = (-3,1)$.

(2) Show that the value of the integral $\int (xy^2dy - x^2ydx)$ taken in the counter-clockwise sense along the cardiode $r = a(1 + \cos\theta)$ is $\dfrac{35a^4\pi}{16}$.

(3) Compute $\displaystyle\int_C \frac{dx}{x+y}$, where C is the curve $x = at^2, y = 2at, 0 \le t \le 2$.

(4) Show that $\int_C (y\,dx - x\,dy) = 24\pi$, where C is the arc of the cycloid $x = 2(t - \sin t)$, $y = 2(1 - \cos t)$, joining the point $(0,0)$ and $(4\pi, 0)$.

(5) Show that

(i) $\iint xy(x+y)dx\,dy = 3/56$ over the area between $y = x^2$ and $y = x$.

(ii) $\iint \sqrt{x^2+y^2}\,dx\,dy = 14\pi/3$ over the area in the XY-plane bounded by $x^2 + y^2 = 1$ and $x^2 + y^2 = 4$.

(6) Express $\iint f(x,y)dx\,dy$ as a repeated integral, taken over the interior of the quadrilateral bounded by the lines $x + y = 0$, $x - y = 0$, $2x - y = 1$, $2x - 3y + 5 = 0$ taken in order.

(7) Evaluate $\displaystyle\iint\limits_R \sin(x+y)dx\,dy,\ R : \left\{0 \le x \le \frac{\pi}{2}, 0 \le y \le \frac{\pi}{2}\right\}$.

(8) Represent $\iint f(x,y)\,dx\,dy$ as a repeated integral (iterated integral) over the trapizium whose vertices are $(4,3), (0,1), (3,3), (2,5)$ taken in order.

(9) Evaluate $\displaystyle\iint\limits_R \frac{2x^2+y^2}{xy}\,dx\,dy$, where R is region in the (+)ve quadrant of XY-plane bounded by the curves $x^2 + y^2 = 4, x^2 + y^2 = 16, y^2 = 4x$ and $x^2 = 16y$.

(10) Change the order of integration in $\int_0^{a\cos\alpha}\int_{\tan\alpha}^{\sqrt{a^2-x^2}} f(x,y)\,dx\,dy$ and verify that when $f(x,y)=1$.

(11) Show that $\int_a^b\int_{a^2/x}^{x} F(x,y)\,dx\,dy = \int_{a^2/b}^{a}\int_{a^2/y}^{b} F\,dx\,dy + \int_a^b\int_y^b F\,dx\,dy.$

(12) Evaluate $\iint\limits_R (x^4-y^4)\,dx\,dy$ over the part of the positive quadrant in which $1 \le x^2-y^2 \le 2,\ 1 \le xy \le 2.$

(13) Let $f : [0,2;\ 0,3] \to R$ be defined by

$$f(x,y)=\begin{cases}2, & \text{when } x \text{ is rational}\\ y^2, & \text{when } x \text{ is irrational}\end{cases}$$

State with reasons whether $\int_0^2\left\{\int_0^3 f(x,y)dy\right\}dx$ and $\int_0^3\left\{\int_0^2 f(x,y)dx\right\}dy$ exist or not.

(14) Evaluate $\iint \dfrac{dx\,dy}{(1+x^2+y^2)^2}$ over a loop of lemniscate $(x^2+y^2)^2-(x^2-y^2)=0.$

(15) Let $F : R\times R \to R$ be defined by

$$F(x,y)=\begin{cases}\dfrac{1}{y^2}, & 0<x<y<1\\ -\dfrac{1}{x^2}, & 0<y<x<1\\ 0, & \text{elsewhere.}\end{cases}$$

Show that the repeated integrals $\int_0^1\left\{\int_0^1 F(x,y)dy\right\}dx$ and $\int_0^1\left\{\int_0^1 F(x,y)dy\right\}dx$ both exist but the double integrals $\int_0^1\int_0^1 F(x,y)\,dx\,dy$ does not exist.

(16) Verify Green's theorem in the plane for $\int_C (3x^2-8y^2)\,dx+(4y-6xy)dy$, where C is the boundary of the region defined by $y=\sqrt{x}$ and $y=x^2$.

(17) Evaluate $\int_C (x^2y\,dx+xy^2dy)$ by Green's theorem, where C denotes the closed path formed by $y=x$ and $y^3=x^2$ in the 1st quadrant described in the positive sense.

(18) Using Green's theorem prove that $\int_\Gamma \dfrac{x\,dy-y\,dx}{x^2+y^2}=2\pi$, if Γ be a positively described closed contour round the origin.

(19) Prove that the integral $\int_C (2xy-y)dx+x^2dy$, where C is a closed contour symmetrical with respect to the axes is equal to the area of the domain bounded by C.

(20) Show the following integrals taken round any closed contour C are equal to zero:

(i) $\int_C f\left(\dfrac{y}{x}\right)\dfrac{x\,dy-y\,dx}{x^2}$

(ii) $\int_C [f(x+y)+f(x-y)]dx+[f(x+y)-f(x-y)]dy.$

(21) Evaluate $\int_\Gamma \dfrac{x\,dy-y\,dx}{x^2+4y^2}$ round the circle $x^2+y^2=1$ in the (+)ve direction.

(22) Evaluate the surface integral $\iint_S (x^3 dy\, dz + y^3 dz\, dx + z^3 dx\, dy)$, where S is the surface of the sphere $x^2 + y^2 + z^2 = 1$. [Apply Gauss's theorem]

(23) Evaluate $\iint_S (x\, dy\, dz + y\, dz\, dx + z\, dx\, dy)$, where S is the outer surface of the part of the ellipsoid $\dfrac{x^2}{a^2} + \dfrac{y^2}{b^2} + \dfrac{z^2}{c^2} = 1$.

(24) Verify Stoke's theorem in evaluating $\int_C 3y\, dx - zx\, dy + yz^2 dz$, where S is the surface of the paraboloid $x^2 + y^2 = 2z$ bounded by $z = 2$ and C is its boundary.

(25) Use Gauss's theorem to show that the surface integral $\iint_S \left[(x^3 - yz)\, dy\, dz - 2x^2 y\, dz\, dx + 2\, dx\, dy\right]$ taken over the outer surface of the cube bounded by the planes $x = 0, x = a, y = 0, y = a$, $z = 0, z = a$ is $\dfrac{a^5}{3}$.

CHAPTER 14

BETA FUNCTION AND GAMMA FUNCTION

We have already introduced the two improper integrals known as Beta function and Gamma function in Ch-8 and discussed their convergence in Th. [8.5] and Th. [8.11].

Beta function and Gamma function denoted by $\beta(m,n)$ and $\Gamma(n)$ respectively and defined by

$$\beta(m,n) = \int_0^1 x^{m-1}(1-x)^{n-1}dx,\ m>0,\ n>0$$

$$\text{and } \Gamma(n) = \int_0^\alpha e^{-x}x^{n-1}dx,\ n>0,$$

since $\beta(m,n)$ converges for $m>0, n>0$ and $\Gamma(n)$ converges for $n>0$.

Relationship between Beta Function and Gamma Function

Theorem [14.1] $\beta(m,n) = \dfrac{\Gamma(m)\cdot\Gamma(n)}{\Gamma(m+n)}$.

Proof. We have $\Gamma(n) = \int_0^\infty e^{-x}x^{n-1}dx$ replacing $x = px,\ p>0$,

$$\Gamma(n) = \int_0^\infty e^{-px}p^{n-1}x^{n-1}p\,dx = \int_0^\infty e^{-px}\cdot p^n\cdot x^{n-1}dx.$$

Multiplying both sides by $e^{-p}\cdot p^{m-1}$, we have

$$\Gamma(n)e^{-p}\cdot p^{m-1} = \int_0^\infty e^{-px}\cdot p^{m-1}\cdot e^{-p}\cdot p^n\cdot x^{n-1}dx = \int_0^\infty e^{-p(x+1)}\cdot p^{m+n-1}\cdot x^{n-1}dx.$$

Now integrating both sides w.r.t. p from 0 to ∞, we get

$$\Gamma(n)\int_0^\infty e^{-p}\cdot p^{m-1}dp = \int_0^\infty\left[\int_0^\infty e^{-p(1+x)}\cdot p^{m+n-1}dp\right]x^{n-1}dx$$

$$\text{or, } \Gamma(n)\cdot\Gamma(m) = \int_0^\infty\left[\int_0^\infty e^{-z}z^{m+n-1}\cdot dz\right]\frac{x^{n-1}}{(1+x)^{m+n}}dx \text{ for putting } p(1+x)=z$$

$$\text{or, } \Gamma(n)\cdot\Gamma(m) = \int_0^\infty \Gamma(m+n)\cdot\frac{x^{n-1}}{(1+x)^{m+n}}dx$$

$$\text{or, } \Gamma(m)\cdot\Gamma(n) = \Gamma(m+n)\int_0^\infty\frac{x^{n-1}}{(1+x)^{m+n}}dx. \tag{i}$$

Again, putting $x = \frac{1}{1+y}$ in $\beta(m,n)$, we get

$$\beta(m,n) = \int_0^1 x^{m-1}(1-x)^{n-1}dx$$
$$= \int_\infty^0 \frac{1}{(1+y)^{m-1}} \cdot \frac{y^{n-1}}{(1+y)^{n-1}} \left[-\frac{1}{(1+y)^2}\right] dy = \int_0^\infty \frac{y^{n-1}}{(1+y)^{m+n}} dy, \quad \text{(ii)}$$

which is same as $\int_0^\infty \frac{x^{n-1}}{(1+x)^{m+n}} dx$.

Therefore, from (1) by (2), we have $\Gamma(m) \cdot \Gamma(n) = \Gamma(m+n) \cdot \beta(m,n)$.

That is, $\beta(m,n) = \frac{\Gamma(m)\Gamma(n)}{\Gamma(m+n)}$.

Remark. (1) Alternative form of Beta Function is

$$\beta(m,n) = \int_0^\infty \frac{x^{n-1}}{(1+x)^{m+n}} dx = \int_0^\infty \frac{x^{m-1}}{(1+x)^{m+n}} dx \left[\begin{array}{l}\text{2nd form is obtained by} \\ \text{putting } x = y/(1+y)\end{array}\right]$$

(2) $\beta(m,n) = \int_0^1 x^{m-1}(1-x)^{n-1}dx$.

For $x = \sin^2\theta$, $dx = 2\sin\theta\cos\theta d\theta$, $\theta : 0 \to \pi/2$.

$$\text{Thus } \beta(m,n) = \int_0^{\pi/2} \sin^{2m-2}\theta \cos^{2n-2}\theta \cdot 2\sin\theta\cos\theta d\theta$$
$$= 2\int_0^{\pi/2} \sin^{2m-1}\theta \cdot \cos^{2m-1}\theta \cdot d\theta.$$

$$\text{So } \beta(m,n) = \frac{\Gamma(m)\cdot\Gamma(n)}{\Gamma(m+n)}$$
$$\Rightarrow \int_0^{\pi/2} \sin^{2m-1}\theta \cos^{2n-1}\theta \, d\theta = \frac{\Gamma(m)\cdot\Gamma(n)}{2\Gamma(m+n)}$$
$$\Rightarrow \int_0^{\pi/2} \sin^p\theta \cos^q\theta \, d\theta = \frac{\Gamma\left(\frac{p+1}{2}\right)\cdot\Gamma\left(\frac{q+1}{2}\right)}{2\Gamma\left(\frac{p+q+2}{2}\right)}.$$

Theorem [14.2] [Duplication formula] $\Gamma m \cdot \Gamma\left(m + \frac{1}{2}\right) = \frac{\sqrt{\pi}\cdot\Gamma(2m)}{2^{2m-1}}$.

Proof. We have $\beta(m,n) = 2\int_0^{\pi/2} \sin^{2m-1}\theta \cdot \cos^{2n-1}\theta d\theta$ [by Remark (2)].

Now for $n = 1/2$, $\beta\left(m, \frac{1}{2}\right) = 2\int_0^{\pi/2} \sin^{2m-1}\theta \, d\theta$.

That is

$$\frac{\Gamma(m)\cdot\Gamma\left(\frac{1}{2}\right)}{\Gamma\left(m+\frac{1}{2}\right)} = 2\int_0^{\pi/2} \sin^{2m-1}\theta d\theta \quad \text{(i)}$$

Again

$$\beta(m,m) = 2\int_0^{\pi/2} \sin^{2m-1}\theta \cdot \cos^{2m-1}\theta \, d\theta \; [\because \; m = n]$$
$$= 2\int_0^{\pi/2} (\sin\theta\cos\theta)^{2m-1}\theta \, d\theta = \frac{2}{2^{2m-1}}\int_0^{\pi/2} \sin^{2m-1} 2\theta \, d\theta.$$

Putting $2\theta = \phi$,

$$\beta(m,m) = \frac{2}{2^{2m-1}} \times \frac{1}{2}\int_0^{\pi} \sin^{2m-1}\phi\, d\phi = \frac{1}{2^{2m-1}}\int_0^{\pi} \sin^{2m-1}\phi\, d\phi.$$

That is

$$\frac{(\Gamma m)^2}{\Gamma(2m)} = \frac{2}{2^{2m-1}}\int_0^{\pi/2} \sin^{2m-1}\phi\, d\phi \quad [\because \sin^{2m-1}(\pi-\phi) = \sin^{2m-1}\phi]$$

$$\text{or, } 2^{2m-1}\cdot\frac{(\Gamma m)^2}{\Gamma(2m)} = 2\int_0^{\pi/2} \sin^{2m-1}\phi\, d\phi. \tag{ii}$$

So from (i) and (ii),

$$\frac{\Gamma(m)\cdot\Gamma\left(\frac{1}{2}\right)}{\Gamma\left(m+\frac{1}{2}\right)} = 2^{2m-1}\frac{(\Gamma m)^2}{\Gamma(2m)}$$

$$\text{or, } \Gamma(m)\cdot\Gamma\left(m+\frac{1}{2}\right) = \frac{1}{2^{2m-1}}\cdot\Gamma\left(\frac{1}{2}\right)\cdot\Gamma(2m) \text{ [by cancelling } \Gamma(m)] \tag{iii}$$

Now we shall show that $\Gamma\left(\frac{1}{2}\right) = \sqrt{\pi}$.

For $m = 1/2,\ n = 1/2$,

$$\beta\left(\frac{1}{2},\frac{1}{2}\right) = 2\int_0^{\pi/2} \sin^{2\cdot\frac{1}{2}-1}\cos^{2\cdot\frac{1}{2}-1}\theta\, d\theta = 2\int_0^{\pi/2} d\theta = \pi \Rightarrow \frac{\Gamma\left(\frac{1}{2}\right)^2}{\Gamma(1)} = \pi$$

$$\Rightarrow \Gamma\left(\frac{1}{2}\right)^2 = \pi\cdot\Gamma(1) = \pi \quad \left[\because \Gamma(1) = \int_0^{\infty} e^{-x}\cdot dx = 1\right]$$

$$\Rightarrow \Gamma\left(\frac{1}{2}\right) = \sqrt{\pi}. \tag{iv}$$

Thus we have, $\Gamma(m)\cdot\Gamma\left(m+\frac{1}{2}\right) = \frac{\sqrt{\pi}\cdot\Gamma(2m)}{2^{2m-1}}$ [from (iii) by (iv)].

Theorem [14.3] $\Gamma(m)\cdot\Gamma(1-m) = \frac{\pi}{\sin m\pi}$ for $0 < m < 1$.

Proof. We have $\beta(m,n) = \dfrac{\Gamma(m)\cdot\Gamma(n)}{\Gamma(m+n)}$.

Then

$$\beta(m,1-m) = \frac{\Gamma(m)\cdot\Gamma(1-m)}{\Gamma(1)} = \Gamma(m)\cdot\Gamma(1-m)\ \ 0 < m < 1$$

$$= \int_0^{\infty}\frac{x^{m-1}}{(1+x)}dx \text{ [Remark (1)]} = \int_0^1 \frac{x^{m-1}}{(1+x)}dx + \int_1^{\infty}\frac{x^{m-1}}{(1+x)}dx.$$

Both these integral converge for $0 < m < 1$.

For the 2nd part, we put $x = 1/y$, then

$$\beta(m,1-m) = \int_0^1 \frac{x^{m-1}dx}{1+x} + \int_0^1 \frac{y^{-m}}{1+y}dy = \int_0^1 \frac{x^{m-1}dx}{1+x} + \int_0^1 \frac{x^{-m}}{1+x}dx$$

$$= \int_0^1 \frac{x^{m-1}+x^{-m}}{1+x}dx = \int_0^1 (x^{m-1}+x^{-m})\left[1-\frac{x}{1+x}\right]dx$$

$$= \int_0^1 (x^{m-1}+x^{-m})dx - \int_0^1 \frac{x^m + x^{1-m}}{1+x}dx. \tag{i}$$

1st Part : $\lim_{\epsilon\to 0}\int_\epsilon^1 \left(x^{m-1}+x^{-m}\right)dx$

$$= \lim_{\epsilon\to 0}\left[\frac{x^m}{m}+\frac{x^{1-m}}{1-m}\right]_\epsilon^1$$

$$= \lim_{\epsilon\to 0}\left[\left(\frac{1}{m}+\frac{1}{1-m}\right)-\left(\frac{\epsilon^m}{m}+\frac{\epsilon^{1-m}}{1-m}\right)\right]=\frac{1}{m}+\frac{1}{1-m}. \tag{ii}$$

2nd Part : $\int_0^1 \frac{x^m+x^{1-m}}{1+x}dx$ (a proper integral, since $0<m<1$)

$$= \int_0^1 \left(x^m+x^{1-m}\right)(1+x)^{-1}dx$$

$$= \int_0^1 \left(x^m+x^{1-m}\right)\left(1-x+x^2-x^3+\cdots\right)dx$$

$$= \int_0^1 \left[x^m-x^{m+1}+x^{2+m}-x^{3+m}+x^{1-m}-x^{2-m}+\cdots\right]dx.$$

We see that the infinite series in the bracket is uniformly convergent and so term by term integration is possible in $0\le x\le 1$.

$$= \left[\frac{x^{m+1}}{m+1}-\frac{x^{m+2}}{m+2}+\frac{x^{m+3}}{m+3}-\frac{x^{m+4}}{m+4}+\frac{x^{2-m}}{2-m}+\frac{x^{3-m}}{3-m}\right]_0^1$$

$$= \left[\frac{1}{m+1}-\frac{1}{m+2}+\frac{1}{m+3}-\frac{1}{m+4}+\frac{1}{2-m}-\frac{1}{3-m}+\cdots\right]$$

$$= \frac{1}{m+1}+\frac{1}{2-m}-\frac{1}{m+2}-\frac{1}{3-m}-\cdots \tag{iii}$$

Therefore, from (i) by (ii) and (iii),

$$\beta(m,1-m)=\left(\frac{1}{m}+\frac{1}{1-m}\right)-\left(\frac{1}{m+1}+\frac{1}{2-m}-\frac{1}{m+2}-\frac{1}{3-m}-\cdots\right)$$

$$= \sum_{k=0}^{\infty}(-1)^k\left(\frac{1}{k+m}+\frac{1}{k+1-m}\right).$$

That is,

$$\Gamma(m)\cdot\Gamma(1-m)=\sum_{k=0}^{\infty}(-1)^k\left(\frac{1}{k+m}+\frac{1}{k+1-m}\right). \tag{iv}$$

Again, we have

$$\frac{\pi}{\sin m\pi}=\sum_{k=0}^{\infty}(-1)^k\left(\frac{1}{k+m}+\frac{1}{k+1-m}\right). \tag{v}$$

[See Ch-11, Problem (11.6)—(3)]

Thus from (iv) and (v), we get $\Gamma(m)\cdot\Gamma(1-m)=\pi/\sin m\pi,\ 0<m<1$.

Example. If $\int_0^\infty \frac{x^{m-1}}{1+x}dx=\pi/\sin m\pi$, show that $\Gamma(m)\cdot\Gamma(1-m)=\pi/\sin m\pi,\ 0<m<1$.

Put $x=\frac{y}{1-y}$. Then $y=\frac{x}{1+x}$

$$x:0\to\infty \Rightarrow y:0\to 1.$$

Then the given integral $\int_0^\infty \frac{x^{m-1}}{1+x}dx$ is reduced to the form

$$\int_0^1 \left(\frac{y}{1-y}\right)^{m-1} \Big/ \left(1+\frac{y}{1-y}\right) \cdot \left[\frac{1(1-y)+y}{(1-y)^2}\right] dy$$

$$= \int_0^1 \frac{y^{m-1}}{(1-y)^{m-1}}(1-y) \cdot \frac{1}{(1-y)^2}dy = \int_0^1 y^{m-1} \cdot (1-y)^{-m} dy$$

$$= \int_0^1 y^{m-1}(1-y)^{(1-m)-1}dy = \beta(m, 1-m).$$

Then $\beta(m, 1-m) = \frac{\pi}{\sin m\pi}$

$$\Rightarrow \frac{\Gamma(m)\cdot\Gamma(1-m)}{\Gamma(m+1-m)} = \frac{\pi}{\sin m\pi} \Rightarrow \Gamma(m)\cdot\Gamma(1-m) = \frac{\pi}{\sin m\pi} \quad [\because \Gamma(1) = 1]$$

Example. Show that $\int_0^{\pi/2} \frac{dx}{\sqrt{\sin x}} \times \int_0^{\pi/2} \sqrt{\sin x}\, dx = \pi.$

$$\int_0^{\pi/2} \frac{dx}{\sqrt{\sin x}} \times \int_0^{\pi/2} \sqrt{\sin x}\, dx$$

$$= \int_0^{\pi/2} \sin^{-1/2} x \cos^0 x\, dx \times \int_0^{\pi/2} \sin^{1/2} x \cos^0 x\, dx$$

$$= \int_0^{\pi/2} \sin^{2\cdot\frac{1}{4}-1} x \cos^{2\cdot\frac{1}{2}-1} x\, dx \times \int_0^{\pi/2} \sin^{2\cdot\frac{3}{4}-1} x \cos^{2\cdot\frac{1}{2}-1} x\, dx$$

$$= \frac{1}{2}\beta\left(\frac{1}{4},\frac{1}{2}\right) \times \frac{1}{2}\beta\left(\frac{3}{4},\frac{1}{2}\right) \quad \text{[by Remark (2)]}$$

$$= \frac{1}{4}\cdot\frac{\Gamma\left(\frac{1}{4}\right)\Gamma\left(\frac{1}{2}\right)}{\Gamma\left(\frac{1}{4}+\frac{1}{2}\right)}\cdot\frac{\Gamma\left(\frac{3}{4}\right)\cdot\Gamma\left(\frac{1}{2}\right)}{\Gamma\left(\frac{3}{4}+\frac{1}{2}\right)} = \frac{1}{4}\cdot\frac{\left(\Gamma\left(\frac{1}{2}\right)\right)^2}{\Gamma\left(\frac{3}{4}\right)}\cdot\frac{\Gamma\left(\frac{1}{4}\right)\cdot\Gamma\left(\frac{3}{4}\right)}{\Gamma\left(\frac{5}{4}\right)}$$

$$= \frac{1}{4}\cdot\frac{\pi\cdot\Gamma\left(\frac{1}{4}\right)}{\frac{1}{4}\cdot\Gamma\left(\frac{1}{4}\right)} = \pi.$$

Elementary properties of Beta function and Gamma function

1. Beta function is symmetrical. That is, $\beta(m, n) = \beta(n, m)$.

Proof. $\beta(m, n) = \int_0^1 x^{m-1}(1-x)^{n-1}dx, \; m, n > 0$

Putting $1 - x = y, \; y : 1 \to 0.$

$$\beta(m, n) = \int_1^0 (1-y)^{m-1}\cdot y^{n-1}(-dy) = \int_0^1 y^{n-1}(1-y)^{m-1}dy, \; n, m > 0$$

$$= \beta(n, m).$$

2. $\beta(m, n)$ is continuous for $m > 0, n > 0$.

Proof. $\beta(m, n)$ converges for $m > 0, n > 0 \;\forall\; x \in [0, 1]$.

Then for every $\epsilon(> 0)$, $\exists$ a (+)ve number δ such that $\left|\int_{\epsilon_1}^{\epsilon_2} f(x)dx\right| < \epsilon, \; 0 < \epsilon_1, \epsilon_2 < \delta$, by considering the convergence at $x = 0$, where

$$f(x) = x^{m-1}(1-x)^{n-1} \qquad \text{(i)}$$

Let $\phi(x) = \int_0^x f(x)dx$ and $c \in [0,1]$.

Then $|\phi(c+h) - \phi(c)| = \left|\int_c^{c+h} f(x)dx\right|,$

$< \epsilon$ by (i), $|h| < \delta$.

Then $\phi(x)$ is continuous $\forall\ x \in [0,1]$ as c is an arbitrary point.

The same is true if we consider the convergence at 1.

Thus $\beta(m,n)$ is continuous for $m > 0,\ n > 0,\ x \in [0,1]$.

3. $\beta\left(\frac{1}{2}, \frac{1}{2}\right) = \pi.$

Proof. We have $\beta(m,n) = 2\int_0^{\pi/2} \sin^{2m-1}\theta \cos^{2m-1}\theta\, d\theta.$

For $m = 1/2,\ n = 1/2.$

$$\beta\left(\frac{1}{2}, \frac{1}{2}\right) = 2\int_0^{\pi/2} d\theta = 2\cdot\frac{\pi}{2} = \pi.$$

4. Gamma function is continuous $\forall\ x > 0.$

Proof. $\Gamma(n) = \int_0^\infty e^{-x}\cdot x^{n-1}dx,\ n > 0.$

It converges for $n > 0,\ \forall\ x \in [0,\infty)$. Now

$$\Gamma(n) = \int_0^1 e^{-x}\cdot x^{n-1}dx + \int_1^\infty e^{-x}x^{n-1}dx. \tag{i}$$

The first part converges at $x = 0$.

So for every $\epsilon(> 0)$, $\exists$ a (+)ve number δ such that $\left|\int_{\epsilon_1}^{\epsilon_2} f(x)dx\right| < \epsilon$, for $0 < \epsilon_1, \epsilon_2 < \delta$, where

$$f(x) = e^{-x}\cdot x^{n-1}. \tag{ii}$$

Let $\phi(x) = \int_0^x e^{-x}\cdot x^{n-1}dx$ and let $c \in [0,1]$.

Then $|\phi(c+h) - \phi(c)| = \left|\int_c^{c+h} e^{-x}\cdot x^{n-1}dx\right| = \left|\int_c^{c+h} f(x)dx\right| < \epsilon,\ |h| < \delta$ [by (ii)].

The first part is continuous $\forall\ x \in [0,1]$.

Again, considering the convergence of the 2nd part $\int_1^\infty e^{-x}x^{n-1}dx$ at ∞, we have, for every $\epsilon(> 0)$, $\exists$ a (+)ve number k such that

$$\left|\int_{t_1}^{t_2} e^{-x}x^{n-1}dx\right| < \epsilon\ \forall\ t_1, t_2 \geq k \tag{iii}$$

Let $\psi(t) = \int_1^t e^{-x}\cdot x^{n-1}dx$ and $t > 1$.

$$\text{Then } |\psi(t+h) - \psi(t)| = \left|\int_t^{t+h} e^{-x}x^{n-1}dx\right| < \epsilon,\ t,\ t+h > 1 \text{ by (3)}$$

$$\Rightarrow \psi(t+h) - \psi(t)| < \epsilon,\ |h| < \delta,\ t > 1$$

$$\Rightarrow \text{2nd part is continuous } \forall\ x \in [1, \infty].$$

Hence $\Gamma(n)$ is continuous for $n > 0$, $\forall\ x \in [0, \infty)$ as both parts in (i) are continuous.

5. (i) $\Gamma(n+1) = n\Gamma(n),\ n > 0$

(ii) $\Gamma(n+1) = n!,\ n \in N$

(iii) $\Gamma(n+m) = (n+m-1)(n+m-2)\cdots(n+1)\cdot n\Gamma(n)$, $\forall$ integers $m \geq 1$, $\forall\ n > 0$.

Proof. (i) We have

$$\Gamma(n+1) = \int_0^\infty e^{-x}\cdot x^{(n+1)-1}dx = \int_0^\infty e^{-x}\cdot x^n dx$$

$$= \left[-e^{-x}x^n\right]_0^\infty + \int_0^\infty e^{-x}nx^{n-1}dx \text{ [integrating by parts]}$$

$$= 0 + n\int_0^\infty e^{-x}x^{n-1}dx = n\cdot\Gamma(n).$$

(ii) $\Gamma(n+1) = n\Gamma(n) = n\cdot(n-1)\Gamma(n-1)$ [by (i)]

$= n(n-1)(n-2)\Gamma(n-2)$

$= n(n-1)(n-2)\cdots 3\cdot 2\cdot 1$ [repeating the process up to $(n-1)$ times]

$= n!,\ n \in N.$

(iii) $\Gamma(n+m) = (n+m-1)\Gamma(n+m-1),\ m \geq 1$ $\quad [m-1 \in N]$

$= (n+m-1)(n+m-2)\Gamma(n+m-2)$ $\quad [m-2 \in N]$.

Here n is fixed, m is changing. So continuing the process $m-1$ times, we get

$$\Gamma(n+m) = (n+m-1)(n+m-2)\cdots(n+1)\Gamma(n+1)$$

$$= (n+m-1)(n+m-2)\cdots(n+1)n\Gamma(n).$$

Note. (i) $\Gamma(0) = \infty$, (ii) $\Gamma(-n) = \infty$, when $n \in N$.

Problem [14.1]: (1) Show that $\int_a^b (x-a)^{m-1}(b-x)^{n-1}dx = (b-a)^{m+n-1}\beta(m,n),\ m,\ n > 0$ and hence show that $\int_a^b (x-a)^3(b-x)^2dx = \dfrac{(b-a)^6}{60}$.

(2) Show that $\int_0^\infty x^m e^{-ax^n}dx = \dfrac{1}{na^{\frac{(m+1)}{n}}}\Gamma\left(\dfrac{m+1}{n}\right)$, where m, n, a are (+)ve integers.

(3) Show that $\int_0^\infty xe^{-x^8}dx \times \int_0^\infty x^2e^{-x^4}dx = \dfrac{\pi}{16\sqrt{2}}$.

(4) $\int_0^\infty e^{-x^2}x^\alpha dx = \dfrac{1}{2}\Gamma\left(\dfrac{\alpha+1}{2}\right),\ \alpha > -1.$

(5) Find the value of $\Gamma\left(\dfrac{1}{n}\right)\cdot\Gamma\left(\dfrac{2}{n}\right)\cdot\Gamma\left(\dfrac{3}{n}\right)\cdots\Gamma\left(\dfrac{n-1}{n}\right)$, when n is an integer.

Solution. (1) Let $x = \alpha y + \beta$, where α, β are so chosen that when $x = a$, $y = 0$ and when $x = b$, $y = 1$. Then we get $a = \beta$ and $\alpha = b - \beta = b - a$ and $dx = \alpha dy = (b-a)dy$. Therefore, $x = (b-a)y + a$ and given integral reduces to the form

$$\int_0^1 \{(b-a)y + a - a)^{m-1}(b - (b-a)y - a)^{n-1}(b-a)dy$$
$$= \int_0^1 (b-a)^{m-1}y^{m-1}(b-a)^{n-1}(1-y)^{n-1} \cdot (b-a)dy$$
$$= \int_0^1 (b-a)^{m+n-1} \cdot y^{m-1} \cdot (1-y)^{n-1}dy$$
$$= (b-a)^{m+n-1} \int_0^1 y^{m-1}(1-y)^{n-1}dy = (b-a)^{m+n-1}\beta(m,n).$$

Taking $m = 4, n = 3$

$$\int_a^b (x-a)^{4-1}(b-x)^{3-1}dx = \int_a^b (x-a)^3(b-x)^2dx = (b-a)^{4+3-1}\beta(4,3)$$
$$= (b-a)^6 \frac{\Gamma(4)\Gamma(3)}{\Gamma(4+3)} = \frac{(b-a)^6}{60}.$$

(2) Given integral $= \int_0^\infty x^m e^{-ax^n}dx$. Put $ax^n = z$. Then $nax^{n-1}dx = dz$ and $x = (z/a)^{1/n}$.

Therefore, $dx = \dfrac{dz}{na} \cdot x^{1-n} = \dfrac{1}{na}\left(\dfrac{z}{a}\right)^{\frac{1-n}{n}} dz.$

$$x : 0 \to \infty \Rightarrow z : 0 \to \infty \; [\because \; a > 0]$$

So

$$\int_0^\infty x^m e^{-ax^n}dx = \int_0^\infty \left(\frac{z}{a}\right)^{\frac{m}{n}} \cdot e^{-z}\frac{1}{na}\left(\frac{z}{a}\right)^{\frac{1-n}{n}} dz$$
$$= \frac{1}{na \cdot a^{\frac{m}{n}} \cdot a^{\frac{1-n}{n}}} \int_0^\infty e^{-z} \cdot z^{\frac{m}{n} + \frac{1-n}{n}} dz$$
$$= \frac{1}{na^{1+\frac{m}{n}+\frac{1-n}{n}}} \int_0^\infty e^{-z} \cdot z^{\frac{m+1}{n}-1}dz = \frac{1}{na^{\frac{m+1}{n}}}\Gamma\left(\frac{m+1}{n}\right).$$

(3) $\displaystyle\int_0^\infty x \cdot e^{-x^8}dx \times \int_0^\infty x^2 \cdot e^{-x^4}dx = I$ (say)

For the 1st term, we put $x^8 = z$.

Then $x = z^{1/8}$ and $8x^7dx = dz \Rightarrow x\,dx = \dfrac{dz}{8x^6} = \dfrac{1}{8}z^{-6/8}dz = \dfrac{1}{8}z^{-3/4}dz.$

Then $\int_0^\infty xe^{-x^8}dx = \int_0^\infty e^{-z} \cdot \frac{1}{8}z^{-3/4}dz = \frac{1}{8}\int_0^\infty e^{-z} \cdot z^{\frac{1}{4}-1}dz = \frac{1}{8}\Gamma\left(\frac{1}{4}\right).$

For the 2nd term, we put $x^4 = t$.

Then $x = t^{1/4}$ and $4x^3dx = dt \Rightarrow x^2dx = \frac{dt}{4x} = \frac{1}{4}t^{-1/4}dt.$

Then $\int_0^\infty x^2e^{-x^4}dx = \frac{1}{4}\int_0^\infty e^{-t} \cdot t^{3/4-1}dt = \frac{1}{4}\Gamma\left(\frac{3}{4}\right).$

Hence $I = \frac{1}{8}\Gamma\left(\frac{1}{4}\right) \cdot \frac{1}{4}\Gamma\left(\frac{3}{4}\right) = \frac{1}{32}\Gamma\left(\frac{1}{4}\right) \cdot \Gamma\left(\frac{3}{4}\right).$

We know $\Gamma(m) \cdot \Gamma(1-m) = \dfrac{\pi}{\sin m\pi}$ for $0 < m < 1$ [Th. 14.3]

For $m = 1/4$,

$$\Gamma\left(\frac{1}{4}\right)\cdot\Gamma\left(\frac{3}{4}\right) = \frac{\pi}{\sin\frac{\pi}{4}} = \sqrt{2}\pi. \tag{i}$$

Therefore $I = \frac{1}{32}\cdot\sqrt{2}\pi = \frac{\pi}{16\sqrt{2}}$ [by (i)]

(4) Let $I = \int_0^\infty e^{-x^2}x^\alpha dx$. Put $x^2 = z$. Then $x = z^{1/2}$

$$2x\,dx = dz \Rightarrow dx = \frac{dz}{2\cdot z^{1/2}} = \frac{1}{2}z^{-\frac{1}{2}}dz.$$

Therefore,

$$\begin{aligned} I &= \int_0^\infty e^{-z}.z^{\frac{\alpha}{2}}\cdot\frac{1}{2}z^{-\frac{1}{2}}dz = \frac{1}{2}\int_0^\infty e^{-z}\cdot z^{\frac{\alpha-1}{2}}dz \\ &= \frac{1}{2}\int_0^\infty e^{-z}\cdot z^{\frac{\alpha+1}{2}-1}dz = \frac{1}{2}\Gamma\left(\frac{\alpha+1}{2}\right),\ \alpha > -1. \end{aligned}$$

(5) Let

$$\lambda = \Gamma\left(\frac{1}{n}\right)\cdot\Gamma\left(\frac{2}{n}\right)\cdot\Gamma\left(\frac{3}{n}\right)\cdots\Gamma\left(\frac{n-1}{n}\right) \tag{i}$$

Writing the expression in the reversed order,

$$\begin{aligned} \lambda &= \Gamma\left(\frac{n-1}{n}\right)\cdot\Gamma\left(\frac{n-2}{n}\right)\cdots\Gamma\left(\frac{3}{n}\right)\cdot\Gamma\left(\frac{2}{n}\right)\cdot\Gamma\left(\frac{1}{n}\right) \\ &= \Gamma\left(1-\frac{1}{n}\right)\cdot\Gamma\left(1-\frac{2}{n}\right)\cdots\Gamma\left(\frac{3}{n}\right)\cdot\Gamma\left(\frac{2}{n}\right)\cdot\Gamma\left(\frac{1}{n}\right). \end{aligned} \tag{ii}$$

Multiplying (i) and (ii),

$$\begin{aligned} \lambda^2 &= \Gamma\left(\frac{1}{n}\right)\cdot\Gamma\left(1-\frac{1}{n}\right)\cdot\Gamma\left(\frac{2}{n}\right)\cdot\Gamma\left(1-\frac{2}{n}\right) \\ &\quad\cdots\Gamma\left(\frac{n-1}{n}\right)\cdot\Gamma\left(1-\frac{n-1}{n}\right) \quad \left[\because \frac{1}{n} = 1-\frac{n-1}{n}\right] \\ &= \frac{\pi}{\sin\frac{\pi}{n}}\cdot\frac{\pi}{\sin\frac{2\pi}{n}}\cdots\frac{\pi}{\sin\frac{n-1}{n}\pi} \quad \left[\because \Gamma m\cdot\Gamma(1-m) = \frac{\pi}{\sin m\pi}\right] \\ &= \frac{\pi^{n-1}}{\sin\frac{\pi}{n}\cdot\sin\frac{2\pi}{n} - \cdots\sin(n-1)\frac{\pi}{n}}. \end{aligned} \tag{iii}$$

To evaluate the denominator of (iii) we consider the equation $1 - x^{2n} = 0$.

Roots of this equation are given by $x = \pm 1$ (real roots) and $x = \cos\frac{k\pi}{n} \pm i\sin k\frac{\pi}{n}, k = 1, 2, \cdots, n-1$.

Then

$$1 - x^{2n} = (1-x^2)\prod_{k=1}^{n-1}\left(x^2 - 2x\cos\frac{k\pi}{n} + 1\right) \text{ or, } \frac{1-x^{2n}}{1-x^2} = \prod_{k=1}^{n-1}\left(x^2 - 2x\cos\frac{k\pi}{n} + 1\right)$$

$$\text{or, } 1 + x^2 + (x^2)^2 + \cdots + (x^2)^{n-1} = \left(1 - 2x\cos\frac{\pi}{n} + x^2\right)\cdot\left(1 - 2x\cos\frac{2\pi}{n} + x^2\right)$$
$$\cdots\left(1 - 2x\cos\frac{n-1}{n}\pi + x^2\right).$$

Putting $x = 1$ and $x = -1$ respectively, we get

$$n = \left(2 - 2\cos\frac{\pi}{n}\right)\left(2 - 2\cos\frac{2\pi}{n}\right)\cdots\left(2 - 2\cos\frac{n-1}{n}\pi\right)$$

$$\text{and } n = \left(2 + 2\cos\frac{\pi}{n}\right)\left(2 + 2\cos\frac{2\pi}{n}\right)\cdots\left(2 + 2\cos\frac{n-1}{n}\pi\right).$$

Multiplying these, we have

$$n^2 = 2^{2n-2}\left(1 - \cos^2\frac{\pi}{n}\right)\left(1 - \cos^2\frac{2\pi}{n}\right)\cdots\left(1 - \cos^2\frac{n-1}{n}\pi\right)$$

$$= 2^{2n-2}\sin^2\frac{\pi}{n}\cdot\sin^2\frac{2\pi}{n}\cdots\sin^2\frac{(n-1)\pi}{n} \Rightarrow \sin\frac{\pi}{n}\cdot\sin\frac{2\pi}{n}\cdots\sin\frac{n-1}{n}\pi$$

$$= \frac{n}{2^{n-1}}. \tag{iv}$$

Therefore, from (iii) by (iv),

$$\lambda^2 = \frac{\pi^{n-1}}{n}2^{n-1} = \frac{(2\pi)^{n-1}}{n}$$

$$\text{or, } \lambda = \frac{(2\pi)^{\frac{n-1}{2}}}{n^{1/2}} = \left[\frac{(2\pi)^{n-1}}{n}\right]^{1/2}.$$

Hence the value of

$$\Gamma\left(\frac{1}{n}\right)\cdot\Gamma\left(\frac{2}{n}\right)\cdots\Gamma\left(\frac{n-1}{n}\right) = \left[\frac{(2\pi)^{n-1}}{n}\right]^{1/2}.$$

Application of Beta function and Gamma function in Multiple integrals

Example. Show that $\iint\limits_E x^{m-1}\cdot y^{n-1}dxdy = \dfrac{\Gamma(m)\cdot\Gamma(n)}{\Gamma(m+n+1)}h^{m+n}$, where the region E is bounded by $x \geq 0$, $y \geq 0$ and $x + y \leq h$.

Putting $x = Xh, y = Yh, dx\,dy = |J|\,dX\,dY = h^2 dX\,dY$, where

$$J = \frac{\partial(x,y)}{\partial(X,Y)} = \begin{vmatrix} \dfrac{\partial x}{\partial X} & \dfrac{\partial y}{\partial X} \\ \dfrac{\partial x}{\partial Y} & \dfrac{\partial y}{\partial Y} \end{vmatrix}, \text{ where } E'[X \geq 0, Y \geq 0, X + Y \leq 1].$$

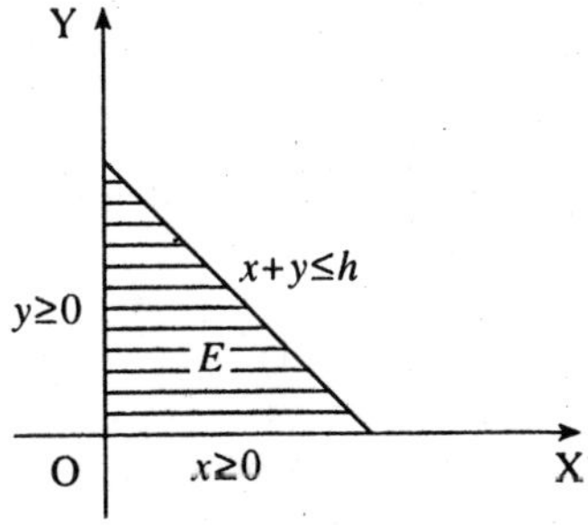

Fig. 14.1(i)

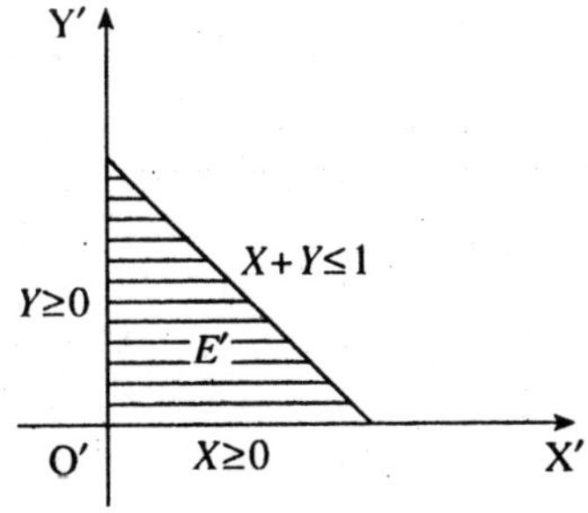

Fig. 14.1(ii)

Then by virtue of transformation

$$\iint_E x^{m-1}y^{n-1}dx\,dy = h^{m+n}\int_0^1\int_0^{1-X} X^{m-1}Y^{n-1}dX\,dY$$

$$= h^{m+n}\int_0^1 X^{m-1}\left[\int_0^{1-X} Y^{n-1}dY\right]dX$$

$$= h^{m+n}\int_0^1 X^{m-1}\left[\frac{Y^n}{n}\right]^{1-X} dX = \frac{h^{m+n}}{n}\int_0^1 X^{m-1}(1-X)^n dX$$

$$= \frac{h^{m+n}}{n}\int_0^1 X^{m-1}(1-X)^{(n+1)-1}dX$$

$$= \frac{h^{m+n}}{n}\beta(m, n+1) = \frac{h^{m+n}}{n}\frac{\Gamma(m)\cdot\Gamma(n+1)}{\Gamma(m+n+1)}$$

$$= \frac{h^{m+n}\cdot\Gamma(m)\cdot n\Gamma(n)}{n\cdot\Gamma(m+n+1)} = h^{m+n}\frac{\Gamma(m)\cdot\Gamma(n)}{\Gamma(m+n+1)}.$$

Theorem [14.4] [Dirichlet's Theorem] $\iiint_V x^{l-1}y^{m-1}z^{n-1}dx\,dy\,dz = \dfrac{\Gamma(l)\cdot\Gamma(m)\cdot\Gamma(n)}{\Gamma(l+m+n+1)}$,

where V is the region bounded by $x \geq 0, y \geq 0, z \geq 0$ and $x+y+z \leq 1$.

Proof.

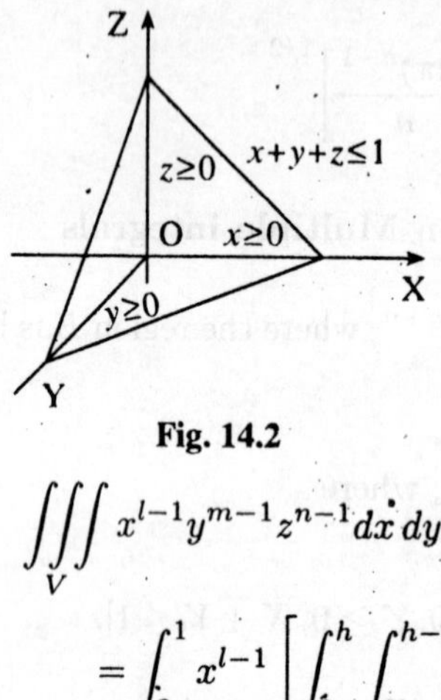

Fig. 14.2

Putting $1 - x = h$, we get
$y + z \leq h,\ z \leq h - y.$

$$\iiint_V x^{l-1}y^{m-1}z^{n-1}dx\,dy\,dz = \int_0^1 x^{l-1}\int_0^{1-x} y^{m-1}dy\int_0^{1-x-y} z^{n-1}dz$$

$$= \int_0^1 x^{l-1}\left[\int_0^h\int_0^{h-y} y^{m-1}z^{n-1}dy\,dz\right]dx$$

$$= \int_0^1 x^{l-1}\left[\frac{\Gamma(m)\cdot\Gamma(n)}{\Gamma(m+n+1)}\cdot h^{m+n}\right]dx \text{ [by Ex. (1)]}$$

$$= \frac{\Gamma(m)\cdot\Gamma(n)}{\Gamma(m+n+1)}\int_0^1 x^{l-1}(1-x)^{m+n}dx = \frac{\Gamma(m)\cdot\Gamma(n)}{\Gamma(m+n+1)}\cdot\beta(l, m+n+1)$$

$$= \frac{\Gamma(m)\cdot\Gamma(n)}{\Gamma(m+n+1)}\cdot\frac{\Gamma(l)\cdot\Gamma(m+n+1)}{\Gamma(l+m+n+1)} = \frac{\Gamma(l)\cdot\Gamma(m)\cdot\Gamma(n)}{\Gamma(l+m+n+1)}.$$

Example. Prove that the value of $\iint_E \sqrt{xy(1-x-y)}\,dx\,dy$ taken over the interior of the triangle bounded by the lines $x=0,\ y=0,\ x+y-1=0$ is $2\pi/105$.

We substitute $x + y = u$ and $x = uv$.

Then $x = uv \Rightarrow y = u(1-v) \Rightarrow |J| = \left|\dfrac{\partial(x,y)}{\partial(u,v)}\right| = u.$

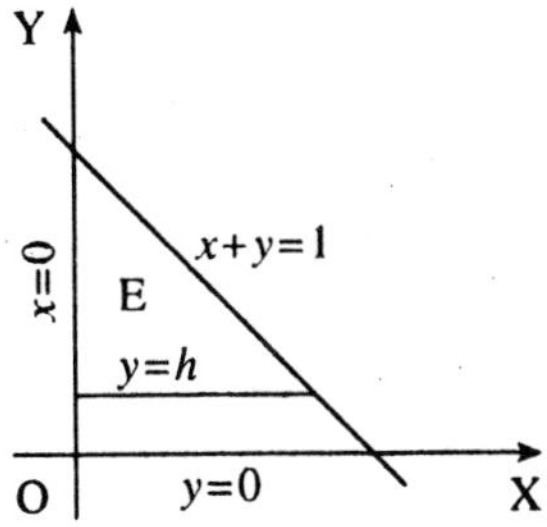

XY-plane

Fig. 14.3(i)

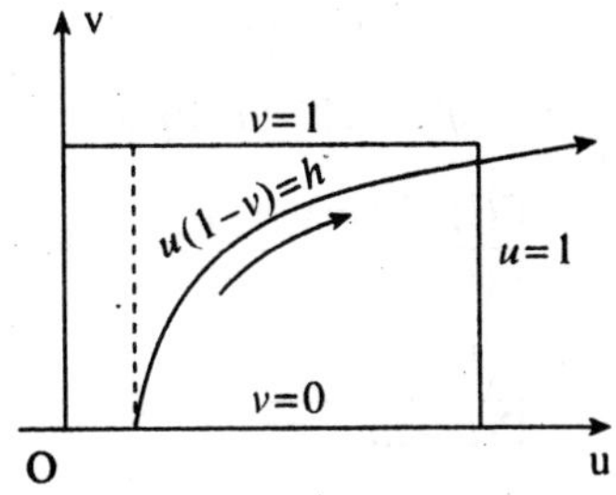

uv-plane

Fig. 14.3(ii)

The integral over the region bounded by $x + y = 1$, $x = 0$, $y = h$, exists if $h \to 0$.

The transformed region is then bounded by the lines $u = 1, v = 0, u(1-v) = h$ corresponding to the three boundaries of the region in XY-plane as the correspondence is one-one. Thus when $h \to 0$, the region in the uv-plane tends to a square bounded by the lines $u = 0$, $v = 0$, $u = 1$, $v = 1$.

Therefore

$$\iint_E \sqrt{xy(1-x-y)}\, dx\, dy$$

$$= \int_0^1 \int_0^1 \sqrt{uv \cdot u(1-v)(1-u)}\, |J|\, du\, dv = \int_0^1 \int_0^1 u\sqrt{v(1-v)(1-u)} \cdot u \cdot du\, dv$$

$$= \int_0^1 u^2(1-u)^{1/2} du \cdot \int_0^1 v^{1/2}(1-v)^{1/2} dv$$

$$= \int_0^1 u^{3-1}(1-u)^{\frac{3}{2}-1} du \cdot \int_0^1 v^{\frac{3}{2}-1}(1-v)^{\frac{3}{2}-1} dv$$

$$= \beta\left(3, \frac{3}{2}\right) \cdot \beta\left(\frac{3}{2}, \frac{3}{2}\right)$$

$$= \frac{\Gamma(3) \cdot \Gamma\left(\frac{3}{2}\right)}{\Gamma\frac{9}{2}} \cdot \frac{\Gamma\left(\frac{3}{2}\right) \cdot \Gamma\left(\frac{3}{2}\right)}{\Gamma(3)} = \frac{\left[\frac{1}{2} \cdot \Gamma\left(\frac{1}{2}\right)\right]^2 \cdot \Gamma\left(\frac{3}{2}\right)}{\frac{7}{2} \cdot \frac{5}{2} \cdot \frac{3}{2}\Gamma\left(\frac{3}{2}\right)}$$

$$= \frac{\frac{1}{4} \cdot \pi}{\frac{105}{8}} = \frac{2\pi}{105} \quad \left[\because \Gamma\left(\frac{1}{2}\right) = \Gamma\pi\right]$$

Theorem [14.5] [Liouville's Theorem (Extension of Dirichlet's Theorem)] If the variables x, y, z are all positive such that $h_1 < x + y + z < h_2$.

Then $$\iiint f(x+y+z) x^{l-1} y^{m-1} z^{n-1} dx\, dy\, dz = \frac{\Gamma(l) \cdot \Gamma(m) \cdot \Gamma(n)}{\Gamma(l+m+n+1)} \int_{h_1}^{h_2} f(u) \cdot u^{l+m+n-1} du.$$

Proof. Let $x + y + z \leq u$. Then under this condition by Dirichlet's theorem [14.5]

$$I = \iiint x^{l-1} y^{m-1} z^{n-1}\, dx\, dy\, dz = u^{l+m+n} \frac{\Gamma(l) \cdot \Gamma(m) \cdot \Gamma(n)}{\Gamma(l+m+n+1)}. \qquad \text{(i)}$$

Now if $x+y+z \le u+\delta u$, then

$$I = (u+\delta u)^{l+m+n}\frac{\Gamma(l)\cdot\Gamma(m)\cdot\Gamma(n)}{\Gamma(l+m+n+1)}. \tag{ii}$$

So for $u < x+y+z < u+\delta u$

$$\iiint x^{l-1}y^{m-1}z^{n-1}dx\,dy\,dz$$

$$= \frac{\Gamma(l)\cdot\Gamma(m)\cdot\Gamma(n)}{\Gamma(l+m+n+1)}\cdot\left[(u+\delta u)^{l+m+n} - u^{l+m+n}\right]$$

$$= \frac{\Gamma(l)\cdot\Gamma(m)\cdot\Gamma(n)}{\Gamma(l+m+n+1)}u^{l+m+n}\left[\left\{1+\left(\frac{\delta u}{u}\right)\right\}^{l+m+n} - 1\right]$$

$$= \frac{\Gamma(l)\cdot\Gamma(m)\cdot\Gamma(n)}{\Gamma(l+m+n+1)}u^{l+m+n}\left[1+(l+m+n)\frac{\delta u}{u}+\cdots-1\right]$$

$$= \frac{\Gamma(l)\cdot\Gamma(m)\cdot\Gamma(n)}{\Gamma(l+m+n+1)}\cdot u^{l+m+n}\cdot(l+m+n)\cdot\frac{\delta u}{u}$$

$$= \frac{\Gamma(l)\cdot\Gamma(m)\cdot\Gamma(n)}{\Gamma(l+m+n)}\cdot u^{l+m+n-1}\delta u \text{ [neglecting higher power of } \delta u].$$

Therefore, $\iiint f(x+y+z)x^{l-1}y^{m-1}z^{n-1}dx\,dy\,dz$ under the condition $h_1 \le x+y+z \le h_2$,

$$= \frac{\Gamma(l)\cdot\Gamma(m)\cdot\Gamma(n)}{\Gamma(l+m+n)}\int_{h_1}^{h_2} f(u)u^{l+m+n-1}du.$$

Since when $x+y+z$ lies between u and $u+\delta u$, the value of $f(x+y+z)$ differs from $f(u)$ by a small quantity as higher power of δu is neglected.

Example. Find the value of $\iiint \log(x+y+z)dx\,dy\,dz$, the integral is extended over all positive values of x, y, z subject to the condition $x+y+z<1$.

Applying Liouville's theorem when $0 < x+y+z < 1$.

$$\iint \log(x+y+z)dx\,dy\,dz$$

$$= \iiint \log(x+y+z)\cdot x^{1-1}\cdot y^{1-1}\cdot z^{1-1}dx\,dy\,dz$$

$$= \frac{\Gamma(1)\cdot\Gamma(1)\cdot\Gamma(1)}{\Gamma(3)}\int_0^1 \log(u)\,u^{1+1+1-1}du = \frac{1}{2\Gamma(1)}\int_0^1 u^2\log u\,du$$

$$= \frac{1}{2}\left[\log u\cdot\frac{u^3}{3} - \frac{1}{3}\cdot\frac{u^3}{3}\right]_0^1 = \frac{1}{2}\left[-\frac{1}{9}\right] = -\frac{1}{18}.$$

Example. Prove that $\iint\limits_E x^{m-1}y^{n-1}f(x+y)dx\,dy = \frac{\Gamma(m)\cdot\Gamma(n)}{\Gamma(m+n)}\int_0^1 u^{m+n-1}f(u)du$, where D is $\{(x,y), x, y \ge 0,\ x+y \le 1\}$.

By Lionville's theorem, when $0 < x+y < 1$, we get

$$\iint\limits_E x^{m-1}y^{n-1}f(x+y)dx\,dy = \frac{\Gamma(m)\cdot\Gamma(n)}{\Gamma(m+n)}\int_0^1 f(u)\cdot u^{m+n-1}du.$$

Theorem [14.6] For $m > 0, n > 0, \beta(m,n) = \dfrac{\Gamma(m)\cdot\Gamma(n)}{\Gamma(m+n)}$.

[An alternative proof of Th. (14.1) with geometrical approach.]

Proof. We have

$$\beta(m,n) = \int_0^1 x^{m-1}(1-x)^{n-1}dx,\ m,n>0 \tag{i}$$

$$\text{and } \Gamma(m) = \int_0^\infty e^{-t}\cdot t^{m-1}dt,\ m>0. \tag{ii}$$

Putting $x = \cos^2\theta$ and $t = x^2$ in (i) and (ii) respectively, we get

$$\beta(m,n) = 2\int_0^{\pi/2} \cos^{2m-1}\theta \sin^{2n-1}\theta\, d\theta \text{ and } \Gamma(m) = 2\int_0^\infty x^{2m-1}e^{-x^2}dx.$$

Similarly, $\Gamma(n) = 2\int_0^\infty y^{2n-1}e^{-y^2}dy$ [for $t = y^2$]

$$\begin{aligned}\Gamma(m)\cdot\Gamma(n) &= 4\int_0^\infty x^{2m-1}e^{-x^2}dx\cdot\int_0^\infty y^{2n-1}e^{-y^2}dy\\ &= 4\int_0^\infty\int_0^\infty x^{2m-1}\cdot y^{2n-1}\cdot e^{-(x^2+y^2)}dx\,dy\\ &= 4\iint_E x^{2m-1}y^{2n-1}\cdot e^{-(x^2+y^2)}dx\,dy, \qquad\text{(iii)}\end{aligned}$$

where E is the square $[0,R;\ 0,R]$ tends to its limit $\Gamma(m)\cdot\Gamma(n)$ as $R\to\infty$.

ADC is the arc of a circle $x^2+y^2=R^2$ and LBM is the arc of a circle $x^2+y^2=2R^2$, both lying in the 1st quadrant. We denote the parts of the circle $x^2+y^2=R^2$ and $x^2+y^2=2R^2$ in the 1st quadrant by E_1 and E_2 respectively. Here we see that $E_1\subset E\subset E_2$.

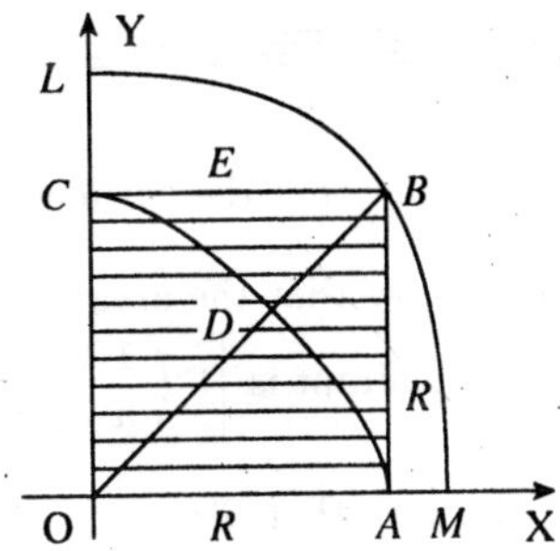

Fig. 14.4

Then it follows that

$$\begin{aligned}&4\iint_{E_1} x^{2m-1}y^{2n-1}e^{-(x^2+y^2)}dx\,dy\\ &\le 4\iint_{E} x^{2m-1}y^{2n-1}e^{-(x^2+y^2)}dx\,dy\\ &\le \iint_{E_2} x^{2m-1}y^{2n-1}e^{-(x^2+y^2)}dx\,dy. \qquad\text{(iv)}\end{aligned}$$

Changing to polar coordinates, i.e., $x = r\cos\theta$, $y = r\sin\theta$,

$$\begin{aligned}4\iint_{E_1} x^{2m-1}y^{2n-1}e^{-(x^2+y^2)}dx\,dy &= 4\int_0^{\pi/2}\int_0^R \cos^{2m-1}\theta\sin^{2n-1}\theta r^{2(m+n)-1}\cdot e^{-r^2}dr\,d\theta\\ &= 4\int_0^{\pi/2}\cos^{2m-1}\theta\sin^{2n-1}\theta d\theta\cdot\int_0^R r^{2(m+n)-1}\cdot e^{-r^2}dr\\ &= 2\beta(m,n)\int_0^R r^{2(m+n)-1}\cdot e^{-r^2}dr.\end{aligned}$$

Similarly,

$$4\iint_{E_2} x^{2m-1}\cdot y^{2n-1}e^{-(x^2+y^2)}dx\,dy = 2\beta(m,n)\int_0^{\sqrt{2}R} r^{2(m+n)-1}\cdot e^{-r^2}dr.$$

From (iv), we get

$$2\beta(m,n)\int_0^R e^{-r^2}\cdot r^{2(m+n)-1}dr \le 4\iint_E x^{2m-1}\cdot y^{2n-1}\cdot e^{-(x^2+y^2)}dx\,dy$$

$$\le 2\beta(m,n)\int_0^{\sqrt{2}R} e^{-r^2}\cdot r^{2(m+n)-1}dr.$$

Now taking $R\to\infty$, we have

$$\beta(m,n)\Gamma(m+n) \le \Gamma(m)\cdot\Gamma(n) \le \beta(m,n)\Gamma(m+n) \text{ [by (iii) and (iv)]}$$

$$\Rightarrow \beta(m,n)\Gamma(m+n) = \Gamma(m)\cdot\Gamma(n)$$

$$\Rightarrow \beta(m,n) = \frac{\Gamma(m)\cdot\Gamma(n)}{\Gamma(m+n)}.$$

Note. $2\int_0^R e^{-r^2}\cdot r^{2(m+n)-1}dr$

$$= \int_0^{R^2} e^{-z}z^{m+n}\cdot\frac{1}{z^{1/2}}\cdot\frac{dz}{z^{1/2}}, \text{ putting } r^2 = z, 2r\,dr = dz, 2dr = \frac{dz}{z^{1/2}}$$

$$= \int_0^{R^2} e^{-z}\cdot z^{m+n-1}dz \Rightarrow \int_0^\infty e^{-z}\cdot z^{m+n-1}dz = \Gamma(m+n),$$

as $R\to\infty$. That is, $2\int_0^R e^{-r^2}r^{2(m+n)-1}dr \Rightarrow \Gamma(m+n)$, when $R\to\infty$.

EXERCISE

(1) Show that the perimeter of a loop of the curve $r^n = a^n\cos n\theta$ is $\frac{a}{n}\cdot 2^{\frac{1}{n}-1}\left[\Gamma\left(\frac{1}{2n}\right)\right]^2\Big/\Gamma\left(\frac{1}{n}\right)$.

(2) Show that the area bounded by the curve $x^m + y^m = a^m$ and the coordinate axes in the first quadrant is $\left[\Gamma\left(\frac{1}{n}\right)\right]^2\Big/2n\Gamma\left(\frac{2}{n}\right)$.

(3) Prove that $\beta(m,n) = \int_0^1 \frac{x^{m-1}+x^{n-1}}{(1+x)^{m+n}}dx, m, n > 0$.

(4) Show that $\int_0^{\pi/2}\sin^n x\,dx = \frac{\sqrt{\pi}}{2}\frac{\Gamma\left(\frac{n+1}{2}\right)}{\Gamma\left(\frac{n+2}{2}\right)}, n > -1$.

(5) Show that $\int_0^\infty x^m e^{-ax^n}dx = \frac{1}{na^{\frac{m+1}{n}}}\Gamma\left(\frac{m+1}{n}\right)$, where m, n, a are (+)ve integers.

(6) Show that

(i) $\int_0^{\pi/2}\sin^p x\,dx \times \int_0^{\pi/2}\sin^{p+1}x\,dx = \frac{\pi}{2(p+1)}$,

(ii) $\int_0^\infty 5^{-x^2}dx = \frac{1}{2\sqrt{\log 5}}\cdot\sqrt{\pi}$.

(7) Prove that the 'volume' and 'area' of n-dimensional sphere of radius a are respectively

$$\frac{\frac{1}{2^n}a^n\left[\Gamma\left(\frac{1}{2}\right)\right]^n}{\Gamma\left(\frac{n}{2}+1\right)} \text{ and } 2a^{n-1}\left[\Gamma\left(\frac{1}{2}\right)^n \Big/ \Gamma\left(\frac{n}{2}\right)\right].$$

(8) If $m \geq 0$, then show that

$$\iint \left(1-x^2-y^2\right)^m f(Ax+By)dx\,xy = \beta\left(\frac{1}{2}, m+1\right)\int_{-1}^{1}\left(1-x^2\right)^{m+\frac{1}{2}} f(kx)dx,$$

where $k^2 = A^2 + B^2$ and the integral on the left is taken over the circle $x^2 + y^2 = 1$.

(9) Show that $\displaystyle\int_0^1 \log\Gamma(x)dx = \frac{1}{2}\log(2\pi)$.

(10) Prove that $\Gamma(x)\cdot\Gamma(2x) = 2^{2x-1}\Gamma(x)\cdot\Gamma\left(x+\frac{1}{2}\right)$.

CHAPTER 15

SOLUTION OF PROBLEMS WITH SOME NEW DEFINITIONS AND THEOREMS

Problem [15.1]: Let $\phi : [a,b] \to \mathbb{R}$ be R-integrable and $f : [a,b] \to \mathbb{R}$ be defined as $f(x) = \int_a^x \phi(t)dt \; \forall \; x \in [a,b] \subseteq \mathbb{R}$. Then show that f is a function of bounded variation [f.b.v.] on $[a,b]$.

Solution. Let $P\{a = x_0 < x_1 \cdots < x_{r-1} < x_r < \cdots < x_{n-1} < x_n = b\}$ be a partition of $[a,b]$.

Then $V_f(a,b) = \sup\left\{\sum_{r=1}^{n} |f(x_r) - f(x_{r-1})|\right\}$.

Now $\sum_{r=1}^{n} |f(x_r) - f(x_{r-1})|$

$$= \sum_{r=1}^{n}\left|\int_a^{x_r} \phi(t)dt - \int_a^{x_{r-1}} \phi(t)dt\right| \quad \left[\because f(x) = \int_a^x \phi(t)dt\right]$$

$$= \sum_{r=1}^{n}\left|\int_{x_{r-1}}^{x_r} \phi(t)dt\right| = \left|\int_a^{x_1} \phi(t)dt\right| + \left|\int_{x_1}^{x_2} \phi(t)dt\right| + \cdots + \left|\int_{x_{n-1}}^{x_n} \phi(t)dt\right|$$

$$\leq \int_a^{x_1} |\phi(t)|dt + \int_{x_1}^{x_2} |\phi(t)|dt + \cdots + \int_{x_{n-1}}^{b} |\phi(t)|dt$$

$$= \int_a^b |\phi(t)|dt \Rightarrow V_f[a,b] \leq \int_a^b |\phi(t)|dt. \quad (1)$$

ϕ is integrable on $[a,b] \Rightarrow |\phi|$ is integrable on $[a,b]$

$\Rightarrow \int_a^b |\phi(t)|dt$ exists finitely

$\Rightarrow V_f[a,b] <$ finite number

$\Rightarrow f$ is f.b.v. on $[a,b]$.

Problem [15.2]: Let $f : [a,b] \to [c,d]$, $g : [c,d] \to \mathbb{R}$ be two functions with g is f.b.v. on $[c,d]$, f is monotonic on $[a,b]$. Then show that $g \circ f : [a,b] \to \mathbb{R}$ is a f.b.v. on $[a,b]$.

Solution. Let $P\{a = x_0 < x_1 \cdots < x_{r-1} < x_r \cdots < x_n = b\}$ be partition of $[a, b]$.

Then

$$\sum_{r=1}^{n} |(g \circ f)(x_r) - (g \circ f)(x_{r-1})| = \sum_{r=1}^{n} |g(f(x_r)) - g(f(x_{r-1}))|$$

$$= \sum_{r=1}^{n} |g(y_r) - g(y_{r-1})|, \qquad (1)$$

when $f(x_r) = y_r$, $f(x_{r-1}) = y_{r-1}$ are the points on $[c, d]$.

Then the points y_{r-1}, y_r forms a division of $[c, d]$.

Let f be monotonically increasing on $[a, b]$.

Then $x_r > x_{r-1} \Rightarrow f(x_r) > f(x_{r-1}) \Rightarrow y_r > y_{r-1}$.

Thus $\exists$ a partition $P'\{c = y_0 < y_1 \cdots < y_{r-1} < y_r \cdots < y_n = d\}$ of $[c, d]$.

Therefore, from (1) it follows that

$$\sup \sum_{r=1}^{n} |(g \circ f)(x_r) - (g \circ f)(x_{r-1})| = \sup \sum_{r=1}^{n} |g(y_r) - g(y_{r-1})| \leq \text{ a finite number},$$

since g is f.b.v. on $[c, d]$.

$\Rightarrow g \circ f$ is a f.b.v. on $[a, b]$.

The same can be proved if f is non-increasing on $[a, b]$.

Alternative proof

Let f be an increasing function on $[a, b]$. Since g is f.b.v. on $[c, d]$, by Jordan's Theorem [6.5], $\exists$ two increasing functions u and v such that $g = u - v$.

Then $g \circ f = u \circ f - v \circ f = u_1 - u_2$, where u_1, u_2 are increasing functions. Hence $g \circ f$ is f.b.v. on $[a, b]$.

Definition [15.1] : (i) Total variation as a function x is defined by

$$v(x) = \begin{cases} V_f[a, x], & a < x \leq b \\ 0, & x = 0 \end{cases}$$

and is obtained by $= \displaystyle\int_a^x |f'(x)| dx$.

(ii) Positive variation $p(x)$ is defined by $p(x) = \frac{1}{2}[v(x) + f(x) - f(a)]$, for $a \leq x \leq b$.

(iii) Negative variation $n(x)$ is defined by $n(x) = \frac{1}{2}[v(x) - f(x) + f(a)]$, for $a \leq x \leq b$.

Problem [15.3] : Find the total variation, positive and negative variation of $f(x) = 3x - 2x^2$, $-1 \leq x \leq 2$.

Solution. $f(x) = 3x - 2x^2 \Rightarrow f'(x) = 3 - 4x$.

It is continuous on $[-1, 2]$. Here $V_f[-1, x] = \displaystyle\int_{-1}^{x} |f'(x)| dx$.

Now $|f'(x)| = \begin{cases} 3 - 4x, & \text{for } -1 \leq x \leq 3/4 \\ 4x - 3, & \text{for } 3/4 < x \leq 2. \end{cases}$

Thus for $-1 \leq x < 3/4$,

$$\int_{-1}^{0} |f(x)|dx + \int_{0}^{x} |f'(x)|dx = \int_{-1}^{0} (3-4x)dx + \int_{0}^{x} (3-4x)dx$$
$$= \left[3x - 2x^2\right]_{-1}^{0} + \left[3x - 2x^2\right]_{0}^{x} = 5 + 3x - 2x^2.$$

For $3/4 < x \leq 2$,

$$\int_{-1}^{3/4} (3-4x)dx + \int_{3/4}^{x} (4x-3)dx = \left[3x - 2x^2\right]_{-1}^{3/4} + \left[2x^2 - 3x\right]_{3/4}^{x} = 2x^2 - 3x + \frac{29}{4}.$$

Therefore,

(i) $v(x) = 5 + 3x - 2x^2, \quad -1 < x \leq \dfrac{3}{4}$

$= 2x^2 - 3x + \dfrac{29}{4}, \quad \dfrac{3}{4} < x \leq 2.$

(ii) $p(x) = \dfrac{1}{2}[v(x) + f(x) - f(a)],$ for $a < x \leq b$

$= \dfrac{1}{2}[5 + 3x - 2x^2 + 3x - 2x^2 + 5],$

$= \dfrac{1}{2}[-4x^2 + 6x + 10] = -2x^2 + 3x + 5,$ for $-1 < x \leq \dfrac{3}{4}.$

Also $p(x) = \dfrac{1}{2}\left[2x^2 - 3x + \dfrac{29}{4} + 3x - 2x^2 - \dfrac{9}{8}\right],$

$= \dfrac{49}{16},$ for $\dfrac{3}{4} < x \leq 2$

(iii) $n(x) = \dfrac{1}{2}[v(x) - f(x) + f(a)],$

$= \dfrac{1}{2}[5 + 3x - 2x^2 - 3x + 2x^2 - 5],$ for $-1 < x \leq \dfrac{3}{4}$

$= 0,$ for $-1 < x \leq \dfrac{3}{4}$

and $n(x) = \dfrac{1}{2}\left[2x^2 - 3x + \dfrac{29}{4} - 3x + 2x^2 + \dfrac{9}{8}\right],$

$= \dfrac{1}{2}\left[4x^2 - 6x + \dfrac{67}{8}\right],$ for $\dfrac{3}{4} < x \leq 2.$

Lipschitz function

Definition [15.2]: A function f is defined to be a Lipschitz function if it satisfies the Lipschitz's condition of order α at c.

That is, if $\exists$ a (+)ve number M depending on c and an open sphere $S_c(1)$ such that

$$|f(x) - f(c)| < M|x - c|^{\alpha} \ \forall \ x \in S_c(1), \ x \neq c.$$

(known as Lipschitz's condition of order α)

Theorem [15.1] Let $f, g : [a, b] \to \mathbb{R}$, $k \geq 0$, $c \geq 1$ be with the property $|f(x) - f(y)| \leq k|g(x) - g(y)|^c \ \forall \ x, \ y \in [a, b]$. **If g is a f.b.v. on $[a, b]$, then f is a f.b.v. on $[a, b]$.**

Proof. Let g be a f.b.v. on $[a, b]$. Then g is bounded on $[a, b]$. So, $\exists$ a (+)ve number M such that $|g(x)| \leq M \ \forall \ x \in [a, b]$.

Let $P\{a = x_0 < x_1 < x_2 \cdots < x_{r-1} < x_r \cdots < x_n = b\}$ be a partition of $[a, b]$.

Then $V_f[a,b] = \sup \sum\limits_{r=1}^{n} |f(x_r) - f(x_{r-1})|$.

Now by the given condition,

$$\begin{aligned}\sum_{r=1}^{n} |f(x_r) - f(x_{r-1})| &\le \sum_{r=1}^{n} k|g(x_r) - g(x_{r-1})|^c \ \forall\ x_{r-1}, x_r \in [a,b]\\ &= k\sum_{r=1}^{n} |g(x_r) - g(x_{r-1})|^{c-1}|g(x_r) - g(x_{r-1}|\\ &\le k\sum_{r=1}^{n} \{|g(x_r)| + |g(x_{r-1})|\}^{c-1}|g(x_r) - g(x_{r-1})|\\ &\le k\sum_{r=1}^{n} \{M + M\}^{c-1}|g(x_r) - g(x_{r-1})|\\ &= k\cdot(2M)^{c-1}\sum_{r=1}^{n} |g(x_r) - g(x_{r-1})|.\end{aligned}$$

$$\sup\sum_{r=1}^{n} |f(x_r) - f(x_{r-1})| \le k\cdot(2M)^{c-1}\cdot\sup\sum_{r=1}^{n} |g(x_r) - g(x_{r-1})|$$

$\Rightarrow V_f[a,b] \le k\cdot(2M)^{c-1}V_g[a,b]$ $[\forall\ c \ge 1]$

$\Rightarrow f$ is a f.b.v. on $[a,b]$, since g is a f.b.v. on $[a,b]$.

Uniform Lipschitz condition

Definition [15.3] : A function $f : [a,b] \to \mathbb{R}$ is said to satisfy a uniform Lipschitz condition of order $\alpha > 0$ on $[a,b]$, if $\exists$ a constant $M > 0$ such that $|f(x) - f(y)| < M|x - y|^\alpha\ \forall\ x, y \in [a,b]$.

Definition [15.4] : **[Absolutely continuous function]** A function $f : [a,b] \to \mathbb{R}$ is said to be absolutely continuous on $[a,b]$, if for every $\epsilon(> 0)$, $\exists$ a (+)ve number δ such that $\sum\limits_{k=1}^{n} |f(b_k) - f(a_k)| < \epsilon$ for every pairwise disjoint subintervals $[a_k, b_k]$ of $[a,b]$, $k = 1, 2, \cdots, n$ such that $\sum\limits_{k=1}^{n} (b_k - a_k) < \delta$.

Problem [15.4] : Show that every absolutely continuous function on $[a,b]$ is continuous and of f.b.v. on $[a,b]$.

Solution. Let f be absolutely continuous on $[a,b]$. Then, by Def. [15.4], we have

$$\sum_{k=1}^{n} |f(b_k) - f(a_k)| < \epsilon, \text{ with } \sum_{k=1}^{n} (b_k - a_k) < \delta, \tag{1}$$

where $[a_k, b_k]$ are pairwise disjoint subintervals on $[a,b]$.

So, from (1), $|f(b_k) - f(a_k)| < \epsilon$ with $(b_k - a_k) < \delta$ $[b_k > a_k]$.

Let $b_k = x, a_k = y, x > y,\ x, y \in [a,b]$

$\Rightarrow |f(x) - f(y)| < \epsilon$ with $|x - y| < \delta,\ \forall\ x, y \in [a,b]$

$\Rightarrow f$ is uniformly continuous on $[a,b]$

$\Rightarrow f$ is continuous on $[a,b]$.

For the next part,
we form a partition $P\{a = c_0 < c_1 < c_2 \cdots < c_{r-1} < c_r \cdots < c_n = b\}$ such that

$$c_r - c_{r-1} < \delta \text{ for } r = 1, 2, \cdots, n \tag{2}$$

Now if the interval $[c_{r-1}, c_r]$ be subdivided into k-finite subintervals by the points of division $c_{r-1} = d_0 < d_1 \cdots < d_k = c_r$, then

$$\sum_{r=1}^{k}(d_r - d_{r-1}) = c_r - c_{r-1} < \delta \text{ [by (2)]}.$$

So, by choice of δ, from Def [15.4],

$$\sum_{r=1}^{k}|f(d_r) - f(d_{r-1})| < 1, \text{ when } \epsilon = 1 \text{ with } \sum_{r=1}^{k}(d_r - d_{r-1}) < \delta$$

$$\Rightarrow \sup \sum_{r=1}^{k}|f(d_r) - f(d_{r-1})| \leq 1 \Rightarrow V_f[c_{r-1}, c_r] \leq 1 \text{ for } r = 1, 2, \cdots, k. \tag{3}$$

Now $V_f[a, b] = V_f[c_0, c_1] + V_f[c_1, c_2] + \cdots + V_f[c_{r-1}, c_r]$

$\Rightarrow V_f[a, b]$ is finite and hence f is a f.b.v. on $[a, b]$.

Problem [15.5] : If $f : [a, b] \to \mathbb{R}$ satisfies the uniform Lipschitz condition of order $\alpha = 1$, then f is uniformly continuous on $[a, b]$.

Solution. For $\alpha = 1$, the uniform Lipschitz condition stands as

$$|f(x) - f(y)| < M|x - y|, \ \forall \ x, y \in [a, b].$$

Now, let $|x - y| < \delta$, $\delta > 0$.

Then $|f(x) - f(y)| < M\delta, |x - y| < \delta$

$$\Rightarrow |f(x) - f(y)| < M \cdot \frac{\epsilon}{M} = \epsilon, |x - y| < \delta, \text{ where } \delta = \frac{\epsilon}{M}(> 0).$$

Thus $|f(x) - f(y)| < \epsilon$ for $|x - y| < \delta$, $\forall \ x, \ y \in [a, b]$

$\Rightarrow f$ is uniformly continuous on $[a, b]$.

Problem [15.6] : Every Lipschitz function of order $\alpha = 1$ is continuous at $x = c$.

Solution. From Def. [15.2] of Lipschitz function for order $\alpha = 1$,

$$|f(x) - f(c)| < M|x - c| \ \forall \ x \in S_c(1), \ x \neq c$$

$$\Rightarrow |f(x) - f(c)| < M \text{ whenever } |x - c| < 1, \ x \neq c.$$

M depending on c, we write $M = \dfrac{\epsilon}{c}$, $c > 0$, $\epsilon(> 0)$

$$\Rightarrow |f(x) - f(c)| < \frac{\epsilon}{c} < \epsilon, \text{ whenever } |x - c| < 1, \ x \neq c.$$

In particular, at $x = c$, L.H.S. $= 0 < \epsilon$ is satisfied.

So, $|f(x) - f(c)| < \epsilon \ \forall \ x \in (c - 1, c + 1)$.

Hence f is continuous at $x = c$.

Corollary [15.1] : Show that a Lipschitz function of order α is continuous, if $\alpha > 0$.
[Follows from Problem [15.6]].

Problem [15.7] : If f satisfies uniform Lipschitz condition for order $\alpha = 1$ on $[a, b]$, then show that f is absolutely continuous on $[a, b]$.

Solution. Let f satisfies uniform Lipschitz condition for $\alpha = 1$ on $[a, b]$.

Then $|f(x) - f(y)| < M|x - y| \; \forall \; x, \; y \in [a, b]$. (1)

Let $[x_1, y_1], \; [x_2, y_2], \cdots, [x_k, y_k], \cdots, [x_n, y_n]$ be the n-disjoint subintervals of $[a, b]$, where $y_k > x_k \; \forall \; k = 1, 2, \cdots, n$.

Then by (1), $|f(x_k) - f(y_k)| < M|x_k - y_k| \; \forall \; (x_k, y_k) \in [a, b]$

$$\Rightarrow \sum_{k=1}^{n} |f(x_k) - f(y_k)| < M \sum_{k=1}^{n} |x_k - y_k|. \tag{2}$$

Let

$$\sum_{k=1}^{n} |x_k - y_k| = \sum_{k=1}^{n} (y_k - x_k) < \delta. \tag{3}$$

Then $\sum\limits_{k=1}^{n} |f(x_k) - f(y_k)| < M\delta$ [from (2) by (3)]

Taking $\delta = \epsilon/M \; (> 0)$,

Thus, we have $\sum\limits_{k=1}^{n} |f(x_k) - f(y_k)| < \epsilon$ with $\sum\limits_{k=1}^{n} (y_k - x_k) < \delta$

$\Rightarrow f$ is absolutely continuous on $[a, b]$.

Problem [15.8] : Let $f : [a, b] \to \mathbb{R}$ and $g : [c, d] \to \mathbb{R}$ and $f[a, b] \subseteq [c, d]$ so that $g \circ f : [a, b] \to \mathbb{R}$ can be defined. Now if (i) g satisfies uniform Lipschitz condition on $f[a, b]$ of order $\alpha = 1$ and (ii) f is a f.b.v. on $[a, b]$, then show that $g \circ f$ is a f.b.v. on $[a, b]$.

Solution. Since g satisfies uniform Lipschitz condition of order $\alpha = 1$ on $f[a, b] \subseteq [c, d]$, then $\exists$ a (+)ve number M such that $\forall \; x, y \in [a, b], \; f(x), f(y) \in f[a, b]$,

$$|g(f(x)) - g(f(y))| \leq M|f(x) - f(y)|$$

$$\Rightarrow |(g \circ f)(x) - (g \circ f)(y)| \leq M|f(x) - f(y)|. \tag{1}$$

Let $P\{a = x_0 < x_1 \cdots < x_{r-1} < x_r \cdots < x_n = b\}$ be a partition of $[a, b]$. Then, since f is a f.b.v. on $[a, b]$,

$$\sup \sum_{r=1}^{n} |f(x_r) - f(x_{r-1})| \leq \text{ a finite number.} \tag{2}$$

From (1) it follows that

$$\sum_{r=1}^{n} |(g \circ f)(x_r) - (g \circ f)(x_{r-1})| < M \sum_{r=1}^{n} |f(x_r) - f(x_{r-1})|.$$

Choosing $x = x_r$ and $y = x_{r-1}, (x_{r-1}, x_r) \in [a, b]$.

$$\Rightarrow \sup \sum_{r=1}^{n} |(g \circ f)(x_r) - (g \circ f)(x_{r-1})$$

$$< M \sup \sum_{r=1}^{n} |f(x_r) - f(x_{r-1})| \leq \text{ a finite number [by (2)]}$$

$\Rightarrow g \circ f$ is a f.b.v. on $[a, b]$.

Problem [15.9] : Find the total variation of $\sin x$ on $[0, 2\pi]$.

Solution. Let $f(x) = \sin x$, defined on $[0, 2\pi]$. Then $f'(x) = \cos x$ is continuous on $[0, 2\pi]$.

Here $|f'(x)| = \begin{cases} \cos x, & \text{for } 0 \le x \le \frac{\pi}{2} \\ -\cos x, & \text{for } \frac{\pi}{2} < x \le \frac{3\pi}{2} \\ \cos x, & \text{for } \frac{3\pi}{2} < x \le 2\pi. \end{cases}$

(i) For $0 \le x \le \frac{\pi}{2}$,

$$v(x) = \int_0^x |f'(x)| dx = \int_0^x \cos x dx = \sin x.$$

(ii) For $\frac{\pi}{2} < x \le \frac{3\pi}{2}$,

$$v(x) = \int_0^{\pi/2} \cos x dx + \int_{\pi/2}^{x} -\cos x dx = 1 + \left[-\sin x\right]_{\pi/2}^{x} = -\sin x + 2.$$

(iii) For $\frac{3\pi}{2} < x \le 2\pi$,

$$v(x) = \int_0^{\pi/2} \cos x dx + \int_{\pi/2}^{3\pi/2} -\cos x dx + \int_{3\pi/2}^{x} \cos x dx$$
$$= (1+2) + \sin x - \sin\frac{3\pi}{2} = 3 + \sin x + 1 = 4 + \sin x.$$

Then the total variation of $\sin x$ is given by

$$\begin{aligned} &\sin x, \ 0 \le x \le \frac{\pi}{2} \\ &-\sin x + 2, \ \frac{\pi}{2} < x \le \frac{3\pi}{2} \\ &4 + \sin x, \ \frac{3\pi}{2} < x \le 2\pi. \end{aligned}$$

Problem [15.10] : Show that (i) $\sin x^2$ (ii) $(x+2)(5-x)$ are f.b.v. on $[0, 1]$ and $[0, 5]$ respectively.

Solution. (i) Let $f(x) = \sin x^2$. Then $f'(x) = 2x \cos x^2 \Rightarrow |f'(x)| = 2|x|\,|\cos x^2| \le 2$ on $[0, 1]$.
Thus f' is bounded on $[0, 1]$. f is also continuous in $[0, 1]$.
Therefore, by Theorem [6.1], f is a f.b.v. on $[0, 1]$.
(ii) Let $f(x) = (x+2)(5-x) = 3x - x^2 + 10$. Then f is continuous in $[0, 5]$
$f'(x) = 3 - 2x$
$\Rightarrow |f'(x)| = |3 - 2x| \le 3 + 2|x| \le 3 + 2 \cdot 5 = 13$ on $[0, 5]$.
f' is bounded on $[0, 5]$.
Hence f is a f.b.v. on $[0, 5]$.

Problem [15.11] : Show that a function satisfying a uniform Lipschitz condition for $\alpha = 1$ is a f.b.v. on $[a, b]$.

Solution. Let f satisfy a uniform Lipschitz condition on $[a, b]$.
Then $|f(x) - f(y)| \le M|x - y|^{\alpha}$, $\alpha \ge 1$, $\forall\ x,\ y \in [a, b]$. (1)
Let $P\{a = x_0 < x_1 < x_2 \cdots < x_{r-1} < x_r \cdots < x_n = b\}$ be a partition of $[a, b]$.

Then by (1), $|f(x_r) - f(x_{r-1})| \le M|x_r - x_{r-1}|^\alpha$ since $x_{r-1}, x_r \in [a,b]$.

Also $|x_r - x_{r-1}| < b - a,\ b > a$

$$\Rightarrow |f(x_r) - f(x_{r-1})| \le M(b-a)^\alpha = \text{a finite number for } \alpha = 1$$

$$\Rightarrow \sup \sum_{r=1}^{n} |f(x_r) - f(x_{r-1})| < \text{ a finite number}$$

$$\Rightarrow V_f[a,b] < \text{a finite number}$$

$$\Rightarrow f \text{ is a f.b.v. on } [a,b].$$

Problem [15.12] : A function of bounded variation is R-integrable but the converse is not true.

Solution. Let $f : [0,1] \to \mathbb{R}$ be defined by $f(x) = \begin{cases} x^2 \sin\frac{1}{x}, & \text{for } 0 < x \le 1 \\ 0, & \text{for } x = 0. \end{cases}$

Then f is continuous on $[0,1]$

$$f'(x) = \begin{cases} 2x\sin\frac{1}{x} - \cos\frac{1}{x}, & \text{for } 0 < x \le 1 \\ 0, & \text{for } x = 0. \end{cases}$$

$\Rightarrow f'(x)$ exists in $[0,1]$

$\Rightarrow |f'(x)| = \left|2x\sin\frac{1}{x} - \cos\frac{1}{x}\right| \le 2|x|\ \left|\sin\frac{1}{x}\right| + \left|\cos\frac{1}{x}\right|$

$\le 2\cdot 1\cdot 1 + 1 = 3$ for $0 < x \le 1$ and $f'(0) = 0$

$\Rightarrow f'$ is bounded on $[0,1]$.

$\Rightarrow f$ is a f.b.v. on $[0,1]$.

Also f being continuous on $[0,1]$ is R-integrable on $[0,1]$.

For, the converse, let $f : [0,2] \to \mathbb{R}$ be defined by

$$f(x) = \begin{cases} x\sin\frac{\pi}{x}, & 0 < x \le 2 \\ 0, & x = 0. \end{cases}$$

Then f is continuous. So f is R-integrable on $[0,2]$.

Let $P\left\{0, \frac{2}{(2n+1)}, \frac{2}{(2n-1)}, \cdots, \frac{2}{5}, \frac{2}{3}, \frac{2}{1} = 2\right\}$ be a partition of $[0,2]$.

Here $x_0 = 0, x_1 = \frac{2}{2n+1}, x_2 = \frac{2}{2n-1}, \cdots, x_{n-1} = \frac{2}{3}, x_n = 2$.

Then $\sum\limits_{r=1}^{n} |f(x_r) - f(x_{r-1})|$

$$= |f(x_1) - f(x_0)| + |f(x_2) - f(x_1)| + |f(x_3) - f(x_2)| + \cdots + |f(x_{n-1}) - f(x_{n-2})| + |f(x_n) - f(x_{n-1})|$$

$$= \left|\frac{2}{(2n+1)}\sin(2n+1)\frac{\pi}{2} - 0\right| + \left|\frac{2}{(2n-1)}\sin(2n-1)\frac{\pi}{2} - \frac{2}{2n+1}\sin(2n+1)\frac{\pi}{2}\right| + \cdots + \left|\frac{2}{3}\sin\frac{3\pi}{2} - \frac{2}{5}\sin\frac{5\pi}{2}\right| + \left|2\sin\frac{\pi}{2} - \frac{2}{3}\sin\frac{3\pi}{2}\right|$$

$$= \frac{2}{2n+1} + \left(\frac{2}{2n-1} + \frac{2}{2n+1}\right) + \left(\frac{2}{2n-3} + \frac{2}{2n-1}\right) + \cdots + \left(\frac{2}{3} + \frac{2}{5}\right) + \left(2 + \frac{2}{3}\right)$$

$$= 2 + 4\left(\frac{1}{3} + \frac{1}{5} + \cdots + \frac{1}{2n-1} + \frac{1}{2n+1}\right).$$

Since the infinite series $\sum\limits_{1}^{\infty} \frac{1}{2n+1}$ is divergent, its partial sum $S_n = \sum\limits_{r=1}^{n} \frac{1}{2r+1}$ is not bounded above.

So $V_f[0,2] = \sup \sum\limits_{r=1}^{n} |f(x_r) - f(x_{r-1})| \nless$ a finite number and hence f is not a f.b.v. on $[0,2]$.

Corollary [15.2] : Every continuous function is not a f.b.v.
[Follows from above problem].

Problem [15.13] : If f, g are absolutely continuous on $[a,b]$, then prove that $|f|$, $\lambda f + \mu g$ (λ, μ are constants), $f \cdot g$ are absolutely continuous on $[a,b]$. Also if g is bounded on $[a,b]$ and $g(x) \neq 0 \; \forall \; x \in [a,b]$, then prove that f/g is absolutely continuous on $[a,b]$.

Solution. Let f, g be absolutely continuous on $[a,b]$. Then from Def. [15.4], we have

$$\sum_{k=1}^{n} |f(b_k) - f(a_k)| < \frac{\epsilon}{2} \text{ with } \sum_{k=1}^{n} (b_k - a_k) < \delta \tag{1}$$

$$\text{and } \sum_{k=1}^{n} |g(b_k) - g(a_k)| < \frac{\epsilon}{2} \text{ with } \sum_{k=1}^{n} (b_k - a_k) < \delta. \tag{2}$$

Therefore, (i) $\Big||f(b_k)| - |f(a_k)|\Big| \leq |f(b_k) - f(a_k)|$

$$\Rightarrow \sum_{k=1}^{n} \Big||f(b_k)| - |f(a_k)|\Big| \leq \sum_{k=1}^{n} |f(b_k) - f(a_k)| < \frac{\epsilon}{2} \text{ with } \sum_{k=1}^{n} (b_k - a_k) < \delta$$

$\Rightarrow |f|$ is absolutely continuous on $[a,b]$.

(ii) Also $|(\lambda f + \mu g)(b_k) - (\lambda f + \mu g)(a_k)| = |\lambda\{f(b_k) - f(a_k)\} + \mu\{g(b_k) - g(a_k)\}|$

$$\leq |\lambda||f(b_k) - f(a_k)| + |\mu||g(b_k) - g(a_k)|$$

$$\Rightarrow \sum_{k=1}^{n} \Big|(\lambda f + \mu g)(b_k) - (\lambda f + \mu g)(a_k)\Big|$$

$$\leq |\lambda| \frac{\epsilon}{2|\lambda|} + |\mu| \frac{\epsilon}{2|\mu|} \text{ with } \sum_{k=1}^{n} (b_k - a_k) < \delta$$

$$= \epsilon \text{ with } \sum_{k=1}^{n} (b_k - a_k) < \delta. \quad \left[\begin{array}{l}\text{by (1) and (2), replacing } \frac{\epsilon}{2} \text{ by} \\ \frac{\epsilon}{2|\lambda|} \text{ and } \frac{\epsilon}{2|\mu|} \text{ respectively.}\end{array}\right]$$

Thus $\sum\limits_{k=1}^{n} \{(\lambda f + \mu g)(b_k) - (\lambda f + \mu g)(a_k)\} < \epsilon$ with $\sum\limits_{k=1}^{n} (b_k - a_k) < \delta$

$\Rightarrow \lambda f + \mu g$ is absolutely continuous on $[a,b]$.

(iii) Proofs of $f \cdot g$ and f/g, $g(x) \neq 0$, g being bounded are analoguous with the proofs of continuity as every absolutely continuous function is continuous.

Theorem [15.2] If $f : [a,b] \to \mathbb{R}$ be R-integrable, then it is bounded.

Proof. If possible, let f be unbounded on $[a,b]$. Let $[a,b]$ be divided into n-finite number of subintervals $[x_0, x_1], \ldots, [x_{r-1}, x_r], \ldots, [x_{n-1}, x_n]$ forming a partition

$$P\{a = x_0 < x_1 < \cdots < x_{r-1} < x_r \cdots < x_n = b\} \text{ of } [a,b].$$

Then f is unbounded in at least one of the subintervals $[x_{k-1}, x_k] = I_k$ (say).

For every $\epsilon(>0)$, $\exists$ a (+)ve number δ with the partition P of $[a,b]$ satisfying $||P||<\delta$.
As $(I_k,\xi_k)\in P$, $\xi_k\in I_k$ [if $x_{k-1}<\xi_k<x_k$],
$|f(\xi_k)|>k$ for any finite number k.

Also $\lim\limits_{||P||\to 0}\sum\limits_{r=1}^{n} f(\xi_r)\delta r=\int_a^b f(x)dx=L$ (say).

$$\text{Then } \left|\sum_{r=1}^{n} f(\xi_r)\delta r-L\right|<\epsilon=1 \text{ (say).} \tag{i}$$

$$\text{Now } |f(\xi_k)\delta k|-\left|L-\sum_{\substack{r=1\\ r\neq k}}^{n} f(\xi_r)\delta r\right|$$

$$\leq\left|f(\xi_k)\delta k-L+\sum_{\substack{r=1\\ r\neq k}}^{n} f(\xi_r)\delta r\right|=\left|\sum_{r=1}^{n} f(\xi_r)\delta r-L\right|<1 \text{ [by (i)]}$$

$$\Rightarrow\ |f(\xi_k)|\delta k<1+\left|L-\sum_{\substack{r=1\\ r\neq k}}^{n} f(\xi_r)\delta r\right|\leq 1+|L|+\left|\sum_{\substack{r=1\\ r\neq k}}^{n} f(\xi_r)\delta r\right|$$

$$\Rightarrow\ |f(\xi_k)|\leq\frac{1}{\delta k}\left[1+|L|+\left|\sum_{\substack{r=1\\ r\neq k}}^{n} f(\xi_r)\delta r\right|\right]. \tag{ii}$$

Numbers on the right hand side of (ii) are fixed and on the left hand side of (ii) ξ_k is any point $[x_{k-1},x_k]$.

So, (ii) $\Rightarrow f$ is bounded on $[x_{k-1},x_k]$, which is a contradiction.

Hence the theorem.

Theorem [15.3] [Cauchy Criterion for *R*-integration] A function $f:[a,b]\to\mathbb{R}$ is R-integrable on $[a,b]$, if and only if for every $\epsilon(>0)$, $\exists$ a (+)ve number δ such that $|S(P,f)-S(Q,f)|<\epsilon$ for any two partitions P and Q of $[a,b]$ satisfying $||P||,\ ||Q||<\delta$.

Proof. Let $\int_a^b f(x)dx=L$.

Then for every $\epsilon(>0)$ and for any partition P of $[a,b]$, $\exists$ a (+)ve number δ such that

$$|S(P,f)-L|<\tfrac{\epsilon}{2} \text{ with } ||P||<\delta. \tag{i}$$

Again, for any partition Q of $[a,b]$ with $||Q||<\delta$, we have $|S(Q,f)-L|<\frac{\epsilon}{2}$. (ii)

So, from (i) and (ii),

$$\begin{aligned}|S(P,f)-S(Q,f)|&=|S(P,f)-L+L-S(Q,f)|\\ &\leq|S(P,f)-L|+|S(Q,f)-L|<\frac{\epsilon}{2}+\frac{\epsilon}{2}=\epsilon \text{ with } ||P||,\ ||Q||<\delta.\end{aligned}$$

Conversely, let $|S(P,f)-S(Q,f)|<\epsilon$ with $||P||,\ ||Q||<\delta$. (iii)

Then for any $n\in N$, we consider $\delta_n>0$ such that $||P||,\ ||Q||<\delta_n$ for $\forall\ P,Q$ of $[a,b]$ and

$$|S(P,f)-S(Q,f)|<\frac{1}{n} \text{ [by (iii)]}.$$

Without loss of generality we can assume $\delta_m>\delta_n$, where $m<n,\ m,n\in N$.

Then $\delta_1 > \delta_2 > \cdots > \delta_n \Rightarrow \{\delta_n, \delta_n > 0, n \in N\}$ forms a decreasing sequence of (+)ve numbers.

For δ_n, let P_n be a partition of $[a, b]$ with $||P_n|| < \delta_n$ and for δ_m, let P_m be a partition of $[a, b]$ with $||P_m|| < \delta_m$, $n > m$.

Then by (iii), $|S(P_m, f) - S(P_n, f)| < \frac{1}{n} = \epsilon \ \forall \ m, n \geq k$

$\Rightarrow \{S(P_n, f)\}$ is a Cauchy sequence in $\mathbb{R}$ and it has a limit L (say).

That is, $\lim\limits_{n \to \infty} S(P_n, f) = L$

$\Rightarrow$ for every $\epsilon(> 0)$, $\exists$ a (+)ve number δ_n such that $|S(P_n, f) - L| < \epsilon$ with $||P_n|| < \delta_n$

$\Rightarrow f$ is R-integrable on $[a, b]$.

Problem [15.14] : **Give example of function in each of the following cases:**

(1) Riemann integrable on $[a, b]$ but not continuous.

(2) Riemann integrable on $[a, b]$ but not monotonic on $[a, b]$.

(3) Riemann integrable on $[a, b]$ but neither continuous nor monotone on $[a, b]$.

(4) f and g are not integrable but $f \cdot g$ and $f + g$ are integrable.

Solution. (1) We define a function f on $[0, 1]$ by

$$f(x) = \begin{cases} 1, & \text{if } x \in [0,\ 1/2] \\ 0, & \text{if } x \in [1/2,\ 1]. \end{cases}$$

Here f is discontinuous at $x = 1/2$.

We consider a partition $P\{0 = x_0 < x_1 \cdots < x_{k-1} < x_k \cdots < x_n = 1\}$ be a partition of $[0, 1]$ so that $\frac{1}{2} \in [x_{k-1}, x_k]$.

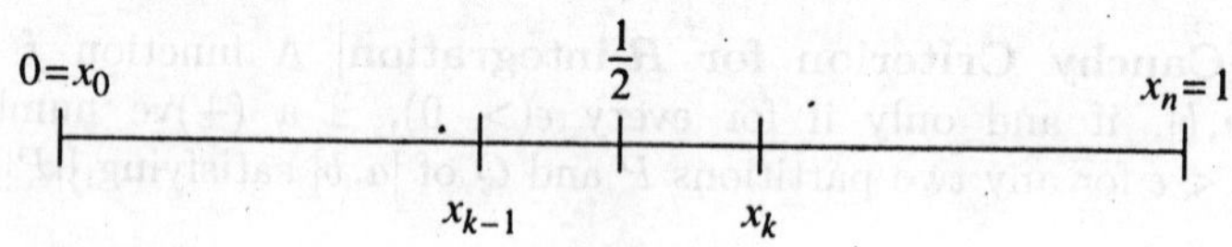

Fig. 15.1

We have $S(P, f) = \sum\limits_{r=1}^{n} M_r \delta_r$ and $s(P, f) = \sum\limits_{1}^{n} m_r \delta_r$.

In $[x_{k-1}, x_k]$, $M_k = 1$ and $m_k = 0$ by definition of f.

Also $M_r = 1$, $m_r = 1$, for $r = 1, 2, \ldots, k - 1$

and $M_r = m_r = 0$, for $r = k, k + 1, \ldots, n$

$$\text{Then } S(P, f) = \sum_{r=1}^{k-1} M_r \delta_r + M_k(x_k - x_{k-1}) + \sum_{r=k}^{n} M_r \delta_r$$

$$= \sum_{r=1}^{k-1} 1 \cdot \delta_r + 1 \cdot (x_k - x_{k-1} + \sum_{r=k}^{n} 0 \cdot \delta_r$$

$$= (x_1 - x_0) + (x_2 - x_1) + \cdots + (x_{k-1} - x_{k-2}) + x_k - x_{k-1}$$

$$= x_k - x_0 = x_k.$$

$$s(P,f)=\sum_{r=1}^{n} m_r\delta_r=\sum_{r=1}^{k-1} m_r\delta_r+(x_k-x_{k-1})\cdot 0+\sum_{r=k}^{n} m_r\delta_r$$
$$=\sum_{r=1}^{k-1} 1\cdot\delta_r+0+0=x_{k-1}-x_0=x_{k-1}.$$

So $\overline{\int_0^1} f(x)dx=\inf[S(P,f)]=1$ and $\underline{\int_0^1} f(x)dx=\sup[s(P,f)]=1$.

Therefore, $\overline{\int_0^1} f(x)dx=\underline{\int_0^1} f(x)dx$

$\Rightarrow f$ is R-integrable on $[0,1]$.

Alternatively. f has finite number of points of discontinuity. So f is R-integrable.

(2) Let $f(x)=\cos x\ \forall\ x\in[-\pi/2,\pi/2]$.

Then f is not monotonic on $[-\pi/2,\pi/2]$ as it is monotonically increasing in $[-\pi/2,0]$ but it is monotonically decreasing on $[0,\pi/2]$. Here f is continuous on $[-\pi/2,\pi/2]$.

So f is R-integrable on $[-\pi/2,\pi/2]$.

Thus f is R-integrable but not monotonic.

(3) Let $f:[1,3]\to\mathbb{R}$ be defined by

$$f(x)=\begin{cases}1, & \text{for } x\in[1,1/2]\\ 0, & \text{for } x\in(1/2,2]\\ 2, & \text{for } x\in(2,3].\end{cases}$$

Then f is not monotonic.

Also f is discontinuous at $x=1/2$ and $x=2$. f having finite number of points of discontinuity in $[1,3]$ is R-integrable on $[1,3]$.

Thus f is R-integrable but neither monotonic nor continuous.

(4) Let $f:[a,b]\to\mathbb{R}$ and $g:[a,b]\to\mathbb{R}$ be defined by

$$f(x)=\begin{cases}0, & \forall \text{ rational values of } x\\ 1, & \forall \text{ irrational values of } x.\end{cases}$$

$$g(x)=\begin{cases}1, & \forall \text{ rational values of } x\\ 0, & \forall \text{ irrational values of } x.\end{cases}$$

Then f,g are not R-integrable.

But $f\cdot g$ is R-integrable on $[a,b]$ as $(f\cdot g)(x)=0\ \forall\ x\in[a,b]$.

Again, $(f+g)(x)=1\ \forall\ x\in[a,b]$ and hence $(f+g)(x)$ is R-integrable on $[a,b]$.

Problem [15.15]: Examine whether f is R-integrable on $[0,1]$. Justify your answer and evaluate $\int_0^1 f(x)dx$ in each case, if it exists.

(1) $f(x)=[10x]\ \forall\ x\in[0,1]$

(2) $f(x)=\begin{cases}0, & \text{if } x \text{ is rational}\\ \sqrt{x}, & \text{if } x \text{ is irrational on } [0,1].\end{cases}$

Solution. (1) $0\le x\le 1\Rightarrow 0\le 10x\le 10$.

Let $10x=y$. Then $0\le y\le 10$.

Therefore, $[y] = 0$ for $0 \le y < 1$
$= 1$ for $1 \le y < 2$
$= 2$ for $2 \le y < 3$
$= 3$ for $3 \le y < 4$
$= 4$ for $4 \le y < 5$
$= 5$ for $5 \le y < 6$
$= 6$ for $6 \le y < 7$
$= 7$ for $7 \le y < 8$
$= 8$ for $8 \le y < 9$
$= 9$ for $9 \le y < 10$
$= 10$ for $y = 10$.

Then $\phi(y) = [y]$ has finite number of points of discontinuity at $y = 1, 2, 3, 4, 5, 6, 7, 8, 9, 10$. So ϕ is R-integrable on $[0, 10]$

$\Rightarrow f(x) = [10x]$ is R-integrable on $[0, 1]$.

Also

$$\int_0^1 [10x]dx = \frac{1}{10}\int_0^{10} [y]dy$$
$$= \frac{1}{10}\left[\int_0^1 0 \cdot dy + \int_1^2 1 \cdot dy + \int_2^3 2 \cdot dy + \cdots + \int_9^{10} 9 \cdot dy\right]$$
$$= \frac{1}{10}[(2-1) + 2(3-2) + 3(4-3) + \cdots + 9(10-9)]$$
$$= \frac{1}{10}[1 + 2 + 3 + \cdots + 9] = \frac{9(9+1)}{2 \cdot 10} = \frac{9}{2}.$$

Alternatively. $M_r = m_r \ \forall\ [x_{r-1}, x_r],\ r = 1, 2, \ldots, 10$. When $[0, 1]$ is divided into 10 equal length of subintervals $[x_{r-1}, x_r]$, where $x_r - x_{r-1} = 1/10\ \forall\ r = 1, 2, \ldots, 10$.

Then $S(P, f) = \sum_{r=1}^{10} M_r\delta_r = M_1\delta_1 + M_2\delta_2 + \cdots + M_{10}\delta_{10}$

$$= 0 \cdot \frac{1}{10} + 1 \cdot \frac{1}{10} + 2 \cdot \frac{1}{10} + \cdots + 9 \cdot \frac{1}{10}$$
$$= \frac{1}{10}(1 + 2 + \cdots + 9) = \frac{9(9+1)}{2 \cdot 10} = \frac{9}{2}.$$

Also $s(P, f) = \sum_{r=1}^{10} m_r\delta_r = \frac{9}{2}$ $[\because M_r = m_r]$.

Therefore $\overline{\int_0^1} f(x)dx = \inf S(P, f) = \frac{9}{2}$

and $\underline{\int_0^1} f(x)dx = \sup s(P, f) = \frac{9}{2}$

$\Rightarrow \int_0^1 f(x)dx$ exists and is equal to $\frac{9}{2}$.

(2) [*Hints.* $\overline{\int_0^1} f(x)dx = \int_0^1 \sqrt{x}dx = \frac{2}{3}$ and $\underline{\int_0^1} f(x)dx = \int_0^1 0 \cdot dx = 0$. So $\overline{\int_0^1} f(x)dx \ne \underline{\int_0^1} f(x)dx \Rightarrow f$ is not R-integrable on $[0, 1]$.]

Problem [15.16]: Let $f, g : [a, b] \to \mathbb{R}$ be both bounded on $[a, b]$ such that $f(x) = g(x)$ except for finite number of points on $[a, b]$. Then show that if f is R-integrable on $[a, b]$, g is also R-integrable on $[a, b]$.

Solution. Let $f(x) \neq g(x)$ for finite number of points $x_1, x_2, \dots, x_p$ on $[a, b]$.

We enclose the points $x_1, x_2, \dots, x_p$ by finite number of non-overlapping subintervals

$$\left[x_1 - \frac{\delta_1}{2},\ x_1 + \frac{\delta_1}{2}\right], \left[x_2 - \frac{\delta_2}{2},\ x_2 + \frac{\delta_2}{2}\right], \dots, \left[x_p - \frac{\delta_p}{2},\ x_p + \frac{\delta_p}{2}\right]$$

such that the sum of their lengths $\delta_1 + \delta_2 + \dots + \delta_p < \frac{\epsilon}{2(M-m)}$, $M,\ m$ be the bounds of g on $[a, b]$.

Let $M_k = \sup \cdot g$ and $m_k = \inf \cdot g$ on $\left[x_k - \frac{\delta_k}{2}, x_k + \frac{\delta_k}{2}\right]$, $k = 1, 2, \dots, p$.

Then the oscillatory sum arising from these subintervals,

$$\sum_{k=1}^{p} (M_k - m_k)\delta_k < \sum_{k=1}^{p} (M - m)\delta_k = (M - m) \sum_{k=1}^{p} \delta_k < (M - m)\frac{\epsilon}{2(M - m)} = \frac{\epsilon}{2}. \quad (1)$$

In the remaining $(p + 1)$ subintervals g is R-integrable as f is so for $g(x) = f(x)\ \forall\ x$ in $(p + 1)$ subintervals.

So each of $(p + 1)$ subintervals can be further subdivided into finite number of non-overlapping subintervals so that the oscillator sum arising from each of the

$$(p + 1) \text{ subintervals} < \tfrac{\epsilon}{2(p+1)}. \quad (2)$$

Hence there exists a partition P of $[a, b]$ so that the resultant oscillatory sum

$$\omega(P, g) < \tfrac{\epsilon}{2} + \tfrac{\epsilon}{2(p+1)}(p + 1) = \tfrac{\epsilon}{2} + \tfrac{\epsilon}{2} = \epsilon.$$

Hence g is R-integrable on $[a, b]$.

Problem [15.17]: Let $f : [a, b] \to \mathbb{R}$ be R-integrable on $[a, b]$ and $f(x) > 0\ \forall\ x \in [a, b]$ with $(+)$ve infimum, then $h(x) = \sqrt{f(x)}$ is R-integrable on $[a, b]$.

Solution. Let $P\{a = x_0 < x_1 \cdots < x_{r-1} < x_r \cdots < x_n = b\}$ be a partition of $[a, b]$. Let M_r, m_r be the $\sup f$ and $\inf f$ in $[x_{r-1}, x_r]$ and M'_r, m'_r be those of $h(x)$ in $[x_{r-1}, x_r]$.

f is bounded, so $\exists$ a $(+)$ve real number m such that $f(x) > m\ \forall\ x \in [a, b]$.

f is R-integrable on $[a, b]$

$$\Rightarrow \omega(P, f) = \sum_{r=1}^{n} (M_r - m_r)\delta_r < \frac{\epsilon\sqrt{m}}{2}. \quad (1)$$

Also $h(x) = \sqrt{f(x)}\ \forall\ x \in [a, b],\ f(x) > 0$

$$\Rightarrow M'_r - m'_r = \sqrt{M_r} - \sqrt{m_r} = \frac{M_r - m_r}{\sqrt{M_r} + \sqrt{m_r}} < \frac{M_r - m_r}{2\sqrt{m}}, \quad (2)$$

since $m \leq m_r \leq M_r$

$$\Rightarrow \sqrt{m} \leq \sqrt{m_r} \leq \sqrt{M_r} \Rightarrow \sqrt{M_r} + \sqrt{m_r} \geq 2\sqrt{m} \Rightarrow \frac{1}{\sqrt{M_r} + \sqrt{m_r}} \leq \frac{1}{2\sqrt{m}}.$$

Therefore, from (2),

$$\omega(h, P) = \sum_{r=1}^{n} (M'_r - m'_r)\delta_r < \sum_{r=1}^{n} \frac{(M_r - m_r)}{2\sqrt{m}}\delta_r \text{ [by (2)]}$$

$$< \frac{\epsilon\sqrt{m}}{2} \cdot \frac{1}{2\sqrt{m}} = \frac{\epsilon}{4} < \epsilon \text{ [by (1)]}$$

$\Rightarrow h$ is R-integrable on $[a, b]$.

Problem [15.18] : Let $f : [a, b] \to \mathbb{R}$ be continuous on $[a, b]$ and $g : [c, d] \to \mathbb{R}$ be differentiable on $[c, d]$, where $g[c, d] \subseteq [a, b]$. If $G : [c, d] \to \mathbb{R}$ be defined by $G(x) = \int_a^{G(x)} f(t)dt \ \forall \ x \in [c, d]$, then show that $G'(x)$ exists and $G'(x) = (f \circ g)(x)g'(x) \ \forall \ x \in [c, d]$.

Solution. $G(x) = \int_a^{g(x)} f(t)dt$.

Then $G(x + h) = \int_a^{g(x+h)} f(t)dt$, when h is so chosen that $x + h \in [c, d]$.

Therefore,

$$G(x + h) - G(x) = \int_{g(x)}^{g(x+h)} f(t)dt = \{g(x + h) - g(x)\}f(\xi), g(x) \leq \xi \leq g(x + h),$$

since f is continuous on $[a, b]$ and $g(x), g(x + h) \in [a, b]$ as $g[c, d] \subseteq [a, b]$

$$\Rightarrow \frac{G(x + h) - G(x)}{h} \cdot \frac{h}{g(x + h) - g(x)} = f(\xi) = f(g(x) + \theta h), \ 0 < \theta < 1$$

$$\Rightarrow \lim_{h \to 0} \frac{G(x + h) - G(x)}{h} \cdot \lim_{h \to 0} \frac{1}{\frac{g(x + h) - g(x)}{h}} = \lim_{h \to 0} f(g(x) + \theta h). \quad (1)$$

Since g is differentiable on $[c, d]$, $\lim\limits_{h \to 0} \frac{g(x+h)-g(x)}{h}$ exists and equal to $g'(x)$ and f is continuous on $[a, b]$, $\lim\limits_{h \to 0} f[g(x) + \theta h]$ exists and equal to $f(g(x))$, it follows that $\lim\limits_{h \to 0} \frac{G(x+h)-G(x)}{h}$ exists.

That is, $G'(x)$ exists on $[c, d]$ and from (1), we get

$$G'(x) \cdot \frac{1}{g'(x)} = f(g(x)) = (f \circ g)(x)$$

$$\Rightarrow G'(x) = (f \circ g)(x) \cdot g'(x) \ \forall \ x \in [c, d].$$

Problem [15.19] : Let $f : [a, b] \to \mathbb{R}$ be R-integrable on $[a, b]$ and $m \leq f(x) \leq M \ \forall \ x \in [a, b]$. Let ϕ be continuous on $[m, M]$ and $h(x) = (\phi \circ f)(x)$ in $[a, b]$. Then show that h is R-integrable on $[a, b]$.

Solution. Continuity of ϕ on $[m, M]$.

$\Rightarrow$ uniformly continuity of ϕ on $[m, M]$.

$\Rightarrow$ for every $\epsilon(> 0)$, $\exists$ a (+)ve number δ such that

$$|\phi(\alpha) - \phi(\beta)| < \epsilon, \ \forall \ \alpha, \beta \in [m, M] \text{ satisfying } |\alpha - \beta| < \delta. \quad (1)$$

R-integrability of f on $[a, b]$.

$\Rightarrow \exists$ a partition $P\{a = x_0 < x_1 \cdots < x_{r-1} < x_r \cdots < x_n = b\}$ of $[a, b]$ with

$$||P|| < \delta \text{ such that } S(P, f) - s(P, f) < \delta^2. \quad (2)$$

Let M_r, m_r, M'_r, m'_r be the sup and inf of f and h respectively in $[x_{r-1}, x_r]$, $r = 1, 2, \ldots, n$. We divide the numbers $1, 2, 3, \ldots, n$ into two classes A and B such that $M_r - m_r < \delta$ for $r \in A$ and $M_r - m_r \geq \delta$ for $r \in B$.

Now for $r \in A$ and $x_{r-1} \leq x < y < x_r$, $|f(x) - f(y)| \leq M_r - m_r < \delta$ and hence

$$|\phi(f(x)) - \phi(f(y))| < \frac{\epsilon}{2(b - a)} \text{ [by (1)]}, \ b > a$$

$$\Rightarrow |(\phi \circ f)(x) - (\phi \circ f)(y)| < \frac{\epsilon}{2(b - a)}$$

$$\Rightarrow |h(x) - h(y)| < \frac{\epsilon}{2(b - a)} \ \forall \ x, y \in [x_{r-1}, x_r] \ r \in A. \quad (3)$$

Next, for $r \in B$, $M'_r - m'_r \leq K$,

$$\text{where } K = \sup |\phi(t)| \; m \leq t \leq M \tag{4}$$

$$\begin{aligned} \text{and } \delta \sum_{r\in B} \delta_r &\leq \sum_{r\in B} (M_r - m_r)\delta_r \text{ [since for } r \in B,\ M_r - m_r \geq \delta] \\ &\leq \sum_{r=1}^{n} (M_r - m_r)\delta_r \; [\because B \subset \{1, 2, \dots, n\}] \\ &= S(P, f) - s(P, f) < \delta^2 \text{ [by (2)]} \end{aligned}$$

$$\Rightarrow \sum_{r\in B} \delta_r < \delta. \tag{5}$$

Therefore, $S(P, h) - s(P, h)$

$$\begin{aligned} &= \sum_{r\in A} (M'_r - m'_r)\delta_r + \sum_{r\in B} (M'_r - m'_r)\delta_r < \frac{\epsilon}{2(b-a)}(b-a) + K\delta \text{ [by (3), (4), (5)]} \\ &< \frac{\epsilon}{2} + K \cdot \frac{\epsilon}{2K} \; \left[\text{taking } \delta = \tfrac{\epsilon}{2K}\right] \\ &= \epsilon \end{aligned}$$

$\Rightarrow h$ is R-integrable on $[a, b]$.

Problem [15.20] : Let $\lim_{n\to\infty} f_n(t) = f(t)$ and g is R-integrable on $[a, b]$. If

$$h(x) = \int_a^x f(t)g(t)dt,\ h_n(x) = \int_a^x f_n(t)g(t)dt$$

for $a \leq x \leq b$. Then show that h_n converges to h uniformly on $[a, b]$.

Solution. $g \in R[a, b] \Rightarrow |g|^2 \in R[a, b] \Rightarrow \int_a^b [g(t)]^2 dt$ exists and say

$$\int_a^b |g(t)|^2 dt = L\ (> 0). \tag{1}$$

$\lim_{n\to\infty} f_n(t) = f(t)$ on $[a, b]$, $b > a$

$\Rightarrow$ for every $\epsilon (> 0)$, $\exists$ a (+)ve integer m such that $|f_n(t) - f(t)| < \frac{\epsilon}{\sqrt{L(b-a)}}$, $\forall\ n \geq m$

$$\Rightarrow \int_a^b |f_n(t) - f(t)|^2 dt < \int_a^b \frac{\epsilon^2}{L(b-a)} dt = \frac{\epsilon^2}{L} \tag{2}$$

$$\Rightarrow \int_a^x |f_n(t) - f(t)|^2 dt < \frac{\epsilon^2}{L}, \text{ if } x \in [a, b].$$

By Cauchy Schwarz's inequality, we have

$$0 \leq \left(\int_a^b |f_n(t) - f(t)|\ |g(t)|dt\right)^2 \leq \left(\int_a^b |f_n(t) - f(t)|^2 dt\right) \times \int_a^b |g(t)|^2 dt$$

$$\Rightarrow 0 \leq \left(\int_a^b |f_n(t)g(t) - f(t)g(t)|dt\right)^2 \leq \frac{\epsilon^2}{L} \cdot L = \epsilon^2 \text{ [by (1) and (2)]}$$

$$\Rightarrow \int_a^b |f_n(t)g(t) - f(t)g(t)|dt < \epsilon\ \forall\ n \geq m$$

$$\Rightarrow \int_a^x |f_n(t)g(t) - f(t)g(t)|dt \le \int_a^b |f_n(t)g(t) - f(t)g(t)|dt < \epsilon,\ a \le x \le b,\ \forall\, n \ge m$$
$$\Rightarrow |h_n(x) - h(x)| < \epsilon,\ \forall\, n \ge m,\ \forall\, x \in [a,b]$$
$$\Rightarrow h_n \to h \text{ uniformly on } [a,b].$$

Problem [15.21]: Let $\lim_{n\to\infty} f_n(t) = f(t)$ and $\lim_{n\to\infty} g_n(t) = g(t)$ on $[a,b]$. If

$$h(x) = \int_a^x f(t)g(t)dt \text{ and } h_n(x) = \int_a^x f_n(t)g_n(t)dt$$

for $x \in [a,b]$. Then show that $h_n \to h$ uniformly on $[a,b]$.

Solution. We have $|f_n(t) - f(t)| < \dfrac{\epsilon}{2\sqrt{b-a}}\ \forall\, n \ge m_1$

$$\text{and } |g_n(t) - g(t)| < \frac{\epsilon}{2\sqrt{b-a}}\ \forall\, n \ge m_2,\ b > a. \qquad (1)$$

$$\text{Now } h_n(x) - h(x) = \int_a^x \{f_n(t)g_n(t) - f(t)g(t)\}dt$$
$$= \int_a^x (f_n(t) - f(t))(g_n(t) - g(t))dt + \left(\int_a^x f_n(t)g(t)dt - \int_a^x f(t)g(t)dt\right)$$
$$+ \left(\int_a^x f(t)g_n(t)dt - \int_a^x f(t)g(t)dt\right)$$

$$\Rightarrow |h_n(x) - h(x)|$$
$$\le \int_a^x |f_n(t) - f(t)||g_n(t) - g(t)|dt + \int_a^x |f_n(t) - f(t)||g(t)|dt$$
$$+ \int_a^x |g_n(t) - g(t)||f(t)|dt$$
$$\le \int_a^x |f_n(t) - f(t)||g_n(t) - g(t)|dt + \epsilon\left[\int_a^x |g(t)|dt + \int_a^x |f(t)|dt\right] \qquad (2)$$
$$\le \int_a^b |f_n(t) - f(t)||g_n(t) - g(t)|dt + \epsilon\left[\int_a^b [|g(t) + |f(t)||]\right] dt\ \forall\, x \in [a,b]$$
$$< \int_a^b \frac{\epsilon}{2\sqrt{b-a}} \cdot \frac{\epsilon}{2\sqrt{b-a}} dt + \epsilon \cdot L,\ \forall\, n \ge m,\ \forall\, x \in [a,b],\ m = \max(m_1, m_2).$$

Here $f, g \in R[a,b] \Rightarrow |f| + |g| \in R[a,b]$

$$\Rightarrow \int_a^b [|f(t)| + |g(t)|]dt \text{ exists and equal to } L, \text{ say}$$

$$\Rightarrow |h_n(x) - h(x)| < \frac{\epsilon^2}{4} + \epsilon L\ \forall\, n \ge m,\ \forall\, x \in [a,b]$$
$$< \epsilon + \epsilon L = \epsilon(1 + L).$$

Choosing $\epsilon(>0) = \frac{\epsilon'}{1+L}$, ϵ' is also an arbitrary small (+)ve number

$$\Rightarrow |h_n(x) - h(x)| < \frac{\epsilon'}{1+L} \cdot (1 + L) = \epsilon'\ \forall\, n \ge m,\ \forall\, x \in [a,b]$$
$$\Rightarrow h_n \to h \text{ uniformly on } [a,b].$$

Problem [15.22] : Every sequence of bounded functions being uniformly convergent is uniformly bounded.

Solution. Let $\{f_n(x)\}$ converge uniformly to f on $[a,b]$, where each $f_n(x)$ is bounded on $[a,b]$.

Then, we have

$$|f_n(x) - f(x)| < \epsilon \ \forall \ n \geq m \text{ and } \forall \ x \in [a,b],$$

we choose $\epsilon = 1$ so that

$$|f_n(x) - f(x)| < 1 \ \forall \ n \in N \text{ and } \forall \ x \in [a,b] \tag{1}$$

$$\text{and } |f_n(x)| \leq M_n \ \forall \ n \in N, \ M_n \text{ is a finite (+)ve real number.} \tag{2}$$

$\Rightarrow |f_1(x)| \leq M_1, \ |f_2(x)| \leq M_2, \cdots, |f_n(x)| \leq M_n \ \forall \ x \in [a,b]$.

Let $M = \max[M_1, M_2, \ldots, M_n]$.

Then

$$|f_n(x)| \leq M \ \forall \ x \in [a,b] \text{ and } \forall \ n \in N. \tag{3}$$

$$\begin{aligned} \text{Now } |f(x)| &= |f(x) - f_n(x) + f_n(x)| \\ &\leq |f(x) - f_n(x)| + |f_n(x)| \\ &< 1 + M \ \forall \ x \in [a,b] \quad \text{[by (1) and (3)]} \end{aligned}$$

$\Rightarrow f$ is uniformly bounded.

Remark. (1) Converse of the above problem is not true.

We consider a sequence of functions $\{f_n(x)\}$, where $f_n(x) = x^n, \ x \in [0,1]$.

$$\begin{aligned} \text{Then } & f_1(x) = x \Rightarrow |f_1(x)| \leq 1 \\ & f_2(x) = x^2 \Rightarrow |f_2(x)| \leq 1 \\ & \cdots\cdots\cdots\cdots\cdots\cdots \\ & f_n(x) = x^n \Rightarrow |f_n(x)| \leq 1. \end{aligned}$$

So, each of the terms of sequence is bounded on $[0,1]$.

$$\lim_{n\to\infty} f_n(x) = \begin{cases} 0, & \text{for } 0 \leq x < 1 \\ 1, & \text{for } x = 1 \end{cases}$$

$\Rightarrow f(x)$ is bounded on $[0,1]$.

$\Rightarrow \{f_n(x)\}$ is uniformly bounded on $[0,1]$.

But $\{f_n(x)\}$ is not uniformly convergent on $[0,1]$ since f is discontinuous at $x = 1$.

(2) It also suggests from the above example that the condition of uniform convergence of $\{f_n(x)\}$ is only a sufficient condition for the boundedness of f.

Problem [15.23] : Show that the pointwise limit of a sequence $\{f_n(x)\}$, where $f_n(x) = (\cos \pi x)^{2n}$, $x \in Z$ of continuous functions is discontinuous while the convergent is not uniform.

Solution. $f_n(x) = (\cos \pi x)^{2n}, \ x \in Z$.

Each $f_n(x)$ is continuous $\forall \ x \in Z$.

$$f(x) = \lim_{n\to\infty} f_n(x) - \begin{cases} 1, & \text{when } x \in Z \\ 0, & \text{when } x \notin Z. \end{cases}$$

So f is discontinuous $\forall \ x \in Z$.

Thus the pointwise limit is a discontinuous function.

Also, for $x \in Z$, $|f_n(x) - f(x)| = |(\cos(\pi x))^{2n} - 1| \le |(\cos(\pi x)^{2n}| + 1$

$$\le 2 \not< \epsilon \ \forall \ n \ge m.$$

So $\{f_n(x)\}$ is not uniformly convergent on Z.

Problem [15.24] : Let $\{f_n\}$ be defined by $f_n(x) = 1 - |1 - x^2|^n$ on E, where

$$E = \{x;\ |1 - x^2| \le 1\} = [-\sqrt{2}, \sqrt{2}].$$

Examine the uniform convergence of $\{f_n(x)\}$ in $[-\sqrt{2}, \sqrt{2}]$.

Solution. Each $f_n(x)$ is continuous on E. Hence each $f_n(x)$ is bounded on E.

$$f(x) = \lim_{n\to\infty} f_n(x) = \begin{cases} 1, & \text{when } |1 - x^2| < 1, \text{ i.e., } 0 < |x| < \sqrt{2} \\ 0, & \text{when } |1 - x^2| = 1 \text{ or, } x = 0, \pm\sqrt{2}. \end{cases}$$

Hence the pointwise limit f is not continuous for $x = 0, \pm\sqrt{2}$.

If possible, let $\{f_n(x)\}$ be uniformly convergent on E so that f is the uniform limit.

Then $|f_n(x) - f(x)| < \epsilon \ \forall \ n \ge m, \ \forall \ x \in [-\sqrt{2}, \sqrt{2}]$

$$\Rightarrow \big|1 - |1 - x^2|^n - 1\big| < \epsilon \text{ for } x \in \left[-\sqrt{2}, \sqrt{2}\right] \Rightarrow |1 - x^2|^n < \epsilon$$

$$\Rightarrow n \log \frac{1}{|1 - x^2|} > \log \frac{1}{\epsilon}$$

$$\Rightarrow n > \frac{\log \frac{1}{\epsilon}}{\log \frac{1}{|1-x^2|}} = m(\epsilon, x).$$

Now as $x \to 0$, $\log \dfrac{1}{|1 - x^2|} \to \log 1 = 0 \Rightarrow m(\epsilon, x) \to \infty$.

The same is true as $x \to +\sqrt{2}$ or $x \to -\sqrt{2}$. So there does not exist an m such that

$$|f_n(x) - f(x) < \epsilon \ \forall \ n \ge m \text{ and } \forall \ x \in \left[-\sqrt{2}, \sqrt{2}\right].$$

Therefore, it is not uniformly convergent in $[-\sqrt{2}, \sqrt{2}]$. The sequence $\{f_n(x)\}$ is pointwise convergent.

It is to be noted that $\{f_n(x)\}$ is uniformly convergent $(-\sqrt{2}, 0)$ and $(0, \sqrt{2})$.

In particular, at $x = 0, \pm\sqrt{2}$, $|f_n(x) - f(x)| < \epsilon$

$$\Rightarrow \big|1 - |1 - x^2|^n\big| < \epsilon \Rightarrow |1 - x^2|^n > 1 - \epsilon$$

$$\Rightarrow n > \log \frac{1 - \epsilon}{\log |1 - x^2|} = m(\epsilon, x)$$

$$m(\epsilon, x) \to \infty \text{ for } x \to 0, \pm\sqrt{2}.$$

Hence $\not\exists\ m$ so that $|f_n(x) - f(x)| < \epsilon \ \forall \ n \ge m$ and $x = 0, \pm\sqrt{2}$.

Theorem [15.4] Let $\{f_n(x)\}$ converge uniformly to $f(x)$ on $[a, b] \subseteq \mathbb{R}$, where $f_n(x) \ne 0 \ \forall \ x \in [a, b]$ and $\forall \ n \in N$, with a property that $f(x) \ge \delta (> 0) \ \forall \ x \in [a, b]$. Then $\left\{\frac{1}{f_n(x)}\right\}$ converges uniformly to $\frac{1}{f(x)}$ on $[a, b]$.

Proof. We have

$$|f_n(x) - f(x)| < \epsilon \;\forall\; x \in [a,b] \text{ and } \forall\; n \geq m. \tag{i}$$

Now $|f(x)| - |f_n(x)| < |f_n(x) - f(x)| < \frac{\delta}{2}$ [taking $\epsilon = \delta/2$]

$$\Rightarrow |f_n(x)| > |f(x)| - \frac{\delta}{2} \geq \delta - \frac{\delta}{2} = \frac{\delta}{2} \;[\because f(x) \geq \delta]$$

$$\Rightarrow |f_n(x)| > \frac{\delta}{2} \;\forall\; n \geq m \text{ and } \forall\; x \in [a,b]. \tag{ii}$$

$$\left|\frac{1}{f_n(x)} - \frac{1}{f(x)}\right| = \frac{|f(x) - f_n(x)|}{|f_n(x)||f(x)|} < \epsilon \cdot \frac{2}{\delta} \cdot \frac{1}{\delta} \quad \begin{bmatrix}\text{by (i), (ii) and} \\ |f(x)| > \delta\end{bmatrix}$$

$$= \frac{2\epsilon}{\delta^2} = \epsilon' \text{ (say)},$$

where ϵ' is an arbitrary small (+)ve number corresponding to ϵ.

Thus $\left|\frac{1}{f_n(x)} - \frac{1}{f(x)}\right| < \epsilon' \;\forall\; x \in [a,b]$ and $\forall\; n \geq m$

$\Rightarrow \left\{\frac{1}{f_n(x)}\right\}$ converges uniformly to $\frac{1}{f(x)}$.

Theorem [15.5] Let $f_n[a,b] \to \mathbb{R}$ be defined $\forall\; n \in N$ and $f_n \to f$ uniformly on $[a,b]$ and let $x_0 \in [a,b]$ so that $\lim_{x \to x_0} f_n(x) = a_n$ for $n = 1, 2, \ldots$. Then

(i) $\{a_n\}$, the sequence of real numbers, is convergent and

(ii) $\lim_{x \to x_0} f(x) = \lim_{n \to \infty} a_n \Rightarrow \lim_{x \to x_0} \lim_{n \to \infty} f_n(x) = \lim_{n \to \infty} \lim_{x \to x_0} f_n(x)$.

Proof. Since $f_n \to f$ uniformly on $[a,b]$, by Cauchy's general principle of uniform convergence, we have

$$|f_m(x) - f_n(x)| < \frac{\epsilon}{3} \;\forall\; m,n \geq k \text{ and } \forall\; x \in [a,b]. \tag{i}$$

Also for $x_0 \in [a,b]$, $\lim_{x \to x0} f_n(x) = a_n \;\forall\; n \in N$

$$\Rightarrow |f_n(x) - a_n| < \frac{\epsilon}{3} \;\forall\; x \in (x_0 - \delta, x_0 + \delta) \subset [a,b] \tag{ii}$$

$$\Rightarrow |f_m(x) - a_m| < \frac{\epsilon}{3} \;\forall\; x \in (x_0 - \delta, x_0 + \delta). \tag{iii}$$

Then $|a_m - a_n| = |a_m - f_m(x) + f_m(x) - f_n(x) + f_n(x) - a_m|$

$\leq |a_m - f_m(x)| + |f_m(x) - f_n(x)| + |f_n(x) - a_n|$

$< \frac{\epsilon}{3} + \frac{\epsilon}{3} + \frac{\epsilon}{3} = \epsilon$ [by (i), (ii) and (iii)] $\forall\; m,n \geq k$

$\Rightarrow \{a_n\}$ is a Cauchy sequence of real numbers.

$\Rightarrow \{a_n\}$ is convergent.

Let $\lim_{n \to \infty} a_n = L$. Then $|a_n - L| < \frac{\epsilon}{3} \;\forall\; n \geq k$. (iv)

Also $\{f_n(x)\} \to f$ uniformly on $[a,b]$.

We have $|f_n(x) - f| < \frac{\epsilon}{3} \;\forall\; x \in [a,b],\; \forall\; n \geq k$. (v)

Therefore, $|f(x) - L| = |f(x) - f_n(x) + f_n(x) - a_n + a_n - L|$

$\leq |f(x) - f_n(x)| + |f_n(x) - a_n| + |a_n - L|$

$< \frac{\epsilon}{3} + \frac{\epsilon}{3} + \frac{\epsilon}{3} = \epsilon \;\forall\; x \in (x_0 - \delta, x_0 + \delta)$ [by (ii), (iv) and (v)]

$$\Rightarrow \lim_{x\to x_0} f(x) = L = \lim_{n\to\infty} a_n$$

$$\Rightarrow \lim_{x\to x_0}\left\{\lim_{x\to\infty} f_n(x)\right\} = \lim_{n\to\infty}\left\{\lim_{x\to x_0} f_n(x)\right\}.$$

Corollary [15.3] : Let $f_n(x) : [a,b] \to \mathbb{R}$ be defined on $[a,b]$ $\forall\, n \in N$ and let $\lim_{x\to x_0} f_n(x) = a_n$ $\forall\, n \in N$, $x_0 \in [a,b]$. If the series $\sum\limits_1^\infty f_n(x)$ converges uniformly to $S(x)$ on $[a,b]$, then (i) $\sum\limits_1^\infty a_n$ converges and (ii) $\lim_{x\to x_0} S(x) = \sum\limits_{n=1}^\infty a_n$. That is, $\lim_{x\to x_0}\left\{\sum\limits_{n=1}^\infty f_n(x)\right\} = \sum\limits_{n=1}^\infty\left\{\lim_{x\to x_0} f_n(x)\right\}$.

[*Hints.* Let $S_n(x) = \sum\limits_{r=1}^n f_n(x)$ and consider that the $\{S_n(x)\}$ converges uniformly to $S(x)$ and then proceed as in Theorem [15.5].]

Example. Show that, for a sequence of functions $\{f_n(x)\}$, where $f_n(x) = x^n$, $0 \le x \le 1$,

$$\lim_{n\to\infty}\lim_{x\to 1-0} f_n(x) \ne \lim_{x\to 1-0}\lim_{n\to\infty} f_n(x).$$

Here $f(x) = \lim_{n\to\infty} f_n(x) = 0$, $0 \le x < 1$

$= 1$ for $x = 1$

f is discontinuous at $x = 1$ on $[0,1]$. So $\{f_n(x)\}$ is not uniformly convergent on $[0,1]$.

For $x = 1$,

$$\lim_{x\to 1-0} f_n(x) = \lim_{x\to 1-0} x^n = 1 \Rightarrow \lim_{n\to\infty}\left\{\lim_{x\to 1-0} f_n(x)\right\} = \lim_{n\to\infty}\{1\} = 1. \tag{1}$$

Again, $\lim_{n\to\infty} f_n(x) = 0$ for $0 \le x < 1$

$$\Rightarrow \lim_{x\to 1-0}\left\{\lim_{n\to\infty} f_n(x)\right\} = \lim_{x\to 1-0}(0) = 0. \tag{2}$$

So, from (1) and (2), $\lim_{n\to\infty}\left(\lim_{x\to 1-0} f_n(x)\right) \ne \lim_{x\to 1-0}\left(\lim_{n\to\infty} f_n(x)\right)$.

Note. For the existence of equality the condition of uniform convergence is sufficient.

Problem [15.25] : Let f be R-integrable on $[a,b]$ and let there be a (+)ve constant M such that $\int_a^b |f(x)|dx \le M$, $\forall\, b > a$. Then show that both $\int_a^\infty f(x)dx$ and $\int_a^\infty |f(x)|dx$ converge.

Solution. Let $F(t) = \displaystyle\int_a^t |f(x)|dx$, $t \ge a$.

It can be defined as $|f|$ is integrable on $[a,b]$ by virtue of the integrability of f on $[a,b]$.

Now by definition of convergence, $\displaystyle\int_a^\infty |f(x)|dx$ is convergent, if and only if $\lim_{t\to\infty} F(t)$ exists as a finite, definite number.

Here $F(t_2) - F(t_1) = \displaystyle\int_{t_1}^{t_2} |f(x)|dx \ge 0\ \forall\, t_2 > t_1 \ge a$

$\Rightarrow F(t_2) \ge F(t_1)$ for $t_2 > t_1$.

$\Rightarrow F$ is monotonically increasing function of t in $[a,\infty]$.

Then it converges, if and only if F is bounded above on $[a,\infty)$. Since $\int_a^b |f(x)|dx \le M$, it follows that $F(t) \le M\ \forall\, t \ge a$.

$$\Rightarrow \lim_{t\to\infty} F(t) = \int_a^\infty |f(x)|dx \text{ exists finitely and hence } \int_a^\infty |f(x)|dx \text{ converges.}$$

Also $\left|\int_a^t f(x)dx\right| \le \int_a^t |f(x)|dx \ \forall\, t \ge a$

$\Rightarrow \int_a^\infty f(x)dx$ is convergent.

Problem [15.26] : (1) Let f be continuous in $[a,\infty), a > 0$ and $\lim_{x\to\infty} x^p f(x) = A$. Then prove that (i) If $p > 1$, $\int_a^\infty f(x)dx$ converges. (ii) If $p = 1$ and $A \ne 0$, $\int_a^\infty f(x)dx$ diverges. (iii) Consider the integrals $\int_1^\infty \frac{dx}{x^2}$ and $\int_2^\infty \frac{dx}{x\log x}$ and state what will be your conclusion in case $A = 0$ in (ii).

(2) If $\int_1^\infty f(x)dx$ converges and $\lim_{x\to\infty} f(x) = l$, then show that $l = 0$.

(3) A function $f(x)$ is unbounded on $[0,1]$ but is R-integrable in every interval of the form $[\epsilon, 1]$, where $0 < \epsilon < 1$.

The improper integral $\int_0^1 f(x)dx$ diverges. Is it true that $\int_0^1 \{f(x)\}^2 dx$ will also diverge? Give reasons.

(4) Show that $\int_{-1}^1 \frac{dx}{x^3}$ is not convergent but its Cauchy principal value (CPV) exists.

Solution. (1) Here f is continuous on $[a,\infty)$. So f is integrable on $[a,\infty)$.

(i) $\lim_{x\to\infty} x^p f(x) = A$

$$\Rightarrow |x^p f(x) - A| < \epsilon \ \forall\, x \ge a,\ a > 0$$
$$\Rightarrow A - \epsilon < x^p f(x) < A + \epsilon,\ \forall\, x \ge a. \tag{1}$$

Also $|x^p f(x)| - |A| < |x^p f(x) - A| < \epsilon \ \forall\, x \ge a$

$$\Rightarrow |x^p f(x)| < |A| + \epsilon = k(> 0)$$
$$\Rightarrow |f(x)| < \frac{k}{|x^p|} = \frac{k}{x^p} \Rightarrow \int_a^\infty |f(x)|dx < k\int_a^\infty \frac{1}{x^p}dx. \tag{2}$$

For $p > 1$, $\int_a^\infty \frac{1}{x^p}dx$ is convergent.

So by (2), $\int_a^\infty |f(x)|dx$ being less than a convergent integral is convergent.

Therefore, $\int_a^\infty f(x)dx$ is absolutely convergent and consequently it is convergent.

(ii) Let $A > 0$ and $\epsilon < A$.

Then from (1), $x^p f(x) > A - \epsilon = k\ (> 0) \Rightarrow f(x) > \frac{k}{x^p}\ [x > 0]$

$$\Rightarrow \int_a^\infty f(x)dx > k\int_a^\infty \frac{1}{x^p}dx.$$

For $p = 1$, $\int_a^\infty \frac{1}{x}dx$ is divergent.

Then $\int_a^\infty f(x)dx >$ a divergent integral $\Rightarrow \int_a^\infty f(x)dx$ diverges.

Next let $A < 0$ and $\epsilon < -A$. Then from (1), $x^p f(x) < A + \epsilon$

$$\Rightarrow -x^p f(x) > -A - \epsilon = A' - \epsilon = k(>0) \ [\because A' > 0]$$

$$\Rightarrow -f(x) > \frac{k}{x^p}.$$

$$\Rightarrow \int_a^\infty -f(x)dx > k\int_a^\infty \frac{1}{x^p}dx.$$

For $p = 1$, $\int_a^\infty \frac{1}{x^p}dx$ diverges.

So $\int_a^\infty -f(x)dx$ diverges and hence $\int_a^\infty f(x)dx$ diverges.

(iii) Let $A = 0$. Then $\lim_{x\to\infty} x^p f(x) = 0 \Rightarrow |x^p f(x)| < \epsilon$.

$\Rightarrow \int_a^\infty f(x)dx$ is convergent or divergent according as $p > 1$ or $p = 1$. (1)

We consider the integrals $\int_1^\infty \frac{dx}{x^2}$ and $\int_2^\infty \frac{dx}{x\log x}$.

The improper integral $\int_1^\infty \frac{dx}{x^2}$ is convergent for $p = 2 > 1$. (2)

But the improper integral $\int_2^\infty \frac{dx}{x\log x}$ diverges.

For $\log x = z$, $\int_2^\infty \frac{dx}{x\log x} = \int_{\log 2}^\infty \frac{dz}{z}$, which is divergent for $p = 1$. (3)

However, if $p = 1$,

$$\lim_{x\to\infty} x^p \frac{1}{x^2} = \lim_{x\to\infty} \frac{1}{x} = 0 \text{ and } \lim_{x\to\infty} x^p \cdot \frac{1}{x\log x} = \lim_{x\to\infty} \frac{1}{\log x} = 0.$$

Thus, $A = 0$ in each case.

(2) $\int_1^\infty f(x)$ is convergent, iff $\left|\int_{t_1}^{t_2} f(x)dx\right| < \epsilon \ \forall\, t_1, t_2 \geq k$,

$$\text{i.e.,} -\epsilon < \int_{t_1}^{t_2} f(x)dx < \epsilon, \ \forall\, t_1, t_2 \geq k. \quad (1)$$

Again $\lim_{x\to\infty} f(x) = l \Rightarrow |f(x) - l| < \frac{\epsilon}{t_2 - t_1} \ \forall\, x \geq k; t_2 > t_1$

$$\Rightarrow l - \frac{\epsilon}{t_2 - t_1} < f(x) < l + \frac{\epsilon}{t_2 - t_1}.$$

Then $l - \frac{\epsilon}{t_2 - t_1} < f(x) < l + \frac{\epsilon}{t_2 - t_1}$

$$\Rightarrow \left(l - \frac{\epsilon}{t_2 - t_1}\right)\int_{t_1}^{t_2} dx < \int_{t_1}^{t_2} f(x)dx < \left(l + \frac{\epsilon}{t_2 - t_1}\right)\int_{t_1}^{t_2} dx$$

$$\Rightarrow l(t_2 - t_1) - \epsilon < \int_{t_1}^{t_2} f(x)dx < l(t_2 - t_1) + \epsilon, \quad (2)$$

which is eqivalent to (1), if $l = 0$.

Hence the result.

(3) f is unbounded on $[0,1]$. So $f(x) > M \ \forall \ x \in [0,1]$. But $\int_\epsilon^1 f(x)dx$ is R-integrable.

That is, $\int_\epsilon^1 f(x)dx$ exists.

Therefore, $\int_\epsilon^1 f(x)dx > \int_\epsilon^1 M\,dx = M(1-\epsilon)$

$$\Rightarrow \lim_{\epsilon\to 0}\int_\epsilon^1 f(x)dx > M$$

$$\Rightarrow \int_0^1 f(x)dx \text{ is divergent.}$$

f is unbounded on $[0,1] \Rightarrow f^2$ is unbounded on $[0,1]$ and so $\int_0^1 f^2dx$ also diverges.

(4) $$\int_{-1}^1 \frac{dx}{x^3} = \int_{-1}^0 \frac{dx}{x^3} + \int_0^1 \frac{dx}{x^3}$$

$$= \lim_{\epsilon\to 0}\int_{-1}^\epsilon \frac{dx}{x^3} + \lim_{\epsilon'\to 0}\int_{\epsilon'}^1 \frac{dx}{x^3} \quad \left[\begin{array}{l}0 \text{ is a point of infinite discontinuity}\\ \text{in both parts.}\end{array}\right]$$

$$= \lim_{\epsilon\to 0}\left[-\frac{1}{2x^2}\right]_{-1}^{\epsilon} + \lim_{\epsilon'\to 0}\left[-\frac{1}{2x^2}\right]_{\epsilon'}^{1}$$

$$= \lim_{\epsilon\to 0}\left[-\frac{1}{2\epsilon^2}+\frac{1}{2}\right] + \lim_{\epsilon'\to 0}\left[-\frac{1}{2}+\frac{1}{2\epsilon'^2}\right] \to \infty. \qquad (1).$$

So $\int_{-1}^1 \frac{dx}{x^3}$ does not converges. But if we choose $\epsilon = \epsilon'$, then

$$\int_{-1}^1 \frac{dx}{x^3} = \lim_{\epsilon\to 0}\left[-\frac{1}{2\epsilon^2}+\frac{1}{2}-\frac{1}{2}+\frac{1}{2\epsilon^2}\right] = 0 \text{ [from (1)]}$$

Thus C.P.V. exists.

Theorem [15.6] [Integral test of convergence or divergence of an infinite series]
If f be a non-increasing function on $[1,\infty)$ such that $f(x) \geq 0 \ \forall \ x \in [1,\infty)$, then $\sum_1^\infty f(n)$ will converge or diverge according as $\int_1^\infty f(x)dx$ converges or diverges.

Proof. We have $n \leq x \leq n+1$ for any $n \in \mathbb{N}$. Then $f(n) \geq f(x) \geq f(n+1)$ since f is non-increasing function.

Integrating from n to $n+1$, we get

$$\int_n^{n+1} f(n)dx \geq \int_n^{n+1} f(x)dx \geq \int_n^{n+1} f(n+1)dx \Rightarrow f(n) \geq \int_n^{n+1} f(x)dx \geq f(n+1).$$

For any $m \in \mathbb{N}$,

$$\sum_{n=1}^{m-1} f(n) \geq \int_1^m f(x)dx \geq \sum_{n=1}^{m-1} f(n+1) = \sum_{k=2}^{m} f(k) \qquad \text{(i)}$$

Let $\int_1^\infty f(x)dx$ converge.

Then $\exists$ a (+)ve number L such that $\int_1^m f(x)dx < L \ \forall \ x \geq 1$.

Then from (i), $\sum_{k=2}^{m} f(k) \leq \int_1^m f(x)dx < L$

$\Rightarrow$ partial sums of the infinite series $\sum_{k=2}^{\infty} f(k)$ are bounded above.

$\Rightarrow \sum_{k=2}^{\infty} f(k)$ converges

$\Rightarrow \sum_{k=1}^{\infty} f(k)$ converges.

That is, $\sum_{n=1}^{\infty} f(n)$ converge.

Next, let $\int_1^{\infty} f(x)dx$ diverge.

Then $\lim_{m\to\infty} \int_1^m f(x)dx = \infty \Rightarrow \int_1^m f(x)dx$ is not bounded above

$$\Rightarrow \int_1^m f(x)dx > K \ \forall\ x \geq 1.$$

So from (i), $\sum_{n=1}^{m-1} f(n) \geq \int_1^m f(x)dx > K$

$\Rightarrow$ partial sums of $\sum_{n=1}^{\infty} f(n)$ are unbounded

$\Rightarrow \sum_{n=1}^{\infty} f(n)$ diverges.

Problem [15.27]: (1) Examine the convergence of the series $\sum_{n=2}^{\infty} \frac{1}{n(\log n)^\alpha}$, where $\alpha > 0$.

(2) If $f(x) \geq 0\ \forall\ x \in [1, \infty)$ and f is non-increasing on $[1, \infty)$ and if $\int_1^{\infty} f(x)dx$ converges, then show that $\lim_{x\to\infty} xf(x) = 0$.

Solution. (1) Let $f(n) = \dfrac{1}{n(\log n)^\alpha}$. Then $f(x) = \dfrac{1}{x(\log x)^\alpha}$.

For $\alpha > 0$, $f(x) > 0$ for every $x \geq 2$.

Also f is monotonically decreasing $\forall\ x \in [2, \infty)$.

Then by Theorem [15.6], $\sum_{n=2}^{\infty} \frac{1}{n(\log n)^\alpha}$ and $\int_2^{\infty} f(x)dx$ converge or diverge together.

$$\text{Now } \int_2^t f(x)dx = \int_2^t \frac{1}{x(\log x)^\alpha}dx,\ \alpha > 0 \text{ and } \alpha \neq 1$$

$$= \int_{\log 2}^{\log t} \frac{1}{z^\alpha}dz \text{ [putting } \log x = z]$$

$$= \left[\frac{z^{-\alpha+1}}{-\alpha+1}\right]_{\log 2}^{\log t} = \frac{1}{1-\alpha}\left[\frac{1}{z^{\alpha-1}}\right]_{\log 2}^{\log t}$$

$$= \frac{1}{1-\alpha}\left[\frac{1}{(\log t)^{\alpha-1}} - \frac{1}{(\log 2)^{\alpha-1}}\right].$$

Then

$$\begin{aligned}\lim_{t\to\infty}\int_2^t f(x)dx &= \lim_{t\to\alpha}\frac{1}{1-\alpha}\left[\frac{1}{(\log t)^{\alpha-1}}-\frac{1}{(\log 2)^{\alpha-1}}\right]\\ &= -\frac{1}{1-\alpha}\cdot\frac{1}{(\log 2)^{\alpha-1}},\ \text{if } \alpha>1\\ &= \infty, \text{if } \alpha<1\end{aligned} \tag{1}$$

Also

$$\begin{aligned}\int_2^t f(x)dx &= \int_2^t \frac{1}{x\log x}dx \text{ for } \alpha=1\\ &= \int_{\log 2}^{\log t}\frac{1}{z}dz \text{ for } \log x = z\\ &= [\log(\log t)-\log(\log 2)].\end{aligned}$$

$$\lim_{t\to\infty}\int_2^t f(x)dx = \lim_{t\to\infty}[\log(\log t)-\log(\log 2)] = \infty. \tag{2}$$

So from (1) and (2), it follows that $\int_2^\infty f(x)dx$ converges when $\alpha>1$ and diverges, when $\alpha\le 1$.

Hence, $\sum_2^\infty \frac{1}{n(\log n)^\alpha}$ converges when $\alpha>1$ and diverges when $0<\alpha\le 1$.

(2) By Theorem [15.6], it follows that $\sum_1^\infty f(n)$ converges, since $\int_1^\infty f(x)dx$ converges.

Here $f(n)$ is a decreasing sequence of (+)ve numbers and $\sum f(n)$ converges, then

$$\lim_{n\to\infty} nf(n) = 0 \text{ [Pringsheim Theorem]}.$$

Hence, $\lim_{x\to\infty} xf(x) = 0$.

Problem [15.28] : Let $g:[0,1]\to\mathbb{R}$ be a continuous function with $g(1)=0$. Show that the sequence of functions $\{f_n(x)\}$, where $f_n(x)=x^ng(x)$, $n\in\mathbb{N}$, converges uniformly to a constant function zero.

Solution. g is continuous on $[0,1]\Rightarrow g$ is bounded on $[0,1]\Rightarrow\exists$ a (+)ve real number M s.t.

$$|g(x)|\le M\ \forall\ x\in[0,1]. \tag{1}$$

Here $f(x)=\lim_{n\to\infty}x^ng(x)=0$ for $0\le x<1$.

For $x=1$, $\lim_{n\to\infty}x^ng(x)=g(1)=0$.

Thus $f(x)=0\ \forall\ x\in[0,1]$.

Now $|x^ng(x)|=|x^n|\,|g(x)|<\frac{\epsilon}{M}\cdot M\ \left[\lim_{n\to\infty}x^n=0 \text{ for } 0\le x<1\Rightarrow|x^n|<\frac{\epsilon}{M}\ \forall\ n\ge N\right]$

$$\Rightarrow |x^ng(x)|<\epsilon\ \forall\ n\ge N \text{ and } \forall\ x\in[0,1). \tag{2}$$

Again, at $x=1$,

$$|x^ng(x)|=|g(1)|=0<\epsilon\ \forall\ n. \tag{3}$$

So from (2) and (3), it follows that $|x^ng(x)-0|<\epsilon\ \forall\ n\ge N\ \forall\ x\in[0,1]$.

$\Rightarrow\{x^ng(x)\}$ converges uniformly to a constant function zero.

Problem [15.29]: (1) Examine the uniform convergence of $\{f_n(x)\}$, where $f_n(x) = x^n$ in $(0, 2)$.

(2) $f(x) = x$, if $x \neq 1/n$ and $f(1/n) = 1 \; \forall \; n \in \mathbb{N}$. Is f R-integrable on $[0, 1]$? Justify your answer.

(3) If $f(x) = 2$ when x is rational and $f(x) = 3$ when x is irrational. Find the upper R-integrable in $[1, 5]$.

(4) If $f_n(x) = \frac{3}{x+n}, 0 \leq x \leq 2$, state with reasons whether $\{f_n(x)\}$ converges uniformly on $[0, 2]$.

Solution. (1) $f_n(x) = x^n$ on $0 < x < 2$

$$f(x) = \lim_{n\to\infty} x^n = \begin{cases} 0, & \text{for } 0 < x < 1 \\ 1, & \text{for } x = 1 \\ \infty, & \text{for } 1 < x < 2. \end{cases}$$

f is discontinuous at $x = 1$ and limit does not exist on $1 < x < 2$.

$\{f_n(x)\}$ is pointwise convergent but not uniformly on $(0, 2)$.

(2) $\left.\begin{aligned} f(x) &= x, \text{ for } x \neq 1/n \\ &= 1, \text{ for } x = 1/n \end{aligned}\right\} n \in \mathbb{N}.$

f has infinite number of points of discontinuities.

S = Set of points of discontinuities $= \{\frac{1}{2}, \frac{1}{3}, \ldots, \frac{1}{n}, \ldots\}$.

S has a limit point 0 only. So f is R-integrable on $[0, 1]$.

(3) Every interval contains both rational and irrational points.

So, $M_r = 3$ in $[x_{r-1}, x_r] \; \forall \; r = 1, 2, \ldots, n$

$$U(f, P) = \sum_{r=1}^{n} M_r \delta_r = 3 \sum_{r=1}^{n} \delta_r = 3(5-1) = 3 \cdot 4 = 12.$$

Therefore, $\overline{\int_1^5} f(x)dx = \inf\,[U(f, P),$ for any partition $P]$
$= 12.$

(4) $f_n(x) = \dfrac{3}{x+n}, \quad 0 \leq x \leq 2.$

$f(x) = \lim_{n\to\infty} f_n(x) = 0 \; \forall \; x \in [0, 2].$

Now $|f_n(x) - f(x)| < \epsilon$, iff $\left|\frac{3}{x+n}\right| < \epsilon$,

$$\text{i.e., if } x + n > \frac{3}{\epsilon}$$

$$\text{i.e., if } n > \left(\frac{3}{\epsilon} - x\right) = m(\epsilon, x).$$

$m(\epsilon, x) \to$ finite number $\forall \; x \in [0, 2]$.

So $\exists \; m$ such that $|f_n(x)| < \epsilon \; \forall \; n \geq m$ and $\forall \; x \in [0, 2]$.

So $\{f_n(x)\}$ converges uniformly to zero on $[0, 2]$.

Theorem [15.7] Let f, g defined on $[a, \infty)$ be integrable on $[a, t] \; \forall \; a < t < \infty$ and $0 \leq f(x) \leq g(x) \; \forall \; x \in [a, \infty)$. Then $\int_a^\infty f(x)dx$ converges, if $\int_a^\infty g(x)dx$ converges and if $\int_a^\infty f(x)dx$ diverges, then $\int_a^\infty g(x)dx$ diverges.

Proof. Let $M = \int_a^\infty g(x)dx$

$$G(t) = \int_a^t g(x)dx,\ F(t) = \int_a^t f(x)dx \ \forall\ t \geq a. \tag{i}$$

For any $t \geq a$,

$$M = \int_a^\infty g(x)dx = \int_a^t g(x)dx + \int_t^\infty g(x)dx = G(t) + \int_t^\infty g(x)dx. \tag{ii}$$

Since $g(x) \geq 0\ \forall\ x \geq a$, $\int_t^\infty g(x)dx \geq 0$.

So from (ii),

$$G(t) = M - \int_t^\infty g(x)dx \leq M,\ \forall\ t \geq a. \tag{iii}$$

Again, $f(x) \leq g(x)\ \forall\ x \geq a$, so $F(t) = \int_a^t f(x)dx \leq \int_a^t g(x)dx = G(t)$.

Therefore,

$$F(t) \leq G(t) \Rightarrow F(t) \leq M \text{ [by (iii)]}. \tag{iv}$$

Furthermore, for $t_1 \geq t \geq a$,

$$\begin{aligned} F(t_1) = \int_a^{t_1} f(x)dx &= \int_a^t f(x)dx + \int_t^{t_1} f(x)dx \\ &\geq \int_a^t f(x)dx = F(t) \quad \left[\begin{array}{l}\text{since } f(x) \geq 0\ \forall\ x \in [a,\infty) \\ \Rightarrow \int_t^{t_1} f(x) \geq 0\end{array}\right] \end{aligned} \tag{v}$$

From (v), we conclude that F is a non-decreasing function. Also from (iv), it follows that F is bounded above by M. A non-decreasing sequence bounded above is convergent.

Therefore, $\lim_{t\to\infty} F(t) = \lim_{t\to\infty} \int_a^t f(x)dx$ exists.

Therefore, $\int_a^\infty f(x)dx$ converges.

Let $\int_a^\infty f(x)dx$ diverges.

$F(t)$ and $G(t)$ are defined as above in (i) $\forall\ t,\ a \leq t < \infty$.

Then $F(t)$ is non-decreasing with a limit as $t \to \infty$. That is, $\lim_{t\to\infty} F(t) = \infty$.

Since as above $G(t) \geq F(t)\ \forall\ t \geq a$, it follows $\lim_{t\to\infty} \int_a^t g(x)dx = \lim_{t\to\infty} G(t) = \infty$.

Hence, $\int_a^\infty g(x)dx$ diverges.

Note. Results hold for the integrals of the form $(-\infty, t]$ and $(-\infty, \infty)$.

Example. Let $f(x) = \dfrac{1}{1+x^2}$ be defined on $[0, \infty)$.

Examine the convergence of $\int_0^\infty f(x)dx$.

$$x^2 < 1 + x^2\ \forall\ x \in [0, \infty) \Rightarrow 0 < \frac{1}{1+x^2} < \frac{1}{x^2}\ \forall\ x > 0.$$

We have $\displaystyle\int_1^\infty \frac{1}{x^2}dx = 1.$

Therefore, by above theorem,

$$\int_1^\infty \frac{1}{1+x^2}dx \text{ converges with } \int_1^\infty \frac{1}{1+x^2}dx \leq 1. \tag{i}$$

Again, $1+x^2 \geq 1 \; \forall \, x \geq 0$

$$\Rightarrow \frac{1}{1+x^2} \leq 1 \; \forall \, x \geq 0 \Rightarrow \int_0^1 \frac{1}{1+x^2}dx \leq \int_0^1 dx = 1. \tag{ii}$$

So $\displaystyle\int_0^\infty \frac{1}{1+x^2}dx = \int_0^1 \frac{1}{1+x^2}dx + \int_1^\infty \frac{1}{1+x^2}dx \leq 1+1 = 2$ [by (i) and (ii)].

Hence, $\displaystyle\int_0^\infty \frac{1}{1+x^2}dx$ is convergent.

Example. Examine the divergence of $\displaystyle\int_2^\infty \frac{1}{\sqrt{x}-1}dx$.

We have $\sqrt{x}-1 < \sqrt{x} \; \forall \, x \geq 1$.

Therefore, $\dfrac{1}{\sqrt{x}-1} > \dfrac{1}{\sqrt{x}}$

$$\Rightarrow \int_2^\infty \frac{1}{\sqrt{x}-1}dx > \int_2^\infty \frac{1}{\sqrt{x}}dx$$

$$\Rightarrow \int_2^\infty \frac{1}{\sqrt{x}-1}dx \text{ diverges as } \int_2^\infty \frac{1}{\sqrt{x}} \text{ diverges.}$$

Problem [15.30]: Prove that $\displaystyle\lim_{K\to\infty} \int_0^\infty \frac{dx}{1+Kx^{10}} = 0$.

Solution. We write,

$$\int_0^\infty \frac{dx}{1+Kx^{10}} = \int_0^\epsilon \frac{dx}{1+Kx^{10}} + \int_\epsilon^\infty \frac{dx}{1+Kx^{10}}, \tag{1}$$

where ϵ is an arbitrary small (+)ve number.

Here

$$\int_0^\epsilon \frac{dx}{1+Kx^{10}} < \int_0^\epsilon dx = \epsilon \tag{2}$$

$$\left[\text{since } 1+Kx^{10} > 1, \; \forall \, x > 0 \Rightarrow \frac{1}{1+Kx^{10}} < 1\right]$$

Again

$$\int_\epsilon^\infty \frac{dx}{1+Kx^{10}} < \int_\epsilon^\infty \frac{dx}{Kx^{10}} \left[\begin{array}{l}\text{since } 1+Kx^{10} > Kx^{10} \\ \Rightarrow \frac{1}{1+Kx^{10}} < \frac{1}{Kx^{10}}\end{array}\right] = \left[-\frac{1}{9K}\cdot\frac{1}{x^9}\right]_\epsilon^\infty = \frac{1}{9K\epsilon^9}.$$

Now $K \to \infty$,

$$\frac{1}{9K\epsilon^9} \to 0 \Rightarrow \frac{1}{9K\epsilon^9} < \epsilon \; \forall \, K > M, \tag{3}$$

where M is any (+)ve real number.

So from (1), we have by (2) and (3)

$$0 < \int_0^\infty \frac{dx}{1+Kx^{10}} < \epsilon + \epsilon = 2\epsilon \; \forall \, K > M \Rightarrow \lim_{K\to\infty} \int_0^\infty \frac{dx}{1+Kx^{10}} = 0.$$

Problem [15.31]: If $f(x)$ be continuous, non-increasing function on $[a, \infty)$ and $\int_a^\infty f(x)dx$ converges, then show that $\displaystyle\lim_{x\to\infty} f(x) = 0$.

Solution. Since $\int_a^\infty f(x)dx$ converges, we have $\left|\int_{t_1}^{t_2} f(x)dx\right| < \epsilon \ \forall\ t_2 > t_1 \geq a$

$\Rightarrow\ |(t_2 - t_1)f(\xi)| < \epsilon$, since f is continuous on $[t_1, t_2]$, where $[t_1, t_2] \subseteq [a, \infty), \xi \in [t_1, t_2]$

$\Rightarrow\ (t_2 - t_1)|f(\xi)| < \epsilon$

$\Rightarrow\ |f(\xi)| < \dfrac{\epsilon}{t_2 - t_1} < \epsilon.$

Since f is non-increasing and $|f(\xi)| < \epsilon$, it follows that $\lim\limits_{x\to\infty} f(x) = 0$, as for each set of intervals $[t_1, t_2]$, there exists a point ξ satisfying $|f(\xi)| < \epsilon \Rightarrow \exists$ an aggregate of points $\xi_1, \xi_2, \ldots$, with $\xi_i < \xi_{i+1}$ satisfying $|f(\xi_i)| < \epsilon,\ i = 1, 2, \ldots$ and

$$f(\xi_1) \geq f(\xi_2) \geq \cdots \geq f(\xi_n), \cdots \Rightarrow \lim_{\xi_n\to\infty} f(\xi_n) = 0.$$

Problem [15.32] : (1) Let f be continuous on $[2, \infty)$ and $\lim\limits_{x\to\infty} x(\log x)^2 f(x) = l(\neq 0)$, then prove that $\int_2^\infty f(x)dx$ is convergent.

(2) Examine the convergence of $\displaystyle\int_0^\infty \frac{\sin(ax+b)}{x^n}dx$,

when (i) $0 < n < 2,\ a \neq 0,\ b = 0$

(ii) $0 < n < 1,\ a \neq 0,\ b \neq 0$.

Solution. (1) $\lim\limits_{x\to\infty} x(\log x)^2 f(x) = l\ (\neq 0)$.

$$\Rightarrow\ |x(\log x)^2 f(x) - l| < \epsilon\ \forall\ x \geq 2$$

$$\Rightarrow\ l - \epsilon < x(\log x)^2 f(x) < l + \epsilon. \tag{1}$$

Let $l > 0$. Then $x(\log(x)^2 f(x) < l + \epsilon = k\ (> 0)$

$$\Rightarrow\ f(x) < \frac{k}{x(\log x)^2} \quad [\because x(\log x)^2 > 0\ \forall\ x \in [2, \infty)]$$

$$\Rightarrow\ \int_2^\infty f(x)dx < k\int_2^\infty \frac{1}{x(\log x)^2}dx \quad \left[\begin{array}{l} f \text{ is continuous on } [2,\infty) \\ \Rightarrow f \text{ is integrable on } [2,\infty) \end{array}\right]$$

Now $\displaystyle\int_2^\infty \frac{1}{x(\log x)^2}dx = \int_{\log 2}^\infty \frac{1}{z^2}dz$ which is convergent.

So $\int_2^\infty f(x)dx$ is convergent.

Next, let $l < 0$. Then from (1),

$$x(\log x)^2 f(x) > l - \epsilon$$

$$\Rightarrow\ -x(\log x)^2 f(x) < -l + \epsilon = k'(> 0)\ [\because -l > 0]$$

$$\Rightarrow\ \int_2^\infty -f(x)dx < k'\int_2^\infty \frac{1}{x(\log x)^2}dx, \text{ which is convergent.}$$

So $\int_2^\infty -f(x)dx$ is convergent and hence $\int_2^\infty f(x)dx$ is convergent.

(2) Let $I = \displaystyle\int_0^\infty \frac{\sin(ax+b)}{x^n}dx$.

CASE (i) Let $a \neq 0$ and $b = 0$. Then

$$I = \int_0^\infty \frac{\sin ax}{x^n} = \int_0^\infty \frac{\sin ax \cdot a}{ax \cdot x^{n-1}} dx = a\left[\int_0^1 \frac{\frac{\sin ax}{ax}}{x^{n-1}} dx + \int_1^\infty \frac{\frac{\sin ax}{ax}}{x^{n-1}} dx\right]. \quad (1)$$

For the 1st part, let $f(x) = \dfrac{\sin ax}{ax}$ and $\phi(x) = \dfrac{1}{x^{n-1}}$.

0 is a point of infinite discontinuity when $(n-1) > 0$. For, $(n-1) < 0$, the integral is proper.

Now, $\lim_{x \to 0+} \dfrac{f(x)}{\phi(x)} = 1$.

Also $\displaystyle\int_0^1 \phi(x)dx = \int_0^1 \frac{1}{x^{n-1}} dx$.

Converges for $n - 1 < 1$, i.e., if $n < 2$.

For the 2nd part, $a \displaystyle\int_1^\infty \frac{\sin ax}{ax} \cdot \frac{1}{x^{n-1}} dx = \int_1^\infty \frac{\sin ax}{x^n} dx.$

Here $\displaystyle\int_1^t \sin ax dx$ is bounded $\forall\, t \geq 1$ and $\phi(x) = \dfrac{1}{x^n}$ is monotonically decreasing on $[2, \infty]$ and $\to 0$ as $x \to \infty$ if $n > 0$.

Therefore, by Abel's test $\displaystyle\int_1^\infty \frac{\sin ax}{x^n} dx$ is convergent.

Hence, from (1) the given integral is convergent for $0 < n < 2$.

CASE (ii) $I = \displaystyle\int_0^\infty \frac{\sin(ax+b)}{x^n}$, $a \neq 0$, $b \neq 0$.

Put $ax + b = z$. Then $dx = \dfrac{dz}{a}$, $x = \dfrac{z-b}{a}$.

$$I = \int_{z=b}^\infty \frac{a^n \sin z}{(z-b)^n} \cdot \frac{dz}{a} = a^{n-1} \int_b^\infty \frac{\sin z}{(z-b)^n} dz$$

$$= a^{n-1}\left[\int_b^1 \frac{\sin z}{(z-b)^n} dz + \int_1^\infty \frac{\sin z}{(z-b)^n} dz\right].$$

For the 1st part. If $n > 0$, $z = b$ is a point of infinite discontinuity on $f(z) = \dfrac{\sin z}{(z-b)^n}$.

Let $\phi(z) = \dfrac{1}{(z-b)^n}$.

Then $\lim_{z \to b} \dfrac{f(z)}{\phi(z)} = \sin b (\neq 0)$.

Also $\displaystyle\int_b^1 \phi(z)dz$ is convergent, when $n < 1$.

So the first part converges for $n < 1$.

For the 2nd part. $\displaystyle\int_1^t \sin z dz$ is bounded $\forall\, t \geq 1$.

$\phi(z) = \dfrac{1}{(z-b)^n}$ is bounded and monotonically decreases and $\phi(z) \to 0$ as $z \to \infty$ $\forall\, z \in [1, \infty)$. Therefore, by Abel's test the 2nd part converges for $n > 0$.

So the given integral converges for $0 < n < 1$.

If $n < 0$, the first is proper but the 2nd part does not converge. So the given integral converges, only when $0 < n < 1$.

Definition [15.5] : Let the Fourier Series $\frac{a_0}{2} + \sum_1^{\infty}(a_n \cos nx + b_n \sin nx)$ converge uniformly to $f(x)$ on $[0, 2\pi]$.

That is, $f(x) \sim \frac{a_0}{2} + \sum_1^{\infty}(a_n \cos nx + b_n \sin nx)$ on $[0, 2\pi]$. (1)

If $s_n(x)$ be the nth partial sum of the Fouries Series (1), then $\{S_n(x)\}$ is a convergent sequence and converges to $\frac{1}{2}[f(x+0) + f(x-0)]$.

When f is continuous, $f(x) =$ sum of the F.S. in (1) and $\{S_n(x)\}$ converges to $f(x)$.

From the sequence of partial sums $\{S_n(x)\}$, we define the sequence of arithmetic mean $\{\sigma_n(x)\}$ by

$$\sigma_n(x) = \frac{S_0(x) + S_1(x) + \cdots + S_{n-1}(x)}{n}, \tag{1}$$

where $S_0(x) = \frac{a_0}{2}$,

$$S_1(x) = \frac{a_0}{2} + (a_1 \cos x + a_1 \sin x),$$

$$S_{n-1}(x) = \frac{a_0}{2} + \sum_{k=1}^{n-1}(a_k \cos kx + b_k \sin kx).$$

Thus, if f is continuous, then $\lim_{n\to\infty} \sigma_n(x) = f(x)$, since each $S_n(x)$ converges to $f(x)$. (2)

Also, by Parseval's theorem,

$$\int_0^{2\pi} [f(x)]^2 dx = a_0^2 \frac{\pi}{2} + \pi \sum_{n=1}^{\infty}(a_n^2 + b_n^2). \tag{3}$$

Problem [15.33] : Let $f(x) \sim \frac{a_0}{2} + \sum_{n=1}^{\infty}(a_n \cos nx + b_n \sin nx)$ and let $\{\{\sigma_n(x)\}$ be the sequence of arithmetic means of partial sums $S_n(x)$ of the series, then show that

(1) $\sigma_n(x) = \frac{a_0}{2} + \sum_{k=1}^{n-1}\left(1 - \frac{k}{n}\right)(a_k \cos kx + b_k \sin kx)$.

(2) $\int_0^{2\pi} \{f(x) - \sigma_n(x)\}^2 dx = \int_0^{2\pi} [f(x)]^2 dx - \frac{\pi}{2}a_0^2 - \pi \sum_{k=1}^{n-1}(a_k^2 + b_k^2) + \frac{\pi}{n^2} \sum_{k=1}^{n-1} k^2(a_k^2 + b_k^2)$.

(3) If f is continuous on $[0, 2\pi]$ and has period 2π, then

$$\lim_{n\to\infty} \frac{\pi}{n^2} \sum_{k=1}^{n-1} k^2(a_k^2 + b_k^2) = 0.$$

Solution. (1) $\sigma_n(x) = \dfrac{S_0(x) + S_1(x) + S_2(x) + \cdots + S_{n-1}(x)}{n}$

$$= \frac{\frac{a_0}{2} + \left\{\frac{a_0}{2} + (a_1 \cos x + b_1 \sin x)\right\} + \left\{\frac{a_0}{2} + (a_1 \cos x + b_1 \sin x) + \{a_2 \cos 2x + b_2 \sin 2x)\right\} + \cdots + \left\{\frac{a_0}{2} + (a_1 \cos x + b_1 \sin x) + (a_2 \cos 2x + b_2 \sin 2x) + \cdots + a_{n-1} \cos(n-1)x + b_{n-1} \sin(n-1)x\right\}}{n}$$

$$= \frac{n\frac{a_0}{2} + (n-1)(a_1 \cos x + b_1 \sin x) + (n-2)(a_2 \cos 2x + b_2 \sin 2x) + \cdots + \{n-(n-1)\}\{a_{n-1} \cos(n-1)x + b_{n-1} \sin(n-1)x\}}{n}$$

$$= \frac{a_0}{2} + \left(1 - \frac{1}{n}\right)(a_1 \cos x + b_1 \sin x) + \left(1 - \frac{2}{n}\right)(a_2 \cos 2x + b_2 \sin 2x) + \cdots + \left(1 - \frac{n-1}{n}\right)\{a_{n-1} \cos(n-1)x + b_{n-1} \sin(n-1)x\}$$

$$= \frac{a_0}{2} + \sum_{k=1}^{n-1}\left(1 - \frac{k}{n}\right)(a_k \cos kx + b_k \sin kx).$$

(2) $$\int_0^{2\pi} \{f(x) - \sigma_n(x)\}^2 dx = \int_0^{2\pi} [f(x)]^2 dx - 2\int_0^{2\pi} f(x)\sigma_n(x) dx + \int_0^{2\pi} \sigma_n^2(x) dx. \qquad \text{(i)}$$

Now $\displaystyle\int_0^{2\pi} f(x)\sigma_n(x) dx$

$$= \int_0^{2\pi}\left[\frac{a_0}{2} + \sum_{n=1}^{\infty}(a_n \cos nx + b_n \sin nx)\right]\left[\frac{a_0}{2} + \sum_{k=1}^{n-1}\left(1 - \frac{k}{n}\right)(a_k \cos kx + b_k \sin kx)\right] dx$$

$$= \left(\frac{a_0}{2}\right)^2 \int_0^{2\pi} dx + \frac{a_0}{2}\sum_{k=1}^{n-1}\int_0^{2\pi}\left[\left(1 - \frac{k}{n}\right)(a_k \cos kx + b_k \sin kx)\right] dx$$

$$+ \sum_{k=1}^{n-1}\int_0^{2\pi}\left(1 - \frac{k}{n}\right)(a_k \cos kx + b_k \sin kx)^2 dx$$

$$+ \sum_{k=1}^{n-1}\int_0^{2\pi}[(a_n \cos nx + b_n \sin nx) + (a_{n+1} \cos(n+1)x + b_{n+1} \sin(n+1)x) + \cdots \infty]$$

$$\times \left[\left(1 - \frac{k}{n}\right)(a_k \cos kx + b_k \sin kx)\right] dx \qquad \text{(ii)}$$

$$= \frac{a_0^2}{2} \cdot \pi + \frac{a_0}{2} \cdot 0 + \sum_{k=1}^{n-1}\int_0^{2\pi}\left(1 - \frac{k}{n}\right)[a_k^2 \cos^2 kx + b_k^2 \sin^2 kx + a_k b_k \sin 2kx] dx + 0$$

$$= \frac{a_0^2}{2} \cdot \pi + \frac{\pi}{n}[(n-1)(a_1^2 + b_1^2) + (n-2)(a_2^2 + b_2^2) + \cdots + (a_{n-1}^2 + b_{n-1}^2)]$$

since $\int_0^{2\pi} \cos^2 kx dx = \int_0^{2\pi} \sin^2 kx dx = \pi$

$$\left.\begin{aligned} \text{and } \int_0^{2\pi} \cos mx \cos nx dx &= \int_0^{2\pi} \sin mx \sin nx dx \\ &= \int_0^{2\pi} \cos mx \sin nx dx = 0 \text{ for } m \neq n \end{aligned}\right\} \quad \text{(iii)}$$

Each term of the expression in the 4th part of (ii) is one of the forms in (iii).

$$\text{Thus } \int_0^{2\pi} f(x)\sigma_n(x)dx = \pi\frac{a_0^2}{2} + \frac{\pi}{n}\sum_{k=1}^{n-1}(n-k)(a_k^2 + b_k^2). \quad \text{(iv)}$$

$$\begin{aligned} \int_0^{2\pi} \sigma_n^2(x)dx &= \int_0^{2\pi} \left[\frac{a_0}{2} + \sum_{k=1}^{n-1}\left(1 - \frac{k}{n}\right)(a_k \cos kx + b_k \sin kx)\right]^2 dx \\ &= \frac{a_0^2}{2}\pi + a_0 \sum_{k=1}^{n-1} \int_0^{2\pi} \left(1 - \frac{k}{n}\right)(a_k \cos kx + b_k \sin kx) \\ &\quad + \sum_{k=1}^{n-1} \int_0^{2\pi} \left(1 - \frac{k}{n}\right)^2 (a_k \cos kx + b_k \sin kx)^2 dx \\ &= \frac{a_0^2}{2}\pi + 0 + \sum_{k=1}^{n-1} \int_0^{2\pi} \left(1 - \frac{k}{n}\right)^2 \left[a_k^2 \cos^2 kx + b_k^2 \sin^2 kx + a_k b_k \sin 2kx\right] dx \\ &= \frac{a_0^2}{2}\pi + \frac{\pi}{n^2}\left[\sum_{k=1}^{n-1}(n-k)^2(a_k^2 + b_k^2)\right]. \quad \text{(v)} \end{aligned}$$

Therefore, from (i), by (ii), (iv) and (v),

$$\begin{aligned} &\int_0^{2\pi} [f(x) - \sigma_n(x)]^2 dx \\ &= \int_0^{2\pi} [f(x)]^2 dx - 2\left[\pi \cdot \frac{a_0^2}{2} + \frac{\pi}{n}\sum_{k=1}^{n-1}(n-k)(a_k^2 + b_k^2)\right] + \frac{a_0^2}{2} \cdot \pi + \frac{\pi}{n^2}\sum_{k=1}^{n-1}(n-k)^2(a_k^2 + b_k^2) \\ &= \int_0^{2\pi} [f(x)]^2 dx - a_0^2\frac{\pi}{2} - 2\pi\sum(a_k^2 + b_k^2) + \frac{2\pi}{n}\sum_{k=1}^{n-1} k(a_k^2 + b_k^2) \\ &\quad + \frac{\pi}{n^2}\sum_{k=1}^{n-1}\{n^2 - 2nk + k^2)(a_k^2 + b_k^2) \\ &= \int_0^{2\pi} [f(x)]^2 dx - a_0^2\frac{\pi}{2} - 2\pi\sum_{k=1}^{n-1}(a_k^2 + b_k^2) + \frac{2\pi}{n}\sum_{k=1}^{n-1} k(a_k^2 + b_k^2) \\ &\quad + \pi\sum_{k=1}^{n-1}(a_k^2 + b_k^2) - \frac{2\pi}{n}\sum_{k=1}^{n-1} k(a_k^2 + b_k^2) + \frac{\pi}{n^2}\sum_{k=1}^{n-1} k^2(a_k^2 + b_k^2) \\ &= \int_0^{2\pi} [f(x)]^2 dx - a_0^2\frac{\pi}{2} - \pi\sum_{k=1}^{n-1}(a_k^2 + b_k^2) + \frac{\pi}{n^2}\sum_{k=1}^{n-1} k^2(a_k^2 + b_k^2). \quad \text{(vi)} \end{aligned}$$

(3) Let f be continuous on $[0, 2\pi]$. Then sum of the series $\frac{a_0}{2} + \sum_{n=1}^{\infty}(a_n \cos nx + b_n \sin nx)$ is $f(x)$ and in that case, by Parseval's theorem,

$$\int_0^{2\pi} [f(x)]^2 dx = \frac{a_0^2}{2}\pi + \pi \sum_{k=1}^{\infty}(a_k^2 + b_k^2).$$

Also $\lim_{n\to\infty} \sigma_n(x) = f(x)$ [by (2) of Def. [15.5].]

Hence taking limit $n \to \infty$ in (vi), we get

$$0 = \frac{a_0^2\pi}{2} + \pi\sum_{k=1}^{\infty}(a_k^2 + b_k^2) - \frac{a_0^2\pi}{2} - \pi\sum_{k=1}^{\infty}(a_k^2 + b_k^2) + \lim_{n\to\infty}\frac{\pi}{n^2}\sum_{1}^{n-1} k^2(a_k^2 + b_k^2)$$

$$0 = \lim_{n\to\infty}\frac{\pi}{n^2}\sum_{k=1}^{n-1} k^2(a_k^2 + b_k^2)$$

Hence the result.

Definition [15.6] : [Harmonic Analysis] If the function is not defined by a formula but the table for the values of a function and the independent variable are given, then the process of finding the Fourier Series is known as *Harmonic Analysis.*

The Fourier co-efficients are evaluated by the following formulae.

(i) $a_0 = \frac{1}{\pi}\int_0^{2\pi} f(x)dx = \frac{2}{2\pi - 0}\int_0^{2\pi} f(x)dx = 2$ [Mean value of $f(x)$ in $(0, 2\pi)$], since by M.V. Theorem $\int_a^b f(x)dx = \frac{1}{b-a}\int_a^b f(x)dx$.

(ii) $a_n = \frac{1}{\pi}\int_0^{2\pi} f(x)\cos nx dx = \frac{2}{2\pi - 0}\int_0^{2\pi} f(x)\cos nx dx$.

So $a_n = 2$ [Mean value of $f(x)\cos nx$ in $[0, 2\pi]$].

(iii) $b_n = \frac{1}{\pi}\int_0^{2\pi} f(x)\sin nx dx = \frac{2}{2\pi - 0}\int_0^{2\pi} f(x)\sin nx dx$

$= 2$ [Mean value of $f(x)\sin nx$ in $(0, 2\pi)$].

Note. The term $(a_1 \cos x + b_1 \sin x)$ in the F.S. is called first harmonic. The term $(a_2 \cos 2x + b_2 \sin 2x)$ in the F.S. is called second harmonic and so on.

Example. Determine the Fourier Series from the given Table up to 2nd harmonic to represent the function $f(x)$.

x	0	60°	120°	180°	240°	300°
$f(x)$	1	1.5	3	3.5	2	2.5

x	$f(x)$	$\sin x$	$\sin 2x$	$\cos x$	$\cos 2x$	$f(x)\sin x$	$f(x)\sin 2x$	$f(x)\cos x$	$f(x)\cos 2x$
0°	1	0	0	1	1	0	0	1	1
60°	1.5	0.87	+0.87	0.50	−0.50	1.30	1.30	0.75	−0.75
120°	3	0.87	−0.87	−0.50	−0.50	2.61	−2.61	−1.50	−1.50
180°	3.5	0	0	−1.00	1	0	0	−3.50	3.50
240°	2	−0.87	0.87	−0.50	−0.50	−1.74	1.74	−1.00	−1.00
300°	2.5	−0.87	−0.87	0.50	−0.50	−2.17	−2.17	1.25	−1.25
	$\Sigma f(x) = 13.5$					$\Sigma f(x)\sin x = 1.82$	$\Sigma f(x)\sin 2x = -1.82$	$\Sigma f(x)\cos x = -1.00$	$\Sigma f(x)\cos 2x$ 3.00

$$\overline{f(x)} = \text{mean of } f(x) = \frac{13.5}{6} = 2.25$$

$$\overline{f(x)\sin x} = \text{mean of } f(x)\sin x = \frac{1.82}{6} = 0.30$$

$$\overline{f(x)\sin 2x} = \text{mean of } f(x)\sin 2x = -\frac{1.82}{6} = -0.30$$

$$\overline{f(x)\cos x} = \text{mean of } f(x)\cos x = -\frac{1.00}{6} = -0.16$$

$$\overline{f(x)\cos 2x} = \text{mean of } f(x)\cos 2x = \frac{3}{6} = 0.50$$

$$a_0 = 2\times \text{ mean of } f(x) = 2 \times 2.25 = 4.50$$

$$a_1 = 2\times \text{ mean of } f(x)\cos x = 2 \times (-0.16) = -0.32$$

$$a_2 = 2\times \text{ mean of } f(x)\cos 2x = 2 \times 0.50 = 1.00$$

$$b_1 = 2\times \text{ mean of } f(x)\sin x = 2 \times 0.30 = 0.60$$

$$b_2 = 2\times \text{ mean of } f(x)\sin 2x = 2 \times (-0.30) = -0.60$$

F.S. corresponding to $f(x)$ up to 2nd harmonic is

$$\begin{aligned} f(x) &= \frac{a_0}{2} + (a_1\cos x + b_1 \sin x) + (a_2\cos 2x + b_2\sin 2x) + \cdots \\ &= \frac{4.50}{2} + (-0.32\cos x + 0.60\sin x) + (\cos 2x - 0.60\sin 2x) + \cdots \\ &= 2.25 - 0.32\cos x + 0.60\sin x + \cos 2x - 0.60\sin 2x + \cdots \\ &= 2.25 - 0.32\cos x + \cos 2x + 0.60\sin x - 0.60\sin 2x. \end{aligned}$$

Fourier Series in Complex Forms

Definition [15.7]: Let f be a function defined in $[-\pi, \pi]$ with period 2π. Then the Fourier Series representing to $f(x)$ is defined by

$$f(x) = \frac{a_0}{2} + \sum_{n=1}^{\infty}(a_n\cos nx + b_n \sin nx),$$

where $a_0 = \frac{1}{\pi}\int_{-\pi}^{\pi} f(x)dx, a_n = \frac{1}{\pi}\int_{-\pi}^{\pi} f(x)\cos nx dx$ and $b_n = \frac{1}{\pi}\int_{-\pi}^{\pi} f(x)\sin nx dx$.

We know that $\cos nx = \dfrac{e^{inx}+e^{-inx}}{2}$ and $\sin x = \dfrac{e^{inx}-e^{-inx}}{2i}$.

Therefore,

$$\begin{aligned} f(x) &= \frac{a_0}{2} + \sum_{n=1}^{\infty}\left[a_n\left(\frac{e^{inx}+e^{-inx}}{2}\right) + b_n\left(\frac{e^{inx}-e^{-inx}}{2i}\right)\right] \\ &= \frac{a_0}{2} + \sum_{n=1}^{\infty}\left[a_n\left(\frac{e^{inx}+e^{-inx}}{2}\right) - ib_n\left(\frac{e^{inx}-e^{-inx}}{2}\right)\right] \\ &= \frac{a_0}{2} + \sum_{n=1}^{\infty}\left[\left(\frac{a_n - ib_n}{2}\right)e^{inx} + \left(\frac{a_n+ib_n}{2}\right)e^{-inx}\right] \\ &= C_0 + \sum_{n=1}^{\infty}\left[C_n e^{inx} + C_{-n}e^{-inx}\right], \end{aligned} \qquad (1)$$

where $C_0 = \dfrac{a_0}{2}$, $C_n = \dfrac{a_n - ib_n}{2}$, $C_{-n} = \dfrac{a_n+ib_n}{2}$.

Thus

$$C_0 = \frac{1}{2\pi}\int_{-\pi}^{\pi} f(x)dx$$

$$C_n = \frac{1}{2\pi}\left[\int_{-\pi}^{\pi} [f(x)]\cos nx dx - i\int_{-\pi}^{\pi} f(x)\sin nx dx\right]$$

$$= \frac{1}{2\pi}\int_{-\pi}^{\pi} f(x)\{\cos nx - i\sin nx\}dx = \frac{1}{2\pi}\int_{-\pi}^{\pi} f(x)e^{-inx}dx.$$

$$C_{-n} = \frac{1}{2\pi}\int_{-\pi}^{\pi} f(x)e^{inx}dx.$$

C_0, C_n, C_{-n} are all complex co-efficients in the Fourier Series (1).

Remarks. (1) Fourier Series in complex form of a function $f(x)$ defined in $[0, 2l]$ with period $2l$ is defined by $f(x) = C_0 + \sum_{n=1}^{\infty}\left[C_n e^{\frac{in\pi x}{l}} + C_{-n}e^{-\frac{in\pi x}{l}}\right]$, where

$$C_0 = \frac{a_0}{2} = \frac{1}{2l}\int_0^{2l} f(x)dx, C_n = \frac{1}{2l}\int_0^{2l} f(x)e^{-\frac{in\pi x}{l}}dx, C_{-n} = \frac{1}{2l}\int_0^{2l} f(x)e^{\frac{in\pi x}{l}}dx.$$

(2) The F.S. in complex form of $f(x)$ in $[-l, l]$ with period $2l$ is of the form

$$C_0 + \sum_1^{\infty}\left\{C_n e^{\frac{in\pi x}{l}} + C_{-n}e^{-\frac{in\pi x}{l}}\right\}, \text{ where } C_n = \frac{1}{l}\int_{-l}^{l} f(x)e^{-\frac{in\pi x}{l}}dx.$$

(3) The F.S. in (1) can also be written as

$$f(x) = \sum_{-\infty}^{\infty} C_n e^{inx}, \text{ where } C_n = \frac{1}{\pi}\int_{-\pi}^{\pi} f(x)e^{-inx}dx.$$

Problem [15.34]: Find the F.S. in complex form in $[-1, 1]$ with period 2 for the function f defined by $f(x) = e^{-x}\ \forall\ x \in [-1, 1]$.

Solution. Let the F.S. of $f(x)$ be $f(x) = \sum_{-\infty}^{\infty} C_n e^{in\pi x}$ in $[-1, 1]$, where

$$C_n = \frac{1}{1}\int_{-1}^{1} e^{-x}\cdot e^{-in\pi x}dx = \int_{-1}^{1} e^{-x(1+in\pi)}dx = \frac{-1}{1+in\pi}\left[e^{-x(1+in\pi)}\right]_{-1}^{1}$$

$$= \frac{1}{1+in\pi}\left\{e^{(1+in\pi)} - e^{-(1+in\pi)}\right\}$$

$$= \frac{1-in\pi}{1+n^2\pi^2}\left\{e(\cos n\pi + i\sin n\pi) - \frac{1}{e}(\cos n\pi - i\sin n\pi)\right\}$$

$$= \frac{1-in\pi}{1+n^2\pi^2}\left\{\left(e - \frac{1}{e}\right)\cos n\pi\right\} \quad [\because \sin n\pi = 0\ \forall\ n \in N]$$

$$= \frac{(1-in\pi)(-1)^n}{1+n^2\pi^2}\left(\frac{e^1 - e^{-1}}{2}\right)\cdot 2 = \frac{2(-1)^n(1-in\pi)}{1+n^2\pi^2}\sinh 1.$$

Therefore, the required F.S. is of the form

$$\sum_{-\infty}^{\infty}\frac{2(-1)^n(1-in\pi)}{1+n^2\pi^2}\sinh 1\cdot e^{in\pi x}.$$

Problem [15.35] : Let $S_n(x)$ denote the nth partial sum of the F.S. of

$$f(x) = \frac{a_0}{2} + \sum_{n=1}^{\infty}(a_n \cos nx + b_n \sin nx) \text{ in } [-\pi, \pi].$$

Then show that the F.S. generated by f in $[-\pi, \pi]$ will converge for a given value of x, if and only if $\lim_{n\to\infty} \frac{1}{2\pi}\int_{-\pi}^{\pi} f(x+t)\frac{\sin nt}{t}dt$ exists. Here f is bounded and R-integrable on $[-\pi, \pi]$.

Solution. We have

$$\begin{aligned}
S_n(x) &= \frac{a_0}{2} + \sum_{k=1}^{n}(a_k \cos kx + b_k \sin kx) \\
&= \frac{1}{2}\frac{1}{\pi}\int_{-\pi}^{\pi} f(t)dt + \sum_{k=1}^{n}\left\{\frac{1}{\pi}\int_{-\pi}^{\pi} f(t)\cos ktdt\right\}\cos kx + \left\{\frac{1}{\pi}\int_{-\pi}^{\pi} f(t)\sin ktdt\right\}\sin kx \\
&= \frac{1}{\pi}\int_{-\pi}^{\pi} f(t)\left[\frac{1}{2} + \sum_{k=1}^{n}\{\cos kt \cos kx + \sin kt \sin kx\}\right] dt \\
&= \frac{1}{\pi}\int_{-\pi}^{\pi} f(t)\left[\frac{1}{2} + \sum_{k=1}^{n}\cos(t-x)k\right] dt \\
&= \frac{1}{\pi}\int_{-\pi}^{\pi} f(x+t)\left[\frac{1}{2} + \sum_{k=1}^{n}\cos kt\right] dt.
\end{aligned}$$

$$\left[\because \int_{-\pi}^{\pi} f(t)dt = \int_{-\pi}^{\pi} f(t+x)dt \text{ for any } x\right]$$

Now,

$$\begin{aligned}
\frac{1}{2} + \sum_{k=1}^{n}\cos kt &= \frac{1}{2} + [\cos t + \cos 2t + \cdots + \cos nt] \\
&= \frac{1}{2} + \frac{\sin(2n+1)\frac{t}{2} - \sin\frac{t}{2}}{2\sin\frac{t}{2}} = \frac{\sin\left(n+\frac{1}{2}\right)t}{2\sin\frac{t}{2}}.
\end{aligned}$$

Therefore,

$$S_n(x) = \frac{1}{\pi}\int_{-\pi}^{\pi} f(x+t)\frac{\sin\left(n+\frac{1}{2}\right)t}{2\sin\frac{t}{2}}dt. \tag{1}$$

Here

$$\begin{aligned}
\frac{\sin\left(n+\frac{1}{2}\right)t}{2\sin\frac{t}{2}} &= \frac{\sin nt\cos\frac{t}{2} + \cos nt\sin\frac{t}{2}}{2\sin\frac{t}{2}} = \sin nt \cdot \frac{1}{2\tan\frac{t}{2}} + \frac{1}{2}\cos nt \\
&= \frac{\sin nt}{t} + \sin nt\left[\frac{1}{2\tan\frac{t}{2}} - \frac{1}{t}\right] + \frac{\cos nt}{2}.
\end{aligned}$$

We define a function $F(t)$ by

$$F(t) = \begin{cases} \frac{1}{2\tan\frac{t}{2}} - \frac{1}{t}, & \text{for } -\pi \le t < 0 \le \pi \\ 0, & \text{for } t = 0 \end{cases}$$

Then $F(t)$ is continuous on $[-\pi, \pi]$ and hence it is bounded and integrable there. From (1), we get

$$S_n(x) = \frac{1}{\pi}\int_{-\pi}^{\pi} f(x+t)\left[\frac{\sin nt}{t} + F(t)\sin nt + \frac{1}{2}\cos nt\right] dt. \tag{2}$$

By Riemann-Lebesgue theorem,

$$\lim_{n\to\infty}\int_{-\pi}^{\pi} f(x+t)F(t)\sin ntdt = 0 \text{ and } \lim_{n\to\infty}\int_{-\pi}^{\pi} f(x+t)\cos ntdt = 0.$$

So from (2), $\lim_{n\to\infty} S_n(x)$ exists, if and only if $\lim_{n\to\infty}\frac{1}{\pi}\int_{-\pi}^{\pi} f(x+t)\frac{\sin nt}{t}dt$ exists.

Hence the F.S. of f converges for any value of x on $[-\pi,\pi]$, if and only if

$$\lim_{n\to\infty}\frac{1}{\pi}\int_{-\pi}^{\pi} f(x+t)\frac{\sin nt}{t}dt \text{ exists.}$$

Corollary [15.4]: Show that the F.S. generated by f on $[-\pi,\pi]$ with period 2π will converge for any value of x in $[-\pi,\pi]$ iff, for some (+)ve number $\delta < \pi$,

$$\lim_{n\to\infty}\frac{1}{\pi}\int_0^{\delta}[f(x+t)+f(x-t)]\frac{\sin nt}{t}dt \text{ exists.}$$

[*Hints.*

$$S_n(x) = \frac{1}{\pi}\int_{-\pi}^{\pi} f(x+t)\frac{\sin\left(n+\frac{1}{2}\right)t}{2\sin\frac{t}{2}}dt \quad \text{[From (1)]}$$

$$= \frac{1}{\pi}\left[\int_{-\pi}^{0} f(x+t)\frac{\sin\left(n+\frac{1}{2}\right)t}{2\sin\frac{t}{2}}dt + \int_0^{\pi} f(x+t)\frac{\sin\left(n+\frac{1}{2}\right)t}{2\sin\frac{t}{2}}dt\right]$$

$$= \frac{1}{\pi}\left[\int_0^{\pi} f(x-t)\frac{\sin\left(n+\frac{1}{2}\right)(-t)}{2\sin\left(-\frac{t}{2}\right)}dt + \int_0^{\pi} f(x+t)\frac{\sin\left(n+\frac{1}{2}\right)t}{2\sin\frac{t}{2}}dt\right]$$

$$= \frac{1}{\pi}\int_0^{\pi}[f(x+t)+f(x-t)]\frac{\sin\left(n+\frac{1}{2}\right)t}{2\sin\frac{t}{2}}. \tag{i}$$

To use *R-L*-integral, we replace the integral $\int_0^{\pi}$ by $\int_0^{\delta}+\int_{\delta}^{\pi}$, where δ is a small (+)ve number $< \pi$ and $\int_{\delta}^{\pi} \to 0$ as $n\to\infty$.

Hence, proceeding as in above example, we say that $\lim_{n\to\infty} S_n(x)$ exists, if and only if

$$\lim_{n\to\infty}\frac{1}{\pi}\int_0^{\delta}[f(x+t)+f(x-t)]\frac{\sin nt}{t}dt \text{ exists.]} \tag{ii}$$

Corollary [15.5]: The F.S. $\frac{a_0}{2}+\sum_{n=1}^{\infty}(a_n\cos nx + b_n\sin nx)$ of f in $[-\pi,\pi]$ converges at x to a real number s, if and only if

$$\lim_{n\to\infty}\frac{1}{\pi}\int_0^{\pi}[f(x+t)+f(x-t)-2s]\frac{\sin nt}{t}dt = 0.$$

Proof. From (ii) of Corollary [15.4], we have

$$\lim_{n\to\infty} S_n(x) = \lim_{n\to\infty}\frac{1}{\pi}\int_0^{\pi}[f(x+t)+f(x-t)]\frac{\sin nt}{t}dt. \tag{i}$$

Now the F.S. converges to s for any value of x, then $\lim_{n\to\infty}[S_n(x)-s] = 0$.

Also

$$\lim_{n\to\infty}\int_0^{\pi}\frac{\sin nt}{t}dt = \frac{\pi}{2} \Rightarrow s = \frac{2}{\pi}\lim_{n\to\infty}\int_0^{\pi} s\cdot\frac{\sin nt}{t}dt. \tag{ii}$$

Therefore,

$$\lim_{n\to\infty}\{S_n(x)-s\}=0 \Rightarrow \lim_{n\to\infty} S_n(x)-s=0$$

$$\Rightarrow \lim_{n\to\infty}\left[\frac{1}{\pi}\int_0^{\pi}[f(x+t)+f(x-t)]\frac{\sin nt}{t}dt-\frac{2}{\pi}\int_0^{\pi}\frac{s\sin nt}{t}dt\right]=0 \text{ [by (i) and (ii)]}$$

$$=\lim_{n\to\infty}\left[\frac{1}{\pi}\int_0^{\pi}[f(x+t)+f(x-t)-2s]\frac{\sin nt}{t}dt\right]=0$$

Problem [15.36]: (1) If the P.S. $f(x)=\sum\limits_0^{\infty}a_nx^n$ converges for $x=1$, then show that the P.S. $f(\alpha)+\dfrac{f'(\alpha)h}{1!}+\dfrac{f''(\alpha)h^2}{2!}+\cdots(0<\alpha<1)$ converges for $h=1-\alpha$.

(2) If $f(x)$ is integrable on $[a,b]$ and vanishes outside $[a,b]$ and $\phi(x)=\dfrac{1}{2h}\displaystyle\int_{x-h}^{x+h}f(t)dt$, then show that $\displaystyle\int_a^b|\phi(x)dx|\le\int_a^b|f(x)|dx$.

(3) Prove that $\displaystyle\int_a^b f(x)dx=(b-a)f\left(\frac{a+b}{2}\right)+\frac{(b-a)^3}{24}f''(\xi)$, where $a<\xi<b$.

Solution. (1) $f(x)=\sum\limits_0^{\infty}a_nx^n$ converges at $x=1$.

So $f(1)=\sum\limits_0^{\infty}a_n$ is a convergent series. (i)

We have by Tailor's expansion,

$$f(\alpha+h)=f(\alpha)+hf'(\alpha)+\frac{h^2}{2!}f''(\alpha)+\cdots.$$

For $h=1-\alpha>0$ $(0<\alpha<1)$,

$$f(1)=f(\alpha)+(1-\alpha)f'(\alpha)+\frac{(1-\alpha)^2}{2!}f''(\alpha)+\cdots=\sum_0^{\infty}a_n \text{ [by (i)]}$$

Hence, the P.S. $f(\alpha)+hf'(\alpha)+\dfrac{h^2}{2!}f''(\alpha)+\cdots$ converges for $h=1-\alpha$.

(2) $\displaystyle\int_a^b f(x)dx$ exists and $f(x)\ne 0\ \forall\ x\in[a,b]$.

Then

$$\phi(x)=\frac{1}{2h}\int_{x-h}^{x+h}f(t)dt=\frac{1}{2h}[(x+h)-(x-h)]\mu \left[\begin{array}{l}\text{by M.V. Th. of}\\ \text{Integral Calculus}\end{array}\right]$$

$\left[\displaystyle\int_{x-h}^{x+h}f(t)dt\right.$ exists; $(x-h,x+h)\in[a,b]\ \forall$ small values of h,

where $m\le\mu\le M, m, M$ are the bounds of f on $\left.[a,b]\right]$

$$\therefore \phi(x)=\frac{1}{2h}\cdot 2h\mu=\mu.$$

Then

$$\int_a^b |\phi(x)|dx = \int_a^b |\mu|dx = (b-a)|\mu|$$

$$= (b-a)\frac{1}{|b-a|}\left|\int_a^b f(x)dx\right| = (b-a)\frac{1}{(b-a)}\left|\int_a^b f(x)dx\right| \quad [\because b > a]$$

$$= \left|\int_a^b f(x)dx\right| \le \int_a^b |f(x)|dx.$$

(3) We have by Tailor's expansion about a point a, up to 2nd order.

$$f(x) = f(a) + (x-a)f'(a) + \frac{(x-a)^2}{2!}f''(\xi) \quad [a < \xi < b] \tag{i}$$

Then

$$\int_a^b f(x)dx = f(a)\int_a^b dx + f'(a)\int_a^b (x-a)dx + f''(\xi)\int_a^b \frac{(x-a)^2}{2!}dx$$

$$= (b-a)f(a) + \frac{(b-a)^2}{2}f'(a) + \frac{(b-a)^3}{6}f''(\xi), \tag{ii}$$

where $\xi = a + \theta h$, $0 < \theta < 1$.

Taking $\theta = 1/2, h = b-a$, $\exists$ a point $x = a + \frac{b-a}{2} = \frac{a+b}{2}$ in $[a, b]$ satisfying (i).

So from (i), for $x = \frac{a+b}{2}$,

$$f\left(\frac{a+b}{2}\right) = f(a) + \frac{b-a}{2}f'(a) + \frac{(b-a)^2}{8}f''(\xi).$$

$$\therefore \; (b-a)f\left(\frac{a+b}{2}\right) = (b-a)f(a) + \frac{(b-a)^2}{2}f'(a) + \frac{(b-a)^3}{8}f''(\xi)$$

$$\Rightarrow (b-a)f(a) + \frac{(b-a)^2}{2}f'(a) = (b-a)f\left(\frac{a+b}{2}\right) - \frac{(b-a)^3}{8}f''(\xi). \tag{iii}$$

From (ii) by (iii),

$$\int_a^b f(x)dx = (b-a)f\left(\frac{a+b}{2}\right) - \frac{(b-a)^3}{8}f''(\xi) + \frac{(b-a)^3}{6}f''(\xi)$$

$$= (b-a)f\left(\frac{a+b}{2}\right) + (b-a)^3 f''(\xi)\left\{\frac{1}{6} - \frac{1}{8}\right\}$$

$$= (b-a)f\left(\frac{a+b}{2}\right) + \frac{(b-a)^3}{24}f''(\xi) \text{ (Proved)}.$$

Theorem [15.8] If f is R-integrable on $[a, b]$ for every $b \ge a$ and if there exists a (+)ve number M such that $\int_a^b |f(x)|dx < M$, then both f and $|f|$ are improper R-integrable on $[a, \infty)$.

Proof. Let $F(b) = \int_a^b |f(x)|dx$.

Let $x \in [a, b]$ and $h > 0$ so that $(x+h) \in [a, b]$.

Then $F(x+h) - F(x) = \int_x^{x+h} |f(x)|dx > 0 \; [\because |f(x)| > 0]$

$\Rightarrow F(x+h) > F(x)$

$\Rightarrow F$ is an increasing function of b.

Also F is bounded above as $F(b) < M \Rightarrow \lim_{b\to\infty} F(b)$ exists $\Rightarrow |f(x)|$ is an improper R-integrable on $[a, \infty)$.

Again $-|f(x)| \le f(x) \le |f(x)| \; \forall \; x \in [a, b]$

$\Rightarrow |f(x)| \ge -f(x) \ge -|f(x)|$

$\Rightarrow 2|f(x)| \ge |f(x)| - f(x) \ge 0.$

That is, $0 \le |f(x)| - f(x) \le 2|f(x)|$. (1)

Then $\displaystyle\int_a^b \{|f(x)| - f(x)\}dx \le 2\int_a^b |f(x)|dx.$

$$\Rightarrow \lim_{b\to\infty} \int_a^b \{|f(x)| - f(x)\}dx \text{ exists since } \lim_{b\to\infty}\int_a^b |f(x)|dx \text{ exists.}$$

$$\Rightarrow \lim_{b\to\infty}\int_a^b f(x)dx \text{ exists.}$$

That is, $f(x)$ is an improper R-integrable on $[a, \infty)$.

Corollary [15.6] : If f is R-integrable on $[a, b]$ for $b \ge a$ and, if there exists a (+)ve number m such that $\int_a^b |f(x)|dx \ge m$, then both f and $|f|$ are improper R-integrable on $(-\infty, a]$.

[*Hints.* Let $F(a) = \int_a^b |f(x)|dx$. Then $F(a) \ge m$. Also $F(x+h) - F(x) = -\int_x^{x+h} |f(x)|dx < 0$.

$\Rightarrow F(a)$ is monotonically decreasing function bounded below.

So $\lim_{a\to-\infty} F(a)$ exists.

$\Rightarrow \int_{-\infty}^a |f(x)|dx$ exists $\Rightarrow |f(x)|$ is an improper R-integrable on $(-\infty, a]$.

Again, by (1) Theorem [15.8], $f(x)$ is an improper R-integrable on $(-\infty, a]$.]

Example. Let $f(x) = e^{-x}x^{p-1}$ for any $p \in \mathbb{R}$.

Now $e^{-x/2}x^{p-1} \to 0$ as $x \to +\infty$.

So there exists a (+)ve constant M such that $e^{-x/2}x^{p-1} \le M \; \forall \; x \ge a (a > 0)$.

Then $e^{-x}x^{p-1} \le Me^{-x/2}$.

Therefore, $\int_a^b |f(x)|dx \; \forall \; b \ge a$

$$= \int_a^b \left|e^{-x}x^{p-1}\right| dx \le M\int_a^b e^{-x/2}dx$$

$$= -2M\left[e^{-b/2} - e^{-a/2}\right] = 2M\left[e^{-a/2} - e^{-b/2}\right] = 2\frac{M}{e^{a/2}}\left[1 - e^{-\left(\frac{b-a}{2}\right)}\right]$$

$$= \frac{2M}{e^{a/2}}\left[1 - \frac{1}{e^{\frac{b-a}{2}}}\right] < 2M \quad \left[\because e^{\frac{b-a}{2}} > 0 \; \forall \; b \ge a\right]$$

$\Rightarrow \lim_{b\to+\infty} \int_a^b |f(x)|dx$ exists for $p \in \mathbb{R}$.

Thus by above theorem both f and $|f|$ are improper R-integrable on $[a, \infty), a > 0$.

Problem [15.37] : A function f is R-integrable on $[a - c, \; a + c]$ and $|f(x)| \le M \; \forall \; x \in [a - c, \; a + c]$, $\int_{a-c}^{a+c} f(x)dx = 0$ and $F(x) = \int_{a-c}^{x} f(t)dt$. Then prove that $\left|\int_{a-c}^{a+c} F(x)dx\right| \le 2Mc^2$.

Solution. $\left|\int_{a-c}^{a+c} F(x)dx\right| \leq \int_{a-c}^{a+c} |F(x)|dx$

$$= \int_{a-c}^{a+c} \left|\int_{a-c}^{x} f(t)dt\right| dx = \int_{a-c}^{a+c} \left|\int_{a-c}^{a+c} f(t)dt + \int_{a+c}^{x} f(t)dt\right| dx$$

$$= \int_{a-c}^{a+c} \left|0 + \int_{a+c}^{x} f(t)dt\right| dx \quad \left[\because \int_{a-c}^{a+c} f(x)dx = 0 \text{ by hypothesis}\right]$$

$$= \int_{a-c}^{a+c} \left|\int_{x}^{a+c} f(t)dt\right| dx \leq \int_{a-c}^{a+c} \left\{\int_{x}^{a+c} |f(t)|dt\right\} dx$$

$$\leq M \int_{a-c}^{a+c} \{(a+c) - x\}dx \ [\because |f(x)| \leq M \ \forall \ x \in [a-c, a+c]]$$

$$= M \left[(a+c)x - \frac{x^2}{2}\right]_{a-c}^{a+c} = M \left[\left\{(a+c)^2 - \frac{(a+c)^2}{2}\right\} - \left\{(a^2 - c^2) - \frac{(a-c)^2}{2}\right\}\right]$$

$$= M \left[\frac{(a+c)^2}{2} + \frac{(a-c)^2}{2} - (a^2 - c^2)\right] = M\left[(a^2 + c^2) - (a^2 - c^2)\right] = 2Mc^2.$$

Thus $\left|\int_{a-c}^{a+c} F(x)dx\right| \leq 2Mc^2.$

Problem [15.38] : If f be bounded and integrable on $[a, b]$, then show that $\int_a^b [f(x)]^2 dx = 0$, if and only if $f(c) = 0$ at every point c, of continuity of f.

Solution. Let $c \in [a, b]$ and f is continuous at $x = c$ such that $f(c) = 0$. (1)

If possible, let $\int_a^b [f(x)]^2 dx \neq 0$.

Then $\int_a^b [f(x)]^2 dx > 0$ $[\because \{f(x)\}^2 > 0 \ \forall \ x \in [a, b]]$

$$\Rightarrow f(x) \neq 0 \ \forall \ x \in [a, b].$$

Therefore, $f(c) \neq 0$, which contradicts (1).

Hence, $\int_a^b [f(x)]^2 dx = 0$.

Conversely, let $\int_a^b [f(x)]^2 dx = 0$. (2)

Let $c \in [a, b]$ and f is continuous at $x = c$. If possible, let $f(c) \neq 0$.

We take $f(c) > 0$. Then $f(x) > 0 \ \forall \ x \in (c - \delta, c + \delta)$.

$$\int_a^b [f(x)]^2 dx = \int_a^{c-\delta} [f(x)]^2 dx + \int_{c-\delta}^{c+\delta} [f(x)]^2 dx + \int_{c+\delta}^{b} [f(x)]^2 dx$$

$$\geq \int_{c-\delta}^{c+\delta} [f(x)]^2 dx > 0,$$

which contradicts our assumption (2). The same can be proved, if $f(c) < 0$.

For, if $\phi(x) = [f(x)]^2$, then $\phi(c) > 0 \ \forall \ x \in (c - \delta, c + \delta)$ and hence

$$\int_a^b \phi(x)dx = \int_a^b [f(x)]^2 dx > 0.$$

Further, if $c = a$ or b, the proof follows immediately from above.

Problem [15.39]: Let $f(x)$ have a derivative $f'(x)$ which is monotonic increasing and satisfies $f'(x) \geq m > 0 \; \forall \; x \in [a, b]$. Then prove that $\left|\int_a^b \cos f(x)dx\right| \leq \frac{2}{m}$.

Solution. We consider

$$\left|\int_a^b \cos f(x)dx\right| = \left|\int_a^b \frac{f'(x)\cos f(x)}{f'(x)}dx\right| \quad [\because f'(x) > 0 \; \forall \; x \in [a, b]] \tag{1}$$

Here $\int_a^b f'(x)\cos f(x)$ and $\int_a^b \frac{1}{f'(x)}dx$ both exist, $\frac{1}{f'(x)}$ is monotonically decreasing and (+)ve in $[a, b]$.

So by applying Bonnet's form in (1),

$$\begin{aligned}\left|\int_a^b \cos f(x)dx\right| &\leq \left|\frac{1}{f'(a)}\int_a^\xi f'(x)\cos f(x)dx\right|, \text{ where } \xi \in [a, b] \\ &\leq \frac{1}{|f'(a)|}\left|\left\{\sin f(x)\right\}_a^\xi\right| \leq \frac{1}{m}|\sin f(\xi) - \sin f(a)| \\ &\leq \frac{1}{m}\{|\sin f(\xi)| + |\sin f(a)|\} \leq \frac{2}{m} \quad \left[\begin{array}{l}\because f'(x) > m \\ \Rightarrow \frac{1}{f'(x)} < \frac{1}{m} \; \forall \; x \in [a, b]\end{array}\right]\end{aligned}$$

Problem [15.40]: Show that $\lim\limits_{x\to\infty} e^{-x^2}\int_0^x e^{t^2}dt = 0$.

Solution. $\lim\limits_{x\to\infty} \dfrac{\int_0^x e^{t^2}dt}{e^{x^2}} \left(\dfrac{\infty}{\infty}\right)$.

For, if $\phi(x) = \int_0^x e^{t^2}dt$, then $\lim\limits_{x\to\infty}\phi(x) = \infty$. Also $\phi'(x) = e^{x^2}$.

$$\begin{aligned}&= \lim_{x\to\infty}\frac{\phi(x)}{e^{x^2}} = \lim_{x\to\infty}\frac{\phi'(x)}{e^{x^2}\cdot 2x} \\ &= \lim_{x\to\infty}\frac{e^{x^2}}{e^{x^2}\cdot 2x} = \lim_{x\to\infty}\frac{1}{2x} = 0. \quad \left[\because e^{x^2} > 0 \; \forall \; x > 0\right]\end{aligned}$$

Problem [15.41]: If $f(x, y) = xy^2e^{-xy} + \frac{x^2y}{1+y}$ and $a, b > 0$, then show that

$$\lim_{y\to\infty}\int_a^b f(x, y)dx = \int_a^b \left[\lim_{y\to\infty} f(x, y)\right]dx.$$

The result does not hold for $a = 0$.

Solution. $\int_a^b f(x, y)dx$

$$\begin{aligned}&= \int_a^b \left(xy^2e^{-xy} + \frac{x^2y}{1+y}\right)dx = y^2\int_a^b xe^{-xy}dx + \frac{y}{1+y}\int_a^b x^2dx \\ &= y^2\left[-\frac{x}{y}e^{-xy} - \frac{1}{y^2}e^{-xy}\right]_a^b + \frac{y}{1+y}\left[\frac{x^3}{3}\right]_a^b = \left[-yxe^{-xy} - e^{-xy}\right]_a^b + \frac{y}{1+y}\left[\frac{b^3 - a^3}{3}\right] \\ &= \left[-ybe^{-by} - e^{-by} + yae^{-ay} + e^{-ay}\right] + \frac{y}{1+y}\left[\frac{b^3 - a^3}{3}\right] \\ &= \left[\frac{(ya+1)}{e^{ay}} - \frac{(yb+1)}{e^{by}}\right] + \frac{y}{1+y}\left[\frac{b^3 - a^3}{3}\right].\end{aligned}$$

$$\lim_{y\to\infty}\int_a^b f(x,y)dx = \lim_{y\to\infty}\left[\frac{(ya+1)}{e^{ay}} - \frac{(yb+1)}{e^{by}}\right] + \lim_{y\to\infty}\frac{y}{1+y}\left(\frac{b^3-a^3}{3}\right)$$

$$= \lim_{y\to\infty}\left[\frac{a}{ae^{ay}} - \frac{b}{be^{by}}\right] + \frac{b^3-a^3}{3}$$

$$= 0 + \frac{b^3-a^3}{3} = \frac{b^3-a^3}{3} \quad [\because a,b>0] \tag{1}$$

Again,

$$\lim_{y\to\infty} f(x,y) = \lim_{y\to\infty}\left[xy^2e^{-xy} + \frac{x^2y}{1+y}\right]$$

$$= x\lim_{y\to\infty}\frac{y^2}{e^{xy}} + x^2\lim_{y\to\infty}\frac{y}{1+y} \quad [x\in[a,b]\Rightarrow x>0]$$

$$= x\lim_{y\to\infty}\frac{2y}{xe^{xy}} + x^2\cdot 1 = x\cdot\lim_{y\to\infty}\frac{2}{x^2e^{xy}} + x^2 = x\cdot 0 + x^2 = x^2.$$

$$\int_a^b\left[\lim_{y\to\infty} f(x,y)\right]dx = \int_a^b x^2dx = \frac{b^3-a^3}{3}. \tag{2}$$

So from (1) and (2) the result follows.

For $a=0$,

$$\lim_{y\to\infty}\int_0^b f(x,y)dx = 1+\frac{b^3}{3}, \text{ while } \int_0^b\left[\lim_{y\to\infty} f(x,y)\right]dx = \frac{b^3}{3}.$$

So equality does not hold for $a=0$.

Problem [15.42]: If $f(x)=0\ \forall\ x\in[0,1]$ except at the set of points $x_1,x_2,\ldots,x_n,\ldots, n\in\mathbb{N}$ and $f(x_n)=\frac{1}{\sqrt{n}}$, then show that f is R-integrable on $[0,1]$.

Solution. f has infinite number of points of discontinuities $x_1,x_2,\ldots,x_n,\ldots$ having a single limit point 0, since $\lim\limits_{n\to\infty} f(x_n) = \lim\limits_{n\to\infty}\frac{1}{\sqrt{n}} = 0$. We enclose the point '0' by an interval $[\delta_1,\delta_2]$ so that $(\delta_2-\delta_1)<\frac{\epsilon}{2}$. Therefore, f has only finite number of points of discontinuities in $[0,\delta_1]$ and $[\delta_2,1]$. Each of them can be further subdivided so that the oscillatory sum arising from each of the subintervals is less than $\frac{\epsilon}{4}$. Hence the resultant oscillatory sum

$$\omega_D < (1-0)\cdot\frac{\epsilon}{2} + \frac{\epsilon}{4}\cdot 2 = \frac{\epsilon}{2} + \frac{\epsilon}{2} = \epsilon. \quad \left[\begin{array}{l}\because \sup f = 1\\ \inf f = 0 \text{ on } [0,1]\end{array}\right]$$

Hence f is R-integrable on $[0,1]$.

Problem [15.43]: Examine the convergence of the improper integral

$$\int_1^\infty \frac{dx}{x\left(\sqrt{\ln(x)}+\ln^2(x)\right)}.$$

Solution. Here $f(x) = \frac{1}{x\left(\sqrt{\ln(x)}+\ln^2(x)\right)}$.

$x=1$ is a point of infinite discontinuity since $f(x)\to\infty$ as $x\to 1$.

So we consider the given integral

$$\int_1^\infty \frac{dx}{x\left(\sqrt{\ln(x)}+\ln^2(x)\right)} = \int_1^2 \frac{dx}{x\left(\sqrt{\ln(x)}+\ln^2(x)\right)} + \int_2^\infty \frac{dx}{x\left(\sqrt{\ln(x)}+\ln^2(x)\right)}. \tag{1}$$

For the 1st part. We have $f(x) = \frac{1}{x\left(\sqrt{\ln(x)}+\ln^2(x)\right)}$.

Let $\phi(x) = \frac{1}{\sqrt{\ln(x)}}$.

Then

$$\lim_{x\to1}\frac{f(x)}{\phi(x)} = \lim_{x\to1}\frac{\sqrt{\ln(x)}}{x\left(\sqrt{\ln(x)}+\ln^2(x)\right)}\left(\frac{0}{0}\right)$$

$$= \lim_{x\to1}\frac{1}{2x\sqrt{\ln(x)}}\times\frac{1}{\left\{\sqrt{\ln(x)}+\ln^2(x)\right\}+x\left\{\frac{1}{2\sqrt{\ln(x)}}\cdot\frac{1}{x}+2\ln(x)\cdot\frac{1}{x}\right\}}$$

$$= \lim_{x\to1}\frac{1}{2x\sqrt{\ln(x)}\left\{\ln(x)+\ln^2(x)\right\}+\left\{x+4(\ln(x))^{3/2}x\right\}} = 1.$$

Again, by Taylor's expansion of $\ln(x)$ about $x = 1$,

$$\ln(x) = \ln(1) + \ln'(1)\cdot(x-1) \text{ [approximately]}$$
$$= (x-1).$$

$\therefore \int_1^2 \frac{1}{\sqrt{\ln(x)}} \sim \int_1^2 \frac{1}{(x-1)^{1/2}}$, which is convergent.

Hence by comparison test, $\int_1^2 f(x)dx = \int_1^2 \frac{dx}{x\left\{\sqrt{\ln(x)}+\ln^2(x)\right\}}$ is convergent.

For the 2nd part. We consider $\int_2^\infty \frac{1}{x\left\{\sqrt{\ln(x)}+\ln^2(x)\right\}}dx < \int_2^\infty \frac{1}{x\ln^2(x)}dx$.

Now for, $\int_2^\infty \frac{1}{x\ln^2(x)}dx$ we put $\ln(x) = z$, $\frac{1}{x}dx = dz$.

Then $\int_2^\infty \frac{1}{x\ln^2(x)}dx = \int_{\log 2}^\infty \frac{1}{z^2}dz$ which is convergent, since $\mu = 2 > 1$.

Therefore, from (1) it follows that the given integral is convergent.

Problem [15.44]: Examine convergence or divergence of

(1) $\int_1^{\pi/2} \frac{\tan x}{x^{3/2}}$, (2) $\int_0^\infty \frac{\sin x}{\sqrt{x+\cos x}}dx$,

(3) $\int_0^{\pi/2} \frac{dx}{1+(\tan x)^a}$, (4) $\int_1^\infty \frac{1}{\log x}dx$.

Solution. (1) $f(x) = \frac{\tan x}{x^{3/2}}$.

$f(x)\to\infty$ as $x\to\frac{\pi}{2}$. $\frac{\pi}{2}$ is a point of infinite discontinuity.

$f(x) = \frac{\sin x}{x^{3/2}\cos x}$.

Let $\phi(x) = \frac{1}{\cos x}$. Then $\lim_{x\to\frac{\pi}{2}}\frac{f(x)}{\phi(x)} = \lim_{x\to\frac{\pi}{2}}\frac{\sin x}{x^{3/2}} = \frac{1}{\left(\frac{\pi}{2}\right)^{3/2}}$.

Again, by Taylor's expansion about $x = \frac{\pi}{2}$,

$$\cos x = \cos\frac{\pi}{2} + \left(-\sin\frac{\pi}{2}\right)\left(x - \frac{\pi}{2}\right) \quad \text{[approximately]}$$
$$= \left(\frac{\pi}{2} - x\right)\sin\frac{\pi}{2} = \frac{\pi}{2} - x.$$

$$\int_1^{\pi/2} \phi(x)dx = \int_1^{\pi/2} \frac{1}{\cos x}dx \sim \int_1^{\pi/2} \frac{1}{\left(\frac{\pi}{2} - x\right)}dx, \text{ which diverges at } x = \frac{\pi}{2}.$$

Therefore, the given integral $\displaystyle\int_1^{\pi/2} \frac{\tan x}{x^{3/2}}dx$ is divergent.

(2) $$\int_0^\infty \frac{\sin x}{\sqrt{x + \cos x}}dx = \int_0^\infty \frac{1 - (1 - \sin x)}{\sqrt{x + \cos x}}dx = \int_0^\infty \frac{dx}{\sqrt{x + \cos x}} + \int_0^\infty \frac{1 - \sin x}{\sqrt{x + \cos x}}dx.$$

1st part. $x = 0$ is not a point of infinite discontinuity. So we shall test the convergence at $x = \infty$, $x + \cos x \le x + 1 \ \forall \ x \in [0, \infty)$.

$$\frac{1}{\sqrt{x + \cos x}} \ge \frac{1}{(x+1)^{1/2}} \Rightarrow \int_0^\infty \frac{1}{\sqrt{x + \cos x}}dx \ge \frac{dx}{(x+1)^{1/2}}$$

which diverges at ∞ since $\mu = \frac{1}{2} < 1$.

2nd part. $\displaystyle\int_0^\infty \frac{1 - \sin x}{\sqrt{x + \cos x}}dx$. Let $x + \cos x = z$.

Then $(1 - \sin x)dx = dz$, $x = 0$, $z = 1$, $x \to \infty$, $z \to \infty$

$\Rightarrow \displaystyle\int_1^\infty \frac{dz}{z^{1/2}}$ which diverges at ∞, since $\mu = \frac{1}{2} < 1$.

Hence the given integral is divergent.

(3) $$I(a) = \int_0^{\pi/2} \frac{dx}{1 + (\tan x)^a} = \int_0^{\pi/4} \frac{dx}{1 + (\tan x)^a} + \int_{\pi/4}^{\pi/2} \frac{dx}{1 + (\tan x)^a}$$

For the 2nd part replacing x by $\frac{\pi}{2} - x$, we get

$$= \int_0^{\pi/4} \frac{dx}{1 + (\tan x)^a} - \int_{\pi/4}^0 \frac{dx}{1 + (\cot x)^a} = \int_0^{\pi/4} \frac{dx}{1 + (\tan x)^a} + \int_0^{\pi/4} \frac{dx}{1 + (\cot x)^a}$$
$$= \int_0^{\pi/4} \left[\frac{1}{1 + (\tan x)^a} + \frac{1}{1 + (\cot x)^a}\right]dx = \int_0^{\pi/4} \left[\frac{1}{1 + (\tan x)^a} + \frac{(\tan x)^a}{1 + (\tan x)^a}\right]dx$$
$$= \int_0^{\pi/4} \frac{1 + (\tan x)^a}{1 + (\tan x)^a}dx = \int_0^{\pi/4} 1 \cdot dx = \frac{\pi}{4}.$$

Therefore, the given integral is convergent.

(4) $\displaystyle\int_2^\infty \frac{1}{\log x}dx$. We consider Taylor's expansion of $\log x$ about $x = 1$ up to 2nd order.

$$\log x = \log 1 + (x - 1)\left[\log(x)\right]'_{x=1} + \frac{(x-1)^2}{2}\left(\log x\right)''_{x=1}$$
$$= (x - 1) \cdot 1 - \frac{1}{2}(x - 1)^2 < x - 1$$
$$\Rightarrow \frac{1}{\log x} > \frac{1}{(x - 1)} \Rightarrow \int_2^\infty \frac{1}{\log x}dx > \int_2^\infty \frac{1}{(x - 1)}dx.$$

Therefore, the given integral is divergent.

Problem [15.45]: Prove that $\sum x^n(1-x)$ converges pointwise but not uniformly on $[0,1]$, whereas $\sum(-1)^n x^n(1-x)$ converges uniformly on $[0,1]$.

Solution.
$$\begin{aligned} S_n(x) &= \sum_1^n x^n(1-x) \\ &= x(1-x)+x^2(1-x)+\cdots+x^n(1-x) \\ &= (1-x)\cdot[x+x^2+\cdots+x^n] = (1-x)\cdot\frac{x(1-x^n)}{1-x} \\ &= x(1-x^n) \text{ for } 0 \le x < 1. \end{aligned}$$

Also $S(x) = \lim_{n\to\infty} S_n(x) = x$ for $0 \le x < 1$.

$S_n(1) = 0+0+\cdots+0 = 0.$

$$\begin{aligned} \Rightarrow S(x) &= x, \quad 0 \le x < 1 \\ &= 0, \quad x = 1 \end{aligned}$$

$S(x)$ is discontinuous at $x = 1$.

$\therefore$ the given series of function is not uniformly convergent on $[0,1]$.

Alternative. Let $|S_n(x) - S(x)| < \epsilon \;\forall\, n \ge m$ and $\forall\, x \in [0,1]$.

Now, $|S_n(x) - S(x)| < \epsilon$, if $|x(1-x^n) - x| < \epsilon$,

i.e., if $|-x^{n+1}| < \epsilon$, i.e., if $x^{n+1} < \epsilon$

i.e., if $(n+1)\log\frac{1}{x} > \log\frac{1}{\epsilon}$, i.e., if $n > \dfrac{\log\frac{1}{\epsilon}}{\log\frac{1}{x}} - 1 = m(\epsilon, x)$.

Now, let $x \to k$, $0 < k < 1$, $m(\epsilon, x) \to$ to finite number. So $\exists$ a (+)ve number m such that

$$\forall\, n > m \text{ and } \forall\, x \in (0,1),\ |S_n(x) - S(x)| < \epsilon. \tag{1}$$

$$\lim_{x\to 0} m(\epsilon,x) = \lim_{x\to 0}\frac{\log\frac{1}{\epsilon}}{\log\frac{1}{x}} - 1 = \lim_{x\to 0}\frac{\log\epsilon}{\log x} - 1 = -1.$$

So $m(\epsilon, x) \to$ a finite number as $x \to 0$. Therefore, $\exists$ a (+)ve integer m such that

$$\forall\, n \ge m,\ \forall\, x \in [0,1),\ |S_n(x) - S(x)| < \epsilon. \tag{2}$$

$x \to 1$, $m(\epsilon, x) \to \infty$; so we cannot have a (+)ve integer m such that

$$|S_n(x) - S(x)| < \epsilon \;\forall\, n \ge m. \tag{3}$$

From (1), (2) and (3) it follows that $\sum x^n(1-x)$ converges pointwise on $[0,1]$. However, it converges uniformly on $[0,1)$.

We now consider the series $\sum(-1)^n x^n(1-x)$.

Let $u_n(x) = (-1)^n$ and $v_n(x) = x^n(1-x)$.

Then $\sum_1^\infty u_n(x)$ oscillates finitely and $\left|\sum_{r=1}^n u_r(x)\right| \le 1$.

So the sequence $\{S_n(x)\}$, where $S_n(x) = \sum_{r=1}^n u_r(x)$ is uniformly bounded.

$\{v_n(x)\}$ is monotonically decreasing sequence and $v_n(x) \to 0$ uniformly on $[0,1]$.

Therefore, by Dirichlet's test $\sum u_n(x)v_n(x) = \sum(-1)^n x^n(1-x)$ converge uniformly on $[0,1]$.

Problem [15.46] : Let $g_{n+1}(x) \le g_n(x)\ \forall\ x \in [a,b]$ and for each $n = 1, 2, \ldots$ and $g_n \to 0$ uniformly on $[a, b]$. Then prove that $\sum(-1)^{n+1} g_n(x)$ converges uniformly on $[a, b]$.

Solution. Let $\sum f_n(x) = \sum(-1)^{n+1}$.

Then, $\sum f_n(x)$ oscillates finitely. Here, $\left|\sum\limits_{r=1}^{n} f_r(x)\right| \le 1$.

That is, the sequence $\{S_n(x)\}$, where $S_n(x)$ is the nth partial sum of $\sum f_n(x)$, is uniformly bounded.

Also $g_n(x)$ is a monotonic decreasing function of x in $[a, b]$ and $g_n(x) \to 0$ uniformly in $[a, b]$. Therefore, by Dirichlet's test [Th. 9.15] $\sum(-1)^{n+1} g_n(x)$ converges uniformly on $[a, b]$.

Problem [15.47] : Let $\{f_n\}$ be a sequence of real valued continuous functions on $[0, 1]$ and let $f_n \to f$ uniformly on $[0,1]$. Then prove or disprove:

$$\lim_{n\to\infty} \int_0^{1-\frac{1}{n}} f_n(x)dx = \int_0^1 f(x)dx.$$

Solution. In support of the proof we first consider examples:

(i) Let $f_n(x) = \dfrac{x}{nx+1}$, $x \in [0, 1]$. Then $f(x) = \lim\limits_{n\to\infty} \dfrac{x}{nx+1} = 0,\ 0 \le x \le 1$.

Each $f_n(x)$ is real valued continuous functions on $[0, 1]$. Also $f_n \to f$ uniformly on $[0, 1]$. $f_n(x), f(x)$ are integrable on $[0, 1]$.

Now

$$\int_0^1 f(x)dx = 0. \tag{1}$$

Also

$$\begin{aligned}
\lim_{n\to\infty} \int_0^{1-\frac{1}{n}} f_n(x)dx &= \lim_{n\to\infty} \int_0^{1-\frac{1}{n}} \frac{x}{nx+1}dx = \lim_{n\to\infty} \frac{1}{n} \int_0^{1-\frac{1}{n}} \frac{nx+1-1}{nx+1}dx \\
&= \lim_{n\to\infty} \frac{1}{n}\left[x - \frac{1}{n}\log(nx+1)\right]_0^{1-\frac{1}{n}} = \lim_{n\to\infty} \frac{1}{n}\left[1 - \frac{1}{n} - \frac{1}{n}\log n\right] \\
&= \lim_{n\to\infty}\left[\frac{1}{n} - \frac{1}{n^2} - \frac{1}{n^2}\log n\right] = -\lim_{n\to\infty} \frac{\log n}{n^2}\left(\frac{\infty}{\infty}\right) \\
&= -\lim_{n\to\infty} \frac{1}{2n^2} = 0.
\end{aligned} \tag{2}$$

Therefore, $\lim\limits_{n\to\infty} \displaystyle\int_0^{1-\frac{1}{n}} f_n(x)dx = \int_0^1 f(x)dx$ [by (1) and (2)].

It is to be noted that $\lim\limits_{n\to\infty} \displaystyle\int_{1-\frac{1}{n}}^1 f_n(x)dx = 0$.

Since

$$\begin{aligned}
\lim_{n\to\infty}\left[x - \frac{1}{n}\log(nx+1)\right]_{1-\frac{1}{n}}^{1} &= \lim_{n\to\infty}\left[1 - \frac{1}{n}\log(n+1) - 1 + \frac{1}{n} + \frac{1}{n}\log n\right] \\
&= \lim_{n\to\infty}\left[\frac{1}{n}\log\frac{n}{n+1}\right] = 0.
\end{aligned}$$

(ii) Let $f_n(x) = \frac{nx^2}{nx+1}$, $x \in [0,1]$

$$f(x) = \lim_{n\to\infty} \frac{nx^2}{nx+1} = \begin{cases} x, & 0 < x < 1 \\ 0, & x = 0 \\ 1, & x = 1. \end{cases}$$

$f(x)$ is continuous on $[0,1]$. $\{f_n(x)\}$ converges uniformly to $f(x)$ on $[0,1]$.
$f_n(x), f(x)$ are integrable on $[0,1]$.

Now $\displaystyle\int_0^1 f(x)dx = \int_0^1 x dx = \frac{1}{2}$. (3)

Again, $\displaystyle\lim_{n\to\infty} \int_0^{1-\frac{1}{n}} f_n(x)dx = \lim_{n\to\infty} \int_0^{1-\frac{1}{n}} \frac{nx^2}{nx+1} dx$

$$= \lim_{n\to\infty} \int_0^{1-\frac{1}{n}} \left[x - \frac{x}{nx+1}\right] dx = \lim_{n\to\infty} \int_0^{1-\frac{1}{n}} \left[x - \frac{1}{n}\frac{nx+1-1}{nx+1}\right] dx$$

$$= \lim_{n\to\infty} \int_0^{1-\frac{1}{n}} \left[x - \frac{1}{n} + \frac{1}{n}\cdot\frac{1}{nx+1}\right] dx = \lim_{n\to\infty} \left[\frac{x^2}{2} - \frac{1}{n}x + \frac{1}{n^2}\log(nx+1)\right]_0^{1-\frac{1}{n}}$$

$$= \lim_{n\to\infty} \left[\frac{1}{2}\left(1-\frac{1}{n}\right)^2 - \frac{1}{n}\left(1-\frac{1}{n}\right) + \frac{1}{n^2}\log n\right] = \frac{1}{2}. \quad (4)$$

So by (3) and (4), $\displaystyle\int_0^1 f(x)dx = \lim_{n\to\infty} \int_0^{1-\frac{1}{n}} f_n(x)dx.$

We can also note here that $\displaystyle\lim_{n\to\infty} \int_{1-\frac{1}{n}}^1 f_n(x)dx = 0.$

Thus the result is true for a uniform convergence of sequence of functions.

Proof. $\{f_n(x)\}$ converges uniformly on $[0,1] \Rightarrow |f_n(x) - f(x)| < \epsilon \; \forall\, n \geq m$ and $\forall\, x \in [0,1]$.

Hence, $\displaystyle\left|\int_0^1 f_n(x) - \int_0^1 f(x)dx\right|$

$$= \left|\int_0^1 [f_n(x) - f(x)]\,dx\right| \leq \int_0^1 |f_n(x) - f(x)|dx < \int_0^1 \epsilon dx = \epsilon \; \forall\, n \geq m$$

$$\Rightarrow \lim_{n\to\infty} \int_0^1 f_n(x)dx = \int_0^1 f(x)dx$$

$$\Rightarrow \lim_{n\to\infty} \left[\int_0^{1-\frac{1}{n}} f_n(x)dx + \int_{1-\frac{1}{n}}^1 f_n(x)dx\right] = \int_0^1 f(x)dx \quad \left[\because 0 < 1 - \frac{1}{n} \leq 1\right]$$

$$\Rightarrow \lim_{n\to\infty} \int_0^{1-\frac{1}{n}} f_n(x)dx = \int_0^1 f(x)dx.$$

Since $n \to \infty$,

$$1 - \frac{1}{n} \to 1 \text{ and consequently } \int_{1-\frac{1}{n}}^1 f_n(x)dx \to 0.$$

Definition [15.8] : **[Affinely extended of Real Numbers]** Let $\mathbb{R}$ be the set of real numbers. Then the set $\mathbb{R} \cup [-\infty, +\infty]$ obtained by adjoining of two improper elements $-\infty$ and $+\infty$ to the set $\mathbb{R}$ is called the set of affinely extended of real numbers and generally we denote it by $\overline{\mathbb{R}}$.

(i) A very important property of $\overline{\mathbb{R}}$ which $\mathbb{R}$ lacks is that every subset S of $\overline{\mathbb{R}}$ has an infimum (g.l.b.) and supremum (l.u.b.).

In particular, $\sup \phi = -\infty$ and if $S \subseteq \overline{\mathbb{R}}$ is unbounded above, then $\sup S = +\infty$. Similarly, $\inf \phi = +\infty$ and if $S \subseteq \overline{\mathbb{R}}$ is unbounded below, then $\inf S = -\infty$.

(ii) Order relation can be extended from $\mathbb{R}$ to $\overline{\mathbb{R}}$. For $x \in \overline{\mathbb{R}}$,

$$-\infty < x < +\infty, \text{ if } x \neq \pm\infty, -\infty < +\infty \quad \text{(i)}$$
$$-(+\infty) = -\infty, -(-\infty) = +\infty \quad \text{(ii)}$$
$$x + (+\infty) = +\infty + x = +\infty, \text{ if } x \neq -\infty \quad \text{(iii)}$$
$$x + (-\infty) = -\infty + x = -\infty, \text{ if } x \neq +\infty \quad \text{(iv)}$$
$$x \cdot (\pm\infty) = \pm\infty \cdot x = \pm\infty, \text{ if } x > 0 \quad \text{(v)}$$
$$x \cdot (\pm\infty) = \pm\infty \cdot x = \mp\infty, \text{ if } x < 0 \quad \text{(vi)}$$
$$\frac{x}{\pm\infty} = 0, \text{ if } x \neq \pm\infty \quad \text{(vii)}$$
$$\left|\frac{x}{0}\right| = +\infty, \text{ if } x \neq 0. \quad \text{(viii)}$$

However, the expressions of the forms $+\infty + (-\infty)$, $-\infty + (+\infty)$ and $\frac{0}{0}$ are undefined.

(iii) Every open integral $(c, +\infty)$ is called a neighbourhood of $+\infty$ or an open sphere with centre at $+\infty$. Every open interval $(-\infty, c)$ is called a neighbourhood of $-\infty$ or an open sphere with centre at $-\infty$.

Here $c \in \mathbb{R}$.

$f(x, y) = x^y$ can be obtained for $x, y \in \overline{\mathbb{R}}$ as

$$(-\infty)^y = \begin{cases} 0, & \text{if } y < 0 \\ +\infty, & \text{if } y > 0 \end{cases}$$

$$x^{+\infty} = \begin{cases} 0, & \text{if } 0 < x < 1 \\ +\infty, & \text{if } x > 1 \end{cases}$$

$$x^{-\infty} = \begin{cases} +\infty, & \text{if } 0 < x < 1 \\ 0, & \text{if } x > 1. \end{cases}$$

The function e^x and $\ln|x|$ can be extended to $\overline{\mathbb{R}}$ with

$$e^{-\infty} = 0$$
$$e^{+\infty} = +\infty$$
$$\ln|0| = -\infty$$
$$\ln|\pm\infty| = +\infty.$$

Some other improper functions, $\tanh(\pm\infty) = \pm 1$, $\tan^{-1}(\pm\infty) = \pm\frac{\pi}{2}$ can be extended to $\overline{\mathbb{R}}$ but not $\sin x, \cos x$.

Definition [15.9] : [Algebraic numbers] A real number which is a solution of non-constant polynomial equation with rational co-efficients is known as algebraic number.

That is, an algebraic number is a real number which is a solution of a polynomial equation $a_0 + a_1 x + a_2 x^2 + \cdots + a_n x^n = 0$, $a_i \in Q$ [or $a_i \in z$].

Example. (i) $x^2 - 2 = 0$, a polynomial equation with rational co-efficients, has a solution $x = \pm\sqrt{2}$. So $\pm\sqrt{2}$ are algebraic number.

(ii) Consider a polynomial equation $\frac{1}{18}x^2 - \frac{2}{3}x + 1 = 0$ with rational co-efficients. The solutions are given by

$$x = \frac{\frac{2}{3} \pm \sqrt{\frac{4}{9} - \frac{4}{18}}}{2 \cdot \frac{1}{18}} = \frac{\frac{2}{3} \pm 2\frac{1}{3\sqrt{2}}}{\frac{1}{9}} = 6 \pm 3\sqrt{2}.$$

So $6 \pm 3\sqrt{2}$ are algebraic numbers.

(iii) $27x^3 - 8 = 0 \Rightarrow x = \frac{2}{3}, \frac{2}{3}\omega, \frac{2}{3}\omega^2$. So $\frac{2}{3} \in \mathbb{R}$ is an algebraic number.

Property [15.1] : [Algebraic numbers are countable] Since the polynomials with integer coefficients are countable and since each polynomial has a finite number of zero, algebraic numbers must also be countable.

Definition [15.10] : [Transcendental Number] A number (possibly a complex number) which is not algebraic is called a transcendental number.

π, e and e^{π} are transcendental numbers. Only a few classes of transcendental numbers are known. However, transcendental numbers are not rare. Indeed almost all real and complex numbers are transcendental, since the algebraic numbers are countable but the sets of reals and complex numbers are uncountable.

All *real transcendental numbers are irrational* since all rational numbers (being countable) are algebraic.

Converse is not true. That is all *irrational numbers* are not transcendental. $\sqrt{2}$ is irrational but not transcendental as it is a solution of a polynomial equation $x^2 - 2 = 0$.

Property [15.2] : (1) The set of transcendental numbers is uncountable infinite. Cantor's diagonal argument proves that the set of real numbers are uncountable, so the set of all transcendental numbers are uncountable.

(2) The name transcendental comes from Leibniz (1682), where he proved that $\sin x$ is not algebraic. Joseph Liouville first proved the existence of transcendental number in 1841, and in 1851 he introduced the first decimal examples such as Liouville's constant

$$\sum_{k=1}^{\infty} 10^{-kt} = 0.110001000000000000000001000\cdots$$

in which kth digit after the decimal point is 1 if k is factorials (like, $1, 2, 6, 24, 120, 720, \ldots$) and 0 otherwise, [i.e., $1!, 2!, 3!, 4!, 5!, 6!, \ldots$]. This number is known as Liouville number. This can be well approximated by a rational number. All Liouville's numbers are *transcendental.*

Golden ratio

Definition [15.11] : If the ratio of sum of the quantities to the larger quantity is equal to the ratio of the larger to the smaller one, then this ratio is defined to the *Golden ratio.*

Golden ratio is also termed as

(i) Extreme and mean ratio (ii) Medial section

(iii) Divine proportion (iv) Golden number.

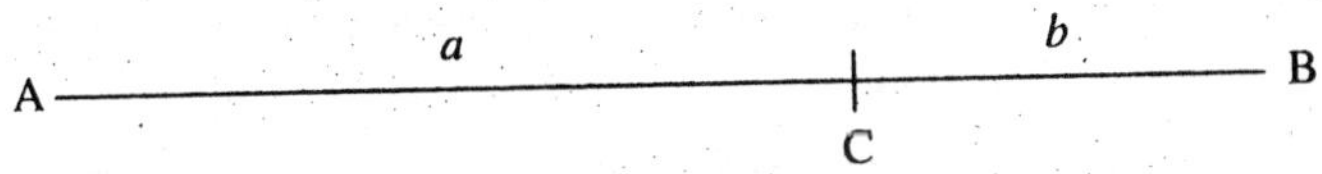

Fig. 15.2.

The line segment AB is divided into two parts.

$$|AC| = a \text{ and } |CB| = b,\ |AC| > |CB| \text{ [i.e., } a > b]$$

Then $\dfrac{a+b}{a} = \dfrac{\text{sum of two quantities } a \text{ and } b \text{ of which } a > b}{\text{larger quantity}}$

$= \dfrac{\text{larger quantity}}{\text{smaller one}} = \dfrac{a}{b}.$

Thus $\dfrac{a+b}{a} = \dfrac{a}{b}$ is a golden ratio, and this is an algebraic expression of it. It is generally denoted by φ.

That is,

$$\frac{a+b}{a} = \frac{a}{b} = \varphi. \tag{1}$$

From (1),

$$\frac{b+a}{a} = \varphi \Rightarrow \frac{b}{a} = \varphi - 1 \Rightarrow \frac{a}{b} = \frac{1}{\varphi - 1} \tag{2}$$

$$\Rightarrow \varphi = \frac{1}{\varphi - 1} \Rightarrow \varphi^2 - \varphi - 1 = 0. \tag{3}$$

This quadratic equation in φ has a unique (+)ve solution $\frac{1+\sqrt{5}}{2}$, an algebraic irrational number.

So $\varphi = \frac{1+\sqrt{5}}{2} = 1.6180339887$ (approx.).

The (−)ve root of the quadratic equation (3) for φ is $1 - \varphi \approx -0.618$.

The absolute value of this quantity, i.e., $|1 - \varphi| \approx 0.618$, corresponds to the length ratio taken in reverse order (i.e., b/a) is sometimes referred as the *golden ratio conjugate* and it is denoted by Φ.

That is, $\Phi = \dfrac{1}{\varphi} = \dfrac{1}{1.6180339887} \approx 0.6180339887.$

This shows the unique property of the *golden ratio* among (+)ve numbers that

$$\frac{1}{\varphi} = \varphi - 1 \text{ [from (3)].}$$

Since $\Phi = \dfrac{1}{\varphi}$, so $\Phi = \dfrac{1}{\varphi} - 1$ or $\dfrac{1}{\varphi} = \Phi + 1$ known as the reverse form of (3). (4)

Property [15.3]: (1) The golden ratio represents an irrational number.

Proof. We have $\frac{a+b}{a} = \frac{a}{b} = \varphi$. (i)

Let $a + b = n$ and $a = m$. Then $b = n - m$.

Then from (i) $\frac{n}{m} = \frac{m}{n-m}$. (ii)

If φ is rational, then φ is a fraction $\frac{n}{m}$, when $n, m \in Z$ and $(n, m) = 1$.

Here we may take $\frac{n}{m}$ to be the lowest terms and n and m are both (+)ve integers. But if $\frac{n}{m}$ is in its lowest terms then (ii) says that $\frac{m}{n-m}$ is still in its lowest term. This is a contradiction from the assumption that φ is rational.

(2) $\varphi = 1 + \frac{1}{\varphi}$ can be expanded recursively to obtain a continued fraction for the *golden ratio.*

That is, $\varphi = 1.6180339887 = 1 + 0.6180339887$

$$= 1 + \cfrac{1}{\cfrac{10000000000}{6180339887}} = 1 + \cfrac{1}{1 + \cfrac{3819660113}{6180339887}}$$

$$= 1 + \cfrac{1}{1 + \cfrac{1}{\cfrac{6180339887}{3819660113}}} = 1 + \cfrac{1}{1 + \cfrac{1}{1 + \cfrac{2360679774}{3819660113}}}$$

Continuing the process we get the continued fraction for

$$\varphi = [1; 1, 1, 1, \ldots] = 1 + \cfrac{1}{1 + \cfrac{1}{1 + \cfrac{1}{1 \cdots}}}$$

$$\text{and } \frac{1}{\varphi} = [0; 1, 1, 1, \ldots] = 0 + \cfrac{1}{1 + \cfrac{1}{1 + \cfrac{1}{1 + \cdots}}}$$

Thus convergents for φ are $1, 2, \frac{3}{2}, \frac{5}{3}, \frac{8}{5}, \frac{13}{8}, \ldots$ (i)

and for $\frac{1}{\varphi}$ are $1, \frac{1}{2}, \frac{2}{3}, \frac{3}{5}, \frac{5}{8}, \frac{8}{13}, \ldots$ (ii)

Fibonacci sequence is $1, 1, 2, 3, 5, 8, 13, 21, 34, 55, 89, 144, \ldots$. (iii)

The convergents defined in (i) and (ii) are the ratios of successive Fibonacci numbers given in (iii). (iv)

Thus golden ratio (φ) and Fibonacci sequence are ultimately interconnected.

φ has also a representation of the form $\varphi = \sqrt{1 + \sqrt{1 + \sqrt{1 +}}} \cdots$ which is equivalent to the recurrence equation $\alpha_n^2 = \alpha_{n-1} + 1$ with $\alpha_1 = 1$, giving $\lim_{n \to \infty} \alpha_n = \varphi$.

When $\varphi = [1; 1, 1, 1, \ldots]$ the nth convergent $x_n = \frac{p_n}{q_n}$ is given by the quadratic recurrence equation $x_n = 1 + \frac{1}{x_{n-1}}$ with $x_1 = 1$, which has a solution $x_n = \frac{F_{n+1}}{F_n}$, where F_n is the nth Fibonacci number followed by (iv).

Therefore, $\varphi = \lim_{n \to \infty} x_n = \lim_{n \to \infty} \frac{F_{n+1}}{F_n}$. (v)

Golden ratio satisfies the recurrence relation

$$\varphi^n = \varphi^{n-1} + \varphi^{n-2} \text{ or } \varphi(n) = \varphi(n-1) + \varphi(n-2), \text{ where } \varphi(n) = \varphi^n.$$

Also

$$\left.\begin{aligned} \varphi &= 1 + 2\sin\left(\frac{\pi}{10}\right) = 1 + 2\sin 18^\circ \\ \varphi &= \frac{1}{2}\operatorname{cosec}\left(\frac{\pi}{10}\right) = \frac{1}{2}\operatorname{cosec} 18^\circ \\ \varphi &= 2\cos\left(\frac{\pi}{5}\right) = 2\cos 36^\circ \end{aligned}\right\} \qquad \text{(vi)}$$

Limit superior and Limit inferior

Definition [15.12]: Let $f : X \to \overline{\mathbb{R}}$ be a real valued function and let $x_0 \in X$. Then limit superior (limit inferior) of $f(x)$ as $x \to x_0$, denoted by $\overline{\lim}_{x\to x_0} f(x) \left(\underline{\lim}_{x\to x_0} f(x)\right)$ are defined by respectively

$$\overline{\lim}_{x\to x_0} f(x) = \sup\{l,\ l \in \overline{\mathbb{R}} \text{ such that } \exists \text{ a sequence } \{x_n\} \in X,\ x_n \neq x_0\ \forall\ n \in N, \text{ converging to } x_0 \text{ and } l = \lim_{n\to\infty} f(x_n)\}$$

$$\underline{\lim}_{x\to x_0} f(x) = \inf\{l,\ l \in \overline{\mathbb{R}} \text{ such that } \exists \text{ a sequence } \{x_n\} \in X,\ x_n \neq x_0\ \forall\ n \in N, \text{ converging to } x_0 \text{ and } l = \lim_{n\to\infty} f(x_n)\}$$

Remarks. (1) If $f : X \to \overline{\mathbb{R}}$ and x_0 is a limit point of X, then

(i) $\overline{\lim}_{x\to x_0} (-f(x)) = -\underline{\lim}_{x\to x_0} f(x)$, (ii) $\underline{\lim}_{x\to x_0} (-f(x)) = -\overline{\lim}_{x\to x_0} f(x)$.

(2) If $f : X \to \overline{\mathbb{R}}$ and x_0 is a limit point of X and if $L = \overline{\lim}_{x\to x_0} f(x)$ and $l = \underline{\lim}_{x\to x_0} f(x)$, then $\exists$ a sequence $\{x_n\}$, $x_n \in X$, $x_n \neq x_0$, converging to x_0 such that $L = \lim_{n\to\infty} f(x_n)$ and $l = \lim_{n\to\infty} f(x_n)$.

(3) If x_0 is a limit of X and $f : X \to \overline{\mathbb{R}}$, then

(i) $\overline{\lim}_{x\to x_0} f(x) = \inf_{\delta>0} \sup\,[f(x),\ x \in X,\ 0 < |x - x_0| < \delta]$

(ii) $\underline{\lim}_{x\to x_0} f(x) = \sup_{\delta>0} \inf\,[f(x),\ x \in X,\ 0 < |x - x_0| < \delta]$.

Semi-continuous functions

Definition [15.13]: Semi-continuity is a property of extended real valued function $f : X \to \overline{\mathbb{R}}$, that is weaker than continuity.

f is upper (lower) semi-continuous at a point $x_0 \in X$, if roughly speaking, the function has a value nearer to x_0, which is less than (greater than) $f(x_0)$.

Thus a real valued function $f : X \to \overline{\mathbb{R}}$ is upper semi-continuous (u.s.c.) at $x_0 \in X$, if for every $\epsilon(> 0)$, $\exists$ a neighbourhood $(x_0 - \delta, x_0 + \delta)$, $\delta > 0$, of x_0 such that $f(x) \leq f(x_0) + \epsilon\ \forall\ x \in (x_0 - \delta, x_0 + \delta)$.

Equivalently, $\overline{\lim}_{x\to x_0} f(x) \leq f(x_0)$.

f is called upper semi-continuous (u.s.c.), if it is upper semi-continuous at every point of its domain X.

Similarly. A real valued function $f : X \to \overline{\mathbb{R}}$ is lower semi-continuous (l.s.c.) at a point $x_0 \in X$, if for every $\epsilon(> 0)$, $\exists$ a neighbourhood $(x_0 - \delta, x_0 + \delta)$, $\delta > 0$, of x_0 such that $f(x) \geq f(x_0) - \epsilon\ \forall\ x \in (x_0 - \delta, x_0 + \delta)$.

Equivalently, $\underline{\lim}_{x\to x_0} f(x) \geq f(x_0)$.

f is lower semi-continuous (l.s.c.), if it is lower semi-continuous at every point of its domain X.

Note. If x_0 is an isolated point of X, then we say that f is both upper and lower semi-continuous at x_0.

Example. (1) We consider a function f, defined by

$$f(x) = \begin{cases} -1, & \text{for } x < 0 \\ 1, & \text{for } x \geq 0. \end{cases}$$

Here $\overline{\lim}_{x \to 0} f(x) = \sup\Big[\Big\{1,\ \exists$ a sequence $\{x_n\} = \Big\{\frac{1}{2}, \frac{1}{3}, \dots, \frac{1}{n} \dots\Big\}$, where $x_n \neq 0\ \forall\ n \in N$ converging to 0 and $\lim_{n\to\infty} f(x_n) = \lim_{n\to\infty}(1) = 1\Big\} \cup \Big\{-1,\ \exists$ a sequence $\{x_n\} = \Big\{-\frac{1}{3}, -\frac{1}{5}, \dots, -\frac{1}{2n+1} \dots\Big\}$, when $x_n \neq 0$, converging to 0 and $\lim_{n\to\infty} f(x_n) = \lim_{n\to\infty}(-1) = -1\Big\}\Big]$ [by Def. [15.9]]

$= \sup\{1, -1\} = 1 = f(0)(= 1)$.

So f is *upper semi-continuous* at $x = 0$

$$\underline{\lim}_{x\to 0} f(x) = \inf\{1, -1\} = -1 \not\geq f(0)(= 1).$$

So f is not lower semi-continuous at $x = 0$.

Example. (2) We consider a function f defined by

$$f(x) = \begin{cases} 1, & \text{for } x < 1 \\ 2, & \text{for } x = 1 \\ \frac{1}{2}, & \text{for } x > 1. \end{cases}$$

We examine semi-continuity of f at $x = 1$.

$\overline{\lim}_{x\to 1} f(x) = \sup\Big[\Big\{\frac{1}{2},\ \exists$ a sequence $\{x_n\} = \Big\{1 + \frac{1}{2}, 1 + \frac{1}{3}, \dots\Big\}$, where $x_n \neq 1\ \forall\ n \in N$ converging to 1 and $\lim_{n\to\infty} f(x_n) = \lim_{n\to\infty} \frac{1}{2} = \frac{1}{2}\Big\} \cup \Big\{1,\ \exists$ a sequence $\{x_n\} = \Big\{1 - \frac{1}{2}, 1 - \frac{1}{3}, \dots\Big\}$, when $x_n \neq 1\ \forall\ n \in N$, converging to 1 and $\lim_{n\to\infty} f(x_n) = \lim_{n\to\infty}(1) = 1\Big\}\Big]$

$= \sup\Big\{\frac{1}{2}, 1\Big\} = 1 < f(1) = 2.$

So f is *upper semi-continuous* at $x = 1$.

$$\underline{\lim}_{x\to 1} f(x) = \inf\Big\{\frac{1}{2}, 1\Big\} = \frac{1}{2} \not> f(1) = 2.$$

So f is not lower semi-continuous at $x = 1$.

Property [15.4]: (1) A function $f : X \to \overline{\mathbb{R}}$ is upper semi-continuous, if and only if $\{x \in X,\ f(x) < \alpha\}$ is an open set for every $\alpha \in \overline{\mathbb{R}}$.

(2) $f : X \to \overline{\mathbb{R}}$ is lower semi-continuous, if and only if $\{x \in X,\ f(x) \geq \alpha\}$ is a closed set for every $\alpha \in \overline{\mathbb{R}}$.

(3) $f : X \to \overline{\mathbb{R}}$ is continuous at $x = x_0 \in X$, if and only if it is upper and lower semi-continuous at x_0.

(4) If $f, g : X \to \overline{\mathbb{R}}$ be upper (lower) semi-continuous at $x = x_0 \in X$, then (i) $f + g$ is so also at x_0, if f, g are both non-negative (ii) $f \cdot g$ will be upper (lower) semi at x_0 if f, g are both non-negative.

(5) If $f : X \to \overline{\mathbb{R}}$ be upper semi-continuous at $x_0 \in X$, then $-f$ is lower semi-continuous at x_0.

(6) Let $f_i : X \to [-\infty, \infty]$ be lower semi-continuous for every i in a non-empty set J, and let f be defined as pointwise supremum, i.e., $f(x) = \sup_{i \in J} f_i(x)$, $x \in X$ (i) then f is lower semi-continuous, (ii) If f_i are continuous at $x = x_0$, f need not be continuous. (iii) Indeed every lower semi-continuous function on a uniform space (i.e., Metric space) arises as the supremum of a sequence of continuous functions.

(7) The limit of monotone inc (dec) sequence of functions which are lower (upper) semi-continuous at a point $x_0 \in X$ is again lower (upper) semi-continuous.

Definitions of some functions

Definition [15.14]:

(1) *Riemann function* or *Thomae's function.* A function $f : \mathbb{R} \to \mathbb{R}$ defined by

$$f(x) = \begin{cases} \frac{1}{q}, & \text{if } x = \frac{p}{q}, \text{ a rational, } (p, q) = 1,\ q \neq 0 \\ 0, & \text{if } x \text{ is irrational.} \end{cases}$$

is known as Riemann function or Thomae's function.

(2) *Dirichlet's function.* A function $f : \mathbb{R} \to \mathbb{R}$ defined by

$$f(x) = \begin{cases} 1, & \text{when } x \text{ is rational} \\ -1, & \text{when } x \text{ is irrational.} \end{cases}$$

is known as Dirichlet's function.

(3) *Characteristic function.* Let $A \subset X$. Then the characteristic function of A, denoted by X_A, is defined by

$$X_A(x) = \begin{cases} 1, & \text{if } x \in A \\ 0, & \text{if } x \in X - A \text{ (i.e., } A^C) \end{cases}$$

If $X_A(x) = 0\ \forall\ x \in X$, then $X_A(x)$ is a characteristic function of null set (ϕ) and if $X_A(x) = 1$ on X, then $X_A(x)$ is a characteristic function of X.

(4) *Riemann function (R-function) for a Fourier Series.* The R-function $F(x)$ for a Fourier Series $\frac{a_0}{2} + \sum_{n=1}^{\infty}(a_n \cos nx + b_n \sin nx)$ is obtained by integrating the F.S. twice term by term. That is, $F(x) = \frac{1}{4}a_0x^2 - \sum_{n=1}^{\infty} \frac{1}{n^2}(a_n \cos nx + b_n \sin nx) + Cx + D$, where C, D are constants of integration (introduced in 1957).

Rectifiable paths and arc length

We want to introduce the concept of arc length of a curve. Since the straight line joining the two points is the shortest distance between the points, the length of a curve showed by an upper bound to the lengths of all inscribed polygon.

Thus the length of a curve = least upper bound (supremum) of the lengths of all possible inscribed polygons.

There are curves for which there is no upper bound to the lengths of the inscribed polygons.

Therefore, the curves are classified in two types: (i) those which have a length and (ii) those which do not have a length.

The type (i) is known as rectifiable and type (ii) is known as non-rectifiable.

Example. (i) Let $f(x) = x\cos\left(\dfrac{\pi}{2x}\right)$ for $x \neq 0$, $f(0) = 0$.

Here f is continuous but not of bounded variation on $[0, 1]$. The curve has no upper bound. So the graph is a non-rectifiable curve. That is of type (ii).

(ii) Let $f(x) = x^2\cos\left(\dfrac{1}{x}\right)$, $x \neq 0$, $f(0) = 0$.

Then f is of B.V. on $[0, 1]$. The curve has an upper bound. So the graph is a rectifiable curve, that is, of type (i).

Definition [15.15] : Let $f : [a, b] \to \mathbb{R}^n$ be a path in $\mathbb{R}^n$ and let $P = \{a = x_0 < x_1 < \cdots < x_{n-1} < x_n = b\}$ be a partition of $[a, b]$.

Let the points $f(x_0), f(x_1), \ldots, f(x_n)$ be the vertices of an inscribed polygon and $L_f(P)$, the length of this polygon be defined by the sum $\sum\limits_{r=1}^{n} |f(x_r) - f(x_{r-1})|$. If the set $S = \{L_f(P)$, P is a partition of $[a, b]\}$ is bounded $\forall$ P of $[a, b]$, then the path f is said to be *rectifiable* and its arc length, denoted by $AL_f[a, b]$ is defined by $AL_f[a, b] = \sup\{L_f(P)$, P is a partition of $[a, b]\}$. If $S = \{L_f(P)$, P is a partition of $[a, b]\}$ is unbounded, f is called *non-rectifiable.*

Follow the figure below:

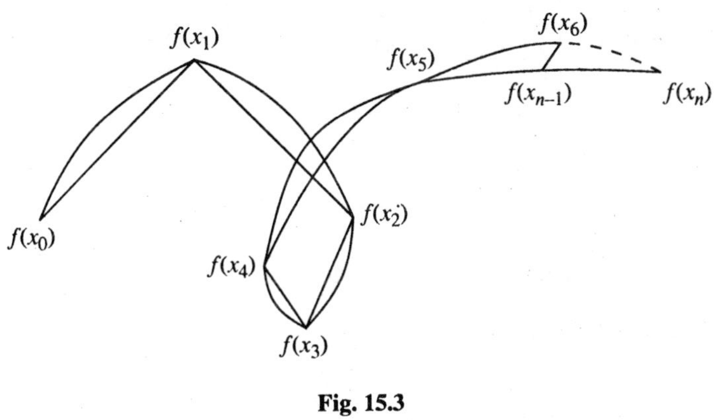

Fig. 15.3

Note. (1) A rectifiable curve is a curve of finite length.

(2) Let Γ be a continuous parametric curve in $\mathbb{R}^3$. That is, $\Gamma\{x, x_1 = x_1(t),\ x_2 = x_2(t),\ x_3 = x_3(t)\}$, $\alpha \le t \le \beta$, where $x_k(t)$, $k = 1, 2, 3$ are continuous functions on $[\alpha, \beta]$.

Let $P = \{\alpha = t_0 < t_1 < \cdots < t_{r-1} < t_r \cdots < t_n = \beta]$ be a partition of $[\alpha, \beta]$ and let

$$A_j = \{x_1(t_j), x_2(t_j), x_3(t_j)\}$$

be the sequence of points on Γ corresponding to P. Also let Γ_P be the polygon arc inscribed in Γ having vertices $A_0, A_1, \ldots, A_j$.

The length of the arc is $S(\Gamma_P) = \sum_{j=1}^{n} m_{j=1}|A_{j-1}, A_j|$, where

$$|A_{j-1}, A_j| = \sqrt{\sum_{j=1}^{n} \{x_k(t_j) - x_k(t_{j-1})^2\}}.$$

Thus $S(\Gamma_P) = \sum_{j=1}^{n} \sqrt{\sum_{k=1}^{3} \{x_k(t_j) - x_k(t_{j-1})\}^2}$.

$S(\Gamma) = \sup_P(S\Gamma_P)$ is defined as the length of Γ. If $S(\Gamma) < \infty$, then Γ is called a rectifiable curve on $\mathbb{R}^3$.

(3) A rectifiable curve Γ has a tangent at most every point $(x_1(t), x_2(t), x_3(t))$ for almost all parametric values of $t \in [\alpha, \beta]$.

(4) The study of rectifiable curve was initiated by L. Schllffer and continued by C. Jordan. Jordan found that, the curve Γ is rectifiable, iff the parametric functions $x_k(t), k = 1, 2, 3$ are of B.V. on $[\alpha, \beta]$.

Theorem [15.9] A path $\Gamma : [a, b] \to \mathbb{R}^3$ with components $(x_1(t), x_2(t), x_3(t))$ is rectifiable, if and only if $x_k(t)$ is of B.V. on $[a, b]$ for $k = 1, 2, 3$.

Proof. Let f be rectifiable. Then $S(\Gamma) < \infty$

$$\Rightarrow \sup_P(S\Gamma_P) < \infty$$

$$\Rightarrow \sum_{j=1}^{n}\left(\sum_{k=1}^{3}\sqrt{\{x_k(t_j) - x_k(t_{j-1})\}^2}\right) < \infty$$

$$\Rightarrow \sup_P\left[\sum_{j=1}^{n}\left(\sum_{k=1}^{3}\sqrt{(x_k(t_j) - x_k(t_{j-1}))}\right)\right] < \infty$$

$$\Rightarrow x_k(t) \text{ is of B.V. on } [a, b], \text{ for } k = 1, 2, 3.$$

Again, let $x_k(t)$ is of B.V. on $[a, b]$.

Then $S(\Gamma_P) = \sum_{j=1}^{n}\left(\sum_{k=1}^{3}\sqrt{(x_k(t_j) - x_k(t_{j-1})^2}\right) < \infty$

$$\Rightarrow S(\Gamma) < \infty \text{ [by Def. } S(\Gamma) = \sup_P(S\Gamma_P)$$

$$\Rightarrow \Gamma \text{ is rectifiable.}$$

Remark. (1) Concept of rectifiable curve defined in Note (2) and Theorem [15.9] can be extended to n-dimensional space $\mathbb{R}^n$ with n-components $(x_1(t), x_2(t), \ldots, x_n(t))$.

(2) Let $f : [a, b] \to \mathbb{R}$ and f' is continuous on $[a, b]$. Then f is reflectifiable and its arc length can be expressed as an integral $AL_f[a, b] = \int_a^b |f'(t)|dt$.

Proof. Let $P = [a = t_0 < t_1 \ldots < t_n = b]$ be any division of $[a, b]$ and f' is continuous on $[a, b]$.

$$\text{Then } \sum_{i=1}^{n}|f(t_r) - f(t_{r-1})| = \sum_{r=1}^{n}\left|\int_{t_{r-1}}^{t_r} f'(t)dt\right| \leq \sum_{r=1}^{n}\int_{t_{r-1}}^{t_r} |f'(t)|dt$$

$$= \int_a^b |f'(t)|dt.$$

Since f' is continuous, it is integrable on $[a, b]$.

So $\int_a^b |f'(t)|dt$ exists.

Hence $\sup_P \left\{ \sum_{r=1}^{n} |f(t_r) - f(t_{r-1})| \right\} <$ finite number.

$\Rightarrow f$ is rectifiable and arc length $AL_f[a, b] = \int_a^b |f'(t)|dt$.

(3) (i) If the curve be non-parametrical, i.e., of form $y = f(x)$, $a \le x \le b$, the arc length is given by

$$\int_a^b \left| \sqrt{1 + \left(\frac{dy}{dx}\right)^2} \right| dx.$$

(ii) If the curve be parametric, i.e., of the form $x = x(t), y = y(t), a \le t \le b$, then the arc length is given by

$$\int_a^b \left| \sqrt{\left(\frac{dx}{dt}\right)^2 + \left(\frac{dy}{dt}\right)^2} \right| dt.$$

(iii) If the curve be given in polar form, i.e., of the form $r = f(\theta)$, $a \le \theta \le \beta$, the arc length

$$= \int_\alpha^\beta \left| \sqrt{r^2 + \left(\frac{dr}{d\theta}\right)^2} \right| d\theta.$$

Problem [15.48] : Find the perimeter of the curve given by $x = \dfrac{1-t^2}{1+t^2}$ and $y = \dfrac{2t}{1+t^2}$.

Solution. Here $x^2 + y^2 = 1$. It is a closed curve, $-1 \le x,\ y \le 1$.

Then $-1 \le t \le 1$. Arc length in the first quadrant

$$= \int_0^1 \sqrt{\left(\frac{dx}{dt}\right)^2 + \left(\frac{dy}{dt}\right)^2}\, dt = \int_0^1 \sqrt{\left(\frac{-4t}{(1+t^2)^2}\right)^2 + \frac{4(1-t^2)^2}{(1+t^2)^2}}\, dt$$

$$= \int_0^1 \sqrt{\frac{4(4t^2 + 1 - 2t^2 + t^4)}{(1+t^2)^4}}\, dt = \int_0^1 2\sqrt{\frac{1}{(1+t^2)^2}}\, dt$$

$$= \int_0^1 2\frac{1}{1+t^2} dt = 2\left[\tan^{-1} t\right]_0^1 = 2\left[\frac{\pi}{4}\right] = \frac{\pi}{2}.$$

Perimeter of the circle $= 4 \cdot \dfrac{\pi}{2} = 2\pi$.

Part II

COMPLEX ANALYSIS

CHAPTER 1

COMPLEX NUMBERS

Introduction

The equations $x^2 + 4 = 0$, $x^2 - x + 3 = 0$ cannot be solved in the domain $\mathbb{R}$ of real numbers. Mathematician *Leonard Euler* (1707-1785) was first to introduce the famous notation $i = \sqrt{-1}$ and later *Care-Friedrich Gauss* (1777–1855) proved that every algebraic equation with real coefficients has a solution of the form $a + ib$, where $a, b \in \mathbb{R}, i = \sqrt{-1}$, termed as a 'complex number'. Thus $\mathbb{R}$ has been extended to the set $\mathbb{C}$ of complex numbers $\mathbb{R} \subseteq \mathbb{C}$. A real root is a particular case of a complex root $a+ib$ for which $b = 0$. *Girolamo Cardon* (1501-1576) solved a cubic equation $ax^3+bx^2+cx+d = 0$, $a, b, c, d \in \mathbb{R}$, by using '$\omega$' as an imaginary root of unity but he did not use the term 'complex number'. *Gasper Wessel* (1745-1818) first represented complex numbers as points on XY-plane by using X-axis as real axis and Y-axis as an imaginary axis. *Jean Robert Argand* (1768-1822) gave his consent of such representation of a complex number and since then this plane is known as Argand plane. The concept of stereographic projection has been introduced by German mathematician *George Friedrich Bernard Riemann* (1836-66). French mathematician *Augustin-Louis Cauchy* (1789-1857) and German mathematician *Weierstrass* (1815-97) have done immense work to the development of complex analysis.

Definition [1.1] : **[Complex number]** A number of the form $a+ib$, associated with an ordered pair (a, b) of real numbers, is defined to be a complex number and denoted by the pair (a, b).

If we write, $z = (x, y), x, y \in \mathbb{R}$, then $z = x + iy$ is called a complex variable.

The real numbers x and y are known as real and imaginary part of z and we write $\operatorname{Re}(z) = x$ and $\operatorname{Im}(z) = y$.

If $x = 0$, then $z = iy$ is called pure imaginary number.

Complex conjugate of $z = x + iy$ is defined by $x - iy$ and is denoted by $\bar{z}$.

That is, $\bar{z} = x - iy$, is conjugate of $z = x + iy$.

Thus $\operatorname{Re}(z) = \dfrac{z + \bar{z}}{2}$ and $\operatorname{Im}(z) = \dfrac{z - \bar{z}}{2}$.

Note. The set $\mathbb{C}$ of complex numbers forms a field under addition and multiplication defined by:

(i) $z_1 + z_2 = (x_1, y_1) + (x_2, y_2) = (x_1 + x_2, y_1 + y_2)$

(ii) $z_1 z_2 = (x_1, y_1) \cdot (x_2, y_2) = (x_1 x_2 - y_1 y_2, x_1 y_2 + y_1 x_2)$.

Representation of a complex number

A complex number $z = a + ib$ is regarded as an ordered pair (a, b) of real numbers. So $z = (a, b)$ can be represented by a point P whose x-co-ordinate is a and y-co-ordinate is b relative to the rectangular axes X and Y.

To each complex number $z = (a, b)$ there corresponds one and only one point in the XY-plane and to each point in the XY-plane, there exists one and only one complex number. This plane is known as complex plane or Argand plane or Gaussian plane.

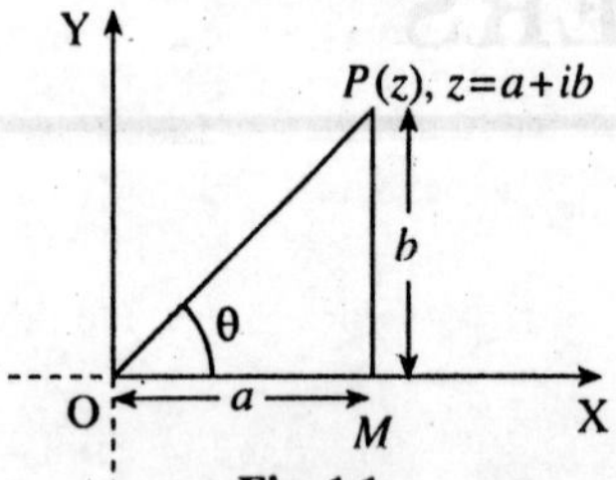

Fig. 1.1

The point $(x, 0)$ on the X-axis corresponds to real number and the $(0, y)$ on the Y-axis corresponds to purely imaginary number. This is the reason for which X-axis is called the real axis and Y-axis is called the imaginary axis in the complex plane.

This distance of the point $z = (a, b)$ from the origin O is given by

$$|OP| = \sqrt{(a-0)^2 + (b-0)^2} = \sqrt{a^2 + b^2}$$

is called the absolute value of z and is denoted by $|z|$. That is, $|z| = \sqrt{a^2 + b^2}$ for $z = (a, b)$.

The distance between the points $P_1(x_1, y_1)$ and $P_2(x_2, y_2)$ is

$$|P_1P_2| = \sqrt{(x_1 - x_2)^2 + (y_1 - y_2)^2} = |z_1 - z_2|,$$

where $z_1 = (x_1, y_1)$ and $z_2 = (x_2, y_2)$.

So the distance between the points $P_1(z_1)$ and $P_2(z_2)$ in the complex plane is denoted by $|z_1 - z_2|$.

Representation of a complex number in polar co-ordinate (r, θ)

Let $\angle POX = \theta$ and $OP = r$, with OX as an initial line. Then P has a polar co-ordinate (r, θ), where $a = r\cos\theta, b = r\sin\theta$.

Therefore, $z = a + ib = r(\cos\theta + i\sin\theta)$

$$|z| = r \text{ and } \theta = \tan^{-1} b/a.$$

Then r is called the absolute value of z and θ is called amplitude (or Argument) of z.

We denote $\theta = \text{amp } z$ (or $\arg z$).

$\tan\theta = b/a$

$\Rightarrow \tan(\theta + 2n\pi) = b/a,$ n is any integer

$\Rightarrow \theta + 2n\pi = \tan^{-1} b/a$

$\Rightarrow \theta = \tan^{-1} b/a - 2n\pi$

$\Rightarrow$ amp z is many valued and they differ from each other by an angle 2π.

If θ(amp z) satisfies $a = r\cos\theta$, $b = r\sin\theta$ and $-\pi \leq \theta \leq \pi$, then θ is called the principal value of amp z.

We remark that

(i) amp (0) is undefined.

(ii) amp of a (+)ve real number is zero. Let $a \in R$ and $a > 0$.

Then

$$\left.\begin{array}{l} a = r\cos\theta \\ 0 = r\sin\theta \end{array}\right\} \Rightarrow \theta = \tan^{-1}\frac{0}{a} = 0.$$

(iii) amp of a $(-)$ve real number is $\pm\pi$.

$$-a = r\cos\theta \qquad [a > 0]$$
$$0 = r\sin\theta.$$

Then $r = a$. Therefore, $\cos\theta = -1 \Rightarrow \theta = \pm\pi$ [which also satisfies $r\sin\theta = 0$]

(iv) amp (ib), $b > 0$ is $\pi/2$.

$$\left.\begin{array}{l} 0 = r\cos\theta \\ b = r\sin\theta \end{array}\right\} \begin{array}{l} \Rightarrow r = b \\ \Rightarrow \sin\theta = 1 \\ \Rightarrow \theta = \pi/2 \quad \text{[which also satisfies } r\cos\theta = 0] \end{array}$$

It is to be noted that in each case mentioned above, amp is a principal value.

1.1 Properties of a Complex Number

(i) If $z = x + iy$, then $z \cdot \bar{z} = |z|^2$

(ii) If z_1 and z_2 be two complex numbers, then $\overline{(z_1 \pm z_2)} = \bar{z}_1 \pm \bar{z}_2$

(iii) $\overline{(z_1 \cdot z_2)} = \bar{z}_1 \cdot \bar{z}_2$

(iv) $\overline{\left(\dfrac{z_1}{z_2}\right)} = \dfrac{\bar{z}_1}{\bar{z}_2}$

(v) $|z_1 \pm z_2| \le |z_1| + |z_2|$

(vi) $|z_1 - z_2| \ge |z_1| - |z_2|$

(vii) $|z_1 z_2| = |z_1||z_2|$

(viii) $\text{amp}\,(z_1 z_2) = \text{amp}\; z_1 + \text{amp}\; z_2$

(ix) $\left|\dfrac{z_1}{z_2}\right| = \dfrac{|z_1|}{|z_2|}$

(x) $\text{amp}\left(\dfrac{z_1}{z_2}\right) = \text{amp}\; z_1 - \text{amp}\; z_2$

(xi) $|z| = |\bar{z}|$ and $\text{amp}\;\bar{z} = -\,\text{amp}\; z$.

1.2 Some Results on Geometry in Complex Plane

(1) Equation of a straight line in a complex plane.

(2) Equation of a circle in a complex plane.

(3) Equation of a circle if the three points $A(z_1), B(z_2)$ and $C(z_3)$ lie on the circle.

(4) Inverse points with respect to a line.

(5) Inverse points with respect to a circle.

Solution. (1) Let $P(z)$ be any point in the complex plane XY. Then the equation of a straight line through $P(x, y)$, $x, y \in \mathbb{R}$ is of the form $ax + by + c = 0$, where $a, b, c \in \mathbb{R}$ and a, b are not both zero.

We have $z = x + iy$ and $\bar{z} = x - iy$.

Then $x = \dfrac{z + \bar{z}}{2}$ and $y = \dfrac{z - \bar{z}}{2i}$.

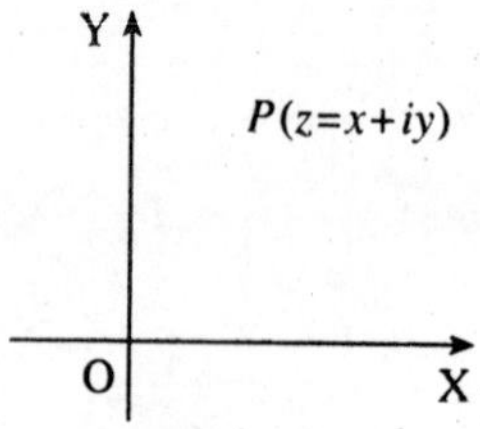

Fig. 1.2

Therefore, the equation of the line in the complex plane becomes

$$a\left(\frac{z+\bar{z}}{2}\right)+b\left(\frac{z-\bar{z}}{2i}\right)+c=0$$

$$\Rightarrow a(z+\bar{z})-ib(z-\bar{z})+2c=0$$

$$\Rightarrow (a-ib)z+(a+ib)\bar{z}+2c=0$$

$$\Rightarrow \bar{\alpha}z+\alpha\bar{z}+2c=0,\ \text{where } \alpha=a+ib.$$

Thus $\bar{\alpha}z+\alpha\bar{z}+2c=0$ is the equation of a line in a complex plane.

(2) Let us consider a circle with centre $C(a)$ and of radius r.

Let $P(z)$ be any point on the circle. Then $|CP|=r$.

$$\Rightarrow |z-a|=r \Rightarrow |z-a|^2=r^2$$

$$\Rightarrow (z-a)\overline{(z-a)}=r^2$$

$$\Rightarrow (z-a)(\bar{z}-\bar{a})=r^2$$

$$\Rightarrow z\cdot\bar{z}-z\bar{a}-a\cdot\bar{z}+a\cdot\bar{a}=r^2$$

$$\Rightarrow z\bar{z}-z\bar{a}-a\bar{z}+\{|a|^2-r^2\}=0$$

$$\Rightarrow z\bar{z}-z\bar{a}-a\bar{z}+c=0,$$

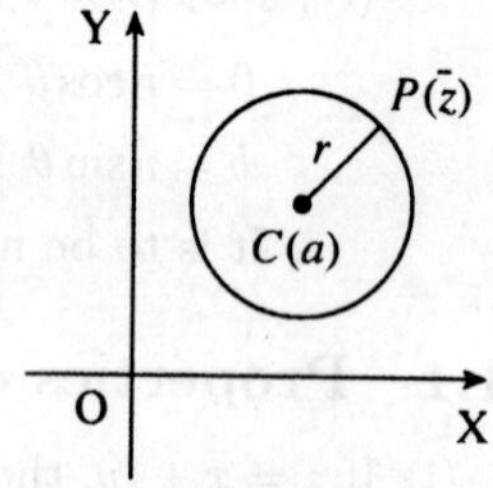

Fig. 1.3

where $c=|a|^2-r^2$ is real, which represents the equation of a circle in the complex plane, a is a complex number.

(3) In Fig 1.4 (i) A, B, C does not lie on the semicircular arc $\widehat{ABC}$.

In Fig 1.4 (ii) A, B, C lie on the semicircular arc $\widehat{ABC}$, where AC passes through the centre O of the circle.

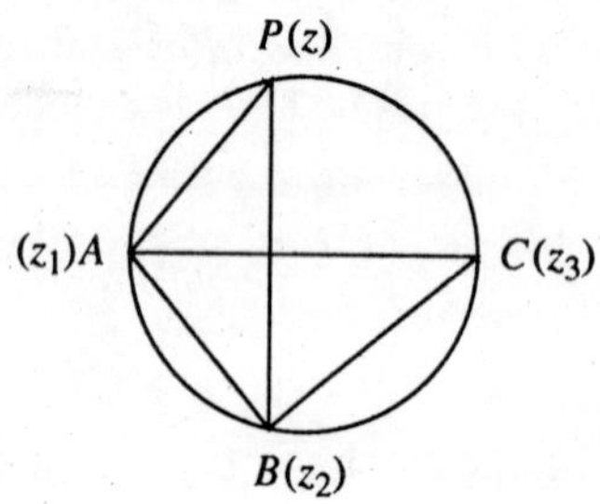

Fig. 1.4(i)

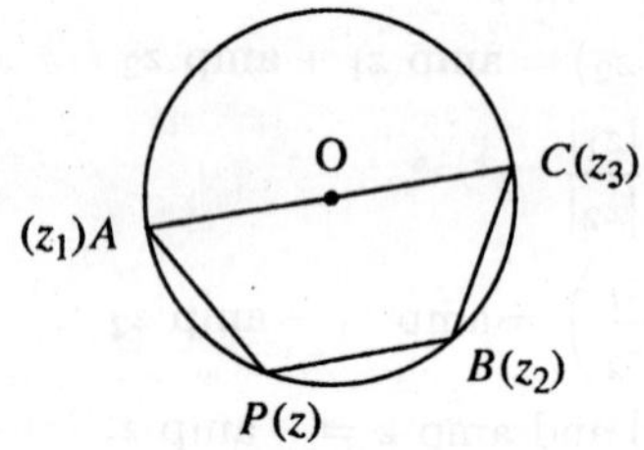

Fig. 1.4(ii)

Let $P(z)$ be any point on the circle shown in Fig 1.4 (i) and Fig 1.4 (ii).

$\angle APB$ and $\angle ACB$ lie on the same arc AB.

Therefore, $\angle APB=\angle ACB$ [Fig 1.4 (i)]

$$\Rightarrow \text{amp}\,\frac{z-z_2}{z-z_1}=\text{amp}\,\frac{z_3-z_2}{z_3-z_1}$$

$$\Rightarrow \text{amp}\,\frac{z-z_2}{z-z_1}-\text{amp}\,\frac{z_3-z_2}{z_3-z_1}=0$$

$$\Rightarrow \text{amp}\left(\frac{z-z_2}{z-z_1}\Big/\frac{z_3-z_2}{z_3-z_1}\right)=0$$

$$\Rightarrow \text{amp}\,\frac{(z-z_2)(z_3-z_1)}{(z-z_1)(z_3-z_2)}=0. \tag{1}$$

In Fig 1.4 (ii), $\angle APB + \angle ACB = \pi$

$$\Rightarrow \text{amp}\,\frac{z-z_2}{z-z_1} + \text{amp}\,\frac{z_3-z_1}{z_3-z_2} = \pi$$

$$\Rightarrow \text{amp}\left[\frac{(z-z_2)(z_3-z_1)}{(z-z_1)(z_3-z_2)}\right] = \pi. \qquad (2)$$

It follows from (1) and (2) that $\dfrac{(z-z_2)(z_3-z_1)}{(z-z_1)(z_3-z_2)}$ is purely real and in that case

$$\frac{(z-z_2)(z_3-z_1)}{(z-z_1)(z_3-z_2)} = \overline{\frac{(z-z_2)(z_3-z_1)}{(z-z_1)(z_3-z_2)}} \quad [z = \bar{z} \Leftrightarrow z \text{ is real}]$$

$$\Rightarrow \frac{(z-z_2)(z_3-z_1)}{(z-z_1)(z_3-z_2)} = \frac{(\bar{z}-\bar{z}_2)(\bar{z}_3-\bar{z}_1)}{(\bar{z}-\bar{z}_1)(\bar{z}_3-\bar{z}_2)}, \quad \left[\because \overline{\left(\frac{z_1}{z_2}\right)} = \frac{\bar{z}_1}{\bar{z}_2}\right]$$

which is the required equation of the circle.

(4) **Definition [1.2] : Inverse points**

(a) **Inverse points with respect to a line.** Let P and Q be two points on a complex plane and l be a line lying on the plane. Then P is said to be inverse of Q (or vice versa) if either of P or Q is the image of the other point with respect to the reflexion by the line l.

Here $PM = QM$. P is the inverse of Q or Q is the inverse of P.

Any points P' on l is its own inverse.

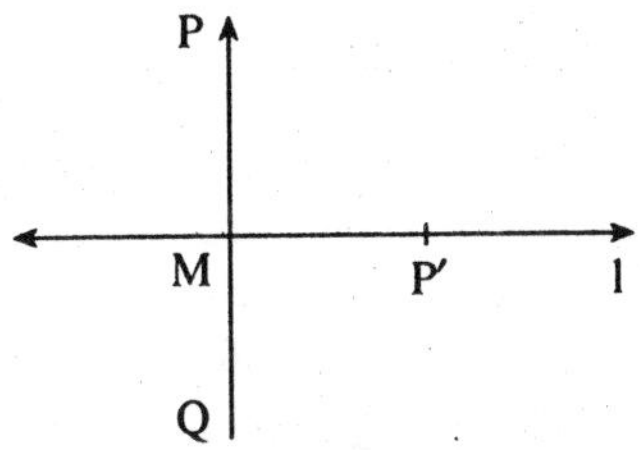

Fig. 1.5

(b) **Inverse points with respect to a circle**

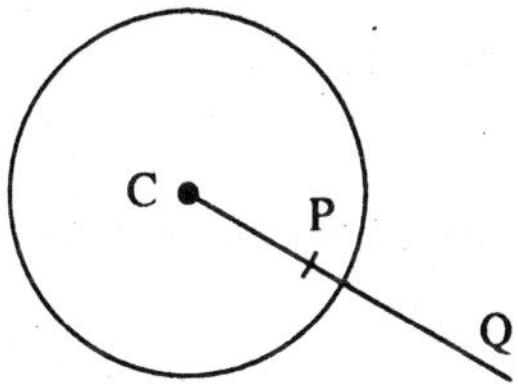

Fig. 1.6

Let us consider a circle with centre at $C(a)$ and radius r. Let $P(z_1)$ and $Q(z_2)$ be points on the same complex plane.

Then $P(z_1)$ and $Q(z_2)$ are said to be inverse points with respect to the circle, if

(i) C, P and Q are collinear and

(ii) $|z_1 - a| \cdot |z_2 - a| = r^2$ [i.e., $CP \cdot CQ = r^2$].

Remarks. (i) If $a = 0$, i.e., when centre C is at origin, P and Q are inverse points, if $|z_1|\,|z_2| = r^2$.

(ii) Since C, P, Q are collinear, $\text{amp}\,(z_1 - a) = \text{amp}\,(z_2 - a) = -\,\text{amp}\,\overline{(z_2 - a)}$.

$\Rightarrow \text{amp}\,(z_1 - a) + \text{amp}\,\overline{(z_2 - a)} = 0$

$\Rightarrow (z_1 - a)\overline{(z_2 - a)}$ is purely real and equal to r^2.

Thus for inverse points, $(z_1 - a)\overline{(z_2 - a)} = r^2$ is the required equation.

Hence, when $P(z_1)$ and $Q(z_2)$ are inverse points with respect to a circle $|z| = r$, then $z_1 \cdot \bar{z}_2 = r^2$.

(iii) z_1 and $\dfrac{1}{\bar{z}_1}$ are inverse points with respect to the circle $|z| = 1$, since

$$z_1 \cdot \overline{\left(\frac{1}{\bar{z}_1}\right)} = z_1 \cdot \frac{1}{z_1} = 1.$$

(iv) $z = 0$ and $z = \infty$ are inverse points of $|z| = 1$.

1.3 Stereographic Projection

Complex numbers are represented by the points on a complex plane. It can be shown that the complex numbers can also be represented by the points on a sphere. Thus we can establish one to one correspondence between the points on a sphere and the points on a plane. This correspondence is established through the projection of a sphere on its tangential plane (Argand plane) at the origin. Then this projection is called 'Stereographic projection'.

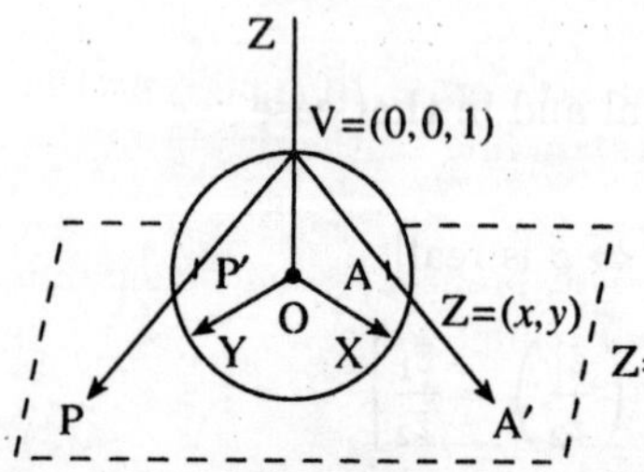

Fig. 1.7

We consider a sphere of radius 1 with centre at the origin. Let $V(0,0,1)$ be a point on sphere as the vertex of projection and the tangential plane $Z=0$ as the plane of projection. Let A be any point on the sphere except the vertex V. Then there corresponds a unique point A' of the plane since VA intersects the plane at A' and conversely, to each point P of the plane, there corresponds a unique point P' on the sphere, since PV intersects the sphere at P'.

The above geometrical approach can be proceeded analytically.

Let the equation of the sphere of unit radius and centre at $(0,0,0)$ be $X^2+Y^2+Z^2=1$ and the plane of projection be $Z=0$. We take the point $V(0,0,1)$ as vertex of projection. Let (X,Y,Z) be the co-ordinate of any point A on the sphere and $(x,y,0)$, the co-ordinate of A', where VA meets the plane of projection $Z=0$.

The points $V(0,0,1)$, $A(X,Y,Z)$ and $A'(x,y,0)$ are lying on a straight line whose equation is given by $\frac{X}{x}=\frac{Y}{y}=\frac{Z-1}{-1}$.

Then $x=\frac{X}{1-Z}$, $y=\frac{Y}{1-Z}$.

Also z (complex point) $= x+iy = \dfrac{X+iY}{1-Z}$, (1)

which shows that the points on the complex plane corresponds to the points on the sphere, except $V(0,0,1)$.

Now

$$x^2+y^2+1=\frac{X^2+Y^2}{(1-Z)^2}+1=\frac{X^2+Y^2+(1-Z)^2}{(1-Z)^2}$$

$$=\frac{1-Z^2+(1-Z)^2}{(1-Z)^2}\quad [\because X^2+Y^2+Z^2=1]$$

$$=\frac{1+Z+1-Z}{1-Z}=\frac{2}{1-Z}.$$

$$\therefore 1-Z=\frac{2}{x^2+y^2+1}.$$

$$\left.\begin{aligned}\therefore X=x(1-Z)&=\frac{2x}{x^2+y^2+1}=\frac{z+\bar{z}}{|z|^2+1}\\ Y=y(1-Z)&=\frac{2y}{x^2+y^2+1}=\frac{z-\bar{z}}{i(|z|^2+1)}\\ Z=1-\frac{2}{x^2+y^2+1}&=\frac{x^2+y^2-1}{x^2+y^2+1}=\frac{|z|^2-1}{|z|^2+1}=\frac{z\cdot\bar{z}-1}{|z|^2+1}\end{aligned}\right\}\quad (2)$$

which shows that the points (X,Y,Z) on the sphere corresponds to the points on the complex plane.

Thus from (1) and (2) we establish one to one correspondence between the points of the complex plane and the points on the sphere except that no points of the complex plane correspondence to the vertex $V(0, 0, 1)$ of the sphere.

$\mathbb{C}$ is the field of complex number. It does not contain the element which we call the point at infinitely or an 'improper complex number', denoted by ∞. By introducing ∞ in C, we extend the system of complex number by $\mathbb{C}_\infty$, where $\mathbb{C}_\infty = \mathbb{C} \cup \{\infty\}$.

It is to be noted that there is no point in the Argand plane adequate to represent the number ∞. Here the point at ∞ is defined to be the point corresponding to the origin of the complex plane (Argand plane) by the transformation $z = 1/z'$.

Again, if the point (X, Y, Z) on the sphere corresponds to the point z, then the point (X, Y, Z) on the sphere corresponds to the point z' in the complex plane, where

$$z' = \frac{X + iY}{1 + Z} = \frac{X^2 + Y^2}{(1 + Z)(X - iY)} = \frac{1 - Z^2}{(1 + Z)(X - iY)} = \frac{1 - Z}{X - iY}. \tag{3}$$

Also from (1), $z = \dfrac{X + iY}{1 - Z}$. Then $\bar{z} = \dfrac{X - iY}{1 - Z}$.

Therefore, $z' = \dfrac{1}{\bar{z}}$ [From (3)]

That is, $z' \cdot \bar{z} = 1$

$\Rightarrow |z'||\bar{z}| = 1$

$\Rightarrow |z'||z| = 1$

and amp $z = -$ amp $\bar{z} =$ amp $\dfrac{1}{\bar{z}} =$ amp z'.

Therefore, z, z' are inverse points with respect to the unit circle $|z| = 1$. Thus we conclude that, if $z = x + iy$ on the complex plane, corresponds to the point (α, β, γ) on the sphere, then z' (inverse of z), the point on the complex plane, corresponds to a point $(\alpha', \beta', \gamma')$ which lies symmetrically on the sphere with respect to the plane $Z = 0$, $(0, 0, 1)$ and $(0, 0, -1)$ are symmetrical points on the sphere.

Now for $z = 0$, $X = 0$, $Y = 0$, $Z = -1$, from (2).

So $z = 0$, the point on the complex plane corresponds to the point $(0, 0, -1)$ on the sphere and consequently $z' = \infty$, inverse point of z, corresponds to $(0, 0, 1)$ on the sphere.

Therefore, on the extended complex plane $\mathbb{C}_\infty$, the point at infinity ∞ corresponds to the point $(0, 0, 1)$ on the sphere.

Remarks. (i) Every straight line will pass through the point at infinity.

(ii) From (2), it follows that the hemisphere $Z < 0$ (called as southern hemisphere) corresponds to the interior of the circular disc $|z| < 1$ and the hemisphere $Z > 0$ (called as northern hemisphere) corresponds to the exterior of $|z| > 1$.

[Here the co-ordinate of the point A lying on the section of the sphere by the plane $Z = 0$ is (X, Y, O), while the co-ordinate of A' on the Argand plane is (x, y). The line VA intersects the plane at A'.

The equation of the line VA is $\dfrac{X}{x} = \dfrac{Y}{y} = \dfrac{0 - 1}{-1} = 1$.

$\therefore x = X,\ y = Y.$

$\Rightarrow x^2 + y^2 = X^2 + Y^2 = 1$ $[\because X^2 + Y^2 + O^2 = 1 \Rightarrow X^2 + Y^2 = 1]$

$\Rightarrow A'(x, y)$ lies on the unit circle $|z| = 1$ in the Argand plane.]

(iii) The sphere discussed above is known as *Riemann sphere.*

(iv) The circle $|z| > 1$ is called as the circle in $\mathbb{C}_\infty$.

Example. Let P be a point on the Argand plane corresponding to $z = (x, y)$ and let the straight line NP intersects the sphere S at P', where S is the Riemann sphere with N as the north pole. Then find the co-ordinate of the point P'.

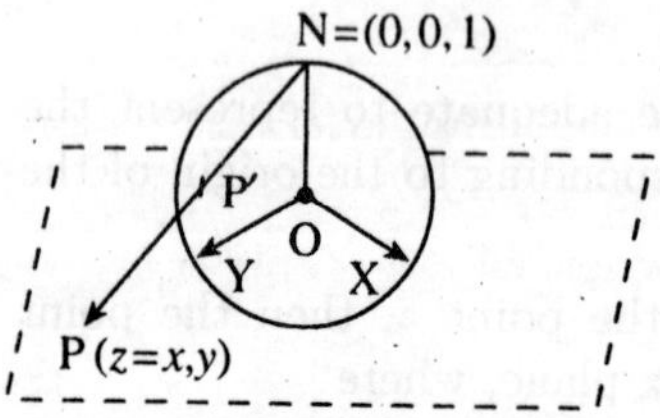

Fig. 1.8

Let the equation of the sphere be $X^2 + Y^2 + Z^2 = 1$ and let (α, β, γ) be the co-ordinates of the point P'.

Then equation of the line NP' is

$$\frac{x}{\alpha} = \frac{y}{\beta} = \frac{0-1}{\gamma - 1} = \frac{1}{1-\gamma}$$

$$\Rightarrow \frac{\alpha}{x} = \frac{\beta}{y} = \frac{1-\gamma}{1}.$$

Then from (2) of Art. 1.3, it follows that

$$\alpha = \frac{2x}{x^2 + y^2 + z^2},\ \beta = \frac{2y}{x^2 + y^2 + z^2} \text{ and } \gamma = \frac{x^2 + y^2 - 1}{x^2 + y^2 + 1}.$$

Example. Let the equation of the sphere be $x^2 + y^2 + \left(z - \frac{1}{2}\right)^2 = \frac{1}{4}$, with centre at $C\left(0, 0, \frac{1}{2}\right)$. $N(0, 0, 1)$ is the vertex of projection and the section of the sphere is considered by the plane $z = \frac{1}{2}$.

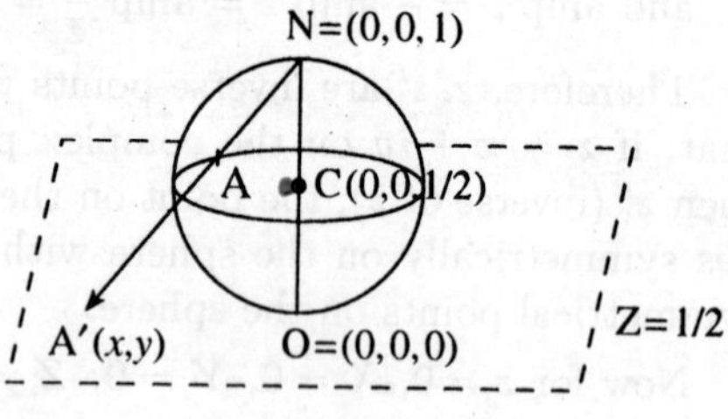

Fig. 1.9

Let the co-ordinate of A be (α, β, γ). Then the equation of the line NA which intersects the plane $Z = 0$ at $A'(x, y)$, the point on the complex plane is given by

$$\frac{x}{\alpha} = \frac{y}{\beta} = \frac{-1}{\gamma - 1}$$

$$x = \frac{\alpha}{1-\gamma},\ y = \frac{\beta}{1-\gamma} \qquad (1)$$

The point $A(\alpha, \beta, \gamma)$ lies on the sphere, we get $\alpha^2 + \beta^2 + \left(\gamma - \frac{1}{2}\right)^2 = \frac{1}{4}$.

So $z = (x, y) = \left(\frac{\alpha}{1-\gamma}, \frac{\beta}{1-\gamma}\right)$.

Now

$$x^2 + y^2 + 1 = \frac{\alpha^2 + \beta^2}{(1-\gamma)^2} + 1 = \frac{\frac{1}{4} - \left(\gamma - \frac{1}{2}\right)^2}{(1-\gamma)^2} + 1$$

$$= \frac{-\gamma^2 + \gamma + (1-\gamma)^2}{(1-\gamma)^2} = \frac{\gamma(1-\gamma) + (1-\gamma)^2}{(1-\gamma)^2} = \frac{1}{(1-\gamma)}.$$

$$\therefore\ 1 - \gamma = \frac{1}{x^2 + y^2 + 1}$$

From (1),

$$\alpha = x(1-\gamma) = \frac{x}{x^2 + y^2 + 1},\ \beta = \frac{y}{x^2 + y^2 + 1},\ \gamma = \frac{x^2 + y^2}{x^2 + y^2 + 1}.$$

EXERCISE

(1) Find $\arg i(x+iy)$, if $\arg(x+iy)=\alpha$.

(2) Prove that $|z_1-z_2|^2+|z_1+z_2|^2=2|z_1|^2+2|z_2|^2$ and hence show that

$$\left|\alpha+\sqrt{\alpha^2-\beta^2}\right|+\left|\alpha-\sqrt{\alpha^2-\beta^2}\right|=|\alpha+\beta|+|\alpha-\beta|.$$

(3) Show that the equation of a straight line in the Argand plane can be put in the form $z\bar{a}+a\bar{z}=c$, where a is a non-zero complex constant and c is real.

(4) Show that the point $Q(\beta)$ is the reflexion of the point $P(\alpha)$ in the line $z\bar{b}+\bar{z}b=c$, if

$$\beta\bar{b}+\bar{\alpha}b=c.$$

(5) Show that, if the points $A(z_1)$, $B(z_2)$, $C(z_3)$, $D(z_4)$ taken in order are concyclic, then $\dfrac{(z_3-z_1)(z_4-z_2)}{(z_3-z_2)(z_4-z_1)}$ is purely real.

(6) If the vertices of a triangle ABC be represented by the complex numbers z_1, z_2, z_3 on the Argand plane, then the triangle is equilateral, if and only if

$$\frac{1}{z_2-z_3}+\frac{1}{z_3-z_1}+\frac{1}{z_1-z_2}=0.$$

That is, if $z_1^2+z_2^2+z_3^2=z_1z_2+z_2z_3+z_3z_1$.

(7) If $|z_1|=|z_2|=|z_3|=1$ and $z_1+z_2+z_3=0$, show that z_1, z_2, z_3 are the vertices of an equilateral triangle inscribed on a unit circle.

(8) For any complex number $z=x+iy$, prove that $|x|+|y|\le\sqrt{2}|z|$. That is, $|\operatorname{Re} z|+|\operatorname{Im} z|\le\sqrt{2}|z|$.

(9) If $az_1+\bar{a}\bar{z}_2=b$, then prove that z_1 and z_2 are inverse points with respect to the line

$$az+\bar{a}\bar{z}=b.$$

(10) Show that the general equation of a circle in the Argand plane can be expressed in the form $z\bar{z}+\alpha\bar{z}+\bar{\alpha}z+r=0$, where r is a real number and $|\alpha|^2-r\ge 0$.

(11) Define extended complex plane and Riemann sphere. Show that the northern hemisphere corresponds to the exterior of the unit circle (i.e., $|z|>1$) and southern hemisphere corresponds to the interior of the unit circle [i.e., $|z|<1$] of the Argand plane.

(12) Prove that (i) z is real, if and only if $z=\bar{z}$, (ii) z is either real or pure imaginary, if and only if $(\bar{z})^2=z^2$.

CHAPTER 2

LIMIT, CONTINUITY AND DIFFERENTIABILITY

Definition [2.1]: *Domain* or *Region*: We define domain D as the set of all complex points. That is, $D = \{z = x + iy,\ x, y \in \mathbb{R}\}$.

Complex variable. If z assumes any one of the values of D, then z is called a complex variable. z stands as the symbol of a complex variable.

Jordan arc. Let $x(t)$ and $y(t)$ be continuous functions defined on a real interval $\alpha \leq t \leq \beta$, $\alpha < \beta$. Then the complex valued function defined by $z(t) = x(t) + iy(t)$ is called a continuous arc Γ. $z(\alpha)$ and $z(\beta)$ are called the end points of the arc Γ.

(i) If $z(\alpha) = z(\beta)$, then Γ is called a closed curve.

(ii) If $z(t)$ is one-one, that is, if $z(t_1) = z(t_2) \Rightarrow t_1 = t_2$, then Γ is called a *Jordan arc* or simple curve.

(iii) A closed curve Γ is called simple if the parametric representation $z(t) = x(t) + iy(t)$, for $\alpha \leq t \leq \beta$, be such that $t_1 < t_2$ and $z(t_1) = z(t_2) \Rightarrow t_1 = \alpha$ and $t_2 = \beta$.
We refer to such curves as *simple-closed Jordan curves.*

Example. $z = t^2, \ -1 \leq t \leq 1.$

$$z(-1) = 1,\ z(1) = 1.$$

So $z(-1) = z(1)$. So the curve is closed.

Let $t_1 = -1/2$ and $t_2 = 1/2$. Then $t_1 < t_2$.

$$z\left(-\frac{1}{2}\right) = \frac{1}{4},\ z\left(\frac{1}{2}\right) = \frac{1}{4}.$$

$z(t_1) = z(t_2)$ but $t_1 = -1/2 \neq -1 (= \alpha)$ and $t_2 = 1/2 \neq 1\ (= \beta)$.

So the curve is not simple.

Example. Let $z = \cos t + i \sin t$ for $0 \leq t \leq 2\pi$.

Then $z(t)$ is a simple closed Jordan curve.

Since (i) $z(0) = 1 = z(2\pi)$, $z(t)$ is closed. But (ii) Let $t_1 < t_2$ so that $z(t_1) = z(t_2)$.

Then $\cos t_1 + i\sin t_1 = \cos t_2 + i\sin t_2$

$$\Rightarrow \cos t_1 - \cos t_2 = i(\sin t_2 - \sin t_1)$$

$$\Rightarrow 2\sin\frac{t_1+t_2}{2}\cdot\sin\frac{t_2-t_1}{2} = i\cdot 2\cos\frac{t_2+t_1}{2}\sin\frac{t_2-t_1}{2}$$

$$\Rightarrow 2\sin\frac{t_2-t_1}{2}\left[\sin\frac{t_1+t_2}{2} - i\cos\frac{t_2+t_1}{2}\right] = 0$$

$$\Rightarrow \sin\frac{t_2-t_1}{2} = 0 \quad \left[\because \sin\frac{t_1+t_2}{2} - i\cos\frac{t_1+t_2}{2} \neq 0\right]$$

$$\Rightarrow t_2 - t_1 = 0 \text{ or, } t_2 - t_1 = 2\pi$$

$\Rightarrow t_2 = t_1$ or, $t_2 = t_1 + 2\pi$, which arises a contradiction in each case.

$\therefore z(t_1) \neq z(t_2)$ for any $t_1, t_2 \in (0, 2\pi)$.

Hence, $z(t_1) = z(t_2) \Rightarrow t_1 = 0$ and $t_2 = 2\pi$.

Therefore, $z(t)$ is a simple closed Jordan curve.

Functions of a complex variable

Let $f : D \to C$ be a function such that for every value of $z \in D$, it is possible to find out one or more complex numbers ω, by some rule, then ω is called a function of z and is written as $\omega = f(z)$.

Since $z = x + iy$, $\omega = f(x+iy) = u(x,y) + iv(x,y)$, where $u(x,y)$ and $v(x,y)$ are functions of two real variables x and y.

$\omega = f(z)$ is called *single valued* if for a given value of z, there corresponds a single value of ω.

$\omega = f(z)$ is called *multiple valued* if for a given value of z, there corresponds many values of ω.

Example. (i) $\omega = z$, a single valued function, (ii) $\omega = z^{1/3}$, a multiple valued function since for every value of z, we get 3 values of ω.

Limit of a function $f(z)$

Definition [2.2] : Let $\omega = f(z)$ be a function of z defined in a domain D. Let $a \in D$. Function $f(z)$ is said to have a limit l as z tends to a, if for every $\epsilon(>0)$, however small, there is a (+)ve number δ such that

$$|f(z) - l| < \epsilon, \text{ whenever } 0 < |z - a| < \delta.$$

We write it symbolically, $\lim_{z\to a} f(z) = l$.

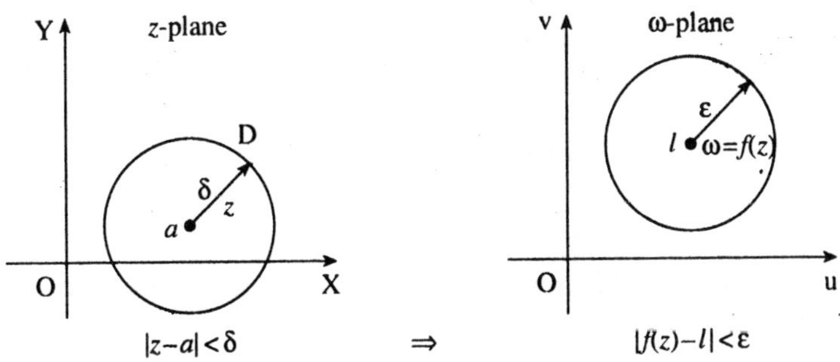

Fig. 2.1

Example. Let $f(z) = \dfrac{iz}{3}$, defined in the open disc $|z| < 1$.

Then $\lim\limits_{z \to 1} f(z) = \dfrac{i}{3}$.

Here $\left|f(z) - \dfrac{i}{3}\right| = \left|\dfrac{iz}{3} - \dfrac{i}{3}\right| = \dfrac{1}{3}|z - 1|$.

Taking $\delta = 3\epsilon$, $\left|f(z) - \dfrac{i}{3}\right| < \epsilon$, for $0 < |z - 1| < \delta$.

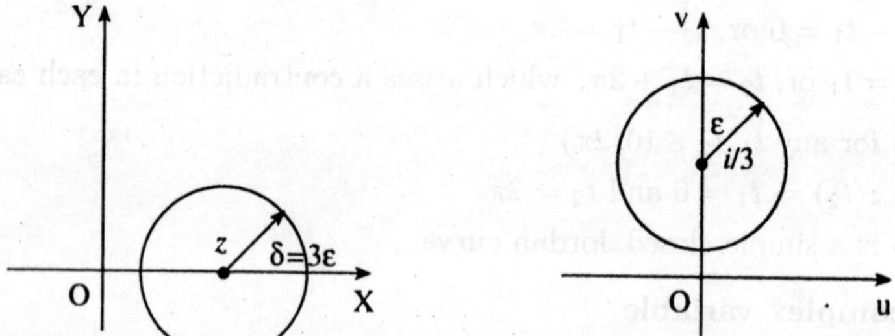

Fig. 2.2

Theorem [2.1] If a limit of function $f(z)$ exists at a point $z = a$, it is unique.

Proof. If possible, let $\lim\limits_{z \to a} f(z) = l_1$ and $\lim\limits_{z \to a} f(z) = l_2$.

Then from definition, $|f(z) - l_1| < \dfrac{\epsilon}{2}$, whenever $0 < |z - a| < \delta_1$ and $|f(z) - l_2| < \dfrac{\epsilon}{2}$, whenever $0 < |z - a| < \delta_2$.

Then
$$|l_1 - l_2| = |l_1 - f(z) + f(z) - l_2|$$
$$\leq |f(z) - l_1| + |f(z) - l_2| < \frac{\epsilon}{2} + \frac{\epsilon}{2} = \epsilon,$$
whenever $0 < |z - a| < \delta$, where $\delta = \min(\delta_1, \delta_2)$.

Here, $|l_1 - l_2|$ is a non-negative number and $\epsilon (> 0)$ can be chosen arbitrarily small.

Hence, $|l_1 - l_2| = 0 \Rightarrow l_1 - l_2 = 0 \Rightarrow l_1 = l_2$.

Therefore, the $\lim\limits_{z \to a} f(z)$ exists uniquely.

Example. (i) If $f(z) = \dfrac{z}{\bar{z}}$, then $\lim\limits_{z \to 0} f(z)$ does not exist.

Here $f(z) = \dfrac{x + iy}{x - iy}$.

Let $z \to 0$ along the real axis. Then $y = 0$.

Therefore, $\lim\limits_{z \to 0} f(z) = \lim\limits_{x \to 0} \dfrac{x}{x} = 1$.

Next, let $z \to 0$ along the imaginary axis. Then $x = 0$.

$$\therefore \lim_{z \to 0} f(z) = \lim_{y \to 0} \frac{iy}{-iy} = -1.$$

Here, limits are different on different approach. But since a limit is unique, we conclude that $\lim\limits_{z \to 0} f(z)$ does not exist.

(ii) $\lim\limits_{z \to z_0} \bar{z} = \bar{z}_0$.

For $0 < |z - z_0| < \delta$, where δ is a (+)ve real number.

We have $|\overline{z - z_0}| < \delta$. (1)

Now $|\bar{z} - \bar{z}_0| = |\overline{z - z_0}| < \epsilon$, taking $\delta = \epsilon$. [by (1)]

Hence $|\bar{z} - \bar{z}_0| < \epsilon$, whenever $0 < |z - z_0| < \delta \Rightarrow \lim_{z \to z_0} \bar{z} = \bar{z}_0$.

(iii) $\lim_{z \to 0} \left(\frac{\bar{z}^2}{z} \right) = 0$

Here $z = x + iy$. $\bar{z} = x - iy$.

Then $\bar{z}^2 = x^2 - y^2 - 2ixy$.

Therefore, $\lim_{z \to 0} \frac{x^2 - y^2 - 2ixy}{x + iy}$ to be evaluated.

Let $z \to 0$ along the real axis. Then $y = 0$.

$$\therefore \lim_{z \to 0} \frac{x^2 - y^2 - 2ixy}{x + iy} = \lim_{x \to 0} \frac{x^2}{x} = 0.$$

Let $z \to 0$ along the imaginary axis. Then $x = 0$.

$$\therefore \lim_{z \to 0} \frac{x^2 - y^2 - 2ixy}{x + iy} = \lim_{y \to 0} \frac{-y^2}{iy} = 0.$$

Also let $z \to 0$ along a line $y = mx$. Then

$$\lim_{z \to 0} \frac{x^2 - y^2 - 2ixy}{x + iy} = \lim_{x \to 0} \frac{x^2 - m^2x^2 - 2imx^2}{x(1 + im)}$$
$$= \lim_{x \to 0} x(1 - m^2 - 2im) = 0.$$

So limit exists uniquely as $z \to 0$ in any manner. Hence, $\lim_{z \to 0} \left(\frac{\bar{z}^2}{z} \right) = 0$.

Alternative method

$$\left| \frac{\bar{z}^2}{z} - 0 \right| = \frac{|\bar{z}|^2}{|z|} = |z| = |z - 0| < \epsilon, \text{ (taking } \delta = \epsilon) \text{ whenever } 0 < |z - 0| < \delta.$$

Hence, $\lim_{z \to 0} \left(\frac{\bar{z}^2}{z} \right) = 0$.

Theorem [2.2] Let $f(z) = u(x, y) + iv(x, y)$, $z = x + iy$, $a = a_0 + ib_0$ and $l = l_0 + im_0$. Then $\lim_{z \to a} f(z) = l$, if and only if

$$\lim_{(x,y) \to (a_0, b_0)} u(x, y) = l_0 \text{ and } \lim_{(x,y) \to (a_0, b_0)} v(x, y) = m_0.$$

Proof. Let $\lim_{z \to a} f(z) = l$.

Then for $\epsilon (> 0)$, $\exists$ a (+)ve number δ, such that $|f(z) - l| < \epsilon$, whenever $0 < |z - a| < \delta$

$\Rightarrow |u(x, y) + iv(x, y) - l_0 - im_0| < \epsilon$, whenever $0 < |x + iy - a_0 - ib_0| < \delta$

$\Rightarrow |\{u(x, y) - l_0\} + i\{v(x, y) - m_0\}| < \epsilon$, whenever $0 < \sqrt{(x - a_0)^2 + (y - b_0)^2} < \delta$. (i)

Now $|u(x, y) - l_0| \leq |\{u(x, y) - l_0\} + i\{v(x, y) - m_0\}| < \epsilon$.

Also $|v(x, y) - m_0| \leq |\{u(x, y) - l_0\} + i\{v(x, y) - m_0\}| < \epsilon$, whenever

$$0 < \sqrt{(x - a_0)^2 + (y - b_0)^2} < \delta \quad \text{[by (1)]}.$$

Therefore, it follows that $\lim_{(x,y)\to(a_0,b_0)} u(x,y) = l_0$ and $\lim_{(x,y)\to(a_0,b_0)} v(x,y) = m_0$.

Next, let $\lim_{(x,y)\to(a_0,b_0)} u(x,y) = l_0$ and $\lim_{(x,y)\to(a_0,b_0')} v(x,y) = m_0$ hold.

Then for each (+)ve number ϵ, $\exists$ (+)ve numbers δ_1 and δ_2 such that

$$|u(x,y), l_0| < \frac{\epsilon}{2}, \text{ whenever } 0 < \sqrt{(x-a_0)^2 + (y-b_0)^2} < \delta_1$$

$$\text{and } |v(x,y) - m_0| < \frac{\epsilon}{2}, \text{ whenever } 0 < \sqrt{(x-a_0)^2 + (y-b_0)^2} < \delta_2. \quad \text{(ii)}$$

Let $\delta = \min(\delta_1, \delta_2)$.

Now $|\{u(x,y) + iv(x,y)\} - (l_0 + im_0)|$

$$= |\{u(x,y) - l_0\} + i\{v(x,y) - m_0\}|$$

$$\leq |u(x,y) - l_0| + |v(x,y) - m_0|$$

$$< \frac{\epsilon}{2} + \frac{\epsilon}{2} = \epsilon, \text{ whenever } 0 < \sqrt{(x-a_0)^2 + (y-b_0)^2} < \delta$$

$\Rightarrow |f(z) - l| < \epsilon$, whenever $0 < |(x-a_0) + i(y-b_0)| < \delta$

i.e., $0 < |(x+iy) - (a_0 + ib_0)| < \delta$

i.e., $0 < |z - a| < \delta$.

$\Rightarrow \lim_{z\to a} f(z) = l$.

Theorem [2.3] If $\lim_{z\to a} f(z) = p$ and $\lim_{z\to a} \phi(z) = q$, where $z = x + iy$, $f(z) = u(x,y) + iv(x,y)$, $p = p_0 + iq_0$, $q = r_0 + is_0$ and $a = a_0 + ib_0$, then

(i) $\lim_{z\to a} \{f(z) \pm \phi(z)\} = p \pm q$

(ii) $\lim_{z\to a} f(z)\phi(z) = pq$

(iii) $\lim_{z\to a} \dfrac{f(z)}{\phi(z)} = \dfrac{p}{q}$, $q \neq 0$

(iv) $\lim_{z\to a} |f(z)| = |p|$

(v) $\lim_{z\to a} z^n = a^n$

(vi) $\lim_{z\to a} p(z) = p(a)$, where $p(z)$ is a polynomial in z of the form $a_0 + a_1 z + a_2 z^2 + \cdots + a_n z^n$.

Proof. (ii) Given that $|f(z) - p| < \epsilon$, whenever $0 < |z - a| < \delta_1$

$$|\phi(z) - q| < \epsilon, \text{ whenever } 0 < |z - a| < \delta_2. \quad \text{(i)}$$

We have $|f(z)| - |p| < |f(z) - p| < \epsilon$

$$\Rightarrow |f(z)| < |p| + \epsilon = p' \quad [p' > 0].$$

Now $|f(z)\phi(z) - pq| = |f(z)\{\phi(z) - q\} + q\{f(z) - p\}|$

$$\leq |f(z)|\,|\phi(z) - q| + |q|\,|f(z) - p| \leq p'|\phi(z) - q| + |q|\,|f(z) - p|$$

$$< p'\frac{\epsilon}{2p'} + |q|\frac{\epsilon}{2|q|} \text{ whenever } 0 < |z - a| < \delta, \text{ where } \delta = \min(\delta_1, \delta_2) \text{ [by (i)]}$$

$$= \frac{\epsilon}{2} + \frac{\epsilon}{2} = \epsilon, \text{ whenever } 0 < |z - a| < \delta$$

$\Rightarrow \lim_{z\to a} f(z)\phi(z) = pq$.

(iv) $|f(z)| - |p| < |f(z) - p| < \epsilon$ whenever $0 < |z - a| < \delta_1$ [by (i)]

Also $|p| - |f(z)| < |f(z) - p| < \epsilon$ whenever $|z - a| < \delta$.

Hence $\lim_{z \to a} |f(z)| = |p|$.

(v) For $n = 1$, $|z - a| < \delta = \epsilon \Rightarrow \lim_{z \to a} z = a$ for $0 < |z - a| < \delta$.

Result holds for $n = 1$. Let it be true for $n = m$.

Then $\lim_{z \to a} z^m = a^m$.

Now $\lim_{z \to a} z^{m+1} = \lim_{z \to a} z \cdot z^m = \lim_{z \to a} \cdot z \lim_{z \to a} z^m = a \cdot a^m = a^{m+1}$.

Hence it is true $\forall\, n \in \mathbb{N}$.

$\therefore \lim_{z \to a} z^n = a^n$.

(vi)
$$\begin{aligned}\lim_{z \to a} p(z) &= \lim_{z \to a} \{a_0 + a_1 z + a_2 z^2 + \cdots + a_n z^n\} \\ &= a_0 + \lim_{z \to a} a_1 z + \lim_{z \to a} a_2 z^2 + \cdots + \lim_{z \to a} a_n z^n \quad \text{[by (1)]} \\ &= a_0 + a_1 a + a_2 a^2 + \cdots + a_n a^n \quad \text{[by (5)]} \\ &= p(a).\end{aligned}$$

Note. Proofs of (i), (iii) are left.

Remarks. For any complex number $a = a_0 + ib_0$, $a \cdot a^n = a^{n+1}$.

Let $a_0 = r\cos\theta$, $b_0 = r\sin\theta$.

Then $a = a_0 + ib_0 = r(\cos\theta + i\sin\theta)$

$$\begin{aligned} a^n &= r^n(\cos n\theta + i\sin n\theta) \\ \therefore\ a \cdot a^n &= r^{n+1}\{\cos(n+1)\theta + i\sin(n+1)\theta\} \\ &= r^{n+1}\{\cos\theta + i\sin\theta\}^{n+1} = \{r\cos\theta + ir\sin\theta\}^{n+1} \\ &= (a_0 + ib_0)^{n+1} = a^{n+1}.\end{aligned}$$

Example. Show that $\lim_{z \to 1-i} \{x + i(2x + y)\} = 1 + i$, where $z = x + iy$.

Here $f(z) = x + i(2x + y) = u(x, y) + iv(x, y)$.

It is sufficient to show that $\lim_{(x,y) \to (1,-1)} x = 1$ and $\lim_{(x,y) \to (1,-1)} 2x + y = 1$.

Now

$$\lim_{(x,y) \to (1,-1)} x = \lim_{x \to 1} x = 1$$

$$\text{and} \quad \lim_{(x,y) \to (1,-1)} 2x + y = \lim_{x \to 1} 2x + \lim_{y \to -1} y = 2 - 1 = 1.$$

Hence $\lim_{z \to 1-i} f(z) = 1 + i$.

Problem [2.1]: (1) Show that $\lim_{z \to 0} \left(\dfrac{z}{\bar{z}}\right)^2$ does not exist.

(2) Find $\lim_{z \to i} \dfrac{iz^3 - 1}{z + i}$.

(3) Let $\Delta z = z - z_0$. Then show that $\lim_{z \to z_0} f(z) = \omega_0$, if and only if $\lim_{\Delta z \to 0} f(z_0 + \Delta z) = \omega_0$.

(4) Show that $\lim_{z \to z_0} f(z)g(z) = 0$, if $\lim_{z \to z_0} g(z) = 0$ and if there exists a (+)ve number M such that $|f(z)| \le M\ \forall\ z$ in the neighbourhood of z_0.

Solution. (1) $\lim\limits_{z\to 0}\left(\dfrac{z}{\bar z}\right)^2 = \lim\limits_{(x,y)\to(0,0)} \dfrac{(x+iy)^2}{(x-iy)^2}$.

Let $(x,y)\to(0,0)$ along $y=mx$.

Then

$$\lim_{(x,y)\to(0,0)} \frac{(x+iy)^2}{(x-iy)^2} = \lim_{x\to 0} \frac{x^2(1+im)^2}{x^2(1-im)^2} = \left(\frac{1+im}{1-im}\right)^2 = \frac{(1+im)^4}{\{1+m^2\}^2}$$

which is different for different values of m.

Therefore, $\lim\limits_{(x,y)\to(0,0)} \dfrac{(x+iy)^2}{(x-iy)^2}$ does not exist.

Hence $\lim\limits_{z\to 0}\left(\dfrac{z}{\bar z}\right)^2$ does not exist.

(2) $\lim\limits_{z\to i} \dfrac{iz^3-1}{z+i} = \dfrac{\lim\limits_{z\to i} iz^3-1}{\lim\limits_{z\to i} z+i} = \dfrac{\lim\limits_{z\to i} i(z^3-i^3)}{2i} = \dfrac{0}{2i} = 0.$

(3) $\Delta z = z - z_0$

$$z\to z_0 \Rightarrow \Delta z\to 0 \text{ and } z = z_0+\Delta z.$$

Then $\lim\limits_{z\to z_0} f(z) = \omega_0 \Rightarrow \lim\limits_{\Delta z\to 0} f(z_0+\Delta z) = \omega_0$.

Conversely, let $\lim\limits_{\Delta z\to 0} f(z_0+\Delta z) = \omega_0$.

Put $z_0+\Delta z = z$. Then $\Delta z = z - z_0$ and as $\Delta z\to 0$, $z\to z_0$.

$$\therefore \lim_{\Delta z\to 0} f(z_0+\Delta z) = \omega_0 \Rightarrow \lim_{z\to z_0} f(z) = \omega_0.$$

(4) $\lim\limits_{z\to z_0} g(z) = 0 \Rightarrow |g(z)| < \frac{\epsilon}{M}$, whenever $0<|z-z_0|<\delta$, where M is a (+)ve number.

Now $|f(z)g(z) - 0| = |f(z)|\,|g(z)| \le M|g(z)|$

$< M\frac{\epsilon}{M} = \epsilon$, whenever $0<|z-z_0|<\delta$.

$\Rightarrow \lim\limits_{z\to z_0} f(z)g(z) = 0.$

In the extended complex plane $\mathbb{C}_\infty$ limits involving ∞

The meaning of the symbol $\lim\limits_{x\to a} f(z) = l$ is given above. Now if a or l or both of these numbers is replaced by the point at infinity, then we shall define the meaning of the symbol at each stage. So the following cases arise, (i) $\lim\limits_{z\to a} f(z) = \infty$ (ii) $\lim\limits_{z\to\infty} f(z) = l$, (iii) $\lim\limits_{z\to\infty} f(z) = \infty$.

Case (i) $\lim\limits_{z\to a} f(z) = \infty$, means for each $\epsilon(>0)$, there exists a (+)ve number δ such that

$$|f(z) > \frac{1}{\epsilon}, \text{ whenever } 0<|z-a|<\delta. \tag{1}$$

The point $\omega = f(z)$ lies in the neighbourhood $|\omega| > \frac{1}{\epsilon}$ of ∞, whenever z lies in the deleted neighbourhood $0<|z-z_0|<\delta$. The statement (1) can be written as $\left|\frac{1}{f(z)} - 0\right| < \epsilon$, whenever $0<|z-a|<\delta$.

Thus $\lim\limits_{z\to a} f(z) = \infty$, if and only if $\lim\limits_{z\to a} \dfrac{1}{f(z)} = 0$.

Example. $\lim\limits_{z\to 1} \dfrac{z+3i}{z-1} = \infty$, since $\lim\limits_{z\to 1} \dfrac{1}{f(z)} = \lim\limits_{z\to 1} \dfrac{z-1}{z+3i} = 0$.

Case (ii) $\lim_{z\to\infty} f(z) = l$, means that for each $\epsilon(>0)$, there exists a (+)ve number δ such that $|f(z)-l| < \epsilon$, whenever $|z| > \frac{1}{\delta}$.

Replacing z by $\frac{1}{z}$, we get

$$\left|f\left(\frac{1}{z}\right) - l\right| < \epsilon, \text{ whenever } 0 < |z-0| < \delta \quad \left[\because \left|\frac{1}{z}\right| > \frac{1}{\delta} \Rightarrow |z| < \delta\right]$$

Thus $\lim_{z\to\infty} f(z) = l$, if and only if $\lim_{z\to 0} f\left(\frac{1}{z}\right) = l$. (2)

Example. $$\lim_{z\to\infty} \frac{4z^2}{(z-1)^2} = \lim_{z\to 0} \frac{4\left(\frac{1}{z^2}\right)}{\left(\frac{1}{z}-1\right)^2} \quad \text{[by (2)]}$$
$$= \lim_{z\to 0} \frac{4}{(1-z)^2} = 4.$$

Case (iii) $\lim_{z\to\infty} f(z) = \infty$ means that for each $\epsilon(>0)$, there exists a (+)ve number δ such that $|f(z)| > \frac{1}{\epsilon}$, whenever $|z| > \frac{1}{\delta}$.

Replacing z by $\frac{1}{z}$, we get

$$\left|f\left(\frac{1}{z}\right)\right| > \frac{1}{\epsilon}, \text{ whenever } \left|\frac{1}{z}\right| > \frac{1}{\delta}$$
$$\Rightarrow \left|\frac{1}{f\left(\frac{1}{z}\right)} - 0\right| < \epsilon, \text{ whenever } 0 < |z-0| < \delta.$$

Thus $\lim_{z\to\infty} f(z) = \infty$ if and only if $\lim_{z\to 0} \frac{1}{f\left(\frac{1}{z}\right)} = 0$.

Example. $$\lim_{z\to\infty} \frac{z^2+1}{z+1} = \infty, \text{ since } \lim_{z\to 0} \frac{\frac{1}{z}+1}{\frac{1}{z^2}+1} = \lim_{z\to 0} \frac{(1+z)z}{z^2+1} = 0.$$

Problem [2.2]: Show that, when $T(z) = \dfrac{az+b}{cz+d}$, $ad \quad bc \neq 0$,

(1) $\lim_{z\to\infty} T(z) = \infty$, if $c = 0$

(2) $\lim_{z\to\infty} T(z) = \frac{a}{c}$ and $\lim_{z\to -\frac{d}{c}} T(z) = \infty$, if $c \neq 0$.

Solution. (1) $$\lim_{z\to 0} \frac{1}{T\left(\frac{1}{z}\right)} = \lim_{z\to 0} \frac{c/z+d}{a/z+b} = \lim_{z\to 0} \frac{c+dz}{a+bz} = \frac{c}{a} = 0 \text{ when } c = 0.$$
$$[a \neq 0 \text{ as } ad - bc \neq 0]$$

Hence $\lim_{z\to\infty} T(z) = \infty$.

(2) $$\lim_{z\to 0} T\left(\frac{1}{z}\right) = \lim_{z\to 0} \frac{a/z+b}{c/z+d} = \lim_{z\to 0} \frac{a+bz}{c+dz} = \frac{a}{c}.$$

$\therefore \lim_{z\to\infty} T(z) = \dfrac{a}{c}$.

$$\lim_{z\to -\frac{d}{c}} \frac{1}{T(z)} = \lim_{z\to -\frac{d}{c}} \frac{cz+d}{az+b} = 0 \quad [\because ad - bc \neq 0]$$

$\therefore \lim_{z\to -\frac{d}{c}} T(z) = \infty \quad [c \neq 0]$.

Continuity of a function $f(z)$

Definition [2.3] : A function $f(z)$ is said to be continuous at a point $a \in D$, if all three of the following conditions hold simultaneously.

(i) $\lim_{z \to a} f(z)$ exists

(ii) $f(a)$ exists

(iii) $\lim_{z \to a} f(z) = f(a)$

We observe that conditions (i) and (ii) are involved in (iii). From (iii) it follows that for every $\epsilon(> 0)$, there exists a (+)ve number δ such that $|f(z) - f(a)| < \epsilon$, whenever $|z - a| < \delta$. A function $f(z)$ is said to be continuous on D, if it is continuous at each point $a \in D$.

Theorem [2.4] The continuity of $f(z) = u(x, y) + iv(x, y)$ at a point $z = a$, $z = x + iy$, $a = a_0 + ib_0 \Rightarrow$ the continuity of $u(x, y)$ and $v(x, y)$ at $(x, y) = (a_0, b_0)$ and conversely.

Proof. Proof follows immediately from the Proof of Th. [2.2].

Theorem [2.5] If f and g be two functions continuous at a point $a \in D$, then

(1) $f \pm g$ is continuous at a

(2) $f \cdot g$ is continuous at a

(3) f/g is continuous at a, provided $g(a) \neq 0$

(4) Compositions of f and g, i.e., $g \circ f$ [or $f \circ g$] is continuous.

Proof. (3) g is continuous at $z = a$. Then for each $\epsilon > 0$, $\exists$ a (+)ve number δ_1 such that

$$|g(z) - g(a)| < \epsilon \text{ for } |z - a| < \delta_1.$$

Now choosing $\epsilon = \frac{|g(a)|}{2}$, $g(a) \neq 0$, we have

$$|g(a)| - |g(z)| \leq |g(z) - g(a)| < \epsilon = \frac{|g(a)|}{2}$$

$$\Rightarrow \frac{1}{2}|g(a)| < |g(z)| \Rightarrow |g(z)| > \frac{1}{2}|g(a)| \text{ for } |z - a| < \delta_1.$$

Now

$$\begin{aligned} \left|\frac{f(z)}{g(z)} - \frac{f(a)}{g(a)}\right| &= \left|\frac{g(a)f(z) - f(a)g(z)}{g(a) \cdot g(z)}\right| \\ &= \left|\frac{g(a)\{f(z) - f(a)\} - f(a)\{g(z) - g(a)\}}{g(a) \cdot g(z)}\right| \\ &\leq \frac{|f(z) - f(a)|}{|g(z)|} + \frac{|f(a)|\,|g(z) - g(a)|}{|g(a)|\,|g(z)|} \\ &< \frac{2}{|g(a)|}|f(z) - f(a)| + \frac{2|f(a)|}{|g(a)|^2}|g(z) - g(a)|. \end{aligned} \quad \text{(i)}$$

Since $f(z)$ and $g(z)$ are continuous at $z = a$, we have

$$|f(z) - f(a)| < \frac{\epsilon|g(a)|}{4}, \text{ for } |z - a| < \delta_2$$

$$\text{and } |g(z) - g(a)| < \frac{\epsilon}{4} \frac{|g(a)|^2}{|f(a)| + 1}, \text{ for } |z - a| < \delta_3.$$

Hence from (i), we get

$$\left|\frac{f(z)}{g(z)} - \frac{f(a)}{g(a)}\right| < \frac{2\cdot\epsilon|g(a)|}{4|g(a)|} + \frac{\epsilon|g(a)^2|\,|f(a)|}{4[|f(a)|+1]|g(a)|^2}$$

$$< \frac{\epsilon}{2} + \frac{\epsilon}{2} = \epsilon \quad \left[\because \frac{|f(a)|}{|f(a)|+1} < 1\right]$$

for $|z-a| < \delta$, where δ is min of $\delta_1, \delta_2, \delta_3$.

$\Rightarrow \frac{f}{g}$ is continuous at $z = a$.

(4) Let $f : A \to B$, $g : C \to E$, where $A, B, C, E \subseteq D$ and $B \subseteq C$ so that $g \circ f$, compositions of f and g, can be defined in the Argand plane D.

Let $a \in A$ and let $\omega = f(z) \in B \subseteq C$. $f(z)$ is defined for all z in the neighbourhood of a and $g(\omega)$ is a function whose domain of definition contains the image of the neighbourhood of a.

Thus $g(w) = g(f(z)) = (g \circ f)(z)$, composition of f and g is defined $\forall$ z satisfying $|z - z_0| < \delta$. g is continuous at $\omega = c \in C$, $c = f(a)$. Then for every $\epsilon(>0)$, $\exists$ a (+)ve number δ_1 such that $|g(\omega) - g(c)| < \epsilon$, for $|\omega - c| < \delta_1$

$$\Rightarrow |g(f(z)) - g(f(a))| < \epsilon, \text{ for } |f(z) - f(a)| < \delta_1$$

$$\Rightarrow |(g \circ f)(z) - (g \circ f)(a)| < \epsilon, \text{ for } |f(z) - f(a)| < \delta_1 \qquad \text{(i)}$$

Again f is continuous at $z = a$.

Therefore, $|f(z) - f(a)| < \epsilon$, for $|z - a| < \delta_2$.

Taking $\epsilon < \delta_1$, $|f(z) - f(a)| < \delta_1$, for $|z - a| < \delta_2$. (ii)

From (i) and (ii), $|(g \circ f)(z) - (g \circ f)(a)| < \epsilon$, for $|z - a| < \delta_2 \Rightarrow (g \circ f)$ is continuous at $z = a$.

[Proofs of (1), (2) are left.]

Theorem [2.6] If $f(z)$ be continuous at $z = a$ and $f(a) \neq 0$, then $f(z) \neq 0$ $\forall$ z in the neighbourhood $|z - a| < \delta$.

Proof. Since $f(z)$ is continuous at $z = a$, for every $\epsilon(>0)$, there exists a (+)ve number δ such that $|f(z) - f(a)| < \epsilon$, whenever $|z - a| < \delta$.

We choose $\epsilon = \frac{|f(a)|}{2}$ $[\because f(a) \neq 0]$.

Then $|f(z) - f(a)| < \frac{|f(a)|}{2}$.

Now if $f(z) = 0$ for some point z in the neighbourhood $|z - a| < \delta$, then $|f(a)| < \frac{|f(a)|}{2}$, which is not possible.

Hence $f(z) \neq 0$ $\forall$ z in the neighbourhood $|z - a| < \delta$.

Theorem [2.7] If $f(z) = u(x, y) + i\,v(x, y)$ be continuous on a closed and bounded domain D, then f is bounded on D. That is, if f is continuous on D, then there exists a (+)ve real number M such that $|f(z)| \leq M$ $\forall$ $z \in D$.

Proof. We have $|f(z)| = \sqrt{[u(x,y)]^2 + [v(x,y)]^2}$.

Continuity of $f(z)$ $\Rightarrow$ continuity of $u(x, y)$ and $v(x, y)$ [Th. 2.4].

$\Rightarrow$ continuity of $\sqrt{[u(x,y)]^2 + [v(x,y)]^2}$ $\forall$ $(x, y) \in D$.

Therefore, $\sqrt{[u(x,y)]^2+[v(x,y)]^2}$ being a real valued continuous function is bounded and hence there exists a (+)ve real number M such that $\sqrt{[u(x,y)]^2+[v(x,y)]^2} \le M \;\forall\; (x,y) \in D$. Therefore, $|f(z)| \le M \;\forall\; z = x+iy \in D$.

Example. Let $f(z) = |z|^2, z = x+iy$. f is defined in a closed and bounded domain D.

Then $f(z) = \left(\sqrt{x^2+y^2}\right)^2 = x^2+y^2$.

$$u(x,y) = x^2+y^2,\; v(x,y) = 0$$

$u(x,y)$ being polynomial in x and y is continuous everywhere.

$\therefore\; f(z)$ is continuous $\forall\; z \in D$.

Example. Let $f(z) = \dfrac{x^3(1+i) - y^3(1-i)}{x^2+y^2}$ for $z \neq 0$ and $f(0) = 0$.

Then f is continuous on D.

Here $f(z) = \dfrac{x^3-y^3}{x^2+y^2} + \dfrac{i(x^3+y^3)}{x^2+y^2} = u(x,y) + i\,v(x,y)$

$$\Rightarrow\; u(x,y) = \frac{x^3-y^3}{x^2+y^2},\; v(x,y) = \frac{x^3+y^3}{x^2+y^2},$$

when $z \neq 0$, $x^2+y^2 \neq 0$.

So $u(x,y)$ are $v(x,y)$ are two rational functions of x and y with non-zero denominator.

Also x^3+y^3, x^2+y^2, being polynomial in x and y, are continuous.

Therefore, for $z \neq 0$, $u(x,y)$ and $v(x,y)$ are continuous.

Now for $z = 0$. Let $z \to 0$ along $y = mx$, then

$$\lim_{(x,y)\to(0,0)} \frac{x^3-y^3}{x^2+y^2} = \lim_{x\to 0} \frac{x^3(1-m^3)}{x^2(1+m^2)} = 0 = f(0)$$

$$\lim_{(x,y)\to(0,0)} \frac{x^3+y^3}{x^2+y^2} = 0 = f(0).$$

So $u(x,y)$ and $v(x,y)$ are also continuous at $(0,0)$.

Hence $f(z)$ is continuous $\forall\; z \in D$.

Differentiability

Let $\omega = f(z)$ be a function defined in a domain D. $f(z)$ is said to be differentiable at $z = a$, if $\displaystyle\lim_{\Delta z \to 0} \frac{f(a+\Delta z) - f(a)}{\Delta z}$ exists finitely and the unique limit is called the derivative of $f(z)$ at $z = a$ and is denoted by $f'(a)$.

That is, $f'(a) = \displaystyle\lim_{\Delta z\to 0} \frac{f(a+\Delta z) - f(a)}{\Delta z} = \lim_{\Delta z \to 0} \frac{\Delta w}{\Delta z}$.

Writing $a + \Delta z = z$ or $\Delta z = z - a$, we have $z \to a$ as $\Delta z \to 0$.

So, $f'(a) = \displaystyle\lim_{z\to a} \frac{f(z) - f(a)}{z-a}$.

Example. (1) Let $f(z) = z^2$.

Then for any point z,

$$f'(z) = \lim_{\Delta z \to 0} \frac{f(z+\Delta z) - f(z)}{\Delta z}$$
$$= \lim_{\Delta z \to 0} \frac{(z+\Delta z)^2 - z^2}{\Delta z} = \lim_{\Delta z \to 0} [2z + \Delta z] = 2z.$$

Example. (2) Let $f(z) = |z|^2$.

Then for any point z,

$$f'(z) = \lim_{\Delta z \to 0} \frac{f(z+\Delta z) - f(z)}{\Delta z} = \lim_{\Delta z \to 0} \frac{|z+\Delta z|^2 - |z|^2}{\Delta z}$$
$$= \lim_{\Delta z \to 0} \frac{(z+\Delta z)(\bar{z} + \overline{\Delta z}) - z \cdot \bar{z}}{\Delta z} = \lim_{\Delta z \to 0} \frac{z \cdot \overline{\Delta z} + \Delta z \cdot \bar{z} + \Delta z \cdot \overline{\Delta z}}{\Delta z}$$
$$= \lim_{\Delta z \to 0} \left[\bar{z} + \overline{\Delta z} + z \frac{\overline{\Delta z}}{\Delta z}\right]. \quad (1)$$

$\Delta z \to 0 \Rightarrow \overline{\Delta z} \to 0.$

Now $\lim_{\Delta z \to 0} \frac{\overline{\Delta z}}{\Delta z}$ does not exist.

For if $\Delta z \to 0$ along real axis through the point $(\Delta x, 0)$, then

$$\lim_{\Delta x \to 0} \frac{\Delta x - i\Delta y}{\Delta x + i\Delta y} = \lim_{\Delta x \to 0} \frac{\Delta x}{\Delta x} = 1$$

and if $\Delta z \to 0$ along imaginary axis through the point $(0, \Delta y)$, then $\lim_{\Delta y \to 0} \frac{-i\Delta y}{i\Delta y} = -1.$

Hence $\lim_{\Delta z \to 0} \frac{\overline{\Delta z}}{\Delta z}$ does not exist.

Therefore, from (1) it follows that $f'(z)$ does not exist, except at $z = 0$.

Theorem [2.8] Existence of the derivative of a function $f(z)$ at a point $z = a \in D \Rightarrow$ the continuity of the function $f(z)$ at $z = a$.

Proof. Let $f'(a)$ exist.

Then $\lim_{z \to a} \frac{f(z) - f(a)}{z - a} = f'(a).$

Now $\lim_{z \to a} [f(z) - f(a)] = \lim_{z \to a} \frac{f(z) - f(a)}{z - a} \cdot \lim_{z \to a} (z - a) = f'(a) \cdot 0 = 0.$

$\Rightarrow \lim_{z \to a} f(z) = f(a).$

$\Rightarrow f$ is continuous at $z = a$.

Theorem [2.9] (1) Let c be a complex constant and f be differentiable at a point z, then $\frac{d}{dz}[cf(z)] = cf'(z)$.

(2) If n be a (+)ve integer, then $\frac{d}{dz} z^n = nz^{n-1}$. The result is also true when n is a negative integer, provided $z \neq 0$.

(3) If f and g be two functions differentiable at a point z, then

(i) $\frac{d}{dz}[(f \pm g)z] = f'(z) \pm g'(z)$

(ii) $\frac{d}{dz}[(f \cdot g)z] = f'(z)g(z) + f(z)g'(z)$

(iii) $\frac{d}{dz}[(f/g)z] = \frac{f'(z)g(z) - f(z)g'(z)}{[g(z)]^2}, \; g(z) \neq 0.$

(iv) **Chain rule for differentiation of composite functions.** If $f(z)$ has a derivative at $z = a$ and g has a derivative at $f(a)$, then the composite function $\phi(z) = g[f(z)]$ has a derivative at $z = a$ and $\phi'(a) = g'[f(a)]f'(a)$.

Proof. (1) Left for students.

(2) Let n be a (+)ve integer, $f(z) = z^n$.

$$f'(z) = \lim_{\Delta z \to 0} \frac{f(z+\Delta z) - f(z)}{\Delta z} = \lim_{\Delta z \to 0} \frac{(z+\Delta z)^n - z^n}{\Delta z}$$

$$= \lim_{\Delta z \to 0} \frac{z^n + {}^nc_1 z^{n-1} \cdot \Delta z + {}^nc_2 z^{n-2}(\Delta z)^2 + \cdots + (\Delta z)^n - z^n}{\Delta z}.$$

[by Binomial theorem which holds for both real and complex numbers]

$$= nz^{n-1} + \lim_{\Delta z \to 0} \text{(terms containing } \Delta z)$$

$$= nz^{n-1}.$$

Next, let n be a (−)ve integer.

Let $n = -m$, $m > 0$. $f(z) = z^n = z^{-m}$. Here $z \neq 0$.

$$\text{Then } f'(z) = \lim_{\Delta z \to 0} \frac{f(z+\Delta z) - f(z)}{\Delta z}$$

$$= \lim_{\Delta z \to 0} \frac{(z+\Delta z)^{-m} - z^{-m}}{\Delta z} = \lim_{\Delta z \to 0} \frac{z^m - (z+\Delta z)^m}{\Delta z \cdot z^m \cdot (z+\Delta z)^m}$$

$$= \lim_{\Delta z \to 0} \left[\frac{-mz^{m-1}}{z^m(z+\Delta z)^m} + \lim_{\Delta z \to 0} \frac{\text{(terms containing } \Delta z)}{z^m(z+\Delta z)^m} \right]$$

$$= \frac{-mz^{m-1}}{z^{2m}} + 0 = -mz^{-m-1} = nz^{n-1}.$$

So the result is also true for a negative integer n, provided $z \neq 0$.

(3) Proofs of (i), (ii) are left for students.

$$\text{(iii) } \frac{d}{dz}\left[\left(\frac{f}{g}\right)z\right] = \lim_{\Delta z \to 0} \frac{\frac{f(z+\Delta z)}{g(z+\Delta z)} - \frac{f(z)}{g(z)}}{\Delta z}$$

$$= \lim_{\Delta z \to 0} \frac{g(z)f(z+\Delta z) - f(z)g(z+\Delta z)}{\Delta z \cdot g(z)g(z+\Delta z)}$$

$$= \lim_{\Delta z \to 0} \frac{g(z)\frac{\{f(z+\Delta z) - f(z)\}}{\Delta z} + \frac{g(z)f(z)}{\Delta z} - \frac{f(z)g(z+\Delta z)}{\Delta z}}{g(z)g(z+\Delta z)}$$

$$= \lim_{\Delta z \to 0} \frac{g(z)\left\{\frac{f(z+\Delta z) - f(z)}{\Delta z}\right\} - f(z)\left\{\frac{g(z+\Delta z) - g(z)}{\Delta z}\right\}}{g(z)g(z+\Delta z)}$$

$$= \frac{g(z) \lim_{\Delta z \to 0} \frac{f(z+\Delta z) - f(z)}{\Delta z} - f(z) \lim_{\Delta z \to 0} \frac{g(z+\Delta z) - g(z)}{\Delta z}}{g(z)g(z+\Delta z)}$$

$$= \frac{g(z)f'(z) - f(z)g'(z)}{[g(z)]^2} \quad \left[\begin{array}{l} g \text{ being differentiale is continuous at } z \in D. \\ \therefore \lim_{\Delta z \to 0} g(z+\Delta z) = g(z). \end{array}\right]$$

(iv) $f(z)$ has a derivative at $z = a$. Therefore, $f'(a)$ exists. Let $c = f(a)$, $a \in D$. We assume that $g'(c)$ exists. For some δ-neighbourhood $|\omega - c| < \delta$ of $c\ \forall\ \omega$ in $|\omega - c| < \delta$, we define a function F such that $F(c) = 0$ and

$$F(\omega) = \frac{g(\omega) - g(c)}{\omega - c} - g'(c), \text{ when } \omega \neq c. \tag{i}$$

$$\lim_{\omega \to c} F(\omega) = \lim_{\omega \to c} \frac{g(\omega) - g(c)}{\omega - c} - g'(c) = g'(c) - g'(c) = 0.$$

Hence F is continuous at c.

Therefore, from (i) $g(\omega) - g(c) = [F(\omega) + g'(c)](\omega - c)$, $|\omega - c| < \delta$

$$\Rightarrow \frac{g(f(z)) - g(f(a))}{z - a} = \{F[f(z)] + g'[f(a)]\}\frac{f(z) - f(a)}{z - a}, \tag{ii}$$

where $\omega = f(z)$, when z is any point in the neighbourhood $|z - a| < \delta_1$.

f is continuous at $z = a$ and F is continuous at $c = f(a)$. Thus the composite function $F(f(z))$ is continuous at a by Th. [2.5] (4).

Since $F(c) = 0$, $\lim\limits_{z \to a} F[f(z)] = 0$.

Therefore, taking limit $z \to a$ in (ii), we get

$$\lim_{z \to a} \frac{\phi(z) - \phi(a)}{z - a} = \lim_{z \to a}[F(f(z)) + g'(f(a))] \lim_{z \to a} \frac{f(z) - f(a)}{z - a},$$

where $\phi(z) = g(f(z))$.

$\Rightarrow \phi'(a) = 0 + g'(f(a))f'(a)$

$\Rightarrow\ \phi(z) = g(f(z))$ has a derivative at a and $\phi'(a) = g'(f(a)) \cdot f'(a)$.

Problem [2.3]: (1) Show that $f'(z)$ does not exist at any point z, when (i) $f(z) = \bar{z}$, (ii) $f(z) = \text{Re } z$, (iii) $f(z) = \text{Im } z$.

(2) Let $f(z) = \begin{cases} \bar{z}^2/z, & \text{when } z \neq 0 \\ 0, & \text{when } z = 0. \end{cases}$

Show that $f'(0)$ does not exist.

(3) Let $f(z_0) = g(z_0) = 0$ and $f'(z_0)$ and $g'(z_0)$ both exist, where $g'(z_0) \neq 0$. Then show that, from the definition of derivative, $\lim\limits_{z \to z_0} \dfrac{f(z)}{g(z)} = \dfrac{f'(z_0)}{g'(z_0)}$.

Solution. (1) (i) $f(z) = \bar{z}$.

Then $$f'(z) = \lim_{\Delta z \to 0} \frac{\overline{(z + \Delta z)} - \bar{z}}{\Delta z} = \lim_{\Delta z \to 0} \frac{\bar{z} + \overline{\Delta z} - \bar{z}}{\Delta z} = \lim_{\Delta z \to 0} \frac{\overline{\Delta z}}{\Delta z}$$

$$= \lim_{(\Delta x, \Delta y) \to (0,0)} \frac{\Delta x - i\Delta y}{\Delta x + i\Delta y}.$$

Let $(\Delta x, \Delta y) \to (0, 0)$ along the real axis through the point $(\Delta x, 0)$.

Then $$f'(z) = \lim_{\Delta x \to 0} \frac{\Delta x}{\Delta x} = 1. \tag{1}$$

Let $(\Delta x, \Delta y) \to (0, 0)$ along the imaginary axis through the $(0, \Delta y)$, then

$$f'(z) = \lim_{\Delta y \to 0} \frac{-i\Delta y}{i\Delta y} = -1. \tag{2}$$

$\therefore$ from (1) and (2), it follows that $f'(z)$ does not exist.

(ii) $f(z) = \text{Re } z = x$, when $z = x + iy$.

$$f(z + \Delta z) = R(x + iy + \Delta x + i\Delta y) = x + \Delta x.$$

Therefore, $f'(z) = \lim\limits_{(\Delta x, \Delta y) \to (0,0)} \dfrac{x + \Delta x - x}{\Delta x + i\Delta y}$ [$\because$ $\Delta z = \Delta x + i\Delta y$].

Let $(\Delta x, \Delta y) \to (0, 0)$ along the real axis through the point $(\Delta x, 0)$, then

$$f'(z) = \lim_{\Delta x \to 0} \frac{\Delta x}{\Delta x} = 1.$$

Let $(\Delta x, \Delta y) \to (0, 0)$ along the imaginary axis through the point $(0, \Delta y)$; then

$$f'(z) = \lim_{\Delta y \to 0} \frac{0}{i\Delta y} = 0.$$

Hence $f'(z)$ does not exist.

(iii) Proceed as in (ii).

(2) $f(z) = \bar{z}^2/z$

$$f'(0) = \lim_{z \to 0} \frac{f(z) - f(0)}{z} = \lim_{z \to 0} \frac{\bar{z}^2}{z^2} = \lim_{z \to 0} \frac{(x - iy)^2}{(x + iy)^2}.$$

Let $z \to 0$ along the line $y = mx$.

Then $f'(0) = \lim\limits_{(x,y) \to (0,0)} \dfrac{(x - imx)^2}{(x + imx)^2} = \lim\limits_{x \to 0} \dfrac{x^2(1 - im)^2}{x^2(1 + im)^2} = \dfrac{(1 - im)^2}{(1 + im)^2}$,

which is different for different values of m and so the limit does not exist. Therefore, $f'(0)$ does not exist.

(3) $\lim\limits_{z \to z_0} \dfrac{f(z)}{g(z)} = \lim\limits_{z \to z_0} \dfrac{f(z) - f(z_0)}{g(z) - g(z_0)}$ [since $f(z_0) = 0, g(z_0) = 0$]

$$= \lim_{z \to z_0} \frac{\frac{f(z) - f(z_0)}{z - z_0}}{\frac{g(z) - g(z_0)}{z - z_0}} \quad [z \neq z_0]$$

$$= \frac{f'(z_0)}{g'(z_0)}, \; g'(z_0) \neq 0 \quad \left[\begin{array}{l} \text{Since } f'(z_0) \text{ and } g'(z_0) \text{ both exist} \\ \text{and } \lim\limits_{z \to a} \dfrac{f(z)}{g(z)} = \dfrac{\lim\limits_{z \to 0} f(z)}{\lim\limits_{z \to a} \phi(z)} \end{array}\right]$$

EXERCISE

(1) Show that

(i) $\lim\limits_{z \to z_0} \text{Re } z = \text{Re } z_0$, $z = x + iy$, $z_0 = x_0 + iy_0$

(ii) $\lim\limits_{z \to 1-i} [x + i(2x + y)] = 1 + i$, $z = x + iy$

(iii) $\lim\limits_{z \to 0} \left(\dfrac{\bar{z}^2}{z}\right) = 0$ (iv) $\lim\limits_{z \to i} \dfrac{iz^3 - 1}{z + i}$

(v) $\lim\limits_{z \to \infty} \dfrac{2z^2}{(z - 1)^2} = 2$ (vi) $\lim\limits_{z \to \infty} \dfrac{2z^2 + 1}{3z - 1} = \infty$.

(2) If $f(z) = \dfrac{az + b}{cz + d}$, $z = x + iy$, $ad - bc \neq 0$, then show that

$$\lim_{z \to \infty} f(z) = \frac{a}{c} \text{ and } \lim_{z \to -\frac{d}{c}} f(z) = \infty, \text{ if } c \neq 0.$$

(3) Show that a polynomial $P(z)$ of degree $n(\geq 1)$, where

$$P(z) = a_n z^n + a_{n-1} z^{n-1} + \cdots + a_1 z + a_0 \ (a_n \neq 0)$$

is differentiable everywhere with $P'(z) = n a_n z^{n-1} + \cdots + 2a_2 z + a_1$ and hence show that

$$\frac{P^n(0)}{n!} = a_n.$$

(4) Show that $f(z) = \cos(x^2 - y^2)\cosh 2xy - i\sin(x^2 - y^2)\sinh 2xy$ is continuous everywhere in the Argand plane.

(5) If $f(z)$ is continuous in a closed and bounded region R, then show that

$$|f(z)| \leq M \ \forall \ z \in R.$$

(6) Let $f(z_0) = g(z_0) = 0$ and let $f'(z)$ and $g'(z)$ both exist at $z = z_0$, where $g'(z_0) \neq 0$, then using the definition of the derivative show that $\lim\limits_{z \to z_0} \dfrac{f(z)}{g(z)} = \dfrac{f'(z_0)}{g'(z_0)}$.

(7) Using definition of the derivative, find $f'(z)$, where $f(z) = \dfrac{2z-1}{3z+1}.\left(z \neq -\dfrac{1}{3}\right)$.

(8) Show that $f(z) = e^{2z}$ is continuous and differentiable for all z.

(9) A function $f(z) = u(x, y) + iv(x, y)$ is differentiable at a point $z = z_0$, iff $u(x, y)$ and $v(x, y)$ are separately differentiable at the points (x_0, y_0).

(10) If $f(z) = u(x, y) + iv(x, y)$ and $\lim\limits_{z \to z_0} f(z) = l_0 + im_0$, $z = x + iy$, $z_0 = x_0 + iy_0$, then show that

$$\lim_{(x,y) \to (x_0,y_0)} u(x, y) = l_0 \text{ and } \lim_{(x,y) \to (x_0,y_0)} v(x, y) = m_0$$

and hence show that $\lim\limits_{z \to z_0} \overline{f(z)} = l_0 - im_0$.

(11) If $f(z)$ is continuous at $z = z_0$, then show that $\overline{f(z)}$ is also continuous at $z = z_0$.

CHAPTER 3

ANALYTIC FUNCTIONS, REGULAR FUNCTIONS AND HARMONIC FUNCTIONS

Definition [3.1] : Let $\omega = f(z)$ be a single valued function defined in a domain D. $f(z)$ is said to be analytic in D, if f is differentiable at each point $z \in D$.

f is said be *analytic at a point* $z = a \in D$, if $f'(z)$ exists $\forall\ z \in |z - a| < \delta$. $|z - a| < \delta$ is called a δ-neighbourhood of a.

If f fails to be analytic at a point $z = a$ but analytic at some point for every neighbourhood of a, then 'a' is called singular point or singularity of f.

Example. The point $z = 0$ is a singular point of the function f defined by $f(z) = \frac{1}{z}$, since $f'(z)$ exists $\forall\ z \in (-\delta, \delta)$ except at $z = 0$.

The function $f(z) = |z|^2$ has no singular point since it is nowhere analytic. That is, $f'(z)$ does not exist for any non-zero value of $z \in D$ [See Example (2), Differentiability].

The terms regular and homographic are sometimes used as synonymous with the term analytic.

Theorem [3.1] The necessary condition that a function $f(z) = u(x, y)+iv(x, y)$ to be analytic in a domain D is that the partial derivatives u_x, v_x, u_y, v_y should exist and satisfy the equations $u_x = v_y$ and $u_y = -v_x$ in D.

The equations $u_x = v_y$ and $u_y = -v_x$ are known as *Cauchy-Riemann* partial differential equations.

Proof. Let $f(z)$ be analytic in a domain D. Then $f'(z)$ exists at every point $z \in D$, where $z = x + iy$.

$$f'(z) = \lim_{\Delta z \to 0} \frac{f(z + \Delta z) - f(z)}{\Delta z}$$

Now $f(z) = u(x, y) + iv(x, y)$

$$\Rightarrow f(z + \Delta z) = u(x + \Delta x, y + \Delta y) + iv(x + \Delta x, y + \Delta y).$$

Also $\Delta z = \Delta x + i\Delta y$ and $\Delta z \to 0 \Rightarrow (\Delta x, \Delta y) \to (0, 0)$ and $\Delta z \to 0$ in any manner.

So,

$$f'(z) = \lim_{(\Delta x,\Delta y)\to(0,0)} \frac{u(x+\Delta x, y+\Delta y) + iv(x+\Delta x, y+\Delta y) - u(x,y) - iv(x,y)}{\Delta x + i\Delta y}.$$

Let $(\Delta x, \Delta y) \to (0,0)$ along the real axis through the point $(\Delta x, 0)$, then

$$\begin{aligned} f'(x) &= \lim_{\Delta x\to 0} \frac{u(x+\Delta x, y) + iv(x+\Delta x, y) - u(x,y) - iv(x,y)}{\Delta x} \\ &= \lim_{\Delta x\to 0}\left[\frac{u(x+\Delta x,y)-u(x,y)}{\Delta x} + i\frac{v(x+\Delta x,y)-v(x,y)}{\Delta x}\right] \\ &= \frac{\partial u}{\partial x} + i\frac{\partial v}{\partial x} = u_x + iv_x. \end{aligned} \quad \text{(i)}$$

Again, if $(\Delta x, \Delta y) \to (0,0)$ along the imaginary axis through the point $(0, \Delta y)$, then

$$\begin{aligned} f'(z) &= \lim_{\Delta y\to 0} \frac{u(x, y+\Delta y) + iv(x, y+\Delta y) - u(x,y) - iv(x,y)}{i\Delta y} \\ \Rightarrow f'(z) &= \lim_{\Delta y\to 0}\left[\frac{u(x,y+\Delta y)-u(x,y)}{i\Delta y} + \frac{v(x,y+\Delta y)-v(x,y)}{\Delta y}\right] \\ &= -i\frac{\partial u}{\partial y} + \frac{\partial v}{\partial y} = -iu_y + v_y \end{aligned} \quad \text{(ii)}$$

Since $f'(z)$ exists at every point $z \in D$, it follows from (i) and (ii) that u_x, v_x, u_y, v_y all exists in D and $u_x + iv_x = -iu_y + v_y = v_y - iu_y \Rightarrow u_x = v_y$ and $v_x = -u_y$ or $u_y = -v_x$.

Thus we have $u_x = v_y$ and $u_y = -v_x$.

Theorem [3.2] The sufficient conditions for a single valued continuous function

$$f(z) = u(x,y) + iv(x,y)$$

to be analytic in a domain D are: (i) the partial derivatives u_x, v_x, u_y, v_y exist and are continuous (ii) they satisfy Cauchy-Riemann equations at each point of D, i.e., $u_x = v_y$ and $u_y = -v_x$.

Proof. Let the conditions (i) and (ii) hold in a domain D and the function $f(z)$ is continuous in D. We have to show that $f(z)$ is analytic in D.

In other words, $f'(z)$ exists at every point $z \in D$.

We write, $u = u(x,y)$ and $u + \Delta u = u(x+\Delta x, y+\Delta y)$ so that

$$\begin{aligned} \Delta u &= u(x+\Delta x, y+\Delta y) - u(x,y) \\ &= u(x+\Delta x, y+\Delta y) - u(x+\Delta x, y) + u(x+\Delta x, y) - u(x,y) \\ \Rightarrow \Delta u &= \Delta y u_y(x+\Delta x, y+\theta\Delta y) + \Delta x u_x(x+\theta'\Delta x, y) \end{aligned}$$

for $0 < \theta < 1,\ 0 < \theta' < 1$ [by Mean Value Theorem]

Now u_x, u_y are continuous in D and, therefore, by the property of uniform continuity, we have $|u_x(x+\theta'\Delta x, y) - u_x(x,y)| < \frac{\epsilon}{4}$ and $|u_y(x+\Delta x, y+\theta\Delta y) - u_y(x,y)| < \frac{\epsilon}{4}$, whenever

$$|\Delta x| < \frac{\delta}{2} \text{ and } |\Delta y| < \frac{\delta}{2}.$$

Taking $\alpha = u_x(x+\theta'\Delta x, y) - u_x(x,y)$ and $\beta = u_y(x+\Delta x, y+\theta'\Delta y) - u_y(x,y)$, we get $|\alpha| < \frac{\epsilon}{4}$, $|\beta| < \frac{\epsilon}{4}$, whenever $|\Delta x| < \frac{\delta}{2}, |\Delta y| < \frac{\delta}{2}$ and consequently from (i),

$$\Delta u = \Delta x\{u_x(x,y) + \alpha\} + \Delta y\{u_y(x,y) + \beta\},\ |\alpha| < \frac{\epsilon}{4}, |\beta| < \frac{\epsilon}{4}.$$

Proceeding as above we can get $\Delta v = \Delta x\{v_x(x,y)+\alpha'\} + \Delta y\{v_y(x,y)+\beta'\}$, where

$$|\alpha'| < \frac{\epsilon'}{4} \text{ and } |\beta'| < \frac{\epsilon'}{4}, \text{ whenever } |\Delta x| < \frac{\delta}{2}, |\Delta y| < \frac{\delta}{2}.$$

Therefore,

$$\frac{\Delta\omega}{\Delta z} = \frac{\Delta u + i\Delta v}{\Delta x + i\Delta y}, \text{ where } \omega = f(z)$$

$$= \frac{[\Delta x\{u_x(x,y)+\alpha\} + \Delta y\{u_y(x,y)+\beta] + i[\Delta x(v_x(x,y)+\alpha') + \Delta y(v_y(x,y)+\beta')]}{\Delta x + i\Delta y}$$

$$= \frac{[\Delta x u_x + \Delta y u_y + \alpha\Delta x + \beta\Delta y] + i[\Delta x v_x + \Delta y v_y + \alpha'\Delta x + \beta'\Delta y]}{\Delta x + i\Delta y}$$

$$= \frac{[u_x\Delta x - v_x\Delta y + iv_x\Delta x + iu_x\Delta y] + [\alpha\Delta x + \beta\Delta y + i\alpha'\Delta x + i\beta'\Delta y]}{\Delta x + i\Delta y}$$

$$[\because u_x = v_y \text{ and } u_y = -v_x]$$

$$= \frac{(u_x+iv_x)\Delta x + i\Delta y(u_x+iv_x) + (\alpha+i\alpha')\Delta x + (\beta+i\beta')\Delta y}{\Delta x + i\Delta y}$$

$$= \frac{(u_x+iv_x)(\Delta x+i\Delta y)}{\Delta x+i\Delta y} + \frac{(\alpha+i\alpha')\Delta x}{\Delta x+i\Delta y} + \frac{(\beta+i\beta')\Delta y}{\Delta x+i\Delta y}$$

$$\Rightarrow \left|\frac{\Delta\omega}{\Delta z} - (u_x+iv_x)\right| = \left|\frac{(\alpha+i\alpha')\Delta x}{\Delta x+i\Delta y} + \frac{(\beta+i\beta')\Delta y}{\Delta x+i\Delta y}\right| \le \frac{|\alpha+i\alpha'||\Delta x|}{|\Delta x+i\Delta y|} + \frac{|\beta+i\beta'||\Delta y|}{|\Delta x+i\Delta y|}$$

$$\le |\alpha| + |\alpha'| + |\beta| + |\beta'| \quad \left[\begin{array}{l}\text{since } |\Delta x| \le |\Delta x+i\Delta y| \\ \text{and } |\Delta y| \le |\Delta x+i\Delta y|\end{array}\right]$$

$$< \frac{\epsilon}{4} + \frac{\epsilon}{4} + \frac{\epsilon}{4} + \frac{\epsilon}{4} = \epsilon.$$

Whenever $|\Delta z| = |\Delta x + i\Delta y| \le |\Delta x| + |\Delta y| < \frac{\delta}{2} + \frac{\delta}{2} = \delta$.

Thus $\left|\frac{\Delta\omega}{\Delta z} - (u_x+iv_x)\right| < \epsilon$, whenever $|\Delta z| < \delta$

$\Rightarrow \lim_{\Delta z\to 0} \frac{\Delta\omega}{\Delta z} = u_x + iv_x = v_y - iu_y$ [by (ii)]

$\Rightarrow f'(z)$ exists in D, since by (i), u_x, v_x, u_y, v_y all exist in D.

$\Rightarrow f(z)$ is analytic in D.

Example. (1) Let $f(z) = (3x+y) + i(3y-x)$.

Then $f(z) = 3(x+iy) - i(x+iy) = (3-i)z,\ z = x+iy$.

$$f'(z) = \lim_{\Delta z\to 0} \frac{f(z+\Delta z) - f(z)}{\Delta z} = \lim_{\Delta z\to 0} \frac{(3-i)(z+\Delta z) - (3-i)z}{\Delta z}$$

$$= (3-i) \lim_{\Delta z\to 0} \frac{\Delta z}{\Delta z} = (3-i).$$

So $f'(z)$ exists $\forall\ z \in D$. Therefore, f is analytic in D.

Here, $u(x,y) = 3x+y,\ v(x,y) = 3y-x$.

$$u_x = 3,\ u_y = 1,\ v_x = -1,\ v_y = 3.$$

$\Rightarrow u_x = v_y,\ u_y = -v_x$.

Thus Cauchy-Riemann equations are satisfied.

Example. (2) Let $f(z) = e^{-y} \sin x - ie^{-y} \cos x$.

Here $u(x, y) = e^{-y} \sin x,\ v(x, y) = -e^{-y} \cos x$

$$u_x = e^{-y} \cos x,\ u_y = -e^{-y} \sin x$$

$$v_x = e^{-y} \sin x,\ v_y = e^{-y} \cos x$$

Thus $u_x = v_y$ and $u_y = -v_x$ everywhere in D.

Also each of the partial derivatives are continuous since $e^{-y}, \sin x, \cos x$ are continuous $\forall\ x, y \in \mathbb{R}$.

Therefore, sufficient conditions for $f(z)$ to be analytic are satisfied. Hence, $f'(z)$ exists $\forall\ z \in D$ and $f'(z) = u_x + iv_x = e^{-y} \cos x + ie^{-y} \sin x = e^{-y}(\cos x + i \sin x)$.

Example. (3) Let $f(z) = |z|^2 = (x^2 + y^2)$

$$u(x, y) = x^2 + y^2,\ v(x, y) = 0.$$

We have shown in [Ex. (2), differentiation] that $f'(z)$ exists at $z = 0$, but $f'(z)$ does not exist for any non-zero point of D.

Here $u_x = 2x,\ u_y = 2y,\ v_x = 0,\ v_y = 0$

$$\Rightarrow u_x(0, 0) = v_y(0, 0), u_y(0, 0) = -v_x(0, 0).$$

But $u_x \neq v_x,\ u_y = -v_x$ for $(x, y) \neq (0, 0)$.

Thus Cauchy-Riemann equations are satisfied at $(0, 0)$ but not satisfied at the non-zero points.

Example. (4) Let $f(z) = z - \bar{z}$.

Then

$$\begin{aligned} f'(z) &= \lim_{\Delta z \to 0} \frac{(z + \Delta z) - \overline{(z + \Delta z)} - (z - \bar{z})}{\Delta z} \\ &= \lim_{\Delta z \to 0} \frac{\Delta z - \overline{\Delta z}}{\Delta z} \\ &= \lim_{\Delta z \to 0} \left[1 - \frac{\overline{\Delta z}}{\Delta z}\right], \end{aligned}$$

which does not exist since $\lim\limits_{\Delta z \to 0} \dfrac{\overline{\Delta z}}{\Delta z}$ does not exist [See Ex. (2), differentiation].

$\therefore f'(z)$ does not exist for any $z \in D$.

Here $f(z) = (x + iy) - (x - iy) = 2iy$

$$\Rightarrow u(x, y) = 0,\ v(x, y) = 2y$$

$$\Rightarrow u_x = 0,\ u_y = 0 \text{ and } v_x = 0,\ v_y = 2.$$

Though $u_y = -v_x$ but $u_x \neq v_y$ for any $x, y \in \mathbb{R}$.

So Cauchy-Riemann equations are not satisfied.

Problem [3.1] : If u and v be the real and imaginary components of $f(z)$ defined by

$$f(z) = \begin{cases} \dfrac{(\bar{z})^2}{z}, & z \neq 0 \\ 0, & z = 0, \end{cases}$$

then verify that Cauchy-Riemann equations are satisfied at $z = (0, 0)$ though $f'(0)$ does not exist.

Solution. It follows from (2) of Problem [2.3], that $f'(0)$ does not exist.

Now $f(z) = \frac{\bar{z}^2}{z} = \frac{(x-iy)^3}{x^2+y^2} = \frac{(x^3-3xy^2)}{x^2+y^2} + \frac{i(y^3-3x^2y)}{x^2+y^2}$, $(x,y) \neq (0,0)$

$$\therefore u(x,y) = \frac{x^3-3xy^2}{x^2+y^2}$$

$$v(x,y) = \frac{y^3-3x^2y}{x^2+y^2}$$

$u(0,0) = 0$, $v(0,0) = 0$ since $f(z) = 0$ at $z = (0,0)$.

Now

$$\frac{\partial u(0,0)}{\partial x} = \lim_{h\to 0} \frac{u(h,0)-u(0,0)}{h} = \lim_{h\to 0} \frac{h^3}{h^2 h} = 1$$

$$\frac{\partial u(0,0)}{\partial y} = \lim_{k\to 0} \frac{u(0,k)-u(0,0)}{k} = \lim_{k\to 0} \frac{0-0}{k} = 0$$

$$\frac{\partial v(0,0)}{\partial x} = \lim_{h\to 0} \frac{v(h,0)-v(0,0)}{h} = \lim_{h\to 0} \frac{0-0}{h} = 0$$

$$\frac{\partial v(0,0)}{\partial y} = \lim_{k\to 0} \frac{v(0,k)-v(0,0)}{k} = \lim_{k\to 0} \frac{k^3-0}{k^2 k} = 1.$$

Thus at the point $(0,0)$, $u_x = v_y$ and $u_y = -v_x$.

So Cauchy-Riemann equations are satisfied at $(0,0)$ though $f'(0)$ does not exist.

Problem [3.2]: (1) Show that $f(z) = e^{-z^{-4}}$ $(z \neq 0)$ and $f(0) = 0$ is not analytic at $z = 0$ although Cauchy-Riemann equations are satisfied at that point.

(2) If $f(z) = \frac{xy^2(x+iy)}{x^2+y^4}(z \neq 0)$, $f(0) = 0$, then prove that $\frac{f(z)-f(0)}{z-0} \to 0$ as $z \to 0$ along any radius vector, but not as $z \to 0$ in any manner. Examine the continuity of $f(z)$ at $z = 0$.

(3) Show that the function $f(z) = \sin x \cosh y + i \cos x \sinh y$ is continuous as well as analytic everywhere.

(4) Show that the functions (i) $f(z) = 2xy + i(x^2 - y^2)$ and (ii) $f(z) = e^y \cdot e^{ix}$ is nowhere analytic.

(5) Determine the singular points of the function and state why the functions are analytic everywhere except at these points, where $f(z) = \frac{(2z+1)}{z(z^2+1)}$.

Solution. (1) $f(z) = e^{-z^{-4}}$ $(z \neq 0)$, $z = x + iy$

$$= e^{-\frac{1}{(x+iy)^4}} = e^{-\frac{(x-iy)^4}{(x^2+y^2)^4}}$$

$$= e^{-\{x^4+y^4-6x^2y^2-4ix^3y+4ixy^3\}/(x^2+y^2)^4}$$

$$= e^{-\{(x^4+y^4-6x^2y^2)+i(4xy^3-4x^3y)\}/(x^2+y^2)^4}$$

$$= e^{-\frac{(x^4+y^4-6x^2y^2)}{(x^2+y^2)^4}}\left[\cos\frac{(4xy^3-4x^3y)}{(x^2+y^2)^4} + i\sin\frac{4xy^3-4x^3y}{(x^2+y^2)^4}\right].$$

$$u(x,y) = e^{-\frac{(x^4+y^4-6x^2y^2)}{(x^2+y^2)^4}}\cos\frac{4xy(y^2-x^2)}{(x^2+y^2)^4}$$

$$\text{and } v(x,y) = e^{-\frac{(x^4+y^4-6x^2y^2)}{(x^2+y^2)^4}}\sin\frac{4xy(y^2-x^2)}{(x^2+y^2)^4}$$

$$\frac{\partial u(0,0)}{\partial x} = \lim_{h\to 0} \frac{u(h,0)-u(0,0)}{h} = \lim_{h\to 0} \frac{e^{-h^{-4}}-0}{h} = \lim_{h\to 0} \frac{e^{-\frac{1}{h^4}}}{h}$$

$$= \lim_{h\to 0} \frac{1}{h\cdot e^{+\frac{1}{h^4}}} = \lim_{h\to 0} \frac{1}{h\left[1+\frac{1}{(h^4)}+\frac{1}{2!}\frac{1}{(h^4)^2}+\cdots\right]}$$

$$= \frac{1}{\lim\limits_{h\to 0} h + \lim\limits_{h\to 0}\frac{1}{h^3} + \lim\limits_{h\to 0}\frac{1}{2h^7}+\cdots} = \frac{1}{\infty} = 0 \quad \left[\begin{array}{l}\text{as } \lim\limits_{h\to 0}\frac{1}{h^r} = \infty. \\ r>0\end{array}\right]$$

$$\frac{\partial u(0,0)}{\partial y} = \lim_{k\to 0} \frac{u(0,k)-u(0,0)}{k} = \lim_{k\to 0} \frac{e^{-\frac{1}{k^4}}-0}{k} = 0 \text{ [by above]}$$

$$\frac{\partial v(0,0)}{\partial x} = \lim_{h\to 0} \frac{v(h,0)-v(0,0)}{h} = \lim_{h\to 0} \frac{0-0}{h} = 0$$

$$\frac{\partial v(0,0)}{\partial y} = \lim_{k\to 0} \frac{v(0,k)-v(0,0)}{k} = \lim_{k\to 0} \frac{0-0}{k} = 0.$$

Hence Cauchy-Riemann equations are satisfied at $(0,0)$.

Here, $f'(0) = \lim\limits_{z\to 0} \frac{f(z)-f(0)}{z-0} = \lim\limits_{z\to 0} \frac{f(z)}{z} = \lim\limits_{z\to 0} \frac{e^{-z^{-4}}}{z}$ which does not exists if $z = re^{\frac{i\pi}{4}}$, since $z \to 0 \Rightarrow r \to 0$

$$\text{and } \lim_{z\to 0} \frac{e^{-z^{-4}}}{z} = \frac{1}{e^{\frac{i\pi}{4}}} \lim_{r\to 0} \frac{e^{\frac{1}{r^4}}}{r} \quad [\because z^{-4} = r^{-4}\cdot e^{-i\pi} = -r^{-4}]$$

$$= \frac{1}{e^{\frac{i\pi}{4}}} \lim_{r\to 0} \left[\frac{1}{r} + \frac{1}{r^5} + \frac{1}{r^9\cdot 21} + \cdots\right] \to \infty$$

$\Rightarrow f'(z)$ does not exist at $z = 0$.

(2) Let $z \to 0$ along any radius vector $y = mx$.

Then

$$\lim_{z\to 0} \frac{f(z)-f(0)}{z} = \lim_{z\to 0} \frac{xy^2(x+iy)}{(x^2+y^4)z} = \lim_{x\to 0} \frac{x\cdot m^2x^2(x+imx)}{\{x^2+m^4x^4\}(x+imx)}$$

$$= \lim_{x\to 0} \frac{m^2x^3}{x^2(1+m^4x^2)} = \lim_{x\to 0} \frac{x^3m^2}{x^2(1+m^4x^2)}$$

$$= \lim_{x\to 0} \frac{xm^2}{(1+m^4x^2)} = 0 \ \forall\ m \in R.$$

Let $z \to 0$ along $x = my^2$, then

$$\lim_{z\to 0} \frac{f(z)-f(0)}{z-0} = \lim_{z\to 0} \frac{xy^2(x+iy)}{(x^2+y^4)z} = \lim_{z\to 0} \frac{xy^2}{x^2+y^4} \quad [\because z = x+iy]$$

$$= \lim_{y\to 0} \frac{my^4}{m^2y^4+y^4} = \frac{m}{m^2+1}, \text{ which does not exist.}$$

Also $\lim\limits_{z\to 0} f(z) = \lim\limits_{z\to 0} \frac{xy^2(x+iy)}{x^2+y^4)}$. Let $z \to 0$ along $y = mx$, then

$$\lim_{z\to 0} f(z) = \lim_{x\to 0} \frac{x\cdot m^2x^2(x+imx)}{x^2+m^4x^4} = \lim_{x\to 0} \frac{x^2m^2(1+im)}{1+m^4x^2} = 0 = f(0).$$

$\therefore f$ is continuous at $z = 0$.

(3) $f(z) = \sin x \cosh y + i \cos x \sinh y$

$\sinh(y - ix) = \sinh y \cos x - i \cosh y \sin x$

$\Rightarrow i \sinh(y - ix) = \sin x \cosh y + i \cos x \sinh y = f(z)$.

Thus $f(z) = i \cdot \left[\dfrac{e^{y-ix} - e^{-(y-ix)}}{2}\right] = i \cdot \left[\dfrac{e^{-i(x-iy)} - e^{i(x+iy)}}{2}\right] = i \cdot \left[\dfrac{e^{-i\bar{z}} - e^{iz}}{2}\right]$.

Let $f_1(z) = \dfrac{i}{2}e^{-i\bar{z}}$ and $f_2(z) = \dfrac{i}{2}e^{iz}$.

Now

$$\begin{aligned}
|f_2(z) - f_2(a)| &= \left|\frac{i}{2}e^{iz} - \frac{i}{2}e^{ia}\right| = \frac{1}{2}|(\cos z - \cos a) + i(\sin z - \sin a)| \\
&\le \frac{1}{2}[|\cos z - \cos a| + |\sin z - \sin a|] \\
&\le \frac{1}{2} \cdot 2\left[\left|\sin\frac{z+a}{2}\right|\left|\sin\frac{z-a}{2}\right| + \left|\cos\frac{z+a}{2}\right|\left|\sin\frac{z-a}{2}\right|\right] \\
&\le 1 \cdot \left|\frac{z-a}{2}\right| + 1 \cdot \left|\frac{z-a}{2}\right| \\
&= |z-a| < \epsilon, \text{ whenever } |z-a| < \delta (= \epsilon)
\end{aligned}$$

$\Rightarrow f_2(z)$ is continuous at $z = a$.

Similarly, $f_1(z) = \dfrac{i}{2}e^{i\bar{z}}$ is continuous at $z = a$, since $|f_1(z) - f_1(a)| \le |\overline{z-a}| = |z-a| < \epsilon$, whenever $|z-a| < \delta\ (= \epsilon)$ (Proceeding as above).

Thus $f_1(z)$ and $f_2(z)$ are continuous at $z = a$.

Therefore, $f(z)$ being the difference of two continuous functions is continuous at $z = a$ in a domain D.

Here $u(x, y) = \sin x \cosh y$ and $v(x, y) = \cos x \sinh y$.

$$\frac{\partial u}{\partial x} = \cos x \cosh y, \quad \frac{\partial v}{\partial x} = -\sin x \sinh y$$

$$\frac{\partial u}{\partial y} = \sin x \sinh y, \quad \frac{\partial v}{\partial y} = \cos x \cosh y.$$

Thus $\dfrac{\partial u}{\partial x} = \dfrac{\partial v}{\partial y}$ and $\dfrac{\partial u}{\partial y} = -\dfrac{\partial v}{\partial x}$.

That is, $u_x = v_y$ and $u_y = -v_x\ \forall\ (x, y) \in D$ and the partial derivatives are all continuous in D.

Hence $f(z)$ is analytic in D.

(4) (i) $f(z) = 2xy + i(x^2 - y^2)$. Here $u(x, y) = 2xy$ and $v(x, y) = x^2 - y^2$.

$$\frac{\partial u}{\partial x} = 2y, \quad \frac{\partial v}{\partial x} = 2x, \quad \frac{\partial u}{\partial y} = 2x, \quad \frac{\partial v}{\partial y} = -2y$$

$$\frac{\partial u}{\partial x} \ne \frac{\partial v}{\partial y} \text{ and } \frac{\partial u}{\partial y} \ne -\frac{\partial v}{\partial x}.$$

That is, $u_x \ne v_y, u_y \ne -v_x\ \forall\ z \in D$.

So Cauchy-Riemann equations are not satisfied.

Hence $f'(z)$ does not exist at every point in D. So $f(z)$ is not analytic in D.

(ii) $f(z) = e^y \cdot e^{ix} = e^y[\cos x + i\sin x]$
$u(x, y) = e^y \cos x$ and $v(x, y) = e^y \sin x$.

Then $u_x = -e^y \sin x$, $v_x = e^y \cos x$
$u_y = e^y \cos x$, $v_y = e^y \sin x$

$u_x \neq v_y$ and $u_y \neq -v_x \ \forall\, z \in D$. Cauchy-Reimann equations are not satisfied; so $f'(z)$ does not exist in D and hence $f(z)$ is not analytic in D.

(5) $f(z) = \dfrac{(2z+1)}{z(z^2+1)}$.

Here $f'(z)$ exists when $z(z^2+1) \neq 0$. That is, $f'(z)$ does not exist when $z = 0$ and $z^2+1 = 0$, i.e., when $z = 0$ and $z = \pm i$. These points are called the singular points. Thus singular points are $z = 0, \ \pm i$. Now we find $f'(z)$ at $z = a$, where $a \neq 0, \pm i$.

$$f'(z) = \frac{2z(z^2+1) - (2z+1)\{(z^2+1) + z\cdot 2z\}}{z^2(z^2+1)^2} = \frac{2z^3 + 2z - (2z+1)(3z^2+1)}{z^2(z^2+1)^2}$$
$$= \frac{2z^3 + 2z - 6z^3 - 2z - 3z^2 - 1}{z^2(z^2+1)^2} = \frac{-4z^3 - 3z^2 - 1}{z^2(z^2+1)^2} = -\frac{(z+1)(4z^2 - z + 1)}{z^2(z^2+1)^2}$$

$$\Rightarrow f'(a) = \frac{-(a+1)(4a^2 - a + 1)}{a^2(a^2+1)^2} \text{ exists } \forall\, a \in D \text{ except } 0 \text{ and } \pm i.$$

For the existence of $f'(0)$ and $f'(\pm i)$,

$$\lim_{z\to 0} \frac{f(z) - f(0)}{z} \text{ and } \lim_{z\to i} \frac{f(z) - f(i)}{z - i} \text{ and } \lim_{z\to -i} \frac{f(z) - f(-i)}{z + i}$$

should exist, but $f(0), f(\pm i)$ are unbounded.

An important observation

$$z = x + iy \text{ and } \bar{z} = x - iy \Rightarrow x = \frac{z+\bar{z}}{2} \quad y = \frac{z-\bar{z}}{2i} \tag{1}$$

$$f(z) = u + iv, \text{ where } u \text{ and } v \text{ are functions of } x, y$$
$$= u\left(\frac{z+\bar{z}}{2}, \frac{z-\bar{z}}{2i}\right) + iv\left(\frac{z+\bar{z}}{2}, \frac{z-\bar{z}}{2i}\right).$$

Here u and v can be considered as the functions of two independent variables z and $\bar{z}$. The conditions that ω is independent of $\bar{z}$ is $\dfrac{\partial\omega}{\partial\bar{z}} = 0$, where $\omega = f(z)$.

$$\Rightarrow \frac{\partial}{\partial\bar{z}}(u + iv) = 0$$
$$\Rightarrow \left(\frac{\partial u}{\partial x}\cdot\frac{\partial x}{\partial\bar{z}} + \frac{\partial u}{\partial y}\cdot\frac{\partial y}{\partial\bar{z}}\right) + i\left(\frac{\partial v}{\partial x}\cdot\frac{\partial x}{\partial\bar{z}} + \frac{\partial v}{\partial y}\cdot\frac{\partial y}{\partial\bar{z}}\right) = 0$$
$$\Rightarrow \frac{1}{2}\cdot\frac{\partial u}{\partial x} - \frac{1}{2i}\frac{\partial u}{\partial y} + \frac{i}{2}\frac{\partial v}{\partial x} - \frac{1}{2}\frac{\partial v}{\partial y} = 0$$
$$\Rightarrow \frac{1}{2}\left[\frac{\partial u}{\partial x} - \frac{\partial v}{\partial y}\right] + \frac{i}{2}\left[\frac{\partial v}{\partial x} + \frac{\partial u}{\partial y}\right] = 0$$
$$\Rightarrow \frac{\partial u}{\partial x} - \frac{\partial v}{\partial y} = 0 \text{ and } \frac{\partial v}{\partial x} + \frac{\partial u}{\partial y} = 0$$
$$\Rightarrow \frac{\partial u}{\partial x} = \frac{\partial v}{\partial y} \text{ and } \frac{\partial u}{\partial y} = -\frac{\partial v}{\partial x},$$

which are Cauchy-Riemann equations. Thus, if $\omega = f(z)$ is analytic and independent of $\bar{z}$, Cauchy-Riemann equations are satisfied.

Note. (1) If $f(z) = u(x, y) + iv(x, y)$ is analytic, then the functions $u(x, y)$ and $v(x, y)$ are called conjugate functions.

Example. Let $f(z) = (3x + y) + i(3y - x)$. Here $f'(z)$ exists $\forall\ z \in D$.

So, $f(z)$ is analytic in D. Therefore, $u(x, y) = 3x + y$ and $v(x, y) = 3y - x$ are conjugate functions.

(2) If $f(z) = u(x, y) + iv(x, y)$ is analytic, then $f'(z) = u_x + iv_x = v_y - iu_y$ [using Cauchy-Riemann equations].

$$\Rightarrow |f'(z)|^2 = u_x^2 + v_x^2 = u_x \cdot u_x + v_x \cdot v_x = u_x \cdot v_y - u_y \cdot v_x = \begin{vmatrix} u_x & u_y \\ v_x & v_y \end{vmatrix}$$

$$\Rightarrow |f'(z)|^2 = \frac{\partial(u, v)}{\partial(x, y)} = J\left(\frac{u, v}{x, y}\right)$$

$\Rightarrow |f'(z)|^2$ is the *Jacobian of u and v* with respect to x and y.

(3) If $\phi(x, y)$ be a function of two variables and if it possesses continuous partial derivatives of 2nd order, then the differential equation $\frac{\partial^2 \phi}{\partial x^2} + \frac{\partial^2 \phi}{\partial y^2} = 0$ is known as *Laplace's differential equation* and $\phi(x, y)$ is called a *Harmonic function.*

(4) Two families of curves $u(x, y) = \alpha$ and $v(x, y) = \beta$ are said to form an *orthogonal system* if they intersect at right angles at each of their point of intersection.

Differentiating $u(x, y) = \alpha$, we get

$$\frac{\partial u}{\partial x} + \frac{\partial u}{\partial y} \cdot \frac{dy}{dx} = 0 \Rightarrow \frac{dy}{dx} = -\frac{\partial u}{\partial x} \Big/ \frac{\partial u}{\partial y} = m_1 \text{ (say)}.$$

Similarly, from $v(x, y) = \beta$, we get

$$\frac{dy}{dx} = -\frac{\partial v}{\partial x} \Big/ \frac{\partial v}{\partial y} = m_2 \text{ (say)}.$$

At each point of intersection, they cut orthogonally, if $m_1 \cdot m_2 = -1$

$$\Rightarrow -\left(\frac{\partial u}{\partial x} \Big/ \frac{\partial u}{\partial y}\right) \cdot -\left(\frac{\partial v}{\partial x} \Big/ \frac{\partial v}{\partial y}\right) = -1 \Rightarrow \frac{\frac{\partial u}{\partial x} \cdot \frac{\partial v}{\partial x}}{\frac{\partial u}{\partial y} \cdot \frac{\partial v}{\partial y}} = -1$$

$$\Rightarrow \frac{\partial u}{\partial x} \cdot \frac{\partial v}{\partial x} + \frac{\partial u}{\partial y} \cdot \frac{\partial v}{\partial y} = 0,$$

which is the condition of orthogonality.

Example. Let $u(x, y) = 3y - x = \alpha$ and $v(x, y) = 3x + y = \beta$.

Then $\frac{\partial u}{\partial x} \cdot \frac{\partial v}{\partial x} + \frac{\partial u}{\partial y} \cdot \frac{\partial v}{\partial y} = (-1) \cdot (3) + (3) \cdot (1) = -3 + 3 = 0.$

Therefore, $u = \alpha$ and $v = \beta$ form an orthogonal system.

Theorem [3.3] (1) Real and imaginary parts of an analytic function satisfy Laplace's equation.

(2) If the harmonic functions u and v satisfy Cauchy-Riemann equations, then $f(z) = u + iv$ is analytic.

(3) If $f(z) = u + iv$ be an analytic function of $z = x + iy$, then the curves $u(x,y) = \alpha$, $v(x,y) = \beta$, α, β are constants, form an orthogonal system.

(4) An analytic function with constant modulus is constant.

(5) If $f'(z) = 0 \; \forall \; z \in D$, then $f(z)$ must be constant in D.

Proof. (1) Let $f(z) = u(x,y) + iv(x,y)$ be analytic in a domain D.

Then Cauchy-Riemann equations are satisfied. That is, $u_x = v_y$ and $u_y = -v_x$.

$$\frac{\partial u}{\partial x} = \frac{\partial v}{\partial y} \text{ and } \frac{\partial u}{\partial y} = -\frac{\partial v}{\partial x} \Rightarrow \frac{\partial^2 u}{\partial x^2} = \frac{\partial^2 v}{\partial x \partial y} \text{ and } \frac{\partial^2 u}{\partial y^2} = -\frac{\partial^2 v}{\partial y \partial x}.$$

We suppose that $u(x,y)$ possesses continuous 2nd order partial derivatives.

Then $\dfrac{\partial^2 v}{\partial x \partial y} = \dfrac{\partial^2 v}{\partial y \partial x}$.

Therefore, $\dfrac{\partial^2 u}{\partial x^2} + \dfrac{\partial^2 u}{\partial y^2} = 0.$ (i)

Similarly, assuming that $v(x,y)$ possesses continuous 2nd order partial derivatives, we get

$$\frac{\partial^2 v}{\partial x^2} + \frac{\partial^2 v}{\partial y^2} = 0. \quad \text{(ii)}$$

So from (i) and (ii) it follows that real and imaginary parts satisfy Laplace's equations.

(2) Here $f(z) = u(x,y) + iv(x,y)$, where u and v are harmonic, i.e.,

$$\frac{\partial^2 u}{\partial x^2} + \frac{\partial^2 u}{\partial y^2} = 0 \text{ and } \frac{\partial^2 v}{\partial x^2} + \frac{\partial^2 v}{\partial y^2} = 0$$

and they satisfy Cauchy-Riemann equations $u_x = v_y$ and $u_y = -v_x$.

Therefore, the four partial derivatives u_x, u_y, v_x, v_y exist and are continuous. Hence by Th. [3.2] the function $f(z)$ is analytic.

(3) Since $f(z) = u + iv$, is analytic in a domain D, Cauchy-Riemann equations are satisfied.

That is, $u_x = v_y$ and $u_y = -v_x$.

Therefore,

$$\frac{\partial u}{\partial x} \cdot \frac{\partial v}{\partial x} + \frac{\partial u}{\partial y} \cdot \frac{\partial v}{\partial y} = \frac{\partial v}{\partial y} \cdot \left(-\frac{\partial u}{\partial y} \right) + \frac{\partial u}{\partial y} \cdot \frac{\partial v}{\partial y} = 0,$$

which is the condition of orthogonality of two curves $u(x,y) = \alpha$ and $v(x,y) = \beta$ by Note (4). Hence the result.

(4) Let $f(z) = u(x,y) + iv(x,y)$.

Then $|f(z)| = \sqrt{u^2 + v^2} = \text{constant} = k \Rightarrow u^2 + v^2 = k^2$ (constant).

Then differentiating w.r.t. x and y respectively, we get

$$2u\frac{\partial u}{\partial x} + 2v\frac{\partial v}{\partial x} = 0 \Rightarrow u\frac{\partial u}{\partial x} + v\frac{\partial v}{\partial x} = 0 \quad \text{(i)}$$

$$\text{and } 2u\frac{\partial u}{\partial y} + 2v\frac{\partial v}{\partial y} = 0 \Rightarrow u\frac{\partial u}{\partial y} + v\frac{\partial v}{\partial y} = 0. \quad \text{(ii)}$$

Since $f(z)$ is analytic, Cauchy-Riemann equations are satisfied. That is,

$$u_x = v_y, \; u_y = -v_x. \quad \text{(iii)}$$

So from (i) and (ii), we get by (iii)

$$u\frac{\partial u}{\partial x} - v\frac{\partial u}{\partial y} = 0 \tag{iv}$$

$$u\frac{\partial u}{\partial y} + v\frac{\partial u}{\partial x} = 0 \tag{v}$$

$u \times$ (iv) $+ v \times$ (v), we get $(u^2 + v^2)\dfrac{\partial u}{\partial x} = 0 \Rightarrow \dfrac{\partial u}{\partial x} = 0$ $[\because u^2 + v^2 = k^2]$

$v \times$ (iv) $- u \times$ (v) gives $-(u^2 + v^2)\dfrac{\partial u}{\partial y} = 0 \Rightarrow \dfrac{\partial u}{\partial y} = 0.$

Similarly, we can deduce that $\dfrac{\partial v}{\partial x} = 0$ and $\dfrac{\partial v}{\partial y} = 0.$

Thus $\dfrac{\partial u}{\partial x} = 0, \dfrac{\partial u}{\partial y} = 0, \dfrac{\partial v}{\partial x} = 0, \dfrac{\partial v}{\partial y} = 0$, where u and v are functions of x and y,

$\Rightarrow u(x, y) =$ constant and $v(x, y) =$ constant.

Hence, $f(z) = u + iv$ is constant.

(5) Let $f(z) = u(x, y) + iv(x, y)$

$f'(z) = u_x + iv_x = v_y - iu_y$ [by Th. (3.1)].

Since $f'(z) = 0 \;\forall\; z \in D$,

$$u_x = 0,\; u_y = 0,\; v_x = 0,\; v_y = 0 \;\forall\; (x, y) \in D$$

$\Rightarrow u(x, y) =$ constant and $v(x, y) =$ constant.

$\Rightarrow f(z)$ is constant.

Theorem [3.4] If $f(z)$ and $g(z)$ be both analytic in a domain D and if $f(z)g(z) = U(x, y) + iV(x, y)$, $z = x + iy$, then U and V are both harmonic in D.

Proof. Let $f(z) = u + iv$ and $g(z) = p + iq$, where u, v, p, q are functions of x and y.

Since $f(z)$ and $g(z)$ are both analytic, Cauchy-Riemann equations are satisfied in both cases. That is,

$$\frac{\partial u}{\partial x} = \frac{\partial v}{\partial y},\; \frac{\partial u}{\partial y} = -\frac{\partial v}{\partial x} \tag{i}$$

$$\text{and } \frac{\partial p}{\partial x} = \frac{\partial q}{\partial y},\; \frac{\partial p}{\partial y} = -\frac{\partial q}{\partial x}. \tag{ii}$$

From (i) and (ii), we get

$$\frac{\partial u}{\partial x}\cdot\frac{\partial p}{\partial x} + \frac{\partial u}{\partial y}\cdot\frac{\partial p}{\partial y} = \frac{\partial v}{\partial y}\cdot\frac{\partial q}{\partial y} + \frac{\partial v}{\partial x}\cdot\frac{\partial q}{\partial x}. \tag{iii}$$

Now $f(z)g(z) = (up - vq) + i(pv + qu) = U + iV$.

Therefore, $U = up - vq$ and $V = pv + qu$.

Then

$$\frac{\partial U}{\partial x} = \frac{\partial u}{\partial x}p + u\frac{\partial p}{\partial x} - q\frac{\partial v}{\partial x} - v\frac{\partial q}{\partial x}$$

$$\frac{\partial^2 U}{\partial x^2} = \left[p\frac{\partial^2 u}{\partial x^2} + 2\frac{\partial u}{\partial x}\frac{\partial p}{\partial x} + u\frac{\partial^2 p}{\partial x^2} - 2\frac{\partial v}{\partial x}\frac{\partial q}{\partial x} - q\frac{\partial^2 v}{\partial x^2} - v\frac{\partial^2 q}{\partial x^2}\right].$$

Similarly, we get

$$\frac{\partial^2 U}{\partial y^2} = \left[p\frac{\partial^2 u}{\partial y^2} + 2\frac{\partial u}{\partial y}\frac{\partial p}{\partial y} + u\frac{\partial^2 p}{\partial y^2} - 2\frac{\partial v}{\partial y}\frac{\partial q}{\partial y} - q\frac{\partial^2 v}{\partial y^2} - v\frac{\partial^2 q}{\partial y^2}\right].$$

Therefore,

$$\begin{aligned}\frac{\partial^2 U}{\partial x^2} + \frac{\partial^2 U}{\partial y^2} &= p\left[\frac{\partial^2 u}{\partial x^2} + \frac{\partial^2 u}{\partial y^2}\right] + 2\left[\frac{\partial u}{\partial x}\frac{\partial p}{\partial x} + \frac{\partial u}{\partial y}\frac{\partial p}{\partial y}\right] + u\left[\frac{\partial^2 p}{\partial x^2} + \frac{\partial^2 p}{\partial y^2}\right] \\ &\quad -2\left[\frac{\partial v}{\partial x}\frac{\partial q}{\partial x} + \frac{\partial v}{\partial y}\frac{\partial q}{\partial y}\right] - q\left[\frac{\partial^2 v}{\partial x^2} + \frac{\partial^2 v}{\partial y^2}\right] - v\left[\frac{\partial^2 q}{\partial x^2} + \frac{\partial^2 q}{\partial y^2}\right] \\ &= 0 \text{ [by (iii) and by harmonic property of } u, v, p \text{ and } q].\end{aligned}$$

Similarly, we can show that $\dfrac{\partial^2 V}{\partial x^2} + \dfrac{\partial^2 V}{\partial y^2} = 0.$

Thus U and V are both harmonic in D.

Example. Let $f(z) = e^{-y}[\sin x - i\cos x] = u + iv$ and

$$g(z) = z^2 = (x+iy)^2 = (x^2 - y^2) + i2xy = p + iq.$$

Then $f(z)$ and $g(z)$ are both analytic in a domain D, since u, v, p, q satisfy Cauchy-Riemann equations.

Here $u_x = e^{-y}\cos x,\ v_x = e^{-y}\sin x$

$u_y = -e^{-y}\sin x,\ v_y = e^{-y}\cos x$

$\Rightarrow u_x = v_y$ and $u_y = -v_x$.

Also for $p = (x^2 - y^2)$ and $q = 2xy$

$$\left.\begin{aligned} p_x &= 2x,\ q_x = 2y \\ p_y &= -2y\ q_y = 2x \end{aligned}\right\} \Rightarrow p_x = q_y \text{ and } p_y = -q_x.$$

Also all the partial derivatives exist and are continuous in D.

The product function

$$\begin{aligned} f(z)g(z) &= e^{-y}\{(x^2 - y^2)\sin x + 2xy\cos x\} + ie^{-y}\{2xy\sin x - (x^2 - y^2)\cos x\} \\ &= U + iV. \end{aligned}$$

Then Re $\{f(z)g(z)\}$ and Im $\{f(z)g(z)\}$ are both harmonic.

Here

$$\begin{aligned} \frac{\partial U}{\partial x} &= e^{-y}[(1-y)2x\sin x + (x^2 - y^2 + 2y)\cos x]. \\ \frac{\partial^2 U}{\partial x^2} &= e^{-y}[2(1-y)\sin x + (1-y)2x\cos x + 2x\cos x - (x^2 - y^2 + 2y)\sin x] \\ &= e^{-y}[(2 - 4y + y^2 - x^2)\sin x + 2(2-y)x\cos x]. \\ \frac{\partial U}{\partial y} &= -e^{-y}[(x^2 - y^2)\sin x + 2xy\cos x] + e^{-y}[-2y\sin x + 2x\cos x] \\ &= e^{-y}[(y^2 - x^2 - 2y)\sin x + 2(1-y)x\cos x]. \end{aligned}$$

$$\begin{aligned}\frac{\partial^2 U}{\partial y^2} &= -e^{-y}[(y^2 - x^2 - 2y)\sin x + 2(1-y)x\cos x] + e^{-y}[(2y-2)\sin x - 2x\cos x] \\ &= e^{-y}[(4y - y^2 + x^2 - 2)\sin x - 2(2-y)x\cos x] \\ &= -e^{-y}[(2 - 4y + y^2 - x^2)\sin x + 2(2-y)x\cos x].\end{aligned}$$

$$\therefore \frac{\partial^2 U}{\partial x^2} + \frac{\partial^2 U}{\partial y^2} = 0$$

Similarly, we can show that $\dfrac{\partial^2 V}{\partial x^2} + \dfrac{\partial^2 V}{\partial y^2} = 0.$

3.1 Methods of Construction of Analytic Function

(1) **Mini-Thomson's method.** We have $z = x + iy$, $\bar{z} = x - iy$.

Then $x = \dfrac{z + \bar{z}}{2}, y = \dfrac{z - \bar{z}}{2i}$, $f(z) = u(x, y) + iv(x, y)$ and let $u(x, y)$ be given.

Then $f(z) = u\left(\dfrac{z+\bar{z}}{2}, \dfrac{z-\bar{z}}{2i}\right) + iv\left(\dfrac{z+\bar{z}}{2}, \dfrac{z-\bar{z}}{2i}\right).$ (i)

(i) can be considered as an identity of two independent variables z and $\bar{z}$. Then for $z = \bar{z}$, we get from (i)

$$f(z) = u(z, 0) + iv(z, 0) \quad \text{(ii)}$$

Again, $f'(z) = u_x + iv_x = u_x - iu_y$ [since $u_y = -v_x$].

Let $u_x(x, y) = \phi_1(x, y)$ and $u_y(x, y) = \phi_2(x, y)$.

Then $f'(z) = \phi_1(x, y) - i\phi_2(x, y) = \phi_1(z, 0) - i\phi_2(z, 0)$.

Integrating, we get

$$f(z) = \int \phi_1(z, 0)dz - i\int \phi_2(z, 0)dz + c,$$

where c is constant of integration.

Similarly, when $v(x, y)$ is given, we have

$$f(z) = \int \psi_1(z, 0)dz + \int \psi_2(z, 0)dz + d,$$

where $f'(z) = u_x + iv_x = v_y + iv_x$ and $v_y = \psi_1(x, y)$ and $v_x = \psi_2(x, y)$.

Example. Find the analytic function $f(z) = u + iv$, where $u = x^3 - 3xy^2 + 3x^2 - 3y^2 + 1$.

$$f'(z) = u_x + iv_x = u_x - iu_y = (3x^2 - 3y^2 + 6x) - i(-6xy - 6y).$$

Let $\phi_1(x, y) = 3x^2 - 3y^2 + 6x$ and $\phi_2(x, y) = (6xy + 6y)$.

Then $\phi_1(z, 0) = 3z^2 + 6z$, $\phi_2(z, 0) = 0$.

Therefore, $f'(z) = (3z^2 + 6z) + i \cdot 0$.

Integrating, we get

$$\begin{aligned}f(z) &= \int (3z^2 + 6z)dz + c \ [c = \text{constant}] \\ &= z^3 + 3z^2 + c = (x + iy)^3 + 3(x + iy)^2 + c \\ &= x^3 + 3ix^2y - 3xy^2 - iy^3 + 3x^2 + 6ixy - 3y^2 + c \\ &= (x^3 - 3xy^2 + 3x^2 - 3y^2) + i(6xy + 3x^2y - y^3) + c.\end{aligned}$$

(2) Using Cauchy-Riemann equations. Let $f(z) = u(x,y) + iv(x,y)$ and let $u(x,y)$ is given. We have to find out $v(x,y)$.

We have $dv = \frac{\partial v}{\partial x}dx + \frac{\partial v}{\partial y}dy$.

By Cauchy-Riemann equations $dv = -\frac{\partial u}{\partial y}dx + \frac{\partial u}{\partial x}dy$, which is in the form

$$dv = Mdx + Ndy. \tag{i}$$

Here $M = -\frac{\partial u}{\partial y}$ and $N = \frac{\partial u}{\partial x}$

$$\Rightarrow \frac{\partial M}{\partial y} - \frac{\partial N}{\partial x} = -\frac{\partial^2 u}{\partial y^2} - \frac{\partial^2 u}{\partial x^2} = 0 \text{ [Since } u \text{ is a harmonic function]}$$

$$\Rightarrow \frac{\partial M}{\partial y} = \frac{\partial N}{\partial x}.$$

So (i) is exact differential equation and hence v can be obtained.

Example. Let $u(x,y) = x^3 - 3xy^2 + 3x^2 - 3y^2 + 1$.

Let $f(z) = u(x,y) + iv(x,y)$ be analytic in D.

$$\begin{aligned} dv &= \frac{\partial v}{\partial x}dx + \frac{\partial v}{\partial y}dy \\ &= -\frac{\partial u}{\partial y}dx + \frac{\partial u}{\partial x}dy \text{ [by Cauchy-Riemann equations]} \\ &= (6xy + 6y)dx + (3x^2 - 3y^2 + 6x)dy. \end{aligned} \tag{i}$$

$$\left.\begin{matrix} M = 6(xy+y) & N = (3x^2 - 3y^2 + 6x) \\ \frac{\partial M}{\partial y} = 6x + 6 & \frac{\partial N}{\partial x} = 6x + 6 \end{matrix}\right\} \Rightarrow \frac{\partial M}{\partial y} = \frac{\partial N}{\partial x}.$$

Equation is exact.

$$\int Mdx = 6\int (xy + y)dx = 6y \cdot \frac{x^2}{2} + 6y \cdot x = 3yx^2 + 6xy. \tag{ii}$$

$$\int Ndy = \int (3x^2 - 3y^2 + 6x)dy = 3x^2y - y^3 + 6xy. \tag{iii}$$

From (i), $v = 3yx^2 + 6xy - y^3 + c$, considering the repeated terms in (ii) and (iii) once only.

Therefore,

$$\begin{aligned} f(z) &= (x^3 - 3xy^2 + 3x^2 - 3y^2 + 1) + i(3yx^2 + 6xy - y^3 + c) \\ &= (x^3 - 3xy^2 + 3x^2 - 3y^2) + i(3yx^2 + 6xy - y^3) + 1 + ic \\ \text{or, } f(z) &= (x+iy)^3 + 3(x+iy)^2 + (1+ic) \\ &= (x+iy)^3 + 3(x+iy)^2 + d, \text{ where } d = 1 + ic = \text{constant.} \end{aligned}$$

Method (3) : Using difference of two conjugate functions u and v.

Let $f(z) = u + iv$ be analytic in a domain D and let $u - v = \phi(x,y)$. Then determine $f(z)$.

$$u - v = \phi(x,y) \Rightarrow \frac{\partial u}{\partial x} - \frac{\partial v}{\partial x} = \phi_x(x,y) \tag{i}$$

$$\text{and } \frac{\partial u}{\partial y} - \frac{\partial v}{\partial y} = \phi_y(x,y) \text{ or, } -\frac{\partial v}{\partial x} - \frac{\partial u}{\partial x} = \phi_y(x,y) \text{ or, } \frac{\partial v}{\partial x} + \frac{\partial u}{\partial x} = -\phi_y(x,y), \tag{ii}$$

(i) + (ii) gives

$$\frac{\partial u}{\partial x} = \frac{1}{2}[\phi_x(x,y) - \phi_y(x,y)] = \phi_1(x,y) \text{ (say)} \tag{iii}$$

and (i) − (ii) gives

$$\frac{\partial v}{\partial x} = -\frac{1}{2}[\phi_x(x,y) + \phi_y(x,y)] = \phi_2(x,y) \text{ (say)}. \tag{iv}$$

Therefore,

$$f'(z) = \frac{\partial u}{\partial x} + i\frac{\partial v}{\partial x} = \phi_1(x,y) + i\phi_2(x,y)$$

$$\Rightarrow f'(z) = \phi_1(z,0) + i\phi_2(z,0). \text{ [replacing } x \text{ by } z,\ y \text{ by } 0]$$

Integrating, we get $f(z) = \int \phi_1(z,0)dz + i\int \phi_2(z,0)dz + c.$

Alternative

From (iii) and (iv) we have $\frac{\partial u}{\partial x} = \phi_1(x,y)$ and $\frac{\partial v}{\partial x} = \phi_2(x,y)$.

On integrating, we get

$$v = \int \phi_2(x,y)dx + f(y), \tag{v}$$

where the integrating constant is a function of y.

Now differentiating w.r.t. y, we get

$$\frac{\partial v}{\partial y} = \int \frac{\partial}{\partial y}\phi_2(x,y)dx + f'(y)$$

$$\Rightarrow \frac{\partial u}{\partial x} = \int \frac{\partial}{\partial y}\phi_2(x,y)dx + f'(y) \quad \left[\because \frac{\partial u}{\partial x} = \frac{\partial v}{\partial y}\right]$$

$$\Rightarrow \phi_1(x,y) = \int \frac{\partial}{\partial y}\phi_2(x,y)dx + f'(y),$$

which will determine $f(y)$.

Therefore, from (v)

$$v = \int \phi_2(x,y)dx + f(y) \text{ and } u = v + \phi(x,y) = \int \phi_2(x,y)dx + \phi(x,y) + f(y)$$

and hence we can determine $f(z) = u + iv$.

Example. Let $f(z) = u + iv$ be analytic in D, and $u - v = e^x(\cos y - \sin y)$. Then find $f(z)$.

Applying alternative method

From $u - v = e^x(\cos y - \sin y)$,

$$\frac{\partial u}{\partial x} - \frac{\partial v}{\partial x} = e^x(\cos y - \sin y) \tag{i}$$

and $\frac{\partial u}{\partial y} - \frac{\partial v}{\partial y} = e^x(-\sin y - \cos y)$ or, $-\frac{\partial v}{\partial x} - \frac{\partial u}{\partial x} = e^x(-\sin y - \cos y)$

$$\text{or, } \frac{\partial v}{\partial x} + \frac{\partial u}{\partial x} = e^x(\sin y + \cos y). \tag{ii}$$

(i) + (ii) gives

$$\frac{\partial u}{\partial x} = e^x \cos y \tag{iii}$$

and (i) – (ii) gives

$$\frac{\partial v}{\partial x} = e^x \sin y. \tag{iv}$$

Now integrating (iv), we get

$$v = \int e^x \sin y dx + \text{constant} = e^x \sin y + f(y), \tag{v}$$

where constant is a function of y.

Differentiating it w.r.t. y, we get

$$\frac{\partial v}{\partial y} = e^x \cos y + f'(y)$$

$$\Rightarrow \frac{\partial u}{\partial x} = e^x \cos y + f'(y) \quad \left[\because \frac{\partial u}{\partial x} = \frac{\partial v}{\partial y}\right]$$

$$\Rightarrow e^x \cos y = e^x \cos y + f'(y) \Rightarrow f'(y) = 0 \Rightarrow f(y) = c.$$

$$\therefore\ v = e^x \sin y + c \text{ [from (v)]}$$

Also,

$$u - v = e^x(\cos y - \sin x)$$

$$\Rightarrow u = v + e^x(\cos y - \sin x) = e^x \sin y + c + e^x(\cos y - \sin y)$$

$$= e^x \cos y + c.$$

Therefore,

$$f(z) = u + iv = e^x \cos y + c + i(e^x \sin y + c)$$

$$= e^x(\cos y + i \sin y) + c + ic = e^x \cdot e^{iy} + c = e^z + d,$$

where $d = (c + ic)$.

Applying direct Method (3)

$$f'(z) = \frac{\partial u}{\partial x} + i\frac{\partial v}{\partial x} = e^x \cos y + ie^x \sin y \text{ [by above (iii) and (iv)]}$$

$$f'(z) = e^z \ [\because x = z, y = 0].$$

Integrating $f(z) = e^z + c$, here c is a constant of integration in the form $a + ib$, where a, b are constant real numbers.

Problem [3.3]: (1) Determine $f(z) = u + iv$ by obtaining a harmonic conjugate of a given harmonic function $u(x, y) = y^3 - 3x^2y$.

(2) Show that $u(x, y) = \sinh x \sin y$ is harmonic in some domain D and find harmonic conjugate $v(x, y)$ of $u(x, y)$.

(3) If $f(z)$ be analytic in a domain D, then show that $f(z)$ must be constant in D, if (i) $f(z)$ is real valued $\forall\ z \in D$, (ii) $\overline{f(z)}$ is analytic in D, (iii) $\arg f(z) =$ constant.

(4) If $u + v = \dfrac{2 \sin 2x}{e^{2y} - e^{-2y} - 2 \cos 2x}$, find the analytic function $f(z) = u + iv$.

(5) If $u = \dfrac{\sin 2x}{\cosh 2y + \cos 2x}$, then find the analytic function $f(z) = u + iv$.

(6) Show that $u(x, y) = \dfrac{1}{2} \log(x^2 + y^2)$ is harmonic and find its harmonic conjugate.

(7) If $f(z)$ be a analytic function of z, then show that $\left(\dfrac{\partial^2}{\partial x^2} + \dfrac{\partial^2}{\partial y^2}\right) |\operatorname{Re} f(z)|^2 = 2|f'(z)|^2$.

(8) If $f(z)$ be analytic functions of $z = x + iy \in D$, and $\bar{z} = x - iy$, then $\dfrac{\partial f}{\partial \bar{z}} = 0$.

(9) Show that (i) if v and V are harmonic conjugates of u in a domain D, then $v(x, y)$ and $V(x, y)$ can differ by at most an additive constant, (ii) if v is harmonic conjugate of u in a domain D and u is a harmonic conjugate of v, then $u(x, y)$ and $v(x, y)$ must be constant in D, (iii) Give an example that v is a harmonic conjugate of u but u cannot be a harmonic conjugate of v

(10) Prove that if $u = x^2 - y^2$, $v = \dfrac{-x}{x^2 + y^2}$ satisfy Laplace's equations, but $f(z) = u + iv$ is not analytic function of z.

Solution. (1) $f(z) = u + iv,\ u = y^3 - 3x^2y$

$$u_x = -6xy,\ u_y = 3y^2 - 3x^2 = 3(y^2 - x^2).$$

By Cauchy-Riemann equations, $u_x = v_y$ and $u_y = -v_x$.

Then $v_y = -6xy$ and $-v_x = 3(y^2 - x^2)$. (i)

Integrating w.r.t. y, we get

$$v = -3xy^2 + \phi(x) \text{ [Here constant of integration is a function of } x] \quad \text{(ii)}$$

$$\Rightarrow v_x = -3y^2 + \phi'(x) \Rightarrow -3(y^2 - x^2) = -3y^2 + \phi'(x) \text{ [by (i)]}$$

$$\Rightarrow \phi'(x) = 3x^2 \Rightarrow \phi(x) = x^3 + c.$$

Hence from (ii), $v = -3xy^2 + x^3 + c$, harmonic conjugate of u.

Therefore,

$$\begin{aligned} f(z) &= u + iv = y^3 - 3x^2y + i(-3xy^2 + x^3) + ic \\ &= i[x^3 - 3xy^2 + i3x^2y - iy^3] + ic \\ &= i[x + iy]^3 + ic = i(z^3 + c). \end{aligned}$$

(2) $u(x, y) = \sinh x \sin y$

$$\frac{\partial u}{\partial x} = \cosh x \sin y,\quad \frac{\partial u}{\partial y} = \sinh x \cos y \quad \text{(i)}$$

$$\frac{\partial^2 u}{\partial x^2} = \sinh x \sin y,\quad \frac{\partial^2 u}{\partial y^2} = -\sinh x \sin y$$

$$\Rightarrow \frac{\partial^2 u}{\partial x^2} + \frac{\partial^2 u}{\partial y^2} = 0$$

$\Rightarrow$ u is harmonic.

We have $\dfrac{\partial v}{\partial y} = \cosh x \sin y$ [by (i)]

Integrating w.r.t. y, we get

$$v = -\cosh x \cos y + \phi(x) \text{ [Here constant of integration is a function of } x] \quad \text{(ii)}$$

$$\Rightarrow \frac{\partial v}{\partial x} = -\sinh x \cos y + \phi'(x)$$

$$\Rightarrow -\sinh x \cos y = -\sinh x \cos y + \phi'(x) \text{ [by Cauchy-Riemann equation, } u_y = -v_x]$$

$$\Rightarrow \phi'(x) = 0 \Rightarrow \phi(x) = c.$$

Therefore, $v = -\cosh x \cos y + c$, a harmonic conjugate of u.

(3) (i) $f(z)$ is analytic. So Cauchy-Riemann equations hold. That is, $u_x = v_y$ and $u_y = -v_x$, $f(z) = u + iv$. Now if $f(z)$ is real valued $\forall\, z \in D$, then $v = 0$.

$$\left.\begin{aligned} &\Rightarrow \frac{\partial v}{\partial x} = 0,\ \frac{\partial v}{\partial y} = 0 \\ &\Rightarrow \frac{\partial u}{\partial y} = 0,\ \frac{\partial u}{\partial x} = 0 \end{aligned}\right\} \qquad (1)$$

Thus, $\dfrac{\partial u}{\partial x} = 0,\ \dfrac{\partial u}{\partial y} = 0,\ \dfrac{\partial v}{\partial x} = 0,\ \dfrac{\partial v}{\partial y} = 0.$

$\Rightarrow u(x, y)$ and $v(x, y)$ are constant.

$\Rightarrow f(z)$ is constant.

(ii) $f(z) = u + iv$. Since $f(z)$ is analytic,

$$u_x = v_y \text{ and } u_y = -v_x. \qquad (1)$$

Let $\overline{f(z)} = u - iv$ be analytic. Then Cauchy-Riemann equations are satisfied.

So $\dfrac{\partial u}{\partial x} = -\dfrac{\partial v}{\partial y}$ and $\dfrac{\partial u}{\partial y} = -\left(-\dfrac{\partial v}{\partial x}\right) = \dfrac{\partial v}{\partial x}$.

Using (1), we get

$$\frac{\partial v}{\partial y} = -\frac{\partial v}{\partial y} \Rightarrow \frac{\partial v}{\partial y} = 0 \text{ and } -\frac{\partial v}{\partial x} = \frac{\partial v}{\partial x} \Rightarrow \frac{\partial v}{\partial x} = 0. \qquad (2)$$

$\therefore v$ is constant.

Also $\dfrac{\partial u}{\partial x} = 0,\ \dfrac{\partial u}{\partial y} = 0$ [by (2) from (1)]

$\Rightarrow u$ is constant.

Hence $f(z)$ is constant.

(iii) $\arg f(z) =$ constant, where $f(z) = u(x, y) + iv(x, y)$

$\Rightarrow \tan^{-1}\dfrac{v}{u} =$ constant $= k$ (say)

$\Rightarrow \dfrac{v}{u} = \tan k \Rightarrow v = u\tan k = cu$, where $c = \tan k$ is a real constant. (1)

Since $f(z)$ is analytic, Cauchy-Riemann equations hold.

That is, $u_x = v_y$ and $u_y = -v_x$. (2)

From (1) we have $v_x = cu_x$ and $v_y = cu_y$ (3)

Thus $cu_x = -u_y$ and $cu_y = u_x$ [by (2) and (3)]

$$\Rightarrow cu_x + u_y = 0 \qquad (4)$$

$$u_x - cu_y = 0. \qquad (5)$$

$(4) \times c + (5) \Rightarrow (c^2 + 1)u_x = 0 \Rightarrow u_x = 0$ [$c^2 + 1 \neq 0$, since c is real].

$(5) \times c - (4) \Rightarrow (c^2 + 1)u_y = 0 \Rightarrow u_y = 0$.

From (3), $v_y = 0,\ v_x = 0$.

Hence $u_x = 0,\ u_y = 0,\ v_x = 0,\ v_y = 0$.

Therefore, u and v are constant and hence $f(z)$ is constant.

(4) Given that

$$u+v=\frac{2\sin 2x}{e^{2y}-e^{-2y}-2\cos 2x}=\frac{2\sin 2x}{2\sinh 2y-2\cos 2x}=\frac{\sin 2x}{\sinh 2y-\cos 2x}. \tag{1}$$

$$f(z)=u+iv.$$

Now from (1),

$$u_x+v_x=\frac{2\cos 2x(\sinh 2y-\cos 2x)+2\sin^2 2x}{(\sinh 2y-\cos 2x)^2}. \tag{2}$$

Also $u_y+v_y=-\dfrac{\sin 2x\cdot 2\cosh 2y}{(\sinh 2y-\cos 2x)^2}$

$$\Rightarrow -v_x+u_x=-\frac{2\sin 2x\cosh 2y}{(\sinh 2y-\cos 2x)^2} \quad \left[\begin{array}{l}\text{Since } f(z) \text{ is analytic, Cauchy-}\\ \text{Riemann equations hold.}\end{array}\right] \tag{3}$$

(2) + (3) gives

$$2u_x=\frac{2\cos 2x(\sinh 2y-\cos 2x)+2\sin^2 2x-2\sin 2x\cosh 2y}{(\sinh 2y-\cos 2x)^2}=\phi_1(x,y).$$

(2) − (3) gives

$$2v_x=\frac{2\cos 2x(\sinh 2y-\cos 2x)+2\sin^2 2x+2\sin 2x\cosh 2y}{(\sinh 2y-\cos 2x)^2}=\phi_2(x,y).$$

Then, $2f'(z)=2u_x+i2v_x=\phi_1(x,y)+i\phi_2(x,y)$.

By Mini-Thomas Method,

$$2\int f'(z)dz=\int\phi_1(z,0)dz+\int\phi_2(z,0)dz$$

$$\Rightarrow 2\int f'(z)dz=\int\frac{-2\cos^2 2z+2\sin^2 2z-2\sin 2z}{\cos^2 2z}dz$$
$$+i\int\frac{-2\cos^2 2z+2\sin^2 2z+2\sin 2z}{\cos^2 2z}dz$$

$$\Rightarrow 2\int f'(z)dz=\int\frac{2-4\cos^2 2z-2\sin 2z}{\cos^2 2z}dz+i\int\frac{2-4\cos^2 2z+2\sin 2z}{\cos^2 2z}dz$$

$$\Rightarrow 2\int f'(z)dz=\int 2\sec^2 2z dz-4\int dz-\int 2\tan 2z\sec 2z dz$$
$$+i\left\{\int 2\sec^2 2z dz-4\int dz+\int 2\tan 2z\sec 2z dz\right\}$$

$$\Rightarrow 2f(z)=(\tan 2z-4z-\sec 2z)+i(\tan 2z-4z+\sec 2z)+c$$

$$\Rightarrow f(z)=\frac{1}{2}[(1+i)\tan 2z-4z(1+i)-(1-i)\sec 2z]+c.$$

Alternative Method

Let $f(z)=u+iv$.

Then $(1+i)f(z)=(u-v)+i(u+v)=U+iV$, where $U=u-v$ and $V=u+v$.

Then we have $V = \dfrac{\sin 2x}{\sinh 2y - \cos 2x}$.

$$\frac{\partial V}{\partial x} = \frac{2\cos 2x(\sinh 2y - \cos 2x) + 2\sin^2 2x}{(\sinh 2y - \cos 2x)^2} = \phi_2(x,y)$$

$$\frac{\partial V}{\partial y} = \frac{-2\sin 2x\cosh 2y}{(\sinh 2y - \cos 2x)^2} = \frac{\partial V}{\partial x} = \phi_1(x,y).$$

$$\left[\because f(z) \text{ is analytic} \Rightarrow (1+i)f(z) \text{ is analytic} \Rightarrow \frac{\partial U}{\partial x} = \frac{\partial V}{\partial y}\right]$$

Then by Mini-Thomson's method, $(1+i)f'(z) = U_x + iV_x$

$$\Rightarrow (1+i)\int f'(z)dz = \int \phi_1(z,0)dz + i\int \phi_2(z,0)dz$$

$$= \int \frac{-2\sin 2z}{\cos^2 2z}dz + i\int \frac{-2\cos^2 2z + 2\sin^2 2z}{\cos^2 2z}dz$$

$$= -2\int \tan 2z \sec 2z dz + i\int -2dz + i\int 2\tan^2 2z dz$$

$$= -\sec 2z - 2iz + i\tan 2z - 2iz + c$$

$$\Rightarrow (1+i)f(z) = -\sec 2z - 4iz + i\tan 2z + c$$

$$\Rightarrow f(z) = -\frac{1}{(1+i)}\sec 2z - \frac{4iz}{1+i} + \frac{i}{1+i}\tan 2z + \frac{c}{1+i}$$

$$= -\frac{1}{2}(1-i)\sec 2z - 2i(1-i)z + \frac{i}{2}(1-i)\tan 2z + \frac{1}{2}(1-i)c$$

$$= -\frac{1}{2}(1-i)\sec 2z - 2(1+i)z + \frac{1}{2}(1+i)\tan 2z + \frac{1}{2}(1-i)c.$$

(5) Let $f(z) = u + iv$ be an analytic function of z, $u = \dfrac{\sin 2x}{\cosh 2y + \cos 2x}$.

Then

$$\frac{\partial u}{\partial x} = \frac{2\cos 2x(\cosh 2y + \cos 2x) + \sin 2x \cdot 2\sin 2x}{(\cosh 2y + \cos 2x)^2}$$

$$= \frac{2 + 2\cosh 2y\cos 2x}{(\cosh 2y + \cos 2x)^2} = \phi_1(x,y)$$

$$\frac{\partial u}{\partial y} = -\frac{\sin 2x \cdot 2\sinh 2y}{(\cosh 2y + \cos 2x)^2}$$

$$\Rightarrow \frac{\partial v}{\partial x} = \frac{2\sinh 2y\sin 2x}{(\cosh 2y + \cos 2x)^2} = \phi_2(x,y) \quad \left[\because \frac{\partial u}{\partial y} = -\frac{\partial v}{\partial x}\right].$$

Therefore, $f'(z) = \dfrac{\partial u}{\partial x} + i\dfrac{\partial v}{\partial x}$.

Then by Mini-Thomson's method,

$$\int f'(z)dz = \int \phi_1(z,0)dz + i\int \phi_2(z,0)dz$$

$$\Rightarrow \int f'(z)dz = \int \frac{2 + 2\cos 2z}{(1+\cos 2z)^2}dz + i\cdot 0$$

$$\Rightarrow f(z) = 2\int \frac{1}{1+\cos 2z}dz = 2\int \frac{1}{2\cos^2 z}dz = \int \sec^2 z dz = \tan z + c.$$

(6) $u = \frac{1}{2}\log(x^2 + y^2)$.

Then $\frac{\partial u}{\partial x} = \frac{x}{x^2+y^2}$, $\frac{\partial^2 u}{\partial x^2} = \frac{1(x^2+y^2) - 2x^2}{(x^2+y^2)^2} = \frac{y^2 - x^2}{(x^2+y^2)^2}$.

Similarly, $\frac{\partial^2 u}{\partial y^2} = \frac{x^2 - y^2}{(x^2+y^2)^2}$.

Therefore, $\frac{\partial^2 u}{\partial x^2} + \frac{\partial^2 u}{\partial y^2} = 0$.

$\Rightarrow u$ is harmonic.

Let $f(z) = u + iv$ be analytic function of z.

$$\frac{\partial u}{\partial x} = \frac{x}{x^2+y^2}$$

$$\frac{\partial u}{\partial y} = \frac{y}{x^2+y^2} = -\frac{\partial v}{\partial x} \quad \left[\because \frac{\partial u}{\partial y} = -\frac{\partial v}{\partial x}\right]$$

$$\Rightarrow \frac{\partial v}{\partial x} = -\frac{y}{(x^2+y^2)}.$$

Integrating w.r.t. x,

$$v = \int -\frac{y}{x^2+y^2}dx + \phi(y) \quad \left[\begin{array}{l}\text{Constant of integration}\\ \text{is a function of } y.\end{array}\right]$$

$$= -y \cdot \frac{1}{y}\tan^{-1}\frac{x}{y} + \phi(y) = -\tan^{-1}\frac{x}{y} + \phi(y). \qquad \text{(i)}$$

$$\frac{\partial v}{\partial y} = -\frac{1}{1+\frac{x^2}{y^2}}\left(-\frac{x}{y^2}\right) + \phi'(y)$$

$$\Rightarrow \frac{\partial u}{\partial x} = \frac{x}{y^2+x^2} + \phi'(y) \Rightarrow \frac{x}{x^2+y^2} = \frac{x}{x^2+y^2} + \phi'(y)$$

$$\Rightarrow \phi'(y) = 0 \Rightarrow \phi(y) = c.$$

Therefore, $v = -\tan^{-1}\frac{x}{y} + c$ from (i) which is the harmonic conjugate of u.

(7) Let $f(z) = u + iv$.

$\text{Re } f(z) = u$, $|\text{Re } f(z)|^2 = u^2$.

Then $\left(\frac{\partial^2}{\partial x^2} + \frac{\partial^2}{\partial y^2}\right)|\text{Re } f(z)|^2$

$$= \left(\frac{\partial^2}{\partial x^2} + \frac{\partial^2}{\partial y^2}\right)u^2 = \frac{\partial}{\partial x}\left[2u\frac{\partial u}{\partial x}\right] + \frac{\partial}{\partial y}\left[2u\frac{\partial u}{\partial y}\right]$$

$$= 2\left(\frac{\partial u}{\partial x}\right)^2 + 2u\frac{\partial^2 u}{\partial x^2} + 2\left(\frac{\partial u}{\partial y}\right)^2 + 2u \cdot \frac{\partial^2 u}{\partial y^2}$$

$$= 2\left[\left(\frac{\partial u}{\partial x}\right)^2 + \left(\frac{\partial u}{\partial y}\right)^2\right] + 2u\left[\frac{\partial^2 u}{\partial x^2} + \frac{\partial^2 u}{\partial y^2}\right] = 2\left[\left(\frac{\partial u}{\partial x}\right)^2 + \left(\frac{\partial u}{\partial y}\right)^2\right] + 0$$

$$= 2\left[\left(\frac{\partial u}{\partial x}\right)^2 + \left(\frac{\partial u}{\partial y}\right)^2\right]. \text{ [Since } f(z) \text{ is analytic, } u \text{ is harmonic.]} \qquad (1)$$

Again,

$$f'(z) = \frac{\partial u}{\partial x} + i\frac{\partial v}{\partial x} = \frac{\partial u}{\partial x} - i\frac{\partial u}{\partial y} \quad \left[\because \frac{\partial u}{\partial y} = -\frac{\partial v}{\partial x}\right]$$

$$|f'(z)|^2 = \left(\frac{\partial u}{\partial x}\right)^2 + \left(\frac{\partial u}{\partial y}\right)^2. \tag{2}$$

So from (1) and (2),

$$\left(\frac{\partial^2}{\partial x^2} + \frac{\partial^2}{\partial y^2}\right) |\text{Re } f(z)|^2 = 2|f'(z)|^2.$$

(8) $f(z) = u(x,y) + iv(x,y)$, where

$$z = x + iy,\ \bar{z} = x - iy \Rightarrow x = \frac{z+\bar{z}}{2},\ y = \frac{z-\bar{z}}{2i}.$$

Then

$$\begin{aligned}\frac{\partial f}{\partial \bar{z}} &= \left(\frac{\partial u}{\partial x}\frac{\partial x}{\partial \bar{z}} + \frac{\partial u}{\partial y}\cdot\frac{\partial y}{\partial \bar{z}}\right) + i\left(\frac{\partial v}{\partial x}\frac{\partial x}{\partial \bar{z}} + \frac{\partial v}{\partial y}\cdot\frac{\partial y}{\partial \bar{z}}\right)\\ &= \left(\frac{1}{2}\frac{\partial u}{\partial x} - \frac{1}{2i}\frac{\partial u}{\partial y}\right) + i\left(\frac{1}{2}\frac{\partial v}{\partial x} - \frac{1}{2i}\frac{\partial v}{\partial y}\right)\\ &= \frac{1}{2}\frac{\partial u}{\partial x} + \frac{i}{2}\frac{\partial u}{\partial y} + \frac{i}{2}\frac{\partial v}{\partial x} - \frac{1}{2}\frac{\partial v}{\partial y}\\ &= \frac{1}{2}\frac{\partial v}{\partial y} - \frac{i}{2}\frac{\partial v}{\partial x} + \frac{i}{2}\frac{\partial v}{\partial x} - \frac{1}{2}\frac{\partial v}{\partial y} = 0.\end{aligned}$$

[Since $f(z)$ is analytic function of z, Cauchy-Riemann's equations are satisfied, that is, $u_x = v_y$, $u_y = -v_x$].

(9) (i) By the given condition, let $f(z) = u + iv$ and $f(z) = u + iV$. Then $f(z)$ is analytic in both cases, where v, V are conjugate to u

$$f'(z) = u_x + iv_x = u_x + iV_x \Rightarrow v_x = V_x. \tag{1}$$

Also

$$f'(z) = v_y - iu_y = V_y - iu_y \Rightarrow v_y = V_y. \tag{2}$$

Integrating (1) w.r.t. x, we get

$$v = V + \phi(y), \quad \left[\begin{array}{l}\text{where constant of integration}\\ \text{is a function of } y\end{array}\right] \tag{3}$$

Then $v_y = V_y + \phi'(y)$

$\Rightarrow \phi'(y) = 0$ [by (2)]

$\Rightarrow \phi(y) = c$ (constant).

Therefore, from (3) $v = V + c$. Hence the result.

(ii) Here $f(z) = u + iv$ and also $f(z) = v + iu$. $f(z)$ is supposed to be analytic in D. Then, in both cases by Cauchy-Riemann equations,

$$u_x = v_y,\ u_y = -v_x \tag{4}$$

$$\text{and } v_x = u_y,\ v_y = -u_x. \tag{5}$$

It follows from (4) and (5) that $u_x = 0$, $u_y = 0$, $v_x = 0$, $v_y = 0$ and hence $u(x,y)$ and $v(x,y)$ are constant.

(iii) Let $f(z) = u + iv$, where $u = x^3 - 3xy^2$, $v = 3yx^2 - y^3$.

Here $f(z)$ is supposed to be analytic in a domain D.

$$u_x = 3x^2 - 3y^2,\ v_x = 6xy$$
$$u_y = -6xy,\ v_y = 3x^2 - 3y^2.$$

Then $u_x = v_y$ and $u_y = -v_x$.

Therefore, v is harmonic conjugate of u. But $v_x \neq u_y$ and $v_y \neq -u_x$. Therefore, u cannot be harmonic conjugate of v.

3.2 Cauchy-Riemann Equations in Polar Form

Let $z = x + iy$. In Polar co-ordinates by means of co-ordinate transformation,

$$x = r\cos\theta,\ y = r\sin\theta. \tag{1}$$

Then $r^2 = x^2 + y^2$ and $\theta = \tan^{-1} y/x$. Let $f(z) = u(x, y) + iv(x, y)$ be analytic in a domain D. Cauchy-Riemann equations are

$$u_x = v_y \text{ and } u_y = -v_x. \tag{2}$$

$u(x, y)$ and $v(x, y)$ become functions of r and θ after transformation of co-ordinates.

Therefore,

$$\frac{\partial u}{\partial r} = \frac{\partial u}{\partial x}\cdot\frac{\partial x}{\partial r} + \frac{\partial u}{\partial y}\cdot\frac{\partial y}{\partial r} \Rightarrow u_r = u_x \cdot \cos\theta + u_y \sin\theta. \tag{3}$$

Also

$$\frac{\partial u}{\partial \theta} = \frac{\partial u}{\partial x}\cdot\frac{\partial x}{\partial \theta} + \frac{\partial u}{\partial y}\cdot\frac{\partial y}{\partial \theta} \Rightarrow u_\theta = -r\sin\theta u_x + r\cos\theta u_y. \tag{4}$$

Similarly,

$$v_r = v_x\cos\theta + v_y\sin\theta \tag{5}$$
$$\text{and } v_\theta = -r\sin\theta v_x + r\cos\theta v_y. \tag{6}$$

By (2), (5) and (6) are reduced to the form

$$v_r = -u_y\cos\theta + u_x\sin\theta = u_x\sin\theta - u_y\cos\theta \tag{7}$$
$$\text{and } v_\theta = r\sin\theta u_y + r\cos\theta u_x. \tag{8}$$

Therefore, $\frac{1}{r}v_\theta = \cos\theta u_x + \sin\theta u_y = u_r$ [by (3)]

Again, $-\frac{1}{r}u_\theta = \sin\theta u_x - \cos\theta u_y$ [from (4)]

$= v_r$. [by (7)]

Thus

$$u_r = \frac{1}{r}v_\theta \tag{9}$$
$$\text{and } v_r = -\frac{1}{r}u_\theta \text{ or, } \frac{1}{r}u_\theta = -v_r. \tag{10}$$

The equations (9) and (10) are known as Cauchy-Riemann equation in polar form at any point $z \in D$.

3.2.1 Some Results

(1) Let $f(z) = u + iv$ be analytic in a domain D.

From (3) and (4) above, we have

$$u_r = u_x \cos\theta + u_y \sin\theta \tag{i}$$

$$u_\theta = -r\sin\theta u_x + r\cos\theta u_y \Rightarrow \frac{1}{r}u_\theta = -\sin\theta u_x + \cos\theta u_y. \tag{ii}$$

(i) $\times \cos\theta$ – (ii) $\times \sin\theta$ gives

$$u_r \cos\theta - \frac{1}{r}\sin\theta u_\theta = u_x \tag{iii}$$

and (i) $\times \sin\theta$ + (ii) $\times \cos\theta$ gives

$$u_r \sin\theta + \frac{1}{r}u_\theta \cos\theta = u_y. \tag{iv}$$

Then by Cauchy-Riemann equations, we get

$$v_y = u_r \cos\theta - \frac{1}{r}\sin\theta u_\theta \tag{v}$$

$$\text{and } v_x = -\left(u_r \sin\theta + \frac{1}{r}u_\theta \cos\theta\right). \tag{vi}$$

(2) Let $f(z)$ be differentiable at a non-zero point $z_0 = x_0 + iy_0 = r_0 e^{i\theta_0}$, where $x_0 = r_0\cos\theta_0$ and $y_0 = r_0 \sin\theta_0$.

Then

$$\begin{aligned} f'(z_0) &= u_x(x_0, y_0) + iv_x(x_0, y_0) \\ &= \left(u_{r_0}\cos\theta_0 - \frac{1}{r_0}\sin\theta_0 u_{\theta_0}\right) + i\left(-u_{r_0}\sin\theta_0 - \frac{1}{r_0}u_{\theta_0}\cos\theta_0\right) \end{aligned}$$

[By (iii) and (vi) above at the point (r_0, θ_0)]

$$\begin{aligned} &= u_{r_0}(\cos\theta_0 - i\sin\theta_0) - \frac{i}{r_0}u_{\theta_0}(\cos\theta_0 - i\sin\theta_0) \\ &= e^{-i\theta_0}\left[u_{r_0} - \frac{i}{r_0}u_{\theta_0}\right] \end{aligned}$$

$$\Rightarrow f'(z_0) = e^{-i\theta_0}\left[u_{r_0} + iv_{r_0}\right] \text{ [by (10) at the point } (r_0, \theta_0)] \tag{vii}$$

Also applying Cauchy-Riemann equations in polar form we can write the alternative form of

$$\begin{aligned} f'(z_0) &= e^{-i\theta_0}\left[\frac{1}{r_0}v_{\theta_0} - i\frac{1}{r_0}u_{\theta_0}\right] \text{ [by (9) and (10)]} \\ &= \frac{-ie^{-i\theta_0}}{r_0}\left[u_{\theta_0} + iv_{\theta_0}\right] = \frac{-i}{r_0 e^{i\theta_0}}\left[u_{\theta_0} + iv_{\theta_0}\right] = \frac{-i}{z_0}\left[u_{\theta_0} + iv_{\theta_0}\right]. \end{aligned} \tag{viii}$$

Problem [3.4] : Let $\omega = f(z) = u + iv$ and let $x = r\cos\theta$, $y = r\sin\theta$. Then show that

$$\frac{d\omega}{dz} = e^{-i\theta}\frac{\partial\omega}{\partial r} = -\frac{i}{r}e^{-i\theta}\frac{\partial\omega}{\partial\theta}.$$

Solution.

$$\frac{d\omega}{dz} = f'(z) = \frac{\partial u}{\partial x} + i\frac{\partial v}{\partial x} = \left(\frac{\partial u}{\partial r}\cdot\frac{\partial r}{\partial x} + \frac{\partial u}{\partial\theta}\cdot\frac{\partial\theta}{\partial x}\right) + i\left(\frac{\partial v}{\partial r}\cdot\frac{\partial r}{\partial x} + \frac{\partial v}{\partial\theta}\cdot\frac{\partial\theta}{\partial x}\right). \tag{1}$$

We have $r^2 = x^2 + y^2$ and $\theta = \tan^{-1}\dfrac{y}{x}$.

$$\frac{\partial r}{\partial x} = \frac{x}{r} = \cos\theta$$

$$\frac{\partial \theta}{\partial x} = \frac{1}{1+\frac{y^2}{x^2}}\left(-\frac{y}{x^2}\right) = -\frac{r\sin\theta}{r^2} = -\frac{\sin\theta}{r}.$$

From (1),

$$\frac{d\omega}{dz} = \left(\frac{\partial u}{\partial r}\cdot\cos\theta - \frac{\partial u}{\partial \theta}\cdot\frac{\sin\theta}{r}\right) + i\left(\frac{\partial v}{\partial r}\cos\theta - \frac{\partial v}{\partial \theta}\cdot\frac{\sin\theta}{r}\right) \tag{2}$$

$$= \cos\theta\left(\frac{\partial u}{\partial r} + i\frac{\partial v}{\partial r}\right) - \frac{\sin\theta}{r}\left(\frac{\partial u}{\partial \theta} + i\frac{\partial v}{\partial \theta}\right)$$

$$= \cos\theta\left(\frac{\partial u}{\partial r} + i\frac{\partial v}{\partial r}\right) - \frac{\sin\theta}{r}\left(-r\frac{\partial v}{\partial r} + ir\frac{\partial u}{\partial r}\right) \text{ [by (9) and (10) of Art. 3.2]}$$

$$= \cos\theta\left(\frac{\partial u}{\partial r} + i\frac{\partial v}{\partial r}\right) - \frac{\sin\theta}{r}\cdot ir\left[\frac{\partial u}{\partial r} + i\frac{\partial v}{\partial r}\right]$$

$$= (\cos\theta - i\sin\theta)\left(\frac{\partial u}{\partial r} + i\frac{\partial v}{\partial r}\right) = e^{-i\theta}\frac{\partial \omega}{\partial r} \left[\begin{array}{c}\because \text{from } \omega = u + iv \\ \dfrac{\partial \omega}{\partial r} = \dfrac{\partial u}{\partial r} + i\dfrac{\partial v}{\partial r}\end{array}\right].$$

Again, from (2)

$$\frac{d\omega}{dz} = \left(\frac{1}{r}\frac{\partial v}{\partial \theta}\cos\theta - \frac{\partial u}{\partial \theta}\frac{\sin\theta}{r}\right) + i\left(-\frac{1}{r}\frac{\partial u}{\partial \theta}\cos\theta - \frac{\partial v}{\partial \theta}\frac{\sin\theta}{r}\right)$$

$$= -\frac{\partial u}{\partial \theta}\cdot\frac{1}{r}(\sin\theta + i\cos\theta) + \frac{1}{r}\frac{\partial v}{\partial \theta}(\cos\theta - i\sin\theta)$$

$$= -\frac{i}{r}\frac{\partial u}{\partial \theta}(\cos\theta - i\sin\theta) + \frac{1}{r}\frac{\partial v}{\partial \theta}(\cos\theta - i\sin\theta)$$

$$= -(\cos\theta - i\sin\theta)\frac{i}{r}\left(\frac{\partial u}{\partial \theta} + i\frac{\partial v}{\partial \theta}\right)$$

$$= \frac{-i}{r}e^{-i\theta}\left(\frac{\partial u}{\partial \theta} + i\frac{\partial v}{\partial \theta}\right) = \frac{-i}{r}e^{-i\theta}\left(\frac{\partial \omega}{\partial \theta}\right) = \frac{-i}{r}e^{-i\theta}\cdot\frac{\partial \omega}{\partial \theta}.$$

Theorem [3.5] If $f(z) = u(r,\theta) + iv(r,\theta)$ be defined in some neighbourhood of a non-zero point $z_0 = r_0e^{i\theta_0}$ and if the 1st order partial derivatives of u and v with respect to r and θ exist in that neighbourhood and they are continuus at (r_0,θ_0), and if the Cauchy-Riemann equations hold in polar form, then $f'(z_0)$ exists and is given by $f'(z_0) = e^{-i\theta_0}\left(u_{r_0} + iv_{r_0}\right)$ at the point (r_0,θ_0) (see (vii), Art. 3.2.1).

Example. Let $f(z) = \dfrac{1}{z} = \dfrac{1}{r\cdot e^{i\theta}} = \dfrac{1}{r}[\cos\theta - i\sin\theta]$.

Then $u(r,\theta) = \dfrac{\cos\theta}{r}$ and $v(r,\theta) = -\dfrac{\sin\theta}{r}$.

Here $f(z)$ can be defined at a non-zero point $z_0 = (r_0,\theta_0)$.

$$\frac{\partial u}{\partial r} = -\frac{\cos\theta}{r^2},\quad \frac{\partial v}{\partial r} = \frac{\sin\theta}{r^2}$$

both exist and are continuous in the neighbourhood of (r_0,θ_0).

Therefore, by above theorem,

$$f'(z) = e^{-i\theta}\left[-\frac{\cos\theta}{r^2} + i\frac{\sin\theta}{r^2}\right] = -\frac{e^{-i\theta}}{r^2}\cdot e^{-i\theta} = -\frac{1}{(re^{i\theta})^2} = -\frac{1}{z^2}$$

exists at a non-zero point $z_0 = (r_0, \theta_0)$.

Problem [3.5]: Let $f(z) = u(r,\theta) + iv(r,\theta)$ be analytic in a domain D which does not contain the origin. Assuming continuous partial derivatives in D and using Cauchy-Riemann equations in polar form show that the function $u(r,\theta)$ satisfies the partial differential equation $r^2u_{rr}(r,\theta)+ru_r(r,\theta)+u_{\theta\theta}(r,\theta) = 0$, which is known as the polar form of the Laplace's equation.

Show that the same is true for the function $v(r,\theta)$.

Solution. Cauchy-Riemann equations in polar form are

$$u_r = \frac{1}{r}v_\theta \tag{1}$$

$$\text{and } v_r = -\frac{1}{r}u_\theta. \tag{2}$$

From (1), $u_{rr} = -\frac{1}{r^2}v_\theta + \frac{1}{r}v_{r\theta}$ [diff. partially w.r.t. r]

$$\Rightarrow r^2u_{rr} = -v_\theta + rv_{r\theta} \Rightarrow r^2u_{rr} = -ru_r + rv_{r\theta} \text{ [by (1)]}$$

$$\Rightarrow r^2u_{rr} + ru_r = rv_{r\theta}. \tag{3}$$

Again, from (2), $v_{\theta r} = -\frac{1}{r}u_{\theta\theta}$ [diff. partially w.r.t. θ]

$$\Rightarrow rv_{\theta r} = -u_{\theta\theta}. \tag{4}$$

Hence from (3) by (4), we get

$$r^2u_{rr} + ru_r = -u_{\theta\theta} \Rightarrow r^2u_{rr} + ru_r + u_{\theta\theta} = 0. \ [\because v_{r\theta} = v_{\theta r}]$$

Proceeding as above, we can prove $r^2v_{rr} + rv_r + v_{\theta\theta} = 0$.

3.3 Reflection Principle

It has been observed that some analytic function $f(z)$ possesses the property $\overline{f(z)} = f(\bar{z}) \ \forall\ z \in D$ while others do not.

For example, $f(z) = z^3$ and $z + 2$, being analytic in D satisfies the property

$$\overline{f(z)} = f(\bar{z}) \ \forall\ z \in D,$$

while $f(z) = z + 2i$ does not satisfy $\overline{f(z)} = f(\bar{z})$, though it is analytic in D. When $\overline{f(z)} = f(\bar{z})$ for some analytic function in a domain D, it is called as 'Reflection Principle'.

Thus the way of predicting when the reflection of $f(z)$ in the real axis corresponds to the reflection of z has been derived from the following theorem.

Theorem [3.6] Let $f(z)$ be analytic in a domain D having a segment of X-axis and is symmetric in that axis. Then $\overline{f(z)} = f(\bar{z}) \ \forall\ z \in D$, if and only if $f(x)$ is real for each x on the segment.

Proof. We first assume that $f(x)$ is real at each point x on the segment.

Let $F(z) = \overline{f(\bar{z})}$. We shall show that $F(z)$ is analytic in D.

We write, $f(z) = u(x, y) + iv(x, y)$ and $F(z) = U(x, y) + iV(x, y)$.

Then $f(\bar{z}) = u(x, -y) + iv(x, -y)$ $\left[\begin{array}{l}\because z = x+iy \text{ associated to } (x, y), \\ \bar{z} \text{ is associated to } (x, -y)\end{array}\right]$

and hence $\overline{f(\bar{z})} = u(x, -y) - iv(x, -y)$.

Therefore, $F(z) = \overline{f(\bar{z})} \Rightarrow U(x, y) = u(x, -y)$ and $V(x, y) = -v(x, -y)$. (i)

$f(z)$ is analytic $\Rightarrow u_x = v_y$ and $u_y = -v_x$. (ii)

From (i) $U_x = u_x$, $V_y = -v_y(-1) = +v_y$.

Also $U_y = -u_y$ and $V_x = -v_x$. (iii)

$\Rightarrow V_x = -v_x = u_y$ [by (ii)]

$\Rightarrow V_x = -U_y$ [by (iii)] i.e., $U_y = -V_x$.

Also $V_y = v_y = u_x = U_x$ [by (i) and (ii)].

Thus $U(x, y)$ and $V(x, y)$ satisfy Cauchy-Riemann equations $U_x = V_y$ and $U_y = -V_x$ and these partial derivatives are continuous as the partial derivatives of $u(x, y)$ and $v(x, y)$ are continuous. Hence $F(z)$ is analytic in D.

Now since $f(x)$ is real on the segment of the real axis lying in D, we have $v(x, 0) = 0$ on the segment and consequently, $f(x) = u(x, 0)$.

Also $F(z) = U(x, 0) + iV(x, 0) = u(x, 0) - iv(x, 0)$ [by (i)]

$= u(x, 0)$ [$\because v(x, 0) = 0$].

Therefore, $F(z) = f(z)$ at each point x on the real axis contained in D.

$$\therefore \overline{f(\bar{z})} = f(z) \Rightarrow f(\bar{z}) = \overline{f(z)}.$$

Next, let the given condition hold. That is, $f(\bar{z}) = \overline{f(z)}\ \forall\ z \in D$.

Then $u(x, -y) + iv(x, -y) = u(x, y) - iv(x, y)$, which is also true for $z = (x, 0)$, a point on the real axis lying in D.

Therefore, $u(x, 0) + iv(x, 0) = u(x, 0) - iv(x, 0) \Rightarrow v(x, 0) = 0$.

$\Rightarrow f(x)$ is real on the segment of real axis lying in D.

Example. We have noted that $f(z) = z^3 = (x^3 - 3xy^2) + i(3x^2y - y^3)$ is analytic and $f(\bar{z}) = \overline{f(z)}$.

Now by preceding theorem $f(x) = x^3$ is real for all real values of x on the segment of X-axis in D so the reflection property holds.

But for

$$f(z) = z + 2i = x + i(2 + y),$$
$$f(x) = x + 2i$$

is not real for all real values of x on the segment of X-axis in D. So the reflection property does not hold though $f(z) = z + 2i$ is analytic in D.

EXERCISE

(1) Show that each of the following functions is nowhere analytic (i) $f(z) = 2xy + i(x^2 - y^2)$, (ii) $f(z) = e^{2y}e^{ix}$.

(2) Determine the singular points of the function and state why the function is analytic everywhere except at those points: (i) $\dfrac{z^2+1}{(z+3)(z^2+3z+1)}$, (ii) $\dfrac{3z+1}{z^2+4}$.

(3) Show that the function $f(z) = x^2y + iy$ is continuous everywhere but is not analytic.

(4) If $f(z)$ and $\phi(z)$ are analytic in a domain D, prove that (i) $f(z) \pm \phi(z)$, (ii) $f(z)\phi(z)$ are also analytic in D.
Also show that $f(z)/\phi(z)$ is analytic in D, if $\phi(z) \neq 0 \ \forall \ z \in D$.

(5) Prove that $f(z)$ and $f(\bar{z})$ are simultaneously harmonic.

(6) Prove that $f(z) = e^{2iz}$ is analytic and find $f'(z)$.

(7) Show that $u(x, y)$ is harmonic in some domain and find a harmonic conjugate $v(x, y)$ when (i) $u(x, y) = 2x - x^3 + 3xy^2$, (ii) $u(x, y) = \sinh x \sin y$.

(8) Show that the function $u(x, y) = \sin x \cosh y + 2\cos x \sinh y + x^2 - y^2 + 4xy$ satisfies Laplace's equation and find the corresponding analytic function $u + iv$.

(9) If $u + v = \dfrac{2\sin 2x}{e^{2y} + e^{-2y} - 2\cos 2x}$ and $f(z) = u + iv$ is an analytic function of z, then find $f(z)$ in terms of z.

(10) Show that the function $f(z) = \dfrac{x^2y^5(x+iy)}{x^4+y^{10}}$, $z \neq 0$, $f(0) = 0$, is not analytic at the origin, although Cauchy-Riemann equations are satisfied at $(0, 0)$.

(11) Examine the continuity and differentiability at $(0, 0)$ and the Cauchy-Riemann equations at $(0, 0)$ for the following function defined by

$$f(z) = \begin{cases} \dfrac{x^5 - y^5}{x^2 + y^2} + i\dfrac{x^5 + y^5}{x^2 + y^2}, & \text{if } (x, y) \neq (0, 0) \\ 0, & \text{if } x = 0,\ y = 0. \end{cases}$$

(12) Show that the function $u(x, y) = \dfrac{1}{2}\log(x^2 + y^2)$ is harmonic and find its harmonic conjugate.

(13) If $u - v = (x - y)(x^2 + 4xy + y^2)$ and $f(z) = u + iv$ is an analytic function of z, find $f(z)$ in terms of z.

(14) Show that when $f(z) = x^3 + i(1-y)^3$, it is legitimate to write $f'(z) = u_x + iv_x = 3x^2$, only when $z = i$.

CHAPTER 4

SEQUENCES AND INFINITE SERIES

Definition [4.1] : A function $f : N \to \mathbb{C}$, where $\mathbb{C}$ is a set of complex numbers, defined by $f(n) = z_n \ \forall \ n \in N$, is said to be a sequence of complex numbers. We denote it by $\langle f(n) \rangle$ or $\langle z_n \rangle$.

Definition [4.2] : A complex sequence $\langle z_n \rangle$ is said to have a limit z_0, written as $\lim\limits_{n\to\infty} z_n = z_0$ if for every $\epsilon (> 0)$, there exists a (+)ve integer m such that $|z_n - z_0| < \epsilon \ \forall \ n \geq m$.

The sequence $\langle z_n \rangle$ can have at most one limit. We say the limit exists if z_0 is unique and in this case the sequence $\langle z_n \rangle$ is said to converge to z_0.

Otherwise, if the limit does not exist, the sequence is said to be divergent.

That is, if $\lim\limits_{n\to\infty} z_n = \infty$, we say that $\langle z_n \rangle$ diverges to infinity.

Theorem [4.1] The limit of a convergent sequence $\langle z_n \rangle$ is unique.

Proof. If possible, let z_0 and z_0' be two limits of a convergent sequence $\langle z_n \rangle$ such that $z_0 \neq z_0'$.

Then for every $\epsilon (> 0)$, $\exists$ (+)ve integers m_1 and m_2 such that

$$|z_n - z_0| < \frac{\epsilon}{2}, \ \forall \ n \geq m_1 \text{ and } |z_n - z_0'| < \frac{\epsilon}{2}, \ \forall \ n \geq m_2.$$

Let $|z_0 - z_0'| = \delta > \epsilon$.

Then $|z_0 - z_0'| = |z_0 - z_n + z_n - z_0'| \leq |z_0 - z_n| + |z_n - z_0'| < \dfrac{\epsilon}{2} + \dfrac{\epsilon}{2} = \epsilon \ \forall \ n > m$, where $m = \max(m_1, m_2) \Rightarrow \delta < \epsilon$, and hence a contradiction arises.

Therefore, the limit of a convergent sequence is unique.

Theorem [4.2] The sequence $\langle z_n \rangle$ converges to z_0, where $z_n = x_n + iy_n$ and $z_0 = x_0 + iy_0$, if and only if the real sequences $\{x_n\}$ and $\{y_n\}$ converge to x_0 and y_0 respectively.

That is, $\lim\limits_{n\to\infty} z_n = z_0$, if and only if $\lim\limits_{n\to\infty} x_n = x_0$ and $\lim\limits_{n\to\infty} y_n = y_0$.

Proof. Let the sequence $\langle z_n \rangle$ converge to z_0. Then for every $\epsilon (> 0)$, $\exists$ a (+)ve integer m such that $|z_n - z_0| < \epsilon, \ \forall \ n \geq m$.

$$\Rightarrow |(x_n + iy_n) - (x_0 + iy_0)| < \epsilon, \ \forall \ n \geq m$$

$$\Rightarrow |(x_n - x_0) + i(y_n - y_0)| < \epsilon, \ \forall \ n \geq m. \qquad \text{(i)}$$

$$\left.\begin{aligned}&\text{Now}\quad |x_n - x_0| \le |(x_n - x_0) + i(y_n - y_0)| < \epsilon,\ \forall\ n \ge m\\&\text{Also}\quad |y_n - y_0| \le |(x_n - x_0) + i(y_n - y_0)| < \epsilon,\ \forall\ n \ge m\end{aligned}\right\}\ \text{[by (1)]}$$

$\Rightarrow \{x_n\}$ converges to x_0 and $\{y_n\}$ converges to y_0.

Conversely, let $\{x_n\}$ converge to x_0 and $\{y_n\}$ converge to y_0.

Then, we have $|x_n - x_0| < \frac{\epsilon}{2}$, $\forall\ n \ge m_1$ and $|y_n - y_0| < \frac{\epsilon}{2}$, $\forall\ n \ge m_2$.

Now, if $m = \max(m_1, m_2)$, then $|x_n - x_0| < \frac{\epsilon}{2}$ and $|y_n - y_0| < \frac{\epsilon}{2}$, $\forall\ n \ge m$.

Then

$$\begin{aligned}|z_n - z_0| &= |(x_n + iy_n) - (x_0 + iy_0)|\\&= |(x_n - x_0) + i(y_n - y_0)|\\&\le |x_n - x_0| + |y_n - y_0| < \frac{\epsilon}{2} + \frac{\epsilon}{2}\ \forall\ n \ge m,\end{aligned}$$

Therefore $|z_n - z_0| < \epsilon$, $\forall\ n \ge m$.

$\Rightarrow \{z_n\}$ converges to z_0.

Definition [4.3] : **[Cauchy sequence]** A sequence $\langle z_n\rangle$ is said to be a Cauchy sequence, if for every $\epsilon(> 0)$ there exists a (+)ve integer m_0 such that $|z_m - z_n| < \epsilon\ \forall\ m, n \ge m_0$ or, $|z_{n+p} - z_n| < \epsilon$, $\forall\ n \ge m_0$ and $p \in N$.

Theorem [4.3] A sequence $\langle z_n\rangle$ is convergent; if and only if it is a Cauchy sequence.

Proof. Let the sequence $\langle z_n\rangle$ converge to z_0. Then, for $\epsilon(> 0)$, $\exists$ a (+)ve integer m_0 such that

$$|z_m - z_0| < \frac{\epsilon}{2}\ \forall\ n \ge m_0.$$

Hence, if $m \ge m_0$, $n \ge m_0$, then $|z_m - z_0| < \frac{\epsilon}{2}$ and $|z_n - z_0| < \frac{\epsilon}{2}$.

Therefore,

$$\begin{aligned}|z_m - z_n| &= |z_m - z_0 + z_0 - z_n| \le |z_m - z_0| + |z_n - z_0|\\&< \frac{\epsilon}{2} + \frac{\epsilon}{2} = \epsilon\ \forall\ m, n \ge m_0.\end{aligned}$$

$\Rightarrow \{z_n\}$ is a Cauchy sequence.

Conversely, let $\{z_n\}$ be a Cauchy sequence.

Therefore, $|z_m - z_n| < \epsilon$, $\forall\ m, n \ge m_0$

$$\begin{aligned}&\Rightarrow |(x_m + iy_m) - (x_n + iy_n)| < \epsilon,\ \forall\ m, n \ge m_0\\&\Rightarrow |(x_m - x_n) + i(y_m - y_n)| < \epsilon,\ \forall\ m, n \ge m_0.\end{aligned}\qquad\text{(i)}$$

Now, $|x_m - x_n| \le |(x_m - x_n) + i(y_m - y_n)| < \epsilon$ and also

$|y_m - y_n| \le |(x_m - x_n) + i(y_m - y_n)| < \epsilon\ \forall\ m, n \ge m_0$ [by (i)]

$\Rightarrow \{x_n\}$ and $\{y_n\}$ are Cauchy sequences of real numbers.

$\Rightarrow \{x_n\}$ and $\{y_n\}$ are both convergent.

Let $\{x_n\}$ converge to x_0 and $\{y_n\}$ converge to y_0.

Then $\langle z_n\rangle = \langle x_n + iy_n\rangle$ converges to $x_0 + iy_0 = z_0$ [by Th. [4.2]]

Hence $\langle z_n\rangle$ is convergent.

Example. (1) Show that the sequence $\langle z_n \rangle$, where

$$z_n = -2 + i\frac{(-1)^n}{n^2} \quad (n = 1, 2, \dots) \text{ converges to } -2.$$

Here $z_n = x_n + iy_n \Rightarrow x_n = -2$ and $y_n = \dfrac{(-1)^n}{n^2}$.

Here, $\lim_{n\to\infty} x_n = -2$ and $\lim_{n\to\infty} y_n = 0$.

Therefore, the sequences $\{x_n\}$ and $\{y_n\}$ are convergent and converges to -2 and 0 respectively.

Hence $\langle z_n \rangle$ is convergent and converges to $-2 + i \cdot 0 = -2$.

Alternative method

$$|z_n - (-2)| = |z_n + 2| = \left|-2 + \tfrac{(-1)^n}{n^2} + 2\right| = \tfrac{1}{n^2} < \epsilon \ \forall\, n \geq m$$

$$\left[\because \tfrac{1}{n^2} < \epsilon \Rightarrow n > \tfrac{1}{\sqrt{\epsilon}} \Rightarrow n > m, \ m \textbf{ is the integral part of } \tfrac{1}{\sqrt{\epsilon}}\right]$$

$\Rightarrow \lim_{n\to\infty} z_n = -2$

$\Rightarrow \langle z_n \rangle$ converges to -2.

Example. (2) Show that, if $\lim_{n\to\infty} z_n = z_0$, then $\lim_{n\to\infty} |z_n| = |z_0|$.

Here, $|z_n - z_0| < \epsilon, \ \forall\, n \geq m$.

Then $|z_n| - |z_0| < |z_n - z_0| < \epsilon, \ \forall\, n \geq m$.

Also $|z_0| - |z_n| < |z_n - z_0| < \epsilon, \ \forall\, n \geq m$.

Therefore, $\big||z_n| - |z_0|\big| < \epsilon, \ \forall\, n \geq m$

$\Rightarrow \lim_{n\to\infty} |z_n| = |z_0|$.

Example. (3) Let the sequence $\langle z_n \rangle$ converge to a number z_0. Then show that there exists a (+)ve number M such that $|z_n| \leq M \ \forall\, n$.

Here, $|z_n - z_0| < \epsilon \ \forall\, n \geq m$.

We take $\epsilon = 1$ so that $|z_n - z_0| < 1 \ \forall\, n$.

Then $|z_n| = |z_n - z_0 + z_0| \leq |z_n - z_0| + |z_0| < 1 + |z_0| = M$ (say) $\forall\, n$.

Thus $\exists$ a (+)ve number M such that $|z_n| \leq M \ \forall\, n$.

Alternative method

The sequence $\langle z_n \rangle$ converges to $z_0 = x_0 + iy_0 \Rightarrow \{x_n\}$ converges to x_0 and $\{y_n\}$ converges to y_0, where $z_n = x_n + iy_n$.

The real convergent sequences are bounded. Hence $\exists$ (+)ve numbers M_1 and M_2 such that $|x_n| \leq M_1$ and $|y_n| \leq M_2, \ \forall\, n$.

Therefore, $|z_n| = |x_n + iy_n| \leq |x_n| + |y_n| \leq M_1 + M_2 = M$ (say).

Hence, $|z_n| \leq M \ \forall\, n$.

Example. (4) If the sequence $\langle z_n \rangle$ converges to z_0, then show that $\langle \bar{z}_n \rangle$ converges to $\bar{z}_0$.

Here $|z_n - z_0| < \epsilon \ \forall\, n \geq m$. (1)

Then $|\bar{z}_n - \bar{z}_0| = |\overline{z_n - z_0}| = |z_n - z_0| < \epsilon \ \forall\, n \geq m$ [by (1)]

$\Rightarrow \langle \bar{z}_n \rangle$ converges to $\bar{z}_0$.

4.1 Infinite Series

Definition [4.4] : A series of the form $z_1 + z_2 + \cdots \infty$, where $z_n \in \mathbb{C} \ \forall \ n \in N$, is called an infinite series of complex numbers.

The series is written as $\sum_1^{\infty} z_n$.

Let $S_n = z_1 + z_2 + \cdots + z_n$.

Then S_n is called the nth partial sum of the series $\sum_1^{\infty} z_n$.

The series $\sum_1^{\infty} z_n$ is said to converge, if and only if the corresponding sequence $\langle S_n \rangle$ of partial sums is convergent.

Also if the sequence $\langle S_n \rangle$ converges to S, then we write $\sum_1^{\infty} z_n = S$.

Theorem [4.4] The necessary and sufficient condition that an infinite series $\sum_1^{\infty} z_n$ to be convergent is that for every $\epsilon (> 0)$, there exists a (+)ve integer m such that

$$|S_{n+p} - S_n| < \epsilon \ n \geq m, \ \forall \ p \geq 0$$

$$\text{i.e., } |z_{n+1} + z_{n+2} + \cdots + z_{n+p}| < \epsilon, \ \forall \ n \geq m, \ \forall \ p \geq 0.$$

Proof. It follows immediately from Th. [4.3].

Theorem [4.5] Let $z_n = x_n + iy_n$ and $S = X + iY$. Then $\sum_1^{\infty} z_n = S$, if and only if

$$\sum_{n=1}^{\infty} x_n = X \text{ and } \sum_1^{\infty} y_n = Y.$$

Proof. Let S_N denote the partial sum consisting of first N terms of the series $\sum_1^{\infty} z_n$.

Then it follows that $S_N = X_N + iY_N$, where

$$X_N = \sum_{n=1}^{N} x_n \text{ and } Y_N = \sum_{n=1}^{N} y_n.$$

Now the series $\sum_1^{\infty} z_n$ converges to S, if and only if $\langle S_N \rangle$ converges to $S = X + iY$.

That is, $\lim\limits_{N \to \infty} S_N = S$, if and only if

$$\lim_{N \to \infty} X_N = X \text{ and } \lim_{N \to \infty} Y_N = Y. \text{ [Proceeding as in Th. [4.2].]}$$

Remarks. (1) In case of convergent series of real numbers, nth term of the series tends to zero as $n \to \infty$.

That is, if $\sum x_n$ and $\sum y_n$ be convergent, then $\lim\limits_{n \to \infty} x_n = 0$ and $\lim\limits_{n \to \infty} y_n = 0$ and hence when $\sum z_n$ is convergent it follows that $\lim\limits_{n \to \infty} z_n = 0$.

Thus the necessary condition for the convergence of the series $\sum_1^{\infty} z_n$ is that $\lim\limits_{n \to \infty} z_n = 0$.

Alternatively

If $\sum_1^\infty z_n$ is convergent, then $|S_m - S_n| < \epsilon, \ \forall \ m, n \geq N_0$.

Taking $m = n-1$, we get $|S_{n-1} - S_n| < \epsilon \ \forall \ n \geq N_0$

$\Rightarrow |z_n| < \epsilon, \ \forall \ n \geq N_0$

$\Rightarrow \lim_{n\to\infty} z_n = 0.$

(2) The terms of a convergent series of complex numbers are, therefore, bounded. That is, there exists a (+)ve number M such that $|z_n| < M \ \forall \ n$.

(3) Let the series $\sum_1^\infty z_n$ be absolutely convergent. Then $\sum_1^\infty |z_n|$ is convergent

$\Rightarrow \sum_{n=1}^\infty \left[x_n^2 + y_n^2\right]^{1/2}$ is convergent, where $z_n = x_n + iy_n$

$\Rightarrow$ a series of real numbers converges.

Now $|x_n| \leq \sqrt{x_n^2 + y_n^2}$ and $|y_n| \leq \sqrt{x_n^2 + y_n^2}$

$\Rightarrow \sum_1^\infty |x_n|$ and $\sum_1^\infty |y_n|$ are convergent by comparison test.

$\Rightarrow \sum_1^\infty x_n$ and $\sum_1^\infty y_n$ are convergent, since absolutely convergent series of real numbers are convergent.

$\Rightarrow \exists$ real numbers X and Y such that $\sum_1^\infty x_n = X$ and $\sum_1^\infty y_n = Y \Rightarrow \sum_1^\infty z_n$ is convergent [by Th. [4.2].]

Thus absolutely convergent series of complex numbers is convergent.

(4) Let the sequence $\langle S_N \rangle$ of the nth partial sum of the $\sum_1^\infty z_n$ be convergent and converges to S. Then we have $\sum_1^\infty z_n = S$, we define R_N, the remainder after n terms by $R_N = S - S_N$.

Now $|S_N - S| < \epsilon, \ \forall \ N \geq m$

$\Rightarrow |R_N| < \epsilon, \ \forall \ N \geq m$

$\Rightarrow R_N \to 0$ as $N \to \infty$.

Thus a series $\sum_1^\infty z_n$ converges to S, if and only if the sequence of remainders tends to zero.

4.2 Power Series

Definition [4.5] : An infinite series of the form $\sum_0^\infty a_n(z-z_0)^n = a_0 + a_1(z-z_0) + a_2(z-z_0)^2 + \cdots$, where z is a complex variable, the constants a_n, z_0 are complex numbers, is called a power series, a_n is independent of $z \ \forall \ n = 0, 1, 2, \ldots$.

By changing of variable $z = t + z_0$, the power series $\sum a_n(z-z_0)^n$ reduces to the form $\sum a_n t^n$. So it is sufficient to consider the power series of the form $\sum a_n z^n$.

Note. As a special case z, a_n, z_0 may be real. But here we shall prove all the results in general case.

4.3 Behaviours of a Power Series

(1) A power series $\sum a_n z^n$ may converge for all z. It follows from the example:

We consider a power series $\sum \frac{z^n}{n!}$. Then, by D'Alembert's ratio test,

$$\lim_{n\to\infty}\left|\frac{a_{n+1}}{a_n}\right| = \lim_{n\to\infty}\left|\frac{z}{n+1}\right| = \lim_{n\to\infty}\frac{|z|}{n+1} = 0 < 1.$$

Therefore, the series converges absolutely $\forall\ z$. Hence it converges $\forall\ z$.

(2) A power series may converge at $z = 0$ only. We consider a power series $\sum nz^n$. Then $\lim\limits_{n\to\infty} nz^n = \infty$ for $z \neq 0$.

Therefore, the series converges for $z = 0$ only.

(3) A power series may converge for some value of z (i.e., in some region) and diverges for other values of z. We consider a power series $\sum\limits_{0}^{\infty} z^n$.

This is a geometrical series and it converges for $|z| < 1$ and diverges for $|z| \geq 1$.

Condition for Convergence of a Power Series

A power series $\sum a_n z^n$ being an infinite series is convergent if $\lim\limits_{n\to\infty} a_n z^n = 0$ [by Remark (1) of Art 4.1]

$\Rightarrow\ |a_n z^n| < \epsilon,\ \forall\ n \geq m$

$\Rightarrow\ |a_n z_0^n| < \epsilon,\ \forall\ n \geq m$, when converges at $z = z_0$.

We can choose a (+)ve number M such that

$$|a_n z_0^n| \leq M\ \forall\ n. \tag{1}$$

Theorem [4.6] [Abel's theorem] If a power series $\sum a_n z^n$ converges for $z = z_0$, then it converges absolutely for all values of z satisfying $|z| < |z_0|$.

Proof. Let the P.S. $\sum a_n z^n$ converges at $z = z_0$.

Then from (1) it follows that $|a_n z_0^n| \leq M,\ \forall\ n$, where M is a (+)ve real number. (i)

Then

$$\sum_{0}^{\infty} |a_n z^n| = |a_0| + |a_1 z| + |a_2 z^2| + \cdots$$

$$= |a_0| + |a_1 z_0|\left|\frac{z}{z_0}\right| + |a_2 z_0^2|\left|\frac{z}{z_0}\right|^2 + \cdots \leq M + M\left|\frac{z}{z_0}\right| + M\left|\frac{z}{z_0}\right|^2 + \cdots \text{ [by (i)]}$$

$$= M\left[1 + \left|\frac{z}{z_0}\right| + \left|\frac{z}{z_0}\right|^2 + \cdots\right]$$

Now for $|z| < |z_0| \Rightarrow \left|\frac{z}{z_0}\right| < 1$. Therefore, the series in the bracket being a geometric series with common ratio < 1, converges. Hence by comparison test, $\sum\limits_{0}^{\infty} |a_n z^n|$ is convergent for all z satisfying $|z| < |z_0|$.

Therefore, the power series $\sum a_n z^n$ converges absolutely for all z satisfying $|z| < |z_0|$.

Corollary [4.1] : If the power series $\sum_{n=0}^{\infty} a_n(z-z_0)^n$ converges, when $z = z_1$ $(z_1 \neq z_0)$, then it is absolutely convergent for all z in the open disk $|z-z_0| < R_1$, where $R_1 = |z_1 - z_0|$.

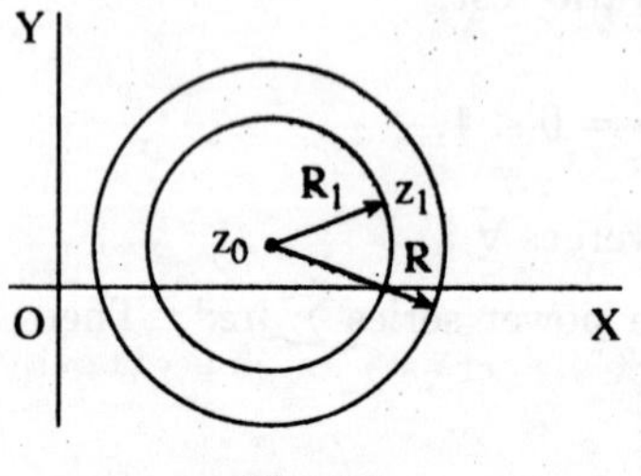

Fig. 4.1

Proof. We have proved the theorem for $z_0 = 0$. Now if z_0 is any non-zero complex number, we assume that the series is convergent at $z = z_1$ $(z_1 \neq z_0)$.

Now if we write $\omega = z - z_0$, then the series $\sum a_n(z-z_0)^n$ becomes $\sum a_n\omega^n$ and it converges at $\omega = z_1 - z_0$. Therefore, $\sum_{0}^{\infty} a_n\omega^n$ converges absolutely for all ω satisfying $|\omega| < |z_1 - z_0|$.

Hence, the P.S. $\sum a_n(z-z_0)^n$ converges absolutely for all z satisfying $|z-z_0| < |z_1 - z_0|$ [replacing ω by $z - z_0$].

Writing $|z_1 - z_0| = R_1$, we say that the P.S. $\sum a_n(z-z_0)^n$ converges absolutely for all z such that $|z - z_0| < R_1$.

4.4 Circle of Convergence and Radius of Convergence

Definition [4.6] : If a power series $\sum_{0}^{\infty} a_n z^n$ converges for all values of z satisfying $|z| < R$, the interior points of a circle $|z| = R$, then the circle $|z| = R$ is called the circle of convergence and R is called the radius of convergence.

Example. The power series $\sum z^n$ converges for all z satisfying $|z| < 1$, the interior points of the circle $|z| = 1$.

Then $|z| = 1$ is called the circle of convergence and 1 is called the radius of convergence.

Working rule for finding the radius of convergence

Let the power series $\sum_{0}^{\infty} a_n z^n = \sum u_n(z)$. Then $\sum u_n(z)$ is convergent, if $\lim_{n\to\infty} |u_n|^{1/n} < 1$.

That is, if $\lim_{n\to\infty} |a_n z^n|^{1/n} < 1$

i.e., if $\lim_{n\to\infty} |a_n^{1/n}||z| < 1$

i.e., if $\dfrac{|z|}{R} < 1$, taking $\lim_{n\to\infty} |a_n|^{1/n} = \dfrac{1}{R}$

i.e., if $|z| < R$.

Thus the power series converges for all z such that $|z| < R$.

Thus corresponding to every power series, there is a non-negative real number R such that $|z| < R$, if the series is convergent and $|z| > R$, if the series is divergent.

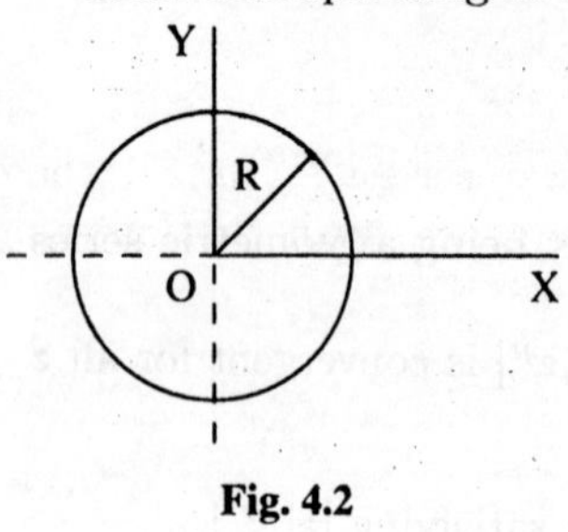

Fig. 4.2

If we draw a circle with centre ($z = 0$) at the origin and of radius R, then

(i) the P.S. $\sum a_n z^n$ is convergent for all z within the circle $|z| = R$

(ii) the P.S. $\sum a_n z^n$ is divergent for all z outside the circle $|z| = R$.

Three possibilities arise for R.

(i) $R = 0$

In this case the series converges only for $z = 0$.

(ii) $R = a$ finite number $= a$ (say)

In this case the series converges for $|z| < a$ and diverges for $|z| > a$.

(iii) $R =$ infinite

In this case the series converges for all z.

Remark. (1) We have shown that the radius of convergence of a power series $\sum a_n z^n$ is given by

$$\frac{1}{R} = \overline{\lim_{n\to\infty}}\ |a_n|^{1/n}.$$

This is known as *Hadamard's formula* for the radius of convergence.

(2) From *Cauchy Limit Theorem–2*, we have $\overline{\lim}\,|a_n|^{1/n} = \lim\limits_{n\to\infty}\left|\dfrac{a_{n+1}}{a_n}\right|$ and, therefore,

$$\frac{1}{R} = \lim_{n\to\infty}\left|\frac{a_{n+1}}{a_n}\right|.$$

This is a simpler form for finding the radius of convergence.

Definition [4.7] : A power series $\sum\limits_{0}^{\infty} a_n z^n$ converges uniformly to $S(z)$ on a set $A \subseteq C$, if and only if the sequence $\{S_n(z)\}$ converges uniformly to $S(z)$, where $S_n(z) = \sum\limits_{r=0}^{n-1} a_r z^n$. That is, if for every $\epsilon(> 0)$ $\exists$ a (+)ve integer m_0 such that $|S_n(z) - S(z)| < \epsilon\ \forall\ n \geq m_0$ and $\forall\ z \in A$.

Definition [4.8] : [**Cauchy criterion for uniform convergence of a P.S.** $\sum a_n z^n$]: A power series $\sum a_n z^n$ converges uniormly on A, if for every $\epsilon(> 0)$, $\exists$ a (+)ve integer m_0 such that $|S_m(z) - S_n(z)| < \epsilon\ \forall\ m, n \geq m_0$ and $\forall\ z \in A$.

Definition [4.9] : [**Weierstrass's Test for uniform convergence of a P.S.** $\sum a_n z^n$]: A power series $\sum a_n z^n$ converges uniformly $\forall\ z \in A$, if (i) $|a_n z^n| \leq M_n\ \forall\ n \in N$, where M_n is a (+)ve constant (independent of z) (ii) $\sum M_n$ is convergent.

Definition [4.10] : [**Dirichlet's Test**]: A power series of the form $\sum\limits_{1}^{\infty} a_n u_n$, where $a_n = f_n(z)$ and $u_n = \phi(n)$, converges uniformly on A if

(i) $|S_n(z)| = |n\text{th partial sum of } f_n(z)| \leq K\ \forall\ z \in A$, i.e., $\langle S_n(z)\rangle$ is bounded.

(ii) $\sum(u_n - u_{n+1})$ is convergent.

(iii) $\lim\limits_{n\to\infty} u_n = 0$.

Theorem [4.7] [**Cauchy-Hadamard Theorem**] For every power series $\sum_0^\infty a_n z^n$, there exists a real number R, $0 \leq R < \infty$, called the radius of convergence, with the following properties:

(1) The series converges absolutely for every $|z| < R$.

(2) The series diverges for $|z| > R$.

(3) The series converges uniformly for $|z| \leq \rho$, where $0 \leq \rho < R$.

Proof. (1) We consider R according to the formula $\frac{1}{R} = \overline{\lim_{n\to\infty}} |a_n|^{1/n}$.

Let $|z| < R$. Then there exists a real number ρ such that $|z| < \rho < R$.

So $\frac{1}{\rho} > \frac{1}{R}$.

From definition of limit superior for every $\epsilon > 0$, there exists a (+)ve integer m such that

$$|a_n|^{1/n} < \frac{1}{R} + \epsilon \ \forall \ n \geq m$$

$$\Rightarrow |a_n|^{1/n} < \frac{1}{\rho} \ \forall \ n \geq m \qquad \left[\frac{1}{\rho} > \frac{1}{R} \Rightarrow \frac{1}{\rho} = \frac{1}{R} + \epsilon\right]$$

$$\Rightarrow |a_n| < \frac{1}{\rho^n}, \ \forall \ n \geq m$$

$$\Rightarrow |a_n z^n| < \left(\frac{|z|}{\rho}\right)^n \text{ for large values of } n.$$

Here the $\sum \left(\frac{|z|}{\rho}\right)^n$ is a geometric series with common ratio $\frac{|z|}{\rho} < 1$, since $|z| < \rho$ and so it converges. Therefore, by comparison test, $\sum |a_n z^n|$ is convergent and hence, $\sum a_n z^n$ is absolutely convergent for $|z| < R$.

(2) For $|z| > R$, we can choose a real number ρ such that $R < \rho < |z|$. So $\frac{1}{R} > \frac{1}{\rho}$.

Again, since $\frac{1}{R} = \overline{\lim_{n\to\infty}} |a_n|^{1/n}$ for every $\epsilon > 0$, there exists a (+)ve integer m such that

$$|a_n|^{1/n} > \frac{1}{R} - \epsilon, \text{ for infinitely many values of } n$$

$$\Rightarrow |a_n|^{1/n} > \frac{1}{\rho} \left[\text{as } \frac{1}{\rho} < \frac{1}{R}, \text{ for every } \epsilon > 0, \text{ we take } \frac{1}{\rho} = \frac{1}{R} - \epsilon\right]$$

$$\Rightarrow |a_n z^n| > \left(\frac{|z|}{\rho}\right)^n \text{ for infinitely many } n$$

$\Rightarrow$ terms of the series $\sum |a_n z^n|$ are unbounded as $|z| > R$

$\Rightarrow$ the series $\sum a_n z^n$ is divergent.

(3) For $0 \leq \rho < R$, we choose a (+)ve real number ρ' such that $\rho < \rho' < R$.

So $\frac{1}{R} < \frac{1}{\rho'} < \frac{1}{\rho}$.

From definition of $\overline{\lim_{n\to\infty}} |a_n|^{1/n} = \frac{1}{R}$

$$\Rightarrow |a_n|^{1/n} \leq \frac{1}{R} + \epsilon \ \forall \ n \geq m$$

$$\Rightarrow |a_n|^{1/n} \leq \frac{1}{\rho'} \quad \left[\text{as } \frac{1}{\rho'} > \frac{1}{R}, \text{ for every } \epsilon > 0, \text{ we can take } \frac{1}{\rho'} = \frac{1}{R} + \epsilon\right]$$

$$\Rightarrow |a_n| \leq \frac{1}{\rho'^n} \Rightarrow |a_n z^n| \leq \frac{|z|^n}{\rho'^n} \leq \left(\frac{\rho}{\rho'}\right)^n \quad \text{for } |z| \leq \rho.$$

Let $M_n = \left(\frac{\rho}{\rho'}\right)^n$.

Then $\sum M_n = \sum \left(\frac{\rho}{\rho'}\right)^n$ is convergent being a G.P. series with common ratio $\frac{\rho}{\rho'} < 1$.

Therefore, by Weierstrass' M-test $\sum a_n z^n$ converges uniformly for $|z| \le \rho' < R$.

Problem [4.1]: (1) Find the radii of convergence of the following power series:

(i) $\sum \left(1+\frac{1}{n}\right)^{n^2} z^n$ (ii) $\sum \frac{(n!)^2}{2n!} z^n$ (iii) $\sum \frac{2^{-n}}{1+in^2} z^n$

(iv) $\sum (4+5i)^n z^n$ (v) $\sum \frac{(-1)^n}{n}(z+2i)^n$.

(2) Show that $1 + \frac{a\cdot b}{1\cdot c}z + \frac{a(a+1)b(b+1)}{1\cdot 2\cdot c(c+1)}z^2 + \cdots$ has unit radius of convergence.

(3) Show that the radius of convergence of the series $\frac{z}{2} + \frac{1\cdot 3}{2\cdot 5}z^2 + \frac{1\cdot 3\cdot 5}{2\cdot 5\cdot 8}z^3 + \cdots$ is $\frac{3}{2}$.

(4) Examine the behaviour of the power series $\sum_0^\infty \frac{z^{4n}}{1+4n}$ on the circle of convergence.

(5) Examine the behaviour of (i) $\sum \frac{z^n}{n}$ and (ii) $\sum_{n=2}^\infty \frac{z^n}{n(\log n)^2}$ on the circle of convergence.

Solution. (1) (i) $\frac{1}{R} = \overline{\lim_{n\to\infty}} |a_n|^{1/n}$. Here $a_n = \left(1+\frac{1}{n}\right)^{n^2}$.

$$\text{Therefore,}\quad \frac{1}{R} = \overline{\lim_{n\to\infty}} \left\{\left(1+\frac{1}{n}\right)^{n^2}\right\}^{1/n} = \overline{\lim_{n\to\infty}} \left(1+\frac{1}{n}\right)^n = e.$$

$$\therefore R = \frac{1}{e}.$$

(ii) We have $\frac{1}{R} = \lim_{n\to\infty}\left|\frac{a_{n+1}}{a_n}\right|$. Here $a_n = \frac{(n!)^2}{2n!}$, $a_{n+1} = \frac{((n+1)!)^2}{(2n+2)!}$.

$$\text{Then}\quad \frac{1}{R} = \lim_{n\to\infty} \frac{\{(n+1)!\}^2}{(2n+2)!} \times \frac{2n!}{(n!)^2} = \lim_{n\to\infty} \frac{(n+1)^2(n!)^2}{(2n+2)(2n+1)2n!}\cdot\frac{2n!}{(n!)^2}$$

$$= \lim_{n\to\infty} \frac{(n+1)^2}{(2n+2)(2n+1)} = \lim_{n\to\infty} \frac{1+\frac{2}{n}+\frac{1}{n^2}}{4+\frac{6}{n}+\frac{2}{n^2}} = \frac{1}{4}.$$

$$\therefore R = 4.$$

(iii) Here
$$a_n = \frac{2^{-n}}{1+in^2} = \frac{2^{-n}(1-in^2)}{1+n^4} = \frac{1-in^2}{2^n(1+n^4)}$$

$$a_{n+1} = \frac{1}{2^{n+1}}\cdot\frac{1-i(n+1)^2}{1+(n+1)^4}.$$

$$\frac{1}{R} = \lim_{n\to\infty}\left|\frac{a_{n+1}}{a_n}\right| = \lim_{n\to\infty} \frac{2^n}{2^{n+1}}\cdot\frac{1-i(n+1)^2}{1+(n+1)^4}\times\frac{(1+n^4)}{1-in^2}$$

$$= \frac{1}{2}\lim_{n\to\infty} \frac{\frac{1}{n^2}-i\left(1+\frac{1}{n}\right)^2}{\frac{1}{n^2}-i}\times\frac{\frac{1}{n^4}+1}{\frac{1}{n^4}+\left(1+\frac{1}{n}\right)^4} = \frac{1}{2}\cdot\frac{-i}{-i}\times\frac{1}{1} = \frac{1}{2}.$$

$$\therefore R = 2.$$

(iv) Here $a_n = (4+5i)^n$

$$\frac{1}{R} = \overline{\lim_{n\to\infty}} |a_n|^{1/n} = \overline{\lim_{n\to\infty}} |4+5i| = \sqrt{16+25} = \sqrt{41}.$$

$$\therefore R = \frac{1}{\sqrt{41}}.$$

(v) $\frac{1}{R} = \overline{\lim_{n\to\infty}} |a_n|^{1/n}$. Here $a_n = \frac{(-1)^n}{n}$, with centre of the circle $z = -2i$.

$$\therefore \frac{1}{R} = \overline{\lim_{n\to\infty}} \left|\frac{1}{n^{1/n}}\right| = 1.$$

$$\therefore R = 1.$$

Thus the radius of the circle of convergence is 1 with centre at $z = -2i$.

(2) Here $a_n = \frac{a(a+1)\cdots(a+n-1)\cdot b(b+1)\cdots(b+n-1)}{1\cdot 2\cdot 3\cdots n\cdot c(c+1)\cdots(c+n-1)}$ with $a_0 = 1$ in the given power series.

Then

$$\frac{1}{R} = \lim_{n\to\infty}\left|\frac{a_{n+1}}{a_n}\right| = \lim_{n\to\infty}\frac{a(a+1)\cdots(a+n-1)(a+n)\cdot b(b+1)\cdot(b+n-1)(b+n)}{1\cdot 2\cdot 3\cdots n(n+1)\cdot c(c+1)\cdots(c+n-1)(c+n)}$$

$$\times\frac{1\cdot 2\cdot 3\cdots n\cdot c(c+1)\cdots(c+n-1)}{a(a+1)\cdots(a+n-1)\cdot b(b+1)\cdots(b+n-1)}$$

$$= \lim_{n\to\infty}\frac{(a+n)(b+n)}{(n+1)(c+n)} = \lim_{n\to\infty}\frac{\left(\frac{a}{n}+1\right)\left(\frac{b}{n}+1\right)}{\left(1+\frac{1}{n}\right)\left(\frac{c}{n}+1\right)} = 1.$$

$$\therefore R = 1.$$

Therefore, the given P.S. has unit radius of convergence.

(3) Here $a_n = \frac{1\cdot 3\cdot 5\cdots(2n-1)}{2\cdot 5\cdot 8\cdots(3n-1)}$.

Then

$$\frac{1}{R} = \lim_{n\to\infty}\left|\frac{a_{n+1}}{a_n}\right| = \lim_{n\to\infty}\frac{1\cdot 3\cdot 5\cdots(2n-1)(2n+1)}{2\cdot 5\cdot 8\cdots(3n-1)(3n+2)}\times\frac{2\cdot 5\cdot 8\cdots(3n-1)}{1\cdot 3\cdot 5\cdots(2n-1)}$$

$$= \lim_{n\to\infty}\frac{2n+1}{3n+2} = \lim_{n\to\infty}\frac{2+\frac{1}{n}}{3+\frac{2}{n}} = \frac{2}{3}.$$

$$\therefore R = \frac{3}{2}.$$

(4) Here $a_n = \frac{1}{1+4n}$.

Then

$$\frac{1}{R} = \lim_{n\to\infty}\left|\frac{a_{n+1}}{a_n}\right| = \lim_{n\to\infty}\frac{1+4n}{1+4(n+1)} = \lim_{n\to\infty}\frac{\frac{1}{n}+4}{\frac{1}{n}+4+\frac{4}{n}} = \frac{4}{4} = 1.$$

$$\therefore R = 1.$$

Thus the radius of circle of convergence is 1 with centre at $z = 0$.

Now at $z = \pm 1$ and $\pm i$, the series becomes $\sum_0^{\infty}\frac{1}{1+4n}$, which is divergent, since it is of the form of a divergent series. Therefore, the series is not convergent at $z = \pm i, \pm 1$ in the circle of convergence. Now to determine the convergence of the series in other points of circle of convergence, we apply Dirichlet's test.

We compare $\sum \frac{z^{4n}}{1+4n}$ with $\sum a_n u_n$. Then $a_n = z^{4n}$ and $u_n = \frac{1}{1+4n}$.

(i) $S_n = n\text{th partial sum of } \sum a_n = \sum_{0}^{n-1} z^{4n}$

$$= 1 + z^4 + z^8 + \cdots + z^{4(n-1)} = \frac{1-(z^4)^n}{1-z^4}$$

$$|S_n| = \frac{|1-z^{4n}|}{|1-z^4|} \leq \frac{1+z^{4n}}{|1-z^4|} \leq \frac{1+1}{|1-z^4|} \text{ for } |z| = 1$$

$\Rightarrow |S_n| \leq \frac{2}{|1-z^4|}$ in the circle of convergence $|z| = 1$

$\Rightarrow$ the sequence $\langle S_n \rangle$ is bounded.

(ii) $$u_n - u_{n+1} = \frac{1}{1+4n} - \frac{1}{1+4(n+1)} = \frac{1}{4n+1} - \frac{1}{4n+5}$$

$$= \frac{4}{(4n+1)(4n+5)} < \frac{4}{16n^2} = \frac{1}{4n^2}$$

$\Rightarrow u_n - u_{n+1} < \frac{1}{4n^2}$.

Also $\sum \frac{1}{n^2}$ is convergent.

$\Rightarrow \sum (u_n - u_{n+1})$ is convergent.

(iii) $\lim_{n\to\infty} u_n = \lim_{n\to\infty} \frac{1}{4n+1} = 0 \Rightarrow \sum a_n u_n = \sum \frac{z^{4n}}{1+4n}$ is convergent $\forall$ z on the circle of convergence except at the points $z = \pm i, \pm 1$.

(5) (i) For the series $\sum_{1}^{\infty} \frac{1}{n} z^n$,

$$\frac{1}{R} = \overline{\lim_{n\to\infty}} |a_n|^{1/n} = \overline{\lim_{n\to\infty}} \left|\frac{1}{n}\right|^{1/n} = \overline{\lim_{n\to\infty}} \frac{1}{n^{1/n}} = 1.$$

So $R = 1$.

Therefore, the radius of the circle of convergence with centre $z = 0$ is 1. At $z = 1$, the power series takes the form $\sum_{1}^{\infty} \frac{1}{n}$, which is a divergent series.

Now to determine the existence of convergence of the power series $\sum \frac{1}{n} z^n$ for the values of z within $|z| < 1$, we apply Dirichlet's test.

Let $\sum \frac{1}{n} z^n = \sum a_n u_n$ so that $a_n = \frac{1}{n}$ and $u_n = z^n$.

(i) $$|S_n| = \left| n\text{th partial sum of } \sum_{1}^{\infty} z^n \right| = |z + z^2 + \cdots + z^n| = \left|\frac{z(1-z^n)}{1-z}\right|$$

$$= \left|\frac{z - z^{n+1}}{1-z}\right| \leq \frac{|z| + |z|^{n+1}}{|1-z|} \leq \frac{1+1}{|1-z|} = \frac{2}{|1-z|}$$

$\Rightarrow \langle S_n \rangle$ is bounded.

(ii) $\lim_{n\to\infty} a_n = \lim_{n\to\infty} \frac{1}{n} = 0$

(iii) $u_n - u_{n+1} = \dfrac{1}{n} - \dfrac{1}{n+1} = \dfrac{1}{n(n+1)} < \dfrac{1}{n^2}$.

Also $\sum \dfrac{1}{n^2}$ is convergent.

Therefore, $\sum (u_n - u_{n+1})$ is convergent.

Hence the $\sum a_n u_n = \sum \dfrac{z^n}{n}$ is convergent for every value of z on the circle of convergence except at $z = 1$.

(ii) We have,

$$\begin{aligned}\frac{1}{R} &= \lim_{n\to\infty} \left|\frac{a_{n+1}}{a_n}\right|, \text{ here } a_n = \frac{1}{n(\log n)^2}\\ &= \lim_{n\to\infty} \left|\frac{n(\log n)^2}{(n+1)\log(n+1)^2}\right| = \lim_{n\to\infty} \frac{1}{1+\frac{1}{n}} \cdot \frac{(\log n)^2}{\log(n+1)^2}\\ &= \lim_{n\to\infty} \frac{(\log n)^2}{\log(n+1)^2} \left(\frac{\infty}{\infty}\right) = \lim_{n\to\infty} \frac{\log n}{\log(n+1)} \cdot \left(1+\frac{1}{n}\right)\\ &= \lim_{n\to\infty} \frac{1/n}{1/(n+1)} = \lim_{n\to\infty} \frac{n+1}{n} = 1.\end{aligned}$$

$\therefore R = 1$.

The radius of the circle of convergence with centre at $z = 0$ is 1.

We now consider the series $\sum \left|\dfrac{z^n}{n(\log n)^2}\right|$.

For $|z| = 1$, $\sum \dfrac{1}{n(\log n)^2}$ is convergent by Cauchy's condensation test, since here

$$f(n) = \frac{1}{n(\log n)^2} \Rightarrow a^n f(a^n) = \frac{a^n}{a^n(\log a^n)^2} = \frac{1}{n^2(\log a)^2},$$

where a is a (+)ve integer greater than 1.

$\Rightarrow$ the series $\sum \dfrac{1}{n^2(\log a)^2}$ is convergent since the factor $\dfrac{1}{(\log a)^2}$ is constant and $\sum \dfrac{1}{n^2}$ is convergent.

Hence the series is absolutely convergent $\forall$ z on $|z| = 1$.

Theorem [4.8] The power series $\sum_{1}^{\infty} n a_n z^{n-1}$, obtained by differentiating term by term the power series $\sum_{0}^{\infty} a_n z^n$ has the same radius of convergence as that of $\sum a_n z^n$.

Proof. Let R, R' be the radius of convergence of the power series $\sum_{0}^{\infty} a_n z^n$ and $\sum_{1}^{\infty} a_n z^{n-1}$ respectively.

Then $\dfrac{1}{R} = \overline{\lim_{n\to\infty}}\, |a_n|^{1/n}$ and $\dfrac{1}{R'} = \overline{\lim_{n\to\infty}}\, |na_n|^{1/n}$.

Therefore,

$$\frac{1}{R'} = \overline{\lim_{n\to\infty}}\, n^{1/n} \cdot |a_n|^{1/n} = \frac{1}{R} \cdot \lim_{n\to\infty} n^{1/n}. \tag{i}$$

Now by Cauchy's limit theorem

$$\lim_{n\to\infty} a_n^{1/n} = \lim_{n\to\infty} \frac{a_{n+1}}{a_n} \text{ and so } \lim_{n\to\infty} n^{1/n} = \lim_{n\to\infty} \frac{n+1}{n} = 1.$$

Hence, from (i), $\dfrac{1}{R'} = \dfrac{1}{R}$, i.e., $R = R'$.

Theorem [4.9] Let $f(z) = \sum\limits_0^\infty a_n z^n$ and R be the radius of convergence of $\sum\limits_0^\infty a_n z^n$. Then (1) $f(z)$ is analytic for $|z| < R$, (2) $f'(z) = \sum\limits_1^\infty n a_n z^{n-1}$.

Proof. Let $\phi(z) = \sum\limits_1^\infty n a_n z^{n-1}$. Then the radius of convergence of this power series is also R, by Theorem [4.8].

Let z be any point within the circle of convergence $|z| = R$, so that $|z| < R$. Then we can choose a (+)ve number r such that $|z| < r < R$.

We write, $|z| = \rho$ and $|h| = \eta$. Then $\rho < r$ and h is so chosen that $\rho + \eta < r$.

Since $\sum a_n z^n$ is convergent for $|z| < R$, $a_n r^n$ is bounded for $0 < r < R$. So there exists a finite (+)ve number M such that $|a_n r^n| < M \;\forall\; r,\; 0 < r < R$. (i)

Now

$$\begin{aligned}
\left|\frac{f(z+h) - f(z)}{h} - \phi(z)\right| &= \left|\frac{\sum a_n (z+h)^n - \sum a_n z^n}{h} - \sum n a_n z^{n-1}\right| \\
&= \left|\sum_{n=0}^\infty \left[a_n \left\{\frac{(z+h)^n - z^n}{h} - n z^{n-1}\right\}\right]\right| \\
&= \left|\sum_{n=0}^\infty \left[a_n \left\{\frac{n z^{n-1} h + \frac{n(n-1)}{2} z^{n-2} h^2 + \cdots + h^n}{h} - n z^{n-1}\right\}\right]\right| \\
&= \left|\sum_{n=0}^\infty \left[a_n \left\{\frac{n(n-1)}{2} z^{n-2} h + \cdots h^{n-1}\right\}\right]\right| \\
&\le \sum_{n=0}^\infty |a_n| \left\{\frac{n(n-1)}{2} |z|^{n-2} |h| + \cdots + |h|^{n-1}\right\} \\
&\le \sum_{n=0}^\infty \frac{M}{r^n} \left\{\frac{n(n-1)}{2} \rho^{n-2} \eta + \cdots + \eta^{n-1}\right\} \quad \text{[by (i)]} \\
&= \sum_{n=0}^\infty \frac{M}{r^n} \left\{\frac{1}{\eta}(\rho+\eta)^n - \frac{\rho^n}{\eta} - \frac{n\rho^{n-1}}{\eta} \cdot \eta\right\} \\
&= \sum_{n=0}^\infty \frac{M}{r^n} \left[\frac{1}{\eta} \left\{(\rho+\eta)^n - \rho^n - n\rho^{n-1}\eta\right\}\right] \\
&= \frac{M}{\eta} \sum_{n=0}^\infty \left[\left(\frac{\rho+\eta}{r}\right)^n - \left(\frac{\rho}{r}\right)^n - \frac{n\eta}{\rho} \cdot \left(\frac{\rho}{r}\right)^n\right].
\end{aligned} \tag{ii}$$

Here

$$\sum_0^\infty \left(\frac{\rho+\eta}{r}\right)^n = 1 + \frac{\rho+\eta}{r} + \left(\frac{\rho+\eta}{r}\right)^2 + \cdots$$

$$= \frac{1}{1-\frac{\rho+\eta}{r}} = \frac{r}{r-\rho-\eta} \quad [\because 1+x+x^2+\cdots = (1-x)^{-1}]$$

$$\sum_0^\infty \left(\frac{\rho}{r}\right)^n = \frac{1}{1-\rho/r} = \frac{r}{r-\rho}.$$

Now let $S = \sum n\left(\frac{\rho}{r}\right)^n$. Then $S = \frac{\rho}{r} + 2\left(\frac{\rho}{r}\right)^2 + 3\left(\frac{\rho}{r}\right)^3 + \cdots$.

Also $S\frac{\rho}{r} = \left(\frac{\rho}{r}\right)^2 + 2\left(\frac{\rho}{r}\right)^3 + 3\left(\frac{\rho}{r}\right)^4 + \cdots$, then $S - S\frac{\rho}{r} = \frac{\rho}{r} + \left(\frac{\rho}{r}\right)^2 + \left(\frac{\rho}{r}\right)^3 + \cdots$

$$\Rightarrow S\left(\frac{r-\rho}{r}\right) = \frac{\rho/r}{1-\rho/r} = \frac{\rho}{r-\rho}$$

$$\Rightarrow S = \frac{\rho_r}{(r-\rho)^2}.$$

From (ii), we get

$$\left|\frac{f(z+h)-f(z)}{h} - \phi(z)\right| \le \frac{M}{\eta}\left\{\frac{r}{r-\rho-\eta} - \frac{r}{r-\rho} - \frac{\eta r}{(r-\rho)^2}\right\}$$

on substituting these values

$$= \frac{M}{\eta}\,\frac{r\cdot\eta^2}{(r-\rho-\eta)(r-\rho)^2} = \frac{Mr\eta}{(r-\rho-\eta)(r-\rho)^2}$$

which tends to zero as $\eta \to 0$.

$$\therefore \lim_{h\to 0} \frac{f(z+h)-f(z)}{h} = \phi(z) \quad [\because |h| = \eta,\ \eta \to 0 \Rightarrow h \to 0]$$

$\Rightarrow f'(z) = \phi(z)$

$\Rightarrow f(z)$ has a derivative $\phi(z)$

$\Rightarrow f(z)$ is differentiable $\forall\ z,\ |z| < R$

$\Rightarrow f(z)$ is analytic within $|z| = R$.

Also $f'(z) = \phi(z) \Rightarrow f'(z) = \sum_1^\infty n a_n z^{n-1}$.

Theorem [4.10] [Abel's Limit Theorem] Let $f(z) = \sum_0^\infty a_n z^n$ and $\sum_0^\infty a_n$ converge. Then $f(z) \to f(1)$ as $z \to 1$ so that $\frac{|1-z|}{(1-|z|)}$ remains bounded.

Proof. Since $\sum_0^\infty a_n$ converges, there is no loss of generality if we suppose that $\sum_0^\infty a_n = 0$.

$$f(z) = \sum_0^\infty a_n z^n \Rightarrow f(1) = \sum_0^\infty a_n = 0.$$

Let $S_n = a_0 + a_1 + \cdots + a_n$.

Then

$$\begin{aligned} S_n(z) &= a_0 + a_1 z + a_2 z^2 + \cdots + a_n z^n \\ &= S_0 + (S_1 - S_0)z + (S_2 - S_1)z^2 + \cdots + (S_n - S_{n-1})z^n \\ &= S_0(1-z) + S_1(z - z^2) + S_2(z^2 - z^3) + \cdots + S_{n-1}(z^{n-1} - z^n) + S_n z^n \\ &= (1-z)(S_0 + S_1 z + S_2 z^2 + \cdots + S_{n-1} z^{n-1}) + S_n z^n. \end{aligned}$$

Making $n \to \infty$, we have

$$\lim_{n\to\infty} S_n(z) = (1-z)\sum_0^\infty S_n z^n + \lim_{n\to\infty} S_n z^n. \tag{i}$$

Now

$$\lim_{n\to\infty} S_n = 0 \Rightarrow \lim_{n\to\infty} S_n z^n = 0 \;\forall\; z.$$

Therefore, from (i),

$$f(z) = (1-z)\sum_0^\infty S_n z^n. \tag{ii}$$

Since $\frac{|1-z|}{1-|z|}$ remains bounded as $z \to 1$, there exists a (+)ve number k such that

$$\left|\frac{1-z}{1-|z|}\right| \le k. \tag{iii}$$

Again, since $\lim_{n\to\infty} S_n = 0$, for every $\epsilon(>0)$, there exists a (+)ve integer m such that

$$|S_n| < \frac{\epsilon}{2k} \;\forall\; n \ge m. \tag{iv}$$

From (ii),

$$\begin{aligned} |f(z)| = \left|(1-z)\sum_0^\infty S_n z^n\right| &= \left|(1-z)\left\{\sum_{n=0}^{m-1} S_n z^n + \sum_{n=m}^\infty S_n z^n\right\}\right| \\ &\le \left|(1-z)\sum_0^{m-1} S_n z^n\right| + \left|(1-z)\sum_{n=m}^\infty S_n z^n\right| \\ &\le \left|(1-z)\sum_0^{m-1} S_n z^n\right| + |1-z|\sum_{n=m}^\infty |S_n||z^n| \\ &\le \left|(1-z)\sum_0^{m-1} S_n z^n\right| + |1-z|\frac{\epsilon}{2k}\sum_{n=m}^\infty |z^n| \text{ [by (iv)]} \\ = \left|(1-z)\sum_0^{m-1} S_n z^n\right| &+ \frac{\epsilon}{2k}|1-z|\frac{|z^m|}{(1-|z|)} \left[\begin{array}{l}\text{summing the G.P. series} \\ \text{of infinite form}\end{array}\right] \\ = \left|(1-z)\sum_0^{m-1} S_n z^n\right| &+ \frac{\epsilon}{2k}k|z^m| \text{ [by (iii)].} \end{aligned}$$

$$\therefore |f(z)| \le |(1-z)|\left|\sum_0^{m-1} S_n z^n\right| + \frac{\epsilon}{2}|z|^m. \tag{v}$$

As $z \to 1$, $|1-z|\left|\sum_{0}^{m-1} S_n z^n\right| \to 0$, so for every $\epsilon(>0)$, there exists a (+)ve integer m_1 such that

$$|1-z|\left|\sum_{0}^{m-1} S_n z^n\right| < \frac{\epsilon}{2}\ \forall\ n \geq m_1.$$

Hence, taking $z \to 1$ in (v), we get

$$\lim_{z\to 1}|f(z)| \leq \lim_{z\to 1}|1-z|\left|\sum_{0}^{m-1} S_n z^n\right| + \lim_{z\to 1}\frac{\epsilon}{2}|z|^m$$

$$< \frac{\epsilon}{2}+\frac{\epsilon}{2} = \epsilon\ \forall\ n \geq p = \max(m, m_1)$$

$$\Rightarrow \lim_{z\to 1} f(z) = 0 = f(1).$$

EXERCISE

(1) Find the radii of convergence of the following power series:
(i) $\sum \frac{in+2}{2^n}z^n$, (ii) $\sum (1+2i)^n z^n$, (iii) $\sum \frac{(-1)^n}{n}(z-2i)^n$.

(2) Show that the domain of convergence of the series $\sum \frac{(iz-1)^n}{(2+i)}$ is given by $|z+1| < \sqrt{5}$.

(3) Show that $\sum_{1}^{\infty}\frac{1}{n}z^n$ converges in $|z \leq 1$ except at $z = 1$.

(4) Find the radius of convergence of $f(z) = \sum_{0}^{\infty}\frac{z^n}{3^n+1}$ and prove that $\{(3-z)f(z)-3\} \to 0$ as $z \to 3$.

(5) If c be any complex number and if $\sum_{n=1}^{\infty} z_n = S$, then show that $\sum_{1}^{\infty} cz_n = cS$.

(6) If $\sum_{n=1}^{\infty} z_n = S$ and $\sum_{1}^{\infty} T_n = T$ then show that $\sum_{1}^{\infty}(\lambda z_n + \mu T_n) = \lambda S + \mu T$.

(7) If $z_n = x_n + iy_n$ and $\{x_n\}$ and $\{y_n\}$ be sequence of real numbers, then show that convergence of $\{x_n\}$ and $\{y_n\} \Rightarrow |x_n| \leq M_1$ and $|y_n| \leq M_2$ for some (+)ve integers M_1 and M_2.

(8) If R_1 and R_2 be the radii of convergence of the power series $\sum_{0}^{\infty} a_n z^n$ and $\sum_{0}^{\infty} b_n z^n$ respectively, then find the radius of convergence of $\sum_{0}^{\infty} a_n b_n z^n$.

(9) Show that a power series $\sum_{0}^{\infty} a_n z^n$ can be integrated term by term within its circle of convergence.

(10) Prove that a power series $\sum_{0}^{\infty} a_n z^n$ represents an analytical function within its circle of convergence.

(11) Let $S_n(z)$ be the nth partial sum of the power series $\sum a_n z^n$ and let $|S_n(z)| \leq M_n$, M_n is independent of z, in a region R and let $\sum M_n$ converge in R. Then show that $\sum S_n(z)$ is uniformly convergent in R.

CHAPTER 5

CONFORMAL MAPPINGS

5.1 Mappings or Transformations

In case of real valued function $f(x)$, $y = f(x)$ represents a curve on the xy-plane. But in case of complex valued function $f(z)$, z is a complex variable, $w = f(z)$ does not represent any convenient graph since separate plane is required to represent each of the complex variable w and z. The plane representing the variable z is called z-plane and the plane representing the variables w is called w-plane. We write $z = x + iy$ and $w = u + iv$, where u and v are functions of x and y. That is, $u = u(x, y)$ and $v = v(x, y)$.

Thus to each point (x, y) in z-plane, in a domain D, there corresponds a point (u, v) in w-plane. Such correspondence is called a *mapping* or *transformation* of the points of z-plane into the points of w-plane and is represented by the transformation $w = f(z)$.

5.2 Jacobian of a Transformation

In general, $w = f(z) = u(x, y) + iv(x, y)$ maps a closed region D_z of z-plane into a closed region D_w of w-plane. Let Δ_z and Δ_w denote the areas of D_z and z_w respectively. If u and v are continuously differentiable functions, then $\lim\limits_{\Delta z \to 0} \dfrac{\Delta w}{\Delta z} = \dfrac{\partial(u, v)}{\partial(x, y)}$, where

$$\frac{\partial(u,v)}{\partial(x,y)} = \begin{vmatrix} \dfrac{\partial u}{\partial x} & \dfrac{\partial u}{\partial y} \\ \dfrac{\partial v}{\partial x} & \dfrac{\partial v}{\partial y} \end{vmatrix} = \frac{\partial u}{\partial x} \cdot \frac{\partial v}{\partial y} - \frac{\partial u}{\partial y} \cdot \frac{\partial v}{\partial x} \tag{1}$$

is called Jacobian of the transformation $u = u(x, y)$ and $v(x, y)$.

If, in particular, $f(z)$ be analytic, then by Cauchy Riemann equations,

$$\frac{\partial u}{\partial x} = \frac{\partial v}{\partial y}, \; \frac{\partial u}{\partial y} = -\frac{\partial v}{\partial x} \tag{2}$$

Therefore, from (1),

$$\begin{aligned} \frac{\partial(u,v)}{\partial(x,y)} &= \frac{\partial u}{\partial x} \cdot \frac{\partial u}{\partial x} - \frac{\partial v}{\partial x}\left(-\frac{\partial v}{\partial x}\right) \\ &= \left(\frac{\partial u}{\partial x}\right)^2 + \left(\frac{\partial v}{\partial x}\right)^2 = \left|\frac{\partial u}{\partial x} + i\frac{\partial v}{\partial x}\right|^2 = |f'(z)|^2. \end{aligned}$$

Problem [5.1] : Let $w = 2z + i$ be a transformation. Then determine the region D_w in the w-plane corresponding to the rectangular region D_z in the z-plane bounded by $x = 0, y = 0$ and $x = 2, y = 3$.

Solution. Here $w = 2z + i = 2(x + iy) + i = 2x + i(2y + 1)$.

Then $u(x, y) = 2x, v(x, y) = 2y + 1$.

The boundary lines of the rectangular region $D_z, x = 0, y = 0, x = 2, y = 3$ are mapped into the lines $u = 0, v = 1, u = 4, v = 7$. The regions D_z and D_w are shown in the diagram.

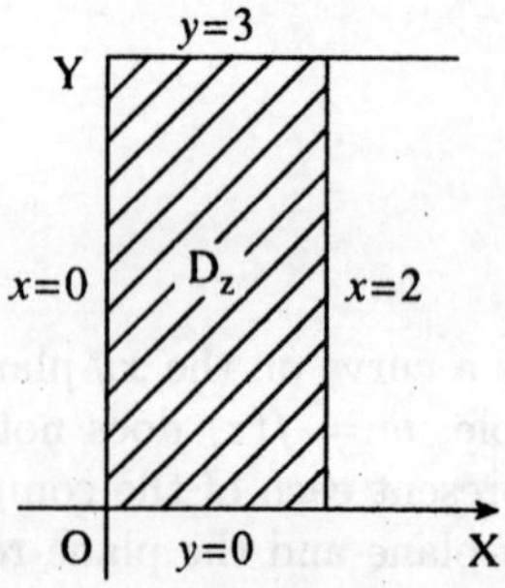

Fig. 5.1(i)

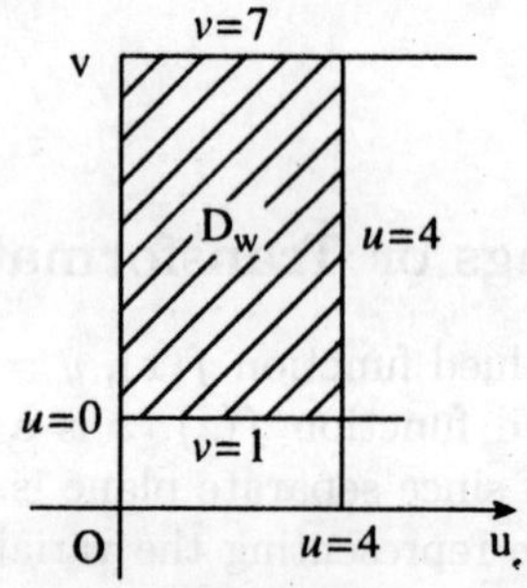

Fig. 5.1(ii)

Here the transformation is one-one, since each point of z-plane is mapped into one and only one point of w-plane and conversely. Also transformation performs a translation of rectangle given by $w = 2z + \alpha$, where $\alpha = i$ and the rectangle is translated in the direction of the vector $\bar{\alpha} = i$.

Problem [5.2] : Let the transformation be $w = ze^{i\pi/4}$. Then determine the region D_w in the w-plane corresponding to the triangular region D_z in z-plane bounded by the lines $x = 0$, $y = 0$ and $x + y = 1$.

Solution. $w = ze^{i\pi/4} = z\left(\dfrac{1}{\sqrt{2}} + i\dfrac{1}{\sqrt{2}}\right)$

$$\Rightarrow u + iv = (x + iy)\left(\frac{1}{\sqrt{2}} + i\frac{1}{\sqrt{2}}\right)$$

$$\Rightarrow u + iv = \frac{x - y}{\sqrt{2}} + i\left(\frac{x + y}{\sqrt{2}}\right) \Rightarrow u = \frac{x - y}{\sqrt{2}} \text{ and } v = \frac{x + y}{\sqrt{2}}.$$

For $x = 0, u = -\dfrac{y}{\sqrt{2}}, v = \dfrac{y}{\sqrt{2}} \Rightarrow v = -u$

$y = 0, u = \dfrac{x}{\sqrt{2}}, \quad v = \dfrac{x}{\sqrt{2}} \Rightarrow v = u$ and $x + y = 1 \Rightarrow v = \dfrac{1}{\sqrt{2}}$.

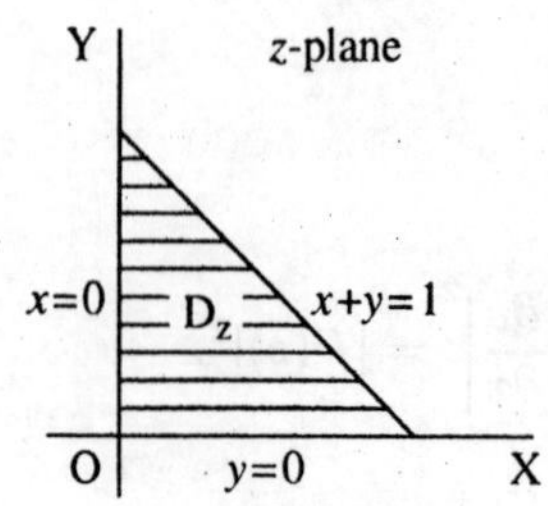

Fig. 5.2(i)

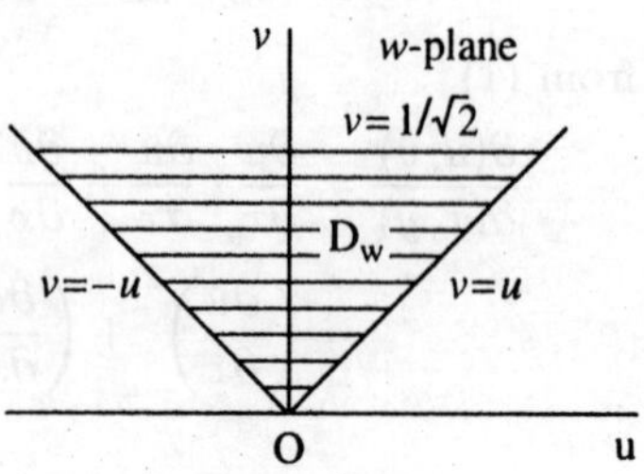

Fig. 5.2(ii)

Hence the triangular region D_z in the z-plane is mapped on a triangular region D_w in the w-plane bounded by the lines $v = u$, $v = -u$ and $v = \frac{1}{\sqrt{2}}$.

The transformation $w = ze^{i\pi/4}$ is a rotation of D_z through an angle $\pi/4$.

Problem [5.3] : Find the image of the infinite strips (i) $\frac{1}{4} < y < \frac{1}{2}$ (ii) $0 < y < \frac{1}{2}$ under the transformation $w = \frac{1}{z}$.

Solution. $w = \frac{1}{z} \Rightarrow u + iv = \frac{x - iy}{x^2 + y^2}$

$$\Rightarrow u = \frac{x}{x^2 + y^2},\ v = -\frac{y}{x^2 + y^2}.$$

Then $\frac{u}{v} = -\frac{x}{y}$, i.e., $x = -\frac{uy}{v}$.

Again, $v = -\frac{y}{x^2 + y^2}$

$$\Rightarrow v = -\frac{y}{\frac{u^2y^2}{v^2} + y^2} = -\frac{v^2}{y(u^2 + v^2)} \Rightarrow 1 = -\frac{v}{y(u^2 + v^2)}$$

$$\Rightarrow y = -\frac{v}{u^2 + v^2}.$$

(i) For the strip, $\frac{1}{4} < y < \frac{1}{2}$

$$y < \frac{1}{2} \Rightarrow -\frac{v}{u^2 + v^2} < \frac{1}{2} \Rightarrow -2v < u^2 + v^2 \Rightarrow u^2 + (v + 1)^2 > 1 \quad (1)$$

$$\text{and } y > \frac{1}{4} \Rightarrow -\frac{v}{u^2 + v^2} > \frac{1}{4} \Rightarrow -4v > u^2 + v^2 \Rightarrow u^2 + (v + 2)^2 < 4. \quad (2)$$

So from (1) and (2) it follows that the strip $\frac{1}{4} < y < \frac{1}{2}$ in z-plane is mapped into a region D_w bounded by the circles $u^2 + (v+1)^2 = 1$ and $u^2 + (v+2)^2 = 4$ in the w-plane having centres at $(0, -1)$ and $(0, -2)$.

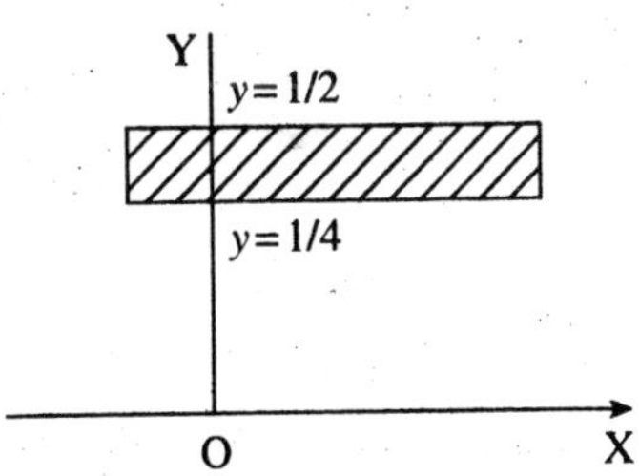

Fig. 5.3(i)

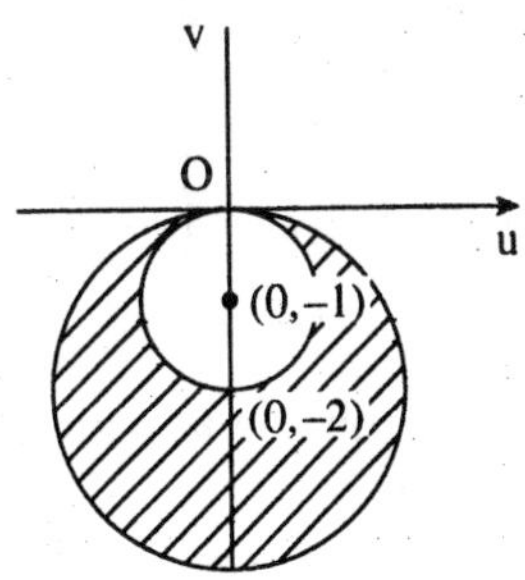

Fig. 5.3(ii)

(ii) $0 < y < \frac{1}{2}$

$$y = -\frac{v}{u^2 + v^2},\ y > 0 \Rightarrow v < 0$$

$$y < \frac{1}{2} \Rightarrow -\frac{v}{u^2 + v^2} < \frac{1}{2} \Rightarrow u^2 + (v + 1)^2 > 1.$$

It follows that as y diminishes from $\frac{1}{2}$ to 0 the circle in w-plane becomes bigger and bigger.

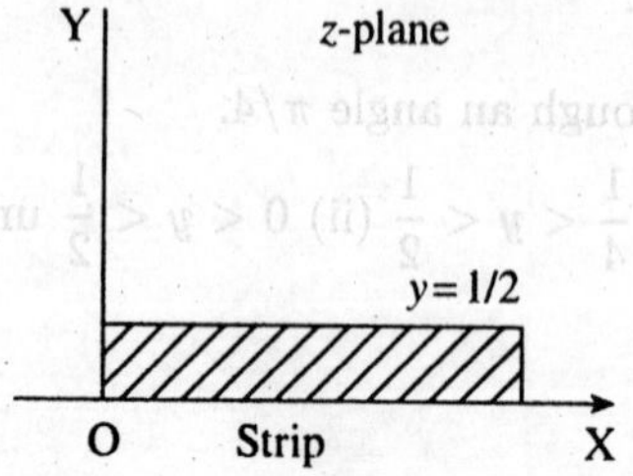

Fig. 5.4(i)

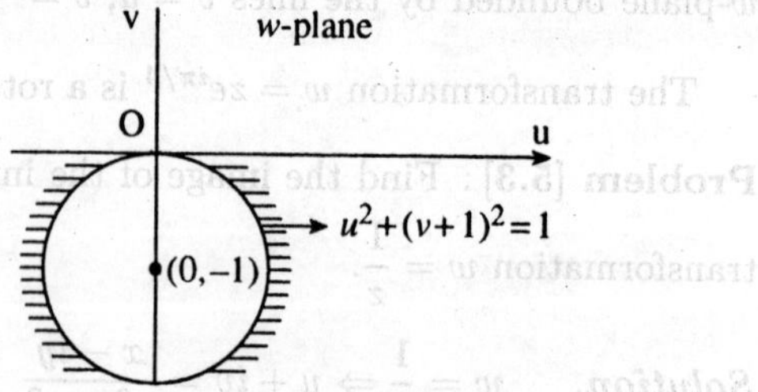

Fig. 5.4(ii)

Hence the infinite strip $0 < y < \frac{1}{2}$ is mapped into the region outside the circle $u^2+(v+1)^2 = 1$ in the lower half of w-plane.

Problem [5.4] : A rectangular region D_z in the z-plane bounded by $x = 0$, $y = 0$, $x = 2$, $y = 1$, determines the region D_w in the w-plane into which D_z is mapped under the transformation $w = \sqrt{2}e^{i\pi/4}z + (2 - i)$.

Solution. $w = \sqrt{2}ze^{i\pi/4} + (2 - i)$

$$\Rightarrow u + iv = \sqrt{2}(x + iy)\left(\frac{1}{\sqrt{2}} + i\frac{1}{\sqrt{2}}\right) + 2 - i$$

$$= (x + iy)(1 + i) + 2 - i$$

$$= (x - y + 2) + i(x + y - 1)$$

$\Rightarrow u = x - y + 2$ and $v = x + y - 1$.

Now for $x = 0$, $u = -y + 2$, $v = y - 1 \Rightarrow u + v = 1$

$x = 2$, $u = 4 - y$, $v = 1 + y \Rightarrow u + v = 5$

$$\Rightarrow \frac{u}{5} + \frac{v}{5} = 1.$$

Also for $y = 0$, $u = x + 2$, $v = x - 1 \Rightarrow u - v = 3$

$$\Rightarrow \frac{u}{3} + \frac{v}{(-3)} = 1$$

and $y = 1$, $u = x + 1$, $v = x \Rightarrow u - v = 1$

$$\Rightarrow \frac{u}{1} + \frac{v}{(-1)} = 1.$$

Transformed lines in the w-plane are

$$u + v = 1$$

$$\frac{u}{5} + \frac{v}{5} = 1$$

$$\frac{u}{3} + \frac{v}{(-3)} = 1$$

$$\frac{u}{1} + \frac{v}{(-1)} = 1.$$

Hence the transformed region D_w in the w-plane is bounded by the above lines. The region D_w in w-plane has been shown in the diagram.

The above transformation is a linear transformation of the type $w = \alpha z + \beta$, where α, β are complex constants. We rotate the region D_z in the z-plane through an angle $\theta = \arg \alpha$ and finally we translate the region D_w so obtained parallel to the vector β.

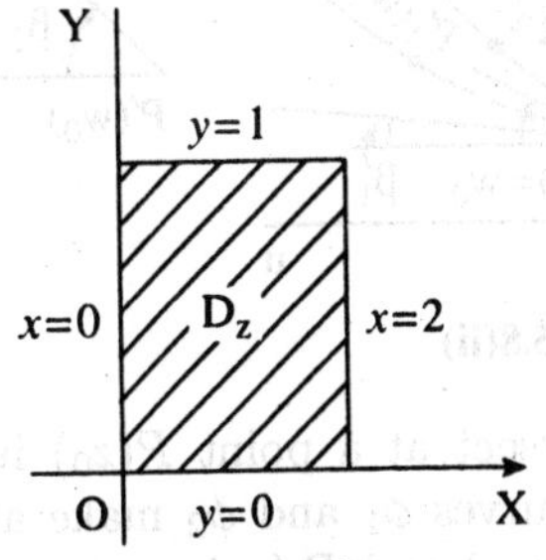

Fig. 5.5(i)

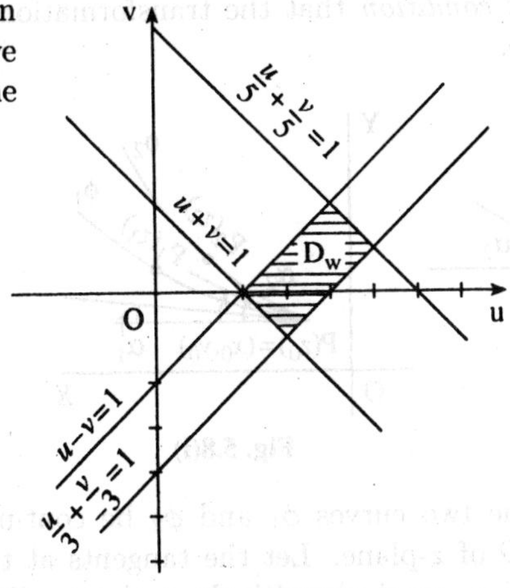

Fig. 5.5(ii)

Here rotational angle $\theta = \arg(1+i) = \pi/4$ and translated in the direction to the vector $\beta = 2 - i$ through a distance $|2 - i| = \sqrt{5}$ unit.

5.3 Conformal Mapping

Definition [5.1] : Let the two curves ϕ_1 and ϕ_2 intersect at a point $P(x_0, y_0)$ in the z-plane and let by the transformation $u = u(x,y)$ and $v = v(x,y)$ the curves mapped respectively into the curves ψ_1 and ψ_2 meeting at $P'(u_0, v_0)$, where (x_0, y_0) is mapped onto (u_0, v_0), in w-plane. Then the transformation is said to be *conformal* if the angle between ϕ_1 and ϕ_2 at (x_0, y_0) is equal in both magnitude and direction to the angle between ψ_1 and ψ_2 at (u_0, v_0).

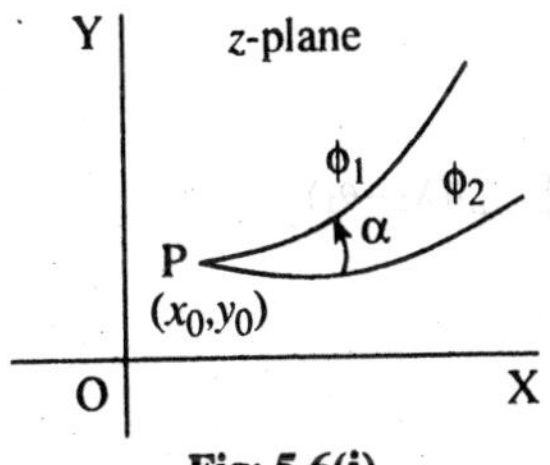

Fig: 5.6(i)

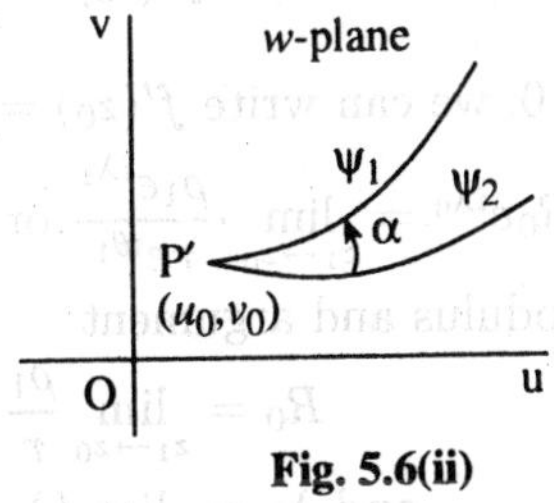

Fig. 5.6(ii)

The transformation is said to be *isogonal* if it preserves the magnitude of the angle but not necessarily the direction.

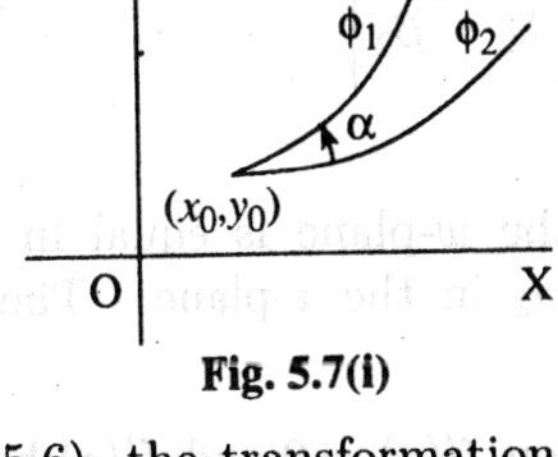

Fig. 5.7(i)

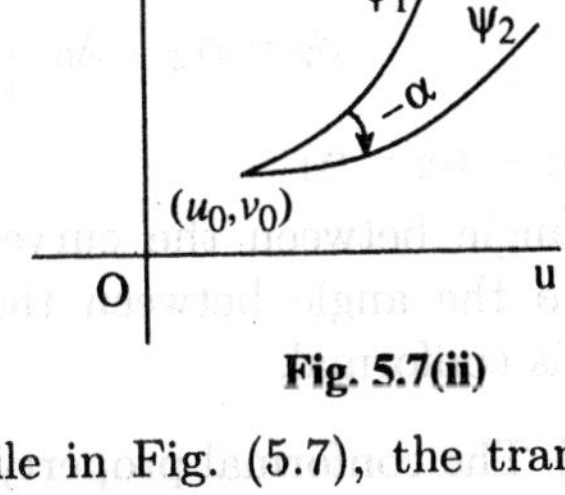

Fig. 5.7(ii)

In Fig. (5.6), the transformation is *conformal* while in Fig. (5.7), the transformation is *isogonal*.

Theorem [5.1] Let $f(z)$ be analytic function of z in a domain D in the z-plane. Then the *sufficient condition* that the transformation $w = f(z)$ to be conformal at the points of D is $f'(z) \neq 0$.

Proof.

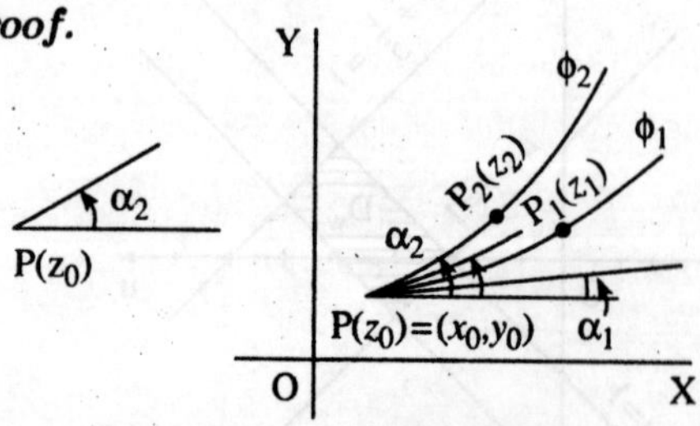

Fig. 5.8(i)

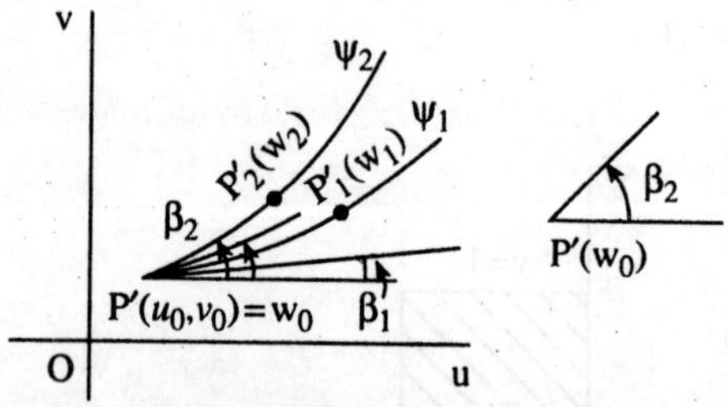

Fig. 5.8(ii)

Let the two curves ϕ_1 and ϕ_2 be continuous and they intersect at a point $P(z_0)$ in the domain D of z-plane. Let the tangents at the point P to the curves ϕ_1 and ϕ_2 make angles α_1 and α_2 respectively with the real axis. We take the points $P_1(z_1)$ and $P_2(z_2)$ on ϕ_1 and ϕ_2 respectively at the same distance r from the point $P(z_0)$, where r is very small.

Then we have $z_1 - z_0 = re^{i\theta_1}$ and $z_2 - z_0 = re^{i\theta_2}$ so that $\theta_1 \to \alpha_1, \theta_2 \to \alpha_2$ as $r \to 0$.

Now as $P(z_0) \to P_1(z_1)$ along ϕ_1, the image point moves along ψ_1 in the w-plane from $P'(w_0)$ to $P_1'(w_1)$. Similarly as the point $P(z_0)$ moves to $p_2(z_2)$ in the z-plane, the image point $P'(w_0)$ moves to $P_2'(w_2)$ in the w-plane.

Let the tangents at $P'(w_0)$ to the curves ψ_1 and ψ_2 make angles β_1 and β_2 with the real axis in the w-plane. We suppose that $w_1 - w_0 = \rho_1 e^{i\lambda_1}$ and $w_2 - w_0 = \rho_2 e^{i\lambda_2}$.

Since $f(z)$ is analytic, $f'(z)$ exists and

$$f'(z_o) = \lim_{z_1 \to z_0} \frac{w_1 - w_0}{z_1 - z_0}, \; z_0 \in D.$$

As $f'(z_0) \neq 0$, we can write $f'(z_0) = R_0 e^{i\lambda_0}$.

Therefore, $R_0 e^{i\lambda_0} = \lim\limits_{z_1 \to z_0} \cdot \dfrac{\rho_1 e^{i\lambda_1}}{re^{i\theta_1}}$ or, $R_0 e^{i\lambda_0} = \lim\limits_{z_1 \to z_0} \dfrac{\rho_1}{r} \cdot e^{i(\lambda_1 - \theta_1)}$.

Equating modulus and argument

$$R_0 = \lim_{z_1 \to z_0} \frac{\rho_1}{r}$$

$$\text{and } \lambda_0 = \lim_{z_1 \to z_0} (\lambda_1 - \theta_1) = \beta_1 - \alpha_1$$

$$\Rightarrow \beta_1 = \alpha_1 + \lambda_0 \left[\text{since } \lim_{z_1 \to z_0} \lambda_1 = \lim_{\rho_1 \to 0} \lambda_1 = \beta_1\right]$$

Similarly, we can show that

$$\beta_2 = \alpha_2 + \lambda_0 \left[\because \lim_{z_2 \to z_0} \lambda_2 = \lim_{\rho_2 \to 0} \lambda_2 = \beta_2\right].$$

Thus $\beta_2 - \beta_1 = \alpha_2 - \alpha_1$.

Hence the angle between the curves ψ_1 and ψ_2 in the w-plane is equal in magnitude and direction to the angle between the curves ϕ_1 and ϕ_2 in the z-plane. Therefore, the transformation is conformal.

Theorem [5.2] The conformal property does not hold when $f'(z) = 0$ and $f'(z)$ has a zero of nth order at $z = z_0$.

Proof. Let $f'(z)$ have a zero of nth order at $z = z_0$.

Then $f'(z) = (z - z_0)^n \phi(z)$, where $\phi(z)$ is analytic and $\phi(z_0) \neq 0$. It immediately follows that

$$f'(z_0) = f''(z_0) = \cdots = f^n(z_0) = 0, \text{ but } f^{n+1}(z_0) \neq 0. \quad \text{(i)}$$

Now expanding $f(z)$ by Taylor's theorem in the neighbourhood of z_0, we get

$$\begin{aligned} f(z) &= f(z_0) + (z - z_0)f'(z_0) + \frac{(z-z_0)^2}{2!}f''(z_0) + \cdots + \frac{(z-z_0)^n}{n!}f^n(z_0) \\ &\quad + \frac{(z-z_0)^{n+1}}{(n+1)!}f^{n+1}(z_0) + \frac{(z-z_0)^{n+2}}{(n+2)!}f^{n+2}(z_0) + \cdots \\ &= f(z_0) + \frac{(z-z_0)^{n+1}}{(n+1)!}f^{n+1}(z_0)\left[1 + \frac{(z-z_0)}{(n+2)}\frac{f^{n+2}(z_0)}{f^{n+1}(z_0)} + \cdots\right] \end{aligned}$$

$$\Rightarrow f(z) - f(z_0) = \frac{(z-z_0)^{n+1}}{(n+1)!}f^{n+1}(z_0)[1 + g(z)],$$

where $g(z)$ is continuous at z_0 and $g(z_0) = 0$

$$\Rightarrow w_1 - w_0 = \frac{f^{n+1}(z_0)}{(n+1)!}(z_1 - z_0)^{n+1}[1 + g(z_1)]. \text{ [for } z = z_1]. \quad \text{(ii)}$$

We now write, $w_1 - w_0 = \rho_1 e^{i\lambda_1}$ and $z_1 - z_0 = re^{i\theta_1}$ by using the notation and assumptions of the previous theorem.

Therefore, $\rho_1 e^{i\lambda_1} = \dfrac{f^{n+1}(z_0)}{(n+1)!}r^{n+1}e^{i(n+1)\theta_1}[1 + g(z_1)] = \dfrac{\rho e^{i\lambda}}{(n+1)!}r^{n+1}e^{i(n+1)\theta_1}[1 + g(z_1)]$

writing $f^{n+1}(z_0) = \rho e^{i\lambda}$ as $f^{n+1}(z_0) \neq 0$,

$$\Rightarrow \rho_1 e^{i\lambda_1} = \frac{1}{(n+1)!}\rho r^{n+1}\left[e^{i\{(n+1)\theta_1 + \lambda\}}\right][1 + g(z_1)]. \quad \text{(iii)}$$

Then

$$\lim_{z_1 \to z_0} \arg(w_1 - w_0) = \lim_{z_1 \to z_0} \arg\left[(z_1 - z_0)^{n+1} \cdot f^{n+1}(z_0)\right] \text{ [from (ii)]}$$

[Since $g(z_1) = 0$ as $z_1 \to z_0$ by virtue of continuity of g.]

Therefore, $\lim\limits_{z_1 \to z_0} \lambda_1 = \lim\limits_{z_1 \to z_0} [(n+1)\theta_1 + \lambda] = (n+1)\alpha_1 + \lambda$ [from (iii)]

That is

$$\lim_{z \to z_0} \lambda_1 = (n+1)\alpha_1 + \lambda \text{ [by Th. [5.1]]}. \quad \text{(iv)}$$

Similarly, we can show that $\lim\limits_{z_2 \to z_0} \lambda_2 = (n+1)\alpha_2 + \lambda$ [following the notation in Th. [5.1]]

That is,

$$\lim_{z \to z_0} \lambda_2 = (n+1)\alpha_2 + \lambda. \quad \text{(v)}$$

Therefore, from (iv) and (v) it follows that the curves ψ_1 and ψ_2 have definite tangent at $w = w_0$.

Also $\lim(\lambda_1 - \lambda_2) = (n+1)(\alpha_1 - \alpha_2)$ shows that the angle between the tangents to the curves ψ_1 and ψ_2 is $(n+1)$ times the angle between the curves ϕ_1 and ϕ_2 passing through z_0. Consequently the transformation is not conformal. That is, the conformal property does not hold.

Theorem [5.3] If the transformation $w = f(z)$ is conformal, then $f(z)$ is analytic.

Proof. $w = f(z) \Rightarrow u + iv = f(z)$.

Then $u = u(x, y)$ and $v = v(x, y)$ are the transformations from z-plane to w-plane.

Also let ds and $d\sigma$ be the elements of an arc in the z-plane and w-plane respectively.

Then $ds^2 = dx^2 + dy^2$ and $d\sigma^2 = du^2 + dv^2$.

Also $du = \frac{\partial u}{\partial x}dx + \frac{\partial u}{\partial y}dy$, $dv = \frac{\partial v}{\partial x}dx + \frac{\partial v}{\partial y}dy$.

Therefore,

$$d\sigma^2 = \left\{\frac{\partial u}{\partial x}dx + \frac{\partial u}{\partial y}dy\right\}^2 + \left\{\frac{\partial v}{\partial x}dx + \frac{\partial v}{\partial y}dy\right\}^2$$

$$= \left[\left(\frac{\partial u}{\partial x}\right)^2 + \left(\frac{\partial v}{\partial x}\right)^2\right]dx^2 + \left[\left(\frac{\partial u}{\partial y}\right)^2 + \left(\frac{\partial v}{\partial y}\right)^2\right]dy^2 + 2\left[\frac{\partial u}{\partial x}\frac{\partial u}{\partial y} + \frac{\partial v}{\partial x}\frac{\partial v}{\partial y}\right]dx\,dy.$$

Since the transformations is conformal, the ratio $d\sigma^2 : ds^2$ is independent of the direction.

$$d\sigma^2 : ds^2 \Rightarrow \frac{\left(\frac{\partial u}{\partial x}\right)^2 + \left(\frac{\partial v}{\partial x}\right)^2}{1} = \frac{\left(\frac{\partial u}{\partial y}\right)^2 + \left(\frac{\partial v}{\partial y}\right)^2}{1} = \frac{\frac{\partial u}{\partial x}\frac{\partial u}{\partial y} + \frac{\partial v}{\partial x}\frac{\partial v}{\partial y}}{0}$$

$$\Rightarrow \left(\frac{\partial u}{\partial x}\right)^2 + \left(\frac{\partial v}{\partial x}\right)^2 = \left(\frac{\partial u}{\partial y}\right)^2 + \left(\frac{\partial v}{\partial y}\right)^2 \qquad \text{(i)}$$

$$\text{and } \frac{\partial u}{\partial x}\cdot\frac{\partial u}{\partial y} + \frac{\partial v}{\partial x}\cdot\frac{\partial v}{\partial y} = 0. \qquad \text{(ii)}$$

From (ii), $\frac{\partial u}{\partial x}\Big/\frac{\partial v}{\partial y} = \frac{\partial v}{\partial x}\Big/ - \frac{\partial u}{\partial y} = \lambda$ (say).

$$\therefore \frac{\partial u}{\partial x} = \lambda\frac{\partial v}{\partial y} \text{ and } \frac{\partial v}{\partial x} = -\lambda\frac{\partial u}{\partial y}. \qquad \text{(iii)}$$

Putting in (i),

$$\lambda^2\left(\frac{\partial v}{\partial y}\right)^2 + \lambda^2\left(\frac{\partial u}{\partial y}\right)^2 = \left(\frac{\partial u}{\partial y}\right)^2 + \left(\frac{\partial v}{\partial y}\right)^2$$

$$\Rightarrow \left(\frac{\partial u}{\partial y}\right)^2 + \left(\frac{\partial v}{\partial y}\right)^2 [\lambda^2 - 1] = 0$$

$$\Rightarrow \lambda^2 - 1 = 0 \quad \left[\because \left(\frac{\partial u}{\partial y}\right)^2 + \left(\frac{\partial v}{\partial y}\right)^2 \neq 0\right]$$

$$\Rightarrow \lambda = \pm 1.$$

For $\lambda = 1$, from (iii),

$$\frac{\partial u}{\partial x} = \frac{\partial v}{\partial y} \text{ and } \frac{\partial v}{\partial x} = -\frac{\partial u}{\partial y}, \qquad \text{(iv)}$$

which are Cauchy-Riemann equations and hence $w = f(z)$ is an analytic function of z. But for $\lambda = -1$,

$$\frac{\partial u}{\partial x} = -\frac{\partial v}{\partial y} \text{ and } \frac{\partial v}{\partial x} = \frac{\partial u}{\partial y}. \qquad \text{(v)}$$

Equation (v) is reduced to equation (iv) if v is replaced by $-v$, that is, if we take the image figure obtained by the reflection in the real axis of w-plane. So the equations (v) corresponds to isogonal but not conformal transformation. Hence, it has been shown that for conformal transformation $f(z)$ is analytic.

Problem [5.5] : Prove that the transformation

$$u = \left(\sqrt{x^2+y^2}+x\right)^{1/2} \text{ and } v = \left(\sqrt{x^2+y^2}-x\right)^{1/2}$$

is conformal of z-plane to w-plane except at $z = 0$. Prove also that the angles at the origin are doubled in passing from z-plane to w-plane.

Solution. Changing into polar system, $x = r\cos\theta,\ y = r\sin\theta$

We get, $u = (r + r\cos\theta)^{1/2},\ v = (r - r\cos\theta)^{1/2}$

$$\Rightarrow u = \sqrt{2r}\cos\frac{\theta}{2} \text{ and } v = \sqrt{2r}\sin\frac{\theta}{2}$$

$$\Rightarrow \frac{\partial u}{\partial r} = \frac{\sqrt{2}}{2\sqrt{r}}\cos\frac{\theta}{2},\ \frac{\partial u}{\partial \theta} = -\sqrt{2r}\cdot\frac{1}{2}\sin\frac{\theta}{2}.$$

That is,

$$\frac{\partial u}{\partial r} = \frac{1}{\sqrt{2r}}\cos\frac{\theta}{2},\ \frac{1}{r}\frac{\partial u}{\partial \theta} = -\frac{1}{\sqrt{2r}}\sin\frac{\theta}{2}. \tag{1}$$

Again, from $v = \sqrt{2r}\sin\dfrac{\theta}{2}$, we get

$$\frac{\partial v}{\partial r} = \frac{1}{\sqrt{2r}}\cdot\sin\frac{\theta}{2}$$

$$\frac{\partial v}{\partial \theta} = \frac{\sqrt{2r}}{2}\cos\frac{\theta}{2}, \text{ i.e., } \frac{1}{r}\frac{\partial v}{\partial \theta} = \frac{1}{\sqrt{2r}}\cos\frac{\theta}{2}. \tag{2}$$

Thus from (1) and (2), we have

$$\left.\begin{aligned} \frac{\partial u}{\partial r} &= \frac{1}{r}\frac{\partial v}{\partial \theta}, \quad r \neq 0 \\ \text{and } \frac{1}{r}\frac{\partial u}{\partial \theta} &= -\frac{\partial v}{\partial r}, \quad r \neq 0 \end{aligned}\right\} \tag{3}$$

Thus the Cauchy-Riemann equations in polar form are satisfied and hence $w = u+iv = f(z)$ is an analytic function of z except at $z = 0$ (since (3) holds for $r \neq 0$ and so $z \neq 0$). Cnsequently, the transformation is conformal everywhere in the z-plane except at $z = 0$.

For the second part of the problem

Let $w = Re^{i\lambda}$ and $z = re^{i\theta}$.

Also $w = f(z) = u + iv = (r + r\cos\theta)^{1/2} + i(r - r\cos\theta)^{1/2}$.

Then we get, $R\cos\lambda = (r + r\cos\theta)^{1/2},\ R\sin\lambda = (r - r\cos\theta)^{1/2}$

$$\Rightarrow \tan\lambda = \frac{(1-\cos\theta)^{1/2}}{(1+\cos\theta)^{1/2}} = \frac{\sin\theta/2}{\cos\theta/2} = \tan\theta/2$$

$$\Rightarrow \lambda = \frac{\theta}{2} \Rightarrow \theta = 2\lambda,$$

which proves that the angles in z-plane is doubled the angle in w-plane at the origin.

Bilinear or Linear Fractional Transformation

Definition [5.2] : A transformation T defined by

$$w = T(z) = \frac{az+b}{cz+d}, \tag{1}$$

where a, b, c, d are complex constants with $ad - bc \neq 0$ is known as *bilinear* or *linear fractional* transformation. $ad - bc$ is called the determinant of the transformation. If $ad - bc = 1$, the transformation (1) is said to be *normalized*.

From (1), we have

$$czw + dw = az + b \text{ or, } czw + dw - az - b = 0, \tag{2}$$

which is linear both in z and w. That is why, it is called bilinear.

Möbius (1770-1868) first studied this type of transformation, so it is sometimes called *Möbius transformation.*

From (2),

$$z(cw - a) = -dw + b \text{ or, } z = \frac{-dw+b}{cw-a} = T^{-1}(w) \tag{3}$$

is called inverse transformation of (1).

Determinant of this transformation is $(-d)(-a) - b \cdot c = ad - bc$, which is same as that of (1).

The transformation (1) corresponds a unique point of w-plane to each point of the z-plane except at $z = -\frac{d}{c}, c \neq 0$.

Also the inverse transformation (3) corresponds a unique point of z-plane to each point of w-plane except at $w = a/c,\ c \neq 0$.

The points $z = -d/c$, and $w = a/c$, $c \neq 0$ are known as exceptional points. These exceptional points are mapped into the point $w = \infty$ and $z = \infty$ respectively. So if the complex plane is closed by the addition of the point ∞, then every bilinear transformation establishes a one-to-one correspondence between the z-plane closed with the point ∞ and the w-plane closed with ∞. That is, $T : z\{a + ib, \infty\} \to w\{c + id, \infty\}$, then $T(z)$ is one-one.

Example. (1) $w = \dfrac{2z+1}{4z+3}$.

Here $ad - bc = 6 - 4 = 2 \neq 0$. So the transformation is bilinear.

(2) $w = \dfrac{(3+2i)z+i}{(-13iz+(3-2i)}$.

Here $ad - bc = (3+2i)(3-2i) + 13i \cdot i = (9+4) - 13 = 0$.

So the transformation is not bilinear.

(3) $w = z = \dfrac{1 \cdot z + 0}{0 \cdot z + 1}$.

Here $ad - bc = 1 - 0 = 1 \neq 0$. So the transformation is bilinear.

Remark. The condition $ad - bc \neq 0$ is essential for a bilinear transformation $w = T(z) = az + b/cz + d$ so that there exists a one-to-one correspondence between the points of the closed z-plane and the closed w-plane.

For, if $ad - bc = 0$, $T(z)$ becomes a constant mapping and in that case, it fails to establish a one-to-one correspondence between the points of z-plane and w-plane. It follows from the following.

Let w_1 and w_2 be the points corresponding to z_1 and z_2 given by (1)

Then $w_1 = \dfrac{az_1+b}{cz_1+d}$ and $w_2 = \dfrac{az_2+b}{cz_2+d}$.

$$w_1 - w_2 = \frac{az_1+b}{cz_1+d} - \frac{az_2+b}{cz_2+d}$$
$$= \frac{(ad-bc)(z_2-z_1)}{(cz_1+d)(cz_2+d)} = 0 \quad \begin{bmatrix} \text{provided } z_1 \neq -d/c \\ \text{and } z_2 \neq -d/c \end{bmatrix}$$

$\Rightarrow w_1 = w_2$

$\Rightarrow w$ is constant.

Critical Points

$$w = \frac{az+b}{cz+d}, \quad ad-bc \neq 0.$$

Differentiaing this equation w.r.t. z, we get

$$\frac{dw}{dz} = \frac{ad-bc}{(cz+d)^2} \Rightarrow \frac{dw}{dz} = \begin{cases} \infty, & \text{if } z = -d/c \\ 0, & \text{if } z = \infty. \end{cases}$$

The points $z = -d/c$, $z = \infty$ are called critical points, where the conformal property does not hold good.

Definition [5.3] : [Invariant or Fixed points] The points which are mapped onto itself by a bilinear transformation (1) are called invariant points or fixed points. That is, a transformation $T(z) = z$ gives fixed points.

Let $T(z) = \dfrac{2z+2}{3z+1}$, where $ad - bc = 2 - 6 \neq 0$.

Fixed points are given by $T(z) = z$.

Theorem [5.4] If there are two distinct fixed points p and q, the transformation may be put in the form

$$\frac{w-p}{w-q} = k\,\frac{z-p}{z-q} \text{ (normal form)}$$

and if there is only one fixed point p, the transformation may be put in the form

$$\frac{1}{w-p} = \frac{1}{z-p} + k \text{ (normal form)}.$$

Proof. We consider a bilinear transformation $w = \dfrac{az+b}{cz+d}$, $ad - bc \neq 0$.

The invariant (fixed) points are given by $w = z$.

That is,

$$z = \frac{az+b}{cz+d} \Rightarrow cz^2 - (a-d)z - b = 0. \tag{i}$$

If p, q be two finite distinct points of (i), then

$$cp^2 - (a-d)p - b = 0 \tag{ii}$$
$$\text{and } cq^2 - (a-d)q - b = 0. \tag{iii}$$

Now

$$\frac{w-p}{w-q}=\frac{\dfrac{az+b}{cz+d}-p}{\dfrac{az+b}{cz+d}-q}=\frac{az+b-p(cz+d)}{az+b-q(cz+d)}$$

$$=\frac{(a-pc)z+b-pd}{(a-cq)z+b-qd}=\frac{(a-pc)z+cp^2-ap}{(a-cq)z+cq^2-aq}\text{ [by (ii) and (iii)]}$$

$$=\frac{(a-pc)z-p(a-pc)}{(a-qc)z-q(a-qc)}=\frac{(a-pc)(z-p)}{(a-qc)(z-q)}=k\,\frac{z-p}{z-q}$$

the required form by taking $k=(a-pc)/(a-qc)$.

(ii) Let there be only one fixed point p (say).

Then $p=\dfrac{a-d}{2c}$ [From (i), z having equal roots p] (iv)

and $cp^2-(a-d)p-b=0.$ (v)

Therefore,

$$\frac{1}{w-p}=\frac{1}{\dfrac{az+b}{cz+d}-p}=\frac{cz+d}{(a-cp)z+b-dp}=\frac{cz-(2pc-a)}{(a-cp)z+cp^2-ap}\text{ [by (iv) and (v)]}$$

$$=\frac{cz-cp+a-cp}{(a-cp)(z-p)}=\frac{c(z-p)+(a-cp)}{(a-cp)(z-p)}=\frac{c}{a-cp}+\frac{1}{z-p}$$

$=k+\dfrac{1}{z-p}$, required form by taking $\dfrac{c}{a-cp}=k.$

Remark. (1) A bilinear transformation with two distinct fixed points p and q has a normal form of the type $\dfrac{w-p}{w-q}=k.\ \dfrac{z-p}{z-q}$, if $p,q\neq\infty$.

If $q=\infty$, then $w-p=k(z-p)$ $\left[w=\dfrac{az+b}{cz+d}\right.$ has a finite fixed point p and $q=\infty$, then $c=0$, $d\neq 0\Rightarrow w=\dfrac{a}{d}z+\dfrac{b}{d}$ and $p=\dfrac{a}{d}p+\dfrac{b}{d}\Rightarrow w-p=\dfrac{a}{d}(z-p)=k(z-p)$, where $\left.k=\dfrac{a}{d}\right]$.

A transformation with two different fixed points is called

(i) elliptic, if $k=e^{i\alpha}$, $\alpha\neq 0$,

(ii) hyperbolic, if $k>0$

(iii) loxodormic, if $k=ae^{i\alpha}$, where $a\neq 1$, $\alpha>0$ α, a are both real numbers.

(2) A bilinear transformation with one fixed point p has a normal form of the type

$$\frac{1}{w-p}=\frac{1}{z-p}+k,\text{ if } p\neq\infty$$

and the transformation is called parabolic.

Problem [5.6]: Determine the fixed points and normal form of the following bilinear transformation (1) $w=\frac{2z}{z-3i}$, (2) $w=\frac{z-4}{z+5}$, (3) $w=\frac{(2+i)z-2}{z+i}$.

Solution. (1) Fixed points are given by $w=z$.

Then $\dfrac{2z}{z-3i}=z\Rightarrow z\left(\dfrac{2}{z-3i}-1\right)=0\Rightarrow z=0$ and $z=2+3i$.

These two are distinct fixed points.

Then

$$\frac{w-0}{w-(2+3i)}=\frac{\frac{2z}{z-3i}}{\frac{2z}{z-3i}-(2+3i)}=\frac{2z}{6i-3iz-9}$$
$$=\frac{2z}{-3i\{z-(2+3i)\}}=\frac{2i}{3}\,\frac{z-0}{z-(2+3i)}=k\cdot\left[\frac{z-0}{z-(2+3i)}\right],$$

which is the normal form, where $k=\frac{2}{3}i=\frac{2}{3}e^{i\pi/2}$.

Here $a=2/3\neq 1,\ \alpha=\pi/2>0$. So the transformation is loxodormic.

(2) Fixed points are given by $w=z$.

Then $\dfrac{z-4}{z+5}=z\Rightarrow z^2+4z+4=0\Rightarrow (z+2)^2=0.$

So there is only one distinct fixed point -2.

Then

$$\frac{1}{w+2}=\frac{1}{\frac{z-4}{z+5}+2}=\frac{z+5}{3z+6}=\frac{(z+2)+3}{3(z+2)}\Rightarrow\frac{1}{w+2}=\frac{1}{3}+\frac{1}{z+2}=\frac{1}{z+2}+k,$$

which is the normal form, where $k=\frac{1}{3}$. The transformation is parabolic.

(3) Fixed points are given by $w=z$.

Then, $\dfrac{(2+i)z-2}{z+i}=z\Rightarrow 2z+iz-2=z^2+iz\Rightarrow z^2-2z+2=0.$

$$\Rightarrow z=\frac{2\pm\sqrt{4-8}}{2}=\frac{2\pm 2i}{2}=1\pm i.$$

Fixed points are $1+i, 1-i$.

Therefore,

$$\frac{w-(1+i)}{w-(1-i)}=\frac{\frac{(2+i)z-2}{z+i}-(1+i)}{\frac{(2+i)z-2}{z+i}-(1-i)}=\frac{(2+i)z-2-z(1+i)-i+1}{(2+i)z-2-z(1-i)-i-1}$$
$$=\frac{z-1-i}{z+2iz-3-i}=\frac{z-(1+i)}{z(1+2i)-(3+i)}$$
$$=\frac{1}{1+2i}\cdot\frac{z-(1+i)}{z-\frac{(3+i)}{1+2i}}=\frac{1}{1+2i}\cdot\frac{z-(1+i)}{z-\frac{(3+i)(1-2i)}{5}}$$
$$=\frac{1}{1+2i}\cdot\frac{z-(1+i)}{z-\frac{5(1-i)}{5}}=\frac{1}{1+2i}\frac{z-(1+i)}{z-(1-i)}$$

$$\text{or,}\quad \frac{w-(1-i)}{w-(1+i)}=(1+2i)\frac{z-(1-i)}{z-(1+i)}=k\cdot\frac{z-(1-i)}{z-(1+i)},$$

which is the normal form, where $k=1+2i$, i.e., $k=\sqrt{5}e^{i\theta}$, where $\tan\theta=2$, i.e., $\theta=\tan^{-1}(2)$. Here the transformation is loxodromic.

Remark. The invariant or fixed points are given by $T(z)=z$

$$\Rightarrow\ \frac{az+b}{cz+d}=z,\ ad-bc\neq 0 \tag{1}$$

$$\Rightarrow\ cz^2-(a-d)z-b=0\Rightarrow z=\frac{(a-d)\pm\sqrt{(a-d)^2+4bc}}{2c}. \tag{2}$$

Now the number of fixed points is one or two according as $(a-d)^2+4bc=0$ or, $\neq 0$.

CASE (1) Let $c = 0$ and $d \neq 0$. Then from (1), we have

$$z = \frac{az+b}{d} = \frac{a}{d}z + \frac{b}{d} \Rightarrow z\left(\frac{d-a}{d}\right) = \frac{b}{d} \Rightarrow z = \frac{b}{d-a}. \tag{3}$$

Now if $a - d \neq 0$, then (2) determines one fixed point ∞ and (3) determines other finite fixed point.

Thus in this case, there are two fixed points one is infinite and the other is finite.

CASE (2) Let $c = 0$ and $a - d = 0$, then there exists only one fixed point ∞ determined by (3).

CASE (3) Let $c \neq 0$ and $(a-d)^2 + 4bc \neq 0$, then there are two finite fixed points.

CASE (4) Let $c \neq 0$ and $(a-d)^2 + 4bc = 0$, there exists only one finite fixed point $\dfrac{a-d}{2c}$.

Theorem [5.5] Every bilinear transformation is the resultant of elementary bilinear transformations with simple geometric properties.

Proof. We consider a bilinear transformation $w = \dfrac{az+b}{cz+d}$, $ad - bc \neq 0$.

(i) Let $c \neq 0$. Then

$$w = \frac{a\left[z + \frac{b}{a}\right]}{c\left[z + \frac{d}{c}\right]} = \frac{a}{c}\cdot\frac{z+\frac{b}{a}}{z+\frac{d}{c}} = \frac{a}{c}\left[\frac{\left(z+\frac{d}{c}\right) + \left(z+\frac{b}{a}\right) - \left(z+\frac{d}{c}\right)}{z+\frac{d}{c}}\right]$$

$$= \frac{a}{c}\left[1 + \frac{\frac{b}{a} - \frac{d}{c}}{z+\frac{d}{c}}\right] = \frac{a}{c} + \frac{a}{c}\cdot\frac{bc-ad}{ac}\cdot\frac{1}{z+\frac{d}{c}} = \frac{a}{c} + \frac{bc-ad}{c^2}\cdot\frac{1}{z+\frac{d}{c}}. \tag{1}$$

Taking $z_1 = z + \dfrac{d}{c}$, $z_2 = \dfrac{1}{z_1}$ and $z_3 = \dfrac{bc-ad}{c^2}\cdot z_2$, then from (1) we have $w = \dfrac{a}{c} + z_3$, which is similar to $z_1 = z + \dfrac{d}{c}$.

The above three auxiliary transformations z_1, z_2, z_3 are of the form

$$w = z + \alpha,\ w = \frac{1}{z},\ w = \beta z.$$

Thus it proves that every general bilinear transformation is the resultant of the elementary bilinear transformations $w = z + \alpha$, $w = \dfrac{1}{z}$, $w = \beta z$, where

(a) $w = z + \alpha$ represents translation.

(b) $w = \dfrac{1}{z}$ represents inversion.

(c) $w = \beta z$ represents rotation and magnification.

Finally the resultant transformation is again a translation.

(ii) Let $c = 0$. Then $w = \dfrac{a}{d}z + \dfrac{b}{d}$, provided $d \neq 0$.

Taking $z_1 = \dfrac{a}{d}z$, we have $w = z_1 + \dfrac{b}{d}$.

The auxiliary transformation z_1 is of the form $w = \beta z$, which represents rotation and magnification.

Thus the given transformation is the resultant of rotation, magnification and translation. Inversion falls out in this case.

Theorem [5.6] Product of two bilinear transformations is a bilinear transformation.

Proof. Let us consider two bilinear transformations defined by

$$w = T_1(z) = \frac{a_1 z + b_1}{c_1 z + d_1},\ a_1 d_1 - b_1 c_1 \neq 0 \tag{i}$$

$$\text{and } \zeta = T_2(w) = \frac{a_2 w + b_2}{c_2 w + d_2},\ a_2 d_2 - b_2 c_2 \neq 0. \tag{ii}$$

The transformation (i) establishes a one-to-one correspondence between the points of z-plane to the points of w-plane, while the transformation (ii) establishes a one-to-one correspondence between the points of w-plane to the points of ζ-plane and hence there defines a transformation from the z-plane to the ζ-plane by the relation

$$\zeta = T_2 T_1(z) = \frac{a_2 \cdot \dfrac{a_1 z + b_1}{c_1 z + d_1} + b_2}{c_2 \cdot \dfrac{a_1 z + b_1}{c_1 z + d_1} + d_2} = \frac{(a_2 a_1 + b_2 c_1)z + a_2 b_1 + b_2 d_1}{(c_2 a_1 + d_2 c_1)z + c_2 b_1 + d_2 d_1} = \frac{az + b}{cz + d}, \tag{iii}$$

where $a = a_2 a_1 + b_2 c_1$, $b = a_2 b_1 + b_2 d_1$, $c = c_2 a_1 + d_2 c_1$, $d = c_2 b_1 + d_2 d_1$.

The determinant of the product transformation $T_2 T_1$ is $ad - bc$

$$\begin{aligned}
&= (a_2 a_1 + b_2 c_1)(c_2 b_1 + d_2 d_1) - (a_2 b_1 + b_2 d_1)(c_2 a_1 + d_2 c_1) \\
&= a_2 a_1 d_2 d_1 + b_2 b_1 c_2 c_1 - b_2 c_2 d_1 a_1 - a_2 b_1 d_2 c_1 \\
&= a_1 d_1 (a_2 d_2 - b_2 c_2) + b_1 c_1 (b_2 c_2 - a_2 d_2) \\
&= a_1 d_1 (a_2 d_2 - b_2 c_2) - b_1 c_1 (a_2 d_2 - b_2 c_2) \\
&= (a_2 d_2 - b_2 c_2)(a_1 d_1 - b_1 c_1) \neq 0 \text{ [by (i) and (ii)]}
\end{aligned}$$

Therefore, $\zeta = T_2(T_1(z))$ is a bilinear transformation.

Example. Let $w = T_1(z) = \dfrac{z+2}{z+3}$ and $w = T_2(z) = \dfrac{z}{z+1}$.

Then $T_1(T_2(z)) = T_1\left(\dfrac{z}{z+1}\right) = \dfrac{\frac{z}{z+1} + 2}{\frac{z}{z+1} + 3} = \dfrac{3z+2}{4z+3}$.

That is, $T_1 T_2(z) = \dfrac{3z+2}{4z+3}$, where $3 \cdot 3 - 4 \cdot 2 = 1 \neq 0$.

T_1, T_2 are bilinear $\Rightarrow T_1 T_2$ is bilinear. Similarly, we can show that $T_2 T_1$ is also bilinear.

Again, from $w = T_1(z) = \dfrac{z+2}{z+3} \Rightarrow z(w-1) = -3w + 2 = 2 - 3w \Rightarrow z = \dfrac{2-3w}{w-1} = T_1^{-1}(w)$, where $2(-1) - (-3) \cdot 1 = -2 + 3 = 1 \neq 0$.

So $T_1^{-1}(w)$ is bilinear. Similarly, we can show that $T_2^{-1}(w)$ is also bilinear.

Again, $T_1 T_1^{-1}(w) = T_1\left(\dfrac{2-3w}{w-1}\right) = \dfrac{\frac{2-3w}{w-1} + 2}{\frac{2-3w}{w-1} + 3} = \dfrac{-w}{-1} = w$

$\Rightarrow T_1 T_1^{-1}$ is an identity transformation, which is bilinear.

Note. In general, if $w = T(z) = \dfrac{az+b}{cz+d}$, $ab - bc \neq 0$, then $TT^{-1}(w)$ is an identity bilinear transformation.

$$w = T(z) = \frac{az+b}{cz+d} \Rightarrow z = \frac{b - dw}{cw - a} = \frac{-dw + b}{cw - a} = T^{-1}(w).$$

Then

$$TT^{-1}(w) = T\left(\frac{-dw+b}{cw-a}\right) = \frac{a\left(\frac{-dw+b}{cw-a}\right)+b}{c\left(\frac{-dw+b}{cw-a}\right)+d} = \frac{(-ad+bc)w}{(bc-ad)} = w\ [\because\ ad-bc \neq 0].$$

Hence the result.

Theorem [5.7] Set of all bilinear transformations form a non-abelian group under the product of transformations.

Proof. Product of two bilinear transformations is a bilinear transformation.

That is, if T_1, T_2 be two bilinear transformations, then $T_1 \cdot T_2$ is a bilinear transformation, which follows from Theorem [5.6].

(i) Thus closure law holds w.r.t. product composition.

(ii) The associative law $(T_1 \cdot T_2)T_3 = T_1 \cdot (T_2 \cdot T_3)$ follows from the fact that this law holds for arbitrary mappings.

(iii) The identity mapping I defined by $w = I(z) = z$ is a bilinear transformation. Hence the identity law holds.

(iv) The inverse of a bilinear transformation $w = T(z) = \frac{az+b}{cz+d}$, $ad - bc \neq 0$, defined by $T^{-1}(w) = \frac{dw-b}{-cw+a}$ is a bilinear transformation as $ad - bc \neq 0$.

So the law of inverse holds.

That is, $TT^{-1}(w) = w,\ T^{-1}T(z) = z \Rightarrow TT^{-1} = I = T^{-1}T$.

Thus the set of all bilinear transformation forms a group under product composition. But in general, $T_1T_2(z) \neq T_2T_1(z)$.

For, if $T_1(z) = \dfrac{z+2}{z+3}$, $T_2(z) = \dfrac{z}{z+1}$.

Both are bilinear transformation.

$$T_1T_2(z) = \frac{3z+2}{4z+3}.$$

$$T_2T_1(z) = T_2\left(\frac{z+2}{z+3}\right) = \frac{\frac{z+2}{z+3}}{\frac{z+2}{z+3}+1} = \frac{z+2}{2z+5}.$$

Therefore, $T_1 \cdot T_2(z) \neq T_2 \cdot T_1(z)\ \forall\ z \in D$.

So the group is non-abelian.

Theorem [5.8] Every bilinear transformation maps circles or straight lines into circles or straight lines.

Proof. The equation of a circle with p and q as inverse points in the z-plane is given by

$$\left|\frac{z-p}{z-q}\right| = k,\ k \neq 1. \tag{1}$$

It stands a straight line, if $k = 1$.

CASE (1) We first consider the transformation of a circle under a bilinear transformation

$$w = \frac{az+b}{cz+d},\ ad - bc \neq 0, \tag{2}$$

$$\text{i.e., } z = \frac{dw-b}{-cw+a}.$$

Then (1) is transformed into

$$\left|\frac{\dfrac{dw-b}{-cw+a}-p}{\dfrac{dw-b}{-cw+a}-q}\right| = k \text{ or, } \left|\frac{(d+pc)w-(ap+b)}{(d+qc)w-(aq+b)}\right| = k$$

$$\text{or, } \left|\frac{w-\dfrac{ap+b}{d+pc}}{w-\dfrac{aq+b}{d+qc}}\right| \frac{|d+pc|}{|d+qc|} = k \text{ or, } \left|\frac{w-p'}{w-q'}\right| = k\left|\frac{d+qc}{d+pc}\right| \quad (3)$$

$$\text{or, } \left|\frac{w-p'}{w-q'}\right| = k',$$

where $p' = \dfrac{(ap+b)}{d+pc}$, $q' = \dfrac{(aq+b)}{d+qc}$ and $k' = k\left|\dfrac{d+qc}{d+pc}\right|$.

If $k' \neq 1$, then the image of circle in (1) under a bilinear transformation is a circle

$$\left|\frac{w-p'}{w-q'}\right| = k', \; k' \neq 1$$

w.r.t. the inverse points p' and q', which are respectively the image point of p and q.

If $k' = 1$, then the transformation $\left|\dfrac{w-p'}{w-q'}\right| = 1$ represents a straight line with p' and q' as symmetrical points.

CASE (2) We now consider the transformation of a straight line given by $\left|\dfrac{z-p}{z-q}\right| = 1$ in the z-plane by a bilinear transformation (2).

Now from

$$\left|\frac{z-p}{z-q}\right| = 1 \Rightarrow |z-p| = |z-q| \Rightarrow |z-p|^2 = |z-q|^2$$

$$\Rightarrow (z-p)(\bar{z}-\bar{p}) = (z-q)(\bar{z}-\bar{q})$$

$$\Rightarrow z(\bar{p}-\bar{q}) + \bar{z}(p-q) + \lambda = 0, \text{ where } \lambda = q\cdot\bar{q} - p\cdot\bar{p} \text{ is real.} \quad (4)$$

$$\Rightarrow A\bar{z} + \bar{A}z + \lambda = 0, \text{ where } A = p - q = \text{complex number},$$

which is the general form of a straight line.

Now under a bilinear transformation (2), where $z = \dfrac{dw-b}{-cw+a}$, the equation of the straight line in (4) is transformed into

$$\frac{dw-b}{-cw+a}(\bar{p}-\bar{q}) + \frac{\bar{d}\bar{w}-\bar{b}}{-\bar{c}\bar{w}+\bar{a}}(p-q) + \lambda = 0$$

$$\text{or, } (dw-b)(-\bar{c}\bar{w}+\bar{a})(\bar{p}-\bar{q}) + (\bar{d}\bar{w}-\bar{b})(-cw+a)(p-q) + \lambda(-cw+a)\left(\overline{-cw+a}\right) = 0$$

$$\text{or, } -\left\{\bar{c}d\left(\overline{p-q}\right) + \bar{d}c(p-q)\right\}w\bar{w} + \left\{d\bar{a}\left(\overline{p-q}\right) + \bar{b}c(p-q)\right\}w$$
$$+\left\{\bar{d}a(p-q) + b\bar{c}\left(\overline{p-q}\right)\right\}\bar{w} + \lambda' = 0, \text{ where } \lambda' = \lambda|-cw+a|^2 = \text{real}$$

$$\text{or, } \bar{B}w + B\bar{w} + k = 0, \; k \text{ is real, since } (z+\bar{z}) = \text{real and } w\cdot\bar{w} = |w|^2 = \text{real}$$

$$\Rightarrow -\{\bar{c}d(\overline{p-q}) + \bar{d}c(p-q)\}w\bar{w} \text{ is real, where } B = \bar{d}a(p-q) + b\bar{c}(\overline{p-q}),$$

which is again a straight line. Thus a straight line is mapped onto a straight line under a bilinear transformation (2).

Also it follows from (3) for $k = 1$,

$$\left|\frac{w-p'}{w-q'}\right| = \left|\frac{d+qc}{d+pc}\right| \neq 1,$$

which is a circle with respect to the inverse points p' and q'. Thus a straight line in the z-plane is mapped into a circle in the w-plane under a bilinear transformation (2).

Cross-Ratio

Definition [5.4] : The cross-ratio of four points z_1, z_2, z_3, z_4 taken in this order is defined by the ratio $\left(\frac{z_1 - z_2}{z_2 - z_3}\right) \Big/ \left(\frac{z_4 - z_1}{z_3 - z_4}\right) = \frac{(z_1 - z_2)(z_3 - z_4)}{(z_2 - z_3)(z_4 - z_1)}$ and is denoted by (z_1, z_2, z_3, z_4).

That is, $(z_1, z_2, z_3, z_4) = \dfrac{(z_1 - z_2)(z_3 - z_4)}{(z_2 - z_3)(z_4 - z_1)}$.

The way of writing the form of cross-ratio of the four points z_1, z_2, z_3, z_4 is the following.

First we take the difference of the points in the cyclic order starting from z_1. Thus we have four elements $z_1 - z_2, z_2 - z_3, z_3 - z_4, z_4 - z_1$, then we form the ratio by taking the first element $z_1 - z_2$ in numerator, 2nd element $z_2 - z_3$ in the denominator, third element $z_3 - z_4$ in the numerator and the fourth element $z_4 - z_1$ in the denominator.

Example. Let $1 - i, i, 1 + i, 3i$ be the four points taken in this order. Then their cross-ratio

$$(1 - i, i, 1 + i, 3i) = \frac{(1 - i - i)}{\{i - (1 + i)\}} \cdot \frac{(1 + i - 3i)}{\{3i - (1 - i)\}}$$

$$= \frac{(1 - 2i)(1 - 2i)}{(-1) \cdot (4i - 1)} = \frac{-3 - 4i}{-(4i - 1)} = \frac{4i + 3}{4i - 1}$$

Remark. (1) Different cross-ratios can be obtained according to the order in which the points are taken.

So permuting the four points z_1, z_2, z_3, z_4 we can obtain $4! = 24$ cross-ratios. But they are equivalent to only six different cross-ratios.

$$(z_1, z_2, z_3, z_4) = (z_2, z_1, z_4, z_3) = (z_3, z_4, z_1, z_2) = (z_4, z_3, z_2, z_1) \quad (1)$$
$$(z_1, z_4, z_2, z_3) = (z_1, z_2, z_4, z_3) = (z_3, z_4, z_2, z_1) = (z_4, z_3, z_1, z_2) \quad (2)$$
$$(z_3, z_2, z_1, z_4) = (z_2, z_3, z_4, z_1) = (z_1, z_4, z_3, z_2) = (z_4, z_1, z_2, z_3) \quad (3)$$
$$(z_4, z_2, z_3, z_1) = (z_2, z_4, z_1, z_3) = (z_3, z_1, z_4, z_2) = (z_1, z_3, z_2, z_4) \quad (4)$$
$$(z_1, z_3, z_2, z_4) = (z_3, z_1, z_4, z_2) = (z_2, z_4, z_1, z_3) = (z_4, z_2, z_3, z_1) \quad (5)$$
$$(z_1, z_2, z_4, z_3) = (z_2, z_1, z_3, z_4) = (z_4, z_3, z_1, z_2) = (z_3, z_4, z_2, z_1) \quad (6)$$

Now, if

$$\lambda = \frac{(z_1 - z_2)}{(z_2 - z_3)}\frac{(z_3 - z_4)}{(z_4 - z_1)} \text{ defined in (1), then}$$

$$1 - \lambda = 1 - \frac{(z_1 - z_2)(z_3 - z_4)}{(z_2 - z_3)(z_4 - z_1)} = \frac{(z_3 - z_1)(z_2 - z_4)}{(z_2 - z_3)(z_4 - z_1)}$$

$$= \frac{(z_4 - z_2)(z_3 - z_1)}{(z_2 - z_3)(z_1 - z_4)} \text{ represents (4).}$$

$$\frac{1}{\lambda} = \frac{(z_2 - z_3)(z_4 - z_1)}{(z_1 - z_2)(z_3 - z_4)} = \frac{(z_3 - z_2)(z_1 - z_4)}{(z_2 - z_1)(z_4 - z_3)} \text{ represents (3).}$$

Now,

$$\lambda - 1 = \frac{(z_3 - z_1)(z_2 - z_4)}{(z_2 - z_3)(z_1 - z_4)}.$$

Then

$$\frac{\lambda}{\lambda - 1} = -\frac{(z_1 - z_2)(z_3 - z_4)}{(z_3 - z_1)(z_2 - z_4)} = \frac{(z_1 - z_2)(z_4 - z_3)}{(z_3 - z_1)(z_2 - z_4)} \text{ represents (6)}$$

$$\frac{1}{1 - \lambda} = \frac{(z_2 - z_3)(z_1 - z_4)}{(z_4 - z_2)(z_3 - z_1)} \text{ represents (2).}$$

Also

$$\frac{\lambda - 1}{\lambda} = \frac{(z_3 - z_1)(z_2 - z_4)}{(z_1 - z_2)(z_4 - z_3)} = \text{represents (5).}$$

Remark. (2) Let z_2, z_3, z_4 be three distinct finite points in the extended z-plane (i.e., the z-plane containing ∞) and $w = T(z) = \dfrac{(z - z_2)(z_3 - z_4)}{(z_2 - z_3)(z_4 - z)}$ be a bilinear transformation.

Then $T(z_2) = 0$, $T(z_3) = 1$ and $T(z_4) = \infty$. Thus the points z_2, z_3, z_4 are mapped onto $0, 1, \infty$ respectively in the extended w-plane. Also the image of any point z_1 under the given bilinear transformation is

$$w = T(z_1) = \frac{(z_1 - z_2)(z_3 - z_4)}{(z_2 - z_3)(z_4 - z_1)} = (z_1, z_2, z_3, z_4).$$

Thus the cross-ratio (z_1, z_2, z_3, z_4) is defined as the image of a bilinear transformation

$$w = T(z) = \frac{(z - z_2)(z_3 - z_4)}{(z_2 - z_3)(z_4 - z)},$$

which transforms z_2, z_3, z_4 into $0, 1, \infty$ respectively.

Preservation of cross-ratio under bilinear transformation

Theorem [5.9] The cross-ratio remains invariant under a bilinear transformation.

Proof. Let w_1, w_2, w_3, w_4 be the images of the four points z_1, z_2, z_3, z_4 respectively under a bilinear transformation

$$w = T(z) = \frac{az + b}{cz + d}, \quad ad - bc \neq 0.$$

We have to show that $(w_1, w_2, w_3, w_4) = (z_1, z_2, z_3, z_4)$.

$$w_1 = \frac{az_1 + b}{cz_1 + d}, \quad w_2 = \frac{az_2 + b}{cz_2 + d}, \quad w_3 = \frac{az_3 + b}{cz_3 + d}, \quad w_4 = \frac{az_4 + b}{cz_4 + d}.$$

Now,

$$w_1 - w_2 = \frac{az_1 + b}{cz_1 + d} - \frac{az_2 + b}{cz_2 + d} = \frac{bc(z_2 - z_1) + ad(z_1 - z_2)}{(cz_1 + d)(cz_2 + d)} = \frac{(ad - bc)(z_1 - z_2)}{(cz_1 + d)(cz_2 + d)}$$

$$w_3 - w_4 = \frac{(ad - bc)(z_3 - z_4)}{(cz_3 + d)(cz_4 + d)}$$

$$w_2 - w_3 = \frac{(ad - bc)(z_2 - z_3)}{(cz_2 + d)(cz_3 + d)}$$

$$w_4 - w_1 = \frac{(ad - bc)(z_4 - z_1)}{(cz_4 + d)(cz_1 + d)}.$$

Then

$$(w_1, w_2, w_3, w_4) = \frac{(w_1 - w_2)}{(w_2 - w_3)} \cdot \frac{(w_3 - w_4)}{(w_4 - w_1)} = \frac{(ad - bc)^2(z_1 - z_2)(z_3 - z_4)}{(cz_1 + d)(cz_2 + d)(cz_3 + d)(cz_4 + d)}$$

$$\times \frac{(cz_1 + d)(cz_2 + d)(cz_3 + d)(cz_4 + d)}{(ad - bc)^2(z_2 - z_3)(z_4 - z_1)}$$

$$= \frac{(z_1 - z_2)(z_3 - z_4)}{(z_2 - z_3)(z_4 - z_1)} = (z_1, z_2, z_3, z_4).$$

Theorem [5.10] The cross-ratio of four points z_1, z_2, z_3, z_4 is real, if and only if the four points z_1, z_2, z_3, z_4 are concyclic or collinear.

Proof.

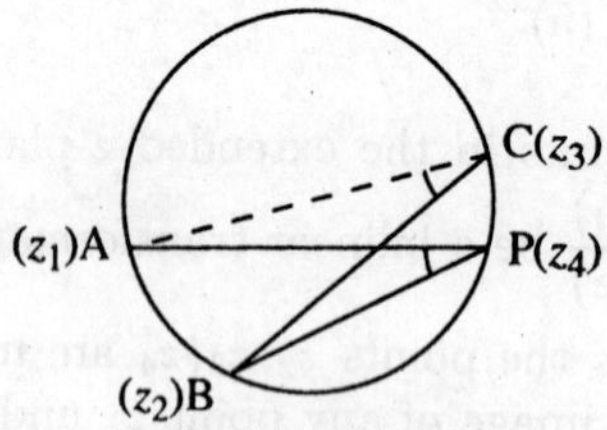

Fig. 5.9(i)

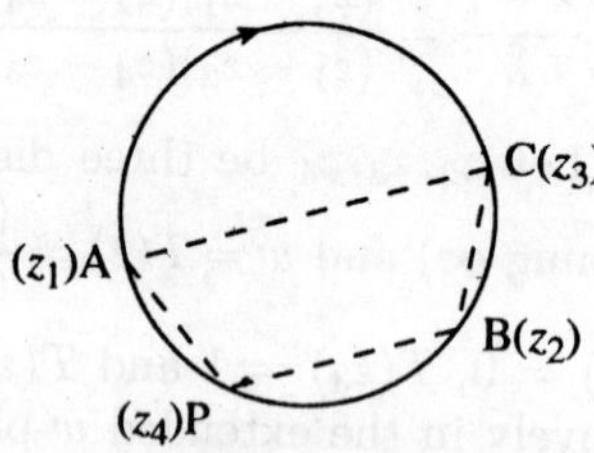

Fig. 5.9(ii)

Let four points $A(z_1), B(z_2), C(z_3)$ and $P(z_4)$ lie on a circle. In Fig. (5.9)(i), $P(z_4)$ lies on the semicircular arc $\widehat{ACB}$ and in Fig.5.9(ii) $P(z_4)$ does not lie on the semicircular arc $\widehat{ACB}$.

So in Fig. 5.9(i), we have

$$\angle ACB = \angle APB \tag{i}$$

and in Fig. 5.9(ii),

$$\angle ACB + \angle APB = \pi. \tag{ii}$$

So from (i),

$$\arg \frac{z_3 - z_2}{z_3 - z_1} = \arg \frac{z_4 - z_2}{z_4 - z_1}$$

$$\Rightarrow \arg \frac{z_3 - z_2}{z_3 - z_1} - \arg \frac{z_4 - z_2}{z_4 - z_1} = 0$$

$$\Rightarrow \arg \left. \frac{z_3 - z_2}{z_3 - z_1} \right/ \frac{z_4 - z_2}{z_4 - z_1} = 0$$

$$\Rightarrow \arg \frac{(z_3 - z_2)(z_4 - z_1)}{(z_3 - z_1)(z_4 - z_2)} = 0$$

$$\Rightarrow \arg(z_2, z_3, z_1, z_4) = 0$$

$\Rightarrow$ cross-ratio of four points is real.

From Fig. 5.9(ii),

$$\arg \frac{z_3 - z_2}{z_3 - z_1} + \arg \frac{z_4 - z_1}{z_4 - z_2} = \pi$$

$$\Rightarrow \arg \frac{(z_3 - z_2)(z_4 - z_1)}{(z_3 - z_1)(z_4 - z_2)} = \pi$$

$$\Rightarrow \arg(z_2, z_3, z_1, z_4) = \pi$$

$\Rightarrow$ cross-ratio of four points is real.

Problem [5.7]: Determine a bilinear transformation which transforms three distinct points in the z-plane into three specified distinct points.

Solution. Let z_1, z_2, z_3 be three distinct finite points in the z-plane and w_1, w_2, w_3 be three specified distinct points in the w-plane.

Let us consider a bilinear transformation

$$\frac{(w-w_1)(w_2-w_3)}{(w_1-w_2)(w_3-w)} = \frac{(z-z_1)(z_2-z_3)}{(z_1-z_2)(z_3-z)} \tag{1}$$

$$\text{or, } (w-w_1)(w_2-w_3)(z_1-z_2)(z_3-z) = (z-z_1)(z_2-z_3)(w_1-w_2)(w_3-w). \tag{2}$$

Now, if we put $z = z_1$ in (2), then

$$(w-w_1)(w_2-w_3)(z_1-z_2)(z_3-z_1) = 0 \Rightarrow w = w_1$$

[since $z_1, z_2, z_3, w_1, w_2, w_3$ are distinct points.]

Again, if we put $z = z_3$ in (2) we have $0 = (z_3-z_1)(z_2-z_3)(w_1-w_2)(w_3-w) \Rightarrow w = w_3$.

For $z = z_2$ in (2), we get

$$(w-w_1)(w_2-w_3)(z_1-z_2)(z_3-z_2) = (z_2-z_1)(z_2-z_3)(w_1-w_2)(w_3-w)$$
$$\Rightarrow (z_1-z_2)(z_3-z_2)\{(w-w_1)(w_2-w_3)-(w_1-w_2)(w_3-w)\} = 0$$
$$\Rightarrow (w-w_1)(w_2-w_3)-(w_1-w_2)(w_3-w) = 0 \; [z_1 \neq z_2,\; z_2 \neq z_3]$$
$$\Rightarrow w\{w_2-w_3+w_1-w_2\} = w_1w_2-w_1w_3+w_1w_3-w_2w_3$$
$$\Rightarrow w(w_1-w_3) = w_2(w_1-w_3)$$
$$\Rightarrow w = w_2 \; [w_1-w_3 \neq 0].$$

Hence (1) is the required bilinear transformation, which transforms z_1, z_2, z_3 into w_1, w_2, w_3 respectively.

Problem [5.8]: (1) Find a bilinear transformation in the following cases:

(a) the points $z = 2, i, -2$ are transformed into $w = 1, i, -1$ respectively,

(b) the points $z = \infty, i, 0$ are transformed into $w = 0, i, \infty$ respectively.

(2) Show that the transformation $w = \frac{z+3}{2z+4}$ transforms the circle $x^2+y^2+2x = 0$ into the straight line $-4u+3 = 0$.

(3) Show that the bilinear transformation which transforms the point $z = i, 0, -i$ into $w = 0, -1, \infty$ maps

(a) the real axis $\text{Im}(z) = 0$ on $|w| = 1$,

(b) the upper half-plane $\text{Im}(z) > 0$ on $|w| < 1$,

(c) the lower half-plane $\text{Im}(z) < 0$ on $|w| > 1$.

(4) Show that the transformation $w = i\frac{1-z}{1+z}$ transforms the circle $|z| = 1$ onto the real axis of w-plane and the interior of the circle $|z| < 1$ into the upper half of the w-plane.

(5) Find all Möbius transformation which transform

(a) the half-plane $\text{Im}(z) \geq 0$ into circles $|w| \leq 1$,

(b) the half-plane $\text{R}(z) \geq 0$ into the unit circle $|w| \leq 1$.

(6) Find the general bilinear transformation which transforms

(a) the circular disc $|z| \leq \rho$ onto the circular disc $|w| \leq \rho'$,

(b) the unit circular disc $|z| \leq 1$ onto the unit circular disc $|w| \leq 1$.

Solution. (1) (a) Let the bilinear transformation be

$$w = \frac{az+b}{cz+d}, \ ad - bc \neq 0. \tag{1}$$

According to the given transformation of points

$$1 = \frac{2a+b}{2c+d} \Rightarrow 2a + b - 2c - d = 0 \tag{2}$$

$$i = \frac{ia+b}{ic+d} \Rightarrow ia + b + c - id = 0 \tag{3}$$

$$-1 = \frac{-2a+b}{-2c+d} \Rightarrow -2a + b - 2c + d = 0. \tag{4}$$

(2) + (4) gives

$$2b - 4c = 0 \Rightarrow c = \frac{b}{2}. \tag{5}$$

(2) − (4) gives

$$4a - 2d = 0 \Rightarrow d = 2a. \tag{6}$$

From (3),

$$ia + b + \frac{b}{2} - 2ai = 0 \Rightarrow -ai + \frac{3b}{2} = 0 \Rightarrow a = \frac{3b}{2i}. \tag{7}$$

Therefore, from (6), $d = \dfrac{3b}{i}$.

Hence the transformation (1) is of the form:

$$w = \frac{\frac{3b}{2i}z + b}{\frac{b}{2}z + \frac{3b}{i}} = \frac{3z+2i}{iz+6}. \left[\begin{array}{l}\text{multiplying numerator and} \\ \text{denominator by } \dfrac{2i}{b}\end{array}\right]$$

(b) $z = \infty \to w = 0$.

So from Eqn. (1), we have

$$0 = \frac{a + \frac{b}{z}}{c + \frac{d}{z}} = \frac{a}{c} \Rightarrow a = 0. \tag{i}$$

$z = i \to w = i$

$$\therefore i = \frac{ai+b}{ci+d} \Rightarrow i = \frac{b}{ci+d} \text{ [by (i)]}$$

$$\Rightarrow b + c - id = 0. \tag{ii}$$

Again, $z = 0 \to w = \infty$.

$$\therefore \infty = \frac{b}{d} \Rightarrow d = \frac{b}{\infty} = 0. \tag{iii}$$

$d = 0 \Rightarrow c = -b$ from (ii)

Thus the transformation (1) is of the form $w = \dfrac{b}{-bz} = -\dfrac{1}{z}$.

(2) Here $w = (z+3)/(2z+4)$

$$\Rightarrow z = \frac{-4w+3}{2w-1}, \text{ then } \bar{z} = \frac{-4\bar{w}+3}{2\bar{w}-1}. \tag{i}$$

Now, $x^2 + y^2 + 2x = 0$

$$\Rightarrow (x+iy)(x-iy) + (x+iy+x-iy) = 0$$

$$\Rightarrow z \cdot \bar{z} + (z+\bar{z}) = 0 \quad [\because z = x+iy, \bar{z} = x-iy]$$

$$\Rightarrow \left(\frac{-4w+3}{2w-1}\right)\left(\frac{-4\bar{w}+3}{2\bar{w}-1}\right) + \left(\frac{-4w+3}{2w-1}\right) + \left(\frac{-4\bar{w}+3}{2\bar{w}-1}\right) = 0$$

$$\Rightarrow (-4w+3)(-4\bar{w}+3) + (-4w+3)(2\bar{w}-1) + (-4\bar{w}+3)(2w-1) = 0$$

$$\Rightarrow 16w\bar{w} - 12w - 12\bar{w} + 9 - 8w\bar{w} + 4w + 6\bar{w} - 3 - 8w\bar{w} + 4\bar{w} + 6w - 3 = 0$$

$$\Rightarrow -2w - 2\bar{w} + 3 = 0 \Rightarrow -2(w+\bar{w}) + 3 = 0$$

$$\Rightarrow -2(2u) + 3 = 0 \Rightarrow -4u + 3 = 0 \quad [\because w = u+iv,\ \bar{w} = u-iv]$$

$\therefore -4u + 3 = 0$ is a straight line in the w-plane.

Thus a circle in z-plane is mapped into a straight line in w-plane.

(3) Let us consider a bilinear transformation

$$w = \frac{az+b}{cz+d},\ ad - bc \neq 0. \tag{i}$$

(a) Transforms $z = i, 0, -i$ into $w = 0, -1, \infty$ respectively, then

$$0 = \frac{ai+b}{ci+d} \Rightarrow ai + b = 0 \Rightarrow b = -ai \tag{ii}$$

$$-1 = \frac{b}{d} \Rightarrow d = -b. \tag{iii}$$

Again,

$$\infty = \frac{-ai+b}{-ci+d} \Rightarrow -ci + d = 0 \Rightarrow c = \frac{d}{i} = -id = ib \text{ [by (iii)]} \tag{iv}$$

Again,

$$a = -\frac{b}{i} = bi \text{ [from (ii)]} \tag{v}$$

Therefore,

$$w = \frac{biz+b}{ibz-b} \text{ [by (iii), (iv) and (v)]}$$

$$= \frac{iz+1}{iz-1} = \frac{z-i}{z+i}. \tag{vi}$$

Any point on the real axis $\text{Im}(z) = 0 \Rightarrow y = 0 \Rightarrow z = x$.

Then its image is

$$w = \frac{x-i}{x+i},\ \text{i.e.,}\ |w| = \left|\frac{x-i}{x+i}\right| = \frac{|x-i|}{|x+i|} = \frac{\sqrt{x^2+1}}{\sqrt{x^2+1}} = 1.$$

Thus the bilinear transformation (i) maps $\text{Im}(z) = 0$ into a unit circle $|w| = 1$.

(b) Any point of the upper half-plane $\text{Im}(z) > 0$ can be taken as $z = x + iy$, $y > 0$.

Then its image is

$$w = \frac{x+iy-i}{x+iy+i} = \frac{x+i(y-1)}{x+i(y+1)} \text{ [by (vi)]}$$

$$\Rightarrow |w| = \frac{\sqrt{x^2+(y-1)^2}}{\sqrt{x^2+(y+1)^2}} < 1. \ [\because y > 0]$$

(c) Any point of lower half-plane $\text{Im}(z) < 0$ is $z = x + iy,\ y < 0$.

Then its image is

$$w = \frac{x + i(y-1)}{x + i(y+1)},\ y < 0 \text{ [by (vi)]}$$

$$\Rightarrow |w| = \frac{\sqrt{x^2 + (y-1)^2}}{\sqrt{x^2 + (y+1)^2}} > 1 \quad \left[\begin{array}{l}\text{for every } y < 0,\ (y-1)^2 > (y+1)^2 \\ \Rightarrow x^2 + (y-1)^2 > x^2 + (y+1)^2\end{array}\right]$$

(4) Given transformation

$$w = i\frac{1-z}{1+z} \Rightarrow w(1+z) = i(1-z) \Rightarrow z(w+i) = i - w \Rightarrow z = \frac{i-w}{w+i}.$$

Now, $|z| = 1 \Rightarrow \dfrac{|i-w|}{|w+i|} = 1$

$$\Rightarrow |i - u - iv| = |u + i(v+1)| \Rightarrow |-u + i(1-v)|^2 = |u + i(v+1)|^2$$

$$\Rightarrow u^2 + (1-v)^2 = u^2 + (v+1)^2 \tag{1}$$

$$\Rightarrow 1 - 2v + v^2 = v^2 + 2v + 1 \Rightarrow 4v = 0 \Rightarrow v = 0 \quad \text{[real axis on } w\text{-plane]}$$

Thus the given transformation transforms $|z| = 1$ maps onto the real axis of w-plane.

Again, $|z| < 1 \Rightarrow u^2 + (1-v)^2 < u^2 + (v+1)^2 \Rightarrow -2v < 2v$

$$\Rightarrow 4v > 0 \Rightarrow v > 0 \Rightarrow \text{ upper half of } w\text{-plane.}$$

Thus the interior of the circle $|z| = 1$ is transformed into the upper half of w-plane.

(5) (a) Let the Möbius transformation be

$$w = \frac{az+b}{cz+d},\ ad - bc \neq 0 = \frac{a}{c} \cdot \frac{z + b/a}{z + d/c}. \tag{1}$$

By this transformation, symmetric points with respect to real axis must be mapped onto inverse points with respect to the points of $|w| = 1$.

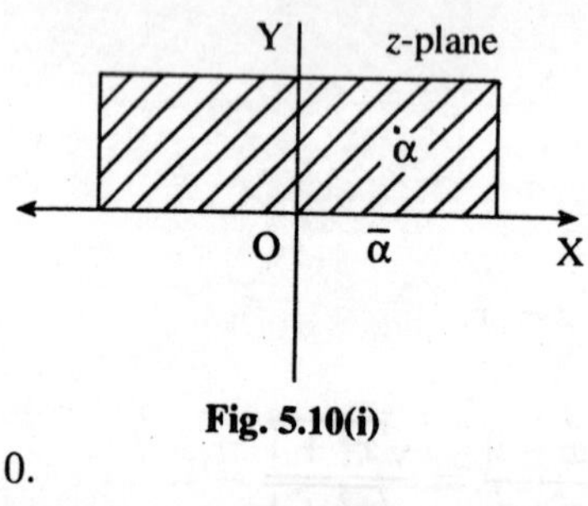

Fig. 5.10(i)

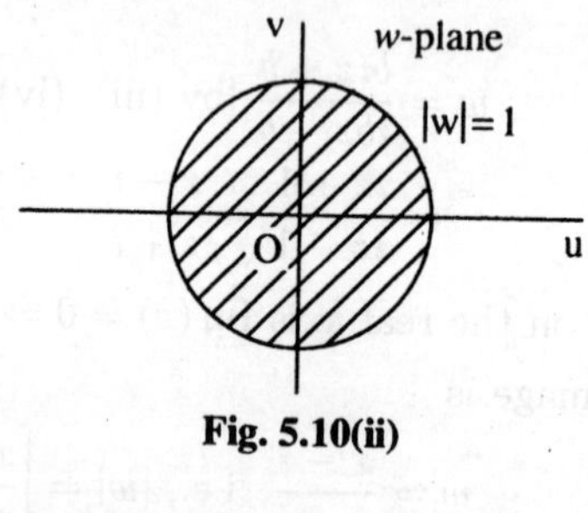

Fig. 5.10(ii)

Let $c \neq 0$.

Then $w = 0$ corresponds to $z = -b/a$ and $w = \infty$ corresponds to $z = -d/c$.

Here 0 and ∞ are inverse points w.r.t. the circle $|w| = 1$ and corresponds to $z = -b/a$ and $z = -d/c$ respectively. Then in this case, $z = -b/a$ and $z = -d/c$ are conjugate w.r.t. the real axis $I(z) = 0$. We may write, $\alpha = -b/a$ and $\bar{\alpha} = -d/c$.

Therefore, from (1),

$$w = \frac{a}{c} \cdot \frac{z - \alpha}{z - \bar{\alpha}}. \tag{2}$$

The point $z = 0$ must correspond to a point $|w| = 1$.

So from (2),

$$|w| = \left|\frac{a}{c}\right|\left|\frac{-\alpha}{\bar{\alpha}}\right| = \left|\frac{a}{c}\right| \Rightarrow 1 = \left|\frac{a}{c}\right|.$$

Hence we may write, $\frac{a}{c} = e^{i\lambda}$, λ is real.

Then we get from (2),

$$w = e^{i\lambda} \cdot \frac{z-\alpha}{z-\bar{\alpha}}. \tag{3}$$

For $z = \alpha$, $w = 0$.

So α corresponds to $w = 0$, an interior point of $|w| = 1$. The point $z = \alpha$ must be a point of the upper half-plane $\text{Im}\,(z) > 0$ and so $\text{Im}\,(\alpha) > 0$. Thus (3) is the desired transformation. It can be verified by the following way,

$$\begin{aligned} w \cdot \bar{w} - 1 &= e^{i\lambda} \cdot \frac{z-\alpha}{z-\bar{\alpha}} \cdot \bar{e}^{i\lambda} \cdot \frac{\bar{z}-\bar{\alpha}}{\bar{z}-\alpha} - 1 \\ &= \frac{(z-\alpha)(\bar{z}-\bar{\alpha}}{(z-\bar{\alpha})(\bar{z}-\alpha)} - 1 = \frac{z\bar{z} - z\bar{\alpha} - \bar{z}\alpha + \alpha\bar{\alpha}}{z\bar{z} - z\alpha - \bar{z}\bar{\alpha} + \alpha\bar{\alpha}} - 1 \\ &= \frac{(z-\bar{z})(\alpha-\bar{\alpha})}{|z-\bar{\alpha}|^2} = \frac{2i\,\text{Im}\,(z) 2i\,\text{Im}\,(\alpha)}{|z-\bar{\alpha}|^2} = \frac{-4\,\text{Im}\,(z) \cdot \text{Im}\,(\alpha)}{|z-\bar{\alpha}|^2}. \end{aligned}$$

$\text{Im}\,(z) = 0$ is mapped on $w \cdot \bar{w} - 1 = 0 \Rightarrow |w|^2 = 1$, i.e., $|w| = 1$. $[\because \text{Im}\,(\alpha) > 0]$

and $\text{Im}\,(z) > 0$ is mapped onto $w \cdot \bar{w} - 1 < 0 \Rightarrow |w|^2 < 1$, i.e., onto $|w| < 1$.

Hence $\text{Im}\,(z) \geq 0$ is mapped onto $|w| \leq 1$.

(b) Here we shall find out a bilinear transformation, which maps the half-plane of $\text{R}(z) \geq 0$ on $|w| = 1$.

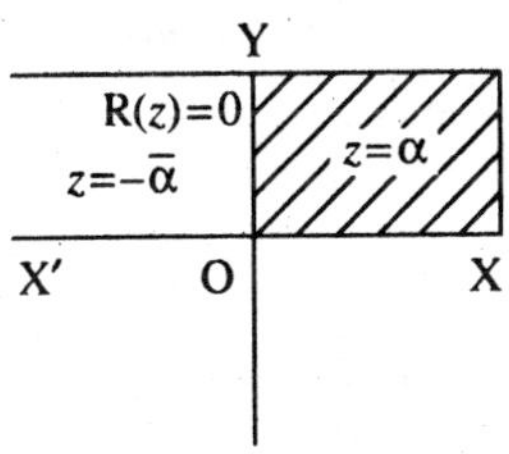

Fig. 5.11(i)

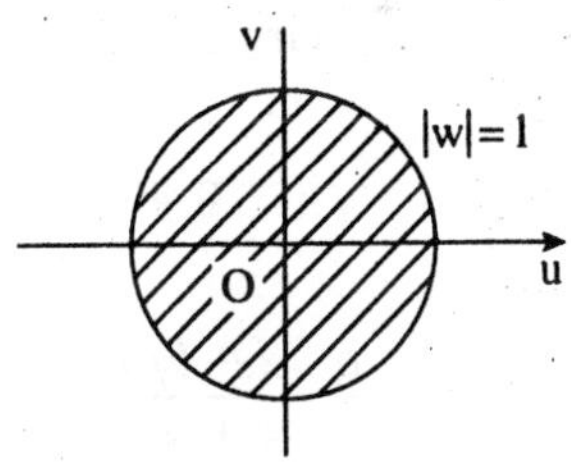

Fig. 5.11(ii)

Let the required bilinear transformation be

$$w = \frac{az+b}{cz+d},\ ad - bc \neq 0 = \frac{a}{c} \cdot \frac{z + b/a}{z + d/c}. \tag{1}$$

Let $c \neq 0$.

The points $w = 0$ and $w = \infty$ are the inverse points with respect to the unit circle $|w| = 1$, corresponding to the points $z = -b/a$ and $z = -d/c$ in the z-plane. These two points must be symmetrical with respect to the imaginary axis $\text{R}(z) = 0$. So we write, $z = -b/a = \alpha$ and $-d/c = -\bar{\alpha}$ [Here symmetrical points are conjugate points.]

Then (1) can be written as $w = \frac{a}{c} \cdot \frac{z-\alpha}{z+\bar{\alpha}}$. $z = 0$ must correspond to a point on $|w| = 1$, as any point on $\text{R}(z) = 0$ corresponds to a point on $|w| = 1$.

Then

$$w = \frac{a}{c}\cdot\frac{-\alpha}{\bar{\alpha}} \Rightarrow |w| = \left|\frac{a}{c}\right|\frac{|-\alpha|}{|\bar{\alpha}|} = \left|\frac{a}{c}\right| \; [\because |-\alpha| = |\bar{\alpha}|] \Rightarrow \left|\frac{a}{c}\right| = 1.$$

Hence we may write, $\frac{a}{c} = e^{i\lambda}$, λ is real.

Therefore, the required transformation is

$$w = e^{i\lambda}\frac{z-\alpha}{z+\bar{\alpha}} \tag{2}$$

since $z = \alpha$ gives $w = 0$, centre of the circle $|w| = 1$, α must be a point on the right half-plane $R(z) > 0$ [i.e., $R(\alpha) > 0$].

It can be easily verified that (4) is the required transformation by the following way,

$$\begin{aligned} w\bar{w} - 1 &= e^{i\lambda}\frac{z-\alpha}{z+\bar{\alpha}}\cdot e^{-i\lambda}\cdot\frac{\bar{z}-\bar{\alpha}}{\bar{z}+\alpha} - 1 = \frac{(z-\alpha)(\bar{z}-\bar{\alpha})}{(z+\bar{\alpha})(\bar{z}+\alpha)} - 1 \\ &= -\frac{(z+\bar{z})(\alpha+\bar{\alpha})}{|z+\bar{\alpha}|^2} = -\frac{4R(z)\cdot R(\alpha)}{|z+\bar{\alpha}|^2}. \end{aligned} \tag{3}$$

Since $R(\alpha) > 0$, (3) shows that $R(z) = 0$ is transformed onto $w\bar{w} - 1 = 0$, i.e., $|w| = 1$ and $R(z) > 0$, (3) is transformed onto $w\bar{w} - 1 < 1$, i.e., onto $|w| < 1$. Hence the right half-plane $R(z) \geq 0$ is mapped onto $|w| \leq 1$.

(6) (a) Let

$$w = \frac{az+b}{cz+d}, \quad ad - bc \neq 0 \tag{1}$$

be the required bilinear transformation.

The points $w = 0, \infty$ corresponds to $z = -b/a, -d/c$ respectively. Since $0, \infty$ are inverse points with respect to a circle $|w| = \rho'$, the points $-b/a$ and $-d/c$ are inverse points with respect to the circle, $|z| = \rho$.

So, we may write, $-b/a = \alpha$ and $-d/c = \rho^2/\bar{\alpha}$, where $|\alpha| < \rho$.

From (1),

$$w = \frac{a}{c}\cdot\frac{z+b/a}{z+d/c} = \frac{a}{c}\cdot\frac{z-\alpha}{z-\rho^2/\bar{\alpha}} = \frac{a}{c}\bar{\alpha}\frac{z-\alpha}{z\bar{\alpha}-\rho^2} \quad [c \neq 0]. \tag{2}$$

Any point on the boundary of $|z| = \rho$ corresponds to a point on the boundary of $|w| = \rho'$.

Therefore,

$$\begin{aligned} |w| &= \left|\frac{a}{c}\bar{\alpha}\right|\frac{|z-\alpha|}{|z\bar{\alpha}-\rho^2|} = \left|\frac{a}{c}\bar{\alpha}\right|\frac{|z-\alpha|}{|z\bar{\alpha}-z\cdot\bar{z}|} \quad [\because z : \bar{z} = |z|^2 = \rho^2] \\ &= \left|\frac{a}{c}\bar{\alpha}\right|\frac{|z-\alpha|}{|z||\bar{\alpha}-\bar{z}|}. \end{aligned}$$

$$|w| = \left|\frac{a}{c}\bar{\alpha}\right|\cdot\frac{1}{\rho} \quad [\because |z-\alpha| = |\bar{\alpha}-\bar{z}| \text{ and } |z| = \rho]$$

$$\Rightarrow \rho' = \left|\frac{a}{c}\bar{\alpha}\right|\frac{1}{\rho}$$

$$\Rightarrow \rho\rho' = \left|\frac{a}{c}\bar{\alpha}\right|.$$

So we may write, $\frac{a}{c}\bar{\alpha} = \rho\rho' e^{i\lambda}$ and hence (2) takes the form

$$w = \rho\rho' e^{i\lambda} \cdot \frac{z-\alpha}{\bar{\alpha}z - \rho^2}. \tag{3}$$

Therefore, (3) is the required transformation which transforms $|z| \le \rho$ onto $|w| \le \rho'$.

Verification is given below:

$$\begin{aligned}
w\bar{w} - \rho'^2 &= \rho^2\rho'^2 \frac{(z-\alpha)(\bar{z}-\bar{\alpha})}{(\bar{\alpha}z-\rho^2)(\alpha\bar{z}-\rho^2)} - \rho'^2 \\
&= \rho'^2 \frac{[\rho^2(z-\alpha)(\bar{z}-\bar{\alpha}) - (\bar{\alpha}z-\rho^2)(\alpha\bar{z}-\rho^2)]}{(\bar{\alpha}z-\rho^2)(\alpha\bar{z}-\rho^2)} \\
&= \rho'^2 \frac{[\rho^2 z\bar{z} - \rho^2 z\bar{\alpha} - \rho^2\alpha\bar{z} + \rho^2\alpha\bar{\alpha} + \bar{\alpha}z\rho^2 - \alpha\bar{\alpha}z\bar{z} + \rho^2\alpha\bar{z} - \rho^4]}{(\bar{\alpha}z-\rho^2)(\alpha\bar{z}-\rho^2)} \\
&= \rho'^2 \frac{[\rho^2 z\bar{z} + \rho^2\alpha\bar{\alpha} - \alpha\bar{\alpha}z\bar{z} - \rho^4]}{(\bar{\alpha}z-\rho^2)(\alpha\bar{z}-\rho^2)} = \rho'^2 \frac{[\rho^2|z|^2 + \rho^2|\alpha|^2 - |\alpha|^2|z|^2 - \rho^4]}{|\bar{\alpha}z-\rho^2|^2} \\
&= \rho'^2 \frac{[(\rho^2-|\alpha|^2)(|z|^2-\rho^2)]}{|\bar{\alpha}z-\rho^2|^2},
\end{aligned}$$

which shows that $|z| < \rho \Rightarrow |w| < \rho'$ as $|\alpha| < \rho$.

(b) Let

$$w = \frac{az+b}{cz+d}, \quad ad - bc \ne 0 \tag{1}$$

be the required bilinear transformation. Here the points $w = 0$ and $w = \infty$ are the inverse points with respect to the unit circle $|w| = 1$, corresponding to the points $z = -b/a$ and $z = -d/c$, $c \ne 0$. Then these two points must be the inverse points with respect to the circle $|z| = 1$.

So we may write, $-b/a = \alpha$ and $-d/c = 1/\bar{\alpha}$, $|\alpha| < 1$,

since, $\alpha = x + iy \Rightarrow \dfrac{1}{\bar{\alpha}} = \dfrac{1}{x-iy} \Rightarrow |\alpha| \cdot \left|\dfrac{1}{\bar{\alpha}}\right| = \sqrt{\dfrac{x^2+y^2}{x^2+y^2}} = 1$, the condition of inverse point w.r.t. a circle $|z| = 1$ is satisfied.

Then (1) can be written as

$$w = \frac{a}{c} \cdot \frac{z + b/a}{z + d/c}, \; c \ne 0 = \frac{a}{c}\frac{z-\alpha}{z - 1/\bar{\alpha}} = \frac{a}{c}\bar{\alpha} \cdot \frac{z-\alpha}{z\bar{\alpha}-1}. \tag{2}$$

The point $z = 1$ on the boundary of $|z| = 1$ must correspond to a point on the boundary of $|w| = 1$ so that

$$1 = |w| = \left|\frac{a}{c}\bar{\alpha}\right| \frac{|1-\alpha|}{|\bar{\alpha}-1|} = \left|\frac{a}{c}\bar{\alpha}\right|. \; [\because |1-\alpha| = |\bar{\alpha}-1|]$$

Thus $\left|\dfrac{a}{c}\bar{\alpha}\right| = 1$. So we can write, $\dfrac{a}{c}\bar{\alpha} = e^{i\lambda}$, λ is real.

Therefore, the transformation (2) takes the form

$$w = e^{i\lambda} \cdot \frac{z-\alpha}{z\bar{\alpha}-1} \tag{3}$$

$z = \alpha$ gives $w = 0$, α must be an interior point of $|z| = 1$, i.e., $|\alpha| < 1$.

Hence $w = e^{i\lambda}\dfrac{z-\alpha}{z \cdot \bar{\alpha} - 1}$ with $|\alpha| < 1$ is the required bilinear transformation.

The above results can be verified from the following:

$$w \cdot \bar{w} - 1 = e^{i\lambda} \cdot \frac{z-\alpha}{z \cdot \bar{\alpha} - 1} \cdot e^{-i\lambda} \frac{\bar{z} - \bar{\alpha}}{\bar{z} \cdot \alpha - 1} - 1 = \frac{(z-\alpha)(\bar{z}-\bar{\alpha})}{(z \cdot \bar{\alpha} - 1)(\bar{z} \cdot \alpha - 1)} - 1$$

$$= \frac{z\bar{z} - z\bar{\alpha} - \alpha\bar{z} + \alpha \cdot \bar{\alpha} - z \cdot \bar{z} \cdot \alpha\bar{\alpha} + z\bar{\alpha} + \bar{z}\alpha - 1}{|z\bar{\alpha} - 1|^2}$$

$$= \frac{z\bar{z} + \alpha\bar{\alpha} - z\bar{z} \cdot \alpha\bar{\alpha} - 1}{|z\bar{\alpha} - 1|^2}$$

$$\text{or, } |w|^2 - 1 = \frac{(z\bar{z} - 1)(1 - \alpha\bar{\alpha})}{|z\bar{\alpha} - 1|^2} = \frac{(|z|^2 - 1)(1 - |\alpha|^2)}{|z\bar{\alpha} - 1|^2} \tag{4}$$

with the condition $|\alpha| < 1$ so that $1 - |\alpha|^2 > 0$.

So from (4), for $|z| = 1$, $|w| = 1$.

That is, a point on the boundary of $|z| = 1$ corresponds to a point on the boundary of $|w| = 1$.

Again for $|z| < 1$, $|w|^2 - 1 < 0 \Rightarrow |w| < 1$.

Hence the transformation (3) shows that $|z| \leq 1$ is mapped onto $|w| \leq 1$, provided $|\alpha| < 1$.

EXERCISE

(1) Let $w = \dfrac{az+b}{cz+d}$, $ad - bc \neq 0$ transform the points $z_1 = -1, z_2 = 0$ and $z_3 = 1$ onto the points $w_1 = -i$, $w_2 = 1$ and $w_3 = i$. Then find w and $T^{-1}(w)$. Also find $T^{-1}(1)$ and $T^{-1}(\infty)$.

(2) Show that the equation

$$\frac{(w - w_1)(w_2 - w_3)}{(w - w_3)(w_2 - w_1)} = \frac{(z - z_1)(z_2 - z_3)}{(z - z_3)(z_2 - z_1)} \tag{1}$$

defines a bilinear transformation which maps three distinct points z_1, z_2, z_3 in the finite z-plane onto distinct points w_1, w_2, w_3 respectively in the finite w-plane.
Also show that (1) can be written as $Azw + Bz + cw + D = 0$, $AD - BC \neq 0$.

(3) Show that the bilinear transformation that maps the points $z_1 = \infty, z_2 = i, z_3 = 0$ onto the points $w_1 = 0, w_2 = i, w_3 = \infty$ is $w = -1/z$.

(4) A fixed point of a transformation $w = f(z)$ is a point z_0 such that $f(z_0) = z_0$.
Show that every bilinear transformation with the exception of identity transformation $w = z$, has at most two fixed points in the extended plane c_∞.

(5) Find the fixed points of the transformation (i) $w = \dfrac{z-1}{z+1}$, (ii) $w = \dfrac{3z-2}{z}$.

(6) Let $T(z) = \dfrac{az+b}{cz+d}$, $ad - bc \neq 0$ be any bilinear transformation other than $T(z) = z$. Show that $T^{-1} = T$, iff $d = -a$.

(7) Show that the transformation $w = \dfrac{z-1}{z+1}$ maps the half-plane $y > 0$ onto the half-plane $v > 0$ and the x-axis onto the u-axis, when $z = x + iy$ and $w = u + iv$.

(8) (a) Finding the inverse of the transformation $w = \dfrac{i-z}{i+z}$, show that the transformation $w = \dfrac{i(1-z)}{1+z}$ maps the disk $|z| \leq 1$ onto the upper half-plane $\text{Im } w \geq 0$.

(b) Show that $w = (z-2)/z$ can be written as $Z = z - 1$, $W = i\dfrac{1-z}{1+z}$, $w = iW$.

Then with the aid of the result in (a), verify that the transformation $w = \dfrac{z-2}{z}$ maps the disk $|z-1| \leq 1$ onto the left half-plane Re $w \leq 0$.

(9) Show that

(a) every bilinear transformation with two finite fixed points α, β can be put in the form $\dfrac{w-\alpha}{w-\beta} = \lambda \cdot \dfrac{z-\alpha}{z-\beta}$, where $\lambda = \dfrac{(\alpha-\delta)(\beta-\gamma)}{(\beta-\delta)(\alpha-\gamma)}$, when a point γ in the z-plane is transformed to a point δ in the w-plane.

(b) every bilinear transformation with one fixed point α can be put in the form $\dfrac{1}{w-\alpha} = \dfrac{1}{z-\alpha} + \lambda$, where λ is constant.

(10) Express the relation $w = \dfrac{13iz+75}{3z-5i}$ in the form $\dfrac{w-\alpha}{w-\beta} = \lambda\dfrac{z-\alpha}{z-\beta}$, where α, β, λ are constants. Show that the circle in the z-plane whose centre is $z = 0$ and radius 5 is transformed into the circle in the w-plane on the line joining $w = \alpha$ and $w = \beta$ as diameters.

(11) Show that the relation $w = \dfrac{5-4z}{4z-2}$ transforms the circle $|z| = 1$ into the circle of unit radius in w-plane and find the centre of this circle.

(12) Let $w = T_1(z) = \dfrac{2z+1}{z+3}$ and $w = T_2(z) = \dfrac{2z}{z+1}$ be the two transformations. Find (a) $T_2T_1(z)$, (b) $T_1 \cdot T_2(z)$, (c) $T_2^{-1}T_1(z)$, (d) $T_1^{-1}(z)$.

Part III

METRIC SPACES

CHAPTER 1

METRIC SPACES AND EXAMPLES

Discussion. In real numbers the distance between two points x and y is given by $|x-y|$. We also know that this definition of distance satisfies the following properties:

(i) $|x-y| \geq 0$, for any x, $y \in \mathbb{R}$ (non-negativity)

(ii) $|x-y| = 0$, iff $x = y$

(iii) $|x-y| = |y-x|$ (symmetry)

(iv) $|x-z| \leq |x-y| + |y-z|$, for any $x, y, z \in \mathbb{R}$.

With the help of the above notion of distance we discuss the continuity of a function, convergence of sequences in $\mathbb{R}$.

We now try to generalise the same in an arbitrary non-empty set.

1.1 Definition of Metric Spaces and Examples

Definition [1.1] : Let X be a non-empty set. Define a function $d : X \times X \to \mathbb{R}$ satisfying the following conditions:

(d.1) $d(x,y) \geq 0, \ \forall \ x, y \in X$

(d.2) $d(x,y) = 0$, iff $x = y$

(d.3) $d(x,y) = d(y,x)$, for all $x, y \in X$

(d.4) $d(x,z) \leq d(x,y) + d(y,z), \ \forall \ x, y, z \in X.$

The function d is called distance function on X or a metric on X and the pair (X,d) is called a metric space.

The last property (d.4) is known as triangle inequality.

Remark (1). In the above definition, the condition (d.1) can be deduced from the rest of the conditions. If we put $x = z$ in (d.4) and use (d.2) and (d.3), we see that

$$\begin{aligned} & d(z,z) \leq d(x,y) + d(y,x) \\ \Rightarrow \ & 0 \leq d(x,y) + d(x,y) \\ \Rightarrow \ & 2 \cdot d(x,y) \geq 0 \\ \Rightarrow \ & d(x,y) \geq 0, \text{ this is true for all } x, y \in X. \end{aligned}$$

Thus the definition of a distance function can be re-stated as follows:

A function $d : X \times X \to \mathbb{R}$ is said to be a distance function if it satisfies the following conditions:

(d.1) $d(x,y) = 0$, iff $x = y$ in X

(d.2) $d(x,y) = d(y,x) \ \forall \ x,y \in X$

(d.3) $d(x,z) \leq d(x,y) + d(y,z), \ \forall \ x,y,z \in X.$

Example [1.1]: Let $X = \mathbb{R}$, the set of all real numbers. Define $d(x,y) = |x-y| \ \forall \ x,y \in \mathbb{R}$.

Then we find that d satisfies all the conditions of a distance function and hence d is a distance function on $\mathbb{R}$. Consequently, $(\mathbb{R}, d)$ is a metric space.

Example [1.2]: Let us take $X = \mathbb{R}^2 = \mathbb{R} \times \mathbb{R}$. Define a function d on $\mathbb{R}^2$ as follows:

For $x = (x_1, x_2)$, $y = (y_1, y_2)$ in $\mathbb{R}^2$, $d(x,y) = \sqrt{(x_1-y_1)^2 + (x_2-y_2)^2}$.

Then d is a distance function on $\mathbb{R}^2$.

This is the usual notion of distance between two points in $\mathbb{R}^2$.

This is known as usual or Euclidean metric on $\mathbb{R}^2$. Thus $(\mathbb{R}^2, d)$ is a metric space.

We now define another metric on $\mathbb{R}^2$ as follows:

For $x = (x_1, x_2)$, $y = (y_1, y_2)$ in $\mathbb{R}^2$, $d^*(x,y) = |x_1 - y_1| + |x_2 - y_2|$.

It is easy to verify that d^* satisfies all the conditions of a distance function and so $(\mathbb{R}^2, d^*)$ is a metric space.

Let us define $\rho : \mathbb{R}^2 \times \mathbb{R}^2 \to \mathbb{R}$ by $\rho(x,y) = \max\{|x_1-y_1|, |x_2-y_2|\}$, where $x = (x_1, x_2)$, $y = (y_1, y_2)$ in $\mathbb{R}^2$. The conditions (d.1), (d.2) and (d.3) are very easy to verify. We now verify the condition (d.4) only.

Let $x = (x_1, x_2)$, $y = (y_1, y_2)$ and $z = (z_1, z_2)$ be three points in $\mathbb{R}^2$. Then for any $i = 1, 2$ we have

$$\begin{aligned} |x_i - z_i| &\leq |x_i - y_i| + |y_i - z_i| \\ &\leq \max\{|x_1-y_1|, |x_2-y_2|\} + \max\{|y_1-z_1|, |y_2-z_2|\} \\ &\leq \rho(x,y) + \rho(y,z). \end{aligned}$$

Since the above relation is true for any $i = 1, 2$, it follows that

$$\max\{|x_1-z_1|, |x_2-z_2|\} \leq \rho(x,y) + \rho(y,z),$$

$$\text{i.e., } \rho(x,z) \leq \rho(x,y) + \rho(y,z).$$

Hence $(\mathbb{R}^2, \rho)$ becomes a metric space.

Remark (2). Thus from the above example we can conclude that metric on a non-empty set is not unique, in general. That means on a non-empty set we can define more than one metric.

Example [1.3]: We now consider the Euclidean n-space $\mathbb{R}^n (n \geq 1)$ which consists all ordered n-tuples $x = (x_1, x_2, \ldots, x_n)$, $x_i \in \mathbb{R}$, $i = 1, 2, \ldots, n$.

Define a function d on $\mathbb{R}^n$ as

$$d(x,y) = \left\{ \sum_{i=1}^{n} (x_i - y_i)^2 \right\}^{1/2},$$

where $x = (x_1, x_2, \ldots, x_n)$, $y = (y_1, y_2, \ldots, y_n)$ in $\mathbb{R}^n$.

We now show that d is a metric on $\mathbb{R}^n$.

(d.1) It is clear from the definition of d that $d(x,y) \geq 0,\ \forall\ x, y \in \mathbb{R}^n$.

(d.2) Now let $d(x,y) = 0$

$$\begin{aligned}
&\Leftrightarrow \left\{\sum_{i=1}^{n}(x_i - y_i)^2\right\}^{1/2} = 0 \\
&\Leftrightarrow \sum_{i=1}^{n}(x_i - y_i)^2 = 0 \\
&\Leftrightarrow (x_i - y_i)^2 = 0,\ \forall\ i = 1, 2, \ldots, n \\
&\Leftrightarrow x_i = y_i,\ \forall\ i = 1, 2, \ldots, n \\
&\Leftrightarrow x = y.
\end{aligned}$$

Thus, $d(x,y) = 0$, iff $x = y$ in $\mathbb{R}^n$.

(d.3) Since $(x_i - y_i)^2 = (y_i - x_i)^2,\ \forall\ i = 1, 2, \ldots, n$, it follows that $d(x,y) = d(y,x),\ \forall\ x, y \in \mathbb{R}^n$.

(d.4) We now verify the triangle inequality. To establish triangle inequality, we take the help of two important inequalities known as Cauchy-Schwarz inequality and Minkowski inequality.

Cauchy-Schwarz Inequality

If $(a_1, a_2, \ldots, a_n)$ and $(b_1, b_2, \ldots, b_n)$ are two sets of reals, then

$$\left(\sum_{i=1}^{n} a_i b_i\right)^2 \leq \sum_{i=1}^{n} a_i^2 \sum_{i=1}^{n} b_i^2.$$

Proof. Let λ be any non-zero real number. Then we always have

$$\begin{aligned}
&\sum_{i=1}^{n}(\lambda a_i + b_i)^2 \geq 0 \\
\text{or,}\quad &\lambda^2 \sum_{i=1}^{n} a_i^2 + 2\lambda \sum_{i=1}^{n} a_i b_i + \sum_{i=1}^{n} b_i^2 \geq 0 \\
\text{or,}\quad &A\lambda^2 + 2B\lambda + C \geq 0,
\end{aligned}$$

where $A = \sum\limits_{i=1}^{n} a_i^2, B = \sum\limits_{i=1}^{n} a_i b_i, C = \sum\limits_{i=1}^{n} b_i^2$ and $A \neq 0$, because otherwise all $a_i = 0$.

Now $A\lambda^2 + 2B\lambda + C \geq 0$ whenever the discriminant

$$4(B^2 - AC) \leq 0 \text{ or } B^2 \leq AC$$

$$\text{or,}\quad \left(\sum_{i=1}^{n} a_i b_i\right)^2 \leq \left(\sum_{i=1}^{n} a_i^2\right)\left(\sum_{i=1}^{n} b_i^2\right)$$ —this is the Cauchy-Schwarz inequality.

Minkowski Inequality

If $(a_1, a_2, \ldots, a_n), (b_1, b_2, \ldots, b_n)$ are two sets of real numbers, then

$$\left\{\sum_{i=1}^{n}(a_i + b_i)^2\right\}^{1/2} \leq \left(\sum_{i=1}^{n} a_i^2\right)^{1/2} + \left(\sum_{i=1}^{n} b_i^2\right)^{1/2}.$$

Proof.

$$\sum_{i=1}^{n}(a_i+b_i)^2=\sum_{i=1}^{n}a_i^2+2\sum_{i=1}^{n}a_ib_i+\sum_{i=1}^{n}b_i^2\le\sum_{i=1}^{n}a_i^2+2\left|\sum_{i=1}^{n}a_ib_i\right|+\sum_{i=1}^{n}b_i^2$$

$$\le\sum_{i=1}^{n}a_i^2+2\left\{\sum_{i=1}^{n}a_i^2\right\}^{1/2}\left\{\sum_{i=1}^{n}b_i^2\right\}^{1/2}+\sum_{i=1}^{n}b_i^2.$$

[by Cauchy-Schwarz inequality]

$$\therefore\sum_{i=1}^{n}(a_i+b_i)^2\le\left(\left\{\sum_{i=1}^{n}a_i^2\right\}^{1/2}\right)^2+\left(\left\{\sum_{i=1}^{n}b_i^2\right\}^{1/2}\right)^2+2\left\{\sum_{i=1}^{n}a_i^2\right\}^{1/2}\left\{\sum_{i=1}^{n}b_i^2\right\}^{1/2}$$

$$=\left(\left\{\sum_{i=1}^{n}a_i^2\right\}^{1/2}+\left\{\sum_{i=1}^{n}b_i^2\right\}^{1/2}\right)^2.$$

This gives $\left\{\sum_{i=1}^{n}(a_i+b_i)^2\right\}^{1/2}\le\left\{\sum_{i=1}^{n}a_i^2\right\}^{1/2}+\left\{\sum_{i=1}^{n}b_i^2\right\}^{1/2}$, which is the Minkowski inequality.

Now let $x=(x_1,x_2,\dots,x_n)$, $y=(y_1,y_2,\dots,y_n)$, $z=(z_1,z_2,\dots,z_n)$ be three points in $\mathbb{R}^n$. Put $a_i=x_i-y_i,\ b_i=y_i-z_i;\ i=1,2,\dots,n$. Then from Minkowski inequality we get

$$\left\{\sum_{i=1}^{n}(x_i-z_i)^2\right\}^{1/2}\le\left\{\sum_{i=1}^{n}(x_i-y_i\right\}^{1/2}+\left\{\sum_{i=1}^{n}(y_i-z_i)^2\right\}^{1/2}$$

$\Rightarrow d(x,z)\le d(x,y)+d(y,z)$—this is the required triangle inequality.

Hence, $(\mathbb{R}^n,d)$ is a metric space.

Example [1.4]: Denote by $l_p(p\ge1)$ the set of all sequences of real or complex numbers $x=\{x_1,x_2,\dots,x_n,\dots\}$ such that $\sum_{n=1}^{\infty}|x_n|^p<\infty$, i.e., $l_p=\left\{x=\{x_n\}:\sum_{n=1}^{\infty}|x_n|^p<\infty\right\}$.

Then $l_p(p\ge1)$ is a non-empty set (why?).

We now define on l_p a function d as $d(x,y)=\left\{\sum_{n=1}^{\infty}|x_n-y_n|^p\right\}^{1/p}$, $p\ge1$ and $x=\{x_n\}$, $y=\{y_n\}$ are two members of l_p.

We now show that l_p is a metric space with respect to d.

The verification of distance axioms (d.1), (d.2) and (d.3) are very easy and are left as an exercise to the reader. We verify triangle inequality only.

To establish triangle inequality we use Minkowski inequality which is stated below:

Minkowski Inequality

If $a=\{a_1,a_2,\dots\}$, $b=\{b_1,b_2,\dots\}$ are two elements of l_p, then

$$\left\{\sum_{i=1}^{\infty}|a_i+b_i|^p\right\}^{1/p}\le\left\{\sum_{i=1}^{\infty}|a_i|^p\right\}^{1/p}+\left\{\sum_{i=1}^{\infty}|b_i|^p\right\}^{1/p}.$$

Now let $x=\{x_1,x_2,\dots,x_n,\dots\}$, $y=\{y_1,y_2,\dots,y_n,\dots\}$ and $z=\{z_1,z_2,\dots,z_n\dots\}$ be three elements in l_p.

We now put $a_i = x_i - y_i, b_i = y_i - z_i, \ \forall\ i = 1, 2, \ldots$ and use Minkowski inequality to get

$$\left\{\sum_{i=1}^{\infty} |(x_i - y_i + y_i - z_i)|^p\right\}^{1/p} \leq \left\{\sum_{i=1}^{\infty} |x_i - y_i|^p\right\}^{1/p} + \left\{\sum_{i-1}^{\infty} y_i - z_i|^p\right\}^{1/p}$$

or, $\left\{\sum_{i=1}^{\infty} |x_i - z_i|^p\right\}^{1/p} \leq d(x, y) + d(y, z)$

or, $d(x, z) \leq d(x, y) + d(y, z)$—this is the required triangle inequality.

Hence (l_p, d) is a metric space, $1 \leq p < \infty$.

Example [1.5] : Let us denote by l_∞, the set of all bounded sequences in $\mathbb{R}$ (or C), i.e.,

$$l_\infty = \{x = \{x_n\} : \sup_n |x_n| < \infty\}.$$

We now define a function d_∞ on l_∞ by $d_\infty(x, y) = \sup\limits_n |x_n - y_n|$, where $x = \{x_n\}$, $y = \{y_n\}$ are two elements from l_∞.

It is very easy to verify that d_∞ defines a metric on l_∞ and hence (l_∞, d_∞) is a metric space.

Example [1.6] : Let us consider the space $C[a, b]$ consisting of all real-valued continuous functions on $[a, b]$, i.e., $C[a, b] = \{f : [a, b] \to \mathbb{R} : f$ is continuous on $[a, b]\}$. Then $C[a, b]$ is non-empty. We now define a function d on $C[a, b]$ as $d(f, g) = \sup\limits_{a \leq t \leq b} |f(t) - g(t)|$, where f, $g \in C[a, b]$. We now show that d is a metric on $C[a, b]$.

(d.1) Let $f, g \in C[a, b]$. Since f, g are continuous on $[a, b]$, it follows that f, g are bounded on $[a, b]$ and so $f - g$ is bounded on $[a, b]$. Hence $\sup\limits_{a \leq t \leq b} |f(t) - g(t)|$ exists as a finite non-negative real number, i.e., $d(f, g) \geq 0$ for any $f, g \in C[a, b]$.

(d.2) If $f = g$, then from the definition of equality of two functions, it follows that $d(f, g) = 0$. Conversely, let $d(f, g) = 0$

$$\begin{aligned}
&\Rightarrow \sup_{a \leq t \leq b} |f(t) - g(t)| = 0 \\
&\Rightarrow |f(t) - g(t)| \leq \sup_{a \leq t \leq b} |f(t) - g(t)| = 0, \ \forall\ t \in [a, b] \\
&\Rightarrow |f(t) - g(t)| = 0, \ \forall\ t \in [a, b] \\
&\Rightarrow f(t) = g(t), \ \forall\ t \in [a, b] \\
&\Rightarrow f = g.
\end{aligned}$$

(d.3) $d(f, g) = \sup\limits_{a \leq t \leq b} |f(t) - g(t)| = \sup\limits_{a \leq t \leq b} |g(t) - f(t)| = d(g, f)$.

(d.4) Let $f, g, h \in C[a, b]$. Then we have for any $t \in [a, b]$,

$$\begin{aligned}
|f(t) - h(t)| &= |\{f(t) - g(t)\} + \{g(t) - h(t)\}| \\
&\leq |f(t) - g(t)| + |g(t) - h(t)| \leq \sup_{a \leq t \leq b} |f(t) - g(t)| + \sup_{a \leq t \leq b} |g(t) - h(t)| \\
&= d(f, g) + d(g, h).
\end{aligned}$$

The above relation is true for any $t \in [a, b]$. So we have

$$\sup_{a \leq t \leq b} |f(t) - h(t)| \leq d(f, g) + d(g, h), \text{ or, } d(f, h) \leq d(f, g) + d(g, h),$$

which is the required triangle inequality.

Thus $(C[a, b], d)$ is a metric space.

Example [1.7] : On $C[a,b]$ we can define another metric ρ as

$$\rho(f,g)=\int_a^b |f(t)-g(t)|dt,$$

integration being taken in the Riemann sense. It is not difficult to verify that ρ is a metric on $C[a,b]$. So $(C[a,b],\rho)$ is a metric space.

Remark (3). If we replace $C[a,b]$ by $R[a,b]$, the space of all Riemann integrable function and ρ is the same function as defined in example [1.7], then $(R[a,b],\rho)$ is **not** a metric space. That is ρ is not a metric on $R[a,b]$.

Note that ρ satisfies conditions (d.1), (d.3) and (d.4) as in example [1.7].

Also if $f=g$, then $f(t)=g(t),\ \forall\, t\in[a,b]$

$$\Rightarrow |f(t)-g(t)|=0,\ \forall\, t\in[a,b]$$

$$\Rightarrow \int_a^b |f(t)-g(t)|dt=0$$

$$\Rightarrow \rho(f,g)=0.$$

But the converse is not true, i.e., $\rho(f,g)=0$ does not imply $f=g$ in general.

For example, let

$$f(x)=\begin{cases}0, & \text{if } a\le x<b\\ k, & \text{if } x=b \quad (k\ne 0)\end{cases}$$

$$\text{and } g(x)=\begin{cases}0, & \text{if } a\le x<b\\ -k, & \text{if } x=b \quad (k\ne 0).\end{cases}$$

Then $\rho(f,g)=\int_a^b |f(t)-g(t)|dt=0$, but $f\ne g$.

Hence ρ is not a metric on $R[a,b]$.

Example [1.8]: Let $X=\mathbb{R}$. For any $x,y\in X$ define $d\colon X\times X\to\mathbb{R}$ by $d(x,y)=|\sin(x-y)|$.

Then d is **not** a metric on X.

Note that, if $y=x+\pi$, then $y\ne x$ but $d(x,y)=|\sin(-\pi)|=0$.

$\therefore d$ is not a metric on X.

Theorem [1.1] Let (X,d) be a metric space and let $x,y,z\in X$.

Then $|d(x,z)-d(y,z)|\le d(x,y)$.

Proof. Since (X,d) is a metric space, we have from (d.4).

$$d(x,z)\le d(x,y)+d(y,z)\Rightarrow d(x,z)-d(y,z)\le d(x,y). \tag{i}$$

Interchanging the role of x and y in both side of (i), we get

$$d(y,z)-d(x,z)\le d(y,x)\Rightarrow -\{d(x,z)-d(y,z)\}\le d(x,y). \tag{ii}$$

Combining (i) and (ii), we obtain $|d(x,z)-d(y,z)|\le d(x,y)$.

Theorem [1.2] Let (X,d) be a metric space. For $x,y\in X$, define $d_1(x,y)=\dfrac{d(x,y)}{1+d(x,y)}$.

Then d_1 is a metric on X.

Proof. We have $d_1(x,y) = \dfrac{d(x,y)}{1+d(x,y)}$, $\forall\ x, y \in X$.

(d.1) Since $d(x,y) \geq 0$, we have $1 + d(x,y) > 0$.

Hence, $\dfrac{d(x,y)}{1+d(x,y)} \geq 0$, i.e., $d_1(x,y) \geq 0$, $\forall\ x, y \in X$.

(d.2) If $x = y$, then $d(x,y) = 0$ and so $d_1(x,y) = \dfrac{d(x,y)}{1+d(x,y)} = 0$.

Now let $d_1(x,y) = 0 \Rightarrow \dfrac{d(x,y)}{1+d(x,y)} = 0 \Rightarrow d(x,y) = 0 \Rightarrow x = y$.

Thus $d_1(x,y) = 0$, iff $x = y$ in X.

(d.3) $d_1(y,x) = \dfrac{d(y,x)}{1+d(y,x)} = \dfrac{d(x,y)}{1+d(x,y)} = d_1(x,y)$, $\forall\ x, y \in X$.

(d.4) For triangle inequality, we consider a function $f : [0, \infty) \to \mathbb{R}$ defined by

$$f(x) = \frac{x}{1+x},\ \forall\ x \in [0, \infty).$$

Then f is continuous and also since

$$f'(x) = \frac{d}{dx}\left\{1 - \frac{1}{1+x}\right\} = \frac{1}{(1+x)^2} > 0,\ \forall\ x \in [0, \infty),$$

it follows that f is an increasing function on $[0, \infty)$.

Since d is a metric on X, for $x, y, z \in X$, we have $d(x,z) \leq d(x,y) + d(y,z)$

$$\Rightarrow\ f(d(x,z)) \leq f(d(x,y) + d(y,z))$$

$$\Rightarrow\ \frac{d(x,z)}{1+d(x,z)} \leq \frac{d(x,y)+d(y,z)}{1+d(x,y)+d(y,z)} = \frac{d(x,y)}{1+d(x,y)+d(y,z)} + \frac{d(y,z)}{1+d(x,y)+d(y,z)}$$

$$\leq \frac{d(x,y)}{1+d(x,y)} + \frac{d(y,z)}{1+d(y,z)}$$

$$\Rightarrow\ d_1(x,z) \leq d_1(x,y) + d_1(y,z).$$

Hence d_1 is a metric on X.

Theorem [1.3] Let (X, d) be a metric space. Define d_2 on X by $d_2(x,y) = \min\{1, d(x,y)\}$. Then d_2 is also a metric on X.

Proof. (d.1) Since $d(x,y) \geq 0$, it follows that $d_2(x,y) \geq 0$, for all x, y in X.

(d.2) If $x = y$ then $d(x,y) = 0$ and so $\min\{1, d(x,y)\} = 0$, i.e., $d_2(x,y) = 0$.

Now let $d_2(x,y) = 0 \Rightarrow \min\{1, d(x,y)\} = 0$

$\Rightarrow d(x,y) = 0$

$\Rightarrow x = y$.

Thus $d_2(x,y) = 0$, iff $x = y$.

(d.3) Since $d(x,y) = d(y,x)$, we get $d_2(x,y) = d_2(y,x)$.

(d.4) Let $x, y, z \in X$.

CASE (1) If $d(x,y) \geq 1$ or, $d(y,z) \geq 1$ or, both ≥ 1, then we have $1 \leq d_2(x,y) + d_2(y,z)$.

Hence $d_2(x,z) = \min\{1, d(x,z)\} \leq 1 \leq d_2(x,y) + d_2(y,z)$,

i.e., $d_2(x,z) \leq d_2(x,y) + d_2(y,z)$.

CASE (2) Let $d(x, y) < 1$ and $d(y, z) < 1$. Then we have

$$\min\{1, d(x, y)\} = d(x, y) \text{ and } \min\{1, d(y, z)\} = d(y, z)$$

and so $d_2(x, z) = \min\{1, d(x, z)\} \leq d(x, z) \leq d(x, y) + d(y, z) = d_2(x, y) + d_2(y, z)$,

$$\text{i.e., } d_2(x, z) \leq d_2(x, y) + d_2(y, z).$$

Thus in both the cases we find that $d_2(x, z) \leq d_2(x, y) + d_2(y, z)$.

Hence d_2 is a metric on X.

Example [1.9]: [**Trivial Metric**] Let X be a non-empty set. Define a function $d : X \times X \to \mathbb{R}$ by

$$d(x, y) = \begin{cases} 1, & \text{if } x \neq y \\ 0, & \text{if } x = y, \end{cases} \qquad \text{where } x, y \in X.$$

It is not very difficult to verify that d is a metric on X and hence (X, d) is a metric space. This metric d is called trivial metric and (X, d) is called trivial (or discrete) metric space.

Example [1.10]: [**Pseudo-metric Space**] Let X be a non-empty set. A function $d : X \times X \to \mathbb{R}$ satisfying

(i) $d(x, y) \geq 0, \ \forall \ x, y \in X$

(ii) $x = y$ in $X \Rightarrow d(x, y) = 0$

(iii) $d(x, y) = d(y, x), \ \forall \ x, y \in X$

(iv) $d(x, z) \leq d(x, y) + d(y, z), \ \forall \ x, y, z \in X$,

is called a pseudo-metric on X, and (X, d) is called a pseudometric space. From Remark (3), it follows that $(R[a, b], \rho)$ is a pseudo-metric space.

1.2 Open Ball, Closed Ball; Open Set, Closed Set

Definition [1.2]: Let (X, d) be a metric space and let $x_0 \in X$. Let r be a (+)ve real number. Then the *open ball* centred at x_0 and having radius r is denoted by $B_r(x_0)$ (or, $B(x_0, r)$) and is defined as the set $B_r(x_0) = \{x \in X : d(x, x_0) < r\}$.

Note that $B_r(x_0) \neq \phi$ since $x_0 \in B_r(x_0)$ as $0 = d(x_0, x_0) < r$.

Definition [1.3]: Let (X, d) be a metric space and let $x_0 \in X$. Let r be a (+)ve real number. Then the *closed ball* centred at x_0 having radius r is denoted by $\bar{B}_r(x_0)$ or, $\bar{B}(x_0, r)$ and is given by the set $\bar{B}_r(x_0) = \{x \in X : d(x, x_0) \leq r\}$.

By the set $\{x \in X : d(x, x_0) = r\}$ we denote the surface of the ball [closed or open] and is written as $S_r(x_0)$ or, $S(x_0, r)$, i.e., $S_r(x_0) = \{x \in X : d(x, x_0) = r\}$.

Example [1.11]: Consider the metric space $\mathbb{R}$ with the usual metric d, i.e.,

$$d(x, y) = |x - y|, \ \forall \ x, y \in \mathbb{R}.$$

Then an open ball in $(\mathbb{R}, d)$ is given by

$$\begin{aligned} B_r(x_0) &= \{x \in \mathbb{R} : d(x, x_0) < r\} \\ &= \{x \in \mathbb{R} : |x - x_0| < r\} \\ &= \{x \in \mathbb{R} : x_0 - r < x < x_0 + r\} \\ &= (x_0 - r, x_0 + r), \text{ where } x_0 \in \mathbb{R} \text{ and } r(> 0) \in \mathbb{R}. \end{aligned}$$

Thus an open ball centred at x_0 and radius r is the open interval $(x_0 - r, x_0 + r)$.

Similarly, one can check that a closed ball centred at x_0 and radius r is the closed interval $[x_0 - r, x_0 + r]$ in $(\mathbb{R}, d)$.

Example [1.12] : Consider the metric space $\mathbb{R}^2$ with the Euclidean or usual metric

$$d(x,y) = \sqrt{(x_1 - y_1)^2 + (x_2 - y_2)^2},$$

where $x = (x_1, x_2)$ and $y = (y_1, y_2)$ in $\mathbb{R}^2$.

Let $x_0 = (x_1, x_2) \in \mathbb{R}^2$ and $r > 0$ be any real number.

Then the open ball

$$\begin{aligned} B_r(x_0) &= \{y = (y_1, y_2) : d(x_0, y) < r\} \\ &= \{y = (y_1, y_2) \in \mathbb{R}^2 : (x_1 - y_1)^2 \\ &\qquad +(x_2 - y_2)^2 < r^2\} \end{aligned}$$

If we take $x_0 = (0,0)$, then

$$B_r(x_0) = \{y = (y_1, y_2) \in \mathbb{R}^2 : y_1^2 + y_2^2 < r^2\}.$$

This shows that $B_r(x_0)$ is the set of all points in $\mathbb{R}^2$ which are inside the circle $y_1^2 + y_2^2 = r^2$.

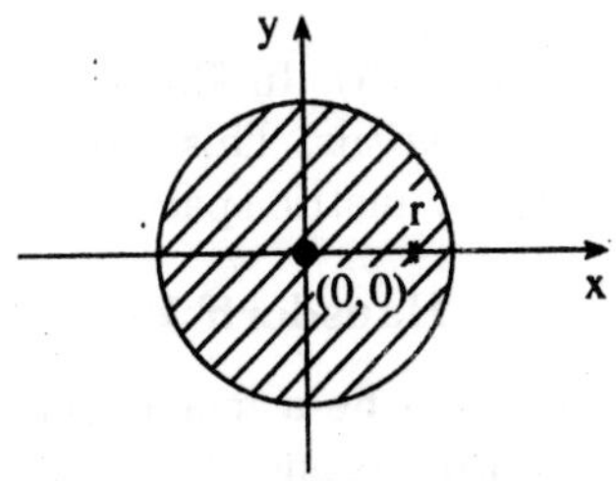

Fig. 1.1 : $B_r(0,0)$ in $(\mathbb{R}^2, d)$

Example [1.13] : We now consider the metric space $(\mathbb{R}^2, d^*)$, where $d^*(x,y) = |x_1 - y_1| + |x_2 - y_2|$, where $x = (x_1, x_2)$, $y = (y_1, y_2)$ in $\mathbb{R}^2$.

Let $\alpha = (\alpha_1, \alpha_2) \in \mathbb{R}^2$ and r is a positive real number.

Then $B_r(\alpha) = \{x = (x_1, x_2) \in \mathbb{R}^2 : d(x, \alpha) < r\}$

$$= \left\{x = (x_1, x_2) \in \mathbb{R}^2 : |x_1 - \alpha_1| + |x_2 - \alpha_2| < r\right\}.$$

If we take $\alpha = (0,0)$, then

$$B_r(0,0) = \{x = (x_1, x_2) \in \mathbb{R}^2 : |x_1| + |x_2| < r\}.$$

This is shown in Fig. 1.2.

Similarly,

$$\bar{B}_r(\alpha) = \{x = (x_1, x_2) \in \mathbb{R}^2 : |x_1 - \alpha_1| + |x_2 - \alpha_2| \leq r\}.$$

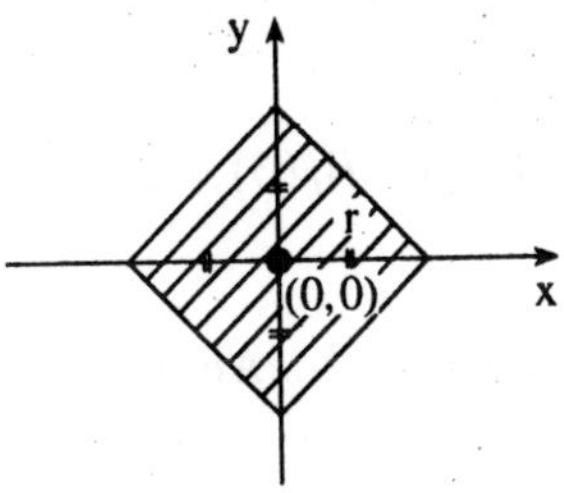

Fig. 1.2 : $B_r(0,0)$ in $(\mathbb{R}^2, d^*)$

Remark (4). From the above two examples [1.12] and [1.13] we see that if we change the metric defined on a non-empty set X, then the characteristic of open ball, closed ball changes with the metric. We will discuss later that it will effect many characteristic of a metric space.

Example [1.14] : Consider the discrete metric space (X, d) as defined in example [1.12].

Let $x_0 \in X$ and r be any real number such that $r > 0$.

$$\text{Then } B_r(x_0) = \{x \in X : d(x, x_0) < r\} = \begin{cases} \{x_0\}, & \text{if } 0 < r \leq 1 \\ X, & \text{if } r > 1. \end{cases}$$

First suppose that $0 < r < 1$. Let $r = \dfrac{1}{2}$.

Then $B_{1/2}(x_0) = \left\{x \in X : d(x, x_0) < \dfrac{1}{2}\right\}$. But we know that $d(x, x_0) = 0$ or 1 according as $x = x_0$ or, $x \neq x_0$ respectively. Since $d(x, x_0) < \dfrac{1}{2}$, so we only have $d(x, x_0) = 0 \Rightarrow x = x_0$.

Hence $B_{1/2}(x_0) = \left\{x \in X : d(x, x_0) < \frac{1}{2}\right\} = \{x_0\}$.

Thus for any r, $0 < r < 1$, we have $B_r(x_0) = \{x_0\}$.

For $r = 1, B_1(x_0) = \{x \in X : d(x, x_0) < 1\} = \{x_0\}$.

If $r > 1$, then from the definition of d it follows that $B_r(x_0) = X$.

Also note that $\bar{B}_r(x_0) = \{x \in X : d(x, x_0) \le r\} = \begin{cases} \{x_0\}, & \text{if } 0 < r < 1 \\ X, & \text{if } r \ge 1. \end{cases}$

It is very important to note that, for any $x \in X$, $\{x\}$ is an open ball in (X, d).

Remark (5). In $\mathbb{R}$, we know that any two distinct points can be separated by two disjoint open intervals. This property is known as *Hausdorff property.* This property is also true in an arbitrary metric space.

1.2.1 Hausdorff Property

Let (X, d) be a metric space and let $\alpha, \beta \in X$. Then we can find two open balls in (X, d) separating α and β.

Proof. If $\alpha = \beta$, then there is nothing to prove. Suppose that $\alpha \ne \beta$. Then $d(\alpha, \beta) > 0$. Let us choose $r < \frac{1}{2}d(\alpha, \beta)$, where $r > 0$.

Consider the open balls $B_r(\alpha)$ and $B_r(\beta)$. We claim that $B_r(\alpha) \cap B_r(\beta) = \phi$. To establish our claim, let $x \in B_r(\alpha) \cap B_r(\beta) \Rightarrow d(x, \alpha) < r$ and $d(x, \beta) < r$.

$\therefore d(\alpha, \beta) \le d(\alpha, x) + d(x, \beta) < r + r = 2r < d(\alpha, \beta)$—a contradiction.

Hence $B_r(\alpha) \cap B_r(\beta) = \phi$.

Thus we find two open balls $B_r(\alpha)$ and $B_r(\beta)$ in (X, d) such that $B_r(\alpha) \cap B_r(\beta) = \phi$.

This shows that every metric space satisfies Hausdorff property.

Definition [1.4]: **[Interior point]** Let (X, d) be a metric space and let $A \subset X$. An element $\alpha \in A$ is called an interior point of A if we find an open ball $B_r(\alpha)$, for some $r > 0$ such that $\alpha \in B_r(\alpha) \subset A$.

Example [1.15]: In the metric space $(\mathbb{R}, d)$ with the usual metric, the point 0 is an interior point of the set $(-1, 1)$.

1.2.2 Interior of a Set

Let (X, d) be a metric space and $A \subset X$. The set of all interior points of A is called the interior of the set A and is written as $\mathring{A}$ or int (A).

Thus, $\mathring{A} = \{x \in A : x \text{ is an interior point of } A\}$.

From the definition, it follows that $\mathring{A} \subset A$.

Example [1.16]: Consider the metric space $(\mathbb{R}, d)$ of all real number with the usual metric.

Let $A = \left\{0, 1, \frac{1}{2}, \frac{1}{3}, \dots, \frac{1}{n}, \dots\right\}$.

Then one can check that $\mathring{A} = \phi$.

Example [1.17] : Consider the trivial metric space (X,d). Let $A \subset X$. Then for any $a \in A$, we have $a \in \{a\} = B_r(a) \subset A,\ 0 < r \leq 1$.

This shows that each $a \in A$ is an interior point of A. Hence $\mathring{A} = A$.

Theorem [1.4] Let (X,d) be a metric space and A, B are two subsets of X. Then

(i) $A \subset B \Rightarrow \mathring{A} \subset \mathring{B}$

(ii) $(A \cap B)^\circ = \mathring{A} \cap \mathring{B}$

(iii) $\mathring{A} \cup \mathring{B} \subset (A \cup B)^\circ$

Proof. (1) Let $x \in \mathring{A}$, i.e., x is an interior point of A

$\Rightarrow \exists\, r > 0$ such that $x \in B_r(x) \subset A \subset B$ as $A \subset B$

$\Rightarrow \exists\, r > 0$ such that $x \in B_r(x) \subset B$

$\Rightarrow x \in \mathring{B}$.

$\therefore\ \mathring{A} \subset \mathring{B}$, if $A \subset B$.

(2) If $A \cap B = \phi$, then from defintion it follows that $(A \cap B)^\circ = \phi$.

On the other hand, $\mathring{A} \subset A, \mathring{B} \subset B$ and so $\mathring{A} \cap \mathring{B} \subset A \cap B = \phi \Rightarrow \mathring{A} \cap \mathring{B} = \phi$

$$\therefore\ (A \cap B)^\circ = \mathring{A} \cap \mathring{B}.$$

Let us now suppose that $A \cap B \neq \phi$.

Let $x \in (A \cap B)^\circ \Rightarrow x$ is an interior point of $A \cap B$.

$\Rightarrow \exists\, r > 0$ such that $x \in B_r(x) \subset A \cap B$

$\Rightarrow \exists\, r > 0$ such that $x \in B_r(x) \subset A$ and $x \in B_r(x) \subset B$

$\Rightarrow x \in A^\circ$ and $x \in B^\circ$

$\Rightarrow x \in A^\circ \cap B^\circ$

$$\therefore (A \cap B)^\circ \subset A^\circ \cap B^\circ. \tag{i}$$

Now take $x \in A^\circ \cap B^\circ$

$\Rightarrow x \in A^\circ$ and $x \in B^\circ$

$\Rightarrow \exists\, r_1, r_2 > 0$ such that $x \in B_{r_1}(x) \subset A$ and $x \in B_{r_2}(x) \subset B$.

Put $r = \min\{r_1, r_2\}$. Consider the open ball $B_r(x)$.

Then for any $y \in B_r(x)$ we have $d(x,y) < r \leq r_1$ and $d(x,y) < r \leq r_2$.

$\Rightarrow y \in B_{r_1}(x)$ and $y \in B_{r_2}(x)$.

$\therefore B_r(x) \subset B_{r_1}(x)$ and also $B_r(x) \subset B_{r_2}(x)$

This shows that $\exists\, r > 0$ such that $x \in B_r(x) \subset A$ and $x \in B_r(x) \subset B$.

$\Rightarrow \exists\, r > 0$ such that $x \in B_r(x) \subset A \cap B$

$\Rightarrow x \in (A \cap B)^\circ$

$$\therefore\ A^\circ \cap B^\circ \subset (A \cap B)^\circ. \tag{ii}$$

Combining (i) and (ii), we get $(A \cap B)^\circ = A^\circ \cap B^\circ$.

(3) Left as an exercise.

Remark (6). Equality in (3) does not hold, in general. For example, consider the metric space $(\mathbb{R}, d)$ of all real numbers with the usual metric.

Take $A = (1, 2)$ and $B = [2, 4)$.

Then we have $A^\circ = (1, 2)$ and $B^\circ = (2, 4)$. Also $A \cup B = (1, 4)$ and so $(A \cup B)^\circ = (1, 4)$.

$\therefore A^\circ \cup B^\circ \neq (A \cup B)^\circ$.

Example [1.18] : In the metric space of all real numbers with the usual metric consider the set $A = (1, 2) \cup \{0\} \cup \{5\}$. Note that every point of $(1, 2)$ is an interior point of A but neither 0 nor 5 can be an interior point of A.

Definition [1.5] : [Open set] Let (X, d) be a metric space and $G \subset X$. Then G is called an open set if each point of G is an interior point of G.

The definition tells us that each $x \in G \Rightarrow x \in G^\circ \Rightarrow G \subset G^\circ$. But we know that $G^\circ \subset G$. Hence $G = G^\circ$.

Thus from definition it follows that if G is an open set, then we have $G = G^\circ$.

On the other hand, if $G = G^\circ$, then each $x \in G \Rightarrow x \in G^\circ$.

$\Rightarrow$ each x in G is an interior point of G

$\Rightarrow$ G is an open set.

Hence we find that

Theorem [1.5] A subset G in a metric space (X, d) is an open set, iff $G = G^\circ$.

Remark (7). From the definition of open set it follows that every open ball is an open set. But the converse is not true. For example, consider the metric space $(\mathbb{R}, d)$ with the usual metric d. Then the set $A = (0, 1) \cup (2, 3)$ is an open set but it is not an open ball in $(\mathbb{R}, d)$.

Remark (8). Consider the trivial metric space (X, d) as defined in example [1.9]. Let $A \subset X$. Then we see in example [1.17] that $A = A^\circ$. Hence by theorem [1.5], A is an open set. Since A is taken arbitrarily, it follows that in the trivial metric space each subset of X is an open set.

Theorem [1.6] Union of any number of open sets in a metric space is an open set. That means, the collection of all open sets in a metric space is closed sender the arbitrary union.

Proof. Let (X, d) be a metric space and let $\{G_\alpha\}\alpha \in I$ be a family of open sets in (X, d), where I is an index set.

Let $G = \bigcup_{\alpha \in I} G_\alpha$.

We now show that G is an open set in (X, d).

Let us take an element x in $G = \bigcup_{\alpha \in I} G_\alpha$.

Then $\exists\, i_1 \in I$ such that $x \in G_{i_1}$. Since G_{i_1} is an open set, $\exists\, r > 0$ such that

$$x \in B_r(x) \subset G_{i_1} \subset \bigcup_{\alpha \in I} G_\alpha = G$$

$\Rightarrow x \in B_r(x) \subset G$ for some $r > 0$.

$\Rightarrow x$ is an interior point of G.

This shows that each element of G is an interior point of G and consequently G is an open set.

Theorem [1.7] Intersection of finite number of open sets in metric space is an open set.

Proof. Let (X, d) be a metric space. Let $\{G_1, G_2, \dots, G_n\}$ be n open sets in (X, d). We show that $\bigcap\limits_{i=1}^{n} G_i$ is an open set in (X, d). For this we first show that $G_1 \cap G_2$ is an open set.

If $G_1 \cap G_2 = \phi$, then there is nothing to prove.

Let $G = G_1 \cap G_2$ and let $x \in G$.

Then $x \in G_1 \cap G_2 \Rightarrow x \in G_1$ and $x \in G_2$.

Since G_1, G_2 are open sets, it follows that x is an interior point of both G_1 and G_2. Then there exists real numbers $r_1 > 0$ and $r_2 > 0$ such that $x \in B_{r_1}(x) \subset G_1$ and $x \in B_{r_2}(x) \subset G_2$.

Let us choose $0 < r < \min\{r_1, r_2\}$. Consider the open ball $B_r(x)$. One can easily verify that $B_r(x) \subset B_{r_1}(x)$ and $B_r(x) \subset B_{r_2}(x)$.

So we have

$$\begin{aligned} & x \in B_r(x) \subset B_{r_1}(x) \subset G_1 \text{ and } x \in B_r(x) \subset B_{r_2}(x) \subset G \\ \Rightarrow\ & x \in B_r(x) \subset G_1 \text{ and } x \in B_r(x) \subset G_2 \\ \Rightarrow\ & x \in B_r(x) \subset G_1 \cap G_2 \\ \Rightarrow\ & x \text{ is an interior point of } G_1 \cap G_2 = G. \end{aligned}$$

Thus each point of G is an interior point of G. Hence G is an open set.

Now $G_1 \cap G_2$ and G_3 are open sets. So by above, $(G_1 \cap G_2) \cap G_3 = G_1 \cap G_2 \cap G_3$ is an open set. Thus by repeated use of the above result we have $\bigcap\limits_{i=1}^{n} G_i$ is an open set in (X, d).

Remark (9). Note that arbitrary intersection of open sets in a metric space need not be an open set. For example, consider the metric space $(\mathbb{R}, d)$ of all real numbers together with the usual metric.

Let $G_n = \left(0, 1 + \dfrac{1}{n}\right)$, for $n \in \mathbb{N}$. Then each G_n, $n \in \mathbb{N}$ is an open set in $(\mathbb{R}, d)$.

Now $G = \bigcap\limits_{n=1}^{\infty} G_n = \bigcap\limits_{n=1}^{\infty} \left(0, 1 + \dfrac{1}{n}\right) = (0, 1]$ (verify) is not an open set.

But if we take $G_n = (-n, n), n \in \mathbb{N}$, then each G_n is an open set in $(\mathbb{R}, d)$.

Then $G = \bigcap\limits_{n=1}^{\infty} G_n = (-1, 1)$ is also an open set.

Thus intersection of any number of open sets may or may not be an open set.

1.2.3 Metric Topology

We first define what is a *topology.* Let X be a non-empty set and let $\tau \subset \mathbb{P}(X)$, the power set of X. If

(i) $\phi, X \in \tau$

(ii) $\{V_i\}_{i \in I} \subset \tau \Rightarrow \bigcup\limits_{i \in I} V_i \in \tau$

(iii) $V_1, V_2 \in \tau \Rightarrow V_1 \cap V_2 \in \tau$,

then τ is called a topology on X and the pair (X, τ) is called a topological space.

Now let (X, d) be a metric space. Let τ_d be the collection of all open sets in (X, d).

Then we see that

(i) $\phi, X \in \tau_d$

(ii) if $\{V_i\}_{i\in I} \subset \tau_d \Rightarrow \bigcup_{i\in I} V_i \in \tau_d$

(iii) if $V_1, V_2 \in \tau_d \Rightarrow V_1 \cap V_2 \in \tau_d$.

That means, τ_d satisfies all the conditions to be a topology. Thus τ_d is a topology on X and consequently (X, τ_d) is a topological space. This topology τ_d is known as *metric topology.* Hence every metric space is a topological space.

Definition [1.6] : A metric space (X, d) is said to be a discrete metric space if $\tau_d = p(X)$, i.e., each subset of X is open.

It follows from remark (8) that a trivial metric space is a discrete metric space. It is very easy to verify that $(\mathbb{N}, d)$ is a discrete metric space, where $d(x, y) = |x - y| \; \forall \, x, y \in \mathbb{N}$, however the metric d is **not** a trivial metric.

1.2.4 Neighbourhood

Let (X, d) be a metric space and let $x \in X$. A subset A of X is said to be a neighbourhood of x if there exists an $r > 0$ such that $x \in B_r(x) \subset A$.

Sometimes by N_x we denote the neighbourhood of a point $x \in X$. Let us denote by $\mathbb{N}_x$ the set of all neighbourhoods of a point $x \in X$.

Theorem [1.8] Let (X, d) be a metric space. A subset G of X is an open set, iff it is a neighbourhood of each of its points.

Proof. First we suppose that G is an open set in (X, d). Let $x \in G$. Since G is open, x is an interior point of G. Then there is a real number $r > 0$ such that $x \in B_r(x) \subset G$.

This implies that G is a neighbourhood of x. Hence G is a neighbourhood of each of its points.

Conversely, let the condition hold. That means G is a neighbourhood of each of its points.

Thus for each $x \in G$, $\exists \, r_x > 0$ such that $x \in B_{r_x}(x) \subset G$.

Then it is easy to verify that $G = \bigcup_{x \in G} B_{r_x}(x)$. We know that each open ball $B_r(x)$ in a metric space is an open set and arbitrary union of open sets is an open set.

Hence G is an open set.

Theorem [1.9] Let (X, d) be a metric space and $x \in X$.

(i) If A is a nbd. of x and $A \subset B$, then B is also a nbd. of x.

(ii) If A, B are two neighbourhoods of x, then $A \cup B$ and $A \cap B$ are nbds of x.

Proof. (1) Let A be a nbd. of x. Then $\exists$ an $r > 0$ such that $x \in B_r(x) \subset A$ and so $x \in B_r(x) \subset B$ as $A \subset B$.

$\Rightarrow B$ is a nbd. of x.

(2) Let A and B be two neighbourhoods of x. Then since $A \subset A \cup B$ and A is a neighbourhood of x, it follows by (1) $A \cup B$ is a neighbourhood of x.

Since A and B are neighbourhoods of x, $\exists \, r_1, r_2 > 0$ such that $x \in B_{r_1}(x) \subset A$ and $x \in B_{r_2}(x) \subset B$.

Let $0 < r < \min\{r_1, r_2\}$. Then we know that $x \in B_r(x) \subset A \cap B$.

$\Rightarrow A \cap B$ is a neighbourhood of x.

1.2.5 Limiting Point of a Set

Definition [1.7] : Let (X,d) be a metric space and $A \subset X$. A point $x \in X$ is said to be a *limit point* of A, if every neighbourhood of x intersects A at a point other than x, if $x \in A$. i.e., x is a limit point of A if $(N_x - \{x\}) \cap A \neq \phi$ for all neighbourhoods N_x of x.

The set of all limiting points of a set A is denoted by A' and is called the *derived set* of A.

Limit point of a set may or may not belong to the set.

Example [1.19] : Consider the metric space $(\mathbb{R}, d)$ of all real numbers together with the usual metric.

Let $A = \left\{1, \frac{1}{2}, \frac{1}{3}, \ldots, \frac{1}{n}, \ldots\right\} \subseteq \mathbb{R}$. Then it is very easy to verify that 0 is a limit point of A. But $0 \notin A$.

Let $B = [0,1] \subseteq \mathbb{R}$. Then every point of B is a limit point of B.

Let $\mathbb{N} = \{1, 2, 3, \ldots, n, \ldots\} \subseteq \mathbb{R}$. Then $\mathbb{N}$ has no limit point.

Remark (10). From definition of limit points it follows that if a set A has limit point α, then every neighbourhood of α contains infinitely many elements of A. Hence A must be an infinite set. From this we can conclude that a finite set F has no limit point, i.e., $F' = \phi$.

Definition [1.8] : Let (X,d) be a metric space. A subset F of X is called a *closed set* if every limiting point of F is a member of F, i.e., $F' \subset F$.

From the above remark we find that if F is a finite set, then $\phi = F' \subset F$ and so a finite set in a metric space is always a closed set.

Definition [1.9] : Let (X,d) be a metric space and let $A \subset X$. Let A' be the set of all limit points of A. Then the set $A \cup A'$ is called the closure of A. We denote it by $\bar{A}$, i.e., $\bar{A} = A \cup A'$.

Note that if A is a finite set, then $A' = \phi$ and so $\bar{A} = A \cup \phi = A$.

Theorem [1.10] Let (X,d) be a metric space and $F \subset X$. Then F is a closed set, iff $F = \bar{F}$.

Proof. First assume that F is a closed set.

Then from definition we get $F' \subset F$. So $F \cup F' \subset F \Rightarrow \bar{F} \subset F$.

Since $\bar{F} = F \cup F'$, we have $F \subset \bar{F}$. Hence we get $F = \bar{F}$.

Conversely, let $F = \bar{F} = F \cup F' \Rightarrow F' \subset F$.

$\Rightarrow F$ is a closed set.

Theorem [1.11] Let (X,d) be a metric space and $F \subset X$. Then F is closed in (X,d), iff $F^c = X|F$ is open in (X,d).

Proof. Let us first assume that F is a closed set. We now show that $X|F$ is an open set. Let $u \in X|F$. Then $u \notin F$. Since F is a closed set, F contains all its limit points. Hence u is not a limit point of F. This imply that we can find a neighbourhood N_u of u such that $N_u \cap F = \phi$. Since N_u is a neighbourhood of u, there is an open ball $B_r(u)$ for some $r > 0$ such that $u \in B_r(u) \subset N_u$. Hence we get $B_r(u) \cap F = \phi$. This imply that $u \in B_r(u) \subset (X|F)$.

$\Rightarrow u$ is an interior point of $X|F$

$\Rightarrow X|F$ is an open set.

Conversely, suppose that $X|F$ is an open set. We now show that F is a closed set. Let α be a limit point of F such that $\alpha \notin F$. Then $\alpha \in (X|F)$. Since $X|F$ is an open set, α is an interior point of $X|F$. So $\exists\, r > 0$ such that $\alpha \in B_r(\alpha) \subset X|F$.

Hence $B_r(\alpha) \cap F = \phi$—a contradiction to the fact that α is a limit point of F.

Thus our assumption $\alpha \notin F$ is wrong. So $\alpha \in F$.

$\Rightarrow F$ is a closed set.

Remark (11). From the above theorem we get an alternative definition of closed sets. A set F in a metric space (X, d) is called a **closed set** if $X|F$ is an open set in (X, d).

Theorem [1.12] Let (X, d) be a metric space and $\{F_i\}_{i \in I}$ be a family of closed sets. Then $\bigcap\limits_{i \in I} F_i$ is a closed set.

Proof. Let $F = \bigcap\limits_{i \in I} F_i$. If $F = \phi$, then F is a closed set. Suppose that $F \neq \phi$.

To show that F is closed set, we show that $X|F$ is an open set.

Note that $X|F = X|\left(\bigcap\limits_{i \in I} F_i\right) = \bigcup\limits_{i \in I} (X|F_i)$, by DeMorgan's law. Since each $X|F_i$ is an open set and arbitrary union of open set is an open set, it follows that $X|F$ is an open set and consequently F is a closed set.

Remark (12). It is easy to show that finite union of closed sets is a closed set. But arbitrary union of closed sets may or may not be a closed set.

For example, consider the metric space $(\mathbb{R}, d)$ of all real numbers with the usual metric.

Let $F_n = \left[\frac{1}{n}, 1\right]$ for all $n \in \mathbb{N}$. Then each F_n is a closed set. But $F = \bigcup\limits_{n=1}^{\infty} F_n = (0, 1]$ is not a closed set. But if we take $P_n = \left[-\frac{1}{n}, \frac{1}{n}\right]$ for all $n \in \mathbb{N}$, then each P_n is a closed set and $P = \bigcup\limits_{n=1}^{\infty} P_n = [-1, 1]$ is also a closed set.

Theorem [1.13] Let (X, d) be a metric space and $A, B \subset X$. Then

(i) $A \subset B \Rightarrow \bar{A} \subset \bar{B}$
(ii) $\overline{A \cup B} = \bar{A} \cup \bar{B}$
(iii) $\overline{A \cap B} \subseteq \bar{A} \cap \bar{B}$
(iv) $\bar{A}$ is the smallest closed set containing A.

Proof. (i) Left as an exercise.

We now prove (iv).

(iv) We first show that $\bar{A}$ is a closed set.

Let x be a limit point of $\bar{A}$. We show that $x \in \bar{A}$. Then for each $r > 0$, $B_r(x) \cap \bar{A} \neq \phi$

$\Rightarrow B_r(x) \cap (A \cup A') \neq \phi$

$\Rightarrow (B_r(x) \cap A) \cup (B_r(x) \cap A') \neq \phi.$

Then either $B_r(x) \cap A \neq \phi$ or $B_r(x)(x) \cap A' \neq \phi$ or both are non-empty.

Let $B_r(x) \cap A \neq \phi$. This is true for any $r > 0$.

$\Rightarrow x$ is a limit point of A

$\Rightarrow x \in A' \subset \bar{A}$

$\Rightarrow x \in \bar{A}$.

Now let $B_r(x) \cap A' \neq \phi$.

Let $y \in B_r(x) \cap A'$. Then $y \in B_r(x)$ and $y \in A'$. Since $y \in A'$, y is a limit point of A and since $B_r(x)$ is an open set and $y \in B_r(x)$, $B_r(x)$ is a nbd of y. So $B_r(x) \cap A \neq \phi \Rightarrow x \in A' \subset \bar{A} \Rightarrow x \in \bar{A}$.

Hence, in any case, we find that $x \in \bar{A}$. So $\bar{A}$ is a closed set.

Since $\bar{A} = A \cup A'$, $A \subset \bar{A}$, i.e., $\bar{A}$ is a closed set containing A. We now show that $\bar{A}$ is the smallest closed set containing A.

Let F be any other closed set containing A. We claim that $\bar{A} \subset F$. Let $x \in \bar{A}$. Then, if $x \in A$, we get $x \in F$. So let $x \in A'$. Then for each $r > 0$, $B_r(x) \cap A \neq \phi \Rightarrow B_r(x) \cap F \neq \phi$ as $F \supset A$.

Hence, $x \in F' \subset \bar{F} \Rightarrow x \in \bar{F} = F$, as F is closed. Thus we see that $\bar{A} \subset F$ [or, since $A \subset F$, from (i) we get $\bar{A} \subset \bar{F} = F$, i.e., $\bar{A} \subset F$].

(ii) Note that $\bar{A} \supset A$, $\bar{B} \supset B$ and so $\bar{A} \cup \bar{B} \supset A \cup B$, i.e., $\bar{A} \cup \bar{B}$ is a closed set containing $A \cup B$. But we have seen from (iv) that $\overline{A \cup B}$ is the smallest closed set containing $A \cup B$; so $\overline{A \cup B} \subset \bar{A} \cup \bar{B}$ [or, $A \cup B \subset \bar{A} \cup \bar{B} \Rightarrow \overline{A \cup B} \subset \overline{\bar{A} \cup \bar{b}} = \bar{A} \cup \bar{B}$].

On the other hand, $A \subset A \cup B, B \subset A \cup B$. So from (i) $\bar{A} \subset \overline{A \cup B}$ and $\bar{B} \subset \overline{A \cup B}$ and hence $\bar{A} \cup \bar{B} \subset \overline{A \cup B}$.

Hence, we get $\overline{A \cup B} = \bar{A} \cup \bar{B}$.

(iii) Left as an exercise.

Remark (13). Equality in (iii) may not hold, in general. Try to find out an example to support this remark.

1.2.6 Exterior Point, Boundary Point

Definition [1.10] : **[Exterior point]** Let (X, d) be a metric space and $A \subseteq X$. A point $x \in X$ is called an exterior point of A if x is an interior point of $X \mid A$.

The set of all exterior points of A is called exterior of A and is denoted by $\text{ext}\,(A)$.

Therefore $\text{ext}\,(A) = [X \mid A]^0$. From definition it follows that $\text{ext}\,(X \mid A) = \overset{\circ}{A}$. It is also clear from definition that $\text{ext}\,(X) = \phi$ and $\text{ext}\,(\phi) = X$.

Theorem [1.14] Let (X, d) be a metric space and $A \subseteq X$. Then $\text{ext}\,(A) = (\bar{A})^c$.

Proof. Let $x \in \text{ext}\,(A)$

$\Rightarrow x$ is an interior point of $X \mid A$

$\Rightarrow \exists r > 0$ such that $B_r(x) \subseteq X \mid A$

$\Rightarrow \exists r > 0$ such that $B_r(x) \cap A = \phi$

$\Rightarrow x \notin \bar{A} \Rightarrow x \in (\bar{A})^c$

$\therefore \text{ext}\,(A) \subseteq (\bar{A})^c$

In a similar way, we can show that $(\bar{A})^c \subseteq \text{ext}\,(A)$. Hence the result follows.

Definition [1.11] : **[Boundary point]** Let (X, d) be a metric space and $A \subseteq X$. A point $x \in X$ is called a boundary point of A if it is neither an interior point of A nor of $X \mid A$, that is, it is neither interior point of A nor exterior point of A.

It follows from definition that a point $x \in X$ is a boundary point of A if every open ball centred at x intersects both A and $X \mid A$. The set of all boundary points of a set A is denoted by Bd (A).

It follows from definition that $\text{Bd}\,(A) = \bar{A} \cap \overline{(X \mid A)}$ and therefore $\text{Bd}\,(X \mid A) = \text{Bd}\,(A)$.

Example [1.20] : Consider the metric space of all real numbers $(\mathbb{R}, d)$ with the usual metric d.

Let $A = (a, b)$. Then $\text{Bd}\,(A) = \{a, b\}$

Similarly, if $A = [a, b]$ or $[a, b)$ or $(a, b]$, then $\text{Bd}\,(A) = \{a, b\}$

If $B = Q$, the set of all rational numbers then $\text{Bd}\,(Q) = \mathbb{R}$

If $B =$ the set of all irrational numbers, then $\text{Bd}\,(B) = \mathbb{R}$.

If $C = \mathbb{N}$ or $\mathbb{Z}$, then $\text{Bd}\,(C) = \mathbb{N}$ or $\mathbb{Z}$ respectively.

If $D = \left\{\frac{1}{n} : n \in \mathbb{N}\right\}$ then $\text{Bd}\,(D) = D \cup \{0\}$.

Example [1.21] : Consider the metric space $(\mathbb{R}^2, d)$ where d is the usual metric of $\mathbb{R}^2$.

Let $A = \{(x, y) \in \mathbb{R}^2 : y = cx, \text{ where } c > 0\}$. Then $\text{Bd}\,(A) = \{(x, y) \in \mathbb{R}^2 : x = 0\}$.

Example [1.22] : Let (X, d) be the trivial metric space and $A \subseteq X$. Then it is clear that $\text{Bd}\,(A) = \phi$.

Remark (14). Let (X, d) be a metric space and $A \subseteq X$.

Then $\overline{(X \mid A)} = X \mid \mathring{A}$.

Let $x \in X \mid \mathring{A} \Leftrightarrow x \notin \mathring{A}$

$\Leftrightarrow$ for any $r > 0, B_r(x) \not\subset A$

$\Leftrightarrow$ for any $r > 0, B_r(x) \cap (X \mid A) \neq \phi$

$\Leftrightarrow x \in \overline{(X \mid A)}$

In view of the remark above, it follows that $\text{Bd}\,(A) = \bar{A} \cap \overline{(X \mid A)} = \bar{A} \cap (X \mid \mathring{A}) = \bar{A} \mid \mathring{A}$.

Theorem [1.15] Let (X, d) be a metric space and $A \subseteq X$. Then (i) A is closed if and only if $\text{Bd}\,(A) \subseteq A$, (ii) A is open if and only if $A \cap \text{Bd}\,(A) = \phi$, (iii) $\text{Bd}\,(A) = \phi$ iff A is both open and closed.

Proof. (i) Let A be closed. Then $\bar{A} = A$. Since $\text{Bd}\,(A) = \bar{A} \cap \overline{(X \mid A)} \subseteq \bar{A}$ we have $\text{Bd}\,(A) \subseteq A$ conversely, let $\text{Bd}\,(A) \subseteq A$. Let $x \in A^c$. Then we must have some open ball centred at x which contains only points of A. Therefore A^c is open and hence A is closed.

(ii) Let A be open.

We prove that $A \cap \text{Bd}\,(A) = \phi$. If possible, let $A \cap \text{Bd}\,(A) \neq \phi$.

Let $x \in A \cap \text{Bd}\,(A) \Rightarrow x \in A$ and $x \in \text{Bd}\,(A)$.

Since A is open, $\exists r > 0$ such that $B_r(x) \subseteq A$ and therefore $B_r(x) \cap A^c = \phi$-a contradiction to $x \in \text{Bd}\,(A)$.

$\therefore A \cap \text{Bd}\,(A) = \phi$.

Conversely, let $A \cap \text{Bd}(A) = \phi$. We claim that A is open.

Let $x \in A$. Clearly $x \notin \text{Bd}(A)$. So $\exists r > 0$ such that $B_r(x) \cap A^c = \phi$ and hence $B_r(x) \subseteq A$.

$\Rightarrow x$ is an interior point of A.

$\therefore A$ is open.

(iii) Let $\text{Bd}(A) = \phi$. Then $\text{Bd}(A^c) = \phi$. Since $\text{Bd}(A) = \phi, A \cap \text{Bd}(A) = \phi$. Hence by (ii) above. A is open.

Similarly, A^c is open and so A is closed.

Theorem [1.16] $\text{Bd}(A) = \bar{A} - \mathring{A}$.

Proof. Let $x \in \text{Bd}(A) \Rightarrow B_r(x) \cap A \neq \phi$ and $B_r(x) \cap A^c \neq \phi$ for any $r > 0$.

We show that $x \in \bar{A}$. If $x \in A$, then clearly $x \in \bar{A}$.

Let $x \notin A$. Then $B_r(x) \cap A \neq \phi$ for any $r > 0$.

$\Rightarrow x \in A'$ and therefore $x \in \bar{A}$.

Clearly $x \notin \mathring{A}$, $\quad \therefore x \in \bar{A} - \mathring{A}$.

Hence,

$$\text{Bd}(A) \subseteq \bar{A} - \mathring{A}. \tag{i}$$

Let $x \in \bar{A} - \mathring{A}$. Then $x \in \bar{A}$ and $x \notin \mathring{A}$.

Let $r > 0$. Consider the open ball $B_r(x)$.

Then $B_r(x) \not\subset A$ which implies that $B_r(x) \cap A^c \neq \phi$.

Since $x \in \bar{A}$, it follows that $B_r(x) \cap A \neq \phi$. Hence $x \in \text{Bd}(A)$. Therefore

$$\bar{A} - \mathring{A} \subseteq \text{Bd}(A) \tag{ii}$$

From (i) and (ii) we have the required result.

Theorem [1.17] Let (X, d) be a metric space and $A \subseteq X$. Then

(i) $\text{Bd}(A) = \text{Bd}(X - A)$ (ii) $\text{Bd}(A) = \bar{A} \cap \overline{X - A}$

(iii) $\text{Bd}(A) = \bar{A} - \mathring{A}$ (iv) $\text{Bd}(A) = (\overline{X - A}) - (X - A)^\circ$

(v) $X - \text{Bd}(A) = \mathring{A} \cup (X - A)^\circ$ (vi) $\bar{A} = \mathring{A} \cup \text{Bd}(A) = A \cup \text{Bd}(A)$

(vii) $\mathring{A} \cap \text{Bd}(A) = \phi$ (viii) $\mathring{A} = A - \text{Bd}(A)$

Proof. Left as an exercise.

1.2.7 Dense Set, Perfect Sets etc

Let (X, d) be a metric space. A subset D of X is said to be dense in X if $\bar{D} = X$.

Example [1.23] :

(i) The set of all rational numbers Q is dense in the metric space $(\mathbb{R}, d)$.

(ii) In a discrete metric space (X, d), a subset D is dense in (X, d) iff $D = X$.

A subset A in a metric space (X, d) is said to be a set dense in itself if $A \subseteq A'$.

A subset A in a metric space (X, d) is said to be perfect if A is closed and A is dense in itself. In other words A is perfect if $A = A'$.

Example [1.24] :

(1) Consider the metric space $(\mathbb{R}, d)$ where d is the usual metric of $\mathbb{R}$.

(i) Q is dense in itself but **not** perfect.

(ii) $[a, b], a, b \in \mathbb{R}$ is a perfect set

(iii) $\mathbb{R}, \phi$ are perfect sets

(2) In a trivial metric space, X and ϕ are the only perfect sets.

1.3 Subspaces of a Metric Space

Definition [1.12] : Let (X, d) be a metric space and $Y \subset X$ a subset. We now consider the function d_Y, the restriction of d on $Y \times Y$. Then d_Y also satisfies all the distance axioms (d.1), (d.2), (d.3) and (d.4) and d_Y is a metric on Y and consequently (Y, d_Y) is a metric space. Note that $d_Y : Y \times Y \to \mathbb{R}$ is defined by $d_Y(x, y) = d(x, y)$, $\forall\ x, y \in Y$.

The metric space (Y, d_Y) is called a subspace of the metric space (X, d).

For example, consider the metric space $(\mathbb{R}, d)$ of all real numbers with the usual metric.

Let $Y = [0, 1]$. Then by the above definition (Y, d_Y) is a subspace of $(\mathbb{R}, d)$.

1.3.1 Open Ball, Open Set in a Subspace

Let (X, d) be a metric space and Y be a subset of X. Then (Y, d_Y) is a subspace of (X, d).

Let $r > 0$ be a real number and let $\alpha \in Y$. Then we denote an open ball centred at α and radius r by $B_r^Y(\alpha)$.

Then $B_r^Y(\alpha) = \{y \in Y : d_Y(\alpha, y) < r\} = \{y \in X : d(\alpha, y) < r\} \cap Y = B_r(\alpha) \cap Y$, i.e., an open ball in Y = (an open ball in X) $\cap Y$.

In a similar way, we can show that an open set G in (Y, d_Y) = (an open set in X) $\cap Y$.

Similar results hold for closed ball and closed sets.

Example [1.25] : Let us consider the metric space $(\mathbb{R}, d)$ of all real numbers with the usual metric.

Let $Y = [0, 1]$. Let $G = (-1, 1/2) \subseteq \mathbb{R}$. Then G is an open set in $\mathbb{R}$.

Now $G \cap Y = (-1, 1/2) \cap [0, 1] = [0, 1/2)$ is an open set in the subspace (Y, d_Y), but is not an open set in $\mathbb{R}$.

Illustrative Examples:

Problem [1.1] : (1) If U and V are subsets of a metric space (X, d) with U an open set, show that (a) $U \cap V = \phi \Rightarrow U \cap \bar{V} = \phi$, (b) $U \cap \bar{V} \subseteq \overline{U \cap V}$.

Solution. (a) Let U and V are subset of (X, d) and U is an open set in (X, d) s.t. $U \cap V = \phi$.

If possible, suppose that $U \cap \bar{V} \neq \phi$.

Let $x \in U \cap \bar{V} \Rightarrow x \in U$ and $x \in \bar{V}$.

Now $x \in \bar{V}$ and U is an open set with $x \in U$

$\Rightarrow x \in \bar{V}$ and U is a neighbourhood of x

$\Rightarrow U \cap V \neq \phi$—a contradiction.

Hence we must have $U \cap V = \phi \Rightarrow U \cap \bar{V} = \phi$.

(b) We now show that $U \cap \bar{V} \subseteq \overline{U \cap V}$.

Let $x \in U \cap \bar{V} \Rightarrow x \in U$ and $x \in \bar{V}$.

Let N_x be a neighbourhood of x. Since U is an open set and $x \in U$, so U is also a neighbourhood of x. Then $U \cap N_x$ is a neighbourhood of x.

Since $x \in \bar{V}$, we have $V \cap (U \cap N_x) \neq \phi \Rightarrow (U \cap V) \cap N_x \neq \phi$.

This is true for all neighbourhood N_x of x. So $(U \cap V) \cap N_x \neq \phi$ for all neighbourhood N_x of x

$\Rightarrow x \in \overline{U \cap V}$.

Hence, $U \cap \bar{V} \subseteq \overline{U \cap V}$.

Problem [1.2] : Let α be a point in a metric space (X, d) and $0 < p < q$. Prove that the set $\{x \in X : p < d(x, \alpha) < q\}$ is open in (X, d).

Solution. Note that

$$\begin{aligned}
&\{x \in X : p < d(x, \alpha) < q\} \\
&= \{x \in X : d(x, \alpha) < q\} - \{x \in X : d(x, \alpha) \leq p\} \\
&= B_q(\alpha) - \bar{B}_p(\alpha) \\
&= B_q(\alpha) \cap \{\bar{B}_p(\alpha)\}'
\end{aligned}$$

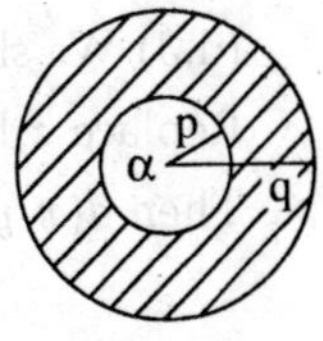

Fig. 1.3

$'$ denotes complement of a set. Since $B_q(\alpha)$ is an open set and $\bar{B}_p(\alpha)$ is a closed set and so $\{\bar{B}_p(\alpha)\}'$ is an open set, it follows that

$B_q(\alpha) \cap \{\bar{B}_p(\alpha)\}'$ is an open set

$\Rightarrow \{x \in X : p < d(x, \alpha) < q\}$ is an open set.

Problem [1.3] : Let (X, d) be a metric space and let $A \subset X$. Let $x \in X$. Then $x \in \bar{A}$, if and only if given any $\epsilon > 0$, there exists a point $a \in A$ and $a \neq x$ such that $d(x, a) < \epsilon$.

Solution. First, let $x \in \bar{A}$. If $x \in A$, then there is nothing to prove. Suppose that $x \notin A$. Then $x \in A'$, i.e., x is a limit point of A. So for any $\epsilon > 0$, $B_\epsilon(x) \cap A \neq \phi$. Let $a \in B_\epsilon(x) \cap A$. Then $d(x, a) < \epsilon$ and $a \in A$ with $a \neq x$.

Conversely, let the condition holds, i.e., for any $\epsilon > 0$ $\exists$ a point $a \in A$ such that $d(x, a) < \epsilon$ and $a \neq x$

$\Rightarrow a \in B_\epsilon(x)$ and $a \neq x$

$\Rightarrow (B_\epsilon(x) - \{x\}) \cap A \neq \phi$

$\Rightarrow x$ is a limit point of A

$\Rightarrow x \in \bar{A}$.

Problem [1.4] : Let (X, d) be a metric space and $x_0 \in X$. Then for any $r > 0$, the set $G = \{x \in X : d(x, x_0) > r\}$ is an open set in X.

Solution. We show that G is an open set, i.e., each point of G is an interior point of G.

Let $x \in G$. Then $d(x, x_0) > r$. Let us put $\delta = d(x, x_0) - r$. Consider the open ball $B_\delta(x)$. Then for any $y \in B_\delta(x)$, we have

$$\begin{aligned}
&d(x, x_0) \leq d(x, y) + d(y, x_0) = d(x, y) + d(x_0, y) \\
&\Rightarrow d(x_0, y) \geq d(x, x_0) - d(x, y) > d(x, x_0) - \delta = r \\
&\Rightarrow y \in G \Rightarrow B_\delta(x) \subset G \Rightarrow x \text{ is an interior point of } G
\end{aligned}$$

Hence G is an open set.

Problem [1.5] : Let X be a non-empty set. Define a function $d : X \times X \to \mathbb{R}$, which satisfies the conditions.

(i) $d(x, y) = 0$, if and only if $x = y$ in X

(ii) $d(x, y) \leq d(x, z) + d(y, z)$ for all $x, y, z \in X$.

Then prove that d is a metric on X.

Solution. Putting $x = y$ in (ii), we get $d(y, y) \leq d(y, z) + d(y, z)$

$\Rightarrow 2d(y, z) \geq 0$, since by (i), $d(y, y) = 0$

$\Rightarrow d(y, z) \geq 0$ for any $y, z \in X$.

$\therefore d(x, y) \geq 0$ for all $x, y \in X$. This is (d.1).

(d.2) follows from condition (i).

(d.3) We show that $d(x, y) = d(y, x)$ for all $x, y \in X$.

Replace z by x in (ii).

Then $d(x, y) \leq d(x, x) + d(y, x) = d(y, x)$ since $d(x, x) = 0$

$$\Rightarrow d(x, y) \leq d(y, x). \tag{1}$$

Now replacing x, y, z by y, x, y respectively, we get from (ii),

$$d(y, x) \leq d(y, y) + d(x, y) = d(x, y) \Rightarrow d(y, x) \leq d(x, y) \tag{2}$$

Hence from, (1) and (2) we get $d(x, y) = d(y, x)$ for all $x, y \in X$.

(d.4) We have $d(x, y) \leq d(x, z) + d(y, z) = d(x, z) + d(z, y)$ by (d.3)

$$\Rightarrow d(x, y) \leq d(x, z) + d(z, y), \ \forall \ x, y, z \in X.$$

Hence, (X, d) is a metric space.

Problem [1.6]: Let (X_1, d_1) and (X_2, d_2) be two metric spaces. Let $X = X_1 \times X_2$. Define $d : X \times X \to \mathbb{R}$ by

$$d(x, y) = \max\{d_1(x_1, y_1), d_2(x_2, y_2)\},$$

where $x = (x_1, x_2)$ and $y = (y_1, y_2) \in X$.

Then show that d is a metric on X.

Solution. (d.1) From definition, it follows that $d(x, y) \geq 0$ for all $x = (x_1, x_2), y = (y_1, y_2)$ in X, since $d_1(x_1, y_1) \geq 0$ and $d_2(x_2, y_2) \geq 0$.

(d.2) Let $d(x, y) = 0$, where $x = (x_1, x_2), y = (y_1, y_2)$ in X

$\Rightarrow \max\{d_1(x_1, y_1), d_2(x_2, y_2)\} = 0$

$\Rightarrow d_1(x_1, y_1) = 0$ and $d_2(x_2, y_2) = 0$,

since $d_1(x_1, y_1) \leq \max\{d_1(x_1, y_1), d_2(x_2, y_2)\} = 0$

$\Rightarrow x_1 = y_1$ and $x_2 = y_2$, since d_1, d_2 are metrics

$\Rightarrow x = y$.

If $x = y$, then it is easy to see that $d(x, y) = 0$.

(d.3) $d(x, y) = d(y, x)$ follows from definition.

(d.4) The triangle inequality follows from $d_k(x_k, y_k) \leq d_k(x_k, z_k) + d_k(z_k, y_k), k = 1, 2$

$\Rightarrow d_k(x_k, y_k) \le \max\limits_{k=1,2} d_k(x_k, z_k) + \max\limits_{k=1,2} d_k(z_k, y_k)$, for any $k = 1, 2$

$\Rightarrow d(x, y) \le d(x, z) + d(z, y)$, where $x = (x_1, x_2)$, $y = (y_1, y_2)$ and $z = (z_1, z_2)$ are in X.

Hence, (X, d) is a metric space.

Problem [1.7]: Let (X_1, d_1) and (X_2, d_2) be two metric spaces. Let $X = X_1 \times X_2$. Define $d^* : X \times X \to \mathbb{R}$ by

$d^*(x, y) = d_1(x_1, y_1) + d_2(x_2, y_2)$, where $x = (x_1, x_2)$ and $y = (y_1, y_2) \in X$.

Then show that d^* is a metric on X.

Solution. (d.1) Since $d_1(x_1, y_1), d_2(x_2, y_2) \ge 0$ for all $x = (x_1, x_2)$, $y = (y_1, y_2)$ in X, it follows that $d(x, y) = d_1(x_1, y_1) + d_2(x_2, y_2) \ge 0$, $\forall\ x, y \in X$.

(d.2) Let $d(x, y) = 0$, where $x = (x_1, x_2)$, $y = (y_1, y_2) \in X$

$\Rightarrow d_1(x_1, y_1) + d_2(x_2, y_2) = 0$

$\Rightarrow d_1(x_1, y_1) = 0 = d_2(x_2, y_2)$, since $d_1(x_1, y_1) \ge 0, d_2(x_2, y_2) \ge 0$

$\Rightarrow x_1 = y_1$ and $x_2 = y_2$

$\Rightarrow x = y$.

Converse part is obvious.

(d.3) $d(y, x) = d_1(y_1, x_1) + d_2(y_2, x_2)$

$= d_1(x_1, y_1) + d_2(x_2, y_2)$, since d_1, d_2 are metric

$= d(x, y)$ for all $x, y \in X$.

(d.4) Let $x = (x_1, y_2), y = (y_1, y_2), z = (z_1, z_2) \in X$.

Then since d_1 and d_2 are metric, we get

$$d_1(x_1, y_1) \le d_1(x_1, z_1) + d_1(z_1, y_1)$$
$$\text{and } d_2(x_2, y_2) \le d_2(x_2, z_2) + d_2(z_2, y_2).$$

Adding both sides, we get $d(x, y) \le d(x, z) + d(y, z)$, which is the required triangle inequality.

$\therefore (X, d)$ is a metric space.

Problem [1.8]: Show that in a trivial metric space (X, d), every subset is open and closed.

Solution. Let X be a non-empty set. From example [1.9] we have

$$d(x, y) = \begin{cases} 0, & \text{iff } x = y \\ 1, & \text{iff } x \ne y. \end{cases}$$

Let $a \in X$ and take $r = 1/2$.

Then $B_{1/2}(a) = \left\{ x \in X : d(x, a) < \dfrac{1}{2} \right\} = \{a\}$, see example [1.14].

Since every open ball is an open set, it follows that $\{a\}$ is an open set for all $a \in X$.

Let $A \subset X$. Then $A = \bigcup\limits_{a \in A} \{a\}$ is union of open sets and so A is an open set.

Thus every subset $A \subset X$ is an open set.

For any $A \subset X$, we have A^c, the complement of A, is a subset of X and hence by above it is an open set.

So its complement $(A^c)^c = A$ is a closed set.

Thus every subset of X is closed.

Problem [1.9] : Show that the closure $\overline{B_r(x_0)}$ of an open ball $B_r(x_0)$ in a metric space (X, d) can differ from the closed ball $\bar{B}_r(x_0)$.

Solution. Let X be a non-empty set where X contains more than one point and d be the trivial metric on X. Then we know that

$$B_r(x_0) = \begin{cases} \{x_0\}, & \text{if } 0 < r \leq 1 \\ X, & \text{if } r > 1 \end{cases} \qquad \text{(see example [1.14])}$$

Take $r = 1$. Then we have $B_1(x_0) = \{x_0\}$ and hence $\overline{B_1(x_0)} = \overline{\{x_0\}} = \{x_0\}$, by (8).

Note that $\bar{B}_r(x_0) = \begin{cases} \{x_0\}, & \text{if } 0 < r < 1 \\ X, & \text{if } r \geq 1. \end{cases}$

Hence, $\bar{B}_1(x_0) = X$.

Thus $\overline{B_1(x_0)} = \{x_0\} \neq \bar{B}_1(x_0)$.

This example shows that closure of an open ball in a metric space may **not** be the closed ball.

Problem [1.10] : Let (X, d) be a metric space. Let G be an open set and F be a closed set. Then show that $G|F$ is an open set and $F|G$ is a closed set.

Solution. Note that $G|F = G \cap F^c$, F^c denotes complement of F. Since F is closed, F^c is an open set and so $G \cap F^c$ is an open set.

Hence $G|F$ is an open set.

Similarly, $F|G$ is a closed set.

Problem [1.11] : Let (X, d) be a metric space and $Y \subset X$. Then show that for any subset F of Y, $\bar{F}^X \cap Y$ is the closure of F in Y, where $\bar{F}^X$ denotes the closure of F in (X, d).

Solution. Let $\bar{F}^Y$ denote the closure of F in (Y, d_Y). Note that $\bar{F}^X \cap Y$ is a closed set in (Y, d_Y). Since $F \subset \bar{F}^X$ and $F \subset Y$, it follows that $F \subset \bar{F}^X \cap Y$. Thus $\bar{F}^X \cap Y$ is a closed set in (Y, d_Y) containing F. So we have $\bar{F}^Y \subset \bar{F}^X \cap Y$, since $\bar{F}^Y$ is the smallest closed set containing F in (Y, d_Y).

Now, since $\bar{F}^Y$ is a closed set in Y, there exists a closed set A in X such that $\bar{F}^Y = A \cap Y$. Then $F \subset \bar{F}^Y \subset A \Rightarrow F \subset A$. So $\bar{F}^X \subset \bar{A}^X = A \Rightarrow \bar{F}^X \cap Y \subseteq A \cap Y = \bar{F}^Y$.

$\therefore \bar{F}^X \cap Y \subset \bar{F}^Y$ and consequently, $\bar{F}^Y = \bar{F}^X \cap Y$.

EXERCISE

(1) Let X be a non-empty set. Let $d : X \times X \to \mathbb{R}$ be a function such that
   (a) $d(x, y) = 0$, if and only if $x = y$ in X
   (b) $d(x, y) \leq d(x, z) + d(y, z)$, $x, y, z \in X$.
   Prove that (X, d) is a metric space.

(2) Show that $(\mathbb{R}, d)$ is a metric space, where d is defined by $d(x, y) = \sqrt{|x - y|}$, $x, y \in \mathbb{R}$.

(3) Let $X = C[0, 1]$. Define d on X as $d(x, y) = \sup\limits_{0 \leq t \leq 1} |x(t) - y(t)|$, where $x, y \in C[0, 1] = X$.
   Let $f, g \in X$ be such that $f(t) = t$ and $g(t) = t^2$, $t \in [0, 1]$. Find $d(f, g)$.

(4) Let (X, d) be a metric space and $x, y \in X$ be such that $x \neq y$. Show that there exist disjoint open balls such that one is centred at x and another at y.

(5) Let (X, d) be a metric space and $x \in X$. Let $0 < r_1 < r_2$ be two real numbers. Prove that $B_{r_1}(x) \subset B_{r_2}(x)$.

(6) Prove that in a metric space (X, d), $\overline{B_r(x)} \subseteq \bar{B}_r(x)$, $x \in X$ and $r > 0$ be a real number. Show by example that the inclusion is proper in the above result.

(7) Show that in a discrete metric space (X, d), every subset A of X is both open and closed.

(8) Let (X, d) be a metric space. Show that the function ρ defined by $\rho(x, y) = \min\{1, d(x, y)\}$, where $x, y \in X$, is also a metric on X.

(9) Let $R[0, 1]$ be the collection of all Riemann-integrable functions in $[0, 1]$. Show that the function d on $R[0, 1]$ defined by $d(f, g) = \int_0^1 |f(t) - g(t)| dt$ is a pseudometric on $R[0, 1]$ but **not** a metric.

(10) Let X be the set of all sequences of real or complex numbers. Define

$$d(x, y) = \sum_{k=1}^{\infty} \frac{1}{2^k} \frac{|x_k - y_k|}{1 + |x_k - y_k|},$$

   where $x = \{x_k\}, y = \{y_k\}$ in X. Show that d is a metric on X.

(11) Show that $(\mathbb{R}^2, d)$ is a metric space where

$$d((x_1, y_1), (x_2, y_2)) = \begin{cases} |x_1 - x_2| & \text{if } y_1 = y_2 \\ |x_1| + |y_1 - y_2| + |x_2| & \text{otherwise.} \end{cases}, (x_1, y_1), (x_2, y_2) \in \mathbb{R}^2$$

(12) Let d_1 and d_2 be two metrics on X. Show that the mapping d defined by $d(x, y) = d_1(x, y) + d_2(x, y)\ \forall\ x, y \in X$ is a metric on X.

(13) Let (X, d) be a metric space and $x_0 \in X$. Let $r > 0$. Prove that the closed ball $\bar{B}_r(x_0)$ is a closed set.

(14) Let (X, d) be a metric space and $A \subseteq X$. Prove that $\left(\mathring{A}\right)^{\circ} = \mathring{A}$.

(15) Consider the metric space $(C[0, 1], d)$ where d is the sup metric. Let $S \subseteq [0, 1]$. Show that the set $F = \{f \in C[0, 1] : f(x) = 0 \text{ for all } x \in S\}$ is a closed set.

(16) Let (X, d) be a metric space. Show that
   (i) for any $A \subseteq X$, $\mathring{A}$ is the largest open set contained in A.
   (ii) for any $B \subseteq X$, $\bar{B}$ is the smallest closed set containing B.

(17) Let A be a subset of a metric space (X, d). Show that $\bar{A} = \bigcap\{F \subseteq X : F \text{ is a closed set and } A \subseteq F\}$

(18) Let (X, d) be a metric space and $A (\neq X)$ be an open set in X. Show that (A, d_A) is a metric space where $d_A : A \times A \to \mathbb{R}$ is defined by

$$d_A(x, y) = d(x, y) + \left| \frac{1}{d(x, X \mid A)} - \frac{1}{d(y, X \mid A)} \right|, x, y \in A.$$

(19) Let (X,d) be a metric space and $x, y \in X$ with $x \neq y$. Show that there exists a real number $r > 0$ such that $B_r(x) \cap B_r(y) = \phi$

(20) Let (X,d) be a metric space and $x, y \in X$. Let $r, s > 0$. If $z \in B_r(x) \cap B_s(y)$, then there exists a real number $t > 0$ such that $B_t(z) \subseteq B_r(x) \cap B_s(y)$.

(21) Examine which of the following are metrics on $\mathbb{R}$?

(i) $\rho_1(x,y) = \left|\dfrac{x}{1+|x|} - \dfrac{y}{1+|y|}\right|$ (ii) $\rho_2(x,y) = |x^3 - y^3|$

(iii) $\rho_3(x,y) = |x^2 - y^2|$ (iv) $\rho_4(x,y) = |\sin x - \sin y|$

(v) $\rho_5(x,y) = |e^{-x} - e^{-y}|$ (vi) $\rho_6(x,y) = |x-y|^2$

for all $x, y \in \mathbb{R}$. Determine which of the above are pseudometrics on $\mathbb{R}$.

(22) In the metric space $(\mathbb{R}, \rho_1)$, ρ_1 is defined as in the previous exercise, find

(a) $B_{\frac{1}{4}}(1)$ (b) $B_{\frac{1}{2}}(1)$ (c) $B_1(0)$ (d) $B_{\frac{1}{2}}(0)$

(23) In the pseudometric space $(\mathbb{R}, \rho_3)$, ρ_3 is defined as in the exercise 23, find $B_{\frac{1}{4}}(1), B_1(0)$.

(24) Find $\overset{\circ}{B}, \text{Bd}\,(B), \bar{B}$ in the metric space $(\mathbb{R}, d)$ where $B = \left\{\dfrac{1}{n} : n \in \mathbb{N}\right\}$.

CHAPTER 2

SEQUENCE AND OTHER PROPERTIES OF METRIC SPACES

2.1 Diameter of a Set

Definition [2.1] : Let (X, d) be a metric space and $B \subset X$ such that $B \neq \phi$. Then diameter of B is denoted by diam (B) [or $D(B)$] and is defined as diam $(B) = \sup\{d(x, y) : x, y \in B\}$.

From definition we see that diameter of a set exists as a non-negative real number, i.e., $0 \leq$ diam $(B) \leq \infty$.

A set B in a metric space is said to be **bounded** if diam $(B) < \infty$. Otherwise, B is called an unbounded set. By convention diam $(\phi) = -\infty$.

Example [2.1] : A closed circular disc or an elliptic disc in a plane is a bounded set with diameter as a finite (+)ve real number. But an infinite straight line in a plane is an unbounded set.

Theorem [2.1] Let (X, d) be a metric space and $B \subset X$. Then diam $(B) =$ diam $(\bar{B})$, $\bar{B}$ denotes the closure of B.

Proof. Since $B \subseteq \bar{B}$, we have $\{d(x, y) : x, y \in B\} \subseteq \{d(x, y) : x, y \in \bar{B}\}$ and hence $\sup\{d(x, y) : x, y \in B\} \leq \sup\{d(x, y) : x, y \in \bar{B}\}$

$$\Rightarrow \text{diam}(B) \leq \text{diam}(\bar{B}). \tag{i}$$

Thus, if diam $(B) = +\infty$, then so is diam $(\bar{B})$. So, let us suppose that diam $(B) < +\infty$.

Let $x, y \in \bar{B}$ and $\epsilon > 0$ be given. Then we can find $u, v \in B$ such that $d(x, u) < \epsilon/2$ and $d(y, v) < \epsilon/2$.

$$\begin{aligned} \text{Now } d(x, y) &\leq d(x, u) + d(u, v) + d(v, y) \\ &< \epsilon/2 + d(u, v) + \epsilon/2 < \epsilon + \text{diam}(B). \end{aligned}$$

The above relation is true for any $x, y \in \bar{B}$ and hence

$$\sup_{x,y \in \bar{B}} d(x, y) \leq \epsilon + \text{diam}(B) \Rightarrow \text{diam}(\bar{B}) \leq \epsilon + \text{diam}(B).$$

Since $\epsilon > 0$ is arbitrary, we get

$$\text{diam}(\bar{B}) \leq \text{diam}(B). \tag{ii}$$

Combining (i) and (ii), we get diam $(B) =$ diam $(\bar{B})$.

Theorem [2.2] Let (X, d) be a metric space. Then

(1) for $A, B \subset X$ with $A \subset B$ we have $\text{diam}(A) \leq \text{diam}(B)$

(2) for $A \subset X$, $\text{diam}(A) = 0$, if and only if A is a singleton set.

Proof. Left as an exercise.

Theorem [2.3] Let (X, d) be a metric space and $A, B \subset X$. Then $\text{diam}(A \cup B) \leq \text{diam}(A) + \text{diam}(B)$, if $A \cap B \neq \phi$.

Proof. Let $A, B \subset X$ be such that $A \cap B \neq \phi$.

If $A \subset B$ or $B \subset A$, then we have $A \cup B = B$ or $A \cup B = A$ and there is nothing to prove.

So suppose that neither $A \subset B$ nor $B \subset A$.

Let us take $x, y \in A \cup B$. If $x, y \in A$, then from definition we have

$$d(x, y) \leq \text{diam}(A) \leq \text{diam}(A) + \text{diam}(B).$$

This is true for any $x, y \in A \cup B$ and so $\text{diam}(A \cup B) \leq \text{diam}(A) + \text{diam}(B)$.

If $x, y \in B$, then we proceed as above to get the required result.

So suppose that, $x \in A$ and $y \in B$. Since $A \cap B \neq \phi$, we can find an element $a \in A \cap B$ such that $d(x, y) \leq d(x, a) + d(a, y) \leq \text{diam}(A) + \text{diam}(B)$.

The above relation is true for any $x, y \in A \cup B$. Hence $\text{diam}(A \cup B) \leq \text{diam}(A) + \text{diam}(B)$.

2.1.1 Distance of a Point from a Set

Definition [2.2]: Let (X, d) be a metric space and $A \subset X$. Let $x \in X$. Then the distance of x from the set A is denoted by $\text{dist}(x, A)$ [or $d(x, A)$] and is given by

$$\text{dist}(x, A) = \inf\{d(x, y) : y \in A\}.$$

From the above definition it follows that if $x \in A$, then $\text{dist}(x, A) = 0$.

2.1.2 Distance between Two Sets

Definition [2.3]: Let (X, d) be a metric space and let $A, B \subset X$. Then distance between A and B is denoted by $d(A, B)$ and is defined as

$$d(A, B) = \inf\{d(x, y) : x \in A, y \in B\}.$$

Note that, $d(A, B) = 0$, if $A \cap B \neq \phi$.

Theorem [2.4] Let (X, d) be a metric space and $A \subset X$. Then $\text{dist}(x, A) = \text{dist}(x, \bar{A})$.

Proof. Since $A \subset \bar{A}$, we get $\{d(x, y) : y \in A\} \subseteq \{d(x, y) : y \in \bar{A}\}$ and hence taking infimum, we get

$$\inf\{d(x, y) : y \in A\} \geq \inf\{d(x, y) : y \in \bar{A}\} \Rightarrow \text{dist}(x, A) \geq \text{dist}(x, \bar{A}). \tag{i}$$

Let $\alpha \in \bar{A}$. Then for $\epsilon > 0$ we find an element $u \in A$ such that $d(\alpha, u) < \epsilon$.

So $\text{dist}(x, A) \leq d(x, u) \leq d(x, \alpha) + d(\alpha, u) < \epsilon + d(x, \alpha)$.

This relation is true for any $\alpha \in \bar{A}$. Therefore, we get $\text{dist}(x, A) \leq \epsilon + \text{dist}(x, \bar{A})$.

Since $\epsilon > 0$ is arbitrary, we get

$$\text{dist}(x, A) \leq \text{dist}(x, \bar{A}). \tag{ii}$$

Combining the relations (i) and (ii), we get $\text{dist}(x, A) = \text{dist}(x, \bar{A})$.

This completes the proof of the theorem.

Theorem [2.5] Let (X, d) be a metric space and A be a subset of X. Then for any $x, y \in X$,

$$|\operatorname{dist}(x, A) - \operatorname{dist}(y, A)| \leq d(x, y).$$

Proof. Let $x, y \in X$ and $\epsilon > 0$ be arbitrary. Then from the definition of infimum, we find an element $a \in A$ such that $d(x, a) < \operatorname{dist}(x, A) + \epsilon$.

Now $\operatorname{dist}(y, A) \leq d(y, a) \leq d(y, x) + d(x, a) < d(x, y) + \operatorname{dist}(x, A) + \epsilon$.

Since $\epsilon > 0$ is arbitrary, we get

$$\operatorname{dist}(y, A) \leq d(x, y) + \operatorname{dist}(x, A) \Rightarrow \operatorname{dist}(y, A) - \operatorname{dist}(x, A) \leq d(x, y). \quad \text{(i)}$$

Interchanging the role of x and y we get from (i),

$$\operatorname{dist}(x, A) - \operatorname{dist}(y, A) \leq d(y, x) = d(x, y). \quad \text{(ii)}$$

Combining (i) and (ii), we get

$$|\operatorname{dist}(x, A) - \operatorname{dist}(y, A)| \leq d(x, y).$$

Theorem [2.6] Let (X, d) be a metric space and $A \subset X$. Then for any $x \in X$,

$$\operatorname{dist}(x, A) = 0, \text{ iff } x \in \bar{A},$$

i.e., $\bar{A} = \{x \in X : \operatorname{dist}(x, A) = 0\}$.

Proof. Let $x \in X$.

First suppose that $\operatorname{dist}(x, A) = 0$. Then if $x \in A$, we get $x \in \bar{A}$. Otherwise, given any $\epsilon > 0$, there must exist an element $u \in A$ such that

$$\begin{aligned} d(x, u) &< \operatorname{dist}(x, A) + \epsilon \text{ [from definition of infimum]} \\ &= 0 + \epsilon, \text{ since } \operatorname{dist}(x, A) = 0 \\ \Rightarrow d(x, u) &< \epsilon \\ \Rightarrow u &\in B_\epsilon(x). \end{aligned}$$

Since $\epsilon > 0$ is arbitrary, we can conclude that x is a limit point of A, i.e., $x \in A' \subset \bar{A}$.

Thus, if $\operatorname{dist}(x, A) = 0$, then we find $x \in \bar{A}$.

Conversely, let $x \in \bar{A} = A \cup A'$. Then if $x \in A$, we get $\operatorname{dist}(x, A) = 0$.

So suppose that $x \in A'$ and $x \notin A$, i.e., x is a limit point of A without being in A.

Since $x \in A'$, so for given any $\epsilon > 0$ we can find an element $u \in A$ such that $d(x, u) < \epsilon$.

Hence $\operatorname{dist}(x, A) \leq d(x, u) < \epsilon \Rightarrow \operatorname{dist}(x, A) < \epsilon$.

As $\epsilon > 0$ is arbitrary, we get $\operatorname{dist}(x, A) = 0$.

This completes the proof.

Theorem [2.7] Let (X, d) be a metric space and let A, B be two subsets of X. Then

$$\operatorname{diam}(A \cup B) \leq \operatorname{diam}(A) + \operatorname{diam}(B) + d(A, B).$$

Proof. Let us assume that

$$\operatorname{diam}(A \cup B) > \operatorname{diam}(A) + \operatorname{diam}(B) + d(A, B).$$

Then from definition of diameter we find x, y in $A \cup B$ such that

$$d(x, y) > \operatorname{diam}(A) + \operatorname{diam}(B) + d(A, B). \quad \text{(i)}$$

Since $x, y \in A \cup B$, so we may have $x, y \in A$ or, $x, y \in B$ or, $x \in A, y \in B$ or, $x \in B, y \in A$.

If $x, y \in A$, then

$$d(x,y) \leq \text{diam}(A) \leq \text{diam}(A) + \text{diam}(B) + d(A,B)\text{—a contradiction to (i)}$$

In the similar way, we can say that $x, y \in B$ is not possible.

So suppose that $x \in A$ and $y \in B$.

Then from $d(A,B) < d(x,y) - \text{diam}(A) - \text{diam}(B)$ and definition of $d(A,B)$, we get $a \in A, b \in B$ such that

$$d(a,b) < d(x,y) - \text{diam}(A) - \text{diam}(B). \tag{ii}$$

But from triangle inequality we have

$$\begin{aligned} d(x,y) &\leq d(x,a) + d(a,b) + d(b,y) \\ &\leq \text{diam}(A) + d(a,b) + \text{diam}(B) \end{aligned}$$

$\Rightarrow d(a,b) \geq d(x,y) - \text{diam}(A) - \text{diam}(B)$—which is a contradiction to (ii).

Thus our assumption is wrong and the result established.

2.2 Countability Axioms

We have seen in Chapter-1 (1·2·3) that every metric space is a topological space. Not only that every metric space is a Hausdorff topological space (1 · 2 · 1). We now discuss the countability properties such as 1st countable, 2nd countable, etc. For this purpose we now define what is known by base, local base in a topological space.

Definition [2.4] : Let (X, τ) be a topological space and $x \in X$. Let $\mathcal{N}_x$ be the collection of all neighbourhoods of x. A collection $B_x \subset \mathcal{N}_x$ is called a **local base** at x if for any $N_x \in \mathcal{N}_x$, there exists a $B \in B_x$ such that $x \in B \subset N_x$.

For example, let us consider the metric space $(\mathbb{R}, d)$ of all real numbers with the usual metric. Then for $x \in \mathbb{R}$, the collection $B_x = \{(x - \epsilon, x + \epsilon) : \epsilon \in \mathbb{R}, \epsilon > 0\}$ is a local base at x.

Definition [2.5] : Let (X, τ) be a topological space. A collection $B \subset \tau$ is called a **base** for the open sets if every open set can be expressed as a union of some members of B.

For example, in $\mathbb{R}$ the collection of all open intervals form a base for the open sets in the usual topology of $\mathbb{R}$.

Definition [2.6] : A topological space (X, τ) is said to be **first countable** if each point of X admits a countable local base.

In $\mathbb{R}$, the collection $\{(x - r, x + r) : r > 0, r \in Q\}$, where $x \in \mathbb{R}$, is countable. Also it is a local base at x in $\mathbb{R}$. Hence $\mathbb{R}$ is first countable.

We now prove that every metric space is first countable.

Theorem [2.8] Let (X, d) be a metric space and let $x \in X$. Then there exists a countable local base $\{B_n^x\}_{n=1}^{\infty}$ at x, i.e., for any neighbourhood N_x of x, $\exists\, k \in \mathbb{N}$ such that $x \in B_k^x \subset N_x$.

Proof. We know that every open ball is an open set and an open set is a neighbourhood of each of its points. So for each $n \in \mathbb{N}$, $B_n^x = B_{1/n}(x)$ is a nbd of x. Hence $\{B_n^x\}_{n \in \mathbb{N}}$ is a countable collection of neighbourhoods of x. Let $N_x \in \mathcal{N}_x$, i.e., N_x is a neighbourhood of x then from definition, there exists $r > 0$ such that $x \in B_r(x) \subset N_x$. Now we can find $k \in \mathbb{N}$ such that $\frac{1}{k} < r$ and so $B_k^x = B_{1/k}(x) \subset B_r(x) \subset N_x$ and $B_{1/k}(x) \in \{B_n^x\}_{n \in \mathbb{N}}$. Thus $\{B_n^x\}_{n \in \mathbb{N}}$ forms a countable local base at x.

Definition [2.7] : A topological space (X, τ) is said to be **second countable** if it has a countable base.

It can be proved that every second countable space is first countable.

For example, we consider $(\mathbb{R}, d)$ when d is the usual metric in $\mathbb{R}$. Then the collection $B = \{(a, b) : a, b \in Q\}$ is a countable base for the open sets in $\mathbb{R}$.

Hence $\mathbb{R}$ is a 2nd countable space.

Example [2.2] : Let X be an uncountable set and let d be the trivial metric on X, i.e., (X, d) is a trivial metric space. Let $\mathcal{B}$ be a base for the open sets in X. Then for each $x \in X$, $B_{1/2}(x) = \{x\}$ [See example [1.14]] is an oepn set in X. So there exists $B \in \mathcal{B}$ such that $x \in B \subset \{x\}$, i.e., $\{x\} \subset B \subset \{x\} \Rightarrow B = \{x\}$.

Thus $\Big\{\{x\} : x \in X\Big\} \subset \mathcal{B}$. Consequently, $\mathcal{B}$ is **not** countable.

Hence (X, d) is not second countable.

However, (X, d) is first countable, since for any $x \in X$, $\{\{x\}\}$ is a countable local base at x.

Definition [2.8] : A metric space (X, d) is said to be separable if X has a countable dense subset, i.e., there exists a subset $C \subset X$ such that $\bar{C} = X$, and C is countable.

Note that $(\mathbb{R}, d)$ with usual metric is separable since $Q \subset \mathbb{R}$, Q is countable and $\bar{Q} = \mathbb{R}$.

Example [2.3] : ($l_p, p \geq 1$ is separable)

Recall that (l_p, d) is a metric space (See 1.4), $1 \leq p < \infty$.

We now show that l_p is separable.

Let C be the set of all sequences y of the form

$$y = \{\eta_1, \eta_2, \ldots, \eta_n, 0, 0, \ldots\},$$

where n is any positive integer and η_i's are rational. Then clearly, C is countable.

We now show that C is dense in l_p, i.e., $\bar{C} = l_p$.

We always have $\bar{C} \subset l_p$, as $C \subset l_p$.

Let $x = \{\xi_i\} \in l_p$.

Then we have from definition $\sum\limits_{n=1}^{\infty} |\xi_n|^p < \infty$, i.e., the series $\sum\limits_{n=1}^{\infty} |\xi_n|^p$ is a convergent series.

So for any $\epsilon > 0$, there exists $n \in \mathbb{N}$ such that

$$\sum_{i=n+1}^{\infty} |\xi_i|^p < \epsilon^p/2$$

since remainder of a convergent series tends to zero in $\mathbb{R}$.

Since $\bar{Q} = \mathbb{R}$, for each ξ_i there is a rational number η_i such that $|\xi_i - \eta_i|^p < \epsilon^p/2n$ and so

$$\sum_{i=1}^{n} |\xi_i - \eta_i|^p < \epsilon^p/2.$$

Therefore, we get

$$[d(x, y)]^p = \sum_{i=1}^{n} |\xi_i - \eta_i|^p + \sum_{i=n+1}^{\infty} |\xi_i - 0|^p < \frac{\epsilon^p}{2} + \frac{\epsilon^p}{2} = \epsilon^p,$$

where $y = \{\eta_1, \eta_2, \ldots, \eta_n, 0, 0, \ldots\} \in C$.

$\Rightarrow d(x,y) < \epsilon$

$\Rightarrow y \in B_\epsilon(x)$

$\Rightarrow B_\epsilon(x) \cap C \neq \phi$, for any $\epsilon > 0$

$\Rightarrow x$ is a limit point of C

$\Rightarrow x \in \bar{C}$.

Hence $l_p \subset \bar{C}$ and consequently, $\bar{C} = l_p$.

Thus l_p is separable $(1 \le p < \infty)$.

Remark (1). (l_∞ is non-separable)

Let l_∞ denote the set of all bounded sequences of complex numbers. Define a function d on l_∞ as $d(x,y) = \sup\limits_{i \in \mathbb{N}} |\xi_i - \eta_i|$, where $x = \{\xi_i\}$, $y = \{\eta_i\} \in l_\infty$.

Then d is a metric on l_∞ (verify) and hence (l_∞, d) is a metric space. The metric space (l_∞, d) is not separable.

Let $y = \{\eta_1, \eta_2, \ldots\}$ be a sequence of zeros and ones. Then y is a bounded sequence and hence $y \in l_\infty$.

Let $\hat{y} = \dfrac{\eta_1}{2^1} + \dfrac{\eta_2}{2^2} + \dfrac{\eta_3}{2^3} + \cdots + \dfrac{\eta_n}{2^n} + \cdots$.

Then $\hat{y} \in [0,1]$ has a binary representation and different $\hat{y}$'s has different binary representations. Hence there are uncountable number of sequences of zeros and ones. Since ξ_i's and η_i's are either 0 or 1, it follows from definition of d that $d(x,y) = 1$, if $x \neq y$. We now consider an open ball $B_{1/3}(x)$, for all sequence x of zeros and ones. Then $\{B_{1/3}(x)\}$ is uncountable and

$$B_{1/3}(x) \cap B_{1/3}(y) = \phi,\ \forall\ x, y (x \neq y).$$

Let A be a dense subset of l_∞.

Then $A \cap B_{1/3}(x) \neq \phi$, $\forall\ x$ as above.

Hence A contains an element from each member of the family $\{B_{1/3}(x)\}$ and consequently A is uncountable. Since A was taken an arbitrary dense subset from l_∞, it follows that l_∞ has no countable dense subset. Thus l_∞ is not separable.

Theorem [2.9] A metric space (X,d) is separable, if and only if X has a countable subset C with the following property: For every $\epsilon > 0$ and every $x \in X$, there is a $y \in C$ such that $d(x,y) < \epsilon$.

Proof. First suppose that (X,d) is a separable metric space. Then we can find a subset C of X such that C is countable and $\bar{C} = X$.

Let $x \in X$ and $\epsilon > 0$ be arbitrary. Since $\bar{C} = X$, we find $x \in \bar{C}$. So $B_\epsilon(x) \cap C \neq \phi$.

Let $y \in B_\epsilon(x) \cap C$.

Then $y \in C$ and $d(x,y) < \epsilon$.

Conversely, suppose that X has a countable subset C such that for any $\epsilon > 0$ and every $x \in X$, $\exists\ y \in C$ with $d(x,y) < \epsilon$. We show that $\bar{C} = X$. If $x \in C$, then there is nothing to prove. So let $x \notin C$. Thus from $d(x,y) < \epsilon$ we get $y \in B_\epsilon(x)$. That means, $B_\epsilon(x) \cap C \neq \phi$ for any $\epsilon > 0$. Hence x is a limit point of $C \Rightarrow x \in C' \subset \bar{C}$ and consequently $\bar{C} = X$. Thus C is a countable dense subset of X, i.e., X is separable.

Theorem [2.10] Every separable metric space is second countable.

Proof. Let (X, d) be a separable metric space. Let D be a countable dense subset of X.

Let $\mathcal{B} = \{B_\delta(a) : a \in D,\ \delta \in Q \cap (0, \infty)\}$.

Clearly $\mathcal{B}$ is countable. We now show that $\mathcal{B}$ is a base for X, i.e., we show that for any open set G in X and for any $p \in G$, $\exists\ B_\delta(a) \in \mathcal{B}$ such that $p \in B_\delta(a) \subset G$.

Since $p \in G$ and G is an open set in (X, d), there exists $\epsilon > 0$ such that $p \in B_\epsilon(p) \subset G$.

Since D is dense in X, there exists $a_0 \in D$ such that $d(p, a_0) < \epsilon/3$, by previous theorem.

Choose a rational number δ_0 in $\frac{1}{3}\epsilon < \delta_0 < \frac{2}{3}\epsilon$. Then $d(p, a_0) < \epsilon/3 < \delta_0 \Rightarrow p \in B_{\delta_0}(a_0)$.

Now let $x \in B_{\delta_0}(a_0)$. Then $d(x, a_0) < \delta_0$.

Also
$$\begin{aligned} d(x,p) &\le d(x,a_0) + d(a_0,p) \\ &< \delta_0 + \epsilon/3 < 2\epsilon/3 + \epsilon/3 = \epsilon \\ &\Rightarrow x \in B_\epsilon(p). \end{aligned}$$

Thus for any $x \in B_{\delta_0}(a_0)$, we get $x \in B_\epsilon(p)$.

So $p \in B_{\delta_0}(a_0) \subset B_\epsilon(p) \subset G \Rightarrow p \in B_{\delta_0}(a_0) \subset G$.

Since $B_{\delta_0}(a_0) \in \mathcal{B}$, it follows that $\mathcal{B}$ is a base for X.

Hence X is second countable.

Theorem [2.11] Every second countable metric space is separable.

Proof. Let (X, d) be a second countable metric space. Let $\mathcal{B}$ be a countable base in X. Then $\mathcal{B}$ can be written as $\mathcal{B} = \{B_1, B_2, \dots, B_n, \dots\}$.

We now show that (X, d) is separable.

Take $x_n \in B_n$ for all $n \in \mathbb{N}$ and Let $D = \{x_n : n \in \mathbb{N}\}$.

Then D is a countable subset of X. We now show that $\bar{D} = X$. For this, let $x \in X$ and let N_x be a neighbourhood of x. Then there exists $r > 0$ such that $x \in B_r(x) \subset N_x$. Since $\mathcal{B}$ is a base, $\exists\ B_K \in \mathcal{B}$ such that $x \in B_K \subset B_r(x)$, for some $K \in \mathbb{N}$. Hence $B_K \cap D \neq \phi$, since $x_K \in B_K \cap D$ and so $N_x \cap D \neq \phi$ since $x \in B_K \subset B_r(x) \subset N_x$.

This is true for any neighbourhood N_x of x.

Thus $x \in \bar{D}$ and so $\bar{D} = X$.

Hence X is separable.

Remark (2). From the above two theorem (Th. [2.10] and Th. [2.11]) we see that

"A metric space is separable, if and only if it is second countable".

We have seen that (l_∞, d) is not separable (Remark 1) so (l_∞, d) is not second countable.

Theorem [2.12] Subspace of a separable metric space is separable.

Proof. Let (X, d) be a separable metric space. Let Y be a non-empty subset of X. Then (Y, d_Y) is a subspace of (X, d) [see definition [1.9]].

Let D be a countable dense subset of X. Since D is countable, we can write
$$D = \{x_1, x_2, \dots, x_n, \dots\}.$$

Then for any $y \in Y$ and any positive integer n, we have $B_{1/n}(y) \cap D \neq \phi$.

Take $\alpha_{i_n} \in B_{1/n}(y) \cap D$. Then we have $B_{1/n}(\alpha_{i_n}) \cap Y \neq \phi$.

We now take an element $y_{i_n} \in B_{1/n}(\alpha_{i_n}) \cap Y$, $\forall\, n \in \mathbb{N}$.

Consider the set $C = \{y_{i_1}, y_{i_2}, \cdots, y_{i_n}, \cdots\}$.

Note that $C \subset Y$ and clearly C is countable.

Let $y \in Y$ and $\epsilon > 0$ be arbitrary. Choose $n \in \mathbb{N}$ such that $1/n < \epsilon/2$. Then

$$d(y, y_{i_n}) \leq d(y, \alpha_{i_n}) + d(\alpha_{i_n}, y_{i_n}) < \frac{1}{n} + \frac{1}{n} < \frac{\epsilon}{2} + \frac{\epsilon}{2} = \epsilon$$

$\Rightarrow d(y, y_{i_n}) < \epsilon$.

Hence (Y, d_Y) is separable [see Th. [2.9]].

This completes the proof.

Example [2.4]: Let (X, d) be a metric space and $Y \subset X$ such that Y is dense in X. If (Y, d_Y) is separable, then (X, d) is separable.

Since Y is separable, there exists a countable dense subset C of Y, i.e., $C \subset Y$ and $\bar{C}^Y = Y$, where $\bar{C}^Y$ denotes the closure of C in (Y, d_Y).

Also it is given that $\bar{Y} = X$.

We now show that $\bar{C} = X$. It is obvious that $\bar{C} \subset X$.

We know from example (11) in Chapter-1 that if $A \subset Y$, then the closure of A in Y is $\bar{A} \cap Y$, i.e., $\bar{A}^Y = \bar{A} \cap Y$, where $\bar{A}$ is the closure of A in X. So we have

$$\bar{C}^Y = \bar{C} \cap Y \Rightarrow Y = \bar{C} \cap Y \subset \bar{C}$$
$$\Rightarrow Y \subset \bar{C} \Rightarrow \bar{Y} \subseteq \bar{\bar{C}} = \bar{C} \Rightarrow X \subseteq \bar{C}.$$

Hence $\bar{C} = X$, i.e., $C \subset Y \subset X$ is countable and $\bar{C} = X \Rightarrow (X, d)$ is a separable metric space.

2.3 Sequence in a Metric Space

We know the convergence and other properties of sequences in $\mathbb{R}$. Let us first recall the definition of convergence of a sequence in $\mathbb{R}$.

A sequence $\{x_n\}$ in $\mathbb{R}$ is said to converge to a point x in $\mathbb{R}$ if for any $\epsilon > 0$, there exists a positive integer N such that

$|x_n - x| < \epsilon$, whenever $n \geq N$

$\Rightarrow d(x_n, x) < \epsilon$, whenever $n \geq N$ (Note that $d(x, y) = |x - y|$ is the usual distance in $\mathbb{R}$)

$\Rightarrow x_n \in B_\epsilon(x)$, whenever $n \geq N$.

Thus the definition of convergence of a sequence in an arbitrary metric space is as follows:

Definition [2.9]: Let (X, d) be a metric space and let $\{x_n\}$ be a sequence in X. Then $\{x_n\}$ is said to converge to a point $x \in X$, if for any $\epsilon > 0$, there exists a positive integer $N \in \mathbb{N}$ such that $d(x_n, x) < \epsilon$, $\forall\, n \geq N$ or, $x_n \in B_\epsilon(x)$ for all $n \geq N$.

Since $\epsilon > 0$ is arbitrary, it follows that every neighbourhood $B_\epsilon(x)$ of the point x contains all but finite number of member of the sequence $\{x_n\}$, i.e., only a finite number of member of the sequence can reside outside $B_\epsilon(x)$.

We write, $\lim\limits_n x_n = x$ or, $x_n \to x$ in X.

We know that, in $\mathbb{R}$, a convergent sequence has unique limit. This property of a convergent sequence also holds in an arbitrary metric space.

Theorem [2.13] A convergent sequence in a metric space has unique limit.

Proof. Let (X, d) be a metric space. Let $\{x_n\}$ be a convergent sequence in X. Let $x_n \to x$ and $x_n \to y$, where $x, y \in X$. If possible, suppose that $x \neq y$. Since X is Hausdorff and $x \neq y$, there exists a real number $\epsilon > 0$ such that $B_\epsilon(x) \cap B_\epsilon(y) = \phi$ (See 1.2.1). Since $x_n \to x$, for this ϵ, there exists a natural number N such that

$$x_n \in B_\epsilon(x), \ \forall\, n \geq N.$$

Thus since $B_\epsilon(x) \cap B_\epsilon(y) = \phi$, $B_\epsilon(y)$ can only contain a finite number of member $\{x_n\}$, say, possibly $x_1, x_2, \ldots, x_{N-1}$—this is a contradiction to the assumption that $x_n \to y$ in X.

Hence we must have $x = y$ and the theorem is proved.

Alternative Proof

Let (X, d) be a metric space and let $\{x_n\}$ be a convergent sequence in X such that $x_n \to x$ and $x_n \to y$ in X.

Since $x_n \to x$, for any $\epsilon > 0$ there exists a positive integer N_1 such that $d(x_n, x) < \epsilon/2$ for all $n \geq N_1$. Again since $x_n \to y$, for this $\epsilon > 0$, there exists a positive integer N_2 such that $d(x, y) < \epsilon/2$ for all $n \geq N_2$.

Let $N = \max\{N_1, N_2\}$. Then for all $n \geq N$, we have $d(x_n, x) < \epsilon/2$ and $d(x_n, y) < \epsilon/2$.

Now for all $n \geq N$, we have

$$d(x, y) \leq d(x, x_n) + d(x_n, y) < \epsilon/2 + \epsilon/2 = \epsilon$$

$\Rightarrow d(x, y) < \epsilon$, for any $\epsilon > 0$.

Since $\epsilon > 0$ is arbitrary, we get $d(x, y) = 0 \Rightarrow x = y$ and the theorem is proved.

Subsequence

Definition [2.10] : Let (X, d) be a metric space and let $\{x_n\}$ be a sequence in X. Then by subsequence of $\{x_n\}$ we mean a subset $\{x_{n_k}\}$ of $\{x_n\}$, where the elements in $\{x_{n_k}\}$, keeps the same order as they are in $\{x_n\}$, i.e., $\{x_{n_k}\} = \{x_{n_1}, x_{n_2}, x_{n_3}, \ldots\}$ is a subsequence of $\{x_n\}$, where $n_1 < n_2 < n_3 < \cdots$.

For example, consider the sequence $\left\{\frac{1}{n}\right\}$ in the metric space of all real numbers $(\mathbb{R}, d)$ with the usual metric. Then $\left\{\frac{1}{2n}\right\} = \left\{\frac{1}{2}, \frac{1}{4}, \frac{1}{6}, \cdots\right\}$ is a subsequence of $\left\{\frac{1}{n}\right\}$.

The set $\left\{1, \frac{1}{3}, \frac{1}{6}, \frac{1}{7}, \frac{1}{9}, \cdots\right\}$ is also a subsequence of $\left\{\frac{1}{n}\right\}$.

Theorem [2.14] Let (X, d) be a metric space and let $\{x_n\}$ be a subsequence in X. If $\{x_n\}$ is convergent, then all subsequences $\{x_{n_k}\}$ of $\{x_n\}$ are convergent in X and converges to the same point at which $\{x_n\}$ is convergent. Converse is also true.

Proof. Left as an exercise.

Theorem [2.15] Let (X, d) be a metric space and $\{x_n\}$ be a sequence in X and let $x_n \to x$ in X. Let $a \in X$. Then the set $\{d(x_n, a)\}$ is bounded and $d(x_n, a) \to d(x, a)$ as $n \to \infty$.

Proof. Using triangle inequality, we get

$$d(x_n, a) \leq d(x_n, x) + d(x, a) \Rightarrow d(x_n, a) - d(x, a) \leq d(x_n, x).$$

Interchanging the role of x_n and x, we get from the above inequality that

$$d(x, a) - d(x_n, a) \leq d(x, x_n) = d(x_n, x).$$

Combining the above inequalities, we get

$$|d(x_n, a) - d(x, a)| \le d(x_n, x).$$

Since $x_n \to x$, for $\epsilon > 0$, there exists a positive integer N such that

$$d(x_n, x) < \epsilon, \ \forall\, n \ge N.$$

Hence $|d(x_n, a) - d(x, a)| \le d(x_n, x) < \epsilon, \ \forall\, n \ge N$

$\Rightarrow |d(x_n, a) - d(x, a)| < \epsilon, \ \forall\, n \ge N$

$\Rightarrow d(x_n, a) \to d(x, a)$ as $n \to \infty$.

Since $\{d(x_n, a)\}$ is a convergent sequence of real numbers and a convergent sequence of real number is bounded, it follows that $\{d(x_n, a)\}$ is bounded in $\mathbb{R}$.

Alternative Proof

The set $\{d(x_n, a)\}$ is bounded.

Since $x_n \to x$ in X, so for any $\epsilon > 0$ there exists a positive integer N such that

$$d(x_n, x) < \epsilon, \ \forall\, n \ge N.$$

Let $M = \max\{d(x_1, x), d(x_2, x), \ldots, d(x_{N-1}, x), \epsilon\}$.

Then we have $d(x_n, x) \le M$ for all $n \in \mathbb{N}$.

Hence $d(x_n, a) \le d(x_n, x) + d(x, a)$

$\le M + d(x, a)$ for all $n \in \mathbb{N}$.

Thus the set $\{d(x_n, a)\}$ is bounded.

Theorem [2.16] Let (X, d) be a metric space. If $x_n \to x$ and $y_n \to y$ in X, then

$$d(x_n, y_n) \to d(x, y) \text{ in } \mathbb{R}.$$

Proof. Let $x_n \to x$ and $y_n \to y$ in X.

Then for $\epsilon > 0$ there exists positive integers N_1, N_2 such that

$$d(x_n, x) < \epsilon/2 \text{ for all } n \ge N_1$$

$$\text{and } d(x_n, y) < \epsilon/2 \text{ for all } n \ge N_2.$$

Let $N = \max\{N_1, N_2\}$.

Then for all $n \ge N$, we get $d(x_n, x) < \epsilon/2$ and $d(y_n, y) < \epsilon/2$.

Now by triangle inequality, we get $d(x_n, y_n) \le d(x_n, x) + d(x, y) + d(y, y_n)$

$$\Rightarrow d(x_n, y_n) - d(x, y) \le d(x_n, x) + d(x_n, y). \tag{i}$$

Interchanging x_n, x and y_n, y, we get

$$d(x, y) - d(x_n, y_n) \le d(x, x_n) + d(y, y_n)$$

$$\Rightarrow -\{d(x_n, y_n) - d(x, y)\} \le d(x_n, x) + d(y_n, y). \tag{ii}$$

Now combining (i) and (ii), we get

$$|d(x_n, y_n) - d(x, y)| \le d(x_n, x) + d(y_n, y)$$

$$< \epsilon/2 + \epsilon/2 \text{ for all } n \ge N$$

$$= \epsilon \text{ for all } n \ge N$$

$\Rightarrow |d(x_n, y_n) - d(x, y)| < \epsilon$ for all $n \ge N$

$\Rightarrow d(x_n, y_n) \to d(x, y)$ as $n \to \infty$.

Theorem [2.17] Let (X, d) be a metric space and let $\{x_n\}$ be a sequence in X such that $x_n \to x$. Then for a sequence $\{y_n\}$, $y_n \to x$, iff $d(x_n, y_n) \to 0$ as $n \to \infty$.

Proof. We have $x_n \to x$ in X.

First suppose that $y_n \to x$. Then by the above theorem (Theorem [2.16]),

$$d(x_n, y_n) \to d(x, x) = 0 \text{ as } n \to \infty$$
$$\Rightarrow d(x_n, y_n) \to 0 \text{ as } n \to \infty.$$

Conversely, suppose that $d(x_n, y_n) \to 0$ as $n \to \infty$. We now show that $y_n \to x$ in X.

Let $\epsilon > 0$ be arbitrary. Then since $x_n \to x$ and $d(x_n, y_n) \to 0$, there exist natural numbers N_1, N_2 such that

$$d(x_n, x) < \frac{\epsilon}{2} \text{ for all } n \geq N_1 \text{ and } d(x_n, y_n) < \frac{\epsilon}{2} \text{ for all } n \geq N_2$$

Let $N = \max\{N_1, N_2\}$. Then for all $n \geq N$,

$$d(x_n, x) < \frac{\epsilon}{2} \text{ and } d(x_n, y_n) < \frac{\epsilon}{2}.$$

Now $d(y_n, x) \leq d(y_n, x_n) + d(x_n, x)$
$$< \frac{\epsilon}{2} + \frac{\epsilon}{2} = \epsilon \text{ for all } n \geq N.$$

This shows that for any $\epsilon > 0$, there exists a positive integer N such that

$$d(y_n, x) < \epsilon \text{ for all } n \geq N$$
$$\Rightarrow y_n \to x \text{ in } X.$$

Theorem [2.18] Let (X, d) be a metric space and let $A \subset X$. Then $x \in \bar{A}$, if and only if there exists a sequence $\{x_n\} \subset A$ such that $x_n \to x$ as $n \to \infty$.

Proof. First suppose that $x \in \bar{A}$. We now show that $x_n \to x$ as $n \to \infty$.

Let $\epsilon > 0$ be arbitrary. Since $x \in \bar{A}$, we have $B_\epsilon(x) \cap A \neq \phi$.

Now taking $\epsilon = 1, \frac{1}{2}, \frac{1}{3}, \ldots, \frac{1}{n}, \ldots$, we get $B_{1/n}(x) \cap A \neq \phi$.

Take a point $x_n \in B_{1/n}(x) \cap A$, for all n.

Then $\{x_n\}$ is a sequence in A and since $x_n \in B_{1/n}(x)$, we have

$$d(x_n, x) \leq \frac{1}{n} \to 0 \text{ as } n \to \infty$$
$$\Rightarrow d(x_n, x) \to 0 \text{ as } n \to \infty$$
$$\Rightarrow x_n \to x \text{ as } n \to \infty.$$

Conversely, let $\{x_n\}$ is a sequence in A such that $x_n \to x$. We show that $x \in \bar{A}$.

Let $\epsilon > 0$ be arbitrary. Then the open ball $B_\epsilon(x)$ contains infinite number of terms of the sequence $\{x_n\}$, except a finite number of terms. Since $\{x_n\}$ is a sequence in A, it follows that the open ball $B_\epsilon(x)$ contains infinite number of elements of A. This is true for any $\epsilon > 0$. Hence $B_\epsilon(x) \cap A \neq \phi$ for any $\epsilon > 0$.

$$\Rightarrow x \text{ is a limit point of } A$$
$$\Rightarrow x \in A' \subset \bar{A}$$
$$\Rightarrow x \in \bar{A} \text{ and the proof is complete.}$$

Corollary [2.1]: Let (X, d) be a metric space and M be a non-empty subset of X. Then M is closed, iff for any sequence $\{x_n\}$ in M such that $x_n \to x$ implies $x \in M$.

Proof. Recall that M is closed, iff $M = \bar{M}$ and the result from the theorem [2.18].

Definition [2.11]: **[Cauchy Sequence]** Let (X, d) be a metric space. A sequence $\{x_n\}$ in X is said to be a Cauchy sequence if for any $\epsilon > 0$ there exists a positive integer N such that

$$d(x_n, x_m) < \epsilon \text{ for all } m, n \geq N.$$

Example [2.5]: Consider the metric space $(\mathbb{R}, d)$ with the usual metric. Let us take a sequence $\{\frac{1}{n}\}$ in $\mathbb{R}$.

We now show that $\{\frac{1}{n}\}$ is a Cauchy sequence in $\mathbb{R}$. Let $\epsilon > 0$ be arbitrary. Choose a natural number N such that $\dfrac{1}{N} < \epsilon/2$.

Then for all $m, n \geq N$ we have $\dfrac{1}{n} \leq \dfrac{1}{N} < \epsilon/2$, $\dfrac{1}{m} \leq \dfrac{1}{N} < \epsilon/2$ and so

$$\left|\frac{1}{m} - \frac{1}{n}\right| \leq \frac{1}{m} + \frac{1}{n} < \epsilon/2 + \epsilon/2 = \epsilon.$$

Thus for any $\epsilon > 0$, we find a positive integer N such that

$$\left|\frac{1}{m} - \frac{1}{n}\right| < \epsilon \text{ for } m, n \geq N$$

$$\Rightarrow d\left(\frac{1}{m}, \frac{1}{n}\right) < \epsilon \text{ for } m, n \geq N$$

$$\Rightarrow \left\{\frac{1}{n}\right\} \text{ is a Cauchy sequence in } \mathbb{R}.$$

Example [2.6]: Consider the sequence $\{(-1)^n\}$ in the metric space $(\mathbb{R}, d)$ with the usual distance. Then note that $|(-1)^n - (-1)^{n+1}| = 2$. So if we take $\epsilon < 2$, then we cannot find a positive integer N such that $|(-1)^n - (-1)^m| < \epsilon$ for all $m, n \geq N$

$\Rightarrow \{(-1)^n\}$ is not a Cauchy sequence.

Theorem [2.19] A Cauchy sequence in a metric space is bounded.

Proof. Let (X, d) be a metric space and let $\{x_n\}$ be a Cauchy sequence in X. Then for $\epsilon = 1 > 0$, there exists a positive integer N such that

$$d(x_n, x_m) < \epsilon = 1 \text{ for all } m, n \geq N$$

$$\Rightarrow d(x_n, x_m) < 1 \text{ for all } m, n \geq N.$$

Let $M = \max\{d(x_n, x_m) : m, n \leq N\}$.

Now if $m < N$ and $n \geq N$ then we have

$$d(x_m, x_n) \leq d(x_m, x_N) + d(x_N, x_n) < M + 1.$$

Hence $d(x_m, x_n) < M + 1$ for all $m, n \in \mathbb{N}$

$\Rightarrow \{x_n\}$ is bounded.

Remark (3). We have seen in theorem [2.19] that a Cauchy sequence in a metric space is bounded. But the converse of this statement is not true in general. Consider the example [2.6]. Note that the sequence $\{(-1)^n\} = \{-1, 1, -1, 1, \dots\}$ is a bounded sequence in $\mathbb{R}$ but it is not a Cauchy sequence.

Theorem [2.20] A convergent sequence in a metric space is a Cauchy sequence.

Proof. Let (X, d) be a metric space and let $\{x_n\}$ be a convergent sequence in X.

Let $x_n \to x$ as $n \to \infty$.

Then for given any $\epsilon > 0$, there is a positive integer N such that

$$d(x_n, x) < \frac{\epsilon}{2} \text{ for all } n \geq N.$$

Hence if $m \geq N$, we have $d(x_m, x) < \frac{\epsilon}{2}$.

Now if $m, n \geq N$ we see that

$$d(x_m, x_n) \leq d(x_m, x) + d(x, x_n) < \frac{\epsilon}{2} + \frac{\epsilon}{2} = \epsilon$$

i.e., $d(x_m, x_n) < \epsilon$ for all $m, n \geq N$

$\Rightarrow \{x_n\}$ is a Cauchy sequence in X.

Remark (4). From the above theorem we find that every convergent sequence in a metric space is a Cauchy sequence. What can we say about the converse of the above theorem?

Consider the real number space $(\mathbb{R}, d)$ with the usual metric. Let $Y = (0, 1]$. Then we know that (Y, d_Y) is a metric space. Consider the sequence $\{1/n\}$ in Y. Then in example [2.5], we see that $\{1/n\}$ is a Cauchy sequence in Y. We know that $1/n \to 0$ as $n \to \infty$ but $0 \notin Y$, that means the sequence $\{1/n\}$ is not convergent in (Y, d_Y).

Thus, we see that a Cauchy sequence in a metric space may not converge. So the converse of the above theorem is not true in general.

However, in $\mathbb{R}$ with the usual metric, the converse statement is also true. For this, recall the Cauchy's general principle of convergence: "A sequence $\{x_n\}$ of real numbers is convergent, if and only if for any $\epsilon > 0$ there exists a natural number N such that $|x_m - x_n| < \epsilon$ for all $m, n \geq N$". That is, a sequence in $\mathbb{R}$ is convergent, if and only if it is a Cauchy sequence in $\mathbb{R}$.

Theorem [2.21] A convergent sequence in a metric space is bounded.

Proof. Let (X, d) be a metric space and let $\{x_n\}$ be a convergent sequence in X. Then by theorem [2.20], $\{x_n\}$ is a Cauchy sequence in X and so by theorem [2.19], $\{x_n\}$ is a bounded sequence.

Theorem [2.22] If a Cauchy sequence in a metric space has a convergent subsequence, then the Cauchy sequence is convergent.

Proof. Let (X, d) be a metric space. Let $\{x_n\}$ be a Cauchy sequence in X and let $\{x_{n_k}\}$ be a sub-sequence of $\{x_n\}$ such that $x_{n_k} \to x$ in X. Since $\{x_n\}$ is a Cauchy sequence, for given any $\epsilon > 0$, there is a natural number N_1 such that $d(x_n, x_m) < \epsilon/2$ for all $m, n \geq N_1$.

Since $x_{n_k} \to x$, for this $\epsilon > 0$, there is a positive integer N_2 such that

$$d(x_{n_k}, x) < \epsilon/2 \text{ for all } n_k \geq N_2.$$

Let $N = \max\{N_1, N_2\}$. Then for all $n \geq N$ we have

$$d(x_n, x) \leq d(x_n, x_{n_k}) + d(x_{n_k}, x) < \epsilon/2 + \epsilon/2 = \epsilon$$

$\Rightarrow d(x_n, x) < \epsilon$ for all $n \geq N$

$\Rightarrow \{x_n\}$ is a convergent sequence and $x_n \to x$ in X.

Remark (5). Let (X, d) be a metric space and $\{x_n\}$ be a convergent sequence in X. Then every subsequence of $\{x_n\}$ is convergent. Thus the converse of the above theorem is also true.

Hence we can say that:

"A Cauchy sequence in a metric space is convergent, if and only it it has a convergent subsequence".

Example [2.7]: If $\{x_n\}$ and $\{y_n\}$ are Cauchy sequences in a metric space (X, d), then $\{d(x_n, y_n)\}$ is a convergent sequence of real numbers.

Let $\epsilon > 0$ be given. Since $\{x_n\}$ is a Cauchy sequence, there is a natural number N_1, such that $d(x_m, x_n) < \epsilon/2$ for all $m, n \geq N_1$.

Again, since $\{y_n\}$ is a Cauchy sequence, there is a natural number N_2, such that $d(y_m, y_n) < \epsilon/2$ for all $m, n \geq N_2$.

Let $N = \max\{N_1, N_2\}$.

Then for all $m, n \geq N$, we have $d(x_m, x_n) < \epsilon/2$ and $d(y_m, y_n) < \epsilon/2$.

Now by using triangle inequality, we get

$$d(x_m, y_m) \leq d(x_m, x_n) + d(x_n, y_n) + d(y_n, y_m)$$
$$\Rightarrow d(x_m, y_m) - d(x_n, y_n) \leq d(x_m, x_n) + d(y_m, y_n). \quad \text{(i)}$$

Similarly, we get

$$d(x_n, y_n) - d(x_m, y_m) \leq d(x_n, x_n) + d(y_m, y_n). \quad \text{(ii)}$$

Combining (i) and (ii), we get

$$\begin{aligned} |d(x_m, y_m) - d(x_n, y_n)| &\leq d(x_m, x_n) + d(y_m, y_n) \\ &< \epsilon/2 + \epsilon/2 \text{ for all } m, n \geq N \\ &= \epsilon \text{ for all } m, n \geq N \end{aligned}$$

$\Rightarrow |d(x_m, y_m) - d(x_n, y_n)| < \epsilon$ for all $m, n \geq N$

$\Rightarrow \{d(x_n, y_n)\}$ is a Cauchy sequence of real number.

Hence by Cauchy's general principle of convergence, we have $\{d(x_n, y_n)\}$ is a convergent sequence of real numbers.

Example [2.8]: Consider the space of all real numbers with the trivial metric d. Show that if a sequence $\{x_n\}$ is a convergent sequence in $(\mathbb{R}, d)$, then there exists $N \in \mathbb{N}$ such that

$$x_N = x_{N+1} = \cdots$$

Let $\{x_n\}$ be a convergent sequence in $(\mathbb{R}, d)$, where d denotes the trivial metric, i.e.,

$$d(x, y) = \begin{cases} 1, & \text{if } x \neq y \\ 0, & \text{if } x = y. \end{cases}$$

Let $x_n \to x \in \mathbb{R}$ as $n \to \infty$.

Then for any $\epsilon > 0$, there exists a natural number N such that $|x_n - x| < \epsilon$ for all $n \geq N$.

Now taking $\epsilon = 1$, we see that

$$d(x_n, x) = |x_n - x| < 1 \text{ for all } n \geq N$$
$$\Rightarrow |x_n - x| = 0 \text{ for all } n \geq N$$
$$\Rightarrow x_n = x \text{ for all } n \geq N$$
$$\Rightarrow x_N = x_{N+1} = x_{N+2} = \cdots = x.$$

Example [2.9] : A sequence $\{x_n\}$ in a metric space (X, d) converges to a number $p \in X$, if and only if given any open set U containing p, there exists a natural number $N \in \mathbb{N}$ such that

$$x_j \in U \text{ for all } j \geq N.$$

Let us first suppose that the sequence $\{x_n\}$ converges to the point p. Let U be an open set containing the point p. Then there exists an $\epsilon > 0$ such that $p \in B_\epsilon(p) \subset U$. Since $x_n \to p$, so for this $\epsilon > 0$, $\exists$ a positive integer N such that

$$d(x_n, p) < \epsilon \text{ for all } n \geq N$$
$$\Rightarrow x_n \in B_\epsilon(p) \text{ for all } n \geq N$$

Since $B_\epsilon(p) \subset U$, we get $x_n \in U$, $\forall\, n \geq N$.

Conversely, suppose the condition holds, i.e., given any open set U containing p, $\exists\, N \in \mathbb{N}$ such that $x_j \in U$ for all $j \geq N$. Let $\epsilon > 0$ be arbitrary. Then we know that $B_\epsilon(p)$ is an open set containing p. So by hypothesis, there exists a natural number N_1 such that $x_n \in B_\epsilon(p)$ for all $n \geq N_1$

$$\Rightarrow d(x_n, p) < \epsilon \text{ for all } n \geq N_1$$
$$\Rightarrow x_n \to p \text{ as } n \to \infty.$$

Remark (6). From the definition [2.3], we find that if A, B are two sets in a metric space with $A \cap B \neq \phi$, then $d(A, B) = 0$. If $A \cap B = \phi$, then is it possible to say that $d(A, B) > 0$.

We see this in the following example:

Example [2.10] : If A, B are disjoint **closed** subsets of a metric space X, then also $d(A, B)$ may be zero.

For example, consider the metric space $(\mathbb{R}^2, d)$, where d denotes the usual metric on $\mathbb{R}^2$.

Let us consider the sets $A = \{(x, 0) \in \mathbb{R}^2 : x \in \mathbb{R}\}$ and $B = \{(x, e^x) \in \mathbb{R}^2 : x \in \mathbb{R}\}$.

Then it is easy to check that A, B are closed disjoint subset of $\mathbb{R}^2$, but $d(A, B) = 0$.

We show later that if one of A and B is compact, and other is closed and disjoint, then $d(A, B) > 0$.

2.4 Equivalent Metrics

Definition [2.12] : Let X be a non-empty set. Let d_1 and d_2 be two metrics defined on X. Then the metrics d_1 and d_2 are said to be equivalent if they generates the same classes of open sets in X. That means, every open set in (X, d_1) is also an open set in (X, d_2) and conversely.

In other words, d_1 and d_2 are equivalent metrics on X if they induce the same metric topology on X.

Example [2.11] : Let X be a non-empty set and d be the trivial metric on X. Then we know that every subset of X is open in (X, d).

Define a metric d^* by

$$d^*(x, y) = \begin{cases} 2, & \text{if } x \neq y \\ 0, & \text{if } x = y \end{cases} \quad \text{for all } x, y \in X.$$

Then the open ball

$$B_1(x) = \{y \in X : d^*(x, y) < 1\}$$
$$= \{x\}, \text{ for all } x \in X.$$

This shows that every singleton set is an open set in (X, d^*) and hence every subset of X is open in (X, d).

Thus the metrics d and d^* are equivalent metrics on X.

Note that on a non-empty set, there are infinitely many equivalent metrics.

Remark (7). Let X be a non-empty set. Then the collection of all metrics forms an equivalence relation, where the relation is "$d\rho d^*$, if and only if d and d^* are equivalent", d and d^* are two metrics on X.

Example [2.12] : Let (X, d) be a metric space. Then by theorem [1.2],

$$d_1(x, y) = \frac{d(x, y)}{1 + d(x, y)} \text{ for all } x, y \in X,$$

is a metric on X. We now show that the metrics d and d_1 are equivalent.

Let G_1 be an open set in the metric space (X, d) and let $x \in G_1$. Then there exists a real number $r > 0$ such that $x \in B_r^d(x) \subset G_1$, where $B_r^d(x)$ is the open ball $B_r(x)$ in (X, d).

Consider the function $f : [0, \infty) \to \mathbb{R}$ by $f(x) = \dfrac{x}{1+x}$, $\forall\, x \in [0, \infty)$.

Then f is continuous and strictly increasing function on $[0, \infty)$.

Hence $d(x, y) < r$ iff $f(d(x, y)) < f(r)$

$$\text{iff } \frac{d(x, y)}{1 + d(x, y)} < \frac{r}{1 + r}$$

$$\text{iff } d_1(x, y) < \frac{r}{1 + r}.$$

If we take $s = \dfrac{r}{1+r}$, then $s > 0$ and $d(x, y) < r$ iff $d_1(x, y) < s$

$$\Rightarrow y \in B_r^d(x) \text{ iff } y \in B_s^{d_1}(x).$$

So we find $s = \dfrac{r}{1+r} > 0$ such that $B_r^d(x) = B_s^{d_1}(x)$ and hence $x \in B_s^{d_1}(x) \subset G_1$. This shows that G_1 is an open set in (X, d_1).

Now let G_2 be an open set in (X, d_1) and let $x \in G_2$. Then there exists $\alpha > 0$ such that

$$x \in B_\alpha^{d_1}(x) \subset G_2.$$

We choose $\alpha < 1$. Then as above, we get $B_\alpha^{d_1}(x) = B_\beta^d(x)$, where

$$\beta = \frac{\alpha}{1 - \alpha} > 0 \text{ and so } x \in B_\beta^d(x) \subset G_2.$$

Hence G_2 is an open set in (X, d).

Consequently, the metrics d and d_1 are equivalent.

Example [2.13] : Let (X, d) be a metric space. Then from theorem [1.3], we see that

$$d_2(x, y) = \min\{1, d(x, y)\}$$

is a metric on X. We now show that d and d_2 are equivalent metrics on X.

Let B_1 be an open set in (X, d) and let $x \in B_1$. Then there exists a real number $r > 0$ such that $x \in B_r^d(x) \subset B_1$, where $B_r^d(x)$ is the open ball $B_r(x)$ in (X, d).

Let $s = \min\{1, r\}$.

Then for any $y \in B_s^{d_2}(x) \Rightarrow d_2(x,y) < s$

$\Rightarrow \min\{1, d(x,y)\} < s$

$\Rightarrow d(x,y) < s$, since if $1 = \min\{1,d(x,y)\} < s \leq 1$—a contradiction.

$\Rightarrow d(x,y) < s \leq r$, since $s = \min\{1,r\}$

$\Rightarrow y \in B_r^d(x)$.

Hence $x \in B_r^{d_2}(x) \subset B_r^d(x) \subset B_1$

$\Rightarrow x \in B_r^{d_2}(x) \subset B_1$

$\Rightarrow B_1$ is an open set in (X, d_2).

Now let B_2 be a open set in (X, d_2). Then for $x \in B_2$, there exists a real number $r > 0$ such that $x \in B_r^{d_2}(x) \subset B_2$.

Consider the open ball $B_r^d(x)$ in (X,d).

Note that, for any $y \in B_r^d(x)$

$\Rightarrow d(x,y) < r$

$\Rightarrow \min\{1, d(x,y)\} \leq d(x,y) < r$

$\Rightarrow d_2(x,y) < r$

$\Rightarrow y \in B_r^{d_2}(x)$.

Hence $x \in B_r^d(x) \subset B_r^{d_2}(x) \subset B_2$

$\Rightarrow x \in B_r^d(x) \subset B_2$

$\Rightarrow B_2$ is an open set in (X,d).

Thus the metrics d and d_2 are equivalent metrics on X.

Remark (8). Let (X,d) be a metric space. Then

$$d_1(x,y) = \frac{d(x,y)}{1+d(x,y)} \text{ and } d_2(x,y) = \min\{1, d(x,y)\}$$

for any $x, y \in X$ are two metrics defined on X.

From example [2.12] we see that the metrics d and d_1 are equivalent and from example [2.13] we see that metrics d and d_2 are equivalent. Since the collection of all metrics on a non-empty set X form an equivalence relation by the relation, two metrics d^* and d^{**} are related, iff d^* and d^{**} are equivalent.

Hence the metrics d_1 and d_2 are equivalent, by transitive property and symmetric property of equivalence relation.

Theorem [2.23] Let X be a non-empty set and d_1, d_2 be two metrics defined on X. Then d_1 and d_2 are equivalent metrics, if and only if for each $x \in X$ and each open ball $B_r^{d_m}(x)$ in (X, d_m) for some $r > 0$, there is an $s > 0$ such that $B_s^{d_n}(x) \subseteq B_r^{d_m}(x)$ for $m, n = 1, 2$ $(m \neq n)$.

Proof. First suppose that the metrics d_1 and d_2 are equivalent. Let $x \in X$ and let $r > 0$. Consider the open ball $B_r^{d_m}(x)$. Then since an open ball is an open set, $B_r^{d_m}(x)$ is an open set in (X, d_m), $m = 1$ or 2. Since d_1 and d_2 are equivalent, $B_r^{d_m}(x)$ is an open set in (X, d_n). So there is a real number $s > 0$ such that $x \in B_s^{d_n}(x) \subset B_r^{d_m}(x)$, where $m, n = 1, 2$ $(m \neq n)$.

Conversely, let the condition holds.

Let G be an open set in (X, d_m), $m = 1$ or 2 and let $x \in G$. Then there exists an $r > 0$ such that $x \in B_r^{d_m}(x) \subset G$.

By hypothesis, there is $s > 0$ such that $x \in B_s^{d_n}(x) \subset B_r^{d_m}(x) \subset G$

$\Rightarrow x \in B_s^{d_n}(x) \subset G$

$\Rightarrow G$ is an open set in (X, d_n); $m, n = 1, 2$ $(m \neq n)$.

This completes the proof.

Theorem [2.24] Let X be a non-empty set and d_1, d_2 be two metrics on X. Then d_1 and d_2 are equivalent if there exists two real number $\alpha, \beta > 0$ such that

$$\alpha\, d_2(x, y) \leq d_1(x, y) \leq \beta\, d_2(x, y) \text{ for all } x, y \in X,$$

or equivalently, $\alpha' d_1(x, y) \leq d_2(x, y) \leq \beta' d_1(x, y)$, $\forall\, x, y \in X$ for two real numbers $\alpha', \beta' > 0$.

Proof. We use the above theorem (theorem [2.23]), we have $\alpha\, d_2(x, y) \leq d_1(x, y) \leq \beta\, d_2(x, y)$ for all $x, y \in X$ and for some $\alpha, \beta > 0$.

Let $x \in X$ and $r > 0$ be a real number. Consider the open ball $B_r^{d_1}(x)$ in (X, d_1).

Let us put $s = \dfrac{r}{\beta}$, then $s > 0$.

Note that for any $y \in B_s^{d_2}(x)$, we have $d_2(x, y) < s = \dfrac{r}{\beta}$

$\Rightarrow \beta\, d_2(x, y) < r$

$\Rightarrow d_1(x, y) \leq \beta\, d_2(x, y) < r$

$\Rightarrow d_1(x, y) < r \Rightarrow y \in B_r^{d_1}(x)$

$\Rightarrow B_s^{d_2}(x) \subset B_r^{d_1}(x)$,

i.e., for any open ball $B_r^{d_1}(x)$ in (X, d_1), there is an open ball $B_s^{d_2}(x)$ such that $B_s^{d_2}(x) \subset B_r^{d_1}(x)$.

Now let $x \in X$ and $r^* > 0$ be a real number. Consider the open ball $B_{r^*}^{d_2}(x)$ in the metric space (X, d_2). Put $s^* = r^* \alpha$. Consider the open ball $B_{s^*}^{d_1}(x)$ in (X, d_1).

Then for any $y \in B_{s^*}^{d_1}(x)$, we get $d_1(x, y) < s^* = r^* \alpha$

$\Rightarrow \dfrac{1}{\alpha} d_1(x, y) < r^*$

$\Rightarrow d_2(x, y) \leq \dfrac{1}{\alpha} d_1(x, y) < r^*$

$\Rightarrow d_2(x, y) < r^* \Rightarrow y \in B_{r^*}^{d_2}(x)$

$\Rightarrow B_{s^*}^{d_1}(x) \subset B_{r^*}^{d_2}(x)$,

i.e., for any open ball $B_{r^*}^{d_2}(x)$ in (X, d_2), we find an open ball $B_{s^*}^{d_1}(x)$ such that $B_{s^*}^{d_1}(x) \subset B_{r^*}^{d_2}(x)$.

Thus by the previous theorem, the metrics d_1 and d_2 are equivalent.

Remark (9). The converse of this theorem is **not** true, in general. For example, consider $(\mathbb{N}, d_u)$, where $d_u(x, y) = |x - y|\ \forall\, x, y \in \mathbb{N}$ and $(\mathbb{N}, d)$, where d is the trivial meric on $\mathbb{N}$. Then d_u and d generates same topology on $\mathbb{N}$ and hence they are equivalent. But we cannot kind real numbers, $\alpha, \beta > 0$ such that $\alpha d_u(x, y) \leq d(x, y) \leq \beta d_u(x, y)\ \forall\, x, y \in \mathbb{N}$.

Example [2.14] : If d_1 and d_2 are two metrics on the same set X and there are positive numbers a, b such that for all $x, y \in X$,

$$a\, d_1(x, y) \leq d_2(x, y) \leq b\, d_1(x, y).$$

Show that the Cauchy sequences in (X, d_1) and (X, d_2) are the same.

Let $\{x_n\}$ be a Cauchy sequence in (X, d_1). Then for $\epsilon > 0$, there exists a positive integer N such that $d_1(x_n, x_m) < \epsilon/b$ for all $m, n \geq N$

$\Rightarrow b\, d_1(x_n, x_m) < \epsilon$ for all $m, n \geq N$.

By the given relation of d_1 and d_2, we get

$$d_2(x_n, x_m) \leq b\, d_1(x_n, x_m) < \epsilon \text{ for all } m, n \geq N$$

$$\Rightarrow d_2(x_n, x_m) < \epsilon \text{ for all } m, n \geq N$$

$\Rightarrow$ $\{x_n\}$ is a Cauchy sequence in (X, d_2).

Conversely, let $\{y_n\}$ be a Cauchy sequence in (X, d_2). Then for any $\epsilon > 0$, there exists a positive integer N_1 such that $d_2(y_n, y_m) < a\epsilon$ for all $m, n \geq N_1$.

By the given relation, we get

$$a\, d_1(y_n, y_m) \leq d_2(y_n, y_m) < a\epsilon \text{ for all } m, n \geq N_1$$

$$\Rightarrow d_1(y_n, y_m) < \epsilon \text{ for all } m, n \geq N_1 \text{ (since } a > 0)$$

$\Rightarrow$ $\{y_n\}$ is a Cauchy sequence in (X, d_1).

Example [2.15] : If A, B are two dense subsets of a metric space (X, d), then $A \cup B$ is dense in (X, d) but $A \cap B$ need not be dense. If one of A, B is, in addition, open then $A \cap B$ is dense in (X, d).

Let A, B be dense in (X, d). Then $\bar{A} = X$ and $\bar{B} = X$.

Now from theorem [1.13] (ii), we have $\overline{A \cup B} = \bar{A} \cup \bar{B} = X \cup X = X$

$\Rightarrow A \cup B$ is dense in (X, d).

Consider the metric space $(\mathbb{R}, d)$ of all real numbers with the usual metric. Let $A = Q$ and $B = I$, where Q denotes the set of all rational numbers and I denotes the set of all irrational numbers. Then we have

$\bar{A} = \bar{Q} = \mathbb{R}$ and $\bar{B} = \bar{I} = \mathbb{R}$, i.e., A and B are dense in $\mathbb{R}$.

But $A \cap B = Q \cap I = \phi$ is **not** dense in $\mathbb{R}$.

Let us now suppose that A and B are dense subsets of X with B an open set. We always have $\overline{A \cap B} \subset X$.

Let $x \in X$. Since $\bar{B} = X$, then we have $x \in \bar{B}$.

Therefore, for all $r > 0$, we have $B_r(x) \cap B \neq \phi$

$\Rightarrow (B_r(x) \cap B) \cap \bar{A} \neq \phi$ (since $\bar{A} = X$)

$\Rightarrow (B_r(x) \cap B) \cap A \neq \phi$ since $B_r(x)$, B are open sets, so $B_r(x) \cap B$ is also an open set and then from example 1 in Chapter 1.

$\Rightarrow (A \cap B) \cap B_r(x) \neq \phi$ for all $r > 0$

$\Rightarrow x \in \overline{A \cap B}$

$\Rightarrow X \subseteq \overline{A \cap B}$.

Consequently, $\overline{A \cap B} = X$, i.e., $A \cap B$ is dense in X.

Note. From this we can say that neither Q nor I is an open set in $(\mathbb{R}, d)$.

Example [2.16] : Let A, B be two subsets in a metric space (X, d). Then

$$d(A, B) = d(\bar{A}, B) = d(\bar{A}, \bar{B}).$$

Since $A \subset \bar{A}$, $B \subset \bar{B}$, then from definition of $d(A, B)$ we always have

$$d(\bar{A}, \bar{B}) \leq d(\bar{A}, B) \leq d(A, B). \tag{1}$$

We now show that $d(A, B) \leq d(\bar{A}, \bar{B})$.

Let $a \in \bar{A}$ and $b \in \bar{B}$. Let $\epsilon > 0$ be given.

Then by example (3), in Chapter 1, we get points $\alpha \in A$ and $\beta \in B$ such that $d(a, \alpha) < \epsilon/2$ and $d(b, \beta) < \epsilon/2$.

Now $d(\alpha, \beta) \leq d(\alpha, a) + d(a, b) + d(b, \beta)$

$< \epsilon/2 + d(a, b) + \epsilon/2$

$= \epsilon + d(a, b)$.

Hence, $d(A, B) \leq d(\alpha, \beta) < \epsilon + d(a, b) \Rightarrow d(A, B) < \epsilon + d(a, b)$.

This is true for all $a \in \bar{A}$, $b \in \bar{B}$.

Thus we get $d(A, B) \leq \epsilon + d(\bar{A}, \bar{B})$. Since $\epsilon > 0$ is arbitrary, we get $d(A, B) \leq d(\bar{A}, \bar{B})$.

Hence, we get $d(A, B) = d(\bar{A}, \bar{B})$.

$\therefore$ from (1), $d(A, B) \leq d(\bar{A}, B) \leq d(A, B) \Rightarrow d(\bar{A}, B) = d(A, B)$.

$\therefore d(A, B) = d(\bar{A}, B) = d(\bar{A}, \bar{B})$.

Example [2.17] : Let $(\mathbb{R}, d)$ be the metric space of all real numbers with the usual metric. Let $A = [0, 1)$, $B = (1, 2]$.

Then $d(A, B) = 0$ since $d(A, B) = d(\bar{A}, \bar{B})$, by the above example.

Note that $A \cap B = \phi$.

If we take the trivial metric d^* (See Def. [1.9]), then $d^*(A, B) = 1$, since $A \cap B = \phi$.

Illustrative Examples:

Problem [2.1] : Let (X, d) be a metric space and $A, B \subseteq X$. Then show that

$$d(A \cap B) \leq d(A) + d(B) \text{ [i.e., } \operatorname{diam}(A \cap B) \leq \operatorname{diam}(A) + \operatorname{diam}(B)].$$

Solution. If $A \cap B = \phi$, then $\operatorname{diam}(A \cap B) = d(\phi) = -\infty$ and the result is obvious.

From theorem [2.2] (1), we find that if $A \subseteq B$, then $\operatorname{diam}(A) \leq \operatorname{diam}(B)$.

Since, $A \cap B \subseteq A$ and $A \cap B \subseteq B$, it follows that

$$\operatorname{diam}(A \cap B) \leq \operatorname{diam}(A) \text{ and } \operatorname{diam}(A \cap B) \leq \operatorname{diam}(B).$$

$$\therefore 2\operatorname{diam}(A \cap B) \leq \operatorname{diam}(A) + \operatorname{diam}(B).$$

Hence, $\operatorname{diam}(A \cap B) \leq 2\operatorname{diam}(A \cap B) \leq \operatorname{diam}(A) + \operatorname{diam}(B)$

$\Rightarrow \operatorname{diam}(A \cap B) \leq \operatorname{diam}(A) + \operatorname{diam}(B)$.

Problem [2.2]: In the metric space of all real numbers $(\mathbb{R}, d)$ with the usual metric, show that $d(A, B) = 0$, where $A = \mathbb{N}$ and $B = \left\{n + \dfrac{1}{2n} : n \in \mathbb{N}\right\}$.

Solution. Recall the definition of $d(A, B)$,

$$d(A, B) = \inf\{d(a, b) : a \in A, b \in B\}.$$

Using this definition, we get

$$d(A,B) \leq d\left(n, n+\frac{1}{2n}\right), \text{ for all } n \in \mathbb{N}$$
$$= \left|n - \left(n+\frac{1}{2n}\right)\right|, \text{ for all } n \in \mathbb{N}$$
$$= \left|-\frac{1}{2n}\right| = \frac{1}{2n}, \text{ for all } n \in \mathbb{N}$$
$$\Rightarrow d(A,B) \leq \frac{1}{2n}, \text{ for all } n \in \mathbb{N}.$$

Hence, we get $0 \leq d(A,B) \leq \frac{1}{2n}$, for all $n \in \mathbb{N}$

$\Rightarrow 0 \leq d(A,B) \leq \lim_n \frac{1}{2n}$, taking limit as $n \to \infty$

$\Rightarrow 0 \leq d(A,B) \leq 0 \Rightarrow d(A,B) = 0.$

Note that both the sets A and B are closed in $(\mathbb{R}, d)$ and $A \cap B = \phi$.

Problem [2.3] : Show that a subset F in a metric space (X,d) is closed, if and only if

$$\{x : \text{dist}\,(x,F) = 0\} \subset F.$$

Solution. In theorem [2.6], we see that if $A \subseteq X$, then $\bar{A} = \{x \in X : \text{dist}\,(x,A) = 0\}$.

Let us now suppose that F is closed in X. Then we have $F = \bar{F}$, by theorem [1.10].

Hence, in particular, $\bar{F} \subset F$, i.e., $\bar{F} = \{x \in X : \text{dist}\,(x,F) = 0\} \subset F$

$\Rightarrow \{x \in X : \text{dist}\,(x,F) = 0\} \subset F.$

Conversely, $\{x \in X : \text{dist}\,(x,F) = 0\} \subset F$

$\Rightarrow \bar{F} \subset F.$

Since $F \subset \bar{F}$ is always true, it follows that $F = \bar{F}$ and hence F is a closed set in (X,d).

Problem [2.4] : Show that a discrete metric space (X,d) is separable, if and only if X is countable.

Solution. Recall that every subset in a discrete metric space (X,d) is both open and closed (See example 8, Chapter 1). Hence, if A is a proper subset of X, then A cannot be dense in X, since $\bar{A} = A \subsetneq X$.

So, the **only** dense subset of X is X itself, since $\bar{X} = X$.

Hence X contains a countable dense subset, if and only if X is countable, i.e., X is separable, if and only if X is countable.

Problem [2.5] : Let X be a non-empty set and d be the trivial metric on X. Then show that every Cauchy sequence in X, is a convergent sequence in X.

Solution. First recall that d is defined by

$$d(a,b) = \begin{cases} 0, & \text{if } a = b \\ 1, & \text{if } a \neq b. \end{cases}$$

Let $\{x_n\}$ be a Cauchy sequence in X.

Let $\epsilon = 1/2$. Then, since $\{x_n\}$ is a Cauchy sequence in X, there exists a natural number N such that $d(x_n, x_m) < \frac{1}{2}$, for all $m, n \geq N$

$\Rightarrow x_n = x_m = p$ (say), for all $m, n \geq N$.

In other words, $\{x_n\}$ is of the form $\{x_1, x_2, \dots, x_{N-1}, p, p, \dots\}$.

Hence, $\{x_n\}$ clearly converges to $p \in X$.

Problem [2.6]: Let $\{a_n\}$ be a sequence in a metric space (X, d) and let

$$A_1 = \{a_1, a_2, \dots,\},\ A_2 = \{a_2, a_3, \dots\},\ A_3 = \{a_3, a_4, \dots\}.$$

Show that $\{a_n\}$ is a Cauchy sequence, if and only if $\lim_n \operatorname{diam}(A_n) = 0$.

Solution. First, suppose that $\{a_n\}$ is a Cauchy sequence in (X, d).

Let $\epsilon > 0$ be arbitrary.

Then there exists a natural number N such that $d(a_n, a_m) < \epsilon$ for $m > n \geq N$.

Hence, $n \geq N \Rightarrow \operatorname{diam}(A_n) < \epsilon \Rightarrow \lim_{n\to\infty} \operatorname{diam}(A_n) = 0$.

Conversely, suppose that $\lim_{n\to\infty} \operatorname{diam}(A_n) = 0$

Let $\epsilon > 0$. Then there exists $n_0 \in \mathbb{N}$ such that

$$\operatorname{diam}(A_n) < \epsilon, \text{ for all } n > n_0 \Rightarrow \operatorname{diam}(A_{n_0+1}) < \epsilon.$$

Hence $m, n > n_0 \Rightarrow a_m, a_n \in A_{n_0+1}$ and so $d(a_m, a_n) \leq \operatorname{diam}(A_{n_0+1}) < \epsilon$ for $m, n > n_0$

$\Rightarrow d(a_m, a_n) < \epsilon$ for all $m, n > n_0$

$\Rightarrow \{a_n\}$ is a Cauchy sequence in (X, d).

Problem [2.7]: Let $\{y_n\}$ be a Cauchy sequence in a metric space (X, d) and let $\{x_n\}$ be a sequence in X such that $d(x_n, y_n) < \frac{1}{n}$ for all $n \in \mathbb{N}$.

(1) Show that $\{x_n\}$ is also a Cauchy sequence in X.

(2) Show that $\{x_n\} \to p \in X$, iff $\{y_n\} \to p$.

Solution. (i) Using triangle inequality, we get

$$d(x_m, x_n) \leq d(x_m, y_m) + d(y_m, y_n) + d(y_n, x_n). \tag{1}$$

Let $\epsilon > 0$ be arbitrary. Then we can find $N_1 \in \mathbb{N}$ such that $1/N_1 < \epsilon/3$. Hence $1/n < \epsilon/3$ and $1/m < \epsilon/3$, for all $m, n \geq N_1$.

Thus $d(x_m, y_m) < \frac{1}{m} < \frac{\epsilon}{3}$ and $d(x_n, y_n) < \frac{1}{n} < \frac{\epsilon}{3}$, for all $m, n \geq N_1$.

Thus from (1), $d(x_m, x_n) < \frac{\epsilon}{3} + d(y_m, y_n) + \frac{\epsilon}{3}$, for all $m, n \geq N_1$.

Since $\{y_n\}$ is a Cauchy sequence in X, there exists $N_2 \in \mathbb{N}$ such that

$$d(y_m, y_n) < \frac{\epsilon}{3}, \text{ for all } m, n \geq N_2$$

Put $N = \max\{N_1, N_2\}$, then $d(x_m, x_n) < \frac{\epsilon}{3} + \frac{\epsilon}{3} + \frac{\epsilon}{3} = \epsilon$, for all $m, n \geq N$

$\Rightarrow d(x_m, x_n) < \epsilon$, for all $m, n \geq N$

$\Rightarrow \{x_n\}$ is a Cauchy sequence in X.

(ii) Again, by triangle inequality, we get $d(y_n, p) \leq d(y_n, x_n) + d(x_n, p)$

$$\Rightarrow \lim_{n\to\infty} d(y_n, p) \leq \lim_{n\to\infty} d(x_n, y_n) + \lim_{n\to\infty} d(x_n, p).$$

But by given condition, $\lim_{n\to\infty} d(x_n, y_n) \leq \lim_{n\to\infty} \frac{1}{n} = 0.$

Hence, if $x_n \to p$, $\lim_{n\to\infty} d(y_n, p) \leq 0 + \lim_{n\to\infty} d(x_n, p) = 0$

$$\Rightarrow \lim_{n\to\infty} d(y_n, p) = 0 \Rightarrow y_n \to p \text{ as } n \to \infty.$$

Similarly, if $y_n \to p$, then $x_n \to p$.

EXERCISE

(1) Let (X, d) be a metric space and $A, B \subset X$ such that $A \cap B \neq \phi$. Show that

$$d(A \cap B) \leq d(A) + d(B).$$

(2) Let A, B be two non-empty subsets of a metric space (X, d). Prove that

$$d(A, B) = d(A, B) = d(\bar{A}, \bar{B}).$$

(3) Find the diameter of the set $S = \{x, x^2, \ldots, x^n, \ldots\}$ in the metric space $(C[0,1], \rho)$, where ρ is defined by

$$\rho(x, y) = \int_0^1 |x(t) - y(t)| dt, \;\; x, y \in C[0,1].$$

(4) If $\{x_n\}$ and $\{y_n\}$ are two Cauchy sequences in a metric space (X, d), then show that $\{d(x_n, y_n)\}$ is a convergent sequence of real numbers.

(5) Let $\{x_n\}$ be a sequence in a metric space (X, d) such that $\sum_{k=1}^{\infty} d(x_k, x_{k+1}) < d$. Show that $\{x_n\}$ is a Cauchy sequence in (X, d).

(6) Let $\{x_n\}$ and $\{y_n\}$ be two sequences in a metric space (X, d) such that $x_n \to x$, $y_n \to y$ as $n \to \infty$ and for some $x, y \in X$. Then show that $d(x_n, y_n) \to d(x, y)$ as $n \to \infty$.

(7) Show that in a trivial metric space (X, d) every Cauchy sequence is convergent.

(8) Give an example of a sequence $\{x_n\}$ in the metric space $(\mathbb{R}, d)$, d denotes the usual metric on the set of all real numbers $\mathbb{R}$, such that $d(x_n, x_{n+1}) \to 0$ as $n \to \infty$ but $\{x_n\}$ is **not** Cauchy.

(9) Let $\{x_1, x_2, \ldots, x_n, \ldots\}$ be a sequence in a metric space (X, d) and let $B_1 = \{x_1, x_2, \ldots\}$, $B_2 = \{x_2, x_3, \ldots\}$ and so on. Show that $\{x_n\}$ is a Cauchy sequence, if and only if $\lim_{n\to\infty} d(B_n) = 0$.

(10) Show that every subset in a trivial metric space is bounded.

(11) Find diameter of a set A in a trivial metric space.

(12) Let (X, d) be a metric space and $A \subseteq X$. Prove that $d(A) = 0$ if and only if A is a singleton set.

(13) Let (X, d) be a metric space and A, B be two non-empty subsets of X. Prove that

(i) if $A \subseteq B$, then $d(A) \leq d(B)$

(ii) for each $x \in A, y \subset B, d(x, y) \leq d(A \cup B)$.

(14) Let (X, d) be a metric space and A, B be two non-empty subsets of X. Prove that

$$d(A \cup B) \leq d(A) + d(B) + d(A, B).$$

(15) Consider the metric space (l_2, d) where $d(x, y) = \left\{\sum_{n=1}^{\infty} |x_n - y_n|^2\right\}^{1/2}$, $x = \{x_n\}, y = \{y_n\} \in l_2$. Let $A = \{e_1, e_2, \ldots, e_n, \ldots\}$ where $e_n = \{0, 0, \ldots, \underset{\text{nth place}}{\underset{\uparrow}{1}}, 0, \ldots\}$. Find $d(A)$.

(16) Consider the metric space $(\mathbb{R}, d)$ where $d(x, y) = \dfrac{|x - y|}{1 + |x - y|}$ for all $x, y \in \mathbb{R}$. Prove that every subset of $\mathbb{R}$ is bounded and $d(\mathbb{R}) = 1$.
[Compare this problem with problem no. 10 and 11]

(17) In $\mathbb{R}^2$, let $A = \{(x, y) \in \mathbb{R}^2 : x^2 + y^2 \leq 1\}$ and $p = (1, 1)$. Find $d(p, A)$ with respect the metric d where

(a) $d((x_1, y_1)(x_2, y_2)) = \sqrt{(x_1 - x_2)^2 + (y_1 - y_2)^2}$.
(b) $d((x_1, y_1), (x_2, y_2)) = |x_1 - x_2| + |y_1 - y_2|$
(c) $d((x_1, y_1), (x_2, y_2)) = \max\{|x_1 - x_2|, |y_1 - y_2|\}$
(d) d is the trivial metric.

(18) Let (X, d) be a metric space and $A \subseteq X$. Prove that $x \in \text{ext}\,(A)$ if and only if $d(x, A) > 0$.

(19) Show that the limit of a convergent sequence in a metric space is a limit point of the range set of the sequence.

(20) Let (X, d) be a metric space and ρ be another metric on X defined as $\rho(x, y) = \min\{1, d(x, y)\}$, $x, y \in X$. Prove that $\{x_n\}$ is a Cauchy sequence in (X, d) if and only if it is a Cauchy sequence in (X, ρ).

(21) Let d_1 and d_2 be two metrics defined on a non-empty set X. Show by example that if d_1 and d_2 are equivalent metrics then the inequality in Theorem 2.24 need not be true.

(22) Show that the metric d_A defined in Exercise 18, Chapter-1 is equivalent to the metric induced by d on A.

(23) Prove that a convergent sequence in a metric space can have atmost one limit. Investigate the case when metric space is replaced by a pseudometric space.

CHAPTER 3

COMPLETE METRIC SPACES

Discussion. In Chapter 2 we have discussed about Cauchy sequences in a metric space and their various properties. We have seen in Theorem [2.20] that a convergent sequence in a metric space is a Cauchy sequence. But from Remark (4), we find that the converse is not true, in general, in a metric space, i.e., a Cauchy sequence in a metric space may not converge.

Definition [3.1] : Let (X, d) be a metric space. Then (X, d) is said to be a **complete metric space** if every Cauchy sequence in X converges to some point in X, i.e., every Cauchy sequence is a convergent sequence in X.

Thus, we find that in a complete metric space a sequence $\{x_n\}$ is convergent, if and only if it is a Cauchy sequence.

Example [3.1] : **[Completeness of $\mathbb{R}$]** Consider the metric space $(\mathbb{R}, d)$ of all real numbers together with the usual metric d. Recall Cauchy's general principle of convergence of a sequence:

"A sequence $\{x_n\}$ of real numbers is convergent, if and only if it is a Cauchy sequence".

That means every Cauchy sequence in $\mathbb{R}$ converges. Hence, $(\mathbb{R}, d)$ is a complete metric space.

Example [3.2] : **[Completeness of $\mathbb{R}^n$) $(n \geq 1)$]** From example [1.3], we find that $\mathbb{R}^n$ is a metric space w.r.t. the metric $d(x, y) = \left\{\sum\limits_{i=1}^{n}(x_i - y_i)^2\right\}^{1/2}$, where $x = (x_1, x_2, \ldots, x_n)$ and $y = (y_1, y_2, \ldots, y_n) \in \mathbb{R}^n$.

We now show that $(\mathbb{R}^n, d)$ is a complete metric space.

Let $\left\{\underset{\sim}{x}^{(m)} = \left(x_1^{(m)}, x_2^{(m)}, \ldots, x_n^{(m)}\right)\right\}$ be a Cauchy sequence in $\mathbb{R}^n$.

Let us look at the terms of the sequence $\{\underset{\sim}{x}^{(m)}\}$.

$$\begin{aligned}
\underset{\sim}{x}^{(1)} &= \left(x_1^{(1)}, x_2^{(1)}, \ldots, x_n^{(1)}\right) \\
\underset{\sim}{x}^{(2)} &= \left(x_1^{(2)}, x_2^{(2)}, \ldots, x_n^{(2)}\right) \\
&\cdots\cdots\cdots\cdots\cdots\cdots \\
\underset{\sim}{x}^{(r)} &= \left(x_1^{(r)}, x_2^{(r)}, \ldots, x_n^{(r)}\right) \\
&\cdots\cdots\cdots\cdots\cdots\cdots \\
\underset{\sim}{x}^{(m)} &= \left(x_1^{(m)}, x_2^{(m)}, \ldots, x_n^{(m)}\right) \\
&\cdots\cdots\cdots\cdots\cdots\cdots
\end{aligned}$$

Since $\{\underset{\sim}{x}^{(m)}\}$ is a Cauchy sequence, we get

$$d\left(\underset{\sim}{x}^{(m)}, \underset{\sim}{x}^{(r)}\right) \to 0 \text{ as } m, r \to \infty$$

$$\Rightarrow \left\{\sum_{i=1}^{n}\left(x_i^{(m)} - x_i^{(r)}\right)^2\right\}^{1/2} \to 0 \text{ as } m, r \to \infty.$$

So for each $i = 1, 2, \ldots, n$, we have

$$\left|x_i^{(m)} - x_i^{(r)}\right|^2 \le \sum_{i=1}^{n}\left(x_i^{(m)} - x_i^{(r)}\right)^2 \to 0 \text{ as } m, r \to \infty$$

$$\Rightarrow \left|x_i^{(m)} - x_i^{(r)}\right| \to 0 \text{ as } m, r \to \infty,\ i = 1, 2, \ldots, n.$$

This shows that for all $i = 1, 2, \ldots, n$, the sequence of real number $\left\{x_i^{(m)}\right\}$ is a Cauchy sequence in $\mathbb{R}$. Hence by Cauchy's general principle of convergence, the sequence $\left\{x_i^{(m)}\right\}$ is a convergent sequence of real numbers. So we can find $x_i^{(0)} \in \mathbb{R}$ such that

$$\lim_{m\to\infty} x_i^{(m)} = x_i^{(0)}, \text{ for each } i = 1, 2, \ldots, n.$$

Since $x_i^{(m)} \to x_i^{(0)}$, given $\epsilon > 0$, we can find a positive integer N such that

$$\left|x_i^{(m)} - x_i^{(0)}\right| < \frac{\epsilon}{\sqrt{n}}, \text{ for all } m \ge N \text{ and for all } i = 1, 2, \ldots, n.$$

Therefore

$$\sum_{i=1}^{n}\left|x_i^{(m)} - x_i^{(0)}\right|^2 < \epsilon^2, \text{ for all } m \ge N$$

$$\Rightarrow \left\{\sum_{i=1}^{n}\left|x_i^{(m)} - x_i^{(0)}\right|^2\right\}^{1/2} < \epsilon, \text{ for all } m \ge N$$

$$\Rightarrow d\left(\underset{\sim}{x}^{(m)}, \underset{\sim}{x}^{(0)}\right) < \epsilon, \text{ for all } m \ge N, \text{ where } \underset{\sim}{x}^{(0)} = \left(x_1^{(0)}, x_2^{(0)}, \ldots, x_n^{(0)}\right) \in \mathbb{R}^n$$

$$\Rightarrow \underset{\sim}{x}^{(m)} \to \underset{\sim}{x}^{(0)} \text{ as } m \to \infty.$$

This shows that the Cauchy sequence $\{\underset{\sim}{x}^{(m)}\}$ in $\mathbb{R}^n$ converges to a point $\underset{\sim}{x}^{(0)} \in \mathbb{R}^n$. Since $\{\underset{\sim}{x}^{(m)}\} \subseteq \mathbb{R}^n$ was taken arbitrarily, we can conclude that every Cauchy sequence in $\mathbb{R}^n$ is a convergent sequence.

Hence, $(\mathbb{R}^n, d)$ is a complete metric space.

Example [3.3] : **[Completeness of $l_p, p \ge 1$]** In example [1.4] we have shown that the set of all sequences $l_p = \left\{x = \{x_n\} : \sum\limits_{n=1}^{\infty} |x_n|^p < \infty\right\}$ is a metric space w.r.t. the metric.

$$d(x, y) = \left\{\sum_{n=1}^{\infty} |x_n - y_n|^p\right\}^{1/p}, p \ge 1 \text{ and } x = \{x_n\}, y = \{y_n\} \text{ are two members of } l_p.$$

We now show that (l_p, d) is a complete metric space.

For this, let $\left\{\underset{\sim}{x}^{(n)} = \left(x_1^{(n)}, x_2^{(n)}, x_3^{(n)}, \ldots, x_r^{(n)} \ldots\right)\right\}$ be a Cauchy sequence in l_p.

Let us look at the terms of the sequence

$$\begin{aligned} \underset{\sim}{x}^{(1)} &= \left(x_1^{(1)}, x_2^{(1)}, \ldots, x_r^{(1)}, \ldots\right) \\ \underset{\sim}{x}^{(2)} &= \left(x_1^{(2)}, x_2^{(2)}, \ldots, x_r^{(2)}, \ldots\right) \\ &\ldots\ldots\ldots\ldots\ldots\ldots\ldots\ldots \\ \underset{\sim}{x}^{(n)} &= \left(x_1^{(n)}, x_2^{(n)}, \ldots, x_r^{(n)}, \ldots\right) \\ &\ldots\ldots\ldots\ldots\ldots\ldots\ldots\ldots \end{aligned}$$

Since $\{\underset{\sim}{x}^{(n)}\}$ is a Cauchy sequence, so for any $\epsilon > 0$ there exists a natural number N such that

$$\begin{aligned} & d\left(\underset{\sim}{x}^{(n)}, \ \underset{\sim}{x}^{(m)}\right) < \epsilon, \text{ for all } m, n \geq N \\ \Rightarrow & \left\{\sum_{i=1}^{\infty} \left|x_i^{(n)} - x_i^{(m)}\right|^p\right\}^{1/p} < \epsilon, \text{ for all } m, n \geq N \\ \Rightarrow & \sum_{i=1}^{\infty} \left|x_i^{(n)} - x_i^{(m)}\right|^p < \epsilon^p, \text{ for all } m, n \geq N. \end{aligned}$$

So for each $i = 1, 2, \ldots$, we get

$$\begin{aligned} & \left|x_i^{(n)} - x_i^{(m)}\right|^p \leq \sum_{i=1}^{\infty} \left|x_i^{(n)} - x_i^{(m)}\right|^p < \epsilon^p, \text{ for all } m, n \geq N \\ \Rightarrow & \left|x_i^{(n)} - x_i^{(m)}\right|^p < \epsilon^p, \text{ for all } m, n \geq N \text{ and } i = 1, 2, 3, \ldots \\ \Rightarrow & \left|x_i^{(n)} - x_i^{(m)}\right| < \epsilon, \text{ for all } m, n \geq N \text{ and } i = 1, 2, 3, \ldots \end{aligned}$$

This shows that $\left\{x_i^{(n)}\right\}$ is a Cauchy sequence of real numbers, i.e., $\left\{x_i^{(n)}\right\}$ is a Cauchy sequence in $(\mathbb{R}, d)$, where $d(x, y) = |x - y|$, $\forall\, x, y \in \mathbb{R}$. From example [3.1], we know that $(\mathbb{R}, d)$ is a complete metric space. Hence the Cauchy sequence $\left\{x_i^{(n)}\right\}$ is a convergent sequence in $\mathbb{R}$.

Let $x_i^{(0)} \in \mathbb{R}$ such that $\lim\limits_{n \to \infty} x_i^{(n)} = x_i^{(0)}$ $(i = 1, 2, 3, \ldots)$.

Let $\underset{\sim}{x}^{(0)} = \left(x_1^{(0)}, x_1^{(0)}, \ldots, x_n^{(0)}, \ldots\right)$.

We now show that $\underset{\sim}{x}^{(0)} \in l_p$ and $d\left(\underset{\sim}{x}^{(n)}, \ \underset{\sim}{x}^{(0)}\right) \to 0$ as $n \to \infty$.

We have $\sum\limits_{i=1}^{\infty} \left|x_i^{(n)} - x_i^{(m)}\right|^p < \epsilon^p$, for all $m, n \geq N$.

From this it follows that $\sum\limits_{i=1}^{K} \left|x_i^{n)} - x_i^{(m)}\right|^p < \epsilon^p$, for all $m, n \geq N$ and for $K \in \mathbb{N}$.

Taking limit as $m \to \infty$, we see that $\sum\limits_{i=1}^{K} \left|x_i^{(n)} - x_i^{(0)}\right|^p \leq \epsilon^p$, for all $n \geq N$.

Since the above relation is true for all $K \in \mathbb{N}$, we get $\sum\limits_{i=1}^{\infty} \left|x_i^{(n)} - x_i^{(0)}\right|^p \leq \epsilon^p$, for all $n \geq N$.

Now from Minkwoski inequality, we get

$$\left(\sum_{i=1}^{\infty}\left|x_i^{(o)}\right|^p\right)^{1/p} \leq \underbrace{\left(\sum_{i=1}^{\infty}\left|x_i^{(0)}-x_i^{(n)}\right|^p\right)^{1/p}+\left(\sum_{i=1}^{\infty}\left|x_i^{(n)}\right|^p\right)^{1/p}}_{<\infty}$$

This implies that $\underset{\sim}{x}^{(0)} \in l_p$.

We also have $\left(\sum\limits_{i=1}^{\infty}\left|x_i^{(n)}-x_i^{(0)}\right|^p\right)^{1/p} < \epsilon$, for all $n \geq N$, i.e., $d\left(\underset{\sim}{x}^{(n)}, \underset{\sim}{x}^{(0)}\right) < \epsilon$, for all $n \geq N$.

Hence, (l_p, d) is a complete metric space.

Example [3.4]: [Completeness of l_∞] We have defined the metric space (l_∞, d_∞), where

$$l_\infty = \left\{x = \{x_n\} : \sup_n |x_n| < \infty\right\} \text{ and } d_\infty(x, y) = \sup_n |x_n - y_n|,$$

where $x = \{x_n\}$, $y = \{y_n\} \in l_\infty$.

We now show that l_∞ is a complete metric space. Let us take a Cauchy sequence $\left\{\underset{\sim}{x}^{(n)}\right\}$ in l_∞, where

$$\left\{\underset{\sim}{x}^{(n)}\right\} = \left\{\left(x_1^{(n)}, x_2^{(n)}, \ldots, x_r^{(n)}, \ldots\right)\right\}.$$

Let us look at the terms of the sequence $\left\{\underset{\sim}{x}^{(n)}\right\}$

$$\begin{aligned}
\underset{\sim}{x}^{(1)} &= \left(x_1^{(1)}, x_2^{(1)}, \ldots, x_r^{(1)}, \ldots\right)\\
\underset{\sim}{x}^{(2)} &= \left(x_1^{(2)}, x_2^{(2)}, \ldots, x_r^{(2)}, \ldots\right)\\
&\cdots\cdots\cdots\cdots\cdots\cdots\\
\underset{\sim}{x}^{(n)} &= \left(x_1^{(n)}, x_2^{(n)}, \ldots, x_r^{(n)}, \ldots\right)\\
&\cdots\cdots\cdots\cdots\cdots\cdots
\end{aligned}$$

By Cauchyness of $\left\{\underset{\sim}{x}^{(n)}\right\}$, for any $\epsilon > 0$, $\exists\, N \in \mathbb{N}$ such that

$$d_\infty\left(\underset{\sim}{x}^{(m)}, \underset{\sim}{x}^{(n)}\right) < \epsilon, \text{ for all } m, n \geq N$$

$$\Rightarrow \sup_r \left|x_r^{(m)} - x_r^{(n)}\right| < \epsilon, \text{ for all } m, n \geq N.$$

Now, for any $r = 1, 2, \ldots$, we get

$$\left|x_r^{(m)} - x^{(n)}\right| \leq \sup_r \left|x_r^{(m)} - x_r^{(n)}\right| < \epsilon, \text{ for all } m, n \geq N$$

$$\Rightarrow \left|x_r^{(m)} - x_r^{(n)}\right| < \epsilon, \text{ for all } m, n \geq N. \tag{1}$$

This shows that $\left\{x_r^{(n)}\right\}$ is a Cauchy sequence in $\mathbb{R}$ (or $\mathbb{C}$), for each $r = 1, 2, 3, \ldots$.

Since $\mathbb{R}$ (or $\mathbb{C}$) is complete, $\exists\, x_r^{(0)} \in \mathbb{R}$ (or $\mathbb{C}$) such that $x_r^{(n)} \to x_r^{(0)}$ as $n \to \infty$, $r = 1, 2, 3, \ldots$.

Let $\underset{\sim}{x}^{(0)} = \left(x_1^{(0)}, x_2^{(0)}, \ldots, x_r^{(0)}, \ldots\right)$.

We now show that $\underset{\sim}{x}^{(0)} \in l_\infty$ and $\underset{\sim}{x}^{(n)} \to \underset{\sim}{x}^{(0)}$ as $n \to \infty$.

Letting $m \to \infty$ and keeping n fixed, we get from (1)

$$\left|x_r^{(0)} - x_r^{(n)}\right| \leq \epsilon, \text{ for all } n \geq N;\ r = 1, 2, 3, \ldots. \tag{2}$$

Hence

$$\left|x_r^{(0)}\right| = \left|x_r^{(0)} - x_r^{(N)} + x_r^{(N)}\right| \leq \left|x_r^{(0)} - x_r^{(N)}\right| + \left|x_r^{(N)}\right| \leq \epsilon + \left|x_r^{(N)}\right|$$

for any $r = 1, 2, 3, \ldots$.

$\Rightarrow \underset{\sim}{x}^{(0)} \in l_\infty$ and from (2), we get

$$\sup_r \left|x_r^{(0)} - x_r^{(n)}\right| \leq \epsilon, \text{ for all } n \geq N$$

$$\Rightarrow d_\infty\left(\underset{\sim}{x}^{(n)}, \underset{\sim}{x}^{(0)}\right) \geq \epsilon, \text{ for all } n \geq N$$

$$\Rightarrow \underset{\sim}{x}^{(n)} \to \underset{\sim}{x}^{(0)} \text{ as } n \to \infty.$$

Thus the metric space (l_∞, d_∞) is complete.

Example [3.5] : [Completeness of the function space $C[a,b]$] From example [1.6] in chapter 1, we see that $C[a,b]$, the collection of all real valued continuous functions on $[a,b]$ is a metric space w.r.t. the metric d defined by

$$d(f,g) = \sup_{a \leq t \leq b} |f(t) - g(t)|, \text{ where } f, g \in C[a,b].$$

We now show that $(C[a,b], d)$ is a complete metric space.

Let us take a Cauchy sequence $\{f_n\}$ in $C[a,b]$. Then we have $d(f_n, f_m) \to 0$ as $m, n \to \infty$.

$\Rightarrow \sup\limits_{a \leq t \leq b} |f_n(t) - f_m(t)| \to 0$ as $m, n \to \infty$.

Thus, $|f_n(t) - f_m(t)| \leq \sup\limits_{a \leq t \leq b} |f_n(t) - f_m(t)| \to 0$ as $m, n \to \infty$, $\forall\, t \in [a,b]$.

$\Rightarrow |f_n(t) - f_m(t)| \to 0$ as $m, n \to \infty$ and $\forall\, t \in [a,b]$.

This shows that $\{f_n(t)\}$ is a Cauchy sequence of real numbers. Hence by Cauchy's general principle of convergence it follows that $\lim\limits_{n \to \infty} f_n(t)$ exists for all $a \leq t \leq b$.

Let us define $f(t) = \lim\limits_n f_n(t)$, $a \leq t \leq b$.

We now show that $f \in C[a,b]$ and $f_n \to f$ as $n \to \infty$. Since $\{f_n\}$ is a Cauchy sequence, for given $\epsilon > 0$ $\exists$ $N \in \mathbb{N}$ such that $d(f_n, f_m) < \epsilon$, for all $m, n \geq N$

$\Rightarrow \sup\limits_{a \leq t \leq b} |f_n(t) - f_m(t)| < \epsilon$, for all $m, n \geq N$.

Keeping $n \geq N$ fixed and making $m \to \infty$, we get

$$\sup_{a \leq t \leq b} |f_n(t) - f(t)| \leq \epsilon, \text{ for all } n \geq N \tag{1}$$

$\Rightarrow |f_n(t) - f(t)| \leq \sup\limits_{a \leq t \leq b} |f_n(t) - f(t)| \leq \epsilon$, $\forall\, n \geq N$ and $\forall\, t \in [a,b]$

$\Rightarrow |f_n(t) - f(t)| \leq \epsilon$, for all $n \geq N$ and $a \leq t \leq b$.

The last inequality shows that $f_n \to f$ uniformly on $[a,b]$. Since each f_n is a continuous function on $[a,b]$ and $f_n \to f$ uniformly on $[a,b]$, it follows that f is also a continuous function on $[a,b]$, i.e., $f \in C[a,b]$.

Now from (1), we get $\sup\limits_{a \leq t \leq b} |f_n(t) - f(t)| \leq \epsilon$, for all $n \geq N$

$\Rightarrow d(f_n, f) \leq \epsilon$, for all $n \geq N$

$\Rightarrow f_n \to f$ as $n \to \infty$.

Thus the metric space $(C[a,b], d)$ is a complete metric space.

Examples of incomplete metric spaces

Example [3.6] : In example [1.7], we have defined another metric ρ on $C[a,b]$, where

$$\rho(f,g) = \int_a^b |f(t) - g(t)|dt, \; f,g \in C[a,b].$$

We now show that $(C[a,b],\rho)$ is **not** a complete metric space. For, let us consider a sequence $\{f_n\}_{n\geq 2}$ in $C[0,1]$ defined as follows:

$$f_n(x) = \begin{cases} 0, & \text{if } 0 \leq x \leq \frac{1}{2} - \frac{1}{n} \\ n\left(x - \frac{1}{2}\right) + 1, & \text{if } \frac{1}{2} - \frac{1}{n} < x \leq \frac{1}{2} \\ 1, & \text{if } \frac{1}{2} < x \leq 1. \end{cases}$$

Now

$$\rho(f_n, f_m) = \int_0^1 |f_n(x) - f_m(x)|dx \leq \int_{\frac{1}{2}-\frac{1}{n}}^{\frac{1}{2}} f_n(x)dx + \int_{\frac{1}{2}-\frac{1}{m}}^{\frac{1}{2}} f_m(x)dx$$

$$= \frac{n}{2}\left(x - \frac{1}{2}\right)^2 + x\Big|_{\frac{1}{2}-\frac{1}{n}}^{\frac{1}{2}} + \frac{m}{2}\left(x - \frac{1}{2}\right)^2 + x\Big|_{\frac{1}{2}-\frac{1}{m}}^{\frac{1}{2}}$$

$$= \frac{1}{2n} + \frac{1}{2m} = \frac{1}{2}\left(\frac{1}{n} + \frac{1}{m}\right).$$

Let $\epsilon > 0$ be arbitrary. Then we can find an integer $N \in \mathbb{N}$ such that $\frac{1}{N} < \epsilon$. Then for all $m, n \geq N$ we have $\frac{1}{n} \leq \frac{1}{N} < \epsilon$ and $\frac{1}{m} \leq \frac{1}{N} < \epsilon$.

So $\rho(f_n, f_m) \leq \frac{1}{2}\left(\frac{1}{n} + \frac{1}{m}\right) < \epsilon, \; \forall \; m, n \geq N.$

Thus, $\{f_n\}_{n\geq 2}$ is a Cauchy sequence in $C[0,1]$.

Suppose now that we can find $f \in C[0,1]$ such that $\rho(f_n, f) \to 0$ as $n \to \infty$.

Then $\rho(f_n, f) = \int_0^{\frac{1}{2}-\frac{1}{n}} |f(x)|dx + \int_{\frac{1}{2}-\frac{1}{n}}^{\frac{1}{2}} |f_n(x) - f(x)|dx + \int_{\frac{1}{2}}^1 |1 - f(x)|dx.$

Since $\rho(f_n, f) \to 0$ as $n \to \infty$, we have

$$\int_0^{\frac{1}{2}} |f(x)|dx = 0, \quad \int_{\frac{1}{2}}^1 |1 - f(x)|dx = 0,$$

i.e., $f(x) = 0$, if $0 \leq x \leq \frac{1}{2}$ and $f(x) = 1$, if $\frac{1}{2} < x \leq 1$—which is a contradiction to the continuity of f.

Hence, $(C[0,1], \rho)$ is **not** complete.

Example [3.7] : Let us consider the metric space of all real numbers $(\mathbb{R}, d)$, where d is the usual metric on $\mathbb{R}$. Let $Y = (0,1]$. Then from the definition of subspace, (Y, d_Y) is also a metric space. Consider a sequence $\left\{\frac{1}{n}\right\}_{n\in\mathbb{N}}$ in (Y, d_Y). Then $\left\{\frac{1}{n}\right\}$ is a Cauchy sequence in (Y, d_Y), but since $\frac{1}{n} \to 0$ as $n \to \infty$ and $0 \notin Y$, so $\left\{\frac{1}{n}\right\}$ is not a convergent sequence in (Y, d_Y). Thus we find a Cauchy sequence $\{x_n\} = \left\{\frac{1}{n}\right\}$ in (Y, d_Y), which does not converge in Y.

Hence, (Y, d_Y) is not a complete metric space.

Example [3.8] : Let us denote by $P\left[0,\frac{1}{2}\right]$, the space of all polynomials over $\left[0,\frac{1}{2}\right]$. Then $P\left[0,\frac{1}{2}\right] \subset C\left[0,\frac{1}{2}\right]$ and again by definition of subspace $P\left[0,\frac{1}{2}\right]$ is a metric space with respect to the sup-metric, i.e., $d(f,g) = \sup_{0\le t\le \frac{1}{2}} |f(t)-g(t)|,\ f,g \in P\left[0,\frac{1}{2}\right]$.

Consider a sequence $\{f_n\}$ of polynomials in $P\left[0,\frac{1}{2}\right]$, where

$$f_n(t) = 1 + t + t^2 + \cdots + t^n,\ 0 \le t \le \frac{1}{2}.$$

Let $n > m$.

Then $|f_n(t) - f_m(t)| = |t^{m+1} + t^{m+2} + \cdots + t^n| = t^{m+1} + t^{m+2} + \cdots + t^n$

$$< \frac{1}{2^{m+1}} + \frac{1}{2^{m+2}} + \cdots + \frac{1}{2^n} \text{ as } 0 \le t \le \frac{1}{2}$$

$$\to 0 \text{ as } m,n \to \infty.$$

This is true for all t in $\left[0,\frac{1}{2}\right]$. Hence, $d(f_n, f_m) \to 0$ as $m,n \to \infty$ $(m < n)$.

The case, when $m > n$, can be similarly proved.

Therefore, $d(f_n, f_m) \to 0$ as $m,n \to \infty$

$\Rightarrow \{f_n\}$ is a Cauchy sequence in $P\left[0,\frac{1}{2}\right]$.

But $\lim_{n\to\infty} f_n(t) = \lim_{n\to\infty} [1 + t + t^2 + \cdots + t^n]$

$$= 1 + t + t^2 + \cdots + t^n + \cdots = \frac{1}{1-t} \text{ is not a polynomial.}$$

Therefore, the space $\left(P\left[0,\frac{1}{2}\right], d\right)$ is not complete.

Remark (1). From the above two examples we see that subspace of a complete metric space need not be complete. In fact, we shall prove later that **closed** subspace of a complete metric space is complete.

Thus the above subspaces are not closed.

Theorem [3.1] Let (X,d) be a metric space and $Y \subset X$. If (Y, d_Y) is complete, then Y is a closed set in (X,d).

Proof. Let (Y, d_Y) be a complete metric space, where d_Y is the restriction of d to Y. We now show that Y is a closed subset of X.

Let y be a limit point of Y in the metric space (X,d). Then for each $\epsilon = \frac{1}{n}$, $n \in \mathbb{N}$, there exists a point $y_n \in Y$ such that $d(y_n, y) < \epsilon = \frac{1}{n}$ [by problem (3) of Chapter 1]. Thus we get a sequence $\{y_n\}$ in Y such that $d(y_n, y) < \frac{1}{n}$, for all n.

This shows that $\{y_n\}$ is a convergent sequence in (X,d) and $y_n \to y$ as $n \to \infty$. Hence $\{y_n\}$ is a Cauchy sequence in (X,d). Since $\{y_n\} \subseteq Y$, it follows that $\{y_n\}$ is a Cauchy sequence in (Y, d_Y). Since (Y, d_Y) is complete, $\{y_n\}$ must converge to an element in Y. Hence $y = \lim_n y_n \in Y$. Thus $\bar{Y} \subseteq Y$ and Y is closed.

Remark (2). (i) From the above theorem it follows that every complete metric space is closed.

(ii) Let (X,d) be a complete metric space and $Y \subseteq X$. Then what can we say about the completeness of (Y, d_Y)? The answer is (Y, d_Y) may or may not be complete. For example, consider the metric space $(\mathbb{R}, d)$ of all real number with the usual metric d.

Let $Y = (0,1]$. Then (Y, d_Y) is not a complete metric space (See example [3.7]). So subspace of a complete metric space may not be complete. But the following result is true:

Theorem [3.2] Let (X,d) be a complete metric space and Y be a **closed** subset of X. Then (Y,d_Y) is also a complete metric space.

Proof. Let $\{y_n\}$ be a Cauchy sequence in (Y,d_Y).

Then $\{y_n\}$ is also a Cauchy sequence in (X,d) (verify!)

Since (X,d) is complete, we can find an element $y \in X$ such that $y_n \to y$ as $n \to \infty$. Hence using theorem [2.18] of Chapter 2, we conclude that $y \in \bar{Y} = Y$, since Y is closed in X.

Since $\{y_n\}$ was taken arbitrarily from Y, it follows that (Y,d_Y) is complete.

Remark (3). From Theorem [3.1] and Theorem [3.2], we can say that

"Let (X,d) be a complete metric space and $Y \subseteq X$. Then Y is complete, iff Y is a closed set in X."

Example [3.9] : Let (X,d) be a metric space. If A and B are complete subsets of (X,d), then $A \cup B$ is also complete.

Let $\{x_n\}$ be a Cauchy sequence in $A \cup B$.

If $\{x_n\} \subset A$ (or, $\{x_n\} \subset B$) then there is nothing to prove.

Let us suppose that this is not the case.

Then either A or B or both must contain infinite number of terms of $\{x_n\}$. In particular, let A contains infinite number of terms of $\{x_n\}$, i.e., A contains a subsequence, say, $\{x_{n_k}\}$ of $\{x_n\}$. Note that $\{x_{n_k}\}$ is a Cauchy sequence in A. Since A is complete, $\exists\ a \in A$ such that $x_{n_k} \to a$ as $k \to \infty$. Thus, the Cauchy sequence $\{x_n\}$ has got a convergent subsequence $\{x_{n_k}\}$ in $A \cup B$, as $a \in A \subset A \cup B$. So by theorem [2.22], $\{x_n\}$ is a convergent sequence in $A \cup B$ and $x_n \to a$ as $n \to \infty$.

Consequently, $A \cup B$ is complete.

Example [3.10] : Let d_1 and d_2 be two metrics on a non-empty set X. Let $\alpha, \beta > 0$ be two real numbers such that $\alpha\, d_2(x,y) \leq d_1(x,y) \leq \beta\, d_2(x,y),\ \forall\ x,y \in X$.

Then (X,d_1) is complete, if and only if (X,d_2) is complete.

First suppose that (X,d_1) is complete. Let $\{x_n\}$ be a Cauchy sequence in (X,d_2), i.e.,

$$d_2(x_n,x_m) \to 0 \text{ as } m,n \to \infty.$$

Therefore, $d_1(x_n,x_m) \leq \beta\, d_2(x_n,x_m) \to 0$ as $m,n \to \infty$

$$\Rightarrow d_1(x_n,x_m) \to 0 \text{ as } m,n \to \infty$$

$$\Rightarrow \{x_n\} \text{ is a Cauchy sequence in } (X,d_1).$$

Since (X,d_1) is complete, $\exists\ x \in X$ such that $d_1(x_n,x) \to 0$ as $n \to \infty$.

$$\Rightarrow \alpha\, d_2(x_n,x) \leq d_1(x_n,x) \to 0 \text{ as } n \to \infty$$

$$\Rightarrow d_2(x_n,x) \to 0 \text{ as } n \to \infty$$

$$\Rightarrow x_n \to x \text{ as } n \to \infty \text{ in } (X,d_2).$$

Hence, (X,d_2) is a complete metric space.

Similarly, we can prove that if (X,d_2) is complete, then (X,d_1) is complete.

Theorem [3.3] [Cantor's Intersection Theorem] A metric space (X,d) is complete, if and only if for any decreasing **chain** of non-empty closed sets $\{F_n\}$ with diam$(F_n) \to 0$ as $n \to \infty$, $\bigcap_{n=1}^{\infty} F_n$ consists of a single element.

Proof. Let (X, d) be complete metric space. Let $\{F_n\}$ be a decreasing chain of non-empty closed sets such that $\text{diam}(F_n) \to 0$ as $n \to \infty$.

Let us take $a_n \in F_n$, for $n = 1, 2, 3, \ldots$.

Then we have $a_{n+p} \in F_{n+p} \subseteq F_n$ [$\{F_n\}$ is decreasing] $p = 1, 2, \ldots$.

Hence, $d(a_n, a_{n+p}) \le \text{diam}(F_n) \to 0$ as $n \to \infty$, $p = 1, 2, \ldots$.

Thus $\{a_n\}$ is a Cauchy sequence in (X, d). Since (X, d) is complete, we can find a point $a \in X$ such that $a_n \to a$ as $n \to \infty$. Since $a_{n+p} \in F_n$, keeping n fixed and letting $p \to \infty$, we get $a = \lim\limits_{p\to\infty} a_{n+p} \in \bar{F}_n = F_n$, since F_n is closed. This is true for all n. Hence, $a \in \bigcap\limits_{n=1}^{\infty} F_n$.

We now show that $\bigcap\limits_{n=1}^{\infty} F_n = \{a\}$.

If $b \in \bigcap\limits_{n=1}^{\infty} F_n$, then $a, b \in F_n$, $\forall\, n$.

So $d(a, b) \le \text{diam}(F_n) \to 0$ as $n \to \infty$

$\Rightarrow d(a, b) = 0$

$\Rightarrow a = b$.

Conversely, suppose that the condition holds, i.e., (X, d) is a metric space and $\{F_n\}$ is a sequence of non-empty closed sets in X such that $F_1 \supseteq F_2 \supseteq F_3 \supseteq \cdots$ and $\text{diam}(F_n) \to 0$ as $n \to \infty$, we have $\bigcap\limits_{n=1}^{\infty} F_n$ consists of a single element. We now prove that (X, d) is complete.

Let $\{x_n\}$ be a Cauchy sequence in (X, d). Put $A_n = \{x_n, x_{n+1}, x_{n+2}, \cdots\}$, $n = 1, 2, \ldots$.

Then $\{A_n\}$ is a decreasing chain of non-empty sets in X. Now since $A_n \supseteq A_{n+1}$, $\forall\, n$, we have $\bar{A}_n \supseteq \bar{A}_{n+1}$, $\forall\, n$. Thus $\{\bar{A}_n\}$ is a decreasing chain of non-empty closed sets in (X, d).

Also $\text{diam}(\bar{A}_n) = \text{diam}(A_n) \to 0$ as $n \to \infty$, since $\{x_n\}$ is a Cauchy sequence.

Hence by assumption, $\bigcap\limits_{n=1}^{\infty} \bar{A}_n$ consists of a single point, say, $u \in X$.

So $u \in \bar{A}_n (\supseteq A_n)$ for all n. Therefore, $d(u, x_n) \le \text{diam}(\bar{A}_n) \to 0$ as $n \to \infty$.

This imply that the Cauchy sequence $\{x_n\}$ converges to the element $u \in X$.

Hence the metric space (X, d) is complete.

This completes the proof of the theorem.

Example [3.11]: Show that $\bigcap\limits_{n=1}^{\infty} [0, \frac{1}{n}] = \{0\}$.

We use Cantor's intersection theorem to prove the above result.

Let us first consider the metric space $(\mathbb{R}, d)$ of all real numbers with the usual metric d.

Let $F_n = \left[0, \dfrac{1}{n}\right]$, $\forall\, n \in \mathbb{N}$.

Then $\{F_n\}_{n=1}^{\infty}$ is a family of non-empty closed sets in $(\mathbb{R}, d)$ such that

$$F_1 \supseteq F_2 \supseteq F_3 \cdots \supseteq F_n \supseteq \cdots.$$

Also note that $\text{diam}(F_n) = \dfrac{1}{n} \to 0$ as $n \to \infty$.

Thus $\{F_n\}$ is a decreasing chain of non-empty closed sets with $\text{diam}(F_n) \to 0$ as $n \to \infty$.

Recall that $(\mathbb{R}, d)$ is a complete metric space.

Hence by Cantor's intersection theorem, $\bigcap_{n=1}^{\infty} F_n = \bigcap_{n=1}^{\infty}\left[0, \frac{1}{n}\right]$ consists of a single point.

Note that $\frac{1}{n} \in F_n, \forall\, n \in \mathbb{N}$. So $\frac{1}{n+p} \in F_n, \forall\, n \in \mathbb{N}$ and $p = 1, 2, 3, \ldots$.

Then $0 = \lim_{p\to\infty} \frac{1}{n+p} \in F_n, \forall\, n \in \mathbb{N}$ and so $0 \in F_n, \forall\, n \in \mathbb{N}$.

Hence, $\bigcap_{n=1}^{\infty} F_n = \bigcap_{n=1}^{\infty}\left[0, \frac{1}{n}\right] = \{0\}$.

3.1 Isometric Spaces and Completion of a Metric Space

Definition [3.2] : Let (X, d) and (Y, ρ) be two metric spaces. A mapping $f : (X, d) \to (Y, \rho)$ is said to be an isometry if for any two points x, y in X,

$$d(x, y) = \rho(f(x), f(y)).$$

The definition of isometry indicates that f preserves the distance between any two points of X into Y.

For example, the rigid motions of the plane $\mathbb{R}^2$ are exactly isometries

$$(\mathbb{R}^2, d_{\text{EUC}}) \to (\mathbb{R}^2, d_{\text{EUC}}),$$

where d_{EUC} denotes Euclidean metric on $\mathbb{R}^2$.

Note that the function $f : (\mathbb{R}, d) \to (\mathbb{R}, d)$ defined by $f(x) = x^2$, $x \in \mathbb{R}$ is not an isometry. For, take $x = 0$ and $y = 2$ in $\mathbb{R}$. Then $d(x, y) = |x - y| = 2$ but $d(f(x), f(y)) = d(0, 4) = 4$, i.e., $d(x, y) \neq d(f(x), f(y))$.

But if one takes the identity mapping $I : (X, d) \to (X, d)$, it is plain that I is an isometry.

Note. It is very easy to check that an isometry is an injective mapping.

Let $f : (X, d) \to (Y, \rho)$ be an isometry.

Let $x, y \in X$ such that $x \neq y$.

Then from properties of a distance function we have $d(x, y) > 0$.

Now from isometry, we get $\rho(f(x), f(y)) = d(x, y) > 0$

$\Rightarrow \rho(f(x), f(y)) > 0$

$\Rightarrow f(x) \neq f(y)$ in (Y, ρ)

$\Rightarrow f$ is injective.

Definition [3.3] : **[Isometric Spaces]** Two metric spaces (X, d) and (Y, ρ) are said to be isometric if there exists a bijective isometry between the spaces (X, d) and (Y, ρ).

Since an isometry is always injective, two metric spaces (X, d) and (Y, ρ) are isometric, if there exists an isometry from (X, d) **onto** (Y, ρ).

Theorem [3.4] **[Completion]** Let (X, d) be a metric space. Then we can find a complete metric space $(\hat{X}, \hat{d})$ which has a subspace W that is isometric with X and is dense in $\hat{X}$. The space $(\hat{X}, \hat{d})$ is unique except for isometrics, that is, if (Y, ρ) is any complete metric space with a dense subspace U isometric with X, then $\hat{X}$ and Y are isometric.

Proof. We divide the proof of this theorem into four steps as follows:

Step (a) : Construction of $(\hat{X}, \hat{d})$.

Step (b) : We construct an isometry $f : X \xrightarrow{\text{onto}} W$, where $\bar{W} = \hat{X}$.

Step (c) : We prove that $(\hat{X}, \hat{d})$ is a complete metric space.

Step (d) : Uniqueness of $\hat{X}$, except for isometrics.

Step (a) : Construction of $(\hat{X}, \hat{d})$

Let $\{x_n\}$ and $\{y_n\}$ be two Cauchy sequences in X.

We say that $\{x_n\}$ to be equivalent to $\{y_n\}$ if

$$\lim_{n\to\infty} d(x_n, y_n) = 0 \tag{1}$$

and we denote it by $\{x_n\} \sim \{y_n\}$.

Then the relation '$\sim$' is an equivalence relation on the set of all Cauchy sequences in X.

Let $\hat{X}$ be the set of all equivalence classes $\hat{x}, \hat{y}, \ldots$ of Cauchy sequences thus obtained [$\hat{x}$ is the equivalence class of the Cauchy sequence $\{x_n\}$].

We now define a function $\hat{d}$ on $\hat{X}$ as

$$\hat{d}(\hat{x}, \hat{y}) = \lim_{n\to\infty} d(x_n, y_n), \tag{2}$$

where $\hat{x}$ = class of $\{x_n\}$, $\hat{y}$ = class of $\{y_n\}$.

We now show that the above definition is well defined. We always have

$$d(x_n, y_n) \le d(x_n, x_m) + d(x_m, y_m) + d(y_m, y_n)$$
$$\Rightarrow d(x_n, y_n) - d(x_m, y_m) \le d(x_n, x_m) + d(y_m, y_n). \tag{3}$$

In a similar way, we can show that

$$d(x_m, y_m) - d(x_n, y_n) \le d(x_n, x_m) + d(y_m, y_n). \tag{4}$$

(3) and (4) together imply,

$$|d(x_n, y_n) - d(x_m, y_m)| \le d(x_n, x_m) + d(y_m, y_n). \tag{5}$$

As $\{x_n\}$ and $\{y_n\}$ are Cauchys sequences it follows that $\{d(x_n, y_n)\}$ is a Cauchy sequence in $\mathbb{R}$ and hence by completeness of $\mathbb{R}$ we say that the limit in (2) exists. We now show that this limit is independent of the particular choice of member from $\hat{x}$ and $\hat{y}$.

For, if $\{x_n\} \sim \{x'_n\}$ and $\{y_n\} \sim \{y'_n\}$, then from (5), we get

$$|d(x_n, y_n) - d(x'_n, y'_n)| \le d(x_n, x'_n) + d(y_n, y'_n) \to 0 \text{ as } n \to \infty$$
$$\Rightarrow \lim_{n\to\infty} d(x_n, y_n) = \lim_n d(x'_n, y'_n).$$

We now prove that $\hat{d}$ is a metric on $\hat{X}$. The metric axioms (d.1) and (d.3) are obvious.

(d.2) From definition it follows that $\hat{d}(\hat{x}, \hat{x}) = 0$. Now let

$$\hat{d}(\hat{x}, \hat{y}) = 0 \Rightarrow \lim_{n\to\infty} d(x_n, y_n) = 0 \Rightarrow \{x_n\} \sim \{y_n\}$$
$$\Rightarrow \text{class of } \{x_n\} = \text{class of } \{y_n\} \text{ [cl(a) = cl(b), iff } a\rho b,$$

where ρ is an equivalence relation on $S; a, b \in S$]

$$\Rightarrow \hat{x} = \hat{y}.$$

(d.3) Follows from definition of 2.

(d.4) Let $\{x_n\} \in \hat{x}, \{y_n\} \in \hat{y}, \{z_n\} \in \hat{z}$.

Then $d(x_n, z_n) \leq d(x_n, y_n) + d(y_n, z_n)$.

$$\Rightarrow \lim_{n\to\infty} d(x_n, z_n) \leq \lim_{n\to\infty} d(x_n, y_n) + \lim_{n\to\infty} d(y_n, z_n)$$

$$\Rightarrow \hat{d}(\hat{x}, \hat{z}) \leq \hat{d}(\hat{x}, \hat{y}) + \hat{d}(\hat{y}, \hat{z}).$$

This is true for any $\hat{x}, \hat{y}, \hat{z} \in \hat{X}$.

Hence, $\hat{d}$ is a metric on $\hat{X}$ and consequently $(\hat{X}, \hat{d})$ is a metric space.

Step (b) : Construction of an isometry $f : X \to W \subset \hat{X}$

Let $a \in X$. Consider the class $\hat{a} \in \hat{X}$, which contains the constant Cauchy sequence $\{a, a, \dots\}$. Define a mapping $f : X \to W \subset \hat{X}$ by $f(a) = \hat{a}$.

Note that $\hat{d}(\hat{a}, \hat{b}) = \lim_{n\to\infty} d(a_n, b_n)$, where $\{a_n\} = \{a, a, \dots\}$, $\{b_n\} = \{b, b, \dots\}$

$= d(a, b)$

$\Rightarrow \hat{d}(f(a), f(b)) = d(a, b)$.

This implies that f is an isometry from $X \to W = f(X)$. Since every isometry is injective, it follows that f is a bijective isometry from $X \to W$.

Hence the spaces X and W are isometric.

Let us now show that $\bar{W} = \hat{X}$. For, let $\hat{x} \in \hat{X}$ and $\{x_n\} \in \hat{x}$. Then for any $\epsilon > 0$, we can find $N \in \mathbb{N}$ such that $d(x_n, x_N) < \frac{\epsilon}{2}$, for all $n > N$.

Let $\{x_N, x_N, \dots\} \in \hat{x}_N$. Then by definition $\hat{x}_N \in W$.

Now $\hat{d}(\hat{x}, \hat{x}_N) = \lim_{n\to\infty} d(x_n, x_N) \leq \frac{\epsilon}{2} < \epsilon$

$\Rightarrow \hat{x}_N \in B_\epsilon(\hat{x})$

$\Rightarrow$ every ϵ-neighbourhood of $\hat{x}$ contains a member of W

$\Rightarrow \hat{x} \in \bar{W}$.

Thus $\bar{W} = \hat{X}$, i.e., W is dense in $\hat{X}$.

Step (c) : Completeness of $(\hat{X}, \hat{d})$

Let $\{\hat{x}_n\} = \{\hat{x}_1, \hat{x}_2, \dots\}$ be a Cauchy sequence in $(\hat{X}, \hat{d})$.

Since $\bar{W} = \hat{X}$, for every $\hat{x}_n \in \hat{X}$, $\exists\ \hat{z}_n \in W$ such that

$$\hat{d}(\hat{x}_n, \hat{z}_n) < \frac{1}{n}. \tag{6}$$

Now by triangle inequality,

$$\hat{d}(\hat{z}_m, \hat{z}_n) \leq \hat{d}(\hat{z}_m, \hat{x}_m) + \hat{d}(\hat{x}_m, \hat{x}_n) + \hat{d}(\hat{x}_m, \hat{z}_n)$$

$$< \frac{1}{m} + \hat{d}(\hat{x}_m, \hat{x}_n) + \frac{1}{n} \to 0 \text{ as } m, n \to \infty.$$

$\Rightarrow \{\hat{z}_m\}$ is a Cauchy sequence in W. Since $f : X \to W$ is onto, for each $\hat{z}_m \in W$, $\exists\ z_m \in X$ such that $f(z_m) = \hat{z}_m$. Now by isometry, $d(z_n, z_m) = \hat{d}(f(z_n), f(z_m)) = \hat{d}(\hat{z}_n, \hat{z}_m)$.

$\Rightarrow \{z_m\}$ is a Cauchy sequence in X, as $\{\hat{z}_n\}$ is a Cauchy sequence in W. Let $\hat{x}$ be the class to which $\{z_m\}$ belongs. We now show that $\lim_n \hat{x}_n = \hat{x}$ in $\hat{X}$.

By (b),

$$\hat{d}(\hat{x}_n, \hat{x}) \leq \hat{d}(\hat{x}_n, \hat{z}_n) + \hat{d}(\hat{z}_n, \hat{x}) < \frac{1}{n} + \hat{d}(\hat{z}_n, \hat{x}). \tag{7}$$

Since $\{z_m\} \in \hat{x}$ and $\hat{z}_n \in W$, so that $\{z_n, z_n, \dots\} \in \hat{z}_n$, the inequality (7) becomes

$$d(\hat{x}_n, \hat{x}) < \frac{1}{n} + \lim_{m\to\infty} d(z_n, z_m)$$

$$\Rightarrow \lim_{n\to\infty} d(\hat{x}_n, \hat{x}) \le \lim_{n\to\infty} \frac{1}{n} + \lim_{m,n\to\infty} d(z_n, z_m) = 0$$

$$\Rightarrow \lim_{n\to\infty} \hat{x}_n = \hat{x} \Rightarrow (\hat{X}, \hat{d}) \text{ is complete.}$$

Step (d) : Uniqueness of completion except for isometries

Let (Y, ρ) and (Z, σ) be two complete metric spaces such that $\exists\ U \subset Y$, $V \subset Z$, $\bar{U} = Y$, $\bar{V} = Z$ and U, V are isometric with X. We now show that Y and Z are isometric.

Since the collection of all isometries forms an equivalence relation, it follows that U, V are isometric. Let $f : U \xrightarrow{\text{onto}} V$ be an isometry.

Let $y \in Y$. Then there is a sequence $\{y_n\}$ in U such that $y_n \to y$ as $n \to \infty$. Therefore $\{y_n\}$ is a Cauchy sequence in U and by isometry, $\{f(y_n)\}$ is a Cauchy sequence in V and hence a Cauchy sequence in Z. Since (Z, σ) is complete, $\exists\ z \in Z$ such that $\lim_n f(y_n) = z$.

We now define a mapping $g : Y \to Z$ as follows:

For $y \in Y, g(y) = \lim_n f(y_n)$, where $\{y_n\} \subseteq U$ and $y_n \to y$ as $n \to \infty$.

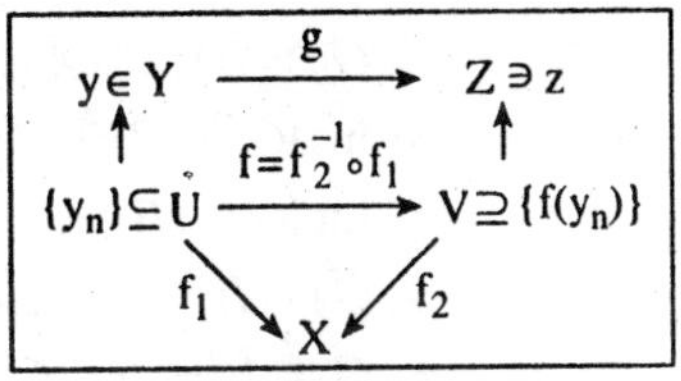

Fig. 3.1

We now show that g is well-defined. For, let $\{y'_n\}$ be another sequence in U such that $y'_n \to y$ as $n \to \infty$.

Then $\lim_{n\to\infty} \sigma(f(y'_n), f(y_n)) = \lim_{n\to\infty} \rho(y_n, y'_n) = 0$.

$\therefore \sigma(f(y'_n), z) \le \sigma(f(y'_n), f(y_n)) + \sigma(f(y_n), z) \to 0$ as $n \to \infty$.

$\Rightarrow \lim_{n\to\infty} f(y'_n) = z = \lim_{n\to\infty} f(y_n)$.

Let $y', y'' \in Y$ and $\{y'_n\}, \{y''_n\}$ be two sequences in U such that $y'_n \to y'$ and $y''_n \to y''$ as $n \to \infty$.

Then $\sigma(g(y'), g(y'')) = \lim_{n\to\infty} \sigma(f(y'_n), f(y''_n)) = \lim_{n\to\infty} \rho(y'_n, y''_n) = \rho(y', y'')$.

It is easy to verify that g is bijective. Hence, (Y, ρ) and (Z, σ) are isometric and the proof is complete.

3.2 Baire Category Theorem

Before going to the Baire category theorem, we need to familiar with some notions in a metric space.

Definition [3.4] : [Nowhere Dense Set] Let (X, d) be a metric space and A be a subset of X. Then A is said to be nowhere dense in (X, d), if $\text{int}(\bar{A}) = \phi$, where $\bar{A}$ denotes the closure of A in (X, d).

Example [3.12] : The set of all integers $\mathbb{Z}$ is nowhere dense subset of the real number space $(\mathbb{R}, d)$, since

$$\text{int}(\bar{\mathbb{Z}}) = \text{int}(\mathbb{Z}), \text{ as } \mathbb{Z} \text{ is a closed set}$$
$$= \phi.$$

Note. It is also very easy to verify that any finite subset of a metric space (X, d) is nowhere dense. The set of natural numbers $\mathbb{N}$ is also nowhere dense in $(\mathbb{R}, d)$.

Example [3.13] : Let N be a nowhere dense set in a metric space (X, d). Then $\bar{N}^c$ is a dense subset of X.

If possible, suppose that $\bar{N}^c$ is **not** dense in X. Then $\overline{\bar{N}^c} \neq X \Rightarrow \overline{\bar{N}^c} \subset X$, i.e., $\exists$ a point $p \in X$ such that $p \notin \overline{\bar{N}^c} \Rightarrow$ there is an open set G in (X, d) such that $p \in G$ and $G \cap \bar{N}^c = \phi \Rightarrow p \in G \subset \bar{N} \Rightarrow p \in \text{int}\,(\bar{N})$—a contradiction to the assumption that N is nowhere dense in (X, d). Thus $\bar{N}^c$ is dense in (X, d).

Remark (4). Since $N \subset \bar{N}$, we always have $N^c \supset \bar{N}^c$. Hence, if $\bar{N}^c$ is dense in X, then so in N^c. Thus it follows from the above example that if N is nowhere dense in X, then N^c is dense in X.

Theorem [3.5] Let A and B be two nowhere dense sets in a metric space (X, d). Then $A \cup B$ is also a nowhere dense set in (X, d).

Proof. To prove that $A \cup B$ is nowhere dense, we must show $\text{int}\,(\overline{A \cup B}) = \phi$.

Let $G = \text{int}\,(\overline{A \cup B})$. Then we have $G \subset \overline{A \cup B} = \bar{A} \cup \bar{B}$.

Now note that

$$
\begin{aligned}
& G \cap (\bar{B})^c \subset (\bar{A} \cup \bar{B}) \cap (\bar{B})^c = [\bar{A} \cap (\bar{B})^c] \cup [\bar{B} \cap (\bar{B})^c] = \bar{A} \cap (\bar{B})^c \subset \bar{A} \\
\Rightarrow\ & \text{int}\,[G \cap (\bar{B})^c] \subset \text{int}\,(\bar{A}) = \phi, \text{ since } A \text{ is nowhere dense in } X \\
\Rightarrow\ & \text{int}\,[G \cap (\bar{B})^c] = \phi \\
\Rightarrow\ & G \cap (\bar{B})^c = \phi, \text{ since } G \cap (\bar{B})^c \text{ is an open set} \\
\Rightarrow\ & G \subset \bar{B} \Rightarrow \text{int}\,(G) \subset \text{int}\,(\bar{B}) = \phi, \text{ since } B \text{ is nowhere dense in } X \\
\Rightarrow\ & \text{int}\,(G) = \phi \Rightarrow G = \phi, \text{ since } G \text{ is an open set} \\
\Rightarrow\ & \text{int}\,(\overline{A \cup B}) = \phi \\
\Rightarrow\ & A \cup B \text{ is nowhere dense in } (X, d).
\end{aligned}
$$

Remark (5). From the above theorem, we can say that

"Finite union of nowhere dense sets in (X, d) is again nowhere dense."

Theorem [3.6] If N is nowhere dense in (X, d), then each open ball $B_r(x)$ contains a closed ball $\bar{B}_s(y)$ such that $\bar{B}_s(y) \cap N = \phi$.

Proof. Let N be nowhere dense in (X, d). Then we have $\bar{N}^c$ is dense in (X, d), by Example [3.13]. That means $\overline{\bar{N}^c} = X$. Hence for each $x \in X$, we have $x \in \overline{\bar{N}^c}$. So for each open ball $B_r(x)$, we get $B_r(x) \cap \bar{N}^c \neq \phi$. Let $y \in B_r(x) \cap \bar{N}^c$.

Since $B_r(x) \cap \bar{N}^c$ is an open set, $\exists\, r_1 > 0$ such that $B_{r_1}(y) \subset B_r(x) \cap \bar{N}^c$, i.e., $B_{r_1}(y) \subset B_r(x)$ and $B_{r_1}(y) \subset \bar{N}^c \Rightarrow B_{r_1}(y) \subset B_r(x)$ and $B_{r_1}(y) \cap N = \phi$.

Let us choose r_2 such that $0 < r_2 < r_1$. Then it is easy to verify that $\bar{B}_{r_2}(y) \subset B_r(x)$ and $\bar{B}_{r_2}(y) \cap N = \phi$.

Thus for each open ball $B_r(x)$, we find a closed ball $\bar{B}_{r_2}(y)$ contained in $B_r(x)$ such that $\bar{B}_{r_2}(y) \cap N = \phi$.

Definition [3.5] : **[First category, second category]** A metric space (X, d) is said to be of **first category** if X can be expressed as a countable union of nowhere dense subsets of X.

Otherwise, (X, d) is said to be of second category.

Example [3.14] : Consider the metric space (Q, d), where Q is the set of all rational numbers and $d(x, y) = |x - y|$, $\forall\, x, y \in Q$.

Note that, for each $x \in Q$, $\overline{\{x\}} = \{x\}$ and so int $\overline{\{x\}}$ = int $\{x\} = \phi \Rightarrow \{x\}$ is a nowhere dense set in (Q, d) for each $x \in Q$.

Also, $Q = \bigcup_{x \in Q} \{x\}$, i.e., Q is a countable union of nowhere dense subsets of $Q \Rightarrow (Q, d)$ is of first category.

To find examples of metric spaces which are of second category, we look at the following theorem which is known as **Baire's category theorem.**

Theorem [3.7] Every complete metric space is of second category.

Proof. Let (X, d) be a complete metric space. If possible, suppose that X is a set of first category and let

$$X = F_1 \cup F_2 \cup \cdots \cup F_n \cup \cdots = \bigcup_{k=1}^{\infty} F_k,$$

where each F_n is nowhere dense in X.

Take an open ball B in (X, d). Since F_1 is nowhere dense, we find a closed ball, say $\bar{B}_{r_1}(x_0)$, such that

$$\bar{B}_{r_1}(x_0) \cap F_1 = \phi \text{ with } 0 < r_1 < 1 \text{ and } \bar{B}_{r_1}(x_0) \subset B \text{ (using Th. [3.6])}$$

Again, since F_2 is nowhere dense in X, we find another closed ball $\bar{B}_{r_2}(x_1) \subset B_{r_1}(x_0)$ such that $\bar{B}_{r_2}(x_1) \cap F_2 = \phi$ with $0 < r_2 < \dfrac{1}{2}$.

Continuing in this way we find a closed ball $\bar{B}_{r_n}(x_{n-1}) \subset B_{r_{n-1}}(x_{n-2})$ such that

$$\bar{B}_{r_n}(x_{n-1}) \cap F_n = \phi \text{ with } 0 < r_n < \frac{1}{n}.$$

Thus we get a decreasing chain of non-empty closed set $\{\bar{B}_{r_n}(x_{n-1})\}$ with diameter going to zero as $n \to \infty$ and, therefore, by Cantor's intersecton theorem, we have

$$\bigcap_{n=1}^{\infty} \bar{B}_{r_n}(x_{n-1}) \text{ is a singleton set, say } \{u\}.$$

Now $u \in \bar{B}_{r_n}(x_{n-1})$, $\forall\, n$

$\Rightarrow u \notin F_n$, $\forall\, n$

$\Rightarrow u \notin \bigcup_{n=1}^{\infty} F_n = X$—a contradiction.

This completes the proof of the theorem.

Note. The above theorem supplies to us a large number of metric spaces which are of second category.

Illustrative Examples:

Problem [3.1] : Consider the metric space [See Example [1.5]] (l_∞, d_∞). Let M be the subset containing of all sequences $x = \{x_n\}$ with atmost finitely many non-zero terms. Prove that (M, d_∞) is not a complete metric space, where d_∞ is the induced metric on M.

Solution. We know from Example [3.4] that (l_∞, d_∞) is a complete metric space. Noting the Remark (3), it only remains to show that M is not closed in (l_∞, d_∞).

Let us consider a sequence $\{x_n\}$ in (M, d_∞). Let x_n be the sequence $\{x_{n,i}\}_{i=1}^{\infty}$ defined by

$$x_{n,i} = \begin{cases} \dfrac{1}{i}, & \text{if } 1 \le i \le n \\ 0, & \text{if } i > n \end{cases}$$

i.e., $x_n = \left\{1, \dfrac{1}{2}, \dfrac{1}{3}, \ldots, \dfrac{1}{n}, 0, \ldots\right\}$. Let x be the sequence whose ith term is $\dfrac{1}{i}$, $\forall\, i$.

That is $x = \left\{1, \dfrac{1}{2}, \dfrac{1}{3}, \ldots, \dfrac{1}{n}, \ldots\right\}$. Then $x \in l_\infty$ but $x \notin M$. Moreover, $\forall\, n \in \mathbb{N}$,

$$d_\infty(x_n, x) = \sup_n \left\{\frac{1}{n+1}, \frac{1}{n+2}, \ldots\right\} \to 0 \text{ as } n \to \infty$$

$\Rightarrow x_n \to x$ as $n \to \infty$

$\Rightarrow x \in \bar{M}$ but $x \notin M$

$\Rightarrow \bar{M} \neq M$

$\Rightarrow M$ is not closed in (l_∞, d_∞).

Problem [3.2] : Let $X = C[0,1]$ and d be defined by $d(f,g) = \displaystyle\int_0^1 |f(x) - g(x)|dx$.

For $n \in \mathbb{N}$, define $f_n \in X$ by $f_n(x) = x^n$, $\forall\, x \in [0,1]$.

(i) Show that the sequence $\{f_n\}$ converges in X and find its limit f.

(ii) Show that the function $f = \lim\limits_n f_n$ is not the pointwise limit of f_n.

Solution. (i) Let f be the zero function, i.e., $f(x) = 0$, $\forall\, x \in [0,1]$. Then $f \in X$ and $\forall\, n \in \mathbb{N}$, we have

$$d(f_n, f) = \int_0^1 |f_n(x) - f(x)|dx = \int_0^1 x^n dx = \frac{1}{n+1} \to 0 \text{ as } n \to \infty \Rightarrow f_n \to f \text{ as } n \to \infty.$$

(ii) It is not true that $\lim\limits_{n\to\infty} f_n(x) = f(x)$, $\forall\, x \in [0,1]$, since

$$f_n(1) = 1^n = 1,\ \forall\, n \in \mathbb{N} \Rightarrow f_n(1) \to 1 \text{ as } n \to \infty, \text{ but } f(1) = 0.$$

So $\lim\limits_{n\to\infty} f_n(1) \neq f(1)$.

Problem [3.3] : Show that the set of all integers $\mathbb{Z}$ with the usual metric $d(m,n) = |m-n|$, $\forall\, m, n \in \mathbb{N}$ is a complete metric.

Solution. We know that $\mathbb{Z} \subset \mathbb{R}$, the set of all real numbers. Also $(\mathbb{R}, d)$ is a metric space, where $d(x,y) = |x-y|$, $\forall\, x, y \in \mathbb{R}$. Hence, $(\mathbb{Z}, d)$ is a subspace of $(\mathbb{R}, d)$ and so $(\mathbb{Z}, d)$ is a metric space. We now show that $\mathbb{Z}$ is a closed subset of $\mathbb{R}$ w.r.to the usual metric.

The complement of $\mathbb{Z}$ in $\mathbb{R}$ is the union of all open intervals of the form $(n, n+1)$, $\forall\, n \in \mathbb{Z}$ and this is an open set, since union of arbitrary number of open set is an open set. Hence $\mathbb{Z}$ is closed in $(\mathbb{R}, d)$.

Since $(\mathbb{R}, d)$ is complete and closed subsets of a complete metric space is complete, it follows that $(\mathbb{Z}, d)$ is complete.

Problem [3.4] : Let $X = (0, 1)$ with the Euclidean metric d. Give an example of a decreasing chain of non-empty closed sets $\{A_n\}$ in X with $\operatorname{diam}(A_n) \to 0$ as $n \to \infty$ but $\bigcap_{n=1}^{\infty} A_n = \phi$.

Solution. Take $A_n = \left(0, \frac{1}{n}\right]$, $\forall\, n \in \mathbb{N}$ with $n \geq 2$. Then we have $A_2 \supset A_3 \supset \cdots$, i.e., $\{A_n\}$ is a decreasing chain of non-empty sets.

Again, since, $A_n = \left[0, \frac{1}{n}\right] \cap X$ and $\left[0, \frac{1}{n}\right]$ is closed in $\mathbb{R}$, it follows that A_n is a closed set in (X, d).

Also $\operatorname{diam}(A_n) = \frac{1}{n} \to 0$ as $n \to \infty$.

However, $\bigcap_{n=2}^{\infty} A_n = \phi$.

Problem [3.5] : Let (X, d) be a metric space and $CS(X)$ denote the collection of all Cauchy sequences in X. For $\{x_n\}, \{y_n\} \in CS(X)$, define $\{x_n\} \sim \{y_n\}$, iff $\lim_{n\to\infty} d(x_n, y_n) = 0$.

Show that $\sim$ is an equivalence relation of $CS(X)$.

Solution. For any Cauchy sequence $\{x_n\}$, we always have

$$\begin{aligned} d(x_n, x_n) = 0,\ \forall\, n &\Rightarrow \lim_{n\to\infty} d(x_n, x_n) = 0 \\ &\Rightarrow \{x_n\} \sim \{x_n\},\ \forall\, \{x_n\} \in CS(X) \\ &\Rightarrow \sim \text{ is reflexive.} \end{aligned}$$

Let $\{x_n\} \sim \{y_n\}$, where $\{x_n\}, \{y_n\} \in CS(X)$

$$\begin{aligned} &\Rightarrow \lim_{n\to\infty} d(x_n, y_n) = 0 \Rightarrow \lim_{n\to\infty} d(y_n, x_n) = 0 \\ &\Rightarrow \{y_n\} \sim \{x_n\} \Rightarrow \sim \text{ is symmetric.} \end{aligned}$$

Now for transitive property, take $\{x_n\}, \{y_n\}, \{z_n\} \in CS(X)$ such that $\{x_n\} \sim \{y_n\}$ and $\{y_n\} \sim \{z_n\}$.

Then $\lim_{n\to\infty} d(x_n, y_n) = 0$ and $\lim_{n\to\infty} d(y_n, z_n) = 0$.

Now, $0 \leq d(x_n, z_n) \leq d(x_n, y_n) + d(y_n, z_n) \to 0$ as $n \to \infty$

$$\begin{aligned} &\Rightarrow \lim_{n\to\infty} d(x_n, z_n) = 0 \Rightarrow \{x_n\} \sim \{z_n\} \\ &\Rightarrow \sim \text{ is transitive.} \end{aligned}$$

Hence $\sim$ is an equivalence relation on $CS(X)$.

Problem [3.6] : Let (X, d) be a complete metric space and let $\{x_n\}$ be a sequence in X such that there is a $\theta \in (0, 1)$ with $d(x_{n+2}, x_{n+1}) \leq \theta \cdot d(x_{n+1}, x_n)$ for $n \in \mathbb{N}$. Show that $\{x_n\}$ is a convergent sequence.

Solution. We first show that $d(x_{n+1}, x_n) \leq \theta^{n-1} d(x_2, x_1)$.

For $n = 1$, the claim is obviously true.

Suppose now that the relation is true for $n = k$, i.e., $d(x_{k+1}, x_k) \leq \theta^{k-1} d(x_2, x_1)$.

Now, $d(x_{k+2}, x_{k+1}) \leq \theta \cdot d(x_{k+1}, x_k) \leq \theta \cdot \theta^{k-1} d(x_2, x_1) = \theta^k d(x_2, x_1)$.

So by induction our claim established.

Now let $m, n \in \mathbb{N}$ and we assume without loss of generality that $m > n$. From the triangle inequality, we have

$$\begin{aligned} d(x_m, x_n) &\le d(x_m, x_{m-1}) + d(x_{m-1}, x_{m-2}) + \cdots + d(x_{n+1}, x_n) \\ &\le (\theta^{m-2} + \theta^{m-3} + \cdots + \theta^{n-1}) d(x_2, x_1) \\ &= \theta^{n-1} \cdot (\theta^{m-n-1} + \theta^{m-n-2} + \cdots + 1) \cdot d(x_2, x_1) \\ &= \theta^{n-1} \cdot \frac{1 - \theta^{m-n}}{1 - \theta} \cdot d(x_2, x_1) \\ &\le \frac{\theta^{n-1}}{1 - \theta} d(x_2, x_1). \end{aligned} \tag{1}$$

Note that $\lim_{n \to \infty} \theta^n = 0$. So from (1), it follows that $\{x_n\}$ is a Cauchy sequence in (X, d) and since (X, d) is complete, $\{x_n\}$ is a convergent sequence in X.

Problem [3.7]: Show that (i) A finite metric space (X, d) is equivalent to (X, d_D), where d_D is the trivial metric. (ii) Any trivial metric space (X, d_D) is complete. (iii) Any finite metric space is complete.

Solution. (i) Let (X, d) be a finite metric space. Then $X \times X$ is also a finite set. Thus the set $\{d(x, y) : x, y \in X, x \neq y\}$ is a finite set and so we can find its maximum and minimum.

Let $c_1 = \min\{d(x, y) : x, y \in X, x \neq y\}$

$c_2 = \max\{d(x, y) : x, y \in X, x \neq y\}$.

Since d is a metric, we have $c_1, c_2 > 0$ and $c_1 \le d(x, y) \le c_2$, $\forall\, x, y (x \neq y)$ in X.

Hence, $c_1 \cdot d_D(x, y) \le d(x, y) \le c_2 \cdot d_D(x, y)$, $\forall\, x, y \in X$ and the result follows.

(ii) Let (X, d_D) be a trivial metric space.

Let $\{x_n\}$ be a Cauchy sequence in (X, d_D). Then for $\epsilon = \frac{1}{2}$, there exists a positive integer N such that

$$\begin{aligned} & d_D(x_n, x_m) < \frac{1}{2}, \ \forall\, m, n \ge N \\ \Rightarrow\ & d_D(x_n, x_m) = 0, \ \forall\, m, n \ge N \\ \Rightarrow\ & x_n = x_m, \ \text{for all } m, n \ge N. \end{aligned}$$

That means $\{x_n\}$ is eventually constant.

Obviously, for an eventually constant sequence $\{x_n\}$, there exists $x \in X$ and $N \in \mathbb{N}$ such that $x_n = x$, $\forall\, n \ge N$ and so $x_n \to x$ as $n \to \infty$.

$\Rightarrow (X, d_D)$ is complete.

(iii) Result follows from Example [3.10].

Remark (6). A discrete metric space may or may not be complete.

For example, consider the metric space $(\mathbb{R}, d)$, where $\mathbb{R}$ denotes the set of all real numbers and d is the usual metric on $\mathbb{R}$. We now take $Y = \left\{1, \frac{1}{2}, \frac{1}{3}, \cdots, \frac{1}{n}, \cdots\right\} \subseteq \mathbb{R}$. Then (Y, d_Y) is a metric space. Not only that (Y, d_Y) is a discrete metric space which is **not** complete.

On the other hand, we have seen in example [3.3] that $(\mathbb{Z}, d)$ is a complete metric space. An easy calculation shows that $(\mathbb{Z}, d)$ is a discrete metric space.

EXERCISE

(1) Show that every trivial metric space (X, d) is complete.

(2) Show that any finite metric space (X, d) is complete.

(3) Let d_1 and d_2 be two metrics on a non-empty set X. Let $\alpha, \beta > 0$ be two real numbers such that

$$\alpha \cdot d_2(x, y) \leq d_1(x, y) \leq \beta \cdot d_2(x, y), \ \forall \ x, y \in X.$$

Show that (X, d_1) is complete, if and only if (X, d_2) is complete.

(4) Show by example that if (X, ρ) is **not** complete then $\bigcap_{n=1}^{\infty} F_n$ may be empty, where $\{F_n\}$ is a decreasing chain of non-empty closed sets in (X, d).

(5) Let (X, d) be a complete metric space and $E \subseteq X$. Then prove that E is complete, if and only if E is closed.

(6) Let $\{x_n\}$ be a sequence in a complete metric space (X, d) with the property that

$$\sum_{k=1}^{\infty} d(x_k, x_{k+1}) < \infty.$$

Show that $\{x_n\}$ is convergent.

(7) Let (X, d) be a metric space. For any sequence $\{x_n\}$ in (X, d) satisfying $\sum_{k=1}^{\infty} d(x_k, x_{k+1}) < \infty$ is convergent. Show that (X, d) is a complete metric space.

(8) Let $X = C[a, b]$. Define $\rho : X \times X \to \mathbb{R}$ by $\rho(x, y) = \sup_{a \leq t \leq b} |x(t) - y(t)|, \ x, y \in X$.

Let $S = \{x \in X : x(a) = x(b)\}$. Show that (S, ρ) is complete.

(9) Let $\{x_n\}$ be a Cauchy sequence in a metric space (X, d). Show that $\{x_n\}$ is convergent, if and only if $\{x_n\}$ has a convergent subsequence in (X, d).

(10) Give an example of a complete metric space (X, d) and a sequence $\{F_n\}$ of non-empty closed sets such that $F_n \supseteq F_{n+1}, \ \forall \ n \geq 1$ but $\bigcap_{n=1}^{\infty} F_n = \phi$. Compare the result with Cantor's intersection theorem.

(11) Let (X, d) and (Y, ρ) be two complete metric spaces. Let $A \subseteq X, B \subseteq Y$ be such that $\bar{A} = X$, $\bar{B} = Y$. If A and B are isometric, then show that X, Y are also isometric.

(12) Let (X, d) be a metric space and ρ be another metric on X defined by $\rho(x, y) = \min\{1, d(x, y)\}$; $x, y \in X$. Prove that (X, d) is complete if and only if (X, ρ) is complete.

(13) Show that the metric space defined in Exercise 10, Chapter-1 is complete.

(14) Let $C^1[0, 1]$ be the set of all continuous functions $f : [0, 1] \to \mathbb{R}$ having continuous f' on $[0, 1]$. Define ρ on $C^1[0, 1]$ by:
for $f, g \in C^1[0, 1], \rho(f, g) = \sup_{0 \leq t \leq 1} |f(t) - g(t)| + \sup_{0 \leq t \leq 1} |f'(t) - g'(t)|$. Prove that $\left(C^1[0, 1], \rho\right)$ is a complete metric space.

(15) Prove that the metric space $(P[a, b], d)$ is not complete, where $P[a, b]$ denote the set of all polynomials on $[a, b]$ and $d(f, g) = \sup_{a \leq t \leq b} |f(t) - g(t)|; f, g \in P[a, b]$.

(16) Let X be a non-empty set and (Y, d) be a complete metric space. Let $B(X, Y)$ denote the set of all bounded functions $f : X \to Y$. Prove that $(B(X, Y), \rho)$ is a complete metric space where $\rho(f, g) = \sup_{t \in X} d(f(t), g(t)); f, g \in B(X, Y)$.

(17) Show that the metric space (Q, d) is not complete, where d is the usual metric of $\mathbb{R}$.

(18) Show that the metric space $(\mathbb{N}, d)$ is not complete where $d(x, y) = \left|\frac{1}{x} - \frac{1}{y}\right| \; \forall \, x, y \in \mathbb{N}$.

(19) Determine whether the following sets in the metric space $(\mathbb{R}^2, d)$ are complete or not:

(a) $\{(x, y) \in \mathbb{R}^2 : x^2 + y^2 < 1\}$

(b) $\{(x, y) \in \mathbb{R}^2 : x^2 + y^2 = 1\}$

(c) $\{(x, y) \in \mathbb{R}^2 : x^2 + y^2 \leq 1\}$

(d) $\{(x, y) \in \mathbb{R}^2 : xy = 0\}$

(20) Let (X, d) be a complete metric space and $a \in X$.
Then for $r > 0$ show that the set $\{x \in X : d(x, a) = r\}$ is complete.

CHAPTER 4

CONTINUITY

Discussion

Let us first recall the continuity of a function from $\mathbb{R}$ into $\mathbb{R}$. Let $f : \mathbb{R} \to \mathbb{R}$ be a function and $c \in \mathbb{R}$. Then f is said to be continuous at c, if for any $\epsilon > 0$ (arbitrarily small), there exists a $\delta > 0$ such that

$$|f(x) - f(c)| < \epsilon, \text{ whenever } |x - c| < \delta \tag{1}$$

$$\Rightarrow f(c) - \epsilon < f(x) < f(c) + \epsilon, \text{ whenever } c - \delta < x < c + \delta$$

$$\Rightarrow f(x) \in (f(c) - \epsilon, f(c) + \epsilon), \text{ whenever } x \in (c - \delta, c + \delta) \tag{2}$$

$$\Rightarrow f(c - \delta, c + \delta) \subset (f(c) - \epsilon, f(c) + \epsilon). \tag{3}$$

Since $\mathbb{R}$ is a metric space with respect to the metric d, where $d(x, y) = |x - y|\ \forall\ x, y \in \mathbb{R}$, the above relations (1), (2), (3) can be written as

$$d(f(x), f(c)) < \epsilon, \text{ whenever } d(x, c) < \delta \tag{1'}$$

$$\Rightarrow f(x) \in B_\epsilon(f(c)), \text{ whenever } x \in B_\delta(c) \tag{2'}$$

$$\Rightarrow f(B_\delta(c)) \subset B_\epsilon(f(c)). \tag{3'}$$

Remembering all these in mind, let us now define the continuity of a function in a metric space.

Definition [4.1] : Let (X, d) and (Y, ρ) be two metric spaces and $f : (X, d) \to (Y, \rho)$ be a function. Let $c \in X$. Then f is said to be continuous at $x = c$, if for any $\epsilon > 0$ (arbitrarily small), there exists a $\delta > 0$ such that

$$\rho(f(x), f(c)) < \epsilon, \text{ whenever } d(x, c) < \delta,\ x \in X.$$

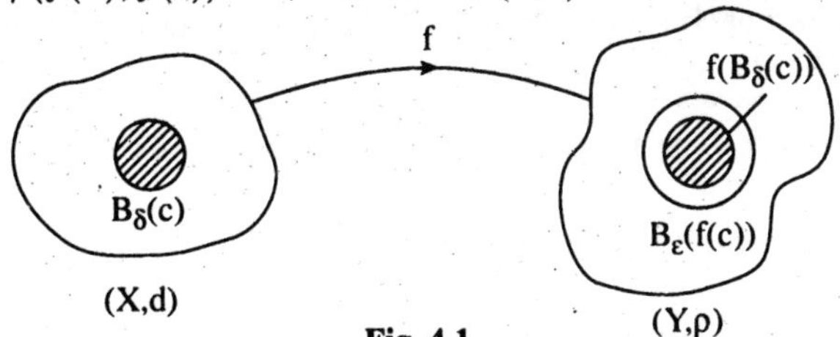

Fig. 4.1

or, equivalently $f(x) \in B_\epsilon(f(c))$, whenever $x \in B_\delta(c)$

or, equivalently $f(B_\delta(c)) \subset B_\epsilon(f(c))$.

If a function f is continuous at each point of (X, d), then f is said to be continuous on X.

Example [4.1]: (i) Let (X, d) be a metric space. Consider the identity mapping $I : X \to X$ defined by $I(x) = x \;\forall\; x \in X$. Let $c \in X$. Let $\epsilon > 0$ be arbitrary. Then $\exists\; \delta > 0$ such that for $d(x, c) < \delta$, we have $d(I(x), I(c)) = d(x, c) < \epsilon = \delta$

$\Rightarrow I$ is continuous at $x = c$.

This is true for any $c \in X$. Hence I is continuous mapping on X.

(ii) A constant function in a metric space (X, d) is continuous.

Theorem [4.1] Let $f : (X, d) \to (Y, \rho)$ be a function and let $c \in X$. Then f is continuous at c, if and only if for any sequence $\{x_n\} \subset X$ with $x_n \to c$ in (X, d), the sequence $f(x_n) \to f(c)$ in (Y, ρ).

Proof. First suppose that f is continuous at c. Now let $\{x_n\}$ be a sequence in (X, d) such that $x_n \to c$. We now show that $f(x_n) \to f(c)$ in (Y, ρ). Let $\epsilon > 0$ be arbitrary. Using the continuity of f at c, we find a number $\delta > 0$ such that

$$\rho(f(x), f(c)) < \epsilon, \text{ whenever } d(x, c) < \delta. \tag{i}$$

Now since $x_n \to c$ in (X, d), for this δ, there exists a positive integer N such that

$$d(x_n, c) < \delta \;\forall\; n \geq N. \tag{ii}$$

From (i) and (ii), we get

$$\rho(f(x_n), f(c)) < \epsilon \;\forall\; n \geq N \Rightarrow f(x_n) \to f(c) \text{ in } (Y, \rho).$$

Conversely, let the condition holds. Then for any sequence $\{x_n\}$ in (X, d) with $x_n \to c$ we have $f(x_n) \to f(c)$ in (Y, ρ). We now show that f is continuous at c. If possible, suppose that f is **not** continuous at c. So there is a positive $\epsilon = \epsilon_0$ (say) such that for any $\delta > 0$, there is a point $x_\delta \in X$ with $d(x_\delta, c) < \delta$ but $\rho(f(x_\delta), f(c)) \geq \epsilon_0$.

Taking $\delta = 1, \frac{1}{2}, \frac{1}{3}, \ldots, \frac{1}{n}, \ldots$ we find a sequence $\{x_n\}$ in (X, d) such that $d(x_n, c) < \frac{1}{n}$ but $\rho(f(x_n), f(c)) \geq \epsilon_0$.

This shows that $x_n \to c$ in (X, d) but $f(x_n) \not\to f(c)$ in (Y, ρ)—which is a contradiction to the assumption. This establishes our claim.

The proof of the theorem is complete.

Corollary [4.1]: Let $f : (X, d) \to (Y, \rho)$ be a mapping. Then f is continuous on X, if and only if to each $x \in X$ and each sequence $\{x_n\} \subset X$ converging to x, the sequence $\{f(x_n)\} \subset Y$ converges to $f(x)$ in (Y, ρ).

Example [4.2]: Let $f : \mathbb{R} \to \mathbb{R}$ be the function given by

$$f(x) = \begin{cases} 1, & x \text{ is rational} \\ 0, & x \text{ is irrational.} \end{cases}$$

Then f is discontinuous at every point of $\mathbb{R}$.

Solution. Here $\mathbb{R}$ is equipped with the usual metric. It is also to be noted that $\bar{Q} = \mathbb{R}$ and $\bar{I} = \mathbb{R}$.

Let $c \in \mathbb{R}$ be a rational number. Then $c \in R = \bar{I}$. Then by using Theorem 2.18 of Chap. 2, we find a sequence $\{x_n\}$ of irrational numbers [i.e., $\{x_n\} \subset I$] such that $x_n \to c$. If f is continuous at $x = c$, then by the above theorem we get $f(x_n) \to f(c)$ as $n \to \infty$.

Note that $x_n \in I \ \forall \ n$, so $f(x_n) = 0 \ \forall \ n$, but since $c \in Q$, we have $f(c) = 1$. Hence, $\{0, 0, \cdots\} \to 1$—a contradiction. This shows that f is **not** continuous at $c \in Q$. Since $c \in Q$ was taken arbitrarily, f is **not** continuous at each point of Q.

In a similar way we can show that f is **not** continuous at each point of I.

Hence f is discontinuous at each point of $\mathbb{R}$.

Remark. The function defined in this example is known as Dirichlet function.

Problem [4.1] : Let us consider $(\mathbb{R}, d)$ where $d(x, y) = |x - y| \ \forall \ x, y \in \mathbb{R}$. Let $f : \mathbb{R} \to \mathbb{R}$ be a continuous function. If $f(x) = 0 \ \forall \ x \in Q$, the set of all rational numbers, then

$$f(x) = 0 \ \forall \ x \in \mathbb{R}.$$

Solution. Let c be an irrational number, i.e., $c \in I$. Then as in the above example, we can find a sequence $\{x_n\}$ of rationals such that $x_n \to c$ as $n \to \infty$. Since f is a continuous function, then by the Theorem 4.1, we get $f(x_n) \to f(c)$ as $n \to \infty$. Since $x_n \in Q \ \forall \ n$, by the given condition $f(x_n) = 0 \ \forall \ n$. Thus we must have $f(x_n) \to 0$ as $n \to \infty$. Consequently, $f(c) = 0$.

Since $c \in I$ was taken arbitrary, we have $f(x) = 0 \ \forall \ x \in I$ and hence $f(x) = 0 \ \forall \ x \in \mathbb{R}$.

Problem [4.2] : Let (X, d) be a metric space and let D be a dense subset of X. Let $f : (X, d) \to (Y, \rho)$ be a continuous function such that $f(x) = 0 \ \forall \ x \in D$. Prove that $f \equiv 0$ in X.

Remark. Theorem 4.1 is very useful to examine whether a function f is continuous or not. However, if f is continuous, then it is **not** in general true that f preserve the Cauchyness of a sequence.

For example, let us consider the function $f : ((0, 1], d_u) \to (\mathbb{R}, d_u)$, where d_u is the usual metric on $\mathbb{R}$, defined by $f(x) = \dfrac{1}{x} \ \forall \ x \in (0, 1] = A$.

Then f is a continuous function on $A \subseteq \mathbb{R}$. Consider the sequence $\left\{\frac{1}{n} : n \in \mathbb{N}\right\} \subseteq A$. Then $\left\{\frac{1}{n}\right\}$ is a Cauchy sequence in A. But $\left\{f\left(\frac{1}{n}\right) : n \in \mathbb{N}\right\} = \{n : n \in \mathbb{N}\}$ is **not** a Cauchy sequence in $(\mathbb{R}, d_u)$.

Theorem [4.2] Let (X, d) and (Y, ρ) be two metric spaces and $f : (X, d) \to (Y, \rho)$ be a function. Then f is continuous, if and only if $f^{-1}(G)$ is an open set in (X, d) for any open set G in (Y, ρ).

Proof. Suppose f is continuous on X. Let G be an open set in (Y, ρ). We now show that $f^{-1}(G)$ is an open set in (X, d). Let us take a point $p \in f^{-1}(G)$. Then $f(p) \in G$. Since G is an open set, it follows that $f(p)$ is an interior point of G. So, for some $\epsilon > 0$, $B_\epsilon(f(p)) \subset G$. Now since f is continuous, for this ϵ we can find a $\delta > 0$ such that $f(B_\delta(p)) \subset B_\epsilon(f(p)) \subset G$

$\Rightarrow f(B_\delta(p)) \subset G$

$\Rightarrow p \in B_\delta(p) \subset f^{-1}(G)$

$\Rightarrow p$ is an interior point of $f^{-1}(G)$

$\Rightarrow f^{-1}(G)$ is an open set in (X, d), since $p \in f^{-1}(G)$ is arbitrary.

Conversely, suppose that the condition holds, i.e., $f^{-1}(G)$ is an open set in (X, d) for any open set G in (Y, ρ).

We now prove that f is a continuous function on X. Let $c \in X$ and let $\epsilon > 0$ be arbitrary. Then $B_\epsilon(f(c))$ is an open set in (Y, ρ). Then by assumption $f^{-1}(B_\epsilon(f(c)))$ is an open set in

(X,d) and $c \in f^{-1}(B_\epsilon(f(c)))$. Then c is an interior point of $f^{-1}(B_\epsilon(f(c)))$. So we can find $\delta > 0$ such that $c \in B_\delta(c) \subset f^{-1}(B_\epsilon(f(c)))$

$\Rightarrow f(B_\delta(c)) \subset B_\epsilon(f(c))$

$\Rightarrow f$ is continuous at $c \in X$

$\Rightarrow f$ is continuous on X since $c \in X$ is arbitrary.

Remark. The theorem can be stated as "A function $f : (X,d) \to (Y,\rho)$ is continuous, if and only if inverse image of an open set in Y is an open set in X".

It is very important to note that if $f : (X,d) \to (Y,\rho)$ be a continuous function and $A \subset X$ is an open set, then $f(A)$ may **not** be an open set in (Y,ρ).

For example, consider the function $f : (\mathbb{R},d) \to (\mathbb{R},d_u)$ defined by $f(x) = x \; \forall \; x \in \mathbb{R}$, where d is the trivial metric, d_u denotes the usual metric on $\mathbb{R}$. Consider an open set G in $(\mathbb{R},d_u)$. Then $f^{-1}(G)$ is obviously an open set in $(\mathbb{R},d)$, since every subset of the discrete metric space $(\mathbb{R},d)$ is open. This shows that f is continuous.

On the other hand, take $A = \{x\}$ in $(\mathbb{R},d)$. Then A is an open set in $(\mathbb{R},d)$. But $f(A) = \{x\}$ is **not** an open set in $(\mathbb{R},d_u)$.

Example [4.3] : Let $f : \mathbb{R} \to \mathbb{R}$ be the constant function. Then f is a continuous function.

Let $f : \mathbb{R} \to \mathbb{R}$ be defined by $f(x) = k \; \forall \; x \in \mathbb{R}$, where k is a constant. Let G be an open set in $\mathbb{R}$. Then

$$f^{-1}(G) = \begin{cases} \mathbb{R}, & \text{if } k \in G \\ \phi, & \text{if } k \notin G. \end{cases}$$

Since ϕ and $\mathbb{R}$ are open, $f^{-1}(G)$ is an open set.

$\Rightarrow f$ is continuous.

Theorem [4.3] Let $f : (X,d) \to (Y,\rho)$ be a function. Then f is continuous, if and only if $f^{-1}(F)$ is a closed set in (X,d) for any closed set F in (Y,ρ).

Proof. First suppose that f is continuous and let F be a closed set in (Y,ρ). Then F^c is an open set in (Y,ρ) [see Theorem 1.11]. Since f is continuous, so by the above theorem, $f^{-1}(F^c)$ is an open set in (X,d).

Note that $f^{-1}(F^c) = [f^{-1}(F)]^c$. Hence $f^{-1}(F)$ is a closed set in (X,d).

Conversely, let $f^{-1}(F)$ is a closed set in (X,d) for any closed set F in (Y,ρ). We now show that f is continuous. Let G be an open set in (Y,ρ). Then G^c is a closed set in (Y,ρ). Thus by assumption, $f^{-1}(G^c)$ is a closed set in (X,d), i.e., $[f^{-1}(G)]^c$ is a closed set in (X,d). Hence, $f^{-1}(G)$ is an open set in (X,d). Consequently, f is continuous on X.

Theorem [4.4] A function $f : (X,d) \to (Y,\rho)$ is continuous, if and only $f(\bar{A}) \subset \overline{f(A)}$, for every subset A of X.

Proof. Suppose that f is continuous and let $A \subset X$. Since $\overline{f(A)}$ is a closed set in (Y,ρ), so by the above theorem $f^{-1}(\overline{f(A)})$ is a closed set in (X,d). Note that $f(A) \subset \overline{f(A)}$.

$\Rightarrow f^{-1}(f(A)) \subset f^{-1}(\overline{f(A)})$ and $A \subset f^{-1}(f(A))$

$\Rightarrow A \subset f^{-1}(\overline{f(A)})$

$\Rightarrow \bar{A} \subset \overline{f^{-1}(\overline{f(A)})} = f^{-1}(\overline{f(A)})$, since $f^{-1}(\overline{f(A)})$ is a closed set

$\Rightarrow f(\bar{A}) \subset \overline{f(A)}$.

Conversely, suppose that the condition holds, i.e., $f(\bar{A}) \subset \overline{f(A)}$ for any $A \subset X$.

Let F be a closed set in (Y, ρ). We now show that $f^{-1}(F)$ is a closed set in (X, d). Note that $f^{-1}(F)$ is a subset of X. Let us now apply the assumption taking $A = f^{-1}(F)$. This gives

$f(\overline{f^{-1}(F)}) \subset \overline{f(f^{-1}(F))} \subset \bar{F}$, since $f(f^{-1}(S)) \subset S, S \subset Y$

$\Rightarrow f(\overline{f^{-1}(F)}) \subset F$, since F is a closed set

$\Rightarrow \overline{f^{-1}(F)} \subset f^{-1}(F)$

$\Rightarrow f^{-1}(F) = \overline{f^{-1}(F)}$, since we always have $f^{-1}(F) \subset \overline{f^{-1}(F)}$

$\Rightarrow f^{-1}(F)$ is a closed set in (X, d)

$\Rightarrow f$ is continuous.

Theorem [4.5] A function $f : (X, d) \to (Y, \rho)$ is continuous, if and only if $\overline{f^{-1}(B)} \subset f^{-1}(\bar{B})$, for any subset B of Y.

Proof. First suppose that f is continuous. Let $B \subset Y$. Then $\bar{B}$ is a closed subset of Y. Hence $f^{-1}(\bar{B})$ is a closed subset in (X, d).

Since $B \subset \bar{B}$, it follows that $f^{-1}(B) \subset f^{-1}(\bar{B})$

$\Rightarrow \overline{f^{-1}(B)} \subset \overline{f^{-1}(\bar{B})} = f^{-1}(\bar{B})$, since $f^{-1}(\bar{B})$ is a closed set in (X, d)

$\Rightarrow \overline{f^{-1}(B)} \subset f^{-1}(\bar{B})$.

Conversely, suppose that the condition holds, i.e., $\overline{f^{-1}(B)} \subset f^{-1}(\bar{B})$ for every subset B of Y.

Let G be an open set in (Y, ρ). Then G^c is a closed set in (Y, ρ). Take $B = G^c$. From the given condition, $\overline{f^{-1}(B)} \subset f^{-1}(\bar{B}) = f^{-1}(B)$

$\Rightarrow f^{-1}(B) = \overline{f^{-1}(B)}$

$\Rightarrow f^{-1}(B)$ is a closed set in (X, d)

$\Rightarrow f^{-1}(G^c) = [f^{-1}(G)]^c$ is a closed set in (X, d)

$\Rightarrow f^{-1}(G)$ is an open set in (X, d)

$\Rightarrow f$ is continuous.

Theorem [4.6] A function $f : (X, d) \to (Y, \rho)$ is continuous, if and only if $f^{-1}(B^0) \subset (f^{-1}(B))^0$, where $B^0 = \text{int}\,(B)$, for $B \subset Y$.

Proof. First assume that f is continuous. Let $B \subset Y$. Then B^0 is an open set in (Y, ρ). So $f^{-1}(B^0)$ is an open set in (X, d).

We know that $B^0 \subset B \Rightarrow f^{-1}(B^0) \subset f^{-1}(B)$

$\Rightarrow (f^{-1}(B^0))^0 \subset (f^{-1}(B))^0$

$\Rightarrow f^{-1}(B^0) \subset (f^{-1}(B))^0$, since $f^{-1}(B^0) = (f^{-1}(B^0))^0$ as $f^{-1}(B^0)$ is an open set.

Conversely, suppose that the condition holds, i.e., for any subset B of Y,

$$f^{-1}(B^0) \subset (f^{-1}(B))^0.$$

Let G be an open set in (Y, ρ). Then $G = G^0$. By hypothesis,

$$f^{-1}(G) = f^{-1}(G^0) \subset (f^{-1}(G))^0.$$

We always have $(f^{-1}(G))^0 \subset f^{-1}(G)$.

Hence we have $f^{-1}(G) = (f^{-1}(G))^0$

$\Rightarrow f^{-1}(G)$ is an open set in (X, d)

$\Rightarrow f$ is continuous.

Theorem [4.7] For a function $f : (X, d) \to (Y, \rho)$, the following statements are equivalent:

(1) f is continuous.

(2) $f^{-1}(G)$ is an open set in (X, d) for any open set G in (Y, ρ).

(3) $f^{-1}(F)$ is a closed set in (X, d) for any closed set F in (Y, ρ).

(4) If $A \subset X, f(\bar{A}) \subset \overline{f(A)}$.

(5) If $B \subset Y$ $\overline{f^{-1}(B)} \subset f^{-1}(\bar{B})$.

Problem [4.3] : Let f, g be two continuous functions from a metric space (X, d) into a metric space (Y, ρ). Then $D = \{x \in X : f(x) = g(x)\}$ is a closed set in (X, d). Also show that if the set D is dense in X, then $f \equiv g$.

Solution. If $D = \phi$, then D is a closed set. Now suppose that $D \neq \phi$. To prove that D is a closed set in (X, d), it is sufficient to show that D^c is an open set in (X, d). Let us first recall the Hausdorff property as defined in Theorem 1.2.1.

"Let (X, d) be a metric space and let $\alpha, \beta \in X$, $\alpha \neq \beta$. Then we can find two open balls in (X, d) separating α and β."

Let $p \in D^c \Rightarrow p \notin D = \{x \in X : f(x) = g(x)\}$

$$\Rightarrow f(p) \neq g(p) \text{ in } (Y, \rho).$$

So by Hausdorff property we can find a positive real number ϵ such that

$$B_\epsilon(f(p)) \cap B_\epsilon(g(p)) = \phi.$$

Since f is continuous at p, we can find $\delta_1 > 0$ such that $f(B_{\delta_1}(p)) \subset B_\epsilon(f(p))$.

Similarly, there exists a $\delta_2 > 0$ such that $g(B_{\delta_2}(p)) \subset B_\epsilon(g(p))$, by continuity of g.

Taking $\delta = \min\{\delta_1, \delta_2\}$, we see that

$$f(B_\delta(p)) \subset B_\epsilon(f(p)) \text{ and } g(B_\delta(p)) \subset B_\epsilon(g(p)).$$

This shows that if $x \in B_\delta(p)$, then $f(x) \neq g(x)$. So we get $p \in B_\delta(p) \subset D^c$

$\Rightarrow p$ is an interior point of D^c

$\Rightarrow D^c$ is an open set in (X, d)

$\Rightarrow D$ is a closed set in (X, d).

Alternative

Let us take a point $x \in \bar{D}$. Then by Theorem 2.18 we find a sequence $\{x_n\} \subset D$ such that $x_n \to x$ as $n \to \infty$. Since f, g are continuous functions, it follows that $f(x_n) \to f(x)$ and $g(x_n) \to g(x)$ as $n \to \infty$. Since $\{x_n\} \subset D$, it follows that $f(x_n) = g(x_n)\ \forall\ n$

$\Rightarrow \lim_n f(x_n) = \lim_n g(x_n)$

$\Rightarrow f(x) = g(x)$

$\Rightarrow x \in D$.

Hence D is a closed set.

2nd part. Suppose that D is dense in (X,d). Then we have $\bar{D} = X \Rightarrow D = X$, since D is a closed set

$$\Rightarrow f(x) = g(x) \ \forall \ x \in X$$

$$\Rightarrow f \equiv g.$$

Problem [4.4] : Let (X_1, d_1) and (X_2, d_2) be two metric spaces and let $f : (X_1, d_1) \to (X_2, d_2)$ be a continuous function. Then graph of f is defined as

$$G = \{(x_1, x_2) \in X_1 \times X_2 : x_2 = f(x_1)\}$$
$$\text{i.e., } G = \{(x_1, f(x_1)) \in X_1 \times X_2 : x_1 \in X_1\}.$$

If d is a metric on $X_1 \times X_2$ given by $d(x, y) = \max\{d_1(x_1, y_1), d_2(x_2, y_2)\}$, $x = (x_1, x_2)$, $y = (y_1, y_2)$ are in $X_1 \times X_2$, prove that G is a closed set in $(X_1 \times X_2, d)$.

Solution. Let $(x, y) \in \bar{G}$. Then by Theorem 2.18, there exists a sequence $\{(x_n, f(x_n)\} \subset G$ such that $(x_n, f(x_n)) \to (x, y)$ as $x \to \infty$

$\Rightarrow d(x_n, f(x_n)), (x, y)) \to 0$ as $n \to \infty$

$\Rightarrow \max\{d_1(x_n, x),\ d_2(f(x_n), y)\} \to 0$ as $n \to \infty$.

Since $d_1(x_n, x) \leq \max\{d_1(x_n, x),\ d_2(f(x_n), y)\}$, it follows that $d_1(x_n, x) \to 0$.

Similarly, $d_2(f(x_n), y) \to 0$

$\Rightarrow x_n \to x$ in (X_1, d_1) and $f(x_n) \to y$ in (X_2, d_2).

Since f is continuous and $x_n \to x$ in (X_1, d_1), we have $f(x_n) \to f(x)$ in (X_2, d_2).

Now $f(x_n) \to f(x)$ and $f(x_n) \to y$ in (X_2, d_2) gives $y = f(x)$ and hence

$$(x, y) = (x, f(x)) \in G.$$

$\Rightarrow G$ is a closed set in $(X_1 \times X_2, d)$.

Theorem [4.8] Let (X, d), (Y, ρ) and (Z, σ) be three metric spaces and let $f : (X, d) \to (Y, \rho)$ and $g : (Y, \rho) \to (Z, \sigma)$ be two continuous functions. Then $g \circ f : (X, d) \to (Z, \sigma)$ is also continuous.

Proof. Let G be an open set in (Z, σ). Then by using continuity of g we say that $g^{-1}(G)$ is an open set in (Y, ρ). Now by continuity of f we get $f^{-1}(g^{-1}(G))$ is an open set in (X, d)

$\Rightarrow (g \circ f)^{-1}(G)$ is an open set in (X, d)

$\Rightarrow g \circ f$ is continuous.

Problem [4.5] : Prove that a function $f : (X, d) \to (Y, \rho)$, where (X, d) is the discrete metric space, is continuous.

Solution. Let us first recall that in the discrete metric space every subset is open.

Let G be an open set in (Y, ρ). Then $f^{-1}(G) \subseteq X$ and hence $f^{-1}(G)$ is an open set in (X, d). This is true for any open set G in (Y, ρ). Thus f is a continuous function on X.

Problem [4.6] : Consider the metric space $(\mathbb{R}, d)$, where $d(x, y) = |x - y| \ \forall \ x, y \in \mathbb{R}$. Define a function $f : (\mathbb{R}, d) \to (\mathbb{R}, d)$ by

$$f(x) = \begin{cases} -5, & x < 0 \\ 5 & x \geq 0. \end{cases}$$

Then f is **not** continuous on $\mathbb{R}$.

Solution. Consider the open set $G = (4,6)$ in $(\mathbb{R}, d)$.

Now $f^{-1}(G) = \{x \in \mathbb{R} : x \geq 0\} = [0, \infty)$—is **not** an open set in $(\mathbb{R}, d)$.

Thus f is **not** continuous on $\mathbb{R}$.

Problem [4.7]: Let $f : (X, d) \to (Y, \rho)$ be a continuous function. Show that if $A \subset Y$ is closed, then $f^{-1}(A)$ is a closed set in (X, d).

Solution. Let x be a limit point of $f^{-1}(A)$. Then by Theorem 2.18, there exists a sequence $\{x_n\} \subset f^{-1}(A)$ such that $x_n \to x \in X$. Since $x_n \to x$ and f is continuous on X, it follows that $f(x_n) \to f(x)$. Now since $x_n \in f^{-1}(A)\ \forall\ n = 1, 2, \ldots$, it follows that $f(x_n) \in A\ \forall\ n$. So in fact, $\{f(x_n)\}$ is a sequence in A such that $f(x_n) \to f(x)$ and so $f(x) \in \bar{A} = A$, since A is a closed set. Therefore $f(x) \in A$ implies that $x \in f^{-1}(A)$. This shows that $\overline{f^{-1}(A)} \subset f^{-1}(A)$. Consequently, $f^{-1}(A) = \overline{f^{-1}(A)}$, i.e., $f^{-1}(A)$ is a closed set in (X, d).

Problem [4.8]: (1) Let $g : (\mathbb{R}, d) \to (\mathbb{R}, d)$ be a continuous function, where d denotes the usual metric on $\mathbb{R}$. Show that the function $G : \mathbb{R} \times \mathbb{R} \to \mathbb{R}$ defined by $G(x, y) = y - g(x)$ is also continuous.

(2) Deduce that $\{(x, y) \in \mathbb{R}^2 : y \geq g(x)\}$ is a closed subset of $\mathbb{R}^2$.

Solution. (1) Let $(x, y) \in \mathbb{R}^2$. Let $\{(x_n, y_n)\}$ be a sequence in $(\mathbb{R}^2, d)$, where d is the usual metric on $\mathbb{R}^2$, such that $(x_n, y_n) \to (x, y)$. Since $(x_n, y_n) \to (x, y)$, it follows that

$$d((x_n, y_n), (x, y)) = \sqrt{(x_n - x)^2 + (y_n - y)^2} \to 0 \text{ as } n \to \infty$$

$$\Rightarrow (x_n - x)^2 + (y_n - y)^2 \to 0 \text{ as } n \to \infty$$

$$\Rightarrow x_n - x \to 0 \text{ and } y_n \to y \text{ as } n \to \infty$$

$$\Rightarrow x_n \to x,\ y_n \to y \text{ as } n \to \infty.$$

Since g is continuous and $x_n \to x$, we get $g(x_n) \to g(x)$.

Hence $G(x_n, y_n) = y_n - g(x_n) \to y - g(x) = G(x, y)$.

Thus $(x_n, y_n) \to (x, y)$ as $n \to \infty$ implies that $G(x_n, y_n) \to G(x, y)$ as $n \to \infty$

$\Rightarrow G$ is continuous on $\mathbb{R}^2$.

(2) Note that

$$\begin{aligned}\{(x, y) \in \mathbb{R}^2 : y \geq g(x)\} &= \{(x, y) \in \mathbb{R}^2 : y - g(x) \geq 0\}\\ &= \{(x, y) \in \mathbb{R}^2 : G(x, y) \geq 0\}\\ &= \{(x, y) \in \mathbb{R}^2 : G(x, y) \in [0, \infty)\}\\ &= G^{-1}([0, \infty)).\end{aligned}$$

Since G is continuous and $[0, \infty)$ is a closed set in $\mathbb{R}$, it follows from Problem 4.7 that $G^{-1}([0, \infty))$ is a closed subset of $\mathbb{R}^2$, i.e., $\{(x, y) \in \mathbb{R}^2 : y \geq g(x)\}$ is a closed subset of $\mathbb{R}^2$.

Problem [4.9]: Let A be any 2×2 matrix of real numbers and regard it as a function $A : \mathbb{R}^2 \to \mathbb{R}^2$, where $\mathbb{R}^2$ has the metric d_∞ defined by $d_\infty((x_1, x_2), (y_1, y_2)) = \max\{|x_1 - y_1|, |x_2 - y_2|\}$. Show that A is a continuous function.

Solution. Let us take $A = \begin{pmatrix} a & b \\ c & d \end{pmatrix}$ where $a, b, c, d \in \mathbb{R}$. Let $(x_1, x_2), (y_1, y_2) \in \mathbb{R}^2$. Then $d_\infty(Ax; Ay) \leq 2M d_\infty(x, y)$ [verify], where $M = \max\{|a|, |b|, |c|, |d|\}$

$$Ax = \begin{pmatrix} a & b \\ c & d \end{pmatrix} \begin{pmatrix} x_1 \\ x_2 \end{pmatrix} = \begin{pmatrix} ax_1 + bx_2 \\ cx_1 + dx_2 \end{pmatrix}.$$

Let $\epsilon > 0$ be arbitrary. Set $\delta = \dfrac{\epsilon}{2M}$ so that when $d_\infty(x, y) < \delta$, we have

$$d_\infty(Ax, Ay) \leq 2Md_\infty(x, y) < 2M\delta = 2M \cdot \frac{\epsilon}{2M} = \epsilon$$

$\Rightarrow d_\infty(Ax, Ay) < \epsilon$

$\Rightarrow A$ is continuous on $\mathbb{R}^2$.

Theorem [4.9] Let (X, d) be a metric space and $A \subset X$. Consider the function $f : X \to \mathbb{R}$ given by $f(x) = \text{dist}\,(x, A)\ \forall\ x \in X$, where $\mathbb{R}$ is equipped with the usual metric. Then f is a continuous function on X.

Proof. Let $x \in X$. We have from Theorem 2.5, $|\,\text{dist}\,(x, A) - \text{dist}\,(y, A)| \leq d(x, y)$ for any x, $y \in X$.

Let $\epsilon > 0$ be arbitrary. Then for $0 < \delta \leq \epsilon$ and $y \in X$ with $d(x, y) < \delta$ we have

$$|\,\text{dist}\,(x, A) - \text{dist}\,(y, A)| \leq d(x, y) < \delta \leq \epsilon$$

$\Rightarrow |f(x) - f(y)| < \epsilon$.

Thus the function f is continuous on X.

Alternative

Let $x \in X$ and $\{x_n\}$ be a sequence in X such that $x_n \to x$ in (X, d), i.e., $d(x_n, x) \to 0$ as $n \to \infty$.

Now $|\,\text{dist}\,(x_n, A) - \text{dist}\,(x, A)| \leq d(x_n, x) \to 0$ as $n \to \infty$.

$\Rightarrow \text{dist}\,(x_n, A) \to \text{dist}\,(x, A)$

$\Rightarrow f(x_n) \to f(x)$ in $\mathbb{R}$.

$\Rightarrow f$ is continuous on X.

Problem [4.10] : Let (X, d) be a metric space and let $x_0 \in X$. Show that the function $f : X \to \mathbb{R}$ defined by $f(x) = d(x, x_0)\ \forall\ x \in X$ is continuous.

Solution. Let $x, y \in X$. Then $d(x, x_0) \leq d(x, y) + d(y, x_0)$

$$\Rightarrow d(x, x_0) - d(y, x_0) \leq d(x, y). \tag{1}$$

Interchanging the role of x, y in (1), we get

$$d(y, x_0) - d(x, x_0) \leq d(y, x) = d(x, y). \tag{2}$$

Combining (1) and (2), we get

$$|d(x, x_0) - d(y, x_0)| \leq d(x, y). \tag{3}$$

Let $a \in X$ and let $\epsilon > 0$ be arbitrary. Choose $\delta > 0$ such that $\delta \leq \epsilon$ and $d(x, a) < \delta$, we get

$$|d(x, x_0) - d(a, x_0)| \leq d(x, a) < \delta \leq \epsilon$$

$\Rightarrow |f(x) - f(a)| < \epsilon$

$\Rightarrow f$ is continuous at a

$\Rightarrow f$ is continuous on X since $a \in X$ is arbitrary.

Theorem [4.10] Let (X, d) be a metric space and F_1, F_2 be two non-empty disjoint closed sets in (X, d). Then there exists a continuous function $f : X \to [0, 1]$ such that $f(x) = 0\ \forall\ x \in F_1$ and $f(x) = 1\ \forall\ x \in F_2$, i.e., $f(F_1) = \{0\}, f(F_2) = \{1\}$.

Proof. We first recall the Theorem 2.18. "Let (X, d) be a metric space and $A \subset X$. Then for any $x \in X$, $\text{dist}\,(x, A) = 0$, if and only if $x \in \bar{A}$".

Keeping this in mind, it is easy to say that $\text{dist}\,(x, F_1) + \text{dist}\,(x, F_2) > 0 \;\forall\; x \in X$, since

$$F_1 \cap F_2 = \phi.$$

Also note that $\dfrac{\text{dist}\,(x, F_1)}{\text{dist}\,(x, F_1) + \text{dist}\,(x, F_2)} \in [0, 1]$, for any $x \in X$.

We now define a function $f : X \to [0, 1]$ by

$$f(x) = \frac{\text{dist}\,(x, F_1)}{\text{dist}\,(x, F_1) + \text{dist}\,(x, F_2)}, \;\forall\; x \in X.$$

Using Theorem 4.9, we can say that f is a continuous function on X. From the definition of f it is clear that

$$f(x) = \begin{cases} 0, & \text{if } x \in F_1 \\ 1, & \text{if } x \in F_2. \end{cases}$$

Note. This theorem is known as **Urysohn's Lemma**.

Example [4.4]: Let (X, d) be a metric space and F_1, F_2 be two non-empty disjoint closed sets in (X, d). Then there exists a continuous function $f : X \to [-1, 1]$ such that $f(F_1) = \{-1\}$ and $f(F_2) = \{1\}$.

[**Hints.** Consider the function $f : X \to [-1, 1]$ by $f(x) = \dfrac{\text{dist}\,(x, F_1) - \text{dist}\,(x, F_2)}{\text{dist}\,(x, F_1) + \text{dist}\,(x, F_2)} \;\forall\; x \in X$.]

4.1 Uniform Continuity

Let us first recall the definition of continuity of a function $f : (X, d) \to (Y, \rho)$. f is said to be continuous at $x = a$, if for any $\epsilon > 0$, there exists $\delta > 0$ such that

$$\rho(f(x), f(a)) < \epsilon, \text{ whenever } d(x, a) < \delta.$$

In this definition, the choice of δ depends not only on ϵ but also on the point a. Thus it is natural to ask whether there is any $\delta > 0$ which is valid for all $a \in X$, i.e., δ depends only on ϵ. If such δ exists, then the function is called uniformly continuous. Let us now give the formal definition of uniform continuity.

Definition [4.2]: Let $f : (X, d) \to (Y, \rho)$ be a function. Then f is said to be uniformly continuous on X, if for given $\epsilon > 0$, there exists $\delta > 0$ such that

$$\rho(f(x), f(y)) < \epsilon, \text{ whenever } d(x, y) < \delta \;\forall\; x, y \in X$$

From the definition above, it is clear that a uniformly continuous function is always continuous. But the converse is **not** true, in general.

Example [4.5]: Consider the metric space $(\mathbb{R}, d)$, $d(x, y) = |x - y| \;\forall\; x, y \in \mathbb{R}$. Consider the function $f : \mathbb{R} \to \mathbb{R}$ by $f(x) = x^2 \;\forall\; x \in \mathbb{R}$. It is very easy to check that f is continuous on $\mathbb{R}$. We now show that f is **not** uniformly continuous on $\mathbb{R}$.

If possible, suppose that f is uniformly continuous on $\mathbb{R}$. Then for $\epsilon = 2$, there exists $\delta > 0$ such that

$$|f(x) - f(a)| < 2, \text{ whenever } |x - a| < \delta \;\forall\; x, a \in \mathbb{R}.$$

We have $f(x) - f(a) = x^2 - a^2 = (x + a)(x - a)$. For $a > 0$ and taking $x = a + \delta/2$, we get

$$|f(x) - f(a)| = |x - a| \cdot |x + a| = \frac{\delta}{2} \cdot \left|2a + \frac{\delta}{2}\right| < 2.$$

This implies that $a\delta < 2$ for all $a > 0$—this is a contradiction [Since δ is small, we neglect $\delta^2/4$].

However, f is uniformly continuous on $A = (0,1]$. First note that

$$|f(x) - f(a)| = |x^2 - a^2| = |x - a| \cdot |x + a| < 2|x - a| \ \forall \ x, a \in A.$$

So if $|x - a| < \delta$, then $|f(x) - f(a)| < 2\delta$. Hence, if $\epsilon > 0$ is given, then we need only to take $\delta \leq \epsilon/2$ so that $|f(x) - f(a)| < \epsilon$.

This shows that f is uniformly continuous on A.

Note. (1) Let $f : (X, d) \to (Y, \rho)$ be a function and let $A \subseteq X$. Then the function $f_A : (A, d_A) \to (Y, \rho)$ defined by $f_A(x) = f(x) \ \forall \ x \in A$, is called restriction of f on A.

On the other hand, f is called extension of f_A from A to X.

(2) Example 4.5 shows that f is not uniformly continuous but its restriction f_A is uniformly continuous.

Example [4.6] : The mapping $f : (0,1] \to \mathbb{R}$ defined by $f(x) = \dfrac{1}{x} \ \forall \ x \in [0,1]$ is **not** uniformly continuous.

Suppose, if possible, f is uniformly continuous on $(0,1]$. Then for given any $\epsilon > 0$, there exists $\delta > 0$ such that

$$\left|\frac{1}{x} - \frac{1}{y}\right| < \epsilon, \text{ whenever } |x - y| < \delta \ \forall \ x, y \in [0,1]. \tag{i}$$

Choose $y = \delta$ itself, then $(y - \delta, y + \delta)$ becomes $(0, 2\delta)$ and $\left|\dfrac{1}{x} - \dfrac{1}{y}\right| < \epsilon$

$$\Rightarrow \left|\frac{1}{x} - \frac{1}{\delta}\right| < \epsilon \ \forall \ x \in (0, 2\delta).$$

But $\dfrac{1}{x} - \dfrac{1}{\delta} \to \infty$ as $x \to 0$—a contradiction.

Thus f is **not** uniformly continuous on $[0,1]$.

Note. We have seen in Theorem 4.1 that a continuous funciton sends a convergent sequence from a metric space (X, d) to a convergent sequence in another metric space (Y, ρ). But this result is not true, if we replace "convergent sequences" by "Cauchy sequence", that means if $f : (X, d) \to (Y, \rho)$ is a continuous function and $\{x_n\}$ is a Cauchy sequence in (X, d), then $\{f(x_n)\}$ may **not** be a Cauchy sequence in (Y, ρ).

For example, consider the function $f : (0,1] \to \mathbb{R}$ defined by $f(x) = \dfrac{1}{x} \ \forall \ x \in (0,1]$, where $\mathbb{R}$ is equipped with the usual metric.

Note that the sequence $\left\{\dfrac{1}{n}\right\}_{n \in \mathbb{N}}$ is a Cauchy sequence in $((0,1], d)$. Also note that f is a continuous function in $(0,1]$. But $\left\{f\left(\dfrac{1}{n}\right)\right\} = \{n\}$ is **not** a Cauchy sequence in $(\mathbb{R}, d)$.

However, we have the following theorem:

Theorem [4.11] Let $f : (X, d) \to (Y, \rho)$ be a uniformly continuous function and $\{x_n\}$ be a Cauchy sequence in (X, d). Then $\{f(x_n)\}$ is a Cauchy sequence in (Y, ρ).

Proof. Let $\{x_n\}$ be a Cauchy sequence in (X, d). Since $f : (X, d) \to (Y, \rho)$ is uniformly continuous, for $\epsilon > 0$, there exists a $\delta > 0$ such that

$$\rho(f(x), f(y)) < \epsilon, \text{ whenever } d(x, y) < \delta \ \forall \ x, y \in X. \quad \text{(i)}$$

Since $\{x_n\}$ is a Cauchy sequence in (X, d), for $\delta > 0$, we can find a natural number N such that

$$d(x_n, x_m) < \delta, \text{ for all } m, n \geq N. \quad \text{(ii)}$$

From (i) and (ii), we get $\rho(f(x_n), f(x_m)) < \epsilon$, for all $m, n \geq N$

$\Rightarrow \{f(x_n)\}$ is a Cauchy sequence in (Y, ρ).

Note. (1) From this theorem it follows that $f : (0, 1] \to \mathbb{R}$ defined by $f(x) = \dfrac{1}{x}$ is **not** uniformly continuous.

(2) However, the uniformly continuous image of a complete metric space need not be complete. For example, consider the metric spaces $(\mathbb{R}, d)$ and $((-1, 1), d)$.

Define a function $f : \mathbb{R} \to (-1, 1)$ by $f(x) = \dfrac{x}{1 + |x|} \ \forall \ x \in \mathbb{R}$.

Then f is a uniformly continuous functions on $\mathbb{R}$ and $f(\mathbb{R}) = (-1, 1)$ but $(-1, 1)$ is not complete.

(3) We will prove later that completeness is preserved under isometry.

Theorem [4.12] [Sequential criterion of uniform continuity]
A function $f : (X, d) \to (Y, \rho)$ is uniformly continuous, iff for any two sequences $\{x_n\}, \{y_n\}$ in (X, d) with $d(x_n, y_n) \to 0$ as $n \to \infty$, we have $\rho(f(x_n), f(y_n)) \to 0$ in (Y, ρ).

Proof. Let us first suppose that f be uniformly continuous on X. Then for any arbitrary $\epsilon > 0$, there is a $\delta > 0$ such that

$$\rho(f(x)), f(y)) < \epsilon, \text{ whenever } d(x, y) < \delta \ \forall \ x, y \in X.$$

Now let $\{x_n\}$ and $\{y_n\}$ be two sequences in (X, d) such that $d(x_n, y_n) \to 0$ as $n \to \infty$. Then for $\delta > 0$, there exists $N \in \mathbb{N}$ such that $d(x_n, y_n) < \delta \ \forall, \ n \geq N$.

Hence for all $n \geq N$, we get $\rho(f(x_n), f(y_n)) < \epsilon \ \forall, \ n \geq N$

$\Rightarrow \rho(f(x_n), f(y_n)) \to 0$ as $n \to \infty$.

Conversely, let the condition holds, i.e., for any two sequences $\{x_n\}, \{y_n\}$ in (X, d) with $d(x_n, y_n) \to 0$ as $n \to \infty$, $\rho(f(x_n), f(y_n)) \to 0$ as $n \to \infty$. Suppose that f fails to be uniformly continuous on X. Then there exists $\epsilon > 0$ such that for any $\delta > 0$ we can find two exceptional points x_δ, y_δ in X such that $d(x_\delta, y_\delta) < \delta$ but $\rho(f(x_\delta), f(y_\delta)) \geq \epsilon$.

Taking $\delta = 1, \dfrac{1}{2}, \dfrac{1}{3}, \ldots, \dfrac{1}{n}, \ldots$ we actually get two sequences $\{x_n\}, \{y_n\}$ in (X, d) with $d(x_n, y_n) < \dfrac{1}{n}$ but $\rho(f(x_n), f(y_n)) \geq \epsilon \ \forall \ n$

$\Rightarrow d(x_n, y_n) \to 0$ as $n \to \infty$ but $\rho(f(x_n), f(y_n)) \not\to 0$—which is a contradiction to the assumption.

Hence the theorem follows.

Example [4.7] : Consider the function $f : (0, 1] \to \mathbb{R}$ defined by $f(x) = \dfrac{1}{x} \ \forall, \ x \in (0, 1]$. We have already shown that f is **not** uniformly continuous on $(0, 1]$. We now use the above theorem to show the same.

Take two sequences $\{x_n\} = \left\{\frac{1}{2n}\right\}$ and $\{y_n\} = \left\{\frac{1}{2n+1}\right\}$ in $(0,1]$. Then note that

$$|x_n - y_n| = \left|\frac{1}{2n} - \frac{1}{2n+1}\right| = \frac{1}{2n(2n+1)} \to 0 \text{ as } n \to \infty.$$

But $|f(x_n) - f(y_n)| = |2n - (2n+1)| = 1 \not\to 0$ as $n \to \infty$

$\Rightarrow f$ is **not** uniformly continuous on $(0,1]$.

Example [4.8]: Define $f : \mathbb{R} \to \mathbb{R}$ by $f(x) = x^2$, where $\mathbb{R}$ is equipped with the usual metric d.

Take two sequences $\{n\}$ and $\left\{n + \frac{1}{n}\right\}$ in $(\mathbb{R}, d)$.

Then we see that $\left|n - \left(n + \frac{1}{n}\right)\right| = \frac{1}{n} \to 0$ as $n \to \infty$.

But $\left|f(n) - f\left(n + \frac{1}{n}\right)\right| = \left|n^2 - n^2 - 2 - \frac{1}{n^2}\right| = 2 + \frac{1}{n^2} \to 2 \neq 0$ as $n \to \infty$.

$\Rightarrow f$ is **not** uniformly continuous on $\mathbb{R}$.

Example [4.9]: The function $f : (0,1) \to \mathbb{R}$ defined by $f(x) = \sin\frac{1}{x}$ is not uniformly continuous.

Let us take two sequences $\{x_n\}, \{y_n\}$ in $(0,1)$, where

$$x_n = \frac{1}{n\pi} \text{ and } y_n = \frac{2}{(2n+1)\pi}, \quad n \in \mathbb{N}.$$

Then we have

$$|x_n - y_n| = \left|\frac{1}{n\pi} - \frac{2}{(2n+1)\pi}\right| = \left|\frac{2n\pi + \pi - 2n\pi}{n\pi(2n+1)\pi}\right| = \frac{1}{n(2n+1)\pi} \to 0 \text{ as } n \to \infty.$$

But

$$|f(x_n) - f(y_n)| = \left|\sin(n\pi) - \sin\frac{(2n+1)\pi}{2}\right| = 1 \not\to 0 \text{ as } n \to \infty.$$

$\Rightarrow f$ is **not** uniformly continuous in $(0,1)$.

Example [4.10]: Let $\alpha, \beta \in \mathbb{R}$. Consider the function $f : \mathbb{R} \to \mathbb{R}$ defined by $f(x) = \alpha x + \beta$, for all $x \in \mathbb{R}$. Then f is uniformly continuous on $\mathbb{R}$.

If $\alpha = 0$, then $f(x) = \beta \; \forall \; x \in \mathbb{R}$ is a constant function and for $\epsilon > 0$, we set

$$|f(x) - f(y)| = |\beta - \beta| = 0 < \epsilon, \text{ for all } x, y \in \mathbb{R}.$$

Suppose that $\alpha \neq 0$.

For $\epsilon > 0$ is given, we set $0 < \delta < \frac{\epsilon}{|\alpha|}$.

Then for all $x, y \in \mathbb{R}$ with $|x - y| < \delta$, we have

$$|f(x) - f(y)| = |(\alpha x + \beta) - (\alpha y + \beta)| = |\alpha||x - y| < |\alpha| \cdot \frac{\epsilon}{|\alpha|} = \epsilon$$

$\Rightarrow f$ is uniformly continuous on $\mathbb{R}$.

Example [4.11] : Let us consider the function $f:\mathbb{R}\to\mathbb{R}$ by $f(x)=\sin x$, for all $x\in\mathbb{R}$. Let $x,y\in\mathbb{R}$. Then

$$\begin{aligned}|f(x)-f(y)| &= |\sin x-\sin y|\\ &= \left|2\cos\frac{x+y}{2}\sin\frac{x-y}{2}\right| = 2\left|\cos\frac{x+y}{2}\right|\left|\sin\frac{x-y}{2}\right|\\ &\le 2\cdot 1\cdot\frac{|x-y|}{2},\ \text{since } |\cos\theta|\le 1,\ |\sin\theta|\le|\theta|, \text{ for all } \theta\in\mathbb{R}\\ &= |x-y|\\ \Rightarrow |f(x)-f(y)| &\le |x-y|.\end{aligned}$$

Let $\epsilon>0$ be arbitrary. Then by setting $0<\delta\le\epsilon$, we have for all $x,y\in\mathbb{R}$ with $|x-y|<\delta$,

$$|f(x)-f(y)|\le|x-y|<\delta<\epsilon\Rightarrow|f(x)-f(y)|<\epsilon$$

$\Rightarrow f$ is uniformly continuous on $\mathbb{R}$.

Note. If we take $f(x)=\cos x$, then f is also uniformly continuous on $\mathbb{R}$.

Definition [4.3] : **[Lipschitz function]** A function $f(X,d)\to(Y,\rho)$ is called a Lipschitz function if there is a constant $M>0$ such that

$$\rho(f(x),f(y))\le Md(x,y)\ \forall,\ x,y\in X.$$

Note. A Lipschitz function $f:(X,d)\to(Y,\rho)$ is uniformly continuous.

Let $\epsilon>0$ be arbitrary.

Taking δ such that $0<\delta<\dfrac{\epsilon}{M+1}$, we get for $d(x,y)<\delta$ and $x,y\in X$,

$$\rho(f(x),f(y))\le Md(x,y)<M\cdot\frac{\epsilon}{M+1}<\epsilon$$

$\Rightarrow \rho(f(x),f(y))<\epsilon\ \forall,\ x,y\in X$

$\Rightarrow f$ is uniformly continuous.

But the converse is **not** true, in general. The function $f:[0,2]\to\mathbb{R}$ defined by $f(x)=\sqrt{x}\ \forall\ x\in[0,2]$ is uniformly continuous but f is **not** a Lipschitz function on $[0,2]$ (verify).

The functions defined in Example [4.10], Example [4.11] are Lipschitz function.

The function in note 2, Theorem [4.11] is also a Lipschitz function (Verify).

Theorem [4.13] Let (X,d) be a metric space and (Y,ρ) be a complete metric space. Let $A\subset X$ such that A is dense in X. Let $f:(A,d_A)\to(Y,\rho)$ be a uniformly continuous function. Then f can be extended uniquely to a uniformly continuous function $g:(X,d)\to(Y,\rho)$.

Proof. If $A=X$, then there is nothing to prove. Suppose that $A\ne X$. Now if $x\in A$, define $g(x)=f(x)$. Let $x\notin A$. Then since $\bar{A}=X$, we can find a sequence $\{x_n\}$ in A such that $x_n\to x$ as $n\to\infty$. Hence $\{x_n\}$ is a convergent sequence in X, and so $\{x_n\}$ is a Cauchy sequence in A. Thus by Theorem [4.11], $\{f(x_n)\}$ is a Cauchy seqeunce in (Y,ρ). Since (Y,ρ) is complete, there exists $y\in Y$ such that $f(x_n)\to y$ as $n\to\infty$. We now show that y is independent of $\{x_n\}$. Let $\{y_n\}$ be another sequence in A such that $y_n\to x$ as $n\to\infty$. Then we get

$$d(x_n,y_n)\le d(x_n,x)+d(x,y_n)\to 0 \text{ as } n\to\infty.$$

Since f is uniformly continuous on A, it follows by Theorem [4.12] that

$$\rho(f(x_n),f(y_n))\to 0 \text{ as } n\to\infty.$$

Therefore, $\rho(f(y_n), y) \le \rho(f(y_n), f(x_n)) + \rho(f(x_n), y) \to 0$ as $n \to \infty$

$$\Rightarrow f(y_n) \to y \text{ as } n \to \infty.$$

This shows that y is independent of the choice of the sequence $\{x_n\}$. Let us define

$$g(x) = y = \lim_n f(x_n),\ x \notin A.$$

Thus we get a function $g : (X, d) \to (Y, \rho)$ defined by

$$g(x) = \begin{cases} f(x), & \text{if } x \in A \\ y = \lim_n f(x_n) & \text{if } x \notin A, \text{ where } x_n \to x. \end{cases}$$

Note that g is an extension of f. We now show that g is uniformly continuous on X.

Let $\epsilon > 0$ be arbitrary. Then there exists a $\delta > 0$ such that $\rho(f(x), f(x')) < \epsilon/2$ whenever $d(x, x') < \delta\ \forall\ x, x' \in A$.

Let $x,\ y \in X$ and let $\{x_n\}$, $\{y_n\}$ be two sequences in A such that $x_n \to x$, $y_n \to y$ as $n \to \infty$. Let $\epsilon > 0$ be arbitrary. Take $\delta > 0$ be such that $d(x, y) < \frac{1}{3}\delta$.

Since $x_n \to x$, $y_n \to y$, for $\delta > 0$, there exists $N \in \mathbb{N}$ such that

$$d(x_n, x) < \frac{1}{3}\delta \text{ and } d(y_n, y) < \frac{1}{3}\delta\ \forall\ n \ge N.$$

Hence $d(x_n, y_n) \le d(x_n, x) + d(x, y) + d(y, y_n) < \frac{1}{3}\delta + \frac{1}{3}\delta + \frac{1}{3}\delta = \delta\ \forall\ n \ge N$

$$\Rightarrow \rho(f(x_n), f(y_n)) < \frac{\epsilon}{2}\ \forall\ n \ge N$$

$$\Rightarrow \rho(g(x), g(y)) = \lim_n \rho(f(x_n), f(y_n)) \le \frac{\epsilon}{2} < \epsilon$$

$\Rightarrow$ g is uniformly continuous on X.

We now prove the uniqueness of the function g. Let $h : (X, d) \to (Y, \rho)$ another extension of f with the requirement. Then $f(x) = h(x) = g(x)\ \forall\ x \in A$.

For $x \notin A$ and $\bar{A} = X$ we get a sequence $\{x_n\} \subset A$ such that $x_n \to x$ as $n \to \infty$. Since a uniformly continuous function is continuous, we get $g(x_n) \to g(x)$, $h(x_n) \to h(x)$ as $n \to \infty$. But we have $g(x_n) = h(x_n)\ \forall\ n$

$$\Rightarrow \lim_n g(x_n) = \lim_n h(x_n)$$

$$\Rightarrow g(x) = h(x).$$

This is true for all $x \notin A$. Hence, $g(x) = h(x)\ \forall\ x \in X \Rightarrow g \equiv h$.

This completes the proof of the theorem.

Example [4.12] : $f(x) = \sin x^2$ is **not** uniformly continuous on $[0, \infty)$. Let us take two sequences $\left\{\sqrt{\frac{n\pi}{2}}\right\}$ and $\left\{\sqrt{\frac{(n+1)\pi}{2}}\right\}$ be two sequences in $[0, \infty)$ such that

$$\left|\sqrt{\frac{n\pi}{2}} - \sqrt{\frac{(n+1)\pi}{2}}\right| \to 0 \text{ as } n \to \infty.$$

But $\left|f\left(\sqrt{\frac{n\pi}{2}}\right) - f\left(\sqrt{\frac{(n+1)\pi}{2}}\right)\right| = \left|\sin\frac{n\pi}{2} - \sin\frac{(n+1)\pi}{2}\right| = 1 \not\to 0$ as $n \to \infty$.

Note. (1) Let f be a real valued function with continuous derivative on $[a,b]$ with usual metric on $\mathbb{R}$, then by mean value theorem, f satisfies Lipschitz condition on $[a,b]$. Hence f is uniformly continuous on $[a,b]$.

(2) The definition of uniform continuity can be written as follows:

Let (X,d) and (Y,ρ) be two metric spaces and $A \subset X$. A function $f:(A,d_A) \to (Y,\rho)$ is uniformly continuous, if for any $\epsilon > 0$, there is $\delta > 0$ such that

$$\rho(f(x),f(y)) < \epsilon, \text{ whenever } d(x,y) < \delta, \text{ for all } x,y \in A$$

$$\Rightarrow \sup\{\rho(f(x),f(y)) : x,y \in A\} < \epsilon, \text{ whenever } d(x,y) < \delta \; \forall \; x,y \in A$$

$$\Rightarrow \operatorname{diam}\{f(A)\} < \epsilon, \text{ whenever } \operatorname{diam}(A) < \delta, \text{ i.e.,}$$

$$\operatorname{diam}(f(A)) < \epsilon, \text{ whenever } \operatorname{diam}(A) < \delta.$$

Example [4.13] : Let (X,d) be a metric space and $A \subset X$. Then the function $f:(X,d) \to (\mathbb{R},d_u)$, where d_u means usual metric of $\mathbb{R}$, defined by $f(x) = \operatorname{dist}(x,A) \; \forall \; x \in X$ is uniformly continuous on X.

Let $x,y \in X$. Then we see from Theorem [2.5] that $|\operatorname{dist}(x,A) - \operatorname{dist}(y,A)| \le d(x,y)$.

Let $\epsilon > 0$ be arbitrary. Then choosing δ such that $0 < \delta \le \epsilon$, we see that, for all $x,y \in X$ with $d(x,y) < \delta$,

$$|f(x) - f(y)| = |\operatorname{dist}(x,A) - \operatorname{dist}(y,A)| \le d(x,y) < \delta \le \epsilon$$

$\Rightarrow |f(x) - f(y)| < \epsilon$

$\Rightarrow d_u(f(x),f(y)) < \epsilon$

$\Rightarrow f$ is uniformly continuous on X.

4.2 Homeomorphism, Isometry

Definition [4.4] : Let (X,d) and (Y,ρ) be two metric spaces. A function $f:(X,d) \to (Y,\rho)$ is called a homeomorphism, if f is bijective and both f and f^{-1} are continuous.

Two spaces (X,d) and (Y,ρ) are called homeomorphic, if there is a homeomorphism from (X,d) **onto** (Y,ρ).

In this case we write, $X \equiv Y$.

Example [4.14] : (i) The function $f:\mathbb{R} \to \mathbb{R}$, where $\mathbb{R}$ is equipped with usual metric, defined by $f(x) = \alpha x + \beta$, where $\alpha, \beta \in \mathbb{R}$ is a homeomorphism, when $\alpha \ne 0$. Note that f is continuous and bijective. Also note that the inverse function $f^{-1}:\mathbb{R} \to \mathbb{R}$ is given by

$$f^{-1}(x) = \alpha^{-1}x - \alpha^{-1}\beta = \frac{x}{\alpha} - \frac{\beta}{\alpha}.$$

(ii) Let us take $X = [a,b]$ and $Y = [c,d]$ with usual metric d. Consider the function $f: X \to Y$ by

$$f(x) = \frac{(c-d)x + (ad-bc)}{a-b} \; \forall \; x \in X.$$

Then f is a continuous bijective function. The inverse of f is given by

$$f^{-1}(x) = \frac{(a-b)x + (bc-ad)}{c-d} \; \forall \; x \in Y,$$

which is also continuous on Y. This shows that f is a homeomorphism from (X,d) **onto** (Y,ρ).

Thus $X \equiv Y$.

Definition [4.5] : Let (X,d) and (Y,ρ) be two metric spaces and $f:(X,d)\to(Y,\rho)$ be a function. Then f is called

(i) an open map, if $f(G)$ is an open set in (Y,ρ) for any open set G in (X,d)

(ii) a closed map, if $f(F)$ is a closed set in (Y,ρ) for any closed set F in (X,d).

Example [4.15] : Consider the mapping $f:\mathbb{R}^2\to\mathbb{R}$ defined by $f(x,y)=x\ \forall\ (x,y)\in\mathbb{R}^2$.

Let D be an open disc in $\mathbb{R}^2$. Then $f(D)$ is an open interval in $\mathbb{R}$. So for any open set G in $\mathbb{R}^2$, $f(G)$ is an open set since any point $f(p)\in f(G)$ belongs to an open interval.

Thus f is an open mapping.

Now let $F=\{(x,y)\in\mathbb{R}^2 : xy\geq 1, x>0\}$. Then F is a closed set. But $f(F)=(0,\infty)$ is **not** closed. Thus f is an open map but **not** closed.

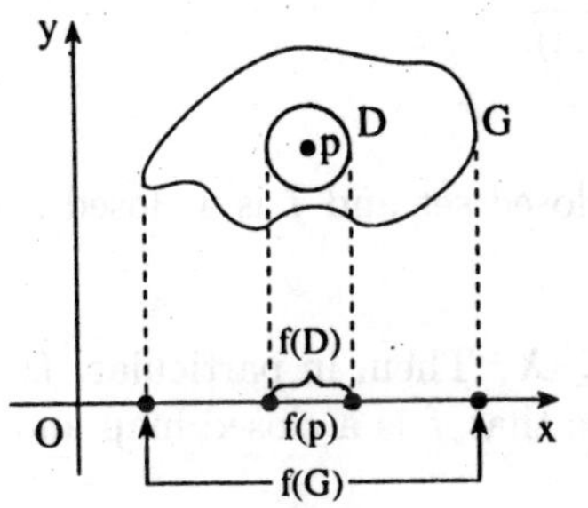

Fig. 4.2

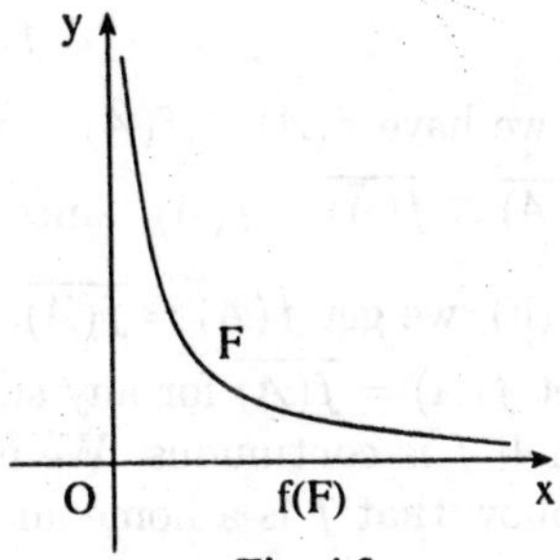

Fig. 4.3

Example [4.16] : We now consider the mapping $f:\mathbb{R}\to\mathbb{R}$ defined by $f(x)=1\ \forall\ x\in\mathbb{R}$.

Note that for any $A\subseteq\mathbb{R}$, $f(A)=\{1\}$. This shows that f is a closed map but f is not an open map.

Example [4.17] : Let (X,d) and (Y,ρ) be two metric spaces with (Y,ρ) a discrete metric space. Since in a discrete metric space every subset is open and closed, it follows that any function $f:(X,d)\to(Y,\rho)$ is both open and closed.

Theorem [4.14] Let (X,d) and (Y,ρ) be two metric spaces and let $f:X\to Y$ be a bijective mapping. Then the follwing conditions are equivalent:

(a) f is a homeomorphism

(b) f is continuous and open

(c) f is continuous and closed.

Proof. (a) $\Rightarrow$ (b) : Let f be a homeomorphism. Then both f and f^{-1} are continuous.

Let G be an open set in (X,d). Then by continuity of $f^{-1}:(Y,\rho)\to(X,d)$, we get $\left(f^{-1}\right)^{-1}(G)=f(G)$ is an open set in (Y,ρ)

$\Rightarrow f$ is an open map. So (b) follows.

(b) $\Rightarrow$ (a) : Now let (b) holds. Then we have f is bijective, continuous and an open map from (X,d) onto (Y,ρ). We now prove that f^{-1} is continuous. Let G be an open set in (X,d). Since $f:(X,d)\to(Y,\rho)$ is an open map, we have $f(G)=\left(f^{-1}\right)^{-1}(G)$ is an open set in (Y,ρ), i.e., $\left(f^{-1}\right)^{-1}(G)$ is an open set in (Y,ρ) for any open set G in (X,d)

$\Rightarrow f^{-1}$ is continuous and (a) follows.

(a) $\Rightarrow$ (c) : Let F be a closed set in (X, d). Then since f^{-1} is continuous, we get by using Theorem [4.3] $(f^{-1})^{-1}(F) = f(F)$ is a closed set in $(Y, \rho) \Rightarrow f$ is a closed map.

(c) $\Rightarrow$ (a) : prove it.

Thus we get (a) $\Leftrightarrow$ (b) and (a) $\Leftrightarrow$ (c)

$\Rightarrow$ (b) $\Leftrightarrow$ (c).

Theorem [4.15] Let (X, d) and (Y, ρ) be two metric spaces. Let $f : X \to Y$ be a bijective mapping. Then f is a homeomorphism, if and only if $f(\bar{A}) = \overline{f(A)}$ for any subset $A \subseteq X$.

Proof. Let us first suppose that f be a homeomorphism. Then f is continuous and closed by previous theorem. We now show that $f(\bar{A}) = \overline{f(A)}$ for any subset $A \subseteq X$.

Since f is continuous, we get from Theorem [4.4],

$$f(\bar{A}) \subset \overline{f(A)}. \tag{i}$$

Since $A \subseteq \bar{A}$, we have $f(A) \subset f(\bar{A})$

$$\Rightarrow \overline{f(A)} \subset \overline{f(\bar{A})} = f(\bar{A}), \text{ since } \bar{A} \text{ is a closed set and } f \text{ is a closed map.} \tag{ii}$$

From (i) and (ii), we get $f(\bar{A}) = \overline{f(A)}$.

Conversely, let $f(\bar{A}) = \overline{f(A)}$ for any subset $A \subseteq X$. Then, in particular, $f(\bar{A}) \subset \overline{f(A)}$ and so by Theorem [4.4] f is continuous. We now prove that f is a closed map and then by above theorem it will follow that f is a homeomorphism.

Let F be a closed set in (X, d). Then $F = \bar{F}$. Now by hypothesis, $f(\bar{F}) = \overline{f(F)}$

$\Rightarrow f(F) = \overline{f(F)}$

$\Rightarrow f(F)$ is a closed set in (Y, ρ).

$\Rightarrow f$ is closed and the theorem follows.

Definition [4.6] : Let (X, d) and (Y, ρ) be two metric spaces. A function $f : X \to Y$ is called an isometry, if $\rho(f(x), f(y)) = d(x, y)$, for all $x, y \in X$, i.e., if it preserves the distances.

Two metric spaces (X, d) and (Y, ρ) are said to be isometric, if there is a bijective isometry from (X, d) **onto** (Y, ρ).

From the definition above it is obvious that an isometry is uniformly continuous.

It is also obvious that an isometry is injective. For, let $f(x) = f(y)$, for any $x, y \in X$

$$\Rightarrow \rho(f(x), f(y)) = 0$$
$$\Rightarrow d(x, y) = 0$$
$$\Rightarrow x = y.$$

Also note that if f is an isometry, then f^{-1} is an isometry.

Let $x, y \in Y$.

Then $f^{-1}(x), f^{-1}(y) \in X$. Since f is an isometry, we get

$$d(f^{-1}(x), f^{-1}(y)) = \rho(ff^{-1}(x), ff^{-1}(y)) = \rho(x, y).$$

Since an isometry is uniformly continuous, it follows that it is continuous.

Thus if $f : (X, d) \to (Y, \rho)$ is a surjective isometry, then f is one-to-one, continuous and f^{-1} is also continuous. Hence, f is a homeomorphism.

We have thus proved the following theorem.

Theorem [4.16] Let (X, d) and (Y, ρ) be two metric spaces and $f : (X, d) \to (Y, \rho)$ be a surjective isometry. Then f is a homeomorphism.

Note. Converse of the above theorem is **not** true, in general. For example, consider the function $f : \mathbb{R} \to \mathbb{R}$ defined by $f(x) = 2x \; \forall \; x \in \mathbb{R}$. It is clear that f is continuous and one-to-one. f^{-1} is given by $f^{-1}(x) = x/2 \; \forall \; x \in \mathbb{R}$. Then. f^{-1} is also continuous. Hence f is a homeomorphism of $\mathbb{R}$ onto itself. But f is **not** an isometry.

Theorem [4.17] Completeness is preserved under isometry.

Proof. Let (X, d) and (Y, ρ) be two metric spaces. Let $f : (X, d) \to (Y, \rho)$ be a surjective isometry. Let (X, d) be a complete metric space. We now prove that (Y, ρ) is complete.

Let $\{y_n\}$ be a Cauchy sequence in (Y, ρ). Then for any $\epsilon > 0$, there exists $N \in \mathbb{N}$ such that $\rho(y_n, y_m) < \epsilon \; \forall \; m, n \geq N$.

Now $d(f^{-1}(y_n), f^{-1}(y_m)) = \rho(y_n, y_m) < \epsilon \; \forall \; m, n \geq \mathrm{N}$

$\Rightarrow \{f^{-1}(y_n)\}$ is a Cauchy sequence in (X, d).

Since (X, d) is complete, there exists $x \in X$ such that $f^{-1}(y_n) \to x$ as $n \to \infty$ in (X, d)

$\Rightarrow y_n \to f(x)$ as $n \to \infty$ in (Y, ρ)

$\Rightarrow (Y, \rho)$ is complete.

EXERCISE

(1) Prove that continuous image of a Cauchy sequence may **not** be a Cauchy sequence.

(2) Let $X = c[a, b]$. Let f be a real valued function on X such that $f(x) = x(t_0)$, $t_0 \in [a, b]$ and $x \in X$. Show that f is continuous.

(3) Let x be a non-empty subset of a metric space (X, d). Let $f(x) = d(x, A) = \inf\{d(x, a) : a \in A\}$, $x \in X$. Show that f is continuous on X. Also show that $\bar{A} = \{x \in X : d(x, A) = 0\}$.

(4) Define a function $f : (0, 1] \to \mathbb{R}^2$ by $f(x) = \left(x, \sin \dfrac{1}{x}\right)$, $x \in (0, 1]$. Show that f is continuous.

(5) Show that continuous image of a complete metric space may **not** be complete.

(6) Let (X, d) be a metric space and $f, g : (X, d) \to (X, d)$ are two continuous functions. Show that the set $\{x \in X : f(x) \neq g(x)\}$ is an open set in (X, d).

(7) Let (X, d) be a separable metric space. Let $f : (X, d) \to (Y, \rho)$ be a continuous mapping, where (Y, ρ) is another metric space. Show that $f(X)$ is separable.

(8) Let d be the usual metric on $\mathbb{R}$. Define $f : (\mathbb{R}, d) \to ((-1, 1), d)$ by $f(x) = \dfrac{x}{1 + |x|}$, $x \in \mathbb{R}$.

Show that f is a homeomorphism. Also show that f is uniformly continuous on $\mathbb{R}$.

(9) Show that image of a complete metric space under homeomorphism need **not** be complete. What can you say about uniformly continuous image of a complete metric space?

(10) Let $f : (X, d) \to (Y, \rho)$ be a surjective isometry. Show that if (X, d) is complete, then so is (Y, ρ).

(11) Show that uniformly continuous image of a Cauchy sequence is a Cauchy sequence.

(12) Show that the function defined in (8) is a Lipschitz function.

(13) Give an example of a continuous function f such that image $f(F)$ of a closed set F in X need not be closed in Y.

(14) Let d_1 and d_2 be two equivalent metrics on a non-empty set X. Let (Y, ρ) be a metric space and $f : X \to Y$ be a mapping. Prove that f is continuous in (X, d_1) if and only if f is continuous in (X, d_2).

(15) Let (X, d) and (Y, ρ) be two metric spaces and $f : X \to Y$ be a function. Prove or disprove: $f : X \to Y$ is continuous if and only if $(f(A))^\circ \subset f(\mathring{A})$ for all subsets A of X.

(16) Let (X, d) be a metric space and $f : X \to \mathbb{R}$ be a continuous mapping. Show that for $r > 0$, the sets $\{x \in X : f(x) \leq r\}$ and $\{x \in X : f(x) \geq r\}$ are closed sets.

(17) Let (X, d) be a metric space and for each $j = 1, 2, \ldots, n$, the function $f_j : X \to \mathbb{R}$ be continuous. Show that the function $f : X \to \mathbb{R}^n$ defined by

$$f(x) = (f_1(x), f_2(x), \ldots, f_n(x)) \text{ for all } x \in X$$

is continuous. Show that the converse is also true.

(18) Let (X, d) be a metric space and $f : X \to X$ be a mapping. Prove that if f is continuous on X, then the set $\{x \in X : f(x) = x\}$ is a closed set in X.

(19) Let A and B be two disjoint closed sets in a metric space (X, d). Show that there exists a real valued continuous function f such that $f(A) = \{1\}$, $f(B) = \{-1\}$ and $-1 < f(x) < 1$ for all $x \in (A \cup B)^c$.

(20) Let (X, d) and (Y, ρ) be homeomorphic metric spaces. Show that (X, d) is discrete if and only if (Y, ρ) is discrete.

(21) Let (X, d) and (Y, ρ) be homeomorphic metric spaces. Prove that (X, d) is separable if and only if (Y, ρ) is separable.

(22) Give an example of a continuous function which is

(a) open but not a homeomorphism

(b) closed but not a homeomorphism

(c) open but not closed

(d) closed but not open.

(23) Let (X, d) and (Y, ρ) be two metric spaces and $f : X \to Y$ be a homeomorphism. Define a function $d^* : X \times X \to \mathbb{R}$ by $d^*(x, y) = \rho(f(x), f(y))$, for all $x, y \in X$. Prove that d^* is a metric on X equivalent to d.

(24) Let (X, d) and (Y, ρ) be two metric spaces and $f : X \to Y$ be a function. Then f is uniformly continuous if and only if for any $\epsilon > 0$, there is a $\delta > 0$ such that for all sets $E \subseteq X$ with $d(E) < \delta$, we have $d(f(E)) < \epsilon$.

(25) Let (X, d) and (Y, ρ) be two metric spaces and $f : X \to Y$ be a surjective continuous mapping such that $d(x, y) \leq \rho(f(x), f(y))$ for all $x, y \in X$. Show that if (X, d) is complete, then so is (Y, ρ).

CHAPTER 5

COMPACT METRIC SPACES

Definition [5.1] : **[Cover and Subcover]** Let X be a non-empty set. A family of subsets $\mathcal{H} = \{A_i : i \in I\}$ of X is said to be a **cover** of X, if $X = \bigcup_{i \in I} A_i$.

Let Y be a subset of X. Then $\mathcal{H} = \{A_i : i \in I\} \subset p(X)$ is said to be a **cover** of Y, if $Y \subset \bigcup_{i \in I} A_i$, where $p(X)$ denotes the power set of X.

$\mathcal{H}$ is said to be a **finite cover** of X, if I is finite.

Let $\mathcal{G} \subset \mathcal{H}$ be such that $X = \bigcup_{A_i \in \mathcal{G}} A_i$. Then $\mathcal{G}$ is called a **subcover** of the cover $\mathcal{H}$ of X.

If $\mathcal{G}$ contains finite number of elements, then $\mathcal{G}$ is called a **finite subcover** of the cover $\mathcal{H}$ of X.

Definition [5.2] : **[Open Cover]** Let (X, d) be a metric space. Let $\mathcal{G} = \{G_i : i \in I\}$ be a family of open sets in (X, d) such that $X = \bigcup_{i \in I} G_i$. Then $\mathcal{G}$ is called an **open cover** of X.

Example [5.1] : Consider the metric space of all real numbers $\mathbb{R}$ with the usual metric d. Then $\{(-n, n) : n \in \mathbb{N}\}$ is a family of open sets in $(\mathbb{R}, d)$ such that $\mathbb{R} = \bigcup_{n \in \mathbb{N}} (-n, n)$.

So $\{(-n, n) : n \in \mathbb{N}\}$ is an open cover of $\mathbb{R}$.

Also note that $\{(-2n, 2n) : n \in \mathbb{N}\}$ is a subcover of the cover $\{(-n, n) : n \in \mathbb{N}\}$ of $\mathbb{R}$.

Example [5.2] : In the metric space $(\mathbb{R}, d)$, take $Y = (2, 1024)$.

Then the collection $\{(i, 2i+1)\}_{i=1}^{\infty}$ is an open cover of $(2, 1024)$.

Now note that the subcollection $\{(2i, 4i+1)\}_{i=1}^{\infty}$ is a **subcover** for Y and the subcollection $\{(2i, 4i+1)\}_{i=1}^{256}$ is a **finite subcover** of the cover $\{(i, 2i+1)\}_{i=1}^{\infty}$ of Y.

Remark. In Example [5.1], the cover $\{(-n, n) : n \in \mathbb{N}\}$ has **no** finite subcover. If possible, let $\{(-n_k, n_k) : k = 1, \ldots, r\}$ be a finite subcover. Let $x = \max\{n_1, n_2, \ldots, n_r\}$. Then $x \in \mathbb{R}$ but

$$x \notin \bigcup_{k=1}^{r} (-n_k, n_k).$$

Definition [5.3] : **[Compact Metric Space]** A metric space (X, d) is said to be compact, if **every** open cover of X admits a **finite** subcover. From Example [5.1] and above Remark, we say that $(\mathbb{R}, d)$ is **not** compact.

Example [5.3]: Consider the discrete metric space (X, d), where X is an infinite set. We know that, in (X, d) every subset of X is open. Then the collection $\{\{x\} : x \in X\}$ forms an open cover of X, i.e., $X = \bigcup_{x \in X} \{x\}$.

But this open cover has **no** finite subcover. Hence the discrete metric space (X, d) is compact, if X is a finite set. From the definition it follows that every finite metric space (X, d) is always compact.

Thus, discrete metric space (X, d) is compact, if and only if X is finite.

Discussion. In Example [1.20] we have seen that, if (X, d) is a metric space and $G \subset Y \subset X$, then G may be open in (X, d) without being open in (Y, d_Y). Similar result is true for closed sets. In this connection it is very important to observe the following:

> "Let (X, d) be a metric space and $Y \subset X$. In order that all open (resp. closed) subsets of Y be open (resp. closed) in X, it is necessary and sufficient that Y be open (resp. closed) in X."

Thus the property of being open (resp. closed) depends on the space in which G is embedded. However, compactness of a set is independent of the space in which the set is embedded.

We see this in the next result.

Theorem [5.1] Let (X, d) be a metric space and $K \subset Y \subset X$. Then K is compact relative to (X, d), if and only if K is compact relative to (Y, d_Y).

Proof. Let us first assume that K is compact relative to (X, d). We now prove that K is compact relative to (Y, d_Y).

Let $\{V_\alpha\}_{\alpha \in I}$ be a family of sets, open relative to Y, such that $K \subset \bigcup_{\alpha \in I} V_\alpha$. Since V_α is an open set in (Y, d_Y) we must find an open set G_α in (X, d) such that $V_\alpha = Y \cap G_\alpha$. This is true, for all $\alpha \in I$.

Hence, $\{G_\alpha\}$ is a family of open sets in (X, d) such that $K \subset \bigcup_{\alpha \in I} G_\alpha$.

Since K is compact relative to (X, d), there exists a finite number of indices $\alpha_1, \alpha_2, \cdots, \alpha_n$ such that $K \subset \bigcup_{i=1}^{n} G_{\alpha_i}$. Since $K \subset Y$, we have

$$\begin{aligned} K \subset \left(\bigcup_{i=1}^{n} G_{\alpha_i}\right) \cap Y &= (G_{\alpha_1} \cap Y) \cup (G_{\alpha_2} \cap Y) \cup \cdots \cup (G_{\alpha_n} \cap Y) \\ &= V_{\alpha_1} \cup V_{\alpha_2} \cup \cdots \cup V_{\alpha_n}. \end{aligned}$$

This proves that K is compact relative to (Y, d_Y).

Conversely, let K be compact relative to (Y, d_Y). Let $\{G_\alpha\}_{\alpha \in I}$ be a family of open sets in (X, d) such that $K \subset \bigcup_{\alpha \in I} G_\alpha$. Put $V_\alpha = Y \cap G_\alpha$, $\forall\, \alpha \in I$. Then $\{V_\alpha\}_{\alpha \in I}$ is a family of open sets in (Y, d_Y) such that $K \subset \bigcup_{\alpha \in I} V_\alpha$ (verify). Since K is compact relative to (Y, d_Y), there exists $\alpha_1, \alpha_2, \ldots, \alpha_r \in I$ such that

$$\begin{aligned} K &\subset V_{\alpha_1} \cup V_{\alpha_2} \cup \cdots \cup V_{\alpha_r} \\ &\subset G_{\alpha_1} \cup G_{\alpha_2} \cup \cdots \cup G_{\alpha_r}. \end{aligned}$$

This shows that K is compact relative to (X, d).

Theorem [5.2] A compact subset K of a metric space (X,d) is closed and bounded.

Proof. To prove that K is closed, we show that K^c is open [K^c denotes complement of K].

Let $p \in X$ be such that $p \notin K$. Then for each $q \in K$, choose ϵ_q such that

$$0 < \epsilon_q < \frac{1}{2}d(p,q).$$

Let $V_q = B_{\epsilon_q}(p)$ and $W_q = B_{\epsilon_q}(q)$. Thus we find a family of open sets $\{W_q\}_{q\in K}$ such that

$$K \subset \bigcup_{q\in K} W_q.$$

By compactness of K, there exists a finite number of points $q_1, q_2, \ldots, q_n$ in K such that

$$K \subset W_{q_1} \cup W_{q_2} \cup \cdots \cup W_{q_n},$$

which proves that K is bounded.

Let $W = W_{q_1} \cup W_{q_2} \cup \cdots \cup W_{q_n}$. Then W is an open set.

Now set $V = V_{q_1} \cap V_{q_2} \cap V_{q_2} \cap \cdots \cap V_{q_n}$. Then V is an open set and

$$V \cap W = \phi \Rightarrow V \cap K = \phi.$$

This shows that $p \in V \subset K^c$, i.e., p is an interior point of K^c and hence K^c is an open set.

Consequently, K is a closed set in (X,d).

Remark. Converse of the above theorem is **not** true, in general. For example, an infinite trivial metric space is closed and bounded but **not** compact.

Theorem [5.3] Closed subsets of a compact metric space are compact.

Proof. Let (X,d) be a metric space which is supposed to be compact and let F be a closed subset of X. We show that F is also compact.

Let $\{G_\alpha\}_{\alpha\in I}$ be a family of open sets in X such that $F \subset \bigcup_{\alpha\in I} G_\alpha$.

Then $\{G_\alpha\}_{\alpha\in I}$ along with $\{F^c\}$ forms an open cover of X. Since (X,d) is a compact metric space, we can find a finite subcover of the cover $\{G_\alpha\}_{\alpha\in I} \cup \{F^c\}$, i.e., there exists $\alpha_1, \alpha_2, \ldots, \alpha_n \in I$ such that $F \subset X = G_{\alpha_1} \cup G_{\alpha_2} \cup \cdots \cup G_{\alpha_n}$.

If F^c is a part of this subcover, we simply remove it to get a finite subcover $\{G_\alpha\}_{\alpha\in I}$ of F.

Hence F is compact.

Remark. Let F_1 and F_2 be two compact subsets of a metric space (X,d), then $F_1 \cap F_2$ is also a compact subset of (X,d). Since F_1 and F_2 are compact, F_1 and F_2 are closed subsets of X and $F_1 \cap F_2 \subset F_1$.

$\Rightarrow F_1 \cap F_2$ is a closed subset of F_1 and F_1 is compact.

$\Rightarrow F_1 \cap F_2$ is compact.

Not only that, if F is a closed subset and K be a compact subset of (X,d), then also $F \cap K$ is a compact subset of (X,d) [Prove it].

From this it follows that, if $\{F_1, F_2, \ldots, F_n\}$ is a family of compact sets in a metric space (X,d), then $\bigcap_{i=1}^{n} F_i$ is also compact.

Theorem [5.4] If $\{K_n\}_{n\in\mathbb{N}}$ is a collection of compact sets in a metric space (X,d) such that the intersection of every **finite** subcollection of $\{K_n\}_{n\in\mathbb{N}}$ is non-empty, then $\bigcap_n K_n \neq \phi$.

Proof. Take a member K_1 from $\{K_n\}_{n\in\mathbb{N}}$ and put $P_n = K_n^c (n \neq 1)$. Assume that no point of K_1 belongs to any $K_n (n \neq 1)$.

Note that, $\{P_n\}$ forms an open cover of K_1. Since it is compact, the open cover $\{P_n\}$ admits a finite subcover for K_1, i.e., there are finitely many indices $\alpha_1, \alpha_2, \ldots, \alpha_n$ such that $K_1 \subset P_{\alpha_1} \cup P_{\alpha_2} \cup \cdots \cup P_{\alpha_n}$.

But this shows that $K_1 \cap K_{\alpha_1} \cap \cdots \cap K_{\alpha_n} = \phi$—a contradiction to our assumption and the theorem is proved.

Corollary [5.1]: If $\{K_n\}$ is a sequence of non-empty compact sets in a metric space (X, d) such that $K_n \supseteq K_{n+1}$ $(n = 1, 2, \ldots)$, then $\bigcap_n K_n \neq \phi$.

Definition [5.4]: **[Bolzano-Weierstrass Property]** A metric space (X, d) is said to have the Bolzano-Weierstrass property (BWP), if every **infinite** subset of X has a limit point in X.

Definition [5.5]: **[Sequentially Compact Metric Space]** A metric space (X, d) is said to be sequentially compact, if every sequence in X has a convergent subsequence.

Theorem [5.5] A metric space is sequentially compact, if and only if it has Bolzano-Weierstrass property.

Proof. Let us first suppose that (X, d) be a sequentially compact metric space. Let A be an infinite subset of X. We show that A has a limit point. Since A is infinite, a sequence $\{x_n\}$ of distinct point can be extracted from A. Since (X, d) is sequentially compact, the sequence $\{x_n\}$ has a subsequence $\{x_{n_k}\}$ converging to a point, say, $x \in X$. Since $\{x_n\}$ is sequence of distinct points, x is a limit point of the set of points of the subsequence $\{x_{n_k}\}$ and since this set is a subset of A, x is a also a limit point of A.

Conversely, let a metric space (X, d) has Bolzano-Weierstrass property. We show that (X, d) is sequentially compact. Let $\{x_n\}$ be an arbitrary sequence in X. If $\{x_n\}$ has a point which is repeated infinitely, then it has a constant subsequence and this subsequence is clearly convergent. Otherwise the set A of distinct points of $\{x_n\}$ is infinite and hence by assumption A has a limit point, i.e., $x \in \bar{A}$. Then by Theorem [2.18] we can find a sequence, say, $\{y_n\} \subset A$ such that $y_n \to x$. Clearly $\{y_n\}$ is a subsequence of $\{x_n\}$. Hence (X, d) is sequentially compact.

This completes the proof of the theorem.

Theorem [5.6] Every compact metric space has Bolzano-Weierstrass property.

Proof. Let (X, d) be a compact metric space and A be an infinite subset of X. We show that A has a limit point.

Suppose that A has no limit point, i.e., no point of X is a limit point of A. Then for every $x \in X$, $\exists\, r_x > 0$ such that $B_{r_x}(x) \cap (A - \{x\}) = \phi$, i.e., $B_{r_x}(x)$ contains no point of A other than (possibly) x. Note that the family $\{B_{r_x}(x) : x \in X\}$ forms an open cover of X and hence by compactness of (X, d), $\exists\, x_1, x_2, \ldots, x_k \in X$ such that

$$X = B_{r_{x_1}}(x_1) \cup B_{r_{x_2}}(x_2) \cup \cdots \cup B_{r_{x_k}}(x_k)$$

and so $A \subset X = B_{r_{x_1}}(x_1) \cup B_{r_{x_2}}(x_2) \cup \cdots \cup B_{r_{x_k}}(x_k)$. Since each $B_{r_{x_i}}$ $(i = 1, 2, \ldots, k)$ contains at most one point of A (possibly each x_i, $i = 1, 2, \ldots, k$), it follows that A has at most k points, i.e., A is a finite set—which is a contradiction to the hypothesis. Hence A must have a limit point in (X, d).

Note. Bolzano-Weierstrass property is also known as **Fréchet compact**.

Definition [5.6] : [**Lebesgue Number**] Let $A = \{G_i\}_{i \in I}$ be an open cover of (X, d). A real number $\delta > 0$ is called a **Lebesgue Number** for the cover, if every subset of X with diameter less than δ is contained in one of the members of the cover A.

Theorem [5.7] In a sequentially compact metric space, every open cover has a Lebesgue number.

Proof. Let (X, d) be a sequentially compact metric space. Let $A = \{G_i\}_{i \in I}$ be an open cover of X. Suppose that A does not have a Lebesgue number. Then for each positive integer n, there exists a subset B_n of X with the property that $0 < \operatorname{diam}(B_n) < \dfrac{1}{n}$ and $B_n \not\subset G_i$ for every G_i in A. For each n, let us take a point $p_n \in B_n$. Thus we get a sequence $\{p_n\}$ from X. Since X is sequentially compact, the sequence $\{p_n\}$ has a convergent subsequence $\{p_{n_k}\}$ and let $p_{n_k} \to p \in X$.

Since $p \in X$, there exists a member $G_p \in A$ such that $p \in G_p$. Hence there exists $\epsilon > 0$ such that $p \in B_\epsilon(p) \subset G_p$.

Since $\{p_{n_k}\} \to p$, there exists a (+)ve integer n_{k_0} with $\dfrac{1}{n_{k_0}} < \dfrac{\epsilon}{2}$ such that

$$d\left(p, p_{n_{k_0}}\right) < \frac{\epsilon}{2}, p_{n_{k_0}} \in B_{n_{k_0}} \text{ and } \operatorname{diam}\left(B_{n_{k_0}}\right) < \frac{\epsilon}{2}.$$

Using triangle inequality, we get $B_{n_{k_0}} \subset B_\epsilon(p) \subset G_p$—a contradiction to the fact that $B_n \not\subset G_i$, $\forall\, i$, in A. This establishes that A admits a Lebesgue number.

Definition [5.7] : [**Totally Bounded**] Let (X, d) be a metric space. If $\epsilon > 0$ be given, a subset A of X is called an ϵ-net, if $X = \bigcup\limits_{a \in A} B_\epsilon(a)$.

If A is **finite**, then A is called a finite ϵ-net of X. A metric space (X, d) is said to be totally bounded, if for each $\epsilon > 0$, it has a **finite** ϵ-net.

Discussion. Every totally bounded subset of a metric space (X, d) is bounded. For, let A be a totally bounded subset of (X, d). Then for $\epsilon > 0$, let $S = \{a_1, a_2, \dots, a_n\}$ be a finite ϵ-net of A, i.e., $A \subset \bigcup\limits_{a \in S} B_\epsilon(a)$.

Let $x, y \in A$. Then there exists $a_i, a_j \in S$ such that $x \in B_\epsilon(a_i)$ and $y \in B_\epsilon(a_j)$ and so $d(x, a_i) < \epsilon$ and $d(y, a_j) < \epsilon$.

$$\therefore d(x, y) \le d(x, a_i) + d(a_i, a_j) + d(a_j, y) < \epsilon + d(a_i, a_j) + \epsilon \le 2\epsilon + M,$$

where $M = \max\limits_{1 \le i, j \le n} d(a_i, a_j)$.

This is true, for all $x, y \in A$

$\Rightarrow \operatorname{diam}(A) \le 2\epsilon + M < \infty$

$\Rightarrow A$ is bounded.

But the converse of the above statement is not true, in general.

For example, let $X = l_2$ and B be the set consisting the points

$$e_1 = (1, 0, 0, \dots),\ e_2 = (0, 1, 0, \dots),\ \dots,\ e_n = (0, 0, \dots, 1, 0, \dots) \text{ and so on.}$$

Note that $d(e_i, e_j) = \sqrt{2}$, if $i \neq j$ [see def. of d from [1.4]] and so

$$\operatorname{diam}(B) = \sup\{d(e_i, e_j) : e_i, e_j \subset B\} = \sqrt{2}.$$

So B is bounded.

We now show that B is **not** totally bounded. Let $\epsilon = 1/2$. We claim that B has **no** finite ϵ-net. If possible, let S be a finite $1/2$-net. Then for each $n \in \mathbb{N}$, $\exists\, x_n \in S$ such that

$$d(e_n, x_n) < \frac{1}{2}.$$

Now for $m, n \in \mathbb{N}$, $m \neq n$ we have $x_n \neq x_m$, since $\sqrt{2} = d(e_n, e_m) \leq d(e_n, x_n) + d(x_n, x_m) + d(x_m, e_m) < \frac{1}{2} + 0 + \frac{1}{2} = 1$—a contradiction. Hence S must be infinite-again a contradiction and our claim established.

Theorem [5.8] Every sequentially compact metric space is totally bounded.

Proof. Let (X, d) be sequentially compact metric space. If possible, suppose that (X, d) is not totally bounded. Then there exists an $\epsilon > 0$ such that X possesses no finite ϵ-net.

Let $p_1 \in X$. Then there exists a point $p_2 \in X$ with $d(p_1, p_2) \geq \epsilon$. If no such points exists, then we get

$$\begin{aligned} & d(p_1, x) < \epsilon,\ \forall\, x \in X \\ \Rightarrow\ & x \in B_\epsilon(p_1),\ \forall\, x \in X \\ \Rightarrow\ & X = B_\epsilon(p_1) \\ \Rightarrow\ & \{p_1\} \text{ is a finite } \epsilon\text{-net of } X. \end{aligned}$$

Hence we must get a point $p_2 \in X$ with $d(p_1, p_2) \geq \epsilon$.

Similarly, there exists a point $p_3 \in X$ with $d(p_1, p_3) \geq \epsilon$ and $d(p_2, p_3) \geq \epsilon$, for otherwise $\{p_1, p_2\}$ would be an ϵ-net for X.

Continuing in this way, we arrive at a sequence $\{p_1, p_2, \ldots\}$ with the property that

$$d(p_i, p_j) \geq \epsilon, \text{ for all } i, j (i \neq j).$$

This reveals that $\{p_n\}$ cannot contain any convergent subsequence—this contradicts the sequential compactness of (X, d). Thus (X, d) is totally bounded.

Theorem [5.9] Every sequentially compact metric space is compact.

Proof. Let (X, d) be a sequentially compact metric space. Let $A = \{G_i\}_{i \in I}$ be an open cover of X. Since (X, d) is sequentially compact, the open cover A has a Lebesgue number [by Theorem [5.7]], say, $\alpha > 0$.

Let us take $\epsilon = \alpha/3$.

By the above theorem (X, d) is totally bounded. Hence X has a finite ϵ-net

$$A = \{p_1, p_2, \ldots, p_n\}.$$

Now for each $k = 1, 2, \ldots, n$, we have $\operatorname{diam}(B_\epsilon(p_k)) \leq 2\epsilon = \frac{2\alpha}{3} < \alpha$.

So by definition of Lebesgue number, for each k we can find a $G_{i_k} \in A$ such that

$$B_\epsilon(p_k) \subseteq G_{i_k}.$$

Thus $X = B_\epsilon(p_1) \cup B_\epsilon(p_2) \cup \cdots \cup B_\epsilon(p_n) \subseteq G_{i_1} \cup G_{i_2} \cup \cdots \cup G_{i_n}$

$\Rightarrow X = G_{i_1} \cup G_{i_2} \cup \cdots \cup G_{i_n}$

$\Rightarrow \{G_{i_1}, G_{i_2}, \ldots, G_{i_n}\}$ is a finite subcover of $\{G_i\}$ for X

$\Rightarrow (X, d)$ is compact.

Theorem [5.10] Let (X, d) be a metric space. Then the following three conditions are equivalent:

(1) (X, d) is compact

(2) (X, d) has the Bolzano-Weierstrass property

(3) (X, d) is sequentially compact.

Proof. (1) $\Rightarrow$ (2) follows from Theorem [5.6].
(2) $\Rightarrow$ (3) follows from Theorem [5.5].
(3) $\Rightarrow$ (1) follows from Theorem [5.9].

We are now in a position to prove the following important theorem.

Theorem [5.11] A metric space (X, d) is **compact**, if and only if it is complete and totally bounded.

Proof. First suppose that (X, d) is a compact metric space. Then (X, d) is sequentially compact and hence totally bounded.

Now, let $\{x_n\}$be a Cauchy sequence in X. Since (X, d) is sequentially compact, the sequence $\{x_n\}$ has a convergent subsequence. Hence by Theorem [2.22], the Cauchy sequence $\{x_n\}$ is also convergent in X. Thus (X, d) is complete.

Conversely, let (X, d) be complete and totally bounded metric space. We now show that (X, d) is sequentially compact and from which it will follow that (X, d) is compact. Let $S_1 = \{x_{11}, x_{12}, x_{13}, \ldots\}$ be a sequence in X. Since X is totally bounded, there exists a finite number of open balls, each with radius $1/2$, whose union is X.

From this we see that S_1 has a subsequence $S_2 = \{x_{21}, x_{22}, x_{23}, \ldots\}$ all of whose points lie in some open ball of radius $1/2$. Another application of totally boundedness of X shows similarly that S_2 has a subsequence $S_3 = \{x_{31}, x_{32}, x_{33}, \ldots\}$ all of whose points lie in some one open ball of radius $1/3$. We continue forming successive subsequences in this manner and let

$$S = \{x_{11}, x_{22}, x_{33}, \ldots\}$$

be the "diagonal" subsequence of S_1. By the nature of this construction, S is clearly a Cauchy subsequence of S_1. Since (X, d) is complete, this Cauchy sequence S is convergent. Thus the sequence $\{x_{11}, x_{12}, \ldots\}$ admits a convergent subsequence.

This completes the proof.

Note. It follows from the above theorem that a compact subset of a metric space is always bounded. Since a closed subset of a complete metric space is complete, it follows the corollary from the above theorem:

Corollary [5.2] : A closed subset of a complete metric space is compact, if and only if it is totally bounded.

Definition [5.8] : **[Countably Compact]** A metric space (X, d) is said to be countably compact, if every countable open cover of X has a finite subcover.

Definition [5.9] : **[Lindelöf Space]** A metric space (X, d) is said to be Lindelöf space, if every open cover of X has a countable subcover.

From definition it follows that every compact metric space is countably compact and Lindelöf.

Theorem [5.12] Every totally bounded metric space is separable.

Proof. Let (X,d) be a totally bounded metric space. Then for each positive integer n, F_n be the finite $1/n$-net (i.e., $\epsilon = 1/n$) of X.

Therefore, $X = \bigcup_{p\in F_n} B_{1/n}(p)$.

Let us take $H = \bigcup_n F_n$. Then H is countable, since each F_n is countable.

We now show that $\bar{H} = X$. Let $x \in X$ and $\epsilon > 0$ be arbitrary. Then there exists a positive integer K such that $\frac{1}{K} < \epsilon$. Then $\exists\ y \in F_K$ such that $x \in B_{1/K}(y) \Rightarrow y \in B_{1/K}(x)$.

$\Rightarrow$ $B_{1/K}(x)$ contains at least one point of F_K

$\Rightarrow$ $B_\epsilon(x)$ contains a point of F_K and hence a point of H

$\Rightarrow$ $B_\epsilon(x) \cap H \neq \phi$, for all ϵ

$\Rightarrow x \in \bar{H}$

$\therefore\ \bar{H} = X$

$\Rightarrow$ X is separable.

Corollary [5.3]: Every compact metric space is separable.

Proof. We have seen that a compact metric space is totally bounded and hence by above theorem every compact metric space is separable.

Corollary [5.4]: Every compact metric space is second countable.

Proof. Follows from Theorem [2.10] and above theorem.

Theorem [5.13] Every separable metric space is a Lindelöf space.

Proof. Let (X,d) be a separable metric space, Let $D = \{x_1, x_2, \ldots\}$ be a countable subset of X such that $\bar{D} = X$. Let $\{G_i\}_{i\in I}$ be an open cover of X, i.e., $X = \bigcup_{i\in I} G_i$.

Let $x \in X$. Then $\exists\ i_x \in I$ such that $x \in G_{i_x}$. Since G_{i_x} is open, $\exists\ r_x > 0$ such that

$$x \in B_{r_x}(x) \subset G_{i_x}.$$

Again, since $x \in X$ and $\bar{D} = X$, $\exists\ x_k \in D$ such that

$$d(x, x_k) < \frac{r_x}{2}, \text{ i.e., } x \in B_{\frac{r_x}{2}}(x_k).$$

Let $y \in B_{\frac{r_x}{2}}(x_k)$. Then $d(y, x_k) < \frac{r_x}{2}$.

Now by triangle inequality we get $d(y,x) \le d(y,x_k) + d(x_k,x) < \frac{r_x}{2} + \frac{r_x}{2} = r_x$

$\Rightarrow y \in B_{r_x}(x)$

Hence, $x \in B_{\frac{r_x}{2}}(x_k) \subset B_{r_x}(x) \subset G_{i_x}$

$$\Rightarrow x \in B_{\frac{r_x}{2}}(x_k) \subset G_{i_x}$$

Thus, for each $x \in X$, $\exists\ x_k \in D$ such that

$$x \in B_{\frac{r_x}{2}}(x_k) \subset G_{i_x} \text{ for some } r_x > 0 \qquad \text{(i)}$$

Note that the collection $\{G_{i_x}\}$ with the property (i) is countable and $\bigcup G_{i_x} = X$, since

$$\bigcup_{x\in X}\{x\} \subset \bigcup G_{i_x} \subset \bigcup_{i\in I} G_i.$$

Hence (X,d) is Lindelöf.

Theorem [5.14] Every countably compact metric space has Bolzano-Weierstrass property.

Proof. Let (X, d) be a countably compact metric space. Let A be an infinite subset of X. Since every infinite set contains a countable subset, let $S = \{x_1, x_2, \dots\}$ be a countably infinite subset of A. Suppose that the set S has no limit point in X, i.e., no $x \in X$ is a limit point of S. Then S is a closed set.

Now for each $n \in \mathbb{N}$, let G_n be an open ball such that $S \cap G_n = \{x_n\}$. Then $\{G_n : n \in \mathbb{N}\} \cup S^c$ is a countable open cover of X. But this open cover does not have a finite subcover—which is a contradiction to the countable compactness of (X, d). Hence S must possess a limit point. Since $S \subset A$, A has a limit point.

Theorem [5.15] A countably compact metric space is totally bounded.

Proof. Let (X, d) be a countably compact metric space. Suppose that (X, d) is **not** totally bounded. Then following the same procedure as in the Theorem [5.8] we get a sequence $\{p_n\}$ in (X, d) with the property that $d(p_i, p_j) \geq \epsilon$, for all i, j $(i \neq j)$. This shows that the sequence $\{p_n\}$ in X does **not** have a convergent subsequence. So (X, d) is **not** sequentially compact and hence by Theorem [5.5], (X, d) does **not** possesses Bolzano-Weierstrass property—this is a contradiction, since every countably compact metric space has Bolzano-Weierstrass property (by previous theorem). Thus (X, d) is totally bounded.

Theorem [5.16] Every countably compact metric space is **compact.**

Proof. Let (X, d) be a countably compact metric space. Then by Theorem [5.15], (X, d) is separable. Since by Theorem [5.13], every separable metric space is Lindelöf, (X, d) is a Lindelöf space.

Now let A be an open cover of X. Then A has a countable subcover B for X, since (X, d) is Lindelöf. Hence by countably compactness, the countable open cover B of X has a finite subcover G, which is also a subcover of A.

Hence (X, d) is compact.

Definition [5.10] : [Finite Intersection Property] Let $\{G_i\}_{i \in I}$ be a family of sets in a metric space (X, d). Then $\{G_i\}_{i \in I}$ is said to have finite intersection property (F.I.P), if each **finite** subfamily of $\{G_i\}_{i \in I}$ has a non-empty intersection, i.e., for any finite subset $I_1 \subset I$,

$$\bigcap_{i \in I_1} G_i \neq \phi.$$

Example. Any decreasing chain of non-empty subsets in a metric space has F.I.P.

Theorem [5.17] A metric space (X, d) is compact, if and only if each family of closed subsets in X with F.I.P has non-empty intersection.

Proof. Let (X, d) be a compact metric space. Let $\{F_i\}_{i \in I}$ be a collection of non-empty closed sets with F.I.P. We show that $\bigcap\limits_{i \in I} F_i \neq \phi$.

If possible, let $\bigcap\limits_{i \in I} F_i = \phi \Rightarrow \left(\bigcap\limits_{i \in I} F_i\right)' = \phi'$.

Then $\bigcup\limits_{i \in I} F_i' = X$.

This shows that $\{F_i'\}_{i \in I}$ is an open cover of X, since F_i are closed sets. Since (X, d) is compact, there exists $i_1, i_2, \dots, i_n \in I$ such that $F_{i_1}' \cup F_{i_2}' \cup \dots \cup F_{i_n}' = X$

$$\Rightarrow \left(\bigcap_{k=1}^{n} F_{i_k}\right)' = X$$

$$\Rightarrow \bigcap_{k=1}^{n} F_{i_k} = \phi \text{—a contradiction to F.I.P.}$$

Hence we must have $\bigcap_{i \in I} F_i \neq \phi$. Converse part of this theorem is the same and left as an exercise.

Theorem [5.18] Every closed bounded intervals of $\mathbb{R}$ is compact.

Proof. Let $[a, b]$ $(a < b)$ be a closed bounded interval of $\mathbb{R}$. Let A be an open cover of $[a, b]$. Let us put $Y = \{x \in [a, b] : [a, x] \text{ has a finite subcover from } A\}$.

Then clearly $Y \neq \phi$, as $a \in Y$. Since $Y \subset [a, b]$, it follows that Y is bounded. Let $\sup Y = M$. So we have $a \leq M \leq b$ and, therefore, there is a member $G_0 \in A$ such that $M \in G_0$. Since M is an interior point of G_0, we get an open intrval $(M - \delta, M + \delta)$, where $\delta > 0$ such that $M \in (M - \delta, M + \delta) \subset G_0$. Applying least upper bound property, there exists $x \in Y$ such that $M - \delta < x < M + \delta$. Since $x \in Y$, $[a, x]$ has a finite subcover from A and $[x, M] \subset (M - \delta, M + \delta) \subset G_0$.

Hence we see that $[a, M]$ has got a finite subcover from A. So $M \in Y$. We now show that $M = b$. Suppose that $M < b$. Take a number p between M and b so that $M < p < M + \delta < b$, then $[a, p] = [a, M] \cup [M, p]$, where $[a, M]$ has a finite subcover from A.

Note that $[M, p] \subset (M - \delta, M + \delta) \subset G_0$. Therefore, $[a, p]$ has a finite subcover from A. Hence, $p \in Y$ and $p > M$—a contradiction to the fact that $M = \sup Y$. Hence we must have $M = b$. Consequently $[a, b]$ has got a finite subcover from A and hence $[a, b]$ is compact.

Theorem [5.19] Continuous image of a compact set is compact.

Proof. Let $f : (X, d) \to (Y, \rho)$ be a continuous function and let (X, d) be compact. We now show that $f(X)$ is a compact subset of Y. Let A be an open cover of $f(X)$. Then for each $G \in A$, there exists an open set K in (Y, ρ) such that $G = K \cap f(X)$. Since f is continuous,

$$\{f^{-1}(K) : G = K \cap f(X)\}$$

forms an open cover of X. Since (X, d) is compact, $\exists\ i_1, i_2, \ldots, i_n$ such that

$$\begin{aligned} X &= f^{-1}(K_{i_1}) \cup f^{-1}(K_{i_2}) \cup \cdots \cup f^{-1}(K_{i_n}) \\ \Rightarrow\ & f(X) \subset K_{i_1} \cup K_{i_2} \cup \cdots \cup K_{i_n} \\ \Rightarrow\ & f(X) = (K_{i_1} \cap f(X)) \cup (K_{i_2} \cap f(X)) \cup \cdots \cup (K_{i_n} \cap f(X)) \\ &= G_{i_1} \cup G_{i_2} \cup \cdots \cup G_{i_n} \\ \Rightarrow\ & f(X) \text{ is compact.} \end{aligned}$$

Theorem [5.20] Let (X_1, d_1) and (X_2, d_2) be two compact metric spaces.

Then $X = X_1 \times X_2$ is compact with respect to the metric d given by

$$d(x, y) = \sqrt{d_1^2(x_1, x_2) + d_2^2(y_1, y_2)} \quad [x = (x_1, y_1),\ y = (x_2, y_2)]$$

Proof. Let $\{x_n\}$ and $\{y_n\}$ are sequences from X_1 and X_2 respectively. Then $\{(x_n, y_n)\}$ is a sequence from X. We show that $\{(x_n, y_n)\}$ has a convergent subsequence in X. Since X_1 is compact, $\{x_n\}$ has a convergent subsequence, say $\{x_{i_n}\} \to x \in X_1$.

Similarly, let $\{y_{k_n}\}$ be a convergent subsequence of $\{y_n\}$ in X_2 and let $y_{k_n} \to y$ in X_2 and the corresponding subsequence $\{x_{k_n}\}$ of $\{x_n\}$ converges to x. Observe that $k_n > i_n \geq n$.

Now for any $\epsilon > 0$, we can find a suitable positive integer N such that

$$d_1(x_{k_n}, x) < \frac{\epsilon}{\sqrt{2}} \text{ and } d_2(y_{k_n}, y) < \frac{\epsilon}{\sqrt{2}}, \ \forall\, n \geq N.$$

Now $d^2((x_{k_n}, y_{k_n}), (x, y)) = d_1^2(x_{k_n}, x) + d_2^2(y_{k_n}, y) < \dfrac{\epsilon^2}{2} + \dfrac{\epsilon^2}{2} = \epsilon^2.$

$\Rightarrow d((x_{k_n}, y_{k_n}), (x, y)) < \epsilon$

$\Rightarrow (x_{k_n}, y_{k_n}) \to (x, y)$

Thus the metric space (X, d) is sequentially compact and hence compact.

Corollary [5.5]: Let $(X_1, d_1), (X_2, d_2), \ldots, (X_n, d_n)$ be n compact metric spaces, then $X = X_1 \times X_2 \times \cdots \times X_n$ is compact with respect to the metric d, where

$$d(x, y) = \sqrt{\sum_{k=1}^{n} d_k^2(x_k, y_k)} \quad \begin{bmatrix} x = (x_1, x_2, \ldots, x_n) \\ y = (y_1, y_2, \ldots, y_n) \in X \end{bmatrix}$$

Note. (1) Let $(X_1, d_1), \ldots, (X_n, d_n)$ be n metric spaces. Then $X = X_1 \times X_2 \times \cdots \times X_n$ is a metric space with respect to the metric d given by

$$d(x, y) = \sqrt{\sum_{k=1}^{n} d_k^2(x_k, y_k)} \quad \begin{bmatrix} x = (x_1, x_2, \ldots, x_n) \\ y = (y_1, y_2, \ldots, y_n) \in X \end{bmatrix}$$

We define the k-th projection map $\Pi_k : X \to X_k$ by

$$\Pi_k(x_1, x_2, \ldots, x_k, \ldots, x_n) = x_k, 1 \leq k \leq n; (x_1, x_2, \ldots, x_n) \in X.$$

Applying sequential definition of continuity it is very easy to show that each Π_k is continuous.

(2) Converse of Corollary [5.5] is also true, i.e., if $(X_1, d_1), (X_2, d_2), \ldots, (X_n, d_n)$ are metric spaces such that (X, d) is compact, $X = X_1 \times X_2 \times \cdots \times X_n$.

Then each X_k is compact, $1 \leq k \leq n$.

Since $\Pi_k : X \to X_k$ is continuous and X is compact, by using Theorem [5.19] we see that $\Pi_k(X) = X_k$ is compact, $1 \leq k \leq n$.

Theorem [5.21] [Heine-Borel Theorem] A subset K of $\mathbb{R}^n$ is compact, if and only if it is closed and bounded.

Proof. Let $K \subset \mathbb{R}^n$ is compact. Then by Theorem [5.2], K is closed. On the other hand, $\Pi_i(K) \subset \mathbb{R}$ is a compact subset of $\mathbb{R}$, for each i and so we have $\Pi_i(K) \subset [a_i, b_i]$ for some $a_i, b_i \in \mathbb{R}$ with $a_i \leq b_i$, $1 \leq i \leq n$. Therefore, we have

$$K \subset [a_1, b_1] \times [a_2, b_2] \times \cdots \times [a_n, b_n]$$

and hence K is bounded, since the R.H.S. set is bounded.

Conversely, let K is closed and bounded. Then as just we saw, K is a closed subset of a product $[a_1, b_1] \times [a_2, b_2] \times \cdots \times [a_n, b_n]$. Since each $[a_i, b_i]$ is compact, it follows by the Corollary [5.4] $[a_1, b_1] \times \cdots \times [a_n, b_n]$ is compact. Since K is a closed subset of $[a_1, b_1] \times \cdots \times [a_n, b_n]$ it follows that K is compact.

Theorem [5.22] Let $f : (X, d) \to (Y, \rho)$ be a continuous function. If (X, d) is compact, then f is uniformly continuous.

Proof. Let us first recall the definition of uniform continuity.

"For any $\epsilon > 0$, there is a $\delta > 0$ such that $\rho(f(x), f(y)) < \epsilon$, whenever $d(x, y) < \delta$, $\forall\, x, y \in X$."

Let $\epsilon > 0$ be given. Then for each $x \in X$, f is continuous at x and so there exists $\delta_x > 0$ such that $\rho(f(x), f(x')) < \frac{\epsilon}{2}$, whenever $d(x, x') < \delta_x$.

Now note that the family of open ball $\left\{B_{\frac{1}{2}\delta x}(x) : x \in X\right\}$ is an open cover of the metric space (X, d). Now by compactness of (X, d) we get a finite subcover from the cover

$$\left\{B_{\frac{1}{2}\delta x}(x) : x \in X\right\}, \text{ say, } \left\{B_{\frac{1}{2}\delta x_1}(x_1), \ldots, B_{\frac{1}{2}\delta x_r}(x_r)\right\}.$$

Let us take $\delta = \min\left\{\frac{1}{2}\delta x_1, \frac{1}{2}\delta x_2, \ldots, \frac{1}{2}\delta x_r\right\}$.

Now for $x, x' \in X$ with $d(x, x') < \delta$, we get $x \in B_{\frac{1}{2}\delta x_i}(x_i)$ for some i, $1 \le i \le r$.

$$\text{So } d(x', x_i) \le d(x', x) + d(x, x_i) < \delta + \frac{1}{2}\delta x_i$$
$$\le \frac{1}{2}\delta x_i + \frac{1}{2}\delta x_i = \delta x_i$$

and hence $\rho(f(x), f(x')) \le \rho(f(x), f(x_i)) + \rho(f(x_i), f(x')) < \frac{\epsilon}{2} + \frac{\epsilon}{2} = \epsilon$.

This is true, for all $x, x' \in X$.

Thus f is uniformly a continuous function on X.

Note. Since $[a, b]$ is a compact subset of $\mathbb{R}$, by above theorem every continuous function $f : [a, b] \to \mathbb{R}$ is uniformly continuous.

Illustrative Examples:

Problem [5.1] : Let (X, d) be a metric space. A subset Y of X is compact, if and only if every cover of Y by the open sets in (X, d) has a finite subcover.

Solution. Let us first suppose that Y is compact, i.e., (Y, d_Y) is a compact metric space.

Let $\{V_i : i \in I\}$ be a collection of open sets in (X, d) such that

$$Y \subset \bigcup_{i\in I} V_i \Rightarrow Y = \left(\bigcup_{i\in I} V_i\right) \cap Y = \bigcup_{i\in I} (V_i \cap Y).$$

Since V_i is an open set in (X, d), the set $V_i \cap Y$ is an open set in (Y, d_Y), for all $i \in I$. So $\{V_i \cap Y : i \in I\}$ is an open cover of Y. Since (Y, d_Y) is compact, there exists $i_1, i_2, \ldots, i_r \in I$ such that

$$Y = (V_{i_1} \cap Y) \cup (V_{i_2} \cap Y) \cup \cdots \cup (V_{i_r} \cap Y)$$
$$= (V_{i_1} \cup V_{i_2} \cup \cdots \cup V_{i_r}) \cap Y$$
$$\Rightarrow Y \subset V_{i_1} \cup V_{i_2} \cup \cdots \cup V_{i_r}.$$

Conversely, let the condition hold. We show that (Y, d_Y) is compact. Let $\{W_i : i \in I\}$ be an open cover of (Y, d_Y). So for each $i \in I$, we can find an open set U_i in (X, d) such that $W_i = U_i \cap Y$. Then $\{U_i : i \in I\}$ is a family of open sets in (X, d) such that $Y \subset \bigcup_{i\in I} U_i$.

By assumption, there exists $i_1, i_2, \dots, i_k \in I$ such that

$$\begin{aligned} Y &\subset U_{i_1} \cup U_{i_2} \cup \cdots \cup U_{i_k} \\ \Rightarrow Y &= (U_{i_1} \cup U_{i_2} \cup \cdots \cup U_{i_k}) \cap Y \\ &= (U_{i_1} \cap Y) \cup (U_{i_2} \cap Y) \cup \cdots \cup (Ui_k \cap Y) \\ &= W_{i_1} \cup W_{i_2} \cup \cdots \cup W_{i_k}. \end{aligned}$$

Hence Y is compact.

Problem [5.2] : Let (X, d) be a metric space and $\{x_n\}$ be a sequence in X such that $x_n \to x$ as $n \to \infty$ for some $x \in X$. Then the set $\{x, x_n : n \in \mathbb{N}\}$ is compact.

Solution. Let $Y = \{x, x_n : n \in \mathbb{N}\}$.

Let $\{G_i\}_{i \in I}$ be a family of open sets in (X, d) such that $Y \subset \bigcup_{i \in I} G_i$.

Then there exists $i_0 \in I$ such that $x \in G_{i_0}$. Since G_{i_0} is an open set, it follows that G_{i_0} is a neighbourhood of x. Hence there exists $N \in \mathbb{N}$ such that $x_n \in G_{i_0}$, for all $n \geq N$.

Now for each $k = 1, 2, \dots, N-1$, there exists $i_k \in I$ such that $x_k \in G_{i_k}$.

Hence $Y \subset G_{i_0} \cup G_{i_1} \cup \cdots \cup G_{i_{N-1}}$.

$\Rightarrow Y$ is compact [by Problem [5.1]].

Alternative

It is clear that Y is complete. Now let $\epsilon > 0$ be arbitrary. Consider the open ball $B_\epsilon(x)$. Since $x_n \to x$ as $n \to \infty$, there exists $N \in \mathbb{N}$ such that $x_n \in B_\epsilon(x)$, for all $n \geq N$.

Note that $Y \subset B_\epsilon(x) \cup B_\epsilon(x_1) \cup \cdots \cup B_\epsilon(x_{N-1})$

This implies that $\{x, x_1, \dots, x_{N-1}\}$ is a finite ϵ-net of Y.

Thus, Y is totally bounded. Hence Y is compact.

Problem [5.3] : Let (X, d) be a metric space and $A \subset X$. Show that A is totally bounded, if and only if $\bar{A}$ is totally bounded.

Solution. It is very easy to show that subset of a totally bounded set is totally bounded. Since $A \subset \bar{A}$, then if $\bar{A}$ is totally bounded, so is A.

Now assume that A is totally bounded. Then for $\epsilon > 0$, there exists $x_1, x_2, \dots, x_n \in A$ such that

$$A \subset B_{\epsilon/2}(x_1) \cup B_{\epsilon/2}(x_2) \cup \cdots \cup B_{\epsilon/2}(x_n). \tag{1}$$

Let $x \in \bar{A}$. Then there exists $y \in A$ such that $d(x, y) < \epsilon/2$. Since $y \in A$, from (1) there exists $1 \leq j \leq n$ such that

$$y \in B_{\epsilon/2}(x_j), \text{ i.e., } d(x_j, y) < \frac{\epsilon}{2}.$$

Now $d(x, x_j) \leq d(x, y) + d(y, x_j) < \frac{\epsilon}{2} + \frac{\epsilon}{2} = \epsilon$

$\Rightarrow x \in B_\epsilon(x_j),\ 1 \leq j \leq n.$

Thus $\bar{A} \subset B_\epsilon(x_1) \cup B_\epsilon(x_2) \cup \cdots \cup B_\epsilon(x_n)$

$\Rightarrow \bar{A}$ is totally bounded.

Problem [5.4] : Let $f : (X, d) \to (Y, \rho)$ be a continuous function, where (X, d) is compact. Show that $f(\bar{A}) = \overline{f(A)},\ A \subset X$.

Solution. Note that $\overline{f(A)}$ is a closed set in (Y, ρ) and

$$f(A) \subset \overline{f(A)} \Rightarrow f^{-1}(f(A)) \subset f^{-1}(\overline{f(A)})$$
$$\Rightarrow A \subset f^{-1}(\overline{f(A)}) \; [\because A \subset f^{-1}(f(A)), \forall A \subset X]$$
$$\Rightarrow \bar{A} \subset f^{-1}(\overline{f(A)}) \; [\text{since } f \text{ is continuous and } \overline{f(A)} \text{ is a closed set, we have } f^{-1}(\overline{f(A)}) \text{ is a closed set.}]$$
$$\Rightarrow f(\bar{A}) \subset \overline{f(A)} \; [\text{since } f(f^{-1}(B)) \subset B, \forall B \subset Y]. \quad (1)$$

Since (X, d) is compact and $\bar{A}$ is a closed subset of X, it follows that $\bar{A}$ is compact. So $f(\bar{A})$ is compact in (Y, ρ) and hence $f(\bar{A})$ is a closed set in (Y, ρ).

We have $A \subset \bar{A} \Rightarrow f(A) \subset f(\bar{A}) \Rightarrow \overline{f(A)} \subset \overline{f(\bar{A})} = f(\bar{A})$.

$$\therefore \overline{f(A)} \subset f(\bar{A}). \quad (2)$$

From (1) and (2), we get $f(\bar{A} = \overline{f(A)}$, for $A \subset X$.

Problem [5.5]: If A and B are two disjoint closed subsets in (X, d), give an example to show that $d(A, B)$ may be zero.

Solution. Remember Example 2 in Chapter 2. In Example 2, we find that if $A = \mathbb{N}$ and

$$B = \left\{ n + \frac{1}{2n} : n \in \mathbb{N} \right\},$$

then $d(A, B) = 0$.

Note that A, B are closed sets in $(\mathbb{R}, d)$, where d is the usual metric of $\mathbb{R}$ and $A \cap B = \phi$.

Problem [5.6]: Let (X, d) be a metric space and A is a compact subset of X and $x \in X$. Then show that there is a point a in A such that $d(x, a) = \text{dist}(x, A)$. Hence show that, if A and B are both compact, then there exists points $a \in A$ and $b \in B$ such that $d(a, b) = d(A, B)$.

Solution. Define $f : X \to \mathbb{R}$ by $f(y) = d(x, y)$, for all $y \in X$.

Then f is a continuous function on X. Since A is a compact subset of X, $f(A)$ is compact in $\mathbb{R}$. Hence $f(A)$ is closed and bounded and thus attains its bounds. So there is a point $a \in A$ such that $f(a) = \inf f(A) = \inf\{d(x, y) : y \in A\} \Rightarrow d(x, a) = \text{dist}(x, A)$.

Now define $g : A \to \mathbb{R}$ by $g(x) = \text{dist}(x, B)$, $\forall\, x \in A$. Then by Theorem [4.9], g is continuous and by the same argument as above, there exists a point $a \in A$ such that

$$g(a) = \inf g(A) = \inf\{\text{dist}(x, B) : x \in A\}$$
$$\text{dist}(a, B) = d(A, B)$$

Since B is compact, by the above case, there exists a point $b \in B$ such that $d(a, b) = \text{dist}(a, B)$.

Hence we find $a \in A, b \in B$ such that $d(a, b) = d(A, B)$.

Problem [5.7]: Let A be a compact set in the metric space (X, d) and let B be a closed set in (X, d) such that $A \cap B = \phi$. Show that $d(A, B) > 0$.

Solution. If possible, suppose that $d(A, B) = 0$. Then by the above problem, there exists $a \in A$ such that $\text{dist}(a, B) = d(A, B) = 0 \Rightarrow \text{dist}(a, B) = 0$

$\Rightarrow a \in \bar{B} = B$, since B is a closed set [From Theorem [2.6]].

Hence $a \in A \cap B \neq \phi$—a contradiction.

Thus $d(A, B) > 0$.

Problem [5.8] : Let (X, d) be a metric space and $A, B \subset X$ be non-empty and B is compact. Prove that $d(A, B) = 0$, if and only if $\bar{A} \cap B \neq \phi$.

Solution. Suppose that $d(A, B) = 0$. Then by Problem [5.7], there exists a point $b \in \bar{A} \cap B$ and so $\bar{A} \cap B \neq \phi$.

Conversely, let $\bar{A} \cap B \neq \phi$. Then from definition, we get $d(\bar{A}, B) = 0$

$\Rightarrow d(A, B) = 0$, from the result $d(A, B) = d(\bar{A}, B)$.

Thus $d(A, B) = 0$, if and only if $\bar{A} \cap B \neq \phi$.

Problem [5.9] : Let A be a compact subset of a metric space (X, d). Show that there is a pair of points $a, b \in A$ such that $d(a, b) = \text{diam}\,(A)$.

Solution. Since A is a compact subset of X, by Theorem [5.20], it follows that $A \times A$ is a compact subset of $X \times X$.

Since $d : X \times X \to \mathbb{R}$ is continuous, we have $d(A \times A)$ is compact and hence closed and bounded subset of $\mathbb{R}$.

So we can find points $a, b \in A$ such that

$$d(a, b) = \sup d(A \times A) = \sup\{d(x, y) : x, y \in A\} = \text{diam}\,(A).$$

EXERCISE

(1) Let (X, d) be a discrete metric space, where X is an infinite set. Find totally bounded subsets of (X, d).

(2) Give an example of a separable metric space which is **not** totally bounded.

(3) Show that continuous image of a compact metric space is compact.

(4) Let (X, d) be a metric space and $\{x_n\}$ be sequence in (X, d) such that $x_n \to x \in X$ as $n \to \infty$. Let us take $S - \{x\} \cup \{x_n : x \in \mathbb{N}\}$. Show that

(a) S is totally bounded.

(b) S is compact.

(5) Let A and B be two compact subsets of a metric space (X, d). Let $d(A, B) = \inf\{d(x, y) : x \in A, y \in B\}$. Show that there are points $a \in A, b \in B$ such that $d(a, b) = d(A, B)$.

(6) If A and B are two non-empty disjoint closed sets in a metric space and if one of which is compact, then show that $d(A, B) > 0$.

(7) Give examples of two non-empty closed sets A, B in a metric space (X, d) such that $A \cap B = \phi$ but $d(A, B) = 0$.

(8) Show that subset of a totally bounded metric space is totally bounded.

(9) Let A be a compact subset of a metric space (X, d). Show that for any $B \subseteq X$, there is a point $a \in A$ such that $d(a, B) = d(a, B)$.

(10) Let A be a compact subset of a metric space (X, d). Show that the derived set A' of A is compact.

(11) Let us consider the metric space $(\mathbb{R}^2, d)$ where d is the usual metric of $\mathbb{R}^2$.
Let $A = \{(x, y) \in \mathbb{R}^2 : x^2 + y^2 < 4\}$. Show that the set $\{(1, -1), (1, 0), (1, 1), (0, -1), (0, 0), (0, 1), (-1, -1), (-1, 0), (-1, 1)\}$ is a finite $\frac{3}{2}$-net of A but not finite $\frac{1}{2}$-net of A.

(12) Prove that finitc union of compact sets is compact.

(13) Let (X, d) be a metric space. Let F_1 and F_2 be two subsets of X such that one is compact and another is closed. Show that $F_1 \cap F_2$ is compact.

(14) Show that the open ball $B_1(0)$ having centre at origin and radius 1 is not compact.

(15) Prove that if every closed ball $\bar{B}_r(x)$ in a metric space (X, d) is compact ($r > 0$ and $x \in X$), then X is complete.

(16) If closed and bounded subsets of a metric space (X, d) are compact, then prove that (X, d) is complete.

(17) Let (X, d) be a complete metric space and A be a totally bounded subset of X. Prove that $\bar{A}$ is compact.

(18) Determine whether the following sets are compact or not:

(a) $\{(x, y) \in \mathbb{R}^2 : x \geq 0, y \geq 0\}$

(b) $\{(x, y) \in \mathbb{R}^2 : x^2 + y^2 = 1\}$

(c) for $0 < a < b, \{(x, y) \in \mathbb{R}^2 : a \leq x^2 + y^2 \leq b\}$

(d) $\{(x, y) \in \mathbb{R}^2 : y = \sin \frac{1}{x}, x \in (0, 1]\}$

(e) $\{(x, y) \in \mathbb{R}^2 : |xy| \leq 1\}$

(f) $\{(x, y) \in \mathbb{R}^2 : x^2 + y^2 \leq 1\}$

(g) $\{(x, y(\in \mathbb{R}^2 : x^2 + y^2 < 1\}$

(h) $\left\{(x, y) \in \mathbb{R}^2 : x \geq 1, 0 \leq y \leq \frac{1}{x}\right\}$

In all of the above cases, $\mathbb{R}^2$ is equipped with the usual metric.

CHAPTER 6

CONNECTED METRIC SPACE

6.1 Separated Sets

Definition [6.1] : Let (X, d) be a metric space. Two non-empty subsets A and B of X are said to be separated if $A \cap \bar{B} = \phi$ and $\bar{A} \cap B = \phi$, $\bar{A}$ denotes closure of A in (X, d).

Since $A \subset \bar{A}$, it follows from definition that

$$A \cap B \subset \bar{A} \cap B = \phi \Rightarrow A \cap B = \phi,$$

i.e., if A and B are two separated sets, then the sets A and B are disjoint.

However, the converse is **not** true, in general. For example, take $A = Q$, the set of all rationals and $B = I$, the set of all irrationals in the metric space $(\mathbb{R}, d)$, where d is the usual metric on $\mathbb{R}$. Then we have $\bar{A} = \bar{Q} = \mathbb{R}, \bar{B} = \bar{I} = \mathbb{R}$ and

$$\bar{A} \cap B = \bar{Q} \cap I = \mathbb{R} \cap I = I \neq \phi,$$

i.e., A and B are **not** separated though $A \cap B = \phi$.

Example [6.1] : In the metric space $(\mathbb{R}, d)$ [d is the usual metric of $\mathbb{R}$], let us take

$$A = (0, 1), B = (2, 3).$$

Then we see that $\bar{A} \cap B = \phi$ and $A \cap \bar{B} = \phi$, i.e., A and B are separated sets.

If we take $C = [1, 2]$, then A and C are **not** separated, since

$$\bar{A} \cap C = [0, 1] \cap [1, 2] = \{1\} \neq \phi.$$

Theorem [6.1] Let (X, d) be a metric space and $Y \subseteq X$. Let $A, B \subseteq Y$. Then A and B are separated in (X, d), if and only if A and B are separated in (Y, d_Y).

Proof. Remembering Problem 11 of Chapter 1, we can write $\bar{A}^Y = \bar{A}^X \cap Y$, $\bar{A}^Y$ is the closure of A in (Y, d_Y) and $\bar{A}^X$ denotes the closure of A in (X, d).

Now note that

$$\bar{A}^Y \cap B = \bar{A}^X \cap Y \cap B = \bar{A}^X \cap B \; [\because B \subset Y].$$

Similarly, $A \cap \bar{B}^Y = A \cap \bar{B}^X \; [\because A \subset Y]$.

Hence, $\bar{A}^X \cap B = \phi, A \cap \bar{B}^X = \phi$, if and only if $\bar{A}^Y \cap B = \phi$ and $A \cap \bar{B}^Y = \phi$.

Thus A, B are separated in (X, d), if and only if A, B are separated in (Y, d_Y).

Remark. Before going to our next result, let us first recall an important result of Chapter-1, Example 1.

"If U and V are subsets of a metric space (X,d) with U an open set, show that

(a) $U \cap V = \phi \Rightarrow U \cap \bar{V} = \phi$

(b) $U \cap \bar{V} \subseteq \overline{U \cap V}$."

From this result we can say that if U and V are two open sets in a metric space, then

$$U \cap V = \phi \Rightarrow U \cap \bar{V} \subseteq \overline{U \cap V} = \bar{\phi} = \phi \text{ and } \bar{U} \cap V \subseteq \overline{U \cap V} = \bar{\phi} = \phi$$
$$\Rightarrow U \cap \bar{V} = \phi \text{ and } \bar{U} \cap V = \phi$$
$$\Rightarrow U \text{ and } V \text{ are separated.}$$

On the other hand, if U and V are two separated sets in a metric space (X,d), then

$$U \cap V = \phi.$$

Thus we have proved the following result.

Theorem [6.2] Let (X,d) be a metric space and U, V are two **open sets** in (X,d). Then U, V are separated, if and only if $U \cap V = \phi$.

Similar result holds, if we replace the word "open" by "closed" which is our next result.

Theorem [6.3] Let (X,d) be a metric space and U, V are two **closed sets** in (X,d). Then U, V are separated, if and only if $U \cap V = \phi$.

Proof. If U, V are separated, then $U \cap V = \phi$. So it only remains to show that if U, V are two disjoint closed sets then they are separated. Since U, V are closed sets, we have $\bar{U} = U$ and $\bar{V} = V$. Thus $U \cap \bar{V} = U \cap V = \phi$ and $\bar{U} \cap V = U \cap V = \phi$.

This completed the proof.

Note. Let A, B be two separated sets in a metric space (X,d), i.e., $A \cap \bar{B} = \phi$ and $\bar{A} \cap B = \phi$. Let $P \subset A$ and $Q \subset B$. So $\bar{P} \subset \bar{A}$ and $\bar{Q} \subset \bar{B}$. Then it follows that

$$P \cap \bar{Q} \subset A \cap \bar{B} = \phi \text{ and } \bar{P} \cap Q \subset \bar{A} \cap B = \phi$$
$$\text{i.e., } P \cap \bar{Q} = \phi \text{ and } \bar{P} \cap Q = \phi.$$

This shows that P and Q are separated. Thus we have the following theorem.

Theorem [6.4] Let (X,d) be a metric space and A, B be two separated sets in (X,d). If $P \subset A$ and $Q \subset B$, then P, Q are also separated.

Theorem [6.5] Let (X,d) be a metric space and A, B bee two subsets of X. If both A and B are closed or open, then $A - B$, $B - A$ are separated.

Proof. First suppose that A and B are closed sets. Let $P = A - B, Q = B - A$.

Then
$$\begin{aligned} P \cap \bar{Q} &= (A - B) \cap \overline{B - A} \\ &= (A \cap B^c) \cap \overline{B \cap A^c} \\ &\subset (A \cap B^c) \cap (\bar{B} \cap \bar{A}^c) \quad [\because \overline{A \cap B} \subset \bar{A} \cap \bar{B}] \\ &= (A \cap \bar{A}^c) \cap (B \cap B^c) \quad [\because B \text{ is closed}] \\ &= \phi \\ \Rightarrow P \cap \bar{Q} &= \phi. \end{aligned}$$

Similarly, $\bar{P} \cap Q = \phi$.

Let us now suppose that A, B are open sets.

Then $\quad P \cap \bar{Q} = (A - B) \cap \overline{B - A}$

$$= (A \cap B^c) \cap \overline{B \cap A^c}$$
$$\subset (A \cap B^c) \cap (\bar{B} \cap \bar{A^c})$$
$$= (A \cap \bar{A^c}) \cap (B^c \cap \bar{B})$$
$$\subset \overline{(A \cap A^c)} \cap (B^c \cap \bar{B})$$

[$\because$ A is open and using (b) of Remark before Theorem [6.2].]

$$= \phi$$

$\Rightarrow P \cap \bar{Q} = \phi$.

Similarly, $\bar{P} \cap Q = \phi$.

Hence, in both the cases, P, Q are separated.

Note. The above result is **not** true if we remove "both open or closed" condition. For example, take $A = (0, 1]$, $B = (1, 2)$ in $(\mathbb{R}, d)$, d is the usual metric of $\mathbb{R}$.

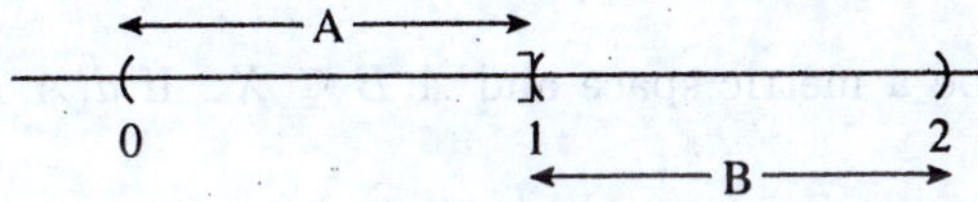

Fig. 6.1

Then $P = (0, 1]$ and $Q = (1, 2)$, where $P = A - B, Q = B - A$.

Now $P \cap \bar{Q} = (0, 1] \cap [1, 2] = \{1\} \neq \phi$

$\Rightarrow P, Q$ are **not** separated.

Note. Let (X, d) be a metric space and A, B are two open sets in (X, d). Then we know that $A \cup B$ is open in (X, d). What can we say about the converse, i.e., if $A \cup B$ is open in (X, d), then is it possible to claim that A, B are open sets? Answer is 'no', in general. For example, let $A = (0, 1)$, $B = [1, 2)$, then $A \cup B = (0, 1) \cup [1, 2) = (0, 2)$ is an open set in $(\mathbb{R}, d)$, but B is **not** open. It is very important to observe that the sets A, B are **not** separated, since

$$\bar{A} \cap B = [0, 1] \cap [1, 2) = \{1\} \neq \phi.$$

Our next theorem is about the converse of the above statement.

Theorem [6.6] Let (X, d) be a metric space and let A, B are two separated sets. If $A \cup B$ is open in (X, d), then A, B are also open sets in (X, d).

Proof. Let $A \cup B$ be open. It is given that A, B are separated sets. Then $A \cap \bar{B} = \phi$ and $\bar{A} \cap B = \phi$.

$\Rightarrow A \subset X - \bar{B}$

$\Rightarrow A \cap (X - \bar{B}) = A$.

Since $B \subset \bar{B}$, we have $B \cap (X - \bar{B}) = \phi$.

$\therefore$ we can write $A = [A \cap (X - \bar{B})] \cup [B \cap (X - \bar{B})] = (A \cup B) \cap (X - \bar{B})$.

Since $\bar{B}$ is a closed set, $X - \bar{B}$ is an open set.

So $A = (A \cup B) \cap (X - \bar{B})$ is an open set.

In a similar way we can show that $B = (A \cup B) \cap (X - \bar{A})$ is an open set.

Note. Above theorem is also valid if we replace the word "open" by "closed".

Theorem [6.7] Let (X, d) be a metric space and A, B are two separated sets in (X, d). If $A \cup B$ is closed, then A, B are closed in (X, d).

Proof. Let $A \cup B$ be a closed set such that $A \cap \bar{B} = \phi$ and $\bar{A} \cap B = \phi$. Since $A \cup B$ is closed, we have $A \cup B = \overline{A \cup B} = \bar{A} \cup \bar{B}$. Since $\bar{A} \subset \bar{A} \cup \bar{B}$, we get

$$\begin{aligned}\bar{A} &= \bar{A} \cap (\bar{A} \cup \bar{B}) = \bar{A} \cap (A \cup B) \\ &= (\bar{A} \cap A) \cup (\bar{A} \cap B) = A \cup \phi \; [\because \bar{A} \cap B = \phi] \\ &= A.\end{aligned}$$

$\therefore$ A is a closed set.

Similarly, B is a closed set.

Note. From Theorem [6.6] and Theorem [6.7], we get "If A, B are two sets in a metric space such that A, B are separated. Then

(i) $A \cup B$ is open, if and only if A, B are open.

(ii) $A \cup B$ is closed, if and only if A, B are closed.

Theorem [6.8] Let (X, d) be a metric space and $A, B \subseteq X$. If $d(A, B) > 0$, then A, B are separated sets in (X, d).

Proof. Suppose that $d(A, B) > 0$. Let $d(A, B) = r > 0$.

Recall the definition of $d(A, B)$,

$$d(A, B) = \inf\{d(x, y) : x \in A, y \in B\}.$$

Then we get $d(x, y) \geq r$, $\forall\, x \in A$ and $\forall\, y \in B$. Clearly, $A \cap B = \phi$.

Let $x \in A$, then $x \notin B'$, i.e., x is **not** a limit point of B, since $B_{r/2}(x) \cap B = \phi$.

Hence $A \cap \bar{B} = \phi$.

Similarly, $\bar{A} \cap B = \phi$.

Alternative

It is very important to observe that if $d(A, B) > 0$, then we must have $A \cap B = \phi$. Recall the result from Chapter-2, Example [2.16]

$$d(A, B) = d(\bar{A}, B) = d(A, \bar{B}) = d(\bar{A}, \bar{B}).$$

Since $d(A, B) > 0$, it follows that $d(\bar{A}, B) > 0$ and $d(A, \bar{B}) > 0$ and hence we get $\bar{A} \cap B = \phi$ and $A \cap \bar{B} = \phi$. Thus A and B are separated sets.

Note. But the converse of the above result is **not** true. For example, let $A = [0, 1)$ and $B = (1, 2]$ in $(\mathbb{R}, d)$, where d is the usual metric on $\mathbb{R}$. Then

$$A \cap \bar{B} = [0, 1) \cap [1, 2] = \phi$$
$$\text{and } \bar{A} \cap B = [0, 1] \cap (1, 2] = \phi$$

i.e., A and B are separated sets.

But $d(A, B) = d(\bar{A}, \bar{B}) = 0$, since $\bar{A} \cap \bar{B} = [0, 1] \cap [1, 2] = \{1\} \neq \phi$.

Take another example in $(\mathbb{R}, d)$.

Let $A = \{x \in \mathbb{R} : x < 0\} = (-\infty, 0)$ and $B = \{x \in \mathbb{R} : x > 0\} = (0, \infty)$. Then A, B are disjoint open sets and so by Theorem [6.2], it follows that A, B are separated sets in $(\mathbb{R}, d)$.

But since $\bar{A} = (-\infty, 0]$ and $\bar{B} = [0, \infty)$, we have $\bar{A} \cap \bar{B} = (-\infty, 0] \cap [0, \infty) = \{0\} \neq \phi$ and hence $d(A, B) = d(\bar{A}, \bar{B}) = 0$.

Problem [6.1] : Show that two sets A, B in a metric space (X, d) are separated, if and only if there exist open sets U, V in (X, d) such that

$$A \subseteq U,\ B \subseteq V,\ A \cap V = \phi = B \cap U.$$

Solution. First suppose that A, B are separated. Then $A \cap \bar{B} = \phi$ and $\bar{A} \cap B = \phi$.

Then $A \subseteq X - \bar{B}$ and $B \subseteq X - \bar{A}$.

Let us take $U = X - \bar{B}$ and $V = X - \bar{A}$. Since $\bar{A}, \bar{B}$ are closed sets, it follows that U, V are open sets in (X, d).

Now $A \cap V = A \cap (X - \bar{A}) = \phi$ [since $A \subset \bar{A}$]

Similarly, $B \cap U = \phi$.

Thus we obtain two open sets U, V in (X, d) such that $A \subseteq U, B \subseteq V, A \cap V = \phi = B \cap U$.

Conversely, suppose that the condition holds, i.e., there exist two open sets U, V in (X, d) such that $A \subseteq U, B \subseteq V, A \cap V = \phi = B \cap U$.

We now show that A, B are separated, i.e., $A \cap \bar{B} = \phi$ and $\bar{A} \cap B = \phi$.

Suppose that $A \cap \bar{B} \neq \phi$ and let $x \in A \cap \bar{B}$

$\Rightarrow x \in A$ and $x \in \bar{B}$

$\Rightarrow x \in U$ and $x \in \bar{B}$ [$\because A \subset U$]

$\Rightarrow B \cap U \neq \phi$ [since U is an open set]—a contradiction.

Thus $A \cap \bar{B} = \phi$.

Similarly, $\bar{A} \cap B = \phi$.

Note. The above proof in the converse part can also be shown as follows:

Recall the result as given in the Remark after Theorem [6.1], keeping this result in mind, note that $U \cap \bar{B} \subseteq \overline{U \cap B} = \phi$, since U is an open set.

$\Rightarrow U \cap \bar{B} = \phi$

$\Rightarrow A \cap \bar{B} = \phi$, since $A \subseteq U$.

Similarly, $\bar{A} \cap V \subset \overline{A \cap V} = \phi$

$\Rightarrow \bar{A} \cap V = \phi$

$\Rightarrow \bar{A} \cap B = \phi$, since $B \subseteq V$.

Problem [6.2] : Let (X, d) be a metric space and A, B are two non-empty separated sets in (X, d). Show that there exist open sets U, V in (X, d) such that $A \subseteq U, B \subseteq V, U \cap V = \phi$.

Solution. Let us consider the function $f : X \to \mathbb{R}$ defined by

$$f(x) = \text{dist}\,(x, A) - \text{dist}\,(x, B),\ \forall\, x \in X.$$

Then f is a continuous function.

Let us take $U = \{x \in X : f(x) < 0\}$ and $V = \{x \in X : f(x) > 0\}$.

Now let $x \in A$. Then $\text{dist}\,(x, A) = 0$.

Since $A \cap \bar{B} = \phi$ and $\text{dist}\,(x, B) = \text{dist}\,(x, \bar{B})$, it follows that $d(x, B) > 0$.

$$\therefore\ f(x) < 0, \text{ if } x \in A \Rightarrow A \subset U.$$

Similarly, $B \subset V$. Also note that $U = f^{-1}\{(-\infty, 0)\}$ and $V = f^{-1}\{(0, \infty)\}$.

Then U, V are open sets in (X, d), since f is continuous and $(-\infty, 0), (0, \infty)$ are open sets in $\mathbb{R}$ with respect to the usual metric.

Clearly, $U \cap V = \phi$.

Thus we get open sets U, V in (X, d) such that $A \subseteq U, B \subseteq V$ and $U \cap V = \phi$.

6.2 Connectedness and Disconnectedness

Definition [6.2] : A metric space (X, d) is said to be **disconnected** if X can be expressed as the union of two **non-empty** separated sets, that means there exist two sets A, B in (X, d) such that

$$X = A \cup B, A \cap \bar{B} = \phi \text{ and } \bar{A} \cap B = \phi,$$
$$A \neq \phi, B \neq \phi.$$

A metric space (X, d) is said to be **connected** if it is **not** disconnected. Thus a metric space is connected if it is **not** possible to find two non-empty separated sets A, B in (X, d) such that $X = A \cup B$.

Note. We have seen in Theorem [6.2] and Theorem [6.3] that two non-empty disjoint open (or closed) sets and separated sets coincide in a metric space. With the help of these theorems we can say that

Theorem [6.9] In a metric space (X, d) the following conditions are equivalent:

(1) (X, d) is disconnected.

(2) There exists two non-empty open sets A, B in (X, d) such that $X = A \cup B, A \cap B = \phi$.

(3) There exists two non-empty closed sets A, B in (X, d) such that $X = A \cup B, A \cap B = \phi$.

(4) There exists a non-empty set U with $U \neq X$ in (X, d) such that U is both open and closed in (X, d).

Proof. **(1) ⇒ (2)**

Let (X, d) be disconnected.

Then by definition there exist two non-empty separated sets A, B in (X, d) such that $X = A \cup B$.

Since X is open, we get $A \cup B$ is open, where A, B are non-empty separated sets. Then by Theorem [6.6], A and B are open in (X, d).

(2) ⇒ (1)

Suppose that there exist two non-empty open sets A, B in (X, d) such that

$$X = A \cup B, A \cap B = \phi.$$

Since A, B are non-empty disjoint open sets, it follows by Theorem [6.2] that A and B are separated sets. Thus **(1) ⇔ (2)**.

Similarly we can show that **(1) ⇔ (3)**. We now prove (3) ⇔ (4).

(3) ⇒ (4)

Suppose that there exist two non-empty closed sets A, B in (X, d) such that

$$X = A \cup B, A \cap B = \phi.$$

Then since $A \cap B = \phi$, $U = X - B = A$. Since A is closed, U is a closed set.

Again, since B is closed, $X - B$ is open, i.e., U is an open set.

$\therefore U$ is both open and closed.

(4) ⇒ (3)

Let us assume that U be a non-empty set in (X, d) which is both open and closed.

Since U is non-empty and $U \neq X$, it follows that $X - U$ is non-empty. Also we have $X = U \cup (X - U)$ and $U \cap (X - U) = \phi$. Since U is open, $X - U$ is closed and (3) follows.

Note. In view of the above theorem we can say that a metric space (X, d) is disconnected, if and only if we can find two non-empty open (or closed) sets in (X, d) such that

$$X = A \cup B, A \cap B = \phi.$$

Definition [6.3] : Let (X, d) be a metric space and $Y \subseteq X$. Then the subset Y of X is said to be disconnected if the subspace (Y, d_Y) is disconnected.

Discussion. Let (X, d) be a metric space and Y be a subset of X. Suppose that Y is disconnected. Then by definition above, the subspace (Y, d_Y) is disconnected. So there exist two non-empty open sets G, H in (Y, d_Y) such that

$$Y = G \cup H, G \cap H = \phi.$$

Since G, H are open in (Y, d_Y), there exist open sets A, B in (X, d) such that

$$G = A \cap Y, H = B \cap Y.$$

So we have $Y = (A \cap Y) \cup (B \cap Y) = (A \cup B) \cap Y \Rightarrow Y \subseteq A \cup B$.

Now $G \cap H = \phi \Rightarrow (A \cap Y) \cap (B \cap Y) = \phi \Rightarrow (A \cap B) \cap Y = \phi$.

Thus we find two non-empty open sets A, B in (X, d) such that

$$Y \subseteq A \cup B \text{ and } A \cap Y \neq \phi, B \cap Y \neq \phi \text{ and } A \cap B \cap Y = \phi.$$

Conversely, let there exist two open sets A, B in (X, d) such that

$$Y \subseteq A \cup B \text{ and } A \cap Y \neq \phi, \; B \cap Y \neq \phi \text{ and } A \cap B \cap Y = \phi.$$

Then $Y = (A \cup B) \cap Y = (A \cap Y) \cup (B \cap Y) = G \cup H$, where

$$G = A \cap Y, H = B \cap Y \text{ (say)}.$$

Since A, B are open sets in (X, d), the sets G and H are open in (Y, d_Y). Also

$$G = A \cap Y \neq \phi, H = B \cap Y \neq \phi$$

$$\text{and } G \cap H = (A \cap Y) \cap (B \cap Y) = (A \cap B) \cap Y = \phi.$$

Thus we find two non-empty open sets G, H in (Y, d_Y) such that $Y = G \cup H, G \cap H = \phi$.

$\Rightarrow (Y, d_Y)$ is disconnected.

In view of the above discussion we have the following theorem.

Theorem [6.10] Let (X, d) be a metric space and Y be a subset of X. Then Y is disconnected, if and only if there exist two non-empty open sets A, B in (X, d) such that

$$Y \subseteq A \cup B, \; A \cap Y \neq \phi, B \cap Y \neq \phi, A \cap B \cap Y = \phi.$$

Example [6.2] : Consider the metric space $(\mathbb{R}, d)$, d is the usual metric of $\mathbb{R}$.

Let $Y = (0, 1) \cup (2, 3)$.

Then $Y \subseteq \mathbb{R}$.

Note that Y is disconnected, since $(-1,1),(2,4)$ are open sets in $(\mathbb{R},d)$ and

$$Y \subseteq (-1,1)\cup(2,4).$$

$(-1,1)\cap Y \neq \phi, (2,4)\cap Y \neq \phi, (-1,1)\cap(2,4)\cap Y = \phi.$

Examples of Connected and Disconnected Sets

(i) The empty set ϕ in any metric space (X,d) is connected.

(ii) Every singleton set in a metric space (X,d) is connected.

(iii) Let (X,d) denotes the **trivial metric space**. In Chapter-1, Illustrative Example (8) we have shown that every subset of X is both open and closed and hence by Theorem [6.9], the metric space (X,d) is **disconnected**, provided that X contains more than one point.

(iv) In $(\mathbb{R},d)$, where d is the usual metric of $\mathbb{R}$, the subset $Q\subseteq\mathbb{R}$ is disconnected, since

$$Q\subseteq(-\infty,a)\cup(a,\infty),$$

where a is an irrational number $[(-\infty,a),(a,\infty)$ are open in $(\mathbb{R},d)]$.

Note. Theorem [6.10] remains valid if we replace the word "open" by the word "closed" (Prove it).

Theorem [6.11] Let (X,d) be metric space and Y be a connected subset of (X,d). If $Y\subseteq G\cup H$, where G,H are separated sets in (X,d), then either $Y\subseteq G$, or $Y\subseteq H$.

Proof. Let $Y\subseteq G\cup H$, where G,H are separated sets in (X,d). We claim that $Y\subseteq G$, or $Y\subseteq H$.

Since $Y\subseteq G\cup H$, we have $Y=(G\cup H)\cap Y=(G\cap Y)\cup(H\cap Y)$.

Our claim will be established, if either $G\cap Y=\phi$, or $H\cap Y=\phi$.

If possible, suppose that both $G\cap Y\neq\phi, H\cap Y\neq\phi$. Since $G\cap Y\subseteq G$, $H\cap Y\subseteq H$ and G,H are separated sets, it follows by Theorem [6.4] the sets $G\cap Y$ and $H\cap Y$ are separated.

Thus $Y=(G\cap Y)\cup(H\cap Y)$ and $G\cap Y, H\cap Y$ are separated sets—a contradiction to the connectedness of Y.

Hence we must have either $G\cap Y=\phi$, or $H\cap Y=\phi$.

If $G\cap Y=\phi$, then $Y=H\cap Y$ and so $Y\subseteq H$.

If $H\cap Y=\phi$, then $Y=G\cap Y$ and hence $Y\subseteq G$ and the proof is complete.

Theorem [6.12] Let (X,d) be a metric space and $S\subset Y\subset X$. Then S is connected in (X,d), if and only if S is connected in (Y,d_Y).

Proof. We know from Illustrative Example 11, Chapter-1, that $\bar{F}^Y=\bar{F}^X\cap Y$ for $F\subseteq Y$, where $\bar{F}^X$ is the closure of F relative to (X,d), $\bar{F}^Y$ is the closure of F in (Y,d_Y).

Note that, if $A,B\subseteq S$, then $A\cap\bar{B}^Y=A\cap\bar{B}^X\cap Y=A\cap Y\cap\bar{B}^X=A\cap\bar{B}^X$.

Similarly, $\bar{A}^Y\cap B=\bar{A}^X\cap B$.

Let S be connected in (X,d). We show that S is connected in (Y,d_Y). Suppose that S is **not** connected in (Y,d_Y). Then there exist two subsets A,B in (Y,d_Y) such that $S=A\cup B$ and $A\cap\bar{B}^Y=\phi$, $\bar{A}^Y\cap B=\phi$.

This implies that $S=A\cup B$ and $A\cap\bar{B}^X=\phi$, $\bar{A}^X\cap B=\phi$—a contradiction to the connectedness of S in (X,d).

Hence S is connected in (Y,d_Y).

Converse part is similar.

Theorem [6.13] Let (X, d) be a metric space and S be a connected subset of X. Then for any subset E of X such that $S \subset E \subset \bar{S}$, E is connected.

Proof. If possible, suppose that E is disconnected. Then there exist non-empty sets A, B of E such that $E = A \cup B, \bar{A} \cap B = \phi, A \cap \bar{B} = \phi$.

$\therefore$ we have $S \subset E = A \cup B,\ \bar{A} \cap B = \phi,\ A \cap \bar{B} = \phi$.

Since S is connected we have by Theorem [6.11] that either $S \subset A$ or $S \subset B$.

Let $S \subset A \Rightarrow \bar{S} \subset \bar{A}$

$$\Rightarrow \bar{S} \cap B \subset \bar{A} \cap B = \phi$$

$$\Rightarrow \bar{S} \cap B = \phi.$$

Again, since $E = A \cup B$ and $E \subset \bar{S}$, we have

$$B \subset A \cup B = E \subset \bar{S} \Rightarrow B \subset \bar{S} \Rightarrow B \cap \bar{S} = B.$$

$\therefore B = \bar{S} \cap B = \phi$—a contradiction.

If $S \subset B$, then we can show that $A = \phi$—which is again a contradiction.

Hence E is connected.

Corollary [6.1] : Let (X, d) be a metric space and E be a connected subset of X. Then $\bar{E}$ is connected.

Note. Converse of the above corollary does **not** hold in general.

For example, let $E = (0, 1) \cup (1, 2)$ in $(\mathbb{R}, d)$, where d is the usual metric of $\mathbb{R}$.

Then $\bar{E} = [0, 2]$.

We shall show later in a theorem that the intervals and connected subsets of $\mathbb{R}$ coincide.

Hence $\bar{E}$ is connected. But E is **not** connected.

Theorem [6.14] Let (X, d) be a metric space and $\{A_\alpha\}_{\alpha \in I}$ be a family of connected subsets of (X, d) such that no two of them are separated. Then $\bigcup\limits_{\alpha \in I} A_\alpha$ is connected.

Proof. Let $E = \bigcup\limits_{\alpha \in I} A_\alpha$ be not connected. Then we can find two non-empty separated sets G, H such that $E = G \cup H$.

Since G, H are non-empty, there exist $i, j \in I$ such that $i \neq j$ and $A_i \subset G$ and $A_j \subset H$.

Since G, H and separated sets, it follows that A_i, A_J are also separated—a contradiction to the assumption.

Hence E is connected.

Theorem [6.15] Let $\{A_i\}_{i \in I}$ be a non-empty family of connected subsets of a metric space (X, d) such that $\bigcap\limits_{i \in I} A_i \neq \phi$. Then $\bigcup\limits_{i \in I} A_i$ is connected.

Proof. Let $S = \bigcup\limits_{i \in I} A_i$ be **not** connected. Then there exists non-empty separated sets A, B such that $S = A \cup B$.

Since $\bigcap\limits_{i \in I} A_i \neq \phi$, let $x \in \bigcap\limits_{i \in I} A_i$. Then $x \in S = A \cup B$. Then either $x \in A$, or $x \in B$. Suppose that $x \in A$. Since $x \in A_i,\ \forall\, i \in I$, we have $A_i \cap A \neq \phi,\ \forall\, i \in I$.

Note that $A_i \subset \bigcup_{i \in I} A_i = A \cup B$, for all $i \in I$. Since A_i is connected, we have from Theorem [6.11] that either $A_i \subset A$, or $A_i \subset B$. Again, since $A_i \cap A \neq \phi$ and $A \cap B = \phi$, we get $A_i \subset A, \forall i \in I$

$$\Rightarrow S = \bigcup_{i \in I} A_i \subset A$$

$$\Rightarrow S \subset A$$

$\Rightarrow B = \phi$—this is a contradiction.

Hence S must be connected.

Theorem [6.16] Let (X, d) be a metric space and $Y \subseteq X$. If every two point of Y are contained in a connected subset of Y, then Y is connected.

Proof. Suppose that Y is **not** connected. Let U, V be two separated sets such that

$$Y = U \cup V,\ U \neq \phi,\ V \neq \phi.$$

Take $x \in U$, $y \in V$. Then $x, y \in Y$ and so by hypothesis there exist a connected subset S of Y such that $x, y \in S$.

$$\therefore S \subset Y = U \cup V.$$

Since S is connected we must have either $S \subset U$ or $S \subset V$. Then $U \cap V \neq \phi$—this is a contradiction.

Hence Y is connected.

6.3 Components

Definition [6.4]: A maximal connected subset in a metric space (X, d) is called a component. From definition it follows that a subset A of a metric space (X, d) is a component, if

(i) A is a connected subset in (X, d).

(ii) $B (\neq X)$ is another connected subset in (X, d) such that $A \subseteq B$, then $A = B$.

Example [6.3]: If a metric space (X, d) is connected, then X is the only component.

Example [6.4]: Let (X, d) be the discrete metric space. Note that if $X = \{x\}$, then X is connected and X is the only component.

If X contains more than one point, then the components are $\{x\}, \forall x \in X$, since any subset of (X, d) containing more than one point is **not** connected.

Example [6.5]: Consider the space Q of all rational numbers as a subspace $(\mathbb{R}, d)$, where d is the usual metric on $\mathbb{R}$.

Let $A = \{x, y\} \subseteq Q$. We know that between any two rational numbers there exists an irrational number. So take an irrational number p such that $x < p < y$.

Then $A \subseteq (-\infty, p) \cup (p, \infty)$, $A \cap (-\infty, p) \neq \phi$

$A \cap (p, \infty) \neq \phi$ and $A \cap (-\infty, p) \cap (p, \infty) = \phi$.

This shows that A is disconnected. Hence the only connected subsets of Q are singleton sets and hence these are the only components of Q.

Remark. Note that (Q, d) is **not** a discrete metric space.

Theorem [6.17] A component of a metric space is closed.

Proof. Let (X, d) be a metric space and A be a component in (X, d). We always have $A \subseteq \bar{A}$. Since A is connected, so is $\bar{A}$. So by maximality of A, it follows that $A = \bar{A}$, i.e., A is a closed set.

Note. If a metric space (X, d) has a connected subset A such that A is dense in (X, d), then (X, d) is connected.

Theorem [6.18] Components of a metric space are either identical or disjoint.

Proof. Let A and B be two components of a metric space (X, d). Suppose that $A \cap B \neq \phi$. Then from Theorem [6.15], $A \cup B$ is connected, since A, B are connected. Now note that

$$A \subseteq A \cup B \text{ and } B \subseteq A \cup B.$$

Hence by maximality of A, B we get $A = A \cup B, B = A \cup B$.

Thus $A = B$.

Discussion (Components forms a partition)

Let (X, d) be a metric space. Define a relation ρ on X as follows:

for any $x, y \in X$, $x\rho y$, if and only if there exist a connected subset A in (X, d) such that $x, y \in A$.

Now for all $x \in X, x, x \in \{x\}$ and $\{x\}$ is connected. Hence $x\rho x$ holds, $\forall\, x \in X$.

Let $x\rho y$ holds for $x, y \in X$. Then by definition there is a connected subset A of X such that $x, y \in A$. We can then say that $y, x \in A$.

$\Rightarrow y\rho x$

$\Rightarrow \rho$ is symmetric.

Let $x\rho y$ and $y\rho z, x, y, z \in X$. Then there exist connected subsets A and B of X such that $x, y \in A$ and $y, z \in B$. Since $A \cap B \neq \phi$, $A \cup B$ is a connected subset of X and $x, z \in A \cup B$

$\Rightarrow x\rho z$

$\Rightarrow \rho$ is transitive.

Hence ρ is an equivalence relation. We now find the equivalence classes. For $x \in X$, let ρ_x denote the equivalence class of x.

So $\rho_x = \{y \in X : x\rho y \text{ holds}\}$.

We now show that $\rho_x = \bigcup\{A \subseteq X : A \text{ is connected and } x \in A\}$.

Let $y \in \rho_x$. Then $x\rho y$ holds and so there exist a connected subset of X, say G, such that $x, y \in G$. Hence $y \in \bigcup\{A : A \text{ is connected and } x \in A\}$.

Now let $y \in \bigcup\{A \subseteq X : A \text{ is connected and } x \in A\}$.

Then we can find a member G in $\{A \subseteq X : A \text{ is connected and } x \in A\}$ such that $y \in G$. That is, G is a connected subset of X with $x, y \in G$. Consequently $x\rho y$ holds, i.e., $y \in \rho_x$.

Thus $\rho_x = \bigcup\{A \subseteq X : A \text{ is connected and } x \in A\}$.

Clearly, ρ_x is a connected subset of X. We show that ρ_x is a component of X. For let G be a connected subset of X such that $\rho_x \subseteq G$.

Since $x \in \rho_x$, it follows that $x \in G$, i.e., G is a connected subset of X such that $x \in G$.

So $G \in \{A \subseteq X : A$ is connected and $x \in A\}$.

$\Rightarrow G \subseteq \bigcup\{A \subseteq X : A$ is connected and $x \in A\} = \rho_x$

$\Rightarrow G \subseteq \rho_x$.

Hence, $\rho_x = G$, i.e., ρ_x is a component of X. Now we show that $\{\rho_x : x \in X\}$ are the only component of X. Let c be a component of X. Then $c \neq \phi$. Take $x \in c$ and so $\rho_x \cap c \neq \phi$

$\Rightarrow c \cup \rho_x$ is connected and $\rho_x \subset c \cup \rho_x, c \subset c \cup \rho_x$

$\Rightarrow \rho_x = c \cup \rho_x, c = c \cup \rho_x$ and so $c = \rho_x$.

Thus $\{\rho_x : x \in X\}$ are the only components of X, i.e., collection of equivalence classes and components coincides in a metric space with respect to the above equivalence relation.

Hence the family of all components forms a partition of the metric space (X, d).

Problem [6.3]: If a metric space (X, d) has only a finite number of components, prove that each component is both open and closed.

Solution. Let $A_1, A_2, \ldots, A_n$ be n components of a metric space (X, d). Since components form a partition of the space (X, d), we have

$$X = A_1 \cup A_2 \cup \cdots \cup A_n \text{ with } A_i \cap A_j = \phi, \text{ for all } i, j (i \neq j).$$

Now for all $i \in \{1, 2, \ldots, n\}$

$$A_i = X - (A_1 \cup A_2 \cup \cdots \cup A_{i-1} \cup A_{i+1} \cup \cdots \cup A_n)$$

Since each component is closed and union of finite number of closed set is closed, it follows that $A_1 \cup A_2 \cup \cdots \cup A_{i-1} \cup A_{i+1} \cup \cdots \cup A_n$ is closed and hence A_i is open, for all i.

So each component is open and closed.

Problem [6.4]: Let (X, d) be a compact metric space. If the components of (X, d) are all open, prove that there are at most a finite number of component.

Solution. If (X, d) is connected, then X is the only component and the result follows:

So, suppose that (X, d) is **not** connected.

Let $\{A_i\}_{i \in I}$ be the family of all components of X. Then by Discussion after Theorem [6.18], we get

$$X = \bigcup_{i \in I} A_i \text{ and } A_i \cap A_j = \phi, \ \forall\ i, j \in I (i \neq j)$$

By hypothesis, A_i is an open set, for all $i \in I$. Thus $\{A_i\}_{i \in I}$ forms an open cover of (X, d). Since (X, d) is compact, the above open cover must have a finite subcover. Since components are either identical or pairwise disjoint, we must have $\{A_i\}_{i \in I}$ is the required finite subcover of $\{A_i\}_{i \in I}$, i.e., I is finite. Hence the result follows.

Theorem [6.19] Every connected subset of a metric space (X, d) is contained in a component of X.

Proof. Let P be a connected subset of (X, d). Let us consider the family $S = \{A_i \subseteq X : A_i$ is connected and $P \subseteq A_i\}$.

Clearly, $S \neq \phi$, since $P \in S$ [P is connected and $P \subseteq P$]

Let $B = \bigcup_i A_i$. Since $P \subseteq A_i$, it follows that $P \subseteq B$. We now show that B is a component of (X, d).

Note that B is connected, since $\bigcap_i A_i = P \neq \phi$ and each A_i is connected. Let G be a connected subset of X such that $B \subseteq G$. Then $P \subseteq B \subseteq G$, i.e., $P \subseteq G$.

Hence $G \in S \Rightarrow G \subseteq B$.

$$\therefore\ B = G.$$

This shows that B is a component of X containing P.

Since the components are pairwise disjoint, it follows that B is the only component of X containing P.

Corollary [6.2] : Each element of a metric space (X, d) is contained in exactly one component of X.

Proof. Let $x \in X$. Then $A = \{x\}$ is a connected subset of X. Hence by above theorem, A is contained in exactly one component B (say), i.e., $A \subset B$, where B is a component of X.

Theorem [6.20] Any non-empty connected subset of a metric space (X, d) is a component, if it is both open and closed.

Proof. Let U be a non-empty connected subset of (X, d) which is both open and closed. We show that U is a component of (X, d). Suppose that U is **not** a component of (X, d).

Since U is a connected set, so by Theorem [6.19] there exist a component V of (X, d) such that $U \subset V$, i.e., $(X - U) \cap V \neq \phi$.

Note that

$V \subset U \cup (X - U)$ and

$$U \cap V \neq \phi, (X - U) \cap V \neq \phi, U \cap (X - U) \cap V = \phi$$

$\Rightarrow V$ is **not** connected—a contradiction.

Hence U is a component of (X, d).

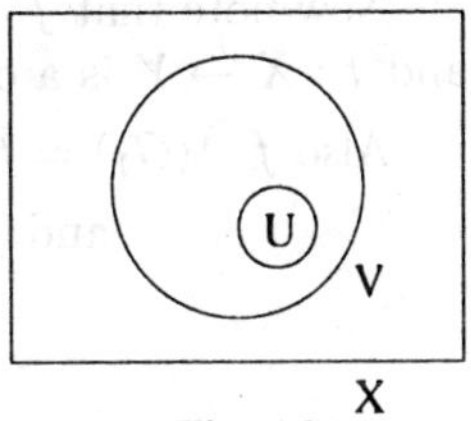

Fig. 6.2

Definition [6.5] : A subset E of $\mathbb{R}$ is an interval if, for all $x, y \in E$ and $x < y$, we have

$$[x, y] \subset E.$$

Theorem [6.21] A subset E of $\mathbb{R}$ is connected, if and only if E is an interval in $\mathbb{R}$.

Proof. First suppose that E is **not** an interval in $\mathbb{R}$. Then by the above definition there exists $x, y \in E$ and $x < y$ but $[x, y] \not\subset E$. So we can find an element $p \in (x, y)$ such that $p \notin E$.

Now note that

$$E \subset (-\infty, p) \cup (p, \infty), E \cap (-\infty, p) \neq \phi, E \cap (p, \infty) \neq \phi$$
$$E \cap (-\infty, p) \cap (p, \infty) = \phi.$$

This shows that E is **not** connected.

This proves that if E is connected, then E is an interval.

Conversely, suppose that E is an interval. We claim that E is connected. If E is **not** connected, then there exist open sets G_1, G_2 in $\mathbb{R}$ such that

$$E \subset G_1 \cup G_2 \qquad E \cap G_1 \neq \phi, E \cap G_2 \neq \phi, E \cap G_1 \cap G_2 = \phi.$$

We now take $x \in E \cap G_1$, $y \in E \cap G_2$. Then $x, y \in E$ such that $x \neq y$.

Suppose that $x < y$.

Then, we have $[x, y] \subset E \subset G_1 \cup G_2$, since E is an interval.

Let $a = \sup([x, y] \cap G_1)$.

Then $a \in \overline{[x, y] \cap G_1} \subset [x, y] \cap \bar{G}_1$. Clearly $a \notin G_2$.

Hence $x \leq a < y$. Then $a \in [x, y) \subset E \subset G_1 \cup G_2$.

So $a \in G_1$. Since G_1 is open, there exist $\epsilon > 0$ such that $a + \epsilon \in G_1 \cap [x, y]$—a contradiction.

Hence E is connected.

Corollary [6.3] : $(\mathbb{R}, d)$ is connected.

Proof. Since $\mathbb{R}$ itself is an interval, so by above theorem $\mathbb{R}$ is connected.

Theorem [6.22] Let (X, d) and (Y, ρ) be two metric spaces and $f : (X, d) \to (Y, \rho)$ a continuous function. If X is connected, then so is $f(X)$.

Proof. Let (X, d) be a connected metric space. We show that $f(X)$ is connected.

Suppose that $f(X)$ is disconnected subset of (Y, ρ). Then there exist open sets G_1, G_2 in (Y, ρ) such that

$$f(X) \subset G_1 \cup G_2,\ f(X) \cap G_1 \neq \phi,\ f(X) \cap G_2 \neq \phi,\ f(X) \cap G_1 \cap G_2 = \phi$$

$\Rightarrow X = f^{-1}(G_1) \cup f^{-1}(G_2)$.

Now note that $f^{-1}(G_1)$ and $f^{-1}(G_2)$ are open sets in (X, d) since G_1, G_2 are open in (Y, ρ) and $f : X \to Y$ is a continuous function.

Also $f^{-1}(G_1) = f^{-1}(G_1 \cap f(X)) \neq \phi,\ f^{-1}(G_2) = f^{-1}(G_2 \cap f(X)) \neq \phi$

$$\begin{aligned} \text{and } f^{-1}(G_1) \cap f^{-1}(G_2) &= f^{-1}(G_1 \cap f(X)) \cap f^{-1}(G_2 \cap f(X)) \\ &= f^{-1}(G_1 \cap G_2 \cap f(X)) \\ &= \phi. \end{aligned}$$

Hence $X = f^{-1}(G_1) \cup f^{-1}(G_2),\ f^{-1}(G_1) \neq \phi,\ f^{-1}(G_2) \neq \phi,\ f^{-1}(G_1) \cap f^{-1}(G_2) = \phi$ and $f^{-1}(G_1), f^{-1}(G_2)$ are open sets in (X, d)—a contradiction to the fact that X is connected.

Hence $f(X)$ is connected.

Alternative

If possible, suppose that $f(X)$ is **not** connected. Then by Theorem [6.9] (4), there exist a non-empty proper subset G of $f(X)$, which is both open and closed. Since f is continuous, $f^{-1}(G)$ is a non-empty subset of X, which is both open and closed—a contradiction to the connectedness of (X, d).

Hence $f(X)$ is connected.

Remark. Let $f : (X, d) \to (Y, \rho)$ be a continuous function. Suppose that f is surjective. Then $Y = f(X)$. Now if X is connected, then by the above theorem (Y, ρ) is connected.

From the above theorem we can say that

"continuous image of a connected metric space is connected".

Problem [6.5] : Let $f : (\mathbb{R}, d) \to (\mathbb{Z}, d)$ be a continuous function. Then show that f is constant.

Solution. Suppose that f is **not** constant.

Then we can find $x_1, x_2 \in \mathbb{R}$ such that $f(x_1) \neq f(x_2)$, where $x_1 \neq x_2$.

Let $f(x_1) < f(x_2)$. Since $\mathbb{R}$ is connected, so is $f(\mathbb{R})$. Hence $f(\mathbb{R})$ is an interval.

So $[f(x_1), f(x_2)] \subseteq f(\mathbb{R}) \subseteq \mathbb{Z}$—a contradiction.

Hence f is a constant function.

Problem [6.6] : Let $f : [a,b] \to \mathbb{R}$ be a continuous function. Then show that $f([a,b])$ is a closed bounded interval.

Solution. We know that $[a,b]$ is a compact subset of $\mathbb{R}$, since $[a,b]$ is closed and bounded. Since continuous image of a compact set is compact, $f([a,b])$ is a compact subset of $\mathbb{R}$ and hence $f([a,b])$ is closed and bounded subset of $\mathbb{R}$.

Also we know that $[a,b]$ is a connected subset of $\mathbb{R}$, so $f([a,b])$ is connected and hence $f([a,b])$ is an interval in $\mathbb{R}$.

Thus $f([a,b])$ is a closed, bounded interval in $\mathbb{R}$.

Theorem [6.23] [Intermediate Value Theorem] Let (X,d) be a connected metric space and $f : X \to \mathbb{R}$ be a continuous function. Then f has intermediate value property.

Proof. Since (X,d) is a connected metric space and $f : X \to \mathbb{R}$ a continuous function, $f(X)$ is a connected subset of $\mathbb{R}$. Hence either $f(X)$ is a singleton set or an interval.

Suppose that $f(X)$ is an interval. Let $x, y \in X$ be such that $f(x) \neq f(y)$. Suppose that $f(x) < f(y)$. Let $f(x) < c < f(y)$. Since $f(X)$ is an interval, we have $[f(x), f(y)] \subseteq f(X)$ and so $c \in f(X)$. Hence we can find $x_0 \in X$ such that $f(x_0) = c$.

Corollary [6.4] : If $f : [a,b] \to \mathbb{R}$ be a continuous function and $f(a) < c < f(b)$, then there exist a real number $x_0 \in (a,b)$ such that $c = f(x_0)$.

Corollary [6.5] : If $f : [a,b] \to \mathbb{R}$ be a continuous function and $f(a) \cdot f(b) < 0$, then there exist a real number x_0 in (a,b) such that $f(x_0) = 0$.

Remark. Converse of the above theorem is also true. This is our next theorem.

Theorem [6.24] Let (X,d) be a metric space. Suppose that every real valued continuous function on (X,d) has the intermediate value property, then (X,d) is connected.

Proof. Let us suppose that (X,d) be **not** connected. Our task is to find a real-valued continuous function on (X,d) without satisfying intermediate value property and this will lead to us a contradiction to our assumption and the proof will be completed.

Since by our assumption (X,d) is not connected, so we can find two non-empty open sets G_1, G_2 in (X,d) such that $X = G_1 \cup G_2$ and $G_1 \cap G_2 = \phi$.

Define a function $f : X \to \mathbb{R}$ by

$$f(x) = \begin{cases} 0, & \text{if } x \in G_1 \\ 1, & \text{if } x \in G_2. \end{cases}$$

Let A be any open set in $\{0,1\}$. Then

$$f^{-1}(A) = \begin{cases} \phi, & \text{if } A = \phi \\ G_1, & \text{if } A \cap \{0,1\} = \{0\} \\ G_2, & \text{if } A \cap \{0,1\} = \{1\} \\ X, & \text{otherwise.} \end{cases}$$

This shows that f is a real-valued continuus function on (X, d). Note that f does **not** have the intermediate value property, since $0 < \frac{1}{3} < 1$, but there is **no** point $x \in X$ such that $f(x) = \frac{1}{3}$.

Thus we have done our job.

Corollary [6.6]: Let (X, d) be a metric space. Then **(X, d) is connected, if and only if every real-valued continuous function $f : X \to \{0, 1\}$ is constant.**

Proof. Follows from Theorem [6.23] and Theorem [6.24].

EXERCISE

(1) If (X, d) is a connected metric space, prove that either X in a singleton set or an infinite set.

(2) Prove that a discrete metric space containing more than one point is **not** connected.

(3) Show that continuous image of a connected metric space is connected.

(4) Let $f : [a, b] \to \mathbb{R}$ be a continuous mapping. Show that $f([a, b])$ is a closed bounded interval.

(5) Show that a continuous function $f : (\mathbb{R}, d) \to (\mathbb{Z}, d)$ is constant, where d denotes the usual metric on $\mathbb{R}$.

(6) Let A and B be two separated subsets in a metric space (X, d). If $A \cup B$ is open, then show that A, B are open sets in (X, d).

(7) Let (X, d) be a metric space and $A, B \subseteq X$. If $d(A, B) > 0$, show that A, B are separated sets.

(8) Let A be a connected subset in a metric space (X, d) and let $A \subset B \subset \bar{A}$. Show that B is connected and hence, in particular, $\bar{A}$ is connected.

(9) Determine the components of a discrete metric space.

(10) Show that any pair of components in a metric space (X, d) are separated.

(11) Prove that if a metric space (X, d) has a finite number of components, then each component is both open and closed.

(12) Show that the set $S = \{(x, y) \in \mathbb{R}^2 : x^2 - y^2 \geq 4\}$ in $(\mathbb{R}^2, d)$ is not connected.

(13) Give an example of a (i) connected set which is not compact and (ii) compact set which is not connected.

(14) Determine whether the following sets are connected or not:

(a) $\{(x, y) \in \mathbb{R}^2 : x^2 + y^2 = 1\}$

(b) $\{(x, y) \in \mathbb{R}^2 : xy \neq 0\}$

(c) $\{(x, y) \in \mathbb{R}^2 : x^2 + y^2 \leq 1\}$

(d) $\{(x, y) \in \mathbb{R}^2 : x > 0, x^2 + y^2 \leq 1\} \cup \{(x, y) \in \mathbb{R}^2 : x < 0, x^2 + y^2 \leq 1\}$

(e) $\{(x, y) \in \mathbb{R}^2 : x = 0, 0 < y < 1\} \cup \{(x, y) \in \mathbb{R}^2 : 0 < x < 1, y = 0\}$

(f) $\{(x, y) \in \mathbb{R}^2 : 0 \leq x \leq \frac{\pi}{4}, y = \cos x\} \cup \{(x, y) \in \mathbb{R}^2, 0 \leq x \leq \frac{\pi}{4}, y = \sin x\}$

(15) Let a metric space has only a finite number of components. Prove that each component is both closed and open.

(16) Let (X, d) be a compact metric space. If the components of X are open, then prove that they are only a finite in number.

(17) Prove that if a connected metric space (X, d) contains atleast two points, then X has uncountably many points.

Index

Some exciting gifts are up for grabs with this book!!

MATHEMATICAL ANALYSIS

Dr Ajay Kr Chaudhuri and Pratikshan Mondal

FREE Gifts Rs. 1000/-

Dear Readers

By now you must have gone through this book in the entirety. So we invite your assessment on this title with absolute impartiality and candid observations. After all, we publishers faithfully depend on your benchmarking. Friends, please fill up the open spaces below with CAPITAL LETTERS. You may add some extra sheets of paper if the space provided is not sufficient for you.

i. What are the important topics of your syllabus which are not adequately covered in this book?

ii. In which chapters and on what topics the treatment of the subject-matter by the author is not satisfactory and why?

iii. Have you come across the typos, factual inaccuracies, inconsistencies, and other forms of mistakes in the book? If yes, then please specify the mistakes and page numbers.

iv. Name top three books on the same subject in your consideration. Could you provide some reasons?

Your 1st Preference: ....

Your 2nd Preference: ....

Your 3rd Preference: ....

v. Any suggestion to improve the quality of illustrations, diagrams, etc.?

vi. How did you like the cover design? Any suggestion on colours?

vii. Suggestions for further improvement of production-standard:

viii. Who recommended you this book:

☐ Teacher ☐ Friend ☐ Parents ☐ Bookseller ☐ Librarian

ix. Name of the person who has recommended:

Designation: ....

Address: ....

Phone No.: ....

e-mail: ....

x. Name and address of the bookseller from whom you have purchased this book:

....

....

....

xi. Name and address of the library where you would like to find the book:

....

....

xii. Name and address of your institution (Please mention the University or Board):

....

....

xiii. Your name, e-mail, and complete postal address:

....

....

....

xiv. A few other titles published by us that you would like to read:

....

....

The best assessment will be rewarded half-yearly. The award will be in the form of our publications, as decided by the Editorial and Marketing Operations Division, amounting to Rs. 1000/- (Rupees one thousand) only.

....

Please mail the filled up coupon at your earliest to :

Editorial & Marketing Operations Division

NEW CENTRAL BOOK AGENCY (P) LTD

8/1 Chintamoni Das Lane, Kolkata 700 009, West Bengal, INDIA

Ph. No.: 033 + 22418411, 22418573, 22419094

Fax: 033 + 22419573, 22413852

e-mail: ncbapvtltd@eth.net ◆ ***Website:*** www.ncbabooks.com